Ultrafast Imaging of Photochemical Dynamics

Royal Society of Edinburgh, UK

31 August – 2 September 2016

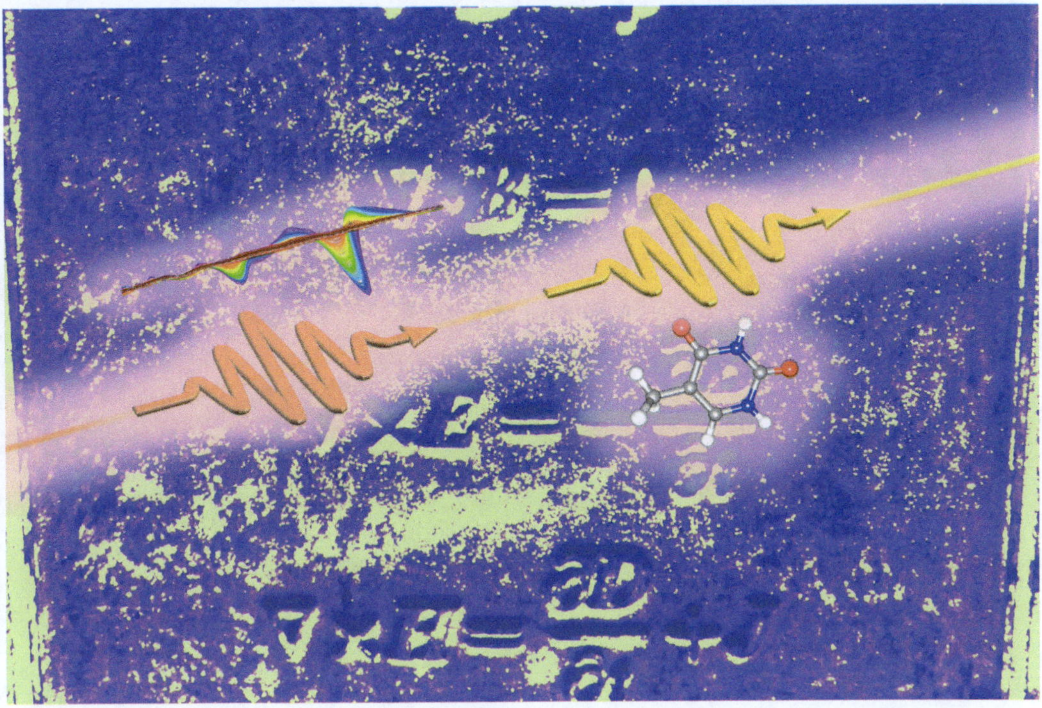

FARADAY DISCUSSIONS

Volume 194, 2016

ROYAL SOCIETY OF CHEMISTRY

The Faraday Division of the Royal Society of Chemistry, previously the Faraday Society, founded in 1903 to promote the study of sciences lying between Chemistry, Physics and Biology.

Faraday Discussions (print ISSN 1359-6640, electronic ISSN 1364-5498) is published 10 times a year by the Royal Society of Chemistry, Thomas Graham House, Science Park, Milton Road, Cambridge, UK CB4 0WF.

Volume 194 ISBN 13: 978-1-78262-479-0

2016 annual subscription price: print+electronic £929 US $1635; electronic only £883, US $1553. Customers in Canada will be subject to a surcharge to cover GST. Customers in the EU subscribing to the electronic version only will be charged VAT. All orders, with cheques made payable to the Royal Society of Chemistry, should be sent to RSC Order Department, Royal Society of Chemistry, Thomas Graham House, Science Park, Milton Road, Cambridge, CB4 0WF, UK.

Tel +44 (0) 1223 432398; E-mail **orders@rsc.org**

If you take an institutional subscription to any RSC journal you are entitled to free, site-wide web access to that journal. You can arrange access via Internet Protocol (IP) address at **www.rsc.org/ip**. Customers should make payments by cheque in sterling payable on a UK clearing bank or in US dollars payable on a US clearing bank.

Printed in the UK

MIX
Paper from responsible sources
FSC® C013604

Faraday Discussions are unique international discussion meetings that focus on rapidly developing areas of chemistry and its interfaces with other scientific disciplines.

Ultrafast Imaging of Photochemical Dynamics

Faraday Discussions

www.rsc.org/faraday_d

General Discussion on Ultrafast Imaging of Photochemical Dynamics was held in Edinburgh, UK on the 31st August, and 1st and 2nd of September 2016.

CONTENTS

ISSN 1359-6640; ISBN 13: 978-1-78262-479-0

Cover

See Orr-Ewing et al., *Faraday Discuss.*, 2016, **194**, 681–706.

Maxwell's Light Fantastic: Pyrimidine nucleoside dynamics probed with ultrafast time-resolved infrared spectroscopy.

Front cover image courtesy of Dr Gareth M. Roberts.

INTRODUCTORY LECTURE

PAPERS AND DISCUSSIONS

CONCLUDING REMARKS

ADDITIONAL INFORMATION

Faraday Discussions

EDITORIAL

Preface

Russell S. Minns*[a] and Adam Kirrander*[b]

DOI: 10.1039/c6fd90074g

It is hard to overstate the scientific and technological importance of photochemical processes and reactions. The fundamental changes that molecules undergo following absorption of light drives important processes in wide ranging fields, from biology to catalysis, photodynamic therapies to molecular machinery, covering the entire spectrum of the physical and life sciences. It is a pillar of chemical reaction dynamics, which itself has a long history that includes developments in spectroscopy, crossed-beam techniques, femtochemistry, and extends to present efforts in ultrafast imaging, with many of these topics reflected in past *Faraday Discussions*. The overarching challenge facing photochemistry is that it is predominantly a non-statistical process, strongly influenced by the quantum dynamics of the photoexcited system. Complications include exponential scaling with the number of degrees of freedom, strong coupling of nuclear and electronic motion through nonadiabatic couplings and conical intersections, tunneling, interference effects, and ultrafast time-scales. All of this means that a detailed comprehension remains elusive.

In recent years, new experimental techniques capable of monitoring photochemical processes in unprecedented temporal, spatial, chemical, and energetic detail have appeared. This is in large part due to the development of intense-laser techniques, the construction of free-electron lasers such as the XFEL in Europe and the LCLS in the USA, new sources of pulsed electrons, and advanced detection techniques. The challenges associated with making such measurements are manyfold, and analysis of the signals produced is equally difficult due to the often complex interactions of the pump and probe pulses, meaning that the signals cannot always be directly related back to the dynamics measured. This amplifies the complexity, and means experiment alone is not enough to uncover the true dynamics. Progress in experiments is therefore combined with new approaches and important advances in theoretical modeling of quantum dynamics. The scope and accuracy of the theoretical models is now such that quantitative comparisons between experiment and theory can be made in ever more complex systems and for increasingly complex processes.

[a]*Chemistry, University of Southampton, Highfield, Southampton, SO17 1BJ, UK. E-mail: r.s.minns@soton.ac.uk*

[b]*EaStCHEM, School of Chemistry, University of Edinburgh, David Brewster Road, Edinburgh EH9 3FJ, UK. E-mail: Adam.Kirrander@ed.ac.uk*

Many of these advances are developed by research communities not traditionally concerned with photochemistry. The purpose of this meeting was therefore to gather key participants representing the full scientific scope of the topic and to set the agenda for future research. The response from the community was overwhelming, and nearly 120 delegates from 16 countries gathered for this Faraday Discussion on *Ultrafast Imaging of Photochemical Dynamics* at the historical Royal Society of Edinburgh on August 31st, 2016. The format of a Faraday Discussion meeting is ideal for discussions that transgress the many emerging experimental technologies and theoretical techniques, and provides an opportunity to identify the overlap and complementarity of different approaches, while also addressing specific areas of contention and controversy. When reading this volume, it is important to remember that key insights and discussion points are often contained within the discussion segments, and we strongly encourage readers to not only explore the articles but also the associated discussions contained herein.

Looking through this volume, a number of important trends emerge clearly. The development of high harmonic sources and free-electron lasers is beginning to give access to pulsed X-ray radiation which opens the door for a wholesale import into the ultrafast regime of an astonishing array of powerful techniques, known for a long time in the synchrotron community. One readily apparent advantage is that X-ray spectra are comparatively uncongested compared to the valence bands. Furthermore, the coherence of lasers increasingly makes nonlinear and multidimensional spectroscopies a viable proposition, even in the X-ray domain. Various spectroscopies are also increasingly complemented by ultrafast scattering, *via* X-rays or electrons, providing a powerful complementary view and more direct access to the time-evolution of molecular geometry. At the same time, the ultrashort pulse durations and high intensities attainable provide alternative routes to structural information and allow us to monitor some of the fastest processes in chemistry through Coulomb explosion imaging and high harmonic spectroscopy.

With such an arsenal of methods, each providing a different view of the dynamics, a combination of techniques is increasingly used in order to attain a more complete picture of the dynamics. Ultrafast imaging techniques are also used to target increasingly complex systems, often including solvents or solid state elements, and dynamic processes that evolve over many different timescales. Theory remains important for interpreting experiments but is also used to charter new opportunities, including for instance the potential of light–matter interactions to alter photochemical outcomes. Interestingly, amidst the sustained and ongoing effort towards ever more complete and inclusive quantum molecular dynamics simulations, reduced dimensionality models still remain useful. A point made during the discussion was that the community should benchmark one theory against another in order to clearly delineate the strengths and weaknesses of different approximations and approaches.

The experimental and theoretical techniques that are now available are transformative and provide a view of dynamics structure that has previously been inaccessible, with new detection methods only just starting to be explored. Such is the pace of development that it is important that we maintain a collaborative international effort, combining multiple experimental and theoretical approaches to obtain a consistent picture and benchmark their relative efficacy.

While the continued development will allow us to follow bond rearrangements directly, in larger molecules and bimolecular reactions, in complex chemical and biological systems and advanced materials, it is crucial that we maintain our efforts to study small molecular systems in an ever more complete manner. The level of detail and precision attainable in small molecules provides the clearest opportunity to develop the general rules and structural patterns that will provide a "bigger picture" understanding, allowing us to predict the key dynamic processes for wide ranging molecular systems, even if the fine details and absolute timescales differ. Ultimately, one may envision a point where it becomes possible to design molecular systems and light fields for a particular function.

We hope this volume serves as an enthusiastic call to arms for the community to not only continue to innovate ever better experiments and theory for the study of ultrafast chemical dynamics, but also to uncover any new simplicity that emerges from the increasingly detailed data obtained.

Acknowledgements

The process of organising and arranging this meeting would have been impossible without the enthusiastic and competent assistance of the staff at the Royal Society of Chemistry (RSC). We would like to thank, in no particular order, Alisa Becker, Marie Cote, Claire Springett, Helen Lunn, and Jeremy Allen of the RSC. Please spare an extra moment of appreciation for the editors Helen Lunn and Jeremy Allen, and the effort involved in producing the beautiful volume you hold in your hand. Shaping the scientific content of the meeting, and selecting among the many strong abstract submissions, would have been difficult without our excellent scientific committee: Jon Marangos (Imperial), Nina Rohringer (CFEL), Olga Smirnova (MBI), and Peter Weber (Brown). Quick-thinking and engaged session chairs are essential for good discussions, and we were superbly served by the members of the scientific committee together with Dan Neumark (Berkeley), Dwayne Miller (MPSD), Hans Jakob Wörner (ETH), Misha Ivanov (MBI), and Eleanor Campbell (Edinburgh). Possibly the two most difficult and important tasks at a Faraday Discussion are to deliver the opening and closing lectures. This was admirably undertaken by our opening speaker Dan Neumark (Berkeley) and closing speaker Dwayne Miller (MPSD). And obviously, without interesting papers of the highest quality, presented by top-notch scientists, and without the active and enthusiastic participation of all delegates, many themselves leaders in the field, this meeting would not have been possible. *Thank you*. Finally, we wish to thank the poster prize committee, Vas Stavros (Warwick), Martin Centurion (Nebraska), Carla Figueira de Morisson Faria (UCL), Dmitry Shalashilin (Leeds) and Nirit Dudovich (Weizmann), for selecting the winner (João Pedro Figueira Nunes, York) among many excellent posters, and Nikola Zotev (Edinburgh) and Andrea Lidberg (Edinburgh) for ensuring that everyone was heard during the meeting. We thank the Faraday Division for their support, and our exhibitor Laser Quantum, and Andrew Turner from the EPCC for an exhilarating exhibit. Finally, we thank the Royal Society of Edinburgh (RSE) for providing an inspiring venue. The RSE counts among its past members James Clerk Maxwell, Paul Dirac, Max Born, and Albert Einstein, and this served as a stark reminder that, as always, we stand on the shoulders of giants, yet look forward to an ever more fascinating future.

Acknowledgements

Faraday Discussions

PAPER

Attosecond science in atomic, molecular, and condensed matter physics

Stephen R. Leone[abc] and Daniel M. Neumark*[ac]

Received 29th July 2016, Accepted 16th August 2016
DOI: 10.1039/c6fd00174b

Attosecond science represents a new frontier in atomic, molecular, and condensed matter physics, enabling one to probe the exceedingly fast dynamics associated with purely electronic dynamics in a wide range of systems. This paper presents a brief discussion of the technology required to generate attosecond light pulses and gives representative examples of attosecond science carried out in several laboratories. Attosecond transient absorption, a very powerful method in attosecond science, is then reviewed and several examples of gas phase and condensed phase experiments that have been carried out in the Leone/Neumark laboratories are described.

Introduction

Physical chemistry has benefited tremendously over the last several decades by the evolution of time-resolved experimental techniques. Advances up to the year 2000 were elegantly described in a review article by Zewail,[1] which discusses how our ability to understand fundamental chemical processes is inextricably linked to the development of methods ranging from flash photolysis to femtosecond lasers. It is now almost routine (albeit expensive!) in chemical physics laboratories to perform experiments that follow detailed and complex molecular motions in real time, using femtosecond lasers that can monitor rotational and vibrational motion, dissociation on bound and repulsive potential energy surfaces, and passage through conical intersections and over transition states. Experiments based on femtosecond methodology can in principle track even the highest frequency vibrations, such as the 7.5 fs vibrational period of the H–H fundamental in H_2.

The next frontier in time-resolved chemical dynamics is the study of electronic dynamics.[2,3] While femtosecond experiments often explore the effects of electron dynamics on nuclear motion, following the electrons themselves requires the development of light sources offering attosecond (as) time resolution; as an example, the classical period of an electron in its ground state orbiting the

[a]Department of Chemistry, University of California, Berkeley, CA 94720, USA. E-mail: dneumark@berkeley.edu
[b]Department of Physics, University of California, Berkeley, CA 94720, USA
[c]Chemical Sciences Division, Lawrence Berkeley National Laboratory, Berkeley, CA 94720, USA

hydrogen atom is ~150 as. Since 2000, developments in laser technology, optics, and atomic and molecular physics have made it possible to generate isolated attosecond pulses as short as 67 as.[4] These technical advances have enabled efforts in many laboratories throughout the world to perform attosecond science experiments that reveal new dynamical time scales in atoms, molecules, liquids, and solids, as discussed in numerous recent reviews.[5–16] This article provides a brief overview of key developments in attosecond science and technology, followed by a discussion of examples that have been carried out in our laboratories at Berkeley over the last several years.[16]

Generation and characterization of attosecond light pulses

Commercial femtosecond laser systems typically employ Ti:sapphire crystals as the active medium. They operate at a central wavelength of around 800 nm, and high power systems involving either multipass or regenerative amplifiers can generate pulses as short as 20 fs. Using a combination of self-phase modulation in optical fibers to increase the bandwidth of the pulse and chirped mirrors to temporally compress the spectrally broadened pulse, one can obtain "few-cycle" near-infrared (NIR) pulses with pulse durations as short as 3–5 fs. The optical period of 2.6 fs for an 800 nm pulse places a lower bound on the pulse duration attainable at this wavelength. Very recently, pulses spanning the visible region of the spectrum with sub-fs peak electric fields have been synthesized.[17] Nonetheless, generation of the shortest attosecond pulses generally requires upconversion to the vacuum ultraviolet/soft X-ray region where the optical period is substantially shorter (*i.e.* 41 as at 100 eV).

Owing to these considerations, attosecond pulse generation is strongly linked to the phenomenon of high harmonic generation (HHG).[18–24] This process involves strong-field manipulation of electrons by focusing a femtosecond laser pulse to high intensities in a target material, typically inert gas in a cell, gas jet, or waveguide, and generates light pulses covering the extreme ultraviolet (XUV) and soft X-ray regime of the electromagnetic spectrum. HHG is typically interpreted using a three-step model.[20,21] First, the high field intensities around the maximum of the electric field cycles of the focused driver pulse distort the electronic potential, resulting in tunneling ionization of the gas and the release of electron wave packets into the continuum. The ejected electrons are accelerated away and then toward the parent ion as the laser electric field changes sign, gaining energy from the field. Finally, the inelastic recollision of these accelerated electrons with the parent ion releases excess kinetic energy in brief attosecond light bursts each half-cycle of the driving field. The upper limit for the emitted energy, called the cut-off energy,[20,22,25] is determined by the maximum amount of kinetic energy gained in the second step and is given by

$$E_{\max} = I_{\mathrm{p}} + 3.17U_{\mathrm{p}}, \tag{1}$$

where I_{p} is the ionization potential of the target gas, and U_{p} is the ponderomotive energy of the photoelectrons in the applied electric field, $U_{\mathrm{p}} = E_{\mathrm{laser}}{}^2/4\omega_{\mathrm{laser}}{}^2$ (in atomic units).

Typical cut-off energies at 800 nm range from ~30 eV in Xe to >100 eV in Ne. HHG out to 1 keV has been observed in He but with a very low flux. At longer wavelengths, the ponderomotive energy is larger for the same laser intensity, resulting in a higher cut-off energy.[26] There has been considerable progress in recent years in experimentally realizing higher cut-off energies with longer wavelength driving pulses, with an important goal being to obtain useable intensities from HHG in the "water window" between 300 and 500 eV;[27–34] this interval lies above the carbon K-edge but below the oxygen K-edge. Efficient HHG out to 280 eV has also been demonstrated recently using ultraviolet driving pulses.[35]

In general, if a driving pulse comprising multiple optical cycles is used for HHG, then the temporal envelope of the HHG pulse is similar to that of the driving pulse. However, around the year 2000,[36–38] it was realized theoretically and shown experimentally that under the right conditions, the HHG pulse comprises an attosecond pulse train with pulses occurring at twice the frequency of the driving field. The interference between these pulses yields a discrete harmonic spectrum comprising odd harmonics of the driving field up to the cut-off energy. Attosecond pulse trains have been used successfully in a series of elegant interferometry experiments that probe important features of electron dynamics.[39–44] However, in many applications involving direct measurements of lifetimes and dynamics, isolated attosecond pulses are more desirable.

The generation of isolated attosecond pulses, first reported by the Krausz group in 2001,[45] required the development of novel advances in laser and optical technology, as well as new techniques for characterizing such pulses. These methods are generally based on isolating the XUV radiation produced by a single half-cycle of the driving field, or by suppressing XUV generation from all but a single half cycle. In amplitude gating,[45,46] one combines appropriate energy filtering and carrier-envelope (C-E) phase stabilization of the driver pulse[47] to obtain an isolated attosecond pulse from a few-cycle driver pulse by reflecting (or transmitting) only the highest energy XUV radiation produced by the most intense half cycle of the driver pulse. This yields an isolated attosecond pulse comprising a continuous spectrum of photon energies close to the cut-off region. Amplitude gating has been used to generate pulses as short as 72 as centered at 105 eV, using Ne as the nonlinear medium.[48] The isolated attosecond pulses and the NIR pulses used to generate them are characterized by attosecond streaking,[49,50] in which the

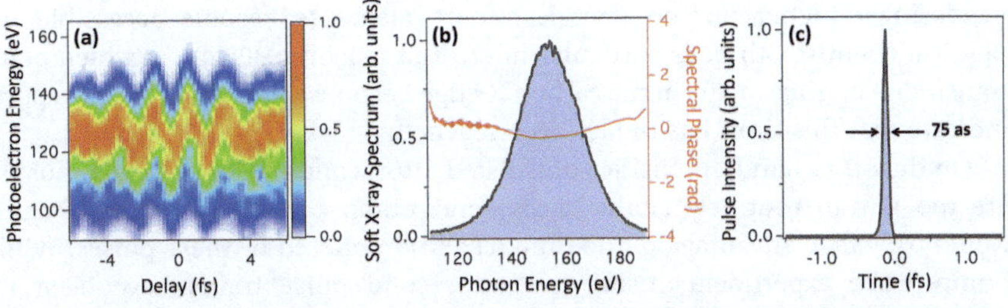

Fig. 1 Isolated attosecond pulse centered at 160 eV generated by amplitude gating. Panel (a) shows an attosecond streaking photoelectron spectrum, from which one extracts the spectral profile and phase (b) and the pulse duration (c).

momentum shifts by a C-E phase stabilized NIR pulse on the photoelectrons generated by the XUV pulse is determined as a function of delay between the two pulses. Fig. 1 shows an example of an isolated attosecond pulse centered at 160 eV along with its streaking trace and reconstruction, generated in our laboratory with amplitude gating using He as the gas medium.[51]

Amplitude gating requires exceedingly short (<4 fs) driver pulses and is thus quite technically demanding. Other gating methods have been developed in which isolated attosecond pulses can be generated from somewhat longer pulses and at lower XUV photon energies. These include polarization gating (PG),[52,53] double optical gating (DOG),[54–56] and ionization gating.[57–59] PG and DOG rely on overlapping two counter-rotating circularly polarized pulses so that only a single half-cycle is linearly polarized and thus produces XUV light. In ionization gating, nearly isolated attosecond pulses are obtained by using ionization to terminate HHG after a particular half-cycle cut-off;[60] the central photon energy can be tuned by varying the C-E phase. A conceptually different method, the attosecond lighthouse, utilizes wavefront-rotated driver pulses to obtain high harmonic spectra consisting of spatially separated, isolated attosecond pulses.[61] Our group has recently developed a gating scheme combining aspects of amplitude and polarization gating (PASSAGE, or Polarization ASSisted Amplitude GatE) that yields isolated attosecond pulses whose central frequency can be varied over energies ranging from 50–130 eV, and which can work with somewhat longer driving pulses than pure amplitude gating.[62] Pulse energies from these methods are typically within an order of magnitude of one nJ. Several laboratories are aiming to generate μJ isolated attosecond pulses using very high power driving lasers,[63,64] with the ultimate goal of performing attosecond pump–probe experiments (see later).

Representative examples of attosecond science

Time-resolved experiments using attosecond pulses offer superior time resolution combined with the capability to either initiate or probe dynamics in the soft X-ray region of the electromagnetic spectrum. Attosecond pulses in the range of 15–30 eV can probe Rydberg and high lying valence electronic states lying near or above the ionization potential of the target species. At higher energies, core ionization and core-to-valence transitions become accessible. The energies of these transitions are highly element-specific, enabling one to track the electronic dynamics associated with particular atoms in the gas phase and condensed phase targets. As technical advances push the useable cut-off energy of HHG to increasingly higher photon energies, more elements become accessible. As physical chemists, there is particular interest in reaching 300 eV, at which point transitions originating from the carbon K-edge become accessible. Recent reports indicate that this limit has in fact been reached.[33]

On the other hand, intensities of isolated attosecond pulses achieved thus far are too low to routinely enable attosecond pump–probe experiments, which would truly take advantage of the time-resolution offered by these pulses. While pump–probe experiments using two attosecond pulse trains have been reported,[39,65] as yet all dynamics experiments involving isolated attosecond pulses vary the time delay between an attosecond pulse and a much stronger, few-cycle NIR pulse. In many of these experiments, the temporal resolution is determined by the cross correlation of the two pulses and thus corresponds to the intensity envelope of the NIR pulse, which can be as short as 3–5 fs. However, if the phase of the electric field of the NIR pulse relative to the attosecond pulse can be controlled, then one can observe dynamics associated with the NIR field and, in favorable cases, resolve dynamics on the attosecond time scale.

One of the first time-domain experiments using isolated attosecond pulses measured the lifetime of the 3d core hole in krypton created by an XUV pulse, which is filled by Auger recombination, emitting electrons from the 4p shell.[66] The lifetime of this process was determined using attosecond streaking, in which the emitted electrons interact with a few-cycle near-infrared pulse, and the evolution of the electron kinetic energy from the $M_{4,5}N_1N_{2,3}$ Auger line as a function of the relative delay between the two pulses reveals a 3d core hole lifetime of 7.9 fs.

Attosecond streaking has also been used to track the birth of photoelectrons in atoms and solids. In the first such experiment, an isolated sub-200-as pulse at 106 eV ionized neon atoms. Attosecond streaking yielded a delay of 21 ± 5 as in the photoemission of a 2s electron relative to a 2p electron in neon.[67] Theoretical efforts to understand the origin of this result have thus far underpredicted the measured value by approximately a factor of two.[68,69] In a similar vein, attosecond photoelectron streaking spectroscopy was applied to study dynamics of electron escape from core and delocalized electronic states of a single crystalline tungsten $\langle 110 \rangle$ surface.[70] The experiment employed isolated XUV pulses with a pulse duration of ~300 as centered at 91 eV to simultaneously excite the 4f core level and the 5d and 6s conduction band electrons to upper conduction band states, leading to the creation of photoelectrons at the surface. The resulting attosecond streaking spectrogram yielded a delay of 110 ± 70 as for the emergence of 4f electrons relative to the conduction band electrons from the tungsten surface. More recently, attosecond streaking and coincidence techniques were combined to measure the difference in photoemission time delays between Ne and Ar, finding longer delays by up to 50 as in Ar that were attributed to resonance effects.[71]

Attosecond XUV pulses have been used to ionize a gas phase target that is then further ionized and/or dissociated by a few-cycle near-infrared pulse. Ions generated using this pulse sequence are detected by a time-of-flight spectrometer as a function of the time delay between the two pulses. This general type of experiment has been applied to systems ranging from inert gas atoms to polyatomic molecules. Electron tunneling ionization in neon was measured in real time using this technique.[72] An attosecond pulse at 91 eV ejected a 2p electron and produced Ne^+ ions in several excited states ($2p^{-2}3p$, for example) *via* shake up. The Ne^{2+} yield was measured as a function of time-delay with respect to a NIR pulse. Steps in the Ne^{2+} yield in synchronicity with the electric field cycles of the NIR pulse were observed at delays near $\Delta t = 0$, where the most intense NIR field cycles overlap the XUV pulse. This step-like structure was attributed to field-induced tunneling of the $2p^{-2}3p$ and $2p^{-2}3s$ states. The ~380 as rise time for the steps was interpreted as an upper bound for the time to populate and tunnel from these states.

In a molecular application, an experiment combining isolated attosecond pulses, NIR excitation, and ion detection was performed to follow charge localization in H_2 and D_2 upon dissociative ionization.[73] This study measured the ultrafast charge redistribution of these molecules following dissociative ionization to explore the coupling between electronic and nuclear motion on sub-femtosecond timescales. Here, an isolated attosecond XUV pulse ranging from 20 eV to 40 eV excited the molecule to doubly-excited electronic states that, left on their own, undergo dissociative ionization. The attosecond pulse was followed by a few-cycle near-infrared pulse that coupled the various neutral and ionic electronic states to one another as dissociation proceeded. The velocity and angular distributions of the resulting H^+ or D^+ photofragments were analyzed as a function of the delay between the two pulses. These experiments provide a sensitive measure of how the electron in dissociating H_2^+ or D_2^+ localizes onto a single atom *en route* to dissociation.

In a related experiment on a much more complex target, the phenylalanine molecule was ionized by an attosecond pulse covering 15–30 eV and then further ionized and fragmented by a few-cycle NIR pulse.[74] The yield of the immonium dication ($R–CH–NH_2^{2+}$) was measured as a function of pump–probe delay and found to oscillate with an average period of 4.3 fs. This time scale is considerably faster than the highest frequency X–H vibrations in the target. This difference in time scale was interpreted in terms of ultrafast hole migration to and from the NH_2 group in a superposition of excited cationic states created by the XUV pulse, an effect that had been predicted theoretically[75–77] but not previously observed.

Finally, given the composition of this Faraday conference, it is appropriate to discuss how attosecond dynamics can be determined from high harmonic imaging/spectroscopy experiments. In 2004, Itatani *et al.*[78] demonstrated that by performing HHG experiments in laser-aligned molecules, one could analyze the spectrum of the high harmonics and obtain "tomographic images" of the Dyson molecular orbitals from which electric field induced ionization occurred. This very elegant method images electrons but is not explicitly time-resolved. Nonetheless, several laboratories[79–83] have shown how one can extract attosecond dynamics from these experiments; two examples are summarized below.

In 2006, Baker *et al.*[79] performed an HHG experiment on H_2, CH_4, and their deuterated isotopologs in which an intense NIR pulse launches dynamics on an ionic potential energy surface. These dynamics evolve as the electron wavepacket, whose trajectory is controlled by the NIR laser field, recombines with the ion, producing high harmonic emission that is spectrally resolved using a diffraction grating. Ionization accesses the repulsive region on the H_2^+ potential energy curve, so the atoms separate during the electron trajectory. As a result, the higher harmonics, which are produced from longer-time trajectories (*i.e.* higher recombination energies), are depleted. To calibrate this effect in terms of molecular motion, HHG signals were compared for H_2 and D_2; the higher harmonics were indeed less depleted for D_2. Hence, nuclear dynamics are resolved on the time scale of recombination, from 0.5 to 1.6 fs for an 800 nm pulse under the reported conditions. Similar experiments on CH_4 and CD_4 probed the initial complex vibrational dynamics that occur upon ionization to the Jahn–Teller-active ground state of CH_4^+, whereupon the ion undergoes distortion from the initial tetrahedral geometry of the neutral.

More recently, high harmonic spectroscopy has been used to follow hole-migration dynamics in iodoacetylene (HCCI).[83] The HCCI molecules were impulsively aligned and then subjected to a few-cycle pulse at either 800 or 1300 nm with the target molecules aligned either perpendicular or parallel to the polarization of the newly created HHG pulse. The HHG driving pulse creates a superposition of the X and A states of $HCCI^+$, and the HHG spectrum upon recombination tracks the hole migration resulting from this superposition. In the perpendicular alignment configuration, the HHG pulse measures but does not affect the hole migration dynamics, *i.e.* the hole dynamics are "quasi-free". Analysis at the two driving wavelengths shows that the hole is initially created on the I atom and oscillates to and from the C–C π molecular orbital with a period of 1.85 fs. The results for parallel alignment are strikingly different, illustrating that the HHG driving laser causes the tunneling electron to chase the charge migration dynamics in this configuration. For example, the population amplitudes of the X and A states in the perpendicular alignment are approximately equal, whereas in the parallel configuration at 800 nm, the population of the X state is significantly depleted at a recombination time of 1 fs. The difference in population amplitudes between these alignment configurations indicate that the HHG driving laser manipulates the A–X ionic transition, affecting both the amplitude and the relative phase of the two ionic states as the hole migration evolves.

Attosecond transient absorption

One of the most powerful methods in attosecond science is attosecond transient absorption (ATA) spectroscopy.[84,85] In this section, experiments based on attosecond transient absorption and its variants are discussed, with particular emphasis on work that has been carried out at Berkeley. An example of an ATA apparatus[84] is shown in Fig. 2.

Here, an NIR pulse passes through a beamsplitter. The transmitted beam passes through optics appropriate for double optical gating and thus produces an isolated attosecond pulse in the HHG cell. The HHG output is focused by means of a toroidal mirror through the hole mirror (HM) into the sample, then dispersed by a grating onto an X-ray CCD camera. The reflected portion of the NIR beam passes through a variable delay stage and is combined with the XUV pulse *via* reflection by the hole mirror to overlap with the XUV beam at the sample. The metal filter downstream of the sample transmits the XUV beam but blocks the NIR beam. In an ATA experiment, one determines how the absorption spectrum of the broadband XUV pulse, as measured by the grating/camera combination, varies with the delay time between the XUV and NIR pulses. Remarkably, in ATA, one can obtain high resolution absorption spectra (∼10–30 meV, depending on the grating and central XUV photon energy) with no loss in temporal resolution. Hence, the broad bandwidth of the isolated attosecond pulse does not have any

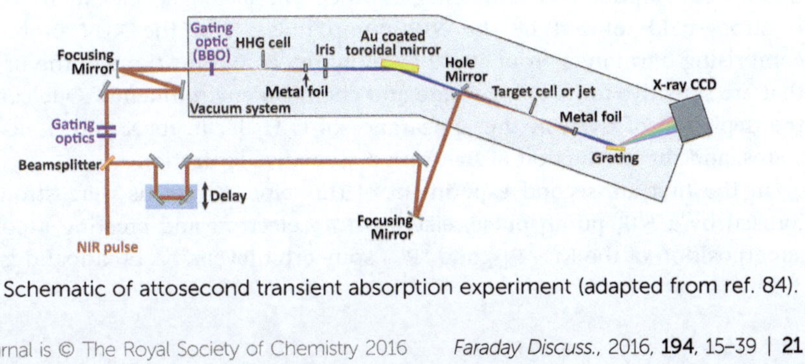

Fig. 2 Schematic of attosecond transient absorption experiment (adapted from ref. 84).

adverse consequences on the spectral resolution of the experiment. In fact, using such a pulse allows one to examine transient phenomena over a wide spectral range, resulting in a highly multiplexed experiment. These considerations make ATA an extremely powerful and versatile method in attosecond science.

ATA experiments can be performed with the NIR pulse preceding the XUV pulse, in which case the NIR and XUV pulses act as the pump and probe pulses, respectively. This is the "conventional" pulse sequence in which one uses a weak probe pulse to investigate the dynamics induced by a strong pump pulse. Additionally, a "counterintuitive" pulse sequence, in which the NIR pulse either overlaps or follows the XUV pulse has been employed in many ATA investigations. Examples of both pulse sequences are given below.

Transient absorption is generally treated in the framework of the third-order polarization in the interaction picture, where an intense pulse (usually the pump) represents a second-order interaction and the weaker pulse (usually the probe) represents a first-order interaction, all in the perturbative limit.[87,88] More generally, the intense few-cycle near-infrared pulses commonly used in attosecond measurements impart strong fields to the system and require the inclusion of higher-order response functions.[89,90] The approach to model these experiments is to calculate the single atom response,

$$\tilde{S}(\omega) = -2\ \mathrm{Im}[\tilde{d}(\omega)\tilde{E}^*(\omega)]. \tag{2}$$

This approach treats all nonlinearities in one calculation. In eqn (2), $\tilde{d}(\omega)$ is the Fourier transform of the time-dependent one-electron single-atom dipole moment, and the corresponding macroscopic polarization is $\tilde{P}(\omega) = 2\rho\tilde{d}(\omega)$, where ρ is the density of atoms in the interaction region. $\tilde{E}^*(\omega)$ is the Fourier transform of the electric field. The sign of the single-atom response determines whether light is absorbed (positive) or emitted (negative). The absorption cross section is then written as

$$\sigma(\omega) = 8\pi\alpha\omega\ \mathrm{Im}\left[\frac{\tilde{d}(\omega)}{\tilde{E}(\omega)}\right], \tag{3}$$

where α is the fine-structure constant. Propagation effects through the sample can also be accounted for as necessary.[91]

In the conventional pulse ordering for transient absorption, the near-infrared excitation serves as the pump pulse and is followed by the attosecond XUV probe pulse. The XUV spectrum is detected as a function of pump–probe delay. This pulse sequence interrogates population dynamics of the system initiated by the near-infrared pump pulse. It has been used to great effect in femtosecond XUV transient absorption measurements, in which a molecular species such as C_2H_3Br is strong-field ionized by the NIR pump pulse, and the XUV probe pulse comprising harmonics from 60–72 eV monitors core transitions in the Br atoms that are sensitive to its charge state and chemical environment.[92] One can track the depletion of C_2H_3Br, the appearance of $C_2H_3Br^+$ in its X and A electronic states, and the production of Br^+ from dissociative ionization.

In the first attosecond experiment of this type, Kr atoms were strong-field ionized by a NIR pump pulse, ejecting a 4p electron and creating a coherent superposition of the Kr^+ $^2P_{1/2}$ and $^2P_{3/2}$ spin–orbit levels.[93] A broadband isolated attosecond pulse probed the set of 4p–3d transitions near 80 eV, and the resulting absorption spectrum was measured as a function of pump–probe delay time. The transitions in this spectral region show clear oscillatory structure with a period of 6.3 fs, corresponding to the energy spacing of 0.67 eV between the two Kr^+ spin-orbit levels. The underlying theory relating the coherent superposition created by the NIR pulse to the transient absorption signal has also been developed.[94] This experiment represents a real-time observation of valence-shell electron wave-packet motion, fulfilling one of the primary goals of attosecond science.

This very general method can also be applied to solid samples.[95] ATA experiments on Si have been carried out according to the scheme shown in Fig. 3.[96] Here, attosecond pulses tuned to the silicon $L_{2,3}$ band edge near 99 eV are used to follow the dynamics of electrons injected into the conduction band of silicon by few-cycle NIR pulses. The attosecond and NIR pulse durations are less than 100 as and 4 fs, respectively, as characterized by attosecond streaking, and the sample thickness is about 140–250 nm to ensure a reasonable optical density at the L-edge and to minimize group velocity mismatch between the NIR and XUV pulses.

In order to observe more clearly time-dependent effects, it is useful to differentiate the observed ATA spectra, turning band edges into peaks, as shown in Fig. 4. Clearly, many of the peaks in the differentiated spectrum shift and broaden in the presence of the NIR pulse. More detailed analysis shows that some of these effects persist for many hundreds of fs, while others are transient and disappear once the NIR pulse is gone.

The most remarkable result is shown in Fig. 5 (left), where the ATA spectrum at 100.35 eV exhibits a resolved step structure at intervals of 1.3 fs, which is half the period of the NIR driving pulse. This step structure signifies that the electric field, not the pulse envelope, of the NIR pulse is driving population from the valence to conduction band, and that this population transfer occurs *via* a tunneling mechanism, similar to the strong-field induced tunneling to the continuum that forms the basis of HHG in the gas phase.[21] Fig. 5 (right) presents an intuitive view of this process;[97] the valence and conduction bands are distorted by the NIR

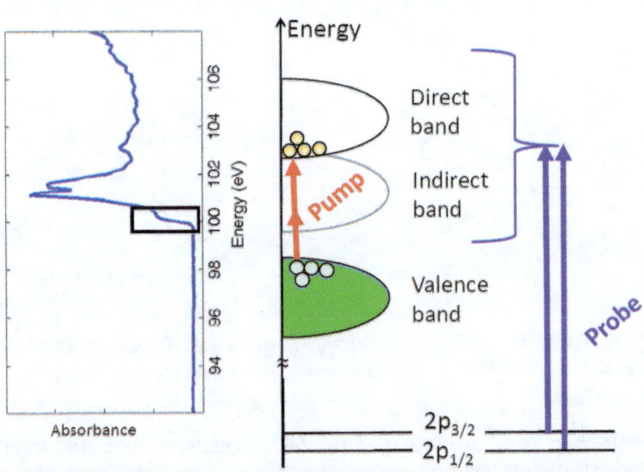

Fig. 3 Principle of ATA experiment in solid Si.

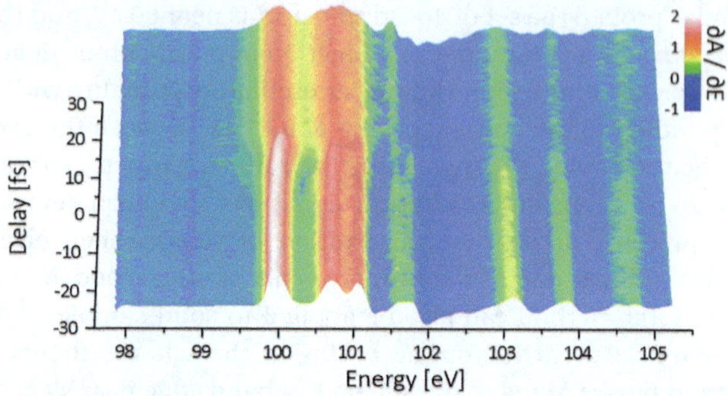

Fig. 4 Derivative of Si ATA spectrum (adapted from ref. 96).

electric field, enabling valence → conduction band tunneling near the maxima of the electric field every half period.

A series of gas phase experiments at Berkeley[98] and elsewhere[43,44,99–102] have made use of a counterintuitive pulse sequence, in which the XUV pulse overlaps or precedes the NIR pulse. These experiments show that the NIR pulse affects the measured XUV absorption spectrum at delays that are considerably longer than attosecond pulse duration. This effect can be understood with reference to Fig. 6.[84] If a single transition at energy $E = \hbar\omega_{\text{XUV}}$ is excited by the XUV pulse, the polarization oscillates with frequency ω_{XUV} and exhibits free induction decay (FID) reflecting the lifetime of the excited state. A NIR pulse applied prior to or during the FID can affect the polarization by, for example, ionizing the excited state or coupling it to a manifold of excited states. Taking the Fourier transforms of the polarization signals in Fig. 6 and inserting them into eqn (2) yields the absorption spectra in the right panels of the figure, which clearly show the effect of the NIR pulse. One application of this pulse sequence is the direct

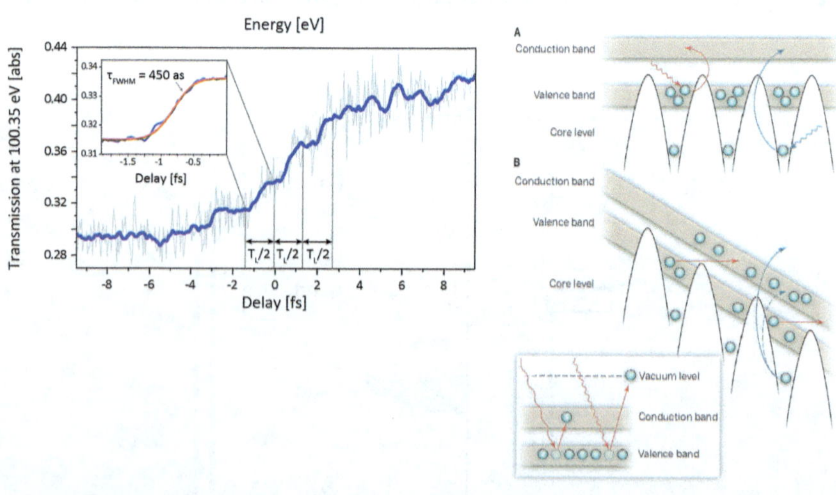

Fig. 5 Sub-cycle step structure in experimental Si spectrum (left) and interpretation in terms of tunneling between valence and conduction bands (right). From references 96 and 97. Reprinted with permission from AAAS.

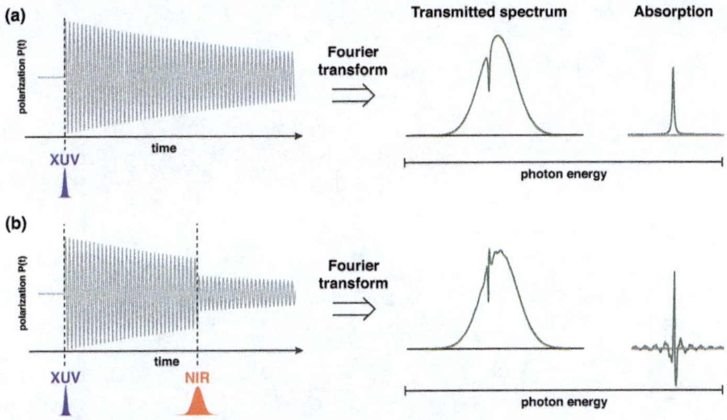

Fig. 6 Principle of ATA experiment with "counterintuitive" pulse sequence. Reprinted from ref. 84 with permission from Elsevier.

measurement of autoionization lifetimes, as demonstrated for Ar and Xe atoms.[103–105] In one example, illustrated in Fig. 7,[104] an attosecond pulse covering 21–25 eV excites a series of $5s5p^6np$ autoionizing states of Xe. The sharp window resonances corresponding to these transitions are clearly perturbed at short positive delay times, where the NIR pulse arrives after the XUV pulse. Lineouts at the center frequencies of the $n = 6$ and 7 resonances, shown in the right-most panel, exhibit depletion at early times and recovery on time scales of 43.8 and 96.8 fs, respectively. These lifetimes reflect decay times of the excited state wavefunctions owing to autoionization, which are readily shown to be twice the excited state population decay times. Hence, the autoionization lifetimes of the $n = 6$ and 7 states are found to be 21.9 and 48.4 fs, in good agreement with those obtained by linewidth analysis of frequency-domain measurements.

We have also investigated the ATA of Rydberg states of Ar below its ionization potential with a similar pulse sequence.[106] The results are shown in Fig. 8. The XUV pulse promotes a 3p valence electron into the Rydberg series $3s^23p^6 \rightarrow 3s^23p^5[^2P_{1/2,3/2}]nd(s)$, populating two groups of ns/nd manifolds that converge to different spin–orbit coupled ion core limits, $I_{P[3/2]} = 15.76$ eV and $I_{P[1/2]} = 15.94$ eV. The valence electronic wavepacket initiated by the XUV pulse

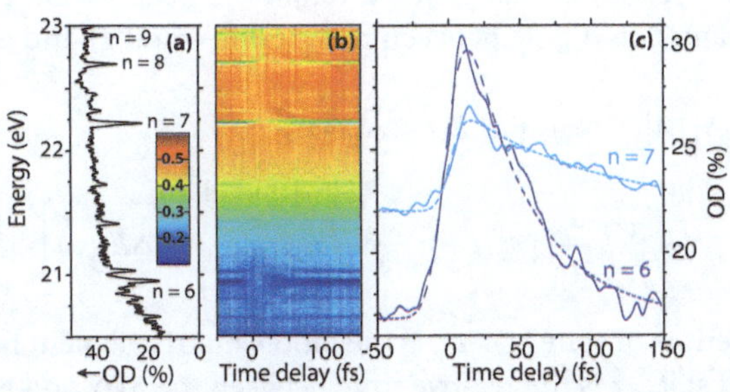

Fig. 7 XUV absorption of Xe (left), ATA spectrum (center), and line-outs showing decay of ATA signals for $n = 6$ and 7 Rydberg states. Reprinted from ref. 104 with permission of the American Physical Society.

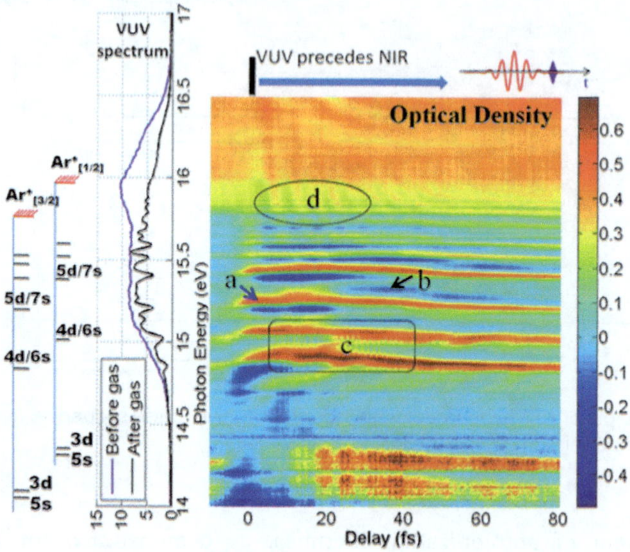

Fig. 8 XUV spectrum of Ar (left) and ATA spectrum (right) from 14–16.5 eV. Reprinted from ref. 106.

experiences an electromagnetic interaction upon the arrival of the NIR pulse. There are four distinct effects attributable to the NIR pulse, each of which is labelled in Fig. 8: (a) broadening and shifting of individual absorption lines at small positive delays, (b) horizontal sideband structures between two adjacent absorption resonances at large positive delays, (c) fast modulations with a period of 1.3 fs on the two 4d states lying around 15 eV, and (d) slow modulations with a period of 5–10 fs on the states approaching the ionization limits (~15.75 eV).

Further insight into the oscillatory structures is gained by Fourier analysis of the experimental spectra as shown in Fig. 9, where one can see the relatively slow oscillations near 15.75 eV and the much faster oscillations between 14.75 and 15 eV. The latter are termed "sub-cycle" oscillations since their characteristic period of 1.3 fs is approximately half the period of the NIR field. Likewise, the corresponding energy of 3.2 eV indicated in Fig. 9 corresponds to two NIR photons.

Features (a)–(d) can all be interpreted by means of a simplified many-level nonperturbative treatment that yields the following expression for the $d_1(t)$, the time varying transition dipole between the ground state $|g\rangle$ and excited state $|1\rangle$:

$$d_1(t) \propto e^{-t/t_1} {\mu_{1g}}^2 \Bigg[\left|A_1^1(t-\tau)\right| \sin\left(E_1 t + \varphi_{A_1^1}\right) + \sum_{n,n\neq 1} \frac{\mu_{ng}}{\mu_{1g}} \left|A_n^1(t-\tau)\right| \sin\left(E_1 t + \varphi_{A_n^1} + \Delta E_{n1}\tau\right) \Bigg] \quad (4)$$

here t_1 is the lifetime of state $|1\rangle$, μ_{ng} is the dipole matrix element between state $|n\rangle$ and the ground state, τ is the relative time between the XUV and NIR pulses, E_n is the energy of state n, and $\Delta E_{n1} = E_n - E_1$. Atomic units are used throughout. The first term represents NIR laser induced transitions where the initial and final

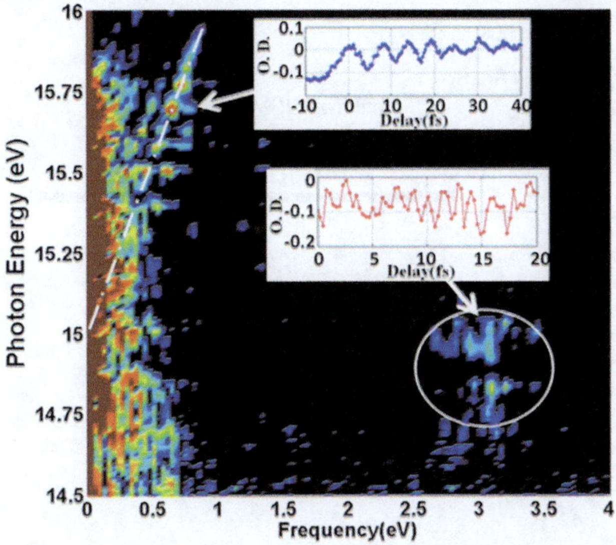

Fig. 9 Fourier analysis of Ar ATA spectrum. Insets show Fourier transform of signals labeled d (upper) and c (lower) in Fig. 8. Reprinted from ref. 106.

states are the same, with $|A_1^1(t-\tau)|$ expressing the laser induced depopulation of state 1 by processes such as ionization and $\varphi_{A_n^1}$ the phase corresponding to any energy shift of the state, such as would be caused by the AC Stark effect. The second term describes situations where the final and initial states do not coincide, *i.e.* population transfer processes between states $|1\rangle$ and$|n\rangle$ induced by the NIR pulse. In this case, $|A_n^1(t-\tau)|$ is the magnitude of mutual population transfer and $\Delta E_{n1}\tau$ is the relative phase accumulated by state n during the different time evolution of the two coherently populated states.

Simulations of the spectra using eqn (4) show that features (a) and (b) are due to the AC Stark shift, as they can be reproduced using only the term proportional to $|A_1^1(t-\tau)|$. Feature (a) is a consequence of the impulsive phase shift induced by the short NIR pulse.[102,107] Feature (b) stems from the interplay between the hyperbolic lines (perturbed free induction decays) originating from two adjacent resonance states[108] and can be used to quantify the AC Stark shift. Features (c) and (d) are due to population transfer involving ladder-type[43,100,109] and vee-type[108] transition schemes, respectively, as shown in Fig. 10. Essentially, these oscillations reflect interference between two excitation pathways to the same final state: absorption of a single XUV photon, and a three-photon pathway involving absorption of on XUV photon and the interaction with two NIR photons. In ladder-type transitions, the oscillation energy in the Fourier analysis corresponds to the sum of the two NIR photon energies, while in vee-type transitions it corresponds to the difference. The broadband nature of the XUV and NIR pulses make it possible to satisfy the resonance conditions for both types of transitions.

As shown in Fig. 10, the vee and ladder transitions involve the dark Ar[$^2P_{1/2,3/2}$] 4p Rydberg level as the intermediate state. Moreover, in the Fourier analysis, one can draw a straight line of unity slope through a series of vee-type transitions; the y-intercept yields the energy of one of the states (15 eV, for the white dashed line

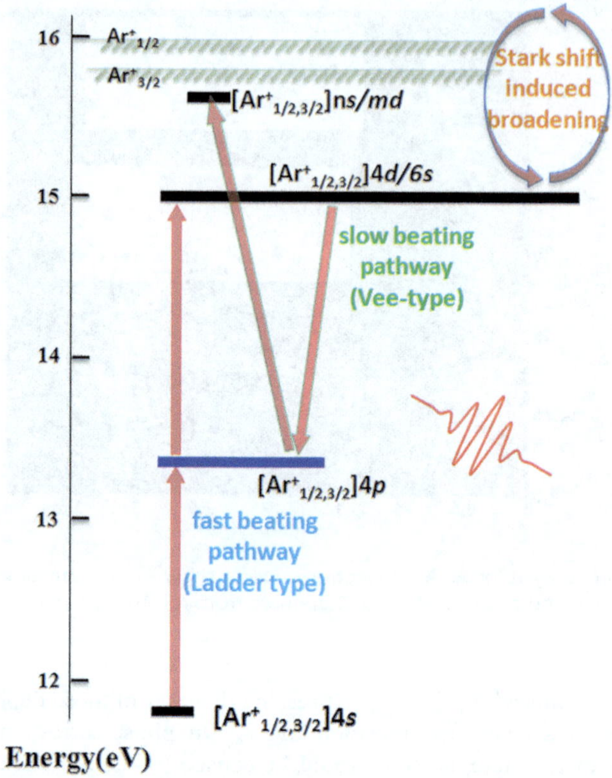

Fig. 10 Origin of interferences between one- and three-photon processes in Ar ATA spectrum.

in Fig. 9) that is coupled *via* the dark state to all the others lying on this line. By the same token, a line of unity slope drawn through the points around 3.2 eV (not shown in Fig. 9) yields the energy of the lower lying state coupled to each upper state by a ladder transition.

Related experiments have recently been reported for electronic states of N_2 near its ionization potential at 15.559 eV.[86,110] The ATA spectrum of N_2 is complicated by Rydberg series and vibrational progressions lying within this spectral region. Nonetheless, many of the electronic effects seen in rare gas atoms can be discerned in N_2, in addition to vibrational coherences that are also produced by interference between one- and three-photon processes. Fig. 11 shows results from our laboratory in the region from 12.5–16.7 eV.[86] The static spectrum in the left panel shows that transitions to the valence b and b′ states of N_2 dominate below 15 eV,[111,112] while the upper energy range is dominated by transitions to Rydberg states converging to the A and B states of N_2^+.[113,114]

The density of transitions is clearly higher than in Ar, as expected, but one can still see broadening and shifting of lines, along with sub-cycle and slow oscillations. For example, Fourier analysis of the sub-cycle oscillations shows clusters of states around 15.8 and 13 eV that are coupled by ladder transitions. Schema of the actual electronic states involved are shown in Fig. 12; we believe that the lower

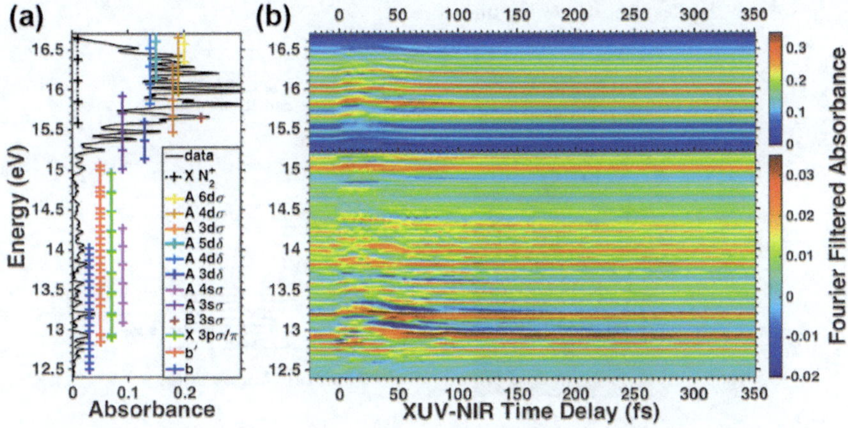

Fig. 11 Static XUV spectrum of N_2 with assignments (left) and ATA spectrum (right). Reprinted from ref. 86 with permission of the American Chemical Society.

and upper states are the b valence state of N_2 and a Rydberg state based on the N_2^+ A state core, with the dark state being another Rydberg state.

Examination of Fig. 11 around 13 eV shows relatively slow oscillations persisting to 350 fs, the longest delays probed in this experiment. Fig. 13 shows Fourier analysis of this spectral region from 12.4–13.3 eV, starting at a delay time of 100 fs. This procedure excludes the fast dynamics caused by electronic coupling near $t = 0$, and allows us to isolate effects associated with vibrational motion. We indeed observe progressions of evenly spaced peaks in the vertical direction that correspond to the $\nu = 1$–8 levels of the N_2 b state;[111] the peak spacing of 0.081 eV corresponds to a period of 51 fs. This value matches the vibrational period of the b state and the oscillations are thus attributed to vibrational levels of the b state

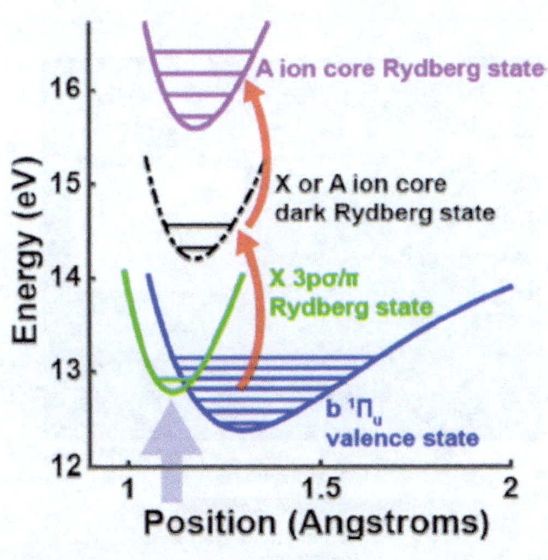

Fig. 12 Schema of electronic states responsible for sub-cycle oscillations in N_2 ATA spectrum near 16 eV. Reprinted from ref. 86 with permission of the American Chemical Society.

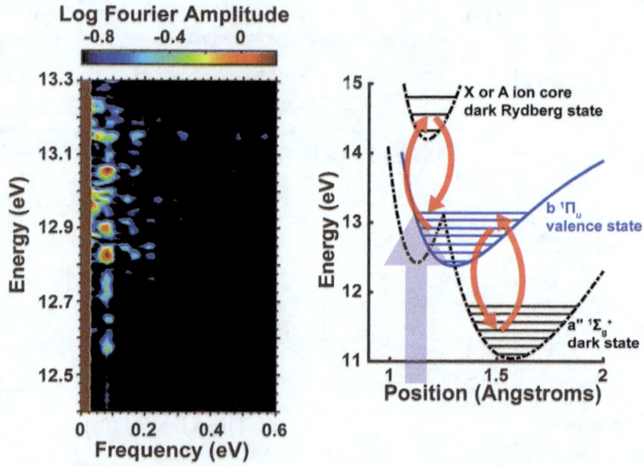

Fig. 13 Fourier analysis of N_2 ATA spectrum from 12.5–14.5 eV, starting at 100 fs in order to emphasize vibrational coherences. Right panel shows possible coupling schemes leading to vibrational coherences. Reprinted from ref. 86 with permission of the American Chemical Society.

separated by $\Delta\nu = \pm1$ that are coupled by the near IR field. Couplings involving larger values of $\Delta\nu$ can also be seen at some energies. For example, the $\nu = 7$ level at 13.26 eV shows peaks indicating couplings to vibrational levels of the b state down to $\nu = 1$. Overall, these vibrational coherences are attributed to vee-type transitions involving either a higher or lower electronic state as illustrated schematically in Fig. 13.

We close this section by considering two experiments performed on Ne. Fig. 14 shows a section of the transient absorption spectrum of Ne using an isolated attosecond pulse.[108] The two main features in the ATA spectrum at 20.04 eV and 20.14 eV are from transitions to the spin–orbit split ($^2P_{3/2}$)3d and ($^2P_{1/2}$)3d′ Rydberg states of Ne. Clear oscillatory structures are seen with a period of 40 fs, which

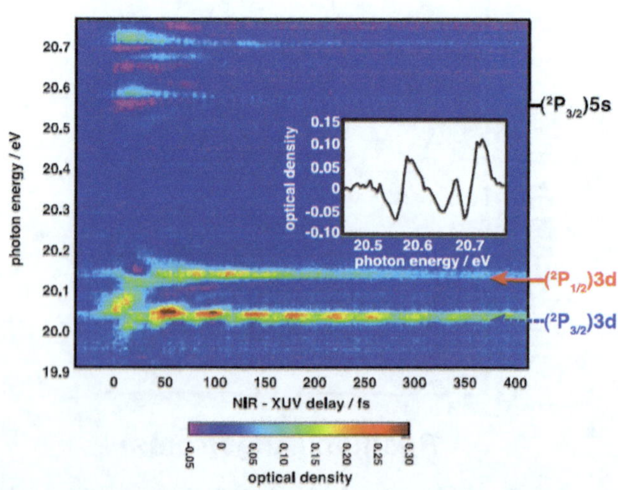

Fig. 14 ATA spectrum of Ne from 19.9–20.8 eV. Reprinted from ref. 108.

corresponds to the spin–orbit splitting of the two Rydberg states. Hence, this beating is attributed to vee-type coupling through an intermediate state, similar to that seen in Ar and N_2.

Recently, we performed a similar experiment on Ne using an attosecond pulse train comprising discrete harmonics in which the 13th harmonic around 20 eV was blocked by an indium filter.[115] Fig. 15 shows the dispersed XUV signal with XUV only (black line) and XUV + NIR (red and blue lines). Remarkably, we observe sharp, spatially coherent emission features around 20 eV in the presence of the two pulses. This result can be understood in terms of resonant four-wave mixing (FWM) with reference to Fig. 15. Here, the 11th harmonic at 16.5 eV is resonant with the transition from the ground state to the spin–orbit split ($^2P_{3/2}$, $^2P_{1/2}$)3s state (the 3s and 3s′ states). The NIR pulse resonantly couples these upper states to the higher lying 4s/4s′ and 3d/3d′ levels *via* the intermediate Ne 3p state. The combination of one XUV and two NIR photons induces emission back to the ground state (ω_4 in Fig. 15), resulting in a series of sharp peaks around 20 eV.

These features appear at the same energies as in the ATA spectrum of Ne, but their time-dependence, as shown in Fig. 16, is distinctly different. The oscillation period is 25 fs, corresponding to the energy splitting of 0.17 eV between the 3s and 3s′ states. In this experiment, we are thus observing quantum beating between the coherent superposition of states formed by the XUV pulse, rather than interferences between the 3d and 3d′ final state by vee-type coupling.

One way to understand the different time-dependences is *via* eqn (5). In this regime, the polarizability $P(\omega)$ is given by

$$P(\omega) = \varepsilon_0\chi^{(1)}E_{XUV}(\omega) + \varepsilon_0\chi^{(3)}E_{HH}(\omega_1)E_{NIR}(\omega_2)E_{XUV}(\omega_3) \quad (5)$$

The first order term represents linear absorption at ω, and the second term is the third-order contribution to the polarizability. When broadband XUV (and NIR) pulses are used, the frequency condition with $\omega = \omega_1 \pm \omega_2 \pm \omega_3$ is easily satisfied, leading to interference between the first and third order terms; this interference is the origin of much of the time-dependent structure. In contrast, under the experimental conditions of Fig. 15, the first order contribution to the signal around 20 eV is largely absent and the third order term dominates;

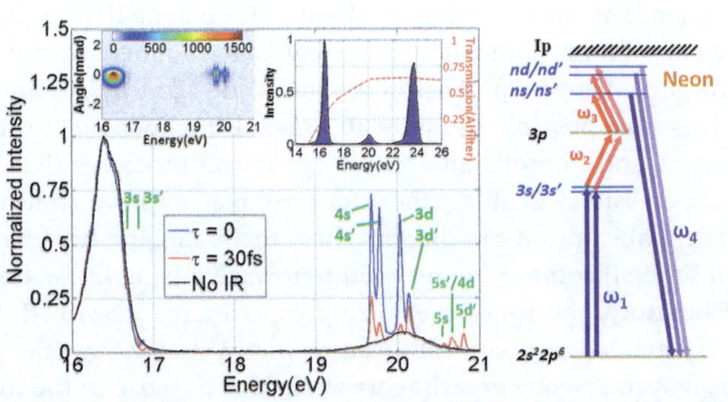

Fig. 15 Left panel shows coherent emission from Ne around 20 eV when pumped by 11th harmonic and two NIR photons. Right panel shows underlying four-wave mixing scheme. Reprinted from ref. 115 with permission of the American Physical Society.

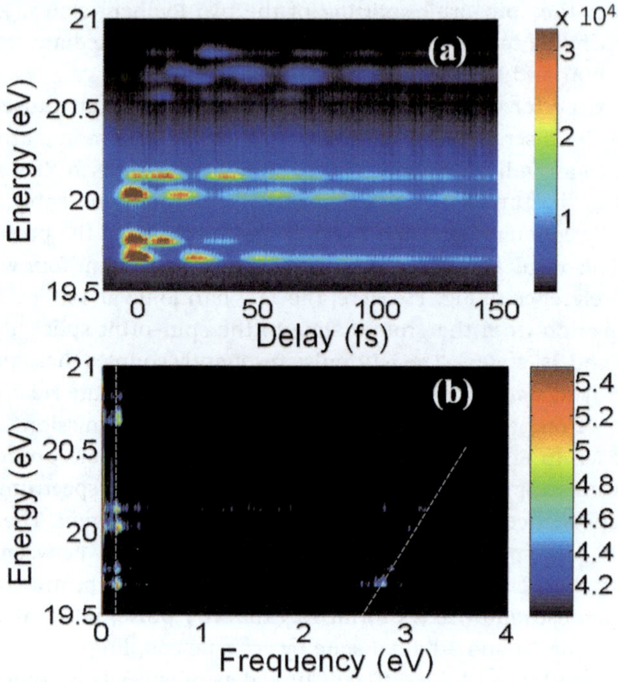

Fig. 16 Time-dependent coherent emission signal from Ne (a) and Fourier analysis of this signal (b). Reprinted from ref. 115 with permission of the American Physical Society.

a description in the language of FWM is more appropriate. We also note that the Ne signals in the FWM configuration (and their Fourier transforms, as shown in the bottom panel of Fig. 16) are considerably simpler than in the ATA experiment. These experiments hint at the potential for carrying out time-resolved non-linear experiments in the XUV, a topic that has received considerable theoretical consideration[116,117] but little experimental attention up to now.

Outlook and conclusions

Attosecond science is an emerging field with the potential to address fundamental problems of interest in atomic, molecular, and condensed matter physics. While technical challenges in generating attosecond light pulses are still being addressed, it is now possible to apply attosecond methodology to increasingly complex systems and to probe an array of dynamical processes that have previously been inaccessible. In this paper, for example, we have shown how attosecond transient absorption can discern phenomena ranging from field-induced tunneling in solid silicon to time-resolved four-wave mixing in Ne atoms.

In our laboratory, the four-wave mixing experiments described above have recently been extended to a non-collinear geometry,[118] allowing one to carry out XUV nonlinear spectroscopy experiments analogous to those in the infrared and visible regions of the spectrum but with sub-fs temporal resolution. The non-adiabatic dynamics of dissociating polyatomic molecules are being investigated with attosecond transient absorption. The broadband probe pulse inherent to these experiments enables one to follow the chemical shift of a core-electron absorption (*i.e.* the Br atom in CH_3Br) as a molecule dissociates. ATA experiments on VO_2 are providing new information on the earliest stages of the photoinduced insulator–metal transition. We have also carried out attosecond transient reflection experiments on solid Ge samples which are highly complementary to ATA experiments on the same system.

Future advances in attosecond science will involve a combination of technique development and new applications. As the upper limit of accessible photon energies is raised, it becomes possible to follow element-specific dynamics with attosecond time resolution. For example, there is much interest (and some progress) in extending the energy range of isolated attosecond pulses beyond 300 eV,[33,119] so that one can follow carbon K-edge chemical shifts in real time either by photoemission or transient absorption. Such a development will enable the study of carbon-containing molecules with attosecond methods and will represent a major extension of attosecond science into chemistry. Advanced optical frequency synthesis methods now allow the generation of nearly single-cycle pulses in the visible and ultraviolet region,[17,120] enabling vis-UV/XUV pump–probe experiments with substantially improved resolution compared to those with few-cycle pump pulses. Accelerator-based light sources will soon be producing attosecond light pulses with significantly higher fluxes and photon energies than laser-based systems, opening up many new directions in attosecond science.

Acknowledgements

This research has been supported by the Director, Office of Science, Office of Basic Energy Sciences, of the US Department of Energy under contract DE-AC02-05CH11231, the Wm. H. Keck Foundation, the Defense Advanced Research Projects Agency PULSE program through grant W31P4Q-13-1-0017, and the Multidisciplinary University Research Initiatives from the Army Research Office (WN911NF-14-1-0383) and the Air Force Office of Scientific Research (FA9550-15-1-0037). S. R. L. acknowledges support from a National Security Science and Engineering Faculty Fellowship (NSSEFF) and from the National Science Foundation under grant CHE-1361226.

References

1 A. H. Zewail, *J. Phys. Chem. A*, 2000, **104**, 5660–5694.
2 P. B. Corkum and F. Krausz, *Nat. Phys.*, 2007, **3**, 381–387.
3 S. R. Leone, C. W. McCurdy, J. Burgdoerfer, L. S. Cederbaum, Z. Chang, N. Dudovich, J. Feist, C. H. Greene, M. Ivanov, R. Kienberger, U. Keller, M. F. Kling, Z. H. Loh, T. Pfeifer, A. N. Pfeiffer, R. Santra, K. Schafer, A. Stolow, U. Thumm and M. J. J. Vrakking, *Nat. Photonics*, 2014, **8**, 162–166.
4 K. Zhao, Q. Zhang, M. Chini, Y. Wu, X. W. Wang and Z. H. Chang, *Opt. Lett.*, 2012, **37**, 3891–3893.
5 P. Agostini and L. F. DiMauro, *Rep. Prog. Phys.*, 2004, **67**, 813–855.
6 M. F. Kling and M. J. J. Vrakking, *Annu. Rev. Phys. Chem.*, 2008, **59**, 463–492.
7 T. Pfeifer, M. J. Abel, P. M. Nagel, A. Jullien, Z. H. Loh, M. J. Bell, D. M. Neumark and S. R. Leone, *Chem. Phys. Lett.*, 2008, **463**, 11–24.
8 F. Krausz and M. Ivanov, *Rev. Mod. Phys.*, 2009, **81**, 163–234.

9 F. Lepine, M. Y. Ivanov and M. J. J. Vrakking, *Nat. Photonics*, 2014, **8**, 195–204.
10 L. Gallmann, C. Cirelli and U. Keller, *Annu. Rev. Phys. Chem.*, 2012, **63**, 447–469.
11 F. Frank, C. Arrell, T. Witting, W. A. Okell, J. McKenna, J. S. Robinson, C. A. Haworth, D. Austin, H. Teng, I. A. Walmsley, J. P. Marangos and J. W. G. Tisch, *Rev. Sci. Instrum.*, 2012, **83**, 071101.
12 M. Chini, K. Zhao and Z. H. Chang, *Nat. Photonics*, 2014, **8**, 178–186.
13 F. Krausz and M. I. Stockman, *Nat. Photonics*, 2014, **8**, 205–213.
14 K. T. Kim, D. M. Villeneuve and P. B. Corkum, *Nat. Photonics*, 2014, **8**, 187–194.
15 F. Calegari, G. Sansone, S. Stagira, C. Vozzi and M. Nisoli, *J. Phys. B: At., Mol. Opt. Phys.*, 2016, **49**, 142001.
16 K. Ramasesha, S. R. Leone and D. M. Neumark, *Annu. Rev. Phys. Chem.*, 2016, **67**, 41–63.
17 M. T. Hassan, T. T. Luu, A. Moulet, O. Raskazovskaya, P. Zhokhov, M. Garg, N. Karpowicz, A. M. Zheltikov, V. Pervak, F. Krausz and E. Goulielmakis, *Nature*, 2016, **530**, 66–70.
18 A. McPherson, G. Gibson, H. Jara, U. Johann, T. S. Luk, I. A. McIntyre, K. Boyer and C. K. Rhodes, *J. Opt. Soc. Am. B*, 1987, **4**, 595–601.
19 M. Ferray, A. L. Huillier, X. F. Li, L. A. Lompre, G. Mainfray and C. Manus, *J. Phys. B: At., Mol. Opt. Phys.*, 1988, **21**, L31.
20 J. L. Krause, K. J. Schafer and K. C. Kulander, *Phys. Rev. Lett.*, 1992, **68**, 3535–3538.
21 P. B. Corkum, *Phys. Rev. Lett.*, 1993, **71**, 1994–1997.
22 M. Lewenstein, P. Balcou, M. Y. Ivanov, A. Lhuillier and P. B. Corkum, *Phys. Rev. A*, 1994, **49**, 2117–2132.
23 A. Rundquist, C. G. Durfee, Z. H. Chang, C. Herne, S. Backus, M. M. Murnane and H. C. Kapteyn, *Science*, 1998, **280**, 1412–1415.
24 M. Bellini, C. Lynga, A. Tozzi, M. B. Gaarde, T. W. Hansch, A. L'Huillier and C. G. Wahlstrom, *Phys. Rev. Lett.*, 1998, **81**, 297–300.
25 I. P. Christov, M. M. Murnane and H. C. Kapteyn, *Phys. Rev. Lett.*, 1997, **78**, 1251–1254.
26 J. Tate, T. Auguste, H. G. Muller, P. Salieres, P. Agostini and L. F. DiMauro, *Phys. Rev. Lett.*, 2007, **98**, 013901.
27 E. J. Takahashi, T. Kanai, K. L. Ishikawa, Y. Nabekawa and K. Midorikawa, *Phys. Rev. Lett.*, 2008, **101**, 253901.
28 P. Colosimo, G. Doumy, C. I. Blaga, J. Wheeler, C. Hauri, F. Catoire, J. Tate, R. Chirla, A. M. March, G. G. Paulus, H. G. Muller, P. Agostini and L. F. Dimauro, *Nat. Phys.*, 2008, **4**, 386–389.
29 G. Doumy, J. Wheeler, C. Roedig, R. Chirla, P. Agostini and L. F. DiMauro, *Phys. Rev. Lett.*, 2009, **102**, 093002.
30 B. E. Schmidt, A. D. Shiner, M. Giguere, P. Lassonde, C. A. Trallero-Herrero, J. C. Kieffer, P. B. Corkum, D. M. Villeneuve and F. Legare, *J. Phys. B: At., Mol. Opt. Phys.*, 2012, **45**, 074008.
31 T. Popmintchev, M. C. Chen, D. Popmintchev, P. Arpin, S. Brown, S. Alisauskas, G. Andriukaitis, T. Balciunas, O. D. Mucke, A. Pugzlys, A. Baltuska, B. Shim, S. E. Schrauth, A. Gaeta, C. Hernandez-Garcia, L. Plaja, A. Becker, A. Jaron-Becker, M. M. Murnane and H. C. Kapteyn, *Science*, 2012, **336**, 1287–1291.

32 M. C. Chen, C. Mancuso, C. Hernandez-Garcia, F. Dollar, B. Galloway, D. Popmintchev, P. C. Huang, B. Walker, L. Plaja, A. A. Jaron-Becker, A. Becker, M. M. Murnane, H. C. Kapteyn and T. Popmintchev, *Proc. Natl. Acad. Sci. U. S. A.*, 2014, **111**, E2361–E2367.
33 F. Silva, S. M. Teichmann, S. L. Cousin, M. Hemmer and J. Biegert, *Nat. Commun.*, 2015, **6**, 6611.
34 C. Spielmann, N. H. Burnett, S. Sartania, R. Koppitsch, M. Schnürer, C. Kan, M. Lenzner, P. Wobrauschek and F. Krausz, *Science*, 1997, **278**, 661–664.
35 D. Popmintchev, C. Hernández-García, F. Dollar, C. Mancuso, J. A. Pérez-Hernández, M.-C. Chen, A. Hankla, X. Gao, B. Shim, A. L. Gaeta, M. Tarazkar, D. A. Romanov, R. J. Levis, J. A. Gaffney, M. Foord, S. B. Libby, A. Jaron-Becker, A. Becker, L. Plaja, M. M. Murnane, H. C. Kapteyn and T. Popmintchev, *Science*, 2015, **350**, 1225–1231.
36 N. A. Papadogiannis, B. Witzel, C. Kalpouzos and D. Charalambidis, *Phys. Rev. Lett.*, 1999, **83**, 4289–4292.
37 P. M. Paul, E. S. Toma, P. Breger, G. Mullot, F. Auge, P. Balcou, H. G. Muller and P. Agostini, *Science*, 2001, **292**, 1689–1692.
38 R. López-Martens, K. Varju, P. Johnsson, J. Mauritsson, Y. Mairesse, P. Salieres, M. B. Gaarde, K. J. Schafer, A. Persson, S. Svanberg, C. G. Wahlstrom and A. L'Huillier, *Phys. Rev. Lett.*, 2005, **94**, 033001.
39 Y. Nabekawa, T. Shimizu, T. Okino, K. Furusawa, H. Hasegawa, K. Yamanouchi and K. Midorikawa, *Phys. Rev. Lett.*, 2006, **97**, 153904.
40 T. Remetter, P. Johnsson, J. Mauritsson, K. Varju, Y. Ni, F. Lepine, E. Gustafsson, M. Kling, J. Khan, R. Lopez-Martens, K. J. Schafer, M. J. J. Vrakking and A. L'Huillier, *Nat. Phys.*, 2006, **2**, 323–326.
41 P. Johnsson, J. Mauritsson, T. Remetter, A. L'Huillier and K. J. Schafer, *Phys. Rev. Lett.*, 2007, **99**, 233001.
42 M. Holler, F. Schapper, L. Gallmann and U. Keller, *Phys. Rev. Lett.*, 2011, **106**, 123601.
43 M. Lucchini, J. Herrmann, A. Ludwig, R. Locher, M. Sabbar, L. Gallmann and U. Keller, *New J. Phys.*, 2013, **15**, 103010.
44 C. Ott, A. Kaldun, L. Argenti, P. Raith, K. Meyer, M. Laux, Y. Z. Zhang, A. Blattermann, S. Hagstotz, T. Ding, R. Heck, J. Madronero, F. Martin and T. Pfeifer, *Nature*, 2014, **516**, 374–381.
45 M. Hentschel, R. Kienberger, C. Spielmann, G. A. Reider, N. Milosevic, T. Brabec, P. Corkum, U. Heinzmann, M. Drescher and F. Krausz, *Nature*, 2001, **414**, 509–513.
46 E. Goulielmakis, M. Schultze, M. Hofstetter, V. S. Yakovlev, J. Gagnon, M. Uiberacker, A. L. Aquila, E. M. Gullikson, D. T. Attwood, R. Kienberger, F. Krausz and U. Kleineberg, *Science*, 2008, **320**, 1614–1617.
47 D. J. Jones, S. A. Diddams, J. K. Ranka, A. Stentz, R. S. Windeler, J. L. Hall and S. T. Cundiff, *Science*, 2000, **288**, 635–639.
48 M. Schultze, E. M. Bothschafter, A. Sommer, S. Holzner, W. Schweinberger, M. Fiess, M. Hofstetter, R. Kienberger, V. Apalkov, V. S. Yakovlev, M. I. Stockman and F. Krausz, *Nature*, 2013, **493**, 75–78.
49 R. Kienberger, E. Goulielmakis, M. Uiberacker, A. Baltuska, V. Yakovlev, F. Bammer, A. Scrinzi, T. Westerwalbesloh, U. Kleineberg, U. Heinzmann, M. Drescher and F. Krausz, *Nature*, 2004, **427**, 817–821.

50 J. Itatani, F. Quere, G. L. Yudin, M. Y. Ivanov, F. Krausz and P. B. Corkum, *Phys. Rev. Lett.*, 2002, **88**, 173903.
51 H. Timmers, private communication.
52 M. Kovacev, Y. Mairesse, E. Priori, H. Merdji, O. Tcherbakoff, P. Monchicourt, P. Breger, E. Mevel, E. Constant, P. Salieres, B. Carre and P. Agostini, *Eur. Phys. J. D*, 2003, **26**, 79–82.
53 G. Sansone, E. Benedetti, F. Calegari, C. Vozzi, L. Avaldi, R. Flammini, L. Poletto, P. Villoresi, C. Altucci, R. Velotta, S. Stagira, S. De Silvestri and M. Nisoli, *Science*, 2006, **314**, 443–446.
54 Z. H. Chang, *Phys. Rev. A*, 2007, **76**, 051403.
55 H. Mashiko, S. Gilbertson, C. Q. Li, S. D. Khan, M. M. Shakya, E. Moon and Z. H. Chang, *Phys. Rev. Lett.*, 2008, **100**, 103906.
56 X. M. Feng, S. Gilbertson, H. Mashiko, H. Wang, S. D. Khan, M. Chini, Y. Wu, K. Zhao and Z. H. Chang, *Phys. Rev. Lett.*, 2009, **103**, 183901.
57 T. Pfeifer, A. Jullien, M. J. Abel, P. M. Nagel, L. Gallmann, D. M. Neumark and S. R. Leone, *Opt. Express*, 2007, **15**, 17120–17128.
58 A. Jullien, T. Pfeifer, M. J. Abel, P. M. Nagel, M. J. Bell, D. M. Neumark and S. R. Leone, *Appl. Phys. B*, 2008, **93**, 433–442.
59 M. J. Abel, T. Pfeifer, P. M. Nagel, W. Boutu, M. J. Bell, C. P. Steiner, D. M. Neumark and S. R. Leone, *Chem. Phys.*, 2009, **366**, 9–14.
60 C. A. Haworth, L. E. Chipperfield, J. S. Robinson, P. L. Knight, J. P. Marangos and J. W. G. Tisch, *Nat. Phys.*, 2007, **3**, 52–57.
61 K. T. Kim, C. Zhang, T. Ruchon, J.-F. Hergott, T. Auguste, D. M. Villeneuve, P. B. Corkum and F. Quere, *Nat. Photonics*, 2013, **7**, 651–656.
62 H. Timmers, M. Sabbar, J. Hellwagner, Y. Kobayashi, D. M. Neumark and S. R. Leone, *Optica*, 2016, **3**, 707–710.
63 E. J. Takahashi, P. F. Lan, O. D. Mucke, Y. Nabekawa and K. Midorikawa, *Nat. Commun.*, 2013, **4**, 2691.
64 E. Cunningham, Y. Wu and Z. Chang, *Appl. Phys. Lett.*, 2015, **107**, 201108.
65 P. Tzallas, E. Skantzakis, L. A. A. Nikolopoulos, G. D. Tsakiris and D. Charalambidis, *Nat. Phys.*, 2011, **7**, 781–784.
66 M. Drescher, M. Hentschel, R. Kienberger, M. Uiberacker, V. Yakovlev, A. Scrinzi, T. Westerwalbesloh, U. Kleineberg, U. Heinzmann and F. Krausz, *Nature*, 2002, **419**, 803–807.
67 M. Schultze, M. Fiess, N. Karpowicz, J. Gagnon, M. Korbman, M. Hofstetter, S. Neppl, A. L. Cavalieri, Y. Komninos, T. Mercouris, C. A. Nicolaides, R. Pazourek, S. Nagele, J. Feist, J. Burgdorfer, A. M. Azzeer, R. Ernstorfer, R. Kienberger, U. Kleineberg, E. Goulielmakis, F. Krausz and V. S. Yakovlev, *Science*, 2010, **328**, 1658–1662.
68 L. R. Moore, M. A. Lysaght, J. S. Parker, H. W. van der Hart and K. T. Taylor, *Phys. Rev. A*, 2011, **84**, 061404.
69 J. Feist, O. Zatsarinny, S. Nagele, R. Pazourek, J. Burgdorfer, X. X. Guan, K. Bartschat and B. I. Schneider, *Phys. Rev. A*, 2014, **89**, 033417.
70 A. L. Cavalieri, N. Muller, T. Uphues, V. S. Yakovlev, A. Baltuska, B. Horvath, B. Schmidt, L. Blumel, R. Holzwarth, S. Hendel, M. Drescher, U. Kleineberg, P. M. Echenique, R. Kienberger, F. Krausz and U. Heinzmann, *Nature*, 2007, **449**, 1029–1032.
71 M. Sabbar, S. Heuser, R. Boge, M. Lucchini, T. Carette, E. Lindroth, L. Gallmann, C. Cirelli and U. Keller, *Phys. Rev. Lett.*, 2015, **115**, 133001.

72 M. Uiberacker, T. Uphues, M. Schultze, A. J. Verhoef, V. Yakovlev, M. F. Kling, J. Rauschenberger, N. M. Kabachnik, H. Schroder, M. Lezius, K. L. Kompa, H. G. Muller, M. J. J. Vrakking, S. Hendel, U. Kleineberg, U. Heinzmann, M. Drescher and F. Krausz, *Nature*, 2007, **446**, 627–632.

73 G. Sansone, F. Kelkensberg, J. F. Perez-Torres, F. Morales, M. F. Kling, W. Siu, O. Ghafur, P. Johnsson, M. Swoboda, E. Benedetti, F. Ferrari, F. Lepine, J. L. Sanz-Vicario, S. Zherebtsov, I. Znakovskaya, A. L'Huillier, M. Y. Ivanov, M. Nisoli, F. Martin and M. J. J. Vrakking, *Nature*, 2010, **465**, 763–767.

74 F. Calegari, D. Ayuso, A. Trabattoni, L. Belshaw, S. De Camillis, S. Anumula, F. Frassetto, L. Poletto, A. Palacios, P. Decleva, J. B. Greenwood, F. Martin and M. Nisoli, *Science*, 2014, **346**, 336–339.

75 L. S. Cederbaum and J. Zobeley, *Chem. Phys. Lett.*, 1999, **307**, 205–210.

76 A. I. Kuleff, S. Lunnemann and L. S. Cederbaum, *J. Phys. Chem. A*, 2010, **114**, 8676–8679.

77 F. Remacle and R. D. Levine, *Proc. Natl. Acad. Sci. U. S. A.*, 2006, **103**, 6793–6798.

78 J. Itatani, J. Levesque, D. Zeidler, H. Niikura, H. Pepin, J. C. Kieffer, P. B. Corkum and D. M. Villeneuve, *Nature*, 2004, **432**, 867–871.

79 S. Baker, J. S. Robinson, C. A. Haworth, H. Teng, R. A. Smith, C. C. Chirila, M. Lein, J. W. G. Tisch and J. P. Marangos, *Science*, 2006, **312**, 424–427.

80 O. Smirnova, Y. Mairesse, S. Patchkovskii, N. Dudovich, D. Villeneuve, P. Corkum and M. Y. Ivanov, *Nature*, 2009, **460**, 972–977.

81 S. Haessler, J. Caillat, W. Boutu, C. Giovanetti-Teixeira, T. Ruchon, T. Auguste, Z. Diveki, P. Breger, A. Maquet, B. Carre, R. Taieb and P. Salieres, *Nat. Phys.*, 2010, **6**, 200–206.

82 Y. Mairesse, J. Higuet, N. Dudovich, D. Shafir, B. Fabre, E. Mével, E. Constant, S. Patchkovskii, Z. Walters, M. Y. Ivanov and O. Smirnova, *Phys. Rev. Lett.*, 2010, **104**, 213601.

83 P. M. Kraus, B. Mignolet, D. Baykusheva, A. Rupenyan, L. Horný, E. F. Penka, G. Grassi, O. I. Tolstikhin, J. Schneider, F. Jensen, L. B. Madsen, A. D. Bandrauk, F. Remacle and H. J. Wörner, *Science*, 2015, **350**, 790–795.

84 A. R. Beck, D. M. Neumark and S. R. Leone, *Chem. Phys. Lett.*, 2015, **624**, 119–130.

85 L. Gallmann, J. Herrmann, R. Locher, M. Sabbar, A. Ludwig, M. Lucchini and U. Keller, *Mol. Phys.*, 2013, **111**, 2243–2250.

86 E. R. Warrick, W. Cao, D. M. Neumark and S. R. Leone, *J. Phys. Chem. A*, 2016, **120**, 3165–3174.

87 W. T. Pollard and R. A. Mathies, *Annu. Rev. Phys. Chem.*, 1992, **43**, 497–523.

88 S. Mukamel, *Principles of Nonlinear Optical Spectropscopy*, Oxford University Press, New York, 1995.

89 M. B. Gaarde, C. Buth, J. L. Tate and K. J. Schafer, *Phys. Rev. A*, 2011, **83**, 013419.

90 W. C. Chu and C. D. Lin, *Phys. Rev. A*, 2012, **85**, 013419.

91 A. N. Pfeiffer and S. R. Leone, *Phys. Rev. A*, 2012, **85**, 053422.

92 M. F. Lin, D. M. Neumark, O. Gessner and S. R. Leone, *J. Chem. Phys.*, 2014, **140**, 064311.

93 E. Goulielmakis, Z. H. Loh, A. Wirth, R. Santra, N. Rohringer, V. S. Yakovlev, S. Zherebtsov, T. Pfeifer, A. M. Azzeer, M. F. Kling, S. R. Leone and F. Krausz, *Nature*, 2010, **466**, 739–744.

94 R. Santra, V. S. Yakovlev, T. Pfeifer and Z. H. Loh, *Phys. Rev. A*, 2011, **83**, 033405.

95 L. J. Borja, M. Zürch, C. D. Pemmaraju, M. Schultze, K. Ramasesha, A. Gandman, J. S. Prell, D. Prendergast, D. M. Neumark and S. R. Leone, *J. Opt. Soc. Am. B*, 2016, **33**, C57–C64.

96 M. Schultze, K. Ramasesha, C. D. Pemmaraju, S. A. Sato, D. Whitmore, A. Gandman, J. S. Prell, L. J. Borja, D. Prendergast, K. Yabana, D. M. Neumark and S. R. Leone, *Science*, 2014, **346**, 1348–1352.

97 C. Spielmann, *Science*, 2014, **346**, 1293–1294.

98 S. H. Chen, M. J. Bell, A. R. Beck, H. Mashiko, M. X. Wu, A. N. Pfeiffer, M. B. Gaarde, D. M. Neumark, S. R. Leone and K. J. Schafer, *Phys. Rev. A*, 2012, **86**, 063408.

99 S. Gilbertson, M. Chini, X. M. Feng, S. Khan, Y. Wu and Z. H. Chang, *Phys. Rev. Lett.*, 2010, **105**, 263003.

100 X. W. Wang, M. Chini, Y. Cheng, Y. Wu, X. M. Tong and Z. H. Chang, *Phys. Rev. A*, 2013, **87**, 063413.

101 F. Martin, Y. Cheng, M. Chini, X. W. Wang, A. Gonzalez-Castrillo, A. Palacios, L. Argenti and Z. H. Chang, in *Xxix International Conference on Photonic, Electronic, and Atomic Collisions*, ed. C. Diaz, I. Rabadan, G. Garcia, L. Mendez and F. Martin, 2015, vol. 635, p. 112070.

102 C. Ott, A. Kaldun, P. Raith, K. Meyer, M. Laux, J. Evers, C. H. Keitel, C. H. Greene and T. Pfeifer, *Science*, 2013, **340**, 716–720.

103 H. Wang, M. Chini, S. Y. Chen, C. H. Zhang, F. He, Y. Cheng, Y. Wu, U. Thumm and Z. H. Chang, *Phys. Rev. Lett.*, 2010, **105**, 143002.

104 B. Bernhardt, A. R. Beck, X. Li, E. R. Warrick, M. J. Bell, D. J. Haxton, C. W. McCurdy, D. M. Neumark and S. R. Leone, *Phys. Rev. A*, 2014, **89**, 023408.

105 X. Li, B. Bernhardt, A. R. Beck, E. R. Warrick, A. N. Pfeiffer, M. J. Bell, D. J. Haxton, C. W. McCurdy, D. M. Neumark and S. R. Leone, *J. Phys. B: At., Mol. Opt. Phys.*, 2015, **48**, 125601.

106 W. Cao, E. R. Warrick, D. M. Neumark and S. R. Leone, *New J. Phys.*, 2016, **18**, 013041.

107 S. Chen, M. Wu, M. B. Gaarde and K. J. Schafer, *Phys. Rev. A*, 2013, **88**, 033409.

108 A. R. Beck, B. Bernhardt, E. R. Warrick, M. X. Wu, S. Chen, M. B. Gaarde, K. J. Schafer, D. M. Neumark and S. R. Leone, *New J. Phys.*, 2014, **16**, 113016.

109 S. H. Chen, M. Wu, M. B. Gaarde and K. J. Schafer, *Phys. Rev. A*, 2013, **87**, 033408.

110 M. Reduzzi, W. C. Chu, C. Feng, A. Dubrouil, J. Hummert, F. Calegari, F. Frassetto, L. Poletto, O. Kornilov, M. Nisoli, C. D. Lin and G. Sansone, *J. Phys. B: At., Mol. Opt. Phys.*, 2016, **49**, 065102.

111 P. Gürtler, V. Saile and E. E. Koch, *Chem. Phys. Lett.*, 1977, **48**, 245–250.

112 A. Lofthus and P. H. Krupenie, *J. Phys. Chem. Ref. Data*, 1977, **6**, 113–307.

113 H. Lefebvre-Brion and K. Yoshino, *J. Mol. Spectrosc.*, 1993, **158**, 140–146.

114 H. Lefebvre-Brion, *J. Chem. Phys.*, 2005, **122**, 144315.

115 W. Cao, E. R. Warrick, A. Fidler, S. R. Leone and D. M. Neumark, *Phys. Rev. A*, 2016, **94**, 021802.

116 I. V. Schweigert and S. Mukamel, *Phys. Rev. Lett.*, 2007, **99**, 163001.

117 S. Mukamel, D. Healion, Y. Zhang and J. D. Biggs, *Annu. Rev. Phys. Chem.*, 2013, **64**, 101–127.

118 W. Cao, E. R. Warrick, A. Fidler, D. M. Neumark and S. R. Leone, *Phys. Rev. A*, submitted.

119 J. Li, X. Ren, Y. Yin, Y. Cheng, E. Cunningham, Y. Wu and Z. Chang, *Appl. Phys. Lett.*, 2016, **108**, 231102.

120 H.-T. Chang, M. Zurch, P. M. Kraus, L. J. Borja, D. M. Neumark and S. R. Leone, *Opt. Lett.*, submitted.

Faraday Discussions

PAPER

Decoherence, control and attosecond probing of XUV-induced charge migration in biomolecules. A theoretical outlook

Manuel Lara-Astiaso,[a] David Ayuso,[a] Ivano Tavernelli,[b] Piero Decleva,[c] Alicia Palacios[a] and Fernando Martín*[ade]

Received 11th April 2016, Accepted 1st June 2016
DOI: 10.1039/c6fd00074f

The sudden ionization of a molecule by an attosecond pulse is followed by charge redistribution on a time scale from a few femtoseconds down to hundreds of attoseconds. This ultrafast redistribution is the result of the coherent superposition of electronic continua associated with the ionization thresholds that are reached by the broadband attosecond pulse. Thus, a correct theoretical description of the time evolution of the ensuing wave packet requires the knowledge of the actual ionization amplitudes associated with all open ionization channels, a real challenge for large and medium-size molecules. Recently, the first calculation of this kind has come to light, allowing for interpretation of ultrafast electron dynamics observed in attosecond pump–probe experiments performed on the amino acid phenylalanine [Calegari *et al.*, Science 2014, **346**, 336]. However, as in most previous theoretical works, the interpretation was based on various simplifying assumptions, namely, the ionized electron was not included in the description of the cation dynamics, the nuclei were fixed at their initial position during the hole migration process, and the effect of the IR probe pulse was ignored. Here we go a step further and discuss the consequences of including these effects in the photoionization of the glycine molecule. We show that (i) the ionized electron does not affect hole dynamics beyond the first femtosecond, and (ii) nuclear dynamics has only a significant effect after approximately 8 fs, but does not destroy the coherent motion of the electronic wave packet during at least few additional tens of fs. As a first step towards understanding the role of the probe pulse, we have considered an XUV probe pulse, instead of a strong IR one, and show that such an XUV probe does not introduce significant distortions in the pump-induced dynamics,

[a]*Departamento de Química, Módulo 13, Universidad Autónoma de Madrid, 28049 Madrid, Spain. E-mail: fernando.martin@uam.es*

[b]*IBM Research GmbH, Zurich Research Laboratory, 8803 Rueschlikon, Switzerland*

[c]*Dipartimento di Scienze Chimiche e Farmaceutiche, Università di Trieste and CNR-Istituto Officina dei Materiali, 34127 Trieste, Italy*

[d]*Instituto Madrileño de Estudios Avanzados en Nanociencia (IMDEA Nano), Campus Universitario de Cantoblanco, 28049 Madrid, Spain*

[e]*Condensed Matter Physics Center (IFIMAC), Universidad Autónoma de Madrid, 28049 Madrid, Spain*

suggesting that pump–probe strategies are suitable for imaging and manipulating charge migration in complex molecules. Furthermore, we show that hole dynamics can be changed by shaping the attosecond pump pulse, thus opening the door to the control of charge dynamics in biomolecules.

1 Introduction

Imaging electron dynamics in biomolecules with few-femtosecond or sub-femtosecond time resolutions is the first step towards achieving control of biologically relevant processes.[1–8] The rapid progress in attosecond technology[9–13] has already permitted the control of electron dynamics in atoms[14,15] and hydrogen molecules,[16,17] and recent experimental efforts aim at doing so in more complex molecules.[4,7] So far two different approaches have been used to visualize electron dynamics in such systems. The first one is high-harmonic spectroscopy, where the light emitted after the round trip of an IR-driven molecular electron carries information about the cation dynamics with sub-fs temporal and sub-Å spatial resolutions.[18–20] The other one is pump–probe spectroscopy,[4,16,17,21] where an attosecond XUV pump pulse ionizes the molecule and a time-delayed fs IR pulse probes the dynamics generated in the molecular cation. The two approaches imply that an electron is ionized, by a strong IR field in the first case or a weak XUV field in the second case, but differ in the way the dynamics is probed in the cation: by the returning an electron itself or by another ultrashort pulse, respectively.

In either case, the sudden ionization of the molecule creates a hole that subsequently migrates on a time scale ranging from a few femtoseconds down to hundreds of attoseconds. The migration is due to a coherent superposition of electronic states generated by the pump pulse. This coherent superposition can be the result of the broad band of the ionizing pulse, which covers several ionization thresholds, or of the appearance of a dense manifold of ionic states due to rearrangements in the electronic structure of the molecular cation induced by electron correlation. In both cases the charge is expected to move from one molecular site to another or to fluctuate all along the molecular backbone. Depending on the initial wave packet that is created, the interplay between the electronic and nuclear dynamics can lead to the final localization of the charge on a specific molecular site.

A fully *ab initio* theoretical description of the correlated motion of electrons and nuclei that follows ionization by laser pulses is still very challenging, even for the simplest molecule H_2,[22] and practically out of reach for many-electron polyatomic molecules. Besides the intrinsic limitations associated with the large number of electronic and nuclear degrees of freedom, which are also applicable to calculations involving molecular bound states, there is the additional limitation of dealing with the electronic continuum. For this reason, most theoretical approaches to large molecules have ignored the ionization step and have only described the cation dynamics that arises from arbitrarily chosen initial wave packets.[1–3] Notably a “sudden ionization” ansatz, corresponding to the creation of a Koopman's hole,[23,24] *i.e.* the removal of an electron from a ground state Hartree–Fock orbital, generating a wave packet through electron correlation in the final ion, has been largely employed in earlier studies. Alternatively, the evaluation of ionization amplitudes employing Coulomb waves located in the molecular center of mass[25–28] have permitted the gap to be reduced between completely ignoring ionization and fully describing it, but with variable success.[29] The first theoretical calculation that fully accounts for the ionization induced by an attosecond pulse in a large molecule, namely phenylalanine, has been reported very recently.[4] At variance with earlier calculations, in this work the initial wave packet was built from actual ionization amplitudes, which were obtained from the solution of the scattering equations in the framework of the static exchange DFT method.[30–32]

The first experimental evidence of ultrafast charge migration has been reported by Calegari *et al.*[4] for the case of the amino acid phenylalanine. In this work, a sub-300-as XUV pulse was used to ionize the molecule and a controlled time-delayed 4 fs IR pulse to probe the dynamics induced by the former. The measured yields for the production of doubly-charged immonium fragments exhibit few-fs oscillations superimposed on a typical smooth decreasing background associated with the much slower dissociation dynamics. The unusually fast oscillations were interpreted as the signature of the coherent electron dynamics generated by the attosecond pulse on the remaining molecular cation. This interpretation was supported by the good agreement between the observed frequencies and those resulting from the above-mentioned theoretical calculations,[4,32] in which the actual transition amplitudes associated with the different ionization channels were accurately evaluated. Evaluation of these amplitudes is essential to obtain a realistic description of the electronic wave packet motion, which is dictated by the relative value of the corresponding moduli and phases associated with the different ionization channels.

In spite of the significant advance provided by such calculations, the comparison with the actual experimental measurements was somewhat indirect. First, following common practice for the subject,[1,3,8,29,33,34] the evolution of the charge density in the cation was described by assuming that the ionized electron leaves the system instantaneously (sudden ionization), so that it does not interact with the remaining electrons at all. In other words, as soon as the interaction of the attosecond pulse with the neutral molecule is over, the theory described the evolution of an $(N-1)$-electron system instead of that of the original N-electron one. This looks like a reasonable approximation when the ejected electron is far away from the remaining cation, *i.e.*, long after the interaction with the attosecond pulse. But what about the early stages of the wave packet evolution when the ionized electron is still close to the cation? Second, due to the complexity of the chosen molecule, charge dynamics was studied by assuming that the nuclei remain fixed during the whole process. As in existing literature,[1,3,4,29,33,34] it was argued that nuclear motion should have a negligible effect at the early stages of this evolution, typically during the first tens of fs, since the shortest vibrational periods in this kind of molecules are around 10 fs and usually correspond to peripheral X–H bonds (X = C, N, O), which are barely involved in charge delocalization processes. However, recent theoretical studies have shown that coherence between two electronic states can be lost in less than 10 fs when the nuclei move across a conical intersection.[35] Furthermore, the quantum distribution of geometries inherent to any nuclear wave packet introduces additional dephasing in a similar time scale.[8] Since conical intersections are a common feature in large molecules, can we really assume that nuclear motion does not play any role in the early stages of the charge migration process? Third, as mentioned above, the charge dynamics is mostly determined by the ionization amplitudes associated with all open channels, *i.e.* by the initial electron wave packet created in the cation. So, in principle, one would expect that by changing those amplitudes one should also change the electron wave packet dynamics. But is this feasible in practice, *e.g.*, by modifying the pump pulse characteristics? The latter question is relevant because, if answered positively, it would open the way to control the charge dynamics in biomolecules. The fourth and final remark is that the effect of the probe pulse was completely ignored. In principle, the role of the probe pulse is to project, at a given time (the pump–probe delay), the dynamics induced by the attosecond pump onto specific observables. Ideally, the probe pulse should not distort the dynamics induced by the pump pulse, but in practice it does. As explained in earlier work,[36] the ideal probe pulses are those containing UV or XUV photons, since, due to their short wavelength, distortion of the molecular potential, and hence of the pump induced dynamics, is reduced to a minimum. Nevertheless, even such a "gentle" probe pulse will further ionize the system, thus leading to a molecular dication that can subsequently break leading to different neutral, singly- and doubly-charged fragments. In fact, the mass and energy of these ions are the actual observables in most pump–probe experiments. The same occurs with IR probe pulses, with the additional complication that, due to their longer wavelengths, they also affect the pump-induced dynamics.[16] The question is then: how does the dynamics generated by the XUV attosecond pulse reflect in the actual pump–probe signal?

In this paper we partially address the previous questions for the case of the glycine molecule. First, we have included the ionized electron in the description of the wave packet dynamics that follows the interaction with the attosecond pulse. We show that the ionized electron only leaves its trace during approximately the first femtosecond. Second, we have made use of Ehrenfest non-adiabatic dynamics to include the effect of nuclear motion on the charge migration process. We show that the latter process is not affected by the nuclear motion during the first 8 fs, but has a significant effect at later times. Nevertheless, modifications of electron dynamics at these later times do not destroy the coherent motion of the electronic wave packet during a few tens of fs. Third, we have employed three different XUV attosecond pulses, all of them generated in recent experiments,[4,37] covering different spectral ranges and with different envelopes, to show that the charge dynamics induced in each case is substantially different. Finally, by means of a simple model that makes use of realistic ionization amplitudes within the fixed-nuclei approximation, we show how the XUV-induced dynamics manifests in the computed double ionization yields after interaction with a "gentle" probe pulse, namely, another XUV pulse, which apart from further ionizing the molecular cation, barely perturbs the molecular potential. This is a first step towards understanding the role of less "gentle" probe pulses, such as the IR one used in earlier experiments.

2 Theoretical methods

A theoretical description of laser–matter interactions using attosecond light sources requires of a time-dependent treatment. We work with two different approaches, both based on the density functional theory (DFT) formalism: (i) the static-exchange DFT method plus time-dependent perturbation theory, and (ii) the time-dependent Kohn–Sham (TDKS) method combined with Ehrenfest molecular dynamics (TDDFT-MD). The former will be used to evaluate the ionization amplitudes and the second one to describe the evolution of the ionic system when nuclear motion is included. The static-exchange DFT method[30,31,38,39] has been successfully applied to describe ionization of many molecules, from the simple N_2 or CO molecules[40] to biomolecules such as the amino acid phenylalanine.[4] The strength of this method lies in the accurate description of the continuum electron wave function, which allows for an accurate evaluation of photoionization amplitudes and cross sections of molecular systems. The combination of this method with first-order time-dependent perturbation theory has recently provided the first reliable description of the electronic wave packet created in the ionization of a biomolecule by a broadband attosecond pulse.[4,32] Previous applications of this method have either ignored the nuclear motion (fixed-nuclei approximation, FNA) or have included at most a single nuclear degree of freedom, which is appropriate for diatomic molecules[40,41] or for small polyatomic molecules in problems where a single vibrational mode is active.[42,43] Here we go a step further and include nuclear motion by means of the TDDFT-MD method, which combines Ehrenfest molecular dynamics (MD) for the nuclear motion with a time-dependent DFT (TDDFT) description of the electronic dynamics as implemented in the CPMD package.[44] In this section, we will first summarize the essence of the static-exchange DFT method and its implementation in the framework of first-order time-dependent perturbation theory, and then briefly describe the basic equations employed to describe the evolution of the N and $N-1$ electron densities associated, respectively, with the full and the cationic system. Finally, we will explain our implementation of the TDDFT-MD method to account for the nuclear motion during the evolution of the electronic wave packet created in the ionization step.

2.1 The initial electronic wave packet

Attosecond pulses are mainly produced from high-harmonic generation. These pulses are typically of low intensity and contain frequencies in the XUV spectral domain. In this context, one can make use of first-order time-dependent perturbation theory to describe their interaction with matter. Thus, the wave packet created upon the interaction of the molecule with such pulses can be written in the basis of molecular eigenstates as

$$\Psi\left(\vec{r}_1,\ldots,\vec{r}_N,t\right) = c_0(T)\mathrm{e}^{-iE_0t}\Psi_0\left(\vec{r}_1,\ldots,\vec{r}_N\right) + \sum_{\alpha,l}\int c_{\alpha,l}(\varepsilon,T)\mathrm{e}^{-i(E_\alpha+\varepsilon)t}\Psi_\alpha{}^l\left(\vec{r}_1,\ldots,\vec{r}_N,\varepsilon\right)\mathrm{d}\varepsilon \tag{1}$$

where $\Psi_0(\vec{r}_1,\ldots,\vec{r}_N)$ is the ground state of the neutral molecule and $\Psi_\alpha{}^l(\vec{r}_1,\ldots,\vec{r}_N,\varepsilon)$ describes the N-electron system where one electron is in the continuum with energy ε and angular momentum l, and the ion is left behind with a hole in the α orbital. The time evolution of the wave packet is given by the stationary phases in the absence of the field ($t > T$, where T is the pulse duration) and the corresponding amplitudes at the end of the pulse:

$$\begin{aligned} c_0(T) &\approx 1 \\ c_{\alpha,l}(\varepsilon,T) &= -i\langle\Psi_\alpha{}^{\varepsilon,l}|\hat{\mu}|\Psi_0\rangle\int_{-\infty}^{\infty}\vec{\mathscr{E}}(t)\mathrm{e}^{i(E_\alpha+\varepsilon-E_0)t}\mathrm{d}t, \end{aligned} \tag{2}$$

where $\langle \Psi_\alpha^{\varepsilon,l} | \hat{\mu} | \Psi_0 \rangle$ is the dipole transition matrix element connecting the ground and a continuum state, $\hat{\mu}$ stands for the dipole operator (here we have used the length gauge form of this operator), the time integral is the Fourier transform of the electric field, $\vec{\mathscr{E}}(t)$, and $(E_\alpha + \varepsilon - E_0)$ is the energy absorbed by the system. The field is defined from 0 to T using experimental temporal envelopes, $F(t)$, in the form $\vec{\mathscr{E}}(t) = F(t)\sin(\omega t + \phi)\varepsilon$, with $\vec{\varepsilon}$ being the polarization vector and ϕ the carrier-envelope phase, which is fixed to zero. Due to our definition of the field, the integral in eqn (2) can be identically evaluated in the interval $[0, T]$.

The final continuum state is obtained by using the static-exchange DFT method as described in ref. 30, 31 and 39. In brief, we make use of the DFT formalism as implemented in the Amsterdam Density Functional (ADF) package[45,46] to compute the ground state of the neutral molecule, employing the LB94 functional with a TZP basis. The first ionization potential of the molecule is computed using the outer-valence Green's function method and the DFT-LB94 eigenvalues are accordingly shifted. The calculated Kohn–Sham density and orbitals are used to represent the exchange and correlation potentials needed to solve the multichannel scattering equations that describe a continuum electron in the presence of an $(N-1)$-electron molecular cation represented by a Slater determinant that includes the above-mentioned orbitals with the exception of that from which the electron has been ionized. Both the original Kohn–Sham orbitals and the continuum electron are written as a multicenter expansion in a basis of B-splines, which is particularly suitable to represent the continuum functions at any desired distance from the molecular center of mass. Evaluation of a continuum state associated with a particular electron energy is made by using the Galerkin approach.[47,48] The multicenter expansion consists of a basis placed at the center of mass of the molecule, containing 111 B-splines enclosed in a sphere of 30 a.u. for the radial components and spherical harmonics with a maximum angular momentum of $l_{\max} = 20$, and a smaller basis located at each atomic center containing B-splines in spheres ranging from 0.2 to 1.6 a.u. and with $l_{\max} = 2$.

In this paper, we will present results for charge migration dynamics originating from two different ways: (i) by removing an electron from a given orbital of the neutral molecule, as it has been done in most extant theoretical works,[1,3,29,49] and (ii) by creating a coherent superposition of one-hole states, as in the case of ionization by an attosecond pulse. It is in the second context where one cannot avoid evaluating the ionization amplitudes.[4,32] The common feature in both scenarios, however, is the formation of a non-stationary cationic state, which induces the hole dynamics along the molecular skeleton.

2.2 *N*-Electron and $(N-1)$-electron densities

A common theoretical practice to visualize charge migration in large molecules is to compute the electron density in the remaining ion, assuming that the dynamics of the photoelectron and that of the cation can be decoupled. This approximation will be valid as long as the interaction of the leaving photoelectron and the ion is negligible, for instance, if the electron is rapidly ejected. To validate this approximation within our DFT-based formalism, we have compared the electron densities of the N and $(N-1)$-electron systems. In the first case, the effect of the continuum electron is explicitly included. In the second case, a common procedure is to make use of the reduced density matrix formalism to get rid of the ejected electron.

We compute the time-dependent one-particle electron density from the perturbed part of the N-electron wave function, as

$$\rho_N(\vec{r}_1,t) = N\int \mathrm{d}\vec{r}_2 \ldots \int \mathrm{d}\vec{r}_N \left|\Psi_N(\vec{r}_1,\ldots,\vec{r}_N,t)\right|^2 \tag{3}$$

where $\Psi_N(\vec{r}_1,\ldots,\vec{r}_N,\ t)$ is the N-electron wave packet created by the XUV pulse, which can be written, as in eqn (1), as an expansion over N-electron continuum eigenstates:

$$\Psi_N(\vec{r}_1,\ldots,\vec{r}_N,t) = \mathcal{Z}\sum_{\alpha,l}\int \mathrm{d}\varepsilon c_{\alpha,l}(\varepsilon,T)\Psi_\alpha{}^l(\vec{r}_1,\ldots,\vec{r}_N,\varepsilon) \tag{4}$$

with $\mathcal{Z}$ being a normalization factor. The corresponding expression for the one-particle density in the basis of spin-orbitals is given by

$$\begin{aligned}\rho_N(\vec{r}_1,t) = \sum_{\alpha,l}\int \mathrm{d}\varepsilon\Bigg[&\left(\sum_{\alpha'\neq\alpha}\left|\varphi_{\alpha'}(\vec{r}_1)\right|^2 + \left|\varphi_l(\vec{r}_1,\varepsilon)\right|^2\right)\left|c_{\alpha}{}^l(\varepsilon)\right|^2 \\ &- \sum_{\alpha'\neq\alpha} c_\alpha^{l*}(\varepsilon)c_{\alpha'}{}^l(\varepsilon)\varphi_\alpha(\vec{r}_1)\varphi_{\alpha'}(\vec{r}_1)\mathrm{e}^{-i(E_{\alpha'}-E_\alpha)t} \\ &+ \sum_{l'}\int \mathrm{d}\varepsilon' c_\alpha^{l*}(\varepsilon)c_{\alpha'}{}^l(\varepsilon')\varphi_l(\vec{r}_1,\varepsilon)\varphi_{l'}(\vec{r}_1,\varepsilon')\mathrm{e}^{-i(\varepsilon'-\varepsilon)t}\Bigg],\end{aligned} \tag{5}$$

where $\varphi_\alpha(\vec{r}_1)$ is a Kohn–Sham spin-orbital and $\varphi_l(\vec{r}_1,\ \varepsilon)$ is a continuum spin orbital. The above expression has a stationary part, the first term, and two other terms that lead to variations of the electron density with time. We will see oscillations in the density with a periodicity dictated by the energy differences between cationic states (second term) and between continuum orbitals themselves (third term). The former are possible because electrons can be ejected with the same energy and angular momentum from different orbitals. The third term is expected to vanish in the neighborhood of the molecule if one waits long enough until the ejected electron is far away from the molecule. When this limit is reached, it is legitimate to ignore this electron and define a one-particle electron density from an $(N-1)$-electron system. For this purpose, we rely on the reduced density matrix formalism, where the N-electron wave function is first projected into the continuum wave functions and then integrated over all continuum energies, thus leading to an $N-1$ ionic electron density

$$\rho_{N-1}^{\mathrm{ion}}(\vec{r}_1,\vec{r}_2,\ldots,\vec{r}_{N-1},t) = \sum_{l'}\mathrm{d}\varepsilon'\left\langle\varphi_l(\vec{r}_N,\varepsilon)\mathrm{e}^{i\varepsilon t}\middle|\Psi_N\right\rangle_{\vec{r}_N}\left\langle\Psi_N\middle|\varphi_l(\vec{r}_N,\varepsilon)\mathrm{e}^{-i\varepsilon t}\right\rangle_{\vec{r}_N}, \tag{6}$$

in which the brackets indicate an integral over the N-th spatial coordinate. We then define the ion one-particle electron density as

$$\rho^{\mathrm{ion}}(\vec{r},t) = \int \rho_{N-1}^{\mathrm{ion}}(\vec{r},\vec{r}_2,\ldots,\vec{r}_{N-1},t)\mathrm{d}\vec{r}_2,\ldots,\mathrm{d}\vec{r}_{N-1} \tag{7}$$

where one now has an integral over the spatial coordinates of $N-2$ instead of $N-1$ electrons. Introducing (4) in the above equation results in the following expression for the electron density of the ion:

$$\rho_{\text{ion}}(\vec{r},t)=\sum_{\alpha}\gamma_{\alpha\alpha}^{(\text{ion})}\sum_{\alpha'\neq\alpha}\left|\varphi_{\alpha'}(\vec{r})\right|^2-\sum_{\alpha}\sum_{\alpha'\neq\alpha}\gamma_{\alpha\alpha'}^{(\text{ion})}\mathrm{e}^{-i\left(E_\alpha-E_{\alpha'}\right)t}\varphi_\alpha(\vec{r})\varphi_{\alpha'}(\vec{r}),\tag{8}$$

where we have introduced the reduced density matrix, $\gamma_{\alpha\alpha'}^{(\text{ion})}$, in the form

$$\gamma_{\alpha\alpha'}^{(\text{ion})}=\sum_{l}\int \mathrm{d}\varepsilon c_{\alpha\varepsilon l}(T)c_{\alpha'\varepsilon l}^{*}(T).\tag{9}$$

For illustrative purposes, it is common to define the hole density as the difference between the electron density that corresponds to the initial state of the neutral target and that of the cation, $\rho_{\text{hole}}(\vec{r},t)=\rho_{\text{neutral}}(\vec{r})-\rho_{\text{ion}}(\vec{r},t)$. For consistency, in the case of the N-electron system we will use a similar expression, where $\rho_{\text{ion}}(\vec{r},t)$ is now replaced by $\rho_{\text{N}}(\vec{r},t)$ given in eqn (5). The electron density of the neutral molecule is simply given by the sum of the squares of the occupied Kohn–Sham orbitals, $\rho_{\text{neutral}}=\sum_{\alpha'}\left|\varphi_{\alpha'}(\vec{r})\right|^2$, and thus the hole density is simply given by

$$\rho_{\text{hole}}(\vec{r},t)=\sum_{\alpha}\left[1-\gamma_{\alpha\alpha}^{(\text{ion})}\right]\sum_{\alpha'\neq\alpha}\left|\varphi_{\alpha'}(\vec{r})\right|^2+\sum_{\alpha}\sum_{\alpha'\neq\alpha}\gamma_{\alpha\alpha'}^{(\text{ion})}\mathrm{e}^{-i\left(E_\alpha-E_{\alpha'}\right)t}\varphi_\alpha(\vec{r})\varphi_{\alpha'}(\vec{r}).\tag{10}$$

2.3 Classical Ehrenfest nuclear dynamics

To account for non-adiabatic effects in the evolution of inner-shell vacancies created by XUV light, we have performed first-principles calculations within the so-called Ehrenfest molecular dynamics (MD) approximation where the mean field potential energy surface driving the nuclear dynamics is computed at the TDDFT level. In earlier work,[35,49] Ehrenfest MD has also been combined with an *ab initio* CASSCF methodology to study charge migration processes involving a reduced number of electronic states. A detailed description of our TDDFT MD method can be found in ref. 44, 50 and 51. Briefly, an isolated glycine molecule was placed in a tetragonal box of dimensions 30 × 20 × 20 Å. Both the Kohn–Sham orbitals and the electronic density were expanded in a plane wave basis set using an energy cutoff of 70 Ry. Core electrons were replaced by pseudo-potentials (PP) of the standard Troullier–Martins form[52] and PP energies and forces were evaluated using the Kleinman–Bylander scheme.[53] The exchange correlation energy was calculated using the PBE functional.[54] The orbitals of the neutral glycine molecule were optimized using a convergence threshold of 10^{-8}. In TDDFT-based Ehrenfest MD, the electronic density evolves according to the time-dependent Kohn–Sham (TDKS) equations.[55] The propagation of the electronic degrees of freedom was started from a non-equilibrium electronic structure in which an electron is removed from a previously occupied KS orbital of neutral glycine. This initial electron density, which does not correspond to a specific state of Gly^+, was then propagated by numerical integration of the unrestricted TDKS equations using an iterative Crank–Nicolson algorithm combined with a two-step Runge–Kutta scheme to maintain the δt^3 order accuracy (for more details on the implementation, see a recent review[56]). A time-step of 0.01 a.u. was chosen to ensure energy conservation within 1% over the entire simulation length (25 fs). The initial atomic configuration was taken from a geometry optimization of the uncharged system at 0 K. The forces on the nuclei were computed according to the Hellmann–Feynman theorem as the analytic derivative of the expectation value of the energy. Due to the fact that the number of α and β electrons is unbalanced in the cation, the hole density at each instant of the propagation is given by the difference between the α and β spin densities.[44]

3 Effect of the ionized electron on the cation dynamics

We explore the signature of the emitted photoelectron on the cationic dynamics right after the ionization event. As mentioned before, most extant works in this field consider a sudden electron emission and simply capture the dynamics of the remaining cation. Alternatively, we calculate here the expected value of the one-particle density operator for the N-electron system in order to trace in time the photoelectron ejection and assess its effect in the remaining cationic density. To that end, we have chosen a realistic case in which the most stable conformer of the glycine molecule is ionized upon the interaction with the experimentally available sub-300-as pulse reported in,[4] with a spectral energy bandwidth covering photon energies from 18 to 35 eV. The selected glycine conformer is the most abundant one at room temperature ($\simeq$60% [57]), so in line with the results of ref. 32 for the phenylalanine molecule, one can expect that the hole dynamics observed in a real experiment should not differ significantly from the one reported here. For consistency, the same conformer has been used in all calculations reported in the forthcoming sections. To avoid spurious reflections of the photoelectron at the box boundaries, we have used a rather large simulation box of 100 a.u. for the radial grid. The XUV pulse is sufficiently energetic to eject an electron from the 15 outer orbitals of glycine, leading to the formation of a coherent superposition of ionic states, as described in eqn (1), including all those 15 states with their associated electronic continua. One can then extract the one-particle electron density as given in eqn (8) for the $(N-1)$-electron system, *i.e.* by first projecting into the continuum states as shown in expression (6), or instead, extract it directly from the N-electron target as defined in eqn (3).

In Fig. 1 we show the early evolution of the hole density, $\rho_{\text{hole}}(\vec{r},t) = \rho_{\text{neutral}}(\vec{r}) - \rho_{\text{ion}}(\vec{r},t)$, in both scenarios, *i.e.*, by including (top row) and excluding (bottom row) the ionized electron, respectively. As can be seen, the ejected electron spreads around the whole molecule when $t < 0.2$ fs, so that one cannot ignore its presence in the evaluation of the hole density. Later on, the ionized electron vanishes from the vicinity of the molecular skeleton, but the correlation effects induced on the dynamics of the remaining cation are still visible, up to $t \sim 0.7$ fs. At longer times, the evolution of the hole density is practically the same as that resulting from ignoring the ionized electron, as shown in Fig. 2. This figure provides a more quantitative information about the role of the ionized electron, by comparing the actual electron densities integrated around different molecular sites. Fig. 2 shows again that the effect of the ionized electron is only appreciable during the first half of femtosecond. The precise value of this elapsed time will strongly depend on the kinetic energy of the photoelectron and, consequently, it will depend on the energy spectrum of the pulse employed to ionize the molecule. In the present case, the broad pulse bandwidth ranges from ~15–35 eV, leading to the emission

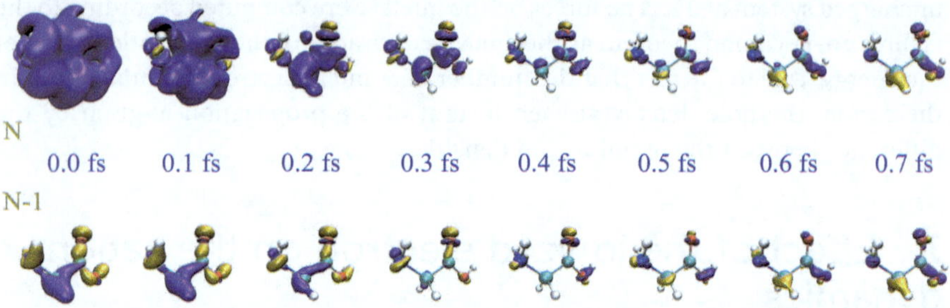

Fig. 1 Comparison of hole densities corresponding to N and $N-1$ electron systems right after the action of a sub-300-as XUV pulse. For the sake of clarity, the hole density has been referenced to its time-averaged value, so that yellow and purple colors indicate, respectively, excess and default of charge with respect to the average hole density. Upper rows: N-electron system (see text). Lower rows: $(N-1)$-electron system. Isosurfaces with values 8×10^{-4} and -8×10^{-4} a.u. are plotted in yellow and purple, respectively. Zero time is taken right at the end of the XUV pulse.

of relatively slow electrons, since the ionization potential (IP) from the highest-occupied molecular orbital (HOMO) of glycine is around 11 eV, while for the lowest valence orbital reached by the pulse (6a′) it is close to 32 eV. The resulting electronic wave packet is thus built from components of relatively low kinetic energy, giving a lower bound for this elapsed time. The general conclusion is that in the very early stages of the process, the observed dynamics will be exactly the same as that of the residual cation in the absence of the photoelectron, thus proving that, although one has to account for ionization in order to obtain a realistic description of the electronic wave packet generated by the XUV pulse, one can safely ignore the ionized electron in the subsequent evolution of the electronic wave packet in the cation.

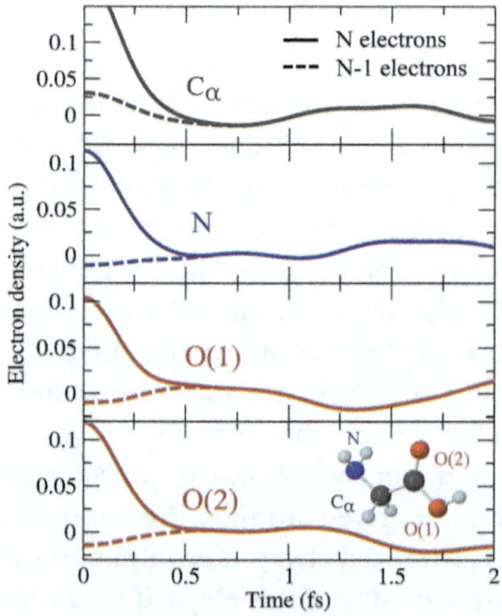

Fig. 2 Variation of the hole density obtained around different molecular sites as a function of time with respect to its time-averaged value. Full lines: N-electron system. Dashed lines: $(N-1)$-electron system. Zero time is taken right at the end of the XUV pulse.

4 Effect of nuclear motion on the cation dynamics

To investigate the effect of the nuclear motion on the XUV induced dynamics, we have compared the results of the TDDFT-MD method described in section 2.3, in which the nuclear motion is described in the framework of a semiclassical Ehrenfest approach, with those obtained from TDDFT calculations in which the nuclear positions remain fixed (fixed-nuclei approximation, FNA). For a meaningful comparison, both types of calculations have been performed by using the same code and the same set of parameters (initial wave packet, basis set, integration steps, exchange-correlation functional, box, *etc.*), the only difference being the turning on or off of the nuclear dynamics, respectively. To simplify the analysis, we have considered one-hole initial states resulting from single-electron removal from specific Kohn–Sham orbitals, instead of initial states that result from the coherent superposition of one-hole states as those arising in ionization by attosecond pulses. As an illustration, we have considered electron removal from the 4a″ and 14a′ orbitals. Fig. 3 and 4 show the corresponding hole densities as functions of time. As can be seen, the effect of nuclear motion is negligible within the first 6–8 fs, whereas it is noticeable at longer times, up to the point that it leads to an entirely different dynamics. The time interval within which nuclear motion can be ignored is significantly shorter than the typical periods of the glycine vibrational modes. This is likely due to the relatively rapid non-adiabatic transitions between different electronic states. This result is compatible with the time scales found by Vacher *et al.*[8,35] for benzene and its derivatives. Since our TDDFT calculations account for non-adiabatic effects involving a large manifold of molecular states, the conclusion is that nuclear motion starts to affect electron dynamics right after the first non-adiabatic transition.

As shown in previous work,[4] coherent hole motion can lead to rapid oscillations in the measured ion yields as a function of the pump–probe delay. These oscillations reflect the typical frequencies of such coherent motion. To analyze the effect of nuclear motion on the measured frequencies, we have calculated the Fourier transform of the hole density around different molecular sites, namely the C_α atom, the two oxygen atoms, and the N atom. Since, in current experiments, ultrafast oscillations have only been observed for time delays shorter than 20 or

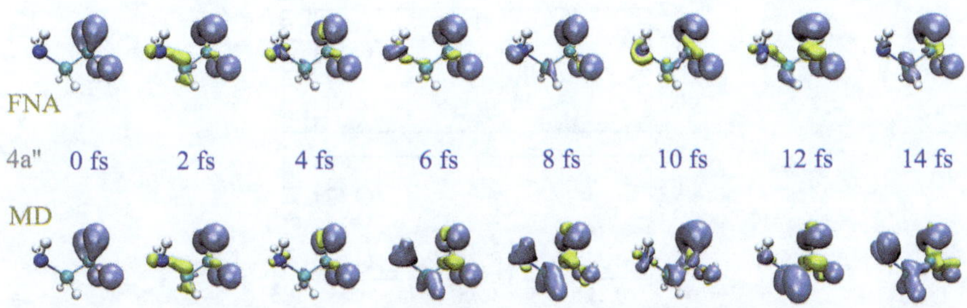

Fig. 3 Time evolution of the spin density, defined as $\rho_\alpha - \rho_\beta$, after removing an electron with α-spin from the 4a″ orbital. The isosurfaces enclose charge densities larger than 0.005 a.u. of β (blue) and α (yellow) electrons.

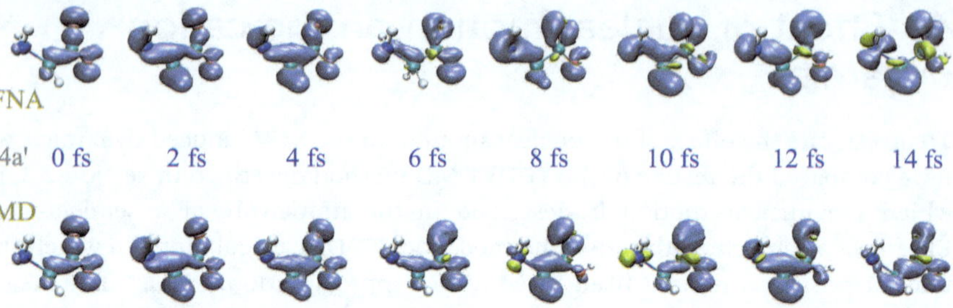

Fig. 4 Same as Fig. 3 for the 14a′ orbital.

30 fs, we have limited our study to the time interval 0–25 fs. The results are shown in Fig. 5 and 6. As a result of the limited time interval used in our analysis, the peaks associated with the different frequencies exhibit a width. As in the case of the phenylalanine molecule,[4] a striking feature of the calculated frequency spectra is the appearance of very few peaks, typically two or three in the interval from 0.1 to 1.5 PHz. These frequencies correspond to energy differences between the cationic states that compose the electronic wave packet generated in the ionization process. However, among all possible energy differences, only a few ones are actually observed, not necessarily involving the state in which the vacancy was initially created. Inclusion or not of the nuclear motion does not alter this general picture. However, the actual values of the frequencies can be significantly affected by the presence of the nuclear motion. As can be seen, some frequencies are slightly shifted, and others can even appear or disappear. This is the consequence of non-adiabatic transitions between the initially populated

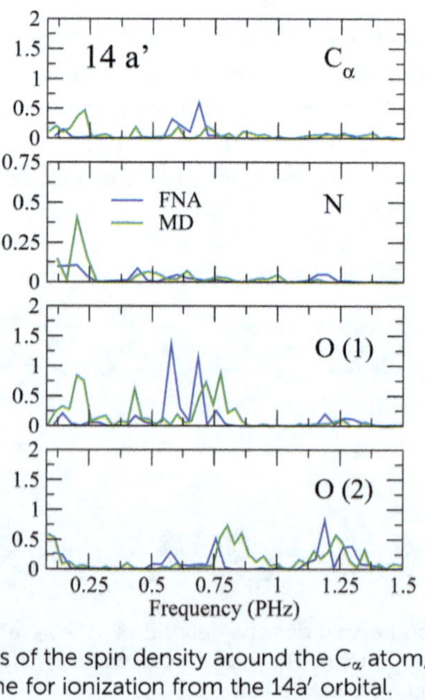

Fig. 5 Fourier transforms of the spin density around the C_α atom, the two oxygen atoms, and the N atom of glycine for ionization from the 14a′ orbital.

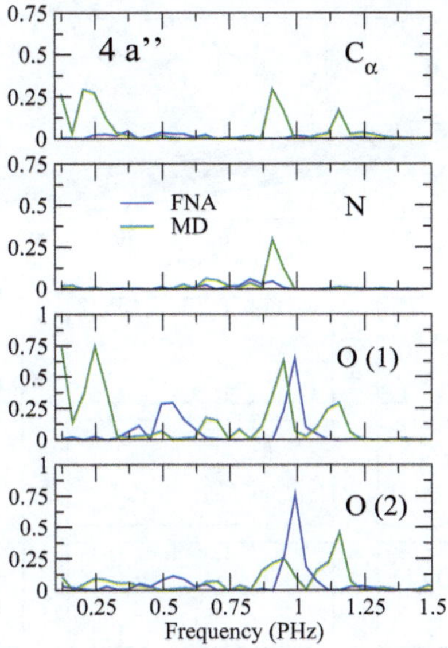

Fig. 6 Same as Fig. 5 for the 4a″ orbital.

cationic states and cationic states that were initially empty. The fact that the values of the frequencies do not differ too much from those obtained in the fixed-nuclei approximation would suggest that non-adiabatic transitions are likely to occur between neighboring states.

5 Control of hole dynamics through attosecond pulse shaping

Hole dynamics reflects the time evolution of the initially created electronic wave packet, which is a coherent superposition of cationic states. Thus different initial electronic wave packets must evolve differently. Here we show that one can indeed modify such dynamics by changing the pulse characteristics. To prove it, we have chosen three different pulses whose frequency spectra are identical to three attosecond pulses produced in recent experiments.[4,37] These frequency spectra are shown in Fig. 7. Two of the pulses cover a similar frequency region, around 20 eV, but have very different envelopes. The third one has a very simple spectral shape but covers a range of higher frequencies, around 90 eV. Fig. 7 also shows the photoionization cross sections of the glycine molecule associated with electron removal from different molecular orbitals. As can be seen, for all pulses, there is a substantial number of ionization channels that are open. Obviously this number is larger for the pulse containing the higher photon energies. All ionization probabilities are comparable in magnitude. However their relative values change significantly with photon energy. Similarly, the relative phases between the different ionization amplitudes change with photon energy. Therefore, one can expect that, by using pulses with different spectral shapes or covering different frequency ranges, different coherent superpositions of cationic states will be generated. Fig. 7 shows the frequency spectra associated with the hole

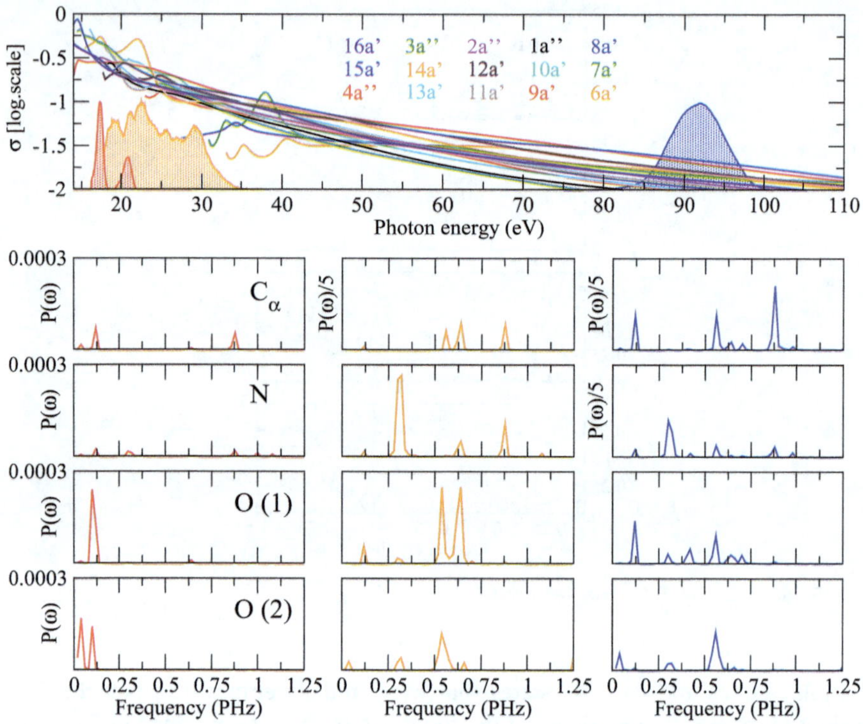

Fig. 7 Upper panel: photoionization cross sections of glycine from the molecular orbitals indicated in the colored labels. The shadowed lines correspond to the experimental energy distributions of three different attosecond pulses centered at 20 eV (red), 18–35 eV (orange) and 90 eV (blue), respectively. Lower panels: Fourier power spectra of the time-dependent spin density integrated around various molecular centers, computed for the three pulses (left column: 20 eV; middle column: 18–35 eV; right column: 90 eV). Each row corresponds to a given atom as labeled in the first column.

dynamics around the C_α atom, the two oxygen atoms, and the N atom of glycine. As can be seen, the frequency spectra, or in other words the hole dynamics, can indeed be very different, but not always. Surprisingly the largest differences are observed for the two pulses that cover a similar range of photon energies. These two pulses, however, have a very different spectral shape. Since the ionization probabilities exhibit the largest relative variations at low photon energies, it is thus clear that the actual shape of the pulse matters. This is because spectral peaks strategically placed at specific photon energies may favor ionization in specific channels, in detriment of ionization in the other channels. Selectivity is much less pronounced at higher energies, where ionization probabilities and their corresponding relative values vary slowly with photon energy. In this case, the calculated frequency spectra are different from those obtained by using the other two pulses, although they exhibit some similarities to those obtained from the pulse with the largest spectral width.

These results show that, in general, a large degree of control can only be achieved in regions of the ionization spectrum where the relative ionization probabilities change significantly with photon energies. In large molecules, this usually occurs close to the ionization threshold, within the first 10 to 40 eV, where specific tuning of shape resonances or broad autoionizing states can play an important role. In this region, the optimum way to exert such control is by introducing peaks in different regions of the pulse spectrum.

6 Role of the probe pulse on the cation dynamics

In pump–probe experiments, common observables employed to capture ultrafast electron dynamics in molecules are total or partial ionization yields as functions of the time delay between the pulses.[4,16] A complete calculation that includes both electron and nuclear dynamics, in conjunction with ionization of a second electron and the eventual dissociation induced by a probe pulse is currently out of reach. For this reason, as a first step to explore the role of the probe pulse, we have ignored the nuclear motion and we have assumed that the second ionization step can be described by using the same Kohn–Sham orbitals as for ionization of the pump pulse. This is equivalent to assume that electron correlation in our Kohn–Sham description of the neutral, singly- and doubly-charged species is the same. Therefore, our analysis can only have a qualitative value. Nevertheless, it will already allow us to understand if the hole dynamics initiated by the pump pulse is captured after the additional ionization produced by the probe pulse. The general expression for the two-photon double ionization amplitude can be written as:

$$c_{\alpha,\beta}(\varepsilon_\alpha,\varepsilon_\beta,\tau) = i\int_0^T \mathrm{d}t' \vec{\mathscr{E}}_{\text{probe}}(t')\mathrm{e}^{i(E_\beta+\varepsilon_\beta-E_\alpha-\varepsilon_\alpha)t'} \times \int_0^{t'} \mathrm{d}t\vec{\mathscr{E}}_{\text{pump}}(t)\mathrm{e}^{i(E_\alpha+\varepsilon_\alpha-E_0)t}\left\langle \Psi_{\alpha,\beta}{}^{\varepsilon_\alpha,l_\alpha,\varepsilon_\beta,l_\beta}|\hat{\mu}|\Psi_\alpha{}^{\varepsilon_\alpha,l_\alpha}\right\rangle\left\langle\Psi_\alpha{}^{\varepsilon_\alpha,l_\alpha}|\hat{\mu}|\Psi_0\right\rangle \tag{11}$$

where the two indices α and β are used to denote the final dicationic states, indicating the Kohn–Sham orbitals from which the two electrons have been ejected, first by the pump pulse, leading to the α vacancy, and then by the probe pulse, leading to the β vacancy. For the reasons explained in the introduction, we consider an XUV probe pulse identical to the pump probe, $\vec{\mathscr{E}}_{\text{probe}}(t) = \vec{\mathscr{E}}_{\text{pump}}(t)$, with a time delay of τ between them such that they never overlap. Hence the double integral in time is separable and the time-dependence can be factored out. Eqn (11) has been evaluated for the sub-300-as pulse of ref. 4. The resulting total double ionization yield as a function of the pump–probe delay is shown in Fig. 8 together with its Fourier transform. We observe that the double ionization yield exhibits rapid oscillations superimposed to a slower one of period ~10 fs. These are the main frequencies seen in Fig. 7 (central panels) for this same pulse. Notice, however, that beatings corresponding to different atoms appear all at once in the double ionization yield, although their relative intensities are no longer modulated by the orbitals' overlap as for the hole densities shown in Fig. 7 [see also eqn (10)]. These findings are in agreement with the conclusion of ref. 36 for the case of the XUV-pump/XUV-probe ionization of H_2, although, in the latter case, only the single ionization channel is open due to the large energy separation between the electronic states of the molecule. The present results show that, as in ref. 36, the use of a "gentle" XUV pulse allows us to uncover the pump-induced

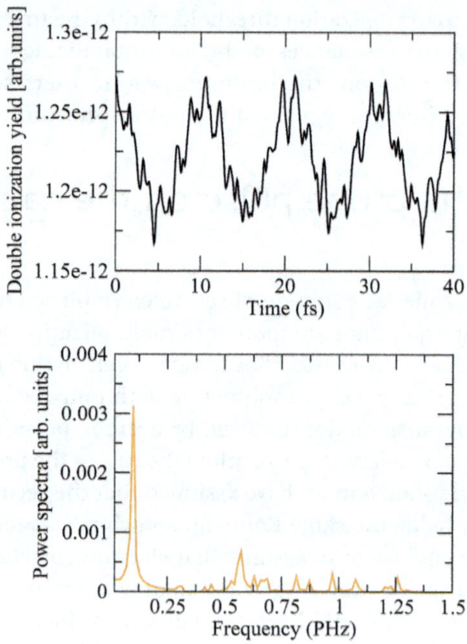

Fig. 8 (Upper panel) Double ionization yields as a function of the pump–probe time delay, computed from eqn (11) using the static exchange DFT method. (Lower panel) Power spectrum of the above signal.

dynamics without introducing additional distortions. That such a conclusion is still valid when IR probe pulses are used or when the nuclear degrees of freedom are taken into account remains to be seen.

7 Conclusion

We have theoretically investigated, by means of the static-exchange DFT and TDDFT-MD methods, the effect of different factors that have hitherto been ignored in the description of charge migration induced by attosecond XUV pulses in biologically relevant molecules, namely, (i) the role of the ionized electron in the electron wave packet dynamics that follows the interaction with a XUV attosecond pulse, (ii) the effect of nuclear motion on the charge migration process, (iii) the possibility to control such dynamics by shaping the attosecond pulse that ionizes the molecule, and (iv) the role of the probe pulse on the pump-induced electron dynamics. By using the glycine molecule as a benchmark, we have shown that (i) the ionized electron only affects these dynamics during approximately the first femtosecond, (ii) nuclear dynamics has a significant effect after approximately 8 fs, but does not destroy the coherent motion of the electronic wave packet during a few tens of fs, (iii) charge dynamics can indeed be changed by modifying the pump pulse characteristics, thus opening the door to the control of charge dynamics in biomolecules, and (iv) a weak XUV probe pulse does not introduce significant distortions in the pump-induced dynamics, suggesting that pump–probe strategies are suitable for imaging and manipulating charge migration in complex molecules.

Acknowledgements

We acknowledge allocation of computer time at Mare Nostrum BSC and CCC-UAM. This work was supported by the European Research Council Advanced Grant No. XCHEM 290853, the MINECO Project No. FIS2013-42002-R, the European COST Action XLIC CM1204, and the CAM project NANOFRONTMAG. A. P. gratefully acknowledges the "Ramón y Cajal" program from the Spanish Ministry of Economy and Competitiveness.

References

1 L. Cederbaum and J. Zobeley, *Chem. Phys. Lett.*, 1999, **307**, 205–210.
2 H. Hennig, J. Breidbach and L. S. Cederbaum, *J. Phys. Chem. A*, 2005, **109**, 409–414.
3 F. Remacle and R. D. Levine, *Proc. Natl. Acad. Sci. U. S. A.*, 2006, **103**, 6793–6798.
4 F. Calegari, D. Ayuso, A. Trabattoni, L. Belshaw, S. De Camillis, S. Anumula, F. Frassetto, L. Poletto, A. Palacios, P. Decleva, J. B. Greenwood, F. Martín and M. Nisoli, *Science*, 2014, **346**, 336–339.
5 F. Lépine, M. Y. Ivanov and M. J. J. Vrakking, *Nat. Photonics*, 2014, **8**, 195–204.
6 S. R. Leone, C. W. McCurdy, J. Burgdörfer, L. S. Cederbaum, Z. Chang, N. Dudovich, J. Feist, C. H. Greene, M. Ivanov, R. Kienberger, U. Keller, M. F. Kling, Z.-h. Loh, T. Pfeifer, A. N. Pfeiffer, R. Santra, K. Schafer, A. Stolow, U. Thumm and M. J. J. Vrakking, *Nat. Photonics*, 2014, **8**, 162–166.
7 P. M. Kraus, B. Mignolet, D. Baykusheva, A. Rupenyan, L. Horny, E. F. Penka, G. Grassi, O. I. Tolstikhin, J. Schneider, F. Jensen, L. B. Madsen, A. D. Bandrauk, F. Remacle and H. J. Worner, *Science*, 2015, **350**, 790–795.
8 M. Vacher, L. Steinberg, A. J. Jenkins, M. J. Bearpark and M. A. Robb, *Phys. Rev. A*, 2015, **92**, 040502.
9 M. Hentschel, R. Kienberger, C. Spielmann, G. A. Reider, N. Milosevic, T. Brabec, P. B. Corkum, U. Heinzmann, M. Drescher and F. Krausz, *Nature*, 2001, **414**, 509–513.
10 P. M. Paul, E. S. Toma, P. Breger, G. Mullot, F. Auge, P. Balcou, H. G. Muller and P. Agostini, *Science*, 2001, **292**, 1689–1692.
11 M. Drescher, M. Hentschel, R. Kienberger, M. Uiberacker, V. Yakovlev, A. Scrinzi, T. Westerwalbesloh, U. Kleineberg, U. Heinzmann and F. Krausz, *Nature*, 2002, **419**, 803–807.
12 M. F. Kling and M. J. Vrakking, *Annu. Rev. Phys. Chem.*, 2008, **59**, 463–492.
13 F. Krausz and M. Ivanov, *Rev. Mod. Phys.*, 2009, **81**, 163–234.
14 M. Uiberacker, T. Uphues, M. Schultze, a. J. Verhoef, V. Yakovlev, M. F. Kling, J. Rauschenberger, N. M. Kabachnik, H. Schröder, M. Lezius, K. L. Kompa, H.-G. Muller, M. J. J. Vrakking, S. Hendel, U. Kleineberg, U. Heinzmann, M. Drescher and F. Krausz, *Nature*, 2007, **446**, 627–632.
15 M. Schultze, M. Fiess, N. Karpowicz, J. Gagnon, M. Korbman, M. Hofstetter, S. Neppl, A. L. Cavalieri, Y. Komninos, T. Mercouris, C. A. Nicolaides, R. Pazourek, S. Nagele, J. Feist, J. Burgdorfer, A. M. Azzeer, R. Ernstorfer,

R. Kienberger, U. Kleineberg, E. Goulielmakis, F. Krausz and V. S. Yakovlev, *Science*, 2010, **328**, 1658–1662.

16 G. Sansone, F. Kelkensberg, J. F. Pérez-Torres, F. Morales, M. F. Kling, W. Siu, O. Ghafur, P. Johnsson, M. Swoboda, E. Benedetti, F. Ferrari, F. Lépine, J. L. Sanz-Vicario, S. Zherebtsov, I. Znakovskaya, A. L'Huillier, M. Y. Ivanov, M. Nisoli, F. Martín and M. J. J. Vrakking, *Nature*, 2010, **465**, 763–766.

17 P. Ranitovic, C. W. Hogle, P. Rivière, A. Palacios, X.-M. Tong, N. Toshima, A. González-Castrillo, L. Martin, F. Martín, M. M. Murnane and H. Kapteyn, *Proc. Natl. Acad. Sci. U. S. A.*, 2014, **111**, 912–917.

18 O. Smirnova, Y. Mairesse, S. Patchkovskii, N. Dudovich, D. Villeneuve, P. Corkum and M. Y. Ivanov, *Nature*, 2009, **460**, 972–977.

19 A. Ferré, D. Staedter, F. Burgy, M. Dagan, D. Descamps, N. Dudovich, S. Petit, H. Soifer, V. Blanchet and Y. Mairesse, *J. Phys. B: At., Mol. Opt. Phys.*, 2014, **47**, 124023.

20 A. Ferré, C. Handschin, M. Dumergue, F. Burgy, A. Comby, D. Descamps, B. Fabre, G. A. Garcia, R. Géneaux, L. Merceron, E. Mével, L. Nahon, S. Petit, B. Pons, D. Staedter, S. Weber, T. Ruchon, V. Blanchet and Y. Mairesse, *Nat. Photonics*, 2015, **9**, 93–98.

21 L. Belshaw, F. Calegari, M. J. Duffy, A. Trabattoni, L. Poletto, M. Nisoli and J. B. Greenwood, *J. Phys. Chem. Lett.*, 2012, **3**, 3751–3754.

22 A. Palacios, J. L. Sanz-Vicario and F. Martín, *J. Phys. B: At., Mol. Opt. Phys.*, 2015, **48**, 242001.

23 B. T. Pickup, *Chem. Phys.*, 1977, **19**, 193–208.

24 Y. Öhrn and G. Born, *Adv. Quantum Chem.*, 1981, **13**, 1–88.

25 B. Mignolet, R. D. Levine and F. Remacle, *J. Phys. B: At., Mol. Opt. Phys.*, 2014, **47**, 124011.

26 N. V. Golubev and A. I. Kuleff, *Phys. Rev. A*, 2015, **91**, 051401.

27 A. Marciniak, V. Despré, T. Barillot, A. Rouzée, M. Galbraith, J. Klei, C.-H. Yang, C. Smeenk, V. Loriot, S. N. Reddy, A. Tielens, S. Mahapatra, a. I. Kuleff, M. Vrakking and F. Lépine, *Nat. Commun.*, 2015, **6**, 7909.

28 A. Trabattoni, M. Klinker, J. González-Vázquez, C. Liu, G. Sansone, R. Linguerri, M. Hochlaf, J. Klei, M. J. J. Vrakking, F. Martín, M. Nisoli and F. Calegari, *Phys. Rev. X*, 2015, **5**, 041053.

29 A. I. Kuleff and L. S. Cederbaum, *J. Phys. B: At., Mol. Opt. Phys.*, 2014, **47**, 124002.

30 M. Stener and P. Decleva, *J. Chem. Phys.*, 2000, **112**, 10871–10879.

31 G. F. D. Toffoli, M. Stener and P. Decleva, *Chem. Phys.*, 2002, **276**, 25–43.

32 F. Calegari, D. Ayuso, A. Trabattoni, L. Belshaw, S. De Camillis, F. Frassetto, L. Poletto, A. Palacios, P. Decleva, J. B. Greenwood, F. Martin and M. Nisoli, *IEEE J. Sel. Top. Quantum Electron.*, 2015, **21**, 1–12.

33 J. Breidbach and L. S. Cederbaum, *J. Chem. Phys.*, 2003, **118**, 3983.

34 J. Breidbach and L. S. Cederbaum, *Phys. Rev. Lett.*, 2005, **94**, 033901.

35 M. Vacher, D. Mendive-Tapia, M. J. Bearpark and M. A. Robb, *J. Chem. Phys.*, 2015, **142**, 094105.

36 A. Palacios, A. González-Castrillo and F. Martín, *Proc. Natl. Acad. Sci. U. S. A.*, 2014, **111**, 3973–3978.

37 D. Fabris, T. Witting, W. Okell, D. Walke, P. Matia-Hernando, J. Henkel, T. Barillot, M. Lein, J. Marangos and J. Tisch, *Nat. Photonics*, 2015, **9**, 383.

38 M. Stener, A. Lisini and P. Decleva, *Int. J. Quantum Chem.*, 1995, **53**, 229–244.

39 M. Stener, G. Fronzoni and P. Decleva, *J. Chem. Phys.*, 2005, **122**, 234301.
40 S. E. Canton, E. Plesiat, J. D. Bozek, B. S. Rude, P. Decleva and F. Martin, *Proc. Natl. Acad. Sci. U. S. A.*, 2011, **108**, 7302–7306.
41 E. Kukk, D. Ayuso, T. D. Thomas, P. Decleva, M. Patanen, L. Argenti, E. Plésiat, A. Palacios, K. Kooser, O. Travnikova, S. Mondal, M. Kimura, K. Sakai, C. Miron, F. Martín and K. Ueda, *Phys. Rev. A*, 2013, **88**, 033412.
42 K. Ueda, C. Miron, E. Plésiat, L. Argenti, M. Patanen, K. Kooser, D. Ayuso, S. Mondal, M. Kimura, K. Sakai, O. Travnikova, A. Palacios, P. Decleva, E. Kukk and F. Martín, *J. Chem. Phys.*, 2013, **139**, 124306.
43 D. Ayuso, M. Kimura, K. Kooser, M. Patanen, E. Plésiat, L. Argenti, S. Mondal, O. Travnikova, K. Sakai, A. Palacios, E. Kukk, P. Decleva, K. Ueda, F. Martín and C. Miron, *J. Phys. Chem. A*, 2015, **119**, 5971–5978.
44 CPMD, *Copyright IBM Corp 1990-2015, Copyright MPI für Festkörperforschung Stuttgart 1997-2001*, http://www.cpmd.org/, 2014.
45 *A. S. Computing and Modeling*, Vrije Universiteit, Amsterdam, Netherlands, http://www.scm.com.
46 G. te Velde, F. M. Bickelhaupt, E. J. Baerends, C. Fonseca Guerra, S. J. A. van Gisbergen, J. G. Snijders and T. Ziegler, *J. Comput. Chem.*, 2001, **22**, 931–967.
47 C. F. Fischer and M. Idrees, *J. Phys. B: At., Mol. Opt. Phys.*, 1990, **23**, 679.
48 H. Bachau, E. Cormier, P. Decleva, J. E. Hansen and F. Martín, *Rep. Prog. Phys.*, 2001, **64**, 1815.
49 D. Mendive-Tapia, M. Vacher, M. J. Bearpark and M. A. Robb, *J. Chem. Phys.*, 2013, **139**, 044110.
50 I. Tavernelli, U. F. Röhrig and U. Rothlisberger, *Mol. Phys.*, 2005, **103**, 963–981.
51 I. Tavernelli, *Phys. Rev. B: Condens. Matter Mater. Phys.*, 2006, **73**, 094204.
52 N. Troullier and J. L. Martins, *Phys. Rev. B: Condens. Matter Mater. Phys.*, 1991, **43**, 1993–2006.
53 L. Kleinman and D. M. Bylander, *Phys. Rev. Lett.*, 1982, **48**, 1425–1428.
54 J. P. Perdew, K. Burke and M. Ernzerhof, *Phys. Rev. Lett.*, 1996, **77**, 3865–3868.
55 E. Runge and E. K. U. Gross, *Phys. Rev. Lett.*, 1984, **52**, 997–1000.
56 B. F. E. Curchod, U. Rothlisberger and I. Tavernelli, *ChemPhysChem*, 2013, **14**, 1314–1340.
57 T. F. Miller III and D. C. Clary, *Phys. Chem. Chem. Phys.*, 2004, **6**, 2563–2571.

Faraday Discussions

ROYAL SOCIETY OF CHEMISTRY

PAPER

Dissipative dynamics at conical intersections: simulations with the hierarchy equations of motion method

Lipeng Chen,[a] Maxim F. Gelin,[b] Vladimir Y. Chernyak,[c] Wolfgang Domcke[b] and Yang Zhao[a]

Received 22nd April 2016, Accepted 29th April 2016
DOI: 10.1039/c6fd00088f

The effect of a dissipative environment on the ultrafast nonadiabatic dynamics at conical intersections is analyzed for a two-state two-mode model chosen to represent the $S_2(\pi\pi^*)$–$S_1(n\pi^*)$ conical intersection in pyrazine (the system) which is bilinearly coupled to infinitely many harmonic oscillators in thermal equilibrium (the bath). The system–bath coupling is modeled by the Drude spectral function. The equation of motion for the reduced density matrix of the system is solved numerically exactly with the hierarchy equation of motion method using graphics-processor-unit (GPU) technology. The simulations are valid for arbitrary strength of the system–bath coupling and arbitrary bath memory relaxation time. The present computational studies overcome the limitations of weak system–bath coupling and short memory relaxation time inherent in previous simulations based on multi-level Redfield theory [A. Kühl and W. Domcke, *J. Chem. Phys.* 2002, **116**, 263]. Time evolutions of electronic state populations and time-dependent reduced probability densities of the coupling and tuning modes of the conical intersection have been obtained. It is found that even weak coupling to the bath effectively suppresses the irregular fluctuations of the electronic populations of the isolated two-mode conical intersection. While the population of the upper adiabatic electronic state (S_2) is very efficiently quenched by the system–bath coupling, the population of the diabatic $\pi\pi^*$ electronic state exhibits long-lived oscillations driven by coherent motion of the tuning mode. Counterintuitively, the coupling to the bath can lead to an enhanced lifetime of the coherence of the tuning mode as a result of effective damping of the highly excited coupling mode, which reduces the strong mode–mode coupling inherent to the conical intersection. The present results extend previous studies of the dissipative dynamics at conical intersections to the nonperturbative regime of system–bath coupling. They pave the way for future first-principles simulations of femtosecond time-resolved four-wave-mixing spectra of chromophores in condensed phases which are nonperturbative in the system dynamics, the system–bath coupling as well as the field-matter coupling.

[a]*Division of Materials Science, Nanyang Technological University, 50 Nanyang Avenue, Singapore 639798*
[b]*Department of Chemistry, Technische Universität München, Garching D-85747, Germany*
[c]*Department of Chemistry, Wayne State University, Detroit, USA*

I. Introduction

The nonequilibrium dynamics of quantum systems which are interacting with a large (macroscopic) environment in thermal equilibrium is a problem with a long history in condensed-matter physics and chemistry, see, *e.g.*,[1–7] and references therein. In the present work, we are addressing the question of how the complex nonadiabatic dynamics occurring at conical intersections of electronic potential-energy (PE) surfaces of polyatomic molecules[8–11] is affected by weak to moderate coupling to a condensed-phase environment. Conical intersections are hypersurfaces of exact degeneracy of adiabatic PE surfaces and are universally encountered in polyatomic molecules.[12–15] The simplest model of a conical intersection involves two electronic states and two vibrational modes which span the so-called branching space of the conical intersection.[16] In the case of a symmetry-allowed conical intersection, these vibrational modes can be classified as coupling and tuning modes.[17] The singularity of the non-Born–Oppenheimer coupling elements at the seam of intersection and the pronounced local anharmonicity of the adiabatic PE surfaces cause exceptionally strong electronic inter-state couplings as well as vibrational mode–mode couplings. The interplay of these couplings gives rise to ultrafast radiationless relaxation processes following preparation of the upper electronic state by optical excitation.[8–11]

In most applications explored in the physics literature, the quantum system of interest is simple and exhibits trivial dynamics when being isolated, such as a nuclear spin[1,2] or a harmonic oscillator.[3] Conical intersections, on the other hand, are multi-level systems with a complex internal dynamics. Even for minimal two-state two-mode models of conical intersections, the Hilbert space required for the description of the photoinduced dynamics can be quite large and the numerical diagonalization of the Hamiltonian can be challenging. Due to the intricate internal dynamics of conical intersections, the physics of energy and phase relaxation due to coupling to an environment can be much richer than the physics of a few-level system or a harmonic oscillator coupled to the same environment.

Early studies of the effects of an external bath on the photoinduced dynamics of two-mode and three-mode conical intersections were performed by Gerdts and Manthe,[18] Kühl and Domcke[19,20] and Baltzer and Stock,[21,22] assuming bilinear coupling of the active modes of the conical intersection to a harmonic bath with ohmic spectral density. The dissipative dynamics of these systems was numerically simulated with multi-level Redfield theory[23,24] which is valid within the limits of weak system–bath coupling and short memory relaxation time of the bath. Kosloff, Ratner and coworkers performed similar computational studies of conical intersections coupled to a bath, considering Lindblad-type relaxation operators or a bath of two-level systems.[25,26] Starting from multi-mode conical intersection Hamiltonians in the diabatic representation, Gindensperger, Burghardt and Cederbaum developed an effective-mode formalism which leads to a hierarchy of system–bath couplings.[27–31] In this approach, the short-time dynamics of the multi-mode conical intersection can be accurately reproduced with the inclusion of just a few effective bath modes. Burghardt and Hynes[32] and Malhado and Hynes[33,34] studied the effect of a dissipative solvent mode, described by a generalized Langevin equation, on the dynamics of a conical intersection related to retinal photoisomerization.

Herein, we employed the hierarchy equation of motion (HEOM) method developed by Tanimura and coworkers[35–38] to simulate the dynamics of a conical intersection in a dissipative environment. The HEOM method is a numerical technique for the solution of the Liouville equation of motion for the reduced density matrix of system–bath models. It is based on the construction and numerical solution of a hierarchy of master equations.[38] An alternative formulation of the HEOM method is based on Fokker–Planck equations.[39] The HEOM method is the most powerful method currently available for the numerically exact solution of master equations for a certain type of system–bath models.[38] In practical applications of the HEOM method, it is convenient to model the bath and the system–bath coupling by the Drude spectral function[7] or linear combination thereof.[40] The Drude bath is characterized by a system–bath coupling parameter λ and a memory relaxation time γ^{-1}. We will focus here on models with weak to moderate system–bath coupling, assuming that the strongly active modes of the conical intersection are included in the system Hamiltonian. Duan and Thorwart recently adopted an alternative strategy to simulate the dynamics of a two-state two-mode conical intersection coupled to a Drude bath.[41] By transforming the coupling and tuning modes of the conical intersection into the bath, they obtained an electronic two-level system which is coupled to a structured bath. The resulting master equation was solved with the HEOM method.

II. Theoretical framework

A. Hamiltonian

We partition the total Hamiltonian into the system Hamiltonian, the bath Hamiltonian, and their coupling

$$H = H^{(\mathrm{S})} + H^{(\mathrm{B})} + H^{(\mathrm{SB})} \tag{1}$$

We adopt a minimal model of a conical intersection which involves two excited electronic states $|\varphi_1\rangle$, $|\varphi_2\rangle$ and two vibrational modes, the coupling mode and the tuning mode. Keeping only the leading (linear) terms in the electronic–vibrational couplings, the system Hamiltonian is written in the diabatic representation as[17]

$$\begin{aligned} H^{(\mathrm{S})} &= \sum_{k=1,2} |\varphi_k\rangle h_k \langle\varphi_k| + (|\varphi_1\rangle\langle\varphi_2| + |\varphi_2\rangle\langle\varphi_1|)\lambda_\mathrm{s} Q_\mathrm{c} \\ h_k &= h_0 + E_k + \kappa_k Q_\mathrm{t} \\ h_0 &= h_\mathrm{c} + h_\mathrm{t} = \frac{1}{2}\sum_{j=\mathrm{c,t}} \hbar\Omega_j \left\{P_j^{\,2} + Q_j^{\,2}\right\} \end{aligned} \tag{2}$$

here, Q_j, P_j and Ω_j are the dimensionless coordinate, momentum and frequency of the coupling mode ($j = \mathrm{c}$) and the tuning mode ($j = \mathrm{t}$), respectively. E_k and κ_k (k = 1,2) are the vertical excitation energies and the intra-state electron-vibrational coupling constants, respectively, of the kth electronic state. λ_s denotes the inter-state coupling constant.

We assume that the coupling mode Q_c and the tuning mode Q_t are bilinearly coupled to an external bath. Due to the different symmetry of the two system modes in the present model of a symmetry-allowed conical intersection, it makes sense to assume that each of the two modes is coupled to its own bath. The Hamiltonian of the environment is thus written as

$$H^{(\mathrm{B})} = \{|\varphi_1\rangle\langle\varphi_1| + |\varphi_2\rangle\langle\varphi_2|\} \sum_{\alpha} \sum_{j=\{\mathrm{c,t}\}} \frac{1}{2}\hbar\omega_{\alpha,j}\left\{p_{\alpha,j}{}^2 + q_{\alpha,j}{}^2\right\} \tag{3}$$

where $p_{\alpha,j}$, $q_{\alpha,j}$ and $\omega_{\alpha,j}$ are the dimensionless momentum, coordinate, and frequency of the αth oscillator of the bath of the coupling mode (j = c) and the tuning mode (j = t), respectively.

We assume that the coupling and tuning vibrational modes are bilinearly coupled to the respective bath. The corresponding system–bath coupling Hamiltonian reads

$$H^{(\mathrm{SB})} = \{|\varphi_1\rangle\langle\varphi_1| + |\varphi_2\rangle\langle\varphi_2|\} \sum_{\alpha} \sum_{j=\{\mathrm{c,t}\}} \{c_{\alpha,j} q_{\alpha,j} Q_j\} \tag{4}$$

where $c_{\alpha,j}$ are the system–bath coupling constants. Alternative choices of system–bath coupling are discussed in ref. 20. The influence of the bath on the system dynamics is determined by the spectral density

$$J_j(\omega) = \sum_{\alpha} c_{\alpha,j}{}^2 \delta(\omega - \omega_{\alpha,j}), \quad j = \mathrm{c,t} \tag{5}$$

The system–bath Hamiltonians specified as above can be used for deriving master equations in two different ways.[39–43] One can either retain the strongly coupled high-frequency vibrational modes in the system Hamiltonian or move the modes to the bath, resulting in a structured bath. The two descriptions are, in principle, equivalent and connected *via* a canonical transformation.[39–44] In the context of the HEOM method, the inclusion of all vibrational modes in the bath is the common choice.[40,41,45] In the present study we prefer to keep the strongly coupled high-frequency vibrational modes in the system Hamiltonian (*cf.* ref. 46). This way, we can treat fairly general system Hamiltonians (*e.g.*, beyond the harmonic approximation for the system vibrational modes).

Introducing a basis set $|n_\mathrm{c}\rangle$, $|n_\mathrm{t}\rangle$ for the coupling mode and the tuning mode, respectively, we can construct the matrix representation of the total Hamiltonian

$$\begin{aligned} H = \sum_{k,k'=1,2} \sum_{n_\mathrm{c}n_\mathrm{t},n'_\mathrm{c}n'_\mathrm{t}} |\varphi_k\rangle|n_\mathrm{c}\rangle|n_\mathrm{t}\rangle \Big\{ H^{(\mathrm{S})}_{kn_\mathrm{c}n_\mathrm{t},k'n'_\mathrm{c}n'_\mathrm{t}} + H^{(\mathrm{SB})}_{kn_\mathrm{c}n_\mathrm{t},k'n'_\mathrm{c}n'_\mathrm{t}} \\ + H^{(\mathrm{B})}_{kn_\mathrm{c}n_\mathrm{t},k'n'_\mathrm{c}n'_\mathrm{t}} \Big\} \langle\varphi_{k'}|\langle n'_\mathrm{c}|\langle n'_\mathrm{t}| \end{aligned} \tag{6}$$

where

$$\begin{aligned} H^{(\mathrm{S})}_{kn_\mathrm{c}n_\mathrm{t},k'n'_\mathrm{c}n'_\mathrm{t}} = \delta_{kk'}\delta_{n_\mathrm{c}n'_\mathrm{c}}\delta_{n_\mathrm{t}n'_\mathrm{t}}(E_k + \hbar\Omega_\mathrm{c}(n_\mathrm{c} + 1/2) + \hbar\Omega_\mathrm{t}(n_\mathrm{t} + 1/2)) + \delta_{kk'}\delta_{n_\mathrm{c}n'_\mathrm{c}}\kappa_k A_{n_\mathrm{t}n'_\mathrm{t}} \\ + \left(\delta_{k,k'-1} + \delta_{k,k'+1}\right)\delta_{n_\mathrm{t}n'_\mathrm{t}}\lambda_\mathrm{s} A_{n_\mathrm{c}n'_\mathrm{c}} \end{aligned} \tag{7}$$

$$H^{(\mathrm{SB})}_{kn_\mathrm{c}n_\mathrm{t},k'n'_\mathrm{c}n'_\mathrm{t}} = \delta_{kk'} \sum_{\alpha} \left\{ \delta_{n_\mathrm{t}n'_\mathrm{t}} A_{n_\mathrm{c}n'_\mathrm{c}} c_{\alpha,\mathrm{c}} q_{\alpha,\mathrm{c}} + \delta_{n_\mathrm{c}n'_\mathrm{c}} A_{n_\mathrm{t}n'_\mathrm{t}} c_{\alpha,\mathrm{t}} q_{\alpha,\mathrm{t}} \right\} \tag{8}$$

$$H^{(\mathrm{B})}_{kn_\mathrm{c}n_\mathrm{t},k'n'_\mathrm{c}n'_\mathrm{t}} = \delta_{kk'}\delta_{n_\mathrm{c}n_\mathrm{c'}}\delta_{n_\mathrm{t}n_\mathrm{t'}} \sum_{\alpha} \sum_{j=\{\mathrm{c,t}\}} \frac{1}{2}\hbar\omega_{\alpha,j}\left\{p_{\alpha,j}{}^2 + q_{\alpha,j}{}^2\right\}$$

with

$$A_{nn'} = \left\langle n|Q|n'\right\rangle = \delta_{n,n'+1}\sqrt{\frac{n'+1}{2}} + \delta_{n,n'-1}\sqrt{\frac{n'}{2}} \tag{9}$$

In matrix form:

$$\mathbf{H} = \mathbf{H}^{(S)} + \sum_{\alpha}\{\mathbf{V}^{c}c_{\alpha,c}q_{\alpha,c} + \mathbf{V}^{t}c_{\alpha,t}q_{\alpha,t}\} + \mathbf{1}H^{(B)} \tag{10}$$

where

$$V^{c}_{kn_{c}n_{t},k'n'_{c}n'_{t}} = \delta_{kk'}\delta_{n_{t}n'_{t}}A_{n_{c}n'_{c}}$$

$$V^{t}_{kn_{c}n_{t},k'n'_{c}n'_{t}} = \delta_{kk'}\delta_{n_{c}n'_{c}}A_{n_{t}n'_{t}}$$

The total (system + bath) dynamics is described by the Liouville–von Neumann equation

$$\frac{\partial}{\partial t}\rho_{\mathrm{tot}}(t) = -\frac{i}{\hbar}[H, \rho_{\mathrm{tot}}] \tag{11}$$

The dynamics of the relevant system is described by the reduced density matrix $\rho(t)$, defined as the trace over the bath degrees of freedom of the full density matrix $\rho_{\mathrm{tot}}(t)$

$$\rho(t) = \mathrm{tr_B}\{\rho_{\mathrm{tot}}(t)\} \tag{12}$$

We assume instantaneous electronic excitation of the system from the electronic ground state $|\varphi_0\rangle$ to the diabatic state $|\varphi_2\rangle$ by an ideally short laser pulse. This corresponds to the preparation of the density matrix

$$\rho(0) = |\varphi_2\rangle|\mathbf{0}\rangle\langle\mathbf{0}|\langle\varphi_2| \tag{13}$$

where $|\mathbf{0}\rangle = |0_c\rangle|0_t\rangle$ denotes the vibrational ground state of the coupling and tuning modes in the electronic ground state.

We model the bath by the Drude spectral density,[7]

$$J_j(\omega) = 2\lambda_j\gamma_j\omega/(\omega^2 + \gamma_j^2),\ j = \mathrm{c, t} \tag{14}$$

here, λ_j determines the coupling strength of the system modes (coupling mode and tuning mode) to the respective bath, and γ_j is the Drude decay constant, so that $1/\gamma_j$ is the bath correlation (memory) time. This choice of the spectral density facilitates the application of the HEOM method.[38]

B. HEOM method

The HEOM method is a numerically exact method which allows the computation of the time evolution of the reduced density matrix $\rho(t)$ beyond the Markovian approximation and for arbitrary system–bath coupling strength.[35–38]

Using the matrix representation of the Hamiltonian (6) and adopting the high-temperature approximation ($\beta\hbar\gamma_j < 1$, where $\beta = (k_B T)^{-1}$ is inverse temperature) which is fulfilled in our simulations, one obtains the following hierarchy of coupled equations of motion for the reduced density matrix

$$\frac{\partial}{\partial t}\rho_{l_c,l_t}(t) = -\left(i\mathscr{L}^{(s)} + l_c\gamma_c + l_t\gamma_t\right)\rho_{l_c,l_t}(t) + \Phi_c\rho_{l_c+1,l_t}(t) + \Phi_t\rho_{l_c,l_t+1}(t) + l_c\gamma_c\Theta_c\rho_{l_c-1,l_t}(t) + l_t\gamma_t\Theta_t\rho_{l_c,l_t-1}(t) \quad (15)$$

where l_c, l_t are non-negative integers. In eqn (15), the density operator with all indexes equal to zero, $\rho_{00}(t)$, is the reduced density operator $\rho(t)$. All other density operators ρ_{l_c,l_t} are auxiliary density operators which are introduced to describe nonperturbative and non-Markovian effects. The Liouvillian super operator corresponding to the system Hamiltonian $H^{(S)}$ is denoted by $\mathscr{L}^{(s)}$, and the relaxation operators Φ and Θ are defined as

$$\Phi_j = iV_j^{\times} \quad (16)$$

$$\Theta_j = i\left(\frac{2\lambda_j}{\beta\hbar^2}V_j^{\times} - i\frac{\lambda_j}{\hbar}\gamma_j V_j^{\circ}\right) \quad (17)$$

where $A^{\circ}\rho = A\rho + \rho A$ and $A^{\times}\rho = A\rho - \rho A$ for any operator A. Note that the HEOM method does not require the construction of the eigenstate representation for the system Hamiltonian, as multi-level Redfield theory does.[20]

The system of eqn (15) represents an infinite number of coupled master equations. It can be shown that the condition for the termination of the hierarchy is $N = l_c + l_t \gg \omega^{(S)}/\min(\gamma_1, \gamma_2)$, where $\omega^{(S)}$ is a characteristic frequency of $\mathscr{L}^{(s)}$.[47] In this case, the last term in the infinite hierarchy of eqn (15) can be replaced by

$$\frac{\partial}{\partial t}\rho_{l_c,l_t}(t) = -i\mathscr{L}^{(s)}\rho_{l_c,l_t}(t) \quad (18)$$

Low-temperature correction terms have to be included in eqn (15) when the condition $\beta\hbar\gamma_j < 1$ is not satisfied (see ref. 37 and 38). The low-temperature correction results in the replacements

$$\begin{aligned} \mathscr{L}^{(s)} &\rightarrow \mathscr{L}^{(s)} - \sum_{j=c,t}\frac{2\lambda_j}{\beta\hbar^2}\frac{2\gamma_j}{\nu_1^2 - \gamma_j^2}V_j^{\times}V_j^{\times} \\ \Theta_j &\rightarrow \Theta_j - \frac{2\lambda_j}{\beta\hbar^2}\frac{2\gamma_j^2}{\nu_1^2 - \gamma_j^2}V_j^{\times} \end{aligned} \quad (19)$$

where $\nu_1 = 2\pi/\beta\hbar$ is the first bosonic Matsubara frequency.[36] Higher-order Matsubara terms can be included to further improve the accuracy of low-temperature correction.

C. Physical observables

Once the reduced density matrix $\rho(t)$ is given, the time evolution of system observables can straightforwardly be calculated. The most important observables for monitoring radiationless decay dynamics are the time-dependent populations of the electronic states. The population of the kth diabatic electronic state is defined as[10]

$$P_k^{\text{di}}(t) = \text{tr}_S\{|\varphi_k\rangle\langle\varphi_k|\rho(t)\} \tag{20}$$

The population of the adiabatic electronic states is given by

$$P_k^{\text{ad}}(t) = \text{tr}_S\{|\tilde{\varphi}_k\rangle\langle\tilde{\varphi}_k|\rho(t)\} \tag{21}$$

The adiabatic electronic states are related to the diabatic states *via*

$$|\tilde{\varphi}_k\rangle = \sum_{k'=1,2} \mathbf{S}(Q_c,\ Q_t)_{kk'}|\varphi_{k'}\rangle, \quad k = 1,2 \tag{22}$$

where

$$\mathbf{S}(Q_c, Q_t) = \begin{pmatrix} \cos\alpha(Q_c, Q_t) & -\sin\alpha(Q_c, Q_t) \\ \sin\alpha(Q_c, Q_t) & \cos\alpha(Q_c, Q_t) \end{pmatrix}$$

and $\alpha(Q_c, Q_t)$ is the diabatic-to-adiabatic mixing angle. The adiabatic population P_2^{ad} is the observable which directly reflects the radiationless electronic decay dynamics from the upper to the lower adiabatic energy surface. The diabatic electronic population P_2^{di}, on the other hand, is directly related to spectroscopic signals, *e.g.* to the total (frequency-integrated) spontaneous emission signal.[48] Note that the system modes Q_c and Q_t are traced out in the definition of P_2^{ad} and P_2^{di} and therefore can be considered as part of the bath.

The coherence of vibrational motion of the system is monitored by the expectation values of the position and momentum operators of the system vibrational modes, which are defined as[10]

$$\langle Q_j\rangle_t = \text{tr}_S\{Q_j\rho(t)\} \tag{23}$$

$$\langle P_j\rangle_t = \text{tr}_S\{P_j\rho(t)\} \tag{24}$$

For the nontotally symmetric coupling mode ν_c of the conical intersection, the expectation values of position and momentum operators vanish by symmetry. To visualize the coupled electronic-vibrational dynamics of the system, we consider the reduced probability densities of the coupling and tuning modes in the adiabatic electronic representation, defined as[10]

$$\begin{aligned} P_k^{\text{ad}}(Q_c, t) &= \int \text{d}Q_t\langle Q_c|\langle Q_t|\langle\tilde{\varphi}_k|\rho(t)|\tilde{\varphi}_k\rangle|Q_t\rangle|Q_c\rangle \\ P_k^{\text{ad}}(Q_t, t) &= \int \text{d}Q_c\langle Q_c|\langle Q_t|\langle\tilde{\varphi}_k|\rho(t)|\tilde{\varphi}_k\rangle|Q_t\rangle|Q_c\rangle \end{aligned} \tag{25}$$

Under certain conditions, these probability densities are related to the time- and frequency-resolved fluorescence spectrum.[49]

D. Parametrization of the Hamiltonian and computational details

The $S_2[^1B_{2u}(\pi\pi^*)]$–$S_1[^1B_{3u}(n\pi^*)]$ conical intersection in pyrazine is a well-known example for ultrafast electronic relaxation *via* a conical intersection.[50–53] Various models of this conical intersection, comprising from 2 to 24 vibrational modes, have been constructed.[54–59] These models have served as testbeds for the benchmarking of novel methods for the theoretical description of electronically nonadiabatic dynamics.[60–72] The out-of-plane normal mode ν_{10a} of B_{1g} symmetry is the only normal mode which can couple the $^1B_{2u}$ and $^1B_{3u}$ electronic states in first order. The mode ν_{6a} of A_{1g} symmetry is the dominant tuning mode of this conical intersection. For the present purposes, a two-mode model of the S_2–S_1 conical intersection including the normal modes ν_{10a} and ν_{6a} is suitable as a simple and generic model of a conical intersection.

The numerical values of the parameters of the model Hamiltonian $H^{(S)}$ of eqn (2) are taken from ref. 55. The frequencies of the coupling and tuning modes are $\hbar\omega_c = 952$ cm^{-1} ($2\pi/\omega_c = 35$ fs) and $\hbar\omega_t = 597$ cm^{-1} ($2\pi/\omega_t = 56$ fs). The intra-state electron-vibration coupling constants are $\kappa_1 = -847$ cm^{-1} and $\kappa_2 = 1202$ cm^{-1}. The inter-state vibronic coupling constant is $\lambda_s = 2110$ cm^{-1}. The vertical excitation energies are $E_1 = 31\,800$ cm^{-1} and $E_2 = 39\,000$ cm^{-1}. Cuts through the potential-energy surfaces along the normal coordinates Q_t and Q_c are shown in Fig. 1. The bath temperature is set to 300 K.

The numerical calculation of the system dynamics is based on the representation of the Hamiltonian $H^{(S)}$ in a direct product basis of diabatic electronic states and harmonic-oscillator basis functions for the coupling and tuning modes. We have employed a basis set of 20 harmonic-oscillator functions for each mode, resulting in a dimension of 800 of the system Hamiltonian matrix $\mathbf{H}^{(S)}$. The numerical evaluation of $\rho(t)$ *via* the HEOM method requires the propagation of a large number of coupled equations of motion for auxiliary matrices of large dimension, which is computationally demanding. We overcame the computational challenge by using a GPU (Graphic Processing Unit) implementation of the ArrayFire package which provides an easy-to-use API (application programming interface) and an array-based function set that facilitates GPU programming.[73] The numerical integration of the equations of motion for the auxiliary density matrices are performed on a NVIDIA Tesla M2050 GPU with 448 processors and 2 GB ECC-protected on-board memory. The HEOM eqn (15) has been solved *via* a fourth-order Runge–Kutta method with a time step of 0.2 fs. Depending on the value of system–bath coupling strength λ_j, j = c, t, the truncation threshold N of the hierarchy was varied from 6 to 12, which was found to be sufficient for

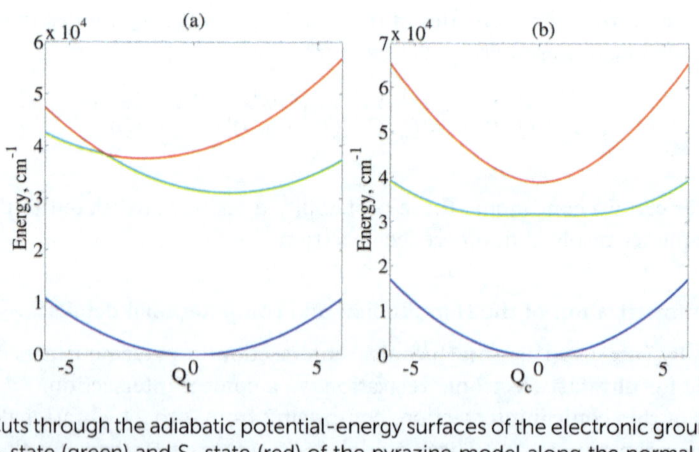

Fig. 1 Cuts through the adiabatic potential-energy surfaces of the electronic ground state (blue), S_1 state (green) and S_2 state (red) of the pyrazine model along the normal coordinates Q_t (a) and Q_c (b).

obtaining converged results. With this efficient algorithm implemented on GPU hardware, the propagation of $\rho(t)$ up to 1000 fs on a single GPU requires from several hours to 10 hours, depending on the truncation number N. The matrix-matrix multiplications in eqn (15) are carried out by the ArrayFire function matmul. The adiabatic populations $P_k^{\rm ad}$ are obtained by evaluating the expectation values of the projection operators on the lower and upper adiabatic electronic states in the diabatic representation, as described in ref. 74. The vibrational probability densities (25) are calculated by using a discrete variable representation (DVR) scheme. In this DVR scheme, a transformation between the harmonic-oscillator basis and a grid in coordinate space is performed to represent operators in coordinate space (see ref. 75 for details).

III. Results

A. Electronic population probabilities and vibrational coherences

Fig. 2(a) shows the time-dependent population $P_2^{\rm di}(t)$ of the diabatic $\pi\pi^*$ state for different system–bath coupling strengths on a time scale of 1 ps. The time evolution of $P_2^{\rm di}(t)$ for the isolated system ($\lambda_c = \lambda_t = 0$, black line) exhibits an initial decay on a timescale of $\approx$30 fs, followed by a few quasi-periodic recurrences with the period of the tuning mode. Beyond about 400 fs, the $\pi\pi^*$ population

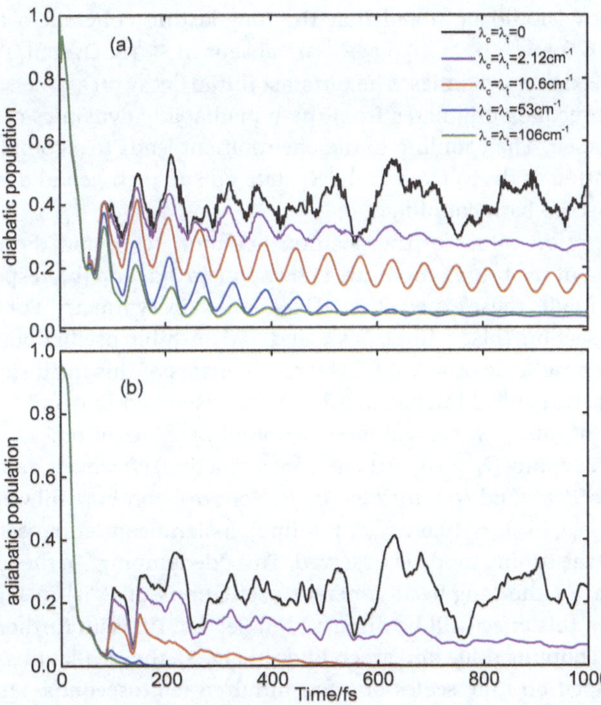

Fig. 2 Diabatic (a) and adiabatic (b) population probability of the upper electronic state of the pyrazine model for bath relaxation times $\gamma_c^{-1} = \gamma_t^{-1} = 50$ fs and different system–bath coupling strengths: $\lambda_c = \lambda_t = 0$ (black), 2.1 (magenta), 10.6 (red), 53.0 (blue), and 106.0 cm^{-1} (green).

probability exhibits irregular fluctuations. The long-time average of $P_2^{\rm di}(t)$ is ≈ 0.4. For the system–bath coupling strengths considered here, the coupling to the bath does not affect the fast initial decay of $P_2^{\rm di}(t)$ which is driven by the strong nonadiabatic coupling at the conical intersection. On the other hand, the irregular fluctuations of $P_2^{\rm di}(t)$ for $t \geq 400$ fs are wiped out by even very weak system–bath coupling ($\lambda_c = \lambda_t = 2.1$ cm^{-1}, magenta line). With increasing, but still weak, system–bath coupling ($\lambda_c = \lambda_t = 10.6$ cm^{-1}, red line), $P_2^{\rm di}(t)$ exhibits, counterintuitively, a revival of the coherent oscillations driven by the tuning mode which extends now well beyond 1 ps. With further increasing system–bath coupling, the coherent oscillations of $P_2^{\rm di}(t)$ are damped out on a time scale of a few hundred femtoseconds, see Fig. 2(a), blue and green lines. The long-time limit of $P_2^{\rm di}(t)$ decreases with increasing system–bath coupling and is seen to converge to a small, but nonzero, value.

The time evolution of the population $P_2^{\rm ad}(t)$ of the adiabatic S_2 state is shown in Fig. 2(b). The initial decay of $P_2^{\rm ad}(t)$ of the isolated conical intersection (black line) occurs on a time scale of ≈ 20 fs and is somewhat faster than the initial decay of $P_2^{\rm di}(t)$. $P_2^{\rm ad}(t)$ approaches an asymptotic value of ≈ 0.2 at 1 ps. The system–bath coupling does not affect the short-time dynamics within the first ≈ 100 fs. As observed for the diabatic population probability, the irregular fluctuations of $P_2^{\rm ad}(t)$ are efficiently damped out by even very weak system–bath coupling ($\lambda_c = \lambda_t = 2.1$ cm^{-1}, magenta line). With increasing system–bath coupling, the population of the S_2 state quickly drops to zero, see Fig. 2(b). For the strongest system–bath coupling considered ($\lambda_c = \lambda_t = 106.0$ cm^{-1}), all recurrences of $P_2^{\rm ad}(t)$ are quenched (green line). It should be noted that the long-lasting coherent oscillations of $P_2^{\rm di}(t)$ observed for $\lambda_c = \lambda_t \geq 10.6$ cm^{-1} are absent in $P_2^{\rm ad}(t)$. Overall, $P_2^{\rm ad}(t)$ reveals a bimodal relaxation dynamics. The ultrafast initial decay on a time scale of a few tens of femtoseconds originates from the nonadiabatic dynamics of the conical intersection itself. The coupling to the environment leads to a complete transfer of the population of the S_2 state to the S_1 state on sub-picosecond time scales for even weak system–bath coupling.

The expectation values of the position operator, $\langle Q_t \rangle_t$, and the momentum operator, $\langle P_t \rangle_t$, of the tuning mode are shown in Fig. 3(a) and (b), respectively. For the coupling mode, these expectation values vanish by symmetry. For the isolated conical intersection (black line), $\langle Q_t \rangle_t$ and $\langle P_t \rangle_t$ exhibit oscillations which are damped on a time scale of ≈ 300 fs. The mechanisms of this intrinsic vibrational dephasing of the conical intersection has been discussed in detail for the three-mode model of the S_2–S_1 conical intersection of pyrazine in ref. 55. A very weak system–bath coupling ($\lambda_c = \lambda_t = 2.1$ cm^{-1} magenta line) effectively damps residual fluctuations of $\langle Q_t \rangle_t$ and $\langle P_t \rangle_t$ for $t \geq 500$ fs. For stronger, but still weak, system–bath coupling ($\lambda_c = \lambda_t = 10.6$ cm^{-1}, red line), a significant enhancement of the amplitude of the tuning mode is observed. This "dedamping" of the tuning mode is the reason for the long-lived coherent oscillations of $P_2^{\rm di}(t)$ in Fig. 2(a). The mechanism of this effect will be discussed in Section IV. With further increasing system–bath coupling (blue and green lines in Fig. 3), the oscillations of $\langle Q_t \rangle_t$ and $\langle P_t \rangle_t$ are damped on time scales of a few hundred femtoseconds. Moreover, the long-time limit of $\langle Q_t \rangle_t$ is seen to shift towards an asymptotic value, which corresponds to the minimum of the adiabatic S_1 PE surface, see Fig. 1. This reflects the flow of excess vibrational energy from the conical intersection into the environment.

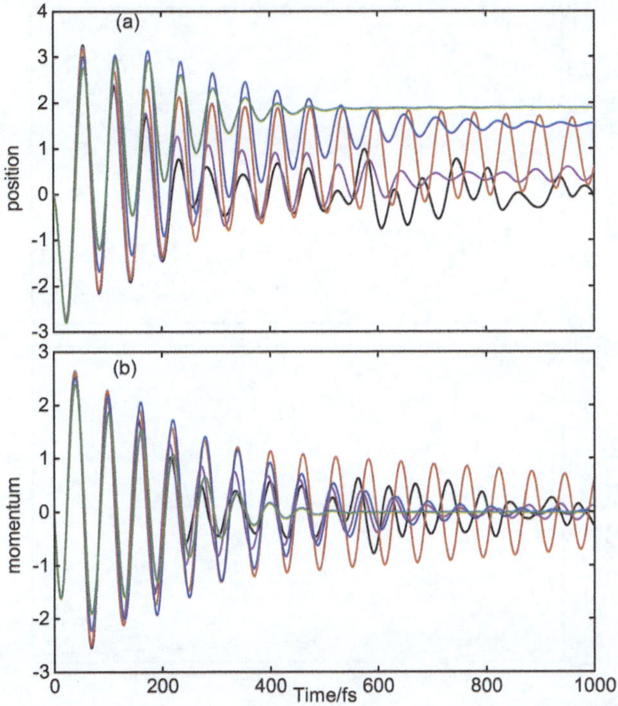

Fig. 3 Expectation values of position (a) and momentum (b) operator of the tuning mode for bath relaxation times $\gamma_c^{-1} = \gamma_t^{-1} = 50$ fs and different system–bath coupling strengths: $\lambda_c = \lambda_t = 0$ (black), 2.1 (magenta), 10.6 (red), 53.0 (blue), and 106.0 cm^{-1} (green).

Next we consider the effect of the bath relaxation time γ^{-1} on the electronic populations and vibrational coherences. Fig. 4 shows the time-dependent population probabilities of the diabatic $\pi\pi^*$ state (a) and the adiabatic S_2 state (b), calculated for a longer bath relaxation time ($\gamma_c^{-1} = \gamma_t^{-1} = 166$ fs) and three different system–bath coupling strengths. Compared to Fig. 2 (bath relaxation time $\gamma_c^{-1} = \gamma_t^{-1} = 50$ fs), the coherent beatings of $P_2^{\mathrm{di}}(t)$ last longer in Fig. 4(a). The adiabatic populations $P_2^{\mathrm{ad}}(t)$ in Fig. 4(b) also decay more slowly than those in Fig. 2(b). The bath with longer relaxation time is less effective in draining coherence and energy from the dynamics of the system.

The corresponding expectation values of $\langle Q_t \rangle_t$ and $\langle P_t \rangle_t$ are shown in Fig. 5(a) and (b), respectively. The damping of the oscillations of $\langle Q_t \rangle_t$ and $\langle P_t \rangle_t$ due to system–bath coupling is less pronounced than in Fig. 3. The effect of dedamping of $\langle Q_t \rangle_t$ and $\langle P_t \rangle_t$ occurs for a higher value of the system–bath coupling strength, $\lambda_c = \lambda_t = 53.0$ cm^{-1} (blue line in Fig. 5). For strong system–bath coupling ($\lambda_c = \lambda_t = 106.0$ cm^{-1} (green line)), the oscillations of $\langle Q_t \rangle_t$ and $\langle P_t \rangle_t$ are completely quenched at $\approx$900 fs, while they were quenched already at $\approx$500 fs for the shorter bath relaxation time (Fig. 3).

B. Vibrational probability densities

The most detailed insight into the dynamics of the ultrafast internal-conversion process in the present model of a dissipative conical intersection is provided by

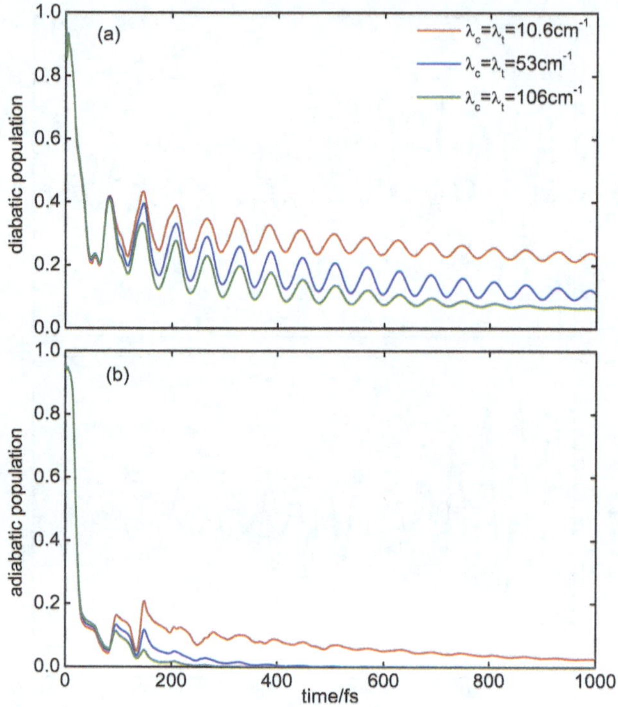

Fig. 4 Diabatic (a) and adiabatic (b) population probability of the upper electronic state of the pyrazine model for bath relaxation times $\gamma_c^{-1} = \gamma_t^{-1} = 166$ fs and different system–bath coupling strengths: $\lambda_c = \lambda_t = 10.6$ (red), 53.0 (blue), and 106.0 cm^{-1} (green).

the time-dependent probability densities $P_k^{ad}(Q_c,t)$, $P_k^{ad}(Q_t,t)$ defined in eqn (25). In what follows we call $P_k^{ad}(Q_j,t)$, j = c, t, probability densities of Q_c and Q_t, but it should be kept in mind that they are reduced probability densities, that is, the respective other vibrational mode is integrated out.

Fig. 6 shows the probability densities $P_2^{ad}(Q_t,t)$ (upper panels) and $P_1^{ad}(Q_t,t)$ (lower panels) of the tuning mode for zero system–bath coupling (left column) and system–bath coupling of intermediate strength ($\lambda_c = \lambda_t = 53.0$ cm^{-1}, right column). Let us consider the isolated conical intersection first (panels (a) and (c)). The localized wave packet prepared at $t = 0$ on the S_2 surface propagates to the right (towards negative Q_t) following the gradient of the S_2 surface (see Fig. 1(a)). Within half of a vibrational period, the wave packet reaches the conical intersection and partially transfers to the S_1 surface, where it experiences a gradient in the opposite direction. This results in coherent wave-packet dynamics in the tuning mode at short times, see panels (a) and (c). For the isolated two-mode conical intersection, the $S_2 \rightarrow S_1$ radiationless transition is incomplete. Beyond ≈200 fs, the structure of the probability density becomes more and more chaotic, especially on the S_1 surface. The coupling of the conical intersection to the environment results in a qualitatively different picture (panels (b) and (d)). The probability densities are smoothed and exhibit rather regular quasi-periodic behavior. The transfer of energy to the bath assists in the relaxation of the wave packet from the upper adiabatic surface to the lower adiabatic

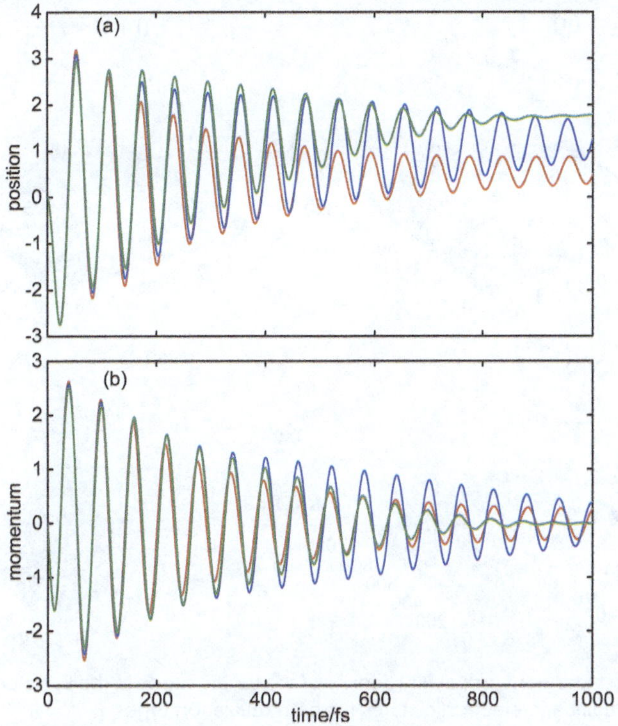

Fig. 5 Expectation values of the position (a) and momentum (b) operator of the tuning mode for bath relaxation times $\gamma_c^{-1} = \gamma_t^{-1} = 166$ fs and different system–bath coupling strengths: $\lambda_c = \lambda_t = 10.6$ (red), 53.0 (blue), and 106.0 cm^{-1} (green).

surface. Recurrences of the probability density to the upper surface are strongly suppressed. For the dissipative conical intersection, the probability density of the tuning mode on the S_1 surface at $t = 1$ ps is peaked at $Q_t \approx 1.8$ (panel (d)), which is not the case for the isolated conical intersection (panel (c)). This illustrates that the vibrational excess energy available after the $S_2 \rightarrow S_1$ radiationless transition is effectively drained into the environment, allowing relaxation of the system to a stationary state.

The probability densities $P_2^{ad}(Q_c,t)$ (upper panels) and $P_1^{ad}(Q_c,t)$ (lower panels) of the coupling mode are shown in Fig. 7. As in Fig. 6, the left column shows the bath-free dynamics, while the right column displays the dynamics of the conical intersection coupled to a bath with $\lambda_c = \lambda_t = 53.0$ cm^{-1}. For the isolated conical intersection (panels (a) and (c)) the probability density in Q_c spreads very rapidly in a symmetric manner. The probability density on the upper surface exhibits three "ridges" at $Q_c = 0$ and $Q_c = \pm 5$ (the outer turning point in Q_c). The wave packet on the lower surface exhibits a symmetry-required node at $Q_c = 0$ and the width of the probability density shrinks with time. Beyond $\approx$200 fs, the probability densities on both surfaces are quite noisy. Panels (b) and (d) of Fig. 7 illustrate the effect of the system–bath coupling on the probability densities $P_2^{ad}(Q_c,t)$ and $P_1^{ad}(Q_c,t)$. The peak of $P_2^{ad}(Q_c,t)$ at $Q_c = 0$ rapidly disappears, while the peaks near the turning points decrease in height and move inwards, reflecting the

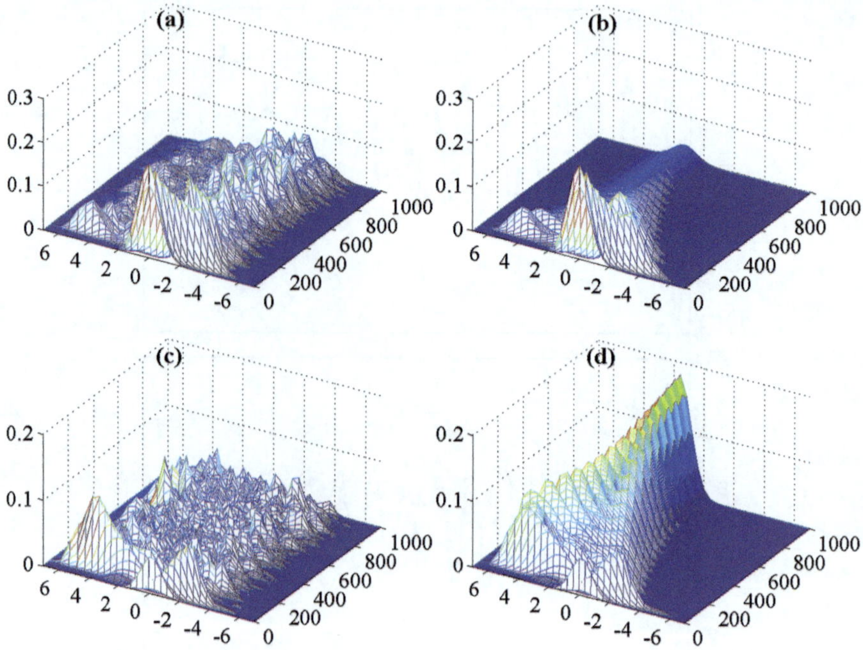

Fig. 6 Time dependent probability density of the tuning mode in the S_2 (upper row) and S_1 (lower row) adiabatic electronic states for bath relaxation times $\gamma_c^{-1} = \gamma_t^{-1} = 50$ fs. Left column: $\lambda_c = \lambda_t = 0$; right column: $\lambda_c = \lambda_t = 53.0$ cm^{-1}.

loss of vibrational energy from the coupling mode. On the lower adiabatic surface, the width of the vibrational distribution narrows even more rapidly and the zero of $P_1^{ad}(Q_c = 0, t = 0)$ develops into a pronounced peak centered at $Q_c = 0$, see Fig. 7(d). As observed for the tuning mode, the probability densities of the coupling mode are much smoother than for the isolated two-mode conical intersection.

IV. Discussion

Considering the population dynamics of the diabatic $\pi\pi^*$ state, it is eye-catching how sensitive the electronic population dynamics is to the coupling of the system modes to an environment (Fig. 2(a) and 4(a)). Even the weakest system–bath coupling considered here ($\lambda_c = \lambda_t = 2.1$ cm^{-1}), completely quenches the irregular fluctuations of $P_2^{di}(t)$ beyond 400 fs (Fig. 2(a), magenta line). With somewhat larger, but still weak, system–bath coupling ($\lambda_c = \lambda_t = 10.6$ cm^{-1}), a qualitatively different behavior of $P_2^{di}(t)$ is observed: the diabatic population probability exhibits completely regular coherent beatings with the frequency of the tuning mode (Fig. 2(a), red line). With further increases in system–bath coupling, these coherent beatings are damped out on time scales of a few hundred femtoseconds. For a longer bath relaxation time ($\gamma_c^{-1} = \gamma_t^{-1} = 166$ fs), the coherent oscillations live longer.

It is interesting to compare these numerically exact results with earlier results obtained with multi-level Redfield theory for the same two-state two-mode conical

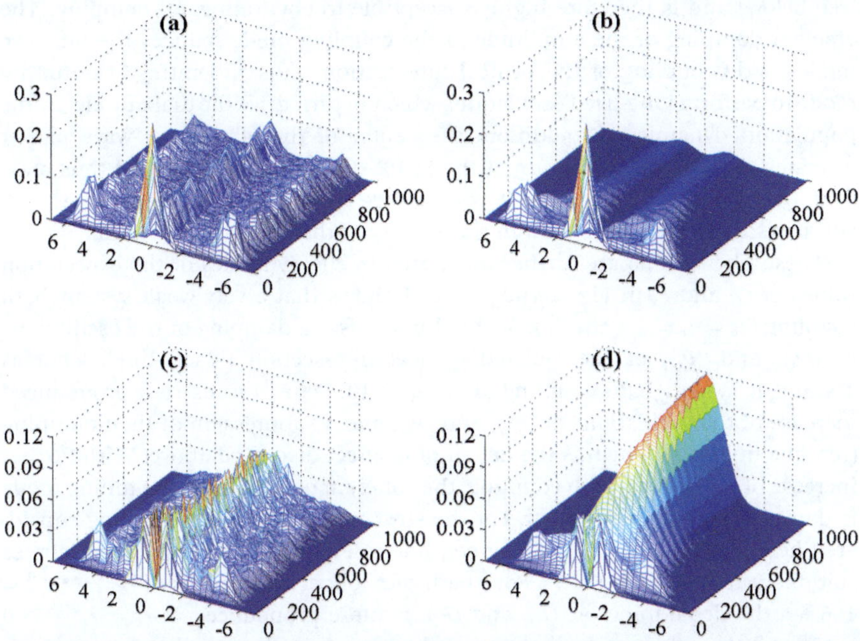

Fig. 7 Time dependent probability density of the coupling mode in the S_2 (upper row) and S_1 (lower row) adiabatic electronic states for bath relaxation times $\gamma_c{}^{-1} = \gamma_t{}^{-1} = 50$ fs. Left column: $\lambda_c = \lambda_t = 0$; right column: $\lambda_c = \lambda_t = 53.0$ cm^{-1}.

intersection coupled in the same manner to an ohmic bath.[19] In this earlier work, the secular approximation[4,24] to multi-level Redfield theory was employed. For a very weak system–bath coupling ($\eta = 0.003$ in ref. 19, $\lambda_c = \lambda_t = 2.1$ cm^{-1} in Fig. 2(a)), the diabatic population probabilities are essentially identical. It can be concluded that for very weak system–bath coupling the spectral function of the bath does not matter and multi-level Redfield theory gives correct results even with the secular approximation. For somewhat stronger system–bath coupling ($\eta = 0.015$ in ref. 19, $\lambda_c = \lambda_t = 10.6$ cm^{-1} in Fig. 2(a)), on the other hand, secular Redfield theory fails dramatically. The pronounced coherent beatings of $P_2^{di}(t)$ (red line in Fig. 2(a)) are completely absent in the simulations with secular Redfield theory.[19] A later study performed with non-secular multi-level Redfield theory for a three-mode model of the S_2–S_1 conical intersection in pyrazine[20] reproduces the coherent oscillations of $P_2^{di}(t)$, albeit somewhat less pronounced and less regular than in Fig. 2(a), which may be a consequence of the additional tuning mode or the different spectral densities employed. A very similar result was obtained by Gelman *et al.*[26] for the two-mode model with a modified (Lindblad-type) Redfield theory. It can be concluded that non-secular multi-level Redfield theory gives a qualitatively correct description of the diabatic electronic population dynamics for weak coupling of a conical intersection to a bath (see also ref. 22). The mechanism of the dedamping of the tuning mode for intermediate system–bath coupling strength, which leads to the pronounced coherent beatings in $P_2^{di}(t)$, was qualitatively explained in ref. 20. The coupling mode becomes particularly highly excited during the ultrafast internal conversion process at the conical intersection (see below) and is therefore highly susceptible to environmental damping. The efficient damping of the amplitude of the coupling mode reduces the intrinsic mode–mode coupling of the conical intersection, thus "liberating" the tuning mode to perform long-lived oscillations which in turn drive the diabatic electronic population dynamics. In addition, the center of the vibrational wave packet migrates towards the minimum of the S_1 PE surface at $Q_t \approx 1.4$ and thus away from the conical intersection at $Q_t \approx -3.5$ (see Fig. 2(a)) to a region where the PE surface is more harmonic than in the vicinity of the conical intersection.

These phenomena are further illustrated by the dynamics of the expectation values of Q_t and P_t in Fig. 3 and 5. Fig. 3 shows that a very weak system–bath coupling ($\lambda_c = \lambda_t = 2.1$ cm^{-1}, magenta line) causes a damping of the oscillations of $\langle Q_t \rangle_t$ and $\langle P_t \rangle_t$ of the isolated conical intersection (black line), whereas a stronger system–bath coupling ($\lambda_c = \lambda_t = 10.6$ cm^{-1}) leads to a pronounced increase of the oscillations of $\langle Q_t \rangle_t$ and $\langle P_t \rangle_t$ and a suppression of their damping (red line in Fig. 3), which is the dedamping effect discussed above. With further increase of the system–bath coupling, the coherent oscillation of the tuning mode is damped by the environment. For the strongest system–bath coupling considered here ($\lambda_c = \lambda_t = 106.0$ cm^{-1}, green line in Fig. 3), the wave packet becomes stationary at $Q_t \approx 2.0$ at 500 fs. For a bath with longer memory time (${\gamma_c}^{-1} = {\gamma_t}^{-1} =$ 166 fs), the dedamping of $\langle Q_t \rangle_t$ and $\langle P_t \rangle_t$ is more pronounced at $\lambda_c = \lambda_t = 53.0$ cm^{-1} and the coherent oscillations of the tuning mode extend to longer times (blue line in Fig. 5).

Let us now consider the time evolution of the population probability $P_2^{ad}(t)$ of the S_2 state. Again, a very weak system–bath coupling, $\lambda_c = \lambda_t = 2.1$ cm^{-1}, is remarkably efficient in suppressing the irregular fluctuations of $P_2^{ad}(t)$ of the isolated conical intersection (Fig. 2(b), magenta line). With increasing system–bath coupling, the decay of $P_2^{ad}(t)$ is significantly enhanced. For $\lambda_c = \lambda_t \geq 53.0$ cm^{-1}, $P_2^{ad}(t)$ decays within less than 200 fs. It is noteworthy that the long-lived coherent oscillations of $P_2^{di}(t)$ are completely absent in $P_2^{ad}(t)$. This observation underlines that the long-lived coherent wave-packet motion occurs exclusively on the lower adiabatic PE surface away from the conical intersection. With longer bath memory time, ${\gamma_c}^{-1} = {\gamma_t}^{-1} = 166$ fs, the quenching of $P_2^{ad}(t)$ by the environment is less pronounced, see Fig. 2(b).

The reduced probability densities of the coupling mode and the tuning mode in Fig. 6 and 7, respectively, provide additional insight. The probability densities in panels (a) and (c) clearly exhibit the irregular, chaotic character of the dynamics of the isolated two-state two-mode conical intersection. These irregularities are the time-domain signatures of the dense eigenvalue spectrum with incommensurable energy-level spacings which arise from the exceptionally strong nonadiabatic coupling at the conical intersection and the pronounced local anharmonicity of the adiabatic PE surfaces. These chaotic features of the dynamics at the conical intersection, which are reflected in various level spacing distributions,[17] are highly sensitive to perturbations and are therefore eliminated by even very weak coupling to a dissipative environment. In contrast, the probability densities obtained for $\lambda_c = \lambda_t = 53.0$ cm^{-1} and shown in panels (b) and (d) of Fig. 6 and 7 are completely smooth and regular. The chaotic character of the dynamics was not observed in previous simulations for a three-mode model of the S_2–S_1 conical intersection in pyrazine,[20,55] presumably because the second tuning mode of the conical intersection (the ν_1 normal mode of pyrazine) already acts as a quencher of the chaotic dynamics.

The vibrational probability densities $P_k^{\mathrm{ad}}(Q_c,t)$ and $P_k^{\mathrm{ad}}(Q_t,t)$ of the dissipative conical intersection in Fig. 6(b) and (d) and 7(b) and (d) illustrate how the large excess energy in the tuning mode and the coupling mode, which is generated within $\approx$30 fs after excitation, is drained out of the system dynamics, resulting in a nearly stationary wave packet at $\approx$1 ps. Comparing panels (a) and (b) of Fig. 7, it can be seen that the ridge of the density surviving on the S_2 surface at $Q_c = 0$ in panel (a) is turned into a valley by the coupling with the bath (panel (b)). The recurrences of the wave packet to the S_2 surface are strongly suppressed and occur near the minimum geometry of the S_1 PE surface in the Q_c coordinate (see Fig. 1(b)), opposed to the isolated conical intersection, where the recurrences to the S_2 surface occur at $Q_c = 0$ and at the outer turning points of the motion in Q_c, see Fig. 7(a).

V. Conclusions

We performed extensive simulations of the electronic and vibrational dynamics of a two-state two-mode conical intersection which is weakly to moderately coupled to a dissipative environment. The bath oscillators are bilinearly coupled to the coupling mode and the tuning mode of the conical intersection. The chaotic features of the intrinsic dynamics of the conical intersection, which are clearly visible in the electronic population dynamics and the reduced vibrational probability densities, are completely wiped out by even very weak system–bath coupling. While the ultrafast nonadiabatic dynamics within the first $\approx$30 fs is unaffected by the coupling to the environment, the population probability of the upper adiabatic state is effectively quenched on sub-picosecond time scales by weak to moderate system–bath coupling. On the other hand, a counterintuitive dedamping of the dynamics of the tuning mode is observed for specific values of the system–bath coupling strength which depend on the bath memory time. This effect is a signature of the exceptionally pronounced nonseparability of the dynamics of the coupling and tuning modes of the conical intersection, to an extent that damping of the coupling mode liberates the dynamics of the tuning mode. Overall, the results confirm that the nonadiabatic wave-packet dynamics at a conical intersection is unusually sensitive to weak coupling to a thermal environment with the exception of the first few tens of femtoseconds. Due to the complexity of the intrinsic dynamics of the conical intersection, the simulations exhibit phenomena which are quite different from the familiar scenarios of the optical Bloch equations or the Brownian oscillator. Our calculations show, in particular, that dissipative conical intersections not only can maintain vibrational wave-packets revealing the tuning modes, but also can create vibrational wave-packets featuring the coupling mode(s). Related effects have been observed in recent time and frequency resolved experiments on β-carotene and rhodopsin.[76,77]

The present study extends earlier simulations of the dynamics of conical intersections coupled to an environment, which were based on multi-level Redfield theory or similar models that are restricted to weak system–bath coupling and short bath relaxation time.[19–22,25,26] With the HEOM method, we were able to explore system–bath couplings in the nonperturbative regime. With this method, system–bath couplings of arbitrary strength can be investigated for complex systems such as conical intersections, as demonstrated recently also by Duan and Thorwart.[41] In future extensions of the present work, the coupling of the material system with femtosecond pulsed laser fields will be included in a nonperturbative manner. While a nonperturbative treatment of the laser-matter interaction may not be a necessity for the laser fields typically employed in spectroscopy, the inclusion of the laser–matter interaction in the equation of motion of the density matrix may be more straightforward and computationally efficient than the calculation of higher-order multi-time response functions and multiple nested integrals over the laser fields, which are required in the perturbative theory of nonlinear optics.[78] The calculation of femtosecond time-resolved four-wave-mixing signals from laser-driven equations of motion has been demonstrated, for example, for pump-probe and photon-echo spectroscopy,[46,79–81] two-dimensional electronic spectroscopy[82] or femtosecond resonant Raman spectroscopy.[83] The combination of nonperturbative HEOM simulations of dissipative nonadiabatic dynamics with the calculation of spectroscopic signals directly from field-driven equations of motion will become a powerful tool for the analysis and/or prediction of state-of-the-art time and frequency resolved nonlinear spectra.

Acknowledgements

Support from the Singapore National Research Foundation through the Competitive Research Programme (CRP) under Project No. NRF-CRP5-2009-04 is gratefully acknowledged. M.F.G. and W.D. acknowledge support from the Deutsche Forschungsgemeinschaft through a research grant and through the DFG-Cluster of Excellence "Munich-Centre for Advanced Photonics" (http://www.munich-photonics.de). M.F.G. thanks the members of the Zhao group for hospitality and numerous stimulating discussions. Fruitful discussions with Akihito Ishizaki are also thankfully acknowledged.

References

1 F. Bloch, *Rev. Mod. Phys.*, 1957, **105**, 1206.
2 A. G. Redfield, *Adv. Magn. Opt. Reson.*, 1965, **1**, 1.
3 G. W. Ford, M. Kac and P. Mazur, *J. Math. Phys.*, 1965, **6**, 504.
4 W. H. Louisell, *Quantum Statistical Properties of Radiation*, Wiley, New York, 1973.
5 A. J. Leggett, S. Chakravarty, A. T. Dorsey, M. P. A. Fischer, A. Garg and W. Zwerger, *Rev. Mod. Phys.*, 1987, **59**, 1.
6 N. G. van Kampen, *Stochastic Processes in Physics and Chemistry*, North-Holland, Amsterdam, 1992.
7 U. Weiss, *Quantum Dissipative Systems*, World Scientific, Singapore, 1993.
8 W. Domcke, D. R. Yarkony and H. Köppel, *Conical Intersections: Electronic Structure, Dynamics and Spectroscopy*, World Scientific, Singapore, 2004.
9 W. Domcke, D. R. Yarkony and H. Köppel, *Conical Intersections: Theory, Computation and Experiment*, World Scientific, Singapore, 2011.
10 W. Domcke and G. Stock, *Adv. Chem. Phys.*, 1997, **100**, 1.
11 G. A. Worth and L. S. Cederbaum, *Annu. Rev. Phys. Chem.*, 2004, **55**, 127.
12 W. Domcke and D. R. Yarkony, *Annu. Rev. Phys. Chem.*, 2012, **63**, 325.

13 M. Klessinger and J. Michl, *Excited States and Photochemistry of Organic Molecules*, VCH, Weinheim, 1995.
14 M. A. Robb, F. Bernardi and M. Olivucci, *Pure Appl. Chem.*, 1995, **67**, 783.
15 D. R. Yarkony, *Rev. Mod. Phys.*, 1996, **68**, 985.
16 D. R. Yarkony, in *Conical Intersections: Electronic Structure, Dynamics and Spectroscopy*, World Scientific, Singapore, 2004, ch. 2, p. 41.
17 H. Köppel, W. Domcke and L. S. Cederbaum, *Adv. Chem. Phys.*, 1984, **57**, 59.
18 T. Gerdts and U. Manthe, *Chem. Phys. Lett.*, 1998, **295**, 167.
19 A. Kühl and W. Domcke, *Chem. Phys.*, 2000, **259**, 227.
20 A. Kühl and W. Domcke, *J. Chem. Phys.*, 2002, **116**, 263.
21 B. Balzer and G. Stock, *J. Phys. Chem. A*, 2004, **108**, 6464.
22 B. Baltzer and G. Stock, *Chem. Phys.*, 2005, **310**, 33.
23 W. T. Pollard, A. K. Felts and R. A. Friesner, *Adv. Chem. Phys.*, 1996, **93**, 77.
24 K. Blum, *Density Matrix Theory and Applications*, Plenum Press, New York, 1981.
25 G. Katz, R. Kosloff and M. A. Ratner, *Isr. J. Chem.*, 2004, **44**, 53.
26 D. Gelman, G. Katz, R. Kosloff and M. Ratner, *J. Chem. Phys.*, 2005, **123**, 134112.
27 L. S. Cederbaum, E. Gindensperger and I. Burghardt, *Phys. Rev. Lett.*, 2005, **94**, 113003.
28 E. Gindensperger, I. Burghardt and L. S. Cederbaum, *J. Chem. Phys.*, 2006, **124**, 144103.
29 E. Gindensperger, I. Burghardt and L. S. Cederbaum, *J. Chem. Phys.*, 2006, **124**, 144104.
30 E. Gindensperger, H. Köppel and L. S. Cederbaum, *J. Chem. Phys.*, 2007, **126**, 034106.
31 E. Gindensperger and L. S. Cederbaum, *J. Chem. Phys.*, 2007, **127**, 124107.
32 I. Burghardt and J. T. Hynes, *J. Phys. Chem. A*, 2006, **110**, 11411.
33 J. P. Malhado, R. Spezia and J. T. Hynes, *J. Phys. Chem. A*, 2011, **115**, 3720.
34 J. P. Malhado and J. T. Hynes, *J. Chem. Phys.*, 2012, **137**, 22A543.
35 Y. Tanimura and R. Kubo, *J. Phys. Soc. Jpn.*, 1989, **58**, 101.
36 Y. Tanimura and P. G. Wolynes, *Phys. Rev. A*, 1991, **43**, 4131.
37 A. Ishizaki and Y. Tanimura, *J. Phys. Soc. Jpn.*, 2005, **74**, 3131.
38 Y. Tanimura, *J. Phys. Soc. Jpn.*, 2006, **75**, 082001.
39 V. Chernyak and S. Mukamel, *J. Chem. Phys.*, 1996, **105**, 4565.
40 K.-B. Zhu, R.-X. Xu, H. Y. Zhang, J. Hu and Y. J. Yang, *J. Phys. Chem. B*, 2011, **115**, 5678.
41 H.-G. Duan and M. Thorwart, *J. Phys. Chem. Lett.*, 2016, **7**, 382.
42 M. F. Gelin, D. Egorova and W. Domcke, *J. Chem. Phys.*, 2012, **136**, 034507.
43 A. W. Chin, J. Prior, R. Rosenbach, F. Caycedo-Soler, S. F. Huelga and M. B. Plenio, *Nat. Phys.*, 2013, **9**, 113.
44 A. Garg, J. N. Onuchic and V. Ambegaokar, *J. Chem. Phys.*, 1985, **83**, 4491.
45 Y. Tanimura, *J. Chem. Phys.*, 2012, **137**, 22A550.
46 M. F. Gelin, Y. Tanimura and W. Domcke, *J. Chem. Phys.*, 2013, **139**, 214302.
47 A. Ishizaki and G. R. Fleming, *J. Chem. Phys.*, 2009, **130**, 234111.
48 A. V. Pisliakov, M. F. Gelin and W. Domcke, *J. Phys. Chem. A*, 2003, **107**, 2657.
49 D. Egorova, M. F. Gelin and W. Domcke, *J. Chem. Phys.*, 2005, **122**, 134504.
50 V. Stert, P. Farmanara and W. Radloff, *J. Chem. Phys.*, 2000, **112**, 4460.
51 T. Horio, T. Fuji, Y.-I. Suzuki and T. Suzuki, *J. Am. Chem. Soc.*, 2009, **131**, 10392.

52 Y.-I. Suzuki, T. Fuji, T. Horio and T. Suzuki, *J. Chem. Phys.*, 2010, **132**, 174302.
53 T. Suzuki, *Molecules*, 2014, **19**, 2410.
54 R. Schneider and W. Domcke, *Chem. Phys. Lett.*, 1988, **150**, 235.
55 R. Schneider, W. Domcke and H. Köppel, *J. Chem. Phys.*, 1990, **92**, 1045.
56 C. Woywod, W. Domcke, A. L. Sobolewski and H. J. Werner, *J. Chem. Phys.*, 1994, **100**, 1400.
57 G. Stock, C. Woywod, W. Domcke, T. Swinney and B. S. Hudson, *J. Chem. Phys.*, 1995, **103**, 6851.
58 S. Krempl, H. Winterstetter and W. Domcke, *J. Chem. Phys.*, 1995, **102**, 6499.
59 A. Raab, G. A. Worth, H. D. Meyer and L. S. Cederbaum, *J. Chem. Phys.*, 1999, **110**, 936.
60 G. S. Latha and M. D. Prasad, *J. Chem. Phys.*, 1996, **105**, 2972.
61 U. Müller and G. Stock, *J. Chem. Phys.*, 1997, **107**, 6230.
62 G. A. Worth, H.-D. Meyer and L. S. Cederbaum, *J. Chem. Phys.*, 1998, **109**, 3518.
63 J. Y. Fang and S. Hammes-Schiffer, *J. Phys. Chem. A*, 1999, **103**, 9399.
64 M. Thoss, W. H. Miller and G. Stock, *J. Chem. Phys.*, 2000, **112**, 10282.
65 M. Ben-Nun and T. J. Martínez, *Adv. Chem. Phys.*, 2002, **121**, 439.
66 D. V. Shalashilin and M. S. Child, *J. Chem. Phys.*, 2004, **121**, 3563.
67 P. Puzari, B. Sarkar and S. Adhikari, *J. Chem. Phys.*, 2006, **125**, 194316.
68 R. Martinazzo, K. H. Hughes, F. Martelli and I. Burghardt, *Chem. Phys.*, 2010, **377**, 21.
69 T. Grinev, M. Shapiro and P. Brumer, *J. Chem. Phys.*, 2012, **137**, 094302.
70 T. Zimmermann and J. Vaníček, *J. Chem. Phys.*, 2014, **141**, 134102.
71 M. Kanno, Y. Ito, N. Shimokura, S. Koseki, H. Kono and Y. Fujimura, *Phys. Chem. Chem. Phys.*, 2015, **17**, 2012.
72 U. Werner, R. Mitrić, T. Suzuki and V. Bonačić-Koutecký, *Chem. Phys.*, 2008, **349**, 319.
73 http://arrayfire.com/.
74 U. Manthe and H. Köppel, *J. Chem. Phys.*, 1990, **93**, 345.
75 W. Domcke, H. Köppel and L. S. Cederbaum, *Mol. Phys.*, 1981, **43**, 851.
76 M. Liebel, C. Schnedermann and P. Kukura, *Phys. Rev. Lett.*, 2014, **112**, 198302.
77 P. J. M. Johnson, A. Halpin, T. Morizumi, V. I. Prokhorenko, O. P. Ernst and R. J. D. Miller, *Nat. Chem.*, 2015, **7**, 980.
78 S. Mukamel, *Principles of Nonlinear Optical Spectroscopy*, Oxford University Press, New York, 1995.
79 L. Seidner, G. Stock and W. Domcke, *J. Chem. Phys.*, 1995, **103**, 3998.
80 M. F. Gelin, D. Egorova and W. Domcke, *J. Phys. Chem. Lett.*, 2011, **2**, 114.
81 M. F. Gelin, D. Egorova and W. Domcke, *Acc. Chem. Res.*, 2009, **42**, 1290.
82 J. Krčmář, M. F. Gelin and W. Domcke, *J. Chem. Phys.*, 2015, **143**, 074308.
83 B. J. Rao, M. F. Gelin and W. Domcke, *J. Phys. Chem. A*, 2016, **120**, 3286.

Faraday Discussions

PAPER

Toward fully quantum modelling of ultrafast photodissociation imaging experiments. Treating tunnelling in the *ab initio* multiple cloning approach

Dmitry V. Makhov,[a] Todd J. Martinez[b] and Dmitrii V. Shalashilin[a]

Received 11th April 2016, Accepted 1st May 2016
DOI: 10.1039/c6fd00073h

We present an account of our recent effort to improve simulation of the photodissociation of small heteroaromatic molecules using the *Ab Initio* Multiple Cloning (AIMC) algorithm. The ultimate goal is to create a quantitative and converged technique for fully quantum simulations which treats both electrons and nuclei on a fully quantum level. We calculate and analyse the total kinetic energy release (TKER) spectra and Velocity Map Images (VMI), and compare the results directly with experimental measurements. In this work, we perform new extensive calculations using an improved AIMC algorithm that now takes into account the tunnelling of hydrogen atoms. This can play an extremely important role in photodissociation dynamics.

I. Introduction

Quantum non-adiabatic molecular dynamics is a powerful tool for understanding the details of the mechanisms of important photo-induced processes, such as the photodissociation of pyrrole and other heteroaromatic molecules. In these processes, quantum effects such as electronically non-adiabatic transitions and tunnelling are important, and an approach that goes beyond surface hopping, such as multiconfigurational time dependent Hartree (MCTDH),[1] for example, is often required. MCTDH can be very accurate, and was recently used to simulate the dissociation of pyrrole.[2] However it needs a parameterized potential energy surface as a starting point, which significantly restricts its practicality. A good alternative is represented by a variety of methods[3–11] based on trajectory-guided Gaussian basis functions (TBF). Despite the fact that such approaches use classical trajectories, they are still fully quantum mechanical because these trajectories are employed only for propagating the basis, while the evolution of their amplitudes and, thus, of the total nuclear wave-function is determined by the

[a]*School of Chemistry, University of Leeds, Leeds, LS2 9JT, UK. E-mail: D.Makhov@leeds.ac.uk; D.Shalashilin@leeds.ac.uk*
[b]*Department of Chemistry, Stanford University, Stanford, CA 94305, USA*

time-dependent Schrödinger equation. An important advantage of trajectory-guided quantum dynamics methods is that they are fully compatible with *direct* or *ab initio* molecular dynamics where excited state energies, gradients, and non-adiabatic coupling terms are evaluated *on the fly* simultaneously with the nuclear propagation. The disadvantage is that trajectory based direct dynamics is very expensive due to the high cost of electronic structure calculations and typically can afford only a limited number of trajectories, which can be an obstacle to full convergence.

Recently, we introduced the *ab initio* multiple cloning (AIMC)[10] method, where TBFs are moving along Ehrenfest trajectories, as in the multiconfigurational Ehrenfest (MCE)[8,9] approach, with bifurcation of the wave-functions taken into account *via* basis function cloning. While leading to the growth of the number of trajectories, the use of cloning helps to adopt the basis set to quantum dynamics significantly better than in the classical MCE approach. AIMC also uses a number of tricks to efficiently sample the trajectory basis and to use the information obtained on the fly: (1) similar to the previously developed trajectory based methods AIMC relies on importance sampling of initial conditions. (2) AIMC uses the so called time displaced or train basis sets,[10,12,13] which increase the basis set size almost without any additional extra cost by reusing the *ab initio* data which has already been obtained. (3) The method calculates quantum amplitudes in a "post-processing technique" after the trajectories of the basis set functions have been found. As a result, the trajectories can be calculated one by one in parallel and good statistics can be accumulated.

In this work, we present a new implementation of the AIMC approach that is improved to take into account the tunnelling of hydrogen atoms by identifying possible tunnelling points and placing additional TBFs of the other side of the barrier. We use this new implementation to simulate the dynamics of the photodissociation of pyrrole, a process where tunnelling can play a very important role. We calculate the TKER spectrum and velocity map image (VMI), and directly compare the results of our calculations with experimental observations.[14]

The paper is organized as follows. In Section II we describe the proposed implementation of the AIMC approach. Section III contains the computational details of our simulations. In Section IV, we present and discuss the results. Conclusions are given in Section V.

II. Theory

II.1 Working equations

The AIMC method[10] is based on the same ansatz as the multiconfigurational Ehrenfest (MCE) approach,[8,9] in which the total wave-function $|\Psi(t)\rangle$ is represented in a trajectory-guided basis $|\psi_n(t)\rangle$:

$$|\Psi(t)\rangle = \sum_n c_n(t)|\psi_n(t)\rangle \tag{1}$$

The basis functions $|\psi_n(t)\rangle$ are composed of nuclear and electronic parts:

$$|\psi_n(t)\rangle = |\chi_n(t)\rangle\left(\sum_I a_I^{(n)}(t)|\phi_I\rangle\right). \tag{2}$$

The nuclear part $|\chi_n(t)\rangle$ is a Gaussian coherent state moving along an Ehrenfest trajectory:

$$\chi_n(\mathbf{R},\overline{\mathbf{R}}_n,\overline{\mathbf{P}}_n) = \left(\frac{2\alpha}{\pi}\right)^{N_{\mathrm{dof}}/4} \exp\left(-\alpha(\mathbf{R}-\overline{\mathbf{R}}_n)^2 + \frac{i}{\hbar}\overline{\mathbf{P}}_n(\mathbf{R}-\overline{\mathbf{R}}_n) + \frac{i}{\hbar}\gamma_n(t)\right), \tag{3}$$

where $\overline{\mathbf{R}}_n(t)$ and $\overline{\mathbf{P}}_n(t)$ are the phase space coordinate and momentum vectors of the basis function centre, $\gamma_n(t)$ is a phase, and the parameter α determines the width of the Gaussians. The electronic part of basis functions $|\psi_n(t)\rangle$ is represented as a superposition of several adiabatic eigenstates $|\phi_I\rangle$ with quantum amplitudes $a_I^{(n)}$.

The time dependence of the Ehrenfest amplitudes $a_I^{(n)}$ is given by the equations

$$\dot{a}_I^{(n)} = -\frac{i}{\hbar}\sum_J H_{IJ}^{\mathrm{el}(n)} a_J^{(n)}, \tag{4}$$

where the matrix elements of electronic Hamiltonian $H_{IJ}^{\mathrm{el}(n)}$ are expressed as:

$$H_{IJ}^{\mathrm{el}(n)} = \begin{cases} V_I(\overline{\mathbf{R}}_n), & I = J \\ -i\hbar\overline{\mathbf{P}}_n\mathbf{M}^{-1}\mathbf{d}_{IJ}(\overline{\mathbf{R}}_n), & I \neq J \end{cases}. \tag{5}$$

here $V_I(\overline{\mathbf{R}}_n)$ is the Ith potential energy surface and $\mathbf{d}_{IJ}(\overline{\mathbf{R}}_n) = \langle\phi_I|\nabla_R|\phi_J\rangle$ is the non-adiabatic coupling matrix element (NACME).

The motion of the centres of the Gaussians follows the standard Newton's equations:

$$\begin{aligned} \dot{\overline{\mathbf{R}}}_n &= \mathbf{M}^{-1}\overline{\mathbf{P}}_n \\ \dot{\overline{\mathbf{P}}}_n &= \overline{\mathbf{F}}_n \end{aligned} \tag{6}$$

where the force $\overline{\mathbf{F}}_n$ is an Ehrenfest force that includes both the usual gradient term and the additional term related to the change of quantum amplitudes as a result of non-adiabatic coupling:

$$\overline{\mathbf{F}}_n = -\sum_I \left|a_I^{(n)}\right|^2 \nabla_R V_I(\overline{\mathbf{R}}_n) + \sum_{I\neq J}\left(a_I^{(n)}\right)^* a_J^{(n)} \mathbf{d}_{IJ}(\overline{\mathbf{R}}_n)[V_I(\overline{\mathbf{R}}_n) - V_J(\overline{\mathbf{R}}_n)] \tag{7}$$

Finally, the phase γ_n evolves as:

$$\frac{\mathrm{d}\gamma_n}{\mathrm{d}t} = \frac{\overline{\mathbf{P}}_n\dot{\overline{\mathbf{R}}}_n}{2}. \tag{8}$$

Eqn (3)–(8) form a complete set, determining the basis and its time evolution.

The evolution of the total wave-function $|\Psi(t)\rangle$ (eqn (1)) is defined by both the evolution of the basis functions $|\psi_n(t)\rangle$ and the evolution of the relevant amplitudes $c_n(t)$. The time dependence of the amplitudes $c_n(t)$ is given by the equation

$$\sum_n \langle\psi_m(t)|\psi_n(t)\rangle\dot{c}_n(t) = -\frac{i}{\hbar}\sum_n\left(H_{mn} - i\hbar\langle\psi_m(t)\left|\frac{\mathrm{d}}{\mathrm{d}t}\right|\psi_n(t)\rangle\right)c_n(t), \tag{9}$$

which can be easily obtained by substituting (1) into the time dependent Schrödinger equation. The Hamiltonian matrix elements H_{mn} can be written as:

$$H_{mn} = \sum_{I,\ J} \left(a_I^{(m)}\right)^* a_J^{(n)} \langle \chi_m \phi_I | \hat{H} | \chi_n \phi_J \rangle \tag{10}$$

Assuming that the second derivative of the electronic wave-function $|\phi_I\rangle$ with respect to $\mathbf{R}$ can be disregarded, we get:

$$\langle \chi_m \phi_I | \hat{H} | \chi_n \phi_J \rangle = \delta_{IJ} \langle \chi_m | -\frac{\hbar^2}{2} \nabla_R \mathbf{M}^{-1} \nabla_R + V_I(\mathbf{R}) | \chi_n \rangle - \hbar^2 \langle \chi_m | \mathbf{d}_{IJ}(\mathbf{R}) \mathbf{M}^{-1} \nabla_R | \chi_n \rangle \tag{11}$$

The matrix elements of the kinetic energy operator $\langle \chi_m | -\frac{\hbar^2}{2} \nabla_R \mathbf{M}^{-1} \nabla_R | \chi_n \rangle$ can be calculated analytically. For potential energy and non-adiabatic coupling matrix elements, we use a simple approximation:[10]

$$\begin{aligned} \langle \chi_m | V_I(\mathbf{R}) | \chi_n \rangle \approx & \langle \chi_m | \chi_n \rangle \left(\frac{V_I(\overline{\mathbf{R}}_n) + V_I(\overline{\mathbf{R}}_m)}{2} \right) \\ & + \left(\frac{\langle \chi_m | (\mathbf{R} - \overline{\mathbf{R}}_n) | \chi_n \rangle \nabla_R V_I(\overline{\mathbf{R}}_n) + \langle \chi_m | (\mathbf{R} - \overline{\mathbf{R}}_m) | \chi_n \rangle \nabla_R V_I(\overline{\mathbf{R}}_m)}{2} \right), \end{aligned} \tag{12}$$

$$\langle \chi_m | \mathbf{d}_{IJ}(\mathbf{R}) \mathbf{M}^{-1} \nabla_R | \chi_n \rangle = \frac{i}{2\hbar} \langle \chi_m | \chi_n \rangle \left(\overline{\mathbf{P}}_m \mathbf{M}^{-1} \mathbf{d}_{IJ}(\overline{\mathbf{R}}_m) + \overline{\mathbf{P}}_n \mathbf{M}^{-1} \mathbf{d}_{IJ}(\overline{\mathbf{R}}_n) \right). \tag{13}$$

The approximation (12) represents a linear interpolation of the potential energy between the two points and can be improved further at the cost of calculating higher derivatives of the potential energy along the trajectories. It has been tested previously,[10] and no visible change of the results was found when this approximation was applied compared to the saddle point approximation which expands around a distinct centroid for each pair of TBFs.[4]

The term $\langle \Psi_m(t) | \frac{\mathrm{d}}{\mathrm{d}t} | \Psi_n(t) \rangle$ in eqn (9), which originates from the time dependence of the basis, can be expressed as:

$$\left\langle \Psi_m \middle| \frac{\mathrm{d}\Psi_n}{\mathrm{d}t} \right\rangle = \left\langle \chi_m \middle| \frac{\mathrm{d}\chi_n}{\mathrm{d}t} \right\rangle \sum_I \left(a_I^{(m)}\right)^* a_J^{(n)} + \langle \chi_m | \chi_n \rangle \sum_I \left(a_I^{(m)}\right)^* \dot{a}_J^{(n)}, \tag{14}$$

where

$$\left\langle \chi_m \middle| \frac{\mathrm{d}\chi_n}{\mathrm{d}t} \right\rangle = \left(\dot{\overline{\mathbf{R}}}_n \langle \chi_m | \frac{\mathrm{d}}{\mathrm{d}\overline{\mathbf{R}}_n} | \chi_n \rangle + \dot{\overline{\mathbf{P}}}_n \langle \chi_m | \frac{\mathrm{d}}{\mathrm{d}\overline{\mathbf{P}}_n} | \chi_n \rangle \right) + \frac{i}{\hbar} \dot{\gamma}_n \langle \chi_m | \chi_n \rangle. \tag{15}$$

Notice that in the AIMC approach, all off-diagonal matrix elements entering eqn (9) are calculated from the electronic structure data at the TBF centres, which is needed for the propagation of the basis. Thus, quantum coupling between the configurations comes at almost no extra cost. Moreover, eqn (9) can be solved *after* the trajectories have been calculated, provided the appropriate electronic structure information has been saved.

The detailed derivation of MCE equations together with the expressions for relevant matrix elements can be found in our previous works.[10,11]

II.2 Basis set sampling and cloning

The Ehrenfest basis set is guided by an average potential, which can be advantageous when quantum transitions are frequent. However, it becomes unphysical in regions of low non-adiabatic coupling when two or more electronic states have significant amplitudes: in this case, the difference of the shapes of potential energy surfaces for different electronic states should lead to branching of the wavepacket.

In order to reproduce the bifurcation of the wave-function after leaving the non-adiabatic coupling region, AIMC methods adopt the cloning procedure,[10] where the appropriate basis function is replaced by two basis functions, each guided (mostly) by a single potential energy surface. After the cloning event, an Ehrenfest configuration $|\psi_n\rangle = |\chi_n\rangle\left(\sum_J a_J^{(n)}|\phi_J\rangle\right)$ yields two configurations:

$$|\psi'_n\rangle = |\chi_n\rangle\left(\frac{a_I^{(n)}}{\left|a_I^{(n)}\right|}\times|\phi_I\rangle + \sum_{J\neq I} 0\times|\phi_J\rangle\right) \tag{16}$$

and

$$\left|\psi'_{n'}\right\rangle = |\chi_n\rangle\left(0\times|\phi_I\rangle + \frac{1}{\sqrt{1-\left|a_I^{(n)}\right|^2}}\sum_{J\neq I} a_J^{(n)}|\phi_J\rangle\right). \tag{17}$$

The first clone configuration has non-zero amplitudes for only one electronic state, and the second clone contains contributions of all other electronic states. The amplitudes of the two new configurations become:

$$c'_n = c_n\left|a_I^{(n)}\right|, \quad c'_{n'} = c_n\sqrt{1-\left|a_I^{(n)}\right|^2}, \tag{18}$$

so that the contribution of the two clones $|\psi'_n\rangle$ and $|\psi'_{n'}\rangle$ to the whole wave-function (1) remains the same as the contribution of original function:

$$c_n|\psi_n\rangle = c'_n|\psi'_n\rangle + c'_{n'}\left|\psi'_{n'}\right\rangle. \tag{19}$$

We apply the cloning procedure shortly after a trajectory passes near a conical intersection, when the non-adiabatic coupling is lower than a threshold, and, at the same time, the so-called breaking force

$$\mathbf{F}_I^{(\mathrm{br})} = |a_I|^2\left(\nabla V_I - \sum_J |a_J|^2\nabla V_J\right), \tag{20}$$

which is the force pulling the Ith state away from the remaining states, is sufficiently strong.

The cloning procedure is very much in spirit of the spawning, used in the *Ab Initio* Multiple Spawning approach (AIMS). Cloning does not require any back-propagation of spawned/cloned basis functions, unlike many[4] (but not all[15,16]) implementations of spawning.

As has been described in our previous work,[7] we rely on importance sampling when generating the initial conditions. Using the linearity of the Schrödinger equation, we first represent the initial wave-function as a superposition of Gaussians and then propagate each of them independently, "bit-by-bit".[7] We use a time-displaced basis set (coherent state trains), where several Gaussian basis functions are moving along the same trajectory but with a time-shift Δt, allowing us to reuse the same electronic structure data for each of the basis functions in the "train." Fig. 1 shows a time displaced basis guided by a trajectory and its bifurcation *via* cloning. The best possible result with AIMC can be achieved when a swarm of trains is used to propagate each "bit" of the initial wave-function.

II.3 Tunnelling

The tunnelling of hydrogen atoms can play an important role in photodissociation processes. As mentioned above, MCE, AIMC and AIMS are fully quantum methods because classical trajectories are used only to propagate the basis, while the amplitudes $c_n(t)$ are found by solving the time dependent Schrödinger equation. When Gaussian basis functions are present on two sides of the potential barrier, the interaction between them can provide quantum tunnelling through the barrier. However, in the case of direct *ab initio* dynamics, the basis is usually very small, far from being complete. As a result, no basis functions normally would be present on the other side, and they must be placed there by hand in order to take tunnelling into account.

In this paper we adopt the ideas[17,18] previously used in the AIMS method to describe tunnelling for use with the AIMC technique. Fig. 2 illustrates the algorithm that we apply. First, we calculate the usual AIMC trajectories and find turning points, where the distance between the hydrogen atom and the radical reaches a local maximum. Then, for each of these turning points, we calculate the shape of the potential barrier: we increase manually the length of the N–H bond keeping all other degrees of freedom frozen, calculate potential energies, and find the point on the other side of the barrier with the same energy as in the turning

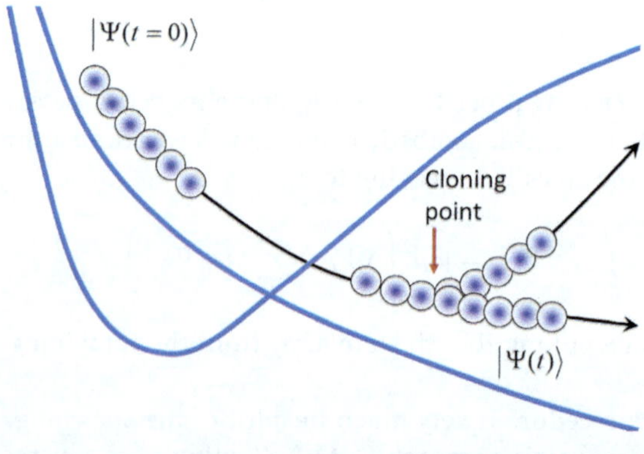

Fig. 1 A sketch of the AIMC propagation scheme. The wave-function is represented as a superposition of Gaussian coherent states, which form a train moving along the trajectory. After passing the intersection, the train branches in the process of cloning. The figure shows a single train with cloning. In the most detailed AIMC calculation, a basis of several cloning trains interacting with each other is used.

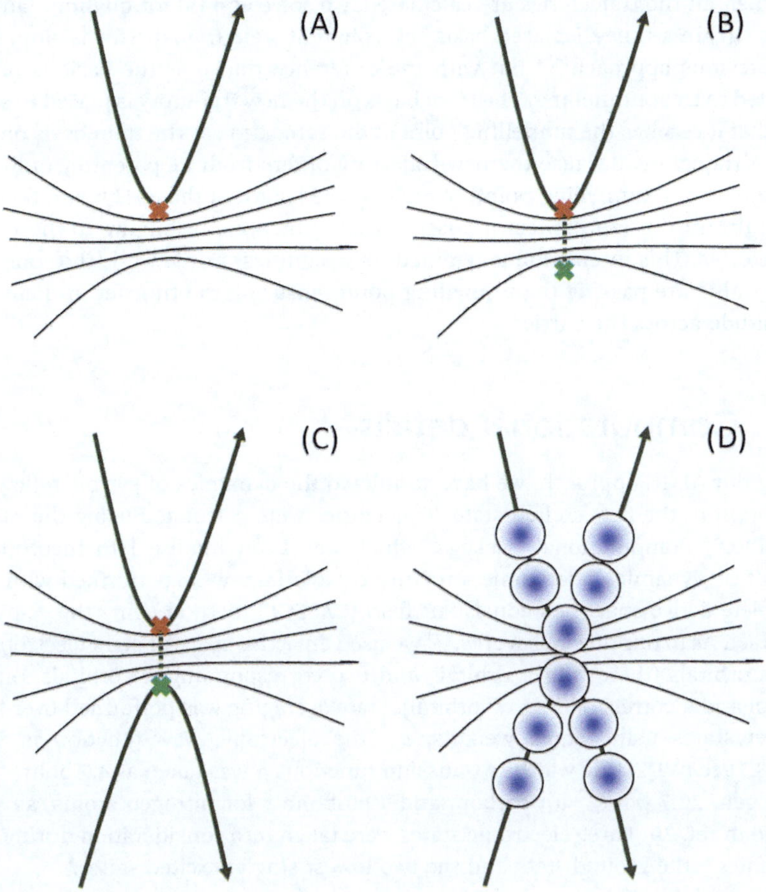

Fig. 2 Illustration of the algorithm used to treat tunnelling in our approach. (A) Identify turning point; (B) find a point with the same potential energy on the opposite side of the barrier; (C) run an additional trajectory through this point; (D) solve time-dependent Schrödinger equation in the basis of a coherent state trains[10] moving along the trajectories on both sides of the barrier.

point. If this point lies further than a set threshold from the turning point, we assume that tunnelling is not possible here, as the potential barrier is too wide. Otherwise, we use it as a starting point for an additional AIMC trajectory. The new trajectory is calculated both forward and backward in time, and the initial momenta are taken as the same as in the turning point, ensuring that new trajectories have the same total classical energies as their parent trajectories. This is exactly the procedure used in the multiple spawning approach, thus our method combines cloning for non-adiabatic events and spawning for tunnelling events. The forward propagation of new trajectories often involves branching as a result of cloning; backward propagation is performed without cloning and for a sufficiently short time, until new and parent trajectories separate in phase space.

When all the trajectories are calculated, we solve eqn (9) for quantum amplitudes $c_n(t)$ in a time-displaced basis set (coherent state trains). This is similar to our previous approach[10,11] but with the difference that now the basis is better adapted to treat tunnelling. The train basis on the new trajectory is placed in such way that it reaches the tunnelling point at the same time as the train basis on the parent trajectory. Because the new trajectory differs from its parent by only one coordinate at a tunnelling point, namely by the length of the N–H bond, there is a significant overlap between Gaussian basis functions belonging to these two trajectories. This interaction is retained for a significant time while the coherent state trains are passing the tunnelling point, ensuring the transfer of quantum amplitude across the barrier.

III. Computational details

Using our AIMC approach, we have simulated the dynamics of pyrrole following excitation to the first excited state. Trajectories were calculated using the AIMS-MOLPRO[19] computational package, which has been modified to incorporate Ehrenfest dynamics. Electronic structure calculations were performed with the complete active space self-consistent field (CASSCF) method using the cc-pVDZ basis set. As in our previous works,[9,11] we used an active space of eight electrons in seven orbitals (three ring π orbitals and two corresponding π^* orbitals, one σ orbital and a corresponding σ^* orbital). State averaging was performed over four singlet states using equal weights, *i.e.* the electronic wave-function is SA4-CAS(8,7)/cc-pVDZ. The width of Gaussian functions α was taken as 4.7 bohr^{-2} for hydrogen, 22.7 bohr^{-2} for carbon, and 19.0 bohr^{-2} for nitrogen atoms, as suggested in ref. 20. Three electronic states were taken into consideration during the dynamics – the ground state and the two lowest singlet excited states.

The initial positions and momenta were randomly sampled from the ground state vibrational Wigner distribution in the harmonic approximation using vibrational frequencies and normal modes were calculated at the same CASSCF level of theory. We approximate the photoexcitation by simply lifting the ground state wavepacket to the excited state, as would be appropriate for an instantaneous excitation pulse within the Condon approximation. Of course, the fine details of the initial photoexcited wavepacket are lost in this approximation, however, we do not expect these details to have much effect on the observables shown in this paper.

We have run 900 initial Ehrenfest trajectories, each propagated with a time-step of $\sim$0.06 fs (2.5 a.u.) for 200 fs or until the dissociation occurred, defined as an N–H distance exceeding 4.0 Å. For a small number of trajectories, simulations exhibiting N–H dissociation were carried out to the full 200 fs in order to investigate the dynamics of the radical. Cloning was applied to TBFs when the breaking acceleration of eqn (20) exceeded a threshold of 5×10^{-6} a.u. and the norm of the non-adiabatic coupling vector was simultaneously less than 2×10^{-3} a.u. For all initial trajectories, as well as for their branches resulting from cloning, we identified turning points for the N–H bond length and calculated the width of the potential barrier. Additional trajectories on the other side of the barrier were placed if the width of the barrier did not exceed 0.5 bohr, which corresponds to an overlap of $\sim$0.3 between Gaussian basis functions. The new trajectories were propagated backward for 20 fs to accommodate the train basis set, and forward until dissociation or until the trajectory time exceeds 200 fs.

For each initial trajectory with all its branches and tunnelling sub-trajectories, we solved eqn (9) using a train basis set of $N = 21$ Gaussians per branch, separated by 10 time steps, which corresponds to an average overlap of ~0.6 between the nearest Gaussians in the train. The total size of the basis is constantly changing because of the inclusion of new branches. The final amplitudes c_n give statistical weights for each of the branches, which are used in the analysis that follows.

IV. Results

As a result of cloning, 900 initial configurations give rise to 1131 trajectory branches. This corresponds to an average of ~0.25 cloning events per initial trajectory. For these branches, we have found 7702 local maxima of N–H bond length, of which 2376 have been identified as possible tunnelling points. For all these points, we run sub-trajectories, which finally gives 3203 additional branches, 4334 branches in total. The majority of these branches undergo N–H dissociation within our computational time of 200 fs: the total statistical weight of dissociative trajectories is 92%, of which 53% is the contribution of tunnelling sub-trajectories.

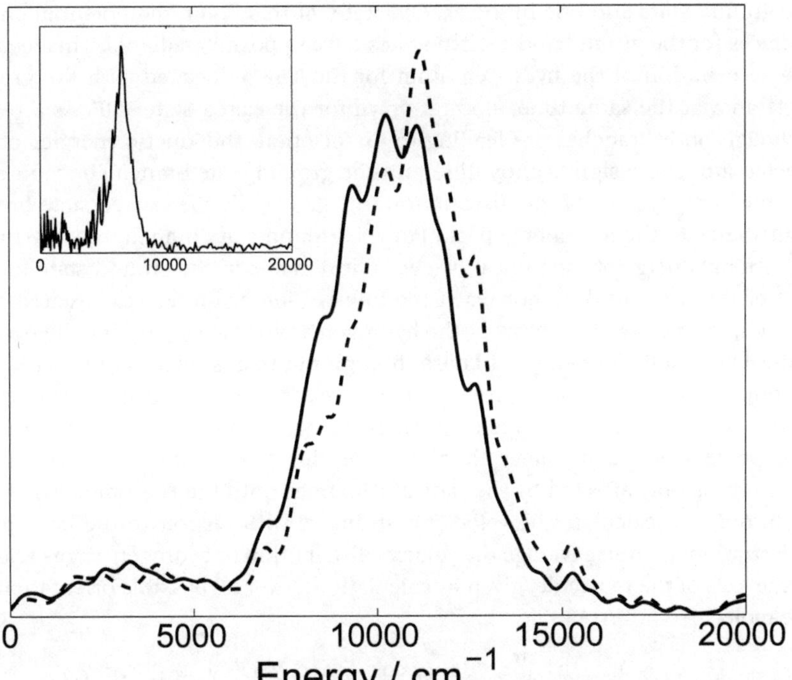

Fig. 3 Total kinetic energy release (TKER) spectrum of hydrogen atoms after dissociation calculated with (solid) and without (dash) taking tunnelling into account. Both spectra are averaged over the same ensemble of initial configuration. The curves are smoothed by replacing delta-functions with Gaussian functions ($\sigma = 200$ cm^{-1}). The inset shows the experimentally measured spectrum.[14]

The kinetic energy distribution of the ejected hydrogen atom is presented in Fig. 3 together with the experimental TKER spectrum.[14] Both distributions clearly exhibit two contributions: a large peak at higher energies, and a small contribution at lower energies. It is important to note that adopting the basis set to tunnelling shifts the high-energy peak of TKER spectrum toward the lower energies by about $\sim$1000 cm^{-1} and makes the low-energy peak slightly more pronounced. While the calculated energies are still on average about 1.5 times higher than experimental values, this difference can be ascribed to the lack of dynamic electron correlation in the CASSCF potential energy surfaces. We previously showed[11] that a more accurate MS-CASPT2 PES would lead to a shift in the kinetic energy peak of approximately $\sim$1800–1900 cm^{-1} towards lower energies, significantly improving the agreement with experiment.

Analysis of the electronic state amplitudes in the Ehrenfest configurations (eqn (2)) shows that the bifurcation of the wave-function while passing through a conical intersection plays an important role in the formation of a two-peak spectrum: the high kinetic energy product is predominantly in the ground state, while the low energy peak is formed by mostly low-weight branches with substantial contribution from excited electronic states. Fig. 4 presents an example of such a bifurcating trajectory. At about 55 fs after photoexcitation, this trajectory reaches an intersection for the first time. After passing the intersection, the ground and first excited states of the original TBF are approximately equally populated, so the cloning procedure is applied creating instead two TBF, one in the ground state and one in the excited state. At this point, the potential energy surfaces for the ground and excited states have opposite gradients. This leads to the acceleration of the hydrogen atom for the TBF associated with the ground state and, at the same time, slows it down for the exited state TBF. As a result, although both branches are leading to dissociation, the kinetic energies of the ejected atoms are significantly different: the ground state branch contributes to the high energy peak of the distribution in Fig. 3, while the excited state branch contributes to the low energy peak. For the ground state branch, the remaining vibrational energy of the radical is low, so it remains in the ground state for the rest of the run and does not reach the intersection again. For the excited state branch, the energy taken away by the hydrogen atom is lower, leaving the pyrrolyl radical with sufficient energy to pass through numerous intersections with population transfer between the ground and both excited states. Naturally, quenching to the ground state will happen eventually for this branch but the time scale of this process is much longer than that for the dissociation, while the TKER spectrum is only affected by the radical dynamics until the H atom is lost.

In order to calculate the velocity map image with respect to the laser pulse polarization, we must average the velocity distribution of hydrogen atoms relative to the axes of the molecule, given by calculations, over all possible orientations of the molecule:

$$I(r,\varphi) \sim \delta(r-|\mathbf{v}|)\iiint \mathrm{d}\alpha\, \mathrm{d}\beta\, \mathrm{d}\gamma\, \cos^2(\xi(\alpha,\beta,\gamma))\delta(\varphi-\phi(\theta,\alpha,\beta,\gamma)), \tag{21}$$

where α, β and γ are Euler angles, θ is the angle between the atom velocity vector $\mathbf{v}$ and the transition dipole of the molecule, $\xi(\alpha,\beta,\gamma)$ is the angle between the transition dipole and light polarization vectors, and $\phi(\theta,\alpha,\beta,\gamma)$ is the angle between the light polarization vector and atom velocity. Here we take into account

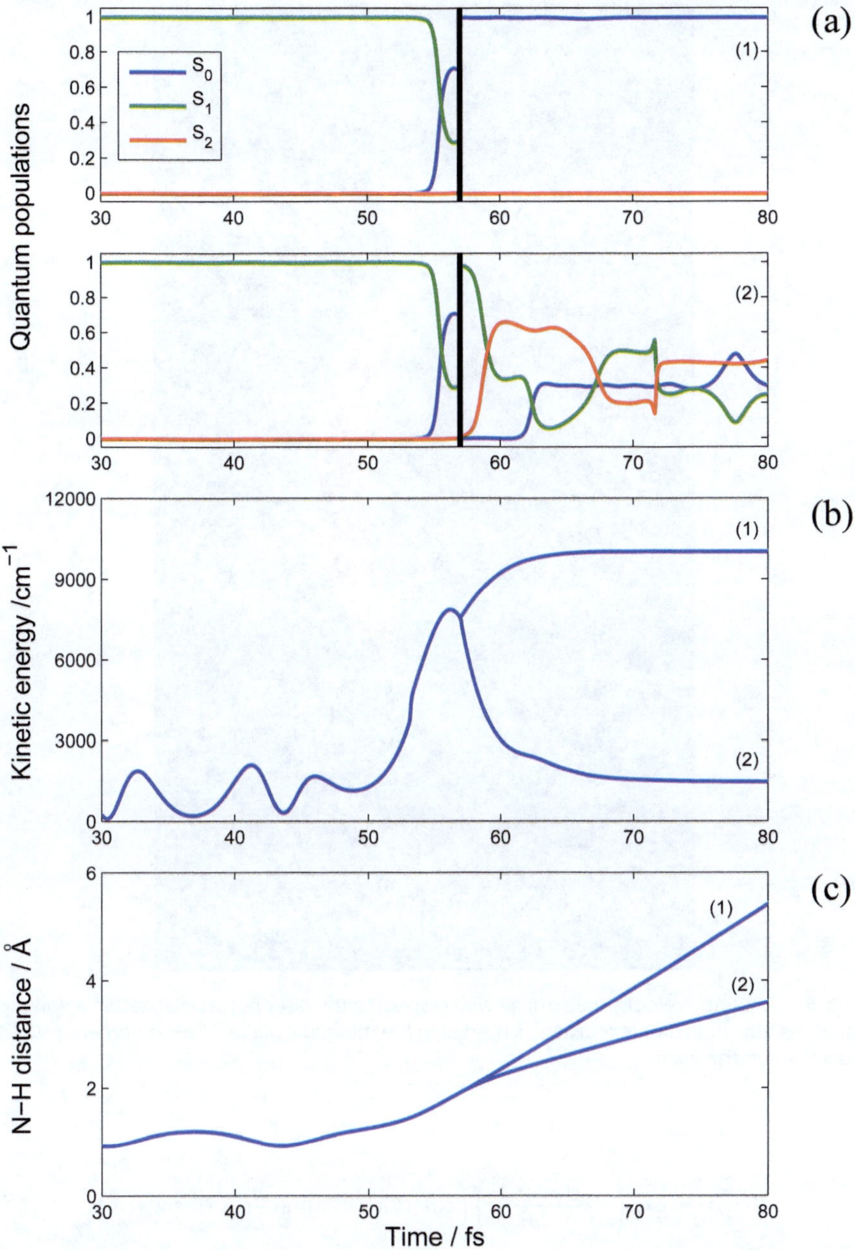

Fig. 4 An example of trajectory bifurcation on conical intersection. Electronic state populations (a), the kinetic energy of the H atom (b) and the N–H distance (c) as a function of time. Fast and slow branches are referred as (1) and (2) respectively. The black vertical line indicates the moment when cloning was applied.

that the probability of excitation is proportional to $\cos^2(\xi)$. Integrating over Euler angles and replacing, as usual, the δ-function for $|\mathbf{v}|$ with a narrow Gaussian function, we obtain

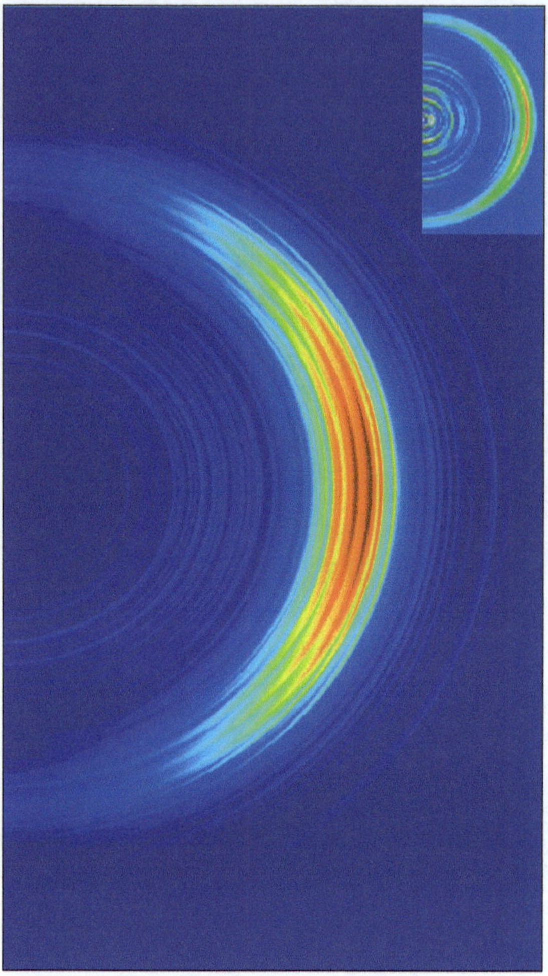

Fig. 5 Simulated velocity map image with respect to the laser pulse polarization assuming that the transition dipole moment is normal to the molecule plane. The experimental VMI[14] is shown in the inset.

$$I(r,\varphi) \sim \exp\left(-\frac{(r-|\mathbf{v}|)^2}{2\sigma^2}\right)\left(\cos^2(\theta)\cos^2(\varphi) + \frac{1}{2}\sin^2(\theta)\sin^2(\varphi)\right). \qquad (22)$$

Fig. 5 shows the simulated velocity map with respect to the laser pulse polarization assuming that the transition dipole is normal to the molecular plane. The simulations reproduce well the main feature of the velocity map image, which is the anisotropy of the intense high energy part. Our results are also consistent with experiment[14] in the low energy region showing an isotropic distribution, although admittedly the statistics of both experiment and simulation are poorer in the region of low energy.

V. Conclusion

We simulated the photodissociation dynamics of pyrrole excited to the lowest singlet excited state ($^1A_1 \rightarrow {}^1A_2$) using a new implementation of the AIMC approach, which now is modified to take into account the tunnelling of hydrogen atoms more accurately. AIMC is a fully quantum technique but its computational cost in our implementation is compatible with classical "on the fly" molecular dynamics, which allows the accumulation of sufficient statistics to clarify the details of photo-induced processes in pyrrole. The treatment of tunnelling in our implementation provides a promising starting point for the further development of fully quantum methods for non-adiabatic dynamics and tunnelling with the ultimate goal of reaching well converged quantitative results. The current version of AIMC is already accurate enough to reproduce features of the experimentally observed TKER spectrum and velocity map images.

Acknowledgements

DM and DS acknowledge the support from EPSRC through grants EP/J001481/1 and EP/N007549/1.

References

1 G. A. Worth, H.-D. Meyer, H. Köppel, L. S. Cederbaum and I. Burghardt, Using the MCTDH wavepacket propagation method to describe multimode non-diabatic dynamics, *Int. Rev. Phys. Chem.*, 2008, **27**, 569–606.

2 G. Wu, S. P. Neville, O. Schalk, T. Sekikawa, M. N. R. Ashfold, G. A. Worth and A. Stolow, Excited state non-adiabatic dynamics of pyrrole: a time-resolved photoelectron spectroscopy and quantum dynamics study, *J. Chem. Phys.*, 2015, **142**, 074302.

3 T. J. Martinez, M. Ben-Nun and G. Ashkenazi, Classical/quantal method for multistate dynamics: a computational study, *J. Chem. Phys.*, 1996, **104**, 2847.

4 M. Ben-Nun and T. J. Martínez, Ab Initio Quantum Molecular Dynamics, *Adv. Chem. Phys.*, 2002, **121**, 439.

5 D. V. Shalashilin, Quantum mechanics with the basis set guided by Ehrenfest trajectories: theory and application to spin-boson model, *J. Chem. Phys.*, 2009, **130**(24), 244101.

6 S. L. Fiedler and J. Eloranta, Nonadiabatic dynamics by mean-field and surface hopping approaches: energy conservation considerations, *Mol. Phys.*, 2010, **108**(11), 1471–1479.

7 D. V. Shalashilin, Nonadiabatic dynamics with the help of multiconfigurational Ehrenfest method: improved theory and fully quantum 24D simulation of pyrazine, *J. Chem. Phys.*, 2010, **132**(24), 244111.

8 D. V. Shalashilin, Multiconfigurational Ehrenfest approach to quantum coherent dynamics in large molecular systems, *Faraday Disc.*, 2011, **153**, 105.

9 K. Saita and D. V. Shalashilin, On-the-fly *ab initio* molecular dynamics with multiconfigurational Ehrenfest method, *J. Chem. Phys.*, 2012, **137**, 8.

10 D. V. Makhov, W. J. Glover, T. J. Martinez and D. V. Shalashilin, *Ab initio* multiple cloning algorithm for quantum nonadiabatic molecular dynamics, *J. Chem. Phys.*, 2014, **141**(5), 054110.

11 D. V. Makhov, K. Saita, T. J. Martinez and D. V. Shalashilin, *Ab initio* multiple cloning simulations of pyrrole photodissociation: TKER spectra and velocity map imaging, *Phys. Chem. Chem. Phys.*, 2015, **17**, 3316.

12 D. V. Shalashilin and M. S. Child, Basis set sampling in the method of coupled coherent states: coherent state swarms, trains and pancakes, *J. Chem. Phys.*, 2008, **128**, 054102.

13 M. Ben-Nun and T. J. Martinez, Exploiting Temporal Non-Locality to Remove Scaling Bottlenecks in Nonadiabatic Quantum Dynamics, *J. Chem. Phys.*, 1999, **110**, 4134–4140.

14 G. M. Roberts, C. A. Williams, H. Yu, A. S. Chatterley, J. D. Young, S. Ullrich and V. G. Stavros, Probing ultrafast dynamics in photoexcited pyrrole: timescales for (1)pi sigma* mediated H-atom elimination, *Faraday Discuss.*, 2013, **163**, 95–116.

15 M. Ben-Nun and T. J. Martínez, A Continuous Spawning Method for Nonadiabatic Dynamics and Validation for the Zero-Temperature Spin-Boson Problem, *Isr. J. Chem.*, 2007, **47**, 75–88.

16 S. Yang, J. D. Coe, B. Kaduk and T. J. Martínez, An "Optimal" Spawning Algorithm for Adaptive Basis Set Expansion in Nonadiabatic Dynamics, *J. Chem. Phys.*, 2009, **130**, 134113.

17 M. Ben-Nun and T. J. Martinez, Semiclassical tunneling rates from *ab initio* molecular dynamics, *J. Phys. Chem. A*, 1999, **103**(31), 6055–6059.

18 M. Ben-Nun and T. J. Martínez, A Multiple Spawning Approach to Tunneling Dynamics, *J. Chem. Phys.*, 2000, **112**, 6113–6121.

19 B. G. Levine, J. D. Coe, A. M. Virshup and T. J. Martinez, Implementation of *ab initio* multiple spawning in the Molpro quantum chemistry package, *Chem. Phys.*, 2008, **347**(1), 3–16.

20 A. L. Thompson, C. Punwong and T. J. Martinez, Optimization of width parameters for quantum dynamics with frozen Gaussian basis sets, *Chem. Phys.*, 2010, **370**, 70–77.

Faraday Discussions

PAPER

Electron and nuclear dynamics following ionisation of modified bismethylene-adamantane†

Morgane Vacher,* Fabio E. A. Albertani, Andrew J. Jenkins, Iakov Polyak, Michael J. Bearpark and Michael A. Robb

Received 6th April 2016, Accepted 13th May 2016
DOI: 10.1039/c6fd00067c

We have simulated the coupled electron and nuclear dynamics using the Ehrenfest method upon valence ionisation of modified bismethylene-adamantane (BMA) molecules where there is an electron transfer between the two π bonds. We have shown that the nuclear motion significantly affects the electron dynamics after a few fs when the electronic states involved are close in energy. We have also demonstrated how the non-stationary electronic wave packet determines the nuclear motion, more precisely the asymmetric stretching of the two π bonds, illustrating "charge-directed reactivity". Taking into account the nuclear wave packet width results in the dephasing of electron dynamics with a half-life of 8 fs; this eventually leads to the equal delocalisation of the hole density over the two methylene groups and thus symmetric bond lengths.

1 Introduction

From 2001, the emergence of tools for attosecond measurement allows the experimental study of the dynamics of electrons and nuclei in molecules, on the attosecond (1 as $= 10^{-18}$ s) time scale.[1,2] The synthesis of attosecond pulses relies on the time–energy uncertainty principle $\Delta E \Delta t \geq \hbar$: collecting together coherent light sources with an energy bandwidth ΔE of several eV gives a pulse with duration $\tau \approx \hbar/\Delta E$ of a few attoseconds. The broad spectral bandwidth of such short pulses leads to the coherent population of several electronic states, thus breaking the Born–Oppenheimer approximation. The system is not confined any more to be in a single stationary electronic state but is now a superposition of electronic states, called an electronic wave packet: this superposition is *non-*

Department of Chemistry, Imperial College London, London SW7 2AZ, UK

† Electronic supplementary information (ESI) available: Convergence check of the spin density average amplitude with respect to the number of sampled geometries in the ensemble and comparison of potential, kinetic and total energies of the Ehrenfest trajectories in BMA[5,5] where Newton's equation of motion is integrated either with the gradient only (without the coupled-perturbed corrections) or with the Hessian-based predictor–corrector algorithm. See DOI: 10.1039/c6fd00067c

stationary, *i.e.* its probability density is time-dependent. Such electron dynamics is often called *charge migration* in the literature.[3]

One key target of attosecond experiments remains the real-time observation of electron dynamics upon ionisation in molecules,[4–8] where changes in both the electronic density and the nuclear geometry are expected. Extended theoretical studies are necessary to help with the understanding of attosecond experiments. Molecular charge migration is still often treated as a purely electronic process solving the electronic time-dependent Schrödinger equation at a single fixed nuclear geometry:[9–14] the effects of the motion of the nuclei and of the natural spatial delocalisation of the nuclei due to the zero-point energy are thus both neglected. Considering a single static molecular geometry, the electronic density oscillates with a period inversely proportional to the energy gap between the populated eigenstates:

$$T = \frac{h}{E_{\rm s} - E_{\rm s'}} \tag{1}$$

The justification given for the *fixed-nuclei approximation* is the difference in time scales of the electron and nuclear dynamics, with the electron distribution changing much faster than the nuclear geometry. Nest and Ulusoy[15] studied electron dynamics after the partial electronic excitation of a diatomic molecule to the first excited state (more than 3 eV above the electronic ground state). Their study suggests that the effect of nuclear motion is very small during the first ten or so femtoseconds, as it takes time for the nuclei to move sufficiently to influence the electronic superposition. Ajay and coworkers have recently shown in another diatomic molecule that the nuclear motion does not destroy the coherent electron dynamics for several tens of fs but it modulates the electronic oscillations.[16] Using the mixed quantum-classical Ehrenfest method,[17] we have previously studied molecules[18,19] where the amplitude and time scale of electron dynamics are affected by the nuclear motion after only a few fs and others[18–21] where the sinusoidal oscillation in the electronic density survives for several tens of fs.

Another effect is that of the nuclear wave packet delocalisation: due the zero-point energy, the nuclei undergo fluctuations in their positions even in the molecular ground state. The various nuclear geometries within the nuclear wave packet lead to a distribution of energy gaps and therefore of oscillation periods in the electronic density (eqn (1)). The oscillations will then dephase with time, calling into question the commonly held picture of long-lived electronic oscillations at a well-defined frequency. Using a simple analytical model, we have shown that it dephases faster when the nuclear wave packet is wider and when the difference of slopes of the potential energy surfaces involved is greater.[22] This effect is neglected in the *single-geometry approximation* used in most simulations until now. We have shown that the intrinsic distribution in nuclear geometries within the nuclear wave packet (sampled from a Wigner distribution) leads to the rapid dephasing of electron dynamics in *para*-xylene, a polycyclic norbornadiene and phenylalanine derivative cations on a few femtosecond time scale.[21,22]

With the population of an electronic wave packet, the convenient picture of a single potential energy surface is lost: the nuclei "feel" the multiple coupled potential energy surfaces and their movement is determined by the gradient of the superposition of electronic states. Experiments suggest that electronic wave packets play a chemical role. For example, Weinkauf *et al.* claimed that

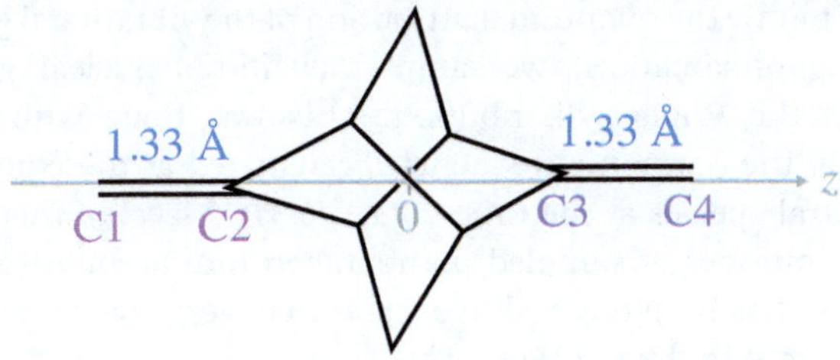

Fig. 1 Structure of the BMA[5,5] molecule: modified bismethylene-adamantane molecule where the cage consists of four connected cyclopentane (instead of cyclohexane) rings. The relevant carbon atoms are numbered. The relevant bond lengths at the equilibrium geometry of the neutral species are indicated.

fragmentation sites in small polypeptide cations were controlled by the dynamics of a superposition of quasi-degenerate electronic states.[23,24] The interpretation and simulation of such coupled electron and nuclear dynamics in molecules, often called *charge-directed reactivity* in the literature, have proven to be challenging. Therefore, several experimental and theoretical studies have since been carried out on diatomic molecules.[25–29] De Vivie-Riedle extended the concept theoretically to larger molecular systems by studying in reduced dimensionality the effect of the initial conditions on the subsequent dynamics:[30,31] it was suggested to control the relative phases and amplitudes of a superposition of electronic states near a conical intersection to "steer" chemical reactions and control the branching ratio of populations for instance. Using the Ehrenfest method, we have previously investigated the nature of the nuclear motion when an electronic wave packet is populated in benzene and toluene cations:[32,33] we have shown that manipulation of the details of the electronic wave packet (relative weight and phase) leads to the control of the initial nuclear motion.

In this article, we study coupled electron and nuclear dynamics upon valence ionisation of modified bismethylene-adamantane (BMA) molecules (Fig. 1), using our implementation of the Ehrenfest method.[17,34] The BMA cation is a model donor–linker–acceptor system where there is an electron transfer between two π bonds connected by the adamantane framework.[35,36] Vertical ionisation of the π system at the equilibrium geometry of the neutral species leads to an exact point of degeneracy between the ground and first excited states of the cation.[37] (This does not induce electron dynamics according to eqn (1).) In a chemically modified BMA molecule, the vertical ionisation may take place at a geometry with a nonzero energy gap between the ground and first excited states of the cation. This is needed to generate a non-stationary electronic wave packet. We present the theoretical methods used in Section 2. In Section 3, we study electron dynamics in modified BMA molecules and the different effects of the nuclei. In Section 4, we study the nuclear motion induced by a non-stationary superposition of the electronic ground and first excited states.

2 Methods and computational details

2.1 Initial conditions

The initial nuclear geometry is the equilibrium geometry of the neutral species, optimised using the complete active space self-consistent field (CASSCF) method with the 4 π orbitals in the active space and a 6-31G* basis set, as implemented in

Gaussian.[34] To mimic the quantum distribution of the vibrational ground state (in the harmonic approximation), we sample 500 initial nuclear geometries and velocities from the Wigner distribution. This was done with the Newton-X program[38] using the harmonic frequencies calculated at the equilibrium geometry of the neutral species at the CASSCF(4,4)/6-31G* level of theory. The results depend on the number of sampled points taken into account; one must make sure convergence has been reached. We test the convergence of the results for the system studied here in the ESI (Fig. S1†).

The initial electronic wave packet is a superposition of the two lowest-energy eigenstates of the cation, ψ_0 and ψ_1,

$$\Psi = \cos(\theta/2)\psi_0 + \sin(\theta/2)e^{i\phi}\psi_1 \tag{2}$$

where $\theta/2$ determines the relative weight between the two electronic states and ϕ the relative phase. We do not aim to simulate a specific experiment: the relative weight $\theta/2$ and phase ϕ are parameters of our simulations, the role of which is worth exploring in its own right.

2.2 Electronic structure and dynamics methods

To simulate coupled electron and nuclear dynamics, we use our CASSCF implementation of the mixed quantum–classical Ehrenfest method in a development version of Gaussian.[34] The derivation and implementation details have been described previously.[17] Here, we only provide a short summary of the method.

Electron dynamics is treated quantum mechanically by integrating the electronic time-dependent Schrödinger equation. The electronic wave packet is expanded in the configuration state function (CSF) basis generated by the active space considered; **A** is the vector that gathers the expansion coefficients. In practice, time is discretised and the electronic Hamiltonian H_e (in the CSF basis) is assumed to be constant over the time step. The electronic wave function at time t_n is calculated from the wave function at time t_{n-1} using the following equation:

$$\mathbf{A}(t_n) = \exp\left(-\frac{i}{\hbar}H_e(t_n) \times (t_n - t_{n-1})\right)\mathbf{A}(t_{n-1}) \tag{3}$$

The nuclear motion is included (if wanted) by solving the classical Newton's equation of motion:

$$\frac{d\mathbf{P}_I}{dt} = -\nabla_I\langle\Psi|\hat{H}_e|\Psi\rangle \tag{4}$$

with $\mathbf{P}_I$ the classical momentum of nucleus I. We integrate eqn (4) using the Hessian-based predictor–corrector algorithm designed by Hase and Schlegel.[17,39] A computationally cheaper alternative is to use only the gradient of the energy (without the coupled-perturbed CASSCF corrections) to integrate eqn (4).[17] We use the latter approach for the ensemble of 500 trajectories.

2.3 Analysis

To follow the evolution of the electronic wave function, its electronic spin density – that allows one to locate the unpaired electron – is computed and partitioned on the atoms (using the standard Mulliken population analysis)[40] at each step of the simulation. This is how we monitor the hole dynamics. We also calculate the time evolution of the molecular dipole moment.

To follow the evolution of the nuclear geometry, the bond lengths of the two methylene groups (forming the two π bonds) are computed at each step of the simulation.

3 Electron dynamics and effects of the nuclei

In this section, we investigate the electron dynamics upon ionisation and the different effects of the nuclei: modification of the nuclear geometry, nuclear motion, and the width of the nuclear wave packet.

3.1 Chemical modification to engineer electron dynamics

The equilibrium geometry of the neutral BMA corresponds to a point of degeneracy between the two lowest-energy electronic states of the cation. These two electronic states correspond to an electron being removed from the π bonds. Since the period of oscillation in the electronic density is inversely proportional to the energy gap (eqn (1)), there is no pure electron dynamics upon valence ionisation of the π system in this molecule. A chemical modification is thus needed to lift the degeneracy and "engineer" electron dynamics.[19] The two nuclear coordinates of the branching space that lift the degeneracy are the antisymmetric stretching of the two π bonds and the deformation of the adamantane cage.[35,36] A distortion along the former coordinate leads to the adiabatic ground state of the cation having the charge localised on the longer methylene bond while the adiabatic first excited state has the charge localised on the shorter bond. A superposition of these two states would result in a delocalised hole over the two methylene groups and not in charge migration between spatially separated parts of the molecule. We therefore choose to lift the degeneracy by "chemically" deforming the adamantane cage.

The original adamantane cage consists of four connected cyclohexane rings (arranged in the "chair" configuration): each ring therefore contains 6 carbon atoms. The chemically modified species are obtained by altering the number of carbon atoms in the rings. By nature of the connection of the rings, two rings must share the same number of atoms. A pair of integers – each indicating the number of carbon atoms in two rings – is therefore enough to define the (modified) adamantane cage. Using this convention, the original BMA molecule corresponds to BMA[6,6]. In this article, we consider the following modified molecules: BMA[5,5] (Fig. 1), BMA[6,5] and BMA[6,7]. The energy gaps between the ground and first excited states calculated at the CASSCF(3,4)/6-31G* level of theory are indicated in Table 1 for the modified molecules studied. For all these molecules, the adiabatic ground and first excited states of the cation have a charge delocalised over the two methylene groups. An equal superposition of the two states results in a localised hole on one methylene group or the other depending on the relative phase.

3.2 Electron dynamics at single fixed nuclear geometries

We start by reporting simulations of pure electron dynamics at a single fixed nuclear geometry: the equilibrium geometry of the neutral species. These results are obtained with eqn (3), keeping the nuclei fixed.

Table 1 Energy gap between the ground and first excited states of the cation at the equilibrium geometry of the neutral species (calculated at the CASSCF(3,4)/6-31G* level with the 4 π orbitals included in the active space) and the corresponding expected period of the oscillations (eqn (1))

Molecule	Energy gap [eV]	Period of oscillations [fs]
BMA[5,5]	0.84	4.9
BMA[6,5]	0.28	14.6
BMA[6,7]	0.11	37.0
BMA[6,6]	0.00	∞

Fig. 2 shows the time evolution of the spin density upon π ionisation of BMA [5,5]. The unpaired electron initially localised on the bond C1–C2 (with a higher density on C1) – see Fig. 1 for atom numbering – migrates to the bond C3–C4: at 1.2 fs, it is delocalised over the two bonds and at 2.4 fs, it is completely localised on the other bond. It then returns to its initial position and the whole process repeats itself. Thus, the hole oscillates back and forth between the two terminal methylene groups, with a period inversely proportional to the energy gap, according to eqn (1).

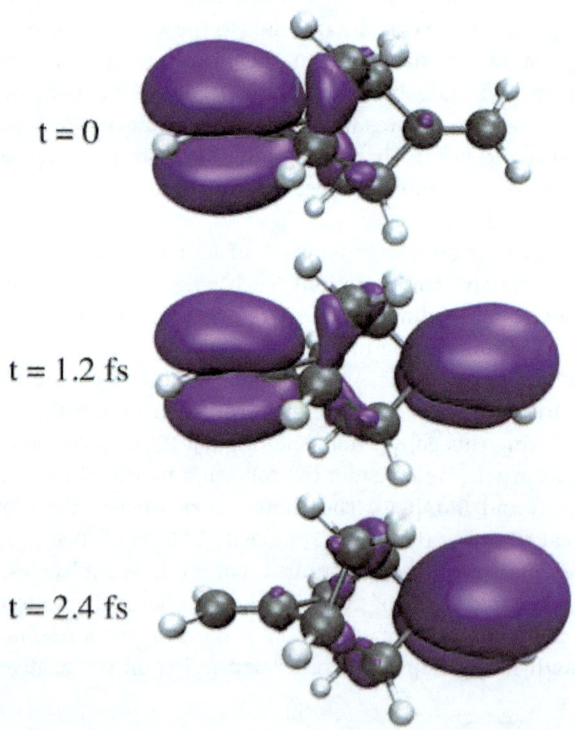

Fig. 2 Snapshots of the spin density upon ionisation of the π system in the BMA[5,5] molecule. The initial electronic wave packet is $\frac{1}{\sqrt{2}}(\psi_0 + \psi_1)$. Simulation with fixed nuclei, at the equilibrium geometry of the neutral species (multimedia view available in the ESI†).

Fig. 3 shows, for the three molecules studied, the Mulliken partitioned spin densities on carbon atoms C1, C2, C3 and C4 as a function of time. The nature and amplitude of the electron dynamics are the same for the three molecules: the unpaired electron oscillates between the two terminal methylene groups, the whole time being mainly located on the outer carbon atom. The time scale of the electron dynamics is different in the three molecules, as expected, because of the different energy gaps between the two electronic states (Table 1).

We have shown that electron dynamics simulated at a single fixed nuclear geometry results in oscillations in the unpaired electron distribution through the spin density. Equivalently, the positive charge distribution can be traced in time by calculating the molecular dipole moment. Fig. 4 shows, for the three molecules studied, the time evolution of the dipole moment component along the z axis (see Fig. 1 for the molecular orientation). (The dipole components along the x and y axes are negligible.) Initially, the unpaired electron and therefore the positive charge is located on the C1–C2 bond: the component of the dipole moment along the z axis is negative and corresponds to a positive charge located approximately 1.6–2 Å away from the origin (here, at the centre of mass). From BMA[5,5] to BMA[6,7], the initial dipole component along the z axis is slightly more negative because the adamantane cage is bigger and the charge is therefore slightly further away from the origin. The dipole moment then oscillates between being positive and negative with the same period as the spin density, as expected. Note that electron dynamics in atoms has been probed experimentally using transient absorption spectroscopy where the electronic transition probability is proportional to the dipole moment.[41,42]

3.3 Effect of the nuclear motion

We have investigated the effect of the nuclear motion on the electron dynamics with simulations of coupled electron and nuclear dynamics started at the equilibrium geometry of the neutral species. The nuclear motion is included according to the Ehrenfest method by integrating eqn (4) with the Hessian-based predictor–corrector algorithm, with a mass-weighted step size of 0.01 amu$^{1/2}$ bohr (corresponding to a time step of approximately 0.1 fs). Fig. 5 shows, for the three molecules studied, the partitioned spin density on the left methylene group (C1 and C2) as a function of time, when the nuclei are kept fixed (solid lines) and when they are allowed to move (dashed lines) for comparison.

In BMA[5,5] (Fig. 5a), the two curves representing the oscillations in the electronic density with fixed and moving nuclei are almost exactly superimposed up to 20 fs. The period of oscillations becomes slightly shorter with time when the nuclei are allowed to move. Overall, the nuclear motion does not have a significant effect on the electron dynamics in this system, when starting from the neutral equilibrium geometry.

In BMA[6,5] (Fig. 5b), the two curves representing the electron dynamics with fixed and moving nuclei are superimposed up to approximately 5 fs. Then, the oscillations get faster when the nuclei are allowed to move: the unpaired electron comes back to its initial position at approximately $t \approx 12$ fs instead of 15 fs. Note that it does not come back completely to its initial position since the amplitude of the oscillations decreases.

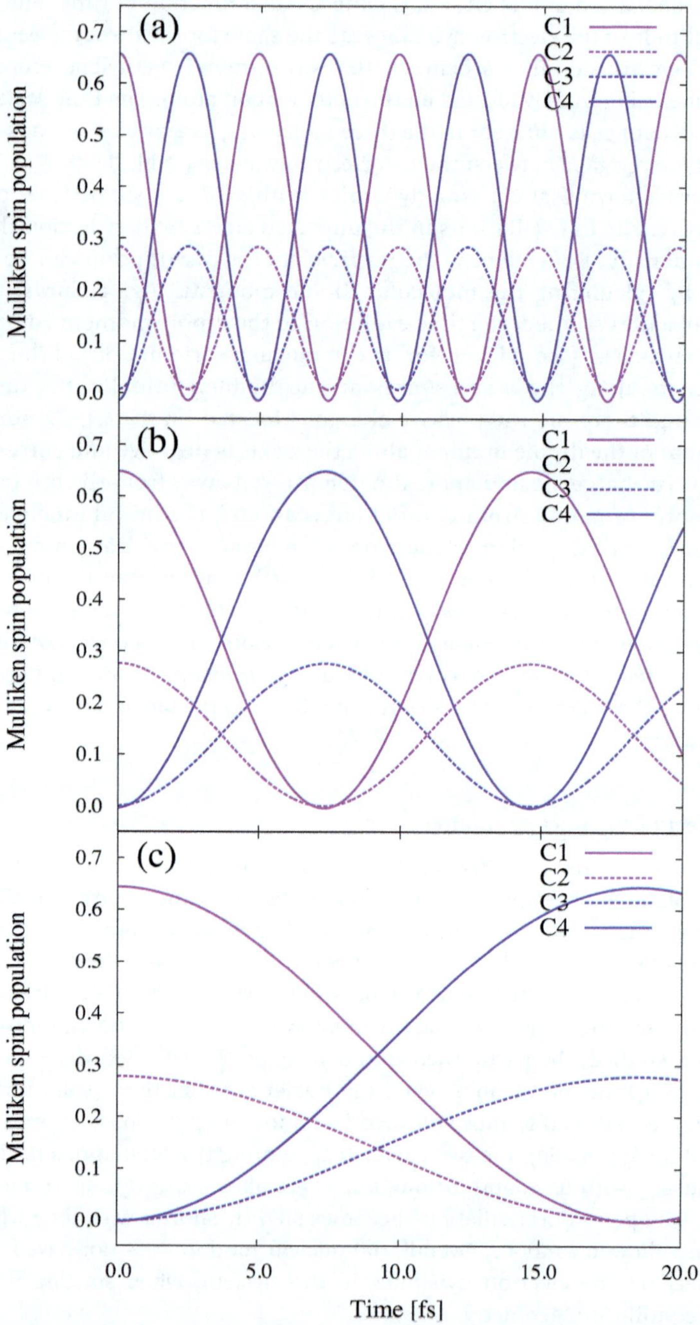

Fig. 3 Time evolution of the partitioned spin densities on the atoms of the two methylene groups, upon ionisation of the π system in the (a) BMA[5,5], (b) BMA[6,5] and (c) BMA[6,7] molecules. The initial electronic wave packet is $\frac{1}{\sqrt{2}}(\psi_0 + \psi_1)$. Simulations with fixed nuclei, at the equilibrium geometries of the neutral species. See Fig. 1 for the atom numbering.

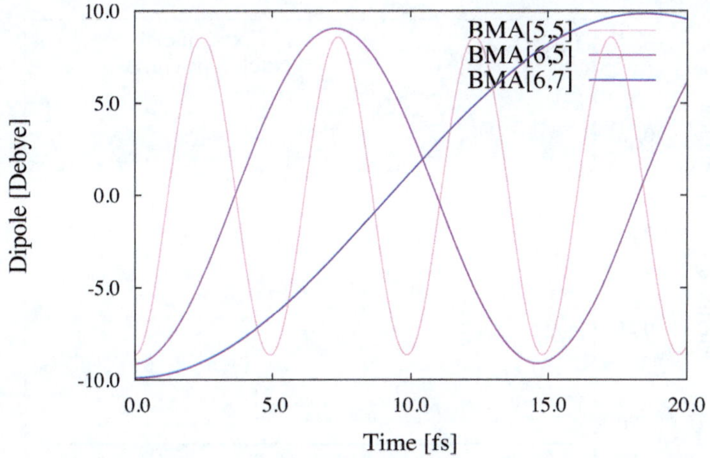

Fig. 4 Time evolution of the dipole moment along the z axis, upon ionisation of the π system in the BMA[5,5], BMA[6,5] and BMA[6,7] molecules. The initial electronic wave packet is $\frac{1}{\sqrt{2}}(\psi_0 + \psi_1)$. Simulations with fixed nuclei, at the equilibrium geometries of the neutral species. See Fig. 1 for the molecular orientation.

In BMA[6,7] (Fig. 5c), the two curves representing the electron dynamics with fixed and moving nuclei are again superimposed until approximately 5 fs. Then, instead of continuing the migration to the other methylene group, the unpaired electron returns to its initial position. By the time it should have migrated completely to the other bond, it is back at the original bond, decreasing the period of oscillations by a factor of two. The amplitude of oscillations is here less than half the one with fixed nuclei.

In summary: the nuclear motion can alter both the period and amplitude of the oscillations in the electronic density. The magnitude of the effect of the nuclear motion is very system-dependent. The smaller the energy gap, the slower the electron dynamics and the more significant the effect of the nuclear motion. This makes sense since the electron dynamics becomes slower where the two electronic states get closer in energy: there, the electron distribution changes on the same time scale as the nuclear geometry, leading to a strong coupling between the electronic and nuclear degrees of freedom.

3.4 Effect of the nuclear wave packet width

Finally, we have studied the effect of the width of the nuclear wave packet on the electron dynamics with simulations in BMA[5,5] started at 500 nuclear geometries sampled from a Wigner distribution to mimic the vibrational ground state nuclear wave packet of the neutral species. The initial electronic wave packet is $\frac{1}{\sqrt{2}}(\psi_0 + \psi_1)$. Fig. 6 shows the sum of spin density on the left methylene group (C1 and C2) as a function of time, for the ensemble of fixed nuclear geometries simulated independently. A close look allows one to distinguish the individual oscillations that dephase with time: the different geometries in the ensemble have different energy gaps and thus result in different periods of oscillation. The

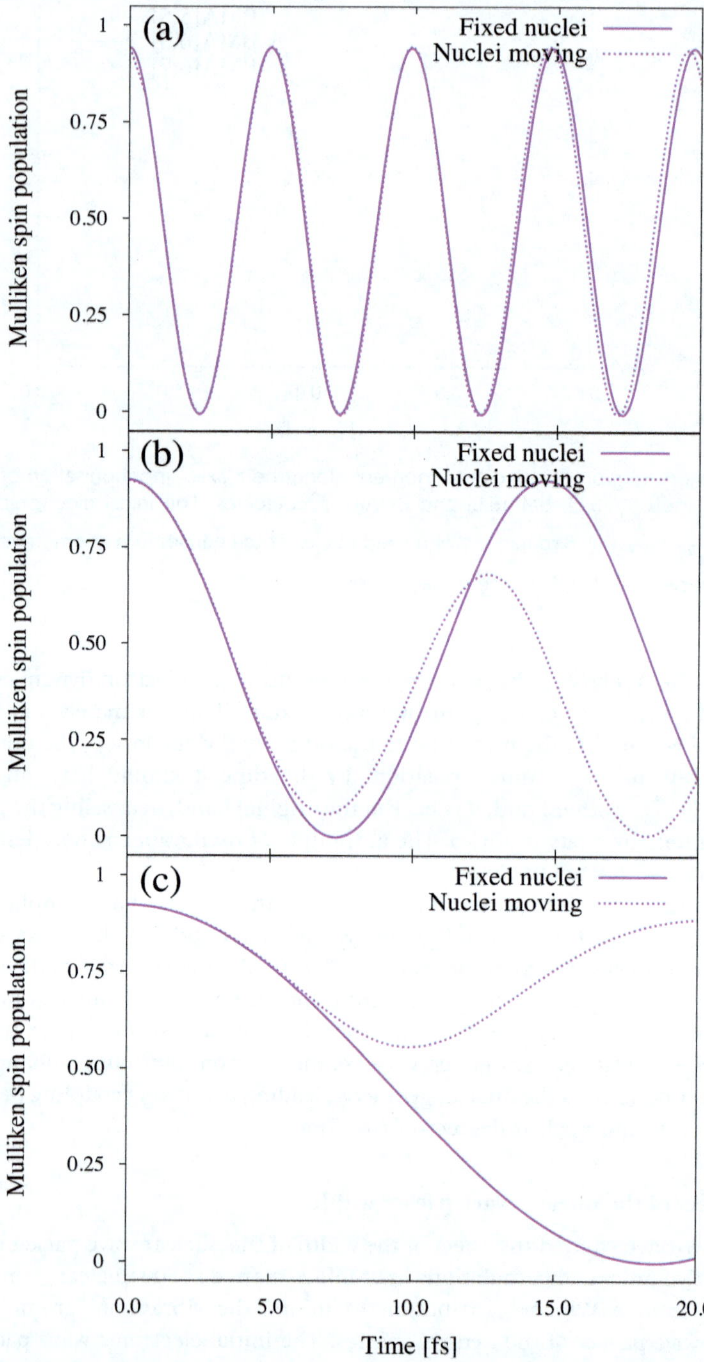

Fig. 5 Time evolution of the sum of partitioned spin densities on the atoms C1 and C2 (left methylene group), upon ionisation of the π system in the (a) BMA[5,5], (b) BMA[6,5] and (c) BMA[6,7] molecules. The initial electronic wave packet is $\frac{1}{\sqrt{2}}(\psi_0 + \psi_1)$. Simulations with fixed nuclei (solid lines) and nuclei moving (dashed lines).

average oscillation amplitude is shown as a solid white line. The amplitude of the average oscillation is quickly damped with time. This results in the spin density being, on average, equally delocalised over the two methylene groups. We can extract a coherence half-life of ≈7 fs by fitting the average spin density as explained in our previous work.[22] As mentioned above, the result depends on the number of sampled geometries taken into account. In the ESI (Fig. S1†), we present the average spin density oscillation amplitude for different sample sizes and demonstrate that 500 geometries corresponds to a converged result.

The dashed white line shows the average spin density for the ensemble when the nuclei are allowed to move according to the Ehrenfest method. Eqn (4) is integrated calculating only the gradient (with a time step of 0.05 fs) so that the 500 trajectories are computationally affordable. We have checked for a few geometries that the results obtained with the gradient only and with the Hessian-based predictor–corrector algorithm are in good agreement (see ESI†). As with fixed nuclei, the average spin density oscillation is quickly damped. Surprisingly, the amplitude of the average spin density oscillation is slightly greater when the nuclei are allowed to move. A coherence half-life of ≈8 fs was extracted in this case. (It is not due to the energy gap distribution becoming narrower; it actually becomes wider.) It could be due to the fact that, when the nuclei move, the amplitude of the oscillations can become lower for some nuclear geometries of the ensemble (see Section 3.3 and Fig. 5). Therefore, the dephasing due to these geometries will be less important, leaving, on average, a higher amplitude oscillation.

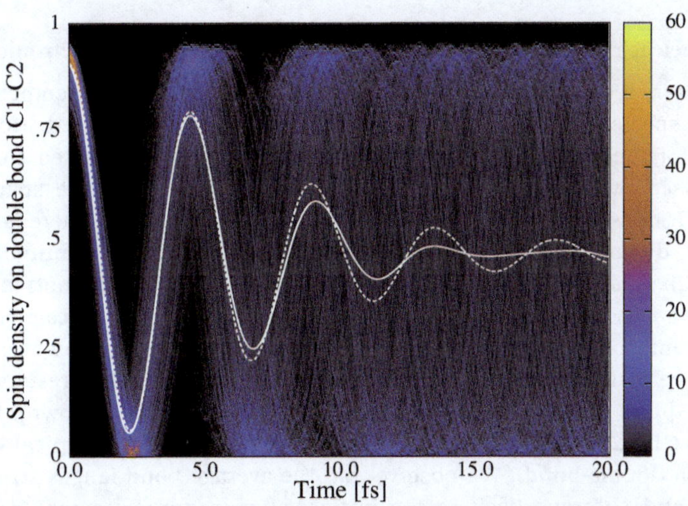

Fig. 6 Time evolution of the sum of partitioned spin densities on the atoms C1 and C2 (left methylene group), upon ionisation of the π system in the BMA[5,5] molecule, for an ensemble of 500 fixed geometries sampled from a Wigner distribution. The colour axis indicates the number of sampled trajectories in each pixel of the heat map. The solid white line indicates the spin density averaged over the ensemble of 500 sampled geometries when the nuclei are kept fixed. Also given is the spin density averaged over the ensemble of 500 trajectories (with initially sampled geometries and velocities) when the nuclei are allowed to move, indicated by the dashed white line. The initial electronic wave packet is $\frac{1}{\sqrt{2}}(\psi_0 + \psi_1)$.

In summary: the effect of the nuclear wave packet width is important: it leads to a fast dephasing of the electron dynamics. In BMA[5,5], it is larger than the effect of the nuclear motion *per se*. The coherence half-life in the studied system here is 7–8 fs which is longer than in other systems we have studied (4 fs in the *para*-xylene cation[22] and 2 fs in the phenylalanine derivative cations[21]). The relevant properties for the dephasing time scale are the width of the nuclear wave packet and the relative slopes of the involved potential energy surfaces:[22] a narrow wave packet and parallel potential energy surfaces result in a narrow energy gap distribution and therefore a slow dephasing. A possible explanation for the "long" coherence half-life in BMA [5,5] is the rigid structure of the cage, which limits the nuclear wave packet width and therefore the energy gap distribution width. Because of the relatively "long" coherence half-life and because of the short period of oscillations (4.9 fs), we observe several oscillations in the electronic density before dephasing occurs.

4 Nuclear motion and electronic coherent control

In this section, we analyse the nuclear motion upon ionisation induced by a single electronic eigenstate and by a superposition of them to investigate the differences. We are particularly interested in the evolution of the bond lengths of the two terminal methylene groups with time.

4.1 Nuclear motion induced by a single stationary adiabatic electronic state

For reference, we report simulations started at the equilibrium geometry of the neutral species with only the electronic ground state populated. Upon population of the adiabatic ground state of the cation, the unpaired electron is delocalised over the two methylene groups (like in the centre snapshot of Fig. 2). It does not evolve significantly with time (not shown): there is no pure electron dynamics since a single stationary adiabatic state is initially populated. (Change in the electronic distribution due to the nuclear motion – called *charge transfer* in the literature – could happen on a longer time scale.) In Fig. 7, we present, for the BMA[5,5] molecule, the time evolution of the average and difference in bond lengths of the two methylene groups. The results for the BMA[6,5] and BMA[6,7] molecules are almost identical (not shown). Initially, the two C1–C2 and C3–C4 bond lengths are equal to 1.33 Å (a typical value for a carbon double bond). We observe that the average bond length stretches to 1.42 Å and shortens back to the initial value in approximately 20 fs. The stretching is expected since an electron has been removed from a bonding π orbital. Note that the frequency of C=C stretching is typically in the range of 1640–1680 cm^{-1}, which corresponds to a period of vibration of approximately 20 fs. Our simulations are in agreement with this value. An important point is that, here, the difference in bond lengths between the two terminal methylene groups is null: the stretching is symmetric, as expected since the unpaired electron is equally, *i.e.* symmetrically, delocalised over the two methylene groups.

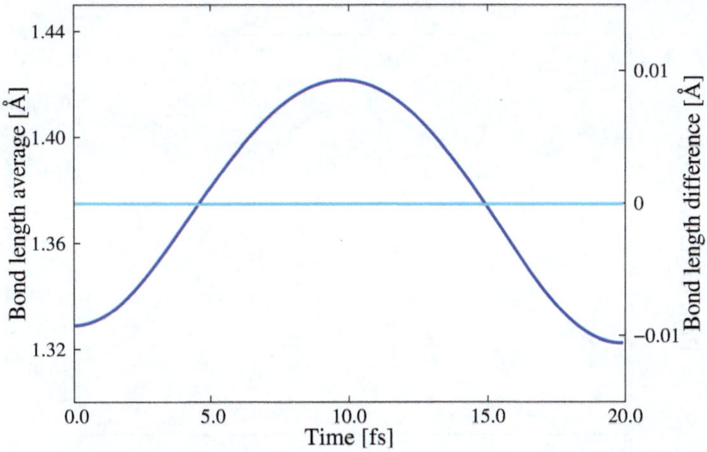

Fig. 7 Time evolution of the bond lengths of the two methylene groups (average length of the C1–C2 and C3–C4 bonds in blue with the left *y* axis and difference in cyan with the right *y* axis), upon ionisation of the π system in the BMA[5,5] molecule. The initial electronic state is the ground state ψ_0.

4.2 Nuclear motion induced by a non-stationary superposition of electronic states

We now analyse the nuclear motion induced when the electronic wave packet $\frac{1}{\sqrt{2}}(\psi_0 + \psi_1)$ is initially populated and we investigate how it differs from the nuclear motion induced by a stationary adiabatic electronic state (Section 4.1). Fig. 8 shows, for the three molecules studied, the time evolution of the average and difference in bond lengths of the two methylene groups in this case. Note that the scale for the bond length difference is the same as for the bond length average except in Fig. 8a for which it is 5 times smaller. The black dashed lines remind the reader of the oscillations in the electronic density (with nuclei moving) from Fig. 5.

In BMA[5,5] (Fig. 8a), the average bond length evolves in a similar way to the case of dynamics induced by the electronic ground state solely (Fig. 7): it stretches and vibrates with a period of 20 fs. The difference in bond lengths between the two methylene groups, however, differs from 0: the stretching is asymmetric. More precisely, we observe oscillations in the difference in bond lengths, the period of which matches the period of the oscillations in the spin density (indicated by the black dashed line). Initially, the electronic density is asymmetric with the unpaired electron (or positive charge) located on the C1–C2 bond; the nuclei thus react asymmetrically *i.e.* the C1–C2 bond stretches more than the C3–C4 bond and the difference in bond length (C3–C4)–(C1–C2) becomes negative. As the unpaired electron migrates to the C3–C4 bond, the process will reverse and the C3–C4 bond will stretch more than the C1–C2 bond. Note that it takes a few fs for the nuclear motion to reverse since it has acquired some kinetic energy. Note that the bond length difference does not just oscillate around 0; there is an overall slower evolution. The oscillations in the bond length difference is an illustration of "charge-directed reactivity": the nuclear and electron dynamics are

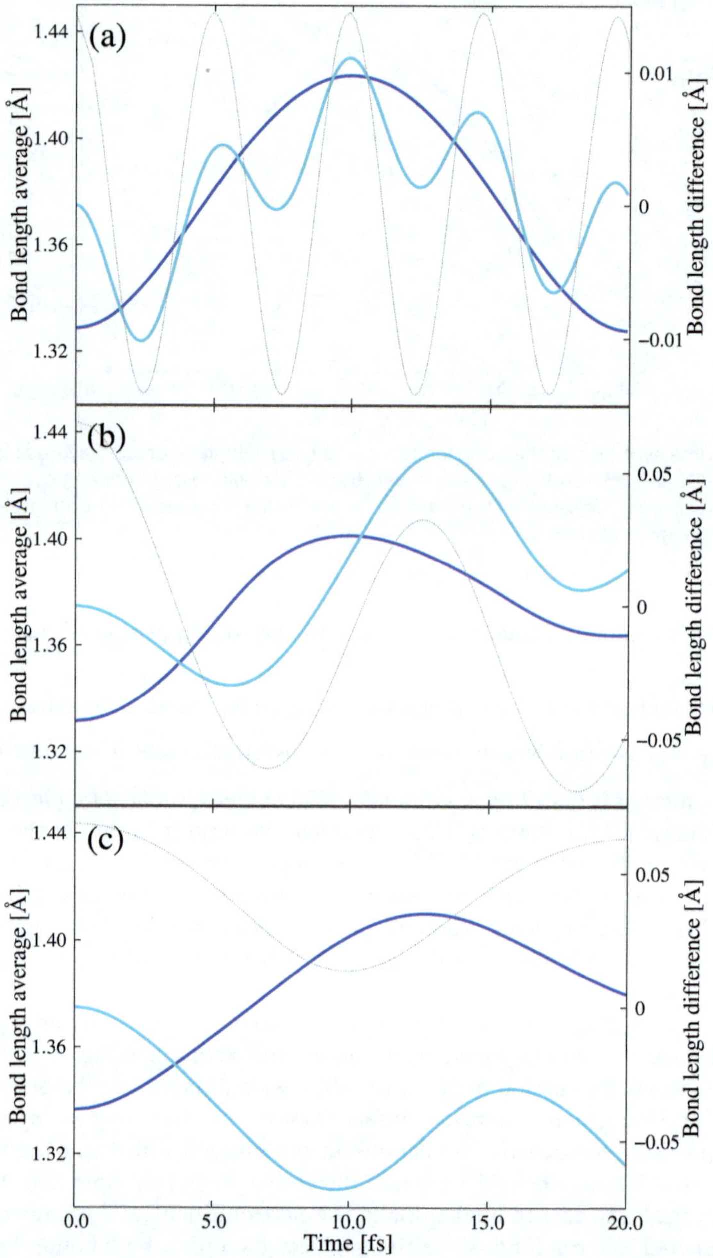

Fig. 8 Time evolution of the bond lengths of the two methylene groups (average length of the C1–C2 and C3–C4 bonds in blue with the left *y* axis and difference (C3–C4)–(C1–C2) in cyan with the right *y* axis), upon ionisation of the π system in the (a) BMA[5,5], (b) BMA[6,5] and (c) BMA[6,7] molecules. The initial electronic wave packet is $\frac{1}{\sqrt{2}}(\psi_0 + \psi_1)$. The scale for the bond length difference is the same as for the bond length average except for BMA[5,5] for which it is 5 times smaller. The black dashed lines remind the reader of the oscillations in the electronic density from Fig. 5.

asynchronous, the nuclei always trying to adapt to the time-dependent electronic distribution.

We observe similar behaviour in the bond length difference for BMA[6,5] (Fig. 8b) and BMA[6,7] (Fig. 8c). Note that in these two molecules, the difference in bond length reaches larger values (the scale being 5 times larger). In these cations the periods of the oscillations in the electronic density are longer, so the bonds stretch asymmetrically for longer. Here, the larger difference in bond lengths affects the time evolution of the average bond length which is not bell-shaped any more.

In summary: the superposition of the two lowest-energy electronic states induces an asymmetric stretching of the two terminal methylene bonds (while each electronic state populated individually induces a symmetric stretching). This is because of the unpaired electron not being equally delocalised over the two terminal groups when a superposition of electronic states is populated. The oscillations with time of the electronic density result in oscillations in the nuclear motion as well, the nuclei continuously trying to adapt to the time-dependent electronic distribution.

To illustrate the "charge-directed reactivity" idea further, we have simulated coupled electron and nuclear dynamics in the BMA[6,5] cation, induced by the electronic wave packets $\frac{1}{\sqrt{2}}(\psi_0 + e^{i\phi}\psi_1)$ with the different relative phases $\phi = 0$, 90, 180 and 270°. The case of $\phi = 0$ corresponds to the equal in-phase superposition as in Fig. 5b and 8b. Fig. 9 shows, for the different initial electronic wave packets, the time evolution of the partitioned spin density on the left methylene group (C1 and C2). The initial relative phase determines the initial position of the unpaired electron: the latter is localised on the C1–C2 bond with $\phi = 0$ (as in Fig. 5b), on the C3–C4 bond with $\phi = 180$, and delocalised over the two bonds with

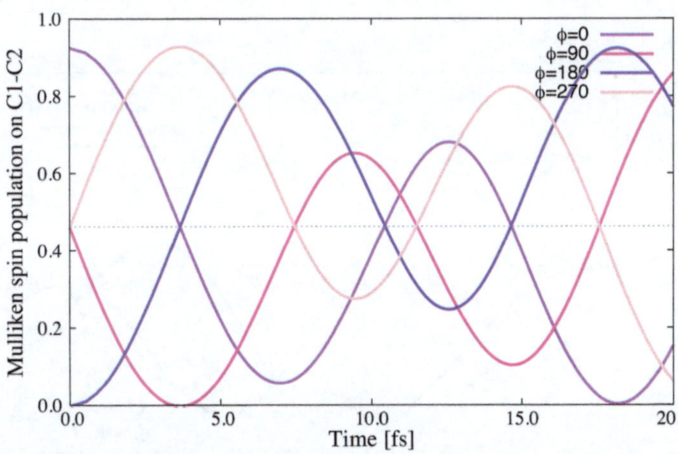

Fig. 9 Time evolution of the sum of partitioned spin densities on the atoms C1 and C2 (left methylene group), upon ionisation of the π system in the BMA[6,5] molecule. The initial electronic wave packets are $\frac{1}{\sqrt{2}}(\psi_0 + e^{i\phi}\psi_1)$ with different relative phases $\phi = 0$, 90, 180 and 270°. Simulations with nuclei moving. The dotted horizontal line at $y = 0.46$ indicates the spin density value when the unpaired electron is equally delocalised over the two terminal methylene groups.

$\phi = 90$ and 270. The spin density then oscillates with time between the two methylene groups. As the initial phase determines the value of the oscillation at $t = 0$, it can be seen as a shift of the time axis. The oscillations are not perfect sinusoids: the amplitude and period vary with time because of the nuclear motion (see Section 3.3). Note that the curves for $\phi = 0$ and 180, and for $\phi = 90$ and 270, are symmetric with respect to $y = 0.46$ (the value corresponding to the spin density equally delocalised over the two methylene groups).

Fig. 10 shows, for the different initial electronic wave packets, the time evolution of the bond length difference. We observe again oscillations in the bond length differences that follow the oscillations in the spin densities (Fig. 9). We notice that the nuclear motions induced by $\phi = 0$ and 180, and by 90 and 270 are symmetric with respect to $y = 0$ (the value corresponding to equal C1–C2 and C3–C4 bond lengths). Indeed, with $\phi = 180$, the C3–C4 bond initially stretches more than C1–C2 since the unpaired electron is this time localised on C3–C4 at $t = 0$. Besides, the absolute difference in bond lengths is initially lower with the relative phases $\phi = 90$ and 270 than with 0 and 180: the unpaired electron being equally delocalised over the two methylene groups at $t = 0$ with $\phi = 90$ or 270, the induced nuclear motions are symmetric stretching. With time, as the unpaired electron migrates to one of the two bonds, it then induces an asymmetric stretching. Note that the unpaired electron initially stays more localised on one bond for 8 fs ($\approx$half a period) with $\phi = 90$ or 270 before migrating to the other bond, while it migrates to the other bond after only 4 fs ($\approx$a quarter of period) with $\phi = 0$ or 180 (because of the "shift" of the time axis). As a result, the (absolute) difference in bond lengths has the time to become greater with $\phi = 90$ or 270: 0.07 Å reached at $\approx$9.0 fs compared to 0.03 Å reached at $\approx$5.6 fs.

In summary: we have illustrated how by manipulating the composition of the initial electronic wave packet (relative weight and phase), one can manipulate the initial nuclear motion, *i.e.* which bond stretches more (if any) and by how much.

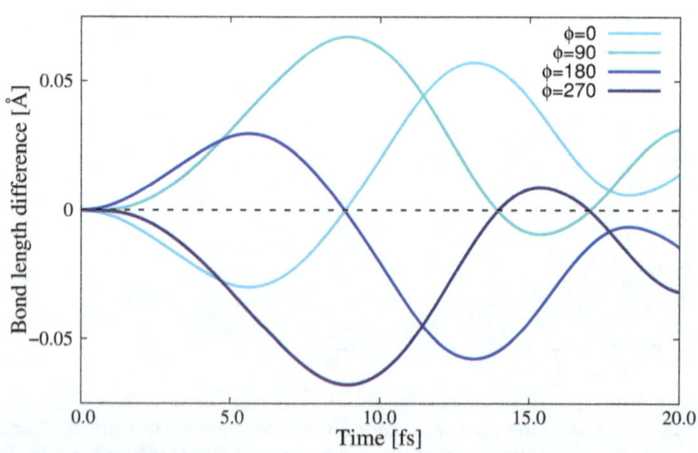

Fig. 10 Time evolution of the bond length difference between the two methylene groups (C3–C4)–(C1–C2), upon ionisation of the π system in the BMA[6,5] molecule. The initial electronic wave packets are $\frac{1}{\sqrt{2}}(\psi_0 + e^{i\phi}\psi_1)$ with different relative phases $\phi = 0$, 90, 180 and 270°. Simulations with nuclei moving.

4.3 Nuclear motion with an ensemble of trajectories to mimic the nuclear wave packet width

Finally, we have studied the effect of the width of the nuclear wave packet on the nuclear motion with simulations in BMA[5,5] started at 500 sampled nuclear geometries and velocities (see Section 3.4). The initial electronic wave packet is $\frac{1}{\sqrt{2}}(\psi_0 + \psi_1)$. Fig. 11 shows the bond length average as a function of time, for the ensemble of trajectories simulated independently. The dashed white line indicates the average value of the ensemble. The initial width of the distribution is characterised by a standard deviation of 0.04 Å. The ensemble of trajectories undergoes a stretching of the average bond length to 1.42 Å with a period of vibration of 20 fs, as the single "un-sampled" trajectory started at the neutral equilibrium geometry with no kinetic energy (Fig. 8a). Note that the time evolution of the average length of the C1–C2 and C3–C4 bonds does not result from the electron dynamics of a non-stationary electronic wave packet; it was also observed with the stationary electronic ground state initially populated (Fig. 7). Therefore, we do not expect a dephasing of this vibrational motion with time.

Fig. 12 shows the time evolution of the *difference* in bond lengths of the two methylene groups, averaged over the ensemble of 500 trajectories. The black dashed line reminds the reader of the oscillations in the electronic density (averaged over the 500 trajectories) from Fig. 6. The bond length difference oscillates with time, as in Fig. 8a initially; then, the oscillations are damped. The oscillations look to be on top of an overall slower evolution, as in Fig. 8a. To help with the interpretation of the damping of the oscillations, Fig. 13 shows the time derivative of the difference in bond lengths of the two methylene groups, averaged over the ensemble of 500

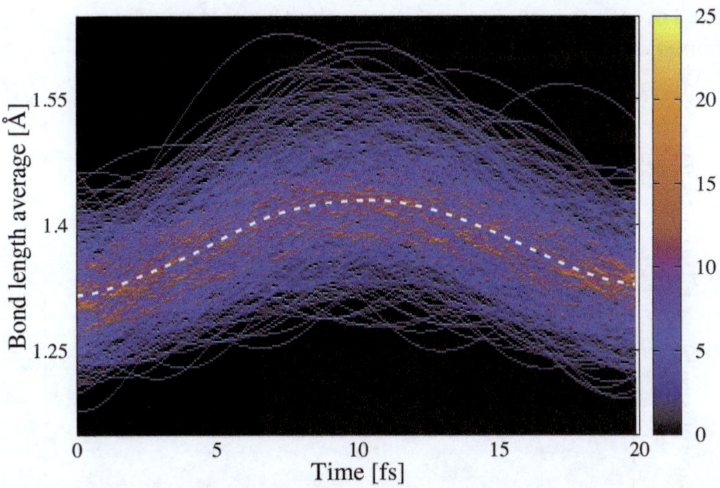

Fig. 11 Time evolution of the average bond length of the C1–C2 and C3–C4 bonds, upon ionisation of the π system in the BMA[5,5] molecule, for an ensemble of 500 geometries and velocities sampled from a Wigner distribution. The colour axis indicates the number of sampled trajectories in each pixel of the heat map. The dashed white line indicates the average value of the ensemble. The initial electronic wave packet is $\frac{1}{\sqrt{2}}(\psi_0 + \psi_1)$.

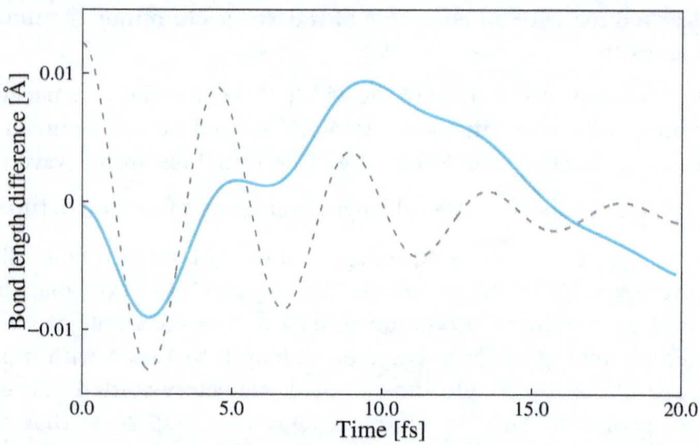

Fig. 12 Time evolution of the difference in bond lengths of the two methylene groups (C3–C4)–(C1–C2), averaged over the ensemble of 500 trajectories (with sampled geometries and velocities), upon ionisation of the π system in BMA[5,5]. The initial electronic wave packet is $\frac{1}{\sqrt{2}}(\psi_0 + \psi_1)$. The black dashed lines remind the reader of the oscillations in the electronic density from Fig. 6.

trajectories. Similarly, the black dashed line is the time derivative of the oscillations in the electronic density (averaged over the 500 trajectories). We clearly see that, when taking into account an ensemble of trajectories to represent the nuclear wave packet, the oscillations in the bond length difference are damped just as the oscillations in the spin density dephase.

In summary: the effect of the nuclear wave packet width is again important. As the electron dynamics dephase and the unpaired electron becomes equally delocalised over the two methylene groups, the asymmetry in the stretching of the two π bonds disappears.

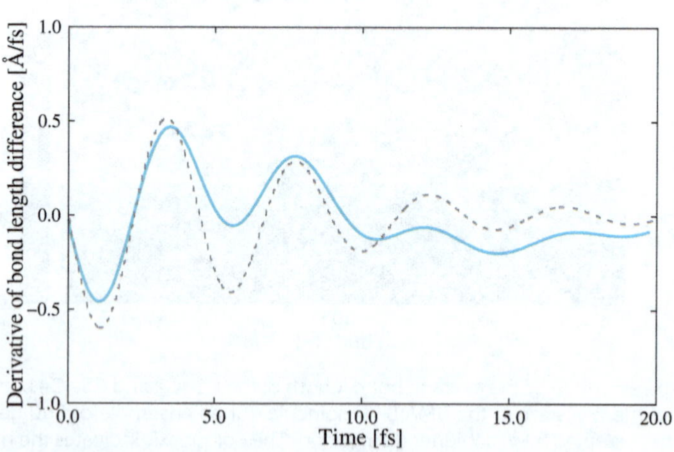

Fig. 13 Time derivative of the curves in Fig. 12.

5 Conclusion

In this article, we have studied the electron and nuclear dynamics upon valence ionisation of the π system in three modified BMA molecules. Chemical modification to the BMA molecule was necessary to lift the energy gap between the ground and first excited cationic states at the neutral equilibrium geometry and "engineer" electron dynamics. Using the Ehrenfest method, we have studied the effect of the nuclear motion on the electron dynamics. The latter is significant in molecules where the energy gap is small (lower than 0.3 eV) since the electron dynamics slows down to the time scale of the nuclear motion. Another important effect is that of the width of the nuclear wave packet. The natural distribution of geometries within the nuclear wave packet leads to the fast dephasing of electron dynamics. In BMA[5,5], we have extracted a coherence half-life of 7 fs which, with the short period of oscillations, leaves enough time for several oscillations in the electronic density before dephasing occurs. Surprisingly, the nuclear motion helps maintaining the coherence for slightly longer by damping the oscillations at some distorted geometries that led to dephasing.

Ionisation of the π system leads to an overall stretching of the two carbon double bonds. We have shown that the population of a single adiabatic state results in a symmetric stretching. On the other hand, the population of a superposition of the ground and first excited state breaks the symmetry by localising the unpaired electron on one methylene group and leads to an asymmetric stretching. This is a demonstration of "charge-directed reactivity". As the unpaired electron oscillates from one methylene group to the other, the difference in bond length of the two bonds oscillates with the same period. The nuclear geometry is continuously trying to adapt to the time-dependent electronic density. By manipulating the composition of the superposition of states, for instance the relative phase, one can manipulate the asymmetry of the nuclear motion, *i.e.* which bond stretches more and by how much. When taking into account an ensemble of sampled trajectories, the asymmetry in stretching eventually disappears as the dephasing of electron dynamics leads to an equal delocalisation of the unpaired electron on the two methylene groups.

Acknowledgements

This work was supported by UK-EPSRC Grant EP/I032517/1. All calculations were run using the Imperial College High Performance Computing service.

References

1 P. M. Paul, E. S. Toma, P. Breger, G. Mullot, F. Augé, P. Balcou, H. G. Muller and P. Agostini, *Science*, 2001, **292**, 1689–1692.
2 M. Hentschel, R. Kienberger, C. Spielmann, G. A. Reider, N. Milosevic, T. Brabec, P. Corkum, U. Heinzmann, M. Drescher and F. Krausz, *Nature*, 2001, **414**, 509–513.
3 J. Breidbach and L. S. Cederbaum, *J. Chem. Phys.*, 2003, **118**, 3983–3996.
4 A. I. Kuleff and L. S. Cederbaum, *Phys. Rev. Lett.*, 2011, **106**, 053001.
5 L. Belshaw, F. Calegari, M. J. Duffy, A. Trabattoni, L. Poletto, M. Nisoli and J. B. Greenwood, *J. Phys. Chem. Lett.*, 2012, **3**, 3751–3754.

6 J. Leeuwenburgh, B. Cooper, V. Averbukh, J. P. Marangos and M. Ivanov, *Phys. Rev. Lett.*, 2013, **111**, 123002.
7 F. Calegari, D. Ayuso, A. Trabattoni, L. Belshaw, S. De Camillis, S. Anumula, F. Frassetto, L. Poletto, A. Palacios, P. Decleva, J. B. Greenwood, F. Martín and M. Nisoli, *Science*, 2014, **346**, 336–339.
8 S. R. Leone, C. W. McCurdy, J. Burgdorfer, L. S. Cederbaum, Z. Chang, N. Dudovich, J. Feist, C. H. Greene, M. Ivanov, R. Kienberger, U. Keller, M. F. Kling, Z.-H. Loh, T. Pfeifer, A. N. Pfeiffer, R. Santra, K. Schafer, A. Stolow, U. Thumm and M. J. J. Vrakking, *Nat. Photonics*, 2014, **8**, 162–166.
9 F. Remacle and R. Levine, *Z. Phys. Chem.*, 2007, **221**, 647–661.
10 A. I. Kuleff and L. S. Cederbaum, *Chem. Phys.*, 2007, **338**, 320–328.
11 S. Lünnemann, A. I. Kuleff and L. S. Cederbaum, *Chem. Phys. Lett.*, 2008, **450**, 232–235.
12 S. Lünnemann, A. I. Kuleff and L. S. Cederbaum, *J. Chem. Phys.*, 2008, **129**, 104305.
13 G. Periyasamy, R. Levine and F. Remacle, *Chem. Phys.*, 2009, **366**, 129–138.
14 A. I. Kuleff, S. Lünnemann and L. S. Cederbaum, *J. Phys. Chem. A*, 2010, **114**, 8676–8679.
15 I. S. Ulusoy and M. Nest, *J. Phys. Chem. A*, 2012, **116**, 11107–11110.
16 J. Ajay, J. Šmydke, F. Remacle and R. D. Levine, *J. Phys. Chem. A*, 2016, **120**, 3335–3342.
17 M. Vacher, D. Mendive-Tapia, M. J. Bearpark and M. A. Robb, *Theor. Chem. Acc.*, 2014, **133**, 1505.
18 D. Mendive-Tapia, M. Vacher, M. J. Bearpark and M. A. Robb, *J. Chem. Phys.*, 2013, **139**, 044110.
19 M. Vacher, D. Mendive-Tapia, M. J. Bearpark and M. A. Robb, *J. Chem. Phys.*, 2015, **142**, 094105.
20 M. Vacher, M. J. Bearpark and M. A. Robb, *J. Chem. Phys.*, 2014, **140**, 201102.
21 A. J. Jenkins, M. Vacher, M. J. Bearpark and M. A. Robb, *J. Chem. Phys.*, 2016, **144**, 104110.
22 M. Vacher, L. Steinberg, A. J. Jenkins, M. J. Bearpark and M. A. Robb, *Phys. Rev. A*, 2015, **92**, 040502.
23 F. Remacle, R. D. Levine, E. W. Schlag and R. Weinkauf, *J. Phys. Chem. A*, 1999, **103**, 10149–10158.
24 R. Weinkauf, E. W. Schlag, T. J. Martinez and R. D. Levine, *J. Phys. Chem. A*, 1997, **101**, 7702–7710.
25 A. D. Bandrauk, S. Chelkowski and H. S. Nguyen, *Int. J. Quantum Chem.*, 2004, **100**, 834–844.
26 V. Roudnev, B. D. Esry and I. Ben-Itzhak, *Phys. Rev. Lett.*, 2004, **93**, 163601.
27 M. F. Kling, C. Siedschlag, A. J. Verhoef, J. I. Khan, M. Schultze, T. Uphues, Y. Ni, M. Uiberacker, M. Drescher, F. Krausz and M. J. J. Vrakking, *Science*, 2006, **312**, 246–248.
28 D. Geppert, P. von den Hoff and R. de Vivie-Riedle, *J. Phys. B: At., Mol. Opt. Phys.*, 2008, **41**, 074006.
29 P. von den Hoff, I. Znakovskaya, M. Kling and R. de Vivie-Riedle, *Chem. Phys.*, 2009, **366**, 139–147.
30 A. Hofmann and R. de Vivie-Riedle, *J. Chem. Phys.*, 2000, **112**, 5054–5059.
31 P. v. d. Hoff, S. Thallmair, M. Kowalewski, R. Siemering and R. de Vivie-Riedle, *Phys. Chem. Chem. Phys.*, 2012, **14**, 14460–14485.

32 J. Meisner, M. Vacher, M. J. Bearpark and M. A. Robb, *J. Chem. Theory Comput.*, 2015, **11**, 3115–3122.

33 M. Vacher, J. Meisner, D. Mendive-Tapia, M. J. Bearpark and M. A. Robb, *J. Phys. Chem. A*, 2015, **119**, 5165–5172.

34 M. J. Frisch, G. W. Trucks, H. B. Schlegel, G. E. Scuseria, M. A. Robb, J. R. Cheeseman, G. Scalmani, V. Barone, B. Mennucci, G. A. Petersson, H. Nakatsuji, M. Caricato, X. Li, H. P. Hratchian, J. Bloino, B. G. Janesko, A. F. Izmaylov, A. Marenich, F. Lipparini, G. Zheng, J. L. Sonnenberg, W. Liang, M. Hada, M. Ehara, K. Toyota, R. Fukuda, J. Hasegawa, M. Ishida, T. Nakajima, Y. Honda, O. Kitao, H. Nakai, T. Vreven, K. Throssell, J. A. Montgomery, J. E. Peralta, F. Ogliaro, M. Bearpark, J. J. Heyd, E. Brothers, K. N. Kudin, V. N. Staroverov, T. Keith, R. Kobayashi, J. Normand, K. Raghavachari, A. Rendell, J. C. Burant, S. S. Iyengar, J. Tomasi, M. Cossi, N. Rega, J. M. Millam, M. Klene, J. E. Knox, J. B. Cross, V. Bakken, C. Adamo, J. Jaramillo, R. Gomperts, R. E. Stratmann, O. Yazyev, A. J. Austin, R. Cammi, C. Pomelli, J. W. Ochterski, R. L. Martin, K. Morokuma, V. G. Zakrzewski, G. A. Voth, P. Salvador, J. J. Dannenberg, S. Dapprich, P. V. Parandekar, N. J. Mayhall, A. D. Daniels, O. Farkas, J. B. Foresman, J. V. Ortiz, J. Cioslowski and D. J. Fox, *Gaussian Development Version, Revision I.04+*, Gaussian, Inc., Wallingford CT, 2014.

35 L. Blancafort, F. Jolibois, M. Olivucci and M. A. Robb, *J. Am. Chem. Soc.*, 2001, **123**, 722–732.

36 L. Blancafort, P. Hunt and M. A. Robb, *J. Am. Chem. Soc.*, 2005, **127**, 3391–3399.

37 I. Tavernelli, *J. Phys. Chem. A*, 2007, **111**, 13528–13536.

38 M. Barbatti, G. Granucci, H. Lischka, M. Persico and M. Ruckenbauer, *Newton-X: a package for Newtonian dynamics close to the crossing seam, version 0.11b*, www.univie.ac.at/newtonx, 2006.

39 J. M. Millam, V. Bakken, W. Chen, W. L. Hase and H. B. Schlegel, *J. Chem. Phys.*, 1999, **111**, 3800–3805.

40 R. S. Mulliken, *J. Chem. Phys.*, 1955, **23**, 1833–1840.

41 E. Goulielmakis, Z.-H. Loh, A. Wirth, R. Santra, N. Rohringer, V. S. Yakovlev, S. Zherebtsov, T. Pfeifer, A. M. Azzeer, M. F. Kling, S. R. Leone and F. Krausz, *Nature*, 2010, **466**, 739–743.

42 A. R. Beck, B. Bernhardt, E. R. Warrick, M. Wu, S. Chen, M. B. Gaarde, K. J. Schafer, D. M. Neumark and S. R. Leone, *New J. Phys.*, 2014, **16**, 113016.

Faraday Discussions

PAPER

Beyond structure: ultrafast X-ray absorption spectroscopy as a probe of non-adiabatic wavepacket dynamics

Simon P. Neville,[a] Vitali Averbukh,[b] Serguei Patchkovskii,[c] Marco Ruberti,[b] Renjie Yun,[b] Majed Chergui,[d] Albert Stolow*[aef] and Michael S. Schuurman[ae]

Received 4th May 2016, Accepted 7th July 2016

DOI: 10.1039/c6fd00117c

The excited state non-adiabatic dynamics of polyatomic molecules, leading to the coupling of structural and electronic dynamics, is a fundamentally important yet challenging problem for both experiment and theory. Ongoing developments in ultrafast extreme vacuum ultraviolet (XUV) and soft X-ray sources present new probes of coupled electronic-structural dynamics because of their novel and desirable characteristics. As one example, inner-shell spectroscopy offers localized, atom-specific probes of evolving electronic structure and bonding (*via* chemical shifts). In this work, we present the first on-the-fly ultrafast X-ray time-resolved absorption spectrum simulations of excited state wavepacket dynamics: photo-excited ethylene. This was achieved by coupling the *ab initio* multiple spawning (AIMS) method, employing on-the-fly dynamics simulations, with high-level algebraic diagrammatic construction (ADC) X-ray absorption cross-section calculations. Using the excited state dynamics of ethylene as a test case, we assessed the ability of X-ray absorption spectroscopy to project out the electronic character of complex wavepacket dynamics, and evaluated the sensitivity of the calculated spectra to large amplitude nuclear motion. In particular, we demonstrate the pronounced sensitivity of the pre-edge region of the X-ray absorption spectrum to the electronic and structural evolution of the excited-state wavepacket. We conclude that ultrafast time-resolved X-ray absorption spectroscopy may become a powerful tool in the interrogation of excited state non-adiabatic molecular dynamics.

[a]*Department of Chemistry, University of Ottawa, 10 Marie Curie, Ottawa, Ontario, K1N 6N5, Canada. E-mail: albert.stolow@nrc-cnrc.gc.ca*

[b]*Department of Physics, Imperial College London, South Kensington Campus, London, SW7 2AZ, UK*

[c]*Max-Born Institute, Max-Born-Strasse 2A, D-12489 Berlin, Germany*

[d]*École Polytechnique Fédérale de Lausanne, Laboratoire de Spectroscopie Ultrarapide and Lausanne Centre for Ultrafast Science (LACUS), Faculté des Sciences de Base, ISIC, Lausanne CH-1015, Switzerland*

[e]*National Research Council of Canada, 100 Sussex Drive, Ottawa, ON K1A 0R6, Canada*

[f]*Department of Physics, University of Ottawa, 150 Louis Pasteur, Ottawa, ON K1N 6N5, Canada*

1 Introduction

Photo-excited polyatomic molecules often exhibit complex dynamics involving the redistribution of both electronic charge and vibrational energy. Their dynamics are dominated by the non-adiabatic coupling of vibrational and electronic degrees of freedom, typically mediated by conical intersections. These ultrafast excited state processes are variously called radiationless transitions, electronic relaxation or internal conversion and are the primary steps in the photochemistry of many polyatomic molecules as well as photobiological process such as vision and photosynthesis. Ultrafast (femtosecond) pump-probe spectroscopies, wherein an excited state wavepacket is projected onto a set of final states as a function of time, provide powerful and intuitive means by which to study these dynamics with a temporal resolution of the order of the natural timescale for nuclear motion. A pertinent question to ask is how to choose a specific method for a particular problem. For the case of excited state non-adiabatic dynamics, the spectroscopic method should be sensitive to the degrees of freedom under investigation – electronic and vibrational – as well as their coupling. Prominent examples of techniques proposed to address this problem include time-resolved photoelectron spectroscopy (TRPES),[1,2] Time-Resolved High Harmonic Generation (TR-HHG) spectroscopy,[3–5] and two-dimensional electronic spectroscopy.[6,7]

Recent years have seen the development of new ultrafast extreme vacuum ultraviolet (XUV) and X-ray sources, both in laboratories and at large facilities such as synchrotrons and free electron lasers.[8–14] Such sources offer new probes of ultrafast excited state dynamics with desirable characteristics.[15–24] Practical advantages of X-ray probes include high photon energies, which enable the molecule to be probed during all aspects of its dynamics (including dissociation), and the ability to directly obtain structural information from the spectrum. Conceptually, the absorption of XUV and X-ray photons by a molecule results in excitation from core-level orbitals which, unlike the valence-shell excitations probed by UV photons, are highly localized in the vicinity of the nuclei. As such, XUV and X-ray photons offer the unique possibility of atom-specific probes of ultrafast electronic and vibrational dynamics.

The X-ray absorption spectrum is naturally split into two distinct regions: the pre-edge region, corresponding to excitation to quasi-bound core-excited states and the post-edge region, corresponding to direct above-threshold ionization to continuum states. The post-edge part of the spectrum offers a wealth of structural information about bond distances and angles, as well as coordination numbers.[25–28] In a similar vein, ultrafast X-ray scattering has also recently been demonstrated as a powerful probe of structural dynamics following photo-excitation.[29–31] The pre-edge region results from transitions between the core-level orbitals and the bound valence-level orbitals. As such, it provides information about the electronic structure of the initial state.[26,32,33] A key aspect of this paper is the focus on the pre-edge part of the spectrum. The principal reason for doing so is that the dominant core-to-valence transitions embedded in this part of the spectrum are found to be particularly sensitive to the electronic character of the excited-state wavepacket. Motivated by this, and a desire for sensitivity to both electronic and structural dynamics, we present here a theoretical study of the use of ultrafast pre-edge X-ray absorption spectroscopy in the study of excited state non-adiabatic molecular dynamics.

In this discussion paper, we report time-resolved X-ray absorption spectra (TRXAS) calculated by coupling the *ab initio* multiple spawning (AIMS) method, employing on-the-fly dynamics simulations, with high-level algebraic diagrammatic construction (ADC) X-ray absorption cross-section calculations. An important aspect of the AIMS approach is that it permits not only 'on-the-fly' wavepacket propagation but also 'on-the-fly' calculation of observables which may be directly compared with experiment. To date, the only existing TRXAS calculation is that reported by Penfold *et al.*,[34] who calculated TRXAS spectra for a Cu(I)–phenanthroline complex using a combination of multiconfigurational time-dependent Hartree wavepacket propagations[35,36] and linear-response time-dependent density functional theory (LR-TDDFT) X-ray absorption cross-section calculations. This work highlighted the sensitivity of the TRXAS to the structural dynamics. However, the LR-TDDFT pre-edge spectrum calculations were restricted to a single electronic state. This was due to the restrictions imposed by the LR-TDDFT methodology, which is rigorously valid only for the lowest-lying state of each spin multiplicity. Importantly, by using the ADC method to calculate X-ray absorption cross-sections, we are able to extend the calculation of pre-edge TRXAS spectra to include both ground and excited states.

To proceed, we calculate the dynamics of the fundamental diene, ethylene, following excitation to its optically bright $S_2(\pi\pi^*)$ state. As well as calculating the TRXAS, we also calculate the TRPES using the same quantum dynamics simulations. By doing so, we can compare and contrast the TRXAS, which probes the core-level electrons, to a well established method (TRPES) which probes the valence electrons, and investigate their respective sensitivities to excited state non-adiabatic dynamics. We find that the calculated TRPES and TRXAS contain, in principle, the same information regarding the electronic and structural dynamics. Importantly, however, this information is predicted to be encoded more clearly in the TRXAS, making its interpretation significantly more straightforward.

2 Theory

2.1 Quantum dynamics simulations

We consider a system composed of f nuclear degrees of freedom R_κ and n electronic degrees of freedom r_i. In the AIMS method[37] the ansatz for the total molecular wavefunction $|\Psi(\mathbf{R}, \mathbf{r}, t)\rangle$ reads

$$|\Psi(\mathbf{R},\mathbf{r},t)\rangle = \sum_{I=1}^{n_s}\left|\chi^{(I)}(\mathbf{R},t)\right\rangle|I(\mathbf{r};\mathbf{R})\rangle. \tag{1}$$

Here, $|\chi^{(I)}\rangle$ denotes the nuclear wavefunction associated with the Ith electronic state $|I\rangle$. The nuclear wavefunctions $|\chi^{(I)}\rangle$ are further expanded in terms of sets of f-dimensional frozen-width Heller-type Gaussian basis functions $\{g_j^{(I)}\}$:

$$|\chi^{(I)}(\mathbf{R},t)\rangle = \sum_{j=1}^{N_I} C_j^{(I)}(t)g_j^{(I)}\left(\mathbf{R};\alpha_j^{(I)},\overline{\mathbf{R}}_j^{(I)}(t),\overline{\mathbf{P}}_j^{(I)}(t),\gamma_j^{(I)}(t)\right), \tag{2}$$

where $\alpha_j^{(I)}$, $\bar{\mathbf{R}}_j^{(I)}$ and $\bar{\mathbf{P}}_j^{(I)}$ denote, respectively, f-dimensional vectors of the Gaussian widths, positions and momenta associated with each nuclear coordinate R_κ. The positions $\bar{\mathbf{R}}_j^{(I)}$ and momenta $\bar{\mathbf{P}}_j^{(I)}$ are taken to evolve according to classical equations of motion, whilst the phases $\gamma_j^{(I)}$ are propagated semi-classically.[37] The equations of motion for the expansion coefficients $C_j^{(I)}$ are derived variationally *via* the insertion of the wavefunction ansatz 1 into the time-dependent Schrödinger equation. The electronic basis $\{|I\rangle\}$ is chosen as the n_s lowest-lying adiabatic electronic states, and the nuclear coordinates $\mathbf{R}$ are taken as the full set of $3N$ Cartesian coordinates.

The set of Gaussian basis functions $\{g_j^{(I)}\}$ is adaptively increased during the course of the AIMS calculation, in a process termed spawning, so as to be able to describe non-adiabatic transitions between electronic states. Briefly, when a Gaussian basis function $g_j^{(I)}$ enters a region of strong non-adiabatic coupling between the electronic states $|I\rangle$ and $|J\rangle$, additional basis functions $g_j^{(J)}$ are created in the Jth electronic state.[37]

2.2 Calculation of time-resolved photoelectron spectra

We consider ionization from a manifold of N-electron states $|I\rangle$ to a manifold of $(N-1)$-electron states $|\alpha\rangle$. In order to simulate the time-resolved photoelectron spectrum for a given system, we approximate the total photoelectron signal $\sigma(E, t)$ at as an incoherent sum of contributions from the relative probabilities of ionization $W_{I\alpha}$ evaluated at the centres $\bar{\mathbf{R}}_j^{(I)}$ of each Gaussian basis function in the corresponding AIMS calculation:

$$\sigma(E,t)=\sum_{I=1}^{n_s^0}\sum_{\alpha=1}^{n_s^+}\sum_{j=1}^{N_I}\left|C_j^{(I)}(t)\right|^2 W_{I\alpha}\left(\bar{\mathbf{R}}_j^{(I)}(t)\right)\delta\left(E-\left(\omega-\Delta E_I{}^{\alpha}\left(\bar{\mathbf{R}}_j^{(I)}(t)\right)\right)\right), \quad (3)$$

Here, ω is the photon energy of the laser pulse, $\Delta E_I{}^{\alpha}$ is the vertical energy difference between $|I\rangle$ and $|\alpha\rangle$, and n_s^0 and n_s^+ denote the number of neutral and cationic electronic states, respectively. The δ-function arises as we assume vertical ionization. In the case of weak field ionization that is well above the threshold, the squares of the norms of the Dyson orbital $\phi_{I\alpha}^{D}$,

$$\phi_{I\alpha}^{D}=\sqrt{N}\langle I|\alpha\rangle, \quad (4)$$

are qualitative estimates of the relative probabilities of ionization $W_{I\alpha}$,[38] and it is this approach that is taken in this work.

The calculated spectrum is broadened *via* convolution in the time domain using a Gaussian function with a full width at half maximum (FWHM) corresponding to the experimental cross-correlation between the pump and probe laser pulses. Furthermore, in order to account for both the finite widths of the Gaussian basis functions used in the AIMS calculations and the inherent resolution of the experimental spectrum, the calculated spectrum is convoluted in the energy domain using a Gaussian function with a FWHM $\tau \simeq O(10^{-1})$ eV.

It should be emphasized that the observable calculated according to eqn (3) and (4) is only a crude, qualitative estimate of the true TRPES. In particular, it ignores the structure of the continuum, the dipole selection rules, and the inter-channel coupling due to electron correlation. Furthermore, since the nuclear wavepacket amplitudes are accumulated incoherently, it cannot describe the vibrational fine structure of the spectrum or vibrational wavepacket revivals. In general, the simulated TRPES spectra discussed below would expected to reproduce the central energies and time delays of the major, short-time, features of an experimental spectrum, but not their relative intensities or the detailed, long-time vibrational dynamics.

2.3 Calculation of time-resolved X-ray absorption spectra

We consider core-excitation from a manifold of initial states $\{|I\rangle; I = 1, \ldots, n_s\}$. We approximate the time-resolved X-ray absorption spectrum (TRXAS) $\sigma(E, t)$ as an incoherent sum of contributions calculated at the centres $\bar{\mathbf{R}}_j^{(I)}$ of each Gaussian basis function in the corresponding AIMS calculation:

$$\sigma(E,t) = \sum_{I=1}^{n_s} \sum_{j=1}^{N_I} \left| C_j^{(I)}(t) \right|^2 \sigma_I\left(E; \bar{\mathbf{R}}_j^{(I)}\right). \tag{5}$$

Here, $\sigma_I(E; \mathbf{R})$ denotes the component XAS for the Ith electronic state evaluated at the nuclear geometry $\mathbf{R}$. The component spectra $\sigma_I(E; \mathbf{R})$ are partitioned as follows:

$$\sigma_I(E; \mathbf{R}) = \sigma_I^{(b)}(E; \mathbf{R}) + \sigma_I^{(c)}(E; \mathbf{R}), \tag{6}$$

where $\sigma_I^{(b)}(E; \mathbf{R})$ denotes the part of the XAS arising from excitation into bound core-excited states, whilst $\sigma_I^{(c)}(E; \mathbf{R})$ corresponds to the part arising from core-ionization. Owing to the different normalizations of the final states contributing to $\sigma_I^{(b)}(E; \mathbf{R})$ and $\sigma_I^{(c)}(E; \mathbf{R})$, different methodologies have to be used in their calculation. For the sake of brevity, we give only a brief description here of the methodologies used to calculate the bound and continuum parts of the XAS. A detailed description is given in Appendix B.

The bound part of the spectrum, $\sigma_I^{(b)}(E; \mathbf{R})$, may be calculated simply from the oscillator strengths and excitation energies between the initial valence state and the core-excited states lying below the threshold for core-ionization. On the other hand, the core-excited states lying above the core-ionization threshold cannot be used directly in the calculation of the continuum oscillator strengths required for the construction of the continuum part of the spectrum, $\sigma_I^{(c)}(E; \mathbf{R})$. This is a result of the square-integrable ($\mathscr{L}^2$) Gaussian basis sets employed in our calculations. However, it has long been recognized that the moments of the photoabsorption cross-sections (also termed the spectral moments), $S_I(k) = \langle I|\hat{D}^\dagger \hat{H}^k \hat{D}|I\rangle$, where $\hat{D}$ is the dipole operator, can be accurately calculated using an $\mathscr{L}^2$ basis. The problem of calculating the continuum oscillator strengths can then be reformulated as that of extracting a function from a finite set of its moments. One commonly used solution to this problem is to use a method known as Stieltjes imaging,[39–43] and it is this approach that is used here.

2.4 Electronic structure calculations

In the AIMS simulations, the requisite adiabatic energies, energy gradients, and non-adiabatic coupling terms were evaluated at the multi-reference first-order configuration interaction (MR-FOCI) level. The underlying reference functions were taken as complete active space self-consistent field (CASSCF) wavefunctions constructed using an active space consisting of the π, π* and Rydberg 3s orbitals, with state averaging over six states. We denote this level of calculation by MR-FOCI(2,3). Generally contracted basis sets of the atomic natural orbital type[44] obtained from the C(10s6p3s)/H(7s3p) primitive sets with the C[3s2p1d]/H[2s1p] contraction schemes were used. In order to satisfactorily describe the low-lying π3s state, a single additional s function contracted from a set of 8s primitive functions was added to the basis, placed at the centre of mass. The additional diffuse s function was constructed using the procedure detailed in ref. 45. We denote this basis by ANO(3s2p1d/2s1p/1s). Further details of the AIMS simulations (including basis information and initial conditions) are given in Appendix E.

The calculation of the TRXAS was performed using the second-order algebraic diagrammatic construction (ADC(2)) method.[46,47] For the calculation of the initial valence excited states $|I\rangle$, the original strict variant of the ADC(2) method was used. For the calculation of the final, core-excited states $|J\rangle$, the extended ADC(2) method was used within the core–valence separation approximation,[48] and this level of theory is denoted here by CVS-ADC(2)x. The bound part of the spectrum was calculated using the 6-311++G** basis. The continuum part of the spectrum was calculated using the cc-pVDZ basis augmented with 3s, 6p and 4d uncontracted continuum-type diffuse functions of the Kauffmann–Baumeister–Jungen (KBJ) type[49] placed at the centre of mass. The spectral moments $S_I(k)$ used in the Stieltjes imaging procedure we generated using the block-Lanczos pseudo-spectrum of the CVS-ADC(2)x Hamiltonian.[41–43] TRPES were calculated using Dyson orbitals and vertical ionization energies calculated at the ADC(2) level of theory using the 6-311++G** basis. The details of these calculations are given in Appendix C. The selection of the initial ADC(2) states used in the TRXAS and TRPES calculations was based on a criterion of maximum overlap with the MR-FOCI states of the AIMS calculations, and details of the procedure used are given in Appendix D.

All MR-FOCI calculations were performed using the COLUMBUS set of programs.[50] The canonical Hartree–Fock orbitals used in the ADC calculations were calculated using the GAMESS program.[51] The AIMS calculations were performed using the FMS90 program.

In order to help identify features in the calculated time-resolved spectra attributable to the population of the π3s Rydberg state, further AIMS simulations were run in which the 3s Rydberg orbital was removed from the CASSCF active space. This eliminates the π3s Rydberg state from the AIMS calculation, and also from any TRXAS and TRPES calculation using the thus obtained nuclear densities. We refer to these calculations as the valence-state-only calculations throughout.

2.5 Analysis of the nuclear wavefunctions

In order to analyze the regions of nuclear configuration space sampled by the evolving wavepacket, a useful quantity to calculate is the one-dimensional reduced density $\rho(s, t)$ defined with respect to a given nuclear degree of freedom s:

$$\rho(s,t) = \int \left|\Psi(\mathbf{R},t)\right|^2 \delta\left(f^{(s)}(\mathbf{R}) - s\right)\mathrm{d}\mathbf{R}, \tag{7}$$

where $f^{(s)}(\mathbf{R})$ denotes the function $f^{(s)}$: $\mathbf{R} \to s$. Following on from the work of Mori *et al.*,[52] we chose to focus on the twisting angle ψ about the C–C bond and the pyramidalization angle ζ defined with respect to the CH_2 groups. The reason for this choice lies in the fact that internal conversion from the $\pi\pi^*$ state to the ground state of ethylene is known to be facilitated predominantly by a conical intersection (CI) involving large amplitude motion about these two degrees of freedom, the so-called twisted-pyramidalized (Tw-Py) CI.[52,53] With reference to the atom numbering shown in Fig. 1, we note that the angles ψ and ζ are defined as follows:

$$\psi = \cos^{-1}[(\mathbf{r}_{12} \times \mathbf{r}_{34}) \cdot (\mathbf{r}_{21} \times \mathbf{r}_{56})] \tag{8}$$

$$\zeta = \max\{\cos^{-1}[(\mathbf{r}_{12} \times \mathbf{r}_{34}) \cdot (\mathbf{r}_{14} \times \mathbf{r}_{13})], \cos^{-1}[(\mathbf{r}_{21} \times \mathbf{r}_{56}) \cdot (\mathbf{r}_{26} \times \mathbf{r}_{25})]\} \tag{9}$$

Additionally, by choosing s to correspond to the distance to a given CI seam (in the Euclidean sense), information regarding the role played by that seam in the excited state dynamics of a molecule may be assessed. To do so, we take s to be defined as the projection onto the branching space of the distance to the minimum energy conical intersection (MECI) geometry $\mathbf{R}_{\mathrm{MECI}}$ between the electronic states $|I\rangle$ and $|J\rangle$ on the seam of interest. We term this quantity the first-order seam distance, and denote it by $D_{IJ}(\mathbf{R}; \mathbf{R}_{\mathrm{MECI}})$. A description of the procedure used to calculate $D_{IJ}(\mathbf{R}; \mathbf{R}_{\mathrm{MECI}})$ is given in Appendix A.

All reduced densities $\rho(s, t)$ were evaluated using a Monte-Carlo procedure, the details of which have been described previously.[54]

3 Results

3.1 Validation of the X-ray absorption cross-section calculations

Before considering the results of the dynamics and spectroscopic calculations, we begin with a validation of our CVS-ADC(2) X-ray absorption cross-section calculations. To do so, we consider the comparison of the calculated ground state X-ray absorption spectrum (XAS) to its experimental counterpart, as shown in Fig. 2. The calculated and experimental spectra are found to be in good qualitative agreement, both in terms of absolute intensities and peak positions. It should be noted that the moderate disagreement between the calculated and experimental cross-sections at higher energies are a result of the relatively small basis set used in the calculation of the continuum part of the spectrum, necessitated by the very large number of cross-section calculations required for the computation of the TRXAS. Nonetheless, the agreement between the experimental and calculated spectra is sufficient for the purposes of the calculation of a qualitatively correct TRXAS, allowing us to evaluate the utility of this measure for probing ultrafast non-adiabatic dynamics.

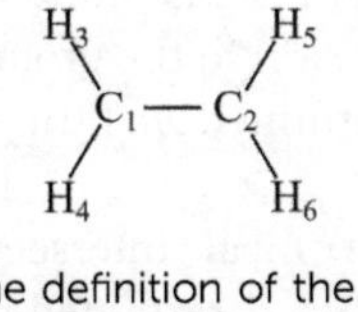

Fig. 1 Atom numbering used in the definition of the twisting angle ψ and the pyramidalization angle ζ.

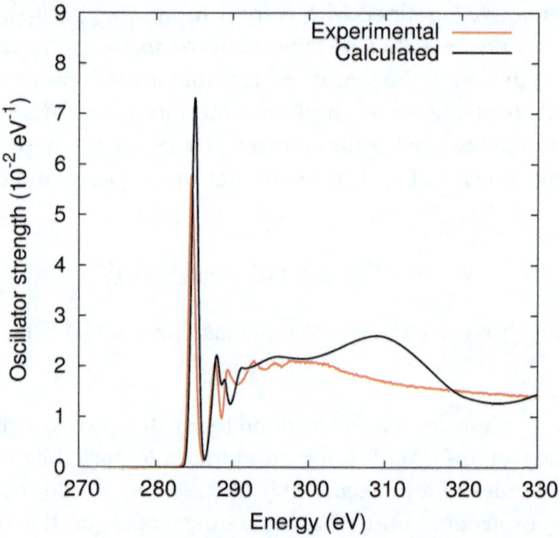

Fig. 2 Experimental[67] and calculated C 1s X-ray absorption spectra for the S_0 state ethylene. The bound part of the calculated spectrum was computed using oscillator strengths and excitation energies calculated at the CVS-ADC(2)x/6-311++G** level of theory. The continuum part was calculated from 20th-order Stieltjes imaging calculation employing spectral moments calculated at the CVS-ADC(2)x/cc_pVDZ/3s6p4d level of theory.

3.2 Dynamics

Previous experimental and computational studies of the dynamics of ethylene following excitation to the optically bright $S_2(\pi\pi^*)$ state exist.[52,53,55–57] The picture emerging from these studies is one of ultrafast internal conversion to the ground electronic state on a sub-100 fs timescale. Previous calculations[52,53] implicated the role of three conical intersections in facilitating internal conversion between the $\pi\pi^*$ and ground states: the twisted-pyramidalized(Tw-Py), ethylidene-type (Et) and hydrogen-bridge (H-br) intersections. The geometries of these conical intersections are shown in Fig. 3. As well as internal conversion to the ground state, a number of previous studies have also suggested that the $S_1(\pi 3s)$ Rydberg state is transiently populated following excitation to the $S_2(\pi\pi^*)$ state: namely, the TRPES study of Champenois *et al.*[56] and the AIMS calculations of Mori *et al.*[52] Our results are found to be in close agreement with this picture and, as such, we present only a brief outline of the most pertinent aspects of our excited state dynamics calculations.

Shown in Fig. 4(a) are the adiabatic state populations calculated following excitation to the $S_2(\pi\pi^*)$ state. Rapid depopulation of the S_2 state is predicted to occur, with 65% of the population of this state being transferred to the S_1 state within 10 fs. Population of the S_1 state is found to reach its maximum at around 25 fs, whereafter internal conversion to the ground state is predicted to occur on a timescale of around 95 fs (determined *via* fitting of a single exponential to the S_0 state population).

In order to reveal which conical intersection seams dominate internal conversion to the ground state, we determined the minimum root mean square

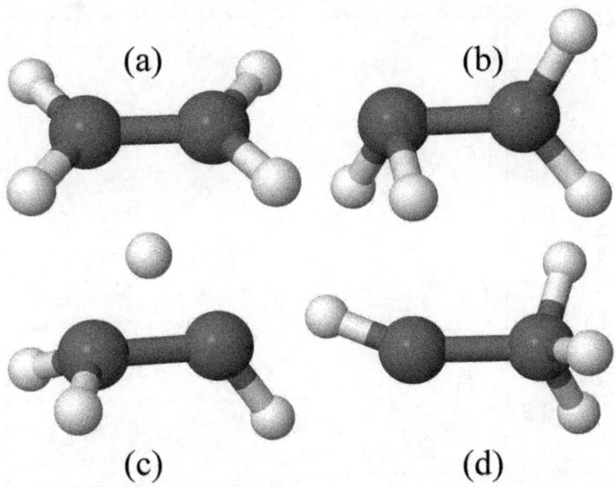

Fig. 3 Geometries of critical points on the S_0/S_1 PESs of ethylene. (a) FC point geometry, (b) Tw-Py MECI, (c) H-br MECI, and (d) Et MECI. All conical intersection geometries were optimized at the MR-FOCI(2,3)/ANO(3s2p1d/2s1p/1s) level of theory.

deviation (RMSD) of each spawn geometry to the ground state from each of the Tw-Py, Et and H-br CI geometries subject to centre of mass translations, rotations and the permutation of identical nuclei. We found that 74% of the spawn geometries map onto the Tw-Py geometry, 4% onto the Et geometry, and 22% onto the H-br geometry. This implies that internal conversion to the ground state *via* the Tw-Py intersection is dominant, in agreement with previous work.[52] As such, we concentrate only on the role played by the Tw-Py conical intersection in the following sections. For reference, the calculated reduced densities for the twisting and pyramidalization angles are shown in Fig. 5(a) and (b), respectively.

Finally, we briefly consider the results of the valence-state-only AIMS calculations, from which the π3s state was excluded. The adiabatic state populations calculated from the valence-state-only AIMS calculations are shown in Fig. 4(b). The time-scale for re-population of the S_0 state is not significantly altered by the exclusion of the π3s state. Additionally, the minimized RMSDs of the spawn-geometries from the Tw-Py, Et and H-br CI geometries are also found to not be significantly altered by the removal of the π3s state, with 78% of the spawn geometries mapping onto the Tw-Py geometry, 1% onto the Et geometry, and 21% onto the H-br geometry. This implies that the removal of the π3s Rydberg state does not significantly affect the excited-state nuclear dynamics. Therefore a meaningful comparison of time-resolved spectra calculated from the AIMS simulations performed with and without the inclusion of the π3s Rydberg state may be made, allowing us to determine the effect of the population of this state on the calculated spectra.

3.3 Time-resolved X-ray absorption spectra

Shown in Fig. 6 is the TRXAS calculated from the AIMS simulations of ethylene following excitation to the $S_2(\pi\pi^*)$ state. For reference, the bound and continuum contributions of the spectrum are shown separately alongside. It is found that the bound part of the TRXAS contains a large amount of structure. In comparison, the

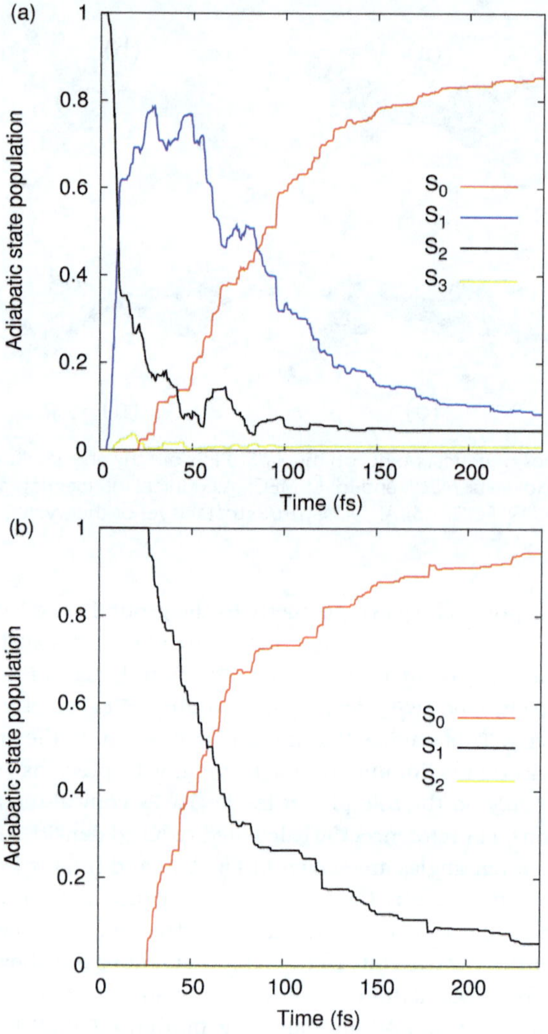

Fig. 4 Adiabatic state populations calculated following excitation to the ππ* state of ethylene. (a) Populations calculated from the all-state AIMS simulation. (b) Populations calculated from the valence-state-only AIMS simulation.

continuum part of the TRXAS is found to be relatively featureless, containing a single, decaying dominant peak. As such, we choose to focus our analysis on the bound part of the TRXAS.

In the following discussion, we consider separately the contributions to the pre-edge TRXAS arising from the dynamics in: (i) the initially excited ππ* state; (ii) internal conversion to the π3s state, and; (iii) the re-population of the ground state.

3.3.1 Dynamics in the ππ* state. At time $t = 0$, the pre-edge part of the TRXAS contains a single, weak peak centred at around 277 eV. At this time, the vertically excited wavepacket is localized around the FC point. In Fig. 7 we show the static X-ray absorption spectrum (XAS) calculated for the S_0, S_1(π3s) and S_2(ππ*) states

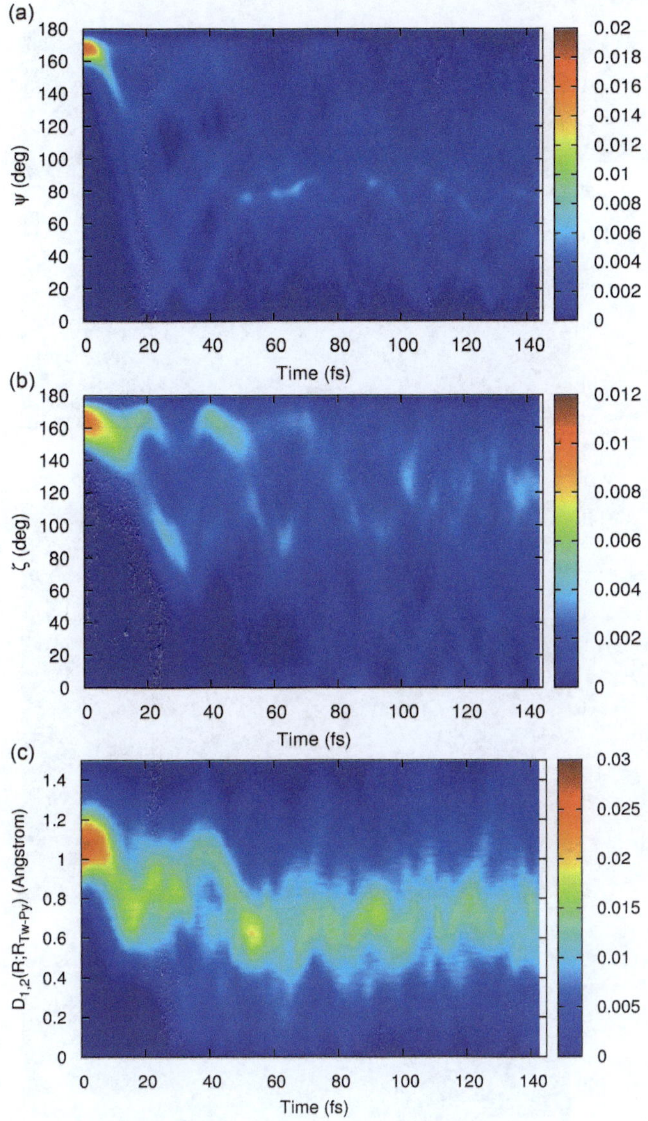

Fig. 5 One-dimensional reduced densities for the twisting angle ψ, the pyramidalization angle ζ, and the distance within the branching space to the Tw-Py conical intersection $D_{1,2}(\mathbf{R};\ \mathbf{R}_{\text{Tw-Py}})$, calculated following excitation to the $\pi\pi^*$ state.

at the FC point. Below the core-ionization threshold, the FC point $S_2(\pi\pi^*)$ state XAS is predicted to be dominated by a single, weak peak at around 277 eV. Analysis of the CVS-ADC(2)x wavefunction of the corresponding final core-excited state reveals that this transition corresponds mainly to excitation of a C 1s electron into the hole in the singly-occupied π orbital of the initial $S_2(\pi\pi^*)$ state, forming the $1s\pi^*$ state. As such, it is trivial to assign the bound part of the TRXAS at $t = 0$ to excitation of a core-electron into the singly-occupied π orbital of the initially excited state. We note that this peak rapidly sweeps to higher energies

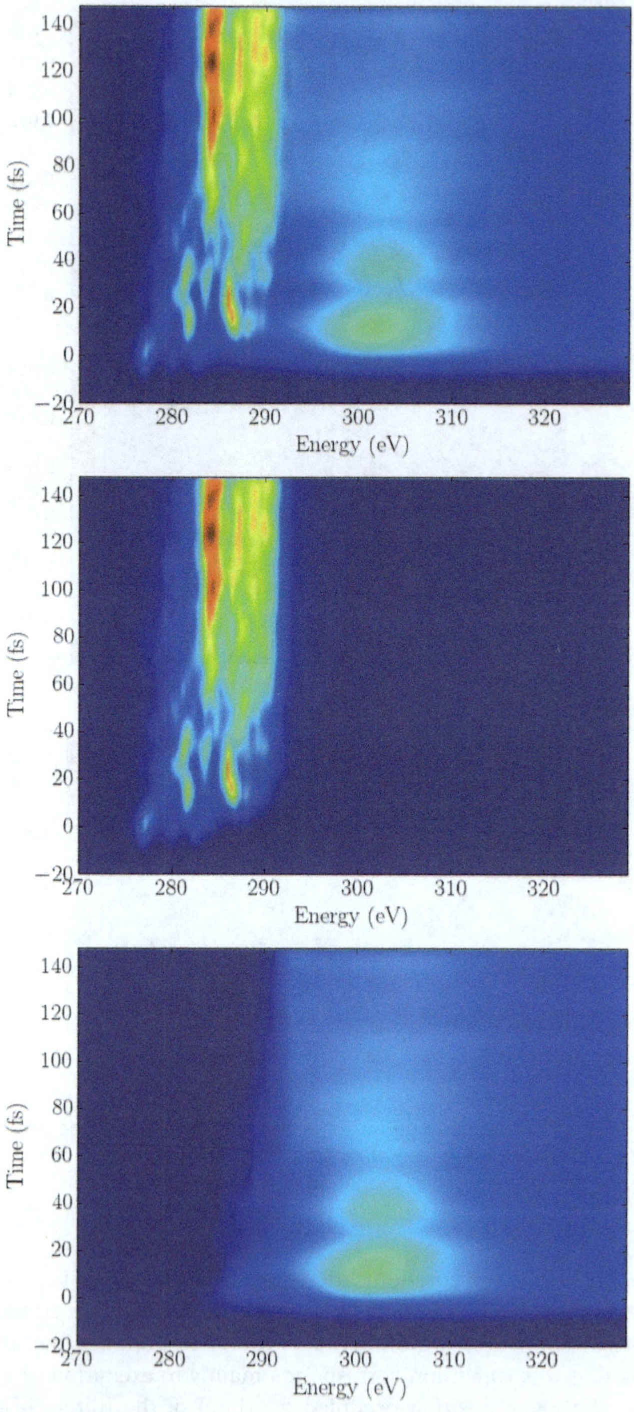

Fig. 6 Calculated TRXAS spectra for ethylene following excitation to the $\pi\pi^*$ state. Top: Total TRXAS spectrum. Middle: Bound part of the TRXAS spectrum. Bottom: Continuum part of the TRXAS spectrum.

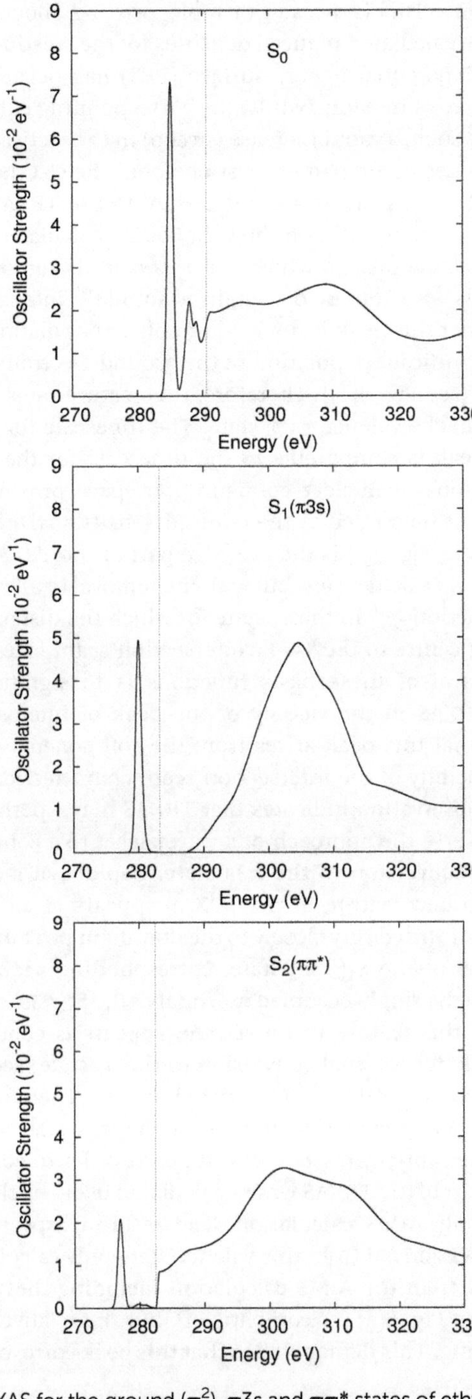

Fig. 7 Calculated XAS for the ground (π^2), π3s and $\pi\pi^*$ states of ethylene at the FC point. The initial valence-excited states were described at the ADC(2) level, and the final core-excited states at the CVS-ADC(2)x level. The dashed lines mark the calculated C 1s ionization energy. The bound-parts of each spectrum were calculated using the 6-311++G** basis. The continuum parts were calculated using the Stieltjes imaging technique with the cc-pVDZ basis augmented with 3s, 6p and 4d continuum-type KBJ functions.

with increasing time. This is a result of rapid twisting about the C–C bond, as corroborated by the calculated reduced densities for the twisting angle ψ shown in Fig. 5(a). The $\pi\pi^*$ potential energy surface (PES) has negative curvature with respect to this degree of freedom, whilst the corresponding curvature of the $1s\pi^*$ state is positive. As such, a rapid increase (sweep) in the vertical excitation energy between the two states accompanies twisting about the C–C bond.

We next consider the peak at around 286 eV in the TRXAS, which develops intensity after ~20 fs. In Fig. 8(b) we show the TRXAS calculated from the valence-state-only AIMS calculations, in which the $\pi3s$ state is removed. This valence-state-only TRXAS is also found to contain a similarly intense peak at around 286 eV, implying that this peak is not attributable to population of the $\pi3s$ state. Additionally, no significant population of the ground electronic state is found to occur at this time (see Fig. 4(a)). Therefore, this feature must be attributable to core-excitation from the valence $\pi\pi^*$ state. The timescale for the growth in the intensity of this peak is comparable to the timescale for the movement of the wavepacket to regions of nuclear configuration space proximate to the Tw-Py intersection seam, as illustrated by the reduced densities $\rho(D_{1,2}(\mathbf{R}; \mathbf{R}_{\text{Tw-Py}}))$ shown in Fig. 5(c). Shown in Fig. 8(c) is the pre-edge part of the TRXAS calculated from the valence-only AIMS calculations, but with the removal of all contributions from Gaussian basis functions $g_j^{(I)}$ in the S_1 state for which the distance $D_{12}(\bar{\mathbf{R}}_j^{(I)}; \mathbf{R}_{\text{Tw-Py}})$ from the Gaussian centre to the Tw-Py intersection seam is less than 0.3 Å. The effect of the removal of these basis functions is to significantly reduce the intensity of the TRXAS in the vicinity of the peak of interest at 286 eV. This strongly suggests that this peak arises from the components of the wavepacket approaching the vicinity of the intersection seam characterized by the Tw-Py CI. Importantly, this observation indicates that TRXAS may – perhaps uniquely – be able to directly observe the approach of a wavepacket to a conical intersection.

3.3.2 Internal conversion to the $\pi3s$ state. Rapidly, at a time of $t = 10$ fs, a relatively intense new feature in the TRXAS appears at an energy of around 281 eV. This peak is centred very closely to the dominant peak in the calculated FC point pre-edge XAS of the $S_1(\pi3s)$ state, corresponding again to excitation of a core-electron into the singly-occupied π orbital of the $S_1(\pi3s)$ state. Additionally, the time at which this feature in the TRXAS appears is concomitant with the calculated timescale for internal conversion to the S_1 state (see Fig. 4(a)), which contains components of both $\pi3s$ and $\pi\pi^*$ character. Therefore, we assign this feature in the TRXAS to core-excitation from the $\pi3s$ state following internal conversion from the initially excited $\pi\pi^*$ state. In order to confirm this assignment, the bound part of the TRXAS was re-calculated using nuclear densities from the valence-state-only AIMS calculation. The resulting spectrum is shown in Fig. 8(b). The TRXAS derived from this valence-state-only calculation is similar to the TRXAS derived from the AIMS calculation including the $\pi3s$ Rydberg state (shown alongside in Fig. 8(a) for comparison), but the relatively intense peak at 281 eV is now missing. This demonstrates that this peak corresponds to excitation from the $\pi3s$ state.

3.3.3 Internal conversion to the ground state. Finally, we consider the broad, intense pre-edge feature in the TRXAS that appears at a time of $t \sim 50$ fs. This part of the TRXAS comprises three dominant peaks of decreasing intensity centred around 284, 287 and 289 eV. With reference to Fig. 7, we note that these positions match those of the dominant pre-edge transitions in the calculated FC point XAS

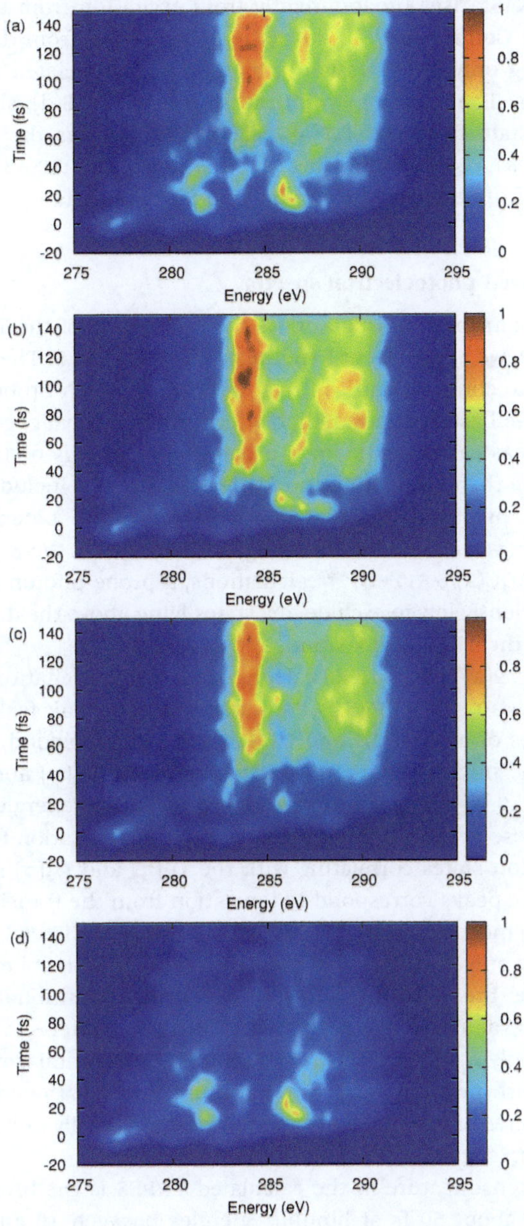

Fig. 8 Pre-edge TRXAS calculated using the results of AIMS simulations of ethylene excited to the $\pi\pi^*$ state. (a) TRXAS calculated with all states included; (b) TRXAS calculated from the valence-state-only AIMS calculation; (c) TRXAS calculated from the valence-state-only AIMS calculation, and with trajectories in the S_1 state close to the Tw-Py seam removed, and; (d) TRXAS calculated from the all-state AIMS calculations with the contributions from the S_0 state trajectories removed from the spectrum calculation.

of the S_0 state. Additionally, the time at which this broad feature appears in the TRXAS is coincident with a significant re-population of the S_0 state, as seen in the calculated adiabatic state populations of Fig. 4(a). This strongly indicates that this part of the TRXAS arises predominantly from excitation from the 'hot' ground electronic state. Confirmation of this assignment comes from the calculation of the TRXAS using only the Gaussian basis functions associated with the excited electronic states. The resulting spectrum, shown in Fig. 8(d), shows that eliminating the S_0 state from the TRXAS calculation removes the majority of the intensity at times $t > 50$ fs, confirming that this part of the TRXAS spectrum arises predominantly from excitation from the 'hot' S_0 ground state.

3.4 Time-resolved photoelectron spectra

Although the main focus of this work is the calculation of the TRXAS of photo-excited ethylene, we also present here the calculated TRPES. The primary reason for doing so is to compare and contrast the TRXAS, which probes the core-level electrons, to a well established method which probes the valence electrons.

The TRPES was calculated using a probe photon energy of 18 eV, chosen so that states above the double ionization threshold are not included in the calculation. Using a previously calculated double ionization potential of 29.5 eV (ref. 58) and the vertical excitation energy of 8.13 eV for the $S_2(\pi\pi^*)$ state furnished by our ADC(2)/6-311++G** calculations, a probe photon energy of 18 eV should be sufficiently low to exclude any states lying above the double ionization threshold from the TRPES calculation.

Shown in Fig. 9(a) is the TRPES calculated following excitation to the $S_2(\pi\pi^*)$ state. At time $t = 0$, the TRPES is dominated by two weak features centred at binding energies of around 2.1 and 15.5 eV. These correspond, respectively, to ionization of the initially excited $S_2(\pi\pi^*)$ state to the $D_0(\pi)$ and $D_9(\pi^*)$ states. Rapidly, within 10 fs, two new features appear at binding energies of around 2.9 and 14.6 eV. These peaks are found to correspond to ionization from the $S_1(\pi 3s)$ state to the cation states correlating with the $D_0(\pi)$ and $D_5(\sigma)$ states at the FC point. That these peaks correspond to ionization from the transiently populated $\pi 3s$ state is seen most clearly in Fig. 9(b), which shows the TRPES calculated from the valence-state-only AIMS calculations. The removal of this state manifests itself in the TRPES as the vanishing of these two peaks. Additionally, the broader feature appearing at a binding energy of around 12.5 eV after ~15 fs is also found to disappear, implying that this also corresponds to ionization from the $\pi 3s$ state. In agreement with previous work,[52] it can thus seen that the calculated TRPES contains signatures of internal conversion from the initially excited $\pi\pi^*$ state to the lower-lying $\pi 3s$ state.

The most intense feature in the calculated TRPES is the broad set of peaks appearing after about 50 fs at binding energies between 10 and 16 eV, corresponding to ionization from the repopulated S_0 ground state. This is confirmed by the TRPES calculated using only Gaussian basis functions associated with the excited electronic states, as shown in Fig. 9(d).

Finally, we investigate whether the calculated TRPES reveals the structural or electronic dynamics upon the approach to the dominant Tw-Py intersection seam in the $\pi\pi^*$ state. To do so, the TRPES was recalculated from AIMS calculation with the removal of the $\pi 3s$ state and all contributions from S_1-state Gaussian basis functions with a first-order distance $D_{12}(\bar{\mathbf{R}}_J^{(I)}; \mathbf{R}_{\text{Tw-Py}})$ from the Gaussian centre to the Tw-Py intersection seam of less than 0.3 Å. The resulting spectrum is shown in Fig. 9(c). This spectrum is compared to that shown in Fig. 9(b), which contains

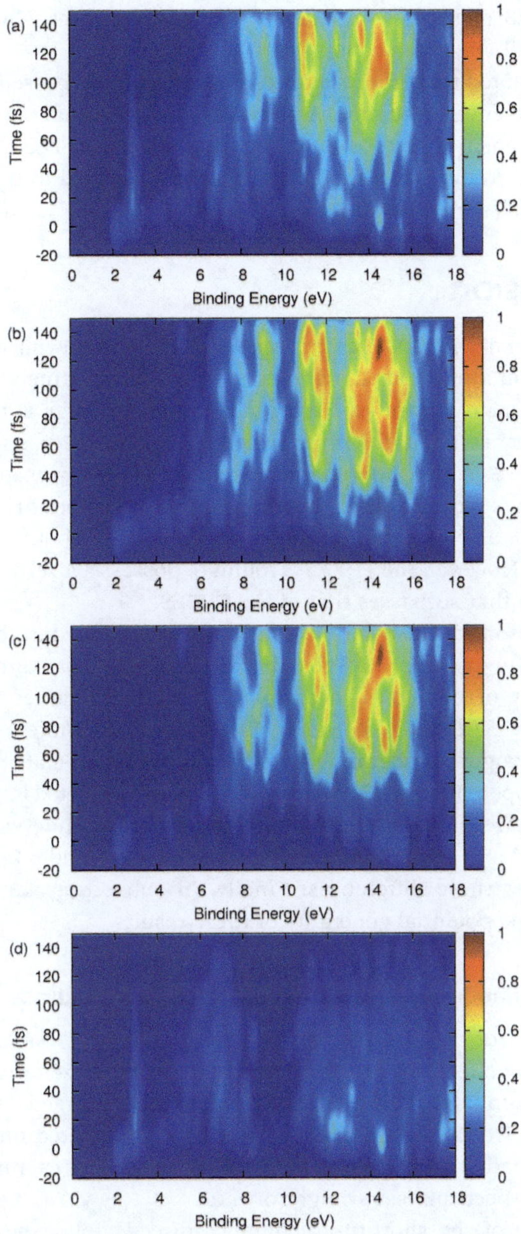

Fig. 9 Main features of TRPES estimated using the results of AIMS simulations of ethylene excited to the $\pi\pi^*$ state and a probe photon energy of 18 eV. (a) TRPES calculated with all states included in the AIMS calculation; (b) TRPES calculated from the valence-state-only AIMS calculation; (c) TRPES calculated using the valence-state-only AIMS calculation, and with trajectories in the S_1 state close to the Tw-Py seam removed, and; (d) TRPES calculated from the all-state AIMS calculations with the contributions from the S_0 state trajectories removed from the spectrum calculation.

contributions from the valence states only. Some small differences are seen. However, in contrast to the TRXAS, there appear to be no clear signatures in the calculated TRPES of those components of the wavepacket which are in close proximity to the Tw-Py intersection seam.

It should be noted that all the features of the TRPES discussed above are well-separated in both energy and time. As a result of this, our analysis does not rely on their relative intensities or the vibrational fine structure. Therefore, it remains valid despite the crudeness of the approximations adopted in the calculation of the model TRPES (see Section 2.2).

4 Discussion

The pre-edge part of the TRXAS calculated for ethylene following excitation to the $S_2(\pi\pi^*)$ state contains clear signatures of the most pertinent aspects of the ensuing excited state dynamics. Namely: (i) the initial large amplitude nuclear motion in the $\pi\pi^*$ state; (ii) the transient population of the π3s state; (iii) the re-population of the S_0 state, and; (iv) the movement of the wavepacket towards the vicinity of the Tw-Py conical intersection seam. It is important to note that the calculated TRPES also contains discernible information about the first three of these processes. However, the TRXAS is found to possess a level of clarity and ease of interpretation that surpasses that of the TRPES.

The main reason for the dynamical information encoded in the TRXAS being so clear is that the pre-edge XAS for the $\pi\pi^*$ and π3s states are dominated by a small number of core-to-valence transitions having rather large energetic separations. The TRPES, in contrast, is significantly more congested, with a comparatively larger number of less well resolved excited-state photoionization channels being open. Remarkably, only the TRXAS is predicted to contain a direct and clear signature of the arrival of the wavepacket at the Tw-Py intersection between the $\pi\pi^*$ and ground states. In subsections 4.1 and 4.2, we discuss the main reasons for these differences. Finally, in subsection 4.3 we offer some perspective on the potential generality of these results.

4.1 Discrimination between the dynamics in the $\pi\pi^*$ and π3s states

The main reason for the distinct clarity of the calculated TRXAS is that the $\pi\pi^*$ and π3s states are effectively projected by core-to-valence excitation onto disjoint sets of small numbers of final states having large energetic separations. In contrast, in the TRPES the excited states are each projected onto larger intersecting (in the mathematical sense) sets of final cation states, making the interpretation of the spectrum less straightforward.

In the course of the short-time dynamics directly following excitation, the valence $\pi\pi^*$ undergoes preferential core-to-valence excitation to the 1sπ^* state, whilst the Rydberg π3s state is excited preferentially to the 1s3s state. This can be interpreted as core-to-valence excitation being dominated in each case by excitation of a C 1s electron into the hole in the singly-occupied π orbital. The vertical energy difference at the FC point between the 1sπ^* and 1s3s states is much greater than that between the $\pi\pi^*$ and π3s states: 2.2 eV *versus* 0.8 eV at the ADC(2)/CVS-ADC(2)x levels of theory using the 6-311++G** basis. Insight into this large energetic separation of the 1sπ^* and 1s3s states may be gained by considering the application of a one-particle, one-hole (1p1h) excitation operator $\hat{C}_{vc} = \hat{c}_v^\dagger \hat{c}_c$ to the correlated ground electronic state $|\Psi_0\rangle$ to yield the singly core-excited configuration $|\Psi_{vc}\rangle = \hat{C}_{vc}|\Psi_0\rangle$, where c indexes a core orbital and v a virtual valence orbital. By applying Rayleigh–Schrödinger perturbation theory through second-order within the core valence separation approximation, the core-excitation energy ΔE_{vc} obtained using the unperturbed 1p1h configuration $|\Psi_{vc}\rangle$ can be written as follows:[46,59]

$$\Delta E_{vc} = \varepsilon_v - \varepsilon_c + [2\langle cv|vc\rangle - \langle cv|cv\rangle] + U_{vc}^{(2)}(1p1h) + U_{vc}^{(2)}(2p2h) + R_{vc}^{(2)}. \qquad (10)$$

Here, ε_p denotes the energy of the canonical Hartree–Fock orbital φ_p. The second-order terms $U_{vc}^{(2)}(1p1h)$, $U_{vc}^{(2)}(2p2h)$ and $R_{vc}^{(2)}$ describe the interaction of $|\Psi_{vc}\rangle$ with other 1p1h and 2p2h configurations, and the partial compensation of the 3p3h contributions with the ground state correlation energy, respectively. Of more interest to us is the first-order Coulomb term $J_{vc} = \langle cv|cv\rangle$, which effectively accounts for the stabilizing Coulomb interaction between the excited electron and the core-hole. Owing to the spatially compact nature of the core-level orbital φ_c, the Coulomb integral J_{vc} can be expected to be smaller for a more diffuse virtual orbital φ_v. It then follows that a core-excited state characterized by excitation into a valence-type orbital may be stabilized to a greater extent relative to a core-excited state characterized by excitation into a Rydberg-type orbital. This, coupled with the propensity for core-to-valence excitation to result in excitation from a core-level orbital into the lowest-lying singly-occupied valence orbital, suggests that pre-edge X-ray absorption may be a particularly useful probe of vibronically coupled dynamics occurring in energetically proximate valence and Rydberg states.

4.2 Signature of the arrival of the wavepacket at the Tw-Py conical intersection seam

A remarkable finding is that the calculated TRXAS contains a peak centred around 286 eV, with a delayed rise, that maps onto the $\pi\pi^*$ state components of the wavepacket which come into close proximity with the Tw-Py intersection seam. To shed more light on the nature of this spectral feature, we consider the calculated XAS for the $\pi\pi^*$ state at the Tw-Py geometry, shown in Fig. 10. The pre-edge part of the spectrum contains two dominant transitions. The lower energy peak corresponds to excitation of a core electron into the hole in the singly-occupied π orbital, forming the $1s\pi^*$ state. This peak thus maps onto the dominant transition from the $\pi\pi^*$ state seen in the FC region spectrum. The appearance of a second, higher-energy peak in the spectrum is linked to the localization – that occurs as the system approaches the Tw-Py seam – of the 1s orbitals about the now non-equivalent carbon atoms. At the FC point, the near-degenerate 1s orbitals are not localized on the carbon atoms, as dictated by the D_{2h} symmetry of the molecule. At the Tw-Py geometry, however, the 1s molecular orbitals become localized on the now non-equivalent carbon atoms. This is accompanied by a large splitting, $\sim$4.5 eV, of the 1s orbital energies. The higher-energy peak in the pre-edge XAS is found to correspond to excitation from the lower-energy 1s orbital localized on the carbon atom of the pyramidalized CH_2 group. This result highlights a unique feature of the pre-edge TRXAS. Namely, that changes in geometry and/or electronic structure localized around specific atomic centres may appear in the spectrum as changes in core-to-valence excitation energies. These in turn

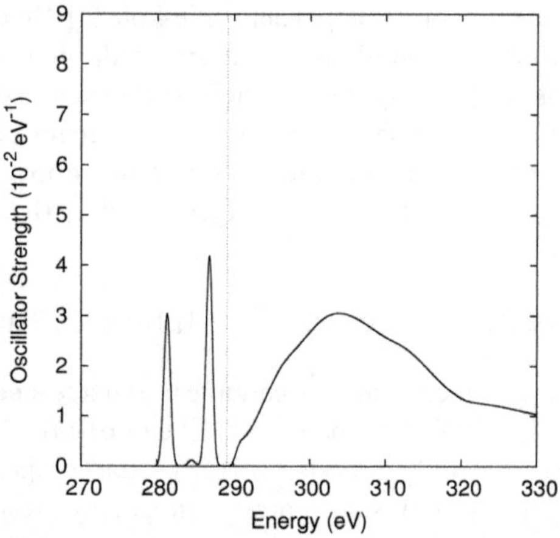

Fig. 10 Calculated XAS for the ππ* state of ethylene at the Tw-Py conical intersection geometry. The initial valence-excited states are described at the ADC(2) level, and the final core-excited states at the CVS-ADC(2)x level. The dashed line marks the calculated C 1s ionization energy. The bound part of the spectrum was calculated using the 6-311++G** basis. The continuum part was calculated using the Stieltjes imaging technique with the cc-pVDZ basis augmented with 3s, 6p and 4d continuum-type KBJ functions.

result from corresponding changes in the energies of the atom-localized core orbitals.

4.3 Generality of the observed trends

We offer, for the purposes of this discussion, our perspective on the generality of the results presented here. Firstly, the excited-state dynamics of ethylene are characteristic of those of a large range of unsaturated hydrocarbons. To emphasize the idea that a chromophore such as the C=C bond acts as a dynamical moiety in larger systems, we have coined the term 'dynamophore'.[60] Accordingly, we expect that the results presented here will apply to related systems containing the C=C dynamophore. Secondly, and more broadly, we consider the predicted propensity for the excited-state, pre-edge XAS to be dominated by those core-to-valence transitions which correspond to the filling of the lowest-lying singly-occupied molecular orbital (SOMO) of the dominant electronic configuration characterizing the state being probed. As dictated by the one-electron nature of the dipole operator, core-to-valence excitation from an initial singly-valence-excited state should exhibit a propensity for excitation to a subset of core-excited states whose dominant configurations are related to that of the initial state by a single pair of orbital occupancies. It then seems reasonable that the lowest energy such transition may in general correspond to excitation of a core-level electron into the hole in the lowest-lying SOMO. Supporting this suggestion, we note that core-excitation into the singly-occupied molecular orbital is recognized as providing a unique fingerprint of radical species.[61] Informed by these two considerations, we propose that there is a propensity for the pre-edge XAS of a singly-valence-excited state to be dominated by the transition which fills the hole in the lowest lying SOMO.

5 Conclusions

Through the coupling of AIMS quantum dynamics calculations with CVS-ADC(2)x X-ray absorption cross-section calculations, a general and tractable method for the calculation of excited-state TRXAS spectra is presented. This methodology was applied to calculation of the TRXAS of ethylene excited to its $S_2(\pi\pi^*)$ state. In doing so, we identified features of TRXAS which make it a useful probe of ultrafast non-adiabatic dynamics. In particular, the pre-edge region of the XAS is predicted to exhibit pronounced sensitivity to the evolving electronic character of the excited-state wavepacket. The principal underlying reasons for this sensitivity are two-fold.

Firstly, we observed that core-to-valence excitation results in transition to a core-excited state which is related to the initial valence-excited state by the filling of the hole in the lowest lying SOMO. Therefore, the different valence-excited states are effectively projected onto distinct (and relatively sparse) sets of final states. As a result, the spectral features in the TRXAS which reflect the electronic dynamics can be well resolved.

Secondly, the high degree of localization of the core orbitals at non-symmetric nuclear geometries should allow for the potential identification of atom-localized changes in geometric and/or electronic structure. This is exemplified in the case of photo-excited ethylene by the prediction of two separate peaks in the TRXAS that map onto the arrival of the wavepacket at the Tw-Py CI seam, corresponding to excitation from the two energetically-split 1s orbitals localized on the now non-equivalent carbon atoms.

For comparison, the qualitative features of the TRPES of ethylene excited to its $S_2(\pi\pi^*)$ state were also estimated. By doing so, an insightful comparison between spectroscopies probing the core- *versus* valence-level electrons could be made. Both the calculated TRXAS and TRPES contain information about the electronic dynamics following excitation. However, this information seems to be encoded more clearly in the TRXAS. The primary reason for this is that, unlike core-to-valence excitation, valence-shell photoionization will generally project a given valence-excited state onto a large number of final (cation) states, leading to a higher density of overlapping spectral features in the TRPES. Additionally, due to the delocalized nature of the valence-level electrons, the atom-localized changes in geometric and electronic structure that appear clearly in the calculated TRXAS are absent from the TRPES.

The results presented here strongly suggest that the pre-edge region of the XAS is sensitive to both electronic and structural dynamics in photoexcited molecules. In particular, core-to-valence excitation provides clear discrimination between the components of the wavepacket associated with the different vibronically-coupled excited electronic states. In conclusion, we suggest that time-resolved pre-edge X-ray absorption spectroscopy will have significant potential as a spectroscopic probe of ultrafast excited state non-adiabatic dynamics.

6 Appendices

A Definition of the first-order seam distance

It is noted that first-order seam distance $D_{IJ}(\mathbf{R}; \mathbf{R}_{\mathrm{MECI}})$ may be written as:

$$D_{IJ}(\mathbf{R};\mathbf{R}_{\text{MECI}}) = \{[\mathbf{P}_{IJ}(\mathbf{R}' - \mathbf{R}_{\text{MECI}})]\cdot[\mathbf{P}_{IJ}(\mathbf{R}' - \mathbf{R}_{\text{MECI}})]\}^{\frac{1}{2}}. \tag{11}$$

Here, $\mathbf{R}'$ denotes the geometry obtained after putting $\mathbf{R}$ into maximum coincidence with $\mathbf{R}_{\text{MECI}}$ (in the least squares sense) *via* centre of mass translations, rotations, and the permutation of identical nuclei. The matrix $\mathbf{P}_{IJ}$ is the orthogonal projector onto the branching space for the MECI of interest:

$$\mathbf{P}_{IJ} = \sum_{i=1,2} \mathbf{x}_{IJ}^{(i)} \otimes \mathbf{x}_{IJ}^{(i)}, \tag{12}$$

where $\otimes$ denotes the outer product, and the $\mathbf{x}_{IJ}^{(i)}$ denote any two orthonormal vectors spanning the branching space. Specifically, we chose

$$\mathbf{x}_{IJ}^{(1)} = \frac{\mathbf{g}_{IJ}}{\|\mathbf{g}_{IJ}\|} \tag{13}$$

$$\mathbf{x}_{IJ}^{(2)} = \frac{\mathbf{h}_{IJ} - \left(\mathbf{h}_{IJ}\cdot\mathbf{x}_{IJ}^{(1)}\right)\mathbf{x}_{IJ}^{(1)}}{\left\|\mathbf{h}_{IJ} - \left(\mathbf{h}_{IJ}\cdot\mathbf{x}_{IJ}^{(1)}\right)\mathbf{x}_{IJ}^{(1)}\right\|}, \tag{14}$$

where $\mathbf{g}_{IJ}$ and $\mathbf{h}_{IJ}$ denote the gradient difference and non-adiabatic coupling vectors, respectively:

$$\mathbf{g}_{IJ} = \frac{\partial(E_J - E_I)}{\partial\mathbf{R}} \tag{15}$$

$$\mathbf{h}_{IJ} = \left\langle I \left| \left(\frac{\partial \hat{H}_{\text{el}}}{\partial \mathbf{R}}\right) \right| J \right\rangle. \tag{16}$$

B Calculation of X-ray absorption spectra

The bound part of the spectrum, $\sigma_I^{(b)}(E;\mathbf{R})$, is calculated from the oscillator strengths $f_{IJ}(\mathbf{R})$ and vertical excitation energies ΔE_I^J between the initial valence-excited state $|I\rangle$ and the core-excited states $|J\rangle$ lying below the threshold for core-ionization:

$$\sigma^{(b)}(E;\mathbf{R}) = \left(\left[\sum_J f_{IJ}(\mathbf{R})\delta\left(E - \Delta E_I^J(\mathbf{R})\right)\right] * g(E;\eta)\right)\Theta(E_{\text{T}}(\mathbf{R}) - E). \tag{17}$$

Here, * denotes the convolution operation, $g(E;\eta)$ corresponding to a Gaussian function with full width at half maximum of η, and $\Theta(E_{\text{T}} - E)$ is the reverse Heaviside step function centred at the core-ionization threshold energy E_{T}. The free parameter η was chosen such that the broadening present in the experimental ground state XAS of a given system is incorporated into the calculated spectra.

Owing to the square integrable ($\mathscr{L}^2$) Gaussian basis sets used in our calculations, the cross-sections above the core-ionization threshold cannot be determined directly from the oscillator strengths for transition to the core-excited states $|J\rangle$ that lie within this domain. This arises from the different normalizations of the $\mathscr{L}^2$ electronic states $|I\rangle$ and the 'true' continuum states $|E\rangle$: the former are normalized to a Kronecker-δ, $\langle I|J\rangle = \delta_{IJ}$, whilst the latter are energy normalized to a Dirac δ-function, $\langle E|E'\rangle = \delta(E - E')$. As such, a direct use of $\mathscr{L}^2$ states in the calculation of the oscillator strengths for excitation to states above the ionization threshold would lead to both incorrect values and incorrect dimensionalities. However, it is known to be possible to indirectly extract oscillator strengths corresponding to photo-ionization using an $\mathscr{L}^2$ basis *via* a procedure pioneered by Langhoff[39,40] and commonly referred to as Stieltjes imaging. The basis of the Stieltjes imaging method is the observation that the spectral moments $S_I(k)$ defined with respect to a given initial electronic state $|I\rangle$,

$$S_I(k) = \langle I|\hat{D}^\dagger \hat{H}^k \hat{D}|I\rangle, \tag{18}$$

can be accurately calculated *via* the insertion of the resolution of the identity expressed in terms of a given set $\{|\alpha\rangle\}$ of N-electron $\mathscr{L}^2$ basis functions:

$$S_I(k) = \sum_\alpha \langle I|\hat{D}^\dagger \hat{H}^k|\alpha\rangle\langle\alpha|\hat{D}|I\rangle = \sum_\alpha E_\alpha{}^k |\langle I|\hat{D}|\alpha\rangle|^2. \tag{19}$$

Equivalently, the resolution of the identity may be expressed in terms of the true bound and continuum states of the Hamiltonian, $\{|\iota\rangle\}$ and $\{|E\rangle\}$, respectively, giving

$$\begin{aligned} S_I(k) &= \sum_\iota E_\iota{}^k |\langle I|\hat{D}|\iota\rangle|^2 + \int_{E_\mathrm{T}}^\infty E^k |\langle I|\hat{D}|E\rangle|^2 \mathrm{d}E \\ &= \sum_i E_\iota{}^k |\langle I|\hat{D}|\iota\rangle|^2 + \frac{3}{2}\int_{E_\mathrm{T}}^\infty E^{k-1} f_I(E)\mathrm{d}E. \end{aligned} \tag{20}$$

From eqn (20), we see that $S_I(k)$ gives the kth moment of the photoabsorption cross-section. Then, by using the techniques of Stieltjes–Chebyshev moment theory, the photoabsorption cross-sections themselves may be extracted from a finite set of calculated spectral moments $S_I(k)$ evaluated using an $\mathscr{L}^2$ basis. In the work presented here, we follow the method outlined by Müller-Plathe and Diercksen.[62]

In the context of the present work, the $\mathscr{L}^2$ basis used in the expansion 19 of the spectral moments would be the CVS-ADC(2)x eigenstates. This approach, however, requires that the entire spectrum of the ADC Hamiltonian to be known, which is possible for only the smallest of systems. It has, however, recently been demonstrated that the spectral moments may be accurately and tractably approximated using the Lanczos pseudo-spectrum of the ADC Hamiltonan,[41–43] and it is this approach that is used in this work. In particular, we use the block variant of the Lanczos method[63,64] to generate the spectral moments $S_I(k)$ used in the Stieltjes imaging procedure.

C Calculation of ionization energies and Dyson orbitals

To calculate core-ionization energies, we make use of an idea from equation-of-motion coupled cluster theory whereby ionization to form an $(N-1)$-electron state is described in an N-electron framework by including an extremely diffuse orbital in the one-electron basis and restricting the N-electron basis to span the subspace for which this 'continuum' orbital is occupied.[65] Accordingly, we calculate the energies of the singly core-ionized states within the N-electron ADC(2) framework by including a single uncontracted s-type orbital in the one-electron basis with an exponent of 10^{-30} and constructing the ADC Hamiltonian matrix in the space spanned by configurations that correspond to excitation of one, and only one core electron and for which only a single electron is excited into the extremely diffuse 'continuum' orbital. We term the resulting state the IP-ADC(2) state.

The Dyson orbital $\phi^{\mathrm{D}}_{I\alpha}$ for ionization from an N-electron state $|I\rangle$ to an $(N-1)$-electron state $|\alpha\rangle$ may be written in the second-quantization form

$$\phi^{\mathrm{D}}_{I\alpha} = \sum_{p} \langle \alpha | \hat{c}_p | I \rangle \varphi_{\mathrm{p}}, \tag{21}$$

where $\{\varphi_{\mathrm{p}}\}$ denotes the set of canonical Hartree–Fock orbitals. We now replace the true $(N-1)$-electron wavefunction $|\alpha\rangle$ in the expansion 21 by $\hat{c}_a|\bar{\alpha}\rangle$, where $|\bar{\alpha}\rangle$ denotes the corresponding N-electron IP-ADC(2) state and a indexes the diffuse 'continuum' orbital used in its construction. The resulting approximate formula for the Dyson orbital $\phi^{\mathrm{D}}_{I\alpha}$ reads

$$\phi^{\mathrm{D}}_{I\alpha} \approx \sum_{p} \langle \bar{\alpha} | \hat{c}^{\dagger}_a \hat{c}_p | I \rangle \varphi_{\mathrm{p}} = \sum_{p} \rho^{(\alpha I)}_{ap} \varphi_{\mathrm{p}}, \tag{22}$$

where $\rho^{(\alpha I)}$ denotes the transition density matrix between the IP-ADC(2) state $|\bar{\alpha}\rangle$, and the ADC(2) state $|I\rangle$. Eqn (22) represents our working equation for the calculation of approximate Dyson orbitals for core-ionization.

D Calculation of approximate overlaps between the ADC(2) and MR-FOCI states

In order to determine the initial ADC(2) state to be used in the calculation of the component spectra $\sigma^{(b/c)}_{I}(E; \bar{\mathbf{R}}^{(I)}_{j}(t))$ that are required for the computation of the TRXAS (see Section 2.3), approximate overlaps between the MR-FOCI states $|\Psi^{(\mathrm{CI})}_{I}\rangle$ of the AIMS calculations and the ADC(2) states $|\Psi^{(\mathrm{ADC})}_{J}\rangle$ at each Gaussian position $\bar{\mathbf{R}}^{(I)}_{j}(t)$ were calculated. The ADC state $|\Psi^{(\mathrm{ADC})}_{J}\rangle$ with the greatest absolute overlap value $|\langle \Psi^{(\mathrm{ADC})}_{J} | \Psi^{(\mathrm{CI})}_{I}\rangle|$ was then selected as the initial state in the X-ray absorption spectrum calculation.

Given the expansion of an MR-FOCI state $|\Psi^{(\mathrm{CI})}_{i}\rangle$ and an ADC(2) state $|\Psi^{(\mathrm{ADC})}_{j}\rangle$ in terms of Slater determinant bases $\{|k^{(\mathrm{CI})}\rangle\}$ and $\{|m^{(\mathrm{ADC})}\rangle\}$,

$$\left|\Psi^{(\mathrm{CI})}_{i}\right\rangle = \sum_{k} C^{(\mathrm{CI})}_{ki} \left|k^{(\mathrm{CI})}\right\rangle, \tag{23}$$

$$\left|\Psi^{(\mathrm{ADC})}_{j}\right\rangle = \sum_{m} C^{(\mathrm{ADC})}_{mj} \left|m^{(\mathrm{ADC})}\right\rangle, \tag{24}$$

the overlap integral $\langle \Psi^{(\mathrm{CI})}_{i} | \Psi^{(\mathrm{ADC})}_{n}\rangle$ may be straightforwardly calculated as

$$\left\langle \Psi^{(\mathrm{CI})}_{i} \middle| \Psi^{(\mathrm{ADC})}_{j} \right\rangle = \sum_{k,m} C^{(\mathrm{CI})*}_{ki} C^{(\mathrm{ADC})}_{mj} \det \mathbf{S}^{k^{(\mathrm{CI})}, m^{(\mathrm{ADC})}}. \tag{25}$$

Here, the $\mathbf{S}^{k(\mathrm{CI}),m(\mathrm{ADC})}$ denotes the matrix of overlaps between the molecular orbitals occupied in the determinants $|k^{(\mathrm{CI})}\rangle$ and $|m^{(\mathrm{ADC})}\rangle$.

The MR-FOCI states, which are expanded originally in a configuration state function basis, are easily re-expressed in the Slater determinant basis. The expression of the ADC(2) states in the Slater determinant basis is, however, non-trivial. Each ADC(2) state $|\Psi_j^{(\mathrm{ADC})}\rangle$ is expanded in terms of a basis $\{|\tilde{\Psi}_J\rangle\}$ of so-called intermediate states,

$$\left|\Psi_j^{(\mathrm{ADC})}\right\rangle = \sum_J X_{jJ}\left|\tilde{\Psi}_J\right\rangle. \tag{26}$$

The intermediate states $|\tilde{\Psi}_J\rangle$ are in turn constructed as orthonormalized linear combinations of the set of so-called correlated excited state $|\Psi_J^0\rangle$ defined by the operation of the physical excitation operators $\hat{C}_J^\dagger$ on the exact ground state wavefunction $|\Psi_0\rangle$:

$$|\Psi_J^0\rangle = \hat{C}_J^\dagger|\Psi_0\rangle, \tag{27}$$

$$\hat{C}_J^\dagger \in \{\hat{c}_a^\dagger\hat{c}_i;\ \hat{c}_a^\dagger\hat{c}_b^\dagger\hat{c}_i\hat{c}_j\ (a < b,\ i < j)\}, \tag{28}$$

where the $\hat{c}_p^\dagger$ and $\hat{c}_p$ denote the elementary creation and annihilation operators. To proceed, Møller–Plesset perturbation theory is used to provide a representation of the ground state wavefunction and energy:

$$|\Psi_0\rangle = |\Psi_0^{\mathrm{HF}}\rangle + |\Psi_0^{[1]}\rangle + |\Psi_0^{[2]}\rangle, \tag{29}$$

where $|\Psi_0^{\mathrm{HF}}\rangle$ denotes the Hartree–Fock ground state wavefunction and $|\Psi_0^{[i]}\rangle$ the ith-order correction to the Hartree–Fock wavefunction. The full expansions of the intermediate states are non-trivial to evaluate due to the appearance of n-particle density matrix elements of high values of n. As such, in the calculation of the overlaps $\langle\Psi_i^{(\mathrm{CI})}|\Psi_j^{(\mathrm{ADC})}\rangle$ we use only the zeroth-order expansions of the intermediate states,

$$|\tilde{\Psi}_{ai}\rangle \rightarrow |\tilde{\Psi}_{ai}^{(0)}\rangle = \hat{c}_a^\dagger\hat{c}_i|\Psi_0^{\mathrm{HF}}\rangle, \tag{30}$$

$$|\tilde{\Psi}_{aibj}\rangle \rightarrow |\tilde{\Psi}_{aibj}^{(0)}\rangle = \hat{c}_a^\dagger\hat{c}_b^\dagger\hat{c}_i\hat{c}_j|\Psi_0^{\mathrm{HF}}\rangle, \tag{31}$$

in conjunction with the expansion coefficients X_{jJ} furnished by the ADC(2) calculation employing the intermediate states to correct order. Although only approximate, this scheme is found to allow for a satisfactory matching of the ADC(2) states to corresponding MR-FOCI states, which is our principle requirement.

E AIMS calculations

In the AIMS simulations, the initial state used corresponds to vertical excitation from the ground state to the $S_2(\pi\pi^*)$ state. The initial positions and momenta were sampled using a Monte Carlo procedure from the ground state Wigner distribution. To do so, the harmonic approximation was employed, with the required frequencies and normal modes calculated at the CCSD(T)/cc-pVTZ level using the CFOUR program[66] program. Propagation times of 240 fs were used, with time steps of 10 a.u. A total of 21 initial basis functions were used, with each being propagated independently, under the so-called independent first generation approximation.[37] Frozen Gaussian widths of 22.5 and 4.5 a.u. were used for the carbon and hydrogen atom coordinates, respectively.

Acknowledgements

AS and MS thank the NSERC Discovery grant program for financial support. VA acknowledges the financial support of the EPSRC through the programme grant on attosecond dynamics (EP/I032517/1).

References

1 A. Stolow, A. E. Bragg and D. M. Neumark, *Chem. Rev.*, 2004, **104**, 1719.
2 A. Stolow and J. G. Underwood, *Adv. Chem. Phys.*, 2008, **139**, 497.
3 J. Itatani, J. Levesque, D. Zeidler, H. Niikura, H. Pépin, J. C. Kieffer, P. B. Corkum and D. M. Villeneuve, *Nature*, 2004, **432**, 867.
4 O. Smirnova, Y. Mairesse, S. Patchkovskii, N. Dudovich, D. Villeneuve, P. Corkum and M. Y. Ivanov, *Nature*, 2009, **460**, 972.
5 W. Li, X. Zhou, R. Lock, S. Patchkovskii, A. Stolow, H. C. Kapteyn and M. M. Murnane, *Science*, 2008, **322**, 1207.
6 D. M. Jonas, *Annu. Rev. Phys. Chem.*, 2003, **54**, 425.
7 T. Brixner, J. Stenger, H. M. Vaswani, M. Cho, R. E. Blankenship and G. R. Fleming, *Nature*, 2005, **434**, 625.
8 M. Chergui, *Faraday Discuss.*, 2014, **171**, 11.
9 G. Sansone, L. Poletto and M. Nisoli, *Nat. Photonics*, 2011, **5**, 655.
10 F. Zamponi, Z. Ansari, M. Woerner and T. Elsaesser, *Opt. Express*, 2010, **18**, 947.
11 S. Khan, K. Holldack, T. Kachel, R. Mitzner and T. Quast, *Phys. Rev. Lett.*, 2006, **97**, 074801.
12 P. Beaud, S. L. Johnson, A. Streun, R. Abela, D. Abramsohn, D. Grolimund, F. Krasniqi, T. Schmidt, V. Schlott and G. Ingold, *Phys. Rev. Lett.*, 2007, **99**, 174801.
13 P. Emma, R. Akre, J. Arthur, R. Bionta, C. Bostedt, J. Bozek, A. Brachmann, P. Bucksbaum, R. Coffee, F. J. Decker, Y. Ding, D. Dowell, S. Edstrom, A. Fisher, J. Frisch, S. Gilevich, J. Hastings, G. Hays, P. Hering, Z. Huang, R. Iverson, H. Loos, M. Messerschmidt, A. Miahnahri, S. Moeller, H. D. Nuhn, G. Pile, D. Ratner, J. Rzepiela, D. Schultz, T. Smith, P. Stefan, H. Tompkins, J. Turner, J. Welch, W. White, J. Wu, G. Yocky and J. Galayda, *Nat. Photonics*, 2010, **4**, 641.
14 T. Ishikawa, H. Aoyagi, T. Asaka, Y. Asano, N. Azumi, T. Bizen, H. Ego, K. Fukami, T. Fukui, Y. Furukawa, S. Goto, H. Hanaki, T. Hara, T. Hasegawa, T. Hatsui, A. Higashiya, T. Hirono, N. Hosoda, M. Ishii, T. Inagaki, Y. Inubushi, T. Itoga, Y. Joti, M. Kago, T. Kameshima, H. Kimura, Y. Kirihara, A. Kiyomichi, T. Kobayashi, C. Kondo, T. Kudo, H. Maesaka, X. M. Marechal, T. Masuda, S. Matsubara, T. Matsumoto, T. Matsushita, S. Matsui, M. Nagasono, N. Nariyama, H. Ohashi, T. Ohata, T. Ohshima, S. Ono, Y. Otake, C. Saji, T. Sakurai, T. Sato, K. Sawada, T. Seike, K. Shirasawa, T. Sugimoto, S. Suzuki, S. Takahashi, H. Takebe,

K. Takeshita, K. Tamasaku, H. Tanaka, R. Tanaka, T. Tanaka, T. Togashi, K. Togawa, A. Tokuhisa, H. Tomizawa, K. Tono, S. K. Wu, M. Yabashi, M. Yamaga, A. Yamashita, K. Yanagida, C. Zhang, T. Shintake, H. Kitamura and N. Kumagai, *Nat. Photonics*, 2012, **6**, 540.
15 C. J. Milne, T. J. Penfold and M. Chergui, *Coord. Chem. Rev.*, 2014, **277–278**, 44.
16 C. Bressler and M. Chergui, *Chem. Rev.*, 2004, **104**, 1781.
17 Z. H. Loh, M. Khalil, R. E. Correa, R. Santra, C. Buth and S. R. Leone, *Phys. Rev. Lett.*, 2007, **98**, 143601.
18 C. La-O-Vorakiat, M. Siemens, M. M. Murnane, H. C. Kapteyn, S. Mathias, M. Aeschlimann, P. Grychtol, R. Adam, C. M. Schneider, J. M. Shaw, H. Nembach and T. J. Silva, *Phys. Rev. Lett.*, 2009, **103**, 257402.
19 L. X. Chen, *Annu. Rev. Phys. Chem.*, 2005, **56**, 221.
20 E. Goulielmakis, Z. H. Loh, A. Wirth, R. Santra, N. Rohringer, V. S. Yakovlev, S. Zherebtsov, T. Pfeifer, A. M. Azzeer, M. F. Kling, S. R. Leone and F. Krausz, *Nature*, 2010, **466**, 739.
21 W. Zhang, R. Alonso-Mori, U. Bergmann, C. Bressler, M. Chollet, A. Galler, W. Gawelda, R. G. Hadt, R. W. Hartsock, T. Kroll, K. S. Kjær, K. Kubiček, H. T. Lemke, H. W. Liang, D. A. Meyer, M. M. Nielsen, C. Purser, J. S. Robinson, E. I. Solomon, Z. Sun, D. Sokaras, T. B. van Driel, G. Vankó, T. C. Weng, D. Zhu and K. J. Gaffney, *Nature*, 2014, **509**, 345.
22 O. Gessner and M. Gühr, *Acc. Chem. Res.*, 2016, **49**, 138.
23 N. Huse, H. Cho, K. Hong, L. Jamula, F. M. F. de Groot, T. K. Kim, J. K. McCusker and R. W. Schoenlein, *J. Phys. Chem. Lett.*, 2011, **2**, 880.
24 B. Cooper, P. Kolorenč, L. J. Frasinski, V. Averbukh and J. P. Marangos, *Faraday Discuss.*, 2014, **171**, 93.
25 C. Bressler and M. Chergui, *Annu. Rev. Phys. Chem.*, 2010, **61**, 263.
26 F. de Groot, *Chem. Rev.*, 2001, **101**, 1779.
27 C. R. Natoli, M. Benfatto, S. D. Longa and K. Hatada, *J. Synchrotron Radiat.*, 2003, **10**, 26.
28 M. Benfatto and S. D. Longa, *J. Synchrotron Radiat.*, 2001, **8**, 1087.
29 T. Northey, N. Zotev and A. Kirrander, *J. Chem. Theory Comput.*, 2014, **10**, 4911.
30 A. Kirrander, K. Saita and D. V. Shalashilin, *J. Chem. Theory Comput.*, 2016, **12**, 957.
31 M. P. Minitti, J. M. Budarz, A. Kirrander, J. S. Robinson, D. Ratner, T. J. Lane, D. Zhu, J. M. Glowina, M. Kozina, H. T. Lemke, M. Sikorski, Y. Feng, S. Nelson, K. Saita, B. Stankus, T. Northey, J. B. Hastings and P. M. Weber, *Phys. Rev. Lett.*, 2015, **114**, 255501.
32 J. Stöhr, *NEXAFS Spectroscopy*, Springer, 1992.
33 A. R. Attar, A. Bhattacherjee and S. R. Leone, *J. Phys. Chem. Lett.*, 2015, **6**, 5072.
34 G. Capano, C. J. Milne, M. Chergui, U. Rothlisberger, I. Tavernelli and T. J. Penfold, *J. Phys. B: At. Mol. Phys.*, 2015, **48**, 214001.
35 G. Capano, M. Chergui, U. Rothlisberger, I. Tavernelli and T. J. Penfold, *J. Phys. Chem. A*, 2014, **118**, 9861.
36 G. Capano, T. J. Penfold, U. Rothlisberger and I. Tavernelli, *Chimia*, 2014, **68**, 227.
37 M. Ben-Nun and T. J. Martínez, *Adv. Chem. Phys.*, 2002, **121**, 439.
38 B. T. Pickup, *Chem. Phys.*, 1977, **19**, 193.
39 P. W. Langhoff, *Chem. Phys. Lett.*, 1973, **22**, 60.

40 P. Langhoff, C. T. Corcoran, J. S. Sims, F. Weinhold and R. M. Glover, *Phys. Rev. A*, 1976, **14**, 1042.
41 K. Gokhberg, V. Vysotskiy, L. S. Cederbaum, L. Storchi, F. Tarantelli and V. Averbukh, *J. Chem. Phys.*, 2009, **130**, 064104.
42 M. Ruberti, R. Yun, K. Gokhberg, S. Kopelke, L. S. Cederbaum, F. Tarantelli and V. Averbukh, *J. Chem. Phys.*, 2013, **139**, 144107.
43 M. Ruberti, R. Yun, K. Gokhberg, S. Kopelke, L. S. Cederbaum, F. Tarantelli and V. Averbukh, *J. Chem. Phys.*, 2014, **140**, 184107.
44 K. Pierloot, B. Dumez, P. O. Widmark and B. O. Roos, *Theor. Chim. Acta*, 1995, **90**, 87.
45 B. O. Roos, M. P. Fülscher, P. A. Malmqvist, M. Merchán and L. Serrano-Andrés, *Quantum Mechanical Electronic Structure Calculations with Chemical Accuracy*, Kluwer Academic Dordrecht, The Netherlands, 1995, p. 357.
46 J. Schirmer, *Phys. Rev. A*, 1982, **26**, 2395.
47 J. Schirmer, *Phys. Rev. A*, 1991, **43**, 4647.
48 J. Schirmer and A. Thiel, *J. Chem. Phys.*, 2001, **115**, 10621.
49 K. Kaufmann, W. Baumeister and M. Jungen, *J. Phys. B: At. Mol. Phys.*, 1989, **22**, 2223.
50 H. Lischka, R. Shepard, I. Shavitt, R. M. Pitzer, M. Dallos, T. MÃijller, P. G. Szalay, F. B. Brown, R. Ahlrichs, H. J. BÃűhm, A. Chang, D. C. Comeau, R. Gdanitz, H. Dachsel, C. Ehrhardt, M. Ernzerhof, P. HÃűchtl, S. Irle, G. Kedziora, T. Kovar, V. Parasuk, M. J. M. Pepper, P. Scharf, H. Schiffer, M. Schindler, M. SchÃijler, M. Seth, E. A. Stahlberg, J.-G. Zhao, S. Yabushita, Z. Zhang, M. Barbatti, S. Matsika, M. Schuurman, D. R. Yarkony, S. R. Brozell, E. V. Beck, J.-P. Blaudeau, M. Ruckenbauer, B. Sellner, F. Plasser and J. J. Szymczak, *Columbus, an ab initio electronic structure program, release 7.0*.
51 M. Schmidt, K. Baldridge, J. Boatz, S. Elbert, M. Gordon, J. Jensen, S. Kosecki, N. Matsunaga, K. Nguyen, S. J. Su, T. Windus, M. Dupuis and J. Montgomery, *J. Comput. Chem.*, 1993, **14**, 1347.
52 T. Mori, W. J. Glover, M. S. Schuurman and T. J. Martínez, *J. Phys. Chem. A*, 2012, **116**, 2808.
53 H. Tao, B. G. Levine and T. J. Martínez, *J. Phys. Chem. A*, 2009, **113**, 13656.
54 J. D. Coe, B. G. Levine and T. J. Martínez, *J. Phys. Chem. A*, 2007, **111**, 11310.
55 T. Kobayashi, T. Horio and T. Suzuki, *J. Phys. Chem. A*, 2015, **119**, 9518.
56 E. G. Champenois, N. H. Shivaram, T. W. Wright, C. S. Yang, A. Belkacem and J. P. Cryan, *J. Chem. Phys.*, 2016, **144**, 014303.
57 H. Tao, T. K. Allison, T. W. Wright, A. M. Stooke, C. Khurmi, J. van Tilborg, Y. Liu, R. W. Falcone, A. Belkacem and T. J. Martínez, *J. Chem. Phys.*, 2011, **134**, 244306.
58 E. Ohrendorf, H. Köppel, L. S. Cederbaum, F. Tarantelli and A. Sgamelloti, *J. Chem. Phys.*, 1989, **91**, 1734.
59 J. Schirmer, M. Braunstein, M. T. Lee and V. McKoy, *VUV and Soft X-Ray Photoionization*, Plenum Press, New York, 1996, ch. 4, p. 105.
60 O. Schalk, A. E. Boguslavskiy, A. Stolow and M. S. Schuurman, *J. Am. Chem. Soc.*, 2011, **133**, 16451.
61 M. Algia, M. Lavollée, R. Richter, U. Ekström, V. Carravetta, D. Stranges, B. Brunetti and S. Stranges, *Phys. Rev. A*, 2007, **76**, 022509.

62 F. Müller-Plathe and G. H. F. Diercksen, *Electronic Structure of Atoms, Molecules and Solids: II Brazilian School on Electronic Structure*, World Scientific, Singapore, 1990.
63 H. D. Meyer and S. Pal, *J. Chem. Phys.*, 1989, **91**, 6195.
64 B. N. Parlett, *The Symmetric Eigenvalue Problem*, Prentice-Hall, 1980, ch. 13.
65 J. F. Stanton and J. Gauss, *J. Chem. Phys.*, 1999, **111**, 8785.
66 J. F. Stanton, J. Gauss, M. E. Harding, P. G. Szalay, A. A. Auer, R. J. Bartlett, U. Benedikt, C. Berger, D. E. Bernholdt, Y. J. Bomble, L. Cheng, O. Christiansen, M. Heckert, O. Heun, C. Huber, T. C. Jagau, D. Jonsson, J. JusÃĽlius, K. Klein, W. J. Lauderdale, D. A. Matthews, T. Metzroth, L. A. MÃijck, D. P. O'Neill, D. R. Price, E. Prochnow, C. Puzzarini, K. Ruud, F. Schiffmann, W. Schwalbach, C. Simmons, S. Stopkowicz, A. Tajti, J. VÃązquez, F. Wang, and J. D. Watts, *Cfour, Coupled-cluster Techniques for Computational Chemistry, a Quantum-chemical Program Package*, see http://www.cfour.de.
67 R. McLaren, S. A. C. Clark, I. Ishii and A. P. Hitchcock, *Phys. Rev. A*, 1987, **36**, 1683.

Faraday Discussions

PAPER

Structural dynamics of photochemical reactions probed by time-resolved photoelectron spectroscopy using high harmonic pulses

Ryo Iikubo,[a] Taro Sekikawa,*[a] Yu Harabuchi[b] and Tetsuya Taketsugu[b]

Received 31st March 2016, Accepted 31st May 2016

DOI: 10.1039/c6fd00063k

Femtosecond ring-opening dynamics of 1,3-cyclohexadiene (CHD) in gas phase upon two-photon excitation at 400 nm (=3.1 eV) was investigated by time-resolved photoelectron spectroscopy using 42 nm (=29.5 eV) high harmonic photons probing the dynamics of the lower-lying occupied molecular orbitals (MOs), which are the fingerprints of the molecular structure. After 500 fs, the photoelectron intensity of the MO constituting the C=C sigma bond ($\sigma_{C=C}$) of CHD was enhanced, while that of the MO forming the C–C sigma bond (σ_{CC}) of CHD was decreased. The changes in the photoelectron spectra suggest that the ring of CHD opens to form a 1,3,5-hexatriene (HT) after 500 fs. The dynamics of the $\sigma_{C=C}$ and σ_{CC} bands between 200 and 500 fs reflects the ring deformation to a conical intersection between the 2^1A and 1^1A potential energy surfaces prior to the ring-opening reaction.

Introduction

Electrocyclic reactions play important roles in biological activities[1] and in photoelectric functions.[2] The stereospecificity of electrocyclic ring-opening and ring-closing reactions is well understood by the Woodward–Hoffmann rules[3] and the dynamics of the electrocyclic reactions has also been investigated experimentally. In particular, 1,3-cyclohexadiene (CHD) is of interest,[4,5] because the ring-opening reaction of CHD to 1,3,5-hexatriene (HT) is the key reaction in the synthesis of vitamin D[6–8] and in photochromic materials.[2] CHD provides us a unique opportunity to investigate the ring-opening dynamics of biological and functional molecules in a simplified manner.

To gain insight into the intrinsic ring-opening dynamics without the interaction with solvent, a gas-phase experiment is suitable. The ring-opening

[a]Department of Applied Physics, Faculty of Engineering, Hokkaido University, Kita 13 Nishi 8, Kita-ku, Sapporo 060-8628, Japan. E-mail: sekikawa@eng.hokudai.ac.jp

[b]Department of Chemistry, Faculty of Science, Hokkaido University, Kita 10 Nishi 8, Kita-ku, Sapporo 060-0810, Japan

dynamics of CHD in gas phase has been extensively investigated by ionization mass and photoelectron spectroscopy,[9–17] electron diffraction,[18] X-ray scattering,[19] theoretical calculations,[20–24] and so forth. These investigations focus on the relaxation processes from the 1^1B state. The configurational coordinate model shown in Fig. 1 is used to understand the processes: upon photoexcitation to 1^1B, the nuclear wavepacket propagates to the 2^1A state *via* a conical intersection (CI). At the second CI with the 1^1A state on the 2^1A potential energy surface, it bifurcates into CHD and HT in the ground state. Experimentally, it was found that the internal conversion from 1^1B to 2^1A took 56 fs by the transient ionization experiment.[12] Time-resolved photoelectron spectroscopy (TRPES) using ultraviolet (UV) light[17] showed that the lifetime of 1^1B was 30 fs and that HT appeared at 142 fs after excitation. This is consistent with other works.[11,12] New light sources were applied to the time-resolved spectroscopy of CHD recently;[16] TRPES probed by 14 eV light gave the similar lifetimes of 70 and 60 fs for 1^1B and 2^1A, respectively, to those obtained by UV light. Structural dynamics observed by X-ray scattering using a free-electron laser (FEL) showed that the bond of CHD broke as the terminal carbon atoms moved perpendicular to the molecular plane.[19] Electron diffraction[18] showed that the CHD ring cleaved on the picosecond timescale including the isomerization processes of HT.

Here, we investigated the ring-opening dynamics pumped by two photons of 400 nm (=3.1 eV) light and probed it by TRPES using high harmonic pulses with a photon energy of 29.5 eV. The present experimental scheme has two features in comparison with the previous studies: first, 3p Rydberg states, found by the resonant-enhanced multiphoton ionization (REMPI),[20] were excited by the two-photon absorption process, while the 1^1B state corresponding to $\pi\pi^*$ transition was excited in the previous works.[9–17] According to ref. 11, the ring opening was observed even at zero delay time under the two-photon excitation by 400 nm photons. This suggests that there are some very fast nonradiative processes from the Rydberg states. Therefore, the ring-opening dynamics might be different from the case of $\pi\pi^*$ excitation.

Second, TRPES using high harmonic pulses enables us to take a snapshot of the molecular orbitals (MOs) under an assumption of Koopmans'

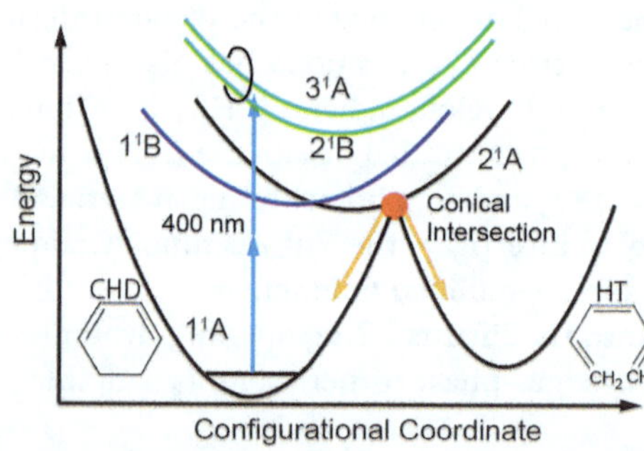

Fig. 1 Configurational coordinate model of the relaxation and the ring-opening processes in 1,3-cyclohexadiene (CHD) pumped by two-photon absorption of 400 nm photons. The arrows at the conical intersection indicate the bifurcation of the relaxation pathway at the conical intersection, leading to CHD or 1,3,5-hexatriene (HT).

theorem.[25] The usage of ultrashort extreme ultraviolet (XUV) pulses by high harmonic generation enables us to probe not only the highest occupied MO (HOMO) but also lower-lying occupied MOs.[26–30] Lower-lying occupied MOs can be the fingerprints of the molecular structure, because each MO can be associated with each bonding nature, *e.g.* single bond, double bond, conjugated double bond, and so forth.[31] The dynamics particular to each MO reflects the local conformational changes of the molecule. Using high harmonic pulses, we can gain insight into both electronic structures and molecular structures, providing us a new perspective on the chemical dynamics. Recently, we observed the ultrafast relaxation of the excited 1,3-butadiene[30] and the C–C bond breaking dynamics of 1,2-butadiene[32] by TRPES using high harmonic pulses.

Experimental

The experimental setup is shown in Fig. 2. We used a Ti:sapphire laser system delivering 800 nm, 1.1 mJ, 30 fs pulses at a repetition rate of 1 kHz for TRPES. After the second harmonic (=3.1 eV) generation by a LiB_3O_5 (LBO) crystal (500 μm thickness) for the pump pulses with a pulse energy of 40 μJ, the remaining fundamental pulses were focused into a Kr pulse gas jet for high harmonic generation. The 19th harmonic (=29.5 eV) for the probe was selected among many harmonics by a time-delay compensated monochromator (TDCM),[33–37] enabling the single harmonic selection with the time duration preserved. A TDCM consists of a pair of toroidal gratings and a slit: the first grating and a slit select the harmonic order and the second grating compensates for the pulse front tilt caused by the diffraction on the first grating. The shortest pulse duration of the selected high harmonics achieved in our system was 11 fs.[35] The pump intensity was reduced to 19 μJ by the attenuator composed of a pair of thin film polarizers and a half-wave plate. Both the pump and probe pulses were focused into the sample gas with a polarization angle of 55° to avoid the time dependence induced by molecular rotation. The photon flux on target was $\sim 10^9$ photons per s, corresponding to a pulse energy of 5 pJ.[37] CHD was purchased from Sigma-Aldrich and was used without any further purification. The sample gas, evaporated at room temperature, was guided to the entrance of a magnetic

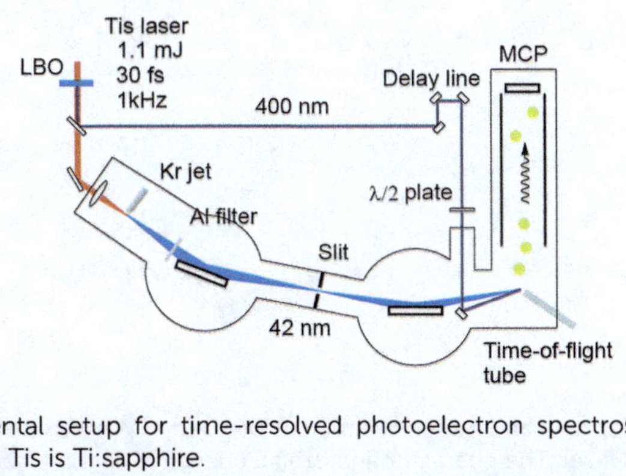

Fig. 2 Experimental setup for time-resolved photoelectron spectroscopy using high harmonic pulses. Tis is Ti:sapphire.

bottle time-of-flight photoelectron spectrometer through a copper pipe. A retardation electric field was applied in the flight tube to improve the spectral resolution of the spectrometer. Hence, the spectral resolution was determined by the spectral width of the XUV light (200 meV). The transient photoelectron spectra were recorded by changing the optical delay between the pump and probe pulses. The temporal evolution of the photoelectrons ejected by the two-photon ionization of Kr gas by pump and probe pulses, *i.e.* the cross-correlation function,[38,39] was employed as the response function of the system, of which the width τ_{FWHM} was 120 fs. Since we have already optimized the diffraction angle of the grating in the TDCM,[35] the temporal width was mainly limited by the pump pulse duration.

Computational details

To understand the photoelectron spectra, the ionization energies of CHD and the isomers of HT were calculated from the orbital energies obtained by long-range corrected density functional theory (LC-DFT) with LC-BOP/cc-pVDZ using the GAMESS-package[25,40] under an assumption of Koopmans' theorem. Also, to investigate the origin of the transient states, the vertical excitation energies and ionization energies were calculated at the Franck–Condon (FC) structure of CHD by 2nd-order multi-reference perturbation theory (CASPT2) with Dunning's basis sets of double-zeta-quality plus Rydberg functions,[41] using MOLPRO2010.[42] Two sets of π and π^* (12a, 10b, π^*-a, π^*-b) orbitals and four Rydberg orbitals (3s, $3p_x$, $3p_y$, $3p_z$) were included in the active space of the reference state-averaged complete active space self-consistent field (SA-CASSCF) wavefunction.

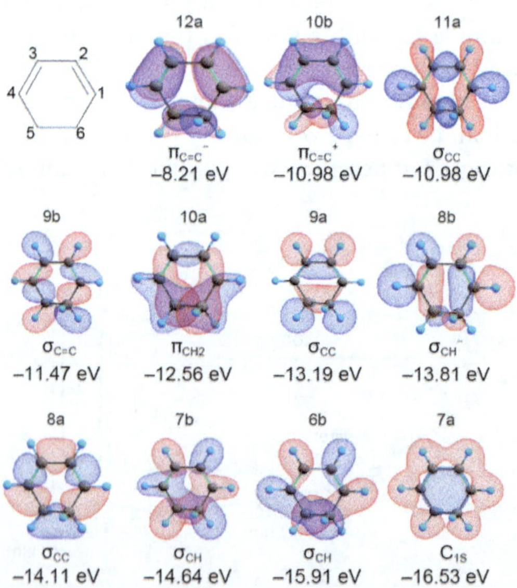

Fig. 3 Structural formula and molecular orbitals of 1,3-cyclohexadiene calculated at the LC-BOP/cc-pVDZ level. The character and orbital energy of each MO are also indicated.

Results and discussion

Molecular orbitals

The orbital energies and the MO shapes of CHD are summarized in Fig. 3. The orbital energies are in good agreement with the experimental ionization energies shown later, and thus, Koopmans' theorem is applicable to CHD. Each MO has a different spatial distribution and the character of the MO is named to signify the distribution in reference to ref. 31. The MO distribution is not directly connected with the classical chemical bond used in the structural formula. However, it is worth mentioning that the MOs with −12.56, −13.19, and −14.11 eV have a high density at the C5–C6 bond that cleaves upon photoexcitation. Thus, we expect that the intensities of photoelectrons or the ionization energies of these MOs are changed by the C5–C6 bond cleavage. These MOs can be the fingerprints of the ring structure of CHD.

Excited states

The calculated and experimental absorption energies of CHD are shown in Table 1. The 3p Rydberg states experimentally have the transition energies of 6.03 and 6.05 eV (ref. 20) and the absorption spectrum has sharp peaks around 6.2 eV,[43] which is a feature of Rydberg states. Since the point group of CHD is C_2, these states are both one- and two-photon transition allowed. Therefore, the excited states by two-photon absorption of 3.1 eV photons were mainly 3p Rydberg states in this experiment. The three-photon energy of the pump laser is larger than the ionization energy of CHD (=8.25 eV)[31] and the second ionization energy should be much larger than the first ionization energy due to the Coulomb interaction. Hence, the photoelectrons from the cations would appear in a totally different spectral region from the neutral species. In this work, therefore, we eliminate the possibility that the dynamics of cations were observed.

Table 1 The calculated and experimental excitation energies [eV] for the neutral and cationic species of CHD. All values are relative to the ground state energy of CHD

	State	CASPT2[a]	CASPT2[b]	exp[b]
Neutral	1^1B(12a → π*-b)	4.88	4.72	4.94
	2^1A(12a → 3s)	5.41	5.49	5.39
	2^1B(12a → 3p$_x$)	5.94	5.98	6.03
	3^1A(12a → 3p$_z$)	6.03	6.12	6.05
	3^1B(12a → 3p$_y$)	6.05	6.04	
Cation	1^2A(GS)	8.09		
	1^2B(10b → 12a)	10.53		
	2^2B(12a → π*-b)	11.78		
	2^2A(12a → π*-a)	13.86		
	3^2A(12a → 3s)	16.00		
	3^2B(12a → 3p$_x$)	17.03		

[a] This work. [b] Ref. 20.

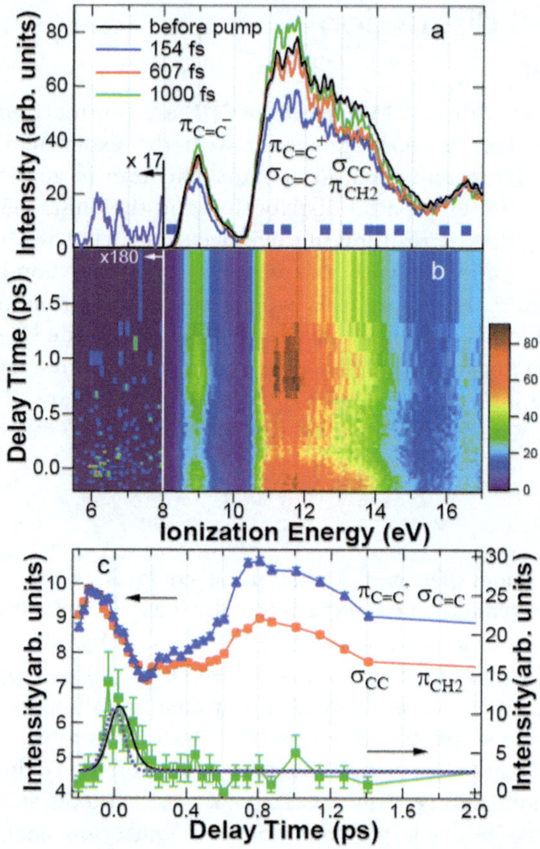

Fig. 4 (a) Photoelectron spectrum before the pump (black) and the spectra at 154 fs (blue), 607 fs (red), and 1000 fs (green). The solid squares indicate the theoretical ionization energies of 1,3-cyclohexadiene. In the assignment to molecular orbitals, notations are taken from Fig. 3. The spectrum of the transient band integrated between −140 and 250 fs is shown by the purple line below 8 eV. (b) Photoelectron spectrogram between −0.2 and 2 ps. (c) Time dependence of the intensity of $\pi_{C=C^+}$ and $\sigma_{C=C}$ (blue triangle), that of σ_{CC} and π_{CH_2} (red circle), and the transient band (green square) integrated between 11.14 and 11.93 eV, 12.20 and 14.00 eV, and 5.69 and 7.75 eV, respectively. The blue, red, and green solid lines are a guide for the eye. The black solid and purple dotted lines are the results fitted to the experimental data and the response function of the system, respectively.

Time-resolved photoelectron spectra

In Fig. 4a, the photoelectron spectrum before the pump is shown by the black line, obtained by averaging the spectra at 5 negative delay times between −200 fs and −86 fs. Two broad photoelectron bands between 10.5 and 12 eV and between 12 and 14 eV are clearly shown. The broad spectrum is due to the spectral width of the probe light. The theoretical ionization energies of CHD indicated by the solid squares are found within the observed spectrum. In particular, the photoelectron band between 12 and 14 eV includes σ_{CC} at 13.19 eV and π_{CH_2} at 12.56 eV constituting the C5–C6 bond, while the band between 10.5 and 12 eV is contributed by $\pi_{C=C^+}$ at 10.98 eV and $\sigma_{C=C}$ at 11.47 eV forming the C=C bond. To characterize each photoelectron band based on the MOs, we denote hereafter the band between 12 and 14 eV as the C–C band, and the band between 10.5 and 12 eV as the C=C band.

Fig. 4a also shows the time-resolved photoelectron spectra at 154 fs (blue), 607 fs (red), and 1000 fs (green) after pump. Fig. 4b shows the photoelectron spectrogram between −200 and 2000 fs. Here, we would like to raise the following two features in Fig. 4a and b. First, the dynamics of the C–C and C=C bands are different to each other: the photoelectron yield was reduced uniformly just after excitation. However, the intensity of the C=C band was almost recovered at 607 fs, while that of the C–C band was still reduced. At 1000 fs, the C=C band was even enhanced, although the C–C band intensity was still reduced. Second, Fig. 4b shows that a transient state appeared around 6–7 eV at 0 fs. The spectrum of the transient band below 8 eV integrated between −140 and 250 fs is shown in Fig. 4a by the purple line. The transient state has a broad spectrum almost reaching to the HOMO band.

To see these MO-dependent dynamics, the intensities of the C=C, C–C, and the transient bands integrated between 11.14 and 11.93 eV, 12 and 14 eV, and 5.69 and 7.75 eV, respectively, were plotted as a function of delay time between −0.2 and 2 ps in Fig. 4c. Here, the intensities of the C–C band and the C=C band are magnified to be on the same scale to verify that their dynamics are quite similar between −200 and 150 fs. The error bars show the statistical errors in the counting experiments. The time dependence of the transient band was fitted by the least squares method to a single exponential function $I(t) = \exp(-t/\tau)$ convoluted with the correlation function of the system $R(t) = \exp\{-4\ln 2\,(t/\tau_{\mathrm{FWHM}})^2\}$, where t, τ, and τ_{FWHM} are the time, decay time, and temporal width of $R(t)$, respectively. The fitting function $f(t)$ is expressed as.

$$f(t) = A\int_{-\infty}^{t} R\left(t'\right) I\left(t - t'\right) \mathrm{d}t' + B, \tag{1}$$

where A and B are the amplitude and the background constant, respectively. The fitting parameters were A, B, and τ. τ_{FWHM} was set to 120 fs. $R(t)$ and the fitting result are shown by the dotted purple line and the solid black line, respectively, in Fig. 4c. The lifetime of the transient band τ from the fitting was 37 ± 13 fs.

Ring-opening dynamics

Fig. 4c shows that, after 150 fs, the C=C band increased gradually until ~500 fs. Then, the intensity was enhanced up to 800 fs, while the C–C band was constant between 200 and 500 fs and was also enhanced after 500 fs. However, the intensity of the C–C band was not fully recovered in the spectrum before the pump. Because the C5–C6 bond mainly comes from the C–C band, the decrease in the relative intensity of the C–C band to the C=C band can be associated with the cleavage of the C–C bond.

To find the intermediates and products of the ring-opening reaction, the difference time-resolved spectra from the photoelectron spectrum before the pump are shown at delay times of 154, 607, and 1000 fs in Fig. 5b. Here, we consider three isomers of HT, all-*cis*-hexatriene (cZc-HT), mono-*s*-*cis*-*Z*-hexatriene (cZt-HT), and di-*s*-*trans*-*Z*-hexatriene (tZt-HT). Their chemical formulas are shown in Fig. 6. The ionization energies of these isomers, calculated by the same method

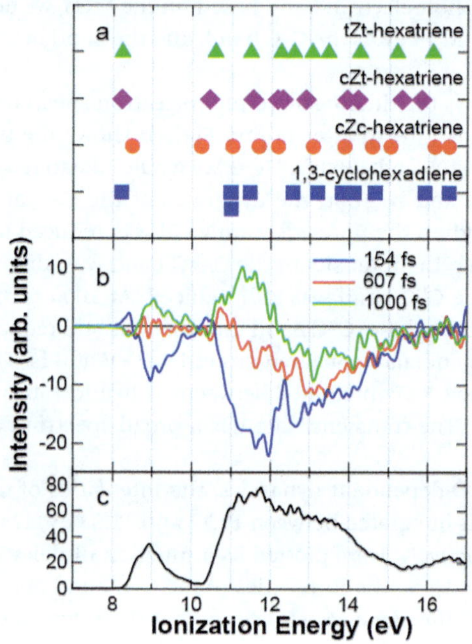

Fig. 5 (a) Ionization energies of tZt-hexatriene (triangle), cZt-hexatriene (diamond), cZc-hexatriene (circle), and 1,3-cyclohexadiene (CHD) (square). (b) Difference photoelectron spectra from the averaged photoelectron spectrum in the negative delay time. (c) Photoelectron spectrum of CHD without optical pump.

as those of CHD, are shown in Fig. 5a together with those of CHD. Fig. 5c shows the photoelectron spectrum of CHD without the pump for comparison. Although we found that the C=C band was enhanced in Fig. 4c, Fig. 5b shows that the temporal evolution of the C=C band was not uniform: the photoelectron band with a lower ionization energy than 11 eV was enhanced faster than the energetically higher region even within the C=C band. As a result, the positive signal appeared below 11 eV at 607 fs in Fig. 5b. This suggests that an MO with an ionization energy lower than 11 eV appeared by ring opening. The theoretical ionization energies of HTs shown in Fig. 5a predict that HTs have smaller ionization energies around 10.2 eV than that of the $\pi_{C=C^+}$ of CHD. The theoretical calculation also predicts the energy shift of the HOMO of cZc-HT to the higher ionization energy, which was experimentally observed at 1000 fs in Fig. 5b. Therefore, we attribute the MO lower than 11 eV observed at 607 fs to $\pi_{C=C}$ of cZc-HT produced just after the ring opening reaction. At almost the same delay time,

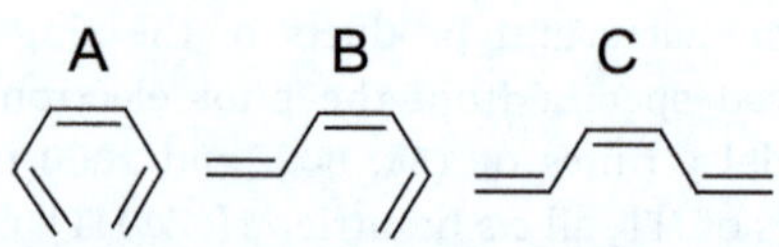

Fig. 6 Chemical formulas of hexatriene isomers. (A) cZc-hexatriene, (B) cZt-hexatriene, and (C) tZt-hexatriene.

we also observed the enhancement of the C–C band in Fig. 4c, suggesting a drastic change in the electronic structure, *i.e.* ring opening, of the molecule. Thus, we conclude that the ring starts to open at ~500 fs.

The enhancement of the C=C band compared to the C–C band upon ring opening is qualitatively understood as the change in the electronic structure from CHD to cZc-HT. If the ionization cross section of each MO is almost the same, the photoelectron intensity is proportional to the number of MOs involved in a selected region: CHD has three MOs in the C=C band between 10 and 12.2 eV, while cZc-HT has four shown in Fig. 5a. Thus, the C=C band is expected to be enhanced by ring opening. On the other hand, CHD has three MOs in the C–C band between 12.2 and 14.0 eV, while cZc-HT has two, also explaining the decrease in the C–C band by ring opening. The enhancement of the C=C and C–C bands ended at 800 fs and then, their band intensities began to be reduced till 1.4 ps. The relative intensity of the C–C band to the C=C band remained smaller after 800 fs, suggesting that the molecule stayed at the HT structures. Therefore, the decrease in the photoelectron intensities of both the C–C and the C=C bands after 800 fs are attributable to subsequent isomerization to cZt-HT or to tZt-HT.

Fig. 7 shows the long-term measurement of the C=C and C–C bands. After 7 ps, the intensities of both bands increased but the C=C band was more enhanced. Then, at 40 ps, the C=C band intensity decreased. Fig. 5a shows that the changes in the ionization energies among the isomers of HT are not so significant. Thus, the assignment of the molecular structure is not conclusive. However, the structural dynamics observed by the electron diffraction revealed the picosecond dynamics and the coexistence of HT isomers.[18] After the ring opening, the rotational inertia around the C–C bond should induce the isomerization of HT. As a result, the long-term dynamics shown in Fig. 7 is attributed to the isomerization among HTs.

TRPES using ultraviolet light also suggested that HTs have the higher vibrational temperature.[17] According to ref. 17, the deconvoluted photoelectron spectrum of HT extended to the lower ionization energies. This suggests that the isomers of HTs coexist and that anharmonic vibrations broaden the spectrum.

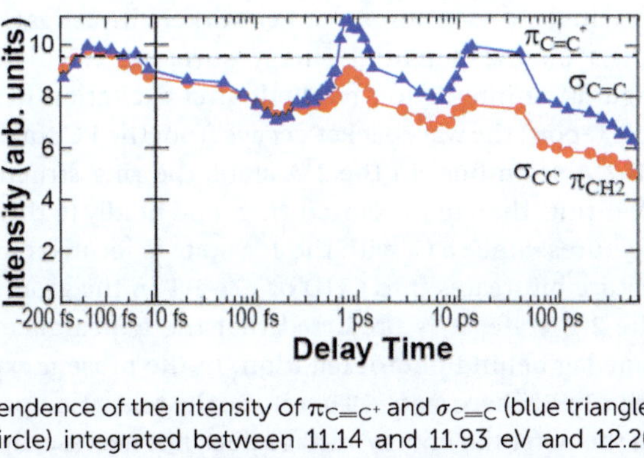

Fig. 7 Time dependence of the intensity of $\pi_{C=C^+}$ and $\sigma_{C=C}$ (blue triangle) and that of σ_{CC} and π_{CH_2} (red circle) integrated between 11.14 and 11.93 eV and 12.20 and 14.00 eV, respectively. Delay times are presented on a logarithmic and on a linear scale in the positive and negative regions, respectively. The blue and red solid lines are a guide for the eye. The dotted line indicates the intensity of $\pi_{C=C^+}$ and $\sigma_{C=C}$ in the averaged spectrum.

Dynamics of transient states

Now, we would like to discuss the processes prior to the ring opening. Just after excitation, we observed the transient state around 6–7 eV. We attribute this transient state to the FC states of 3p Rydberg states, which relax to the lower-lying excited states through CIs with a time constant of 37 ± 13 fs. The corresponding relaxation dynamics of the C=C and C–C bands was also observed: upon photoexcitation, both the C=C and C–C band intensities decreased with the response function of the system, owing to the different ionization cross section of the ground state or FC factors from those of the excited states, and then, the slight increment was observed at 200 fs in Fig. 4c. The relaxation of the transient state took place simultaneously with these increments of the C=C and C–C bands, indicating the relaxation to the lower-lying excited states. The broad spectral width of the transient state over 2 eV shown in Fig. 4a suggests that the transient band is composed of more than a few states.

As shown in Table 1, the present calculations well reproduced the experimental and previously calculated excitation energies for neutral CHD.[20] The first ionization energy was calculated as 8.09 eV which is consistent with the HOMO energy from LC-BOP calculations (8.21 eV). According to the relative energy values in Table 1, the ionization energies from the 3^1A state of the neutral species to the cationic states, 2^2B (12a → π*-b) and 2^2A (12a → π*-a), are 5.75 and 7.83 eV, respectively. Thus, the observed transient photoelectron bands could be assigned to these two states.

We did not detect any signals *via* the transition from the 3p Rydberg states of the neutral species to the 1^2A and 1^2B states of the cation, which are expected to appear around 2.06 and 4.50 eV, respectively. We think that this is due to the lower collection efficiency for the faster photoelectrons of the magnetic bottle spectrometer, because the faster electrons are not well guided to the detector due to the larger Larmor radius of the electron in the magnetic field.[44]

Dynamics on the 2^1A potential energy surface

After the relaxation to the 2^1A or 1^1B state, the temporal evolution of the C=C and C–C bands observed between 200 and 500 fs should reflect the dynamics in the 2^1A or 1^1B excited states. Tamura and coworkers investigated the quantum dynamics of CHD on the potential energy surfaces, calculated by a multi-reference configuration interaction method, after excitation to the 1^1B state.[23] According to this report, the wavepacket decays from the FC state to 2^1A through a CI on the 1^1B-2^1A seam line. In the 2^1A state, the ring structure is distorted firstly to the open-ring, then to the closed-ring, and finally to the five-membered ring (5 MR) structures. Since a CI with the 1^1A state is located close to 5 MR, the molecular structure bifurcates into CHD or cZc-HT in the ground state. Due to dynamics in the 2^1A state, it is predicted that the appearance of cZc-HT has about 180 fs time lag behind photoexcitation. In the present experiment, CHD was excited to the 3p Rydberg states. Therefore, the relaxation pathway from the FC structure on the potential energy surface does not necessarily lead to the 1^1B state and the experimental results might not be directly comparable to the theoretical work by Tamura *et al.*[23] However, the time-resolved photoelectron spectra show that the CHD ring opened not just after photoexcitation but after

500 fs. We would like to ascribe this delayed ring-opening reaction to the ring-deformation dynamics in the 2^1A state after the relaxation from 3p Rydberg states. Still, we do not have any idea why only the C=C band intensity increases gradually during the ring deformation. We suspect that the ionization cross sections of the MOs forming the C=C band are enhanced by the strong vibronic coupling with the symmetry breaking vibrational modes provoking the ring opening reaction.[11,12]

In this work, we found that the excitation to the 3p Rydberg states leads to the longer time for the ring-opening reaction than that to the 1^1B state.[11,12,19] During sliding down to the 2^1A potential energy surface, the absorbed photon energy is redistributed among the vibrational modes in the isolated molecules and the molecule excited to the 3p Rydberg states has more internal energy than that to the 1^1B state. Thus, it takes more time for the energy dissipation to reach the CI on the 2^1A potential energy surface compared with the excitation to the 1^1B state.

Estimation of the ring-opening efficiency

The quantum yield for the ring-opening reaction was estimated as follows: Fig. 4a shows that the band between 13 and 14 eV was not recovered at 1000 fs, which should be due to cZc-HT formation. We focus on the band between 13 and 14 eV, because the intensity at 607 fs, when the ring was not fully opened, was still almost the same as at 154 fs and because it was less modified by the recovery of the C=C band. The reduced area at 154 fs is attributed to the depletion of the ground state of CHD. If the recovered portion at 1000 fs is attributed totally to the ground state of CHD, the quantum yield of cZc-HT is estimated to be 0.46. Here, the ionization cross section between 13 and 14 eV of cZc-HT is assumed to be zero in this estimation. So 0.46 is the minimum quantum yield for the ring-opening reaction after the two-photon excitation to the FC states of the 3p Rydberg states. On the other hand, the efficiency in the solvent[45] and in gas phase by TRPES[16] upon 1^1B excitation were 0.4 and 0.3, respectively. According to the theoretical work,[46] the quantum efficiency upon 2^1A excitation is lower than that upon 1^1B. Since the initial states are different in the present case, the efficiency is not comparable. We suspect that the excess energy by photoexcitation to 3p Rydberg states promotes the ring opening deformation on the 2^1A potential energy surface.

Conclusions

TRPES using high harmonic pulses shed light on the detailed structural dynamics of the ring-opening processes of CHD by correlating the lower-lying MOs with particular spatial electron distributions predicted by quantum chemical calculation. Upon two-photon excitation to the FC state of 3p Rydberg states, the electronic state slides to 2^1A with a time constant of 37 ± 13 fs. On the 2^1A potential energy surface, the ring structure might be deformed to 5MR, where a CI with the ground state is located. Then, the ring-opening reaction to cZc-HT starts at ~500 fs and finishes at ~800 fs. The isomerization to cZt-HT or to tZt-HT proceeds upto 1.4 ps. The long-term measurement suggests further isomerization among HT isomers.

Acknowledgements

This work is supported by CREST, JST. T. S. was also supported by KAKENHI (26600107, 15H03702, and 16K13854) and the Eno Science Foundation. T. T. was also supported by KAKENHI (26288001). Part of the calculations was performed on supercomputers at the Research Center for Computational Science, Okazaki, Japan.

References

1 C. M. Beaudry, J. P. Malerich and D. Trauner, *Chem. Rev.*, 2005, **105**, 4757–4778.
2 M. Irie, *Chem. Rev.*, 2000, **100**, 1683–1684.
3 R. B. Woodward and R. Hoffmann, *Angew. Chem., Int. Ed. Engl.*, 1969, **8**, 781–853.
4 B. C. Arruda and R. J. Sension, *Phys. Chem. Chem. Phys.*, 2014, **16**, 4439–4455.
5 S. Deb and P. M. Weber, *Annu. Rev. Phys. Chem.*, 2011, **62**, 19–39.
6 H. J. C. Jacobs, J. W. J. Gielen and E. Havinga, *Tetrahedron Lett.*, 1981, **22**, 4013–4016.
7 M. Braun, W. Fuß, K. L. Kompa and J. Wolfrum, *J. Photochem. Photobiol., A*, 1991, **61**, 15–26.
8 O. G. Dmitrenko, I. P. Terenetskayab and W. Reischl, *J. Photochem. Photobiol., A*, 1997, **104**, 113–117.
9 S. A. Trushin, W. Fuß, T. Schikarski, W. E. Schmid and K. L. Kompa, *J. Chem. Phys.*, 1997, **106**, 9386.
10 W. Fuß, W. E. Schmid and S. A. Trushin, *J. Chem. Phys.*, 2000, **112**, 8347–8362.
11 N. Kuthirummal, F. M. Rudakov, C. L. Evans and P. M. Weber, *J. Chem. Phys.*, 2006, **125**, 133307.
12 K. Kosma, S. A. Trushin, W. Fuß and W. E. Schmid, *Phys. Chem. Chem. Phys.*, 2009, **11**, 172–181.
13 J. L. White, J. Kim, V. S. Petrovi and P. H. Bucksbaum, *J. Chem. Phys.*, 2012, **136**, 054303.
14 V. S. Petrović, M. Siano, J. L. White, N. Berrah, C. Bostedt, J. D. Bozek, D. Broege, M. Chalfin, R. N. Coffee, J. Cryan, L. Fang, J. P. Farrell, L. J. Frasinski, J. M. Glownia, M. Gühr, M. Hoener, D. M. P. Holland, J. Kim, J. P. Marangos, T. Martinez, B. K. McFarland, R. S. Minns, S. Miyabe, S. Schorb, R. J. Sension, L. S. Spector, R. Squibb, H. Tao, J. G. Underwood and P. H. Bucksbaum, *Phys. Rev. Lett.*, 2012, **108**, 253006.
15 V. S. Petrovic, S. Schorb, J. Kim, J. White, J. P. Cryan, J. M. Glownia, L. Zipp, D. Broege, S. Miyabe, H. Tao, T. Martinez and P. H. Bucksbaum, *J. Chem. Phys.*, 2013, **139**, 184309.
16 S. Adachi, M. Sato and T. Suzuki, *J. Phys. Chem. Lett.*, 2015, **6**, 343–346.
17 C. C. Pemberton, Y. Zhang, K. Saita, A. Kirrander and P. M. Weber, *J. Phys. Chem. A*, 2015, **119**, 8832–8845.
18 C.-Y. Ruan, V. A. Lobastov, R. Srinivasan, B. M. Goodson, H. Ihee and A. H. Zewail, *Proc. Natl. Acad. Sci. U. S. A.*, 2001, **98**, 7117–7122.
19 M. P. Minitti, J. M. Budarz, A. Kirrander, J. S. Robinson, D. Ratner, T. J. Lane, D. Zhu, J. M. Glownia, M. Kozina, H. T. Lemke, M. Sikorski, Y. Feng, S. Nelson, K. Saita, B. Stankus, T. Northey, J. B. Hastings and P. M. Weber, *Phys. Rev. Lett.*, 2015, **114**, 255501.

20 M. Merchán, L. Serrano-Andrés, L. S. Slater, B. O. Roos, R. McDiarmid and X. Xing, *J. Phys. Chem. A*, 1999, **103**, 5468–5476.
21 A. Hofmann, L. Kurtz and R. d. Vivie-Riedle, *Appl. Phys. B*, 2000, **71**, 391–396.
22 M. Garavelli, C. S. Page, P. Celani, M. Olivucci, W. E. Schmid, S. A. Trushin and W. Fuss, *J. Phys. Chem. A*, 2001, **105**, 4458–4469.
23 H. Tamura, S. Nanbu, T. Ishida and H. Nakamura, *J. Chem. Phys.*, 2006, **124**, 084313.
24 A. Nenov, P. Kolle, M. A. Robb and R. d. Vivie-Riedle, *J. Org. Chem.*, 2010, **75**, 123–129.
25 T. Tsuneda, J.-W. Song, S. Suzuki and K. Hirao, *J. Chem. Phys.*, 2010, **133**, 174101.
26 L. Nugent-Glandorf, M. Scheer, D. A. Samuels, A. M. Mulhisen, E. R. Grant, X. Yang, V. M. Bierbaum and S. R. Leone, *Phys. Rev. Lett.*, 2001, **87**, 193002.
27 P. Wernet, M. Odelius, K. Godehusen, J. Gaudin, O. Schwarzkopf and W. Eberhardt, *Phys. Rev. Lett.*, 2009, **103**, 013001.
28 P. Wernet, *Phys. Chem. Chem. Phys.*, 2011, **13**, 16941–16954.
29 M. Fushitani, A. Matsuda and A. Hishikawa, *Opt. Express*, 2011, **19**, 9600–9606.
30 A. Makida, H. Igarashi, T. Fujiwara, T. Sekikawa, Y. Harabuchi and T. Taketsugu, *J. Phys. Chem. Lett.*, 2014, **5**, 1760–1765.
31 *Handbook of HeI Photoelectron Spectra of Fundamental Organic Molecules : Ionization Energies, Ab Initio Assignments, and Valence Electronic Structure for 200 Molecules*, ed. K. Kimura, Japan Scientific Societies Press, Tokyo, 1981.
32 R. Iikubo, T. Fujiwara, T. Sekikawa, Y. Harabuchi, S. Satoh, T. Taketsugu and Y. Kayanuma, *J. Phys. Chem. Lett.*, 2015, **6**, 2463–2468.
33 P. Villoresi, *Appl. Opt.*, 1999, **38**, 6040–6049.
34 L. Poletto, P. Villoresi, E. Benedetti, F. Ferrari, S. Stagira, G. Sansone and M. Nisoli, *Opt. Lett.*, 2007, **32**, 2897–2899.
35 H. Igarashi, A. Makida, M. Ito and T. Sekikawa, *Opt. Express*, 2012, **20**, 3725–3732.
36 H. Igarashi, A. Makida and T. Sekikawa, *Opt. Express*, 2013, **21**, 20632–20640.
37 M. Ito, Y. Kataoka, T. Okamoto, M. Yamashita and T. Sekikawa, *Opt. Express*, 2010, **18**, 6071–6078.
38 T. E. Glover, R. W. Schoenlein, A. H. Chin and C. V. Shank, *Phys. Rev. Lett.*, 1996, **76**, 2468–2471.
39 T. Sekikawa, T. Katsura, S. Miura and S. Watanabe, *Phys. Rev. Lett.*, 2002, **88**, 193902.
40 M. S. Gordon and M. W. Schmidt, in *Theory and Applications of ComputationalChemistry, the First Forty Years*, ed. C. E. Dykstra, G. Frenking, K. S. Kim and G. E. Scuseria, Elsevier, Amsterdam, 2005, ch. 41, pp. 1167–1189.
41 T. H. Dunning Jr., *J. Chem. Phys.*, 1970, **53**, 2823.
42 H.-J. Werner, P. J. Knowles, G. Knizia, F. R. Manby, M. Schütz, P. Celani, T. Korona, R. Lindh, A. Mitrushenkov, G. Rauhut, K. R. Shamasundar, T. B. Adler, R. D. Amos, A. Bernhardsson, A. Berning, D. L. Cooper, M. J. O. Deegan, A. J. Dobbyn, F. Eckert, E. Goll, C. Hampel, A. Hesselmann, G. Hetzer, T. Hrenar, G. Jansen, C. Köppl, Y. Liu, A. W. Lloyd, R. A. Mata, A. J. May, S. J. McNicholas, W. Meyer, M. E. Mura, A. Nicklass, D. P. O'Neill, P. Palmieri,D. Peng, K. Pflüger, R. Pitzer, M. Reiher, T. Shiozaki, H. Stoll, A. J. Stone, R. Tarroni, T. Thorsteinsson and M. Wang, *Molpro, Version 2015.1*, a package of ab initio programs, 2015, see http://www.molpro.net.

43 R. McDiarmid, A. Sabljić and J. P. Doering, *J. Chem. Phys.*, 1985, **83**, 2147–2152.
44 P. Kruit and F. H. Read, *J. Phys. E: Sci. Instrum.*, 1983, **16**, 313–324.
45 S. H. Pullen, N. A. Anderson, L. A. Walker and R. J. Sension, *J. Chem. Phys.*, 1998, **108**, 556–563.
46 A. Ohta, O. Kobayashi, S. O. Danielache and S. Nanbu, *Chem. Phys.*, 2015, **459**, 45–53.

Faraday Discussions

PAPER

Molecular frame photoemission by a comb of elliptical high-order harmonics: a sensitive probe of both photodynamics and harmonic complete polarization state

K. Veyrinas,†*[a] V. Gruson,‡[b] S. J. Weber,§[b] L. Barreau,[b] T. Ruchon,[b] J.-F. Hergott,[b] J.-C. Houver,[a] R. R. Lucchese,[c] P. Salières[b] and D. Dowek*[a]

Received 23rd May 2016, Accepted 6th July 2016

DOI: 10.1039/c6fd00137h

Due to the intimate anisotropic interaction between an XUV light field and a molecule resulting in photoionization (PI), molecular frame photoelectron angular distributions (MFPADs) are most sensitive probes of both electronic/nuclear dynamics and the polarization state of the ionizing light field. Consequently, they encode the complex dipole matrix elements describing the dynamics of the PI transition, as well as the three normalized Stokes parameters s_1, s_2, s_3 characterizing the complete polarization state of the light, operating as molecular polarimetry. The remarkable development of advanced light sources delivering attosecond XUV pulses opens the perspective to visualize the primary steps of photochemical dynamics in time-resolved studies, at the natural attosecond to few femtosecond time-scales of electron dynamics and fast nuclear motion. It is thus timely to investigate the feasibility of measurement of MFPADs when PI is induced *e.g.*, by an attosecond pulse train (APT) corresponding to a comb of discrete high-order harmonics. In the work presented here, we report MFPAD studies based on coincident electron-ion 3D momentum imaging in the context of ultrafast molecular dynamics investigated at the PLFA facility (CEA-SLIC), with two perspectives: (i) using APTs generated in atoms/molecules as a source for MFPAD-resolved PI studies, and (ii) taking advantage of molecular polarimetry to perform a complete polarization analysis of the harmonic emission of molecules, a major challenge of high harmonic spectroscopy. Recent results illustrating both aspects are reported for APTs generated

[a]*Institut des Sciences Moléculaires d'Orsay (ISMO), CNRS, Univ. Paris-Sud, Université Paris-Saclay, F-91405 Orsay, France. E-mail: danielle.dowek@u-psud.fr*

[b]*LIDYL, CEA, CNRS, Université Paris-Saclay, CEA Saclay, 91191 Gif-Sur-Yvette, France*

[c]*Department of Chemistry, Texas A&M University, College Station, Texas 77843, USA*

† Present address: Department of Physics, University of Ottawa, 150 Louis Pasteur, Ottawa, Ontario K1N 6N5, Canada. *E-mail: Kevin.Veyrinas@uottawa.ca*

‡ Present address: INRS, Centre Energie Matériaux et Télécommunications, 1650 Boulevard Lionel-Boulet, Varennes, Quebec, J3X1S2, Canada.

§ Present address: CEMES – CNRS, 29 rue Jeanne Marvig, 31055 Toulouse, France.

in unaligned SF_6 molecules by an elliptically polarized infrared driving field. The observed fingerprints of the elliptically polarized harmonics include the first direct determination of the complete s_1, s_2, s_3 Stokes vector, equivalent to (ψ, ε, P), the orientation and the signed ellipticity of the polarization ellipse, and the degree of polarization P. They are compared to so far incomplete results of XUV optical polarimetry. We finally discuss the comparison between the outcomes of photoionization and high harmonic spectroscopy for the description of molecular photodynamics.

1. Introduction

Photoionization (PI) is one of the basic processes that allows direct investigation of molecular structure and dynamics; therefore it is often used as a probe of the relaxation of transient excited molecular states involved in gas-phase photochemical dynamics. Resulting from the intimate anisotropic interaction between the light field and a molecule, molecular frame photoelectron angular distributions (MFPADs) are the most sensitive observables of the electronic/nuclear photodynamics induced by photoionization, and they simultaneously encode the polarization state of the ionizing light. Consequently, the determination of MFPADs gives access to the complex dipole matrix elements describing the photoionization transition, as well as to the three normalized Stokes parameters s_1, s_2, s_3 (ref. 1) characterizing the complete polarization state of the ionizing light. In particular, the ability to derive s_3, whose sign reflects the light helicity, is a result of the circular dichroism in the molecular frame, *i.e.*, the difference in photoemission when the molecule is exposed to left- or right-handed circularly polarized light. In this work we discuss two ways that the MFPADs can be combined with the use of ultrafast XUV/VUV pulses to contribute to the tool box of ultrafast imaging of photochemical dynamics.

MFPADs can be accessed using different techniques,[2,3] such as photoelectron spectroscopy (PES) applied to rotationally resolved excited molecules or to laser-aligned and oriented molecules, or electron-ion coincidence momentum spectroscopy. In the work presented here we use the latter approach, taking advantage of dissociative photoionization (DPI) of small molecules induced by XUV radiation to determine fully differential MFPADs. On the one hand, this method has provided a wealth of results on the photoionization dynamics following inner-valence shell[4–6] or inner-shell[7–10] ionization of linear molecules or small polyatomic systems most often induced by synchrotron radiation, whose high repetition rate is appropriate for coincidence measurements, *i.e.*, in the weak field regime. On the other hand, the use of MFPADs as a probe of the complete polarization state of ionizing light, known as molecular polarimetry (MP),[11] has been benchmarked recently by comparing data relying on coincident electron-ion 3D momentum spectroscopy and on VUV optical polarimetry[12] available on the DESIRS beamline (SOLEIL) which delivers fully controlled VUV elliptically polarized light.[13] This development is part of the substantial effort performed, in particular in the XUV and X-ray domains at large scale facilities such as synchrotrons[12,14] and free electron lasers,[15,16] to characterize and control the polarization state of the light source, a key parameter for numerous experiments addressing fundamental topics such as chiral properties of matter,[17,18] ranging from biomolecules to magnetic materials. To our knowledge, with the use of optical polarimeters including a dephazer element, the MP method is the only *in situ* approach providing the full determination of the (s_1, s_2, s_3) Stokes vector, equivalent to that of (ψ, ε, P), the orientation and signed ellipticity of the polarization ellipse, and the degree of polarization P. In particular, for partially polarized light, the MP method enables us to extract the light helicity s_3 separately from the degree of unpolarized light $s_4 = 1 - P$ (sometimes considered as a fourth Stokes parameter[1]). One clear advantage of the method[11] is its validity across the whole VUV to X-ray range that offers a broad tunability for applications.

The remarkable development of advanced light sources delivering ultrafast XUV pulses has created unprecedented possibilities to address and possibly visualize the primary steps of photochemical dynamics in time-resolved studies of ultrafast processes at the natural time-scale of electron dynamics and fast nuclear motion, ranging from attoseconds to a few femtoseconds. Among these ultrafast sources free electron lasers (FELs) stand out, as well as secondary XUV sources based on high-order harmonic generation (HHG), resulting from the non-linear interaction between an intense infrared (IR) radiation and a gas medium.[19] Ultrashort coherent XUV pulses may serve two types of application schemes: (i) an "external probing" scheme in time-resolved pump-probe PI experiments with controllable delay at the subfemtosecond scale,[20–27] and (ii) a "self-probing" scheme in the case of HHG, where the three-step generation process (strong field tunnel ionization, laser driven electronic wave-packet motion, and recollision of the returning electron leading to photorecombination)[28,29] probes the system on the time scale of half a laser cycle, a technique now referred to as high harmonic spectroscopy (HHS).[30–35] In both approaches the molecular frame (MF) observables are a unifying concept for characterizing photochemical dynamics *via* snapshot mapping of the electronic density, either in photoemission or by molecular orbital tomography with femto to attosecond temporal resolution. For both approaches too, the occurrence of specific situations where symmetry breaking in the generation process results in the production of elliptically polarized coherent sub-femtosecond pulses has been emphasized in recent experiments, *e.g.*, HHG emission from aligned molecules[36,37] or induced by an elliptical IR laser,[38] or HHG driven by a two-color field involving photons of opposite helicity.[39] This context motivates the development of methods for a complete characterization of the harmonic polarization state providing (i) sources with optimized ellipticities for time-resolved investigations of dichroisms, and (ii) access to the complex induced dipole vector and insight into the interactions occurring during the HHG process, in particular the dynamics of the electron–parent ion recombination.

In the work presented here, we report MFPAD studies based on electron–ion coincident 3D momentum spectroscopy in the context of attosecond molecular dynamics conducted at SLIC facility of CEA-Saclay, with the two perspectives addressed above: on the one hand, using the generated APT as a light source for PI of target systems characterized at the MFPAD level,[40–42] and on the other hand, taking advantage of MFPAD-resolved PI of simple molecules, to perform a polarization analysis of the HHG emission in high harmonic spectroscopy.

(i) We demonstrate the measurement of complete MFPADs in dissociative photoionization of a chosen molecular target by an attosecond pulse train: this constitutes the first step towards the study of subfemtosecond time-resolved electron/nuclear dynamics in molecules where the reaction launched by the APT is either photoionization, or relaxation of a neutral excited state probed by photoionization, in different pump-probe schemes.[43]

(ii) We use molecular polarimetry to determine the complete polarization state of harmonics generated in SF_6 molecules by an elliptically polarized IR field. SF_6 is a polyatomic molecule where it was found that strong field ionization involves contributions from multiple electronic channels, while the recombination step is influenced by resonant transitions due to the presence of autoionizing states and shape resonances[44] trapping the recolliding electron for typically hundreds of attoseconds before recombination to the ground state occurs. Our MP results are compared with data from optical polarimetry (OP) in the conditions of the MP study.

The paper is organized as follows. In order to ensure a self-contained content, we revisit in Section 2 the general expression of the MFPAD for PI of linear molecules induced by elliptically polarized light, and the subsequent key steps for the extraction of (i) the dynamical parameters of the PI reaction and (ii) the Stokes parameters, providing the grounds for MP. Section 3 describes the experimental methodology, where dissociative photoionization of small molecules induced by an attosecond pulse train is analyzed using 3D-momentum spectroscopy. In Section 4, we report the first complete measurements of MFPADs in PI from an incident APT. In Section 5, we apply the MP method to the complete characterization of the polarization state of the HH comb produced in SF_6 gas by an elliptically polarized driving laser. Conclusions and perspectives are given in Section 6.

2. MFPADs induced by elliptically polarized light: methodology for molecular polarimetry

The molecular polarimetry method, previously described,[11] relies on the remarkable properties of molecular frame photoelectron angular distributions (MFPADs) induced by elliptically polarized light[45,46] as obtained in the study of dissociative photoionization (DPI) for primarily linear molecules using electron-ion coincident 3D momentum spectroscopy.

Briefly, for DPI of a molecule AB producing an ion fragment A^+ and a photoelectron, MFPADs are determined from the correlated ion fragment recoil velocity V_{A^+} and photoelectron velocity V_e measured for each DPI coincident event (A^+, e).[47,48]

$$AB + h\nu(s_1,s_2,s_3) \rightarrow AB^+ + e \rightarrow A^+ + B + e \tag{1}$$

The extension to inner-shell ionization where PI is followed by Auger decay is straightforward.[49] For one-photon PI of a linear molecule, the general expression of the MFPAD in the dipole approximation is a function of three angles, $I(\theta_e,\phi_e,\chi)$, for linear[50] and circular[45] polarizations of the ionizing light, while it is function of an additional angle, $I(\theta_e,\phi_e,\chi,\gamma)$, in the general case where PI is induced by elliptically polarized light.[11,46] (θ_e,ϕ_e) represent the polar and azimuthal electron emission direction in the molecular frame (MF) and (χ,γ) the polar and azimuthal ion fragment emission direction in the field frame (FF) or laboratory frame (LF), as indicated schematically in Fig. 1(a). In the LF, the z_{LF} reference axis for the (χ) polar dependence is parallel to the k

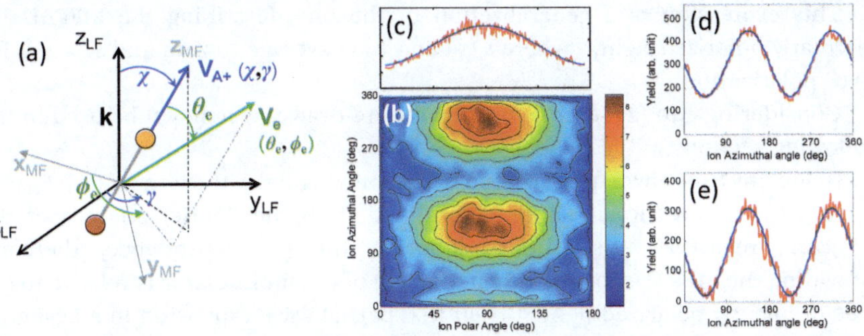

Fig. 1 (a) Schematic of the relevant angles: (θ_e,ϕ_e) represent the polar and azimuthal electron emission direction in the molecular frame (MF) and (χ,γ) the polar and azimuthal ion fragment emission direction in the laboratory frame (LF). (b) Bidimensional histogram of the (N^+, e) events showing the ion fragment polar and azimuthal dependences, for the studied polarization state S at $h\nu = 23.65$ eV; the 1D plots show: (c) the (χ) dependence, (d) and (e) the (γ) dependences according to eqn (4) plotted here as a function of $(\gamma + \pi/2)$, providing two determinations of the normalized s_1, s_2 Stokes parameters, *i.e.*, the polarization ellipse orientation angle (ψ) (see Fig. 2). The example shown leads to $s_1 = -0.05 \pm 0.01$ and $s_2 = -0.87 \pm 0.01$ corresponding to an orientation $\psi = 133.2° \pm 0.3°$.

propagation axis of the ionizing light, where x_{LF} is the reference axis in the polarization plane for the (γ) azimuthal dependence. In the MF, z_{MF}, parallel to the ion fragment recoil velocity, *i.e.*, along the molecular axis in the axial recoil approximation, and x_{MF}, a unitary vector in the plane containing z_{LF} and z_{MF}, are the reference axes for the (θ_e) and (ϕ_e) polar and azimuthal dependences, respectively. The MP method relies, on the one hand, on the strong dependence of the MFPADs induced by linearly polarized light upon the molecular axis orientation relative to the light polarization axis, and, on the other hand, on the circular dichroism in the MF frame[51] which reflects the different responses of the system when exposed to left (LHC) and right handed circularly (RHC) polarized light, respectively. When PI is induced by elliptically polarized light described by the three Stokes parameters s_1, s_2, s_3, defined in the LF,[1] this information is encapsulated in the following analytical form:[11,46]

$$I(\theta_e,\phi_e,\chi,\gamma) = F_{00}(\theta_e) + F_{20}(\theta_e)\left[-\frac{1}{2}P_2^0(\cos\chi) + \frac{1}{4}t_1(\gamma)P_2{}^2(\cos\chi)\right]$$
$$+F_{21}(\theta_e)\left\{\left[-\frac{1}{2}-\frac{1}{2}t_1(\gamma)\right]P_2{}^1(\cos\chi)\cos(\phi_e) - \frac{3}{2}t_2(\gamma)P_1{}^1(\cos\chi)\sin(\phi_e)\right\}$$
$$+F_{22}(\theta_e)\left\{\left[-\frac{1}{2}P_2{}^2(\cos\chi) + t_1(\gamma)\left(2+P_2^0(\cos\chi)\right)\right]\cos(2\phi_e) + 3t_2(\gamma)P_1^0(\cos\chi)\sin(2\phi_e)\right\}$$
$$-s_3F_{11}(\theta_e)P_1{}^1(\cos\chi)\sin(\phi_e)$$

$$\text{with } t_1(\gamma) = s_1\cos(2\gamma) - s_2\sin(2\gamma)$$
$$t_2(\gamma) = s_1\sin(2\gamma) + s_2\cos(2\gamma) \quad (2)$$

This expression is a generalization of the one describing the MFPAD for circularly polarized light,[46] where $t_1(\gamma)$ and $t_2(\gamma)$ average to zero and $s_3 = -1$ for LHC polarization.

Considering eqn (2) several conclusions are drawn, which will be used in the following sections:

(i) For any (unknown) polarization of the ionizing light, the four $F_{00}(\theta_e)$, $F_{20}(\theta_e)$, $F_{21}(\theta_e)$, $F_{22}(\theta_e)$ functions can be determined from the Fourier analysis of the $I(\theta_e,\phi_e,\chi)$ measured distribution, ignoring the (γ) dependence, therefore providing the MFPADs for any orientation (χ) of the molecular axis relative to the axis of linearly polarized light.[50] Their (l,λ) partial wave expansion in a Legendre polynomial basis, where l is the angular momentum of the electron in a one center description of the scattering process and the quantum number λ is its projection on the molecular axis, gives access to the complex transition dipole moments describing the parallel and perpendicular PI transitions, and the amplitude of their relative phases. Knowing the sign of these relative phases requires the determination of the $F_{11}(\theta_e)$ function, which appears as the mixed product $s_3 \times F_{11}(\theta_e)$ in the MFPAD expression (eqn (2)), as discussed below. Access to these complex transition dipole moments is of high value for the advanced study of the photoemission dynamics.

(ii) The s_1 and s_2 Stokes parameters describing the linear component of the polarization are extracted by fitting the $I(\chi,\gamma)$ angular distribution of the ion fragments after integration of the $I(\theta_e,\phi_e,\chi,\gamma)$ angular distribution over (θ_e) and (ϕ_e):

$$I(\chi,\gamma) = C\left\{P_0^0(\cos\chi)\left[1+\frac{\beta}{2}t_1(\gamma)\right] - \frac{\beta}{2}P_2^0(\cos\chi)[1+t_1(\gamma)]\right\} \quad (3)$$

The asymmetry parameter of the ion fragment distribution β is obtained from the (χ) dependence after integration over (γ), while the Fourier analysis in (γ) can be performed in the projections on $P_0^0(\cos\chi)$ and $P_2^0(\cos\chi)$, providing two ways for extracting s_1 and s_2, as illustrated in Fig. 1(b)–(e):

$$\begin{aligned}\mathrm{Proj}_{00}(\gamma) &= C\left\{1+\frac{\beta}{2}s_1\cos(2\gamma) - \frac{\beta}{2}s_2\sin(2\gamma)\right\} \text{ and}\\ \mathrm{Proj}_{20}(\gamma) &= C'\{1+s_1\cos(2\gamma) - s_2\sin(2\gamma)\}\end{aligned} \quad (4)$$

We note that a similar form describes the angular distribution of the photoelectrons in the FF, involving the asymmetry parameter of the electron distribution β_e.

(iii) As for the $F_{11}(\theta_e)$ function which characterizes the circular dichroism in the MF, and the s_3 Stokes parameter which features the helicity of the ionizing radiation, they appear as the product $s_3 \times F_{11}(\theta_e)$ in the expression of the MFPAD $I(\theta_e,\phi_e,\chi,\gamma)$. This implies that determination of $F_{11}(\theta_e)$ requires an independent measurement with a known s_3 helicity, or a calculation. *Vice versa*, if $F_{11}(\theta_e)$ is known independently, the s_3 Stokes parameter is determined from the MFPAD. It is worth noticing that the MF circular dichroism constitutes the dephasing element of molecular polarimetry. It is also featured by the dimensionless CDAD (circular dichroism in electron angular distribution) parameter, proportional to $F_{11}(\theta_e)$, which characterizes the MF left–right emission asymmetry in the polarization plane ($\phi_e = 90°$ or $270°$) when the molecular axis is perpendicular to the light propagation axis ($\chi = 90°$).[51]

The MP method has been benchmarked using the prototype NO DPI reaction corresponding to ionization of the 4σ inner-valence molecular orbital of the NO(X,$^2\Pi$) molecule (eqn (5)), previously studied,[11,45,47] which possesses fingerprint properties to act as an efficient "polarimeter", and is very well adapted to measure the polarization state of HHs composing an APT, as discussed in Section 4.

$$NO(^2\Pi,4\sigma^2 5\sigma^2 1\pi^4 2\pi^1) + h\nu \rightarrow NO^+(c^3\Pi,(4\sigma)^{-1}) + e \rightarrow N^+(^3P) + O(^3P) + e \quad (5)$$

Reaction (5) is the dominant DPI process for the studied XUV photon energies due to a strong shape resonance centered around 30 eV. Dissociation of the $NO^+(c^3\Pi)$ state is prompt relative to the rotational period and satisfies the conditions of axial recoil as validated by the detailed comparison of measured and computed angular anisotropies.[45,47] Both the asymmetry parameter β_{N^+} characterizing the ion fragment emission anisotropy and the MF circular dichroism parameter take significant values ($\beta_{N^+} \approx 1$ and CDAD ranging between 0.5 and 1), which correspond to favorable conditions for the extraction of the s_1, s_2, s_3 Stokes parameters.

Selecting as an example a polarization state, labelled S, similar to those reported in ref. 11 for a photon energy $h\nu = 23.65$ eV, Fig. 1 displays the key features for the extraction of s_1 and s_2, while Fig. 2 illustrates the MF circular dichroism and summarizes the extraction of s_3 leading to the full polarization ellipse.

Beside the three Stokes parameters, the polarization state can also be characterized by the parameters of the polarization ellipse which describes the polarized component of the light, *i.e.*, the orientation (ψ) and ellipticity (ε), and the degree of polarization P.[1] The (ψ, ε, P) quantities are related to the normalized s_1, s_2, s_3 Stokes parameters as follows:

$$\tan 2\psi = \frac{s_2}{s_1}, \quad \varepsilon = \tan\chi \text{ with } \sin 2\chi = \frac{s_3}{\sqrt{{s_1}^2 + {s_2}^2 + {s_3}^2}} \text{ and } P = \sqrt{{s_1}^2 + {s_2}^2 + {s_3}^2}$$

$s_4 = 1 - \sqrt{s_1^2 + s_2^2 + s_3^2} = 1 - P$ represents the degree of unpolarized light.

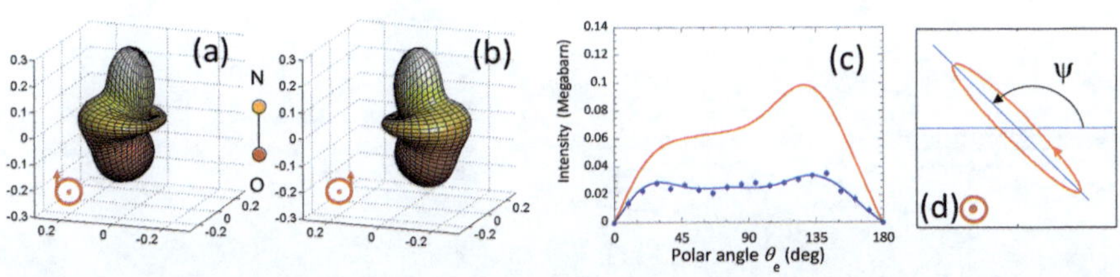

Fig. 2 Illustration of the MF circular dichroism at $h\nu = 23.65$ eV measured on the DESIRS beamline using circularly polarized light:[45] 3D $I_{90°}(\theta_e,\phi_e)$ MFPADs for $\chi = 90°$ induced by (a) RHC ($s_3 = +1$, helicity −1) and (b) LHC ($s_3 = -1$, helicity +1) polarized light. (c) Extraction of the s_3 Stokes parameter for the S polarization state: the measured $-s_3 \times F_{11}(\theta_e)$ function is shown (blue: dots and Legendre polynomial line fit); the $s_3 = -0.35 \pm 0.01$ value is obtained here as the ratio between the blue curve and the reference F_{11} function (red), corresponding to an ellipticity $\varepsilon = -0.2 \pm 0.01$. (d) Full polarization ellipse for the S state, resulting from the measured s_1, s_2, s_3 parameters.

For later comparison with OP, we note that incomplete optical measurements relying on the Malus' law such as those described in Section 3 lead to the determination of the two Stokes parameters (s_1, s_2), equivalent to the ellipse orientation (ψ) and the amplitude of the linear component of the polarization $\sqrt{s_1^2 + s_2^2}$, but do not provide information on the circular and unpolarized components. One obtains only an "upper bound ellipticity" (ε_{ub}) defined as:

$$\varepsilon_{\text{ub}} = \tan \chi_{\text{ub}} \text{ with } \sin 2\chi_{\text{ub}} = s_{3-\text{ub}},\ \ s_{3-\text{ub}} = \sqrt{1 - {s_1}^2 - {s_2}^2} \text{ and } \varepsilon_{\text{ub}} \geq 0$$

ε_{ub} coincides with the absolute value of the ellipticity only for fully polarized light $P = 1$, $\varepsilon_{\text{ub}} = |\varepsilon|$.

As for the HHG emission, due to the absence of practical XUV dephasing elements, only incomplete optical polarimetry measurements have been performed up to now, assuming a totally polarized light ($P = 1$), thus leading to a possible over estimate of the ellipticity. We point out that the Stokes parameters, as well as the parameters of the polarization ellipse and degree of polarization, are quantities averaged in time and space: they provide an effective description of a temporally and spatially variable field – such as the harmonic field – as the sum of steadily polarized and unpolarized parts.

3. Combining 3D-momentum spectrometry with APT generation: experimental methodology

We now turn to the combination of the 3D-momentum spectrometry with APT generation. The experiments have been conducted on the PLFA beamline at the SLIC facility of CEA Saclay.[52] A schematic of the set-up is shown in Fig. 3. The IR laser delivers pulses at 800 nm with up to 8 mJ energy, 50 fs pulse duration at a 1 kHz repetition rate. For the SF_6 experiment, linearly polarized pulses of about 1 mJ are made elliptically polarized through a quarter wave plate whose rotation α determines the signed fundamental ellipticity $\varepsilon_{\text{fun}} = \tan(\alpha)$ as well as the direction of the main axis of the ellipse. The IR beam is focused with a 85 cm lens into an effusive gas jet of SF_6 molecules under vacuum. It is then filtered out of the generated harmonics with a 200 nm aluminum foil (not shown). The harmonics

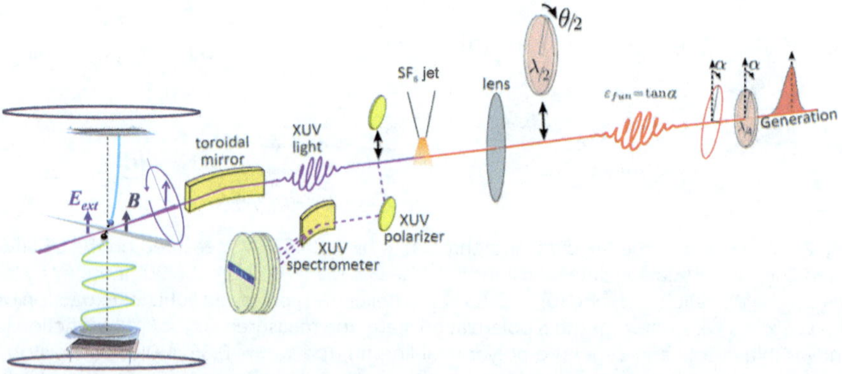

Fig. 3 Schematic of the experimental set-up combining the HHG PLFA beamline and the electron-ion 3D-momentum spectrometer.

are then refocused in the ion-electron coincidence 3D momentum spectrometer CIEL[52] using a 11.5° grazing incidence angle, 22 × 6 cm size 60 cm focal length toroidal gold mirror. Special care has been taken to characterize the transmission of the polarized harmonics by the toroidal mirror: its action has been calibrated using the MP method, providing s_1, s_2, s_3 Stokes parameters for a series of known linear polarizations of the incident harmonics, that is probing the complex reflectivity for the *s* and *p* components for the four studied harmonic energies using Mueller matrix formalism, as briefly discussed in the data analysis of the measured Stokes parameters.

A motorized gold mirror can be inserted upstream from the toroidal mirror. It intersects the light beam at 45° and directs the beam toward a second gold 45°-incidence mirror and an XUV spectrometer (composed of variable groove spacing grating, micro-channel plates and phosphor screen detector). This allows direct HHG monitoring in order to optimize efficiency and beam steering, but also to perform optical polarimetry, *i.e.*, Malus' law type polarization analysis using the two 45° reflections as an analyzer. Indeed, the difference of reflectivity between the *s* and *p* polarization components amounts to a factor 20 in the 15th to 25th harmonic spectral range (23.25–38.75 eV). Instead of turning the analyzer, we use the property that HHG is a field-driven process, the polarization of which in an isotropic target gas is determined by the driving laser. We thus rotate continuously the laser polarization ellipse by an angle θ using a half-wave-plate inserted before the focusing lens, and record the HHG yield $I_q(\theta)$ for the different harmonic orders q on the spectrometer. For a perfect analyzer, this would give the following dependence, also called Malus' law:

$$I_q(\theta) = \frac{1}{2}[S_0 + S_1\cos(2\theta) + S_2\sin(2\theta)] \tag{6}$$

The fit of the HHG yield variation with a Malus' law taking into account the imperfect analyzer provides the values of the two normalized Stokes parameters: $s_1 = S_1/S_0$ and $s_2 = S_2/S_0$, from which one obtains the direction of the harmonic ellipse (ψ), and an upper bound for the magnitude of the ellipticity ε_{ub} (see Section 2). Indeed, this incomplete OP cannot disentangle the circular part of the polarization (measured by s_3) from the unpolarized part. Performing a complete OP would require measuring in addition the variation of the HHG yield $I_q(\theta)$ in the presence of a dephasing element, such as an XUV quarter-wave plate which is not yet available, or using the dephasing induced by reflection on a metallic mirror, that is accompanied with a strong signal attenuation. In most of the published work except for a few studies[53,54] the harmonic emission was considered to be fully polarized, so it is assumed that $\varepsilon_{ub} = |\varepsilon|$.

The XUV light beam directed into the CIEL 3D momentum spectrometer induces PI of the gas target, here mainly NO molecular and He atomic targets, produced by a two skimmer supersonic expansion. The supersonic jet and ultrahigh vacuum chambers originate from a previous version of the COLTRIMS type apparatus[55,56] which combines electric and magnetic fields to guide ions and electrons. The advanced version of the spectrometer based on the two delay line time and position sensitive detectors (DLD PSDs RoentDek), including an electrostatic focusing lens for the ion trajectories,[48] ensures a 4π collection of both particles for the studied DPI processes. An eight-channel time-to-digital converter is used to encode the DLD time signals, providing the position for each particle of the (A^+, e) coincident events as well as the ion time of flight (TOF), while the electron TOF is encoded using a synchronized time to amplitude converter.

One issue for the present measurements was the 1 kHz repetition rate of the laser, which restricts to about 50 c/s the overall coincidence count rate. This imposed a few hours of stability for each measurement on the NO target. The subsequent statistics for these measurements is discussed in the following sections. This situation has motivated complementary measurements relying on PI of the He target which are not discussed here.

4. MFPADs for PI induced by an attosecond pulse train (APT)

In this section we report the first $I(\theta_e,\phi_e,\chi)$ complete MFPADs subsequent to the interaction of an APT with a molecular target, based on the (V_{A^+}, V_e) vector correlation analysis of dissociative photoionization processes. Building on the experience in femtochemistry,[57] two strategies have prevailed so far in the few studies of MFPADs in PI of simple molecular targets by attosecond pulses. The first one relies on the impulsive alignment of small molecules such as O_2, N_2, CO, CO_2 using a near-IR laser and taking advantage of the periodical revival of the rotational wave packet to ionize the molecules at the maximum of field-free alignment or anti-alignment with a properly delayed APT.[40] The energy and $I(\theta_e)$ angular distributions of the photoelectrons were obtained after processing an Abel inversion of the images recorded by Velocity Map Imaging (VMI) and could be assigned to the parallel and perpendicular transitions for the different ionic channels and photon energies in the harmonic comb. Secondly, despite HHG sources usually operate at a rather low repetition rate (1–10 kHz), they allow 3D momentum electron–ion coincidence spectroscopy: first complete measurements of $I(\theta_e,\phi_e,\chi)$ MFPADs were demonstrated in dissociative photoionization of H_2 and D_2 induced by spectrally filtered single high harmonic, following resonant excitation of doubly excited auto-ionizing states,[41] and traced as a function of the kinetic energy release of the atomic fragments. This prototypical reaction features a dynamical interplay between the electronic and nuclear motions at the 1–5 femtosecond time scale, giving rise to quantum interferences between distinct reaction pathways and remarkable symmetry breakings, and it stands as a model process for state-of-the-art calculations.[58,59] A related study, based on a similar electron-ion coincidence technique and using a broad band APT, focused on the dependence of photoelectron emission asymmetries in the MF which also illustrates nicely those symmetry breakings.[42]

The present study is illustrated by DPI of NO molecules as motivated earlier, and more specifically on the prototype reaction (eqn (5)). We use for the ionizing radiation the APT generated by SF_6 molecules driven by elliptically polarized IR laser, the polarization state of which will be characterized using the MP technique in Section 5. Inner-valence ionization of molecules usually involves several DPI processes corresponding to PI into distinct molecular ionic states and dissociation channels, each "reaction pathway" being assigned by the intermediate ionic state and the populated dissociation channel. The first step in the analysis of the (A^+, e) DPI coincident events consists of disentangling the reaction pathways, based on the resolving power of the electron-ion kinetic energy correlation (KEC) featured in KEC diagrams (KECDs),[47,60] and relying on the total energy conservation ($h\nu - E_D = E_e$ + KER), where E_D is the asymptotic potential energy at the dissociation limit, E_e and KER are the photoelectron energy and the kinetic energy release of the two heavy atomic/molecular fragments, respectively. Typical KECDs are displayed in Fig. 4(a) and (b) for one-photon PI corresponding to DPI of the NO molecule induced by (a) a one-photon-energy pulse at SOLEIL, here $h\nu = 23.65$ eV as discussed above, and (b) an APT produced by HHG from the SF_6 gas. In Fig. 4(a) three DPI processes are identified, which correspond (I) to PI into the first $NO^+(c^3\Pi,(4\sigma)^{-1})$ ionic state followed by dissociation into the $N^+(^3P) + O(^3P)$ + e limit and has the largest intensity, and (II) and (III) to PI into the $NO^+(B^1\Pi,(4\sigma)^{-1})$ and $NO^+(B'^1\Sigma^+,(1\pi)^{-1})$ ionic states dissociating into the $N^+(^3P) + O(^3P)$ + e limit (II) and the two close lying ones $N^+(^1D) + O(^3P)$ + e and/or $N^+(^3P) + O(^1D)$ + e (III).[47]

The KECD presented in Fig. 4(b) for DPI induced by the APT shows well resolved peaks corresponding to the same $E_{N^+} \approx 0.4$ eV ion fragment energy and different E_e electron energies, separated by 3.1 eV, *i.e.*, the energy difference between two successive odd harmonics in the APT. These structures are assigned to the dominant DPI process (eqn (5)) produced by the harmonics comb, with energies larger than the ionization potential of NO into the $NO^+(c^3\Pi,\nu = 0)$ ionic state $I_P = 21.7$ eV, namely from harmonic H15 up to H25 in the present experiment. Such a KECD reflects the superposition of as many one-photon-energy KECDs as there are contributing HHs in the APT. To ensure a valid interpretation of the KECD assigned to the APT, as well as to provide reference data for the MF circular dichroism at the H15–H23 photon energies, we have performed a series of complementary experiments on the DESIRS beamline at SOLEIL synchrotron using circularly polarized light at the same energies.[61,62] These results establish that there is no significant overlap between the resolved structures assigned to DPI into the $NO^+(c^3\Pi)$ ionic state, and other DPI processes that might result in comparable (E_{N^+}, E_e) positions in the KECD, so a proper selection of the processes enable us to proceed to the MFPAD analysis for each harmonic. The reason why weaker processes (II) and (III) resolved in Fig. 4(a), or the other DPI processes induced at the higher photon energies, do not provide any significant contribution in Fig. 4(b) is, first, the intrinsically weak probability of process (III) and,

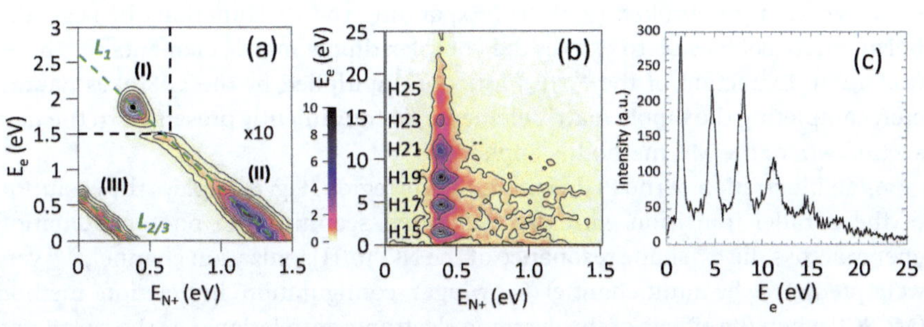

Fig. 4 KECDs characterizing DPI of NO induced by (a) synchrotron radiation at $h\nu = 23.65$ eV (b) an APT generated on the SF_6 gas medium; (c) photoelectron spectrum reflecting the HHG spectrum (H15 to H25) convoluted with the PI cross section for eqn (5) (4π collection of electrons and ions) extracted from the KECD shown in (b) after the 0.25 eV $\leq E_{N^+} \leq$ 0.55 eV ion-fragment energy selection.

second, the significant discrimination of processes leading to more energetic N^+ fragments at the extraction field chosen for the HHG experiment to ensure a 4π collection of process (I) (eqn (5)). Furthermore, scrutiny of the one photon energy KECDs ensures that a unique selection of reaction (eqn (5)) induced by the APT is achieved when the $0.25\ \text{eV} \leq E_{N^+} \leq 0.55\ \text{eV}$ ion fragment selection is performed. As a result, in the photoelectron spectrum displayed in Fig. 4(c) corresponding to this selection, the resolved peaks reflect PI of NO into the $NO^+(c^3\Pi)$ ionic state for each of the HHs composing the APT: it features directly the HHG spectrum for harmonics H15 and higher, convoluted by the total photoionization cross section corresponding to eqn (5). Experimental and/or theoretical PI cross sections can be used as reference to restore the original HHG spectrum.[63,64]

Fig. 5 displays $I(\theta_e,\phi_e,\chi)$ complete MFPADs, subsequent to the interaction of an APT with the NO molecular target, derived from the measured $I(\theta_e,\phi_e,\chi,\gamma)$ angular distribution for selected processes in the KECD, after integration over the (γ) angle. Here the APT was generated in SF_6 gas by an elliptically polarized IR laser $\varepsilon_{fun} = 0.2$, the elliptical polarization of the HHs being unknown at this level. Selecting as an example the peak assigned to PI by the H17 harmonic, at photon energy $h\nu = 26.35$ eV, the complete MFPAD is determined by the extraction of the five $F_{LN}(\theta_e)$ functions implemented in (eqn (2)). The four $F_{00}(\theta_e)$, $F_{20}(\theta_e)$, $F_{21}(\theta_e)$, $F_{22}(\theta_e)$ are displayed in Fig. 5(a) providing the $I(\theta_e,\phi_e)$ MFPAD for any orientation of the molecular axis relative to the axis of linearly polarized light, shown in Fig. 5(b) for three meaningful orientations $\chi = 0°$, $90°$ featuring the parallel and perpendicular transitions, and $\chi = 45°$, where a coherent superposition between parallel and perpendicular orientations occurs. In Fig. 5(a) the $F_{LN}(\theta_e)$ functions are compared with those measured at SOLEIL at photon energy $h\nu = 26.35$ eV. Despite the rather low statistics in the HHG measurements at the 1 kHz laser rep-rate, reflected by the statistical error bars (*e.g.* here 3000 events in the selection), the comparison shows that the main characteristics of the MFPADs are fairly well determined. This is a striking demonstration of the potential of the F_{LN}-based data analysis methodology, since all events collected for a given process are used to generate the MFPAD for each selected orientation of the molecule. The 3D plots of the $I(\theta_e,\phi_e)$ MFPADs for the selected orientations illustrate remarkably well the partial wave expansion different for the parallel and perpendicular transitions resulting from the linear combinations of $F_{00}(\theta_e)$ and $F_{20}(\theta_e)$, and including the azimuthal dependence described by the $F_{21}(\theta_e)$ and $F_{22}(\theta_e)$ as soon as the cylindrical symmetry is broken ($\chi \neq 0°$). Expanding the F_{LN} functions in Legendre polynomials enables us to extract the complex dipole matrix elements for the PI transition. Extraction of the $F_{11}(\theta_e)$ function multiplied by the s_3 Stokes parameter, completing the dipole matrix elements, is conveniently presented in the next section where the MP method is applied.

As an illustration of the PI dynamics in this region, Fig. 6 displays the evolution of the parallel transition with the HH order, scanning the photon excitation energy across the σ^* shape resonance of the $NO^+(c^3\Pi)$ ionization channel,[45,64] very well predicted by multichannel Schwinger configuration interaction method (MCSCI) when the effects of the dynamic electronic correlation, *i.e.*, the correlated motion of the photoelectron and electrons of the target are included.[45,50] The dominance of the *f* partial wave and the π increase of the phases of the matrix elements can therefore be discussed in this context where significant inter-channel coupling occurs. At the maximum of the shape resonance, around 30 eV,

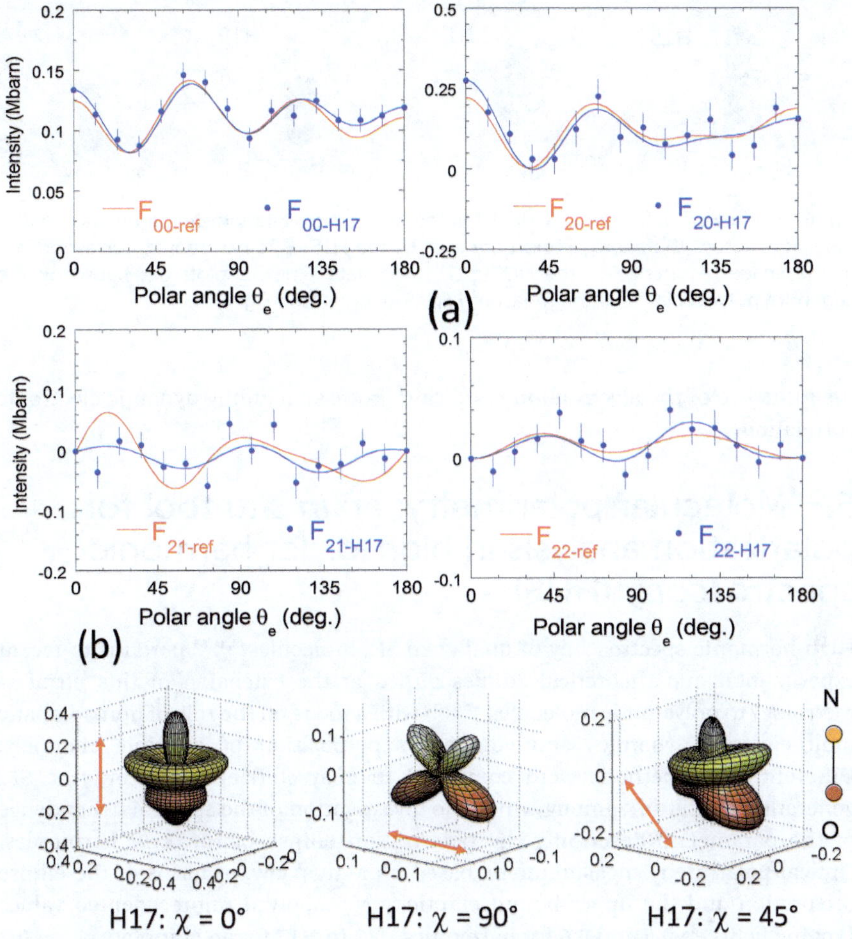

Fig. 5 (a) Measured $F_{00}(\theta_e)$, $F_{20}(\theta_e)$, $F_{21}(\theta_e)$, $F_{22}(\theta_e)$ functions for DPI induced by the harmonic H17 of the APT (blue dots, and Legendre polynomial line fit), compared with the $F_{LN}(\theta_e)$ measured at SOLEIL at $h\nu = 26.35$ eV (red line); both measurements are normalized to the corresponding MCSCI computed cross section for reaction eqn (5), such that the total cross sections $\int_0^{\pi} F_{00}(\theta_e)\sin(\theta_e)d\theta_e$ are identical; (b) $I(\theta_e,\phi_e)$ MFPADs for three meaningful orientations $\chi = 0°, 90°$ featuring the parallel and perpendicular transitions, and $\chi = 45°$ involving their coherent superposition. The emission diagram in the MF results from the interference of the $l\lambda$ partial waves building up the electronic wave function in the continuum for the parallel and perpendicular transitions, described by the complex dipole matrix elements. The 3D plots are based on the Legendre polynomial fit of the F_{LN} functions.

i.e. H19, it leads to a strong electron emission anisotropy, favoring electron ejection along the molecular axis in the direction of the O end of the NO molecule.

These results open perspectives for time-resolved MFPAD studies at the attosecond time scale, with the goal to probe the evolution of *e.g.*, the electronic density in a transient excited molecular state through photoionization. At the same time, the fast and striking evolution of the MFPAD along the shape resonance shown in Fig. 6 illustrates the important role of the PI dynamics, *i.e.*, here

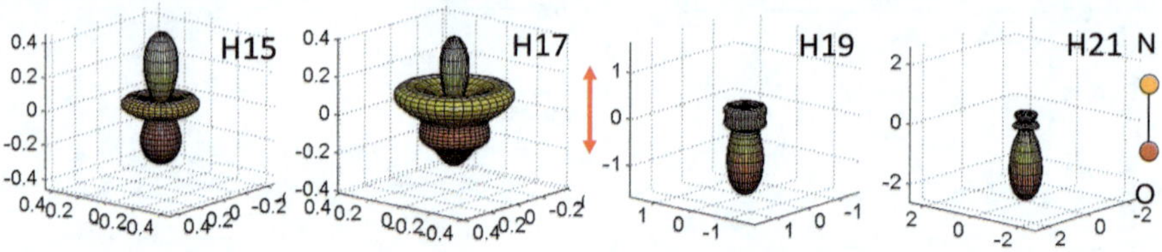

Fig. 6 Evolution of the $I(\theta_e)$ MFPAD for the $\chi = 0°$ orientation, featuring the parallel transition, across the energy region scanned by the H15–H21 harmonics, corresponding to the shape resonance of the $NO^+(c^3\Pi)$ ionic state. The 3D plots are based on the Legendre polynomial fit of the measured F_{LN} functions (see Fig. 5).

the influence of the electron-ion scattering process including dynamic electronic correlation.

5. Molecular polarimetry: an *in situ* tool for polarization analysis in high-order harmonic spectroscopy (HHS)

High-harmonic spectroscopy of unaligned SF_6 molecules[38,44,65] pertains to recent experimental and theoretical studies aimed at the extension of this ultrafast metrology to polyatomic molecules,[32,66–70] with a focus on the role of non-adiabatic multi-electron dynamics and coherent superposition of multiple channels. Different HHS methods were combined to characterize HHG from the SF_6 generation medium,[44] among which the investigation of the spectral dependence of the parameters describing the polarization ellipse of the XUV harmonics. Optical polarimetry measurements based on Malus' law, in terms of the ellipse orientation and the upper bound ellipticity ε_{ub}, showed unprecedented values ranging between 0.8 and 0.6 for harmonics H13 to H17 for an ellipticity $\varepsilon_{fun} = 0.2$ of the 800 nm driving laser.[38] The produced elliptically polarized HHG source was subsequently used to measure photoelectron circular dichroism (PECD)[71] on chiral molecules,[38] *i.e.*, to characterize the electron emission forward/backward asymmetry along the propagation direction of the light. Such measurements should provide the s_3 Stokes parameter, if the PECD is calibrated independently and if the contribution of each HH can be resolved in the photoelectron spectra.

We applied the MP method to the analysis of the APT generated in SF_6 by an elliptically polarized driving IR laser, with the goal of probing the complete polarization state of each HH composing the APT, expressed in terms of the s_1, s_2, s_3 Stokes parameters, or the (ψ, ε, P) quantities characterizing the polarization ellipse and the degree of polarization P. The analysis of the measured $I(\chi,\gamma)$ histograms and the $s_3 \times F_{11}(\theta_e)$ functions for the dominant DPI process (eqn(5)) induced by the H15–H21 major harmonics proceeds along the lines described in Section 2.

For the extraction of the s_3 Stokes parameters, we rely on the reference $F_{11}(\theta_e)$ functions which were measured on the DESIRS beamline at SOLEIL at the same photon energies,[62] and are well predicted by MCSCI calculations:[45] for each harmonic, the s_3 value is obtained as the ratio between the $s_3 \times F_{11}(\theta_e)$ curve and the reference $F_{11}(\theta_e)$ function at the corresponding photon energy. s_1 and s_2 are derived from the analysis of the $I(\chi,\gamma)$ histograms for each harmonic according to the method illustrated in Fig. 1.

However one additional step has to be included in the data analysis, which accounts for the modification of the polarization state of the HHGs which is likely to be introduced by optical elements, in particular the toroidal gold mirror which focuses the XUV pulses at the center of the COLTRIMS spectrometer. As mentioned in the Experimental section, the action of this mirror has been characterized using the MP method, providing s_1, s_2, s_3 Stokes parameters for a series of known orientations of the incident linearly polarized XUV field (*i.e.*, that of the IR laser), and modeled with a designed Mueller matrix which includes transmission parameters for the four relevant harmonics. The other optical element crossed by the XUV pulses is the metallic Al filter which has a flat response in the region of interest and should not influence the polarization state. In the data reported in the following, the Mueller formalism was used to determine the complete polarization state for each harmonic in terms of the normalized Stokes vector upstream from the mirror, based on that directly measured after transmission by the mirror.

Fig. 7 illustrates the extraction of the s_3 Stokes parameters for the four harmonics downstream from the mirror: the MFPADs display significant right–left asymmetries which demonstrate that large positive s_3 values are obtained, in particular for H15 and H17.

The large values determined for the s_3 Stokes parameter, in particular $s_3 \approx 0.8 \pm 0.05$ for H15 and H17, demonstrate unambiguously the large degree of polarization and the quasi-circular character of the polarization state for these harmonics, corresponding to a positive ellipticity $\varepsilon \approx 0.70 \pm 0.06$. For H19 and H21 the sign remains positive while the magnitude of the ellipticity decreases.

In Fig. 8 we present the results obtained with the MP method for two ε_{fun} values of opposite sign close to $\varepsilon_{fun} \approx \pm 0.2$ (which we would rather attribute to

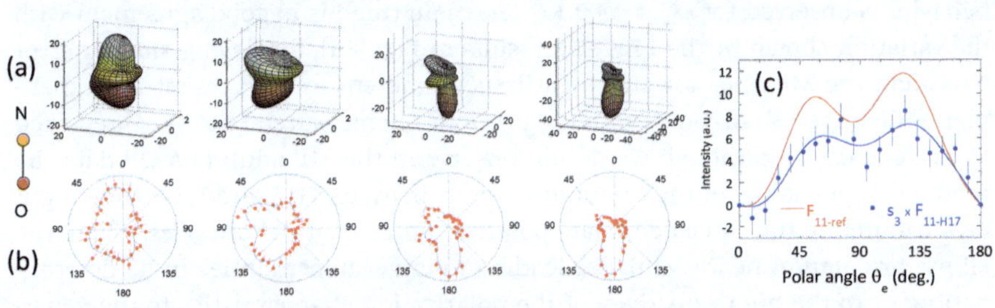

Fig. 7 (a) $I(\theta_e,\phi_e)$ MFPADs for the ($\chi = 90°$) orientation of the molecular axis relative to the propagation axis of the APT generated by $\varepsilon_{fun} \approx 0.23$ of the driving IR laser, for the H15 to H21 harmonics: the significant right-left emission anisotropies reflect the HH dependent s_3 Stokes parameter and F_{11} function (b) cuts of the MFPADs in the polarization plane ($\phi_e = 90°$ or $270°$) displaying the largest CDAD anisotropy (c) $s_3 \times F_{11}(\theta_e)$ for H17 (blue dots and Legendre polynomial fit line) compared to the reference F_{11} function at $h\nu = 26.35$ eV (red line): the $s_3 = 0.76 \pm 0.05$ value is obtained as the ratio between the blue curve and the reference F_{11} function (red); the measured values downstream from the mirror are shown at the bottom of the figure, with ± 0.05 error bars, while the values upstream from the mirror amount to $s_3 = 0.78$, 0.77, 0.51, and 0.22 (± 0.05), respectively, using the Mueller formalism (see text).

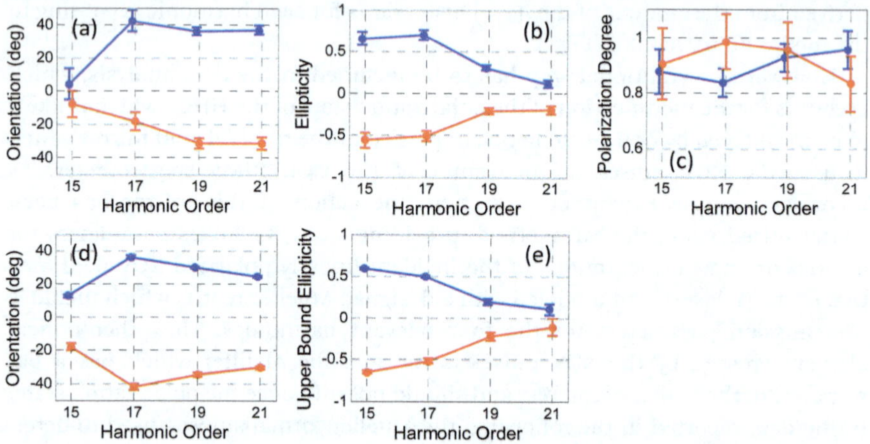

Fig. 8 First row: MP (a) orientation of the polarization ellipse (ψ) relative to the reference direction of the driving IR laser (b) measured ellipticity (ε) (c) polarization degree, for $\varepsilon_{fun} \approx$ 0.23 (blue dots) and $\varepsilon_{fun} \approx -0.17$ (red dots). Second row: OP (d) orientation of the polarization ellipse (ψ) relative to that of the driving IR laser (e) ε_{ub} (affecting artificially a negative sign to facilitate the visualization of the result), for $\varepsilon_{fun} \approx 0.2$ (blue dots) and $\varepsilon_{fun} \approx -0.2$ (red dots).

$\varepsilon_{fun} \approx 0.23$ and $\varepsilon_{fun} \approx -0.17$ in the MP measurement), in terms of the polarization ellipse parameters, orientation (ψ) and ellipticity (ε), as well as the degree of polarization P. Fig. 8 also includes the results of optical polarimetry based on Malus' law obtained at SLIC-PLFA in terms of the orientation (ψ) and the upper bound value ε_{ub}.

The orientation displayed in Fig. 8(a) as a function of the harmonic order, derived from the values of s_1 and s_2, strongly increases from about 5° to 40° from H15 to H17 for $\varepsilon_{fun} \approx 0.23$, and then remains rather stable. A quasi-symmetric behavior is observed for $\varepsilon_{fun} \approx -0.17$. The main trend is in good agreement with the variation shown by the PLFA-OP results in Fig. 8(d), displaying smaller error bars than the MP ones for statistical issues inherent to the present MP experiment. It compares fairly well with the previous OP measurements of (ψ) reported by Ferré *et al.*,[38,44] although we do not observe in the MP and PLFA-OP data the reported sign change of the orientation angle between H15 and H17: this might be attributed to the "quasi-circular" polarization at such HH energies, where the ellipse orientation nearly vanishes leading to larger uncertainties in its determination, or to the high sensitivity of the polarization characteristics to the generation conditions, *e.g.*, a different detuning from resonances in this energy region. As noted above, measuring s_3, in addition to s_1 and s_2, provides an unambiguous determination of the ellipticity of each HH, including its sign, ranging here from $\varepsilon \approx 0.70 \pm 0.06$ for H15 and H17 down to $\varepsilon \approx 0.31 \pm 0.03$ and $\varepsilon \approx 0.12 \pm 0.04$ for H19 and H21, for $\varepsilon_{fun} \approx 0.23$. This demonstrates that the large ε_{ub} found for H15 and H17 in OP reported by Ferré *et al.*,[38,44] and presently measured at SLIC-PLFA, is indeed due to the high degree of ellipticity of the polarization and not to a significant depolarization for these HHs.

The polarization degree shows, for the case $\varepsilon_{fun} \approx 0.23$, some deviation from the maximum value of 1 for the resonant harmonics 15 and 17, although with large error bars. The case $\varepsilon_{fun} \approx -0.17$ does not show this trend, which could possibly be due to an imprecision in the determination of s_1 and s_2, in particular for H17. A higher repetition rate in future experiments will allow increasing the statistics and thus reducing these error bars in order to reach a final conclusion on the presence of depolarization in this case.

These polarization characteristics provide insight into the complex interactions which occur during the HHG process in the SF_6 molecule. It was proposed[44] that the first step, *i.e.* strong field ionization, involves contributions from multiple electronic channels, while the recombination step is influenced by resonant transitions due to the presence of autoionizing states and shape resonances trapping the recolliding electron for typically hundreds of attoseconds before recombination to the ground state occurs. The specific behavior observed for H15 and H17 was related to the presence of the $5t_{1u} \rightarrow \varepsilon t_{2g}$ shape resonance in the A channel that dominates the harmonic emission in this spectral range, as shown by complex Kohn computations of valence PI total cross sections involving strong intercoupling effects.[72] The influence of resonances was then investigated using 2D-TDSE simulations for Ar atoms (see ESI of ref. 38), and it was found that they result in a structuring of the returning electron wavepacket and an increase of the perpendicular component of the dipole in the radiative recombination step of the HHG process. The high ε_{ub} values (assimilated to $\langle\varepsilon\rangle$) measured in SF_6 were then attributed to such an effect induced by the $5t_{1u} \rightarrow \varepsilon t_{2g}$ shape resonance. The reported MP measurements prove that the harmonic emission is indeed highly elliptical, which strengthens the proposed interpretation in terms of resonance. Furthermore, some degree of depolarization is expected in such a case, because the polarization characteristics should be highly dependent on the detuning from the resonance, inducing spectral variations inside the harmonic bandwidth.[38] This calls for more precise MP measurements using a high repetition rate laser. With the direct determination of the signed ellipticity and of the degree of polarization, these results complement those obtained previously for SF_6 and demonstrate the potential of the MP method to access the complete state of polarization of the HHs composing an APT. The interpretation of the striking fingerprints of the ε_{fun} and harmonic order dependent polarization state of the HHs composing the APT generated on SF_6 molecules, in particular the large ellipticities and strong variation of the ellipse orientation in the H13–H17 range, calls for additional theoretical studies.

6. Conclusion

In the work presented we have reported MFPAD studies based on coincident electron-ion 3D momentum imaging in the context of ultrafast molecular dynamics investigated at the SLIC facility of CEA-Saclay, with two perspectives, namely (i) using APT generated in atoms/molecules as a source for PI studies performed at the level of MFPADs, and (ii) taking advantage of MFPAD-resolved PI of simple molecules to perform a polarization analysis of the HHG emission in molecules, a major challenge of high harmonic spectroscopy.

The ability to measure $I(\theta_e,\phi_e,\chi)$ complete MFPADs induced by an APT opens perspectives for two-color pump-probe time-resolved MFPAD studies at the attosecond time scale, with the goal to probe the evolution of *e.g.*, the electronic density in a transient excited molecular state through photoionization. Entangled with the intrinsic properties of the ionized molecular orbital, the PI dynamics at play during the electron-ion scattering process has a major impact on MFPAD. This points to the need for a thorough description of the PI processes, as aimed at by different *ab initio* theoretical approaches.

This direction calls for extensions of the F_{LN} based method for extraction of the MFPADs to a two (multi)-photon PI process (pump-probe) on the one hand, as well as to non-linear small polyatomic molecules, providing then recoil frame photoelectron angular distributions (RFPADs), on the other hand. This type of extension was discussed recently for *e.g.*, the description of DPI of the NO_2 molecule induced by one-photon[73] and multi-photon[74] absorption. The vibrational dynamics at conical intersections, electronic correlations and quantum interferences, or isomerization reactions, are *e.g.*, some of the key processes which can be advantageously studied at the detailed level of time-resolved RFPADs both experimentally and theoretically in small model molecular systems.[43]

In this work, the MP method has been applied to the complete determination of the polarization state of the harmonic comb in APTs generated in SF_6 molecules by an elliptically polarized driving field. We have investigated two other cases where symmetry breaking in the generation process results in the production of elliptically polarized harmonics, well identified in recent literature, namely HHG driven by two color counter rotating circularly polarized fields in Ar atoms, and HHG driven by linearly polarized light in aligned N_2 molecules [in preparation]. One outcome and issue of these first results is the evaluation of depolarization of the generated HHs, in particular in the study of HHG from aligned molecules. It raises interesting questions on the different possible origins of depolarization,[53,54] which will be addressed in future work.

Both above perspectives (i) and (ii) will be pursued, taking advantage of the current development of high repetition rate (>1 kHz) HHG sources for future experiments, such as the 10 kHz FAB10 ATTOLAB source, allowing a significant increase of the statistics.

Finally, the investigation of MFPADs or RFPADs induced by an APT contributes to the ongoing comparison between PI and HHS studies, *i.e.*, the complementary insights on photoinitiated ultrafast electronic and nuclear dynamics in molecules which can be extracted from the two approaches, respectively. In PI, the $I(\theta_e,\phi_e,\chi)$ MFPAD describing photoelectron emission at the microscopic level gives the most complete access to the partial wave resolved complex dipole matrix elements, at each photon energy. On the other hand, in a gas medium HHS observables build up coherently from the microscopic to the macroscopic level, under the condition of phase-matching between the nonlinear polarization and the harmonic field.

Gathering theoretical concepts relevant for HHG and PI, the quantitative rescattering theory (QRS) for linear aligned molecules[75,76] and unaligned polyatomic molecules[66] provides an illustrative theoretical framework for the description of the HHG process, where the photorecombination step is considered as time-reversed PI. Strengthening the fruitful combination of thorough MFPAD-PI and HHS studies, both in experiment and theory, should provide greater insight into the specificities of the strong-field driven dynamics in HHG beyond the QRS theory (see ref. 77 and references therein).

Acknowledgements

We are very grateful to B. Carré and Y. Mairesse for fruitful discussions as well as for a careful reading of the manuscript (BC), and to L. Nahon and G. Garcia for their collaboration on the studies performed on the DESIRS beamline at SOLEIL. The authors gratefully acknowledge F. Lepetit who very efficiently contributed to the PLFA laser operation at the Saclay Laser-Matter Interaction Center (CEA-SLIC), as well as E. Bouisset, J. Guigand, S. Lupone, N. Tournier (ISMO) for their efficient technical support for maintenance of the COLTRIMS set-up. They are also very grateful to P. Billaud, A. Camper, M. Géléoc, N. Lin, B. Manschwetus, S. Marggi Poullain, J. Rothhardt, N. Saquet, who contributed to an earlier stage of this experiment [in preparation]. This work was supported by the FP7 program Laserlab-Europe, Grant No. 284464, the French National Research Agency (ANR) through the "Laboratoire d'Excellence Physics Atoms Light Matter" (ANR10-LABX-0039-PALM), ATTOWAVE (ANR-09-BLAN-0031-01), "Triangle de la Physique" (ATTOCONTROL 2012-046T), and "Equipement d'excellence ATTOLAB" (ANR11-EQPX0005_ATTOLAB). Financial support from Laserlab-Europe (EU-H2020 654148) is also acknowledged. RRL acknowledges support of this work by the United States Department of Energy, Office of Science, Basic Energy Science, Geoscience, and Biological Divisions, under Award No. DE-SC0012198.

Notes and references

1 M. Born and E. Wolf, *Principles of Optics*, Pergamon, New York, 1980.
2 K. L. Reid, *Mol. Phys.*, 2012, **110**, 131–147.
3 Special issue on Molecular-frame Photoelectron Angular Distributions, *J. Phys. B: At., Mol. Opt. Phys.*, 2012, **45**, 19.
4 S. Motoki, J. Adachi, K. Ito, K. Ishii, K. Soejima, A. Yagishita, S. K. Semenov and N. A. Cherepkov, *Phys. Rev. Lett.*, 2002, **88**, 63003.
5 O. Geßner, Y. Hikosaka, B. Zimmermann, A. Hempelmann, R. R. Lucchese, J. H. D. Eland, P.-M. Guyon and U. Becker, *Phys. Rev. Lett.*, 2002, **88**, 193002.
6 M. Lebech, J. C. Houver, D. Dowek and R. R. Lucchese, *Phys. Rev. Lett.*, 2006, **96**, 73001.
7 A. Landers, T. Weber, I. Ali, A. Cassimi, M. Hattass, O. Jagutzki, A. Nauert, T. Osipov, A. Staudte, M. H. Prior, H. Schmidt-Böcking, C. L. Cocke and R. Dörner, *Phys. Rev. Lett.*, 2001, **87**, 13002.
8 T. Jahnke, T. Weber, A. L. Landers, A. Knapp, S. Schössler, J. Nickles, S. Kammer, O. Jagutzki, L. Schmidt, A. Czasch, T. Osipov, E. Arenholz, A. T. Young, R. Díez Muiño, D. Rolles, F. J. García de Abajo, C. S. Fadley, M. A. Van Hove, S. K. Semenov, N. A. Cherepkov, J. Rösch, M. H. Prior, H. Schmidt-Böcking, C. L. Cocke and R. Dörner, *Phys. Rev. Lett.*, 2002, **88**, 73002.
9 T. Osipov, T. N. Rescigno, T. Weber, S. Miyabe, T. Jahnke, A. S. Alnaser, M. P. Hertlein, O. Jagutzki, L. P. H. Schmidt, M. Schöffler, L. Foucar, S. Schössler, T. Havermeier, M. Odenweller, S. Voss, B. Feinberg, A. L. Landers, M. H. Prior, R. Dörner, C. L. Cocke and A. Belkacem, *J. Phys. B: At., Mol. Opt. Phys.*, 2008, **41**, 91001.

10 X.-J. Liu, H. Fukuzawa, T. Teranishi, A. De Fanis, M. Takahashi, H. Yoshida, A. Cassimi, A. Czasch, L. Schmidt, R. Dörner, K. Wang, B. Zimmermann, V. McKoy, I. Koyano, N. Saito and K. Ueda, *Phys. Rev. Lett.*, 2008, **101**, 83001.

11 K. Veyrinas, C. Elkharrat, S. Marggi Poullain, N. Saquet, D. Dowek, R. R. Lucchese, G. A. Garcia and L. Nahon, *Phys. Rev. A*, 2013, **88**, 063411.

12 L. Nahon and C. Alcaraz, *Appl. Opt.*, 2004, **43**, 1024–1037.

13 L. Nahon, N. de Oliveira, G. A. Garcia, J.-F. Gil, B. Pilette, O. Marcouillé, B. Lagarde and F. Polack, *J. Synchrotron Radiat.*, 2012, **19**, 508–520.

14 F. Schäfers, H.-C. Mertins, A. Gaupp, W. Gudat, M. Mertin, I. Packe, F. Schmolla, S. Di Fonzo, G. Soullié, W. Jark, R. Walker, X. Le Cann, R. Nyholm and M. Eriksson, *Appl. Opt.*, 1999, **38**, 4074–4088.

15 E. Allaria, B. Diviacco, C. Callegari, P. Finetti, B. Mahieu, J. Viefhaus, M. Zangrando, G. De Ninno, G. Lambert, E. Ferrari, J. Buck, M. Ilchen, B. Vodungbo, N. Mahne, C. Svetina, C. Spezzani, S. Di Mitri, G. Penco, M. Trovó, W. M. Fawley, P. R. Rebernik, D. Gauthier, C. Grazioli, M. Coreno, B. Ressel, A. Kivimäki, T. Mazza, L. Glaser, F. Scholz, J. Seltmann, P. Gessler, J. Grünert, A. De Fanis, M. Meyer, A. Knie, S. P. Moeller, L. Raimondi, F. Capotondi, E. Pedersoli, O. Plekan, M. B. Danailov, A. Demidovich, I. Nikolov, A. Abrami, J. Gautier, J. Lüning, P. Zeitoun and L. Giannessi, *Phys. Rev. X*, 2014, **4**, 41040.

16 T. Mazza, M. Ilchen, A. J. Rafipoor, C. Callegari, P. Finetti, O. Plekan, K. C. Prince, R. Richter, M. B. Danailov, A. Demidovich, G. De Ninno, C. Grazioli, R. Ivanov, N. Mahne, L. Raimondi, C. Svetina, L. Avaldi, P. Bolognesi, M. Coreno, P. O'Keeffe, M. Di Fraia, M. Devetta, Y. Ovcharenko, T. Möller, V. Lyamayev, F. Stienkemeier, S. Düsterer, K. Ueda, J. T. Costello, A. K. Kazansky, N. M. Kabachnik and M. Meyer, *Nat. Commun.*, 2014, **5**, 3648.

17 L. D. Barron, *Molecular Light Scattering and Optical Activity*, Cambridge University Press, Cambridge, England, 2004.

18 N. Berova, K. Nakanishi and R. Woody, *Circular Dichroism: Principles and Applications*, Wiley-VCH, New York, 2000.

19 F. Krausz and M. Ivanov, *Rev. Mod. Phys.*, 2009, **81**, 163–234.

20 Y. H. Jiang, A. Rudenko, O. Herrwerth, L. Foucar, M. Kurka, K. U. Kühnel, M. Lezius, M. F. Kling, J. van Tilborg, A. Belkacem, K. Ueda, S. Düsterer, R. Treusch, C. D. Schröter, R. Moshammer and J. Ullrich, *Phys. Rev. Lett.*, 2010, **105**, 263002.

21 J. Ullrich, A. Rudenko and R. Moshammer, *Annu. Rev. Phys. Chem.*, 2012, **63**, 635–660.

22 B. Erk, R. Boll, S. Trippel, D. Anielski, L. Foucar, B. Rudek, S. W. Epp, R. Coffee, S. Carron, S. Schorb, K. R. Ferguson, M. Swiggers, J. D. Bozek, M. Simon, T. Marchenko, J. Küpper, I. Schlichting, J. Ullrich, C. Bostedt, D. Rolles and A. Rudenko, *Science*, 2014, **345**, 288–291.

23 R. Boll, A. Rouzée, M. Adolph, D. Anielski, A. Aquila, S. Bari, C. Bomme, C. Bostedt, J. D. Bozek, H. N. Chapman, L. Christensen, R. Coffee, N. Coppola, S. De, P. Decleva, S. W. Epp, B. Erk, F. Filsinger, L. Foucar, T. Gorkhover, L. Gumprecht, A. Hömke, L. Holmegaard, P. Johnsson, J. S. Kienitz, T. Kierspel, F. Krasniqi, K.-U. Kühnel, J. Maurer, M. Messerschmidt, R. Moshammer, N. L. M. Müller, B. Rudek, E. Savelyev, I. Schlichting, C. Schmidt, F. Scholz, S. Schorb, J. Schulz, J. Seltmann,

M. Stener, S. Stern, S. Techert, J. Thøgersen, S. Trippel, J. Viefhaus, M. Vrakking, H. Stapelfeldt, J. Küpper, J. Ullrich, A. Rudenko and D. Rolles, *Faraday Discuss.*, 2014, **171**, 57–80.

24 M. Drescher, M. Hentschel, R. Kienberger, M. Uiberacker, V. Yakovlev, A. Scrinzi, T. Westerwalbesloh, U. Kleineberg, U. Heinzmann and F. Krausz, *Nature*, 2002, **419**, 803–807.

25 S. Gilbertson, S. D. Khan, Y. Wu, M. Chini and Z. Chang, *Phys. Rev. Lett.*, 2010, **105**, 93902.

26 G. Sansone, F. Kelkensberg, J. F. Pérez-Torres, F. Morales, M. F. Kling, W. Siu, O. Ghafur, P. Johnsson, M. Swoboda, E. Benedetti, F. Ferrari, F. Lépine, J. L. Sanz-Vicario, S. Zherebtsov, I. Znakovskaya, A. L'Huillier, M. Y. Ivanov, M. Nisoli, F. Martín and M. J. J. Vrakking, *Nature*, 2010, **465**, 763–766.

27 A. Wirth, M. T. Hassan, I. Grguraš, J. Gagnon, A. Moulet, T. T. Luu, S. Pabst, R. Santra, Z. A. Alahmed, A. M. Azzeer, V. S. Yakovlev, V. Pervak, F. Krausz and E. Goulielmakis, *Science*, 2011, **334**, 195–200.

28 P. B. Corkum, *Phys. Rev. Lett.*, 1993, **71**, 1994–1997.

29 K. J. Schafer, B. Yang, L. F. DiMauro and K. C. Kulander, *Phys. Rev. Lett.*, 1993, **70**, 1599–1602.

30 J. Itatani, J. Levesque, D. Zeidler, H. Niikura, H. Pépin, J. C. Kieffer, P. B. Corkum and D. M. Villeneuve, *Nature*, 2004, **432**, 867–871.

31 S. Haessler, J. Caillat, W. Boutu, C. Giovanetti-Teixeira, T. Ruchon, T. Auguste, Z. Diveki, P. Breger, A. Maquet, B. Carré, R. Taïeb and P. Salières, *Nat. Phys.*, 2010, **6**, 200–206.

32 O. Smirnova, Y. Mairesse, S. Patchkovskii, N. Dudovich, D. Villeneuve, P. Corkum and M. Y. Ivanov, *Nature*, 2009, **460**, 972–977.

33 C. Vozzi, M. Negro, F. Calegari, G. Sansone, M. Nisoli, S. De Silvestri and S. Stagira, *Nat. Phys.*, 2011, **7**, 822–826.

34 N. L. Wagner, A. Wüest, I. P. Christov, T. Popmintchev, X. Zhou, M. M. Murnane and H. C. Kapteyn, *Proc. Natl. Acad. Sci. U. S. A.*, 2006, **103**, 13279–13285.

35 H. J. Wörner, J. B. Bertrand, B. Fabre, J. Higuet, H. Ruf, A. Dubrouil, S. Patchkovskii, M. Spanner, Y. Mairesse, V. Blanchet, E. Mével, E. Constant, P. B. Corkum and D. M. Villeneuve, *Science*, 2011, **334**, 208–212.

36 X. Zhou, R. Lock, N. Wagner, W. Li, H. C. Kapteyn and M. M. Murnane, *Phys. Rev. Lett.*, 2009, **102**, 73902.

37 Y. Mairesse, J. Higuet, N. Dudovich, D. Shafir, B. Fabre, E. Mével, E. Constant, S. Patchkovskii, Z. Walters, M. Y. Ivanov and O. Smirnova, *Phys. Rev. Lett.*, 2010, **104**, 213601.

38 A. Ferré, C. Handschin, M. Dumergue, F. Burgy, A. Comby, D. Descamps, B. Fabre, G. A. Garcia, R. Géneaux, L. Merceron, E. Mével, L. Nahon, S. Petit, B. Pons, D. Staedter, S. Weber, T. Ruchon, V. Blanchet and Y. Mairesse, *Nat. Photonics*, 2015, **9**, 93–98.

39 O. Kfir, P. Grychtol, E. Turgut, R. Knut, D. Zusin, D. Popmintchev, T. Popmintchev, H. Nembach, J. M. Shaw, A. Fleischer, H. Kapteyn, M. Murnane and O. Cohen, *Nat. Photonics*, 2015, **9**, 99–105.

40 A. Rouzée, F. Kelkensberg, W. K. Siu, G. Gademann, R. R. Lucchese and M. J. J. Vrakking, *J. Phys. B: At., Mol. Opt. Phys.*, 2012, **45**, 74016.

41 P. Billaud, M. Géléoc, Y. J. Picard, K. Veyrinas, J. F. Hergott, S. Marggi Poullain, P. Breger, T. Ruchon, M. Roulliay, F. Delmotte, F. Lepetit, A. Huetz, B. Carré and D. Dowek, *J. Phys. B: At., Mol. Opt. Phys.*, 2012, **45**, 194013.
42 A. Fischer, A. Sperl, P. Cörlin, M. Schönwald, H. Rietz, A. Palacios, A. González-Castrillo, F. Martín, T. Pfeifer, J. Ullrich, A. Senftleben and R. Moshammer, *Phys. Rev. Lett.*, 2013, **110**, 213002.
43 A. Stolow, *Faraday Discuss.*, 2013, **163**, 9–32.
44 A. Ferré, A. E. Boguslavskiy, M. Dagan, V. Blanchet, B. D. Bruner, F. Burgy, A. Camper, D. Descamps, B. Fabre, N. Fedorov, J. Gaudin, G. Geoffroy, J. Mikosch, S. Patchkovskii, S. Petit, T. Ruchon, H. Soifer, D. Staedter, I. Wilkinson, A. Stolow, N. Dudovich and Y. Mairesse, *Nat. Commun.*, 2015, **6**, 5952.
45 M. Lebech, J. C. Houver, A. Lafosse, D. Dowek, C. Alcaraz, L. Nahon and R. R. Lucchese, *J. Chem. Phys.*, 2003, **118**, 9653–9663.
46 D. Dowek and R. R. Lucchese, in *Dynamical Processes in Atomic and Molecular Physics*, Bentham, Bussum, The Netherlands, 2012, pp. 57–95.
47 A. Lafosse, M. Lebech, J. C. Brenot, P. M. Guyon, O. Jagutzki, L. Spielberger, M. Vervloet, J. C. Houver and D. Dowek, *Phys. Rev. Lett.*, 2000, **84**, 5987–5990.
48 M. Lebech, J. C. Houver and D. Dowek, *Rev. Sci. Instrum.*, 2002, **73**, 1866–1874.
49 W. B. Li, R. Montuoro, J. C. Houver, L. Journel, A. Haouas, M. Simon, R. R. Lucchese and D. Dowek, *Phys. Rev. A*, 2007, **75**, 52718.
50 R. R. Lucchese, A. Lafosse, J. C. Brenot, P. M. Guyon, J. C. Houver, M. Lebech, G. Raseev and D. Dowek, *Phys. Rev. A*, 2002, **65**, 20702.
51 D. Dowek, M. Lebech, J. C. Houver and R. R. Lucchese, *Mol. Phys.*, 2007, **105**, 1757–1768.
52 S. J. Weber, B. Manschwetus, M. Billon, M. Böttcher, M. Bougeard, P. Breger, M. Géléoc, V. Gruson, A. Huetz, N. Lin, Y. J. Picard, T. Ruchon, P. Salières and B. Carré, *Rev. Sci. Instrum.*, 2015, **86**, 33108.
53 P. Antoine, A. L'Huillier, M. Lewenstein, P. Salières and B. Carré, *Phys. Rev. A*, 1996, **53**, 1725–1745.
54 H. Ruf, C. Handschin, R. Cireasa, N. Thiré, A. Ferré, S. Petit, D. Descamps, E. Mével, E. Constant, V. Blanchet, B. Fabre and Y. Mairesse, *Phys. Rev. Lett.*, 2013, **110**, 83902.
55 M. Gisselbrecht, A. Huetz, M. Lavollée, T. J. Reddish and D. P. Seccombe, *Rev. Sci. Instrum.*, 2005, **76**, 13105.
56 O. Guyétand, M. Gisselbrecht, A. Huetz, P. Agostini, B. Carré, P. Breger, O. Gobert, D. Garzella, J.-F. Hergott, O. Tcherbakoff, H. Merdji, M. Bougeard, H. Rottke, M. Böttcher, Z. Ansari, P. Antoine and L. F. DiMauro, *J. Phys. B: At., Mol. Opt. Phys.*, 2008, **41**, 65601.
57 P. Hockett, C. Z. Bisgaard, O. J. Clarkin and A. Stolow, *Nat. Phys.*, 2011, **7**, 612–615.
58 F. Martin, J. Fernandez, T. Havermeier, L. Foucar, T. Weber, K. Kreidi, M. Schoffler, L. Schmidt, T. Jahnke, O. Jagutzki, A. Czasch, E. P. Benis, T. Osipov, A. L. Landers, A. Belkacem, M. H. Prior, H. Schmidt-Bocking, C. L. Cocke and R. Dorner, *Science*, 2007, **315**, 629–633.
59 D. Dowek, J. F. Pérez-Torres, Y. J. Picard, P. Billaud, C. Elkharrat, J. C. Houver, J. L. Sanz-Vicario and F. Martín, *Phys. Rev. Lett.*, 2010, **104**, 233003.
60 A. Lafosse, J. C. Brenot, A. V. Golovin, P. M. Guyon, K. Hoejrup, J. C. Houver, M. Lebech and D. Dowek, *J. Chem. Phys.*, 2001, **114**, 6605–6617.

61 K. Veyrinas, PhD thesis, Université Paris-Sud 11, U.F.R. Scientifique d'Orsay, 2015.
62 In preparation.
63 Y. Iida, F. Carnovale, S. Daviel and C. E. Brion, *Chem. Phys.*, 1986, **105**, 211–225.
64 R. E. Stratmann, R. W. Zurales and R. R. Lucchese, *J. Chem. Phys.*, 1996, **104**, 8989–9000.
65 B. Manschwetus, N. Lin, J. Rothhardt, R. Guichard, T. Auguste, A. Camper, P. Breger, J. Caillat, M. Géléoc, T. Ruchon, R. Taïeb, B. Carré and P. Salières, *J. Phys. Chem. A*, 2015, **119**, 6111–6122.
66 A.-T. Le, R. R. Lucchese and C. D. Lin, *Phys. Rev. A*, 2013, **87**, 63406.
67 R. Torres, N. Kajumba, J. G. Underwood, J. S. Robinson, S. Baker, J. W. G. Tisch, R. de Nalda, W. A. Bryan, R. Velotta, C. Altucci, I. C. E. Turcu and J. P. Marangos, *Phys. Rev. Lett.*, 2007, **98**, 203007.
68 O. Smirnova, S. Patchkovskii, Y. Mairesse, N. Dudovich, D. Villeneuve, P. Corkum and M. Y. Ivanov, *Phys. Rev. Lett.*, 2009, **102**, 63601.
69 M. C. H. Wong, A.-T. Le, A. F. Alharbi, A. E. Boguslavskiy, R. R. Lucchese, J.-P. Brichta, C. D. Lin and V. R. Bhardwaj, *Phys. Rev. Lett.*, 2013, **110**, 33006.
70 M. Negro, M. Devetta, D. Faccialá, S. D. Silvestri, C. Vozzi and S. Stagira, *Faraday Discuss.*, 2014, **171**, 133–143.
71 L. Nahon, G. A. Garcia and I. Powis, *J. Electron Spectrosc. Relat. Phenom.*, 2015, **204**, 322–334.
72 J. Jose, R. R. Lucchese and T. N. Rescigno, *J. Chem. Phys.*, 2014, **140**, 204305.
73 D. Toffoli, R. R. Lucchese, M. Lebech, J. C. Houver and D. Dowek, *J. Chem. Phys.*, 2007, **126**, 54307.
74 S. Marggi Poullain, C. Elkharrat, W. B. Li, K. Veyrinas, J. C. Houver, C. Cornaggia, T. N. Rescigno, R. R. Lucchese and D. Dowek, *J. Phys. B: At., Mol. Opt. Phys.*, 2014, **47**, 124024.
75 A.-T. Le, R. R. Lucchese, M. T. Lee and C. D. Lin, *Phys. Rev. Lett.*, 2009, **102**, 203001.
76 A.-T. Le, R. R. Lucchese, S. Tonzani, T. Morishita and C. D. Lin, *Phys. Rev. A*, 2009, **80**, 13401.
77 P. Salières, A. Maquet, S. Haessler, J. Caillat and R. Taïeb, *Rep. Prog. Phys.*, 2012, **75**, 62401.

Faraday Discussions

PAPER

Observation of multi-channel non-adiabatic dynamics in aniline derivatives using time-resolved photoelectron imaging

Magdalena M. Zawadzki,[a] Marco Candelaresi,[a] Lisa Saalbach,[a] Stuart W. Crane,[a] Martin J. Paterson[b] and Dave Townsend*[ab]

Received 25th April 2016, Accepted 24th May 2016
DOI: 10.1039/c6fd00092d

We present results from a recent time-resolved photoelectron imaging (TRPEI) study investigating the non-adiabatic relaxation dynamics of *N*,*N*-dimethylaniline (*N*,*N*-DMA) and 3,5-dimethylaniline (3,5-DMA) following excitation at 240 nm. Analysis of the experimental data is supported by *ab initio* coupled-cluster calculations evaluating excited state energies and the evolution of several excited state physical properties as a function of N–H/N–CH_3 bond extension – a critical reaction coordinate. The use of site-selective methylation brings considerable new insight to the existing body of literature concerning photochemical dynamics in the related system aniline at similar excitation wavelengths. The present work also builds on our own previous investigations in the same species at 250 nm. The TRPEI method provides highly differential energy- and angle-resolved data and, in particular, the temporal evolution of the photoelectron angular distributions afforded by the imaging approach offers much of the new dynamical information. In particular, we see no clear evidence of the second excited $2\pi\pi^*$ state non-adiabatically coupling to the lower-lying $S_1(\pi\pi^*)$ state or the mixed Rydberg/valence $S_2(3s/\pi\sigma^*)$ state. This, in turn, potentially raises some unresolved questions about the overall nature of the dynamics operating in these systems, especially in regard to the $2\pi\pi^*$ state's ultimate fate. More generally, the findings for the aromatic systems *N*,*N*-DMA and 3,5-DMA, taken along with our recent TRPEI results for several aliphatic amine species, highlight interesting questions about the nature of electronic character evolution in mixed Rydberg-valence states as a function of certain key bond extensions and the extent of system conjugation. We begin exploring these ideas computationally for a systematically varied series of tertiary amines.

[a]*Institute of Photonics & Quantum Sciences, Heriot-Watt University, Edinburgh, EH14 4AS, UK. E-mail: d.townsend@hw.ac.uk*
[b]*Institute of Chemical Sciences, Heriot-Watt University, Edinburgh, EH14 4AS, UK*

1. Introduction

Amine-based motifs are an integral component within a variety of biological molecules including the DNA bases, amino acids and phenylpropanoids in plants.[1,2] It is therefore of great interest to fully study and understand the physical and chemical role of these sub-units within the larger biological environment. One important characteristic of these species is the relaxation dynamics they exhibit following ultraviolet (UV) irradiation. UV absorption is potentially harmful to living organisms, and the ability of biological chromophores to non-radiatively dissipate this excess energy harmlessly into the surrounding environment *via* efficient ultrafast non-adiabatic processes has attracted much recent study.[3,4] This provides a critical "photostability" that may have been an important evolutionary factor in the development of early life 3.5 billion years ago – particularly as the surface of the Earth was subject to much harsher UV irradiation conditions at that time due to the atmosphere not yet being fully developed.[5]

The aniline molecule is a good example of a system that has been used as a starting model to begin exploring the links between chemical structure, non-adiabatic dynamics and thus the photostability function of larger, biologically relevant molecules. Over the past decade several experimental studies, complemented by in-depth theoretical investigations, have been undertaken using a range of different methodologies and associated observables.[6–14] This body of literature includes our own previously published results, obtained using the time-resolved photoelectron imaging (TRPEI) technique, which interrogated the relaxation dynamics of gas-phase aniline at UV excitation wavelengths in the region 273–266 nm (close to the origin of the S_2 excited state – characterised in more detail below) and also at 250 nm.[6,9] In the latter study we also reported findings for the structurally related species *N*,*N*-dimethylaniline (*N*,*N*-DMA) and 3,5-dimethylaniline (3,5-DMA), which are shown in Fig. 1 along with their UV absorption spectra. This use of site-selective methylation afforded considerable

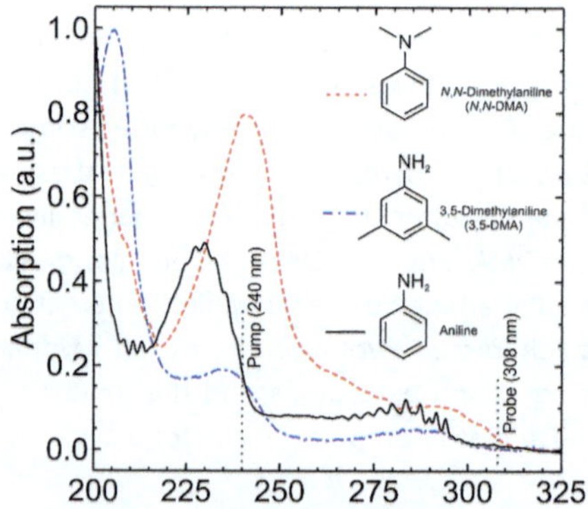

Fig. 1 UV vapour-phase room temperature absorption spectra of *N*,*N*-dimethylaniline (*N*,*N*-DMA) and 3,5-dimethylaniline (3,5-DMA) (schematic structures of which are inset) recorded using a commercial bench-top spectrophotometer (Shimadzu UV-2550). Vertical dashed lines denote the pump and probe wavelengths used in our TRPEI measurements. Equivalent data recorded for aniline is also included for comparison.

new dynamical insight – especially given previous work on aniline had implicated motion along the N–H stretching coordinate and deformations of the aromatic ring structure as being critical to access conical intersections that mediate the overall relaxation process following UV excitation.[8,12,14] The two lowest-lying singlet excited electronic states in aniline, *N,N*-DMA and 3,5-DMA may be designated $S_1(\pi\pi^*)$ and $S_2(3s/\pi\sigma^*)$. Following single-photon excitation at 250 nm, these are the only two states that are energetically accessible and both may be prepared optically. In the Franck–Condon region, the S_2 state is predominantly 3s Rydberg in nature, while at extended N–H bond lengths (N–CH_3 in the case of *N,N*-DMA) this rapidly evolves towards $\pi\sigma^*$ valence character. In aniline and *N,N*-DMA, population loss from the initially prepared $S_2(3s/\pi\sigma^*)$ state was determined to occur *via* two competing ultrafast (<100 fs) mechanisms: internal conversion to the (much longer-lived) $S_1(\pi\pi^*)$ state or, alternatively, direct dissociation along the N–X stretching coordinate (X = H or CH_3). In 3,5-DMA, however, direct N–H dissociation was inferred to be the only significant relaxation pathway due to methylation of the aromatic ring system, significantly slowing down the timescale for the system to reach the $S_2(3s/\pi\sigma^*)/S_1(\pi\pi^*)$ conical intersection (CI) and undergo efficient internal conversion. Of particular importance to our analysis was the highly-differential information revealed in the temporal evolution of the photoelectron angular distributions (PADs) provided by the imaging aspect of the TRPEI approach. We therefore now seek to extend the same methodology to shorter excitation wavelengths and consider the non-adiabatic dynamics that operate in *N,N*-DMA and 3,5-DMA following 240 nm excitation. As explored in some detail previously, these systems exhibit very similar electronic state characters and energies as aniline.[6] As seen in Fig. 1, at 240 nm, there is now a clear onset of strong absorption to a higher-lying state, assigned (in aniline) as having singlet $\pi\pi^*$ character.[8,12,15,16] Several Rydberg states have, however, been predicted to sit slightly below this $\pi\pi^*$ state in energy.[6,17] In order to avoid any confusion relating to state ordering, we therefore simply denote this second $\pi\pi^*$ state as $2\pi\pi^*$ throughout. Literature summaries of the photochemical dynamics studies undertaken in aniline at absorption wavelengths >240 nm have been reported in our previous work.[6,9] As such, only details of direct relevance to the present study (involving $2\pi\pi^*$ excitation) are briefly highlighted below.

Ashfold and co-workers have thoroughly investigated the mechanisms of H atom loss in aniline following UV excitation using photofragment translational spectroscopy.[14] Upon incrementally reducing the excitation wavelength from 250 nm to 193.3 nm, these authors observed distinct changes in the H atom total kinetic energy release spectra, attributed to the onset of excitation to the $2\pi\pi^*$ state at ~240 nm. Subsequent non-adiabatic decay through a $2\pi\pi^*/S_2(3s/\pi\sigma^*)$ CI or *via* successive $2\pi\pi^*/S_1(\pi\pi^*)$ and $S_1(\pi\pi^*)/S_2(3s/\pi\sigma^*)$ CIs was suggested, yielding H atom photoproducts with high kinetic energies formed *via* dissociation on the $S_2(3s/\pi\sigma^*)$ state potential surface. Additionally, competing decay of the $2\pi\pi^*$ population *via* a CI with the S_0 ground state was invoked to account for a lower energy TKER component seen in the data. A more recent time-resolved ion-yield study of aniline by Montero *et al.* has, in agreement with Ashfold and co-workers, concluded that excitation to the $2\pi\pi^*$ state has become an open channel at 240 nm and also suggested similar relaxation pathways to those proposed previously.[13] Stavros and co-workers subsequently explored the ultrafast dynamics in photoexcited aniline using time-resolved H atom imaging.[12] Once again, this work concluded that 240 nm excitation populates the 2ππ* state and supporting calculations predicted a barrierless internal conversion pathway from 2ππ* to S_1(ππ*) that is mediated by out-of-plane molecular deformations. Population then evolves further onto the S_2(3s/πσ*) excited state which undergoes ultrafast N–H bond fission. Finally, Fielding and co-workers have undertaken an intensive TRPEI investigation of the photochemistry of aniline at a range of excitation wavelengths, including the region around 240 nm. The initial findings of these authors concluded that the 2ππ* state decays on a sub 100 fs timescale directly to the S_0 ground state.[10,11] In more recent communications these authors have expanded this interpretation and invoke (based upon extensive supporting theory) a three-state conical intersection involving 2ππ*, S_2(3s/πσ*) and S_1(ππ*).[7,8] This predominantly transfers the 2ππ* population directly to the ground state although some evidence of relaxation *via* the S_1(ππ*) and/or S_2(3s/πσ*) states was also reported. Interestingly, this latter pathway was not directly observed in deuterated d7-aniline.

In addition to providing new insight into the dynamics that operate in aniline-based motifs (and in particular, to further exploring the dynamical role of the 2ππ* state), our present findings, along with those we have reported in several other recent TRPEI studies on related systems,[18–20] begin to raise broader questions about the influence of different chemical substituents attached to the N atom centre on the exact nature of mixed Rydberg/valence development within the electronic states of amines. In addition, the physical evolution of Rydberg-to-valence character along specific molecular coordinates highlights even more general issues relating to intra-state detection sensitivity effects that may potentially arise in many photoionization-based experiments. This also then leads to related considerations for inter-state detection as the population non-adiabatically evolves over multiple potential energy surfaces. The second part of this communication discusses these ideas in more detail and summarizes recent work that begins to explore them computationally for a systematically varied series of tertiary amines.

2. Experimental

Our experimental TRPEI setup has been described in detail elsewhere.[21] The fundamental laser output of a 1 kHz Ti:sapphire laser system (Spectra Physics, Spitfire Pro/Empower) operating at a central wavelength of 800 nm with 30 nm full-width-at-half-maximum (FWHM) bandwidth provided the input for a pair of optical parametric amplifier (OPA) systems (Spectra Physics, OPA-800c and TOPAS Prime-U, Light Conversion – used for the pump and probe generation, respectively). The pump beam (240 nm, ∼0.5 μJ per pulse) was obtained by two sequential non-linear processes: generation of 480 nm radiation through sum frequency mixing a portion of the fundamental 800 nm laser output with the OPA signal beam and then subsequent frequency doubling. The probe beam (308 nm, ∼1.6 μJ per pulse) was obtained by twice frequency doubling the signal beam output of the second OPA. Thin β-barium borate (BBO) crystals were used as the non-linear medium throughout. The UV pump and probe beams then propagated through single-pass prism compressors (fused silica and CaF_2, respectively) and were combined on a dichroic mirror before being focussed into the photoelectron imaging spectrometer using a 25 cm fused silica lens. Temporal delay between the pump and probe was controlled using a motorized linear translation stage running under computer command. As seen in Fig. 1, the choice of a 308 nm probe ensures our data will be free from unwanted "probe–pump" signals evolving towards negative delay times.

Liquid molecular samples were purchased from Sigma-Aldrich (*N*,*N*-DMA 99% and 3,5-DMA 98%) and used without further purification. These were used to soak small pieces of filter paper that were placed within the body of a 1 kHz pulsed molecular beam valve, directly behind the exit nozzle ($\varnothing$ = 150 μm).[22] Helium (1 bar) was used as the carrier gas and the temperature of the valve was regulated at 45 °C using a chiller. After expansion into the source chamber of the spectrometer, samples seeded in the molecular beam passed through a skimmer ($\varnothing$ = 1.0 mm) and into the main interaction chamber. Here they were intersected at 90° by the (co-propagating) UV pump and probe in a region of space between the repeller and extractor electrodes of a magnetically shielded electrostatic lens assembly optimized for velocity-map imaging.[23] Photoelectrons generated from (1 + 1′) resonant multiphoton ionization were then guided along a short flight tube and imaged spatially using a 40 mm MCP/P47 phosphor screen detector in conjunction with a CCD (640 × 480 pixels) camera. Photoelectron images were acquired at a series of repeatedly sampled pump–probe delay positions spanning −300 fs to +500 fs in 25 fs increments and +500 fs to +50 ps in 7 exponentially increasing time steps. At each time step, pump-alone and probe-alone photoelectron images were also recorded for background subtraction. Prior to commencing photoelectron collection, the spectrometer was switched into ion detection mode and the pulsed valve timing conditions tuned to ensure no signals from clusters were observed. Compressor lengths for each beam line were also systematically adjusted to yield a cross correlation of 130 ± 10 fs, as determined from non-resonant two-colour (1 + 1′) multiphoton ionization of pyrrole. This data also provided energy calibration information for the spectrometer.

3. Experimental results

3.1. Time-resolved photoelectron spectra

Fig. 2 presents a series of photoelectron images resulting from (1 + 1′) photoionization of *N*,*N*-DMA and 3,5-DMA at a range of selected temporal delay times between the pump (240 nm) and probe (308 nm) beams. Of particular note is the sharp, and highly anisotropic outer ring that peaks along the direction of laser polarization in the data recorded at very short delay times. Time-resolved photoelectron spectra may be generated from the complete sets of acquired image data (using appropriate energy calibration information and pixel weighting) following application of a rapid matrix inversion approach described in detail elsewhere.[21] These plots are shown in Fig. 3 (note the mixed linear-logarithmic scaling of the time axis). Both systems under study show a long-lived feature spanning the low-kinetic energy region up to ~0.8 eV. An additional feature exhibiting extremely rapid decay is also seen at higher kinetic energies. On first inspection, the overall appearance of these spectra is very similar to that seen previously following excitation at 250 nm,[6] although the high kinetic energy features are somewhat different in shape – a consequence of direct excitation to the $2\pi\pi^*$ state, as expanded upon in more detail later.

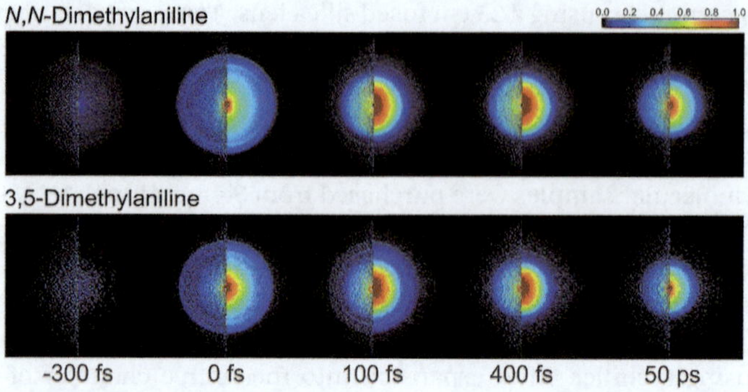

Fig. 2 (1 + 1′) photoelectron images obtained for *N*,*N*-DMA and 3,5-DMA at selected pump–probe delay times following 240 nm excitation and subsequent ionization using 308 nm. Time-invariant pump-alone and probe-alone signals have been subtracted and the images are 4-fold symmetrised. The (linear) polarization direction of the pump and probe beams is vertical with respect to the figure. The left half of the images show the processed data obtained following application of the matrix inversion approach described in ref. 21.

The temporal evolution of the photoelectron data was analysed quantitatively using a global Levenberg–Marquardt fitting routine where the 2D experimental data $S(E, \Delta t)$ are expressed as follows:

$$S(E,\ \Delta t) = \sum_{i} A_i(E) P_i(\Delta t) \otimes g(\Delta t) \tag{1}$$

here $A_i(E)$ is the decay associated photoelectron spectrum of the ith data channel, which has a time-dependent population $P_i(\Delta t)$ described by a series of exponentially decaying profiles (all of which originate from zero pump–probe delay) and

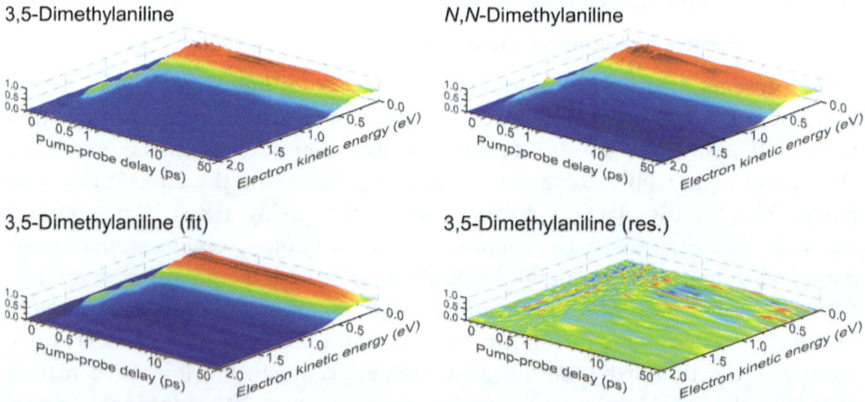

Fig. 3 Time-dependent photoelectron spectra of *N*,*N*-DMA, and 3,5-DMA obtained using a 240 nm pump/308 nm probe. For clear display of the dynamics over all temporal ranges, the time axis is linear to +500 fs and logarithmic between +500 fs and +50 ps. The data are partitioned into 0.025 eV energy bins. Also shown is the fit to the 3,5-DMA data (obtained using the procedure described in the main text) along with the associated residuals (*i.e.*, the fit subtracted from the raw data).

$g(\Delta t)$ represents the (Gaussian) experimental cross-correlation function of the pump and probe pulses. The global nature of the fitting procedure provides a series of decay associated spectra (DAS) that plot the relative amplitude of each exponential component as a function of photoelectron kinetic energy. In order to obtain a satisfactory fit to both the *N,N*-DMA and 3,5-DMA data using eqn (1), a total of four exponentially decaying functions were required. We label these using their respective time constants τ_{1-4} (note that only three such functions were necessary in our equivalent analysis following 250 nm excitation). The DAS obtained are shown in Fig. 4, along with their respective time constants, which are broadly similar to those previously reported by Fielding and co-workers for aniline at similar excitation wavelengths.[7,10,11] To illustrate the good overall quality of the model, Fig. 3 also includes the corresponding global fit and associated residuals for the 3,5-DMA data. In both systems under study, only the τ_1 and τ_2 DAS contribute significant fit amplitude at relatively high photoelectron kinetic energies (>0.8 eV). The τ_1 DAS describes an extremely fast decay that, within the limits of our experimental instrument response, exhibits essentially zero lifetime (*i.e.* the fit component was Gaussian). The τ_2 DAS describes a longer decay (110–120 fs) which shows a sharp, large amplitude feature (at 1.2 eV and 0.95 eV for *N,N*-DMA and 3,5-DMA, respectively) superimposed on a broader background. In regions of lower photoelectron kinetic energy (<0.8 eV), the τ_1 DAS obtained for

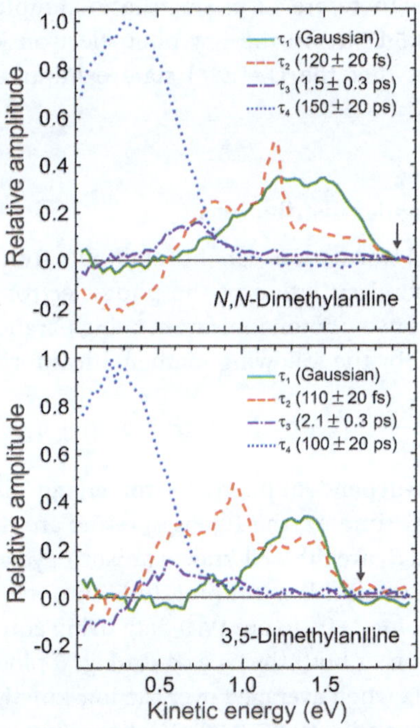

Fig. 4 Decay associated spectra (DAS) obtained from a global multi-exponential fit to the data presented in figure. For additional details, see the main text. Uncertainties quoted are 1σ values and the data are partitioned into 0.025 eV energy bins. Vertical dashed lines denote predicted maximum photoelectron kinetic energy cut-offs based on the central pump and probe wavelengths and the adiabatic ionization potentials (7.26 eV and 7.48 eV for *N,N*-DMA, and 3,5-DMA, respectively, as determined in ref. 6).

N,N-DMA does not contribute significant amplitude to the overall global fit but τ_{2-4} are still significant, with τ_2 having a negative component. This is in contrast to the situation in 3,5-DMA where only τ_3 and τ_4 are effectively non-zero. Overall, the timescales and relative amplitudes of the various DAS associated with τ_{2-4} in the 240 nm excitation data for both N,N-DMA and 3,5-DMA are very similar to those seen previously at 250 nm.[6] The only difference at 240 nm is the inclusion of the additional Gaussian fit component τ_1, which appears to have no significant influence on the behaviour on the other temporal processes observed. Given this observation we may therefore attribute the τ_1 DAS directly to the decay of the $2\pi\pi^*$ state (which is not populated at 250 nm) and assign the τ_{2-4} DAS in accordance with our earlier work: τ_2 describes decay of the $S_2(3s/\pi\sigma^*)$ state, while τ_3 and τ_4 represent intramolecular vibrational redistribution (IVR) within the $S_1(\pi\pi^*)$ state and the ultimate long-time decay of the $S_1(\pi\pi^*)$ state, respectively. This brief summary will be expanded upon further in the Discussion. Finally, the D_1 (π^{-1}) vertical ionisation potentials for N,N-DMA, and 3,5-DMA are 9.00 eV, and 8.55 eV, respectively.[24,25] Our $(1 + 1')$ REMPI scheme at 240/308 nm provides a total photon energy of 9.19 eV, and so ionization to the D_1 (π^{-1}) cation state (in addition to the D_0 (π^{-1}) state) may, in principle, be observed in our TRPEI data. Excitation at 240 nm (5.17 eV) is energetically well above the origins of the $S_1(\pi\pi^*)$ and $S_2(3s/\pi\sigma^*)$ states[6] and so heuristic Franck–Condon arguments lead us to assume the propensity for D_1 (π^{-1}) ionization (with a 308 nm probe) is likely to be small in these instances. In contrast, $2\pi\pi^*$ is prepared closer to its electronic origin and it is perhaps therefore surprising that no amplitude is seen in the τ_1 DAS (particularly for 3,5-DMA) in the low photoelectron kinetic energy region. This possibly indicates that the D_1 (π^{-1}) state exhibits a somewhat different geometry to that of D_0 (π^{-1}).

3.2. Photoelectron angular distributions

Additional dynamical information can be obtained from our TRPEI data by investigating the temporal evolution of the photoelectron angular distribution (PADs). For the case of a two-photon ionization using parallel linear polarizations, the PADs are described by the following mathematical form:[26,27]

$$I(E,\ \Delta t,\ \theta) = \frac{\sigma(E,\ \Delta t)}{4\pi}[1 + \beta_2(E,\ \Delta t)P_2(\cos\theta) + \beta_4(E,\ \Delta t)P_4(\cos\theta)] \quad (2)$$

here $\sigma(E, \Delta t)$ is the time-dependent photoelectron energy distribution acquired at each pump–probe delay time Δt, the $P_n(\cos\theta)$ terms are the nth-order Legendre polynomials and β_2 and β_4 are the well-known anisotropy parameters. The angle $\theta = 180°$ is defined by a vertical line running fully through the images shown in Fig. 2 *via* their centre points. Fits to our PAD data using eqn (2) reveals significant time-dependent variation of both the β_2 and β_4. Fig. 5 plots the temporal evolution of these parameters when averaged over the low kinetic energy region of the photoelectron spectra spanning 0.05–0.70 eV where long-lived ($\gtrsim$100 ps) signals are observed (as modelled by the τ_4 DAS shown in Fig. 4). For both systems under study, the temporal variation of both β_2 and β_4 is consistent within this selected energy region and averaging only serves to improve the statistics in the data presented. Each system exhibits a clear initial rise in β_2 at short times beyond $\Delta t = 0$. This is slightly more rapid in N,N-DMA (~100 fs) than in 3,5-DMA (~250 fs).

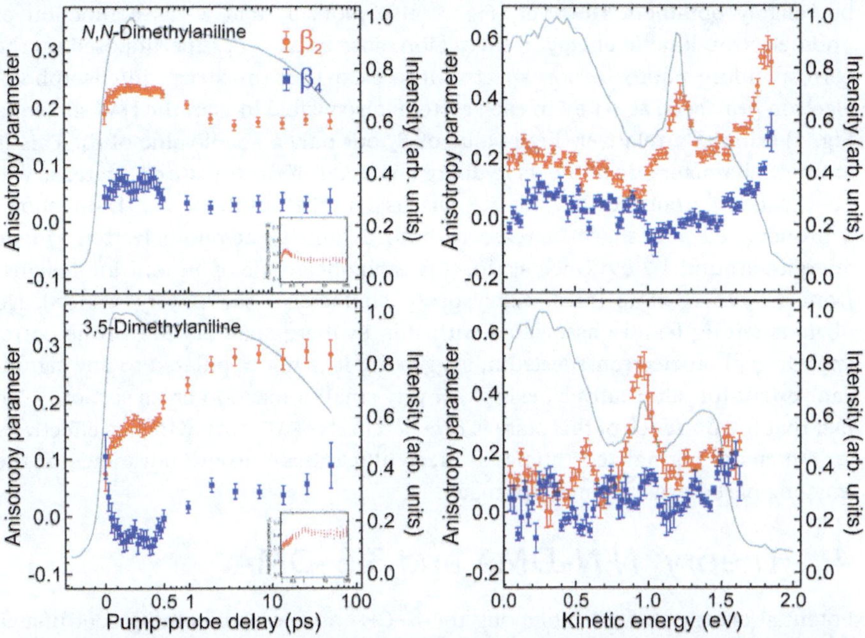

Fig. 5 (Left) Anisotropy parameters β_2 and β_4 as a function of pump–probe delay averaged over the 0.05–0.70 eV photoelectron kinetic energies region following 240 nm excitation/308 nm ionization of *N,N*-DMA and 3,5-DMA. The time axes are linear to +500 fs and logarithmic between +500 fs and +100 ps. The data were partitioned into 0.025 eV energy bins for initial anisotropy fits (before averaging), and the error bars represent one standard deviation. Fits were performed over the angular region $5° \leq \theta \leq 90°$ to eliminate uncertainties from centre-line noise present in the Abel-inverted images (see Fig. 2). Inset panels show the corresponding β_2 data obtained following 250 nm excitation, as previously reported in ref. 6. (Right) Anisotropy parameters β_2 and β_4 as a function of photoelectron kinetic energy at zero pump-probe delay (data points) superimposed over the corresponding photoelectron spectrum (grey line).

Beyond this point in time, the two molecules differ to some extent in their behaviour. In *N,N*-DMA, β_2 plateaus before falling again slightly after ~1 ps. In contrast, 3,5-DMA exhibits an additional increase in β_2 on a more extended timescale, reaching a maximum value in 2 ps and then remaining at this level out to 50 ps. In both systems, the evolution of β_4 occurs in a similar manner to that seen in the corresponding β_2 values, although the initial β_2 rise seen in 3,5-DMA is accompanied by a β_4 fall. In *N,N*-DMA the β_2 plateau region also appears to be resolved into two distinct temporal features in the equivalent β_4 plot. The inset panels included in Fig. 5 permit a direct comparison between the temporal evolution of β_2 in our present 240 nm excitation data and that observed previously at 250 nm. In both systems, the 240 nm data appears to show a feature that is effectively superimposed on top of the behaviour seen at the longer excitation wavelength. As was the case with the DAS data described earlier, we therefore attribute these additional features to a process directly associated with the $2\pi\pi^*$ state.

At higher kinetic energies (<0.8 eV), photoelectron signals in the data are extremely short lived and no information relating to their temporal evolution may be reliably obtained. However, Fig. 5 also plots β_2 and β_4 as a function of photoelectron kinetic energy in the region close to $\Delta t = 0$, superimposed on the corresponding photoelectron spectra. It is clear that the sharp, intense photoelectron peak seen at ~1 eV in each system (also visible in τ_2 in the DAS shown in Fig. 4) exhibits a relatively high value of β_2 but only a small value of β_4. This is strongly characteristic of the 3s Rydberg character of the $S_2(3s/\pi\sigma^*)$ state, as will be expanded upon further in the Discussion. In addition, *N,N*-DMA shows a pronounced and sharp increase in both β_2 and β_4 at photoelectron kinetic energies around 1.7 eV. Once again, this is characteristic of an ionization signal from a Rydberg state (presumably of 3p character – see Theory section). No obvious strong feature associated with this Rydberg state is seen in the corresponding photoelectron spectrum, suggesting it is not populated to any significant extent (or, alternatively, exhibits a very small ionization cross section). The fact that the presence of this state is revealed in the PAD data (which is effectively an "intensity normalized" analysis) nicely illustrates a useful advantage of the imaging-based experimental approach.

4. Theory: *N,N*-DMA and 3,5-DMA

Potential energy surface cuts along the N–CH_3 and N–H stretching coordinates were obtained for *N,N*-DMA and 3,5-DMA, respectively, using equation of motion (EOM) coupled cluster theory with single and double substitutions in conjunction with an aug-cc-pVDZ basis set. These coordinates have been strongly implicated as being key in mediating the overall non-adiabatic dynamics.[8,12,14] Vertical oscillator strengths were computed using fully relaxed linear response transition densities. Equilibrium S_0 geometries were taken from a B3LYP/aug-cc-pVTZ optimisation in C_s symmetry. All calculations were undertaken with the Gaussian 09 program.[28] The results, as summarized in Table 1, are consistent with our previous work.[6] We note that the $2\pi\pi^*$ state is quite diffuse in character and sits energetically above several Rydberg states that have been discussed previously elsewhere.[6,17] The oscillator strengths for transitions to the $2\pi\pi^*$ state are considerably larger than for $S_1(\pi\pi^*)$, in agreement with the absorption spectra in Fig. 1. As seen in Fig. 6, for both *N,N*-DMA and 3,5-DMA all electronic states considered are strongly bound along the N–CH_3/N–H coordinate with the exception of $S_2(3s/\pi\sigma^*)$, which exhibits only a small potential barrier before tending towards a dissociative asymptote and at 240 nm direct N–CH_3/N–H bond fission is expected to be an energetically open pathway. At highly extended bond lengths (beyond the limits of where the coupled cluster approach begins to breakdown) a crossing between the $S_2(3s/\pi\sigma^*)$ state with the S_0 ground state is also expected, as previously shown for the closely related case of aniline.[12]

5. Discussion

As already outlined in Section 3, the DAS and PAD data extracted from our TRPEI measurements point to a situation where the overall 240 nm relaxation dynamics observed in both *N,N*-DMA and 3,5-DMA may be viewed as a combination of decay pathways seen previously at 250 nm and an additional process attributable to the onset of excitation to the $2\pi\pi^*$ state. In the case of the DAS analysis, this was strongly suggested by the similarity of the τ_{2-4} spectra seen in each molecule at

Table 1 EOM-CCSD/aug-cc-pVDZ vertical excitation energies (Δ*E*) and oscillator strengths (*f*) obtained for *N*,*N*-DMA and 3,5-DMA, along with corresponding state assignments. For more details see the main text. Note that for *N*,*N*-DMA the ordering of the first ππ* and 3s/πσ* states depend delicately on electron correlation, with a connected triples correction changing the ordering (see ref. 6). In the absence of any experimental data confirming the true ordering of these states, the $S_1(\pi\pi^*)$ and $S_2(3s/\pi\sigma^*)$ labels will be used throughout for both systems for clarity in subsequent discussions. The p manifold is heavily mixed with orbitals directed away from the *x*, *y* and *z* molecular axes. As such, no specific assignments are given

		Δ*E* (eV) (CCSD)[a]	*f* (CCSD)
***N*,*N*-Dimethylaniline**			
$^1A'$	S_0[b]	0	—
$1\ ^1A'$	$S_2(3s/\pi\sigma^*)$	4.55	0.0231
$1\ ^1A''$	$S_1(\pi\pi^*)$	4.59	0.0406
$2\ ^1A'$	3p	5.14	0.0007
$2\ ^1A''$	3s	5.17	0.0145
$3\ ^1A'$	3p	5.41	0.0119
$4\ ^1A'$	2ππ*	5.49	0.2441
3,5-Dimethylaniline			
$^1A'$	S_0[b]	0	—
$1\ ^1A''$	ππ*	4.68	0.0247
$1\ ^1A'$	3s/πσ*	4.80	0.0034
$2\ ^1A'$	3s	5.49	0.0089
$2\ ^1A''$	3p	5.55	0.0004
$3\ ^1A'$	2ππ*	5.65	0.1135
$4\ ^1A'$	3p	5.92	0.0037

[a] Aug-cc-pVDZ basis set for response properties. [b] S_0 geometries from B3LYP/aug-cc-pVTZ optimisation in C_s symmetry.

the two different excitation wavelengths. In the PADs, the temporal evolution of the β_2 anisotropy parameter following 240 nm excitation appeared to be an effective superposition of features seen at 250 nm with a new, additional component. We therefore structure the Discussion by first summarizing the dynamical processes associated with the $S_2(3s/\pi\sigma^*)$ and $S_1(\pi\pi^*)$ states (described elsewhere in more detail in the context of 250 nm excitation[6]) and then consider the role of the 2ππ* state.

5.1. $S_2(3s/\pi\sigma^*)/S_1(\pi\pi^*)$ dynamics

On the basis of our previous 250 nm excitation study, the τ_{2-4} DAS in both *N*,*N*-DMA and 3,5-DMA may be assigned as follows:

Firstly, the τ_3 and τ_4 DAS are both associated with ionization from the $S_1(\pi\pi^*)$ state. The τ_3 component (1.5 ± 0.3 ps in *N*,*N*-DMA and 2.1 ± 0.3 ps in 3,5-DMA) is assigned to IVR processes, with τ_4 (150 ± 20 ps in *N*,*N*-DMA and 100 ± 20 ps in 3,5-DMA) then describing the overall $S_1(\pi\pi^*)$ decay. The ultimate fate of this state is not apparent in our measurements but several possibilities suggested previously (for aniline) include fluorescence,[29] intersystem crossing to lower-lying triplet states,[30] or internal conversion back to the S_0 ground state.[8,12] Ultrafast IVR processes have previously been implicated as playing a role in the excited states of several other aromatic species in situations where the excitation is well above the

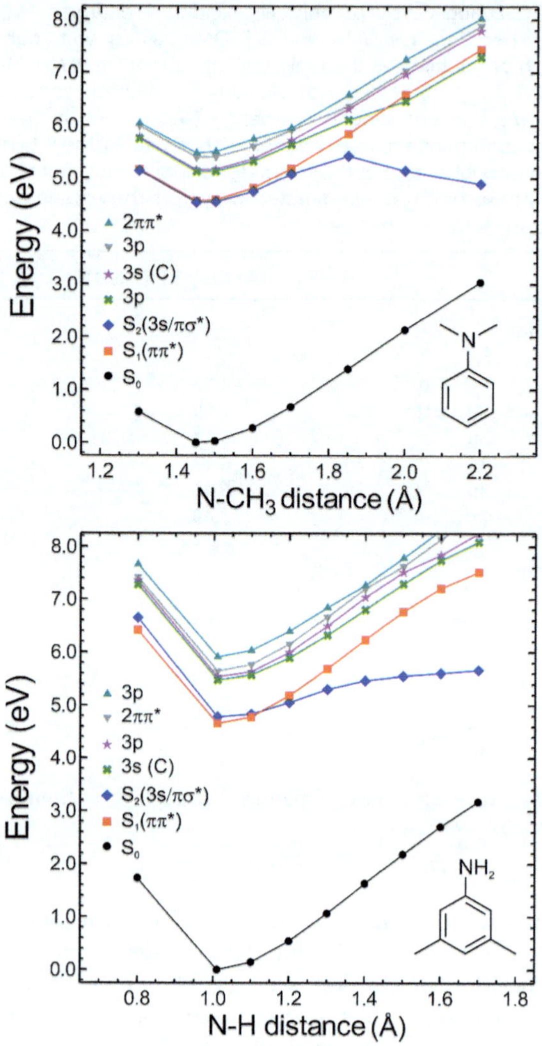

Fig. 6 EOM-CCSD/aug-cc-pVDZ potential energy cuts along the N–H stretching coordinate (3,5-DMA) and the N–CH_3 stretching coordinate (*N*,*N*-DMA) for the S_0 ground state and several low-lying electronically excited singlet states. For additional details, see the main text. The p manifold is heavily mixed with orbitals directed away from the *x*, *y* and *z* molecular axes. As such, no specific assignments are given.

origin, as is the case here for the $S_1(\pi\pi^*)$ state. This has been expanded upon previously.[6,21]

Secondly, the τ_2 DAS (120 ± 20 fs in *N*,*N*-DMA and 110 ± 20 fs in 3,5-DMA) is attributed to the decay of the $S_2(3s/\pi\sigma^*)$ state. At photoelectron kinetic energies >0.8 eV, this appears to have the form of a sharp peak (at 1.2 eV in *N*,*N*-DMA and 0.95 eV in 3,5-DMA) superimposed on a broader background of lower amplitude. As seen in Fig. 5, the sharp peak is associated with a relatively high value of the anisotropy parameter β_2 but a very small β_4 term. In contrast, the broader component exhibits low values of both β_2 and β_4. As confirmed theoretically in our earlier work (see also Section 6), the $S_2(3s/\pi\sigma^*)$ state exhibits considerable 3s Rydberg character in the vertical Franck–Condon region and evolves $\pi\sigma^*$ valence character as the N–H (3,5-DMA) and N–CH_3 (*N*,*N*-DMA) bonds are extended. This evolution will be extremely rapid (*i.e.* it is not temporally resolvable in our present measurements), but we argue that we are able to see this change in the electronic character in a spectrally averaged way – the 3s Rydberg component of the S_2 state gives rise to a narrow photoelectron peak with high β_2 anisotropy – a consequence of a strong propensity for diagonal ionisation and well-defined orbital angular momentum, respectively. Since excited states of predominantly s orbital character possess little or no inherent alignment, the lack of a significant β_4 contribution to the PAD is also expected here. As the $\pi\sigma^*$ valence character of the S_2 state increases (at the expense of the Rydberg character) a reduction in the propensity for diagonal ionisation and less well-defined orbital angular momentum leads to a broader feature in the photoelectron spectra/τ_2 DAS with lower associated PAD anisotropy – as is observed.

At photoelectron kinetic energies <0.8 eV, there is a clear difference between the τ_2 DAS obtained for the two systems under consideration; *N*,*N*-DMA exhibits negative amplitude across this region, whereas 3,5-DMA shows zero amplitude. This negative amplitude in the *N*,*N*-DMA data provides a strong indication of a sequential dynamical process occurring in this system that is absent in 3,5-DMA. As described in Section 3.1, all exponentials used in the global DAS fit originate from zero pump–probe delay ($\Delta t = 0$). The negative amplitude in the *N*,*N*-DMA τ_2 DAS (which describes a short-lived process) is therefore effectively compensating for the positive amplitude at short delay times associated with a sequential feature (*i.e.*, one not truly originating from $\Delta t = 0$) described by a longer-lifetime DAS in the same energy region (which in this case is the τ_3/τ_4 DAS associated with ionization from the $S_1(\pi\pi^*)$ state). Physically, this may be interpreted as a transfer of the $S_2(3s/\pi\sigma^*)$ state population in *N*,*N*-DMA to the $S_1(\pi\pi^*)$ state *via* internal conversion. The apparent absence of this internal conversion process in 3,5-DMA means that a different, but still extremely rapid, mechanism is responsible for the $S_2(3s/\pi\sigma^*)$ decay in this instance. Our TRPEI measurement is effectively "blind" to this alternative pathway, but we may assume that it is direct dissociation along the N–H stretching coordinate within the excited state, as evidenced by observations of H atom release occurring on a very rapid time scale in aniline at similar excitation wavelengths.[12,14] This is also strongly suggested by the dissociative nature of the $S_2(3s/\pi\sigma^*)$ state revealed in our supporting calculations (see Fig. 6) which, additionally, show that an equivalent N–CH_3 fragmentation pathway in *N*,*N*-DMA should be possible (noting that at 240 nm we are exciting well above any small barrier present along this dissociative coordinate). In *N*,*N*-DMA we therefore appear to have the possibility of two competing $S_2(3s/\pi\sigma^*)$ decay pathways; $S_1(\pi\pi^*)$ internal conversion and CH_3 elimination, whereas in 3,5-DMA, H atom elimination is overwhelmingly dominant. Further evidence for this is clearly revealed in the PAD data, as described below.

Non-adiabatic coupling interactions between different states effectively mixes their electronic (or vibronic) character and this may be reflected in the temporal evolution of the PADs observed in a TRPEI measurement.[31–33] As discussed in Section 3.2 (and illustrated in Fig. 5), the temporal evolution of the anisotropy parameters at low photoelectron kinetic energies following 240 nm excitation of

N,N-DMA and 3,5-DMA may be regarded as a superposition of behaviour seen at 250 nm and a new additional component which is attributed to $2\pi\pi^*$ excitation. Here the PADs associated with low kinetic energy photoelectrons directly reflect the electronic character of the $S_1(\pi\pi^*)$ state. Any evolution of β_2 and β_4 may be assumed (as a first approximation[34]) to arise as a consequence of this state sampling nuclear geometries that then significantly change the electronic character (for example, geometries close to conical intersections). Upon comparing the β_2 data in Fig. 5 for 240 nm excitation (main plot) and 250 nm excitation (inset panel), it is apparent that the temporal evolution in the PADs associated with accessing the $S_1(\pi\pi^*)/S_2(3s/\pi\sigma^*)$ CI geometry is considerably slower in 3,5-DMA (~1.5 ps) than in *N,N*-DMA (~150 fs). Such an observation is therefore consistent with *N,N*-DMA exhibiting competing internal conversion and CH_3 elimination decay pathways for the $S_2(3s/\pi\sigma^*)$ state as the timescales for the two processes are likely to be similar. In contrast, methylation of the aromatic ring system in 3,5-DMA slows down the time to access the $S_1(\pi\pi^*)/S_2(3s/\pi\sigma^*)$ CI sufficiently that the faster H atom elimination channel becomes overwhelmingly favoured.

As demonstrated above, being able to use the PAD data afforded by the TRPEI technique to resolve such intramolecular kinetics in more detail provides a useful demonstration of the value the imaging approach brings to the study of ultrafast dynamics. In spite of this, however, we highlight one key piece of information we have so far been unable to comment on: for the case of *N,N*-DMA we are unable to quantify the relative branching of the internal conversion and direct dissociation channels. Upon first inspection of the τ_2 DAS (associated with the decay of the $S_2(3s/\pi\sigma^*)$ state), one may initially conclude that the relative sizes of the (large) positive amplitude feature and the (much smaller) negative amplitude feature implies that internal conversion is a relatively minor pathway. However, the negative amplitude component is effectively associated with ionization from the $S_1(\pi\pi^*)$ state once internal conversion from $S_2(3s/\pi\sigma^*)$ has taken place. As discussed in previous work, we may (conservatively) assume that the relative ionization cross section of the $S_2(3s/\pi\sigma^*)$ state is at least an order of magnitude larger than $S_1(\pi\pi^*)$.[6] As such, the small negative amplitude seen in the low energy region of the τ_2 DAS is therefore likely to describe a relatively large fraction of population transfer from the $S_2(3s/\pi\sigma^*)$ state (*i.e.* we may speculate that internal conversion is likely to be the dominant pathway). We also note, however, that the much smaller absorption cross section exhibited by $S_2(3s/\pi\sigma^*)$ (relative to $S_1(\pi\pi^*)$ – see Table 1) means that the initial population prepared in this state is only a very small contribution to the overall excitation process. A more detailed discussion of issues relating to inter-state detection sensitivity when using ionization detection to track non-adiabatic dynamics evolving over multiple potential energy surfaces is beyond the scope of this communication and the reader is directed to our previous work for a more expanded (albeit exploratory) discussion.[6,35]

5.2. $2\pi\pi^*$ dynamics and comparison with aniline

As already discussed, the τ_1 DAS seen in Fig. 4 is assigned to ionization from the $2\pi\pi^*$ state. The lifetime of this state is extremely rapid, effectively being described by a Gaussian in our global fitting model. The lack of negative amplitude in any region of the τ_1 DAS also appears to suggest that the decay of the $2\pi\pi^*$

state in both *N*,*N*-DMA and 3,5-DMA does not involve population transfer to either the $S_1(\pi\pi^*)$ or $S_2(3s/\pi\sigma^*)$ states (see the argument already outlined in the previous section regarding negative DAS). Initially, this is perhaps a surprising outcome as in the related species aniline, a CI connecting all three states has been predicted and direct evidence for internal conversion from $2\pi\pi^*$ to the $S_1(\pi\pi^*)$/$S_2(3s/\pi\sigma^*)$ states has been reported from experimental TRPEI data following 238 nm excitation (where negative amplitude is seen in the low energy region of the DAS corresponding to that labelled τ_1 here).[7] As was the case for the decay of the $S_2(3s/\pi\sigma^*)$ state, however, this may be rationalized through consideration of the PAD data in Fig. 5. Following 240 nm excitation in both *N*,*N*-DMA and 3,5-DMA, the temporal evolution of β_2 and β_4 show two distinct features, one of which we have already associated with accessing the $S_1(\pi\pi^*)/S_2(3s/\pi\sigma^*)$ CI geometry. We therefore suggest that the second feature is reflecting a different geometry within the $S_1(\pi\pi^*)$ state associated with an additional CI connecting to $2\pi\pi^*$. Although the timescale associated with the evolution of this second feature is relatively fast for both species under consideration (~250 fs and ~350 fs for *N*,*N*-DMA and 3,5-DMA, respectively) it is still much longer than the observed $2\pi\pi^*$ lifetime (as described by the Gaussian τ_1 DAS). In both systems we may therefore infer that (i) although decay of the $2\pi\pi^*$ state *via* the $S_1(\pi\pi^*)$ and/or $S_2(3s/\pi\sigma^*)$ states may be possible in principle, such a process is unable to compete kinetically with an alternative decay pathway (discussed below) and (ii) the $S_1(\pi\pi^*)$ and $S_2(3s/\pi\sigma^*)$ states must therefore be populated to some extent in the initial 240 nm excitation (an assumption already implicitly assumed in the discussion presented in Section 5.1). As mentioned above, this dynamical behaviour is in contrast to the situation in aniline following excitation at similar wavelengths, where internal conversion from $2\pi\pi^*$ to $S_1(\pi\pi^*)$ and/or $S_2(3s/\pi\sigma^*)$ does operate to some extent. Furthermore, it is also interesting to note that fully deuterated d7-aniline appears (on the basis of the DAS data reported by Fielding and co-workers[7]) to revert back to overall dynamical behaviour that is very similar to that seen in *N*,*N*-DMA (*i.e.* population transfer between the $S_2(3s/\pi\sigma^*)$ and $S_1(\pi\pi^*)$ states but *no* population transfer to either of these from $2\pi\pi^*$). This suggests that competition between different $2\pi\pi^*$ decay pathways is extremely sensitive to the vibrational timescales operating within these systems. Finally, it is worth highlighting here that in aniline, β_2 has been previously shown to undergo no temporal evolution in the low photoelectron kinetic energy region (0.2–0.4 eV) following 240 nm excitation.[36] This provides additional support to our kinetic arguments, as we assume this is a consequence of aniline being able to access the CI geometry connecting the $2\pi\pi^*$ state to the $S_1(\pi\pi^*)$ and/or $S_2(3s/\pi\sigma^*)$ states on a timescale that is too fast to observe within the temporal experimental resolution – and so internal conversion to these states from $2\pi\pi^*$ is able to now compete with other decay mechanisms. This is also seen to be the case for the CI connecting the $S_1(\pi\pi^*)$ and $S_2(3s/\pi\sigma^*)$ in aniline following 250 nm excitation.[6]

In both *N*,*N*-DMA and 3,5-DMA, the kinetic picture outlined above for predominant $2\pi\pi^*$ decay *via* a pathway other than internal conversion to $S_1(\pi\pi^*)$ and/or $S_2(3s/\pi\sigma^*)$ appears consistent with the geometry of a 3-state $S_1(\pi\pi^*)$/$S_2(3s/\pi\sigma^*)/2\pi\pi^*$ CI predicted theoretically in aniline by Sala *et al.*[8] and also, alternatively, the $S_1(\pi\pi^*)/2\pi\pi^*$ CI proposed by Stavros and co-workers.[12] Both require significant motion of the NH_2 group out of the ring plane and some additional distortion of the ring system itself. The work of Sala *et al.* also suggests that decay of the $2\pi\pi^*$ state may also proceed *via* two other possible (barrierless) CI pathways, both of which connect the $2\pi\pi^*$ and S_0 ground states. One of these is suggested to first pass through the aforementioned 3-state CI (and so may possibly be ruled out in *N,N*-DMA and 3,5-DMA on the basis of our previous kinetic arguments) and the second samples a geometry similar to the Dewar form of benzene. This latter process once again requires considerable out of (ring) plane motion of the NH_2 group and also significant distortion of the aromatic system. In light of this, it is perhaps surprising that we are unable to resolve any significant differences in the $2\pi\pi^*$ lifetime between *N,N*-DMA, 3,5-DMA and aniline (which has been recently reported as 50 ± 10 fs (ref. 7)). While such lifetime differences (beyond the temporal resolution of our present experiment) may exist, we also tentatively suggest that our findings open up the possibility that a $2\pi\pi^*$ decay pathway other than internal conversion to S_0 may be operating. Since our TRPEI experiment is effectively blind to this pathway, this remains an open question, but we note that our excited state calculations for *N,N*-DMA and 3,5-DMA (see Table 1) and also previously for aniline[6,17] appear to suggest the presence of several Rydberg states sitting energetically just below the $2\pi\pi^*$ state (in the Franck–Condon region). As suggested theoretically by Worth and co-workers,[17] a consideration of the coupling interactions between these Rydberg states and $2\pi\pi^*$ is necessary to model the aniline absorption spectrum satisfactorily. It therefore seems entirely possible that this interaction may also play a role in the excited state dynamics more generally. As seen in the plots of β_2 and β_4 *vs.* photoelectron kinetic energy at $\Delta t = 0$ presented in Fig. 5 (see also Section 3.2), some spectral signatures of a p-Rydberg state do appear to be evident (at least in *N,N*-DMA) and so this is an avenue that warrants further investigation. Since our present experimental findings clearly illustrate the value in using highly differential imaging methods in conjunction with systematic, site-selective substitution to study complex ultrafast non-adiabatic dynamics, we anticipate that further studies in a similar vein will prove instructive in this regard.

Finally, we once again return to the issue of relative detection sensitivity when using ionization detection to track dynamical processes involving multiple potential energy surfaces. On the basis of both the absorption spectra presented in Fig. 1 and also the oscillator strength data presented in Table 1, it is extremely surprising that photoelectron signals associated with the $2\pi\pi^*$ state ionization in both *N,N*-DMA and 3,5-DMA (τ_1 DAS in Fig. 4) are similar/smaller in size to those attributed to the $S_2(3s/\pi\sigma^*)$ and $S_1(\pi\pi^*)$ states (τ_2 DAS and τ_4 DAS, respectively). Since the $2\pi\pi^*$ state clearly exhibits a vastly stronger absorption, this initially suggests that it must therefore have (i) a greatly reduced (electronic) ionization cross section (compared to the $S_2(3s/\pi\sigma^*)$ and $S_1(\pi\pi^*)$ states) and/or (ii) very small Franck–Condon factors (within the limits imposed by 308 nm ionization). In regard to point (i), we have previously employed correlations between photoionization cross sections and the isotropic polarizability volume $\langle\bar{\alpha}\rangle$ as a useful heuristic tool in assisting the interpretation of TRPEI data.[6,35] Although a somewhat speculative approach, we note that evaluation of $\langle\bar{\alpha}\rangle$ for the $2\pi\pi^*$ and $S_1(\pi\pi^*)$ states of *N,N*-DMA (at the equilibrium geometry) suggests that the ionization cross section of the $2\pi\pi^*$ state is, in fact, likely to be somewhat larger than $S_1(\pi\pi^*)$.[37] In considering point (ii), we note that the τ_1 DAS shown in Fig. 4 does not appear to extend all the way to the predicted $(1 + 1')$ energy cut-off, indicating that we are at least sampling the full Franck–Condon progression. We therefore suggest that an alternative explanation is possibly responsible for much of the huge reduction seen in the relative amplitude of the τ_1 DAS over what might reasonably be expected. Specifically, this is the extremely short lifetime of the $2\pi\pi^*$ state relative to the temporal duration of the ionizing laser pulse (~80–100 fs), which leads to a significant reduction in the effective ionization efficiency relative to the $S_2(3s/\pi\sigma^*)$ and $S_1(\pi\pi^*)$ states (which have lifetimes comparable to or greatly in excess of the laser pulse duration). This conclusion (which may obviously be confirmed by conducting experiments with temporally shorter laser pulses) potentially highlights an important additional convoluting factor that may, in some cases, need be taken into consideration when attempting to interpret relative branching ratios in time-resolved experiments using photoionization detection.

6. General functionality effects in amine Rydberg/valence evolution

As discussed in detail above, the evolution of the 3s Rydberg to σ^* valence character within the S_2 states of both *N*,*N*-DMA and 3,5-DMA is clearly an important process in mediating some aspects of the overall relaxation dynamics in these systems following UV excitation. More generally, similar Rydberg-to-valence behaviour has also been strongly implicated in playing a key role in the excited state photochemistry observed in numerous small hetero-atom containing systems. The study of such processes has therefore attracted a great deal of attention over the last decade.[3,4] Recent TRPEI measurements conducted on a series of tertiary aliphatic (*i.e.* saturated) amines have, however, suggested that the very long (>10 ps) overall excited state lifetimes seen in these systems are a consequence of the 3s state being unable to evolve σ^* valence character and therefore they undergo rapid decay (*via* N–C bond fission).[18,19] In contrast, secondary and primary aliphatic amine species exhibit ultrafast ($\ll$1 ps) decay of the 3s state as the σ^* valence character is able to develop in this instance.[20] This difference in dynamical timescales is illustrated in Fig. 7 for the case of piperidine (a secondary aliphatic amine) and *N*-methylpyrrolidine (a tertiary aliphatic amine) following 200 nm excitation. In both cases, a member of the 3p Rydberg manifold is initially populated. This undergoes rapid (<400 fs) internal conversion to the 3s state, which then exhibits a decay lifetime of <200 fs in piperidine and 160 ps in *N*-methylpyrrolidine. We stress that this profound difference in 3s lifetime does not appear to be a simple tunnelling rate effect that may arise due to the presence of any small barrier along the N–H/N–CH_3 stretch of the excited state potential surface. This is initially suggested by the fact that, following 200 nm excitation, the 3s states of both molecules are prepared (conservatively) at least 0.6 eV above their electronic origins. Exploring this in more detail computationally, calculations obtained using both the EOM-CCSD and CASSCF levels of theory indicate that there are clear general differences in the way that tertiary and secondary/primary aliphatic amines evolve valence character within their Rydberg states.[18–20] In both classes of molecule, members of the 3p manifold (but not the 3s state) appear to develop σ^* valence character along N–C bond extensions (a factor responsible for mediating the 3p to 3s internal conversion mentioned above). In contrast, 3s to σ^* evolution is seen along N–H coordinates. Although an

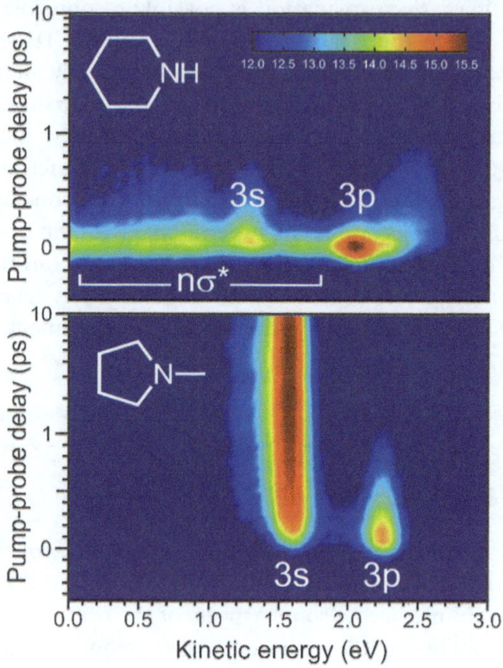

Fig. 7 2D time-dependent photoelectron spectra of piperidine (top) and *N*-methylpyrrolidine (bottom) obtained using a similar TRPEI setup to that described in the Experimental section using a 200 nm pump and 267 nm probe. The time axis is linear to +1 ps and then logarithmic beyond this point. The intensity colour map is presented on a natural logarithmic scale based on the output directly obtained from the CCD camera used to record the raw photoelectron image data. The *N*-methylpyrrolidine and piperidine data are adapted from ref. 18 and 20, respectively.

interesting observation in its own right, this situation is made somewhat more confusing by the fact that the aromatic amines *N*,*N*-DMA (a tertiary system) and 3,5-DMA (a primary system) clearly both exhibit dynamics consistent with the development of σ* character within the 3s Rydberg state (note also here that in Fig. 7 piperidine appears to show the same spectrally averaged ionization signatures of 3s to σ* evolution). This suggests that the presence of the aromatic ring system in *N*,*N*-DMA may exert an additional influence over the electronic state character as a function of nuclear coordinates, in turn modifying the UV relaxation dynamics.

Given the observations highlighted above, we have undertaken a preliminary theoretical investigation that begins to explore the effect of ππ* states associated with unsaturated chemical functionality on the nature of valence character evolution within the Rydberg states of amines. Fig. 8 presents a series of selected tertiary amines that systematically introduce C═C double bonds in close proximity to the amine N atom centre. For all 4 systems considered, PES cuts were evaluated along N–C bond extensions for several low-lying singlet excited states. Rigid scans were performed for the lowest energy conformer of each species. The computational approach was essentially identical to that already applied to *N*,*N*-DMA and 3,5-DMA (see Section 4) and the key results are summarized in

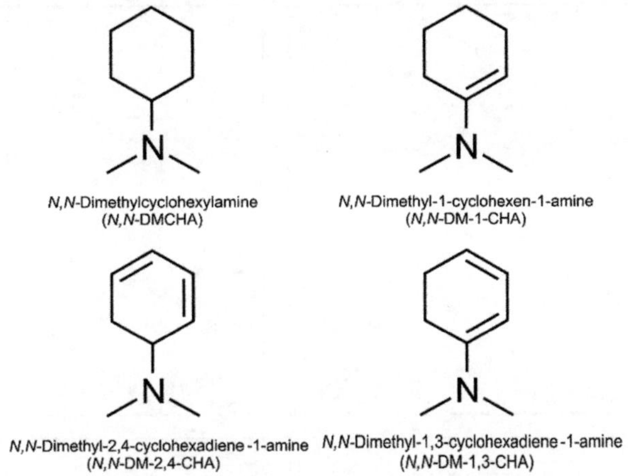

Fig. 8 Schematic depictions of *N,N*-dimethylcyclohexylamine (*N,N*-DMCHA), *N,N*-dimethyl-1-cyclohexen-1-amine (*N,N*-DM-1-CHA), *N,N*-dimethyl-2,4-cyclohexadiene-1-amine (*N,N*-DM-2,4-CHA) and *N,N*-dimethyl-1,3-cyclohexadiene-1-amine (*N,N*-DM-1,3-CHA).

Fig. 9. In the fully saturated system *N,N*-dimethylcyclohexylamine (*N,N*-DMCHA) we find that along the N–CH_3 coordinate evolution of Rydberg to (nσ^*) valence character is observed from a heavily mixed p-Rydberg state lying in the quasi-molecular plane of the cyclohexane ring. At an N–C separation of ~1.9 Å this state crosses with the 3s Rydberg state, which remains strongly bound as the bond is extended. A similar scan along the N–C bond connecting to the ring system (not shown in Fig. 9) yields the same overall behaviour. Upon introducing one C=C double bond into *N,N*-DMCHA, three isomers are possible, and we have investigated the specific species *N,N*-dimethyl-1-cyclohexen-1-amine (*N,N*-DM-1-CHA) – chosen as it places the unsaturated functionality closest to the N atom centre. Here we once again find p-Rydberg to valence evolution, similar to the fully saturated case – although we note that the Rydberg states are extremely mixed in this system and tracking their (adiabatic) evolution as the N–C bond extends is particularly challenging. Energetically, the lowest-lying $\pi\pi^*$ state associated with the C=C double bond sits well above the 3s and 3p Rydberg manifold (and is not shown in the Fig. 9 plot). Increasing the complexity further, for diene systems several isomers are again possible, and we have investigated the conjugated 1,3 species *N,N*-dimethyl-1,3-cyclohexadien-1-amine (*N,N*-DM-1,3-CHA) that is additionally conjugated with the nitrogen lone pair, and the corresponding 2,4 species *N,N*-dimethyl-2,4-cyclohexadien-1-amine (*N,N*-DM-2,4-CHA) where the N atom interaction is absent. For both systems there are now two $\pi\pi^*$ states, one embedded within the Rydberg manifold and one (that is considerably more diffuse) at much higher energies. Only the former is included in Fig. 9. In terms of the valence character evolution along the N–CH_3 stretch we now observe differing behaviour in these species. The nitrogen conjugated species *N,N*-DM-1,3-CHA exhibits less s/p Rydberg mixing, and here it is the lowest (predominantly 3s) state that evolves valence σ^* character, behaviour that is now similar to the aromatic tertiary system *N,N*-DMA (see Fig. 6). In the case of *N,N*-DM-2,4-CHA, however, the

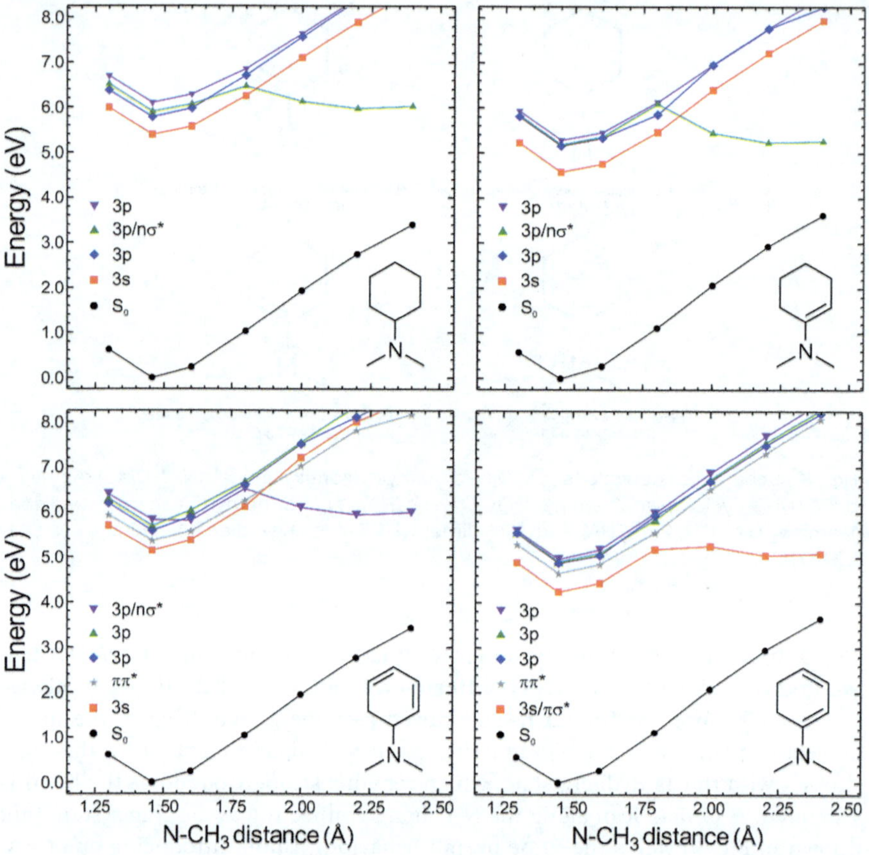

Fig. 9 EOM-CCSD/aug-cc-pVDZ potential energy cuts along the N–CH_3 stretching coordinate of *N*,*N*-DMCHA (top left), *N*,*N*-DM-1-CHA (top right), *N*,*N*-DM-2,4-CHA (bottom left) and *N*,*N*-DM-1,3-CHA (bottom right). For additional details, see the main text. The p manifold is heavily mixed with orbitals directed away from the *x*, *y* and *z* molecular axes. As such, no specific assignments are given.

s/p mixing is greater, and it is a higher notional p-Rydberg state that evolves the valence σ* character in a manner similar to both the species with zero and one double bonds in the ring, and also reminiscent of our previous tertiary aliphatic amines studies.[18,19] Overall these are interesting findings that, although preliminary, reinforce the idea that dynamical behaviour in tertiary amines is strongly influenced by the extent to which the lone pair on the N atom is able to interact with nearby unsaturated functional groups.

To further extend the theoretical characterization of our systematically varied systems we have evaluated the isotropic invariant of the second-moment of charge tensor relative to the S_0 ground state $\Delta\langle r_{\mathrm{iso}}{}^2\rangle$ (a measure of the spatial extent of an electronic state) to compare the two extremes: the aromatic species *N*,*N*-DMA, and the fully saturated tertiary amine *N*,*N*-DMCHA. For the second moments we have used a slightly reduced basis set of aug-cc-pVDZ on N and C atoms, and the cc-pVDZ basis on hydrogens. As demonstrated previously, $\Delta\langle r_{\mathrm{iso}}{}^2\rangle$ provides useful additional information for parametrizing the nature (and

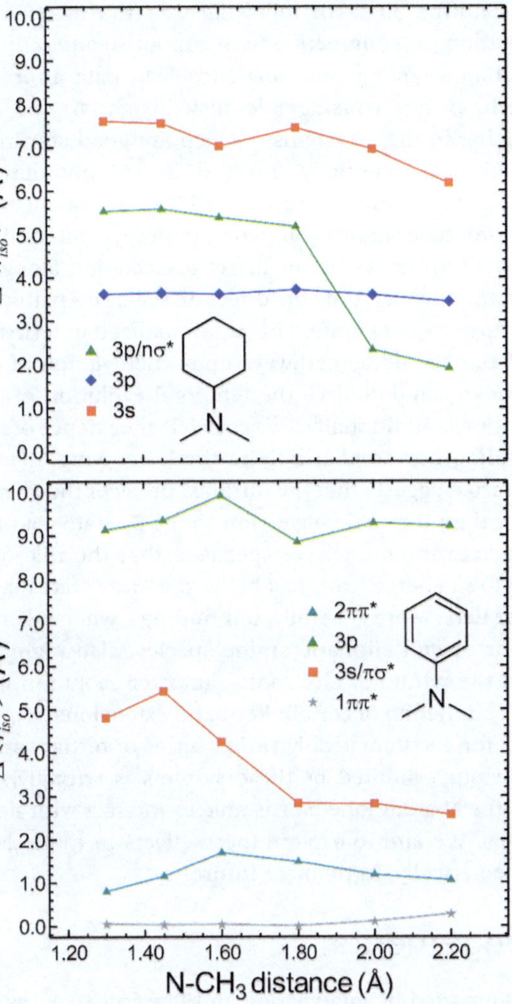

Fig. 10 Evolution of the excited-state second-moment of the electronic charge distribution with respect to the ground state $\Delta\langle r_{\mathrm{iso}}{}^2\rangle$ in *N*,*N*-DMCHA and *N*,*N*-DMA as a function of the N–CH_3 bond extension. For additional details see the main text. Data for only one of the 3p states showing no significant valence evolution is included.

evolution) of electronic character, with Rydberg states typically exhibiting values that are somewhat larger than valence states (often by an order of magnitude or more).[6,35,38] We find here that the Rydberg-to-valence character evolution along the N–CH_3 stretch seen in Fig. 9 is also strongly reflected in the $\Delta\langle r_{\mathrm{iso}}{}^2\rangle$ values, as illustrated in Fig. 10. In *N*,*N*-DMCHA it is a higher-lying (notionally) p-Rydberg state that undergoes a significant orbital size contraction but in *N*,*N*-DMA it is the 3s state that, instead, exhibits this behaviour.

7. Conclusion

Time-resolved photoelectron imaging readily reveals the presence of multiple non-adiabatic pathways in the aniline derivatives *N*,*N*-dimethylaniline (*N*,*N*-DMA) and 3,5-dimethylaniline (3,5-DMA) following excitation at 240 nm. The use of site-selective methylation, in conjunction with supporting *ab initio* calculations and the highly differential energy- and angle-resolved data afforded by our experimental approach, brings considerable new insight to the overall relaxation dynamics operating in these systems – which are good starting models for biological chromophores potentially involved in UV photoprotection. Detailed analysis of our data reveals decay of the mixed Rydberg/valence $S_2(3s/\pi\sigma^*)$ state in *N,N*-DMA likely proceeds *via* two competing pathways: internal conversion to the (much longer-lived) $S_1(\pi\pi^*)$ state or direct dissociation along the N–CH_3 coordinate. In 3,5-DMA, however, only the direct dissociation pathway (in this case N–H dissociation) appears to operate. This is rationalized in terms of changes in the relative rates of the two decay pathways upon methylation of the aromatic ring system – as strongly supported by the temporal evolution of the photoelectron angular distributions. Additionally, we see no clear evidence of the second excited $2\pi\pi^*$ state coupling non-adiabatically to the lower-lying $S_1(\pi\pi^*)$ or $S_2(3s/\pi\sigma^*)$ states. This, in turn, suggests that the ultrafast decay of the $2\pi\pi^*$ state may occur *via* a route other than internal conversion to the S_0 state – a pathway previously suggested to operate in aniline. We speculate that the role of several Rydberg states predicted to sit energetically just below the $2\pi\pi^*$ state may ultimately prove to be a key factor here. More generally, our findings, when taken along with other recent results for several aliphatic amine species, also highlight more general questions about the nature of electronic character evolution in mixed Rydberg-valence states as a function of certain key bond extensions. Exploring these ideas computationally for a systematically varied series of tertiary amines reveals that dynamical behaviour exhibited by these systems is strongly influenced by the extent to which the N atom lone pair is able to interact with nearby unsaturated functional groups. We aim to explore these effects in more detail, both experimentally and theoretically, in the near future.

Acknowledgements

This work was supported by Engineering and Physical Sciences Research Council (EPSRC) Grant No. EP/K021052/1. M. J. P. is supported by the European Research Council under the European Union's Seventh Framework Programme (FP7/2007–2013)/ERC Grant No. 258990. M. M. Z. and S. W. C thank Heriot-Watt University for PhD funding. We also acknowledge several collaborators who have contributed in numerous ways to various aspects of the work discussed here; J. O. F Thompson, E. A Burgess, L. B. Klein and T. I. Sølling.

References

1 C. E. Crespo-Hernández, B. Cohen, P. M. Hare and B. Kohler, *Chem. Rev.*, 2004, **104**, 1977.
2 C. S. Cockell and J. Knowland, *Biol. Rev. Cambridge Philos. Soc.*, 1999, **74**, 311.
3 M. N. R. Ashfold, G. A. King, D. Murdock, M. G. D. Nix, T. A. A. Oliver and A. G. Sage, *Phys. Chem. Chem. Phys.*, 2010, **12**, 1218.
4 G. M. Roberts and V. G. Stavros, *Chem. Sci.*, 2014, **5**, 1698.
5 I. Cnossen, J. Sanz-Forcada, F. Favata, O. Witasse, T. Zegers and N. F. Arnold, *J. Geophys. Res.*, 2007, **112**, E02008.

6 J. O. F. Thompson, L. Saalbach, S. W. Crane, M. J. Paterson and D. Townsend, *J. Chem. Phys.*, 2015, **142**, 114309.
7 O. M. Kirkby, M. Sala, G. Balerdi, R. de Nalda, L. Bañares, S. Guérin and H. H. Fielding, *Phys. Chem. Chem. Phys.*, 2015, **17**, 16270.
8 M. Sala, O. M. Kirkby, S. Guérin and H. H. Fielding, *Phys. Chem. Chem. Phys.*, 2014, **16**, 3122.
9 J. O. F. Thompson, R. A. Livingstone and D. Townsend, *J. Chem. Phys.*, 2013, **139**, 034316.
10 R. Spesyvtsev, O. M. Kirkby, M. Vacher and H. H. Fielding, *Phys. Chem. Chem. Phys.*, 2012, **14**, 9942.
11 R. Spesyvtsev, O. M. Kirkby and H. H. Fielding, *Faraday Discuss.*, 2012, **157**, 165.
12 G. M. Roberts, C. A. Williams, J. D. Young, S. Ullrich, M. J. Paterson and V. G. Stavros, *J. Am. Chem. Soc.*, 2012, **134**, 12578.
13 R. Montero, A. Peralta-Conde, V. Ovejas, R. Martínez, F. Castaño and A. Longarte, *J. Chem. Phys.*, 2011, **135**, 054308.
14 G. A. King, T. A. A. Oliver and M. N. R. Ashfold, *J. Chem. Phys.*, 2010, **132**, 214307.
15 X.-J. Hou, P. Quan, T. Höltzl, T. Veszprémi and M. T. Nguyen, *J. Phys. Chem. A*, 2005, **109**, 10396.
16 Y. Honda, M. Hada, M. Ehara and H. Nakatsuji, *J. Chem. Phys.*, 2002, **117**, 2045.
17 F. Wang, S. P. Neville, R. Wang and G. A. Worth, *J. Phys. Chem. A*, 2013, **117**, 7298.
18 J. O. F. Thompson, L. B. Klein, T. I. Sølling, M. J. Paterson and D. Townsend, *Chem. Sci.*, 2016, **7**, 1826.
19 L. B. Klein, T. J. Morsing, R. A. Livingstone, D. Townsend and T. I. Sølling, *Phys. Chem. Chem. Phys.*, 2016, **18**, 9715.
20 L. B. Klein, J. O. F. Thompson, S. W. Crane, L. Saalbach, T. I. Sølling, M. J. Paterson and D. Townsend, *Phys. Chem. Chem. Phys.*, 2016, **18**, 25070.
21 R. A. Livingstone, J. O. F. Thompson, M. Iljina, R. J. Donaldson, B. J. Sussman, M. J. Paterson and D. Townsend, *J. Chem. Phys.*, 2012, **137**, 184304.
22 U. Even, J. Jortner, D. Noy, N. Lavie and C. Cossart-Magos, *J. Chem. Phys.*, 2000, **112**, 8068.
23 A. T. J. B. Eppink and D. H. Parker, *Rev. Sci. Instrum.*, 1997, **68**, 3477.
24 L. Szepes, G. Distefano and S. Pignataro, *Ann. Chim.*, 1974, **64**, 159.
25 J. P. Maier and D. W. Turner, *J. Chem. Soc., Faraday Trans. 2*, 1973, **69**, 521.
26 T. Suzuki, *Annu. Rev. Phys. Chem.*, 2006, **57**, 555.
27 K. L. Reid, *Annu. Rev. Phys. Chem.*, 2003, **54**, 397.
28 M. J. Frisch, G. W. Trucks, H. B. Schlegel, G. E. Scuseria, M. A. Robb, J. R. Cheeseman, G. Scalmani, V. Barone, B. Mennucci, G. A. Petersson, H. Nakatsuji, M. Caricato, X. Li, H. P. Hratchian, A. F. Izmaylov, J. Bloino, G. Zheng, J. L. Sonnenberg, M. Hada, M. Ehara, K. Toyota, R. Fukuda, J. Hasegawa, M. Ishida, T. Nakajima, Y. Honda, O. Kitao, H. Nakai, T. Vreven, J. A. Montgomery Jr, J. E. Peralta, F. Ogliaro, M. Bearpark, J. J. Heyd, E. Brothers, K. N. Kudin, V. N. Staroverov, R. Kobayashi, J. Normand, K. Raghavachari, A. Rendell, J. C. Burant, S. S. Iyengar, J. Tomasi, M. Cossi, N. Rega, J. M. Millam, M. Klene, J. E. Knox, J. B. Cross, V. Bakken, C. Adamo, J. Jaramillo, R. Gomperts, R. E. Stratmann, O. Yazyev, A. J. Austin, R. Cammi, C. Pomelli, J. W. Ochterski, R. L. Martin,

K. Morokuma, V. G. Zakrzewski, G. A. Voth, P. Salvador, J. J. Dannenberg, S. Dapprich, A. D. Daniels, Ö. Farkas, J. B. Foresman, J. V. Ortiz, J. Cioslowski and D. J. Fox, *Gaussian 09, Revision D.01*, Gaussian, Inc., Wallingford CT, 2009.
29 R. Scheps, D. Florida and S. A. Rice, *J. Chem. Phys.*, 1974, **61**, 1730.
30 B. Kim, C. P. Schick and P. M. Weber, *J. Chem. Phys.*, 1995, **103**, 6903.
31 G. Wu, P. Hockett and A. Stolow, *Phys. Chem. Chem. Phys.*, 2011, **13**, 18447.
32 Y.-I. Suzuki, T. Horio, T. Fuji and T. Suzuki, *J. Chem. Phys.*, 2011, **134**, 184313.
33 C. Z. Bisgaard, O. J. Clarkin, G. Wu, A. M. D. Lee, O. Gessner, C. C. Hayden and A. Stolow, *Science*, 2009, **323**, 1464.
34 We note that additional dynamical factors (other than non-adiabatic coupling) may also influence/convolute the observed form of the PADs. Primarily these are (i) changes in the relative phases and amplitudes of the outgoing photoelectron partial waves due to conformational rearrangement modifying the scattering dynamics off the non-spherical ion core potential and (ii) major changes in orbital character that may occur as a function of nuclear coordinates (for example, the significant Rydberg-to-valence evolution seen here in the $S_2(3s/\pi\sigma^*)$ state). Point (i) is generally expected to be relatively small and point (ii) is not expected to be significant for the predominantly valence $S_1(\pi\pi^*)$ state considered here.
35 M. M. Zawadzki, J. O. F. Thompson, E. A. Burgess, M. J. Paterson and D. Townsend, *Phys. Chem. Chem. Phys.*, 2015, **17**, 26659.
36 R. Spesyvtsev, PhD Thesis, University College London, 2013.
37 Using linear response CCSD with a mixed basis (aug-cc-pVDZ on C and N, and cc-pVDZ on H). Polarizability volumes at ground state equilibrium geometries are 2.04 Å^3 for $S_1(\pi\pi^*)$ and 26.08 Å^3 for $2\pi\pi^*$.
38 H. Reisler and A. I. Krylov, *Int. Rev. Phys. Chem.*, 2009, **28**, 267.

Faraday Discussions

DISCUSSIONS

Electronic and non-adiabatic dynamics: general discussion

Andrew J. Orr-Ewing, Jan R. R. Verlet, Tom J. Penfold, Russell S. Minns, Michael P. Minitti, Theis I. Sølling, Oliver Schalk, Markus Kowalewski, Jon P. Marangos, Michael A. Robb, Allan S. Johnson, Hans Jakob Wörner, Dmitrii V. Shalashilin, R. J. Dwayne Miller, Wolfgang Domcke, Kiyoshi Ueda, Peter M. Weber, Raluca Cireasa, Morgane Vacher, Gareth M. Roberts, Piero Decleva, Filippo Bencivenga, Daniel M. Neumark, Oliver Gessner, Albert Stolow, Pankaj Kumar Mishra, Iakov Polyak, Kyoung Koo Baeck, Adam Kirrander, Danielle Dowek, Álvaro Jiménez-Galán, Fernando Martín, Shaul Mukamel, Taro Sekikawa, Maxim F. Gelin, Dave Townsend, Dmitry V. Makhov and Simon P. Neville

DOI: 10.1039/c6fd90070d

Jon Marangos opened a general discussion of the paper by Fernando Martín: The work was in part motivated by IR XUV probes and strong IR field experiments. Couplings to cation resonances are clearly very important in the IR/vis region – what do you anticipate to be the role of these strong field couplings and how can you unpick this effect in the experiment?

Fernando Martín responded: In the experiment, the molecule is ionized by using a single XUV attosecond pulse. This process is clean and it is not contaminated by the presence of an IR pulse. This second pulse is used to probe the dynamics induced by the attosecond pulse. Therefore it is at this stage where the IR field can induce additional couplings between the cationic states that might complicate the analysis of the dynamics induced by the attosecond pulse or even blur it. Our calculations were able to describe the dynamics induced by the pump pulse, but did not take into account the probing step. Surprisingly, the results of these calculations catch most of the features that are observed in the experiment after probing the system with the IR pulse. Why is this so? We do not know yet. Preliminary results shown during my talk suggest that, if the probe pulse is not very intense, the additional features introduced by this latter pulse do not destroy the original dynamics generated by the pump pulse, although of course the analysis is more cumbersome because the signature of this dynamics is superimposed to that induced by the probe pulse. If this is a general conclusion for all systems, I cannot say. We should develop more sophisticated theoretical tools that allow us to accurately describe the probing step in such complicated systems. In parallel, additional experiments performed with weak IR probe pulses or, even better, XUV probe pulses that lead to further ionization of the system without introducing additional couplings between cationic states, should be carried out.

Kiyoshi Ueda asked: You pointed out that the electron migration is not affected by the nuclear motion in the first 8 fs but has a significant effect later. Could you explain what happens at about 8 fs? Is the nuclear motion that affects the charge migration related to the vibrational motion of the hydrogen atoms? Does the diabatic transition at the conical intersection occur at about 8 fs?

Fernando Martín answered: It is a combination of all the above. In 8 fs, light nuclei have enough time to move significantly, but also the corresponding nuclear wave packets have enough time to pass through a conical intersection. The precise number, 8 fs, is only meaningful for the particular example that was discussed during my presentation. We have other examples where the effect of nuclear dynamics occurs earlier or later. It depends very much on the characteristics of the conical intersection and of how far away they are from the initial position of the nuclear wavepacket that is created after ionization.

Daniel Neumark remarked: Can you offer an intuitive explanation as to why the temporal oscillations associated with charge migration appear specifically in the immonium dication signal?

Fernando Martín responded: We cannot say from the present simulations because they do not include the effect of the probe pulse, which is ultimately responsible for the observed fragmentation. Therefore, I will answer your question based on my intuition. Our calculations show that the observed oscillations in phenylalanine are mainly associated with charge migration from the amino and the carboxyl groups to the other end of the molecule, the phenyl ring. Hole states mainly localized in this ring actively participate in the migration process. So, all fragmentation channels associated with the break up of the ring are unlikely to exhibit any oscillations, since the complex dynamics associated with fragmentation of the ring will be superimposed to the charge migration dynamics initiated by the attosecond pulse and can obscure it. In contrast, the channel leading to immonium fragments respects the ring and is associated to the loss of the small carboxyl group. Intuitively, I would say that the dynamics associated with this fragmentation channel is much simpler than in the former case and less perturbed by the IR probe pulse, therefore, there is a larger chance to observe the original charge migration dynamics. Also, previous experimental work[1] suggests that the probe pulse preferentially removes electrons from the amino group, thus giving additional support to the preferential role played by this group in displaying the XUV-induced dynamics. In any case, these are only speculations. A definite conclusion can only be obtained by including the nuclear dynamics and the probe pulse in the theoretical simulations. We are still far from doing so accurately, but our work shows possible directions to take.

1 L. Belshaw, F. Calegari, M. J. Duffy, A. Trabattoni, L. Poletto, M. Nisoli and J. B. Greenwood, *J. Phys. Chem. Lett.*, 2012, **3**, 3751–3754.

Albert Stolow commented: Could you comment on the role of a selected observable (*e.g.*, the photoelectron spectrum) in elucidating complex wavepacket dynamics? How does the choice of observable "filter" the wavepacket dynamics?

Fernando Martín answered: Indeed, the choice of a given observable filters the wavepacket dynamics. So, by just looking at a single observable one loses a lot of information. In the experiments with which we were comparing the results of our theoretical calculations, only charged molecular fragments were detected. In my opinion, this is not the best observable to choose when investigating electron dynamics induced by attosecond pulses because a lot of other things must happen before fragmentation occurs. Detection of the photoelectron and, in particular, of the corresponding energy and angularly resolved photoelectron yields, would be a much better option, because it would give direct access to dynamics induced by the pump pulse, *i.e.*, well before fragmentation occurs. I know that this is not so easy when working with attosecond pulses because of their low intensity and hence poor statistics in the measured yields. However, I encourage experimentalists to make efforts in this direction.

Even better would be to measure electrons and ions in coincidence, which would give access to a more complete description of the dynamics of the system, but this is even more challenging for the reasons mentioned above.

Danielle Dowek remarked: This question is related to the description of the first step in the studied ultrafast electron dynamics taking place in an attosecond pump–probe process, *i.e.*, ionization of the molecule by the attosecond pulse. including the influence of the emitted electron, while it still interacts with the remaining electrons of the cation long after the interaction of the system with the attosecond pulse. From the study of photoionization dynamics in small molecules, we know that this dynamical electronic correlation may induce electron emission anisotropies in the molecular frame (asymptotically), which *a priori* leave a residual footprint in the electronic state in which the cation is left. In the reported time resolved studies, one might then consider that the "initial state" of the cation mirrors the dynamics of the emitted electron and therefore may be influenced accordingly. The results of the calculations describing the effect of the ionized electron on the cation dynamics in terms of hole density, presented Fig.1 and 2 in the paper, show that after 1 femtosecond the effect of correlation induced by the ionized electron vanishes, in the sense that the N-electron system and (N-1)-electron system calculations lead to similar hole densities: so that the conclusion of this interesting section of the paper is that the description of the electronic wavepacket generated by the XUV pulse must account for ionization, however that it does not mean that it also accounts for this part of the ionization process. The question is why is it so? Is this expected to be general or could we imagine some situations where the effect could be durable?

Fernando Martín responded: I would tentatively say that this is a general conclusion when the only ionized electrons are the photoelectrons, *i.e.*, those that are ejected by directly absorbing the photon, and the energy of these photoelectrons is not too low. Under this circumstance, the electron escapes very rapidly from the molecule, so that 1 fs later is too far away to have any significant influence on the remaining electrons. The situation would be completely different when Auger electrons are produced. In this case, the electrons will take much longer to be emitted and therefore one can expect that they will have a significant effect on the cation dynamics at much longer times. This is reinforced by the fact that Auger electrons may be slow, so that they will stay longer around the molecule. However, in this case, there will be an additional factor that will seriously affect the cation dynamics: by the time the Auger electron is produced, the nuclei may have moved significantly, so that the cation electron dynamics will no longer correspond to the original geometry of the molecule.

When Auger electrons are not produced, there is still a possibility to see a significant effect of the photoelectron on the cation electron dynamics. Imagine a large 3D molecule and an ionization process that selectively removes an electron from the central region of the molecule. Under this circumstance, the photoelectron will certainly have the chance to interact with the molecular periphery in its way towards the vacuum. Thus, any charge migration process involving this peripheral region will be affected by the photoelectron until the latter has completely escaped from the molecular environment.

Hans Jakob Wörner asked: What is the role of electronic decoherence of the ionic states, caused by the departing photoelectron (see Ref. 1)? How is this included in your calculations?

1 S. Pabst and R. Santra, *Phys. Rev. Lett.*, 2011, **106**, 053003.

Fernando Martín replied: The effect of the departing electron is included in our calculations. Our theoretical method allows us to calculate the actual continuum state that is generated by an attosecond pulse in a large molecule. This state represents both the electrons that remain bound to the cation and the electron that is ejected into the continuum. The details of the method can be found in the Faraday Discussion paper related to this meeting and in the references therein (DOI: 10.1039/C6FD00074F). To quantify the actual role of the departing electron, we have compared the electron wavepacket dynamics resulting from this full calculation with that resulting from a calculation in which the departing electron is supposed to reach the asymptotic region instantaneously. For the latter calculation, we use a density matrix formalism similar to that described in the reference you mention. From this comparison we conclude that the departing electron only plays a significant role during approximately the first 0.5 fs, After this, the ejected electron does not affect the cation dynamics any more.

Adam Kirrander asked: As far as I understand, the calculations in your paper account for direct ionisation into the flat continuum of the molecule. Have you considered investigating the effect of closed ionisation channels (*i.e.* the structured continuum) on the charge migration? Presumably the rescattering of the autoionising electron could lead to interesting dynamics, along the lines of what has been examined for Rydberg wave packets in simple molecules (see *e.g.*, Ref. 1).

1 A. Kirrander, Ch. Jungen and H. H. Fielding, *Phys. Chem. Chem. Phys.*, 2010, **12** 8948–8952.

Fernando Martín responded: Several years ago, we did this in detail for the hydrogen molecule (see Refs. 1–3 and references therein). We found that the dynamics associated with autoionization is indeed very interesting, but also more challenging to describe theoretically, mainly because autoionization is a "slow" process and the nuclei can move significantly in the meantime. As a result, the theoretical treatment must account for the coupling between electronic and nuclear degrees of freedom. A similar treatment is still possible for other diatoms, but not for large molecules, mainly because of the difficulty to describe electronic and nuclear dynamics in the molecular continuum on an equal footing. In our group, we are taking the first steps towards achieving this goal.

1 I. Sánchez and F. Martín, *Phys. Rev. Lett.*, 1997, **79**, 1654–1657.
2 F. Martín, *J. Phys. B*, 1999, **32**, R197–R231.
3 A. Palacios, J. L. Sanz-Vicario and F. Martín, *J. Phys. B: At., Mol. Opt. Phys.*, 2015, **48**, 242001.

Morgane Vacher remarked: The exact composition of the initial electronic wavepacket, *i.e.* the relative weights and phases of the electronic states, will depend on the experimental setup. By using the relative weights and phases of the electronic states as free parameters in our own simulations, we have learnt how important the initial composition of the electronic wavepacket is for the subsequent electron and nuclear dynamics.

One of the challenges of theory is to be able to predict the exact composition of the electronic wavepacket in a given experiment. My understanding is that the amplitude and phase of each frequency component within the pulse bandwidth are imprinted on the weight and phase of the electronic state excited by that frequency component. The electric field polarisation also plays an important role. Having access to all these experimental parameters is not always possible.

It seems like in the present work, you use only the temporal envelope of experimental pulses to determine the initial wavepacket; for instance, the optical phase F is fixed to zero (see section 2.1). Can you comment on the potential errors made in the predicted initial electronic wave packet by such assumptions?

Fernando Martín replied: You are right. In order to evaluate the actual electronic wavepacket that is created in a real experiment, one has to use the actual wavepacket used in that experiment. As you say, very often only the spectral shape of the pulse is known, and this is not so easy to measure! In the experiment with which we compare, the optical phase is not known. However, since the attosecond pulse used in that experiment contains many cycles, the effect of the phase is expected to be small. Indeed, we have explicitly checked that the results do not change significantly when we use a pulse with a phase different from zero. For shorter pulses, however, this conclusion is not valid and knowledge of the optical phase is essential.

Theis Sølling asked: You mentioned the need for a wide variety of parameters when modelling dynamics. How are the coupling between electronic surfaces obtained and how accurate are they?

Fernando Martín responded: For small systems, *e.g.*, diatomic molecules, the quality of the calculated couplings is similar to that provided by standard multi-reference quantum chemistry packages. The latter can only provide such information for electronically bound states. When electronic continuum states are involved, as it is always the case when molecules are irradiated by attosecond pulses, we use codes that we have developed in our lab along the years, which provide a similar level of accuracy. For large molecules, like phenylalanine, the accuracy of these couplings is substantially reduced. However, it is still good enough to provide photoelectron spectra in good agreement with those obtained in synchrotron radiation experiments. Hence we can assume that they are also good enough to describe ionization by attosecond pulses, since the dipole couplings are essentially the same as those involved in synchrotron radiation experiments.

Shaul Mukamel said: Fragmentation patterns result from complex multi-step dynamics. A much clearer probe of elementary electron and hole migration may be provided by ultrafast non-linear spectroscopy that can observe electronic coherence and electron and nuclear correlation functions

Fernando Martín replied: I fully agree with this comment. If instead of ionic fragments, ejected electrons were detected in a similar pump–probe scheme, one would have direct access to the dynamics occurring at the probing step without the distortions introduced by nuclear motion and fragmentation, which occur at a much later step. Even better would be the use of XUV pulses for the probing step instead of IR pulses, as discussed in the manuscript.

Raluca Cireasa asked: Your calculations based on single-electron excitations show that the charge migration in glycine is influenced by the ejected electron for less than 1 fs and by the nuclear dynamics after 8 fs, but without destroying the coherences created by the attosecond pulse for several tens of fs. Electron correlations occur on similar timescales and can lead to the migration of a localised hole through the whole system. Could you comment on the possibility of including the electron correlations in the calculations and their effect on the coherences?

Fernando Martín responded: There is absolutely no problem in including electron correlations during the migration dynamics initiated by a coherent superposition of single-electron excitations. Although electron correlations were not taken into account during the charge migration dynamics in our original paper[1] because they did not play a significant role, the TDDFT results on glycine presented in the present manuscript do include the effect of electron correlations (at the DFT level, of course). A much better way to do it would be to use a fully *ab initio* description of the molecular states involved in the migration process, which is feasible by using multi-reference configuration interaction quantum chemistry methods if one neglects the effect of the ejected electron (which, as shown in the manuscript, is a very good approximation). Here, I want to emphasize once again that inclusion of the ejected electron is essential for an accurate description of the initial wavepacket created by the attosecond pulse, but that, once the electron is ejected, it does not play a significant role in the charge migration dynamics. Concerning the role of electron correlation in the evolution of the wavepacket that results from the coherent superposition of single-excitation states, I do not expect significant changes during the first few femtoseconds of the evolution since the early stages of this dynamics is mainly controlled by the initial superposition of singly excited states of the cation. However, at longer times, it will have an important effect, certainly comparable to that due to the coupling with nuclear dynamics.

1 F. Calegari, D. Ayuso, A. Trabattoni, L. Belshaw, S. De Camillis, S. Anumula, F. Frassetto, L. Poletto, A. Palacios, P. Decleva, J. B. Greenwood, F. Martín and M. Nisoli, *Science*, 2014, **346**, 336–339.

R. J. Dwayne Miller opened a general discussion of the paper by Wolfgang Domcke:† It is quite unusual to increase coherence times by increasing damping. Normally damping leads to faster decoherence. I gather you are referring to coherence of the excited nuclear wavepacket and are not considering the electronic decoherence of the excited state. Is the reason for the increased coherence time that this coupling mode has less chance to drive the crossing of the conical intersection, which opens up other relaxation pathways and faster decoherence?

Maxim Gelin answered: The reason for this counter-intuitive phenomenon is the fact that the isolated conical intersection intrinsically exhibits dissipative behavior in the sense that coarse-grained observables such as population probabilities of electronic states or coherences of vibrational modes exhibit irreversible dynamics, see Ref. 1. This is a consequence of the exceptionally strong non-adiabatic coupling of the electronic states and the very pronounced local anharmonicity of the adiabatic electronic potentials near the conical intersection. In the model considered here, the coupling mode becomes particularly highly excited during the ultrafast internal conversion process at the conical intersection and is therefore particularly susceptible to damping by the bath (the damping rate of a damped harmonic oscillator increases linearly with the vibrational quantum number). The damping of the coupling mode by the bath reduces the intrinsic damping of the tuning mode. In the example considered, this effect is stronger than the direct damping of the tuning mode by the bath, which results in the counter-intuitive "de-damping" of the tuning mode.

1 R. Schneider *et al.*, *J. Chem. Phys.*, 1990, **92**, 1045.

R. J. Dwayne Miller remarked: How did you damp the mode? Was the parameter based on physically reasonable damping, presumably to the bath (system–bath coupling). The reason I ask is that normally vibrational damping times for relatively high frequency coupling modes are on the order of ps relative to much faster motions to the CI region. This mechanism may not be operative.

Maxim Gelin responded: We considered a minimal model of a so-called dissipative conical intersection. Namely, two coupled electronic states and two vibrational modes are considered as the system, which is bilinearly coupled to an environment. The results of our studies demonstrate that the conical intersection induces highly non-linear dynamics which is very sensitive to even weak coupling

† Wolfgang Domcke's paper was presented by Maxim Gelin, Technische Universität München, Garching, Germany.

to the bath. If the electronic coupling at the conical intersection were set to zero, the weakest system–bath couplings considered in the present work would virtually produce no effect on the wavepacket dynamics on a picosecond timescale. The dynamics at the conical intersection, on the contrary, is highly non-linear and chaotic and hence very sensitive to even weak system–bath couplings.

Andrew Orr-Ewing asked: Can your results for a conical intersection which is bound along both the tuning and coupling coordinates provide some insights about the role of a dissipative medium on the dynamics of passage through a conical intersection in which the tuning coordinate is dissociative? For example, can we learn something about how a solvent affects the non-adiabatic dissociation of heteroaromatic molecules excited to repulsive $\pi\sigma^*$ states?

Maxim Gelin answered: The role of vibrational energy relaxation due to the coupling to a vibrational bath in photodissociation reactions driven by repulsive $\pi\sigma^*$ states was explored for a model of the photodissociation of pyrrole by Lan and Domcke.[1] It was shown that sufficiently strong system–bath coupling reduces the dissociation probability and enhances the probability for internal conversion to the electronic ground state. This finding is in qualitative agreement with the experimental observation that H-atom photodetachment *via* dissociative $\pi\sigma^*$ states is suppressed in larger chromophores with more vibrational degrees of freedom, see *e.g.*, M.-F. Lin *et al.*[2] While H-atom photodetachment is the dominant photochemical channel in pyrrole and indole, it is completely quenched in tryptophan and larger chromophores.

1 Z. Lan and W. Domcke, *Chem. Phys.*, 2008, **350**, 125.
2 M.-F. Lin *et al.*, *J. Chem. Phys.*, 2007, **126**, 241104.

Dmitrii Shalashilin commented: Is the computational approach you are using generic or can it only be used for specific types of Hamiltonian?

Maxim Gelin replied: The approach assumes that the bath is described by a collection of harmonic oscillators and the system bath coupling is linear in the bath coordinates. The system Hamiltonian may be quite general.

Adam Kirrander remarked: Could one adapt your model of a bath to describe a specific system, for instance, a particular solvent or an anisotropic environment such as a protein embedding a chromophore? Would it make sense to tackle such a problem with this type of formalism?

Maxim Gelin responded: The present model of the bath can be applied to specific systems. The procedure goes like this: the harmonic (Gaussian) bath is fully specified by its spectral density function (in the frequency domain) or, equivalently, by its memory function (in the time domain). For specific systems, the functional form of the memory function can be retrieved, *e.g.*, from molecular dynamics simulations or optical measurements. The next step is to fit the bath memory function by a linear combination of exponentials. The so-represented memory function can readily be incorporated into the hierarchy equation of motion scheme, see Tanimura *et al.*[1] In many applications, the system–bath coupling can be considered as weak and the bath memory is short on the system dynamics timescale (Markovian bath). Then, simpler descriptions (*e.g.*, Redfield theory) are valid.

1 Y. Tanimura, *J. Phys. Soc. Jpn.*, 2006, 75, 082001.

Pankaj Kumar Mishra asked: The authors have very nicely discussed the effect of a dissipative environment on the ultrafast non-adiabatic dynamics at conical intersections in pyrazine (the system studied) coupled with infinitely many harmonic oscillators in thermal equilibrium. If one tries to mimic the effect of the solvent by harmonic oscillators, how can the interaction of the oscillators with themselves be included within this hierarchy of equations of motion? Can that authors also comment on achieving the thermal equilibrium at a particular (desired) temperature?

Maxim Gelin answered: The interaction of harmonic oscillators with themselves can readily be included up to the second order in vibrational displacements, but the so-obtained Hamiltonian can be diagonalized by the introduction of normal modes, resulting in a collection of independent harmonic oscillators. The dynamics of any system in a harmonic bath obeys the so-called fluctuation–dissipation theorem, which ensures the relaxation of the system to equilibrium at the bath temperature. The temperature of the bath is a given external parameter. In many applications, the consideration of the fine characteristics of the bath is not necessary; it is sufficient to consider the strength of the system–bath coupling and the bath relaxation time, see, *e.g.*, Ref. 1.

1 U. Weiss, *Quantum Dissipative Systems*, World Scientific, Singapore, 1993.

Wolfgang Domcke opened a general discussion of the paper by Dmitrii Shalashilin: In the paper, you mention that the second-derivative non-adiabatic coupling operator is neglected. This has two consequences: (i) the Hamiltonian including only the first-derivative operator is not Hermitian; (ii) the second-derivative term is singularly repulsive at the conical intersection and repels the wave packet from the intersection. The non-Hermiticity of the Hamiltonian and the lack of repulsion could have significant consequences for the dynamics at the conical intersection.

Dmitrii Shalashilin responded: Our interpolation for the matrix element is of course an approximation, which helps to decrease the computational cost. It fails when the trajectory hits the conical intersection. However, in a multidimensional case this never happens in practice. The interpolation can be improved at the expense of more calculations. Also I would like to add that no approximations are used when running individual Ehrenfest trajectories, which together with cloning, accounts for the majority of the dynamics. The approximation for the matrix elements is only needed for coupling the amplitudes of the Ehrenfest configurations, which provides only a small correction.

Michael Robb commented: Why do you need cloning? I would have thought that the set of Gausssian basis functions floated on Ehrenfest trajectories was complete. Or is it more efficient to do cloning than to run more Ehrenfest trajectories? If so why?

Dmitrii Shalashilin replied: Although in principle cloning is not necessary, in practice it speeds up the convergence of the results putting the basis in the right place. Too many Ehrenfest trajectories would be required without it.

Iakov Polyak said: In your work you run Ehrenfest dynamics on the basis of 3 adiabatic electronic states. May it be that restriction to just 3 adiabatic states could lead to some of the numeric issues that you have to account for by using "cloning"? Would it be beneficial to use a diabatic basis (configuration state functions) spanning all adiabatic states instead?

Dmitrii Shalashilin responded: An adiabatic basis is used in our direct dynamics because most of electronic structure codes yield adiabatic states and non-adiabatic coupling matrix elements (NACME). Using a limited set of adiabatic states can represent a problem and one always need to check whether the number of states is sufficient. In principle, a diabatic basis can be an alternative, but defining efficient sampling techniques, such as cloning for example, in a diabatic basis is not trivial.

Andrew Orr-Ewing remarked: The simulated image in Fig. 5 of your paper shows a radial structure in the high TKER component. Is this structure associated with the production of the pyrrolyl radical in different vibrational modes? If so, can the outcomes be compared to the high-resolution TKER measurements of Ashfold and coworkers[1] in which such a vibrational structure is observed and assigned?

1 B. Cronin, M. G. D. Nix, R. H. Qadiri and M. N. R. Ashfold, *Phys. Chem. Chem. Phys.*, 2004, **6**, 5031.

Dmitrii Shalashilin responded: No, the structure is not related with the vibrational levels. We believe that it is simply noise.

Wolfgang Domcke asked: While the minimum energy reaction path is linear for H-atom detachment in pyrrole, it is generally strongly curved in H-atom transfer reactions. The tunneling path will then not follow the reaction path and you will not know where the tunneling path emerges on the other side of the barrier. How are you going to deal with this problem?

Dmitrii Shalashilin responded: On the other side of the barrier we put a train of the basis functions, not just one Gaussian. Our hope is that if the train is long enough it will cover significant space in the tunnelling region, which will overlap with the end point of the curved path.

Adam Kirrander commented: Looking at Fig. 3 in your paper, the low-energy part of the TKER spectrum does not change very much with the inclusion of tunneling, despite the fact that one might naively expect that it should. Can you comment? Also, how sensitive are the calculated tunneling rates to the value of the Gaussian width-parameter? We have seen in other calculations that these can influence results, at least until calculations are fully converged.[1–2]

1 A. Kirrander, K. Saita and D. Shalashilin, *J. Chem. Theory Comp.*, 2016, **12** 957–967.
2 A. Kirrander and D. Shalashilin, *Phys. Rev. A*, 2011, **84**, 033406.

Dmitrii Shalashilin responded: The inclusion of tunneling certainly increases the low energy contribution, but in this particular case the low energy contribution is small. We hope that for other molecules we are considering the influence of tunneling will be stronger. As we use train basis sets here, the influence of the width parameter should be smaller than in the previous works, where a simpler version of the approach has been used. However, admittedly we have not yet checked this here.

Markus Kowalewski addressed Dmitrii Shalashilin and Dmitry Makhov: Does the *ab initio* multiple cloning approach describe the time evolution of the electronic coherence created upon cloning correctly? Is the decay of the electronic coherence in the vicinity properly described by the method?

Dmitrii Shalashilin responded: In principle our approach is a fully quantum one. In practice, of course, full convergence is hard to achieve. We tested the approach on model systems and found that full convergence can be achieved for a multi-configurational Ehrenfest method. This will be reported elsewhere.

Dmitry Makhov answered: The electronic coherence is not created upon cloning: cloning does not change the wave function, it changes only the ensuing dynamics. The decay of the coherence is always present in the Ehrenfest configurations: the phases of all electronic states are propagated in the dynamics and, because of the energy difference, they become essentially independent after some time. Cloning does not change much here.

Jon Marangos opened a general discussion of the paper by Morgane Vacher: The importance of the zero point geometry distribution in the dephasing of the coherent distribution is clear. What I am curious about is the fact that when the Ehrenfest dynamics was also included, the dephasing time was somewhat increased in the data shown. Is there a simple/general explanation for that behaviour?

Morgane Vacher replied: The zero point energy alone (with fixed nuclear geometries) leads to the fast dephasing of electron dynamics: in BMA[5,5], studied in the present work, the predicted coherence half-life is about 7 fs. This coherence half-life is slightly increased to 8 fs when the nuclear motion is taken into account (with the Ehrenfest method). We were also surprised by this result, expecting the nuclear motion to further damp the electron dynamics oscillations. According to our model presented in a previous article,[1] the coherence half-life is determined by the width of the energy gap distribution across the ensemble of geometries used to represent the nuclear wavepacket. We checked the time evolution of the energy gap distribution width but the latter was not becoming narrower with time. Our current hypothesis to explain this somewhat surprising result is the following. In section 3.3 of the present work and in previous work,[2] we have shown that the nuclear motion can decrease the amplitude of the oscillations in the electronic density. By decreasing the amplitude of the oscillations for some nuclear geometries of the ensemble, the dephasing caused by these geometries will be less important, leaving on average a higher amplitude oscillation.

1 M. Vacher, L. Steinberg, A. J. Jenkins, M. J. Bearpark and M. A. Robb, *Phys. Rev. A*, 2015, **92**, 040502.
2 M. Vacher, D. Mendive-Tapia, M. J. Bearpark and M. A. Robb, *J. Chem. Phys.*, 2015, **142**, 094105.

Jon Marangos addressed Morgane Vacher and Fernando Martín: The dephasing timescales due to the zero point geometry spread is clearly a general phenomenon that will affect all "charge migration" experiments. Do you anticipate that it will be important for the situation considered in Fernando Martín's work?

Morgane Vacher replied: We have now looked at the effect of the zero point energy on the electron dynamics in several molecules, including some phenylalanine derivatives.[1] In each case, we observe the fast dephasing of oscillations in the electronic density: this result calls into question the commonly held picture of long-lived electronic oscillations at a well-defined frequency (obtained with the single-geometry approximation).[2] We believe that the dephasing of electron dynamics due to the natural width of the nuclear wavepacket is general and will be common to all molecular systems. We therefore anticipate it to be important for Fernando Martín's work, for all electron dynamics simulations, and we also expect it to be particularly relevant for the interpretation of "charge migration" experiments.

The time scale of the oscillations dephasing is system-dependent. In the systems we have studied so far, the coherence half-lives range from 2 fs to 10 fs. The half-life obtained here for BMA[5,5] is therefore one of the longest times we predicted. We can rationalise the different coherence half-lives using our simple analytical model.[3] One of the challenges for theory may be to suggest molecular systems in which the oscillations in the electronic density survive for longer. But for this, one must go beyond the single-geometry approximation and describe realistically the nuclear wavepacket width.

1 A. J. Jenkins, M. Vacher, M. J. Bearpark and M. A. Robb, *J. Chem. Phys.*, 2016, **144**, 104110.
2 F. Calegari, D. Ayuso, A. Trabattoni, L. Belshaw, S. De Camillis, S. Anumula, F. Frassetto, L. Poletto, A. Palacios, P. Decleva, J. Greenwood, F. Martin, M. Nisoli, *Science*, 2014, **346**, 336–339.
3 M. Vacher, L. Steinberg, A. J. Jenkins, M. J. Bearpark and M. A. Robb, *Phys. Rev. A*, 2015, **92**, 040502.

Fernando Martín replied: It will certainly play a role, especially for molecules containing light nuclei, for which the geometrical spread is larger. Even more importantly, the intrinsic quantum nature of the nuclear dynamics, specially near conical intersections, further contributes to the dephasing of the electron wavepacket dynamics. Both effects will play a crucial role in determining the molecular fragments that are produced in XUV pump/IR probe experiments. Since it is precisely these fragments that are actually detected in a real experiment, accounting for the above dephasing effects is crucial for a direct comparison with experiment.

Adam Kirrander asked: In the paper you discuss different factors that influence the dispersion of the electronic wavepacket, including nuclear motion, the initial nuclear wavepacket, and the slopes of the potential energy surfaces. Given that Ehrenfest wavepackets may over-emphasize coherence, in particular when surfaces have different gradients, have you investigated this using a different method, such as DD-vMCG? Also, have you considered using an implementation of Ehrenfest that includes a *breaking force* such as proposed by D. Shalashilin and collaborators?[1–2] This would potentially make it possible to disentangle the influence of the spread of oscillators at different geometries *versus* the motion of the wavepacket on different electronic states.

D. V. Makhov *et al.*, *J. Chem. Phys.*, 2014, **141**, 054110.
D. V. Makhov *et al.*, *Phys. Chem. Chem. Phys.*, 2015, **17**, 3316.

Morgane Vacher answered: For the Ehrenfest method, coming from a single-configuration ansatz for the molecular wave function, the nuclear motion is bound to be the same for all electronic states. It cannot describe the nuclear wavepackets on the different electronic states moving in different directions, which would add another origin for decoherence of electron dynamics. We are currently working on approaches going beyond the Ehrenfest method to check whether more elaborate methods are needed to describe these processes. The first step to go beyond an ensemble of independent Ehrenfest trajectories (without using an *ad hoc* procedure) would be to couple the amplitudes of the trajectories,[1] as proposed by Shalashilin and collaborators.[2–3] Then, if one uses quantum mechanics to guide the coupled trajectories (instead of classical mechanics), one arrives at the DD-vMCG method.[1] There, a multi-configurational ansatz for the molecular wave function is used and the nuclear wavepackets could move in different directions on the different electronic states, according to quantum mechanics.[5] Although such quantum dynamics calculations would include all effects at the same time, some post-analysis of the time-dependent molecular wave function might allow us to disentangle the importance of each factor in the decoherence of electron dynamics.

1 J. Shang, T. Yuan and H. Zhu, *Theo. Chem. Acc.*, 2016, **135**, 1.
2 D. V. Shalashilin and M. S. Child, *Chem. Phys.*, 2004, **304**, 103.
3 D. V. Shalashilin, *J. Chem. Phys.*, 2009, **130**, 244101.
4 G. W. Richings, I. Polyak, K. E. Spinlove, G. A. Worth, I. Burghardt and B. L. Lasorne, *Int. Rev. Phys. Chem.*, 2015, **34**, 269.

Peter M. Weber said: In our experiments on tetramethyl diamine and other molecules, there is a high degree of structural flexibility that can affect the electron dynamics. We have interpreted the observed long timescales, about 1 ps, as due to the fact that the nuclear geometry must find the right place in its multi-dimensional space before charge transfer can take place. Listening to the talk I wonder if, in addition, the effect observed here – that the dephasing time becomes slower with the inclusion of nuclear motion – might play a role as well. In the flexible molecules there are several low frequency, large amplitude vibrations. Is it possible for large amplitude zero point motions to push the timescales significantly toward the picosecond regime? See, amongst others, Refs. 1-3.

1 X. X. Cheng, Y. Zhang, Y. Gao, H. Jónsson and P. M. Weber, *J. Phys. Chem. A*, 2015, **119**, 2813–2818.
2 X. X. Cheng, Y. Zhang, S. Deb, M. P. Minitti, Y. Gao, H. Jónsson and P. M. Weber, *Chem. Sci.*, 2014, 5, 4394–4403.
3 X. X. Cheng, Y. Gao, F. Rudakov and P. M. Weber, *Chem. Sci.*, 2016, 7, 619–627.

Morgane Vacher responded: It is helpful to distinguish between electron dynamics induced by nuclear motion, called *charge transfer* in the literature, and electron dynamics induced by an electronic wavepacket (coherent superposition of electronic states) that happens even with fixed nuclei, called *charge migration*. The two mechanisms are quite different and particularly the role of the nuclear degrees of freedom is different.

In the first case, upon electronic excitation, the electronic state is stationary (a single electronic eigenstate is populated); the nuclei will first respond to the change in electronic distribution upon excitation and then, as the nuclei explores new regions of the potential energy surfaces where the electronic character may change, we observe charge transfer, *i.e.* immediate rearrangement of the electronic distribution to the current position of the nuclei. In this context, low frequency modes sound consistent with a timescale of about 1 ps.

In the second case, a coherent superposition of electronic states is populated. This superposition is non-stationary, *i.e.* its probability density oscillates with time at a period inversely proportional to the energy gap between the populated electronic states. The nuclei are not "active" in this type of electron dynamics. In the present work and in earlier works,[1–2] we have shown that the distribution of geometries within the nuclear wavepacket width at the zero-point energy leads to a distribution of energy gaps, and therefore to a distribution of oscillation periods in the electronic density. The oscillations at individual geometries will dephase with time leading to, on average, a damped oscillation with a half-life of a few fs. In that context, low frequency modes mean larger amplitude motions, *i.e.* a larger nuclear wavepacket, and therefore a larger distribution of energy gaps and faster dephasing.[1] It looks therefore *a priori* difficult for large amplitude zero-point motions to push the dephasing timescale of charge migration toward the picosecond regime.

1 M. Vacher, L. Steinberg, A. J. Jenkins, M. J. Bearpark and M. A. Robb, *Phys. Rev. A*, 2015, **92**, 040502.
2 A. J. Jenkins, M. Vacher, M. J. Bearpark and M. A. Robb, *J. Chem. Phys.*, 2016, **144**, 104110.

Daniel Neumark remarked: Why did you choose this particular molecule for your calculations, and what experiments have previously been performed on it?

Morgane Vacher replied: The real reason we chose to study the BMA cation is that, in order to understand the nuclear motion induced by an electronic wavepacket, we were searching for a molecular system where the induced changes in the nuclear geometry would be simple, localised and therefore straightforward to follow and analyse. Some of the other molecular systems we have looked at so far involve phenyl rings: there, the nuclear distortions become delocalised, complicated and it is harder to get some physical insights. Here, in BMA, removal of an electron from the bonding π orbitals simply results in the stretching of the methylene bonds. With such a system, we were able to demonstrate that the population of a single adiabatic state results in a symmetric stretching while the population of a superposition of the ground and first excited state breaks the symmetry by localising the unpaired electron on one methylene group and leads to an asymmetric stretching. This is a demonstration of "charge-directed reactivity". When taking into account an ensemble of trajectories, the asymmetry in stretching eventually disappears as the dephasing of electron dynamics leads to an equal delocalisation of the unpaired electron on the two methylene groups. The BMA cation had already been studied theoretically in the past.[1–2] There, the focus was not on the dynamics upon photoionisation though. Adamantane itself is a colourless, crystalline chemical compound; it is also the simplest diamondoid. Photoionization energies have been determined for adamantane (as well as for several bigger diamondoids).[3] Adamantane itself has few applications since it is merely an unfunctionalised hydrocarbon. The BMA cation is a model donor–linker–acceptor system where there is an intramolecular electron transfer between the two methylene groups connected by the adamantane framework. Such chemical structures may be used as basic units to make oligomers to produce new organic semiconductor materials or to make new molecular parts in the building of molecular computers.

1 L. Blancafort, F. Jolibois, M. Olivucci and M. A. Robb, *J. Am. Chem. Soc.*, 2001, **123**, 722–732.
2 L. Blancafort, P. Hunt and M. A. Robb, *J. Am. Chem. Soc.*, 2005, **127**, 3391–3399.
3 K. Lenzke, L. Landt, M. Hoener, H. Thomas, J. E. Dahl, S. G. Liu, R. M. K. Carlson, T. Möller and C. Bostedt, *J. Chem. Phys.*, 2007, **127**, 084320.

Hans Jakob Wörner asked: In this paper and in earlier work you showed that the relative phase of an electronic superposition state can influence nuclear dynamics. Do you expect this result to be general for sufficiently close-lying electronic states or are there additional conditions, *e.g.*, concerning the strength of non-adiabatic couplings *etc.*?

Morgane Vacher answered: First, we note that different electronic eigenstates, having different electronic characters, lead to different nuclear motions. This happens even in the Born–Oppenheimer approximation without involving any non-adiabatic coupling. Then, an equal superposition of delocalised eigenstates (typically "left+right" and "left−right") leads to a localised hole density ("left" or "right" – or a more complicated density according to the relative phase). This is a general fact that does not rely on any interaction or coupling or condition on the energy gap. A nice illustration of this is the series of BMA[5,5], BMA[6,5] and BMA [6,7] cations, studied in the present work: these molecules exhibit the same charge migration but with different energy gaps between the two electronic eigenstates (and therefore different periods of oscillation). The same was observed in benzene, toluene and *para*-xylene cations, with the extreme situation where the period of oscillations is infinite in the benzene cation because of the zero energy gap.[1] Now, a given superposition of states leads to a nuclear motion determined by its own electronic character (different from those of the individual electronic eigenstates). The electronic character of the superposition is itself determined by the relative phase (and weight) of the states. In this case, the nuclear motion occurs on an *effective* potential energy surface, rather than on an *adiabatic* potential energy surface. We have shown in previous works,[2–3] the importance of the derivative coupling in such an effective potential energy surface, to describe the nuclear motion induced by a superposition of states. In practice, such delocalised "left+right" and "left−right" electronic eigenstates often occur in the vicinity of avoided crossings where we expect the non-adiabatic coupling to be non-negligible.

1 M. Vacher, D. Mendive-Tapia, M. J. Bearpark and M. A. Robb, *J. Chem. Phys.*, 2015, **142**, 094105.
2 J. Meisner, M. Vacher, M. J. Bearpark and M. A. Robb, *J. Chem. Theory Comput.*, 2015, **11**, 3115–3122.
3 M. Vacher, J. Meisner, D. Mendive-Tapia, M. J. Bearpark and M. A. Robb, *J. Phys. Chem. A*, 2015, **119**, 5165–5172.

Shaul Mukamel asked: Extensive work has been carried out on energy transfer (rather than electron transfer) in donor–bridge–acceptor systems. Could you comment on the comparison of your work to those studies?

Morgane Vacher replied: We believe that energy transfer can proceed by the same mechanism as charge migration (or electron dynamics). The same methods that we use to study coupled electron and nuclear dynamics can thus be used to investigate these energy transfer processes. We are actually looking at such chemical problems now.

Adam Kirrander said: When investigating the charge migration in your paper, you generally stop at 20 fs, with, for instance, Fig. 6 in the paper showing strong damping at that point. Have you explored the dynamics at longer times? Would you expect that any form of revivals could occur, for instance, related to vibrational beatings? Ultimately, I suspect that this would depend on whether the spectrum of states is a broad continuum or if it has some structure. Certainly in smaller molecules beatings related to rotational, vibrational, and electronic motion are common (see *e.g.*, Ref. 1 and 2). Also, have you investigated to what degree the preparation of the initial state, and in particular the initial phase relationship between the electronic states, can influence the observed charge dynamics? In much simpler molecules we have found that phase can influence the dynamics quite significantly (see *e.g.*, Ref. 3 and 4).

1 H. H. Fielding, *Annu. Rev. Phys. Chem.*, 2005, **56**, 91–117.
2 A. Kirrander, H. H. Fielding and Ch. Jungen, *J. Chem. Phys.*, 2007, **127**, 164301.
3 A. Kirrander, H. H. Fielding and Ch. Jungen, *J. Chem. Phys.*, 2010, **132**, 024313.
4 A. Kirrander, Ch. Jungen and H. H. Fielding, *J. Phys. B*, 2008, **41**, 074002.

Morgane Vacher replied: We have not explored the dynamics at longer times. For this, a multi-configurational approach would be recommended to give reliable results. Let me try to explain why we do not *a priori* expect the occurrence of revivals. In general, a coherent superposition of (non-degenerate) eigenstates evolves with time with each eigenstate acquiring a time-dependent phase proportional to its energy. With only two eigenstates, as time passes, the interference between the two states alternates between constructive and destructive. The oscillation period is inversely proportional to the energy difference between the two states. With more than two states in the coherent superposition, the time evolution is more complicated. The revival period is governed by the least common multiple of all the pairwise oscillation periods between its component eigenstates. While this revival time is long, it always exists if a finite number of eigenstates are populated. The decoherence studied in the present and previous works,[1–2] is due to the dephasing of the continuous distribution of oscillation frequencies within the intrinsic width of the nuclear wavepacket; in other words, there are only two electronic states but an infinite number of oscillation frequencies (which can be totally in phase only at one moment in time) because of the geometry-dependence of the energies. For this reason, we believe revivals are not relevant in this context, unless one is able to select a few discrete nuclear geometries during the ionisation step. Besides, we have investigated the importance of the initial relative phase between the two electronic states on the subsequent coupled electron and nuclear dynamics in the current article (Fig. 9 and 10, section 4.2) and in previous works.[3–4] We found that indeed it significantly influences the dynamics: by determining the initial hole localisation, the initial relative phase controls the initial nuclear motion. This opens up the possibility to control chemical reactions *via* a coherent control technique.

1 M. Vacher, L. Steinberg, A. J. Jenkins, M. J. Bearpark and M. A. Robb, *Phys. Rev. A*, 2015, **92**, 040502.
2 A. J. Jenkins, M. Vacher, M. J. Bearpark and M. A. Robb, *J. Chem. Phys.*, 2016, **144**, 104110.
3 J. Meisner, M. Vacher, M. J. Bearpark and M. A. Robb, *J. Chem. Theory Comput.*, 2015, **11**, 3115–3122.
4 M. Vacher, J. Meisner, D. Mendive-Tapia, M. J. Bearpark and M. A. Robb, *J. Phys. Chem. A*, 2015, **119**, 5165–5172.

R. J. Dwayne Miller commented: It is very difficult to optically direct momentum *via* optical excitation. Normally some form of coherent control is needed. The absorbed photon has no momentum or better cannot conserve momentum so that quantum interference in the superposition states with some phase relation picks out the specific momentum (see work by van Driel and Sipe on the control of electron motion in semiconductors).[1] The idea of a charge-directed reactivity using the preparation of an excited superposition state is interesting in this context. It seems, however, when you studied a distribution the effect vanished. I gather this was from the random motions in the ground state and FC projection to the excited state surface that washed out the idea of asymmetric excited state preparation to direct electron distributions through the coupling of nuclear motions to electron distribution. Did you extend your calculations in time to see if the same effect of averaging over an ensemble would occur if the process is averaged over time. This point is raised with respect to similar ideas for coherent control of electron motion.

1 J. Hübner, W. W. Rühle, M. Klude, D. Hommel, R. D. R. Bhat, J. E. Sipe and H. M. van Driel, *Phys. Rev. Lett.*, 2003, **90**, 216601.

Morgane Vacher answered: Our simulations are typically up to 20 fs, including all vibrational modes. Preparing an equal superposition of the two lowest energy electronic states in modified-BMA cationic species (studied in the present work) initially localises the hole on one methylene group. Considering a single fixed nuclear geometry, the hole oscillates back and forth between the two methylene groups with a constant period. Considering an ensemble of fixed nuclear geometries to describe the vibrational ground state of the neutral species and projecting it into the cation, the hole is still initially localised on one methylene group but the oscillations at each individual geometry dephase with time (because of different periods), leading to a hole symmetrically delocalised onto the two methylene groups at about 15 fs. We have shown that considering a single nuclear trajectory (*i.e.* with the nuclei allowed to move along all vibrational modes), the oscillations in the electronic density are not perfect anymore: the period and the amplitude of the oscillations are affected after only a few fs (this time is of course system-dependent). As a consequence, already within the first 20 fs, the "symmetry" in the oscillations of the asymmetrically localised hole is lost. This effect could become larger with time as charge transfer comes into play. A good description of coupled electron and nuclear dynamics at longer times requires using a multi-configurational method for the nuclear motion (beyond the Ehrenfest method).

Adam Kirrander opened the discussion of the papers by Fernando Martín, Wolfgang Domcke, Dmitrii Shalashilin and Morgane Vacher: Given the number of theoreticians in the room, I would like to open the general discussion with a very broad question. In your view, how far away are we from a truly reliable theoretical description of photochemical reactions? What are the biggest obstacles and hurdles, as well as areas of opportunity, for theory and quantum molecular dynamics? Likewise, are there any particular (types of) experiments that you wish that experimentalists took upon themselves to perform, and which might be particularly useful from the perspective of advancing theory?

Dmitrii Shalashilin responded: The complexity of quantum dynamics is in my opinion the main challenge for theory. However, we are now at the stage that fully quantum simulations are becoming possible, albeit for very short times only. Luckily many photochemical processes are very fast and we should thank experimentalists for moving towards faster and faster processes providing a playground for theory.

Fernando Martín answered: Theoretical description of photochemical reactions, occurring either in the femtosecond and sub-femtoseond timescales, is accurate for small molecules, mainly diatomics and very simple polyatomics. There has been a lot of theoretical progress in the last few years to treat more and more complex molecules rather accurately, however, as we heard during the conference, experimentalists can always choose an even larger molecule for which accurate theoretical modeling is not possible. So, in my opinion, the key question is if theory can reach the necessary accuracy to treat systems with the minimum level of complexity that would allow us to obtain some general rules and patterns that can be extrapolated to those molecular systems that will remain forever inaccessible to theory. I think that we have not yet reached this point, mainly because of the difficulty to describe the coupling between electronic and nuclear motions in systems with a large number of degrees of freedom. Also, we are even further away when the dynamics implies ionization of the system, as *e.g.*, when irradiated with XUV pulses or when Auger processes play an important role. Theoreticians still need the support of detailed experiments performed in benchmark molecules (triatomics and small polyatomics), so that the different approaches can be rigorously tested. Theory also needs to be contrasted with other theories, even simplified theories, to check the validity of simplified models, the only alternative when dealing with very large systems.

Michael Robb responded: There is no doubt that quantum dynamics is becoming essential for the treatment of photochemistry. It is like the early days of electronic structure methods; quantum dynamics methods and computer programs are becoming "general" and "user friendly". "On the fly" methods mean that one only computes those parts of the potential surface that are important. Quantum dynamics means that all the reaction pathways are explored simultaneously and the "importance" is thus ascertained from quantum mechanics. However, the accuracy and cost of the electronic structure part of the computation will continue to be an issue.

Concerning the experiments, it is clear that theory needs to be able to "simulate experiment". This raises big theoretical challenges such as including the electric field and an outgoing electron in an ionisation experiment. The influence of the probe on the intrinsic dynamics of the system and the fact that there is very little knowledge and control over the initially populated electronic states (initial conditions of the dynamics) are outstanding questions. I am not sure resolving these would "advance" theory but it would help with testing it.

Wolfgang Domcke addressed Dmitrii Shalashilin and Fernando Martín: In my opinion, it is essential to test theory against theory, rather than theory against experiment. We need non-trivial and representative systems and models for which we can exactly solve the Schrödiger equation numerically. Against these solutions we can then benchmark approximate, but computationally less expensive, methods. In the comparison of theory with experiment for complex systems and/or dynamics, too many approximations and error cancellations are involved. It is also essential to explicitly include the photon field to compute the actually measured observables.

Dmitrii Shalashilin responded: I completely agree with this. Our multi-configurational Ehrenfest method has been extensively tested on a number of models, such as the spin–boson model for example, where good benchmark calculations are available. We currently are preparing a review article which describes the details of such tests. The conclusion there is that, in the case of models, the techniques we are using in our current direct dynamics "on the fly" calculations lead to good convergence with a similar number of the Ehrenfest trajectories.

Fernando Martín answered: I also fully agree with this point of view. However, this is not in contradiction with the fact that theory, even at an approximate level, is usually indispensable to shed some light on complicated experiments, either by identifying the dominant channels leading to a particular observation or by providing the basic mechanisms that operate under certain circumstances. Both aspects are extremely important to develop a know-how applicable to those cases where theoretical simulations are out of reach.

Daniel Neumark addressed Fernando Martín: In the time-resolved charge migration experiments that you have simulated, it seems that using near-infrared strong field ionization as the probe pulse is far from ideal, as its effect on the system under study is difficult to characterize. What in your view would be the ideal means of probing attosecond-induced charge migration dynamics?

Fernando Martín replied: In my opinion, the best way to probe the system is by using another XUV attosecond pulse. This pulse will further ionize the molecular cation created by the pump pulse, but it does so by absorption of just one photon. In this way, one would avoid introducing additional couplings between cationic states and, consequently, the original charge migration dynamics would be directly projected on the double ionization continuum. In other words, one would avoid all the possible resonant ionization transitions that take place in multiphoton ionization processes as those induced by an IR probe pulse (REMPI). Furthermore, due to their short wavelength (*i.e.*, a Keldysh parameter much larger than one), XUV pulses barely perturb the system, so that one can expect that the original dynamics induced by the pump pulse is barely affected by the probe pulse.

These expectations have been recently confirmed by XUV-pump/XUV-probe numerical experiments performed on the hydrogen molecule,[1] for which theory can provide nearly exact results. The results of that work show that the unperturbed molecular wavepacket created by the pump pulse can be reconstructed if the required energy and time resolutions are available. In our current Faraday Discussions paper, we also provide an example for the glycine molecule (although in this case, nuclear motion was ignored). For this molecule we show that the actual beating associated with the electron dynamics generated by the pump pulse remains intact after interaction with the probe pulse and are relatively easy to identify (notice that the beatings shown in the four central panels of Fig. 7 are the same as those appearing the bottom panel of Fig. 8).

1 A. Palacios, A. González-Castrillo and F. Martín, *Proc. Natl. Acad. Sci. U. S. A.*, 2014, **111**, 3973–3978.

Albert Stolow addressed Dmitrii Shalashilin and Fernando Martín: Could you offer any general comments as to the strengths and limitations of semi-classical methods in excited state molecular dynamics simulations?

Dmitrii Shalashilin answered: First of all, I would like to point out that our approach is not semi-classical. We use classical trajectories to speed up fully quantum simulations, but the *ab initio* multiple cloning technique is fully quantum. However, of course various semi-classical methods exist and in many cases they represent the most efficient way of simulating non-adiabatic dynamics. Their main strength of trajectory-based methods is that they explore only the most relevant regions of space, and can be implemented in the form of direct dynamics. The treatment of tunnelling and zero point energy effects is difficult though.

Fernando Martín responded: By using a classical description of the nuclear dynamics in combination with a quantum mechanical description of the electron dynamics, we are losing coherences that may show up in a real experiment, *e.g.*, around conical intersections. However, a full quantum mechanical description is out of reach for very large systems, so one should rely on semi-classical methods to get some qualitative understanding.

Fernando Martín remarked to Dmitrii Shalashilin: Semi-classical methods in which electronic motion is treated quantum mechanically and nuclear motion classically are probably the only ones that are applicable to large molecules. However, in doing so, one is losing part of the coherence due to the coupling between electronic and nuclear degrees of freedom. Your method seems to overcome this limitation by associating quantum phases to the nuclear classical trajectories. Do you think that such an approach would be workable for very large systems?

Dmitrii Shalashilin replied: Yes, the techniques we are using are formally exact and they associate proper quantum phases with trajectory-guided functions. We are now pushing this techniques towards larger and larger systems. Our hope is that in large systems decoherence occurs very quickly and the required simulation time will be short. If there is a longer quantum coherence then typically a small quantum subsystem can be identified and a fully quantum approach may work also.

R. J. Dwayne Miller addressed Wolfgang Domcke, Michael Robb, Fernando Martín and Dmitrii Shalashilin: I am wondering if the level of theory needed to understand photochemical reactions could not be reduced by focusing on the structural changes, the actual chemistry, as opposed to calculating spectroscopic observables. The level of accuracy is really high if one wants to calculate electronic states and changes with structure as probes of the underlying dynamics. The energy differences are on the order of 10's of wavenumbers or better than 0.1 eV accuracy and this then requires very high-level treatment and convergence. However, I think we are missing a bit of physics that makes the problem much simpler if we focus on the structural dynamics to cast out the dominant reaction modes. Synthetic chemists have not been worried about such nuances in electronic states and detailed issues of vibrational non-adiabatic coupling. They typically use "arrow pushing" to conceptualize how the electrons will redistribute to give the lowest energy product or intermediates. If you reflect on this exercise, this approach involves a highly localized basis to rationalize the chemistry. Similar statements can be made for the concepts used to infer transition state structures to guide synthesis. It involves a very localized basis in which the most direct connection from reactant to product is typically invoked. They have not been held back by the details – and this strategy has worked. Over time, there are hundreds of named reactions that can be employed on nearly every molecular target to get the desired chemistry and make nearly any conceivable structure within near biological scales. These ideas are oversimplified but there is something in this view that transcends to different molecules independent of the significant differences in structure and many body potential/properties. There must be a reduction in dimensionality during barrier crossing. This process only has to be faster than all other competing processes to get drive the desired chemistry. We only need to get relative barrier heights and the most strongly coupled modes right to understand the observed chemistry.

With the latest advances in real time imaging of chemical reaction dynamics and inversion methods of femtosecond diffraction data, it is possible to obtain full atom resolved chemical reactions (see Ishakawa *et al.*).[1] Is there any computational saving or means to scale theoretical calculations to try to cast out the key modes that direct the chemistry and to compare to the structural dynamics? New approaches that could be formulated, such as coarse grain sampling or imposed correlations? The level of accuracy needed should be far less than calculating electronic states to properly model shifts in X-ray absorption spectroscopy, electronic, or vibrational spectroscopy as indirect probes of the chemical dynamics. It is this detail that is needed to enable the development of simple rules/guidelines based on reaction modes by which one would be able to better approximate what the structure is a the critical point/barrier and better design means to control barrier heights. This would give a dynamic structural basis to understand chemistry and would be a major advance over "arrow pushing". It would also scale to arbitrarily large systems as the synthetic chemists have admirably shown with synthetic-based approaches to solving chemical problems.

1 T. Ishikawa, S. A. Hayes, S. Keskin, G. Corthey, M. Hada, K. Pichugin, A. Marx, J. Hirscht, K. Shionuma, K. Onda, Y. Okimoto, S. Y. Koshihara, T. Yamamoto, H. Cui, M. Nomura, Y. Oshima, M. Abdel-Jawad, R. Kato and R. J. D. Miller, *Science*, 2015, **350**, 1501–1505.

Wolfgang Domcke responded: These arguments are intriguing and challenging for chemical theory. Simple and generic guidelines for the prediction of photochemical dynamics would certainly be welcomed by both experimentalists and theoreticians. However, it may be worthwhile to point out that several empirical rules developed in the past and widely used for the interpretation of the spectroscopy of polyatomic molecules and photochemistry, such as the "energy gap law" or the "proximity effect" in aromatic molecules, were thrashed as soon as reasonably accurate first-principles calculations of excited-state potential energy surfaces and non-adiabatic chemical dynamics became available. These early empirically derived rules have been superseded by the concept of non-adiabatic chemical dynamics at conical intersections. As a counter-example, the Woodward–Hoffmann rules should be mentioned, which remain a useful concept for polyene photochemistry. There are fundamental differences between ground-state chemical dynamics and photoinduced chemical dynamics. Thermal reaction dynamics is determined by rare events, the crossing of relatively high energy barriers, which is the reason why transition-state theory works so well. Ultrafast (sub-picosecond) photochemical dynamics, on the other hand, is typically not dominated by barriers on excited-state potential-energy surfaces (otherwise, it would not be ultrafast). It is rather determined by usually just a few reaction coordinates leading to conical intersections and it is essential to identify and follow these excited-state reaction coordinates. Moreover, excited states are numerous, often densely spaced and difficult to calculate accurately from first principles. From this viewpoint, it appears unlikely that simple and generically applicable rules governing excited-state chemical dynamics can be found.

Michael Robb replied: Valence Bond theory may be a framework in which the problem posed by R. J. Dwayne Miller can be resolved. Excited states have a simple qualitative representation in this theory. For example, we have recently shown how one may rationalize the occurrence of conical intersections with the aid a simple qualitative valence bond interpretation.[1] This type of qualitative theory is as accessible as "arrow pushing" for understanding excited state reactivity. Further, VB methods can be easily parameterized[2–4] for excited states. Here one is "fitting" an effective VB Hamiltonian rather than many potential energy surfaces.

The parameterization has a very simple form[2] and can be fitted to high accuracy computational data (and even with some experimental input). The crucial point is that the chemical nature of the excited states is clearly differentiated (*e.g.* zwitterionic states, the B state of polyenes have a completely different representation from the covalent states 2A/1A). Finally, the VB basis is inherently diabatic because its localized nature.

1 M. A. Robb, *Adv. Phys. Org. Chem.*, 2014, **48**. 190–239.
2 F. Bernardi, M. Olivucci, J. J. McDouall and M. A. Robb, *J. Chem. Phys.*, 1988, **89**, 6365–6375.
3 F. Bernardi, M. Olivucci and M. A. Robb, *J. Am. Chem. Soc.*, 1992, **114**, 1606–1616.
4 M. Garavelli, F. Ruggeri, F. Ogliaro, M. J. Bearpark, F. Bernardi, M. Olivucci and M. A. Robb, *J. Comput. Chem.*, 2003, **24**, 1357–1363.

Fernando Martín answered: I have little to add to what Michael Robb has explained, except that in the case of chemical phenomena occurring in the attosecond or few-fs timescales, as *e.g.*, charge migration, experimental investigation of the corresponding dynamics implies the use of light sources that always lead to ionization of the system, either at the pump step or at the probe step, or at both! Furthermore, ionization induced by these light sources can also occur from inner molecular orbitals that lead to Auger electron emission on a temporal scale similar to that of the nuclear motion. If, as we have seen, the concepts developed for chemistry in the ground state are difficult to extrapolate to photodynamics in excited states, any tentative approaches to do it for electronic continuum states, as those generated in attosecond experiments, are hopeless unless we first understand the electron dynamics that is generated at the ionization step. As we have heard during this Faraday Discussion, this is not a trivial task, even for simple molecules. Therefore, significant theoretical efforts are necessary in order to understand the chemistry resulting from ionization before trying to look for general rules similar to those used in chemistry performed at thermal energies.

Dmitrii Shalashilin responded: Of course modern theory is always about simplifications and the ideas of developing some key models of key processes is very natural, but the question is always whether these qualitative theories are right (at least qualitatively)

Michael Minitti said: looking at model systems, direct imaging wavepackets, and experiments with 300–400 eV X-ray pulses – is there a model system? What type of model systems should experimentalists focus on and run to best inform theory/computation efforts in calculating electron wavepacket dynamics? FEL sources such as LCLS and European XFEL are capable of producing very intense edge-specific soft X-rays having pulse widths on the order of 2–5 fs. Could these capabilities be used?

Fernando Martín answered: XFELs have a great advantage with respect to HHG sources: the pulse intensity is very large and, therefore, XUV-pump/XUV-probe or X-ray pump/X-ray probe experiments should be perfectly feasible. Earlier attempts performed at FLASH (Hamburg) several years ago worked nicely, but they did not have the necessary time resolution to observe electron dynamics. Definitely, the advent of shorter pulses will allow one to overcome this situation. A combination of pulses of let's say 2 fs duration should be good enough to observe such electron dynamics if the time delay between the pump and the probe pulses can be controlled with sub-fs time resolution. This seems to be possible nowadays. My recommendation would be to start with a simple triatomic or tetraatomic molecule, so that theoretical calculations can be carried out with high precision and, therefore, provide the necessary input to understand what is going on in such experiments. In my opinion, this is necessary before moving to very large systems, because, as explained in my talk, the electron dynamics induced by the pump pulse reflects in the fragments that are detected experimentally after probing the system with the second pulse, and this is very difficult to model in large systems.

Allan Johnson opened a general discussion of the paper by Albert Stolow:‡ In the plot of X-ray absorption, could you clarify what the color axis represents? What variable exactly is being plotted against time and energy here? Also, could you comment about the parameters of the pulses used in this simulation?

Albert Stolow and **Simon Neville** answered: The colour axis corresponds to normalised oscillator strengths. The initial wavepacket is taken to correspond to the instantaneous vertical excitation of the ground state to the $S_2(\pi\pi^*)$ state. That is, the pump pulse in effect corresponds to a delta function. The probe pulse parameters enter into the TRXAS simulation phenomenologically *via* the convolution of the spectrum in the time domain with a Gaussian of full width at half maximum corresponding to the pump–probe cross correlation, which we take as 10 fs.

Oliver Gessner addressed Albert Stolow and Simon Neville: Have you compared your TRXAS results with estimates for time-resolved X-ray/inner-shell photoelectron spectroscopy (TRXPS) measurements? For example, the ~5 eV splitting of the C 1s-pre-edge features due to the onset of distinguishability of the two carbon atoms ~20 fs after excitation (Fig. 8a and d), should also affect the corresponding C 1s inner-shell photolines in XPS, provided the effect is at least in part caused by a difference in the C1s core levels. Comparing TRXAS and TRXPS predictions may highlight the strengths/weaknesses of both methods and their complementarity.

Albert Stolow and **Simon Neville** replied: We agree that it can be anticipated that the splitting of the two C 1s orbital energies that accompanies twisting and pyramidalisation should also manifest itself in the TRXPS. However, a key difference between the pre-edge TRXAS and the TRXPS is the number and energetic density of accessible final states. In the case of the pre-edge TRXAS, the dominant final states are very small in number and well energetically separated. In contrast, in the case of the TRXPS, it may be anticipated that there will be a much larger number of overlapping accessible core-ionized states. Indeed, recent calculations support this idea.[1]

1 S. Neville *et al.*, *J. Chem. Phys.*, 2016, **145**, 144307.

Tom Penfold addressed Albert Stolow and Simon Neville: During the simulations, did you observe any significant difference in the convergence of the

‡ Albert Stolow's paper was co-presented by Simon Neville, of University of Ottawa, Ottawa, Ontario, Canada.

spectroscopic observables with respect to the number of basis functions compared to that of the population dynamics?

Albert Stolow and **Simon Neville** responded: In the case of the time-resolved X-ray absorption spectrum (TRXAS) of photoexcited ethylene, we observe a rapid convergence of the qualitative features of the spectrum with respect to the number of initial Gaussian basis functions. This appears to be due to the vertically excited initial wavefunction being quite tightly localized about the Franck–Condon point, a consequence of the absence of low-frequency normal modes. As such, the preponderance of classically evolving basis functions are effectively funnelled towards the twisted-pyramidalised (Tw-Py) conical intersection along similar pathways.

Daniel Neumark asked: Looking at Fig. 6 in your paper, it appears that the total photoionization cross-section at early times (<10 fs) is much lower than at later times (around 100 fs). Is that correct and if so, can you explain why it is the case?

Albert Stolow responded: This is a consequence of the dominant $\pi\pi^* \rightarrow 1s\pi^*$ transition in the initially excited $\pi\pi^*$ state having a relatively small oscillator strength in the vicinity of the FC point. It should be noted that the dominant feature in the calculated TRXAS around t ~10 fs is actually around 3 times weaker than the dominant transitions after 100 fs, at which point the S_0 state is populated. This is in agreement with the ratios of the calculated oscillator strengths of the dominant pre-edge transitions for the $\pi\pi^*$ and S_0 states at the FC point, as shown in Fig. 7 of our paper (DOI: 10.1039/C6FD00117C).

Daniel Neumark asked: One of your primary conclusions is that time-resolved carbon-edge X-ray absorption spectroscopy (XAS) is a more incisive probe of dynamics in photoexcited ethylene than time-resolved valence photoelectron spectroscopy. It seems that a more relevant comparison would be between the XAS calculations and time-resolved core-level photoelectron spectroscopy at photon energies above the carbon K-edge. Can you comment on what such a comparison would reveal?

Albert Stolow responded: Our primary reason for comparing the calculated TRXAS and TRPES spectra was to contrast methods that probe the core *vs.* valence level electrons, and their respective sensitivities to non-adiabatic dynamics. Our calculations predict that the significant variation of the core-level electronic structure with nuclear geometry gives rise to much of the clear and distinguishable temporal evolution in the calculated TRXAS. As such, it may be anticipated that time-resolved X-ray photoelectron spectroscopy (TRXPS), which probes the core-level electrons, will also provide a good probe of the excited state dynamics. One potential caveat, however, is that the number of core-ionization channels predicted to be open within a few tens of eV above the carbon K-edge in the excited states of ethylene is rather large.[1] As such, the TRXPS may be anticipated to be somewhat more congested that the TRXAS.

1 S. Neville *et al.*, *J. Chem. Phys.*, 2016, **145**, 144307.

Oliver Schalk said: In Fig. 6, you are showing your calculated TRXAS traces which cover excited state as well as ground state dynamics. In the ground state, you see time-dependent changes. Are these changes subject to a limited number of sampled trajectories or do they reflect "real" dynamics induced in the sample (*i.e.* would you expect these features when converging towards an infinite amount of calculated trajectories)?

Simon Neville answered: In the short time limit, when statistical processes have not yet become dominant, we expect that our description of the re-populated ground state dynamics should be somewhat meaningful. As such, even with a moderate number of basis functions, we do expect that the calculated ground state TRXAS will capture the qualitative features of the "real" dynamics.

Oliver Schalk remarked: In your paper, you simulated the TRXAS spectrum of excited ethylene up to 150 fs (see Fig. 6). It has been shown previously that on that timescale, H_2 formation on the ground state surface is observed.[1] Do you observe dissociation of ethylene in trajectories and if not, how would this affect the spectrum? In general, is this technique sensitive for such a bond cleavage?

1 T. K. Allison *et al.*, *Opt. Lett.*, 2010, **35**, 3664.

Simon Neville responded: We do not observe a significant amount of H_2 elimination within the timescale of our AIMS simulations. In general, the AIMS method should be amenable to the description of such dissociation processes provided that they are not statistical in nature. It should be noted, however, that the ADC(2) method, which we use for the X-ray cross-section calculations, is based on a correction to a single determinant reference and, as such, is not well suited to giving a good description of dissociation processes. It is also reasonable to question whether the paper cited[1] can definitively tell us whether the H_2 elimination channel is a significant reaction pathway at short times, as it does not yield information about quantum yields.

1 T. K. Allison *et al.*, *Opt. Lett.*, 2010, **35**, 3664.

Hans Jakob Wörner addressed Albert Stolow and Simon Neville: Can you decompose the ~4.5 eV splitting of the two pre-edge absorption features at early delays into chemical shifts of the core levels and energy shifts of unoccupied orbitals? Which one dominates?

Albert Stolow and **Simon Neville** responded: Within the initially excited $\pi\pi^*$ state, the ~4.5 eV splitting of the two dominant pre-edge peaks is almost entirely due to the splitting of the two C 1s orbitals. This splitting is found to arise due to the arrival of the wavepacket at the Tw-Py conical intersection seam, in the vicinity of which the C 1s orbitals are localized about the non-equivalent carbon atoms. In fact, the splitting of the pre-edge peaks maps directly onto the splitting of the energies of the C 1s canonical Hartree–Fock orbitals at the Tw-Py conical intersection geometry: 4.51 eV using the 6-311++G** basis.

Hans Jakob Wörner asked: Have you included Auger-lifetime effects and the vibronic structures of the final state in your calculations?

Simon Neville replied: In effect, all sources of broadening are included phenomenologically *via* the convolution of the spectrum in the time domain with a Gaussian function, the FWHM of which is determined so as to reproduce the ground state FC point experimental spectrum.

Hans Jakob Wörner asked Albert Stolow and Simon Neville: This is also a comment on an earlier question by Daniel Neumark. Although carbon K-edge XANES and valence-shell TRPES will, in general, provide complementary information, they should both probe the dynamics of partially-occupied orbitals because, at least within the molecular-orbital approximation. Do your calculations validate this picture?

Albert Stolow and **Simon Neville** answered: It is true that the evolution of both the valence- and core-level electronic structure will contribute to the evolution of the X-ray absorption spectrum. However, we find that the temporal evolution of the calculated TRXAS is, in effect, dominated by the changes in the core-level orbitals. This is in particular exemplified by the large splitting of the C 1s orbital energies, by ~4.5 eV, upon the approach to the Tw-Py conical intersection.

Wolfgang Domcke addressed Albert Stolow and Simon Neville: The excited electronic states of ethylene have been a nightmare in quantum chemistry for decades. Did you check whether the rather simple ADC(2) method is reliable for the excited states of ethylene? Did you check the ADC(2) results against more accurate multi-configuration multi-reference methods? Is ADC(2) applicable at the conical intersection of the excited states with the ground state?

Albert Stolow and **Simon Neville** replied: We did check the ability of the ADC(2) method, used to calculate the X-ray absorption cross-sections, to give a satisfactory description of the MR-FOCI states, used in the AIMS dynamics calculations. As a suitable metric, we considered the overlap of the ADC(2) and MR-FOCI states at all geometries used in the calculation of the TRXAS. Those geometries for which the maximum overlap between the MR-FOCI state of the AIMS calculation and the ADC(2) states was not greater than 0.6 were excluded from the TRXAS calculation. For both the bound and continuum spectrum calculation, which employed different basis sets, around 90% of the geometries were retained. This indicates that, in general, the ADC(2) method is capable of giving a reasonable description of the excited states. Moreover, we did not observe a bias in the discarded geometries towards any particular region of nuclear configuration space.

Wolfgang Domcke commented to Albert Stolow and Simon Neville: My compliments on the results shown in Fig. 4(a). Your time-dependent electronic populations show recurrences driven by vibrational motion. These must generally occur in non-adiabatic electronic decay at such short timescales, when the vibrational motion is inertially driven. While many non-adiabatic classical dynamics simulations have been published recently for similar or other systems, the all exhibit just structureless decays. Your calculation is much better.

Oliver Gessner addressed Albert Stolow and Simon Neville: Is the splitting of the C1s pre-edge feature at ~20 fs delay (Fig. 8a and d) dominated by core or valence level shifts? In case of a strong core level shift contribution, the shifts should be related to a difference in core-hole screening on the different carbon atoms as a result of the pyramidalization. The relative positions of the split TRXAS features compared to the ground state pre-edge feature would then suggest that the pyramidalization leads to a charge transfer between the two carbon sites of the molecule. Is that the physics behind the predicted splitting?

Albert Stolow and **Simon Neville** answered: From the calculated CVS-ADC(2)x permanent dipole moments for the final core-excited states involved in these transitions, it is predicted that in both cases the final core-excited state exhibits a strong polarization, resulting in the pyramidalized CH_2 group acquiring a partial negative charge. As such, it seems reasonable that the predicted splitting of the pre-edge $\pi\pi^*$ state signal is intimately linked to the polarization of the molecule at geometries close to the Tw-Py seam. This could conceivably be due to differential core-hole screening between the two types of carbon atoms associated with the differential charge of the two carbon atoms. However, the relaxation of the valence orbitals induced by the core-hole should also be taken into account here. Again, the differential charge of the two carbon atoms may reasonably be expected to result in a difference in orbital relaxation effects induced by the creation of a core-hole on each of the two carbon atoms. In short, it seems likely that the polarization across the C-C bond will induce differential core-hole screening and relaxation effects. However, it is not straightforward to predict the relative magnitudes of these two interrelated effects.

Piero Decleva asked: The adequacy of the CVS-ADC(2) approach for the calculation of core-excited spectra may be questionable in the case of excited initial states, as the approach uses the ground state as a reference, and core singly excited configurations which dominate the spectrum are in general doubly excited with respect to the ground state, apart from the valence-hole filling (1s $\rightarrow$ π) which is prominent in this case. I suggest benchmarking with convergent MRCI calculations which should be affordable on ethylene.

Albert Stolow answered: We acknowledge that benchmarking the calculated excited-state X-ray absorption spectra against a higher-level multi-reference approach would be desirable, and we are in fact working on such an approach. However, it should be noted that, at the CVS-ADC(2)x level of theory, doubly-excited configurations are in fact included in the description of the excited state wavefunctions.

Adam Kirrander said: In your calculations you assigned the lines a fixed Lorentzian width, presumably based on information from other sources such as experiments. Do you have any feeling for how sensitive the lifetimes are to molecular geometry? Could the shifts in line position and width be sufficiently large to complicate the interpretation of the experiment? Also, how computationally expensive are these calculations, for instance, compared to (valence) photoionization cross-sections?

Simon Neville answered: For a C 1s core-hole, which is relatively deep, it is not anticipated that the line width will vary significantly with the nuclear geometry. One caveat is that for dissociation processes that remove some valence electron density from the vicinity of the core-hole, it can be expected that the width may decrease significantly. However, dissociation processes negligibly contribute to our calculated AIMS dynamics of ethylene. As for the computational expense of our method, the pre-edge cross-section calculations are relatively simple, involving only the calculation of the initial valence-excited states, the final core-excited states and the transition dipole moments between the two. The calculation of the continuum part of the spectrum is somewhat more involved, requiring the generation of the Lanczos pseudospectrum of the CVS-ADC(2) Hamiltonian of sufficiently large dimension (typically of the order of 10^3) to converge the spectral moments. The main bottleneck in either case, however, is the $O(N^6)$ scaling of the CVS-ADC(2)x method.

Adam Kirrander opened a general discussion of the paper by Taro Sekikawa: The timescales for the ring-opening reaction of 1.3-cyclohexadiene reported in your paper are significantly slower than the timescales reported in several recent experiments and theoretical simulations.[1–3] Can you comment?

1 O. Schalk *et al.*, *J. Phys. Chem. A*, 2016, **120**, 2320–2329.
2 M. P. Minitti *et al.*, *Phys. Rev. Lett.*, 2015, **114**, 255501.
3 C. C. Pemberton *et al.*, *J. Phys. Chem. A*. 2015. **119**, 8832–8845.

Taro Sekikawa replied: I would like to point out that not all the previous works reported fast ring-opening reaction. Electron diffraction showed that hexatriene (HT) is formed with a time constant of 32 ps (Ref. 1) and it takes about 500 fs to complete the formation of HT according to photoelectron spectroscopy by VUV light.[2] In the latter case, the HT formation time is faster, but the difference seems to be acceptable, taking account of the different initial excited states.

In comparison with the raised references, I would like to describe two points: i) The dynamics might be changed due to the difference in the initial states, as described in later replies. ii) In photoelectron spectroscopy using ultraviolet light, the lifetimes of the excited states are short. However, HT formation was not observed because of the small photon energy of the probe light. The smaller probe photon energy might miss some excited states or transition states on the way to the conical intersection. If so, the observed lifetime of the excited state does not necessarily correspond to the ring-opening time.

1 C.-Y. Ruan *et al.*, *Proc. Natl. Acad. Sci. U. S. A.*, 2001, **98**, 7117.
2 S. Adachi *et al.*, *J. Phys. Chem. Lett.*, 2015, **6**, 343.

Peter M. Weber asked: In our 2011 paper,[1] we saw little evidence for crossing to 1B or 2A after excitation to the 3p Rydberg states. Do you have any evidence that such a curve crossing happens?

1 C. C. Bühler, M. P. Minitti, S. Deb, J. Bao and P. M. Weber, *J. At., Mol., Opt. Phys.*, 2011, 637593, DOI: 10.1155/2011/637593.

Taro Sekikawa answered: No, we do not. According to my collaborators, it is not easy to calculate the potential energy surfaces from the 3p Rydberg states down to 1B states, because many states have to be included. The reference shows that the efficiency of the ring-opening is not zero but low. Therefore, we do not think that the relaxation pathway from the 3p Rydberg states to 1B or 2A is completely closed.

Michael Minitti commented: In your experiment with 1,3-CHD, you used 400 nm to excite *via* two-photon absorption. What was your percentage of excitation in order to guarantee only two photons were absorbed? In the manuscript, you mentioned your team used on the order of 40–50 μJ of pulse energy. This pulse energy, coupled with a presumably small focus, would be nearly impossible to control just two photons being absorbed. What focal spot size did you use?

Taro Sekikawa answered: In the middle of the same paragraph, it is described that the excitation energy was 19 μJ. We did not estimate the peak intensity, because the beam profile was far from the Gaussian or the uniform circular profiles. We tried to excite the molecules at the minimum energy so that we could observe some signals. In addition, as described in the subsection "Excited states" in the Results and discussion, the three-photon absorption ionizes the molecules. The ionization energy of the cation is usually larger than that of the neutral species due to the Coulomb interaction. Therefore, the spectrum of the cation did not overlap with that of the neutral species. The ionization did not deteriorate the observed spectra.

Peter M. Weber commented: A two-photon excitation with significant population transfer, as is the case here, requires a significant laser pulse intensity. Have you explored whether there are further absorption steps? The absorption of multiple further photons could result in fragmentation of the molecule. Have you looked for fragments in the mass spectrum?

Taro Sekikawa responded: Yes, we observed the photoelectrons only by the pump pulses. However, their kinetic energies were low and were energetically well separated from those by high harmonic pulses. We have never measured the mass spectra, but are worrying about the photofragments. We would like to investigate the dynamics under one-photon excitation for comparison.

Oliver Schalk commented: In 2015, Adachi *et al.* submitted data on a 267 nm pump, a HHG probe experiment on 1,3-cyclohexadiene (Fig. 2 in Ref. 1). In this paper, the authors showed formation of hexatriene which was strongly indicated by a peak appearing in the region between 10 and 10.5 eV binding energy, a region where 1,3-cyclohexadiene does not show ionization but hexatriene does. In the present contribution, the authors claim formation of hexatriene upon excitation with two photons of 400 nm (predominantly to the π3p-Rydberg manifold); however, no changes of the spectrum in the region between 10 and 10.5 eV can be observed. Is there an explanation for such a behavior or does that mean that hexatriene is not formed under the present experimental conditions?

1 S. Adachi *et al.*, *J. Phys. Chem. Lett.*, 2015, **6**, 343–346.

Taro Sekikawa responded: We think HT was observed. For simplicity, we call the photoelectron band between 10.5 and 12 eV the C=C band, although it consists of several molecular orbitals (MOs). After excitation, the C=C band increased at 607 and 1000 fs, as shown in Fig. 5. However, please note that the whole band was not increased: the photoelectron band with a lower ionization energy than 11 eV was enhanced faster than the energetically higher region. This indicates that a new MO is formed around 11 eV, shown in Fig. 5b. This should correspond to cZc-HT. In addition, our calculation predicts that the HOMO of cZc-HT is shifted to higher ionization energy than that of 1,3-CHD, it was also observed experimentally at 1000 fs as shown in Fig. 5b. These are described in the subsection of "Ring-opening dynamics" in the manuscript.

Oliver Schalk commented: As a reference to this comment, please see Fig. 1, taken from Schalk *et al.*,[1] where the distance between the two C atoms that break during the photodynamics of 1,3-cyclohexadiene is plotted against the propagation time.

In their paper, the authors claim the following mechanism:

a) The molecule is excited by two photons of 400 nm from the ground state to a π3p-Rydberg state.

b) From there, the dynamics flows to the valence state within 37 fs from where it accesses the ground state after 500 fs.

c) In the ground state, the cZc conformation is formed, which rearranges to cZt after 800 fs before it finally ends up in the tZt conformation after 1.4 ps.

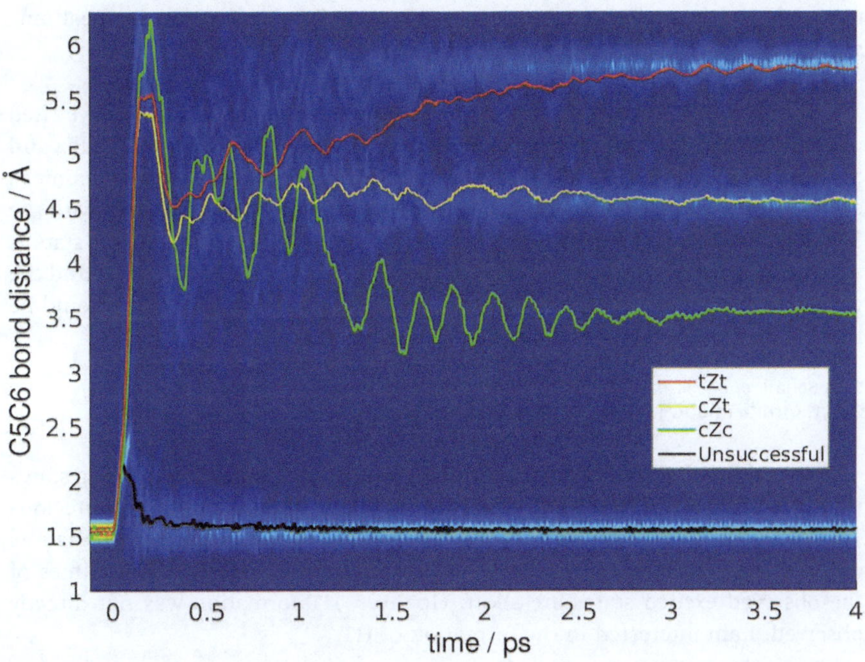

Fig. 1 Time evolution of the bond distance between C5 and C6 for all trajectories. The average values of the trajectories that lead to a certain rotational isomer are indicated by the colored lines. Adopted from Fig. 5 from Schalk *et al.*[1], Fig. 5. Reprinted with permission from Schalk *et al.*, *J. Phys. Chem. A*, 2016, **120**, 2320. Copyright 2016 American Chemical Society.

Regarding this mechanism, could the authors explain the following: in the region below the IP, the signal must originate from excited state dynamics and is always a positive differential signal (*i.e.* no counts before time zero). If the molecule stays in the excited state for more than 500 fs, why don't you see a signal in this region?

1 O. Schalk *et al.*, *J. Phys. Chem. A*, 2016, **120**, 2320.

Taro Sekikawa replied: First of all, b) and c) are slightly different from what we described. The ring opens when relaxing to the ground state of HT. The ring begins to open to form cZc at about 500 fs and ends at about 800 fs. This experimentally observed timescale is determined by the temporal durations of the pump and probe pulses. We did not observe excited states explicitly between 200 and 500 fs. Experimentally, the collection efficiency of the magnetic bottle spectrometer for the faster photoelectrons is lower, because the faster electrons are not well guided to the detector due to the larger Larmor radius of an electron in the magnetic field. This makes it difficult to observe the fast electrons with lower ionization energies. As described in the manuscript, we had difficulty observing even the Franck–Condon 31A state just after excitation, shown in Fig. 4b, with a good S/N ratio in comparison with relatively larger spectral changes above 10 eV. Please make sure that the transient band observed in photoelectron spectra corresponds to the transition not to the ground state of the cation but to the excited states. The transition to the ground state of the cation should appear in the region of the lower ionization energy. Another reason for the low photoelectron yields of the excited states is that the Franck–Condon factors to the cationic states might be smaller under the distorted molecular structure.

Oliver Schalk asked: In previous time resolved studies (see *e.g.* the above cited paper of Schalk *et al.*[1] and papers cited therein, especially the review of Aruda and Sension[2] – there are 10+ studies agreeing on the rough timescale of the dynamics), excited state dynamics of sub 150 fs were observed. How does this agree with the 500 fs lifetime in this paper? One possible reason might be that another state is accessed initially, but passage to the valence state(s) is claimed to be fast and the gradients are supposed to be steep such that the conical intersection should be readily accessed once you are on the valence manifold.

1 O. Schalk *et al.*, *J. Phys. Chem. A*, 2016, **120**, 2320.
2 B. C. Aruda and R. J. Sension, *Phys. Chem. Chem. Phys.*, 2014, **16**, 4439–4455.

Taro Sekikawa responded: In the present work, we excited to 3p Rydberg states in a two-photon absorption scheme, which is not the case with the previous works, except for that by Peter Weber's group. Since the initial excited state is different, the relaxation processes could be different. Certainly, the lifetimes of the observed excited states are short. However, HT formation was not directly observed. I am interested in the formation of HT.

Oliver Schalk remarked: The authors claim that the transformation between cZc, cZt and tZt conformers occurs between 300–500 fs. This significantly contradicts recent theoretical results (see Fig. 1 but also Ref. 1 and 2) which show that the tZt conformer is formed directly after ring opening and that interconversion between these conformers occurs on a sub 100 fs timescale. I would like the authors to also comment on the following mechanism that might be consistent with the data presented in the paper: in accordance with previous results, return to the ground state is fast and occurs on a sub 100 fs timescale which is observed in the region below the IP (sub 8eV in Fig. 4b). There, you observe an extended bleach of the photoelectron bands which might be caused by a weak Franck–Condon overlap between the ring-opened hexatriene and the ionic manifold as emphasized by the "floppiness" suggested by the calculations shown in Fig. 1. This floppiness comes from the momentum stored in the motions along the ring-opening variable which enforces a fast interconversion between the three conformers. After 1 to 1.5 ps, IVR redistributes energy to other modes and the conformers show little interchange (see Fig. 1). This might increase the Franck–Condon overlap which leads to an increase in the ionization yields. The time-scales of the dynamics of theory and experiment match nicely.

1 Y. Lei *et al.*, *J. Photochem. Photobiol., A*, 2016, **317**, 39–49.
2 M. P. Minitti *et al.*, *Phys. Rev. Lett.*, 2015, **114**, 255501.

Taro Sekikawa answered: This could also be attributed to the initial excited states. In the references, the initial exited states are not 3p Rydberg states. Therefore, the dynamics is not necessarily the same. We agree that a molecule might rotate to form tZt after the ring breaks. It is difficult to distinguish among cZc, cZt and tZt experimentally from the photoelectron spectra as described in the paper. However, the HT isomers have lower MO energies of around 10 eV than CHD and the photoelectron band being less than 10 eV appears at 607 fs in Fig. 5b. This indicates the formation of HT. We also think that the formed HT transforms among the isomers coherently because of the impulsive excitation. We suspect that the intensity enhancement of around 10 ps in Fig. 7 might reflect the coherent motion. In our case, the molecules are excited by the pump pulses with a pulse duration of ~100 fs. Although the 100 fs oscillation is predicted by the theoretical calculation in Ref. 1, the oscillation would be washed out because the starting point of the oscillation has an ambiguity of ~100 fs. Therefore, we are not sure that the observed intensity revival is due to the predicted oscillation.

1 O. Schalk *et al.*, *J. Phys. Chem. A*, 2016, **120**, 2320.

Michael Robb said: Fig. 1 is a bit misleading. In reference 25 of your paper it is shown that the 2A/1A conical intersection seam lies parallel to the ring-opening co-ordinate (*i.e.* the reaction path). It follows that the 3s Rydberg state and the 1B state may intersect this conical intersection in different places resulting in different observed dynamics.

Taro Sekikawa responded: We appreciate your comments. We expect that the relation pathway from the 3s Rydberg state is different. The schematic diagram will be improved.

Daniel Neumark remarked: It seems that one source of your disagreement with the previous results of Prof. Weber is that he performed one-photon excitation of CHD at 200 nm, whereas you do two-photon excitation at 400 nm. Can you repeat your measurements at 200 nm to investigate whether these discrepancies in interpretation disappear?

Taro Sekikawa replied: Yes. We are planning the experiment in which 200 nm light is used for the pump. I have one comment on the disagreement with the previous result. In the discussion with Peter Weber, we agreed that their data do not necessarily indicate non-ring opening because the photoelectron spectra in their paper showed some features of the bond cleavage. However, the efficiency of the ring-opening reaction was low. I would like to investigate CHD using 200 nm light.

Hans Jakob Wörner asked: I think that 1,3-cyclohexadiene does not have inversion symmetry, therefore there is no electronic parity (g *vs.* u). Consequently, one- and two-photon transitions may access the same electronic states. How can you reconcile your results with the results obtained following one-photon excitation at 200 nm by Peter Weber's group?

Taro Sekikawa replied: The comment on the symmetry is correct. So both one-photon and two-photon transitions are allowed. In Bühler *et al.*,[1] the final conclusion the authors described in the abstract is that "excitation to 3p does not induce ring opening". Experimentally, they observed the resonant ionization *via* the Rydberg states of CHD, occurring after the relaxation to the 2A state. The disappearance of the Rydberg states was attributed to the ring-opening reaction. In Bühler *et al.*,[1] upon excitation at 207 nm, the intensities of these Rydberg states were weaker than those upon excitation at 268 nm, indicating that the relaxation to the 2A state happens less upon excitation at 207 nm. Consequently, it was concluded that the ring-opening reaction does not happen.

I would like to raise two points. 1) The experimental results indicate that the Rydberg states appeared clearly, even by the excitation at 207 nm. The ring-opening process can take place. Therefore, I think that their final conclusion is too simplified. In fact, the authors described in the text that the ring-opening is not vanishingly small. During the Faraday Discussion, Peter Weber and I reached a consensus on this issue. 2) In Fig. 1 of Bühler *et al.*,[1] the intensities of the Rydberg states by 207 nm light look much weaker at a glance. However, the color scales indicate that the intensities of the Rydberg states less than 2 eV, both in Fig. 1a (207 nm excitation) and 1b (268 nm excitation), are on the order of 16~17. I wonder if they are comparable or not. Since the excitation peak intensity was not identical, it is difficult to estimate the absolute intensity of the Rydberg states only from the presented data. However, if comparable, the efficiency by 207 nm excitation might not be so low. Totally, since the one-photon excitation also opens the ring, I do not think there is a serious discrepancy. I will investigate the relaxation processes under the irradiation of 200 nm photons.

1 C. C. Bühler, M. P. Minitti, S. Deb, J. Bao and P. M. Weber, *J. At., Mol., Opt. Phys.*, 2011, 637593, DOI: 10.1155/2011/637593.

Russell Minns commented: For the signal going to negative times in Fig. 4, there appears to be an initial increase to 10 arb. units before it falls again to 8.5 arb. units giving a total change of 1.5 arb. units. The signal at this point still seems to be falling such that we do not know how far the signal would continue to fall, what a true baseline at negative times is, or if these changes are due to probe–pump or pump–probe processes that are currently unaccounted for. The change in signal after time zero does seem to fall further, down to a minimum of about 7 arb. units, but it is unclear how significant this is without a true baseline to compare the changes with, or a measure of the uncertainty in the measurement itself. I wonder if you can therefore expand on the origin of the changes in the signal occurring at and before time zero seen in Fig. 4 and 7 of the manuscript. I believe you are using the dashed line in Fig. 7 to try to convey this baseline level, my understanding from the manuscript is that this is the average signal for negative times. I am therefore left wondering, depending on your answer to my initial point above, how good a comparison this is? If the changes in signal at negative times are affected by a probe–pump (or even pump–probe) contribution, then the dashed line seems inappropriate as a measure of the changes in signal and the plot needs to extend to much more negative times to obtain a reasonable measure. This would also mean the assignment of the spectrum at these early times would not be particularly accurate. If no such contributions are present, and the changes are due to uncertainties in the experiment then this is currently not captured in the error bars on the plots. Could you please expand on these points.

Taro Sekikawa responded: I suspect that the definition of "true baseline" is different for each of us. The true baseline to us is the average of the negative temporal time. We understand you are worrying about the pump effects by XUV pulses. In the beginning of this research, we compared the spectra without the pump with those at negative delay times longer than 1 ps and made sure that the spectra were almost identical except for the intensity fluctuation. So, at the present stage, we do not think that the pump effects in the negative delay times should be taken into account. In the research presented here, we decided not to measure the spectra in the negative temporal region longer than 200 fs. Although you worry that "this would also mean the assignment of the spectrum at these early times would not be particularly accurate", we think there are no problems in the assignment because the spectral features between with the pump and without the pump were the same. On the other hand, at around 800 fs, the intensity ratio between the C–C and C=C bands changed. This is the evidence of the structural change of the molecule upon photoexcitation. Because high harmonic generation is the highly nonlinear process, we sometimes suffer from the stability of the light source. If the temporal behavior of only one particular band is focused, it is difficult to discuss the dynamics under the worse stability of the light source as you claim. However, we observe spectra covering multiple molecular orbitals and can see the appearance and disappearance of the energy levels simultaneously. Even using the unstable light source, it is not difficult to find new energy levels in comparison with other peaks. The different temporal behaviors among MOs are less distorted by the stability of the light source because the photoelectrons are ejected by the one-photon process and the spectral shape itself does not depend on the high harmonic intensity. This is one of the advantages of the broadband measurement of photoelectron spectra using high harmonics. Please note that the temporal behaviors of the C-C and C=C bands in the negative temporal range were almost identical, although they are different in the positive range with larger amplitude. The dynamics in the negative temporal range might be induced by the

XUV pump as you claim. However, we are not confident about that, which might come from the intensity fluctuation of high harmonics. So, in this paper, we focused on the dynamics after the pump with a larger amplitude. We can see the moment of the structural change without ambiguity, when the spectral shape was altered. We think that the application of high harmonics to spectroscopy has just started. We are neutral about the dynamics in the negative temporal range. We will be able to improve the experimental methods and conditions to investigate the dynamics in the negative temporal range in the near future.

Kyoung Koo Baeck said: Though the major emphasis of this paper was focused on the ring-opening mechanism between 150 fs and 800 fs after the initial photo-excitation, the observation of the lifetime of the initially excited transient 3p Rydberg state (R_{3p}), ~37 fs, seems a new exploration of the early-time dynamics not studied before. Moreover, I think, the longer time for the ring-opening reaction by the initial excitation to the R_{3p} than to the 1^1B state was resulted in *not simply because* the internal energy of the initially excited R_{3p} is larger than that of the initially excited 1^1B *but mainly because* there is a significant difference between the geometries of the $R_{3p}/1^1B$ and the $1^1B/2^1A$ conical intersections (CIs). Therefore, we need more information about the geometrical difference between the two CIs in order to get better understanding of the ring-opening mechanism. The $1^1B/2^1A$ CI was studied before theoretically, but it seems the $R_{3p}/1^1B$ CI has not been studied yet. Unfortunately, the theoretical study of CIs involving such high-lying Rydberg states is more difficult than CIs between valence excited states. Regarding this, I suggest to the authors to try the pre-excitation of vibration levels of the ground electronic state of CHD by IR laser just before the action of the two-photon excitation at 400 nm. The pre-excitation of some vibrational levels could make a noticeable change to the lifetime of the transient band reported in the present work. Then the characteristics of the vibrational mode could be very helpful for a future theoretical study of the $R_{3p}/1^1B$ CI.

Taro Sekikawa replied: The idea to excite some vibrational modes is interesting. We wonder what the necessary amplitude of the vibrational modes is to observe the difference. Still, we appreciate the suggestion of the new experimental scheme. We will consider the scheme further.

Fernando Martín opened a general discussion of the paper by Danielle Dowek: The harmonics that are produced in the lab have a bandwidth. Do you know how much the polarization state of those harmonics varies within the bandwidth and from harmonic to harmonic? Experiments show a significant variation when resonances are involved. But what about those cases where only the smooth non-resonant continuum is playing a role? Since the determination of the polarization of emitted harmonics ignores this fact, it would be nice to have an estimation of the error that is made in making this assumption.

Danielle Dowek responded: To most generally answer your question about the possible influence of resonances in the determination of the polarization state of harmonics in an attosecond pulse train (APT), I think one may distinguish two aspects in the polarization analysis presented in this work.

Firstly, considering the high harmonic generation process, in particular the recombination step, the existence in the HH medium of narrow resonances relative to the bandwidth of the harmonics may induce a spectral dependence of the polarization state of the emitted radiation within the bandwidth of the harmonic (as well as from harmonic to harmonic). Such a variation of the polarization state within the harmonic bandwidth is not resolved by the molecular polarimetry method in the present conditions, due to the low statistics at 1 kHz repetition rate, therefore the polarization state is determined for each harmonic from the analysis of the MFPAD induced by the integrated peak attached to the selected harmonic. This situation is even invoked as a possible origin of the production of a measured "depolarization", *i.e.* a non-zero s_4 parameter. Note that the evolution of the polarization state from harmonic to harmonic is well resolved, as reported in the paper.

The second aspect lies in the way the complete polarization state of a given harmonic is measured: here the method relies on the analysis of the MFPAD resulting from photoionization of the NO molecule by the harmonics and, in particular for the determination of the s_3 Stokes parameter, it requires the predetermination of a reference F_{11} function. Although each harmonic has a given bandwidth (of a few hundreds of meV in the reported experiments) the reference F_{11} functions used in the work were measured using synchrotron radiation at the well-defined central energies of the relevant harmonics. Then an uncertainty could result from the use of this single calibration, for example if the F_{11} function would vary for photon energies sampling the harmonic bandwidth: such a situation might occur if the chosen NO inner-valence shell photoionization reaction would be affected by narrow resonances. This is not the case for the photon energy region addressed in this work: the direct photoionization continuum between 25 eV and 40 eV is affected by the shape resonance of the NO^+(c) state, as shown in the paper, however the related variation of the MFPAD, including that of the F_{11} function, is smooth at the scale of the harmonic bandwidth. This is illustrated *e.g.* by the fact that the MFPADs measured for the overall harmonics are very similar to those measured at the central energies using synchrotron radiation. Finally this issue can also be checked experimentally by tuning the photon excitation energy around the central value of the synchrotron radiation. For the reported results, the estimated error on the determination of the polarization state of the harmonic due to the use of the central energies for F_{11} calibration can be neglected.

Albert Stolow remarked: In your polarimetry analysis, did you need to make use of the s_4 Stokes parameter and, if so, how may this be interpreted?

Danielle Dowek replied: For partially polarized light, the s_4 "fourth" Stokes parameter measures the degree of unpolarized light of the analyzed radiation. The reported molecular polarimetry method enables us to determine its value, since the s_1, s_2 and s_3 three Stokes parameters, which determine the polarization ellipse of the polarized component of the light, are measured and $s_4 = 1 - \sqrt{s_1^2 + s_2^2 + s_3^2}$. The precision in the determination of s_4 is derived from that of the measured s_1, s_2 and s_3.

For the study reported in this paper, where high order harmonics are generated by the interaction of elliptically polarized IR driving laser with SF_6 molecules, the obtained polarization degree presented ($P = 1 - s_4$) shows some deviation from the maximum value of 1, which corresponds to non-zero values of s_4, in particular for the H15 and H17 harmonics, although with rather large error bars. If non-zero values of s_4 for such harmonics would be confirmed in future measurements with higher statistics and precision, these could be attributed *e.g.*, to the role of narrow resonances which are thought to influence the HHG process in SF_6 for such harmonics[1] and might therefore induce a spectral dependence of the polarization state of the light within the harmonic bandwidth analyzed as a whole. Other possible causes of depolarization may occur, in particular depending on the way symmetry breaking is performed in the generation process to produce elliptically polarized harmonics. We have investigated two other cases where HHG is driven by two color counter rotating circularly polarized fields in Ar atoms, and HHG is driven by linearly polarized light in aligned N_2 molecules. We found indications of depolarization for the generated HHs that will be discussed in forthcoming work. The detailed polarization state of harmonic radiation is a rich observable in high harmonic spectroscopy, and its description, including the degree of depolarization, raises interesting questions to be addressed in the future.

1 A. Ferré, A. E. Boguslavskiy, M. Dagan, V. Blanchet, B. D. Bruner, F. Burgy, A. Camper, D. Descamps, B. Fabre, N. Fedorov, J. Gaudin, G. Geoffroy, J. Mikosch, S. Patchkovskii, S. Petit, T. Ruchon, H. Soifer, D. Staedter, I. Wilkinson, A. Stolow, N. Dudovich and Y. Mairesse, *Nat. Commun.*, 2015, **6**, 5952.

Álvaro Jiménez-Galán asked: Your method can characterize beautifully the ellipticity and polarization state of a harmonic in the train. However, would it work for broadband pulses, *i.e.*, attosecond pulses? As far as I understand, the problems that may arise are due to the use of mirrors that have different reflectivities for different frequencies. Also, your method relies on comparison with synchrotron measurements at specific photon energies. Could you clarify where are the limitations of this technique when dealing with attosecond pulses?

Danielle Dowek responded: For broadband pulses such as single attosecond pulses (SAP), there are different issues in the analysis of the polarization state of elliptically polarized light due to the quasi-continuum of photon energies involved on a several eV bandwidth.

The first one, as pointed to in your question, is the need to characterize the transmission function of optical elements which transport the XUV pulses from the HHG source to the center of the end-station, such as the focusing toroidal mirror, in terms of complex reflectivities for numerous photon energies. This requires some effort, although the smooth evolution of their amplitudes and phases also opens the possibility for some extrapolation between discrete energies to be studied.

The second aspect is the photoionization analysis required to determine MFPADs, using reference F_{11} functions that must be recorded in a rather extended energy range corresponding to the bandwidth of the SAP. In the NO inner-valence photoionization region, after selection based on the ion fragment energy in the overall kinetic energy correlation diagram (KECD), the quasi-continuous photoelectron spectrum will be reflecting photoionization through the $NO^+(c)$ state at different photon energies. The method presented can *a priori* be extended by selecting successive ΔE slices centred at different photoelectron energies in the spectrum and extracting the corresponding MFPADs. This extension of the method should provide the polarization state as a function of the photon energy in the SAP.

Filippo Bencivenga asked: I'm wondering if another pair of insertable gold-coated mirrors at a 45° angle of incidence (as the two ones shown in Fig. 3) could be placed downstream the toroidal mirror to route the XUV light into the spectrometer. This is in order to get information on the effects of the optics on the polarization of the XUV radiation at the sample.

Danielle Dowek replied: The pair of insertable gold-coated mirrors you refer to in Fig. 3, combined with a rotatable halfwave plate, constitutes an optical polarimeter based on Malus' law: as described in the paper it provides the s_1 and s_2 Stokes parameters, however it does not contain a dephazer element so it does not give access to the ellipticity of the light. So such a couple of mirrors might be used for the analysis of the light transmitted by the mirror, however only for the orientation of the polarization ellipse, characterized by s_1 and s_2. Here instead, we are using the molecular polarimetry technique itself in order to characterize the transmission function of the mirror (amplitudes and phases), for each harmonic, using incoming linearly polarized light at different orientations (this is required for the further analysis of HHG elliptically polarized light).

Russell Minns asked: You mentioned in your presentation that the requirements of coincidence imaging mean that it requires a large number of laser shots to obtain good signal levels and reasonable statistics. I wonder if you could expand on this and outline how long each measurement took and what would be necessary to perform time-resolved experiments with this technique?

Danielle Dowek answered: It is indeed challenging to measure doubly differential observables such as molecular frame photoelectron angular distributions (MFPADs) induced by an attosecond pulse train (APT), relying on coincidence electron-ion 3D momentum spectroscopy, by employing a 1 kHz laser source. In the reported experiments, providing MFPADs for each harmonic in an APT composed of 4 to 5 relevant harmonics, an acquisition lasting about three hours was required. As illustrated in our paper, this achievement benefited from the efficient data analysis that we developed earlier, taking advantage of the general analytical form of the MFPAD. For future pump–probe experiments aimed at measuring time-resolved MFPADs, the potential applicability of this method will critically rely on the development of HHG sources at a higher repetition rate: this is in progress since HHG sources running at 10 to 100 kHz are being developed by different groups. In Saclay, we will use the FAB10 HHG source of the ATTOLAB facility which will be operational soon. Nevertheless, these experiments will remain quite demanding, in particular in terms of the stability of the XUV light source for each studied delay.

Raluca Cireasa asked: The method proposed for determining the polarisation of the harmonics through measurements of MFPAD is very powerful, but it is an experiment in itself that implies long acquisition time and complex data analysis.

Could it then be envisaged to set it up and use it just as an *in situ* molecular polarimeter? In this context, please, comment on the importance of the choice of the molecular systems to be photo-ionised by the harmonics.

Danielle Dowek responded: Using the molecular polarimetry method, as demonstrated in our paper, as a standard diagnostic of the complete polarization state (three Stokes parameters) of an XUV light source requires one to measure MFPADs: so it intrinsically calls for a coincidence experiment and related data analysis, even if a dedicated more compact instrument and standard analysis routines may be optimized. The choice of a prototype molecular system as a target for that purpose is influenced by some criteria, such as a large dissociative photoionization (DPI) cross-section, the identification of one DPI reaction characterized by (i) a significant asymmetry parameter in the ion fragment angular distribution (which also ensures validity of the axial recoil approximation), and (ii) a significant circular dichroism in the molecular frame, which acts as the dephasing element in the polarization analyzer. When the XUV light is composed of harmonics, it is interesting to favor molecular systems which minimize the overlap of DPI processes induced by different harmonics. The inner-valence shell of the NO molecule considered in the paper is quite favorable with respect to most of the criteria listed above.

Raluca Cireasa asked: The energy of the harmonics used for the experiments presented in the paper is 20–40 eV. Is it possible to extend this method to accurately determine the Stokes parameters, defining the polarisation of harmonics of much higher energy?

Danielle Dowek replied: At much higher photon energies, now becoming available in HHG XUV sources, the molecular polarimetry method can be applied relying on MFPADs measured for other photoionization (PI) reactions, such as inner-shell PI of simple molecules, *e.g.*, PI of the CO molecule at photon excitation energies above the C (1s) threshold ($\sim$293 eV). MFPADs for O (1s) PI of CO ($I_p \approx$ 542.5 eV) induced by elliptically polarized synchrotron radiation was reported as an example in Refs 1–2. The fact that the inner-shell PI of such molecular targets is followed by subsequent Auger decay results in dominant dissociative photoionization (DPI) processes. This is quite favorable for molecular polarimetry coincidence experiments based on MFPADs, since a single PI reaction dominates above each inner-shell ionization threshold, and most of the PI events are dissociative.

1 D. Dowek and R. R. Lucchese, in *Dynamical Processes in Atomic and Molecular Physics*, Bentham, Bussum, Netherlands, 2012, pp. 57–95.
2. K. Veyrinas, PhD Thesis, Université Paris-Sud 11, U. F. R. Scientifique d'Orsay, 2015.

Oliver Schalk opened a general discussion of the paper by Dave Townsend: In section 5.2, you write: "The lack of negative amplitude in any region of the τ_1 DAS also appears to suggest that the decay of the $2\pi\pi^*$ state in both *N*,*N*-DMA and 3,5-DMA does not involve population transfer to either the $S_1(\pi\pi^*)$ or $S_2(3s/\pi\sigma^*)$ states ...". In Fig. 4, you show that τ_1 is within the time resolution of the experiment and that you essentially fit a Gaussian to account for the amplitude of the

$2\pi\pi^*$ state. In this case, there is no way that you would see a delayed rise (or a negative amplitude) that would indicate passage from $2\pi\pi^*$ to a lower lying state, even if you had 100% population transfer. Therefore, I do not see how your argument leads to the claim that you do not see any population transfer from $2\pi\pi^*$ to a lower lying state and, hence, why the photoinduced dynamics in *N,N*-DMA and 3,5-DMA would be different to aniline in this respect.

Dave Townsend answered: If internal conversion between the $2\pi\pi^*$ state and the lower-lying states were taking place, we believe this should be resolvable in the DAS analysis we have undertaken. This is illustrated, for example, in related TRPEI work on aniline using 238 nm excitation with a similar time resolution to our present measurements – as published recently by the Fielding group.[1] In this work, the aniline DAS analysis does indeed reveal signatures of ultrafast internal conversion between the $2\pi\pi^*$ state and the singlet states that sit energetically below it. The absence of such features in our *N,N*-DMA and 3,5-DMA data therefore appears to suggest that in these cases, the $2\pi\pi^*$ state decays (extremely rapidly) *via* another pathway. More critically, this assertion is also supported by the temporal evolution of the anisotropy parameters that we report (see Fig. 5), which indicates that the timescale to access the correct geometry for internal conversion from the $2\pi\pi^*$ state to either the $S_1(\pi\pi^*)$ state or the $S_2(3s/\pi\sigma^*)$ state is too slow (> 250 fs) to effectively compete with this much faster alternative pathway.

1 O. M. Kirkby, M. Sala, G. Balerdi, R. de Nalda, L. Bañares, S. Guérin and H. H. Fielding, *Phys. Chem. Chem. Phys.*, 2015, **17**, 16270.

Michael Robb asked: What is the chemical nature of the two $\pi\pi^*$ states? One should be locally excited (LE), one should be CT (from the lone pair to the benzene), and one a LE benzene (lowest). In addition there is the role on the NH_2 twist (for a theoretical discussion see Gómez *et al.*).[1]

1 I. Gómez, M. Reguero, M. Boggio-Pasqua and M. A. Robb, *J. Am. Chem. Soc.*, 2005, **127**, 7119-7129.

Gareth Roberts responded: Regarding the nature of the second $^1\pi\pi^*$ state ($2^1\pi\pi^*$) in aniline and whether it shares any electronic characteristics with the $2^1\pi\pi^*$ state of 4-(dimethylamino)benzonitrile (DMABN), I can confirm that this indeed appears to be the case. Prior excited state valence bonding analysis using the CASSCF method by Paterson, Stavros and co-workers[1] has indicated that in the vertical Franck–Condon (vFC) region, aniline's $2^1\pi\pi^*$ state exhibits partial intramolecular charge-transfer (ICT) character, with ~0.2 of an electron transferred from the lone pair in the *n*-orbital of the NH_2 group to the 4-position of the phenyl ring (see the valence bond structure in Fig. 1). This is akin to DMABN, but in the absence of any electron accepting/withdrawing group in the 4-position, it means that aniline's $2^1\pi\pi^*$ surface possesses no stable adiabatic minimum, leading to its ultrafast ballistic decay from the vFC region, the latter of which is discussed in Dave Townsend's paper. However, there is no obvious evidence to date that any so-called twisted intramolecular charge-transfer (TICT) process plays a role in aniline. As previously reported by Michael Robb,[2] this may not be the case for DMABN and the addition of the electron withdrawing CN group *para*

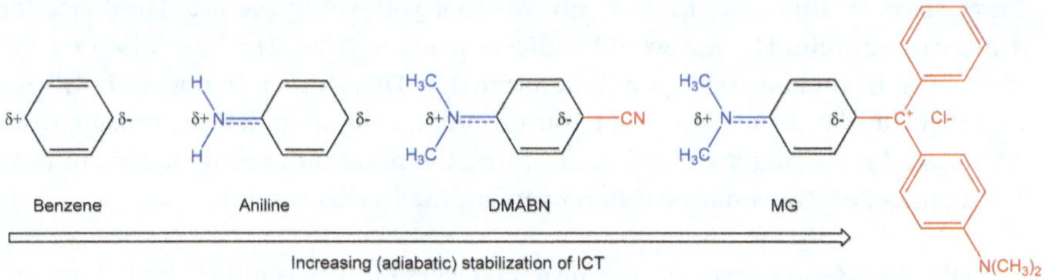

Fig. 2 Scheme showing the zwitterionic ICT valence bonding structure of benzene's $2^1\pi\pi^*$ (1^1B_{1u}) state in the vertical Franck–Condon region and how the character of this state can be stabilized/manipulated through the addition of electron 'donor' and 'acceptor' groups.

to the $N(CH_3)_2$ moiety can lead to the stabilization of ICT in the $2^1\pi\pi^*$ manifold through the twisting of the $N(CH_3)_2$ group relative to the phenyl ring, causing DMABN's classic dual-colour fluorescence.[3] Yet further 'tuning' of this ICT can be seen in the more complex malachite green (MG) system, where a carbocation centre lying *para* to the $N(CH_3)_2$ group leads to dramatic ICT stabilization, effectively deactivating the fluorescence and opening the rapid internal conversion to S_0 though conical intersections.[4,5] This seemingly inherent ICT character of the $2^1\pi\pi^*$ state does not have its origins in the aniline motif, but actually originates from the more fundamental benzene chromophore, whose $2^1\pi\pi^*$ (1^1B_{1u}) state exhibits the same zwitterionic ICT valence bonding structure in the vFC region (see Fig. 2).[6]

These comments/observations feedback more generally into the debate raised throughout this Faraday Discussion of 'predictive rules for photochemical dynamics'. While there may not be a unified set of 'silver bullet' rules on the horizon (as there isn't in organic synthesis), with appropriate theory and experiment, in certain cases trends can clearly be identified, understood and (to some degree) manipulated using ideas borrowed from organic chemistry (*i.e.* donors, acceptors, electron-withdrawing *versus* electron-donating *etc.*). Naturally, there are nuances to this simplified approach, but it can serve as an excellent heuristic starting point, as demonstrated here by a trend series which aims to manipulate an inherent characteristic of a fundamental UV chromophore.[6] This tuning of the ICT character in benzene's $2^1\pi\pi^*$ (1^1B_{1u}) manifold is summarized in Fig. 2, but is by no means an isolated example and other apt demonstrations of tuning organic photochemistry in this manner also exist in the recent literature (see, for example, Ref. 7 and 8).

1 G. M. Roberts, C. A. Williams, J. D. Young, S. Ullrich, M. J. Paterson and V. G. Stavros, *J. Am. Chem. Soc.*, 2012, **134**, 12578–12589.
2 I. Gómez, M. Reguero, M. Boggio-Pasqua and M. A. Robb, *J. Am. Chem. Soc.*, 2005, **127**, 7119–7129.
3 Z. R. Grabowski, K. Rotkiewicz and W. Rettig, *Chem. Rev.*, 2003, **103**, 3889–4031.
4 E. Riedle, M. Bradler, M. Wenninger, C. F. Sailer and I. Pugliesi, *Faraday Discuss.*, 2013, **163**, 139–158.
5 G. M. Roberts, *Faraday Discuss.*, 2013, **163**, 244–245.
6 S. Zilberg and Y. Haas, *J. Phys. Chem. A*, 2002, **106**, 1–11.
7 T. N. V. Karsili, A. M. Wenge, S. J. Harris, D. Murdock, J. N. Harvey, R. N. Dixon and M. N. R. Ashfold, *Chem. Sci.*, 2013, **4**, 2434–2446.

8 G. Bassolino, T. Sovdat, M. Liebel, C. Schnedermann, B. Odell, T. D. W. Claridge, P. Kukura and S. P. Fletcher, *J. Am. Chem. Soc.*, 2014, **136**, 2650–2658.

Michael A. Robb answered: In addition there is the role of the Rydberg states.

Daniel Neumark commented: One of the interesting aspects in the systems you are studying is the competition between dissociation and internal conversion from the $2\pi\pi^*$ state. My understanding is that you think CH_3 loss is competitive in *N,N*-DMA but that dissociation is not important in 3,5-DMA. Since your experiment is blind to this channel, what is the basis for this conclusion?

Dave Townsend answered: The basis for this interpretation stems from the fact that, in contrast to *N,N*-DMA, the $S_2(3s/\pi\sigma^*)$ state of 3,5-DMA exhibits no signatures of internal conversion to the lower-lying $S_1(\pi\pi^*)$ state (*i.e.* there is no negative amplitude in the DAS assigned to the decay of the $S_2(3s/\pi\sigma^*)$ state). The $S_2(3s/\pi\sigma^*)$ state of 3,5-DMA does, however, still exhibit an extremely short lifetime. Population within this state must therefore be rapidly decaying *via* another route. The $S_2(3s/\pi\sigma^*)$ state is only very weakly bound along the N–H coordinate (see Fig. 6) and so direct dissociation to yield H atom photoproducts is the obvious alternative – especially since this pathway has been observed to occur in the related species aniline (see, for example, Ref. 1 and 2). We suggest that, in principle, following 240 nm excitation both the internal conversion and direct dissociation pathways are energetically possible in both *N,N*-DMA and 3,5-DMA and, in the former, both are actively taking place. In the latter case, however, dissociation kinetically out-competes internal conversion to an overwhelming extent due to methylation of the aromatic ring system slowing the non-adiabatic decay pathway (relative to the dissociation pathway) to a much greater extent than methylation of the amino group in *N,N*-DMA. The temporal evolution of the anisotropy parameters presented in Fig. 5 also appears to be consistent with this picture, providing a particularly illustrative example of the interplay between structure and dynamics, as well as the power of the angle resolved data afforded by the TRPEI approach to reveal these effects.

1 G. A. King, T. A. A. Oliver and M. N. R. Ashfold, *J. Chem. Phys.*, 2010, **132**, 214307.
2 G. M. Roberts, C. A. Williams, J. D. Young, S. Ullrich, M. J. Paterson and V. G. Stavros, *J. Am. Chem. Soc.*, 2012, **134**, 12578–12589.

Andrew Orr-Ewing asked: Are there reported quantum yields for photofragments from aniline and the methylated anilines you have studied with which you might compare your deductions about competition between dissociative and non-dissociative pathways? Might a transient absorption spectroscopy measurement of the efficiency of parent recovery following UV excitation of the molecules dissolved in a weakly interacting solvent be instructive because it can quantify fragmentation pathway yields?

Dave Townsend replied: As highlighted in the response to Daniel Neumark's question, experimental schemes interrogating excited state dynamics using energy- and angle-resolved measurements of time-resolved photoreactant ionization provide a particularly powerful approach for revealing very subtle details of specific non-adiabatic pathways that may (or may not) operate within a given molecular system. One inherent problem, however, is that in instances where various competing pathways are present, techniques employing photoreactant ionization are not able to reliably differentiate which pathway is dominant. This issue with determining branching fractions stems from changes in relative detection sensitivity as the population evolves non-adiabatically over multiple potential energy surfaces and is an effect that is a convolution of several factors. These include changes in relative (electronic) photoionization cross-sections, variation in associated Franck–Condon factors for ionization (a particular problem for seeing any ground state recovery), excited state alignment effects and also the relative excited state lifetimes. Although some of these can be accounted for fairly easily, the issue of the electronic photoionization cross-section (which may vary between different excited states by an order of magnitude or more) is, at present, a notoriously challenging problem to address either experimentally or computationally. We have previously presented some speculative thoughts on how one may begin to address this issue,[1–2] although this only (potentially) serves as a rough guide for qualitative discussion. Any complementary measurement that may be able to shed light on this issue (such as the one Andrew Orr-Ewing suggests) would therefore be extremely valuable. This is not only important in terms of direct relevance to the specific problem under consideration here, but also more generally in providing data that may ultimately be used, in conjunction with that from ionization-based studies, to provide useful benchmarks for theoretical developments related to the accurate determination of excited state photoionization cross-sections.

1 J. O. Thompson, L. Saalbach, S. W. Crane, M. J. Paterson and D. Townsend, *J. Chem. Phys.*, 2015, **142**, 114309.
2 M. M. Zawadzki, J. O. F. Thompson, E. A. Burgess, M. J. Paterson and D. Townsend, *Phys. Chem. Chem. Phys.*, 2015, **17**, 26659.

Oliver Schalk remarked: In Fig. 5 (left panels), you show the time-dependent behavior of the β-parameters. It looks like there would be a change on a 1.5 ps timescale which would agree with the timescale you assign to IVR. Why would the β-parameters change if IVR is present?

Dave Townsend responded: The timescales for the evolution of the β-parameters in Fig. 5 are not really that well-matched to the timescales assigned as IVR in Fig. 4. In principle, IVR could lead to changes in the β-parameters but, without this process then inducing additional associated changes in the electronic state character through induced electronic coupling, it seems unlikely this would lead to the increase in the β-parameters that is observed. We therefore assume that Fig 5 is predominantly reflecting a change in the electronic character of the $S_1(\pi\pi^*)$ state as the molecule assumes the required geometry to couple efficiently to other low-lying states and (at least in principle) facilitate non-adiabatic population transfer – although as discussed in the paper, other energetically open pathways may (and in some cases do) effectively out-compete this mechanism if they occur on a faster timescale.

Jan Verlet remarked: In general, β-parameters derived from reconstructed images in time-resolved photoelectron imaging are convoluted. In the time domain, changes in β_2 often only probe changes in population of two states with differing photoelectron angular distributions (PADs).[1] In the energy domain, photoelectron peaks tend to be broad and overlapping so that the PADs are a convolution of the detachment from these overlapping bands. Therefore, although the global fitting of the reconstructed images provides insight into the dynamics through the decay-associated spectra, it does not provide reliable information about PADs. To overcome this limitation, we have shown that the global fitting can be performed directly on the raw time-resolved photoelectron images. This yields decay-associated images. Reconstruction of the decay-associated image provides identical decay-associated spectra, but also provides the PAD associated with a decay-associated spectrum.

1 J. Lecointre, G. M. Roberts, D. A. Horke and J. R. R. Verlet, *J. Phys. Chem. A*, 2010, **114**, 11216.

Dave Townsend replied: This is a very interesting approach. Although we do not currently employ such a scheme in our present data evaluation, it would certainly be an interesting avenue for future investigation – especially in instances where photoelectron bands strongly overlap, as Jan Verlet highlights. In the case of the *N*,*N*-DMA and 3,5-DMA work presented here, the photoelectron signals in the low kinetic energy region originating from ionization of the $S_1(\pi\pi^*)$ state are, however, reasonably well separated energetically from those due to the other states participating in the overall dynamics. This is important to stress, as it is the (consistent) temporal evolution of the anisotropy parameters within this region that provides strong support to our conclusions about the relative timescales of internal conversion processes and the impact this then has on the overall multi-pathway dynamics.

Russell Minns asked: Picking up on R. J. Dwayne Miller's earlier comments, having performed a systematic study of photochemical dynamics in similar systems, do you find any generalities in their photochemistry and do you think such generalities are likely in broader chemical systems?

Dave Townsend replied: Using site-selective methylation to systematically alter a basic molecular motif such as aniline can reveal a great deal of photochemical insight relating to specific relaxation pathways. Methylation (as opposed to substitution with strongly electron-withdrawing or -donating groups) tends not to alter the relative electronic state energies too significantly, although it can profoundly alter dynamical timescales. As such, it is an approach that we would advocate, where possible, as a route to gaining more mechanistic information. In particular, it is the dynamical changes that such an approach may induce that often serve to reveal additional information about the starting motif itself. This is certainly the case in aniline, where the dynamics are slightly too fast (given our present experimental time resolution) to pull out all of the information we would like to access. This is especially true in regard to the photoelectron angular distributions. By studying *N*,*N*-DMA and 3,5-DMA, we are able to slow the dynamics down sufficiently to learn a great deal of information about not only these specific systems, but also about aniline itself. There are some clear similarities between the dynamics operating in these systems, but also some important differences. Critically, however, some of these observed differences are not what might be predicted prior to conducting the experiments – a good example of this would be the apparent lack of significant change in the decay lifetimes of the $2\pi\pi^*$ state within *N*,*N*-DMA and 3,5-DMA when compared to aniline, even though the out-of-plane motion of the amino group and ring distortions have been implicated theoretically as key coordinates in potentially facilitating relevant non-adiabatic decay pathways in this system.[1–2] More generally, there has been considerable discourse at this Faraday Discussion regarding the desire to ultimately develop generalized predictive rules for excited state photochemical dynamics. Although others may disagree, my feeling is that as a community we are still a very long way from achieving this goal. The excited states of polyatomic molecules are extraordinarily complex and, in my opinion, each is essentially unique in terms of dynamical subtleties. New experimental and theoretical studies on seemingly related species consistently yield unexpected and surprising results, clearly reflecting the present lack of predictive insight. This, of course, should serve as a strong impetus to continue pushing at the boundaries of our understanding in the hope that a clearer picture will materialize, but it seems unlikely to me that this will be forthcoming in the near future. A major part of the problem here is that, in order for rules to be useful, they have to predict a measurable photochemical outcome – *i.e.* product branching ratios – and correlate this with specific key features of any given starting species. Although there are a wide variety of experimental strategies (with different associated observables) available to study the excited state dynamics operating in molecular systems, each only reveals a small piece of the overall "photochemical jigsaw", making the task of developing such rules rather incoherent and therefore extremely challenging. Although a somewhat obvious statement, it would therefore be highly desirable to develop a universal approach that is able to fully map the entire reaction coordinate connecting photoreactants to all photoproducts in a manner that is fully quantitative and yet also sufficiently differential to reliably extract such correlations for a wide range of systems. This still represents one of the major challenges to this community.

1 G. M. Roberts, C. A. Williams, J. D. Young, S. Ullrich, M. J. Paterson and V. G. Stavros, *J. Am. Chem. Soc.*, 2012, **134**, 12578–12589.
2 M. Sala, O. M. Kirkby, S. Guérin and H. H. Fielding, *Phys. Chem. Chem. Phys.*, 2014, **16**, 3122.

Russell Minns asked: The advantage the organic chemists have is that they are often dealing with rules about systems in equilibrium. In photochemistry this is not the case, excitation leads to non-equilibrium systems with very different electronic structure and levels of vibrational excitation dependent on the particularly wavelength used. It seems that making generalities in these situations would be particularly challenging.

Dave Townsend responded: I would strongly agree with this statement. Even very subtle changes in molecular composition (for example, simple deuteration of H atom sites) can have a profound effect on the dynamics exhibited by a specific system at the same excitation wavelength. In some cases this may be in a manner that can be predicted based on prior knowledge of dynamics operating in the system before any changes are introduced – but even then, perhaps not always. More generally, and as already outlined in my previous answer, it does not therefore seem that at present we are close to a more general set of useful predictive rules for excited state photochemistry. Being slightly more tongue-in-cheek, I might also argue that in some respects the analogy that is often made with rules in organic chemistry is something of a distraction and, as a community, we perhaps shouldn't start to obsess over this too much – my own experience of organic chemistry was always that "rules" get introduced in the first couple of lectures and the rest of the course was then devoted studying all of the "exceptions" – and as Russell Minns points out, this is for systems in equilibrium.

R. J. Dwayne Miller commented: I agree with Dave Townsend regarding whether or not general guidelines/rules might be found for photochemistry using the aniline derivatives as a case in point. The specific details and the exact timescales for non-radiative relaxation do depend exquisitely on the specific molecular structure. This implies that we are dealing with molecular "snow flakes" in terms of trying to find common features. However, if one focuses on the nuclear motions that are involved in coupling the electronic surfaces, there are great similarities even if the dynamics are very different. The case in point involves motion along the N–C coordinate for the methylated aniline systems. There is strong coupling along this coordinate as a common feature even though the magnitudes vary. The prospect for finding common, dominant modes, in the reaction dynamics is best exemplified with simple photoisomerization. We saw in Albert Stolow's paper (DOI: 10.1039/C6FD00117C) that the isomerization of ethylene involves a bond softening of the C=C mode, followed by torsional motions along this axis, and pyridimalization involving out-of-plane motions. These are clearly localized modes as we are using a very small system without any prospect of alternative pathways or spatial delocalization of the reaction forces driving isomerization. How curious then is the photoisomerization of rhodopsin and bacteriorhodopsin? We just showed[1] that the photoisomerization of rhodopsin involves the localized bond softening, torsions, and out-of-plane motions at the C_{11}–C_{12} location, effectively the exact same localized motions as in ethylene. The isomerization occurs within 30 fs (difficult to assign a precise number as we are dealing with timescales of half periods of the key modes, <50 fs) such that even the dynamics are not so different than ethylene. The primary vision in relation to rhodopsin involves a huge biological molecule where there are innumerable ways to spatially delocalize or distribute the forces. Yet, there is enormous reduction in dimensionality during motion through the conical intersection where the strongly anharmonic potential leads to localized motions of primarily 3 key modes - the very same modes as in ethylene as the targeted location for isomerization. Similarly bacteriorhodopsin (bR) undergoes photoisomerization but an order of magnitude slower. The process is not *cis* to *trans* at the C_{11}–C_{12} position but rather *trans* to *cis* at the C_{13}=C_{14} position. The same highly localized motions are involved in the same class of reaction. Nature found two different solutions for the same class of reaction in which the potential is clearly different in the bR and rhodopsin case. The details regarding dynamics vary by an order of magnitude but the key reaction modes do not. From a theory standpoint, these issues are understood on a case-by-case basis but the focus is on differences not on general relationships.

This point expands upon my comment that there may be a means to cast out the key modes to provide a dynamic reaction mode basis for understanding chemistry. The details are critical to a detailed understanding of the photophysics. However, there are clearly localized motions involving the most strongly coupled modes to the reaction coordinate and these likely are general for a given reaction type. This concept would scale to large systems as can be nicely traced in the family of rhodopsin proteins.

1 P. J. M. Johnson *et al.*, *Nature Chem.*, 2015, 7, 980–986.

Hans Jakob Wörner presented a comment: I presented the results of our recent work on time-resolved X-ray absorption spectroscopy with a water-window high-harmonic source.[1] Using high-harmonic generation from a 1.8 μm driver, we have generated soft-X-ray spectra covering the range of 100 to 350 eV. We have used this source to observe the photodissociation dynamics of CF_4^+ to CF_3^+ + F, induced by strong-field ionization from neutral CF_4, by monitoring transient absorption at the K-edge of carbon. The single resonance that dominates the carbon K-edge absorption spectrum of CF_4 was found to split into multiple lines as a consequence of symmetry lowering during the dissociation. We have performed similar measurements at the L-edges of sulfur, where we have studied the photodissociation of SF_6^+ to SF_5^+ and F.

1 Y. Pertot, C. Schmidt, M. Matthews, A. Chauvet, M. Huppert, V. Svoboda, A. von Conta, A. Tehlar, D. Baykusheva, J.-P. Wolf and H. J. Wörner, to be published, 2016.

Kiyoshi Ueda answered: A very impressive work! As we discussed in person, you should be able to map the delay time to the distances between F and the rest, referring to kinetic energy release. Also you should be able to extract the vibrational excitation of the fragment CF_3^+ (less for SF_5^+) since off-plane vibrational motion should be highly excited *via* recoil between F and C.

Jon Marangos commented: How many cycles are there in the HHG drive pulse? If it is multicycle then it would be a remarkable combination of parameters to lead to an isolated pulse generation (*e.g.* through ionisation gating). Can the spectrum alone (without careful characterisation of the spectrometer resolution) be enough to confirm an IAP?

Hans Jakob Wörner replied: The number of cycles was 5–6 as in Ref. 1, but we worked in free-focusing conditions similar to Teichmann *et al.*[2] Chen *et al.*[1] performed a linear auto-correlation measurement and concluded from this that an isolated attosecond pulse had been generated. Whereas this conclusion cannot be made directly without any information about the spectral phase of the pulse, the linear auto-correlation measurement establishes the generation of a continuous spectrum. The existence of "phase-matching gating" invoked by Chen *et al.*[1] has been confirmed theoretically by Teichmann *et al.*,[2] where it was additionally demonstrated experimentally. Teichmann *et al.*[2] observed continuous spectra for all values of the CEP, instead of the harmonic structure expected for a CEP close to π when two identical attosecond pulses should be generated. They attributed this effect to phase-matching gating. Hence, there is both experimental and theoretical evidence for the existence of phase-matching gating. Nevertheless, it is clear that a measurement sensitive to the spectral phase must be performed to definitely establish the generation of an isolated pulse. Such experiments relying on the attosecond streak camera have been done in our laboratory.

1 M. C. Chen *et al.*, *Proc. Natl. Acad. Sci. U. S. A.*, 2014, **111**, 8329.
2 S. Teichmann *et al.*, *Nat. Comm.*, 2016, 7, 11493.

Allan Johnson commented: I think the supercontinuum above 200 eV in your carbon K-edge spanning HHG spectrum may be due to resolution limits. The small amount of carbon contamination should show a clear π^* signature before the main edge if the resolution is sufficient to resolve harmonics with spacing around 1.4 eV.

Hans Jakob Wörner replied: We can exclude this possibility. We have experimentally determined the resolving power $E/\Delta E$ of our spectrometer to be 308, *i.e.* a resolution of better than 1 eV at 300 eV. This would be sufficient to resolve the harmonic structure beyond 300 eV.

Michael Minitti commented: Knowing how hard it is to perform absorption measurements in the soft X-ray regime, and how critical it is to know your pulse intensity before and after it interacts with the sample. How do you normalize your intensities in the XUV regime?

Hans Jakob Wörner replied: Our soft-X-ray source is based on high-harmonic generation of a 1.8-micron pulse in neon. It appears to have a few advantageous properties that simplify the task of recording absorption spectra. These are (i) a high pulse-to-pulse stability, (ii) a good long-term stability over a few hours and (iii) a high photon flux that enables us to record spectra with $S/N > 100$ in 5 s integration time and detailed pump–probe scans in 45 min. The presented data were obtained by dividing the transmitted spectra at each delay by the average of the spectra without absorption before and after the measurement.

Faraday Discussions

PAPER

Novel photochemistry of molecular polaritons in optical cavities

Kochise Bennett,* Markus Kowalewski and Shaul Mukamel

Received 26th April 2016, Accepted 17th May 2016
DOI: 10.1039/c6fd00095a

Violations of the Born-Oppenheimer approximation (BOA) and the consequent nonadiabatic dynamics have long been an object of intense study. Recently, such dynamics have been induced *via* strong coupling of the molecule to a high-amplitude (spatially confined) mode of the electromagnetic field in optical cavities. However, the effects of a cavity on a pre-existing avoided crossing or conical intersection are relatively unexplored. The dynamics of molecules dressed by cavity modes are usually calculated by invoking the rotating wave approximation (RWA), which greatly simplifies the calculation but breaks down when the cavity mode frequency is higher than the relevant material frequencies. We develop a protocol for computing curve crossing dynamics in an optical cavity by exploiting a recently-developed method of solving the quantum Rabi model without invoking the RWA. The method is demonstrated for sodium iodide.

I. Introduction

The photochemistry of molecules can be significantly influenced by specially-tailored electromagnetic fields.[1–7] While control is usually described in terms of classical laser fields, coupling with the vacuum state of a cavity mode requires a quantum field description. The underlying theoretical framework, cavity quantum electrodynamics, has been well studied in theory and experiment with atoms.[8–10] Its application to molecular systems has been demonstrated recently with great success,[11,12] the dynamics being described using joint photon–matter states called polaritons. However, the inclusion of the internal nuclear degrees of freedom, not present in atomic systems, complicates the problem and leads to nonadiabatic dynamics.[13,14] The electric field mode alters the nuclear dynamics through its interaction with the electronic degrees of freedom, creating avoided curve crossings and conical intersections between the molecular potential energy surfaces. This can in principle be achieved by a strong laser field.[2,3] By placing the molecule in a localized cavity mode, a laser system is not necessary and the field can even be in the vacuum state. The former scenario can be described in terms of

Chemistry Department, University of California, Irvine, California 92697-2025, USA. E-mail: kcbennet@uci.edu

a classical field, while the latter requires a quantum field description. A major difference between the two scenarios is the number of photons available: a strong laser field may give rise to multi-photon pathways that may interfere with the intended control scheme.

A broad variety of photochemical and photophysical dynamics are controlled by naturally-occurring conical intersections (CIs) in the bare electronic-energy landscape, opening up fast, non-radiative relaxation pathways by which the molecule is funneled back to the ground state. In this contribution, we theoretically investigate how a molecule which already possesses a curve crossing is affected by coupling to a nano-scale cavity.

The Hamiltonian describing a two-level system linearly coupled to a quantized radiation field is given by the quantum Rabi model:

$$\hat{H} = \frac{\omega_0}{2}\left(2\hat{\sigma}^\dagger\hat{\sigma} - 1\right) + \omega_c\hat{a}^\dagger\hat{a} + \tilde{g}\left(\hat{a}^\dagger + \hat{a}\right)\left(\hat{\sigma}^\dagger + \hat{\sigma}\right) \tag{1}$$

where ω_0 is the energy difference between the exited state $|e\rangle$ and the ground state $|g\rangle$, $\hat{\sigma} = |g\rangle\langle e|$ is the annihilation operator for bare molecular excitations, and $\hat{a}^{(\dagger)}$ annihilates (creates) cavity-mode photons. The vacuum Rabi-frequency $\tilde{g} = \mu_{eg}\varepsilon_c\sqrt{n}$ determines the coupling strength between the photon mode and the molecular degrees of freedom. The vacuum field amplitude

$$\varepsilon_c = \sqrt{\frac{\omega_c}{V\varepsilon_0}} \tag{2}$$

is determined by the active mode volume V of the cavity mode, making nano-scale cavities favorable candidates to achieve strong coupling. The cavity coupling is further enhanced by a factor of $\sqrt{n}$ if an ensemble of n molecules contributes in a coherent fashion.

The approximations which can be used to find the eigenvalues of eqn (1) depend on the magnitude of $\tilde{g}$. Perturbative solutions are only possible for very small values of $\tilde{g}$, *i.e.*, much smaller than all other system frequencies. If the time scale of $\tilde{g}^{-1}$ is faster than the possible decay processes, we enter the strong coupling regime where the Jaynes–Cummings (JC) model[9] provides approximate solutions to the Hamiltonian (eqn (1)) in the form of dressed states. Formally, the JC model neglects the counter rotating terms ($\hat{\sigma}^\dagger\hat{a}^\dagger$ and $\hat{\sigma}\hat{a}$) in the Hamiltonian, a simplification called the Rotating Wave Approximation (RWA) which thus keeps only the $\hat{\sigma}^\dagger\hat{a} + \hat{\sigma}\hat{a}^\dagger$ term in the cavity–molecule coupling. In this case, the total excitation number $\hat{N} \equiv \hat{n}_c + \hat{\sigma}^\dagger\hat{\sigma}$ is a conserved quantity ($[\hat{N}, \hat{H}] = 0$), reducing the Hamiltonian to a 2-by-2 block-diagonal structure (see Fig. 1 upper panel). The bare states contributing to the dressed state basis are the ground state with n_c and $n_c + 1$ photons and the excited state with n_c photons. An extension to the JC model to include the nuclear degrees of freedom within the RWA has been reported recently.[13]

The RWA is justified in the regime of a weakly-driven system ($\tilde{g} \ll \omega_c$) and a reasonably-large transition frequency ($\omega_0/\omega_c > 0$). Both conditions are violated if we consider ultrastrong coupling to a molecule with a bare avoided crossing or CI. This case is the main focus of the present paper. As a result of this new parameter regime, the Hamiltonian acquires a band-diagonal structure as compared to a block diagonal structure (Fig. 1 upper panel). This Hamiltonian has been well-studied. Analytic expressions in terms of continued fractions were found by Judd

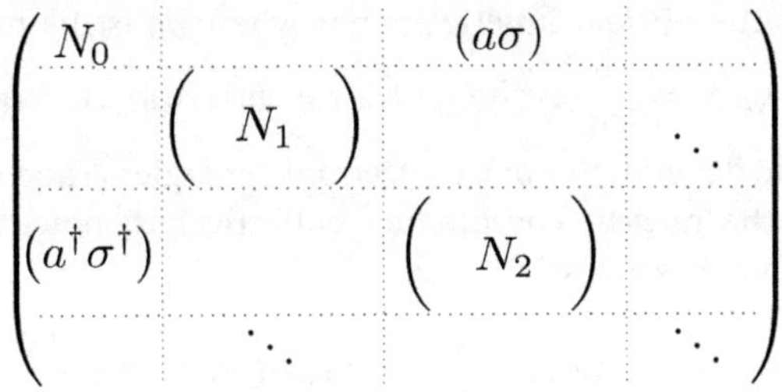

$$\pi = +1 : |g,0\rangle \to |e,1\rangle, |g,2\rangle \to |e,3\rangle, |g,4\rangle \to \ldots$$

$$\pi = -1 : |e,0\rangle, |g,1\rangle \to |e,2\rangle, |g,3\rangle \to \ldots$$

Fig. 1 (Upper) Schematic depiction of the Hamiltonian. Under the RWA, $\hat{H}$ has a block-diagonal structure. Each block with a nonzero total excitation number is 2-by-2 and composed, in this basis, of $|e, n_c\rangle$, $|g, n_c + 1\rangle$. Adding back in the counter rotating terms couples states that differ by 2 in the total excitation number $\hat{N}$. (Lower) The simultaneous eigenstates of parity and excitation number. States with an even (odd) excitation number have even (odd) parity.

for an isolated set of eigenstates and the approach was further explicated in terms of the Bargmann representation and extended.[15–17] While a purely analytic, complete solution without the RWA is not possible, the problem has been treated in Fock space to allow for a numerically exact solution.[18,19] Operator methods have been discussed as well.[20,21]

The paper is organized as follows. In section B, we describe a semi-analytic method for solving the Schrödinger equation in the ultrastrong coupling regime. This method yields the potential energy surfaces of the coupled light-matter system and the nonadiabatic couplings between them. These will be used for the qualitative interpretation of the wavepacket simulations which are carried out using a numerical grid representation of the cavity mode as described recently.[22] In section IV, we discuss wavepacket simulation for sodium iodide in a cavity and show how the time-dependent branching ratio of the ionic and covalent configurations is manipulated in the presence of the cavity mode.

II. Nuclear motion through a nonadiabatic coupling region

We have recently discussed nuclear motion through a CI or avoided crossing (a region of nonadiabatic coupling) and the ways one may track this motion using spectroscopic signals.[13] We shall explore what occurs when such an experiment is performed in an electromagnetic cavity resonant with a material frequency (for some value of the nuclear coordinates). The system Hamiltonian is given by the sum of the nuclear kinetic energy operator $\hat{T}$ and the electronic Hamiltonian,

$$\hat{H} = \hat{T} + \hat{H}_{\mathrm{el}} = -\frac{1}{2}\sum_{i=1}^{3N}\frac{1}{m_i}\nabla_i^{\,2} + \sum_{l=1}^{l_{\mathrm{M}}}\varepsilon_l(\mathbf{q})|\varepsilon_l(\mathbf{q})\rangle\langle\varepsilon_l(\mathbf{q})|. \tag{3}$$

The first term is the nuclear kinetic energy, where m_i is the mass of the i-th nuclear coordinate $\mathbf{q}_i$, $\nabla_i = \dfrac{\partial}{\partial \mathbf{q}_i}$ is the derivative with respect to $\mathbf{q}_i$, and we have explicitly notated the dependence of the electronic energies ε_l and corresponding eigenstates $|\varepsilon_l\rangle$ on the nuclear coordinates, collectively denoted $\mathbf{q}$. Taking the electronic matrix elements gives[23,24]

$$\hat{H}_{kl} = \delta_{kl}\left(\hat{T} + \varepsilon_l(\mathbf{q})\right) + \sum_i \frac{1}{m_i}\left(f_{kli}(\mathbf{q})\nabla_i + \frac{1}{2}h_{kli}(\mathbf{q})\right) \tag{4}$$

with

$$\begin{aligned} f_{kli}(\mathbf{q}) &= \langle \varepsilon_k(\mathbf{q})|\nabla_i|\varepsilon_l(\mathbf{q})\rangle \\ h_{kli}(\mathbf{q}) &= \langle \varepsilon_k(\mathbf{q})|\nabla_i^{\,2}|\varepsilon_l(\mathbf{q})\rangle. \end{aligned} \tag{5}$$

Given the electronic energies $\varepsilon_l(\mathbf{q})$, and the nonadiabatic couplings (NAC) $f_{kli}(\mathbf{q})$ and $h_{kli}(\mathbf{q})$, the total wavefunction may be written as a product of nuclear and electronic wavefunctions $|\Psi\rangle = |\phi\rangle|\psi\rangle = \sum_l c_l|\phi_l\rangle|\varepsilon_l\rangle$ where $|\phi_l\rangle$ is the nuclear wavepacket on the $|\varepsilon_l\rangle$ electronic surface and we have expanded the electronic wavefunction as a sum over eigenstates weighted by coefficients c_l. Any given initial state $|\Psi(t_0)\rangle$ can be propagated using the Hamiltonian eqn (4) to obtain the expectation values and correlation functions of material operators that are related to spectroscopic observables.

III. The quantum Rabi Hamiltonian

In Section II, we introduced the ingredients to the general Hamiltonian (eqn (4)) that we will use to propagate the nuclei. We now examine the quantum Rabi Hamiltonian (eqn (1)) in more detail. Taking matrix elements in the $\{|e\rangle, |g\rangle\}$ basis gives

$$H = \begin{pmatrix} \omega_c\hat{a}^\dagger\hat{a} - \dfrac{\omega_0}{2} & \tilde{g}(\hat{a}^\dagger + \hat{a}) \\ \tilde{g}(\hat{a}^\dagger + \hat{a}) & \omega_c\hat{a}^\dagger\hat{a} + \dfrac{\omega_0}{2} \end{pmatrix} \tag{6}$$

Switching to the basis $\left\{|+\rangle \equiv \dfrac{1}{\sqrt{2}}(|g\rangle + |e\rangle),\ |-\rangle \equiv \dfrac{1}{\sqrt{2}}(|g\rangle - |e\rangle)\right\}$ and dividing through by the cavity frequency (re-defining the unit of energy to be ω_c) gives

$$H = \begin{pmatrix} \hat{a}^\dagger\hat{a} + \dfrac{\tilde{g}}{\omega_c}(\hat{a}^\dagger + \hat{a}) & \dfrac{-\omega_0}{2\omega_c} \\ \dfrac{-\omega_0}{2\omega_c} & \hat{a}^\dagger\hat{a} - \dfrac{\tilde{g}}{\omega_c}(\hat{a}^\dagger + \hat{a}) \end{pmatrix} \tag{7}$$

which is a special (zero-bias) case of the model studied by[25] with parameters $g = \dfrac{\tilde{g}}{\omega_c}$, $\Delta = \dfrac{\omega_0}{\omega_c}$. This eigenvalue problem can be solved in a variety of ways. Numerical diagonalization using cavity field Fock states is possible but, for high values of g, requires that many states be taken into account (up to 10^5 at ultrastrong coupling depending on the convergence criteria[25]). Since our goal is to do non-BOA dynamics simulations, it is beneficial to use as few states as possible (this is particularly true for the calculation of the NAC). Using a Bargmann transformation and analyticity, Braak obtained a transcendental equation, the roots of which yield the energies, and this solution was later re-derived using extended coherent states.[21,26] The eigenstates and energies had also been derived earlier using tunable coherent states.[25] Both of these representations have an advantage over the Fock-space representations in that they automatically include many-body correlations amongst infinitely-many photons in the cavity mode. This is particularly useful in the strong coupling regime. Both of these representations have numerical difficulties when $g \ll 1$ but this regime is solvable within the RWA and can be smoothly connected provided we choose an appropriate separation point. More importantly, the former representation (extended coherent states) involves expressions that have periodic divergences, complicating numerical evaluation. For this reason, we choose to work in the basis of tunable, photon-added coherent states (*i.e.*, the basis of photon-added coherent states with a tunable coherence parameter) and make the ansatz

$$
\begin{aligned}
|\varepsilon\rangle_l &= |+\rangle \otimes \sum_{n=0}^{M} c_{nl} \left(\hat{a}^\dagger\right)^n e^{\alpha_l \hat{a}^\dagger} |0\rangle + \pi_l |-\rangle \otimes \sum_{n=0}^{M} c_{nl} \left(-\hat{a}^\dagger\right)^n e^{-\alpha_l \hat{a}^\dagger} |0\rangle \\
&= \begin{pmatrix} \sum_{n=0}^{M} c_{nl} (\hat{a}^\dagger)^n e^{\alpha_l \hat{a}^\dagger} |0\rangle \\ \pi_l \sum_{n=0}^{M} c_{nl} (-\hat{a}^\dagger)^n e^{-\alpha_l \hat{a}^\dagger} |0\rangle \end{pmatrix}
\end{aligned}
\tag{8}
$$

where the c_{nl} coefficients represent the l-th eigenstate in the basis of n-photon-added tunable coherent states, α_l is the tunable coherent state parameter corresponding to the l-th eigenvalue, and $\pi_l = \pm 1$ is the parity of the l-th eigenstate. Since the Hamiltonian conserves the parity, we can separately consider the even and odd parity manifolds and write π_l explicitly rather than simply $\pm$ because we will later need to keep track of the two parities in the matrix elements $\langle \varepsilon_k | \hat{\mathcal{O}} | \varepsilon_l \rangle$ succinctly. In order that the π_l not be functions of the nuclear coordinates, we do not simply order the eigenstates by energy but rather list first the even and then the odd in ascending order of energy. In this ansatz, M is the truncation number. In practice, we must make M large enough to ensure convergence. This will be addressed in more detail later. Following the analysis of ref. 25 and inserting this wavefunction into the Schrödinger equation yields a connection between the energy and α_l

$$
\varepsilon_l = \alpha_l g - \pi_l \frac{\Delta}{2}
\tag{9}
$$

and a recurrence relation for the coefficients

$$
c_{n+1} = \frac{-c_0}{(n+1)g} \left[\left(n + \pi_l \frac{\Delta}{2} \right) c_n + (\alpha_l + g) c_{n-1} - \pi_l (-1)^n \frac{\Delta}{2} \sum_{j=0}^{n} \frac{(2\alpha_l)^j}{j!} c_{n-j} \right].
\tag{10}
$$

Note that we have used $c_1 = 0$ (permitted since α_l can completely determine the 1-photon contribution to $|\varepsilon_l\rangle$) and that c_0 remains an arbitrary constant that can be used to normalize the states. This normalization and its consequences are discussed in detail in Appendix C but we will otherwise formally work with the normalized coefficients c_{nl}. Setting $c_{M+1} = 0$ (required for convergence) gives a degree M polynomial equation for α_l

$$\left(M + \pi_l \frac{\Delta}{2}\right) c_M + (\alpha_l + g) c_{M-1} - \pi_l (-1)^M \frac{\Delta}{2} \sum_{j=0}^{M} \frac{(2\alpha_l)^j}{j!} c_{M-j} = 0. \tag{11}$$

Chen *et al.*[25] had examined this recurrence relation for various regimes of the parameters Δ and g and we now briefly summarize their findings. For $0.1 < g < 0.5$ and $0.5 < \Delta < 1.5$, $M = 60$ is sufficient to ensure convergence (in the sense of a relative difference of 10^{-8} or less on increasing M further) for the first 20 or more eigenstates and the energies match those obtained *via* Fock state diagonalization (at a much higher truncation order). Moreover, the coefficients are found to die down by around $n = 40$ for $\Delta = 1, g = 0.1$ and this improves to $n = 10$ for $\Delta = 1, g = 1$. This shows the strength of this representation for handling the strong coupling regime. In the simulation section, we will discuss the parameters we used and the truncation.

Given a pair g, Δ, we can solve eqn (11) for the first N roots such that ε_N is sufficiently large to account for the system energetics. The details of this truncation will depend on the way in which the system is probed and the signal being examined.

It is worth repeating the findings of Braak[26] regarding the distribution of the zeroes of his G-function (not shown here for brevity) and thus, the energies. The zeroes of $G^{\pm}(x)$ are distributed relatively regularly, with the property that the number of eigenenergies in the range $[n\omega_c, (n+1)\omega_c]$ is restricted to be 0, 1, or 2. Additionally, two intervals each containing two eigenenergies will never be adjacent to each other (nor will two intervals each containing no eigenenergies). Since the G-function gives the exact energies as well, we are guaranteed that, for reasonably large ω_c, we will not have to include too many states $|\varepsilon_l\rangle$ to obtain converged calculations. This is trivially true for weak coupling at resonance ($g \ll 1$, $\Delta = 1$) but is good to confirm it for all parameter regimes.

In order to propagate an initial wavefunction so as to simulate spectra, we must obtain all the ingredients to compute $H_{kl}(\mathbf{q})$ in eqn (4). Besides the energies, we also need the nonadiabatic couplings $f_{kli}(\mathbf{q})$, and $h_{kli}(\mathbf{q})$. Computing these quantities is tedious but it is fairly straightforward. We outline the, rather lengthy, derivation in Appendix B and make frequent use of auxilliary quantities defined in Appendix A to arrive at

$$f_{kli} = f_{egi}(\pi_l - \pi_k) B_{kl}^{(1)} + (1 + \pi_k \pi_l)[A_{kl}^{(2)} + A_{kl}^{(3)} \nabla_i \alpha_l] \tag{12}$$

for the first-derivative couplings and

$$\begin{aligned} h_{kli} &= \nabla_i f_{kli} - F_{kli} \\ F_{kli} &= \frac{F_{gg} + F_{ee}}{2} \delta_{kl} + \frac{F_{gg} - F_{ee}}{2} (\pi_l + \pi_k) B_{kl}^{(1)} + F'_{eg} (1 - \pi_k \pi_l) A_{kl}^{(1)} \\ &\quad + \left[\left\{ f_{egi} (\pi_k - \pi_l) \left[B_{kl}^{(2)} - B_{kl}^{(3)} \nabla_i \alpha_l \right] \right\} + k \leftrightarrow l \right] \\ &\quad + (1 + \pi_k \pi_l) \left(A_{kl}^{(4)} \nabla_i \alpha_k + A_{lk}^{(4)} \nabla_i \alpha_l + A_{kl}^{(5)} \nabla_i \alpha_k \nabla_i \alpha_l + A_{kl}^{(6)} \right) \end{aligned} \tag{13}$$

for the second (we have suppressed the explicit $\mathbf{q}$-dependence for brevity).

IV. Application to avoided crossing in sodium iodide

The influence of ultrastrong cavity coupling is demonstrated on the nonadiabatic dynamics of photoexcited sodium iodide, a well known example for femtochemistry.[27] The adiabatic electronic states of the bare sodium iodide experience an avoided crossing near $\mathbf{q} \sim 8$ Å (Fig. 2(a)), leading to dissociation into the neutral products through the ^{1}X state upon photoexcitation. The bare nonadiabatic coupling matrix element f_{eg} is shown in Fig. 2(b). The branching ratio, *i.e.*, the amount of population transferred to the ^{1}X state, is determined by the acquired momentum of the wavepacket, which depends on the excitation laser frequency. We will investigate two scenarios: the cavity is set in resonance at $\approx$ 6 Å (case 1), *i.e.*, before the wavepacket in the *A* state reaches the bare avoided crossing, (ω_c = 815 meV) and a low frequency cavity (ω_c = 52.6 meV) resonant directly at the crossing (case 2). Both cases are illustrated in Fig. 3 by plotting the ^{1}X-state curve shifted by the photon energy of the cavity.

For case 1, eqn (8) is used to calculate the resulting potential energy curves in the basis of the photon-added, tunable coherent states. Fig. 4 shows the lowest ten potential energy curves for four values of the coupling strength *g*. For small *g*,

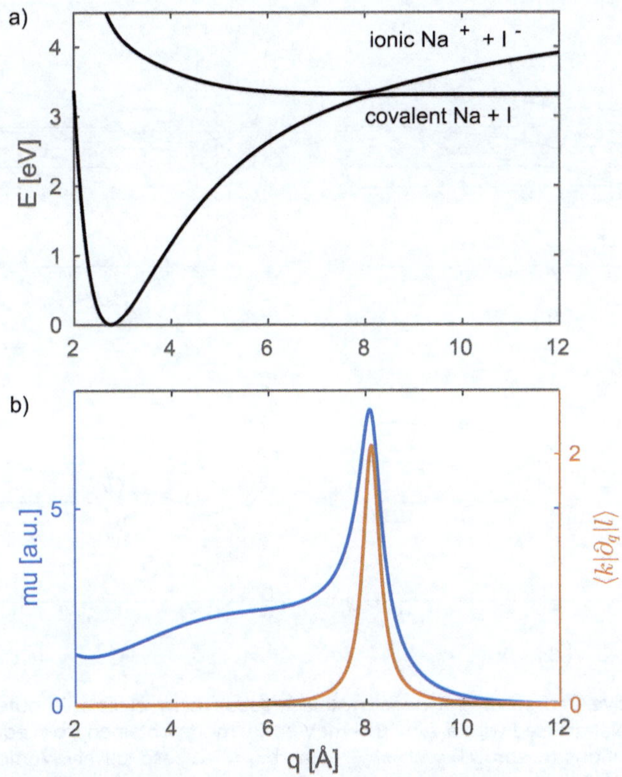

Fig. 2 (a) Bare ground (^{1}X) and excited (^{1}A) electronic potential energy surfaces of NaI. Dissociation on the ^{1}X potential leads to the neutral reaction products. (b) Transition dipole moment μ_{eg} (blue) and the derivative coupling matrix element f_{eg} (red).

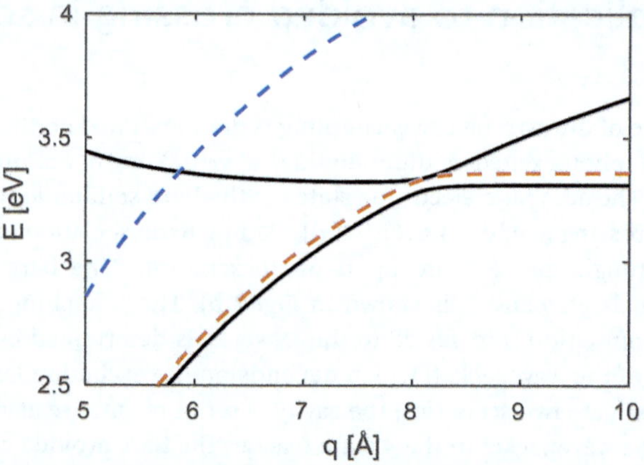

Fig. 3 Potential curve crossings for two different cavity detunings: ground state potential wavefunction shifted by the cavity frequency ($V_g + \omega_c$), for $\omega_c = 815$ meV (blue, dashed) and $\omega_c = 52.6$ meV (red, dashed).

the result converges to curves which coincide with a molecular Jaynes–Cummings model[13] (Fig. 4(a)). New avoided crossings are created in the region at 6 Å, where the cavity mode is in resonance with the bare electronic states of NaI ($\Delta = 1$). New

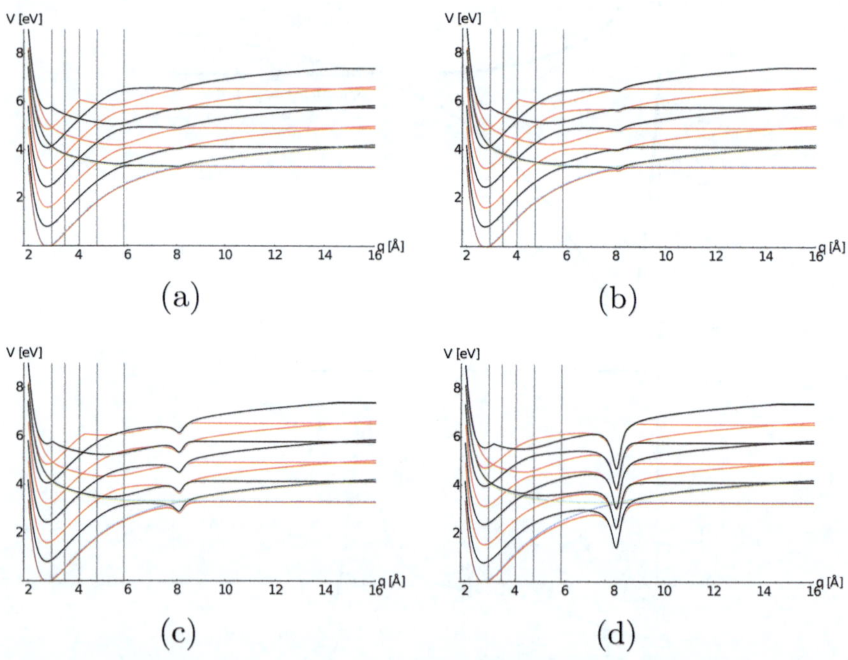

Fig. 4 Positive (red solid) and negative (black solid) parity electronic potential energy surfaces of NaI dressed with a $\omega_c = 815$ meV cavity mode obtained from eqn (9) and the first five solutions to eqn (11) with $\pi_l = +1$ and $\pi_l = -1$ respectively. Vertical gray lines indicate points where the bare states are resonant with integer multiples of the cavity frequency ($\omega_0 = n\omega_c$ for n an integer, $n = 1$ for the rightmost line and counts up going to the left). (a) Peak coupling of 200 meV. (b) Peak coupling of 300 meV. (c) Peak coupling of 600 meV. (d) Peak coupling of 1240 meV.

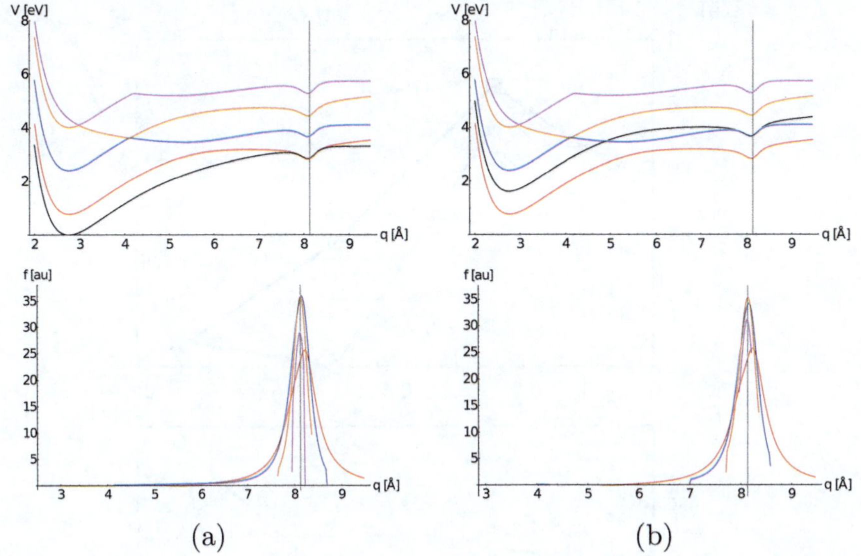

Fig. 5 (a) Potential curves for the lowest energy positive-parity eigenstate (black) and the 4 lowest energy negative-parity eigenstates (upper) and their corresponding NACs (lower) (all couplings are between the "black" state and one of the others and therefore labeled by the color of the latter state). (b) Potential curves for the second-lowest positive-parity state (black) and the 4 lowest energy negative-parity eigenstates (upper) and their corresponding NACs (lower).

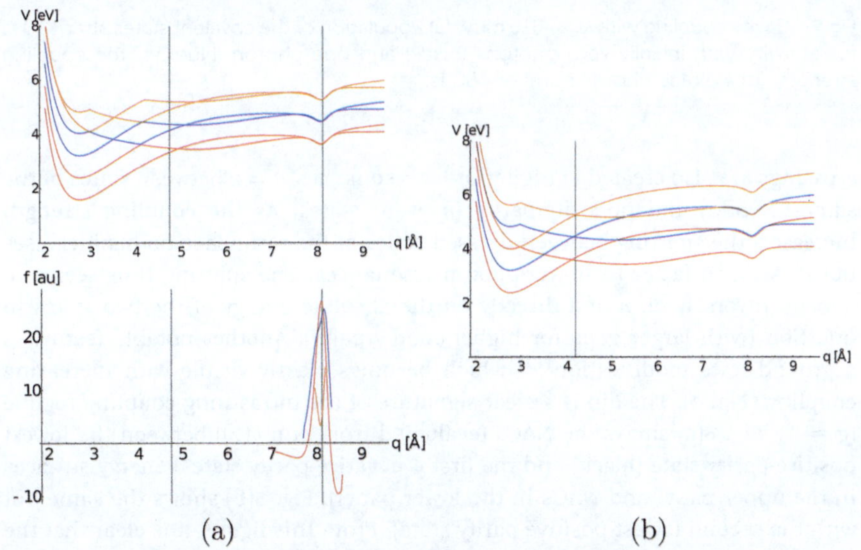

Fig. 6 (a) Potential curves (upper) and corresponding NACs (lower) for 3 different pairs of opposite-parity eigenstates. (b) Potential curves for 3 different pairs of same-parity eigenstates. Corresponding NACs are sharply peaked at the avoided crossings near the gray line and are otherwise negligible.

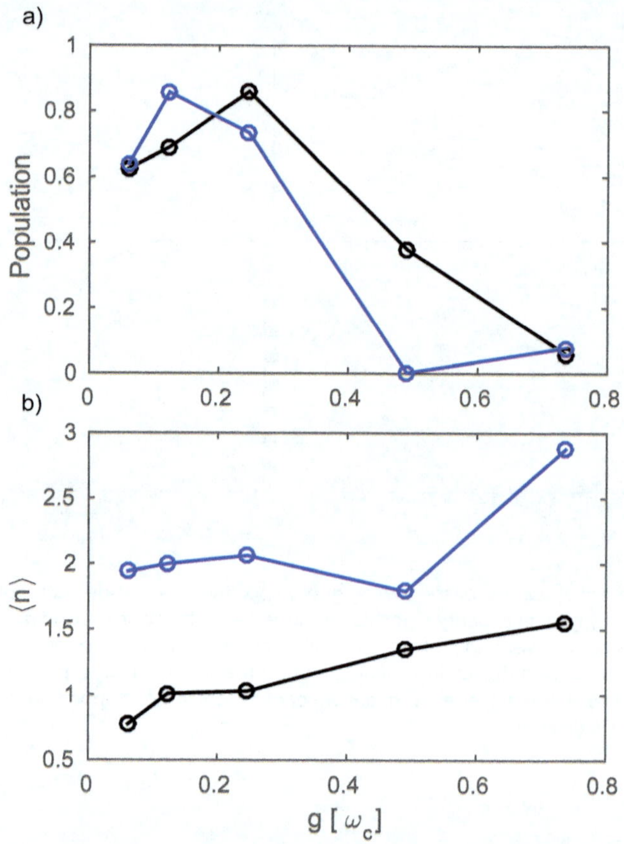

Fig. 7 Cavity coupling with $\omega_c = 815$ meV. (a) Population of the covalent states after 480 fs for a cavity with initially zero photons (black) and one photon (blue) *vs.* the coupling strength. (b) Average photon number 480 fs.

crossings are also created at higher-order resonances $\Delta = n$ between states of the same- (n odd) and opposite-parity (n even) states. As the coupling strength increases, the splitting between dressed states at the resonance points increases but it is much faster for lower-photon resonances. The splitting thus seems to depend inversely on n and directly on the absolute energy of the two states in question (with larger gaps for higher-energy pairs). Another notable feature is a ground-state modification[14,28] which becomes clearly visible with increasing coupling (Fig. 4). The dip is a clear signature of the ultrastrong coupling regime ($g \approx 1$). Fig. 5(a) shows the NACs (evaluated from eqn (12)) between the lowest positive-parity state (black) and the first 4 negative-parity states (energy surfaces in the upper panel and NACs in the lower panel). Fig. 5(b) shows the same but with the second-lowest positive-parity state). From this figure, it is clear that the NACs in the region of the original avoided crossing are strengthened relative to their bare values (Fig. 2 lower panel). Moreover, the coupling strength remains of the same order even for states separated by a large distance in energy. Fig. 6(a) shows the NACs between 3 pairs of opposite-parity states. Note that no NAC is induced in the region of crossing between such pairs (only the coupling in the

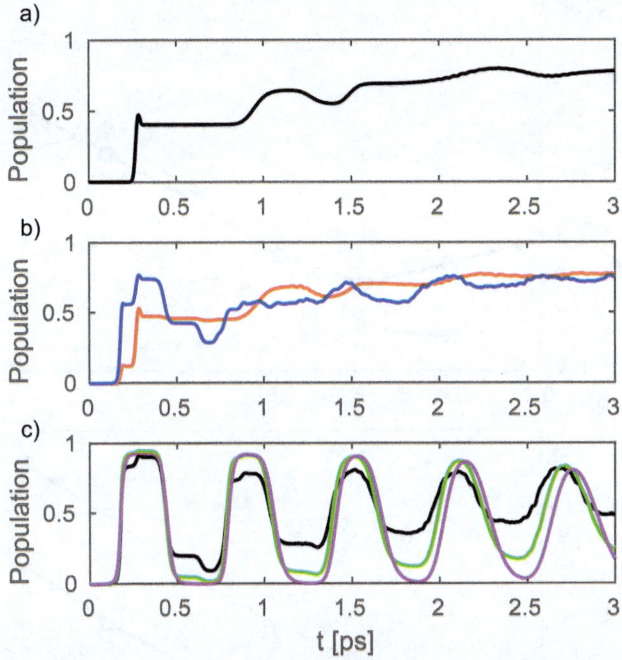

Fig. 8 Selected time traces of the ^{1}X state population ($\omega_c = 815$ meV). No cavity (black line), $g = (0.0502, 0.1256, 0.1884, 0.02512, 0.3793)$ red, blue, black, green, and magenta respectively.

region of the original avoided crossing remains). We can see that crossings between states of opposite parity do not generate any NACs. In contrast, the avoided crossings between states of the same parity correspond to peaks in the NAC curves. For a given resonance point (*e.g.*, $\omega_0 = 3\omega_c$), these peaks become smaller as one considers dressed states of higher energy (corresponding to larger separations at the avoided crossings).

To quantify the possible modifications of the dynamics under the influence of the cavity coupling, wavepacket calculations were carried out (for details see Appendix E). The amount of population in the dissociative part of the ^{1}X state after the wavepacket has reached the crossing point for the first time ($\approx$480 fs) serves as an indicator of the effect on the dynamics. Fig. 7(a) depicts the population in the covalent channel for varying coupling strengths. While an intermediate coupling strength ($g < 0.3$) can increase the dissociation probability, the transfer to the ^{1}X state is suppressed for larger values of g. This behavior can be explained by the increased splitting between the dressed states: for larger splittings, the state which corresponds to the ^{1}A state is well-separated from the lower-lying state, effectively suppressing the population transfer.[13,14] The dynamics is mainly influenced by two modifications of the potential energy surfaces: the new avoided crossings created by the cavity and the modified avoided crossing already present in the bare molecule. With increasing cavity coupling, the cavity-created crossing at the 1-photon resonance point generates well-separated states such that a nuclear wavepacket is trapped and a much smaller fraction even reaches the original crossing. This partially coincides with the effect described by Galego

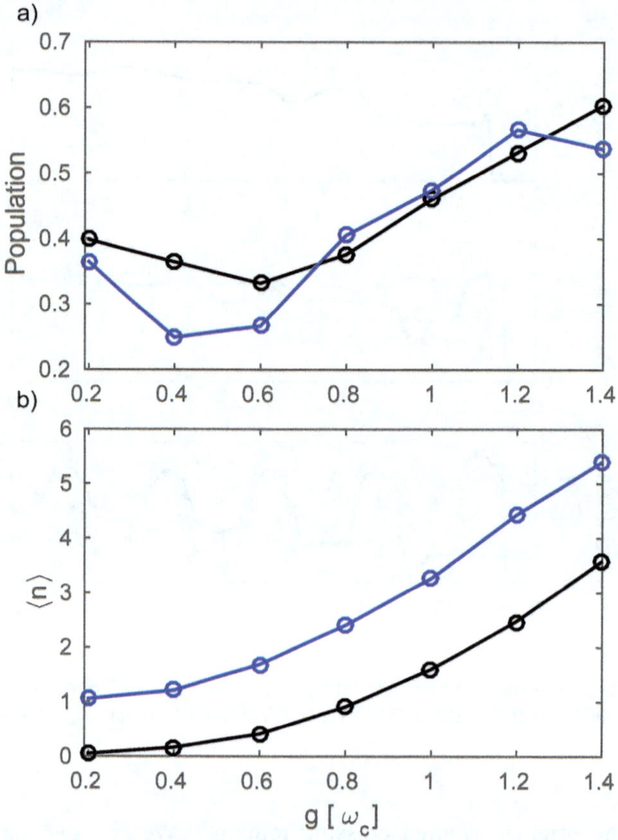

Fig. 9 (a) Population of the covalent states after 480 fs for a cavity with initially zero photons (black) and one photon (blue) *vs.* the coupling strength ($\omega_c = 52.6$ meV). (b) Average photon number after 480 fs.

et al.[14] where for higher coupling strengths the states become well separated and the Born–Oppenheimer approximation is valid again. An alternative view is provided by the time evolution of the population of the ^{1}A state (Fig. 8). With increasing coupling strength, the wavepacket becomes trapped in a tighter effective potential and its recurrence time becomes shorter.

Additionally, both the NAC and the shape of the potential surface at the bare state avoided crossing is affected by the counter-rotating terms, creating a barrier for the nuclear wavepacket from which it reflects and allowing for further suppression of population transfer into the ^{1}X dissociation channel. The dissociation is strongly suppressed for large coupling strengths. The average photon number shown in Fig. 7 is a clear indicator that the system can not be represented by only two Fock states, but requires the present approach.

Case 2 is presented in Fig. 9. This case lacks an artificially-created avoided crossing at the 1-photon resonance point, thus simplifying the dynamics. The population in the X state after one round trip is shown in Fig. 9(a). The branching into the covalent state is slightly suppressed for small coupling strengths ($g < 0.6$) but increases for larger g. The increase in population transfer is accompanied by an increased number of photons (Fig. 9(b)) in the cavity mode, leading to the conclusion that stimulated emission into the cavity is the main mechanism for the control.

V. Discussion

Molecular cavity QED offers many exciting possibilities from enhanced probing techniques[12,29–33] to molecular cooling[34] and cavity-modified photochemistry.[13,35] The underlying Quantum Rabi model Hamiltonian is well-studied in the atomic context and has been solved in a number of ways.[9,15–18,20,21,25,26,36] In this work, we apply the solution in terms of tunable coherent states provided by Chen *et al.*[25] to the case of a molecular Hamiltonian with electronic parameters dependent on the nuclear configuration. We derive transformation formulas for electronic operators as well as expressions for the nonadiabatic coupling due to the nuclear kinetic-energy operator. Combined with the eigenstates, this is all the information necessary to carry-out and interpret dynamical simulations. By reducing the problem to familiar adiabatic energy surfaces and corresponding NACs, the dynamics can be understood even in non-intuitive parameter regimes. In particular, when a molecule possessing a CI or avoided crossing is dressed with an ultrastrong cavity field, the potential surfaces and NACs can be strongly modified from their bare or Jaynes–Cummings (RWA) shape.

Although the dynamics of a molecule interacting with a quantum field can be numerically solved in other ways, they possess limitations or do not provide the intuitive picture in terms of potential surfaces and NACs accessible here. Expansion in Fock-states (or the JC-polaritons) is the most straightforward approach but convergence is slow in the ultrastrong coupling regime ($g \sim 1$) and physical interpretation of the dynamics is complicated by the strong-mixing induced by the non-rotating terms in the Hamiltonian. Dynamical quantities can also be evaluated *via* discretization of the photon space as demonstrated recently[22] but the physical quantities that determine the evolution, the potential energy surfaces and NACs are never calculated in such a procedure and are thus inaccessible.

The method has been used to evaluate the photodissociation dynamics of sodium iodide as reproduced in Fig. 8. An analysis of the dynamics shows that, for small coupling, the population of the ^{1}X state increases in a step-wise fashion with time. Upon increasing the coupling, oscillations between the states are induced. These oscillations eventually die down to a nearly-even population. Examining the potential energy surfaces for increasing g (Fig. 4) reveals the opening of a gap at the 1-photon resonance ($\Delta \sim 1$) and the creation of a dip in the potential surface at the original avoided crossing while the NACs (Fig. 5) connect all pairs of opposite-parity states, providing decay paths to lower-energy pairs. For a nuclear wavepacket prepared in the ^{1}A state, these effects combine to create oscillations in the electronic populations, suppressing the final dissociation of the molecule.

The single-molecule coupling g is proportional to the vacuum field strength of the cavity mode given by

$$\varepsilon_c = \sqrt{\frac{\hbar\omega_c}{V\varepsilon_0}} = \sqrt{\frac{2\hbar\pi c}{\lambda_c V\varepsilon_0}} \tag{14}$$

where V is the mode volume of the cavity. The effective mode volumes used for the sodium iodide simulations range from $0.002\lambda_c^3$ to for $0.04\lambda_c^3$ for $\omega_c =$ 815 meV and from $10^{-7}\lambda_c^3$ to for $10^{-5}\lambda_c^3$ for $\omega_c = 52.6$ meV for couplings present in Fig. 7 and 9 under the assumption that $\approx 10^5$ particles contribute collectively.[37] Nano-cavities,[38,39] nano-plasmon antennas,[40] or nano-guides[41] may provide a solution for the realization of such small effective mode volumes. Additionally, the collective enhancement effects must be leveraged to reach the ultrastrong coupling regime. In molecular systems however, these effects are subject to vibrational dephasing. Collective enhancement effects in molecules is a subject that needs further study and will be part of future investigations.

Appendix A

Working formulas for the tunable coherent-state basis

It is a known fact that eigenvectors corresponding to distinct eigenvalues are orthogonal. Thus, working with the normalized coefficients c_{nl}

$$\langle\varepsilon_k|\varepsilon_l\rangle = \delta_{kl}\langle\varepsilon_l|\varepsilon_l\rangle = \sum_{mn=0}^{M} c_{mk}c_{nl}\Big(\langle\alpha_k, m|\alpha_l, n\rangle + (-1)^{n+m}\pi_k\pi_l\langle -\alpha_k, m|-\alpha_l, n\rangle\Big) \tag{A1}$$

where

$$|\alpha_l,\, n\rangle \equiv (\hat{a}^\dagger)^n e^{\alpha_l \hat{a}^\dagger}|0\rangle \tag{A2}$$

is the (un-normalized) n-photon-added coherent state with parameter α_l. For this orthogonality relation and for other purposes, we will need the quantity

$$\langle\alpha_k, m|\alpha_l, n\rangle = \sum_{pq=0}^{\infty}\left\langle q+m\left|\frac{\alpha_k{}^q\alpha_l{}^p\sqrt{q+m}\sqrt{p+n}}{q!p!}\right|p+n\right\rangle = \sum_{p=0}^{\infty}\frac{\alpha_l{}^p\alpha_k{}^{p+n-m}(p+n)!}{p!(p+n-m)!} \tag{A3}$$

This last formula is only valid for $n > m$ but, since all quantities in this overlap are real, it is symmetric with respect to the simultaneous exchange $k \leftrightarrow l$ and $n \leftrightarrow m$ so that this presents no difficulty. This form is convenient for seeing symmetries such as

$$\begin{aligned}\langle -\alpha_k, m|-\alpha_l, n\rangle &= (-1)^{n+m}\langle\alpha_k, m|\alpha_l, n\rangle\\ \langle\alpha_k, m|-\alpha_l, n\rangle &= (-1)^{n+m}\langle -\alpha_k, m|\alpha_l, n\rangle\end{aligned} \tag{A4}$$

and others that arise later in calculating the non-BOA couplings. For now, we use these relations to reduce the orthonormality condition to

$$\delta_{kl} = (1+\pi_k\pi_l)\sum_{mn=0}^{M} c_{mk}c_{nl}\langle\alpha_k, m|\alpha_l, n\rangle = (1+\pi_k\pi_l)A_{kl}^{(1)}(\mathbf{q}) \tag{A5}$$

where the last equality defines $A_{kl}^{(1)}$ which depends on $\mathbf{q}$ indirectly *via* the g and Δ. It is clear from the pre-factor that this condition is automatically satisfied for $\pi_k\pi_l = -1$, corresponding to states of opposite parity, while considering states of the same parity restricts $A_{kl}^{(1)}$ to vanish for $k \neq l$. The orthonormality condition also directly determines the initial coefficients c_{0l}. From eqn (10), (11), (A3) and

(A5), it is clear that it is not practical to solve it directly for the normalized coefficients c_{nl}. In practice, one must first solve it for a set of un-normalized coefficients using some other initial coefficient $\tilde{c}_{0l}$ (which we take to be 1 for all states) and then normalize after the fact. This will be discussed further in section C.

In calculating signals, we will be interested in evaluating expectation values or correlation functions of electronic operators such as the dipole or polarizability. The propagators in such an expression are easily evaluated *via* diagonalizing the truncated non-BOA Hamiltonian or a real-time propagation scheme. However, we will also need the operator matrix elements in the eigenbasis

$$\begin{aligned}\hat{\mathcal{O}}_{kl}(\mathbf{q}) &= \langle \varepsilon_k | \hat{\mathcal{O}}(\mathbf{q}) | \varepsilon_l \rangle \\ &= \hat{\mathcal{O}}_{++}(\mathbf{q}) \sum_{mn=0}^{M} c_{mk} c_{nl} \langle \alpha_k, m | \alpha_l, n \rangle + \hat{\mathcal{O}}_{--}(\mathbf{q}) \pi_k \pi_l \sum_{mn=0}^{M} c_{mk} c_{nl} \langle \alpha_k, m | \alpha_l, n \rangle \\ &\quad + \hat{\mathcal{O}}_{+-}(\mathbf{q}) \pi_l \sum_{mn=0}^{M} (-1)^n c_{mk} c_{nl} \langle \alpha_k, m | -\alpha_l, n \rangle \\ &\quad + \hat{\mathcal{O}}_{-+}(\mathbf{q}) \pi_k \sum_{mn=0}^{M} (-1)^m c_{mk} c_{nl} \langle -\alpha_k, m | \alpha_l, n \rangle \end{aligned} \tag{A6}$$

where we have written the $\mathbf{q}$-dependence for $\hat{\mathcal{O}}$ explicitly while omitting this dependence from the other variables (g and Δ and through them, the coefficients c_n as well as the α) for brevity. Defining the following summation

$$B_{kl}^{(1)}(\mathbf{q}) \equiv \sum_{mn=0}^{M} (-1)^n c_{mk} c_{nl} \langle \alpha_k, m | -\alpha_l, n \rangle, \tag{A7}$$

we obtain

$$\begin{aligned}\hat{\mathcal{O}}_{kl} = &\frac{\hat{\mathcal{O}}_{gg} + \hat{\mathcal{O}}_{ee}}{2}(1 + \pi_k \pi_l) A_{kl}^{(1)} + \frac{\hat{\mathcal{O}}_{gg} - \hat{\mathcal{O}}_{ee}}{2}(\pi_k + \pi_l) B_{kl}^{(1)} + \Re\{\hat{\mathcal{O}}\}_{eg}(1 - \pi_k \pi_l) A_{kl}^{(1)} \\ &+ i\Im\{\hat{\mathcal{O}}_{eg}\}(\pi_l - \pi_k) B_{kl}^{(1)}\end{aligned} \tag{A8}$$

where we have suppressed the $\mathbf{q}$-dependence for brevity. The $B_{kl}^{(1)}$ are complementary to the $A_{kl}^{(1)}$, differing only by the placement of the minus signs, and arise from cross terms between $|+\rangle$ and $|-\rangle$. Two particular operators are relevant for our purposes. The ionic state projection operator $|e\rangle\langle e| \equiv \hat{P}_e$ is given by

$$\left(\hat{P}_e\right)_{kl} = \frac{\delta_{kl} - (\pi_k + \pi_l) B_{kl}^{(1)}(\mathbf{q})}{2} \tag{A9}$$

and the transition dipole operator (or any operator with vanishing diagonal elements in the bare space) is given by

$$\begin{aligned}\hat{\mu}_{kl}(\mathbf{q}) = &\Re\{\hat{\mu}\}_{eg}(\mathbf{q})(1 - \pi_k \pi_l) A_{kl}^{(1)}(\mathbf{q}) + i\Im\{\hat{\mu}\}_{eg}(\mathbf{q})(\pi_l - \pi_k) B_{kl}^{(1)}(\mathbf{q}) \rightarrow \hat{\mu}_{eg}(\mathbf{q}) \\ &(1 - \pi_k \pi_l) A_{kl}^{(1)}\end{aligned} \tag{A10}$$

where the last relation is on the assumption of real transition dipole moments. We notice that this expression, whether real or complex, clearly vanishes for states of the same parity (as would be expected by the symmetry of the dipole operator). These two special cases correspond to the observables considered in the main text.

Appendix B

Derivation of the non-BOA couplings

To propagate nuclear wavepackets along the energy surfaces will require the non-BOA couplings f_{kli}, h_{kli}, which we obtain in this section. Since we will have frequent occasion to consider the derivative by the i-th nuclear coordinate, we notate such derivatives by a prime for brevity (*i.e.*, $\nabla_i[\ldots] \equiv [\ldots]'$ and there is no ambiguity since no terms mix nuclear derivatives) in the appendix. Using the expansion for the eigenstates (eqn (8)) and the product rules, we have

$$\begin{aligned} f_{kli} &= \langle \varepsilon_k | \nabla_i | \varepsilon_l \rangle \\ &= \langle + | \nabla_i | + \rangle A_{kl}^{(1)} + \langle - | \nabla_i | - \rangle \pi_k \pi_l A_{kl}^{(1)} + \langle + | \nabla_i | - \rangle \pi_l B_{kl}^{(1)} + \langle - | \nabla_i | + \rangle \pi_k B_{kl}^{(1)} \\ &\quad + (1 + \pi_k \pi_l) \sum_{mn=0}^{M} c_{mk} c'_{nl} \langle \alpha_k, m | \alpha_l, n \rangle + \alpha'_l (1 + \pi_k \pi_l) \sum_{mn=0}^{M} c_{mk} c_{nl} \langle \alpha_k, m | \alpha_l, n+1 \rangle \end{aligned} \tag{B1}$$

where we have used $\nabla_i|\pm\alpha, n\rangle = \pm\alpha'|\pm\alpha, n+1\rangle$ and $\langle -\alpha_k, m| -\alpha_l, n+1\rangle = (-1)^{n+m+1}\langle \alpha_k, m|\alpha_l, n+1\rangle$, easily obtained from eqn (A2) and (A3). From basic considerations, $\langle e|\nabla_i|g\rangle = -\langle g|\nabla_i|e\rangle$ so that $\langle +|\partial_q|+\rangle = 0 = \langle -|\partial_q|-\rangle$ and the first two terms in the above vanish. The derivative coupling can then be easily written as

$$f_{kli} = \langle e|\nabla_i|g\rangle(\pi_l - \pi_k)B_{kl}^{(1)} + (1 + \pi_k\pi_l)[A_{kl}^{(2)} + \alpha'_l A_{kl}^{(3)}] \tag{B2}$$

where, for brevity, we have defined shorthand for two additional summations

$$\begin{aligned} A_{kl}^{(2)}(\mathbf{q}) &\equiv \sum_{mn=0}^{M} c_{mk} c'_{nl} \langle \alpha_k, m | \alpha_l, n \rangle \\ A_{kl}^{(3)}(\mathbf{q}) &\equiv \sum_{mn=0}^{M} c_{mk} c_{nl} \langle \alpha_k, m | \alpha_l, n+1 \rangle. \end{aligned} \tag{B3}$$

The first term in eqn (B2) is simply the result of transforming a purely molecular-electronic operator with the appropriate symmetries and is thus proportional to the derivative coupling in the bare basis. The second term contains the two summations $A^{(2)}$ and $A^{(3)}$ which originate from the $\mathbf{q}$-derivative acting on the coefficients c_{nl} and the photonic state $|\alpha_l, n\rangle$ respectively.

To calculate the second-derivative coupling is somewhat more difficult but we can use the separation

$$h_{kli} = \langle \varepsilon_k|\nabla_i^2|\varepsilon_l\rangle = \nabla_i\langle \varepsilon_k|\nabla_i|\varepsilon_l\rangle - \langle \nabla_i\varepsilon_k|\nabla_i|\varepsilon_l\rangle = \nabla_i f_{kli} - F_{kli} \tag{B4}$$

where the last equality defines F_{kli} and the first term can be calculated numerically from the f_{kli} already obtained above. From

$$\begin{aligned} \nabla_i|\varepsilon_l\rangle = \sum_{n=0}^{M} \Big[&\nabla_i|+\rangle \otimes c_{nl}|\alpha_l, n\rangle + \pi_l \nabla_i|-\rangle \otimes (-1)^n c_{nl}|-\alpha_l, n\rangle + |+\rangle \otimes c'_{nl}|\alpha_l, n\rangle \\ &+ \alpha'_l|+\rangle \otimes c_{nl}|\alpha_l, n+1\rangle + \pi_l|-\rangle \otimes (-1)^n c'_{nl}|-\alpha_l, n\rangle \\ &- \alpha'_l \pi_l|-\rangle \otimes (-1)^n c_{nl}|-\alpha_l, n+1\rangle \Big] \end{aligned} \tag{B5}$$

it is straightforward, if somewhat tedious, to calculate the F_{kli}. It is convenient to separate the result into three terms

$$F_{kli} = \sum_{j=1}^{3} F_{kli}^{(j)} \tag{B6}$$

based on where the **q**-derivatives (the ∇_i) act. The first term arises from transforming F from the $|e\rangle$, $|g\rangle$ basis as though it were an ordinary electronic operator

$$F_{kli}^{(1)} = \frac{F_{gg} + F_{ee}}{2}(1 + \pi_k\pi_l)A_{kl}^{(1)} + \frac{F_{gg} - F_{ee}}{2}(\pi_l + \pi_k)B_{kl}^{(1)} + \Re\{F_{eg}\}(1 - \pi_k\pi_l)A_{kl}^{(1)} + i\Im\{F_{eg}\}(\pi_l - \pi_k)B_{kl}^{(1)}. \tag{B7}$$

This is an unsurprising result and easily calculated from the bare F and the already-obtained $A^{(1)}$, $B^{(1)}$. The $F^{(1)}$ contribution can be thought of as collecting the terms in which both (of the two available) **q**-derivatives act on the bare molecule states $|+\rangle$, $|-\rangle$ (due exclusively to the first two terms in eqn (B5)). The second term contributing to F_{kl} arises when only one **q**-derivative acts on the bare molecular states, while the other acts on the coefficients c_{nl} or the photonic state $|\alpha_l, n\rangle$ (*i.e.*, cross terms between the first two and latter four elements of eqn (B5)). These contributions come as

$$F_{kli}^{(2)} = \{f_{egi}(\pi_k - \pi_l)[B_{kl}^{(2)} - \alpha'_l B_{kl}^{(3)}]\} + k \leftrightarrow l, \tag{B8}$$

where we have defined two more associated summations

$$\begin{aligned} B_{kl}^{(2)}(\mathbf{q}) &\equiv \sum_{mn=0}^{M} (-1)^n c_{mk} c'_{nl} \langle \alpha_k, m | -\alpha_l, n \rangle \\ B_{kl}^{(3)}(\mathbf{q}) &\equiv \sum_{mn=0}^{M} (-1)^n c_{mk} c_{nl} \langle \alpha_k, m | -\alpha_l, n+1 \rangle \end{aligned} \tag{B9}$$

that again differ from their respective A summations only by the minus signs. These again arise as cross terms between $|+\rangle$ and $|-\rangle$ states with derivative operators intervening (directly leading to the matrix element f_{egi}). The final contribution to F_{kli} arises when both **q**-derivatives act on coefficients or photonic states. There are several ways to distribute the derivative operations leading to

$$F_{kli}^{(3)} = (1 - \delta_{kl})(1 + \pi_k\pi_l)(\alpha'_l A_{kl}^{(4)} + \alpha'_k A_{lk}^{(4)} + \alpha'_k \alpha'_l A_{kl}^{(5)} + A_{kl}^{(6)}) \tag{B10}$$

where

$$\begin{aligned} A_{kl}^{(4)}(\mathbf{q}) &\equiv \sum_{mn=0}^{M} c'_{mk} c_{nl} \langle \alpha_k, m | \alpha_l, n+1 \rangle \\ A_{kl}^{(5)}(\mathbf{q}) &\equiv \sum_{mn=0}^{M} c_{mk} c_{nl} \langle \alpha_k, m+1 | \alpha_l, n+1 \rangle \\ A_{kl}^{(6)}(\mathbf{q}) &\equiv \sum_{mn=0}^{M} c'_{mk} c'_{nl} \langle \alpha_k, m | \alpha_l, n \rangle \end{aligned} \tag{B11}$$

are the three relevant ways to distribute the two **q**-derivatives between the coefficients and photon modes (the remaining conceivable combination does not contribute due to orthogonality) and the factor $(1 - \delta_{kl})$ is introduced to make clear that $F^{(3)}$ vanishes for diagonal elements (this can be shown analytically using the results of the following section).

Appendix C

Normalization

As mentioned briefly in Appendix A above, it is not practical to solve directly for the normalized coefficients c_{nl}. Instead, we recursively determine un-normalized coefficients $\tilde{c}_{nl}$ using the initial condition $\tilde{c}_{0l} = 1$ and then normalize afterwards. Thus, the normalization constant is

$$c_{0l} = \frac{1}{\sqrt{2\tilde{A}_{ll}^{(1)}(\mathbf{q})}} \tag{C1}$$

where the "tilde" indicates that the un-normalized coefficients $\tilde{c}_{nl}$ were used rather than the normalized c_{nl} as implied by eqn (A5) and we can clearly obtain

$$A_{kl}^{(1)} = \frac{\tilde{A}_{kl}^{(1)}}{2\sqrt{\tilde{A}_{kk}^{(1)}\tilde{A}_{ll}^{(1)}}} \equiv \overline{A}_{kl}^{(1)}. \tag{C2}$$

This last definition is superfluous for the $A^{(1)}$ summation but the consequences of a **q**-dependent normalization will not always be quite so straightforward and defining the quantities

$$\overline{A}_{kl}^{(j)} = \frac{\tilde{A}_{kl}^{(j)}}{2\sqrt{\tilde{A}_{kk}^{(1)}\tilde{A}_{ll}^{(1)}}} \quad \overline{B}_{kl}^{(j)} = \frac{\tilde{B}_{kl}^{(j)}}{2\sqrt{\tilde{A}_{kk}^{(1)}\tilde{A}_{ll}^{(1)}}} \tag{C3}$$

facilitates compact expression of the correction terms. This straightforward normalization procedure only runs into difficulties when the **q**-derivative of the coefficients c_{nl} are called for. As will become apparent from numerical implementation, the relative size of these coefficients spans many orders of magnitude, complicating numerical differentiation of the normalized coefficients. However, as we will show below, they can be obtained *via* a recurrence relation. We thus summarize the connections between normalized and un-normalized quantities

$$\begin{aligned}
A_{kl}^{(j)} &= \overline{A}_{kl}^{(j)}, \quad j = 1, 3, 5 \\
B_{kl}^{(j)} &= \overline{B}_{kl}^{(j)}, \quad j = 1, 3 \\
A_{kl}^{(2)} &= \overline{A}_{kl}^{(2)} - \left(\overline{A}_{ll}^{(2)} + \alpha'_l \overline{A}_{ll}^{(3)}\right)\overline{A}_{kl}^{(1)} \\
B_{kl}^{(2)} &= \overline{B}_{kl}^{(2)} - \left(\overline{A}_{ll}^{(2)} + \alpha'_l \overline{A}_{ll}^{(3)}\right)\overline{B}_{kl}^{(1)} \\
A_{kl}^{(4)} &= \overline{A}_{kl}^{(4)} - \left(\overline{A}_{kk}^{(2)} + \alpha'_k \overline{A}_{kk}^{(3)}\right)\overline{A}_{kl}^{(3)} \\
A_{kl}^{(6)} &= \overline{A}_{kl}^{(6)} - \left[\left(\overline{A}_{ll}^{(2)} + \alpha'_l \overline{A}_{ll}^{(3)}\right)\overline{A}_{lk}^{(2)} + k \leftrightarrow l\right] + \left(\overline{A}_{kk}^{(2)} + \alpha'_k \overline{A}_{kk}^{(3)}\right)\left(\overline{A}_{ll}^{(2)} + \alpha'_l \overline{A}_{ll}^{(3)}\right)\overline{A}_{kl}^{(1)}
\end{aligned} \tag{C4}$$

where we have used the readily-confirmed identity

$$\nabla_i \tilde{A}_{ll}^{(1)} = 2(\tilde{A}_{ll}^{(2)} + \alpha'_l \tilde{A}_{ll}^{(3)}) \tag{C5}$$

One can thus begin with the un-normalized coefficients $\tilde{c}_{nl}$ obtained from the initial condition $\tilde{c}_{0l} = 1$ and eqn (10) and (11). One can then normalize the coefficients c_{nl} *via* eqn (C1) and the amplitudes $\tilde{A}_{ll}^{(1)}$ before calculating all auxilliary summations directly from the c_{nl}. Alternatively, one can work with the un-normalized coefficients $\tilde{c}_{nl}$ and the un-normalized summations $\tilde{A}$ and $\tilde{B}$ can then be easily calculated and normalized afterwards *via* eqn (C4). However, to do this requires the nuclear derivative of the un-normalized coefficients, $\tilde{c}'_{nl}$. These can of course be obtained numerically but the better way is to take the derivative of the original recurrence relation to obtain

$$\begin{aligned}-g(n+1)\tilde{c}'_{n+1} = &-\left[(n+\pi_l)\frac{\Delta}{2}\tilde{c}'_n + (\alpha_l + g)\tilde{c}'_{n-1} - \pi_l(-1)^n\frac{\Delta}{2}\sum_{j=0}^{n}\frac{(2\alpha_l)^j}{j!}\tilde{c}'_{n-j}\right] \\ &+ (n+1)\tilde{c}_{n+1}\left(g' - g\frac{\Delta'}{\Delta}\right) + n\tilde{c}_n\left(2g\alpha'_l - \frac{\Delta}{\Delta}\right) + \tilde{c}_{n-1} \\ &\times\left(g' + \alpha'_l(2n - 1 + \Delta\pi_l) - \frac{\Delta'}{\Delta}(\alpha_l + g)\right) + 2\alpha_l\tilde{c}_{n-2}(\alpha_l + g)\end{aligned} \tag{C6}$$

where we have used the original recursion relation to simplify the result. The derivatives of coefficients then depend only on the g, Δ, α, which are already available, and their derivatives g', Δ', α', the former two of which can be obtained numerically while the latter can be obtained self-consistently from the truncation relation (*i.e.*, setting $\tilde{c}'_M = 0$). In this paper, we take this latter approach to avoid numerically differentiating the coefficients.

Appendix D

Summary of formulas

In this section, we summarize the final formulas obtained for all key quantities. Namely, the following table summarizes the various summations necessary for the transformation of electronic operators and the calculation of the derivative couplings. For brevity, we have omitted the summation symbol $\sum_{mn}^{M}$ from the

Table 1 Summary of direct ($A_{kl}^{(i)}$) and complementary cross term ($B_{kl}^{(i)}$) summations used as auxilliary quantities to obtain operator matrix elements $\mathcal{O}_{kl}$ and non-BOA coupling f_{kli} and F_{kli} coming from first- and second-derivative couplings respectively

i	$A_{kl}^{(i)}$	$B_{kl}^{(i)}$	Type of terms
1	$c_{mk}c_{nl}\langle\alpha_k, m\|\alpha_l, n\rangle$	$(-1)^n c_{mk}c_{nl}\langle\alpha_k, m\|-\alpha_l, n\rangle$	Zero-derivative direct and cross terms
2	$c_{mk}c'_{nl}\langle\alpha_k, m\|\alpha_l, n\rangle$	$(-1)^n c_{mk}c'_{nl}\langle\alpha_k, m\|-\alpha_l, n\rangle$	One-derivative direct and cross terms
3	$c_{mk}c_{nl}\langle\alpha_k, m\|\alpha_l, n+1\rangle$	$(-1)^n c_{mk}c_{nl}\langle\alpha_k, m\|-\alpha_l, n+1\rangle$	
4	$c'_{mk}c_{nl}\langle\alpha_k, m\|\alpha_l, n+1\rangle$		Two-derivative direct terms
5	$c_{mk}c_{nl}\langle\alpha_k, m+1\|\alpha_l, n+1\rangle$		
6	$c'_{mk}c'_{nl}\langle\alpha_k, m\|\alpha_l, n\rangle$		

entries of columns two and three. The photon-added coherent state overlaps are given by eqn (A3) which may be recast as

$$\langle \alpha_k, m|\alpha_l, n\rangle = {}_1F_1(1+n, 1+n-m, \alpha_k\alpha_l)\alpha_l^{n-m}(n)_m \tag{D1}$$

where $(n)_m = n(n-1)(n-(m-1))$ is the falling factorial and ${}_1F_1(a, b, c)$ is the Kummer confluent hypergeometric function. This form is particularly useful for numeric implementation. The un-normalized coefficients $\tilde{c}_{nl}$ are given as polynomials in α by the recurrence relation eqn (10) with $\tilde{c}_{0l} = 1$. After solving eqn (11), we obtain the α_l, and thus the $\tilde{c}_{nl}$, numerically. The sums given in Table 1 can then be calculated and used to evaluate the three key quantities necessary for evaluation of signals (transformed electronic operators, and first and second derivative couplings), which we write here in terms of the easily calculated "tilde" and "bar" quantities for completion

$$\begin{aligned}
\hat{\mathcal{O}}_{kl} &= \frac{\hat{\mathcal{O}}_{gg}+\hat{\mathcal{O}}_{ee}}{2}(1+\pi_k\pi_l)\overline{A}^{(1)}_{kl} + \frac{\hat{\mathcal{O}}_{gg}-\hat{\mathcal{O}}_{ee}}{2}(\pi_k+\pi_l)\overline{B}^{(1)}_{kl} + \Re\{\hat{\mathcal{O}}_{eg}\}(1-\pi_k\pi_l)\overline{A}^{(1)}_{kl} \\
&\quad + i\Im\{\hat{\mathcal{O}}_{eg}\}(\pi_l-\pi_k)\overline{B}^{(1)}_{kl} \\
f_{kli} &= f_{egi}(\pi_l-\pi_k)\overline{B}^{(1)}_{kl} + (1+\pi_k\pi_l)\left[\overline{A}^{(2)}_{kl} - \overline{A}^{(2)}_{ll}\overline{A}^{(1)}_{kl} + \alpha'_l\left(\overline{A}^{(3)}_{kl} - \overline{A}^{(3)}_{ll}\overline{A}^{(1)}_{kl}\right)\right] \\
F_{kli} &= \frac{F_{gg}+F_{ee}}{2}\delta_{kl} + \frac{F_{gg}-F_{ee}}{2}(\pi_l+\pi_k)\overline{B}^{(1)}_{kl} + F'_{eg}(1-\pi_k\pi_l)\overline{A}^{(1)}_{kl} + f_{egi}(\pi_k-\pi_l) \\
&\quad \times\left[\left\{\left[\overline{B}^{(2)}_{kl} - \alpha'_l\overline{B}^{(3)}_{kl}\right] - \left(\overline{A}^{(2)}_{ll} + \alpha'_l\overline{A}^{(3)}_{ll}\right)\overline{B}^{(1)}_{kl}\right\} + k\leftrightarrow l\right] + (1-\delta_{kl})(1+\pi_k\pi_l) \\
&\quad \times\left[\left(\alpha'_l\overline{A}^{(4)}_{kl} + \alpha'_k\overline{A}^{(4)}_{lk} + \alpha'_k\alpha'_l\overline{A}^{(5)}_{kl} + \overline{A}^{(6)}_{kl}\right)\right. \\
&\quad \left. + \left(\left(\overline{A}^{(2)}_{kk} + \alpha'_k\overline{A}^{(3)}_{kk}\right)\left(\overline{A}^{(2)}_{kl} + \overline{A}^{(3)}_{kl}\right) + k\leftrightarrow l\right) + A^{(1)}_{kl}\left(\overline{A}^{(2)}_{ll} + \alpha'_l\overline{A}^{(3)}_{ll}\right)^2\right]
\end{aligned} \tag{D2}$$

Appendix E

Direct space quantization of the photonic degrees of freedom

The analytical diagonalization of the matter-field Hamiltonian as described in this paper yields a set of new potential energy surfaces and nonadiabatic couplings. This representation provides an intuitive picture and insight into the modified structure of the coupled Hamiltonian but might become intractable for a quantum dynamics time propagation. An alternative efficient computational scheme is obtained by the direct treatment of the cavity mode[22] in analogy to the quantum harmonic oscillator. The annihilation and creation operators of the cavity mode can be expressed in terms their quadrature coordinates x and p:

$$a = \sqrt{\frac{\omega_c}{2\hbar}}\left(\hat{x} + \frac{i}{\omega_c}\hat{p}\right) \tag{E1}$$

The coupled light-molecule Hamiltonian from eqn (1) then reads:

$$H = \frac{\omega_0}{2}(2\hat{\sigma}^\dagger\hat{\sigma} - 1) - \frac{\hbar}{2}\frac{\partial^2}{\partial x^2} + \frac{1}{2}\omega_c{}^2\hat{x}^2 + \tilde{g}\sqrt{2\hbar\omega_c}\hat{x}(\hat{\sigma}^\dagger + \hat{\sigma}) \tag{E2}$$

The coordinate X is treated as a spatial coordinate, putting the nuclear coordinates and the cavity mode on an equal footing in a numerical simulation. Here x is discretized on a numerical grid and the second derivative with respect to x can then be conveniently calculated by a discrete Fourier transform.[42] The nuclear degrees of freedom $\mathbf{q}$ are accounted for by treating ω_0 and $\tilde{g}$ as functions of $\mathbf{q}$ (*i.e.* the potential energy curves and transition dipole curves respectively).

The wavepacket simulations have been carried out by propagating the photonic-nuclear wave packet on the potential energy curves of NaI (^{1}X and ^{1}A) using the Hamiltonian from eqn (E2). The time stepping was done with a Chebychev propagation scheme.[43] The potential energy curves NaI, as well as the nonadiabatic coupling matrix element and the transition dipole moment has been calculated with the program package MOLPRO[44] at the MRCI/CAS(6/7)/aug-cc-VQZ level of theory with an effective core potential for iodine (ECP46MWB).

Appendix F

Simulation protocol

Below, we outline the necessary steps to implementing the calculation of a time-dependent signal that tracks the evolving nuclear wavepacket on the multitude of electronic energy surfaces.

1. Obtain eigenenergies ε_l of the electronic Hamiltonian

(a) Obtain $g(\mathbf{q})$, $\Delta(\mathbf{q})$ from quantum chemistry and definitions.

(b) Discretize $\mathbf{q}$-space on a multidimensional grid.

(c) For each point in $\mathbf{q}$-space, obtain the first N solutions, $\alpha_1 \cdots \alpha_N$, of eqn (11) with $\varepsilon = \alpha g - \pi \frac{\Delta}{2}$ sufficiently large for the problem (this depends on the type of signal being discussed).

2. Transform electronic operators to eigenbasis

(a) For each $\varepsilon_l(\mathbf{q})$, use eqn (10) to obtain the coefficients $c_{nl}(\mathbf{q})$ of the eigenstate expansions.

(b) For each $\mathbf{q}$-point, use eqn (A2) and the α_l from step 1(c) above to obtain the photon-state overlaps $\langle \alpha_k, m | \alpha_l, n \rangle$ and $\langle \alpha_k, m | -\alpha_l, n \rangle$ for all choices of $k, l \in \{1,\ldots, N\}$ and $m, n \in \{1,\ldots, M\}$. This is two sets of $(NM)^2$ quantities.

(c) Use the coefficients c_{nl} and the photon-state overlaps $\langle \alpha_k, m | \alpha_l, n \rangle$ and $\langle -\alpha_k, m | \alpha_l, n \rangle$ to calculate the summations $A_{kl}^{(1)}(\mathbf{q})$, $B_{kl}^{(1)}(\mathbf{q})$ *via* eqn (A7)

(d) Use eqn (A8) to transform any electronic operator. Special cases include the ionic-state population operator (eqn (A9)) and the transition dipole operator (eqn (A10)).

3. Obtain the derivative couplings $f_{kli}(\mathbf{q})$

(a) Obtain the derivative couplings for the bare molecule, $f_{egi}(\mathbf{q})$, from quantum chemistry calculations.

(b) Take numeric derivatives of the $\alpha_l(\mathbf{q})$ and $c_{nl}(\mathbf{q})$ obtained above with respect to the nuclear coordinates to obtain the quantities $\alpha'_l(\mathbf{q})$ and $c'_{nl}(\mathbf{q})$.

(c) Using $c'_{nl}(\mathbf{q})$ and eqn (B3), calculate the summations $A_{kl}^{(2,3)}(\mathbf{q})$.

(d) Using $\alpha'_l(\mathbf{q})$, $f_{egi}(\mathbf{q})$ and the auxiliary summations, calculate $f_{kli}(\mathbf{q})$ *via* eqn (12).

4. Obtain second-derivative couplings $h_{kli}(\mathbf{q})$

(a) Use eqn (B4) to recast the problem into finding $F_{kli}(\mathbf{q}) \equiv \langle \nabla_i \varepsilon_k | \nabla_i \varepsilon_l \rangle(\mathbf{q})$.

(b) Obtain matrix elements of F for the bare molecule (F_{gg}, F_{ee}, F_{eg}) from quantum chemistry calculations.

(c) Calculate auxiliary summations $A_{kl}^{(j)}, j \in \{4, 5, 6\}$.

(d) Assemble F_{kli} *via* eqn (13).

5. Calculate desired signals

(a) Write the desired signal in terms of expectation values or correlation functions of electronic operators.

(b) Write the Hamiltonian in the form of eqn (4) with $\varepsilon_l(\mathbf{q})$, $f_{kli}(\mathbf{q})$ and $h_{kli}(\mathbf{q})$ are given by steps 1, 3, and 4 above.

(c) Using this Hamiltonian and some time propagation scheme (such as the Short-Iterative Lanczos time-stepping procedure), any initial state can be propagated in time.

(d) The action of electronic operators in this basis can be computed using the results of step 2 above.

(e) Combining these two ingredients, we can calculate any expectation value or correlation function of electronic operators.

(f) Convolutions of these expectation values or correlation functions with external field envelopes then give the observable signal of interest.

Acknowledgements

The support of the National Science Foundation (grant CHE-1361516) as well as from the Chemical Sciences, Geosciences, and Biosciences division, Office of Basic Energy Sciences, Office of Science, U.S. Department of Energy through award No. DE-FG02-04ER15571, is gratefully acknowledged. The computational resources and support for K. B. was provided by DOE. M. K. gratefully acknowledges support from the Alexander von Humboldt foundation through the Feodor Lynen program. We would like to thank the green planet cluster (NSF Grant CHE-0840513) for allocation of compute resources.

References

1 P. V. Demekhin and L. S. Cederbaum, *J. Chem. Phys.*, 2013, **139**, 154314.
2 J. Kim, H. Tao, J. L. White, V. S. Petrović, T. J. Martinez and P. H. Bucksbaum, *J. Phys. Chem. A*, 2012, **116**, 2758.
3 P. von den Hoff, M. Kowalewski and R. de Vivie-Riedle, *Faraday Discuss.*, 2011, **153**, 159.
4 P. von den Hoff, S. Thallmair, M. Kowalewski, R. Siemering and R. de Vivie-Riedle, *Phys. Chem. Chem. Phys.*, 2012, **14**, 14460.
5 P. Gross, D. Neuhauser and H. Rabitz, *J. Chem. Phys.*, 1992, **96**, 2834.
6 P. Brumer and M. Shapiro, *Chem. Phys. Lett.*, 1986, **126**, 541.
7 D. J. Tannor, R. Kosloff and S. A. Rice, *J. Chem. Phys.*, 1986, **85**, 5805.
8 E. M. Purcell, *Phys. Rev.*, 1946, **69**, 674.
9 E. T. Jaynes and F. W. Cummings, *Proc. IEEE*, 1963, **51**, 89.
10 S. Haroche and J.-M. Raimond, *Exploring the Quantum: Atoms, Cavities, and Photons (Oxford Graduate Texts)*, 1st edn, Oxford University Press, 2006.
11 J. A. Hutchison, T. Schwartz, C. Genet, E. Devaux and T. W. Ebbesen, *Angew. Chem., Int. Ed.*, 2012, **51**, 1592.

12 A. Shalabney, J. George, H. Hiura, J. A. Hutchison, C. Genet, P. Hellwig and T. W. Ebbesen, *Angew. Chem., Int. Ed.*, 2015, **54**, 7971.
13 M. Kowalewski, K. Bennett and S. Mukamel, *J. Chem. Phys.*, 2016, **144**, 054309.
14 J. Galego, F. J. Garcia-Vidal and J. Feist, *Phys. Rev. X*, 2015, **5**, 041022.
15 B. Judd, *J. Phys. C: Solid State Phys.*, 1979, **12**, 1685.
16 H. Reik, H. Nusser and L. A. Ribeiro, *J. Phys. A: Math. Gen.*, 1982, **15**, 3491.
17 M. Kus and M. Lewenstein, *J. Phys. A: Math. Gen.*, 1986, **19**, 305.
18 S. Swain, *J. Phys. A: Math. Gen.*, 1973, **6**, 1919.
19 S. He, C. Wang, Q.-H. Chen, X.-Z. Ren, T. Liu and K.-L. Wang, *Phys. Rev. A*, 2012, **86**, 033837.
20 I. D. Feranchuk, L. I. Komarov and A. P. Ulyanenkov, *J. Phys. A: Math. Gen.*, 1996, **29**, 4035.
21 Q.-H. Chen, C. Wang, S. He, T. Liu and K.-L. Wang, *Phys. Rev. A*, 2012, **86**, 023822.
22 M. Kowalewski, K. Bennett and S. Mukamel, *J. Phys. Chem. Lett.*, 2016, **7**, 2050–2054.
23 A. Hofmann and R. de Vivie-Riedle, *Chem. Phys. Lett.*, 2001, **346**, 299.
24 W. Domcke, D. R. Yarkony and H. Köppel, *Conical Intersections*, WORLD SCIENTIFIC, 2011, vol. 17.
25 Q.-H. Chen, T. Liu, Y.-Y. Zhang and K.-L. Wang, *Europhys. Lett.*, 2011, **96**, 14003.
26 D. Braak, *Phys. Rev. Lett.*, 2011, **107**, 100401.
27 A. Mokhtari, P. Cong, J. L. Herek and A. H. Zewail, *Nature*, 1990, **348**, 225.
28 T. Liu, K.-L. Wang and M. Feng, *J. Phys. B: At., Mol. Opt. Phys.*, 2007, **40**, 1967.
29 F. C. Spano, *J. Chem. Phys.*, 2015, **142**, 184707.
30 F. Caruso, S. K. Saikin, E. Solano, S. F. Huelga, A. A. Guzik and M. B. Plenio, *Phys. Rev. B*, 2012, **85**, 125424.
31 P. Saurabh and S. Mukamel, *J. Chem. Phys.*, 2016, **144**, 124115.
32 J. del Pino, J. Feist and F. J. Garcia-Vidal, *J. Phys. Chem. C*, 2015, **119**, 29132.
33 B. S. Simpkins, K. P. Fears, W. J. Dressick, B. T. Spann, A. D. Dunkelberger and J. C. Owrutsky, *ACS Photonics*, 2015, **2**, 1460.
34 M. Kowalewski, G. Morigi, P. W. H. Pinkse and R. de Vivie Riedle, *Phys. Rev. A*, 2011, **84**, 033408.
35 F. Herrera and F. C. Spano, *Cavity-controlled Chemistry in Molecular Ensembles*, 2015, arXiv:1512.05017.
36 J. Casanova, G. Romero, I. Lizuain, J. J. García-Ripoll and E. Solano, *Phys. Rev. Lett.*, 2010, **105**, 263603.
37 T. Schwartz, J. A. Hutchison, J. Léonard, C. Genet, S. Haacke and T. W. Ebbesen, *ChemPhysChem*, 2013, **14**, 125.
38 S. Tomljenovic-Hanic, M. J. Steel, C. M. de Sterke and J. Salzman, *Opt. Express*, 2006, **14**, 3556–3562.
39 A. D. Greentree, C. Tahan, J. H. Cole and L. C. L. Hollenberg, *Nat. Phys.*, 2006, **2**, 856, arXiv:cond-mat/0609050.
40 M.-K. Kim, H. Sim, S. J. Yoon, S.-H. Gong, C. W. Ahn, Y.-H. Cho and Y.-H. Lee, *Nano Lett.*, 2015, **15**, 4102.
41 S. Faez, P. Türschmann, H. R. Haakh, S. Götzinger and V. Sandoghdar, *Phys. Rev. Lett.*, 2014, **113**, 213601.
42 D. J. Tannor, *Introduction to Quantum Mechanics: a Time-dependent Perspective*, University Science Books.

43 H. T. Ezer and R. Kosloff, *J. Chem. Phys.*, 1984, **81**, 3967.
44 H.-J. Werner, P. J. Knowles, G. Knizia, F. R. Manby and M. Schütz, *et al.*, *Molpro, Version 2015.1*, a package of ab initio programs, 2015, see http://www.molpro.net.

Faraday Discussions

PAPER

Four-wave-mixing experiments with seeded free electron lasers

F. Bencivenga,*[a] A. Calvi,[b] F. Capotondi,[a] R. Cucini,[c] R. Mincigrucci,[a] A. Simoncig,[a] M. Manfredda,[a] E. Pedersoli,[a] E. Principi,[a] F. Dallari,[d] R. A. Duncan,[e] M. G. Izzo,[a] G. Knopp,[f] A. A. Maznev,[e] G. Monaco,[d] S. Di Mitri,[a] A. Gessini,[a] L. Giannessi,[ag] N. Mahne,[a] I. P. Nikolov,[a] R. Passuello,[a] L. Raimondi,[a] M. Zangrando[ac] and C. Masciovecchio[a]

Received 22nd April 2016, Accepted 23rd May 2016

DOI: 10.1039/c6fd00089d

The development of free electron laser (FEL) sources has provided an unprecedented bridge between the scientific communities working with ultrafast lasers and extreme ultraviolet (XUV) and X-ray radiation. Indeed, in recent years an increasing number of FEL-based applications have exploited methods and concepts typical of advanced optical approaches. In this context, we recently used a seeded FEL to demonstrate a four-wave-mixing (FWM) process stimulated by coherent XUV radiation, namely the XUV transient grating (X-TG). We hereby report on X-TG measurements carried out on a sample of silicon nitride (Si_3N_4). The recorded data bears evidence for two distinct signal decay mechanisms: one occurring on a sub-ps timescale and one following slower dynamics extending throughout and beyond the probed timescale range (100 ps). The latter is compatible with a slower relaxation (time decay > ns), that may be interpreted as the signature of thermal diffusion modes. From the peak intensity of the X-TG signal we could estimate a value of the effective third-order susceptibility which is substantially larger than that found in SiO_2, so far the only sample with available X-TG data. Furthermore, the intensity of the time-coincidence peak shows a linear dependence on the intensity of the three input beams, indicating that the measurements were performed in the weak field regime. However, the timescale of the ultrafast relaxation exhibits a dependence on the intensity of the XUV radiation. We interpreted the observed behaviour as the generation of a population grating of free-electrons and holes that, on the sub-ps timescale, relaxes to generate lattice

[a]*Elettra-Sincrotrone Trieste S.C.p.A., S.S. 14 km 163.5 in AREA Science Park, 34149 Basovizza, Italy. E-mail: filippo.bencivenga@elettra.eu*

[b]*Department of Physics, University of Trieste, Via A.Valerio 2, 34127 Trieste, Italy*

[c]*IOM-CNR, Strada Statale 14-km 163.5, 34149 Basovizza, Trieste, Italy*

[d]*Department of Physics, University of Trento, Via Sommarive 14, Povo, TN, Italy*

[e]*Department of Chemistry, Massachusetts Institute of Technology, 77 Massachusetts Avenue, Cambridge, MA, USA*

[f]*Paul Scherrer Institute, Villigen 5232, Switzerland*

[g]*ENEA CR Frascati, Via E. Fermi 45, 00044 Frascati, Rome, Italy*

excitations. The background free detection inherent to the X-TG approach allowed the determination of FEL-induced electron dynamics with a sensitivity largely exceeding that of transient reflectivity and transmissivity measurements, usually employed for this purpose.

Introduction

The discovery of the non-linear optical response of materials has represented a milestone for theoretical and experimental physics.[1] Among the numerous non-linear optical processes occurring in nature, those based on third-order interactions—often termed four-wave-mixing (FWM), the theory of which is comprehensively described in the literature[2]—are the most widely used in applications. In FWM three (input) photons interact with the sample giving rise to a fourth (signal) photon, which carries information on sample excitations. In a four-photon experiment a large number of combinations of photon parameters (frequency, polarization, bandwidth, incidence angle, time delay, *etc.*) can be experimentally controlled and exploited to access different kinds of processes with high selectivity and in a broad timescale range. For instance, FWM methods have been used to monitor in real-time the dynamics of slow (μs) structural changes in macromolecules, picosecond spin and electronic relaxations, and ultrafast (sub-ps) chemical reactions.[3–5] Furthermore, the multi-wave nature of FWM enables probing processes and correlations involving different states of the system, capability at the basis of the multi-dimensional approach.[6] The intriguing idea of displaying spectral features in two dimensions spanned by two parameters was developed in the field of nuclear magnetic resonance about 30 years ago[7] and carried over to coherent optical spectroscopy.[8] Such methods allow the direct measurement of couplings between vibrational and low-lying (valence) electronic excitations. Their extension to shorter wavelengths, in the extreme ultraviolet (XUV) and soft X-ray ranges, has been theoretically explored by the group of Shaul Mukamel. It has been shown that XUV/soft X-ray FWM signals can contain information on the couplings between the aforementioned low energy (electronic and/or vibrational) excitations and high-energy core-excited states.[9,10] This may result in the unprecedented capability to monitor in real-time the dynamics and correlations between selected atomic species in a sample. In addition, the short wavelength of XUV/soft X-ray radiation provides sensitivity to local environments and molecular structures, and enables probing short-wavelength excitations, *e.g.*, in transport phenomena in disordered systems and nanostructures.[11–13]

In spite of such bright perspectives, until a few years ago experiments based on non-linear optics (besides pioneering studies on X-ray parametric down-conversion in crystals and EUV probing of optically induced excitations[14]) were mostly carried out using table-top optical lasers and the exploitation of the FWM concept at sub-optical wavelengths has been left to theory. This situation was essentially dictated by the lack of XUV and soft X-ray sources bright enough to stimulate non-linear interactions. Indeed, the strength of non-linear processes dramatically drops with increasing photon frequency. The availability of FELs dramatically improved the prospects in the field of non-linear XUV/X-ray optics and has already led to achievements such as the emission of coherent X-ray photons in gases due to stimulated radiative decay of core-excited atoms.[15]

Further investigations provided evidence for stimulated XUV emission in condensed matter and electronic inelastic scattering.[16] X-ray FELs were also employed to observe non-linear Compton scattering as well as second order wave-mixing processes, such as second harmonic and sum-frequency generation.[17]

The possibility to exploit highly coherent photon pulses delivered by the seeded FEL FERMI (Elettra, Italy; see Fig. 1 for a schematic layout)[18] to carry out XUV/soft X-ray transient grating (X-TG) experiments in the 0.1–1 nm^{-1} wavevector (**q**) range[11] motivated our efforts in realizing (user-dedicated) experimental infrastructure for FEL-based FWM experiments.[11,19,20] The TG approach is a specific kind of FWM process commonly used in the optical domain to probe transport phenomena, vibrational and thermal properties of condensed matter. However, the long wavelengths of optical photons limits its application to **q**-values lower than a few μm^{-1}. The access to such "mesoscopic" (0.1–1 nm^{-1}) **q**-range could provide relevant insights into, for example, the vibrational and thermal properties of amorphous solids, structural relaxation processes in supercooled liquids and soft matter in general, elasticity and thermodynamics of nanostructures, and the diffusive-to-ballistic transition of the heat transport mechanisms at the nanoscale.[11,13,21] In order to develop the TG approach in the XUV and soft X-ray domain we devised a set-up ("mini-Timer"; sketched in Fig. 2a)[20] that was used to demonstrate FEL-based FWM.[22] The polarizability modulation induced in a SiO_2 sample by two crossed pump-beams stimulates the radiation of a FWM signal when illuminated by an optical probe, having the form of a well-defined beam propagating downstream of the sample along the phase matching direction. The X-TG signal showed a dependence on the probe pulse delay which could be monitored over extended timescales (*i.e.* up to ~100 ps), demonstrating our capability to observe the time evolution of the third order non-linear response driven by XUV radiation.[17] We recently improved this set-up with the aim to extend this new approach to other systems, refine the methodology, and provide reference data for forthcoming measurements. In this paper we report X-TG experiments on a silicon nitride (Si_3N_4) sample. Si_3N_4 is a relevant reference sample, since it has been widely studied in FEL-pump/optical-probe

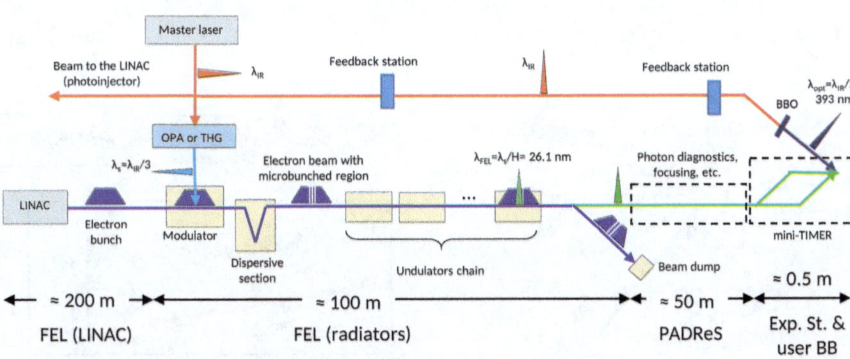

Fig. 1 Layout of the FERMI facility. The names of the main FEL devices and the lengths of the LINAC, FEL radiators, PADReS photon transport system and the experimental area are indicated. The purple trapezium sketches the electron bunch (with the central region flat in the phase space) while the red, blue and green triangles are the master laser, the seed and the FEL photon pulses, respectively.

reflectivity and transmissivity measurements ranging from XUV to hard X-rays[23,24] and the mechanisms responsible for the optical response to ultrafast XUV illumination are therefore well understood.[25]

In a more general sense, the experimental advances in the FEL-based X-TG approach provides a solid foundation for other FEL-based FWM applications such as XUV/soft X-ray coherent Raman scattering (X-CRS).[9,26] The CRS approach can be used to probe excitations (*e.g.* ultrafast molecular vibrations and valence-band electronic excitations) with energies largely exceeding the range typically accessed by TG experiments, while XUV/soft X-ray wavelengths can provide atomic-selectivity and nm spatial resolution. These features can potentially result in major breakthroughs, and the outlook for X-CRS is already motivating interest in both FEL-based non-linear experiments and advanced FEL techniques such as the multi-colour FEL emission (nowadays available at most FEL facilities),[27,28] as well as proposals for FEL sources purposely designed for X-CRS.[12,29]

Experimental

The two main features of the experimental set-up are the seeded FEL source FERMI and the mini-TIMER system, which is able to split and recombine the FEL pulses at the sample at a given crossing angle (2θ).

The FERMI FEL

The FEL at FERMI uses the high-gain-at-high-harmonic approach. In this mechanism the central role is played by the interaction between an ultrafast seed laser pulse and a portion (flat in the phase space) of a longer electron bunch. The wavelength (λ_S) of the seed laser is in the UV range (typically 230–270 nm), generated by frequency up-conversion of the near IR output ($\lambda_{IR} \sim 785$ nm) of a Ti:sapphire laser, referred to as "master laser" in Fig. 1. The up-conversion is achieved either through third harmonic crystals (THG, fixed seed wavelength $\lambda_S = \lambda_{IR}/3$–261.1 nm) or by using an optical parametric amplifier device (OPA, tunable seed wavelength $\lambda_S \sim 230$–270 nm). The former solution provides

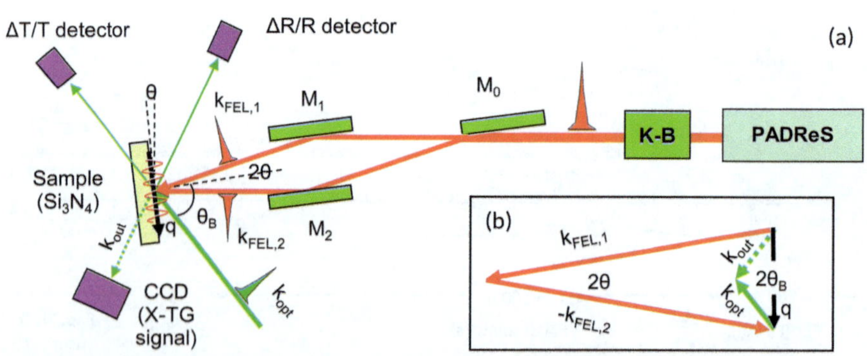

Fig. 2 (a) Sketch of the mini-TIMER set-up: PADReS is the photon transport and analysis system of the facility, KB is the Kirkpatrick-Baez focusing system, $M_{0,1,2}$ are the three plane mirrors that represent the core of the set-up, **q** is the grating vector, $k_{FEL,1}$, $k_{FEL,2}$ and k_{opt} are the wavevectors of the input beams, and k_{out} is that of the signal beam. (b) Phase matching for TG; see eqn (1).

a more stable FEL output, in terms of intensity and spectral fluctuations, while the latter is used when continuous tunability in the FEL output wavelength is required.[30] The interaction between the seed pulse and the electrons takes place in the modulator (a short undulator tuned at λ_s) and results in a modulation of the electron beam energy in the portion of the electron bunch that experienced interaction with the seed laser. The energy modulation is then converted into a spatial modulation (microbunching) by a magnetic chicane (dispersive section). The FEL emission at a high harmonic of λ_s is finally radiated by an undulator chain tuned at $\lambda_{FEL} = \lambda_s/H$, where H is the harmonic number (typically in the 3–15 range). Broad and reliable tunability of λ_{FEL} is obtained by controlling both λ_s and H.[30] The polarization vector of the FEL radiation can be independently set by acting on the undulators.[31] The time duration of the FEL pulses (Δt_{FEL}) is linked to that of the seed laser (Δt_s, typically 100 fs FWHM) by the relation $\Delta t_{FEL} \sim \Delta t_s/H^{1/3}$ and it is independent of the temporal profile of the electron bunch.[32] Thus short FEL pulses can be achieved either by exploiting large H's or by reducing Δt_s. In addition to the straightforward possibility to lock the FEL output to the (easily controllable) seed laser, the seeding scheme also preserves the coherence properties of the seed laser on the FEL radiation, which hence essentially consists of a single photon mode in both the transverse and longitudinal (*i.e.* time) coordinates. Furthermore, the laser seeding permits the stabilization of the FEL output, which otherwise is characterized by multiple longitudinal modes that give rise to a spiky spectrum with large fluctuations in the spectral width and central wavelength.

An additional inherent advantage of laser seeding is the possibility to propagate a fraction of the master laser pulse to the experimental stations for pump–probe experiments. This conceptually simple idea can mitigate the issue of the timing jitter (*i.e.* the stochastic fluctuations in the arrival time difference). In this context, the technical bottleneck of combining FEL and optical pulses is the synchronization between the FEL source and optical source (*i.e.* an external laser)—which are typically separated in space by some hundreds of meters—down to an accuracy comparable with the time duration of the pulses (tens of femtoseconds). The accuracy in the synchronization cannot be reduced below 100 fs r.m.s. with the current stage technology, thus preventing the exploitation of the sub-100 fs time-resolution in principle possible with these sources. To mitigate jitter issues, different kinds of "timing-tools" which are often quite complex and based on demanding post-processing data sorting have been proposed and are routinely used at all working FEL facilities.[25,33] Despite these efforts, however, time resolutions as low as 10 fs have been rarely reported.[25,33] The use of a fraction of the master laser (wavelength λ_{IR}) as a source for the optical radiation has been implemented at FERMI (see Fig. 1), and through the use of active feedbacks[34] timing jitters as low as 6 fs have been demonstrated.[35] This number corresponds to stochastic fluctuations in the optical path difference of about 2 μm, *i.e.* just a few IR wavelengths over more than 100 meters of optical path. The experience gained in successive experiments has confirmed that such a low jitter can be maintained over the typical data collection time, while timing drifts on the order of hours per day can be tracked and compensated by a slow feedback.

In the present experiment we used an optical pulse from the master laser after frequency up-conversion by a second harmonic generation (SHG) crystal, to probe an X-TG generated by two crossed FEL pulses. Since the FEL wavelength was set to the 10th harmonic of the third harmonic seed (see Fig. 1), one may regard the performed experiment as a two-color experiment from the same source, in which we used both the 2nd and 30th harmonic of λ_{IR} and where the whole FEL device can be naively seen as a frequency up-conversion system, not much different from an SHG crystal. The possibility to regard FERMI as a common piece of laboratory equipment is probably one of the most effective ways to state that a seeded FEL has the outstanding performance of a "true" laser.

The mini-TIMER set-up

A sketch of the mini-TIMER set-up that was used to generate the XUV TG's in the sample is shown in Fig. 2a (top-view) and has been described in more detail elsewhere.[19,20,22] However, we will briefly report on its main features and on recent upgrades.

General features

The system consists of three carbon-coated XUV plane mirrors: the first one (M_0) splits the incoming FEL beam into two halves, which are then recombined at the sample at an angle 2θ by two other plane XUV mirrors (M_1 and M_2). The main limitations of the mini-TIMER are: (i) the absence of curved mirrors to focus the FEL radiation, and (ii) the capability of handling only two out of the three input beams needed in a TG experiment. Other issues concerning splitting and recombination of the beams (*e.g.*, the tilted wavefronts, diffraction from the mirror edges, *etc.*) have been investigated in TG experiments carried out in the optical regime using a similar set-up[36] and were found to be of minor relevance. To compensate for the absence of a focusing system, the mini-TIMER set-up is integrated into the DiProI end-station,[37] which is equipped with Kirkpatrick-Baez adaptive optics. These are able to focus the FEL beam at the sample to a spot size ranging from ~10–15 μm to a few mm diameter. The optical probe pulse can be delayed with respect to the FEL pulses by an amount of time (Δt) up to a few hundred ps. The phase matching condition in TG experiments is given by:[2]

$$\mathbf{k}_{\mathrm{FEL},1} - \mathbf{k}_{\mathrm{FEL},2} = \mathbf{k}_{\mathrm{opt}} - \mathbf{k}_{\mathrm{out}} = \pm\mathbf{q} \tag{1}$$

where $\mathbf{k}_{\mathrm{FEL},1}$ and $\mathbf{k}_{\mathrm{FEL},2}$ (with $|\mathbf{k}_{\mathrm{FEL},1}| = |\mathbf{k}_{\mathrm{FEL},2}|$) are the wavevectors of the photons generating the TG, $\mathbf{q}$ is the grating vector, and $\mathbf{k}_{\mathrm{opt}}$ and $\mathbf{k}_{\mathrm{out}}$ (with $|\mathbf{k}_{\mathrm{opt}}| = |\mathbf{k}_{\mathrm{out}}|$) are the wavevectors of the probing photons and output (signal) photons, respectively (see Fig. 2b). Assuming that the sample is oriented with the surface normal perpendicular to $\mathbf{q}$, eqn (1) leads to the scalar relations:

$$|\mathbf{q}| = \frac{2\pi}{L_{\mathrm{TG}}} = \frac{4\pi}{\lambda_{\mathrm{FEL}}}\sin\left(\frac{2\theta}{2}\right) \quad \text{and} \quad \frac{\lambda_{\mathrm{FEL}}}{\lambda_{\mathrm{opt}}} = \frac{\sin(\theta)}{\sin(\theta_{\mathrm{B}})} \tag{2}$$

where θ_B, L_{TG}, λ_{FEL} and λ_{opt} are, respectively, the incident angle of the probe beam at the sample, the spatial period of the TG, the wavelength of the FEL photons that generate the TG (pump), and the wavelength of the optical (probing) photons. Since $\lambda_{opt} \gg \lambda_{FEL}$, eqn (2) can be fulfilled only for $\theta_B \gg \theta$ and does not admit solutions when $\lambda_{FEL} < \lambda_{opt}\sin(2\theta/2)$. Mechanical constraints limit the tunability of

θ_B to just a couple of degrees (°) around 45 + θ_{sam}, where $\theta_{sam} = 2\theta/2$ is the sample angle with respect to the FEL beam coming from M_2 (see Fig. 2a).

Recent and forthcoming improvements

With new actuators for the translational motion of the mirrors in the plane containing the FEL beams (horizontal plane), the 2θ-range has been extended to a range of 3–15°. It is also now possible to vary (at fixed 2θ) the time-delay (Δt_{F-F}) between the two crossed FEL beams over a range from 0 to 0.4/6.5 ps (at $2\theta = 3°/15°$). The increased 2θ range enables the fulfilment of the phase matching condition (see eqn (1), (2) and Fig. 2b) down to $\lambda_{FEL} \sim 13.8$ (9.2) nm, using the 2nd (3rd) harmonic of the Ti:sapphire laser as a probe. The capability of reliably controlling Δt_{F-F} facilitates finding the temporal overlap (*i.e.* $\Delta t_{F-F} = 0$) between the two FEL-pulses, which is necessary for X-TG experiments. To set $\Delta t_{F-F} = 0$, transient reflectivity measurements on a 0.5 μm thick Si_3N_4 membrane (*i.e.* the same sample subsequently used for the X-TG measurements) were carried out by varying the length of the "M_0–M_2–sample" optical path while keeping the "M_0–M_1–sample" path length fixed. The set-up for these measurements is essentially the same as that reported in ref. 24 and has been integrated into the mini-TIMER system. In principle this allows the simultaneous collection of transient reflectivity and transmissivity data with X-TG measurements, a functionality that will be implemented in the near future. The possibility of changing Δt_{F-F} can also be used to compensate for the intrinsic time delay between the two FEL-pulses generated in the twin-seed mode, which enables the use of the two-color XUV pulses for X-CRS.[28,38]

X-TG also requires spatial overlap of the input beams and precise control of 2θ and θ_{sam}. Spatial superposition is achieved by visual inspection of a fluorescent screen, while the value of 2θ is set by using reference marks. At the alignment stage a tracking laser (TL) that mimics the FEL trajectory into the experimental chamber is employed. The TL is also used to optimize the transverse positions (*i.e.* along the bisectors between the incident and reflected beams) of the three XUV mirrors. The fully encoded stages for adjusting the sample angles (accuracy 4×10^{-5} degrees) allows one to use the back reflection of the TL from the sample surface (or from a reference mirror) to precisely cross-check the actual 2θ-value and also to set $\theta_{sam} = 2\theta/2$ accordingly. The actuators of the mirror angles will soon be replaced by more accurate and fully encoded ones.

The in-vacuum CCD sensor (Princeton Instruments, PI-MTE) used to detect the X-TG signal is now equipped with motorized stages. This permits one to center the X-TG signal into the portion of the CCD with a flat response, as well as to move it into a region with low background from spurious scattering of the optical probing pulse not blocked by the pinholes and beamstops in the set-up (see the sub-section "Data extraction").

Experimental settings

In the present study we used the THG seed with the FEL emission set to $H = 10$, yielding an FEL wavelength λ_{FEL} of 26.21 nm, while the probe pulse was set to the second harmonic ($\lambda_{opt} = 393.1$ nm) of the "jitter-free" Ti:sapphire laser pulse. The focal spot area of the FEL and laser beams were $A_{FEL} \sim 0.04$ mm^2 and $A_{opt} \sim 0.01$ mm^2, respectively. The pulse energy of each of the crossed FEL pulses at the sample position was estimated as $I_{FEL} = TI_0/2$, where $T = 0.7$ is the beamline transmission, the factor 1/2 accounts for the assumption of equal intensity in the two branches of the mini-TIMER, and I_0 is the FEL pulse energy at the source as measured on a shot-to-shot basis by the gas ionization chamber (I_0-monitor) of the PADReS system.[39] The FEL intensity was controlled by the gas attenuation chamber of PADReS and kept constant during an X-TG Δt-scan (typical I_0 fluctuations were 7–10%), while it was varied from scan to scan from 0.2 μJ to 1.4 μJ per pulse. The pulse energy of the probing laser was $I_{opt} \sim 1$ μJ with fluctuations less than 1%. The time duration of the FEL pulses was estimated (by the scaling law reported in ref. 32) to be $\Delta t_{FEL} \sim 70$ fs, while that of the probe pulse was $\Delta t_{opt} \sim 100$ fs as measured by an autocorrelator (before the SHG stage). The spectrum of the FEL pulses was also measured and recorded on a shot-by-shot basis by the PADReS spectrometer. The spectral shape was a single narrowband Gaussian line (bandwidth $\Delta\omega_{FEL} \sim 26$ meV; fluctuations $\sim 3\%$) centered on λ_{FEL} (fluctuations $\sim$0.002%), thus indicating a single mode emission. The trends of I_{opt}, I_{FEL}, λ_{FEL}

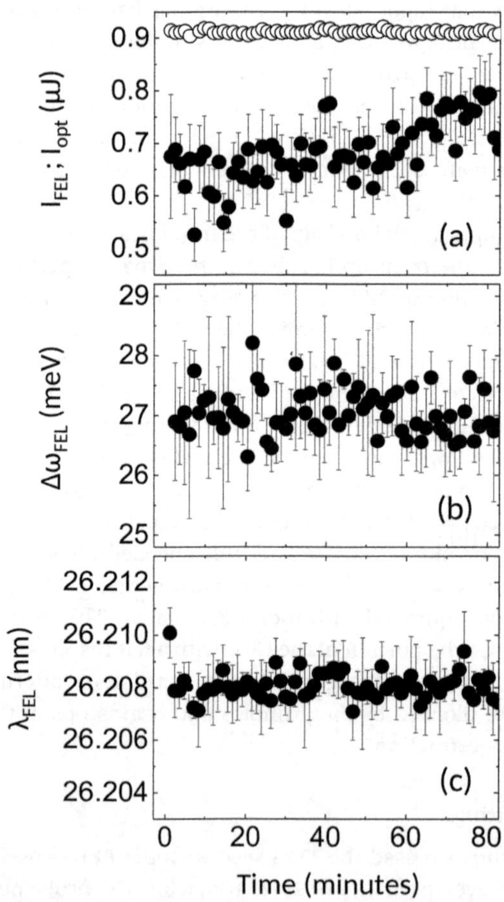

Fig. 3 (a) Example of fluctuations in I_{FEL} (filled circles) and I_{opt} (open circles) over the typical acquisition time for a Δt-dependent X-TG signal. (b) and (c) are the same as (a) for the $\Delta\omega_{FEL}$ and λ_{FEL} parameters, respectively.

and $\Delta\omega_{FEL}$ over the typical data collection time are displayed in Fig. 3. Each point is the average over the shots (25 in this case) that contributed to a single Δt-point in the X-TG scan and error-bars are the corresponding standard deviations. All data were collected with a thin Al filter (200 nm thickness) inserted along the beam transport to remove the residual UV radiation originating from the seed laser. All input beams were vertically polarized. The repetition rate of the FEL was 10 Hz.

To fulfil the phase matching condition, the crossing angle was set to $2\theta = 5.7°$ and $\theta_{sam} = 45° + \theta$. The phase matching relation dramatically affects the TG signal intensity (I_{TG}), which, on order-of-magnitude grounds, can be estimated as:[2,12,22]

$$I_{TG} \propto |\mathbf{E}_{out}|^2 \approx |\chi^{(3)}_{eff}|^2 |\mathbf{E}_{FEL}|^4 |\mathbf{E}_{opt}|^2 \,\mathrm{sinc}^2(|\mathbf{\Delta k}|L_{int}), \tag{3}$$

where $\mathbf{E}_{out}$ is the signal field radiated by the sample, $\mathbf{\Delta k} = \mathbf{k}_{out} - \mathbf{k}_{FEL,1} + \mathbf{k}_{FEL,2} - \mathbf{k}_{opt}$ is the wavevector mismatch, L_{int} is the characteristic length of the interaction region, $\mathbf{E}_{FEL}$ and $\mathbf{E}_{opt}$ are the electric field amplitudes of the FEL (pump) and optical (probe) beams in the interaction region, and $|\chi^{(3)}_{eff}|$ is an effective value for the 3rd-order susceptibility. However, it is worth stressing that the phase matching is relaxed in the present case, since the penetration depth of the exciting FEL radiation ($L_{abs} \sim 33$ nm) is much shorter than the spatial modulation of the X-TG, $L_{TG} = \lambda_{FEL}/2\sin(2\theta/2) \sim 263$ nm (thin grating conditions). This also implies that the observed FWM signal is generated within a small sample depth ($\approx L_{abs}$), which, however, is still much larger than the typical thickness of a surface layer in solids (*i.e.* a few atomic planes, ~1 nm). Under this condition the contribution from the surface-specific response is expectedly lower than that from the bulk. It is worth mentioning that surface-specific nonlinear responses can be studied using 2nd order processes (such as second harmonic, sum or difference frequency generation) which, in centrosymmetric systems, are only associated with the surface response.

Data extraction

As a preliminary step we acquired CCD images without any radiation entering the experimental chamber, using different settings (binning, exposure time, readout speed, *etc.*). This allowed us to evaluate the background level due to residual light leaking into the experimental chamber and also to test the uniformity of the CCD response over the full chip. We then collected background images, either with only FEL or optical radiation entering the experimental chamber. As expected, the dominant contribution to the background arose from optical radiation, which led to well-defined and reproducible structures in the image which were identifiable as shadows of the sample and mirror stages. The CCD sensor and the sample stage were thus positioned to ensure that the X-TG signal was located in a low-background region of the CCD. Raw images were acquired with both the FEL and optical beams impinging onto the sample, while the final signal images were obtained by subtracting the background image taken without the FEL beam. All images were acquired using a 2 × 2 hardware bin (effective pixel size 27 × 27 μm^2).

A typical X-TG Δt-scan consisted of ~70 points in the −0.5 + 2 ps range, sampled with different Δt-steps. In one case the Δt-scan was extended to 130 ps, with a coarse sampling. For each value of Δt we acquired 3 raw and 1 background images. The exposure time was kept fixed during a single scan but was changed (in the 1.5–10 s range) from scan to scan. The number (N) of FEL shots that reached the sample during the acquisition of a given image was ~10 times the exposure time in seconds, given the 10 Hz repetition rate and the few "empty" FEL shots that occasionally occurred (<1%, detected and tagged by the PADReS diagnostics). Overall, a single X-TG scan lasted between 0.5 and 2 hours, including a preliminary signal optimization and the substantial amount of downtime for the CCD readout.† The background subtraction procedure was performed by subtracting equal regions of interest (ROI's) of the two images. The subtracted images were further normalized by N (in order to obtain an X-TG signal proportional to a single FEL shot) as well as by the mean values of I_{opt} and ${I_{FEL}}^2$ calculated over the N shots that reached the sample during the acquisition of a given Δt-point. The latter scaling implies a reasonable assumption that the intensity ratio between the crossed FEL pulses has negligible fluctuation with respect to the uncertainty in the determination of I_{FEL} (see Fig. 3a), which is the predominant source of error. Representative background, raw, and signal images are displayed in Fig. 4. Signal images show a main spot with Gaussian profile and dimensions of about 2×0.3 mm^2 FWHM (horizontal × vertical), a small tilt angle ($\alpha \sim 2°$) and a set of fringes in the vertical direction, *i.e.* orthogonal to the dispersive plane of the X-TG. The horizontal dimension of the main spot is compatible with the dispersion of the probe radiation (bandwidth $\Delta\omega_{opt} \sim 18$ meV) from X-TG generated by FEL pulses with ~26 meV bandwidth (sample-to-CCD distance ~160 mm). The tilt angle and fringes are likely due to a slight tilt of the CCD holder and to diffractive effects arising from the finite dimensions of the interaction region and an inhomogeneous distribution of the FEL intensity. Since we are primarily interested in the determination of I_{TG}, we limited the analysis to a ROI centered on the X-TG signal, highlighted by the box in Fig. 4c. A more detailed analysis of the X-TG signal lineshape will be performed in a future study. The value of I_{TG} from a given signal image was extracted by fitting the main spot of the X-TG signal to a tilted 2D-Gaussian curve centered at (H_0, V_0) with variances σ_H and σ_V plus a residual background term (Bg), where H and V are the horizontal and vertical coordinates of the CCD:

$$\frac{I_{TG}}{2\pi N I_{opt} {I_{FEL}}^2 \sigma_H \sigma_V} \exp\left\{ -\frac{[(H-H_0)\cos(\alpha)-(V-V_0)\sin(\alpha)]^2}{2{\sigma_H}^2} - \frac{[(H_0-H)\sin(\alpha)-(V-V_0)\cos(\alpha)]^2}{2{\sigma_V}^2} \right\} + \mathrm{Bg} \tag{4}$$

Data corresponding to signal images acquired at the same value of Δt were averaged, and data belonging to the same X-TG scan were multiplied by the average value of ${I_{FEL}}^2$ over the entire scan and by a factor of 8.5 (accounting

† X-TG measurements are still quite time consuming compared to many FEL-based experiments but in this work we note almost an order of magnitude increase in the overall throughput of our system with respect to our previous experiments.[17] We believe that this is mostly due to the better procedures for optimizing d the time-space-angular matching of the input beams.

for the efficiency of the detector) to convert the CCD intensities into photon counts.

Results and discussion

Fig. 5 compares the Δt-dependence of the X-TG signal from Si_3N_4 with that of SiO_2;‡ the latter data are scaled to match the peak intensity of the Si_3N_4. Error bars are dominated by the I_0 fluctuations, which yield an almost constant relative error (see Fig. 3) evident from the fairly large scattering of the data around the time coincidence peak where the X-TG signal has the largest intensity. In analogy with that found in SiO_2, the main feature is the large time-coincidence peak. The rise of the X-TG signal from $\Delta t < 0$ can be accounted for by a Gaussian function with a FWHM of 160 fs (shaded area in Fig. 5), which is compatible with the cross-correlation profile of the three input beams in analogy with the optical TG experiments. In the case of SiO_2, the signal is dominated by the time-coincidence peak and $\Delta t > 0$ contributions have a lower intensity with respect to Si_3N_4. In this case the data displays a strong contribution with a sub-ps relaxation time, and the signal asymptotically approaches a plateau in the 1–2 ps interval. The inset of Fig. 5 shows an additional X-TG scan from Si_3N_4 extended up to 130 ps.§ This long delay scan indicates that the X-TG signal slowly decays beyond the probed timescale range. This results in a rather large X-TG signal intensity in comparison to SiO_2, over timescales as large as several hundred ps.

Time dependence of the X-TG signal

In order to describe the Δt-dependence of the X-TG signal a best fit procedure based on a standard χ^2 minimization was used, in which the signal is modeled as a single exponential decay plus a constant term:

$$I_{TG}(\Delta t) = R(\Delta t) \otimes [(A_1 e^{-\Delta t^*/\tau_1} + A_2)\theta(\Delta t^*)]^2, \quad (5)$$

where $R(\Delta t)$ is a unit-area resolution function assumed to have the same shape as the cross-correlation peak (*i.e.* the Gaussian profile with 160 fs FWHM shown in Fig. 5), $\otimes$ is the convolution operator, A_1 and τ_1 are the relaxation amplitude and decay time, respectively, A_2 is a constant accounting for contributions with decay times largely exceeding the probed Δt interval, θ is the Heaviside step function with $\Delta t^* = \Delta t - t_0$, and t_0 is a rigid shift of the time-delay axis. A_1 and A_2 terms are summed and squared because the signal contributions add coherently before the detection with an intensity detector ($I_{det} \propto |\mathbf{E}_{signal}|^2$). The direct determination of $\mathbf{E}_{signal}$ is possible by using heterodyne detection schemes.[40] We are also considering the possibility of using permanent gratings for heterodyning, as proposed in ref. 41. The fitting results are shown in Fig. 6a

‡ SiO_2 data are from ref. 17 and from recent experiments performed a few weeks before the experiments on Si_3N_4 using different values of λ_{FEL} and 2θ but the same value of L_{TG}.

§ It is worth mentioning that for this additional scan we scaled the X-TG signal by a factor of ~3.5 in order to compensate for a drop in the intensity (due to a misalignment of the set-up) and match the trend of $I_{TG,peak}$ *vs.* I_{FEL} observed in the other scans (Fig. 8a), which were performed sequentially with identical experimental conditions.

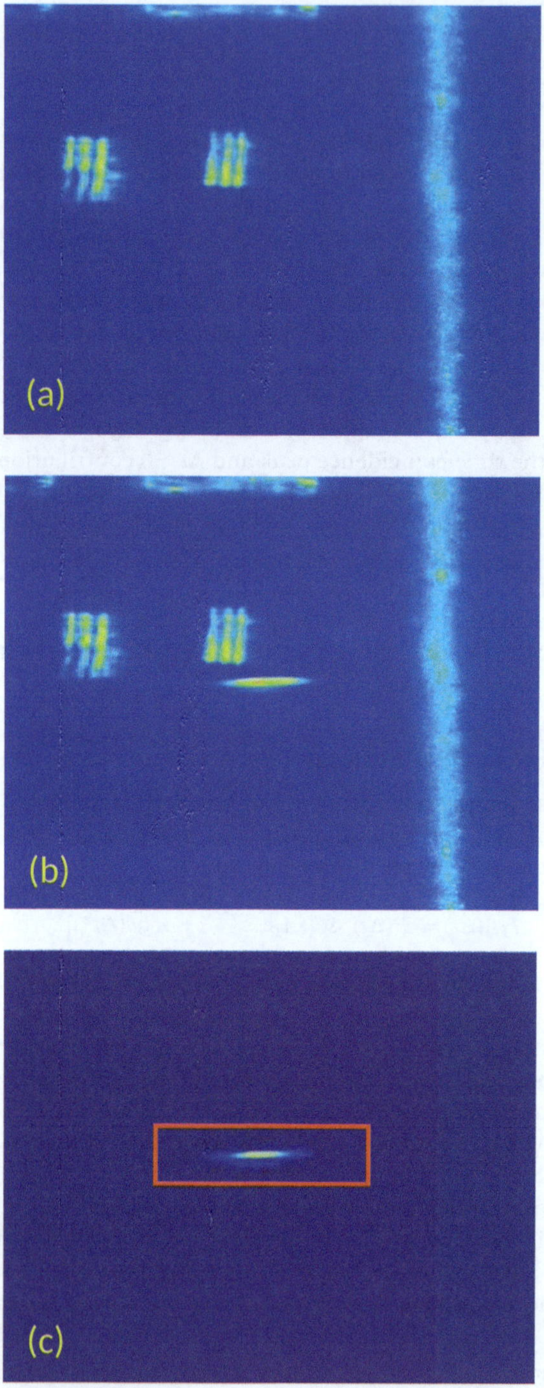

Fig. 4 Representative background (a), raw (b) and signal images (c). The red hatched rectangle in the latter panel is the ROI where the best fit procedure with the function reported in eqn (4) was performed.

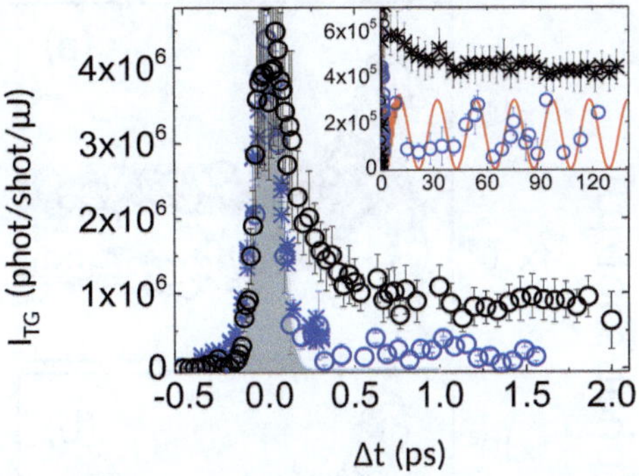

Fig. 5 Comparison between the Δt-dependence of the X-TG signals of SiO_2 (blue circles are from ref. 17, blue asterisks are from the present work) and Si_3N_4 (black circles; $I_{FEL} = 0.7$ μJ). The shaded peak is a Gaussian function with 160 fs FWHM, assumed to be equal to $R(\Delta t)$ in eqn (5). The black asterisks in the inset are the additional X-TG scan collected on Si_3N_4 in the 2–130 ps range while the red line through the SiO_2 data are the signal modulations reported in ref. 22.

and b; within the accuracy of the measurements we found a satisfactory agreement between the experimental data and the single exponential decay model function.

X-TG peak intensity

The dependence of the signal peak intensity, $I_{TG,peak} = (A_1 + A_2)^2$, on I_{FEL} is reported in Fig. 7a and shows a clear quadratic dependence on I_{FEL} as expected from eqn (4). The FWM response in the weak field regime is in general characterized by a linear dependence on the intensity of each of the three beams.[2] In our case this implies a quadratic dependence on I_{FEL} because the two crossed FEL beams are equal in intensity and cannot be varied independently. The largest value of I_{FEL} used in the present work (*i.e.* 1.4 μJ per pulse) corresponds to $\sim 1.9 \times 10^{11}$ XUV photons per pulse. Considering a spot area and absorption length of 0.04 mm^2 and 33 nm, respectively, one may estimate a photon density in the interaction region of $\sim 2.8 \times 10^{20}$ photons per cm^3. Given a sample density of ~ 3.3 g cm^{-3}, there are on average less than 0.003 XUV photons per atom in the interaction region. This value of "photons per atom" is lower than what is used in many FEL-based experiments claimed to be performed in the weak field regime.[24,25] Assuming $\text{sinc}^2(|\mathbf{\Delta k}|L_{int}) = 1$ and considering $|\mathbf{E}_{FEL}|^2 = 2Z_0 I_{FEL}/\Delta t_{FEL} A_{FEL}$, where $Z_0 = 377$ Ohm is the vacuum impedance, eqn (3) can be used to estimate an effective magnitude of the 3rd order susceptibility: $|\chi^{(3)}_{eff}| \approx 1.7 \times 10^{-20}$ m^2 V^{-2} ($|\mathbf{E}_{FEL}|$ is in the 0.16–0.43 V nm^{-1} range). This value is a factor ≈ 30 larger than that estimated for SiO_2 (typical values of $|\chi^{(3)}|$ are in the 10^{-22} m^2 V^{-2} range, but very disparate values have been found in different samples). However, an accurate determination of $|\chi^{(3)}|$ in such kinds of experiments is rather difficult due to alignment issues such as angular matching and temporal and spatial overlap. This likely results in the underestimation of $|\chi^{(3)}|$.

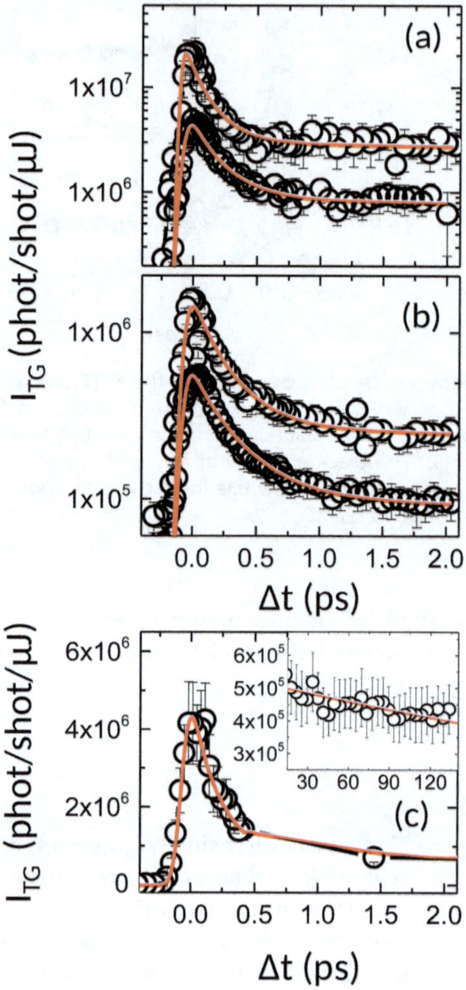

Fig. 6 Experimental data (circles) and best fit results (red lines) obtained using eqn (5), shown on a semi-logarithmic scale; I_{FEL} values are (a) 1.4 μJ (top data) and 0.7 μJ (bottom data), (b) $I_{FEL} = 0.3$ μJ (top) and 0.2 μJ (bottom). (c) shows data for the additional X-TG scan collected in the larger Δt range; the long Δt range is shown in the inset.

Relaxation processes

Fig. 7b displays the dependence on I_{FEL} of the relaxation amplitudes, scaled by the total amplitude ($A_0 = A_1 + A_2$) of the TG signal at $\Delta t = 0$. This figure shows that the relative weights of the two contributions are essentially independent of I_{FEL}, as one may expect in the weak field regime. However, the parameter τ_1 shows a clear dependence on I_{FEL}, as shown in Fig. 7c. The dependence of the decay on I_{FEL} is already observable from the data in Fig. 6. A qualitative explanation for this behavior can be given by following the same arguments used to describe FEL-pump/optical-probe reflectivity measurements on Si_3N_4.[24–26] The crossed FEL pulses generate a spatially modulated free electron density, N_e, that mimics the intensity profile of the X-TG, *i.e.*:

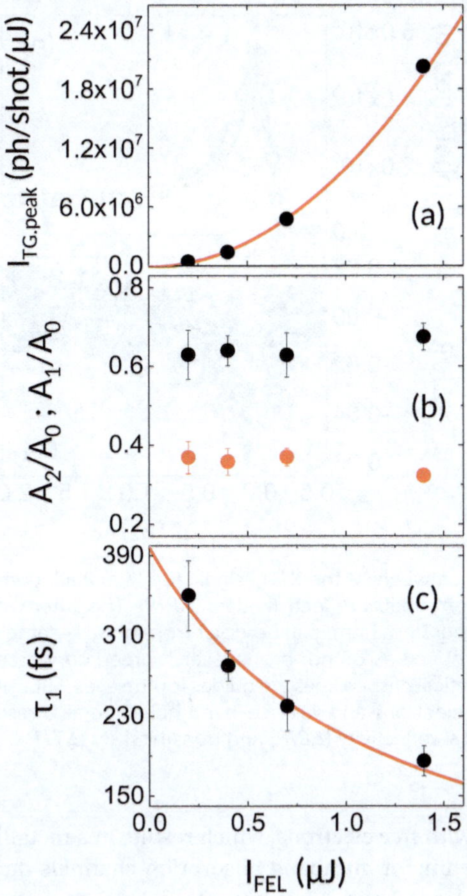

Fig. 7 (a) I_{FEL}-Dependence of the peak intensity of the X-TG signal; the red line is the I_{FEL}^2 trend. (b) I_{FEL}-dependence of A_1/A_0 (black dots) and A_2/A_0 (red dots). (c) I_{FEL}-dependence of τ_1; here the red line is given by the empirical expression $\tau_1 = \tau_1^0/(1 + I_{FEL}/\alpha)$, where $\tau_1^0 = 0.4$ ps and $\alpha = 1.1$ μJ.

$$N_e = \frac{4\beta}{\omega_{FEL} A_{FEL} L_{abs}} G(x,\ y) \exp\left\{-\frac{z}{L_{abs}}\right\} \cos^2\left(\frac{2\pi x}{L_{TG}}\right), \quad (6)$$

where $\omega_{ph} = 47$ eV is the FEL photon energy, $G(x, y)$ is a unit volume function that describes the spatial profile of the two (assumed to be identical) FEL beams in the interaction region, and β is an empirical parameter that quantifies the average number of free electrons generated by a single XUV photon; x is the coordinate parallel to the grating vector and z is the one orthogonal to the sample surface. One may roughly estimate the value of N_e (*e.g.* by assuming $\beta = 1$ and $G(x, y)$ as a constant function in eqn (6)) as on the order of 10^{-20} cm^{-3} in the bright fringes of the TG while, ideally, $N_e = 0$ in the dark fringes. The spatial periodicity of the N_e modulations is hence the same as that of the X-TG (*i.e.* ~263 nm) and results in a periodic modulation of the complex refractive index (n^*). In our case the photon energy of the probe, ~3.1 eV, is lower than the bandgap of the material, ~4.5 eV.[42] Thus the main mechanisms leading to changes in n^* are the filling of the

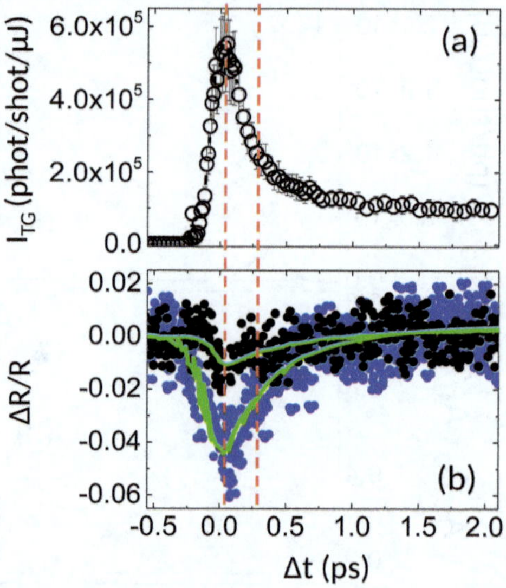

Fig. 8 Comparison between (a) the X-TG signal ($I_{FEL} = 0.2$ μJ, corresponding to 0.5 mJ cm^{-2}) and (b) transient optical reflectivity data ($\Delta R/R$). The latter were collected at 8 mJ cm^{-2} (black dots) and 35 mJ cm^{-2} (blue dots) from a Si_3N_4 sample using a similar FEL wavelength (25.6 nm) and a 785 nm probe laser;[24] green curves are data interpolations from ref. 24 while vertical dashed lines are guides for the eyes. FEL-induced changes in the populations of free electrons and valence-band holes are also responsible for transient changes in the optical reflectivity ($\Delta R/R$) and transmissivity ($\Delta T/T$).

conduction band with free electrons, which results in a metallic-like contribution to n^*, and the opening of intraband absorption channels due to the generation of holes in the valence band. In this context the X-TG signal can be regarded as being proportional to the charge carrier population changes in the valence and conduction band. Moreover, the large spatial modulation of the X-TG as compared to the short mean free path ($\ll$10 nm)[43] for electrons with energy comparable to ω_{FEL} ensures that the photo-excited free electrons remain around the location where they are created. Such charge carriers can then transfer their energy locally to the lattice. In this way the energy of the XUV radiation is "absorbed" by the sample and a thermal grating is generated from the local heating. The energy transfer from the carriers to the lattice becomes faster on increasing the intensity of the radiation at the sample and is dominated by Auger recombination.[44] Qualitatively, this behavior is evident in Fig. 7c. The I_{FEL}-dependence of τ_1 can be empirically quantified by the relation $\tau_1 = \tau_1^0(1 + I_{FEL}\alpha^{-1})^{-1}$, where $\alpha = 1.1$ μJ and $\tau_1^0 \sim 0.4$ ps. We can then suggest that the fast relaxation process observed in the X-TG signal relates to the decrease in the carrier density as a function of Δt. A quantitative analysis of this relation necessitates an appropriate modeling for the carrier–lattice interactions,[45] and is beyond the scope of this paper.

In Fig. 8 we compare the X-TG data and time resolved reflectivity ($\Delta R/R$) data from Si_3N_4 (from ref. 24). The reflectivity measurements were performed under comparable conditions, using a FEL wavelength of 25.6 nm and a near IR probing laser of 780 nm wavelength. The timescale for the recovery of the reflectivity after the drop at the time coincidence is very similar to the fast relaxation time in the X-TG signal. However, the FEL intensity, I_{FEL}/A_{FEL} (in the 0.5–3.5 mJ cm^{-2} range), and hence the value of N_e used in the present work is lower than what is typically needed to observe $\Delta R/R$ and transient transmissivity ($\Delta T/T$) signals in Si_3N_4 (or in other samples). Furthermore the amplitude of $\Delta R/R$ and $\Delta T/T$ variations substantially drops with shorter wavelengths of the optical probe. Therefore the X-TG experiment provides a much higher sensitivity. This is mainly due to the "background-free" nature of the TG signal, which propagates along a direction different from those of the input beams. In time-resolved FEL-induced optical reflectivity, transmissivity, or scattering measurements, an (often small) time-dependent variation of the signal in the presence of a large background has to be detected.

The limited amount of X-TG data and the low FEL intensities probed in the present work (not overlapped with that used for $\Delta R/R$ measurements) does not allow for data interpretation beyond qualitative statements. We are planning to extend this study to a larger FEL intensity range and to implement the parallel detection of $\Delta R/R$ and $\Delta T/T$ signals, in order to correlate these data.

Slow dynamics

In order to describe the slow signal decay observed in the X-TG scan extended to the 2–130 ps range (inset of Fig. 6c), we replaced the constant term (A_2) in eqn (5) with an exponential decay function ($A_2 \exp(\Delta t/\tau_2)$). We found quite reasonable agreement by assuming $\tau_2 = L_{TG}^2/4\pi^2 D_T = 1.1$ ns, which is the expected value for the thermal diffusion time in Si_3N_4 thin films[2,46] where $D_T \approx 1.5 \times 10^{-6}$ m^2 s^{-1} is the thermal diffusivity.[47] Though it is conceivable that such a slow relaxation can be associated with thermal diffusion, a process that always contributes to the TG response in the presence of tangible optical absorption[2,46] as in the present case, we cannot exclude other effects—*e.g.* the excitation of electronic states related to defects, which are expected in thin Si_3N_4 films and depend on sample-specific parameters such as composition, impurities, manufacturing procedures, *etc.* Such states are essentially equivalent to free electrons and holes and can result in time decays as long as ns or μs.[48] X-TG measurements over longer time delays would definitely be needed to clarify the nature of this slow relaxation process. Though this is not possible with the present set-up, the X-TG approach with optical probing in principle allows signal detection in the ns–μs range—*i.e.* with the implementation of CW probing and electronic sampling of the signal, a detection scheme commonly used in table-top TG experiments.[40,46]

Conclusions

We have reported an X-TG experiment performed on a 0.5 μm thick suspended Si_3N_4 film as a function of the intensity of the FEL beams that generate the TG. To our knowledge this is the only report of the intensity-dependence of the TG response stimulated by photon pulses with wavelengths shorter than the optical range. We found that the peak intensity of the X-TG signal, which corresponds to the time coincidence of the three input beams, follows the expected scaling with the FEL intensity. This indicates that the intensity range (0.5–3.5 mJ cm^{-2}) is moderate enough to describe the non-linear response in the weak field regime. From the data we could estimate the magnitude of the effective third-order non-linear susceptibility to be in the 10^{-20} m^2 V^{-2} range, which is more than a factor of 10 larger than that reported for SiO_2—the only other sample with available X-TG measurements. The time dependence of the X-TG signal in Si_3N_4 reveals a sub-ps exponential decay instead of a sharp (*i.e.* a Gaussian-like cross-correlation profile) intensity drop after the time-coincidence peak, as observed in SiO_2. Furthermore, at long Δt's (1–130 ps) the X-TG signal of Si_3N_4 has a high-intensity component (of about one order of magnitude more than in SiO_2) that slowly decays with a time constant exceeding the probed range. These features make Si_3N_4 an excellent reference sample for X-TG, since an appreciable response is expected both at extremely low FEL intensities and at time delays far from time coincidence.

The time evolution of the X-TG signal has been analyzed using a simple model, accounting for two coherent signal contributions having distinct relaxation times. The relative amplitudes of the two relaxation processes are independent of the FEL intensity. The slow relaxation is supposedly due to thermal diffusion and/or to long-lived defect states. The time decay of the fast relaxation shows a clear dependence on the FEL intensity at the sample, decreasing with increasing FEL intensity. This behavior can be interpreted in analogy to time resolved reflectivity measurements on Si_3N_4. The free electrons and holes, with a density that in this case is spatially modulated by the X-TG, rapidly relax into thermal excitations according to the characteristic timescale for energy transfer processes between carriers and the lattice. Indeed, the timescale for carrier–lattice equilibration in semiconductors is expected to decrease upon increasing the photo-generated carrier density.

The inherent "background-free" nature of the X-TG signal results in a much larger sensitivity with respect to "conventional" pump–probe approaches, such as FEL-induced optical reflectivity and transmissivity measurements. This makes X-TG a powerful experimental tool to determine small variations in the sample response.

The sufficiently large non-linear response, robustness, ease of handling, and well-characterized properties of Si_3N_4 make it an attractive test sample for attempting more advanced XUV FWM approaches, *i.e.* X-CRS (Raman active modes are well characterized for this sample). While the TG approach is traditionally used to probe transport properties related to relatively low-energy "thermal" excitations (*e.g.*: phonons, diffusion and structural relaxation processes), CRS can be exploited to probe excitations at higher energy (*e.g.*: ultrafast molecular vibrations or low-lying electronic excitations). The possibility offered by X-CRS to further extend the excitation energy range to multi-eV valence excitations and to achieve atomic-site selectivity through the exploitation of core resonances can potentially make X-CRS a unique experimental tool for studying ultrafast electronic and vibrational dynamics in solids, liquids, and molecules, with large impacts in both physics and chemistry. The combination of the unique "twin-seed" two-colour operation mode[26,28] with our reliable FWM set-up[12,19,20,22] makes X-CRS essentially within reach.[38]

Acknowledgements

The authors acknowledge the European Research Council through the grant no. 202804-TIMER. M. Svandrlik and the FERMI commissioning team are recognized for their valuable support. The contribution by R. A. D. and A. A. M. was supported by U.S. Department of Energy Grant No. DE-FG02-00ER15087.

Notes and references

1 P. A. Franken, A. E. Hill and C. W. Peters, *Phys. Rev. Lett.*, 1961, **7**, 118.

2 N. Bloembergen, *Rev. Mod. Phys.*, 1982, **54**, 685; L. Dhar, J. A. Rogers and K. A. Nelson, *Chem. Rev.*, 1994, **94**, 157; S. Mukamel, *Principles of Nonlinear Optics and Spectroscopy*, Oxford University Press, Oxford, 1995; R. Boyd, *Nonlinear Optics*, Academic Press, 2008.

3 R. J. D. Miller, *Acc. Chem. Res.*, 1994, **27**, 145; G. Dadusc, J. P. Ogilvie, P. Schulenberg, U. Marvet and R. J. D. Miller, *Proc. Natl. Acad. Sci. U. S. A.*, 2001, **98**, 6110; A. C. Beveridge, B. A. Bench, S. M. Gorun and G. J. Diebold, *J. Phys. Chem. A*, 2003, **107**, 5138.

4 A. R. Cameron, P. Riblet and A. Miller, *Phys. Rev. Lett.*, 1996, **76**, 4793; C. P. Weber, N. Gedik, J. E. Moore, J. Orenstein, J. Stephens and D. D. Awschalom, *Nature*, 2005, **437**, 1330; J. D. Koralek, C. P. Weber, J. Orenstein, B. A. Bernevig, S.-C. Zhang, S. Mack and D. D. Awschalom, *Nature*, 2009, **458**, 610.

5 A. R. Cameron, P. Riblet and A. Miller, *Phys. Rev. Lett.*, 1996, **76**, 4793; T. Sjodin, H. Petek and H.-L. Dai, *Phys. Rev. Lett.*, 1998, **81**, 5664; D. Avisar and D. J. Tannor, *Phys. Rev. Lett.*, 2011, **106**, 170405.

6 S. T. Cundiff and S. Mukamel, *Phys. Today*, 2013, **66**, 44.

7 R. R. Ernst, G. Bodenhausen and A. Wokaun, *Principles of Nuclear Magnetic Resonance in One and Two Dimensions*, Oxford University Press, Oxford, 1987.

8 S. Mukamel, *Annu. Rev. Phys. Chem.*, 2000, **51**, 691; J. C. Wright, *Int. Rev. Phys. Chem.*, 2002, **21**, 185; T. Brixner, *et al.*, *Nature*, 2005, **434**, 625.

9 S. Tanaka and S. Mukamel, *Phys. Rev. Lett.*, 2002, **89**, 043001.

10 S. Tanaka, V. Chernyak and S. Mukamel, *Phys. Rev. A*, 2001, **63**, 063405; S. Tanaka and S. Mukamel, *J. Chem. Phys.*, 2002, **116**, 1877; S. Tanaka and S. Mukamel, *Phys. Rev. Lett.*, 2002, **89**, 043001; S. Tanaka and S. Mukamel, *Phys. Rev. A*, 2003, **67**, 033818; S. Mukamel, *Phys. Rev. B: Condens. Matter Mater. Phys.*, 2005, **72**, 235110; I. Schweigert and S. Mukamel, *J. Chem. Phys.*, 2008, **128**, 184307; O. Roslyak, C. A. Marx and S. Mukamel, *Phys. Rev. A*, 2009, **79**, 638271; S. Mukamel, D. Healion, Y. Zhang and J. D. Biggs, *Annu. Rev. Phys. Chem.*, 2013, **64**, 101; D. Healion, *et al.*, *Struct. Dyn.*, 2014, **1**, 014101.

11 F. Bencivenga and C. Masciovecchio, *Nucl. Instrum. Methods Phys. Res., Sect. A*, 2009, **606**, 785.

12 F. Bencivenga, *et al.*, *New J. Phys.*, 2013, **15**, 123023.

13 M. E. Siemens, *et al.*, *Nat. Mater.*, 2010, **9**, 26; J. A. Johnson, *et al.*, *Phys. Rev. Lett.*, 2013, **110**, 025901; A. J. Minnich, *Phys. Rev. B: Condens. Matter Mater. Phys.*, 2015, **92**, 085203.

14 I. Freund and B. F. Levine, *Phys. Rev. Lett.*, 1969, **23**, 854; P. M. Eisenberger and S. L. McCall, *Phys. Rev. A*, 1971, **3**, 1145; R. I. Tobey, M. E. Siemens, M. M. Murnane, H. C. Kapteyn, D. H. Torchinsky and K. A. Nelson, *Appl. Phys. Lett.*, 2006, **89**, 091108; K. Tamasaku, K. Sawada, E. Nishibori and T. Ishikawa, *Nat. Phys.*, 2011, **7**, 705; T. Ding, *et al.*, *Opt. Lett.*, 2016, **41**, 709.

15 N. Rohringer, *et al.*, *Nature*, 2012, **481**, 488.

16 M. Beye, *et al.*, *Nature*, 2013, **501**, 191; C. Weninger, *et al.*, *Phys. Rev. Lett.*, 2013, **111**, 233902; J.-A. You, N. Rohringer and J. M. Dahlström, *Phys. Rev. A*, 2016, **93**, 033413.
17 T. E. Glover, *et al.*, *Nature*, 2012, **488**, 603; S. Shwartz, *et al.*, *Phys. Rev. Lett.*, 2014, **112**, 163901; M. Fuchs, *et al.*, *Nat. Phys.*, 2015, **11**, 964.
18 E. Allaria, *et al.*, *Nat. Photonics*, 2012, **6**, 699; E. Allaria, *et al.*, *Nat. Photonics*, 2013, **7**, 913.
19 C. Masciovecchio, *et al.*, *J. Synchrotron Radiat.*, 2015, **22**, 553; F. Bencivenga, *et al.*, *J. Synchrotron Radiat.*, 2016, **23**, 132.
20 F. Bencivenga, L. Raimondi, C. Svetina and C. Masciovecchio, *Proc. SPIE*, 2013, **8778**, 877807.
21 R. C. Zeller and R. O. Pohl, *Phys. Rev. B: Solid State*, 1971, **4**, 2029; M. Foret, *et al.*, *Phys. Rev. Lett.*, 1996, **77**, 3831; C. Masciovecchio, F. Bencivenga and A. Gessini, *Condens. Matter Phys.*, 2008, **11**, 47; C. Masciovecchio, *et al.*, *Phys. Rev. Lett.*, 2006, **97**, 035501; G. Baldi, *et al.*, *Phys. Rev. Lett.*, 2010, **104**, 195501.
22 F. Bencivenga, R. Cucini, F. Capotondi, *et al.*, *Nature*, 2015, **520**, 205.
23 O. Krupin, *et al.*, *Opt. Express*, 2012, **20**, 11396; M. Beye, *et al.*, *Appl. Phys. Lett.*, 2012, **100**, 121108; M. Harmand, *et al.*, *Nat. Photonics*, 2013, **7**, 215; R. Riedel, *et al.*, *Nat. Commun.*, 2013, **4**, 1731.
24 F. Casolari, *et al.*, *Appl. Phys. Lett.*, 2014, **104**, 191104.
25 J. P. Callan, A. M.-T. Kim, L. Huang and E. Mazur, *Chem. Phys.*, 2000, **251**, 167; E. G. Gamaly, *Phys. Rep.*, 2011, **508**, 91; S. M. Durbin, *AIP Adv.*, 2012, **2**, 042151.
26 F. Bencivenga, *et al.*, *Faraday Discuss.*, 2014, **171**, 487.
27 A. A. Lutman, *et al.*, *Phys. Rev. Lett.*, 2013, **110**, 134801; E. Allaria, *et al.*, *Nat. Commun.*, 2013, **4**, 2476; T. Hara, *et al.*, *Nat. Commun.*, 2013, **4**, 2919; A. A. Lutman, *et al.*, *Phys. Rev. Lett.*, 2014, **113**, 254801; E. Ferrari, *et al.*, *Nat. Commun.*, 2016, **7**, 10343.
28 E. Allaria, *et al.*, *Nat. Commun.*, 2013, **4**, 2476.
29 G. Marcus, G. Penn and A. A. Zholents, *Phys. Rev. Lett.*, 2014, **113**, 024801.
30 E. Allaria, *et al.*, *New J. Phys.*, 2012, **14**, 113009.
31 E. Allaria, *et al.*, *Phys. Rev. X*, 2014, **4**, 041040; T. Mazza, *et al.*, *Nat. Commun.*, 2014, **5**, 3648.
32 D. Ratner, A. Fry, G. Stupakov and W. White, *Phys. Rev. Accel. Beams*, 2012, **15**, 030702.
33 N. Hartmann, *et al.*, *Nat. Photonics*, 2014, **8**, 706.
34 P. Cinquegrana, *et al.*, *Phys. Rev. Accel. Beams*, 2014, **17**, 040702.
35 M. B. Danailov, *et al.*, *Opt. Express*, 2014, **22**, 12869.
36 R. Cucini, F. Bencivenga and C. Masciovecchio, *Opt. Lett.*, 2011, **36**, 1032; R. Cucini, *et al.*, *Opt. Lett.*, 2014, **39**, 5110.
37 F. Capotondi, *et al.*, *Rev. Sci. Instrum.*, 2013, **84**, 051301; L. Raimondi *et al.*, NIMA, 2013, **710**, 131.
38 F. Bencivenga, *et al.*, *Phys. Rev. Lett.*, in preparation
39 M. Zangrando, *et al.*, *Rev. Sci. Instrum.*, 2009, **80**, 113110; M. Zangrando, *et al.*, *J. Synchrotron Radiat.*, 2015, **22**, 565; C. Svetina, D. Cocco, N. Mahne, L. Raimondi, E. Ferrari and M. Zangrando, *J. Synchrotron Radiat.*, 2016, **23**, 35.
40 A. A. Maznev, K. A. Nelson and J. A. Rogers, *Opt. Lett.*, 1998, **23**, 1319.
41 E. Sistrunk, *et al.*, *Opt. Express*, 2015, **23**, 4340.
42 G. Kresse, M. Marsman, L. E. Hintzsche and E. Flage-Larsen, *Phys. Rev. B: Condens. Matter Mater. Phys.*, 2012, **85**, 045205.

43 S. Tanurna, C. J. Powell and D. R. Penn, *Surf. Interface Anal.*, 1991, **17**, 927; B. Ziaja, R. A. London and J. Hajdu, *J. Appl. Phys.*, 2006, **99**, 033514.
44 Y. Gan and J. K. Chen, *Comput. Phys. Commun.*, 2012, **183**, 278.
45 H. M. van Driel, *Phys. Rev. B: Condens. Matter Mater. Phys.*, 1987, **35**, 8166; J. K. Chen, D. Y. Tzou and J. E. Beraun, *Int. J. Heat Mass Transfer*, 2005, **48**, 501; A. Ramer, O. Osmani and B. Rethfeld, *J. Appl. Phys.*, 2014, **116**, 053508; T. Shin, S. W. Teitelbaum, J. Wolfson, M. Kandyla and K. A. Nelson, *J. Chem. Phys.*, 2015, **143**, 194705.
46 A. Taschin, R. Cucini, P. Bartolini and R. Torre, *Europhys. Lett.*, 2010, **92**, 26005; R. Cucini, A. Taschin, P. Bartolini and R. Torre, *J. Mech. Phys. Solids*, 2010, **58**, 1302.
47 X. Zhang and C. P. Grigoropoulos, *Rev. Sci. Instrum.*, 1995, **66**, 1115.
48 L. Zhong and F. Shimura, *Appl. Phys. Lett.*, 1993, **62**, 615; S. V. Deshpande, E. Gulari, S. W. Brown and S. C. Rand, *J. Appl. Phys.*, 1995, **77**, 6534; F. Giorgis, P. Mandracci, L. Dal Negro, C. Mazzoleni and L. Pavesi, *J. Non-Cryst. Solids*, 2000, **266–269**, 588; K. S. Seol, *et al.*, *Phys. Rev. B: Condens. Matter Mater. Phys.*, 2000, **62**, 1532; L. Dal Negro, *et al.*, *Appl. Phys. Lett.*, 2006, **88**, 233109; V. A. Gritsenko, S. S. Nekrashevich, V. V. Vasilev and A. V. Shaposhnikov, *Microelectron. Eng.*, 2009, **86**, 1866.

Faraday Discussions

PAPER

Stimulated X-ray Raman scattering – a critical assessment of the building block of nonlinear X-ray spectroscopy

Victor Kimberg,*[ab] Alvaro Sanchez-Gonzalez,[c] Laurent Mercadier,[a] Clemens Weninger,[ad] Alberto Lutman,[d] Daniel Ratner,[d] Ryan Coffee,[d] Maximilian Bucher,[df] Melanie Mucke,[e] Marcus Agåker,[e] Conny Såthe,[e] Christoph Bostedt,[f] Joseph Nordgren,[e] Jan Erik Rubensson[e] and Nina Rohringer*[a]

Received 29th April 2016, Accepted 13th June 2016

DOI: 10.1039/c6fd00103c

With the invention of femtosecond X-ray free-electron lasers (XFELs), studies of light-induced chemical reaction dynamics and structural dynamics reach a new era, allowing for time-resolved X-ray diffraction and spectroscopy. To ultimately probe coherent electron and nuclear dynamics on their natural time and length scales, coherent nonlinear X-ray spectroscopy schemes have been proposed. In this contribution, we want to critically assess the experimental realisation of nonlinear X-ray spectroscopy at current-day XFEL sources, by presenting first experimental attempts to demonstrate stimulated resonant X-ray Raman scattering in molecular gas targets.

I. Introduction

Femtosecond stimulated Raman spectroscopy[1,2] is a nonlinear optical pump probe technique that allows following vibrational structural dynamics with high spectral and temporal resolution. This relatively new all optical technique has evolved into a well-established method to study the coherent interplay between different vibrational modes and coherent vibrational wave-packet dynamics ranging from small molecules to large biomolecular complexes. Transferring this technique to the X-ray spectral domain would in principle allow following valence electronic wave packets[3–5] or coupled nuclear electronic motion[6] on attosecond timescales, or studying coherent long-range electronic transfer.[7] The potential

[a]*Max Planck Institute for the Structure and Dynamics of Matter, Hamburg, Germany. E-mail: nina.rohringer@mpsd.mpg.de; kimberg@kth.se*

[b]*Royal Institute of Technology, Stockholm, Sweden*

[c]*Imperial College, Department of Physics, London, UK*

[d]*LCLS, SLAC National Accelerator Laboratory, Menlo Park, California, USA*

[e]*Department of Physics and Astronomy, Uppsala University, Uppsala, Sweden*

[f]*Argonne National Laboratory, Illinois, USA*

impact of this method could, therefore, revolutionise our understanding of electron dynamics in complex systems, light-induced catalysis, long-range electron and hole transport, *etc.* The conceptional generalisation of nonlinear coherent spectroscopy to electronic transitions and the X-ray domain was put forward in several papers by Shaul Mukamel and collaborators[3–8] over the past decade, but present-day X-ray sources, such as X-ray free electron laser (XFEL) facilities or table-top attosecond sources, are far from achieving the required pulse characteristics. Realising the dream of coherent nonlinear electronic X-ray spectroscopy would either require a sequence of well-reproducible attosecond pulses, with well-controlled attosecond to femtosecond time separation, of different, tunable central photon energies, or, in the mixed frequency–time domain spectroscopy, the combination of a long coherent pulse (Raman probe pulse) and an attosecond pulse that can be accurately temporally shifted within the duration of the Raman probe pulse. Neither combination is available at present day sources. Moreover, the underlying basic building block of nonlinear Raman spectroscopy, that is stimulated resonant electronic X-ray Raman scattering, also referred to as stimulated resonant inelastic X-ray scattering (SRIXS), could so far only be demonstrated in isolated atoms.[9,10] SRXIS is more difficult to achieve in molecules,[11] due to the generally stronger competing process of inner-shell photoionisation of other atomic constituents of the molecules and an overall smaller Raman gain-cross section that is distributed over many electronic, vibrational and rotational channels.[12] Here we present two experimental studies, along with theoretical predictions, that aimed at demonstrating SRIXS in CO molecules.

A. Introduction to SRIXS in CO

The level scheme for resonant inelastic X-ray scattering (RIXS) at the oxygen $1s_O \rightarrow 2\pi$ resonance in CO is shown in Fig. 1a. The X-ray "pump" pulse is tuned in resonance with the lowest unoccupied molecular orbital (LUMO) in CO

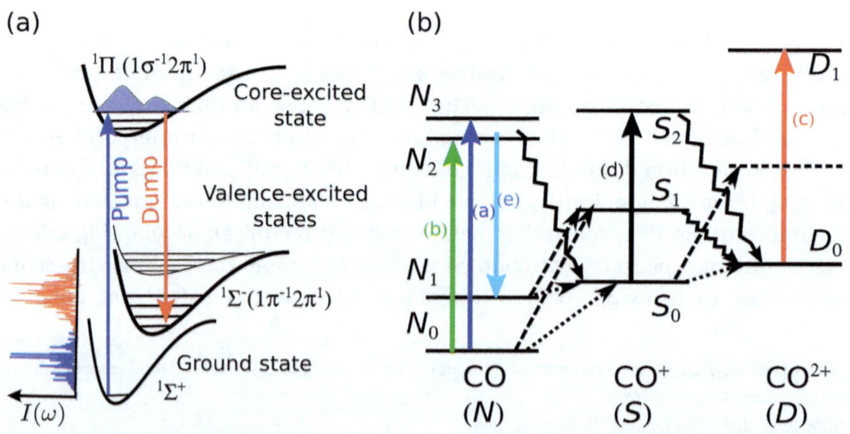

Fig. 1 (a) Schematic illustration of the RIXS transitions in CO stimulated with a two-colour SASE pulse. (b) Scheme of the radiative resonant (solid lines) and non-resonant (dashed lines) X-ray transitions in the CO, CO^+ and CO^{++}; Auger decay channels are shown by zig-zag lines (see text for details).

populating the $1s_O \rightarrow 2\pi$ core-excited state at 534 eV, referred to as the π^* state in the text below. The core-excited π^* state predominantly decays non-radiatively with the emission of an Auger electron, the resonant Auger decay (RAD). With a small probability of 7%, the π^* state can decay radiatively to several different final valence-excited states[13] by the emission of an X-ray photon, the so-called RIXS process. In order to increase the radiative decay channel and the X-ray emission signal, a second X-ray pulse can be applied (the so-called "dump" pulse at Stoke-shifted lower photon energy) to stimulate the resonant scattering process. This process is referred to as stimulated resonant inelastic X-ray scattering (SRIXS). In our experiment, we aimed at stimulating the resonance scattering process to the $^1\Sigma^-(1\pi^{-1}2\pi^1)$, which is the strongest transition in conventional RIXS measurements,[13] with an energy loss of $\approx$9 eV and hence tuning the dump pulse to $\approx$526 eV. The energy of the low-frequency seed pulse is set typically 1–2 orders of magnitude lower than that of the pump pulse. In the course of the SRIXS process, the dump pulse is amplified and changes its spectral shape in the resonance region with the final electronic state.

In a recent theory paper, we discussed SRIXS in a two-colour XFEL set-up,[11] where both XFEL frequencies were produced from a single electron bunch by the self-amplified spontaneous emission (SASE) process. The first experimental attempt and limitations to achieve SRIXS in CO at the oxygen 1s $\rightarrow$ 2π resonance with the two-colour SASE set-up is shortly discussed in ref. 14 and a more detailed analysis of the experimental results is presented in this contribution. Despite the relatively large pulse energies (ranging from 0.07 to 0.17 mJ on target) SRIXS could not be demonstrated in this first attempt. Moreover, the relatively long pulse durations (>50 fs) and large X-ray fluxes ($\sim$1.4 $\times$ 10^{11} photons per μm^2) resulted in multiple ionisation of the CO molecules. These molecular ions, not necessarily in their respective ground states, were shown to have several absorption resonances, also in the 526 eV signal range. Similar absorption of transiently created molecular ions were also recently seen in water droplets.[15] For the predicted small Raman amplification[11] of only a few percent, the observation of the SRIXS signal in averaged spectra can, therefore, be overshadowed by spectral dips resulting from absorption. In Section III we present a second strategy to realise SRIXS, that avoids that problem: the SASE pump pulse, with quite a large pump pulse energy and bandwidth (2 eV), but only a small frequency overlap with the $1s_O \rightarrow 2\pi$ transition is replaced by a self-seeded FEL pulse. Our theory, adapted to the actual experimental conditions shows that we were at the onset of an observable SRIXS signal, but so far, no statistical evidence is seen to confidently report the demonstration of SXRIS in a molecular gas target. We present a critical, comprehensive experimental and theoretical case study of SRXIS in molecules and discuss challenges and the necessary experimental parameters to ultimately reach the conditions for advancing nonlinear X-ray spectroscopy.

II. Stimulated RIXS in two-colour SASE scheme

In a first experimental attempt at the AMO beam line at the LCLS in February 2014,[14] a two-colour SASE scheme was applied to achieve SRIXS in CO. The undulators were set up to produce $\approx$100 fs duration pulses with two SASE bands, each of 2 eV width, centred at 534 and 525 eV. The experimental results for 3000 single SASE shots are reproduced in Fig. 2a. Shown are the recorded spectra (colour coded spectral intensity) as a function of the central photon energy of the pump pulse, that was determined by fitting the linear dependence of electron-beam energy to the measured spectral maximum of a reference measurement without CO. The plot clearly shows the two SASE energy bands at different intensities, separated by ~9 eV and the absorption of the pump radiation at the π^* resonance at 534.5 eV. The width of the absorption feature is defined by the broad vibrational progression excited in the π^* state in agreement with the numerical simulations.[11] One can observe a few weaker absorption lines at the energies below and above the π^* resonance (dashed lines). The absorption features are more apparent in the total averaged spectrum, shown in Fig. 2b for two different SASE pulse combinations. Since multi-photon ionisation is a likely process at the considered pulse energies, we have to take into account some additional excitation and de-excitation channels in the neutral, singly and doubly charged CO molecule for proper assignment of these lines.

A level scheme of the dominant electronic states in CO, CO^+ and CO^{++} excited by the strong X-ray light and decay processes is illustrated in Fig. 1b. Starting from the electronic ground state N_0 resonant absorption of one photon can result in the occupation of the π^* core-excited state N_2 (transition (b) in Fig. 1b, resonant with the pump pulse of 534 eV) or one state of the ensemble of O 1s-Rydberg core-excited states N_3 (transition (a) in Fig. 1b) resonant with the high frequency tails of the SASE pump pulse > 534 eV. Both absorption features, the broad π^* resonance and the Rydberg absorption lines at the higher energy side of the π^* feature are clearly visible in Fig. 2a and b. Resonant excitation of the π^* or Rydberg resonances followed by fast Auger decay (Auger life time of ≈5 fs) results in the creation of the cationic ground state or into the occupation of low-lying valence excited states[16] by resonant Auger decay (RAD), denoted by the ensemble of levels S_0. The ensemble of states S_0 can also be achieved by single photon valence

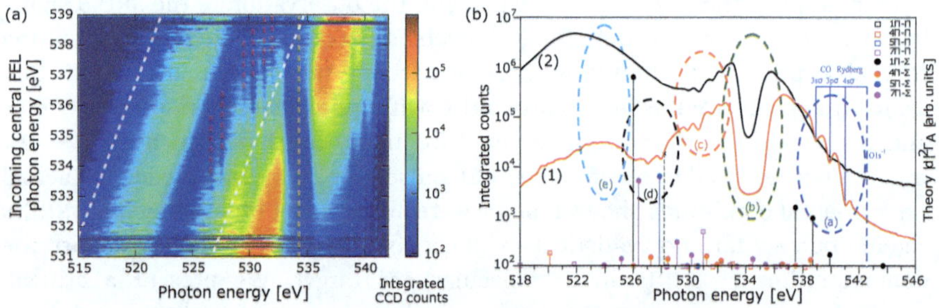

Fig. 2 Experimental spectra (spectral intensity on a logarithmic scale) in the two-colour SASE pulse scheme. (a) Experimental single shot spectra (3000 shots) as a function of the central energy of the incoming pump pulse; colour bar gives the integrated number of counts. The pulse duration was ~100 fs, average total pulse energy on target 0.17 mJ, and the gas pressure 225 mbar. (b) Plot (a) integrated over incoming central FEL photon energy (1); X-ray absorption near π^* resonance in the presence of the strong dressing field on emission channel averaged over 2.2×10^4 SASE shots, gas pressure 85 mbar, total pulse energy on target 0.086 mJ. Vertical lines in (b) show theoretical energies and intensities (arbitrary units) of the strongest resonant X-ray absorption transitions from several $^2\Pi$ states in CO^+ to the Π (empty squares) and Σ (filled circles) symmetry core-excited CO^+ $1s_O^{-1}$ states. The absorption dips observed in areas (a)–(e) are assigned to different channels illustrated in Fig. 1(b) (see text).

ionisation (along with shake-up processes) of neutral CO. Reached by these different single-photon processes, the ensemble S_0 will be generally quite heavily populated during the leading edge of the FEL pulse and can, in a second interaction with the photon fields of the pump or dump pulse, result in resonant O 1s excitations (depicted by the vertical arrow (d) in Fig. 1b). O 1s core excitation of the cationic ground and valence-excited states of the S_0 ensemble will give rise to absorption of photons of the dump pulse (absorption lines summarised in the group (d) in Fig. 2b). A strong non-resonant absorption channel of the neutral molecule is the inner-shell ionisation of the carbon, resulting in C $1s^{-1}$ core-ionized state denoted by S_1 that predominantly decays by the Auger process, resulting in the occupation of the ground state and valence excited states D_0 of the doubly charged molecular ion. The ensemble of states D_0 produced by single photon interaction are also quite strongly occupied, giving rise to resonant absorption lines at slightly higher photon energies as compared to absorption from S_0 and are denoted with (c) in Fig. 1b and 2b.

In order to support the assignment of level groups, we made theoretical estimations for possible absorption lines in the cation: *ab initio* simulations of the O 1s transition energy and intensities of the absorption lines of the CO^+ manifold S_0 (transition group (d) for cationic states) formed in the RAS process are shown in Fig. 2b (legends explain the symmetry of the allowed X-ray absorption transitions). The intensities of the absorption lines are proportional to the partial rates of the formation of the considered ionic states in RAD[17] Γ_A multiplied by the transition probability, proportional to the square of the transition dipole moment, to a correspondent O 1s core-excited state. The transition dipole moments calculations were performed on CASSCF theory level (including up to 7σ, $2\pi_x$ and $2\pi_y$ orbitals) with aug-cc-pvtz basis set using the MOLPRO package.[18] The vibrational degrees of freedom were neglected and the calculations were performed for the internuclear distance 1.153 Å, corresponding to the equilibrium of the π^* core-excited state of the neutral CO. The absolute energy positions of the resonances were corrected by an overall energy shift based on the assignment of the well-known CO^+ O 1s $\rightarrow$ 1π transition at 526 eV observed in non-resonant emission measurements.[16] The transitions from the ground $1^2\Pi$ $(1\sigma^{-1}2\pi^1)$ ionic state to a number of O 1s core-excited states are shown by black lines, where the target states of Π and Σ symmetries are marked with open squares and filled circles, respectively (see legends). The transitions from valence excited $4,5,7^2\Pi$ are shown by red, blue, and magenta lines, respectively. The large absorption probabilities were found for $5,7^2\Pi$ states (main electronic configuration[16] $1\pi^{-2}2\pi^1$) shown by blue and magenta circles in the region (d) of Fig. 2b.

Changing the relative intensities of the dump *versus* pump pulse allows for the suppression or enhancement of certain absorption features and gives supporting evidence for our assignment of different groups of transitions. Fig. 2b shows the recorded spectrum for a strong pump pulse (red line) and the case with a stronger dump pulse (black line), at albeit slightly different CO gas pressures. In the case of a strong pump pulse (0.17 mJ in the pump pulse, ~0.013 mJ in the dump pulse), the O 1s-Rydberg absorption features are clearly visible, but are suppressed in the case for a strong dump pulse (~0.02 mJ in the pump pulse, 0.086 mJ in the dump pulse). The reason for the difference is the ground-state depletion: in the case of a strong pump pulse, only a small fraction (20%) of the total available pulse energy depletes the ground state by the resonant O 1s–π^* transition, so that ample ground-state population is available for O 1s-Rydberg transitions. In the case of a strong dump pulse, depletion of the ground state is driven by off-resonant C 1s ionisation over the whole spectral region of the dump pulse, hence, the total pulse energy that is available for ground-state depletion, so the weaker Rydberg absorption is suppressed. The changes in the relative pump and dump pulse intensity also supports the identification of the O 1s absorption group (d) of cationic valence excited states created in the resonant Auger: the absorption features (d) disappear in the case of a strong dump pulse, since the dump pulse cannot create the necessary valence-excited CO^+ ions summarised as manifold S_0 in Fig. 1b.

Absorption features (c) are clearly visible in a strong pump and a strong dump pulse, corroborating the hypothesis that the double ionised molecules CO^{2+} are predominantly produced by C 1s ionisation followed by Auger decay, a channel that is present in both experimental conditions. SRIXS and RIXS (depicted by the vertical arrow (e) pointing down, resonant with the SASE dump pulse of 525 eV) results in the occupation of one of the valence excited states in the neutral molecule, denoted by N_1 and would result in amplification of the SASE seed radiation. The energy region of the expected peaks is denoted with the region (e) in Fig. 2b. No clear sign of amplification and SRIXS signal is, however, visible in the measured spectra. As it was shown in our numerical simulations,[11] the SRIXS amplification is exponentially sensitive to the applied pump intensity. Our quantitative predictive theory shows that at the present experimental conditions we were at best at the onset of SRIXS. The total transmission of the beam line is unfortunately constantly decreasing at the AMO beam line at LCLS due to damage of the transport optics. Moreover, the achieved focal area of the focusing optics is not readily measurable and only estimated, so that the actually achieved intensity and pulse energy on the CO sample can only be estimated with quite a large uncertainty.

In the following, we present a second strategy for realising SRIXS in molecular CO that should result in more effective coupling to the O 1s–π^* transition by replacing the SASE pump pulse with a self-seeded pulse at albeit smaller total pulse energies, which will reduce the absorption background of transiently created molecular ions in the sample.

III. SRIXS in self-seeded pump and SASE dump scheme

A. Experimental set-up and data acquisition

The experiment was performed at the AMO beamline, using the LAMP chamber. The configuration was very similar to the one described in ref. 9. The Kirkpatrick–Baez mirrors focused the XFEL beam to a spot of about 3.5 to 6 μm diameter with a Rayleigh range of ≈3 mm into a gas cell set to 3 mm length filled with CO gas at a controllable pressure of up to 700 mbar. The transmitted beam was analysed with a spectrometer consisting of a 45 mm long spherical grating with 5 m radius and a CCD camera (Andor, model Newton). The beam impinged on the 10 μm wide entrance slit of the spectrometer at a grazing incidence of 1.88° and diffracted onto the grating to enhance illumination and increase the spectral resolution. The diffracted beam was then imaged onto the CCD camera. Wavelength calibration was achieved by scanning the X-ray photon energy of regular SASE pulses to measure the π^* and 1s-Rydberg absorption lines in CO. We measured a resolution power of $\omega/\Delta\omega \approx 4800$ at 530 eV.

The data acquisition and data analysis were performed in the context of *psana*[19] using the python interface for both online analysis and offline analysis. The main devices involved in the acquisition were standard beam diagnostics on the energy and the trajectory of the electron bunch,[20] two gas monitor detectors to measure the X-ray pulse energy,[21] the pressure reading in the absorption cell, and the X-ray spectrum measured by the CCD camera. Except for the pressure reading, which was only stored at 2 Hz, the information from the other devices was correlated on a shot-to-shot basis at a repetition rate limited by the CCD read-out rate. While storing the full 2048×512 pixel array this was limited to 1.5 Hz, by applying a hardware integration into a 2048×1 array a 60 Hz repetition rate was achieved. Up to 9 datasets were recorded across 12 hours. Each dataset consisted of different 10 minute runs recorded at different gas pressures, including a run at the minimum pressure (<1 mbar) to use as a no-gas reference. Different datasets were taken at different experimental conditions such as varying the position of the cell, beam focusing, and exact tuning of the pump and dump pulses.

B. Setting up of the XFEL

The required two-colour X-ray source in combination of SASE and self-seeded modus to achieve SRIXS in molecules is far from the conventional machine set-up of an XFEL source and a tight collaboration with the FEL scientists is indispensable for a successful experimental realisation. Although set-ups for two-colour SASE generation and self-seeding of the FEL have each been developed previously, the combination of both is a new development that resulted out of this collaboration. A scheme routinely used to generate two-colour pulses in X-ray FELs is the split undulator.[22] There, the same electron bunch generates both X-ray pulses but in two different undulator sections. In the first section, tuned at the undulator strength K_1, a photon beam at energy E_1 is generated while in the second section, tuned at the undulator strength K_2, a second photon beam at energy E_2 is generated. The lasing process in the first section deteriorates the quality of the electron bunch used to generate the second pulse. Particularly if lasing in the first section approaches the saturation point, the electron bunch energy spread growth quickly decreases the performance in the second section, and can prevent lasing when it is too large. Therefore, balancing the colour intensities may prove difficult and is subject to shot-to-shot FEL intensity variations.

The pulse duration can be controlled in several ways, ranging from a shorter electron bunch as operating in low-charge mode, or by using electron bunch collimation,[23] or by using spoiling techniques.[24–26] The bandwidths of each colour generated in this mode are those of a typical SASE pulse, and cannot fulfil the narrowband requirement for the pump pulse. The installation of a grating soft X-ray monochromator and of a dedicated magnetic chicane in the LCLS undulator line recently enabled self-seeding at soft X-ray wavelengths, between 500 and 1000 eV.[27] In a self-seeding scheme, the photon beam generated in a first undulator section is monochromatised and serves as seed in the downstream undulator section. The magnetic chicane is used to provide the required delay to ensure temporal overlap between the monochromatic seed and the electron bunch in the downstream undulator.

The requirement of a strong pump pulse and a weak dump pulse would be easily achievable by generating the weak dump pulse upstream from the intense pump pulse, avoiding the anti-correlation due to the energy spread growth. Unfortunately, the large photon beam divergence after the monochromator mandates to generate the strong seeded pulse prior to the weak SASE pulse. Moreover, aiming for the highest power in the pump pulse prevents adopting solutions similar to the interleaved undulator.[28] Typically, soft X-ray self-seeding is operated with pulses of duration between 70 and 120 fs. To shrink the pulse duration to 20 fs, slotted foil and charge collimation were experimentally tried, with charge collimation providing better performance.

An electron beam of 250 pC charge was collimated in the first bunch compressor down to 80 pC and compressed to a current of 2 kA, generating photon pulses of 25 fs of duration. Five undulator segments were used to generate a 1 μJ SASE pulse to be monochromatised in the first undulator section. The seeding monochromator was set to a photon energy of 534 eV for the pump pulse. The largest ratio between the pump and the dump pulse intensities was achieved by amplifying the monochromatised seed in an eleven undulator segments long section, reaching an average pump energy of 140 μJ, with peak values up to 350 μJ in optimal conditions. However, to match the desired colour separation, the undulator parameter K of that section had to be reduced, decreasing the efficiency, and leaving an average pump energy of 70 μJ, with peak values up to 260 μJ. The weak wideband SASE dump pulse was generated in downstream nine undulator segments at 526 eV and was measurable on the spectrometer. The balance between the intensities was tunable by decreasing the length of the undulator section amplifying the narrowband seed.

C. Data analysis

In the presented combination of self-seeded and SASE operation of the LCLS, the main challenge for data analysis is related to the strong XFEL fluctuations from shot to shot and the correlations between the different parameters. Fig. 3a shows the correlation between the electron-beam energy and the total X-ray pulse energy achieved for an ensemble of XFEL pulses. The first observation is that the high energies above 100 μJ are only reached for a specific region of the electron beam energy centred at 3419 MeV. This is due to the nature of the seeding process: in order to have a good seed, it is necessary to have some SASE emission at the right wavelength, which is only emitted efficiently for a certain electron-beam energy. Furthermore, even for the right electron-beam energy conditions, due to the stochastic nature of the SASE emission, the total X-ray pulse energy can vary between 20 and 260 μJ. As a consequence of these fluctuations, also the averaged spectrum is changing from region to region in the correlation plot. This has to be taken into consideration when trying to observe signals such as SRIXS or transient absorption in spectral averages. Fig. 3b shows the influence of the electron beam energy in the spectrum for the data contained within the red box in Fig. 3a. As expected, the SASE component of the pair of pulses shifts towards higher photon energies as the electron beam energy increases. For the pump spectral component, a much stronger pulse is emitted at the seeding condition of 534.3 eV for

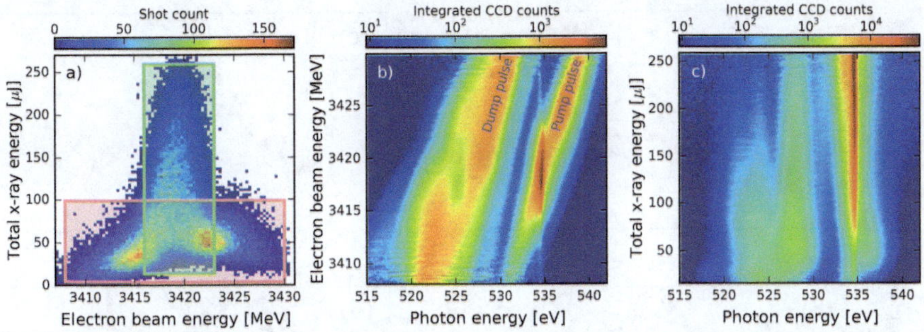

Fig. 3 Fluctuations of the FEL emission. (a) Correlation between the total X-ray pulse energy and the electron-beam energy. The pulse energy was measured at the gas monitor detector that is situated before transport optics to the beam line and focusing optics. The total pulse energy on target is roughly 5% of those values. (b) Average X-ray spectrum as a function of the electron beam energy for the data within the red box in (a). (c) Average X-ray spectrum as a function of the total X-ray energy for the data within the green box in (a).

values of the electron-beam energy between 3415 and 3423 MeV. When seeding becomes strong the weak dump pulse shows a spectral hole. This behaviour becomes more apparent when looking at Fig. 3c, where the spectrum is shown as a function of the total X-ray energy for the narrow green region of Fig. 3a: as the total X-ray energy and seeded contribution at 534.3 eV increases and eventually saturates, the dump-pulse intensity decreases and the spectral hole in the dump becomes more pronounced, to eventually completely split the spectrum. For the highest pulse energies, the spectral intensity of the pump is more than two orders of magnitude larger than that of the dump.

IV. Results and discussion

A. Theoretical simulations of SRIXS

As already discussed, high spectral intensities of the pump pulse at the targeted electronic transition is of crucial importance for SRIXS.[11] In the two-colour SASE scheme (Section II), supplying pulses of relatively broad spectrum (>5 eV, see Fig. 2), only a small number of photons overlap with the $1s \rightarrow \pi^*$ resonance (with a width <1 eV). Moreover, the photons outside of the resonance window drive off-resonant absorption, the creation of transient ionic occupations and finally are responsible for resonant absorption features in the transmitted spectrum of generally both the pump and the dump pulses. In the present section, we numerically test the advantages of using a self-seeded pump pulse in order to optimise the core-excitation and thus enhance the efficiency of SRIXS. We, however, keep SASE radiation on the dump channel, according to the experimental FEL set-up, which provides a broad-band seed pulse to stimulate various RIXS transitions.

Typical averaged spectra of the self-seeded pump and SASE dump pulses measured (without gas) are shown in Fig. 4a. The profiles shown are averaged over about 100 XFEL pulses, resulting in smooth spectra. The quality of the self-seeding process varies greatly from shot to shot, as discussed in the previous sections (see Fig. 3, 7 and 8 for spectral averages and also Fig. 6, that shows

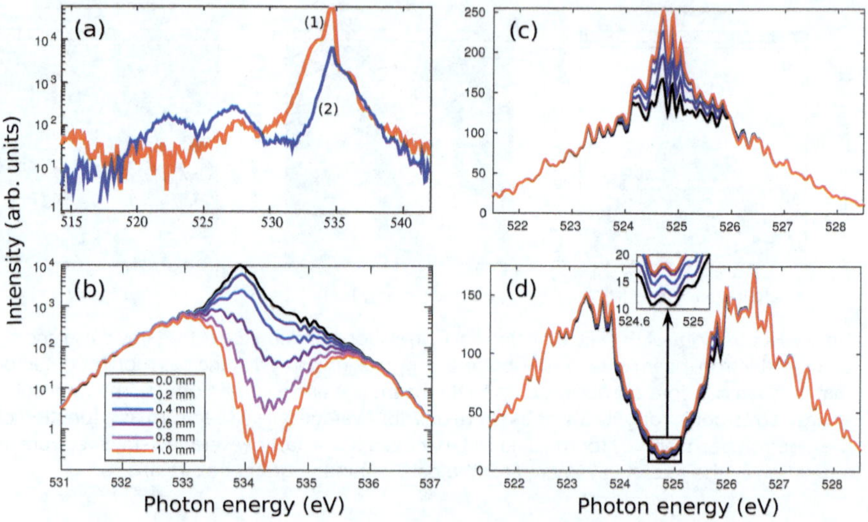

Fig. 4 (a) Measured averaged spectra of the incoming FEL pulses for two typical machine settings: the pulse energies on target are 10 μJ in case (1) and 2 μJ in case (2), supposing a total beamline transmission of 5%. (b) Theoretical estimate of the pump-pulse spectrum at different propagation distances for theoretical spectra mimicking spectrum (1) of the panel (a). Plot (c) and (d): calculated spectra of the dump pulse at various propagation distances (according to the legend of the panel (b)) for a theoretical model of the dump spectra according to the case (1) in (c) and the case (2) in (d). The gas pressure is 500 mbar. All theoretical spectra shown are averages over 128 single shot spectra. The parameters of simulations (b)–(d) are adapted to the high experimental pulse energy of 10 μJ on target: the pump (dump) pulse contains 10^{11} (10^{10}) photons, the pulse duration is 35 fs. The spectra are corrected for off-resonant absorption according to the Beer–Lambert law.

examples for single-shot spectra). Spectrum (1) in Fig. 4a is an example of successful self-seeding with the high pump-pulse energy. Spectrum (2) shows a rather common example of a moderate seeding quality and low pump-pulse energy. As discussed in the previous section, the spectral shape of the dump component develops a spectral hole in case of strong self-seeding (Fig. 3b and c). The characteristics of both dump and pump spectral components are taken into consideration in our numerical simulations presented in Fig. 4b–d: the self-seeded spectrum was modelled as a narrow band (0.5 eV) Gaussian pulse on top of a weak broadband SASE pulse profile of 2.5 eV average width. For the SASE dump component, we simulate two cases: averaged Gaussian spectral profile (Fig. 4c) and a double-peak "m" profile, that corresponds to the low-frequency band SASE emission at strong self-seeding condition (Fig. 4d). All theoretical spectra in Fig. 4 are averaged over 128 simulated XFEL shots. The theory, described in detail in ref. 11, employs a coupled solution of the Liouville–von Neumann equations for the temporal evolution of the electronic-vibrational density matrix equations and Maxwell's equations in the paraxial approximation for the X-ray pulse, which allows tracing the change of the spectral shape of the X-ray fields at different lengths of the gas cell (see legends in Fig. 4b): the pump spectrum (Fig. 4b) shows formation of a strong absorption dip at the π^* resonance. The vibrational structure of the π^* resonance is clearly resolved, in agreement with a conventional X-ray absorption experiment,[29] but difficult to observe in the XFEL experiment due to the large intensity range of about 3 orders of magnitude between the minimum of the absorption dip and the spectral wings of the pump pulse (Fig. 4b). The absorption on the π* resonance is roughly two orders of magnitude passing through 0.5 mm at the gas pressure of 500 mbar. Therefore, a strong core-inversion cannot be maintained over a longer propagation distance, so that SRIXS amplification nearly stops for long propagation distances >0.5 mm (see Fig. 4c and d).

Our numerical simulations mimicking the strongest available set of pump pulses with high seeding component (spectrum (1) in Fig. 4a) predict a SRIXS amplification of up to 50% at the spectral maximum of the emission band (see Fig. 4c). In principle, such amplification should be experimentally detectable. Since the Raman gain is independent of the dump-pulse intensity below saturation,[11] the SASE (Fig. 4c) and m-shaped SASE (Fig. 4d) dump spectra show similar Raman amplification. This also becomes evident in comparing the m-shaped dump pulse in Fig. 5 for several dump intensities. Due to noise, the amplification at the local minimum of the m-shaped pulses (Fig. 4d) will be more difficult to observe experimentally than on top of the SASE spectrum (Fig. 4c). According to our simulations, the population of the core-excited state is ≈20% at the entry of the gas cell. The final state population reached by SRIXS is rather weak (≈1%), according to the small Raman gain cross section. The SRIXS amplification (see Fig. 5, that shows the relative amplification) shows a weak dependence on the number of photons in the pump pulse. A linear increase of the amplification factor is observed for an exponential increase in the number of photons of the pump pulse. In the case of a moderate pump-pulse energy – the more common scenario in our measurements (see Fig. 3b and c) – and assuming 2×10^{10} pump photons in the pump pulse on target (corresponding to a pulse energy of ~2 μJ on target, Fig. 4a(2)), the maximal observable amplification is only 10%. In the next section, we compare these predictions to the experiment.

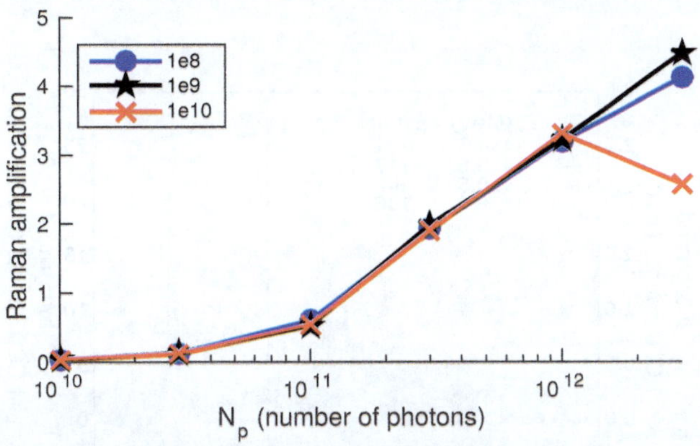

Fig. 5 Raman amplification defined as $[I_d(z, \omega_m)e^{\sigma_{ph}Nz}/I_d(0, \omega_m) - 1]$ at the maximum of the SRIXS signal at $\omega_m = 524.7$ eV, measuring the relative increase of the spectral intensity $I_d(z, \omega_m)$ of the dump field, for different total numbers of photons in the dump pulse (shown in the legend) and as a function of the number of photons in the pump pulse for a propagation distance of $z = 1$ mm. N denotes the molecular concentration and σ_{ph} is the non-resonant absorption cross section. Shown are results corresponding to Fig. 4d, giving Raman amplification of about 0.5.

B. Experimental results

We now present results obtained in the experimental conditions described in Section III-B and in accordance with the parameters of our theoretical discussion of Section IV-A. Examples of single shot spectra without gas of a specific machine set-up (corresponding to the analysis presented in Fig. 7) are presented in Fig. 6, for different combinations of pulse energies and electron-beam energies. The spectra in the dump region show randomly spiked structures, which is a characteristic feature of the SASE process. We can clearly see the anti-correlation between the pump and dump spectral components: when the pump pulse is strong (blue curve), the dump pulse intensity is low, whereas a weaker pump (green curve) allows for a stronger dump pulse. If the electron beam energy is too high (red curve), seeding is no longer achieved and the SASE background of the pump pulse is shifted towards higher photon energies. Even for the right electron-beam energies, a large intensity fluctuation of the seeded spectral component is observed. The average spectrum is also displayed and the standard deviation reflects the highly fluctuating nature of these spectra: it is never less than 70% of the average. These spectral characteristics and fluctuations complicate the detection of a small SRIXS signal. As the signal should appear only when both the pump and dump pulse are at the right energies, we tried making a similar study to that of Fig. 3b with gas. The most promising approach is to compare the average spectra for different pressures to a no-gas reference. To avoid the systematic change of the spectrum for different electron-beam and pulse-energy settings, this average was calculated for different small regions of the correlation plot of electron-beam *versus* pulse energy (see Fig. 7 and 8). In the analysis, the overall absorption by 1s C and valence levels has to be corrected. By making a comparison of the integrated spectral intensity with the exception of the region near the resonant π^* absorption for different pressures, it was possible to fit the Beer–Lambert law, allowing the determination of the transmission for each pressure. As this process assumes a flat transmission, and in reality there are many more resonant and non-linear absorptions, the transmission values obtained are only

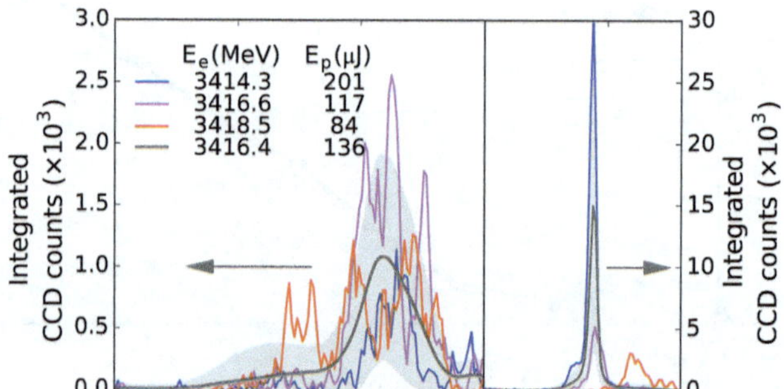

Fig. 6 Single-shot spectra recorded without gas for different combinations of electron-beam energy E_e and pulse energy E_p. As the two vertical scales show, the peak intensity in the seed pump pulse around 534 eV is larger by a factor of 10 compared to that of the dump pulse. The black curve is the mean value over 14 820 pulses and the shaded area is the associated standard deviation.

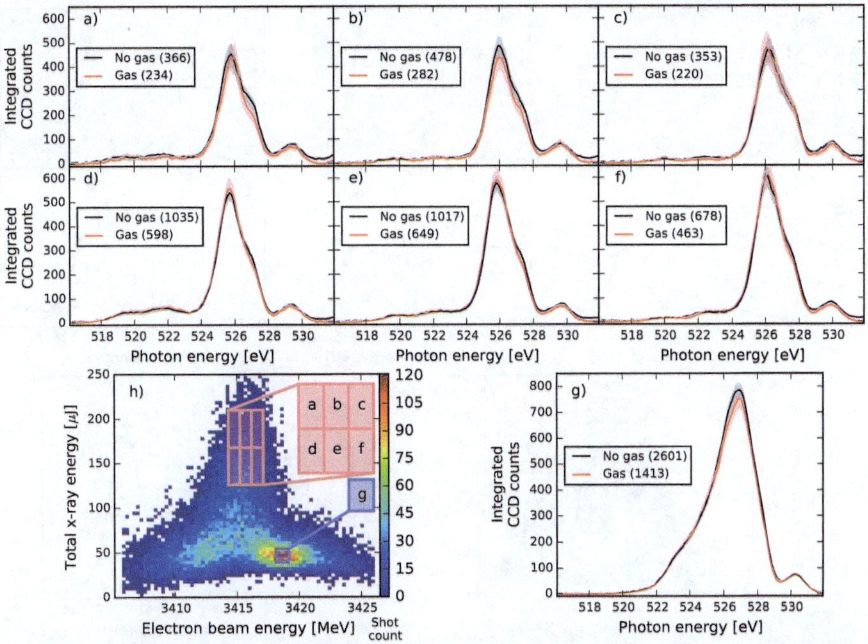

Fig. 7 Comparison of spectra in the dump region with and without gas for a specific setting of the XFEL with a seeded beam centred at 534.43 eV. Each spectrum in plots (a) to (g) is averaged over spectra corresponding to a specific region of pulse energy and electron beam energy that is indicated in the correlation plot (h). The number of spectra used for each region is indicated in the legend for both the gas and no-gas cases. The shaded red and black areas are the 99% confidence intervals (corresponding to 2.58 × standard error). The no-gas spectra are adjusted by a factor of 0.537 (determined by the Beer–Lambert law) to compensate for 1s C and valence absorption (see text). The average gas pressure is 99.57 mbar.

an approximation used mainly to make easier the qualitative comparison across different pressures. Data at each measured pressure was then compared directly to the no-gas reference corrected by the transmission factor (0.537 in Fig. 7 and 0.486 in Fig. 8 and 9) to detect changes in the shape near the 525 eV region.

As can be seen, the averaged spectrum shows systematic changes for different electron beam and pulse energies. In Fig. 7 and 8, the subfigure (g) corresponds to the region with the highest number of recorded shots and with rather low pulse energies and small seeding components (see Fig. 3). In this region, no RIXS signal is expected and therefore the comparison of gas data and reference is a check for a possible drift of machine parameters between the two runs. We can see that except for the very central region of the peak at 527 eV, the overall shape of the dump spectral region is very well reproduced in Fig. 7. In the other regions of the correlation plot (Fig. 7a–f), the wings of the gas and no-gas spectra are also very well superimposed. Slight changes at the peak are observed but they are not statistically relevant: for all ensemble averages presented, we calculated the standard error and the 99% confidence interval that is given by 2.5 times the standard error and is shown as the shaded area in plots 7 and 8. If the confidence intervals of the gas and reference data split apart, it means that the spectral difference of the averages has a good chance to be real and does not reflect the

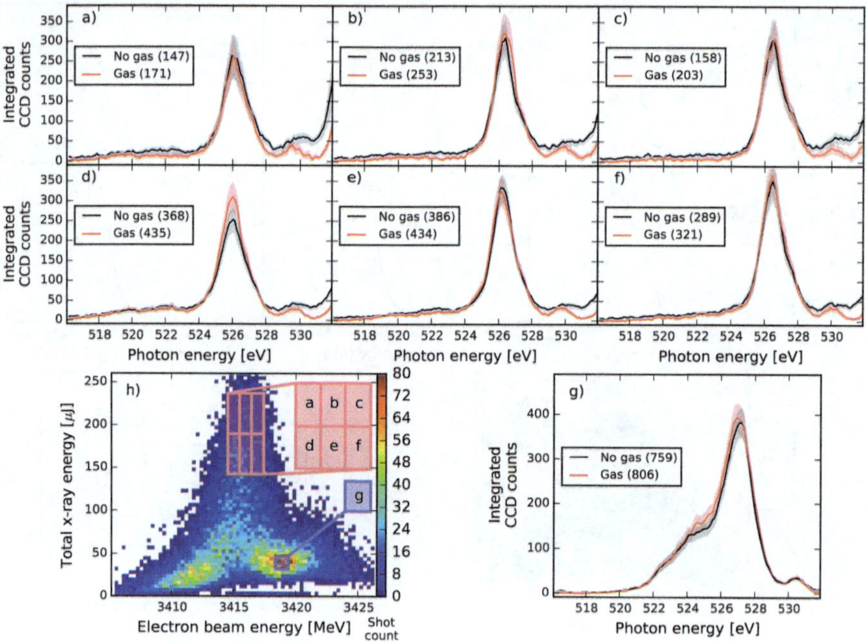

Fig. 8 Same as Fig. 7 at different XFEL settings with a central seeded pulse energy of 534.32 eV. In this case the no-gas spectra are multiplied by a factor of 0.486 to account for 1s C and valence absorption. The average gas pressure is 107.6 mbar.

finite size of the ensemble. The differences observed do not satisfy this criterion for this XFEL configuration. In Fig. 8, slight spectral changes between gas and reference data are observed already in the box (g), which can indicate a drift in the machine settings between the acquisition of the gas and no-gas datasets. An increase of the gas signal is observed at the centre of the peak of the dump in (b) and (d) but the change is not large enough to be statistically relevant. Moreover, the increase in signal does not systematically change when increasing the pulse energy, which would be expected by theory (see Fig. 4c and d). We detect, however, statistically relevant differences at energies of 528 and 531 eV in the spectra shown in Fig. 8. The loss in the spectral intensity in this region can be due to non-linear absorption of transiently created ions, but could also signal a shift in the machine parameters or even be related to not perfectly accurate background subtraction, which becomes significant in these low count regions.

To explore those features we studied the pulse energy dependence of the absorption (Fig. 9). For the same dataset shown in Fig. 4, we selected a series of boxes on increasing intensities (Fig. 9a) and plot the full spectrum for each of the boxes compared to the no-gas reference corrected by the same transmission factor (Fig. 9b–e). Additionally, we plotted the transmission factor for each box calculated as the signal divided by the no-gas reference (Fig. 9f).

We observe absorption dips at 528 and 531 eV equivalent to the absorption resonances measured in the two-colour SASE set-up (see Fig. 2b), but not as well resolved. The absorption dip at 528 eV is not visible at the lowest intensity case (box b, blue curve), but it appears as the pulse energy doubles (box c, cyan curve),

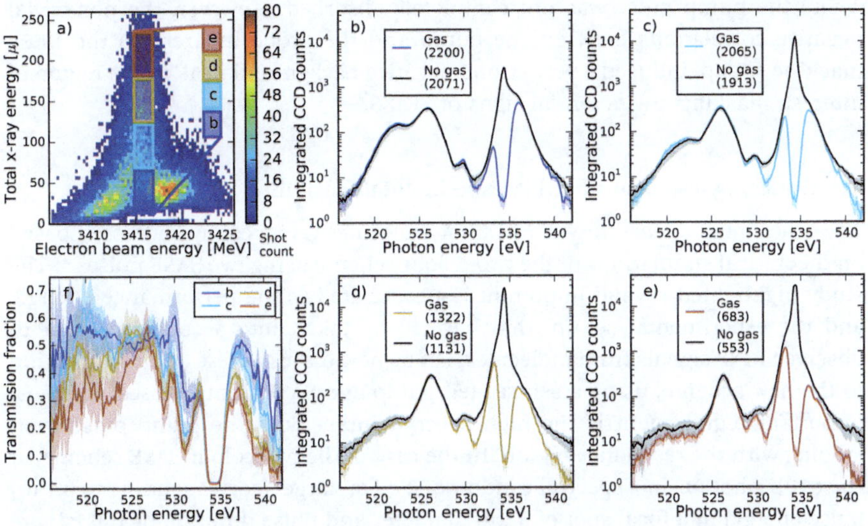

Fig. 9 Comparison of spectra for the whole range (pump and dump, in logarithmic scale) with and without gas for the same XFEL settings as in Fig. 8. Spectra in plots (b–e) are averaged over specific regions of pulse energy and electron beam energy, which are indicated in plot (a), compared to the no gas reference adjusted by a factor of 0.486 as in Fig. 8. Plot (f) illustrates the transmission factor for each of the boxes, calculated as gas spectrum divided by the no gas reference. One can see absorption dips in the region of the Rydberg states of the neutral molecule 539–541 eV (see also Fig. 2b) in the case of small pulse energy, as well as two strong dips at ~528 and ~531 eV. The two dips grow with increase of the pulse energy and thus are tentatively assigned to the absorption by the double ionized molecules, consistent with the two-colour SASE experiment (Fig. 2).

and it continues to increase as the pulse energy triples and quadruples (box d and e, orange and red curves, respectively). Conversely, the absorption dip at 531 eV is already present at the lowest energies (box b), continues to increase when the energy doubles (box c), and reaches saturation when the energy triples (box d), showing almost the same absorption as in box e. This could in principle indicate different degrees of non linearity of the absorption channels. Nevertheless, these features could not be reproduced consistently at all other machine settings, in particular at settings that did not show local minima at these energy positions. This point needs further investigation, as they could be artefacts in the background subtraction at the low signal levels. In fact, we believe the observed apparent non-linear absorption in the region between 516 and 524 eV may just be a consequence of this.

In order to make small changes in the spectral averages more apparent, we also studied the absolute and relative spectral differences of gas and no-gas reference. No consistent evidence of SRIXS amplification could be observed. As our recent theoretical study suggests,[11] a covariance analysis of the recorded single-shot spectra would result in a clear manifestation of the SRIXS process for small SRIXS amplification factors. This analysis is, however, complicated for pulses with systematic spectral changes. We attempted a partial covariance analysis of our data, to find correlations between the spectral pump-pulse region (absorption) and the dump pulse region (emission) as an indication of SRIXS, for low pressures where the pump pulse was not completely absorbed. However, the non-trivial negative correlations between the pump and the probe inherent to the XFEL machine set-up, induced a very strong negative background peak in the region of interest, masking any potential signs of SRIXS.

C. Critical assessment of SRIXS experimental feasibility

The experimental feasibility of SRIXS in molecules was recently addressed based on theoretical simulations of the two-colour scheme using two SASE pulses.[11] The study highlighted several important issues, related to the performance of XFEL and the experimental set-up, that have to be taken into account in order to observe SRIXS signals from molecules. Let us now discuss these points in relation to the new scheme, using a self-seeded pump and SASE dump pulses: the most important requirement is the high spectral intensity of the pump pulse overlapping with the resonance region. In the case of the two-colour SASE scheme, at least 10^{12} photons in a spectral window of 2 eV on target (corresponding to 0.1 mJ pulse energy) in a focal spot of 3 μm diameter and pulse duration of 100 fs[11] are required to obtain vibrationally resolved SRIXS spectra. These relatively high pulse energies, however, result in off-resonant ionisation and the creation of transiently occupied ionic species. This can result in absorption resonances overlapping with the SRIXS signal region. In the present experimental scheme with a self-seeded narrow-band pump pulse, our numerical simulations of Section IV-A demonstrate that much lower total pulse energies are required to efficiently stimulate RIXS. For a pump pulse containing $\sim 10^{11}$ photons on target in a spot of 6 μm diameter and pulse duration of 35 fs the SRIXS signal amplification should be up to be 50%. In the present experiment, these spectral intensities 2×10^8 photons per μm^2 per fs per eV (for 0.5 eV self-seeded pulse of 35 fs duration), however, were on the border of attainability. One of the main limiting factors of our experiment was the detected degradation of the X-ray transport and focusing optics at the AMO beam line at LCLS, resulting in the beam-line transmission drop down to about $\approx 5\%$, as compared to the previous experimental campaign,[14] when the beam-line transmission was estimated to be $\approx 20\%$. We also determined a decrease of the obtainable intensities, comparing amplified spontaneous emission of neon gas obtained in this campaign *versus* measurements in 2014 and 2010. Compared to 2010 the signal outcome of the neon laser was four orders of magnitude lower (with an exponential sensitivity of the signal to the incoming XFEL intensity). Our observation is in accordance with other experimental campaigns, measuring the creation of high charge states in Xe gas by nonlinear sequential photoionisation, pointing to a degradation of the focusing optics, resulting in focal spot sizes roughly twice larger, thus 4 times smaller X-ray intensity on target as compared to previous experiments. Despite the excellent operation of the XFEL and the self-seeding set-up, these deficiencies precluded the observation of SRIXS in this experiment.

The main goal of the present experiment was to observe RIXS amplification as a gross feature shown in Fig. 4c, thus high spectral resolution was not the key interest. In the SASE–SASE pulse configuration, covariance analysis[11] has to be applied to uncover the vibrationally resolved SRIXS maps, showing the spectrum as a function of pump and dump photon energy. To that end, the absorption features have to be properly resolved, requiring a large dynamic range of the spectrometer set-up. In the self-seeded SASE pulse configuration, the hope is that covariance is not necessary, in the case of strong seeding. The averaged spectra for a series of seed photon energies should yield high-resolution SRIXS maps. In the case of a strong superimposed SASE background of the pump spectrum, an additional partial covariance analysis might have to be applied. The partial covariance maps might, however, as seen in this experiment (results not shown in this contribution), be dominated by the XFEL-related correlations in pump and dump frequency components, such as the spectral hole at good seeding conditions. The best configuration for a successful SRIXS experiment is, therefore, to work with only the subset of best seeded pulses, and scanning the seed-energy through the resonance.

As demonstrated by the presented spectra, the averages for a given pulse-energy/e-beam energy region of the correlation plot are fairly stable over a characteristic timescale to record data sets. The detectable SRIXS signal is however limited by the standard error of a given data set. A pure analysis of the statistical error, however, does not allow for detecting drifts of the accelerator. To detect systematic changes of the machine, we monitored the stability of the distributions of all the electron beam parameters and gas detector, and benchmarked our error by comparing the shape of the spectrum for a low energy region of the map, where no SRIXS signal is expected and a good match to the reference is expected. The excellent overlap of the reference and gas spectra in Fig. 7 show that the machine was stable between the two acquisitions. This measurement protocol is applicable for studying all kinds of transient small spectral changes on the base of spectral averages, such as SRIXS and transient absorption spectroscopy at XFEL facilities.

V. Conclusions

We have presented and summarised the results of two experimental campaigns at LCLS (LAMP instrument at the AMO beam line) aimed at a proof of principle observation of the stimulated resonant X-ray Raman scattering in molecules. The high-level numerical simulations of the SRIXS in CO, taking into account vibrational excitation and nonlinear pulse propagation preceded and accompanied the experimental effort. We tested two experimental pump–dump schemes for SRIXS. In the first campaign (February 2014), two-colour SASE emission was used. The high-energy component at 534 eV pump the O 1s → π^* core-excited state in CO, while the lower spectral component at 525 eV stimulates the X-ray emission to valence-excited states. In the second campaign (July 2015), we utilised a narrow band self-seeded pulse for the pump transition together with a broadband SASE pulse on the dump transition.

Our theory predicts that the effect should be observable at the presently attainable parameters of XFEL, though the Raman gain is predicted to be rather moderate: ~50% of relative amplification of the dump radiation near the RIXS resonances. In both experiments, however, we were not able to clearly identify the SRIXS spectral features. The unsuccessful SRIXS observation was explained by the deviation of the critical parameters of the XFEL machine from the optimal parameters used in the simulations. According to the theory, the SRIXS signal drops significantly with decrease of the pump intensity. Among the critical factors of XFEL and beamline performance are: pump-pulse energy (number of photons) in the spectral resonance window; limiting factors of the experiments were the quality of the focusing KB mirrors, losses through the beam transport optics and the potentially missing collinearity of the X-ray beams of pump and dump spectral components in the gas cell.

In spite of the unsuccessful SRIXS observation, our experiments show several new interesting aspects of the interaction of the strong X-rays with molecular targets. With the help of advanced spectral analysis, we were able to register the formation of molecular ions as a function of the X-ray fields intensity of Raman transitions. Moreover, we developed new methods of analysis of the XFEL-molecule interaction by using the statistical averaging of the spectral features, partial averaging by restricting the ensemble to a small region of the XFEL beam parameters (from e-beam energy–pulse energy diagrams), covariance analysis and partial covariance analysis. The methods are implemented in software that now is available for a detailed online data analysis for future experiments. The obstacles of large beam-line losses and degraded focusing mirrors can be overcome in principle, easily pushing up attainable intensities by a factor of 20. The fraction of absorbing and emitting molecules can be drastically increased by alignment of the molecules in the gas phase,[12,30] that, however, requires good synchronisation and timing of an external IR laser pulse to the XFEL beam. SRIXS should, therefore, become observable by overcoming these structural and technical challenges.

Author contributions

V. Kimberg and N. Rohringer developed the theory and conceived the experiment. V. Kimberg performed the numerical simulations of SRIXS. A. Sanchez-Gonzalez performed the data analysis. L. Mercadier was responsible for the overall experimental set-up and supported the experiment. C. Weninger and A. Sanchez-Gonzalez were responsible for the data acquisition and online data analysis. A. Lutman and D. Ratner conceived the ideas and were responsible for the FEL setup in the two-colour SASE/soft X-ray self-seeding scheme. M. Mucke, M. Agåker, Conny Såthe, J.-E. Rubensson and J. Nordgren were responsible for the X-ray spectrometer setup and X-ray spectroscopy. R. Coffee, C. Bostedt and M. Bucher were responsible for the AMO instrument at LCLS and supported the experiment. N. Rohringer, V. Kimberg, A. Sanchez-Gonzalez, L. Mercadier, R. Coffee, C. Bostedt, A. Lutman and J.-E. Rubensson interpreted and discussed the data. V. Kimberg, N. Rohringer, L. Mercadier wrote the paper, with contributions from A. Lutman and A. Sanchez-Gonzalez.

Acknowledgements

We acknowledge financial support from the Max Planck Society; VK also acknowledges financial support from the Knut and Alice Wallenberg Foundation (KAW-2013.0020) and Swedish Research Council (VR). AS-G also acknowledges support from the Science and Technology Facilities Council (STFC). MB and CB acknowledge support from the U.S. Department of Energy, Office of Science, Office of Basic Energy Sciences, Division of Chemical, Geological, and Biological Sciences, under Contract No. DE-AC02-06CH11357. Use of the Linac Coherent Light Source (LCLS), SLAC National Accelerator Laboratory, is supported by the

U.S. Department of Energy, Office of Science, Office of Basic Energy Sciences under Contract No. DE-AC02-76SF00515. The simulations were performed on resources provided by the Swedish National Infrastructure for Computing (SNIC).

References

1 P. Kukura, D. W. McCamant and R. A. Mathies, *Annu. Rev. Phys. Chem.*, 2007, **58**, 461–488.
2 R. R. Frontiera and R. A. Mathies, *Laser Photonics Rev.*, 2011, **5**, 102–113.
3 S. Tanaka and S. Mukamel, *Phys. Rev. Lett.*, 2002, **89**, 043001.
4 I. Schweigert and S. Mukamel, *Phys. Rev. A*, 2007, **76**, 012504.
5 U. Harbola and S. Mukamel, *Phys. Rev. B: Condens. Matter Mater. Phys.*, 2009, **79**, 085108.
6 W. Hua, S. Oesterling, J. D. Biggs, Y. Zhang, H. Ando, R. de Vivie-Riedle, B. P. Fingerhut and S. Mukamel, *Struct. Dyn.*, 2016, **3**, 023601.
7 Y. Zhang, W. Hua, K. Bennett and S. Mukamel, *Top. Curr. Chem.*, 2014, **368**, 273.
8 D. Healion, H. Wang and S. Mukamel, *J. Chem. Phys.*, 2011, **134**, 124101.
9 C. Weninger, M. Purvis, D. Ryan, R. a. London, J. D. Bozek, C. Bostedt, A. Graf, G. Brown, J. J. Rocca and N. Rohringer, *Phys. Rev. Lett.*, 2013, **111**, 233902.
10 C. Weninger and N. Rohringer, *Phys. Rev. A*, 2013, **88**, 053421.
11 V. Kimberg and N. Rohringer, *Struct. Dyn.*, 2016, **3**, 034101.
12 V. Kimberg and N. Rohringer, *Phys. Rev. Lett.*, 2013, **110**, 043901.
13 R. C. Couto, M. Guarise, A. Nicolaou, N. Jaouen, G. S. Chiuzbăian, J. Lüning, V. Ekholm, J.-E. Rubensson, C. Såthe, F. Hennies, F. F. Guimarães, H. Ågren, F. Gel'mukhanov, L. Journel, M. Simon and V. Kimberg, *Phys. Rev. A*, 2016, **93**, 032510.
14 N. Rohringer, V. Kimberg, C. Weninger, A. Sanchez-Gonzalez, A. Lutman, T. Maxwell, C. Bostedt, S. C. Monterro, A. O. Lindahl, M. Ilchen, R. N. Coffee, J. D. Bozek, J. Krzywinski, T. Kierspel, T. Mullins, J. Küpper, B. Erk, D. Rolles, O. D. Mücke, R. A. London, M. Purvis, D. Ryan, J. J. Rocca, R. Feifel, R. Squibb, V. Zhaunerchyk, C. Såthe, M. Agåker, M. Mucke, J. Nordgren and J. E. Rubensson, in *X-ray Lasers 2014. Proceedings of the 14th International Conference on X-Ray Lasers*, ed. J. J. Rocca, C. Meloni and M. Marconi, 2016, vol. 169, p. 201.
15 S. Schreck, M. Beye, J. a. Sellberg, T. McQueen, H. Laksmono, B. Kennedy, S. Eckert, D. Schlesinger, D. Nordlund, H. Ogasawara, R. G. Sierra, V. H. Segtnan, K. Kubicek, W. F. Schlotter, G. L. Dakovski, S. P. Moeller, U. Bergmann, S. Techert, L. G. M. Pettersson, P. Wernet, M. J. Bogan, Y. Harada, A. Nilsson and A. Föhlisch, *Phys. Rev. Lett.*, 2014, **113**, 153002.
16 P. Skytt, P. Glans, K. Gunnelin, J. Guo and J. Nordgren, *Phys. Rev. A*, 1997, **55**, 146–154.
17 M. Piancastelli, M. Neeb, A. Kivimäki, B. Kempgens, H. M. Köppet, K. Maier, A. M. Bradshaw, R. F. Fink and H. M. Köppe, *J. Phys. B: At., Mol. Opt. Phys.*, 1997, **30**, 5677–5692.
18 H. J. Werner and *et al.*, http://www.molpro.net, 2012.
19 D. Damiani, M. Dubrovin, I. Gaponenko, W. Kroeger, T. J. Lane, A. Mitra, C. P. O'Grady, A. Salnikov, A. Sanchez-Gonzalez, D. Schneider and C. H. Yoon, *J. Appl. Crystallogr.*, 2016, **49**, 672–679.

20 S. Smith, S. Hoobler, R. G. Johnson, T. Straumann and A. Young, in *Proceedings of PAC '09*, Vancouver, British Columbia, Canada, p. TU3GRC05.

21 S. P. Hau–Riege, R. M. Bionta, D. D. Ryutov and J. Krzywinski, *J. Appl. Phys.*, 2008, **103**, 053306.

22 A. Lutman, R. Coffee, Y. Ding, Z. Huang, J. Krzywinski, T. Maxwell, M. Messerschmidt and H.-D. Nuhn, *Phys. Rev. Lett.*, 2013, **110**, 134801.

23 P. Emma, R. Akre, J. Arthur, R. Bionta, C. Bostedt, J. Bozek, A. Brachmann, P. Bucksbaum, R. Coffee, F.-J. Decker, Y. Ding, D. Dowell, S. Edstrom, A. Fisher, J. Frisch, S. Gilevich, J. Hastings, G. Hays, P. Hering, Z. Huang, R. Iverson, H. Loos, M. Messerschmidt, A. Miahnahri, S. Moeller, H.-D. Nuhn, G. Pile, D. Ratner, J. Rzepiela, D. Schultz, T. Smith, P. Stefan, H. Tompkins, J. Turner, J. Welch, W. White, J. Wu, G. Yocky and J. Galayda, *Nat. Photonics*, 2010, **4**, 641–647.

24 P. Emma, K. Bane, M. Cornacchia, Z. Huang, H. Schlarb, G. Stupakov and D. Walz, *Phys. Rev. Lett.*, 2004, **92**, 074801.

25 Y. Ding, C. Behrens, R. Coffee, F.-J. Decker, P. Emma, C. Field, W. Helml, Z. Huang, P. Krejcik, J. Krzywinski, H. Loos, A. Lutman, A. Marinelli, T. J. Maxwell and J. Turner, *Appl. Phys. Lett.*, 2015, **107**, 191104.

26 A. Marinelli, R. Coffee, S. Vetter, P. Hering, G. N. West, S. Gilevich, A. A. Lutman, S. Li, T. Maxwell, J. Galayda, A. Fry and Z. Huang, *Phys. Rev. Lett.*, 2016, **116**, 254801.

27 D. Ratner, R. Abela, J. Amann, C. Behrens, D. Bohler, G. Bouchard, C. Bostedt, M. Boyes, K. Chow, D. Cocco, F. J. Decker, Y. Ding, C. Eckman, P. Emma, D. Fairley, Y. Feng, C. Field, U. Flechsig, G. Gassner, J. Hastings, P. Heimann, Z. Huang, N. Kelez, J. Krzywinski, H. Loos, A. Lutman, A. Marinelli, G. Marcus, T. Maxwell, P. Montanez, S. Moeller, D. Morton, H. D. Nuhn, N. Rodes, W. Schlotter, S. Serkez, T. Stevens, J. Turner, D. Walz, J. Welch and J. Wu, *Phys. Rev. Lett.*, 2015, **114**, 054801.

28 A. Marinelli, A. A. Lutman, J. Wu, Y. Ding, J. Krzywinski, H.-D. Nuhn, Y. Feng, R. N. Coffee and C. Pellegrini, *Phys. Rev. Lett.*, 2013, **111**, 134801.

29 M. Coreno, M. de Simone, K. C. Prince, R. Richter, M. Vondracek, L. Avaldi and R. Camilloni, *Chem. Phys. Lett.*, 1999, **306**, 269–274.

30 V. Kimberg, S. Zhang and N. Rohringer, *J. Phys. B: At., Mol. Opt. Phys.*, 2013, **46**, 164017.

Faraday Discussions

PAPER

Probing ultrafast dynamics of chiral molecules using time-resolved photoelectron circular dichroism

Samuel Beaulieu,[ad] Antoine Comby,[a] Baptiste Fabre,[a] Dominique Descamps,[a] Amélie Ferré,[a] Gustavo Garcia,[b] Romain Géneaux,[c] Francois Légaré,[d] Laurent Nahon,[b] Stéphane Petit,[a] Thierry Ruchon,[c] Bernard Pons,[a] Valérie Blanchet[a] and Yann Mairesse*[a]

Received 2nd May 2016, Accepted 8th June 2016
DOI: 10.1039/c6fd00113k

Measuring the ultrafast dynamics of chiral molecules in the gas phase has been a long standing and challenging quest of molecular physics. The main limitation to reach that goal has been the lack of highly sensitive chiroptical measurement. By enabling chiral discrimination with up to several 10% of sensitivity, photoelectron circular dichroism (PECD) offers a solution to this issue. However, tracking ultrafast processes requires measuring PECD with ultrashort light pulses. Here we compare the PECD obtained with different light sources, from the extreme ultraviolet to the mid-infrared range, leading to different ionization regimes: single-photon, resonance-enhanced multiphoton, above-threshold and tunnel ionization. We use single and multiphoton ionization to probe the ultrafast relaxation of fenchone molecules photoexcited in their first Rydberg states. We show that time-resolved PECD enables revealing dynamics much faster than the population decay of the Rydberg states, demonstrating the high sensitivity of this technique to vibronic relaxation.

1 Introduction

Chiral molecules are non-superimposable to their mirror images. The two images, called enantiomers, are defined by a left or right handedness in analogy to human hands. Enantiomers have essentially the same physical and chemical properties and can only be distinguished *via* their interaction with a chiral object, such as circularly polarized light or another chiral molecule.

Terrestrial life is homochiral. All the amino acids and sugars found in living terrestrial organisms have the same handedness (L and D respectively). A

[a]*Université de Bordeaux – CNRS – CEA, CELIA, UMR5107, F33405 Talence, France. E-mail: yann.mairesse@celia.u-bordeaux.fr*
[b]*Synchrotron SOLEIL, Saint Aubin, BP 34, 91192 Gif-sur-Yvette, France*
[c]*Lidyl, CEA/Saclay, 91191 Gif-sur-Yvette Cedex, France*
[d]*Institut Natinal de la Recherche Scientifique, Varennes, Québec, Canada*

consequence of this homochirality is that many biological processes are chiral-sensitive. For instance, two enantiomers of a given chemical species can have different flavors or healing properties. The identification of the left or right character of the molecules by chiral biological sensors is called chiral recognition. The understanding of the interaction of chiral species, and in particular the question of chiral recognition, is a central question in analytical chemistry and catalysis. Chiral reactivity is governed by nuclear rearrangements occurring on picosecond to femtosecond timescales. The nuclear motion is driven by electron dynamics which are typically much faster and at the attosecond scale. On what timescale does chirality imprint from nuclear to electronic structure? What are the dynamical mechanisms for the birth and death of chirality in chemical reactions? What are the fundamental processes governing chiral recognition? Answering these questions requires unraveling the dynamical aspects of the processes, which is only possible using ultrashort light pulses. However, this task is so difficult that, until very recently[1] no ultrafast dynamical chiral process had ever been measured in the gas phase, where isolated molecular systems can be studied with a high level of detail. This is due to the lack of a sensitive enough chiral discrimination technique. In the visible range, the differential response of a chiral molecule to left and right circularly polarized light (Circular Dichroism, CD), is related to the interference of electric and magnetic dipole transitions. Consequently, the CD values are generally weak, on the order of 10^{-5} to 10^{-3} in relative values from the visible to the X-ray range,[2] which is prohibitive for gas-phase measurements where absorption is especially weak. High enantio-sensitivity can be achieved by microwave spectroscopy,[3,4] but this spectral range is not suited for ultrafast measurements on the femtosecond timescale.

Moving from an integral measurement such as absorption or total ionization yield, to a differential measurement such as angle-resolved photoemission enabled a strong increase in the sensitivity of chiral detection in the early 2000s through PhotoElectron Circular Dichroism (PECD).[5,6] Unlike conventional CD, this very intense chiroptical effect is allowed within the electric dipole approximation. The ionization of randomly oriented chiral molecules by circular XUV light leads to a forward/backward asymmetry, with respect to the light propagation axis, in the angular distributions of the released electrons. This asymmetry reverses if the light helicity or the enantiomer is changed (Fig. 1) and can be very large – the highest measured value to date is 35%.[7] PECD gives rise to forward/backward asymmetry because of the intrinsic chirality of the target. It should not be confused with the observed asymmetries which arise in the plane perpendicular to the light propagation axis in photoionization from achiral molecules with a well defined orientation, the so called Circular Dichroism in electron Angular Distribution, CDAD.[8]

PECD possesses a number of unique characteristics (for a review see ref. 9) which make it an ideal probe for molecular structural and dynamical investigations: (i) high signal-to-noise ratio – due to its purely electric-dipole nature, PECD is typically on the order of a few percents; (ii) initial molecular orbital sensitivity;[10] (iii) dynamical final (continuum) state sensitivity depending on the electron kinetic energy; (iv) vibrational sensitivity – PECD was found to change sign between ground and vibrationally excited states, in clear violation of the Franck–Condon approximation.[11] Furthermore, PECD is very sensitive to conformation,[12–14] chemical substitutions,[15,16] isomerism[10,17] and to dimerization and clustering. As

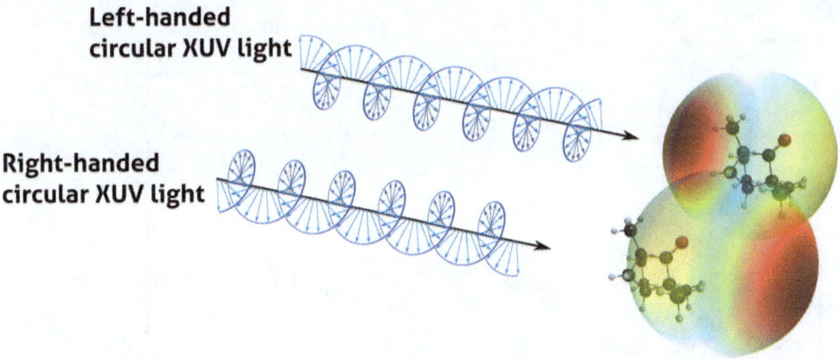

Fig. 1 Principle of photoelectron circular dichroism in XUV photoionization.

a consequence, subtle changes in electronic or nuclear molecular structures are expected to produce large modulations of PECD, which is thus a very good candidate for ultrafast time-resolved measurements on chiral compounds.

Time-resolved measurements of chirality do not only require a sensitive observable but also ultrashort circularly polarized light pulses, which are difficult to produce in the XUV range. One way to circumvent this issue is to use relatively intense circular femtosecond UV-visible pulses to perform multiphoton ionization of the chiral molecules. Several studies based on resonant-enhanced multiphoton ionization have shown that a strong PECD could emerge in this situation, with high sensitivity to isomerism.[18–20] In the first part of the manuscript we compare the PECD obtained using different femtosecond sources, from the extreme ultraviolet to the mid-infrared range, resulting in different ionization regimes: single-photon, resonance-enhanced multiphoton, above-threshold and tunnel ionization. We discuss the advantages and drawbacks of these different cases in the perspective of probing the ultrafast dynamics of chiral molecules. In the second part we study the ultrafast relaxation of the showcase fenchone molecule photoexcited into 3s Rydberg states and show that PECD is a very good probe of the ongoing dynamics, revealing features that are unaccessible in conventional time-resolved photoelectron spectroscopy.

2 Photoelectron circular dichroism using ultrashort light pulses

The vast majority of PECD studies have been carried out in the single-photon XUV ionization regime, using synchrotron radiation. Indeed, synchrotron radiation is bright, tunable over a very large spectral range, and can be fully controlled in polarization – a perfect tool for spectroscopic investigations. However, its long duration prevents ultrafast dynamical studies. Femtosecond lasers can be used to cut temporal slices in the radiation,[22,23] resulting in femtosecond pulses but at the cost of a much lower brightness. Free electron lasers do represent a valuable alternative but so far a single one (FERMI) delivers circular XUV pulses,[24] with

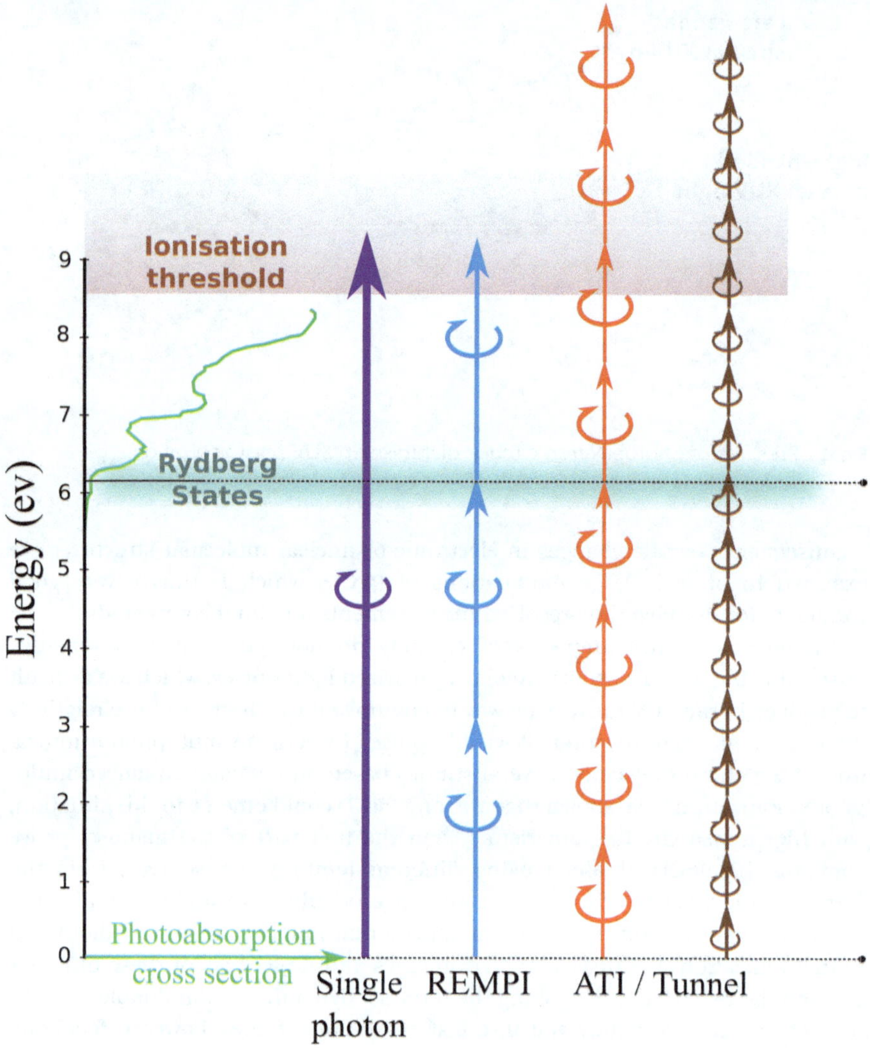

Fig. 2 Ionization schemes of fenchone molecules used in the PECD measurements discussed in the article, by photons at 133, 400, 800 and 1850 nm from left to right. The photoabsorption cross section is adapted from Pulm *et al.*[21] The right panel shows the highest-occupied and 3s Rydberg state molecular orbitals.

a rather low repetition rate not suitable for coincidence imaging. In the following we show that high-order harmonic generation offers an elegant alternative.

Multiphoton PECD has emerged in the past few years as an alternative to XUV PECD, with a natural femtosecond resolution. Using ultrashort ~400 nm laser pulses, Lux *et al.* have demonstrated that a strong PECD could be observed in the 3-photon resonance-enhanced ionization of fenchone and camphor molecules.[25] Fanood *et al.*[19] and Kastner *et al.*[26] have shown that multiphoton PECD has the necessary sensitivity to offer a reliable characterization of enantiomeric excesses. This technique is especially interesting in complex mixtures when combined with coincident mass spectrometry.[19]

In this work, we exclusively use fenchone molecules as a benchmark system to compare different ionization schemes, shown in Fig. 2: single photoionization by high-order harmonics of a 400 nm laser, resonant-enhanced multiphoton ionization by 400 nm pulses, above-threshold ionization by 800 nm pulses, and tunnel ionization by 1850 nm pulses.

2.1 Single-photon ionization from quasi-circular high-harmonics

High-order harmonic generation (HHG) occurs when an intense laser pulse is focused into a gas jet. The highly non-linear interaction of the strong field with the atoms or molecules results in the emission of bright, coherent, ultrashort (femtosecond to attosecond) pulses in the XUV range. HHG has been used for decades as a source of radiation for pump–probe spectroscopy and would be a very good source for PECD studies. Unfortunately, the HHG mechanism has an intrinsic preference for linearly polarized light. The high harmonics are emitted by a three step process.[27–29] When the laser field is maximum, part of the electron wavefunction tunnels into the continuum; the wavepacket is accelerated by the strong field, which drives it back towards the parent ion; lastly, the wavepacket can recombine radiatively with this ion, emitting a short burst of extreme ultraviolet light. This last step is the reverse process of XUV photoionization. If the driving laser field is elliptical, then the minor component of the electric field drives the electron wavepacket away from its parent ion, strongly suppressing recombination. Using 20–30% ellipticity of the driving laser typically lowers the harmonic signal by one order of magnitude.[30] Furthermore, the ellipticity of the harmonics emitted in these conditions remains lower than the driving laser ellipticity.[31]

Converting the polarization state of high-harmonics from linear to circular can be achieved by multiple reflections on metallic surfaces.[32,33] However, this technique is not so easy to implement and results in a strong decrease of the signal, due to the poor reflectivity of the surfaces in this wavelength range. In the past few years, a few solutions have emerged to solve this issue and directly produce bright highly elliptical high-order harmonics. The first one relies on shaping the laser electric field in the time domain. Using a combination of counter-rotating fundamental and second harmonic fields results in the emission of circularly polarized high harmonics with alternatively left or right polarizations.[34] While this configuration would be inadequate for molecular photoionization studies, due to the difficulties of deciphering the spectrum, it is certainly interesting for the condensed phase to investigate X-ray magnetic circular dichroism. The second solution is based on shaping of the laser field in the spatial domain. Generating high-harmonics from two interfering non-collinear counter-rotating circularly polarized fields produces highly elliptical harmonic beams.[35] The third solution, which we use here, is based on shaping the recollision process in the HHG mechanism using resonant high-harmonic generation.[36]

Autoionizing states and shape resonances are known to have a strong influence in XUV photoionization, where they induce structures in the photoelectron spectrum, changes in the photoelectron angular distribution and scattering phase shifts.[37] By performing a complete characterization of the high-order harmonic emission in the vicinity of a shape resonance in SF_6 molecules, we have shown that the resonance had a similarly strong influence on the harmonic spectrum. Besides enhancing the harmonic yield, the resonance creates a strong

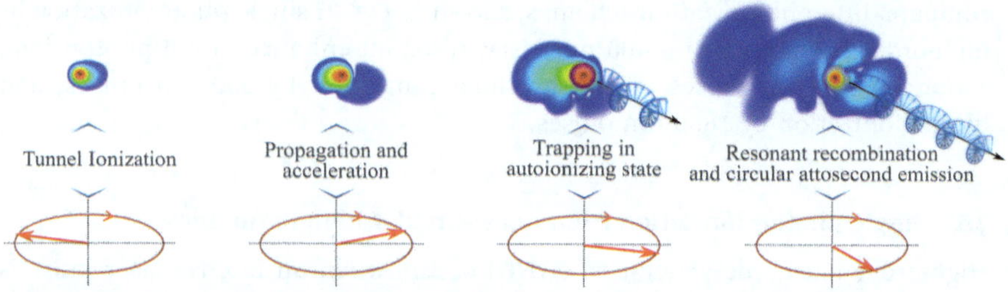

Fig. 3 Mechanism of elliptical resonant high-harmonic generation. The top color plot represents the electronic wavefunction from a 2D quantum calculation, evolving in an elliptical laser field whose instantaneous direction is depicted at the bottom.

orthogonal component to the harmonic electric field when driven by an elliptical laser. This results in the emission of quasi-circularly polarized light. This process is schematized in Fig. 3. An electron wavepacket tunnels out from the molecule and is driven by the elliptical strong field. When the wavepacket returns near its parent ion, it can get trapped into the potential well associated to the resonance before recombining. This adds an additional step to the standard 3-step model of HHG, as proposed by V. Strelkov.[38] Our simulations indicate that the shape resonance modifies the spatial profile of the electron wavepacket as it approaches the ionic core, as illustrated in Fig. 3.[39] Its perpendicular component increases, leading to a higher orthogonal harmonic field. Depending on the exact shape of the potential, this process can lead to the emission of quasi-circular high-order harmonics. This property was experimentally verified in the case of the shape resonance around 22–23 eV in SF_6 molecules, as well as for Rydberg resonances below the ionization threshold of argon.[39]

The PECD measurements were performed using the 1 kHz Aurore laser system at CELIA, which delivers 25 fs, 7 mJ pulses at 800 nm. The laser pulses were frequency doubled in a 200 μm thick BBO crystal and focused by a 50 cm lens in a 3 mm long cell filled with SF_6 molecules. Using 400 nm rather than 800 nm pulses for HHG increases the energy spacing between harmonics from 1.55 to 3.1 eV, easing the assignment of the photoelectron spectra. The laser ellipticity was controlled using a motorized zero order half waveplate in front of a fixed zero order quarter waveplate, in order to keep the orientation of the polarization ellipse fixed in space. The harmonics were directly sent into the interaction region of a Velocity Map Imaging Spectrometer (VMIS), without any focusing optics to avoid modifying the XUV polarization state and perturbing the ionization process with an additional intense 400 nm field. A motorized aperture was tuned to limit the diameter of the XUV beam and achieve a good resolution of the photoelectron images. A sample of enantiopure (+)-fenchone molecules (Sigma Aldrich) at room temperature was connected to a heated metallic nozzle (390 K) with 300 μm diameter located 7 cm away from the interaction zone of the VMIS. The pressure in the interaction region was 2×10^{-6} mbar (with a 5×10^{-8} mbar background pressure).

The velocity map imaging spectrometer records the bidimensional projection of a tridimensional photoelectron angular distribution $S_p(E, \theta)$, which can be decomposed as a sum of Legendre polynomials P_i:

$$S_p(E,\theta) = \sum_{i=0}^{l_{max}=2N} b_i^p(E)P_i(\cos(\theta)) \tag{1}$$

where $p = \pm 1$ is the light helicity, E is the photoelectron energy, θ its ejection angle with respect to the light propagation direction, and N the number of photons involved in the ionization process (see *e.g.* ref. 25). The expansion coefficients $b_i^p(E)$ are extracted by a least-squares fitting of the experimental distributions.[40] The first coefficient $b_0(E)$ is the angle-integrated photoelectron spectrum (PES). The odd coefficients reflect the forward/backward asymmetry along the optical axis in the photoelectron angular distribution. The angle-resolved photoelectron spectrum (ARPES) is calculated by:

$$\text{ARPES}(E,\theta) = \frac{S_{-1} + S_{+1}}{2} \tag{2}$$

The PECD, which quantifies the forward/backward asymmetry relative to the ARPES signal, is defined as

$$\text{PECD}(E,\theta) = \frac{S_{-1} - S_{+1}}{\text{ARPES}(E,\theta)} \tag{3}$$

The normalization procedure can strongly overestimate the PECD in areas where the ARPES is low (and noisy). It also modifies the angular distribution of the asymmetry, and does not directly reflect the shape of the antisymmetric part of the ARPES. To avoid this, in the following we will thus also use an alternative definition to monitor the absolute forward/backward asymmetry:

$$\text{APECD}(E,\theta) = \frac{S_{-1} - S_{+1}}{\max(\text{ARPES})} \tag{4}$$

Last, to evaluate the difference between the number of electrons emitted in the forward and backward hemispheres, we will use the general expression of multiphoton PECD:[18]

$$\text{MPPECD}(E) = \frac{1}{b_0}\left(2b_1 - \frac{1}{2}b_3 + \frac{1}{5}b_5 - \frac{5}{32}b_7 + \frac{6}{64}b_9\ldots\right) \tag{5}$$

which reduces to $2b_1/b_0$ in the single photon case.

Fig. 4 shows the ARPES (a) and PECD (b) obtained by ionizing fenchone by harmonics 3 (9.3 eV photon energy) and 5 (15.5 eV) of the 400 nm laser. This one-photon absorption case corresponds to $N = 1$ in eqn (1). The PES ($b_0(E)$, Fig. 4(c)) presents a sharp peak around 0.5 eV whose height is normalized to 1, followed by a broad bump from ~1 to ~5.5 eV. An additional (tiny) maximum is noticeable around 6.75 eV. In order to assign these features to ionization pathways, we performed Time-Dependent Density Functional Theory (TDDFT) calculations of the electronic spectra of the neutral and cationic fenchone using the GAMESS-US package.[41] The geometry of fenchone has been optimized at the TDDFT level of theory, employing a 6-311++G** basis and CAMB3LYP hybrid exchange–correlation functional. The same geometry has been used for the cation, yielding a vertical ionization potential of 8.74 eV, in very good agreement

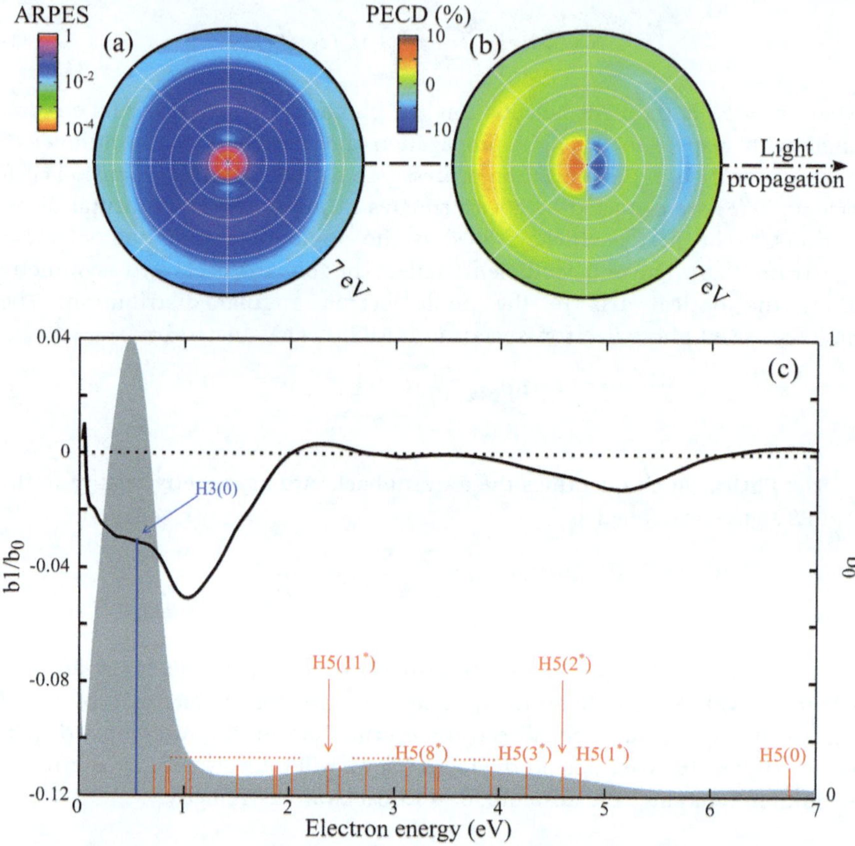

Fig. 4 Single-photon ionization of (+)-fenchone with harmonics 3 and 5 of a 400 nm, ±30% ellipticity laser pulse. ARPES (a) and PECD (b); the light propagation axis is horizontal and the radius extends from 0 to 7 eV. (c): b_0 (gray area) and normalized b_1/b_0 (black line) Legendre coefficients as a function of electron kinetic energy; the positions of PES peaks expected from TDDFT calculations are marked by Hq(i*), which is a shortcut to the Hq(0 → i*) notation employed in the text.

with the experimental one, 8.72 eV. These calculations provide the vertical ionization potentials $I_p^{0\rightarrow i^*}$ from the ground state of fenchone (0) to the ground (0) and excited (i*) states of the associated cation. These ionization potentials (I_p) are used in Fig. 4 to identify the different ionization channels 0 → i* opened by the absorption of the qth harmonic according to energy conservation $E = q\hbar\omega_0 - I_p{}^{0\rightarrow i^*}$. The main and secondary maxima in the PES correspond to the H3(0 → 0) and H5(0 → 0) ionizing processes, respectively. The broad bump results from ionization leaving the cation in excited states, H5(0 → i*). Within the Koopman's approximation, these processes would be understood as ionization of inner molecular orbitals. However for $i > 11$ two-hole-one-particle configurations, *i.e.* ionization + excitation processes, come into play. Interestingly, the excited state contributions vanish in the PECD image which only shows peaks centered at H3(0 → 0) and H5(0 → 1*), with respective magnitudes of 10 and 3%. This indicates that the PECD associated to neighbouring 0 → i* channels, with i in the range 4–12, alternates between positive and negative values so that their overlap leads to an average value close to zero. The dichroism is further presented in Fig. 4 in terms of b_1/b_0 which represents, according to eqn (1), half the PECD along the light propagation axis. Remarkably b_1/b_0 does not maximize at the locations of the PES peaks, but on their falling edges. This could be due to several effects. First, for a given electronic and vibrational state of the cation, b_1 inherently depends on the kinetic energy of the electron.[10] Second, simulations have shown that vibrational excitation of the ion played an important role, with a dramatic b_1 change in the falling edge of the PES associated to vibrations that explore different parts of the potential.[11] Last, b_1 could be modified by the contributions from different ionic states, produced by adjacent H5(0 → i*) channels with $16 \leq i \leq 20$. It should be noted that the harmonics used in the experiment are only elliptical, such that the extracted values of PECD and b_1 do not represent the asymmetry of the response of the molecule to a purely circular radiation as in synchrotron experiments.[17] Nevertheless, the PECD signal is strong, and this measurement demonstrates the possibility of using femtosecond pulses produced by resonant HHG to reveal the chiral response of a molecule.

2.2 Resonant-enhanced multiphoton ionization

The advantage of quasi-circular HHG is that it delivers XUV pulses capable of producing a PECD signal by single-photon absorption. This enables straightforward comparison with spectroscopic studies carried out at synchrotrons, and also

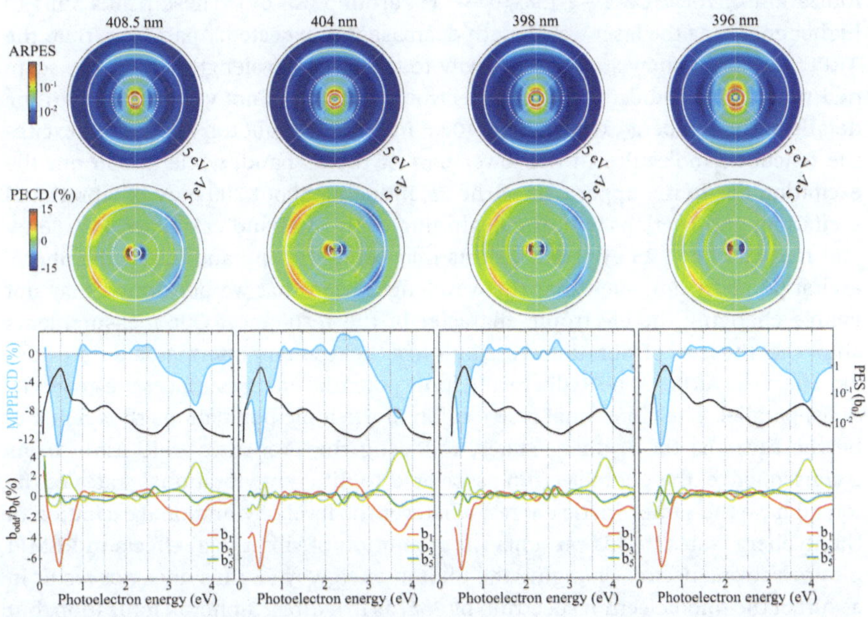

Fig. 5 Resonance-enhanced multiphoton ionization of (+)-fenchone. ARPES (top) and PECD (middle) for 408.5 nm 1 × 10^{13} W cm^{-2}, 404 nm 5 × 10^{12} W cm^{-2}, 398 nm 5 × 10^{12} W cm^{-2} and 396 nm 2 × 10^{13} W cm^{-2} pulses. The light propagation axis is horizontal and the radius extends from 0 to 5 eV. The continuous circle shows the expected electron energies for ionization from the HOMO, and the dashed circle from the HOMO−1. The PECD is set to zero when the ARPES is below 0.25% of its maximum. The associated PES, MPPECD and Legendre coefficients are plotted in the bottom graphs.

offers *a priori* an easier comparison with calculations, avoiding effects due to the transition *via* intermediate states and their possible alignment. On the other hand, multiphoton ionization makes PECD measurements possible directly using femtosecond laser pulses without the need of the intermediate HHG step, offering an interesting alternative. In order to compare the two techniques, we studied the multiphoton ionization of fenchone by the same 400 nm laser pulses as the ones used in HHG. We focused the circularly polarized laser pulses with a 50 cm focal lens into the VMIS. Two-photon absorption resonantly excites the molecules into a 3s-Rydberg state (of energy 6.11 eV according to TDDFT calculations and 5.9 eV according to absorption measurements[21]), facilitating subsequent ionization. The acquisition procedure was similar to that used for XUV ionization. The images were analyzed by setting $N = 5$ in eqn (1) in order to account for absorption of up to 5 photons in above-threshold ionization.

In order to evaluate the influence of the resonance in the PECD signal, we performed measurements at different laser wavelengths. Fig. 5 shows the ARPES and PECD obtained using laser pulses at 408.5 nm (3.04 eV) and 1×10^{13} W cm^{-2}, 404 nm (3.07 eV) and 5×10^{12} W cm^{-2}, 398 nm (3.12 eV) and 5×10^{12} W cm^{-2}, 396 nm (3.14 eV) and 2×10^{13} W cm^{-2}. The laser intensities are slightly different but we have observed that at fixed wavelength, the PECD remains independent of the laser intensity within the 0.5 to 2×10^{13} W cm^{-2} range, in good agreement with previous measurements.[20] The ARPES is constituted of three main peaks, corresponding to 3- and 4-photon ionization of the HOMO ($4\omega_0(0 \rightarrow 0)$ and $4\omega_0(0 \rightarrow 0)$, around 0.5 eV and 3.5 eV electron energy respectively), and 4-photon ionization of the HOMO−1 ($3\omega_0(0 \rightarrow 1^*)$ around 1.5 eV). These peaks shift to higher energy as the laser wavelength decreases, as expected. Apart from that, the ARPES does not show strong sensitivity to the laser wavelength within the scanned range. This indicates that the electron spectrum is not very sensitive to the details of the Rydberg excitation. At 408.5 nm, the two-photon absorption excites the fenchone molecules in the lower part of the 3s band, while at 396 nm the excitation lies in the upper part of the 3s, inducing about 200 meV of vibrational excitation, and overlaps with the beginning of the 3p band calculated at 6.56 eV and measured at 6.25 eV.[21] Stark shifts may bring the 3p band out of the photoexcitation spectrum, such that the wavelength scan that we performed may not enable changing the electronic character of the resonance. Our measurements show that the level of vibrational excitation of the resonant intermediate state does not affect the ARPES. The Rydberg and ionic potential energy surfaces are generally quasi-parallel. The vibrational distribution of the ionic populations thus tends to be the same as the Rydberg states, following the propensity rule $\Delta\nu = 0$. As a consequence, the photoelectron associated to different vibrational states of the ion all have the same kinetic energy, whatever the level of vibrational excitation of the Rydberg states.[42] This prevents the resolution of vibrational effects in REMPI photoelectron spectra. Changing the photon energy by δE thus does not result in a shift of the photoelectron spectrum by $3\delta E$, as in a direct 3-photon ionization, but by δE, since only the last photon of the 2 + 1 process plays a role. The photon energy difference between 396 nm and 408.5 nm is $\delta E = 92$ meV, which agrees reasonably well with the measured energy shift of the 2 + 1 peak ($\delta E^{exp} = 115$ meV) and 2 + 2 peaks ($\delta E^{exp} = 250$ meV). Since PECD is *a priori* very sensitive to vibrational excitation of the cation, one can wonder if it will reveal any feature connected to deviations of the $\Delta\nu = 0$ propensity rule in the REMPI process.

The angular resolved PECD shows a remarkable stability as a function of the laser wavelength. The PECD is strong for the $3\omega_0(0 \rightarrow 0)$ and $4\omega_0(0 \rightarrow 0)$ channels. By contrast, no measurable PECD is detected from the HOMO−1 despite the presence of a significant $4\omega_0(0 \rightarrow 1^*)$ peak in the ARPES. The PECD associated to the 3 and 4 photon ionization of the HOMO have different angular distributions. This is reflected by the evolution of the Legendre coefficients with electron energy (Fig. 5, bottom). The $3\omega_0(0 \rightarrow 0)$ component has a dominant b_1 character (maximum around 0°) while the $4\omega_0(0 \rightarrow 0)$ shows stronger b_3 with an opposite sign (maximum around 45°). These results could be interpreted as an indication that PECD in fenchone is mainly imprinted by the absorption of a single or two photons from the excited state. However, this is not a general effect, since previously reported 2 + 1 REMPI measurements in camphor, an isomer of fenchone with very similar structure of the HOMO, showed a strong b_3 component.[25]

The maximum MPPECD of the $3\omega_0(0 \rightarrow 0)$ transition is around −13% at all wavelengths. It thus has a similar amplitude and the same sign as the one recorded in single photon absorption (see Fig 4). While the photoelectron peak shifts linearly with photon energy, the MPPECD remains maximized at 0.5 eV, with a sign change of b_3 around this energy. This insensitivity to the ionizing photon wavelength indicates that the strong peak of the MPPECD between 0 and 1 eV is the effect of the electron kinetic energy, as established in near threshold single-photon absorption.[17] There is no signature of vibrational excitation with an alternating asymmetry like the ones we recently observed in limonene molecules.[43] This indicates that the $\Delta v = 0$ propensity rule for ionization from the 3s Rydberg states is maintained in the MPPECD of fenchone, and that the magnitude of this MPPECD is mainly fixed by the kinetic energy of the photoelectron and not affected by the variation of the chiral potential induced by the vibrational excitation in the cation.

For the $4\omega_0(0 \rightarrow 0)$ transition, the MPPECD maxima are upshifted by 100–200 meV with respect to the PES peaks. The MPPECD reaches −8% at 404 nm, while it is only −6% at 408.5 nm and 398 nm. These values, which are peak values of the MPPECD, are slightly higher than those previously reported, which were averaged over the FWHM of the electron energy distribution.[44] They can be compared to the +4.78% PECD recorded in single absorption of a 12.4 eV photon ionizing the HOMO.[17] The MPPECD at $4\omega_0$ is stronger, with an opposite sign. Obviously, although the photoelectron has the same kinetic energy and is associated to the same ionic state, it does not experience the same scattering in the chiral potential. In the ATI picture, the absorption of the fourth photon occurs when the electron is already slightly away from the ionic core, when most of the scattering into the molecular potential has happened. Since this scattering, and the associated interference of the outgoing partial waves, is what sets the value of the PECD, the MPPECD is mostly determined by the 2 + 1 process, and only slightly modified by the absorption of one additional photon. We have observed a similar sign change of the MPPECD relative to the single photon PECD in the 2 + 1 and 2 + 2 ATI peak in limonene.[43] Theoretical investigations are required to confirm this hypothesis as well as to establish the sensitivity of MPPECD to vibrational excitation.

2.3 Above-threshold and tunnel ionization

By enabling direct measurements with femtosecond laser pulses, resonance-enhanced multiphoton ionization provides a valuable alternative to single-photon absorption for time-resolved PECD studies. However the necessity to hit a resonance sets a limitation to the generality of the method. Furthermore, the influence of the resonance, and the probable alignment of the corresponding intermediate state, on the measured PECD remain unclear, and the comparison to theory is more complex. We thus investigated the possibility to perform PECD measurements using high-order above threshold ionization, in which the influence of resonances should be much weaker, or in the tunneling regime. The ionization regime can be characterized by the Keldysh parameter, $\gamma = \sqrt{\frac{I_p}{2U_p}} = \omega_0\sqrt{\frac{I_p}{2E^2}}$, where $I_p = 8.72$ eV is the ionization potential of fenchone, U_p the ponderomotive potential, and E the amplitude of the laser electric field. In a semi-classical picture of tunnel ionization, γ is the ratio between the time taken by the electron to go through the potential barrier and the laser period. Low values of $\gamma(<1)$ are characteristic of the tunneling regime, in which the barrier is quasi-static during ionization. By contrast the $\gamma > 1$ regime is the multiphoton ionization regime.

In the REMPI measurements discussed above, the Keldysh parameter was typically 3.8 ($I \sim 2 \times 10^{13}$ W cm^{-2} at 400 nm). In order to decrease γ, we can increase the laser intensity and/or wavelength. Fig. 6 shows the evolution of the ARPES and PECD decreasing γ from 7.6 to 0.6. Increasing the laser intensity at 400 nm increases the cutoff of the ATI spectrum, as seen in the ARPES reaching 12 eV in Fig. 6(b) compared to 5 eV in Fig. 6(a). The ARPES is constituted of two series of peaks, associated to photoionization of the HOMO and HOMO−1. While the $4\omega_0(0 \rightarrow 1^*)$ ionization did not show any PECD in the measurements shown in Fig. 5, a significant one is now observed from the HOMO−1 for higher ATI orders. Multiphoton ionization is thus able to reveal the PECD from multiple orbitals. The spectrum is less congested than in the single-photon ionization case, where much more channels are open. Indeed, multiphoton ionization is

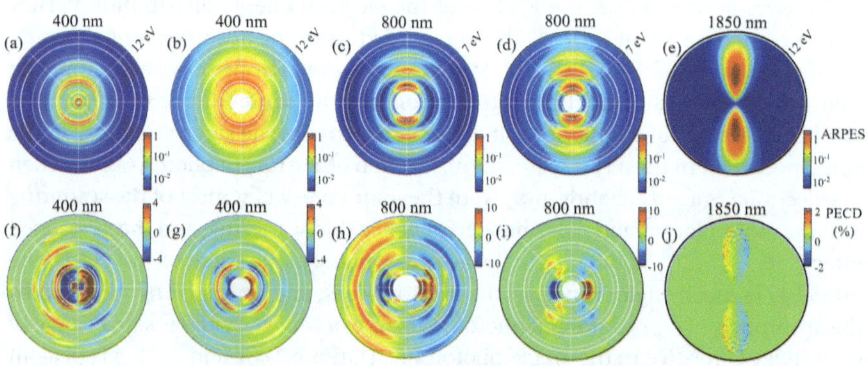

Fig. 6 Photoionization of (+)-fenchone from ATI to the tunneling regime. ARPES and PECD for $\lambda = 402$ nm pulses with $I \sim 5 \times 10^{12}$ (a and f) and $I \sim 4 \times 10^{13}$ (b and g) W cm^{-2}, $\lambda = 800$ nm pulses with $I \sim 9 \times 10^{12}$ (c and h) and $I \sim 1.3 \times 10^{13}$ (d and i) W cm^{-2}. The continuous circle shows the expected electron energies for ionization from the HOMO, and the dashed circle from the HOMO−1. The ponderomotive shift is taken into account. (e and j) Raw projection of the ARPES and PECD for $\lambda = 1850$ nm pulses with $I \sim 4 \times 10^{13}$ W cm^{-2}. The light propagation axis is horizontal and the radius extends from 0 to 12 eV in (a, b and e) and 0 to 7 eV in (c and d).

expected to favor ionization from orbitals that lie close enough to the HOMO, typically within an energy range smaller than the ionizing photon energy.[45]

This trend is confirmed by decreasing the photon energy from 3.1 eV to 1.55 eV (Fig. 6(c) and (h)). A single frequency comb appears in the ARPES and PECD, corresponding to ionization from the HOMO. The PECD remains quite strong, in the $\pm 5\%$ range even for the highest detected peak associated to the absorption of 10 photons. When the laser intensity increases, the ARPES becomes more sharply peaked along 90°, *i.e.* around the laser polarization plane, and the PECD decreases. This decrease is the signature of a stronger influence of the laser field on the outgoing scattering electron compared to the chiral potential.

The pure tunnel-ionization regime is reached by further increasing the laser wavelength to 1850 nm (Fig. 6(e) and (j)). The ARPES is very sharp, and a significant PECD is measured over the whole energy range, up to energies corresponding to the absorption of more than 30 photons. No contribution from the inner orbitals appears, since their ionization probability decays exponentially with the ionization potential. This measurement demonstrates that PECD can be observed even far away from the ionization threshold. This regime is usually described by the strong-field approximation, which consists in neglecting the influence of the ionic potential on the electron dynamics after tunneling. Our results show the failure of this approximation and reveal the influence of the electron scattering in the chiral molecular potential, as recently predicted by Dreissigacker and Lein.[46] We have recently shown that this could be understood within a purely classical picture of ionization through classical trajectory Monte-Carlo simulations.[43]

2.4 Conclusions

Photoelectron circular dichroism can be measured in all ionization regimes, from the single-photon to tunnel ionization, using ultrashort light pulses. The number of open channels depends on the ionization regime, offering complementary pictures of the chiral molecules. All energetically permitted channels are open in XUV one-photon ionization, and the direct transition to the continuum enables the probing of the vibrational state of the ion.[11] In the perspective of dynamical measurements, for instance in the case of a vibrational wavepacket in the ground electronic state, the high number of ionization channels can be considered as an advantage, since it would enable measuring the response of different molecular orbitals to the nuclear dynamics. On the other hand the channel identification in the ARPES and PECD can be complex. Coincidence electron-ion detection is a solution to this issue, enabling (partial) separation of the different channels by their different dissociative ionization pathways. In comparison, REMPI has the advantage to strongly favour resonant channels, even though several resonant channels can co-exist as we have seen in fenchone. The drawback is the complexity of the interpretation: the resonant character of the ionization may be modulated by the ongoing molecular dynamics, introducing an additional parameter to the sensitivity of the experiment. As the driving wavelength and intensity increase towards the tunneling regime, the number of open channels decreases, reflecting the exponential decay of tunnel ionization with increasing ionization potential. In the tunneling regime a weak PECD remains despite the 10^8 V cm^{-1} electric field. This PECD purely originates from the HOMO. This configuration could offer a simple picture for dynamical measurements, but has the disadvantage of showing a rather low sensitivity and of being based on the use of strong laser fields, which can significantly perturb the probed molecule.

3 Time-resolved photoelectron circular dichroism

The resonance-enhanced multiphoton ionization of fenchone studied above was carried out through the excitation of Rydberg states around 6.1 eV. We conducted time-resolved measurements to follow the ultrafast relaxation of these Rydberg states and to investigate the sensitivity of PECD as a probe of the molecular dynamics. After summarizing the results of our first pump–probe study in which we used a single photon process for both the pump and probe steps,[1] we investigate the case of multiphoton absorption.

3.1 Single photon excitation and single-photon ionization

The first (3s) Rydberg states of fenchone lie slightly above 6 eV (Fig. 7). We thus generated the fourth harmonic of our 800 nm pulses to produce 6.17 eV (201 nm) pulses with 30 meV bandwidth and ~2 μJ energy. The pulses were linearly polarized, and were focused around 5×10^{11} W cm^{-2} by a lens into the VMIS.

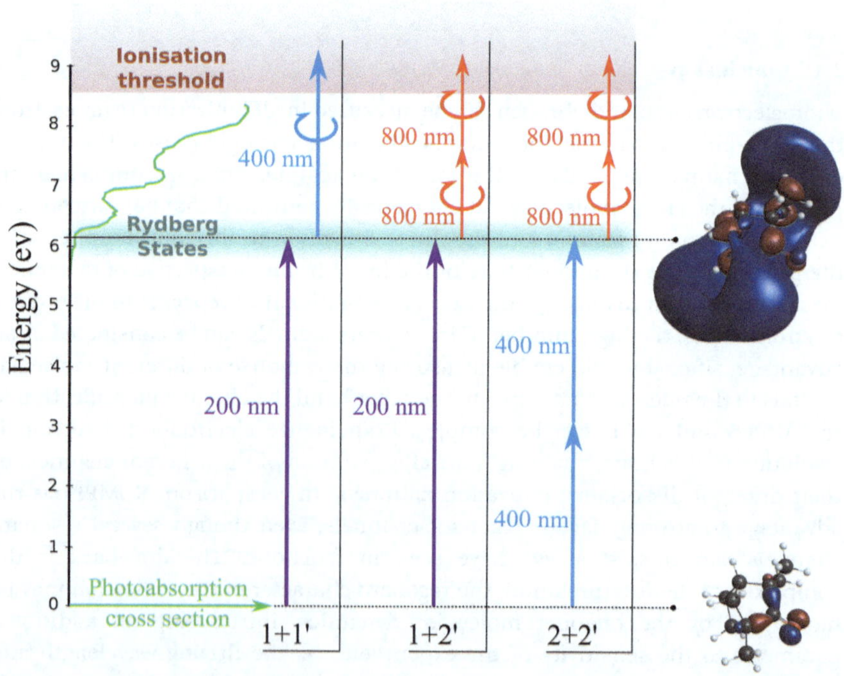

Fig. 7 Pump and probe configuration for time-resolved photoelectron circular dichroism in fenchone molecules. The right panel shows the highest-occupied and 3s Rydberg state molecular orbitals.

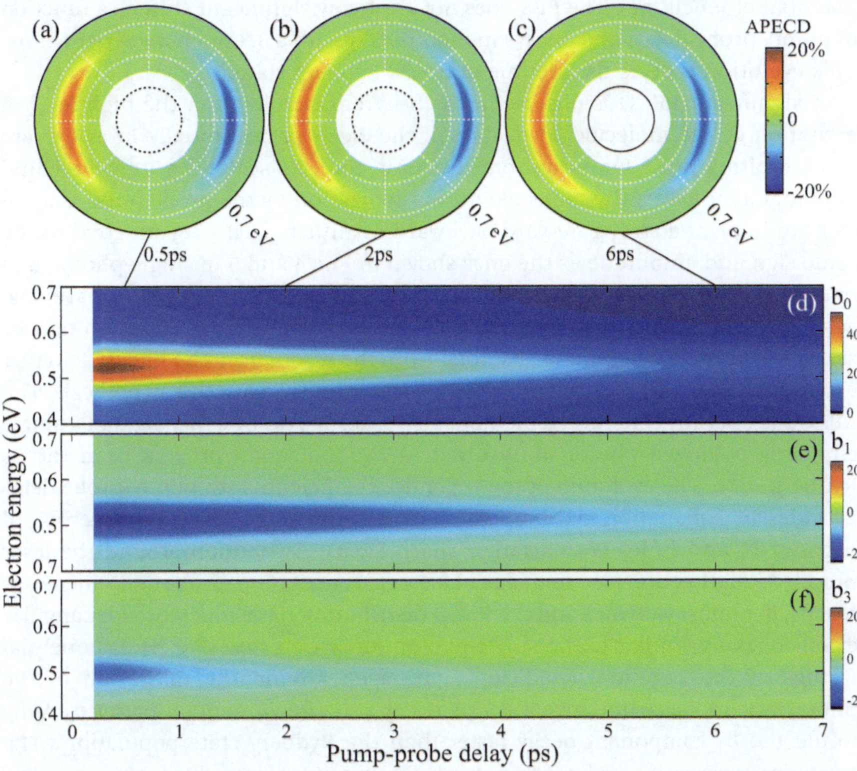

Fig. 8 Time-resolved PECD measurements in fenchone in a 1 + 1′ configuration (201 nm linearly polarized pump and 404 nm circularly polarized probe). The top polar plots show the APECD recorded at three pump probe delays: 0.5 ps (a), 2 ps (b) and 6 ps (c). The light propagation direction is horizontal and the radius extends from 0 to 0.7 eV. The lower contour plots depict the evolution of the Legendre coefficients b_0 (d), b_1 (e) and b_3 (f) as a function of electron kinetic energy and pump probe delay.

Enantiopure fenchone molecules were injected by a pulsed Even–Lavie valve operating at a 1 kHz rate and backed with 12 bars of He. For the probe, we generated 3.08 eV (404 nm) pulses with 35 meV bandwidth by second harmonic generation, and sent them in a zero order quarter waveplate to convert their polarization state to circular. The pulses were focused into the VMIS at $\sim 1 \times 10^{11}$ W cm^{-2} with an independent lens. The temporal resolution of the experiment was limited by the dispersion of the focusing optics and vacuum window, leading to a 170 fs cross-correlation time between the pump and probe. For each pump–probe delay, the probe laser helicity was switched every 45 000 laser shots to record a series of 8 left and 8 right images. The pump–probe delay was then scanned consecutively ten times to smooth out slow drifts in the experiment. Background images corresponding to the probe pulse alone were recorded for each helicity and subtracted from each image.

Fig. 8 shows the evolution of the PES and APECD (eqn (4)) as a function of pump–probe delay. The PES is centered around 0.52 eV, which corresponds to a 1 + 1′ transition in which all the excess energy goes to the electron (KE = $E_{pump} + E_{probe} - I_p = 0.53$ eV, where $I_p = 8.72$ eV is the experimental vertical ionization potential of fenchone). The PES does not show any significant shift as a function of pump–probe delay, and decays monotonically with a 3.3 ps characteristic time. This evolution reflects the lifetime of the 3s Rydberg states.

A significant PECD is observed, which is remarkable given the high level of excitation of the molecule: even though the electron lies initially in a Rydberg state, orbiting quite far away from the ionic core, its scattering into the chiral potential as it leaves the molecule during ionization by the probe pulse is sufficient to imprint a strong forward/backward asymmetry. This asymmetry has the same sign and amplitude as the ones shown in Fig. 4 and 5 in single-photon and one-color REMPI respectively. Our PECD studies on static molecules have suggested that in REMPI, the effective order of the Legendre decomposition may be set by the number of photons absorbed from the resonance. The linear + circular experiment that we perform here gives an opportunity to test this hypothesis. The APECD is not only constituted of b_1 but also b_3 terms, despite the fact that a single circularly polarized photon is involved in the ionization process from the 3s Rydberg state. The evolution of the Legendre coefficients with pump–probe delay provides an explanation to this apparent paradox. At short delays, the interplay of negative b_1 and b_3 leads to a rather sharp PECD distribution around the laser propagation direction (see snapshot at 0.5 ps). As the pump–probe delay increases to 6 ps, b_3 almost vanishes and the PECD distribution has a much broader angular extension, very similar to the single-photon ionization case (Fig. 4). Exponential fittings of the Legendre coefficients (averaged around the maximum of the photoelectron spectrum) give a 3.9 ps decay time for b_1 and 1.4 ps for b_3. Why would the b_3 component decay faster than the Rydberg state population? The excitation process by the linearly polarized pump pulses introduces an anisotropy in the sample, because the excitation probability is maximized for a given molecular orientation. This anisotropy in the molecular excitation means that the ionized molecules are not randomly oriented. Orientation effects have been shown theoretically to strongly modify the angular structure of PECD,[46] and probably cause the appearance of higher order Legendre polynomials in our experiment. The molecules being at non-zero temperature, they rotate with time. This rotation results in a loss of the excitation orientation after a certain time, determined by the moment of inertia of the molecules and rotational temperature of the gas jet.[47]

The PECD can be quantified by integrating the Legendre coefficients around the peak of the PES and calculating the multiphoton MPPECD defined in eqn (5), where the sum is here restricted to its first two terms since the transition is a two-photon process. The MPPECD is found to experience a fast decrease with a minimum around 0.4 ps, followed by an increase to reach the limit value of 14.6%. These results can be found in another publication.[1] If PECD was simply sensitive to the population decay of the Rydberg states, then the MPPECD, which is a normalized quantity, should remain constant with the pump–probe delay. This evolution is thus a first demonstration of the sensitivity of TRPECD to vibronic dynamics.

3.2 Single photon excitation and two-photon ionization

Our measurements on static molecules have illustrated the dependence of the PECD angular distributions on the number of ionizing photons. Following this

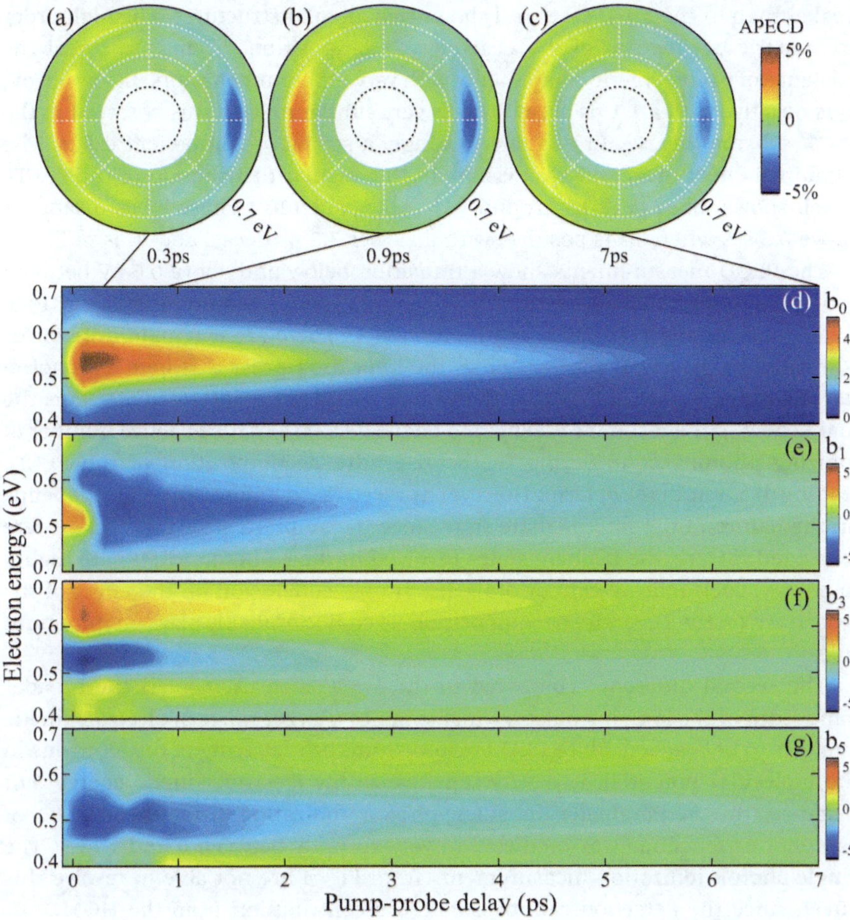

Fig. 9 Time-resolved PECD measurements in fenchone in a 1 + 2′ configuration (201 nm linearly polarized pump and 800 nm circularly polarized probe). The top polar plots show the APECD recorded at three pump–probe delays: 0.3 ps (a), 0.9 ps (b) and 7 ps (c). The light propagation direction is horizontal and the radius extends from 0 to 0.7 eV. The lower contour plots depict the evolution of the Legendre coefficients b_0 (d), b_1 (e), b_3 (f) and b_5 (g) as a function of electron kinetic energy and pump–probe delay.

observation, we have repeated the pump–probe measurements of the relaxation of the 3s Rydberg states using two 800 nm photons as a probe (1 + 2′ configuration). This change is accompanied by an improvement of the temporal resolution of the experiment (~100 fs cross correlation time), since the 800 nm pulses are only 30 fs long. The two-photon ionization here is a REMPI process, with a strong resonance around 7.7 eV to the 3d Rydberg states mixed with the second valence state (Fig. 7).

Fig. 9 shows the temporal evolution of the Legendre coefficients as well as snapshots of the APECD at three pump–probe delays. The PES (b_0) is centered at the same energy as in the 1 + 1′ scheme (0.52 eV) but is significantly broader, reflecting the broader bandwidth of the probe pulses. The signal decays in 3.2 ps, in agreement with the 3s Rydberg lifetime measured above. The APECD is much weaker than in the 1 + 1′ case, and shows rich angular structures. The high-order asymmetric coefficients of the Legendre decomposition, b_3 and b_5, reveal the existence of two energetic components with very different behaviors. Below 0.6 eV, b_3 is negative with a 1.1 ps decay. This is very similar to what was observed in the 1 + 1′ scheme (Fig. 8). In this energy range, a strong b_5 component exists. The combined effect of b_3 and b_5 is visible in the angular distribution of the PECD, which shows lobes peaking around $\theta \approx \pm 60°$ and 120°. These lobes disappear above 0.6 eV, where b_3 is positive, with a slower 3.3 ps decay, and b_5 is null.

The PECD measurements show a transition below and above 0.6 eV between two energy components separated by ~100 meV, which are invisible in the PES. What could be the origin of these components? There are two main differences compared to the 1 + 1′ scheme. First, the ionization process is here a 2-photon REMPI. In the 3-photon REMPI with the 3s Rydberg as the intermediate state, the main effect of the resonance seemed to be the decrease of the effective number of ionizing photons contributing to the Legendre decomposition of the PECD, leading to a dominant b_1 term. However as we stressed before, this effect depends on the nature of the intermediate state since in camphor, strong b_1 and b_3 were measured.[20] Here, the Rydberg states involved in the 2-photon REMPI are higher in energy, such that we cannot draw any simple conclusion on the effect of the resonance on the Legendre decomposition, even if it probably plays an important role.

The second difference compared to the 1 + 1′ measurement is the broader bandwidth of the ionizing pulses, which enables a larger range of electron kinetic energies to be reached. Since PECD results from the scattering of the electrons in the molecular potential, it is very sensitive to the electron kinetic energy. For instance, the b_1 parameter in single-photon ionization from the HOMO of camphor or fenchone was predicted to vary abruptly between 0 and 1 eV.[17] The single-photon ionization measurements from Fig. 4 are not able to resolve this effect, since the detection cannot isolate the contribution from the HOMO. By contrast the 2 + 1 REMPI ionization measurements at 400 nm show a clear electron kinetic energy effect.

The 1 + 2′ scheme selects a well defined ionization channel through the excitation and REMPI detection. It is thus likely that the transition observed at 0.6 eV is a near-threshold kinetic energy effect. In that case, why would the different component show a different time-dependence? The low energy part of the b_3 reflects the decay of excitation anisotropy, meaning that this component is enhanced by the partial selection of molecular orientations operated by the excitation process, as observed in the 1 + 1′ scheme. The high energy component is not affected by this decay, which shows that it is more robust to orientation averaging. These observations provide precious information to understand and to model the electron scattering in the chiral molecular potential, near the ionization threshold.

In addition to the picosecond dynamics, the time-resolved PECD measurements show a faster and spectacular evolution in the first few hundreds of femtoseconds (see Fig. 9(e), with a rapid sign change of the recorded asymmetry). As in the 1 + 1′ case, this demonstrates the high sensitivity of the technique. In order to achieve optimal resolution of these fast dynamics, we performed 2 + 2′ measurements through the same Rydberg resonances using shorter laser pulses.

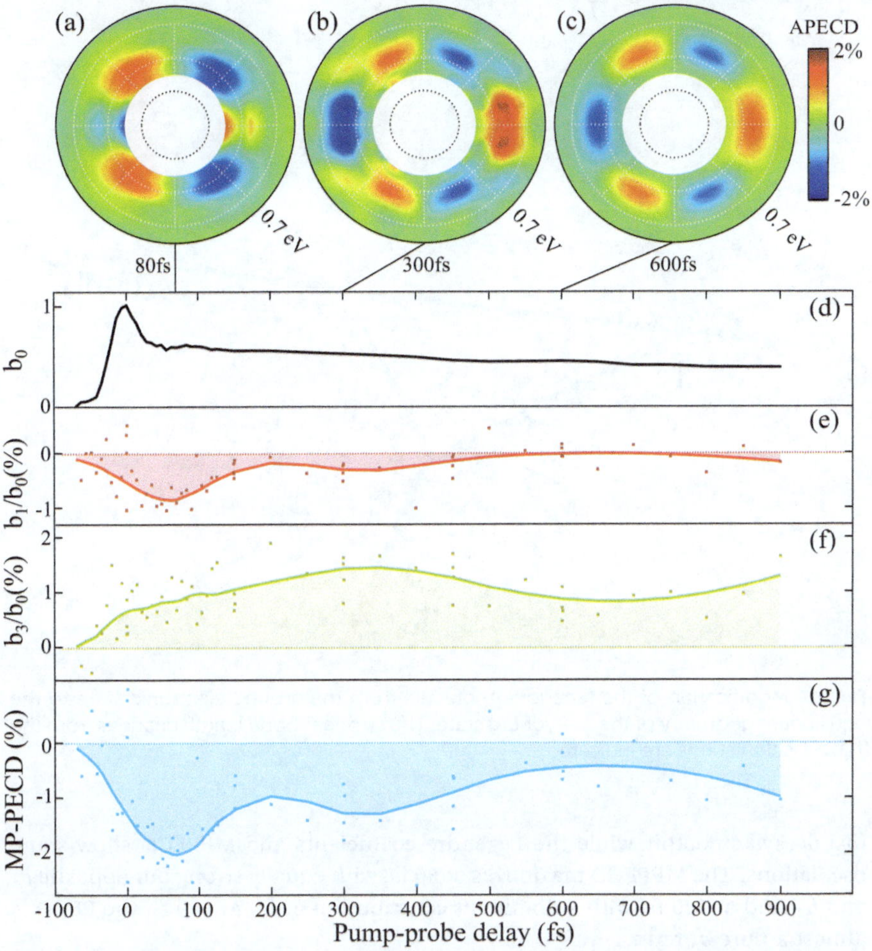

Fig. 10 Time-resolved PECD measurements in fenchone in a 2 + 2′ configuration (400 nm linearly polarized pump and 800 nm circularly polarized probe). (a–c) APECD averaged around three pump–probe delays: 80 ± 20, 300 ± 50 and 600 ± 50 fs. The light propagation direction is horizontal and the radius extends from 0 to 0.7 eV. The lower plots depict the temporal evolution of the PES b_0 (d), normalized Legendre coefficients b_1/b_0 (e), b_3/b_0 (f), and MPPECD (g).

3.3 Two-photon excitation and two-photon ionization

For this last TRPECD experiment, we used linearly polarized 400 nm pulses with 9 nm bandwidth to photoexcite fenchone molecules by resonant two-photon absorption, and ionized with circularly polarized 800 nm pulses. The cross correlation time was below 50 fs in this experiment. The fenchone source here was the continuous nozzle described in the static PECD measurements. The overall PECD is much weaker than in the 1 + 1′ and 1 + 2′ cases, reaching a maximum value of 2%. The angle-resolved APECD show a very fast evolution (Fig. 10(a)–(c)). This evolution can be quantified by monitoring the Legendre coefficients averaged over the width of the PES peak, as a function of pump–probe delay. The PES

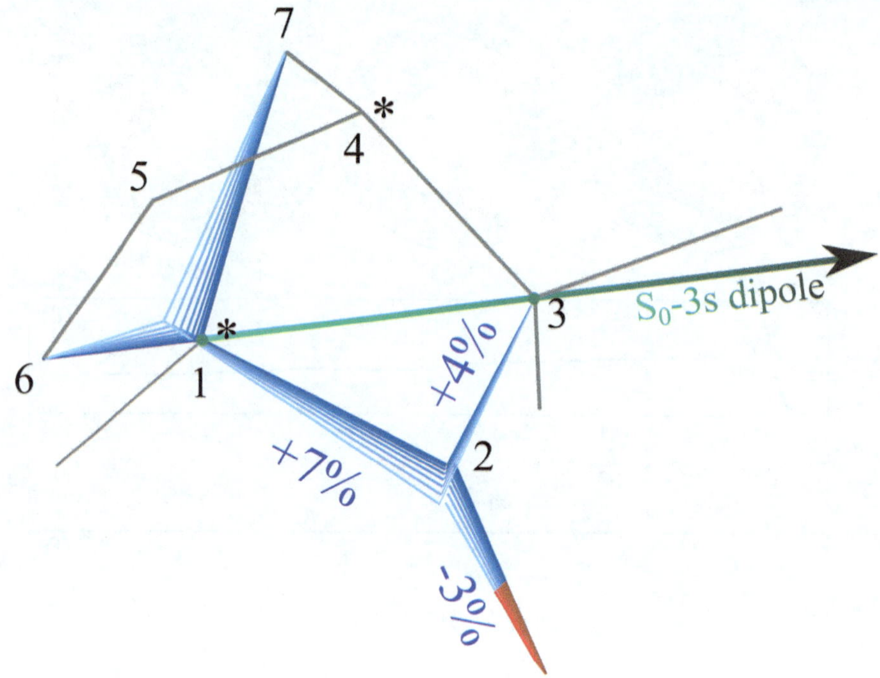

Fig. 11 Modification of the fenchone molecule from the ground electronic state to the equilibrium geometry of the 3s Rydberg state. The stronger bond length changes from the TDDFT calculations are indicated.

(b_0) decays smoothly while the Legendre coefficients and MPPECD show some oscillations. The MPPECD maximizes at 80 fs, with equally strong but opposite b_1 and b_3, and at 320 fs, with a dominant contribution of b_3. At 600 fs, the PECD is almost a pure b_3 term.

The drastic change of the b_1/b_3 ratio between 1 + 2′ and 2 + 2′ is quite fascinating and opens interesting questions that need to be addressed both experimentally and theoretically:

- With a 2-photon excitation, a stronger selectivity is expected in the orientation of the excited molecules. Does this type of alignment effect enhance the b_3 contribution?

- A 2-photon excitation means that antisymmetric vibrational modes can be excited. Does this type of excitation in the 3s Rydberg states enhance the b_3 contribution compared to the fully symmetric vibrational wavepacket created in one-photon excitation?

These fast oscillations of the PECD reflect rapid changes in the molecular geometry. We have performed TD-DFT calculations to determine the dominant motion of the wavepacket launched by the pump pulse on the 3s potential energy surface. The main gradient of this surface corresponds to the C1–C2 stretching, in the vicinity of the carbonyl group. The main modifications of bond lengths as the molecule goes from the ground electronic state to the equilibrium geometry of the 3s Rydberg state are indicated in Fig. 11. We expect that this main vibrational motion will be accompanied by energy transfers to other vibrational modes, as well as to internal conversion onto lower valence states. The detailed understanding of the PECD signal would require a complete calculation of the wavepacket dynamics, which is far beyond the scope of this work. Let us stress that the dynamics revealed by the TRPECD measurements are invisible in the TRPES signal, confirming the relevance of photoelectron circular dichroism as a probe of ultrafast dynamics of chiral molecules.

4 Conclusions

We have performed an extended investigation of photoelectron circular dichroism in the photoionization of enantiopure fenchone molecules in different ionization regimes, and found that PECD is a universal effect. Each regime has its specificities in terms of signal strength and ionization channel sensitivity, offering complementary looks at the chiral molecules. We have used time-resolved PECD to track the ultrafast relaxation of a Rydberg wavepacket, proving that this observable could reveal ultrafast dynamics that is invisible in time-resolved photoelectron spectra. The comparison of different pump and probe schemes shows spectacular differences in the angular distribution of the PECD, which are an additional illustration of its sensitivity. These first results mark the birth of ultrafast time-resolved measurements of chirality in the gas phase. Understanding the outcome of the measurements will require an extensive theoretical work, but the parametric study that we have performed should enable determining what key parameters are at play. A essential aspect of future experiments will be to increase the temporal resolution to track the fastest vibronic dynamics with high contrast.

Several other ultrafast chiral discrimination techniques have been developed very recently, that could provide complementary information on the dynamics of chiral molecules. First, high-order harmonic generation by elliptical laser pulses in chiral molecules was shown to be very sensitive to the handedness of the molecules.[48] This technique, which has intrinsic attosecond resolution, could lead to huge contrasts in pump–probe measurement.[49] Second, we have recently shown that chiral wavepackets could be probed with linearly polarized light. We have measured a forward/backward asymmetry in the photoionization of fenchone molecules excited by circularly polarized light and ionized by linear radiation. This process emerges from a macroscopic asymmetry in the photoexcited wavepacket, an effect we called photo excitation circular dichroism. PXCD is an interferometric process and is a very good probe of ultrafast molecular relaxations.

One question remains open: what will be the sensitivity of these different observables to valence state dynamics? The wavepacket spreading over the reaction coordinates often blurs the photoelectron spectra and prevents resolving the dynamics from TRPES. Will TRPECD, chiral HHG or TRPXCD be able to overcome this issue? The very high sensitivity of these methods to vibrations and electronic states leads us to be optimistic.

Acknowledgements

We thank R. Bouillaud and L. Merzeau for technical assistance and E. Constant and E. Mével for providing key apparatus used in the experiment. We acknowledge financial support of the European Union (LASERLAB-EUROPE 284464), the ANR MISFITS and the ANR Xstase. SB acknowledges the support of a NSERC Vanier Canada Graduate Scholarship.

References

1 A. Comby, S. Beaulieu, M. Boggio-Pasqua, D. Descamps, G. Garcia, F. Légaré, L. Nahon, S. Petit, B. Pons, B. Fabre, Y. Mairesse and V. Blanchet, 2016, submitted.

2 G. Prümper, J. Viefhaus, S. Cvejanović, D. Rolles, O. Gessner, T. Lischke, R. Hentges, C. Wienberg, W. Mahler, U. Becker, B. Langer, T. Prosperi, N. Zema, S. Turchini, B. Zada and F. Senf, *Phys. Rev. A*, 2004, **69**, 062717.

3 D. Patterson, M. Schnell and J. M. Doyle, *Nature*, 2013, **497**, 475–477.

4 C. Medcraft, R. Wolf and M. Schnell, *Angew. Chem., Int. Ed.*, 2014, **53**, 11656–11659.

5 B. Ritchie, *Phys. Rev. A*, 1976, **13**, 1411–1415.

6 N. Böwering, T. Lischke, B. Schmidtke, N. Müller, T. Khalil and U. Heinzmann, *Phys. Rev. Lett.*, 2001, **86**, 1187–1190.

7 S. Daly, I. Powis, G. A. Garcia, H. Soldi-Lose and L. Nahon, *J. Chem. Phys.*, 2011, **134**, 064306.

8 D. Dowek, M. Lebech, J. C. Houver and R. R. Lucchese, *Mol. Phys.*, 2007, **105**, 1757–1768.

9 L. Nahon, G. A. Garcia and I. Powis, *J. Electron Spectrosc. Relat. Phenom.*, 2015, 322–334.

10 I. Powis, C. J. Harding, G. A. Garcia and L. Nahon, *ChemPhysChem*, 2008, **9**, 475–483.

11 G. A. Garcia, L. Nahon, S. Daly and I. Powis, *Nat. Commun.*, 2013, **4**, 2132.

12 G. A. Garcia, L. Nahon, C. J. Harding and I. Powis, *Phys. Chem. Chem. Phys.*, 2008, **10**, 1628–1639.

13 S. Turchini, D. Catone, G. Contini, N. Zema, S. Irrera, M. Stener, D. Di Tommaso, P. Decleva and T. Prosperi, *ChemPhysChem*, 2009, **10**, 1839–1846.

14 M. Tia, B. Cunha de Miranda, S. Daly, F. Gaie-Levrel, G. A. Garcia, L. Nahon and I. Powis, *J. Phys. Chem. A*, 2014, **118**, 2765–2779.

15 M. Stener, G. Fronzoni, D. D. Tommaso and P. Decleva, *J. Chem. Phys.*, 2004, **120**, 3284–3296.

16 G. A. Garcia, H. Dossmann, L. Nahon, S. Daly and I. Powis, *Phys. Chem. Chem. Phys.*, 2014, **16**, 16214–16224.

17 L. Nahon, L. Nag, G. A. Garcia, I. Myrgorodska, U. Meierhenrich, S. Beaulieu, V. Wanie, V. Blanchet, R. Géneaux and I. Powis, *Phys. Chem. Chem. Phys.*, 2016, **18**, 12696.

18 C. S. Lehmann, N. B. Ram, I. Powis and M. H. M. Janssen, *J. Chem. Phys.*, 2013, **139**, 234307.

19 M. M. R. Fanood, M. H. M. Janssen and I. Powis, *Phys. Chem. Chem. Phys.*, 2015, **17**, 8614–8617.

20 C. Lux, M. Wollenhaupt, C. Sarpe and T. Baumert, *ChemPhysChem*, 2015, **16**, 115–137.

21 F. Pulm, J. Schramm, J. Hormes, S. Grimme and S. D. Peyerimhoff, *Chem. Phys.*, 1997, **224**, 143–155.

22 R. W. Schoenlein, S. Chattopadhyay, H. H. W. Chong, T. E. Glover, P. A. Heimann, C. V. Shank, A. A. Zholents and M. S. Zolotorev, *Science*, 2000, **287**, 2237–2240.

23 N. Cutić, F. Lindau, S. Thorin, S. Werin, J. Bahrdt, W. Eberhardt, K. Holldack, C. Erny, A. L'Huillier and E. Mansten, *Phys. Rev. Spec. Top.–Accel. Beams*, 2011, **14**, 030706.

24 E. Allaria, B. Diviacco, C. Callegari, P. Finetti, B. Mahieu, J. Viefhaus, M. Zangrando, G. De Ninno, G. Lambert, E. Ferrari, J. Buck, M. Ilchen, B. Vodungbo, N. Mahne, C. Svetina, C. Spezzani, S. Di Mitri, G. Penco, M. Trovó, W. M. Fawley, P. R. Rebernik, D. Gauthier, C. Grazioli, M. Coreno, B. Ressel, A. Kivimäki, T. Mazza, L. Glaser, F. Scholz, J. Seltmann, P. Gessler, J. Grünert, A. De Fanis, M. Meyer, A. Knie, S. P. Moeller, L. Raimondi, F. Capotondi, E. Pedersoli, O. Plekan, M. B. Danailov, A. Demidovich, I. Nikolov, A. Abrami, J. Gautier, J. Lüning, P. Zeitoun and L. Giannessi, *Phys. Rev. X*, 2014, **4**, 041040.

25 C. Lux, M. Wollenhaupt, T. Bolze, Q. Liang, J. Köhler, C. Sarpe and T. Baumert, *Angew. Chem., Int. Ed.*, 2012, **51**, 1–6.

26 A. Kastner, C. Lux, T. Ring, S. Züllighoven, C. Sarpe, A. Senftleben and T. Baumert, *ChemPhysChem*, 2016, **17**, 1119.

27 M. Y. Kuchiev, *JETP*, 1987, **45**, 404.

28 K. J. Schafer, B. Yang, L. F. DiMauro and K. C. Kulander, *Phys. Rev. Lett.*, 1993, **70**, 1599–1602.

29 P. B. Corkum, *Phys. Rev. Lett.*, 1993, **71**, 1994–1997.

30 K. S. Budil, P. Saliéres, M. D. Perry and A. L'Huillier, *Phys. Rev. A*, 1993, **48**, R3437–R3440.

31 P. Antoine, B. Carré, A. L'Huillier and M. Lewenstein, *Phys. Rev. A*, 1997, **55**, 1314–1324.

32 B. Vodungbo, A. Barszczak Sardinha, J. Gautier, G. Lambert, C. Valentin, M. Lozano, G. Iaquaniello, F. Delmotte, S. Sebban, J. Lüning and P. Zeitoun, *Opt. Express*, 2011, **19**, 4346–4356.

33 F. Willems, C. T. L. Smeenk, N. Zhavoronkov, O. Kornilov, I. Radu, M. Schmidbauer, M. Hanke, C. von Korff Schmising, M. J. J. Vrakking and S. Eisebitt, *Phys. Rev. B*, 2015, **92**, 220405.

34 A. Fleischer, O. Kfir, T. Diskin, P. Sidorenko and O. Cohen, *Nat. Photonics*, 2014, **8**, 543–549.

35 D. D. Hickstein, F. J. Dollar, P. Grychtol, J. L. Ellis, R. Knut, C. Hernández-García, D. Zusin, C. Gentry, J. M. Shaw, T. Fan, K. M. Dorney, A. Becker, A. Jaroń-Becker, H. C. Kapteyn, M. M. Murnane and C. G. Durfee, *Nat. Photonics*, 2015, 743.

36 A. Ferré, C. Handschin, M. Dumergue, F. Burgy, A. Comby, D. Descamps, B. Fabre, G. A. Garcia, R. Géneaux, L. Merceron, E. Mével, L. Nahon, S. Petit, B. Pons, D. Staedter, S. Weber, T. Ruchon, V. Blanchet and Y. Mairesse, *Nat. Photonics*, 2015, **9**, 93–98.

37 M. Piancastelli, *J. Electron Spectrosc. Relat. Phenom.*, 1999, **100**, 167–190.

38 V. Strelkov, *Phys. Rev. Lett.*, 2010, **104**, 123901.

39 A. Ferré, A. E. Boguslavskiy, M. Dagan, V. Blanchet, B. D. Bruner, F. Burgy, A. Camper, D. Descamps, B. Fabre, N. Fedorov, J. Gaudin, G. Geoffroy, J. Mikosch, S. Patchkovskii, S. Petit, T. Ruchon, H. Soifer, D. Staedter,

I. Wilkinson, A. Stolow, N. Dudovich and Y. Mairesse, *Nat. Commun.*, 2015, **6**, 5952.

40 G. A. Garcia, L. Nahon and I. Powis, *Rev. Sci. Instrum.*, 2004, **75**, 4989.

41 M. Schmidt, K. Baldridge, J. Boatz, S. Elbert, M. Gordon, J. Jensen, S. Koseki, N. Matsunaga, K. Nguyen, S. Su, T. Windus, M. Dupuis and J. Montgomery, *J. Comput. Chem.*, 1993, **14**, 1347.

42 V. Blanchet, K. Raffael, G. Turri, B. Chatel, B. Girard, I. A. Garcia, I. Wilkinson and B. J. Whitaker, *J. Chem. Phys.*, 2008, **128**, 164318.

43 S. Beaulieu, A. Ferré, R. Géneaux, R. Canonge, D. Descamps, B. Fabre, N. Fedorov, F. Légaré, S. Petit, T. Ruchon, V. Blanchet, Y. Mairesse and B. Pons, *New J. Phys.*, 2016, **18**, 102002.

44 C. Lux, A. Senftleben, C. Sarpe, M. Wollenhaupt and T. Baumert, *J. Phys. B: At., Mol. Opt. Phys.*, 2016, **49**, 02LT01.

45 A. Zhao, P. Sándor, T. Rozgonyi and T. Weinacht, *J. Phys. B: At., Mol. Opt. Phys.*, 2014, **47**, 204023.

46 I. Dreissigacker and M. Lein, *Phys. Rev. A*, 2014, **89**, 053406.

47 A. Blokhin, M. Gelin, E. Khoroshilov, I. Kryukov and A. Sharkov, *Opt. Spectrosc.*, 2003, **95**, 346–352.

48 R. Cireasa, A. E. Boguslavskiy, B. Pons, M. C. H. Wong, D. Descamps, S. Petit, H. Ruf, N. Thiré, A. Ferré, J. Suarez, J. Higuet, B. E. Schmidt, A. F. Alharbi, F. Légaré, V. Blanchet, B. Fabre, S. Patchkovskii, O. Smirnova, Y. Mairesse and V. R. Bhardwaj, *Nat. Phys.*, 2015, **11**, 654–658.

49 O. Smirnova, Y. Mairesse and S. Patchkovskii, *J. Phys. B: At., Mol. Opt. Phys.*, 2015, **48**, 234005.

Faraday Discussions

PAPER

Role of tunnel ionization in high harmonic generation from substituted benzenes

Dane R. Austin,*[a] Felicity McGrath,[a] Lukas Miseikis,[a] David Wood,[a] Peter Hawkins,[b] Allan S. Johnson,[a] Morgane Vacher,[c] Zdeněk Mašín,[d] Alex Harvey,[d] Misha Ivanov,[d] Olga Smirnova[d] and Jon P. Marangos[a]

Received 3rd May 2016, Accepted 4th July 2016
DOI: 10.1039/c6fd00116e

We theoretically study high-harmonic generation in toluene, *ortho*-xylene and fluorobenzene driven by a 1.8 μm ultrashort pulse. We find that the chemical substitutions have a strong influence on the amplitude and phase of the emission from the highest occupied molecular orbital, despite having a small influence on the orbital itself. We show that this influence is due to the tunnel ionization step, which depends critically on the sign and amplitude of the asymptotic part of the wave function. We discuss how these effects would manifest in phase-sensitive high-harmonic generation spectroscopy experiments.

1 Introduction

In high-harmonic generation (HHG) spectroscopy, strong-field ionization (SFI) creates a free electron that subsequently recombines with the cation. The emitted radiation encodes information about the cation and its evolution in the interval between ionization and recombination which is typically a fraction of a laser cycle. It has been used to probe static structure,[1] nuclear motion,[2,3] charge migration[4,5] and molecular vibrations.[6] Most HHG spectroscopy has used atoms or small molecules (typically di- or tri-atomic) driven by near-IR laser pulses with a corresponding cation evolution duration of ≈ 3 fs. Extended to larger systems, and driven by mid-infrared laser fields, HHG spectroscopy seems promising for studying fundamental processes in photochemistry such as the coupling of electronic and nuclear dynamics.[7,8] However, HHG involves an interplay of ionization, cation evolution and recombination in the presence of a strong laser field, and all three steps must be fully described to interpret experiments.

[a]Blackett Laboratory, Imperial College London, London, UK. E-mail: d.austin@imperial.ac.uk; Fax: +44 (0)20 7594 7714; Tel: +44 (0)20 7594 7520
[b]Department of Physics, Philipps-Universität Marburg, Renthof 5, D-35032 Marburg, Germany
[c]Department of Chemistry, Imperial College London, London, UK
[d]Max-Born Institute for Nonlinear Optics and Short Pulse Spectroscopy, Berlin, Germany

In molecules, SFI can create several cation eigenstates which contribute coherently to the high-harmonic emission. The initial eigenstate populations are determined by tunnel ionization rates, which are affected by the ionization potentials and the molecular geometry and orientation relative to the laser polarization. The various theories of tunnel ionization in molecules, such as MO-ADK[9] and MO-SFA,[10–12] all feature the same strong dependence on field strength and IP found in atomic tunnel ionization, as well as predicting that ionization will be suppressed along nodal planes of the molecular orbitals. The relative phases of the cation eigenstates have been studied in less detail. These phases may depend on electron correlations during ionization[13] and possible tunneling delays,[14,15] and determine the position of the hole left behind in the cation, necessary for predicting its subsequent evolution. In HHG spectroscopy, the relative phases of the cation states after ionization shape the spectrum through interference, but this must be disentangled from the propagation and recombination steps. Experiments[4] in CO_2 found that the relative phase was consistent with a quasi-static picture of the bound-state orbital wavefunctions tunnelling through a barrier. However this view may be complicated by laser coupling of the cation states.[16] In this case a dressed state picture is helpful; studies in iodoacetylene[5] found that the tunnel ionization launches the single lowest eigenstate of the cation dressed by a static electric field. Furthermore, by orienting the molecules, the effects of ionizing along the iodine or hydrogen end of the molecule could be compared and were found to significantly affect the subsequent population evolution driven by the laser field. The ability to selectively ionize along a given axis raises interesting possibilities for larger organic molecules, in which hydrogen atoms may be substituted by a wide range of atoms or groups.

We recently conducted an HHG spectroscopic comparison of benzene, its methyl-substituted forms toluene and the three isomers of xylene, and the singly halo-substituted fluorobenzene, bromobenzene and chlorobenzene. Benzene is particularly attractive for such a systematic scan of the available chemical substitutes because of the two highest occupied molecular orbitals (HOMOs)—which participate most strongly in HHG—only the HOMO is significantly altered by the substitution. The emission from the HOMO−1 is therefore expected to act as a reasonably stable reference. Practically speaking, benzene and its fluoro- and methyl-substituted forms have similar ionization potentials (8.45–9.25 eV), meaning that the appropriate laser intensities (those which cause sufficient ionization to observe harmonics, but do not fully deplete the ground state) are similar between the molecules, enabling comparison of harmonics measured under identical laser conditions. The boiling points of the molecules vary (from 80° to 156°) and hence achieve similar target densities requiring control of the temperature as the vapour is delivered to the vacuum. Our set-up[17] enables rapid switching between molecules and delivery of the vapour into the vacuum through a 200 μm diameter heated nozzle. We did not attempt to align or orient the sample; instead our aim was to explore what may be learned from HHG spectroscopy in unaligned samples. Although alignment and orientation of some benzene substitutes has been demonstrated,[18] for many large molecules, achieving alignment/orientation in a sample of sufficient density for HHG is likely to be unfeasible. One possible workaround would be to use the comparison of HHG spectra from molecules with different chemical substitutions to extract similar information to that obtained from aligned or oriented samples. A full interpretation of our unaligned experimental spectra is currently underway.

To interpret the experimental spectra, we developed a theoretical model incorporating various extensions to the strong-field approximation (SFA) appropriate for the molecules under study. Here, we exploit the ability of the SFA to unpick the steps of HHG to study the role of the substitution as a function of the relative orientation of the electric field and molecular axis. We focus on emission from the HOMO only, since it is more strongly affected by the substitution than the HOMO−1. This work is a necessary step in understanding the differences between the spectra from unaligned samples. We find that, in addition to the well known role of the photorecombination cross section,[19] the tunnelling step can be profoundly altered when tunnelling occurs along a substitution. This is because the substitution introduces additional outer lobes in the orbital which have the opposite sign and greater spatial extent than those of benzene. This can lead to a large phase and amplitude asymmetry between the emission caused by positive and negative half-cycles of the laser. We discuss how these asymmetries would manifest in a phase-sensitive HHG spectroscopy experiment with oriented molecules.

This paper is structured as follows: Section 2 presents the theoretical model, Section 3.1 studies the asymmetries in the HHG from toluene at specific orientations which select emission from the HOMO, and Section 3.2 looks at how these asymmetries manifest once all the orientations are considered. Sections 3.3 and 3.4 look at *ortho*-xylene and fluorobenzene respectively, which provide contrasting examples of how the shape of the HOMO around the substitute group affects HHG. Section 4 provides a summary and conclusion.

2 Numerical model

We use a form of the SFA[20] extended to high-harmonic generation in molecules by summing coherently over multiple cation states.[4,16,21–23] For completeness, our model also includes laser induced dynamics in the cation between ionization and recombination.[24]

In the following, boldface symbols represent spatial vectors, and an overbar over a quantity indicates it is a vector or matrix running over the cation states. For a particular molecular orientation, the high-frequency part of the dipole response at recombination time t_r is

$$\boldsymbol{D}(t_r) = \int_{-\infty}^{t_r} \tau^{-3/2} e^{iS(t_r,t_i)} \overline{\boldsymbol{d}}_r^\dagger(\boldsymbol{v}(t_r)) \overline{U}(t_r, t_i) \boldsymbol{E}(t_i) \cdot \overline{\boldsymbol{d}}_i(\boldsymbol{v}(t_i)) dt_i. \quad (1)$$

The equation is an integral over ionization times t_i and the integrand, a matrix product, can be read from right to left. The factor describing the angular dependence of the ionization rate is a column vector over cation states

$$\bar{\boldsymbol{d}}_i(\boldsymbol{v}) = (\boldsymbol{d}_i^{(1)}(\boldsymbol{v}), \boldsymbol{d}_i^{(2)}(\boldsymbol{v}), \ldots)^T \quad (2)$$

where $\boldsymbol{d}_i^{(n)}(\boldsymbol{v})$ is the ionization dipole of state n for momentum $\boldsymbol{v}$. The usual SFA definitions[20] apply; the kinetic momentum is $\boldsymbol{v}(t) = \boldsymbol{p} + \boldsymbol{A}(t)$ where $\boldsymbol{p}$ is the canonical momentum of the stationary phase

$$\boldsymbol{p} = \frac{-\int_{t_i}^{t_r} \boldsymbol{A}(t)\mathrm{d}t}{\tau}, \tag{3}$$

the trajectory duration $\tau = t_r - t_i$, and $\boldsymbol{E}(t)$ and $\boldsymbol{A}(t)$ are the laser electric field and vector potential respectively. The next term in the integrand of (1) is a matrix describing evolution of the cation states under the influence of the laser field. It is the solution of the Schrödinger equation

$$\frac{\mathrm{d}\overline{U}(t,t_i)}{\mathrm{d}t} = -i\left[\overline{I}_p - \boldsymbol{E}(t)\cdot\Delta\overline{\boldsymbol{\mu}}\right]\overline{U}(t,t_i) \tag{4}$$

$$\bar{U}(t_i, t_i) = \bar{I} \tag{5}$$

where the diagonal matrix $\bar{I}_p$ contains the ionization potentials $I_p^{(1)}$, $I_p^{(2)}$,... on its main diagonal, $\Delta\bar{\boldsymbol{\mu}}$ is a matrix of cation state dipole couplings (the dot product $\boldsymbol{E}(t)\cdot\Delta\bar{\boldsymbol{\mu}}$ runs over the spatial coordinates) and $\bar{I}$ is the identity matrix. The recombination step is described in (1) by $\bar{\boldsymbol{d}}_r(\boldsymbol{v})$, a column vector of recombination dipoles in the same format as $\bar{\boldsymbol{d}}_i(\boldsymbol{v})$, and † denotes the conjugate transpose. Finally the scalar exponential contains the standard SFA action

$$S(t_r, t_i) = -\int_{t_i}^{t_r} \frac{(\boldsymbol{p} + \boldsymbol{A}(t))^2}{2}\mathrm{d}t \tag{6}$$

and the $\tau^{-3/2}$ factor describes spreading of the continuum electron wavepacket.

The ionization dipole amplitude is the length-gauge plane-wave approximation

$$\boldsymbol{d}_i^{(n)}(\boldsymbol{v}) = \langle \boldsymbol{v}|(I - |\psi_n\rangle\langle\psi_n|)\boldsymbol{x}|\psi_n\rangle \tag{7}$$

where $|\psi_n\rangle$ is the Dyson orbital corresponding to cation state n. The projection operator in (7) arises from our use of the dressed form of the SFA, as described by equation (33) of Smirnova *et al.*[25] This is necessary for invariance with respect to translation of the molecule.

For the recombination dipole, we use field-free[26] photorecombination cross sections when these are available; this is consistent with quantitative rescattering theory.[19,27] The cross sections and Dyson orbitals for fluorobenzene and toluene are calculated using the R-matrix method[28] with modifications[29] that enable description of the high energy region crucial for HHG in the long-wavelength regime. We have employed the Static Exchange (*i.e.* Hartree–Fock) model. The calculation included a diffuse basis of continuum orbitals with angular momentum up to $L = 6$. The inclusion of the diffuse continuum functions has proved important for an accurate description of the exponential tail of the Dyson orbitals. For *ortho*-xylene, the photorecombination cross sections were unavailable and so we used the plane wave approximation, and focus on the role played by the tunnel ionization step. For these molecules, we used as approximate Dyson orbitals the natural orbitals (those that diagonalise the 1-electron density matrix) calculated with the complete active space self-consistent field (CASSCF) electronic structure method state-averaging (equally) over the two lowest-energy cationic eigenstates. We used the 6π orbitals as active with the 6-31G* basis set.

Our calculations use the five lowest lying cation states. These are well approximated by a single hole in the neutral (the Koopman picture[30,31]), and therefore throughout this work we equivalently refer to them by their corresponding molecular orbital. For the purposes of HHG, we checked this equivalence directly by comparing spectra in benzene, for which both the Dyson and natural orbitals were available.

The diagonal elements of the cation state dipole couplings $\Delta\bar{\mu}$ lead to dynamic Stark shifts and equal the difference between the dipole moments of the cation states and the neutral.[32–34] The off-diagonal elements lead to laser-driven transitions between the cation states. We calculate $\Delta\bar{\mu}$ directly from the orbitals as

$$\Delta\bar{\mu}_{\mathrm{m,n}} = \langle \psi_{\mathrm{m}} | \boldsymbol{x} | \psi_{\mathrm{n}} \rangle. \tag{8}$$

The results are all within 0.04 au of transition dipoles calculated from cation state wavefunctions.

We have tested our model for invariance with respect to translation of the molecule;[25] to this end inclusion of dynamic Stark shifts is essential. To facilitate comparison of the three steps in HHG between the molecules, we take the centroid of the benzene ring as a common origin in all calculations.

Macroscopic trajectory selection is modelled by integrating coherently across the plane of the laser focus, which is taken as spatially and temporally Gaussian. For the temporal integration we use the adiabatic approximation.[35] The macroscopic harmonic amplitudes are taken as the amplitude of the emission on-axis ($\boldsymbol{k}_\perp = 0$) and at the centre of each harmonic ($\omega = q\omega_1$), which strongly favours the short trajectories. In this approximation the spectra are independent of the laser waist size and pulse duration except for a scale factor.

3 Results and discussion

All simulations in this work use a laser wavelength of 1.8 μm and a peak intensity of 0.3×10^{14} W cm^{-2}, with coherent macroscopic averaging as described in the previous section.

We first introduce the properties of the molecular orbitals under consideration. Their ionization potentials and bonding character are given in Table 1. The HOMO and HOMO−1 of each molecule, denoted X and A, have π character. They are degenerate in benzene and split by 0.2–0.4 eV in the substituted molecules. The remaining three states have ionization potentials several electron volts higher. The HOMO and HOMO−1 dominate the HHG emission and are shown in Fig. 1 for all the molecules we consider. The figure also shows the coordinate system used throughout and the grids indicate the *yz* planes referenced in the discussion below. Note that the contour surfaces in Fig. 1 are drawn at the rather low value of $|\psi| = 0.02$ in order to make visible the additional lobes caused by the substitution. In fact the overall impact of the substitution on the HOMO is minor; this is shown in the left column by the projections of the electron density on to the plane of the benzene ring

$$\rho_{\mathrm{X}}^{\mathrm{proj}}(\boldsymbol{x}_\parallel) = \int \left| \psi_{\mathrm{X}}(x_\perp, \boldsymbol{x}_\parallel) \right|^2 \mathrm{d}x_\perp \tag{9}$$

where $x_\perp$ and $\boldsymbol{x}_\parallel$ are coordinates perpendicular and parallel to the benzene ring plane, respectively. It can be seen that the electron density is concentrated on the

Table 1 Ionization potentials and characters of the relevant cation states. These are taken from photoelectron spectra.[36–40]

Molecule	State/orbital	Character	Ionization pot. (eV)
Fluorobenzene	X	π	9.2
	A	π	9.8
	B	π	13.8
	C	σ	13.7
	D	σ	14.4
Toluene	X	π	8.8
	A	π	9.0
	B	σ	11.2
	C	σ	11.5
	D	π	12.0
ortho-Xylene	X	π	8.6
	A	π	9.0
	B	σ	11.0
	C	σ	11.2
	D	σ	11.8

benzene ring. A quantitative measure of the similarity of the HOMOs of benzene and its substituted forms is given by the cross-correlation (normalized overlap integral). This is 0.96, 0.95 and 0.98 for *ortho*-xylene, fluorobenzene and toluene, respectively.

The dipole coupling matrices $\Delta\bar{\mu}$ are presented in Table 2. The X and A states are only weakly polar (with respect to the neutral) and thus linear Stark shifting plays only a small role, producing a maximum energy shift of 0.3 eV in the methyl substituted species. A lower lying π orbital (B in fluorobenzene and D in toluene) is dipole coupled to X and A. At peaks of the laser field, this causes a dynamic quadratic Stark shift of the energies of these states of magnitude $\approx$0.3 eV. Because the effect is symmetric about the *xy* plane, it does not lead to asymmetry between molecules oriented or anti-oriented along the *z* axis, and so does not play a significant role here.

3.1 Toluene HOMO

We start by studying the behaviour of the emission produced by a single laser half-cycle as the molecule is rotated about an axis running through its centre perpendicular to the carbon–methyl bond. In this way, any asymmetry introduced by the presence of the methyl group can be easily detected. In the molecular frame, the electric field rotates in the *yz* plane indicated in Fig. 1. This is a nodal plane of the HOMO−1, and so emission from that orbital is forbidden. Fig. 2(a) and (b) show the result in polar form; zero degrees is defined to mean the ionizing electric field is antiparallel to the carbon-to-methyl bond *i.e.* the electron leaves through the methyl group and passes by it upon recombination. (We adopt this convention for all monosubstituted forms. For doubly substituted *ortho*-xylene the reference angle is taken as the bisector of the two bonds.) The amplitude and the phase is also shown at 20 eV in panel (c). We first consider the amplitude. The weak emission around the poles ($\theta = 0$ and $\theta = \pi$) is from the C orbital. Although tunnel ionization from this orbital is suppressed by its higher ionization potential, 11.3 eV, this is partially overcome by its extension along the methyl group.

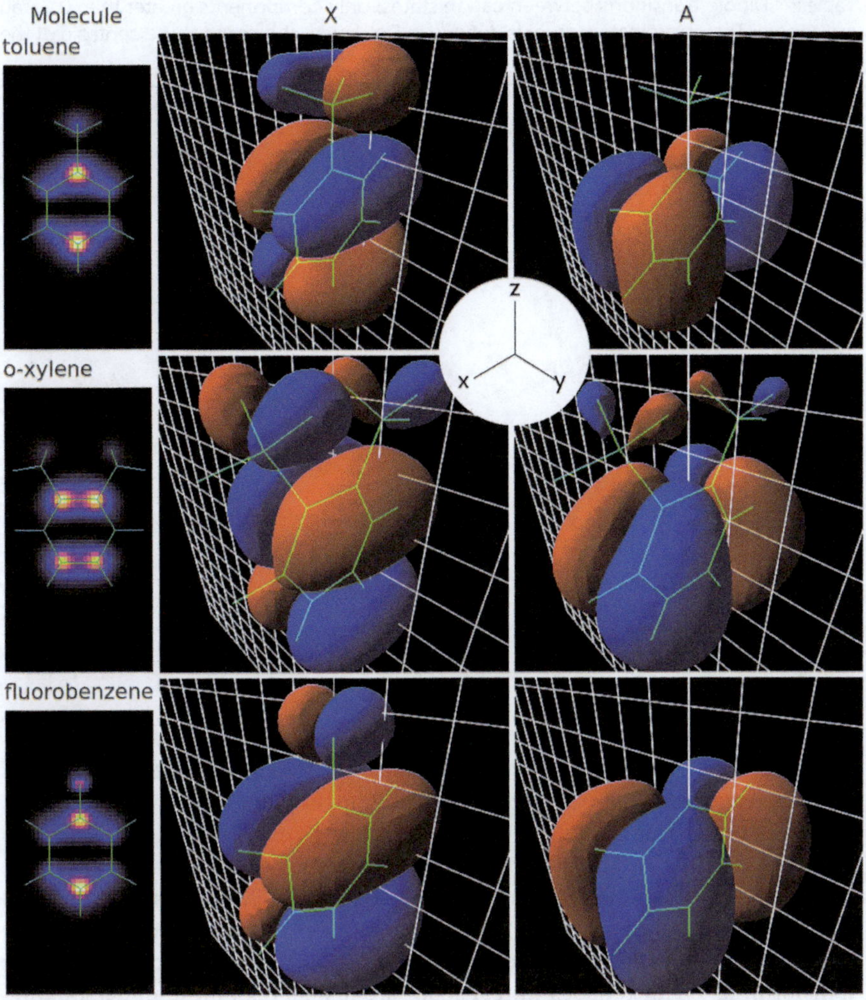

Fig. 1 Contours of $|\psi| = 0.02$ of the highest occupied molecular orbitals (middle column) and next highest orbital (right column) of the molecules considered here (rows). The grids are referred to in the text and show the plane of the laser field for the angle-dependent studies. The left column gives the projection of the electron density onto the benzene ring plane. The coordinate system used throughout is also shown in the middle.

Since the overall contribution from this orbital is minor, we do not consider it further. The four strong lobes centred in each quadrant come from the HOMO. In the upper half-plane, corresponding to the ionization from the methyl side, the amplitude is asymmetric between positive and negative angles. This is due to one hydrogen atom of the methyl group lying in the *yz* plane on the right side, with the other two hydrogens of the methyl group lying above and below this plane on the left side (see Fig. 1). This asymmetry disappears after integrating over the azimuthal angle (*i.e.* around the ring-methyl bond axis).

Table 2 Dipole transitions between cation states; only components greater than 0.05 au are shown. The coordinate system is defined in Fig. 1, with the origin at the centroid of the benzene ring

	X	A	B	C	D
Fluorobenzene					
X	$+0.06\hat{z}$	$+0.12\hat{x}$	$+1.96\hat{z}$		
A		$+0.07\hat{z}$	$-1.84\hat{x}$		
B			$+0.08\hat{z}$		
C				$+0.17\hat{z}$	$-0.07\hat{x}$
D					$-1.33\hat{z}$
Toluene					
X	$+0.28\hat{z}$	$+0.11\hat{x}$		$-0.31\hat{z}$	$+2.10\hat{z}$
A				$+0.27\hat{x}$	$-1.76\hat{x}$
B			$+0.58\hat{z}$	$-0.06\hat{x}$	
C				$+0.78\hat{z}$	$-0.08\hat{z}$
D					$+0.08\hat{y} + 0.38\hat{z}$
***o*-Xylene**					
X	$+0.32\hat{z}$	$+0.19\hat{x}$			$-1.89\hat{z}$
A					$+1.89\hat{x}$
B			$+1.87\hat{z}$	$-0.42\hat{x}$	
C				$+2.10\hat{z}$	
D					$-0.29\hat{z}$

Turning to the phase, shown in Fig. 2(b), there is a striking $\approx 0.6\pi$ phase difference between emission in the lower and upper half-planes. Apart from this the phase is fairly uniform in the regions of strong emission.

To dissect this result we consider the length gauge dipole matrix element, which along with the ionization potential determines the ionization step. This matrix element appears in the SFA in momentum space, but because it appears inside an oscillating integral over ionization times that approximates a Fourier transform, its characteristics in real space are just as relevant to understanding tunnel ionization. In particular, Murray *et al.*[41] showed that the total tunnel ionization rate with the laser electric field lying along an arbitrary direction $\hat{z}$ can be determined by semi-classical propagation starting from the (laboratory frame) molecular orbital $\psi(\boldsymbol{k}_\perp, z_0)$ at a plane $z = z_0$, where the point $z\hat{z}$ lies somewhere in the classically forbidden region. Here $\boldsymbol{k}_\perp$ is momentum in the plane perpendicular to the laser electric field. Tunnel ionization filters out components with high transverse momentum, the acceptance width of filter being proportional to $I_p^{1/2}/E$. In HHG with a linearly polarized field, the filter is even more severe because electrons emitted with non-zero transverse momentum do not recombine. Therefore, for the purposes of HHG, the angular dependence of the amplitude and phase of tunnel ionization should be well described by $\psi(\boldsymbol{k}_\perp = 0, z_0)$, which is simply the length gauge dipole matrix element divided by z_0. It also represents a projection of the wavefunction onto the laser polarization axis, mapping out its spatial extent. The length gauge dipole matrix element is shown in Fig. 3 using the same geometry as in Fig. 2. It is real-valued; this follows from the definition (7) and the fact that the orbitals are real-valued. The π orbitals of the unsubstituted benzene ring appear at $\approx$2–8 au in the lower half plane ($|\theta| > \pi/2$). In the upper

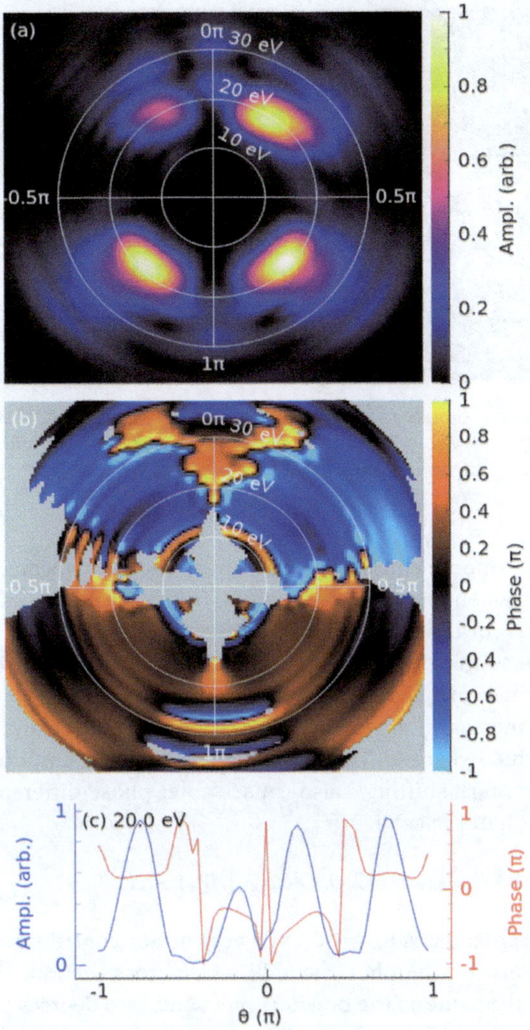

Fig. 2 Spectral amplitude (a) and phase (b) of the attosecond burst from a single laser half-cycle in toluene with the laser field in the HOMO symmetry plane (as shown in Fig. 1); the radial direction corresponds to photon energy and the angle θ is between the ring carbon–methyl carbon bond and the direction of electron motion at ionization. The phase is plotted relative to the average over all angles (removing the angle-independent attochirp), and its color is set to grey where the amplitude is less than 1% of its maximum value. (c) Amplitude (blue, left y-axis) and phase (red, right y-axis) at 20 eV. Both y-axes have the same scale as the respective color scales in (a) and (b).

half plane, the substitution causes these lobes to be concentrated at distances from $\approx$2–5 au. Furthermore, from $\approx$6–8 au, the lobes around the methyl group result in significant extra lobes in the dipole matrix element of opposite sign to the main central lobes. Although these lobes are weaker, they are approximately twice as distant from the centre, leading to a strong ionization enhancement. Furthermore, the electron wavepacket which tunnels from them inherits their π phase shift with respect to the main π orbitals of the ring. One therefore expects

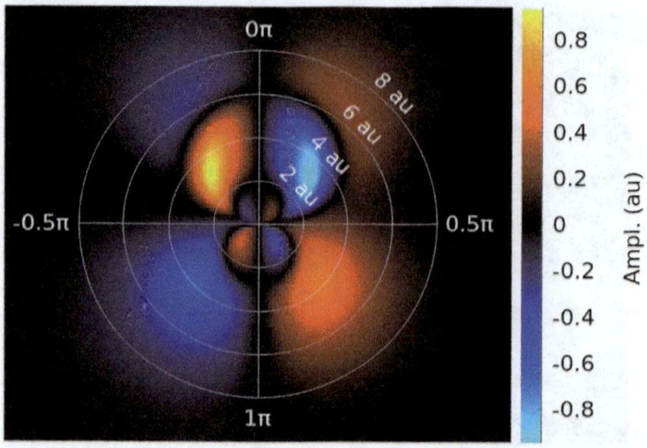

Fig. 3 Length-gauge dipole matrix element of the toluene HOMO in the orbital's symmetry plane (as shown in Fig. 1); the angle is between the direction of a departing electron and the ring carbon-methyl carbon bond.

the sign of the harmonic emission to be correspondingly negated, producing an $\approx\pi$ phase shift when the electron tunnels from the methyl group side. This partially accounts for the observed phase asymmetry.

An additional source of asymmetry is caused by linear dynamic Stark shifting of the HOMO which arises due to its dipole moment of $\Delta\boldsymbol{\mu}_{\mathrm{XX}} = 0.28\hat{z}$ (Table 2). This decreases (increases) the ionization rate when the electron tunnels towards (away from) the methyl group. The resulting change in the amplitude of the HHG is $\approx$10%. Linear Stark shifting[42] also causes a net phase difference between the lower and upper half cycles of

$$\Delta\phi_{\mathrm{Stark}} = -2\cos\theta\Delta\boldsymbol{\mu}_{\mathrm{XX}}[\boldsymbol{A}(t_{\mathrm{r}}) - \boldsymbol{A}(t_{\mathrm{i}})]. \tag{10}$$

At 20 eV, this approximates to $\approx -0.1\pi$, a very minor contribution.

Finally, we consider the photorecombination cross section, shown in Fig. 4. The geometry is the same as the previous plots and zero degrees is defined as the recombining electron approaching the molecule from the substitution side. Therefore Fig. 4 can be directly related to Fig. 2. The lower and upper half-planes are quite similar, with the amplitudes within $\approx$10% and an $\approx -0.2\pi$ phase difference. The methyl substitute also causes weak structures at $\theta = \pm\pi/5$, but these are not evident in the HHG emission because they do not coincide with angles of strong tunnel ionization.

We conclude that the dominant source of the phase and amplitude asymmetry seen in Fig. 2 comes from the tunnel ionization step, and that bound state propagation and photorecombination play only a minor role. We now consider how this asymmetry would manifest once all orientations of the molecule are included.

3.2 Toluene—all orbitals

We now introduce the contribution of the HOMO−1 by performing a similar calculation to that of Fig. 2 but integrating over all rotations of the molecule about the ring-methyl bond axis. In this way, all orbitals can participate, and the

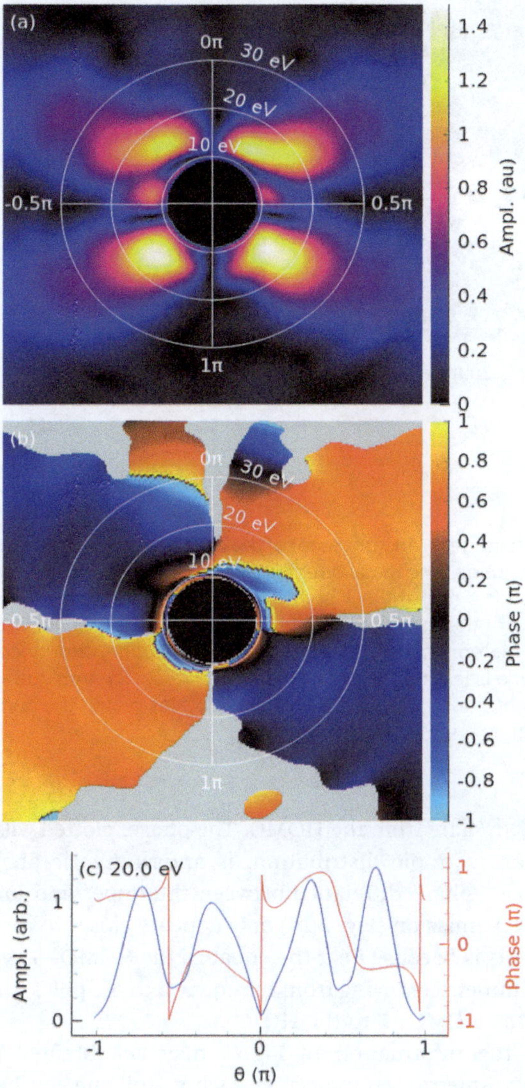

Fig. 4 (a) Amplitude and (b) phase of the photorecombination cross section of the toluene HOMO in orbital's symmetry plane (as shown in Fig. 1); the angle is between the direction from which an electron returns and the carbon–methyl bond. The phase is set to grey where the amplitude is less than 5% of its maximum value. (c) Amplitude (blue, left y-axis) and phase (red, right y-axis) at 20 eV. Both y-axes have the same scale as the respective color scales in (a) and (b).

distribution of molecular angles is closer to what may be achieved using experimental orientation techniques that have been demonstrated with HHG.[43–45] Fig. 5 shows the spectral amplitude (a) and phase (b) of the attosecond burst as a function of the angle θ between the ring-methyl bond and the direction of electron motion at ionization. Only $0 < \theta < \pi$ is shown; the data are symmetric about $\theta = 0$. The positive and negative half cycles of the laser therefore correspond to θ and $\pi - \theta$, respectively. The peak around $\theta = \pi/2$ is from the HOMO−1, and

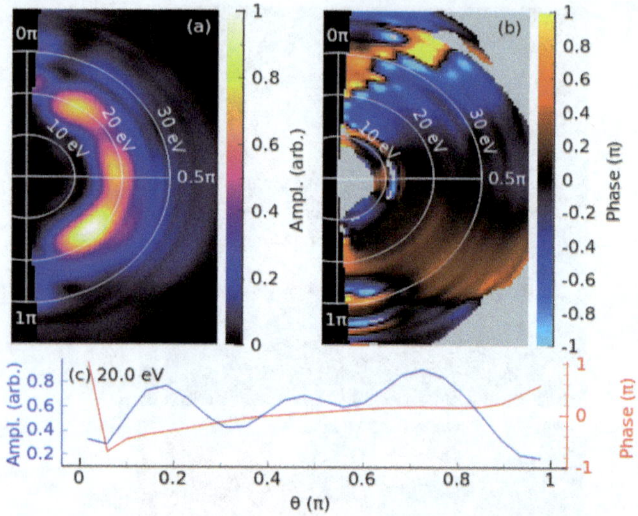

Fig. 5 Spectral amplitude (a) and phase (b) of the attosecond burst from a single laser half-cycle in oriented toluene. The radial direction corresponds to the photon energy and the angle θ is between the ring-methyl bond axis and the direction of electron motion at ionization. Rotation of the molecule about this bond axis is averaged. The phase is plotted relative to that of the emission from an isotropic distribution, and its color is set to grey where the amplitude is less than 1% of its maximum value. (c) Amplitude (blue, left y-axis) and phase (red, right y-axis) at 20 eV. The vertical scales match the corresponding color scales in (a) and (b).

those at $\theta = \pi/4, 3\pi/4$ are from the HOMO. The phase, plotted with respect to the emission from an isotropic distribution, is antisymmetric about the equator ($\theta = \pi/2$). The 0.6π phase difference between the upper and lower half-planes seen in the HOMO emission (Fig. 2(b)) only appears closer to the poles ($\theta < \pi/4$) and ($\theta > 3\pi/4$). This is because near the equator, the HOMO−1 is dominant, and its emission is symmetric. Moving from the equator to the poles, there is a smooth transition from the HOMO−1 to the HOMO.

In principle, the information in Fig. 5 may be obtained experimentally. Orientation of asymmetric top molecules such as toluene can be achieved with the combination of a laser pulse and a static field,[33,46] or a laser pulse and its second harmonic.[47]

Generation of an isolated attosecond burst is possible with a near-single cycle pulse with stable carrier-envelope phase. Phase sensitive detection of the harmonics is possible using two spatially separated sources, which overlap and interfere in the far field.[48,49] One of the sources contains the oriented sample. The other source is a reference and would contain an isotropic distribution.

An experimental alternative to directly producing a single attosecond burst is to examine the even and odd harmonics produced by a multi-cycle laser pulse. The combination of molecular orientation—a macroscopic asymmetry—and an asymmetric dependence on the angle between the molecular axis and the electric field—microscopic asymmetry—leads to the emission of even harmonics.[44,45] This is because the macroscopic emissions from successive half cycles of the laser are no longer identical, rather than being identical except for a π phase shift as is the case when only odd harmonics are present. Measurements of the even harmonics, coupled with a knowledge of the degree of orientation, allows the microscopic asymmetry to be inferred.[44] In general, the asymmetry will be in both amplitude and phase, and therefore the even harmonics must be measured with a phase sensitive technique. This contrasts with previous reconstructions in which only the amplitudes of the even harmonics were measured; additional information was required from theoretical ionization rates and photorecombination cross sections.[5,44] Even harmonics can be produced in the reference source by adding the second harmonic of the drive laser,[50,51] which breaks the odd symmetry of the positive and negative half cycles. Note that although the phase of the reference does not need to be characterized, the relative phase between its even and odd harmonics must be known. Since odd and even harmonics cannot be directly compared, it is worth being precise on the exact meaning of this requirement. The even and odd harmonics, taken separately, each define functions on a common continuous temporal domain *via* the continuous time Fourier transform. These functions have odd or even periodicity with period $T/2$. Provided the functions decay to zero at some point within a period, a temporal window function may be applied to isolate a single period. Fourier transforming the functions yields two continuous spectral amplitudes $D_o(\omega)$ and $D_e(\omega)$, which may be directly compared. Knowledge of their relative phase is equivalent to knowing the relative phase of the even and odd harmonics. The sum and difference of $D_o(\omega)$ and $D_e(\omega)$ correspond to the attosecond bursts from positive and negative half cycles, respectively, which we denote $D_+(\omega)$ and $D_-(\omega)$. An alternative way of stating the requirement is that the (complex-valued) ratio $r = D_+(\omega)/D_-(\omega)$ be known. For weak second harmonic field strengths, this ratio can be obtained theoretically from a first-order perturbation to the action integral,[50,51] and it is likely that combining the perturbative model with a calibration measurement will allow r to be characterized across the spectrum.

Experimental considerations aside, Fig. 6 shows the amplitude and phase of the odd and even harmonics in oriented toluene. These correspond to sums and differences of the attosecond bursts from positive and negative half cycles of the laser. Only $0 < \theta < \pi/2$ is shown; the odd (even) harmonics are symmetric (anti-symmetric) about $\theta = \pi/2$. In the odd harmonics, the contribution from the HOMO−1 (at $\theta = \pi/2$) is slightly stronger than that from the HOMO (at $\theta \approx \pi/4$). The phase of the odd harmonics is nearly constant with angle, except at low $\theta \approx 0$ where there is a contribution from the C orbital. The even harmonics arise solely from the HOMO and are peaked at $\theta \approx \pi/4$. Their phase, relative to that of the odd harmonics, is $\approx -\pi/2$. This reflects the fact that the asymmetry between positive and negative half cycles is in the amplitude and the phase; an amplitude-only asymmetry would produce even harmonics exactly in or out of phase with the odd harmonics.

3.3 *ortho*-Xylene HOMO

We now return to the HOMO-only geometry used in Section 3.1 (*i.e.* with the electric field in the nodal plane of the HOMO−1) and consider some different molecules. Fig. 7 shows the spectral amplitude and phase of the attosecond burst produced in *ortho*-xylene in the same format as Fig. 2. The photorecombination cross section was not available for this molecule so we used the plane wave

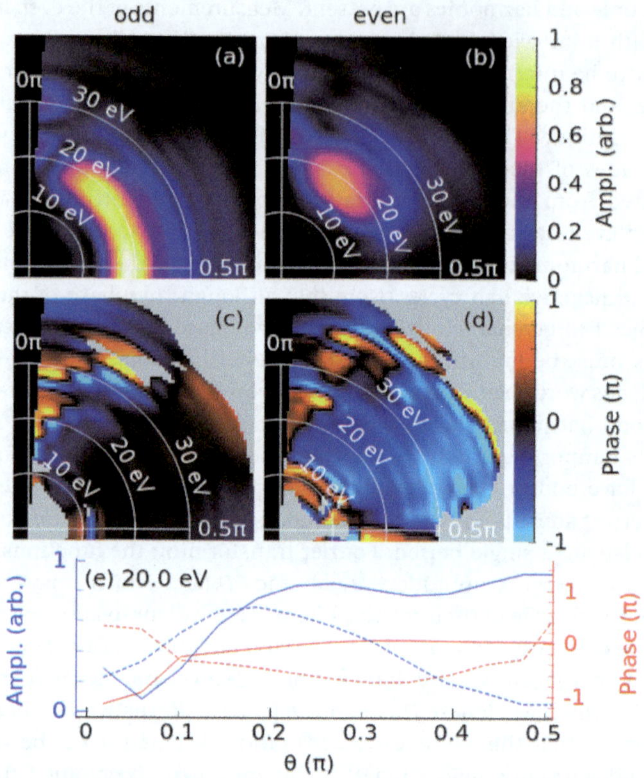

Fig. 6 Spectral amplitude (top row) and phase (bottom row) of the odd (left column) and even (right column) harmonics produced by a many-cycle laser pulse in oriented toluene. The radial direction corresponds to the photon energy and the angle θ is between the ring-methyl bond axis and the direction of electron motion at ionization. Rotation of the molecule about this bond axis is averaged over. The phase of the even harmonics is shown relative to that of the odd harmonics, and the phase of the odd harmonics is shown relative to that produced by an isotropic distribution. In the phase plots, the color is set to grey where the amplitude is less than 1% of its maximum value. Within each row, the color scale is common to both plots. (c) Amplitude (blue, left y-axis) and phase (red, right y-axis) of the odd (solid) and even (dashed) harmonics at 20 eV.

approximation. The amplitude asymmetry is very pronounced, with the emission when tunnelling through the methyl substitutes almost double that through the π orbitals of the benzene ring on the opposite side of the molecule. The phase difference between the half planes $\approx 0.7\pi$.

As with Section 3.1, we dissect this result by examining the three steps. The length gauge dipole for the *ortho*-xylene HOMO (not shown) looks very similar to that of toluene (Fig. 3), except that the lobes produced by the substitution have approximately twice the amplitude. This explains why the emission is stronger in the upper half plane. The HOMO also has ungerade symmetry about xz plane, and this appears as symmetry about $\theta = 0$. Tunnelling from the substitution side also produces $\approx \pi$ phase shift, as with toluene. Linear Stark shifting, eqn (10), contributes a phase difference -0.1π and, as with toluene, decreases (increases) the ionization rate when the electron tunnels towards (away from) the methyl

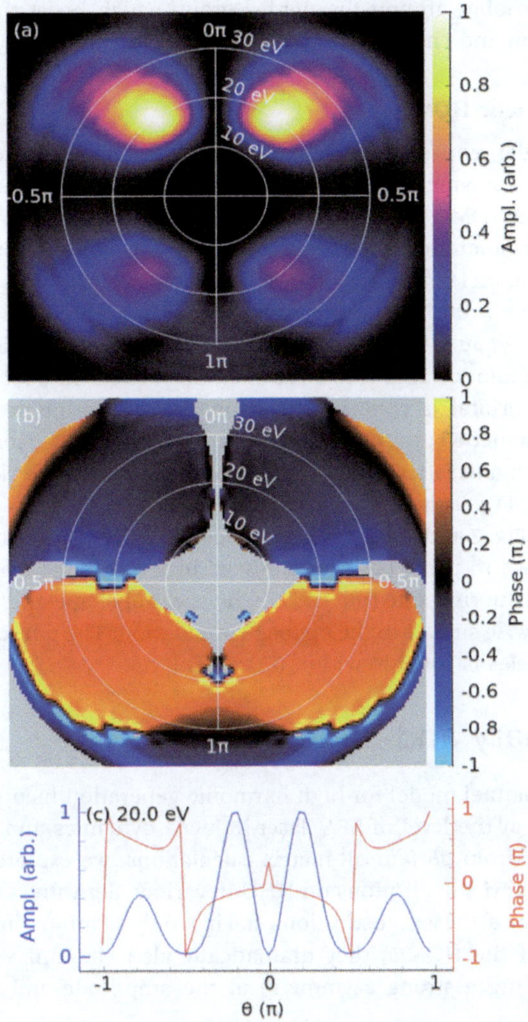

Fig. 7 Spectral amplitude (a) and phase (b) of the attosecond burst from a single laser half-cycle in *ortho*-xylene with the laser field in the HOMO symmetry plane (as shown in Fig. 1); the radial direction corresponds to photon energy and the angle θ is between the bisector of the two ring-methyl bonds and the direction of electron motion at ionization. The phase is plotted relative to the average over all angles (removing the angle-independent attochirp). (c) Amplitude (blue, left y-axis) and phase (red, right y-axis) at 20 eV. Both y-axes have the same scale as the respective color scales in (a) and (b).

group by about 10%. This partially offsets the increased ionization rate when ionizing towards the methyl groups. The plane wave recombination matrix elements are almost entirely unaffected by the substitution, with phase and amplitude differences of less than 0.1π and 5%, respectively between the two directions. Based on the result from toluene and fluorobenzene below, it is likely any asymmetry in the photorecombination matrix element will be similarly small. We tentatively conclude that, as with toluene, the dominant source of amplitude and phase asymmetry in emission from the *ortho*-xylene HOMO comes from the presence of the lobes around the methyl group which boost the amplitude of tunnel ionization and change its sign.

3.4 Fluorobenzene HOMO

We now consider emission from the fluorobenzene HOMO under identical conditions to the previous section. Fluorobenzene is a revealing example because the shape of the HOMO is very similar to that of toluene, except that the lobes around its fluorine atom are more concentrated than those around the methyl group in toluene. Its length gauge dipole is shown in Fig. 8. As with toluene, the main lobes of the π orbitals on the benzene ring are more concentrated on the substitute side (2–4 au in $|\theta| < \pi/2$) compared to the unmodified side (2–8 au in $|\theta| > \pi/2$). Unlike in toluene, there are no additional lobes on the substitute side. This is because the orbital is concentrated around the fluorine atom rather than spread around a methyl group. Following the reasoning in Section 3.1, we expect tunnel ionization and HHG to be suppressed on the substitute side, but without a phase change. Fig. 9 shows the attosecond burst as a function of angle (in the same format as Fig. 2 and 7), and this is indeed what the calculation predicts. The amplitude, shown in panel (a), is suppressed by a factor ≈ 4 when the electron leaves from the fluorine side ($|\theta| < \pi/2$). The spectral shape and angular dependence are otherwise similar to the opposite orientation. The phase of positive and negative half cycles is equal to within 0.1π.

4 Summary and conclusion

Using a multichannel model for high harmonic generation incorporating strong field ionization at the level of SFA, laser-induced dynamics and recombination using the input from *ab initio* R-matrix calculations, we explored the effect of fluorine and methyl substitutions on HHG in various derivatives of benzene. We found that despite these substitutions having only a minor influence on the overall shape of the HOMO, they dramatically alter the emission in oriented molecules, causing a strong asymmetry in the amplitude and phase between

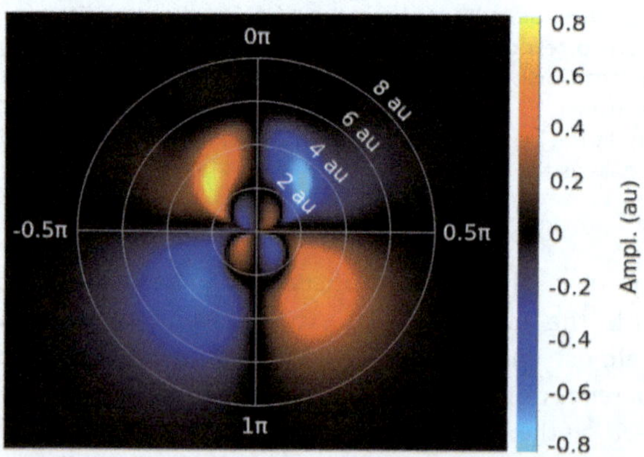

Fig. 8 Length-gauge dipole matrix element of the fluorobenzene HOMO in the orbital's symmetry plane (as shown in Fig. 1); the angle is between the direction of a departing electron and the carbon–fluorine bond.

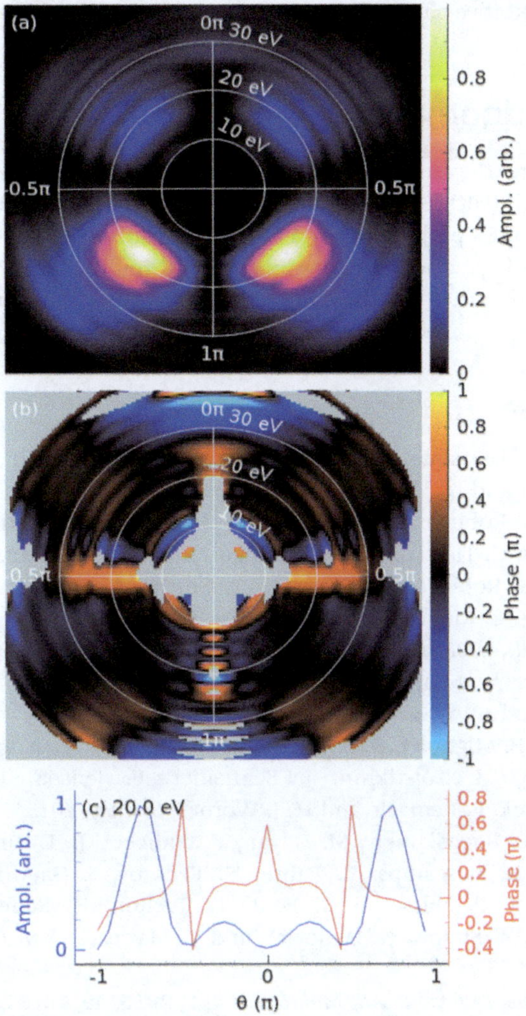

Fig. 9 Amplitude (a) and phase (b) of attosecond burst produced by a single laser half-cycle in fluorobenzene with the laser field in the HOMO symmetry plane (as shown in Fig. 1); the radial direction corresponds to photon energy and the angle θ is between the carbon–fluorine bond and the direction of electron motion at ionization. The phase is plotted relative to the average over all angles (removing the angle-independent attochirp), and its color is set to grey where the amplitude is less than 1% of its maximum value. (c) Amplitude (blue, left y-axis) and phase (red, right y-axis) at 20 eV. Both y-axes have the same scale as the respective color scales in (a) and (b).

molecules oriented and anti-oriented to the laser polarization. In particular, the additional lobes in the orbital wavefunction produced by methyl group substitutions were shown to cause a sign flip in the tunnelled electron wavepacket, which is mapped directly into the high-harmonic emission. We discussed how this asymmetry would manifest itself in a phase- and alignment-sensitive HHG spectroscopy experiment. These insights will also enable us to better understand

and make quantitative comparisons with our existing measurements in isotropic samples.

Acknowledgements

We acknowledge the support from the Engineering and Physical Sciences Research Council (UK) (EPSRC) grant EP/I032517/1 and the European Research Council (ERC) ASTEX project 290467. ZM and OS acknowledge DFG grant Sm 292/5-1. AH and MI gratefully acknowledge DFG grant DFG IV 152/7-1. DRA acknowledges support from the Imperial College Junior Research Fellowship scheme.

References

1 J. Itatani, J. Levesque, D. Zeidler, H. Niikura, H. Pepin, J. C. Kieffer, P. B. Corkum and D. M. Villeneuve, *Nature*, 2004, **432**, 867–871.
2 S. Baker, J. S. Robinson, C. A. Haworth, H. Teng, R. A. Smith, C. C. Chirilă, M. Lein, J. W. G. Tisch and J. P. Marangos, *Science*, 2006, **312**, 424–427.
3 S. Baker, J. S. Robinson, M. Lein, C. C. Chirilă, R. Torres, H. C. Bandulet, D. Comtois, J. C. Kieffer, D. M. Villeneuve, J. W. G. Tisch and J. P. Marangos, *Phys. Rev. Lett.*, 2008, **101**, 053901.
4 O. Smirnova, Y. Mairesse, S. Patchkovskii, N. Dudovich, D. Villeneuve, P. Corkum and M. Y. Ivanov, *Nature*, 2009, **460**, 972–977.
5 P. M. Kraus, B. Mignolet, D. Baykusheva, A. Rupenyan, L. Horný, E. F. Penka, G. Grassi, O. I. Tolstikhin, J. Schneider, F. Jensen, L. B. Madsen, A. D. Bandrauk, F. Remacle and H. J. Wörner, *Science*, 2015, **350**, 790–795.
6 A. Ferre, A. E. Boguslavskiy, M. Dagan, V. Blanchet, B. D. Bruner, F. Burgy, A. Camper, D. Descamps, B. Fabre, N. Fedorov, J. Gaudin, G. Geoffroy, J. Mikosch, S. Patchkovskii, S. Petit, T. Ruchon, H. Soifer, D. Staedter, I. Wilkinson, A. Stolow, N. Dudovich and Y. Mairesse, *Nat. Commun.*, 2015, **6**, 5952.
7 J. P. Marangos, *J. Phys. B: At., Mol. Opt. Phys.*, 2016, **49**, 132001.
8 S. Haessler, J. Caillat and P. Salières, *J. Phys. B: At., Mol. Opt. Phys.*, 2011, **44**, 203001.
9 X. M. Tong, Z. X. Zhao and C. D. Lin, *Phys. Rev. A*, 2002, **66**, 033402.
10 J. Muth-Böhm, A. Becker and F. H. M. Faisal, *Phys. Rev. Lett.*, 2000, **85**, 2280–2283.
11 T. K. Kjeldsen, C. Z. Bisgaard, L. B. Madsen and H. Stapelfeldt, *Phys. Rev. A*, 2003, **68**, 063407.
12 T. K. Kjeldsen and L. B. Madsen, *J. Phys. B: At., Mol. Opt. Phys.*, 2004, **37**, 2033.
13 C. F. de Morisson Faria and X. Liu, *J. Mod. Opt.*, 2011, **58**, 1076–1131.
14 P. Eckle, A. N. Pfeiffer, C. Cirelli, A. Staudte, R. Dörner, H. G. Muller, M. Büttiker and U. Keller, *Science*, 2008, **322**, 1525–1529.
15 L. Torlina, F. Morales, J. Kaushal, I. Ivanov, A. Kheifets, A. Zielinski, A. Scrinzi, H. G. Muller, S. Sukiasyan, M. Ivanov and O. Smirnova, *Nat. Phys.*, 2015, **11**, 503–508.
16 Y. Mairesse, J. Higuet, N. Dudovich, D. Shafir, B. Fabre, E. Mével, E. Constant, S. Patchkovskii, Z. Walters, M. Y. Ivanov and O. Smirnova, *Phys. Rev. Lett.*, 2010, **104**, 213601.

17 F. McGrath, A. S. Johnson, L. Miseikis, D. Wood, P. Hawkins, D. Austin and J. P. Marangos, 2016, Manuscript in preparation.
18 J. L. Hansen, L. Holmegaard, J. H. Nielsen, H. Stapelfeldt, D. Dimitrovski and L. B. Madsen, *J. Phys. B: At., Mol. Opt. Phys.*, 2012, **45**, 015101.
19 A.-T. Le, R. R. Lucchese, S. Tonzani, T. Morishita and C. D. Lin, *Phys. Rev. A*, 2009, **80**, 013401.
20 M. Lewenstein, P. Balcou, M. Y. Ivanov, A. L'Huillier and P. B. Corkum, *Phys. Rev. A*, 1994, **49**, 2117–2132.
21 O. Smirnova, S. Patchkovskii, Y. Mairesse, N. Dudovich, D. Villeneuve, P. Corkum and M. Y. Ivanov, *Phys. Rev. Lett.*, 2009, **102**, 063601.
22 O. Smirnova, S. Patchkovskii, Y. Mairesse, N. Dudovich and M. Y. Ivanov, *Proc. Natl. Acad. Sci. U. S. A.*, 2009, **106**, 16556–16561.
23 O. Smirnova and M. Ivanov, in *Multielectron High Harmonic Generation: Simple Man on a Complex Plane*, Wiley-VCH Verlag GmbH & Co. KGaA, 2014, pp. 201–256.
24 A. Etches and L. B. Madsen, *J. Phys. B: At., Mol. Opt. Phys.*, 2010, **43**, 155602.
25 O. Smirnova, M. Spanner and M. Ivanov, *J. Mod. Opt.*, 2007, **54**, 1019–1038.
26 O. Smirnova, A. S. Mouritzen, S. Patchkovskii and M. Y. Ivanov, *J. Phys. B: At., Mol. Opt. Phys.*, 2007, **40**, F197.
27 C. D. Lin, A.-T. Le, Z. Chen, T. Morishita and R. Lucchese, *J. Phys. B: At., Mol. Opt. Phys.*, 2010, **43**, 122001.
28 A. G. Harvey, D. S. Brambila, F. Morales and O. Smirnova, *J. Phys. B: At., Mol. Opt. Phys.*, 2014, **47**, 215005.
29 D. S. Brambila, A. G. Harvey, Z. MaŽín, J. D. Gorfinkiel and O. Smirnova, *J. Phys. B: At., Mol. Opt. Phys.*, 2015, **48**, 245101.
30 B. H. Bransden and C. J. Joachain, *Physics of atoms and molecules*, Prentice Hall, 2003.
31 T. Koopmans, *Physica*, 1934, **1**, 104–113.
32 D. Dimitrovski, C. P. J. Martiny and L. B. Madsen, *Phys. Rev. A*, 2010, **82**, 053404.
33 L. Holmegaard, J. L. Hansen, L. Kalhoj, S. Louise Kragh, H. Stapelfeldt, F. Filsinger, J. Kupper, G. Meijer, D. Dimitrovski, M. Abu-samha, C. P. J. Martiny and L. Bojer Madsen, *Nat. Phys.*, 2010, **6**, 428–432.
34 H. Li, D. Ray, S. De, I. Znakovskaya, W. Cao, G. Laurent, Z. Wang, M. F. Kling, A. T. Le and C. L. Cocke, *Phys. Rev. A*, 2011, **84**, 043429.
35 K. Varjú, Y. Mairesse, P. Agostini, P. Breger, B. Carré, L. J. Frasinski, E. Gustafsson, P. Johnsson, J. Mauritsson, H. Merdji, P. Monchicourt, A. L'Huillier and P. Salières, *Phys. Rev. Lett.*, 2005, **95**, 243901.
36 K. Kimura, *Handbook of HeI photoelectron spectra of fundamental organic molecules: ionization energies, ab initio assignments, and valence electronic structure for 200 molecules*, Japan Scientific Societies Press, 1982.
37 M. Klessinger, *Angew. Chem., Int. Ed. Engl.*, 1972, **11**, 525–526.
38 T. Koenig, M. Tuttle and R. Wielesek, *Tetrahedron Lett.*, 1974, **15**, 2537–2540.
39 K. Kesper, N. Münzel, W. Pietzuch, H. Specht and A. Schweig, *J. Mol. Struct.: THEOCHEM*, 1989, **200**, 375–400.
40 J. A. Sell and A. Kuppermann, *Chem. Phys.*, 1978, **33**, 367–378.
41 R. Murray, M. Spanner, S. Patchkovskii and M. Y. Ivanov, *Phys. Rev. Lett.*, 2011, **106**, 173001.
42 A. Etches, M. B. Gaarde and L. B. Madsen, *Phys. Rev. A*, 2011, **84**, 023418.

43 E. Frumker, C. T. Hebeisen, N. Kajumba, J. B. Bertrand, H. J. Wörner, M. Spanner, D. M. Villeneuve, A. Naumov and P. B. Corkum, *Phys. Rev. Lett.*, 2012, **109**, 113901.

44 E. Frumker, N. Kajumba, J. B. Bertrand, H. J. Wörner, C. T. Hebeisen, P. Hockett, M. Spanner, S. Patchkovskii, G. G. Paulus, D. M. Villeneuve, A. Naumov and P. B. Corkum, *Phys. Rev. Lett.*, 2012, **109**, 233904.

45 P. M. Kraus, A. Rupenyan and H. J. Wörner, *Phys. Rev. Lett.*, 2012, **109**, 233903.

46 B. Friedrich and D. Herschbach, *J. Chem. Phys.*, 1999, **111**, 6157–6160.

47 S. De, I. Znakovskaya, D. Ray, F. Anis, N. G. Johnson, I. A. Bocharova, M. Magrakvelidze, B. D. Esry, C. L. Cocke, I. V. Litvinyuk and M. F. Kling, *Phys. Rev. Lett.*, 2009, **103**, 153002.

48 R. Zerne, C. Altucci, M. Bellini, M. B. Gaarde, T. W. Hänsch, A. L'Huillier, C. Lyngå and C.-G. Wahlström, *Phys. Rev. Lett.*, 1997, **79**, 1006–1009.

49 M. Bellini, C. Lyngå, A. Tozzi, M. B. Gaarde, T. W. Hänsch, A. L'Huillier and C.-G. Wahlström, *Phys. Rev. Lett.*, 1998, **81**, 297–300.

50 N. Dudovich, O. Smirnova, J. Levesque, Y. Mairesse, M. Y. Ivanov, D. M. Villeneuve and P. B. Corkum, *Nat. Phys.*, 2006, **2**, 781–786.

51 J. M. Dahlström, A. L'Huillier and J. Mauritsson, *J. Phys. B: At., Mol. Opt. Phys.*, 2011, **44**, 095602.

Faraday Discussions

PAPER

Multidimensional high harmonic spectroscopy of polyatomic molecules: detecting sub-cycle laser-driven hole dynamics upon ionization in strong mid-IR laser fields

Barry D. Bruner,*[a] Zdeněk Mašín,[b] Matteo Negro,[c] Felipe Morales,[b] Danilo Brambila,[b] Michele Devetta,[c] Davide Faccialà,[d] Alex G. Harvey,[b] Misha Ivanov,[b] Yann Mairesse,[e] Serguei Patchkovskii,[b] Valeria Serbinenko,[b] Hadas Soifer,[af] Salvatore Stagira,[d] Caterina Vozzi,[c] Nirit Dudovich*[a] and Olga Smirnova*[b]

Received 11th May 2016, Accepted 7th July 2016
DOI: 10.1039/c6fd00130k

High harmonic generation (HHG) spectroscopy has opened up a new frontier in ultrafast science, where electronic dynamics can be measured on an attosecond time scale. The strong laser field that triggers the high harmonic response also opens multiple quantum pathways for multielectron dynamics in molecules, resulting in a complex process of multielectron rearrangement during ionization. Using combined experimental and theoretical approaches, we show how multi-dimensional HHG spectroscopy can be used to detect and follow electronic dynamics of core rearrangement on sub-laser cycle time scales. We detect the signatures of laser-driven hole dynamics upon ionization and reconstruct the relative phases and amplitudes for relevant ionization channels in a CO_2 molecule on a sub-cycle time scale. Reconstruction of channel-resolved complex ionization amplitudes on attosecond time scales has been a long-standing goal of high harmonic spectroscopy. Our study brings us one step closer to fulfilling this initial promise and developing robust schemes for sub-femtosecond imaging of multielectron rearrangement in complex molecular systems.

[a]*Department of Physics of Complex Systems, Weizmann Institute of Science, Rehovot 76100, Israel. E-mail: nirit.dudovich@weizmann.ac.il; Fax: +972 8934 4109; Tel: +972 8934 6290*

[b]*Max Born Institute for Nonlinear Optics and Short Pulse Spectroscopy, Max-Born-Strasse 2 A, 12489 Berlin, Germany*

[c]*Istituto di Fotonica e Nanotecnologie-CNR, Milan 20133, Italy*

[d]*Dipartimento di Fisica, Politecnico di Milano, Milan 20133, Italy*

[e]*CELIA, Université de Bordeaux-CNRS-CEA, UMR5107, F33405 Talence, France*

[f]*Stanford Institute for Materials and Energy Sciences, SLAC National Accelerator Laboratory, Menlo Park, California 94025, USA*

1 Introduction

The ability to manipulate and probe the evolution of electronic wavepackets on their natural time-scale has been the focus of attosecond science for the past two decades. Application of strong (10^{13} to 10^{14} W cm^{-2}) infrared fields in attosecond spectroscopy gives access to phenomena evolving on a sub-laser cycle time scale[1] – well below 2.6 fs for 800 nm wavelength fields.

One of the fastest phenomena is photoionization – electron removal from an atom or a molecule using attosecond XUV pulses[2–4] or strong IR laser fields.[5–8] Much experimental and theoretical effort in the attosecond community has been devoted to understanding and reconstructing ionization times from experimental observations.[2–8] Electron removal creates a hole in an atomic or molecular system, and the evolution of the hole upon ionization is a prominent example of the sub-cycle time-scale dynamics. This evolution can be driven by spin–orbit interaction,[9,10] a laser field,[11,12] or pure electronic correlations.[13] It can also arise from strongly non-equilibrium initial conditions created by ultrashort excitation inducing complex hole dynamics,[14] which may involve electron correlations.[15]

Among the most exciting advances in atto-science is high harmonic generation (HHG) spectroscopy. It allows one to combine sub-Angstrom spatial with attosecond temporal resolution, holding the potential of resolving the structure of electronic wavefunctions as they evolve in time.[11,12,16–21] HHG spectroscopy exploits the built-in pump-probe process. Ionization acts as a pump: it removes an electron and creates a hole in the system, launching non-equilibrium dynamics. Driven by the laser field, the liberated electron returns to the parent ion (electron–ion recollision) and probes it *via* radiative recombination.[22] The emitted attosecond burst of light takes a snapshot of the system, probing the evolved wavefunction. HHG spectroscopy is not limited by the duration of the driving laser pulse. The dynamics only start when the electron is detached from the system, and probed as the electron recollides. The spatial resolution is given by the de-Broglie wavelength of the returning electron.[23,24] High harmonic spectroscopy has been used to resolve ultrafast electronic,[11,12,16–19,25] chiral[26] and nuclear[27–34] dynamics in molecules and to visualize the structure of molecular orbitals.[35–38]

A unique property of high harmonic spectroscopy is its ability to time-resolve the interference between multiple quantum pathways in the response of a multielectron system to the ionizing pulse, encoding their relative phases.[16–18] These phases, in turn, encode the underlying multielectron rearrangement during ionization. However, the high temporal resolution and interferometric sensitivity of HHG spectroscopy come with a clear trade-off. The main parts of the experiment – the pump-probe delay, the fundamental field's intensity and wavelength – cannot be independently controlled. They are all coupled by the light–matter interaction and are dictated by the recollision mechanism.

In conventional multidimensional ultrafast spectroscopy, the coherent evolution of the system is excited and probed by a sequence of pulses with independently controllable time delays between them. The spectrum is spread into multiple dimensions where the full dynamical picture can be more easily understood.[39] In contrast, in conventional HHG spectroscopy, these independent controls are not available, and the complex multielectron response is mapped into a 1D experimental observable – the HHG spectrum.

Disentangling the underlying dynamics requires a multi-dimensional approach, capable of separating the contributions of different ionization pathways, reminiscent of how multidimensional spectroscopy has been used for mapping energy transfer and coupling between electronic and vibrational states.[40,41] In the absence of such controls, HHG cannot advance to be a robust tool for probing complex multielectron dynamics in molecular systems.

In this paper we describe advanced HHG spectroscopy, demonstrating the ability to resolve complex dynamics of multielectron rearrangement in CO_2 molecules. In prior work, we have employed similar schemes for separating the spatial and temporal information in HHG from molecules.[7] In addition, we have applied these schemes to calibrate the intrinsic pump-probe delay in HHG spectroscopy by reconstructing ionization and recombination times.[7,42–44]

Here, we introduce the idea of a laser-driven intra-molecular interferometer and exploit it to detect core rearrangement dynamics during ionization. The interferometer is created by sub-cycle laser driven dynamics inside the molecular ion between ionization and recombination. It is naturally built into molecules interacting with strong IR fields, and is particularly simple for longer wavelengths of the driving laser field.

Laser induced dynamics in the molecular ion result in new HHG pathways (channels), where ionization and recombination involve two different states of the ion.[11] In this work we first show that the 2D HHG spectroscopy is particularly sensitive to such channels. Second, we use these channels to perform reconstruction of sub-cycle ionization dynamics. The reconstruction strategy benefits from the fact that purely laser-driven transitions can be characterized theoretically by solving the Schrödinger equation for laser-driven bound dynamics in the ion, between ionization and recombination. Therefore, the new HHG pathways associated with these dynamics can serve as a reference – the analogue of a local oscillator in heterodyne-style detection. The interference of these "reference" pathways with "signal" pathways evolving towards the same HHG recombination channels records the contribution of these signal pathways in the harmonic spectrum. These contributions can include additional interactions leading to inter-channel coupling, such as electron correlations during strong field ionization,[45–47] which can introduce nontrivial amplitudes and phases between different ionization channels.

For longer wavelengths of the driving field, laser-driven dynamics in the molecular ion is dominated by laser-induced polarization of different ionic states. The "virtual" transitions between the ionic states associated with polarization are phase-locked to the driving field. Thus, the reference pathways of the intra-molecular interferometer become particularly simple. We focus our analysis of the multidimensional HHG spectra on this regime.

The paper is organized as follows. Section 2 describes the essential high harmonic channels and their interference in mid-IR fields. Section 3 provides the background for "reading" the information recorded in 2D HHG spectra. Section 4 describes the experimental results. Section 5 describes the theoretical analysis of 2D HHG spectra, demonstrates the contribution of laser-driven HHG channels and performs the reconstruction of sub-cycle ionization dynamics from experimental 2D spectra.

2 Encoding electronic dynamics in molecular HHG spectra

The CO_2 molecule represents a complex system in which strong field ionization can leave the ion in four different states (ground, $\tilde{X}$, and three excited $\tilde{A}$, $\tilde{B}$, $\tilde{C}$), as shown in Fig. 1. For clarity, in the remainder of the paper we shall omit the tilde in the notations of these states. The strength of each ionization channel depends exponentially on the corresponding ionization potential, but also very strongly on the structure of the Dyson orbital associated with each channel.[17,48,49]

The four ionization channels give rise to at least four recombination channels: XX, AA, BB, CC, where the second letter labels the recombination channel. We call these channels "direct" (or "diagonal"), because the state of the ion does not change between ionization and recombination. As detailed in ref. 11, the laser field can also couple the different states of the ion between ionization and recombination and lead to additional cross-channels. For example, the ionization channel X now leads to four HHG channels XX, XA, XB, XC – one direct and three cross-channels. Note that the three cross-channels arise even in the absence of real excitations from the state X – in this case, they reflect polarization of the ionic core at the moment of recombination.

The observed signal is the result of the coherent addition of all pathways, leading to a particular recombination channel, as shown schematically in Fig. 2(a). *Ab initio* calculations of laser-driven dynamics in the ion and recombination amplitudes open the opportunity to recover the less well-known pathways in the network of channels. The interferometer becomes simpler in the long wavelength limit, as shown in Fig. 2(b). Here we briefly summarize the dominant channel-resolved contributions to the HHG spectra, a more detailed analysis will appear in Section 5 of the paper.

The XX channel usually dominates the HHG signal, since the ionization potential for this channel is significantly lower than for other direct channels, such as AA, BB, CC. The disparity between these channels is further accentuated in mid-IR fields, where the strong field ionization is expected to leave an even smaller fraction of molecular ions in the excited state. Thus, we expect smaller

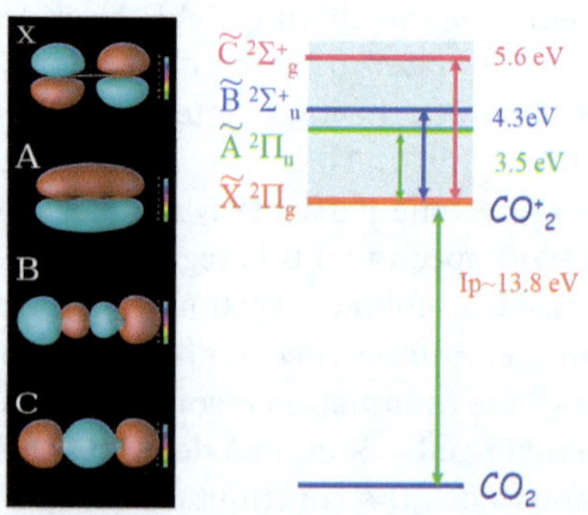

Fig. 1 Orbital structure and energy levels of the CO_2 ion. Left: four key Dyson orbitals of the CO_2 ion: X, A, B, and C. Right: energy level structure of the CO_2 ion. The ionization potentials are 13.8, 17.3, 18.1, and 19.4 eV for channels X, A, B, and C, respectively.

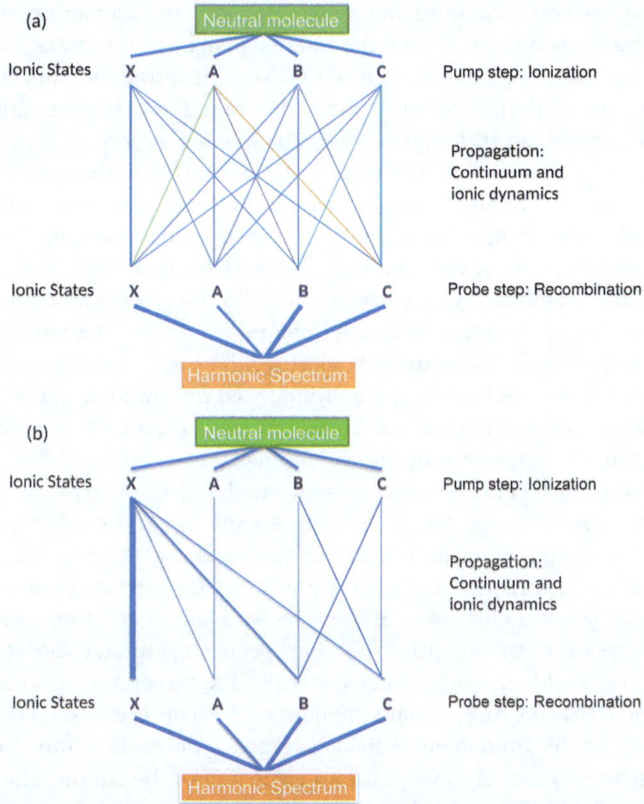

Fig. 2 Schematic representation of HHG channels in the CO_2 molecule. Each ionization channel is defined with respect to a fully correlated spectroscopic state of the ion. However, it gives rise to several pathways due to interchannel couplings, driven by the laser field or electron correlations during ionization. The observed harmonic signal results from coherent addition of all pathways, which form an inter-connected network, see panel (a). The network is simplified in the long-wavelength regime, as shown in panel (b). The thickness of the lines connecting each pathway provides an estimate of their relative contributions to the total signal.

contributions of the AA, BB and CC channels to the HHG spectrum. These contributions interfere with cross-channels XA, XB and XC, which all originate from X and are associated with the laser-induced dynamics between the states X, A, B, C between ionization and recombination. These channels are mostly associated with "virtual" transitions, which reflect the laser-induced polarization of state X. The contributions of these cross-channels to HHG spectra are on par with the contributions of direct AA, BB, CC channels. The cationic states B and C are strongly coupled by the laser field, which leads to significant population transfer between these states and results in relatively important contributions from CB and BC HHG channels. Finally, channels AX, BX, CX are much weaker than XA, XB, XC due to their lower ionization rates.

Thus, the pathways XA, XB, and XC form the analogue of the reference, locked by the driving laser field to the XX channel. They heterodyne the "unknown" signal encoded in the pathways AA, BB, CC. *Ab initio* calculations of the laser-driven dynamics and recombination matrix elements yield an accurate reference and open the possibility to decode the unknown signal – the initial ionic superposition state formed during the ionization step. In principle, this can be performed for each harmonic and thus the dynamics of the initial ionic superposition state can be mapped with sub-cycle resolution.

While strong field interaction with CO_2 molecules is described by multichannel dynamics, its rich complex nature has not yet been fully resolved experimentally. One major limitation is imposed by the interplay between the structural and dynamical information in the HHG spectrum. Sample experimental spectra appear in Fig. 3, where aligned CO_2 molecules were probed using HHG for two laser intensities. The experimental set-up and the molecular alignment mechanism used to obtain such spectra will be discussed in more detail in Section 3. The stand-out feature is the pronounced minimum around 55 eV, with a width of about 15 eV. This is a well known structural feature associated with so-called two center interference in the recombination step of HHG.[37]

Much effort has been directed toward resolving these types of structural signatures in the HHG spectra of CO_2 (ref. 36 and 37) and N_2,[19,35,50,51] aimed at extracting the spatial information of the participating orbitals. The interplay between the spatial and the temporal features in HHG spectra has been studied in CO_2 and N_2O by systematic scans of the laser wavelength and intensity.[16,52,53] For particular experimental conditions, it has been shown that the strong field dynamics scales with the laser intensity while the structural information contained in the HHG spectrum remain unchanged.[16,54] This is also seen in Fig. 3 – as the intensity of the fundamental field increases, the basic features of the 1D spectra appear very similar, except for an extension of the cut-off. The details of the complex multi-channel dynamics remain elusive, their presence is not apparent.

Extracting dynamical information from the HHG spectrum is one of the primary challenges in HHG spectroscopy. Measurements of dynamical

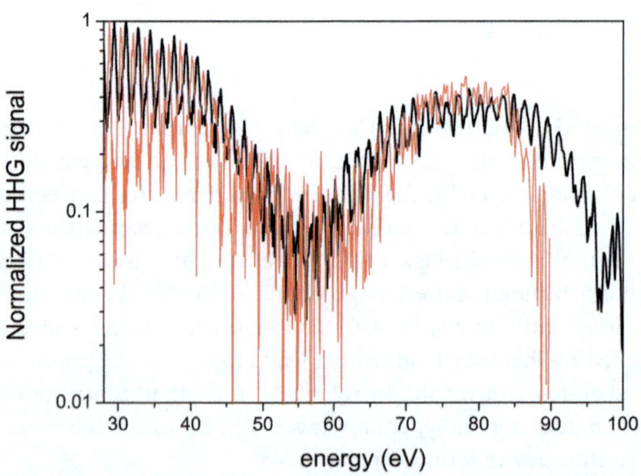

Fig. 3 Experimental 1D spectra of CO_2 for two intensities. The spectra are normalized in order to compare their basic features. Black curve: 1.4×10^{14} W cm^{-2} laser intensity, red curve: 1.0×10^{14} W cm^{-2}. The laser wavelength is 1480 nm.

interference attributed to two independently evolving ionization channels in these molecules have also been reported.[7,54–56] The coupling between different channels can leave its signature in the HHG polarization state[11] and in the interference between HHG signals from long and short trajectories.[25] In the next section we show that applying a two-colour multidimensional scheme allows us to record the complex multielectron dynamics, which leaves clear fingerprints in the HHG spectrum.

3 Increasing the dimensionality of HHG measurements

3.1 Two-colour gating of HHG signals

Our multidimensional method is based upon HHG generated by two-colour, orthogonally polarized fields. The concepts behind these experiments have been described in detail previously.[7,8,42–44,56] For a concise overview of the most pertinent concepts, we refer the reader to ref. 56 as a starting point. Here in Sections 3.1 and 3.2, we provide a brief introduction for the benefit of the reader.

A fundamental, high intensity field dictates the basic properties of the interaction while a weak phase-locked second harmonic field, in an orthogonal polarization, serves as a subcycle gate. The second harmonic field perturbs the dynamics of the continuum electron, controlling the direction and magnitude of the lateral kick imparted to the electron before it recollides. A large kick in the lateral direction means that the probability of electron recombination to the parent ion is lowered and therefore the HHG signal is suppressed.

As we vary the two-colour phase delay ϕ between the fields, we modulate this kick, leading to clear sinusoidal oscillations of the signal at each harmonic order. Optimization of the HHG yield in this two-colour configuration selects a very specific condition for ionization: it remains virtually unaffected by the perturbative second harmonic field, especially in the long-wavelength limit.[43] Thus it is essential that the perturbing field is sufficiently weak. Under these conditions, the overall HHG yield depends weakly on ϕ, such that the amplitude of the ϕ-dependent modulation is small compared to the average HHG yield. This allows us to extract a key observable of the experiment – the optimal two-colour delay, ϕ_{max}, that maximizes the high harmonic yield for each harmonic.

Scanning the two-colour delay increases the dimensionality of the measurement and helps to resolve the dynamical information driven by the interaction. The optimal delay ϕ_{max} depends on the strong field parameters that define each electron trajectory – the ionization time and recollision time. Therefore, the modulation phase of each channel serves as a "label" which identifies the underlying dynamics. When the dynamics evolve along several channels, each channel is modulated according to its characteristic phase ϕ_{max}. The HHG spectrum records the interference of these modulations, encoding the relative phases and amplitudes dictated by the hole dynamics.

3.2 Single channel dynamics

When ionization occurs from a single isolated channel, the 2D measurement provides a universal fingerprint, which offers an important calibration tool. Representative examples of single-channel ionization in Krypton gas are shown in

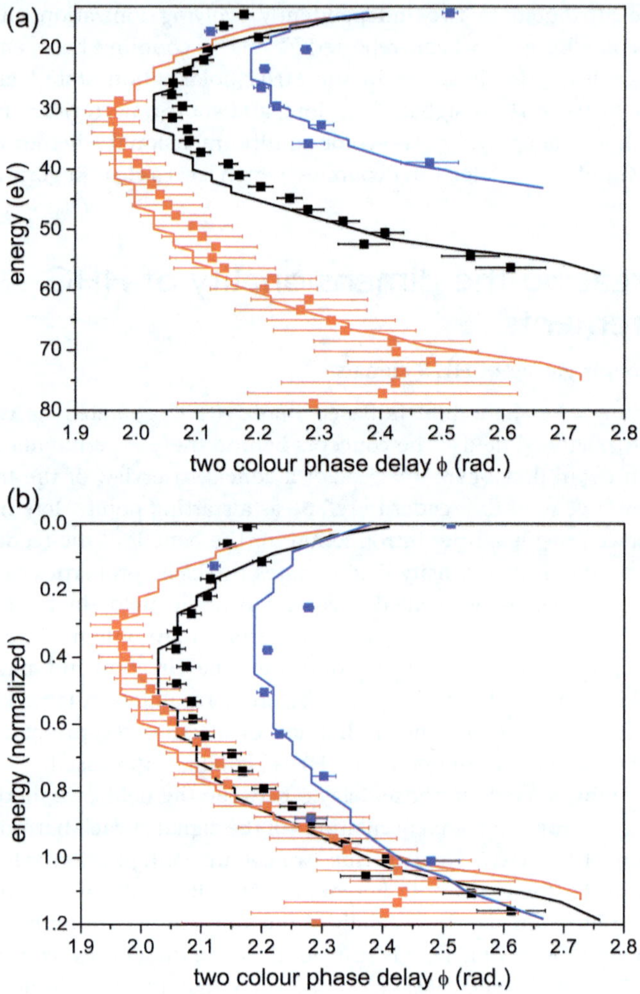

Fig. 4 Single channel ionization in Krypton. (a) Optimal two-colour phase delay ϕ_{max} *vs.* harmonic energy and (b) *vs.* harmonic energy normalized to the classical HHG cut-off. Experimental data are shown as scatter graphs and theoretical calculations are shown as solid lines. Calculated values for ϕ_{max} are found by solving for the times of ionization and recombination within the strong field approximation, propagating the semiclassical electron in the two-colour field and solving for the optimal ϕ. These methods are described in more detail in ref. 44. Red curves: laser wavelength = 1434 nm, intensity = 9 $\times 10^{13}$ W cm^{-2}, corresponding to $U_p = 16.9$ eV and $\gamma = 0.64$. Black curves: laser wavelength = 1300 nm, intensity = 7×10^{13} W cm^{-2}, corresponding to $U_p = 11.5$ eV and $\gamma = 0.78$. Blue curves: laser wavelength = 800 nm, intensity = 1.3×10^{14} W cm^{-2}, corresponding to $U_p = 7.8$ eV and $\gamma = 0.95$.

Fig. 4. The 2D HHG spectra were recorded for different laser wavelengths and intensities, and ϕ_{max} was extracted as a function of harmonic energy for each measurement. The variation of ϕ_{max} with energy follows the same basic curve for all measurements. As the electron energy increases, there is a gradual monotonic increase in ϕ_{max} over nearly the entire range of the HHG spectrum. The exact shape of the curve can be robustly predicted as a function of a single parameter, the well known Keldysh parameter defined as $\gamma = \sqrt{\frac{I_p}{2U_p}}$, where U_p is the electron ponderomotive energy and I_p is the ionization potential. The Keldysh parameter encompasses the important experimental parameters that determine the single channel response, namely the laser wavelength, intensity, and the ionization potential. The relations between the three data sets can be seen more clearly by re-scaling the HHG energy over a range set between zero and unity, where the minimum energy (zero) corresponds to I_p and the maximum to the classical HHG cut-off (Fig. 4(b)). These concepts were confirmed experimentally for a number of other simple atomic systems as well.[44] The single-channel response can be considered as the basic reference for decoding more complex electronic dynamics involving multiple ionization channels.

3.3 Two channel interference

How can we link the 2D measurement with the underlying dynamics and multi-channel contributions? Consider the case of two ionization channels,[7] when the HHG signal can be written as:

$$A_{\text{tot}} = A_1 \exp(i\varphi_1) + A_2 \exp(i\varphi_2), \tag{1}$$

where A_1 and A_2 are the absolute values of the amplitudes of the HHG fields emitted from the two channels and φ_1 and φ_2 are their phases. The total signal is given by $A_{\text{tot}}{}^2 = A_1{}^2 + A_2{}^2 + 2A_1A_2 \cos(\Delta\varphi)$ with $\Delta\varphi$ the phase difference $\varphi_2 - \varphi_1$. The amplitudes of each channel are themselves modulated by the two-colour field, varying sinusoidally with respect to the relative two-colour phase delay ϕ, and labelled by their optimal two-colour phase delays $\phi_{\text{max},1}$ and $\phi_{\text{max},2}$. Such a modulation is then mapped, *via* the interference $\Delta\varphi$ between channels, into the total signal modulation having an optimal phase $\phi_{\text{max}}^{\text{tot}}$.

One mechanism that leads to deviations from the single channel response, as observed in our experiment, arises under two conditions. The first requires a comparable contribution of the two channels – $A_1 \sim A_2$, so that their interference is visible. In contrast, when one of the channels completely dominates the interaction, the total signal will follow the single-channel response.

The second condition requires substantial variation of the relative phase, $\Delta\varphi \sim \pi$. A dramatic shift of $\phi_{\text{max}}^{\text{tot}}$ can originate from a destructive interference between the two channels.[7,56] In this case the interference becomes a differential measurement, mapping small phase differences between $\phi_{\text{max},1}$ and $\phi_{\text{max},2}$ into a significant phase jump observed in $\phi_{\text{max}}^{\text{tot}}$.

Consider the interference between two direct channels, when the total signal is the coherent sum of two independent channels induced separately by the strong laser field. The relative phases between the two channels can be approximated by[17]

$$\Delta\varphi = (I_{p2} - I_{p1})\tau + (\varphi_{r2} - \varphi_{r1}). \tag{2}$$

where I_{p1}, I_{p2} are the ionization energies of the two channels, τ is the excursion time of the continuum electron between ionization and recombination, for the average ionization potential $(I_{p2} + I_{p1})/2$, and φ_{r1}, φ_{r2} are the recombination phases. If ionization is dominated by the X-channel (see Fig. 1), the HHG amplitudes resulting from this channel will dominate over nearly the entire spectrum. The condition $A_1 \sim A_2$ is satisfied near the HHG cut-off or near a spectral minimum, such as the structural minimum around 55 eV in CO_2. Such suppression also manifests itself in the vicinity of nodal planes of the usually dominant channel. In such cases, the X-channel contribution is sufficiently suppressed due to both ionization and recombination. A significant deviation from the single channel optimal phase ϕ_{max} can be observed in these spectral regions when the relative phase between the two channels, $\Delta\varphi$, approaches π.

A different picture could be expected when the two channels are coupled, so that the laser field can induce real (population transfer) or virtual (polarization) transitions between the channels, during the time-delay between ionization and recombination. For a sufficiently strong coupling field, the contribution of the two coupled channels may become comparable anywhere in the HHG spectrum. In addition, the laser-induced A_2 will vary relative to A_1 from one harmonic to another. Therefore, the condition $A_1 \sim A_2$ can be achieved over a large spectral range, not only near the cut-off or a structural minimum. As a result, a smooth deviation from the single channel response over a large spectral range, rather than discrete phase jumps in ϕ_{max}, might be expected.

In Section 5 we show that while all diagonal channels have very similar ϕ_{max}, the laser-driven cross-channels lead to a different optimization of the two-colour delay. Fundamentally, it is related to the entanglement between the electron and the ion. For a given harmonic order, laser-driven "real" or "virtual" transitions in the ion are correlated to changes in the continuum electron trajectory leading to this harmonic. Consider, for example, channels XX and XC. The ionization time is the same for these two channels, but the recombination times are different due to a different final state, C (instead of X), with a different I_p (mathematically, this leads to an extra phase accumulated in the XC channel upon transition to the state C). Since the optimal two-colour delay depends on both ionization and recombination times, the pair of times corresponding to the XX channel will optimize ϕ_{max} for the XX channel, but the shift of the recombination time in the XC channel for the same ionization time as in XX will inevitably lead to a different value of ϕ_{max} for the XC channel. The magnitude of this effect and its appearance in the spectrum depends on the value of ΔI_p between the two coupled channels.

While this general assessment of multichannel interference provides an essential background for our overview of the experimental results in Section 4, Section 5 rigorously analyzes the contributions of the individual direct and cross channels and uses this analysis to reveal the dynamics recorded in experimental 2D spectra.

3.4 Measurement and control of the laser driven dynamics

We explore multiple channels in HHG spectroscopy using a fundamental wavelength in the mid-IR regime.[57] The cut-off frequency scales linearly with $U_p = I/4\omega_0^2$ with I being the laser intensity and ω_0 the laser frequency. Increasing the fundamental wavelength (*i.e.* decreasing the frequency) greatly increases the HHG spectral bandwidth and therefore the available spatial resolution and dynamical range provided in HHG spectroscopy.[58–60]

High harmonics are generated by focusing a 20 fs, 1480 nm, 1.3 mJ self-stabilized tunable optical parametric amplifier (OPA) laser source into a pulsed gas jet. The harmonic spectra were measured with a flat field XUV spectrometer consisting of a spherical variable line space grating focusing onto a micro channel plate coupled to a phosphor screen and imaged with a CCD camera. The second harmonic (SH) gating field is produced using a thin (typical thickness 100 μm) BBO (βBaB_2O_4) nonlinear crystal in a type-I phase matching configuration, which ensures that the SH field is orthogonally polarized with respect to the fundamental field. The group velocity walk off between the beams is compensated using a birefringent crystal (calcite) and IR-coated glass wedges. The sub-cycle delay ϕ of the SH field relative to the fundamental field is controlled by translation of the glass wedges.

The use of the pulsed gas jet at moderate backing pressure (~3–4 bars) produces a supersonic expansion of molecules that are rotationally cooled upon entering the vacuum chamber.[61] For laser-induced molecular alignment,[62] we split off a small portion of the 800 nm Ti:S beam that is used to pump the OPA. The pulses are stretched to 100 fs in order to improve the efficiency of the alignment. The energy in the alignment beam is approximately 250 μJ, with a focused intensity of $\sim 4 \times 10^{13}$ W cm^{-2}. The polarization of the alignment beam relative to the generating beam is set with a zero-order half-wave plate. The alignment and two-colour generating beams are collinearly combined using a dichroic mirror. Spatial filtering and mode matching of all beams is achieved using irises placed near the entrance to the vacuum chamber.

In this work we control the key parameters of our dynamical imaging scheme – the sub-cycle laser driven dynamics between ionization and recombination – by modifying the fundamental laser field intensity and the dipole couplings between different states of the ion. The latter is achieved by variation of the molecular alignment angle.

The alignment pulse excites a rotational wavepacket in the molecular ensemble that undergoes periodic revivals for linear molecules such as CO_2. Changing the polarization of the strong probe pulse relative to the polarization of the aligning pulse allows one to probe the molecular ensemble at different alignment angles. The alignment distribution displays cylindrical symmetry around the axis defined by the polarization direction of the alignment pulse.[37]

Calculations show that the X-channel ionizes most strongly for all alignment angles due to its lower I_p. However, for the channels with higher I_p, the interplay of their electronic structure and the molecular alignment angle plays a significant role in defining the angular dependence of the ionization rate. For example, for molecules aligned perpendicular to the HHG generating field, the A-channel has a higher ionization rate than the B-channel, whereas for parallel alignment, ionization from the B-channel dominates over the A-channel despite its higher I_p.[17,49] Thus, the alignment angle affects the relative contributions of the individual channels, and therefore modulates the phase and amplitude of their interference.

Multidimensional HHG scans were recorded for two fundamental field intensities and two alignment directions. The polarization of the alignment pulse was set parallel or perpendicular to the polarization of the fundamental field that drives HHG, while the fundamental field and its SH are orthogonally polarized. The two-colour field was delayed by 21.2 ps relative to the alignment pulse, which corresponds to the half-revival time for CO_2. Under these alignment conditions, HHG spectra were recorded as a function of the relative phase delay ϕ between the strong fundamental and weak second harmonic fields. The experiment was performed at two intensities of the generating field: $\sim 1.0 \times 10^{14}$ W cm^{-2} and $\sim 1.3 \times 10^{14}$ W cm^{-2}, which we will refer to as the "low" and "high" intensity scans, respectively.

4 Experimental results

To facilitate the visualization of the optimal two-colour delay, which maximises HHG yield for each harmonic, we perform a normalization of 2D HHG spectra. Specifically, for each photon energy, the HHG yield is normalized to its maximal value, achieved at ϕ_{max}. Fig. 5 shows the 2D plots of the normalized HHG energy as a function of the relative two-colour phase delay ϕ for the case of perpendicular alignment. By normalizing the 2D HHG spectrum along the ϕ-axis, such that for each harmonic the magnitude of the signal is equal to one at $\phi = \phi_{max}$, the dependence of ϕ_{max} *versus* photon energy is easier to visualize.

Both the high (Fig. 5(a)) and low (Fig. 5(b)) intensity scans produce similar results, with a slow, monotonic increase in ϕ_{max} with respect to energy. Since the I_p of the X-channel in CO_2 is nearly the same as that of Krypton, we expect them to have the same characteristic 2D HHG spectrum.

As a confirmation, in Fig. 5 and 6 we plot the expected ϕ_{max} for a single channel with the same I_p as the X-channel of CO_2. The single channel calculation is based on a model that relies only on the strong field approximation for predicting ϕ_{max} (see Fig. 4). This model was used for calculating the predicted HHG signal dependence on the two-colour phase delay in our prior work on single channel dynamics in atomic systems.[44] The experiment measures the relative two-colour phase ϕ only. We calibrate the absolute value of ϕ_{max} for CO_2 by assuming that the X-channel dominates the signal for low harmonic energies. This is a reasonable assumption based on the concepts presented in Section 3.3. The absolute value of ϕ_{max} for a single channel (the X-channel) can then be calculated using the model, and used to calibrate the absolute value of ϕ_{max} for the entire harmonic spectrum.[7,44] Once the absolute ϕ_{max} is determined, we can see that for both intensities in perpendicularly aligned CO_2, the measured signal closely follows the single X-channel response over the entire HHG spectral range. A small deviation from the single channel response appears near the cut-off, in particular for high intensity. Contributions from additional channels, in particular the A state of the cation, only appear in this region of the HHG spectrum.

A significantly different response is observed as we rotate the molecular system to be parallel to the strong fundamental field. Fig. 6 shows the normalized HHG energy as a function of the relative two-colour phase delay for the two intensities. For the low intensity, the HHG signal is suppressed almost completely in the structural minimum around 55 eV in this range. In contrast, for the high intensity, the HHG signal is suppressed by over an order of magnitude in this region of the spectrum, but the remaining signal is still well within the sensitivity of the measurement. For both intensities, ϕ_{max} deviates significantly from the expected single channel dynamics, except at very low harmonic energies. Importantly, comparing the high (Fig. 6(a)) and low (Fig. 6(b)) intensities shows a dramatic difference. In particular, at higher energies, we observe a pronounced shift in

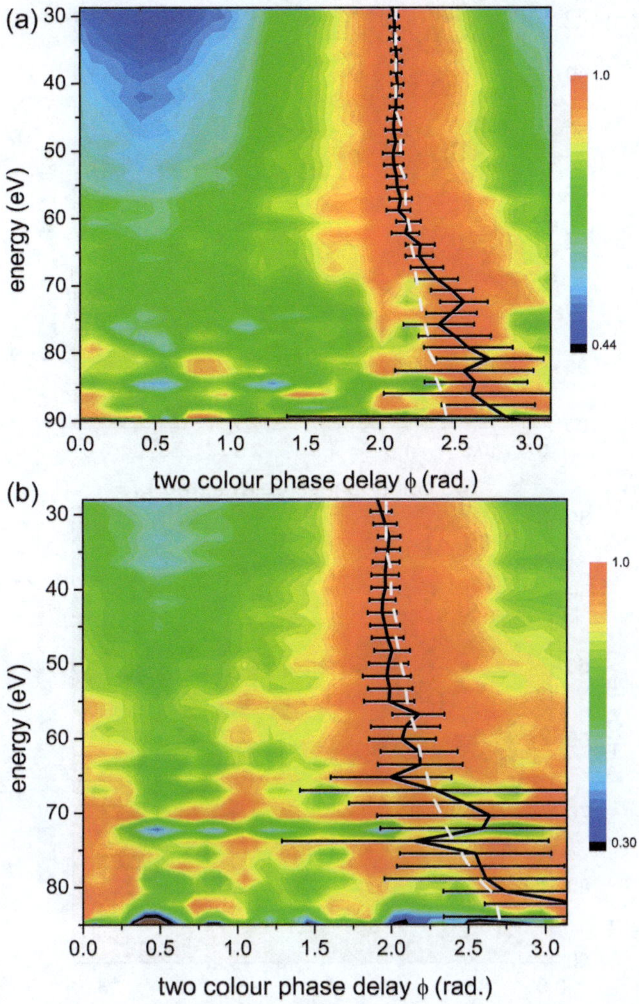

Fig. 5 Normalized experimental 2D colour plots of HHG signal *vs.* relative two-colour phase delay ϕ for high (1.3×10^{14} W cm^{-2}) (a) and low (1.0×10^{14} W cm^{-2}) (b) intensities of the strong 1480 nm generating field for perpendicular alignments (alignment field polarization is perpendicular to that of the generating field). Bold black lines: maximum two-colour phase ϕ_{max}, with experimental error bars. White dashed lines: calculated maximum two-colour phase for a single ionization channel with the same I_p as the X-channel in CO_2 (13.8 eV).

ϕ_{max} of about 1.5 radians near 70 eV for the low intensity, and near 95 eV for the high intensity. This feature is similar to the phase shift observed in our earlier experiments performed at 800 nm.[7] The shift around 75 eV at high intensity has a different origin as will be discussed below.

A more direct comparison between the experimental data and the expected single channel response appears in Fig. 7. Here we plot the comparative phase difference between ϕ_{max} extracted from the high and low intensity measurements and the calculated phase, ϕ^{X}_{max} associated with the single channel (channel X)

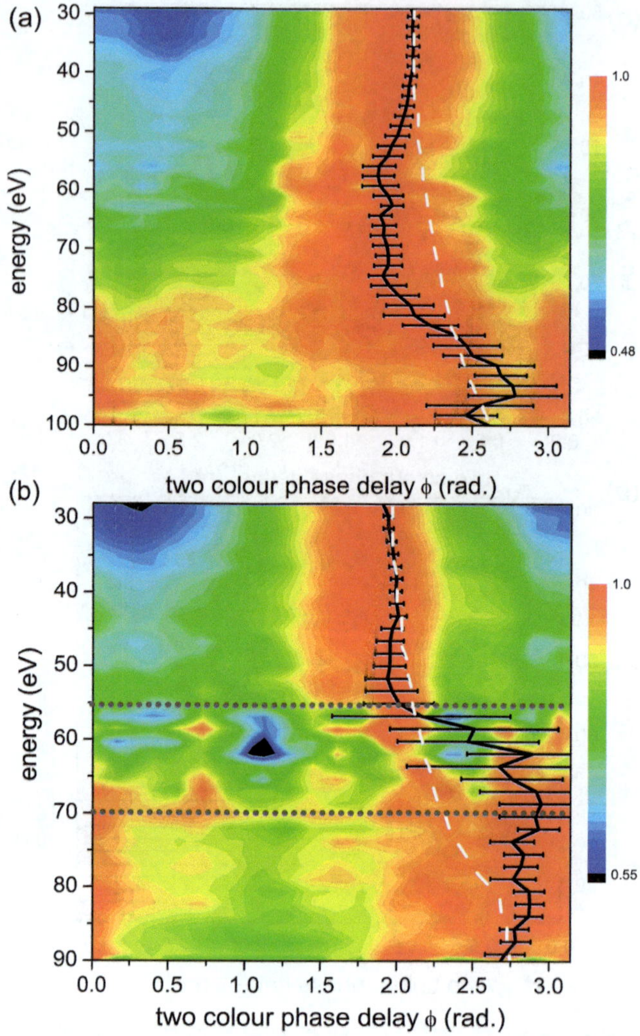

Fig. 6 Normalized experimental 2D colour plots of HHG signal *vs.* relative two-colour phase delay ϕ for high (1.3×10^{14} W cm^{-2}) (a) and low (1.0×10^{14} W cm^{-2}) (b) intensities of the strong 1480 nm generating field for parallel alignments (alignment field polarization is parallel to that of the generating field). Bold black lines: maximum two-colour phase ϕ_{max}, with experimental error bars. White dashed lines: calculated maximum two-colour phase for a single ionization channel with the same I_p as the X-channel in CO_2 (13.8 eV), as in Fig. 5. The horizontal grey dotted lines in panel (b) mark the energy region, from 55 eV to 70 eV, where the non-normalized experimental HHG signal is extremely weak, making reliable assignment of ϕ_{max} in this region very challenging.

response. For low harmonics, ϕ_{max} follows the single channel case for both intensities. Above 45 eV, the two curves sharply diverge. For the low intensity, $\phi_{max}^{X} - \phi_{max}$ shifts toward larger values of ϕ compared with a single channel, having the largest deviation at 70 eV. At high intensity $\phi_{max}^{X} - \phi_{max}$ shifts towards the opposite direction, having the maximal deviation at 75 eV. The comparative phase difference with respect to energy is a key indicator for the multichannel

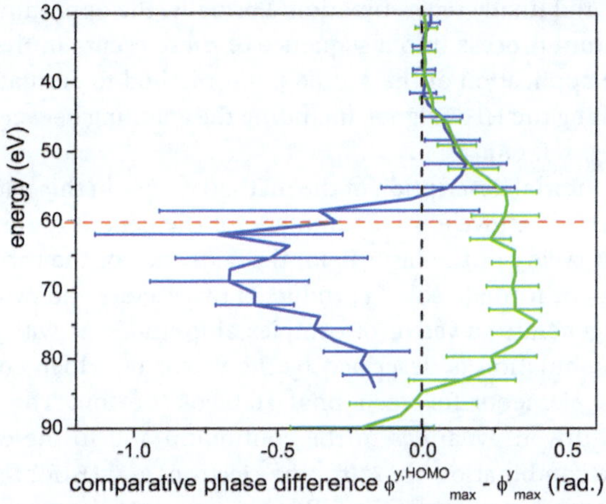

Fig. 7 Optimal two-colour phase ϕ_{max} for parallel alignments, compared with that for a single channel. The comparative phase difference between the measured and calculated single channel ϕ_{max} is shown as a function harmonic energy. Blue curve: low intensity (1.0 $\times$ 10^{14} W cm^{-2}), green curve: high intensity (1.3 $\times$ 10^{14} W cm^{-2}). The dashed black vertical line at zero phase difference corresponds to the expected behaviour of a single channel with the same I_p as the X-channel in CO_2. Red dashed line: centre position of the structural minimum.

dynamics that are encoded in the 2D spectrum. Note that for low intensity and parallel alignment the non-normalized experimental HHG signal is extremely weak in the energy region from 55 eV to 75 eV, thus reliable assignment of ϕ_{max} in this region becomes very challenging. This region is indicated by horizontal lines in Fig. 6(b).

We note that modulations of the HHG spectrum due to structural features, such as the minimum near 55 eV in aligned CO_2, do not hinder the detection of multichannel dynamics in our experiment. In fact, the structurally-based reduction of the signal from the HOMO relative to the deeper orbitals leads to more comparable amplitude contributions between the channels.[17,54] This leads to larger and more easily measurable shifts in ϕ_{max}[56] and increases the sensitivity of the measurement to the channels with higher I_p.

In the following section, we analyze the 2D HHG response and the encoded dynamics in greater detail. We identify the contribution of different channels, analyze their coupling, and describe the approach which allows us to decode the relative phases of the different cationic channels formed at the ionization step, with sub-cycle time-resolution.

5 Theoretical analysis of the experimental 2D HHG spectra

5.1 The AZUR method

In the time domain, the HHG process can be understood as a sequence of several events which happen within a single half-cycle of the fundamental field: ionization, laser-driven dynamics in the ion accompanied by electron propagation in the continuum, and finally recombination. Formally, the opportunity to factorize the whole quantum process into a sequence of these events in the time domain comes from the application of the saddle point method to evaluate the multiple integrals describing the HHG signal, including the coupling between the different channels (see *e.g.* ref. 63).

The detailed tutorial description of the method we use in this paper is available on the arXiv.[64] Here we give a brief summary for the benefit of the reader.

For each half cycle j of the laser field, the sequence of the three steps in the HHG process can be formalized as a product of two vectors and two matrices. The ionization step generates a vector of complex amplitudes $\mathbf{A}^j$, with j labelling the half-cycle. Recombination is described by the vector $\mathbf{R}^j$, which contains recombination matrix elements for each final state of the ion. The two matrices, describing laser-driven dynamics in the continuum and in the cation between ionization and recombination are $\mathbf{Z}^j$ (for the electron) and $\mathbf{U}^j$ (for the ion) (see *e.g.* ref. 63). Note that factorization of the full propagator into the product of $\mathbf{Z}^j$ and $\mathbf{U}^j$ is an approximation. For the recombination step, this approximation is motivated by short duration of the recollision event. The laser field is "frozen" during recombination, so that real transitions between the states of the ion do not occur during this time. The same approximation is, however, questionable during the ionization step. The tunnelling electron departs with low energy and can strongly interact with the ion within the ionization window, leading to correlation-driven processes in strong-field ionization[45,46] and potentially to changes in the amplitudes and phases of the ionization vector $\mathbf{A}^j$. As we will discuss in the next section, the experimental results may allow us to check the validity of this approximation.

The complete signal is given by the time-dependent induced dipole as follows:

$$\mathbf{D}^j(t) = \mathbf{R}^j\mathbf{U}^j\mathbf{Z}^j\mathbf{A}^j \tag{3}$$

where j labels the half-cycle. The ionization vector is a column with entries $\mathbf{A}_\mathrm{n} = a_\mathrm{n}(t_{\mathrm{i,n}}(t))$. These entries depend on the ionization time $t_{\mathrm{i,n}}(t)$, which, in turn, depends on the recombination time t and the ionization channel n = X, A, B, C *via* channel-specific ionization potential. The sub-cycle dynamics of ionization enters into the vector $\mathbf{A}_\mathrm{n}$ *via* the imaginary component of the Volkov action

$$S_\mathrm{n}^\mathrm{V}(\mathscr{R}[t_{\mathrm{i,n}}], t_{\mathrm{i,n}}) = \frac{1}{2}\int_{t_{\mathrm{i,n}}}^{\mathscr{R}[t_{\mathrm{i,n}}]} \mathbf{k}_\mathrm{n}{}^2(\tau)\mathrm{d}\tau, \tag{4}$$

and *via* the additional exponential factor $\exp(-I_{\mathrm{p,n}}\mathscr{I}[t_{\mathrm{i,n}}])$. Here $\mathscr{R}[t_{\mathrm{i,n}}]$, $\mathscr{I}[t_{\mathrm{i,n}}]$ are the real and imaginary parts of the complex ionization time $t_{\mathrm{i,n}}$, $\mathbf{k}_\mathrm{n}(\tau) = \mathbf{k}_{t,\mathrm{n}} + \mathbf{A}(\tau)$, $\mathbf{A}(t)$ is the vector potential of the laser field, and $\mathbf{k}_{t,\mathrm{n}}$ is the complex saddle point solution for the electron canonical momentum, specified for every moment of recombination t.

The recombination vector is a row, with entries given by the corresponding recombination dipoles $\mathbf{R}_\mathrm{n} = \mathbf{r}_\mathrm{n}(\mathbf{k}_\mathrm{n}(t), \theta)$. These depend on the channel n, the channel-specific electron kinetic momentum at the recombination $\mathbf{k}_\mathrm{n}(t)$, and the molecular alignment angle θ. Note that in orthogonally polarized two-colour fields, $\mathbf{k}_\mathrm{n}(t)$ modifies the recombination angle for every half cycle. This is why we keep the information about the molecular alignment angle and the vector $\mathbf{k}_\mathrm{n}(t)$ in the argument of $\mathbf{R}_\mathrm{n}$.

The diagonal continuum propagation matrix $\mathbf{Z}$ has elements $Z_{nn} = \exp(-iS^{V}_{n}(t,\mathscr{R}[t_{i,n}]))a^{prop}_{n}(t,t_{i,n})$, which describe the laser-driven electron dynamics and contain the Volkov action and the pre-factor a^{prop}_{n} describing the spreading of the electron wave-packet. Finally, the matrix $\mathbf{U}$ describes laser-driven dynamics in the ion. Its respective matrix elements are the amplitudes of laser-induced transitions between ionic states m and n: $u_{mn}(t, \mathscr{R}[t_{i,n}])$ (see ref. 63), with the energies of the ionic states calculated from the ground state energy of the neutral. The whole expression, described in eqn (3), has to be read from right to left, following the standard convention of quantum mechanics and the Hebrew tradition, hence the name AZUR.

The full high harmonic dipole comes from summing over all odd and even half-cycles of the fundamental field, $\mathbf{D}(t) = \mathbf{D}(t)^{j_o} + \mathbf{D}(t)^{j_e}$. In linearly polarized fields and for a symmetric ground state, $\mathbf{D}^{j+1}(t) = -\mathbf{D}^{j}(t + \pi/\omega)$. However, in two-colour fields this symmetry between the two consecutive half-cycles of the fundamental is broken, leading to the generation of even harmonics. In molecules, the two successive half cycles $j_o = 2k - 1$ and $j_e = 2k$ may also correspond to different ionization and recombination angles. Note also that all saddle point solutions depend on the two-colour delay ϕ. Thus, the full high harmonic dipole becomes a function of the two-colour phase delay ϕ. We suppress the index describing this dependence to simplify notations.

Now consider the auxiliary dipole vectors including all steps except the recombination: $\mathbf{d}^{o}(t) = \sum_{j_o}\mathbf{d}^{j_o}(t) = \sum_{j_o}\mathbf{U}^{j_o}\mathbf{Z}^{j_o}\mathbf{A}^{j_o}$ and $\mathbf{d}^{e}(t) = \sum_{j_e}\mathbf{d}^{j_e}(t) = \sum_{j_e}\mathbf{U}^{j_e}\mathbf{Z}^{j_e}\mathbf{A}^{j_e}$. In our method, we first calculate the Fourier transforms of the auxiliary dipole vectors separately for even and odd half-cycles, yielding $\mathbf{d}^{o,e}_{\mathscr{F}}(N) = \mathscr{F}[\mathbf{d}^{o,e}(t)]$, and then include the recombination step. To this end, we multiply odd and even components of the auxiliary frequency domain dipole vectors $\mathbf{d}^{o,e}_{\mathscr{F}}(N)$ by the respective recombination vectors for odd $\mathbf{R}^{j_o}(N,\theta^{j_o})$ and even half-cycles $\mathbf{R}^{j_e}(N,\theta^{j_e})$, where $\mathbf{R}^{o,e}$ are taken at the respective different recombination angles $\theta^{o,e}$ and energies $N\omega$. Thus, finally the HHG light is given by

$$\mathbf{D}(N,\theta,\phi) = \mathbf{R}^{o}(N,\theta^{o})\mathbf{d}^{o}_{\mathscr{F}}(N) + \mathbf{R}^{e}(N,\theta^{e})\mathbf{d}^{e}_{\mathscr{F}}(N). \tag{5}$$

To calculate this expression we need to know the respective vectors and matrices.

The ionization vector is obtained using the method described in ref. 49 and describes strong -field ionization for uncoupled X, A, B, and C channels. Both phase and amplitude corrections to these quantities, which may arise *e.g.* due to channel coupling during ionization, are the targets of reconstruction from the experimental data.

The recombination matrix elements are calculated using the *ab initio* R-matrix method (UKRMol), and specifically its recent extension to photoionization from aligned molecules.[65,66] We have developed two photoionization models. The first model (for full details see ref. 65, model 3) uses a sophisticated modelling of correlation/polarization but uses a version of UKRmol which limits the calculation in several parameters. Subsequently, in collaboration with J. Gorfinkiel, we have significantly improved on the codes which has allowed us to refine the original model and to generate more accurate photoionization dipoles. The new model differs from the original one in several aspects.

First, the original model used a description of the continuum based on Gaussian-type (GTO) continuum basis functions with angular momentum up to $L = 5$. In the new model we have been able to include continuum functions with $L = 6$. Inclusion of these functions has significantly extended the upper limit of the energy range for which we are able to generate accurate photoionization dipoles, from approx. 50 eV to approx. 90 eV of photon energy. This extension is crucial for the long wavelength driving laser fields, considered herein. Second, the original model included 96 states in the close-coupling expansion of the scattering wavefunction. In the new model we have included 300 target states spanning the energy range of approx. 30 eV. This has allowed us to further improve the description of polarization/correlation and to further suppress the appearance of unphysical pseudoresonances.

To enable these improvements we have made the following changes to the R-matrix codes. First, we have developed a completely new parallel integral code which is capable of generating the molecular integrals in quadrupole precision. This enabled us to include in our calculations a large number of continuum basis functions without encountering problems with linear dependencies among the diffuse GTO functions. Second, MPI-parallelization of the outer region calculation was implemented. This modification was essential to allow us to include a large number of target states in the calculation. Finally, to generate data for use in HHG calculations, the Dyson orbitals, photoionization dipoles and target transition dipole moments have to be all phase-consistent. Previously, the phase-correction had to be applied manually using a laborious procedure. We have implemented changes to the codes which eliminated a number of arbitrary phases, thus allowing us to automatically generate phase-consistent data.

High quality recombination amplitudes and phases are essential for our HHG calculations due to the interferometric sensitivity of the HHG signal. In the present calculations, we know the recombination dipoles with high accuracy in the region up to 90 eV of photon energy.

The matrix **U** is also modelled with adequate accuracy. To model laser-driven dynamics of bound states in the CO_2 ion we first included around 160 correlated target states (the corresponding dipole couplings and energies were taken from our *ab initio* R-matrix calculations). Finally, we could show that the four essential states X, A, B, and C are sufficient to describe these dynamics with adequate accuracy (see also ref. 67). We have also checked that the coupling dipoles between these essential states resulting from the quantum chemical calculations based on the CASSCF method yield the same dynamics. These (CASSCF) coupling dipoles (with appropriate phase correction) we have used in the present calculations.

The continuum propagation matrix **Z** is standard and has been verified over the years in one-electron studies of HHG in atoms, see *e.g.* ref. 7 and 68.

Having accurate models for the recombination and propagation steps, and using the interferometric sensitivity of the 2D HHG signal allows one to reconstruct the initial conditions dictated by the ionization process. Specifically, we can reconstruct the possibly missing information about the sub-cycle dynamics of multielectron rearrangement during ionization, which occurs in the presence of both correlation-induced and laser-induced coupling between the channels. These dynamics are encoded in the relative phases between different ionic states set up by the ionization process, and also impact the ionization amplitudes. In the following we describe how these quantities can be extracted from the experimental observables of 2D HHG spectroscopy.

5.2 Reconstructing sub-cycle ionization dynamics from 2D HHG spectra

The comparison between theory and experiment allows us to assess the accuracy of the ionization vector used in the calculation. Note that the approximations we have made in describing the ionization step and replacing the exact propagator with **UZ** are formally equivalent to using a different ionization vector **A** in eqn (3); given a very large electron oscillation amplitude in an intense IR field, the factorization **UZ** is appropriate just after the ionization step.

The propagator **UZ** is unitary and performs a well-defined rotation of the initial vector **A** between ionization and recombination. To find the corrected ionization vector **A**, we need to perform an additional rotation of the vector $\mathbf{d} = \mathbf{UZA}$ before applying the recombination operator **R**, yielding the rotated vector $\mathbf{d}'$. This extra rotation should allow us to match the observed HHG spectrum. Then, performing the back propagation $(\mathbf{UZ})^{-1}$ of the rotated vector $\mathbf{d}'$ we can determine the corrected ionization vector $\mathbf{A}'$.

We perform the rotation by introducing a diagonal matrix Φ with entries $\Phi_{\rm nm} = \rho_{\rm n}\exp(-{\rm i}\varphi_{\rm n})$, where n = X, A, B, C, $\varphi_{\rm n} \in [0, 2\pi]$; only three relative phases matter. In the present analysis, we assume $\varphi_{\rm n}$ to be constant as a function of the photon energy. We also allow for varying the strength of the channels X, A, B, C, by varying $\rho_{\rm n}$. Specific reconstructed values of $\varphi_{\rm n}$ and $\rho_{\rm n}$ will be presented and discussed below. In the calculation we ensure that the norm of vector $\mathbf{d}(t)$ is preserved upon multiplication by matrix Φ.

Thus, the 2D HHG spectrum for a given angle of molecular alignment θ becomes:

$$\tilde{\mathbf{D}}(N,\theta,\Phi,\phi) = \sum_{\rm o,e}[\mathbf{R}^{\rm o}(N,\theta^{\rm o})\Phi\mathbf{d}^{\rm o}(N) + R^{\rm e}(N,\theta^{\rm e})\Phi\mathbf{d}^{\rm e}(N)], \tag{6}$$

where $\mathbf{R}^{\rm o}$ and $\mathbf{R}^{\rm e}$ are associated with odd and even half-cycles of the field, respectively. The matrix Φ modifies the elements of the vector **d** as $d'_{\rm n}(t) = \exp(-{\rm i}\varphi_{\rm n})\rho_{\rm n}d_{\rm n}(t)$, which are then used to calculate the 2D harmonic spectrum and compare it with the experimentally measured one.

5.3 Details of the HHG calculations

To connect the theoretical analysis to the experiment, we note that the 1D experimental HHG spectra exhibit a strong minimum around 55 eV, associated with the structure of the recombination matrix element correlated to the X state of the cation. In our *ab initio* calculations of angle-resolved recombination in a CO_2 molecule,[65] the deep minimum at this energy arises for molecules aligned at 40°, see Fig. 8. Therefore, for our calculations, we coherently add the calculated HHG signals arising from molecules aligned at ±35°, ±40°, and ±45°. We note that high harmonic fields corresponding to "positive" and "negative" alignment angles, *e.g.* ±40° are different due to the presence of the perturbative second harmonic field. Due to the very weak second harmonic gate field used in the experiment, almost no even harmonic signal was detected. Nonetheless, phases for the parallel high harmonic light for "positive" and "negative" alignment angles are different and they depend on the two-colour delay.

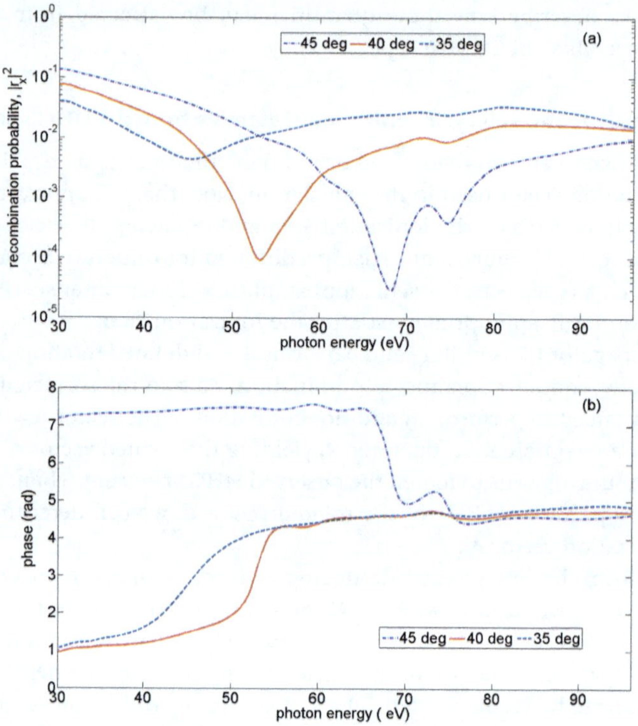

Fig. 8 Recombination strength for X-channel *vs.* the emitted photon energy, for three alignment angles of the CO_2 molecule: 35°, 40°, and 45°. (a) The squared modulus of the recombination amplitude, (b) the recombination phase.

The calculations of 2D spectra were performed for a $\lambda = 1480$ nm, 4-cycle pulse with constant intensity, for laser field strengths ranging from $F = 0.059$ a.u. to $F = 0.042$ a.u. with the step $\Delta F = 0.001$ a.u. Focal averaging is performed using these sets of calculations, for the peak field strengths $F = 0.057$ a.u. ("high" intensity) and $F = 0.051$ a.u. ("low" intensity). These fundamental peak field strengths were chosen to match the experimentally observed cut-offs. The second harmonic field strength was set to $F_2 = 0.14F$.

5.4 Channel-resolved contributions to high harmonic spectrum

One of the most surprising findings of this work is the sensitivity of the 2D HHG spectra to laser-driven cross-channels. To explain the underlying physics, we start with the analysis of high harmonic channels pertinent to our experiment. These channels can be separated into three groups. The first group includes all diagonal channels: XX, AA, BB, CC (the first letter indicates the ionization channel, the second letter indicates the recombination channel), see Fig. 9(a). The second group of channels includes all channels originating from the X-channel: XX, XA, XB, XC. The cross-channels XA, XB, XC originate mainly due to laser-induced (sub-cycle) polarization, *i.e.* they are predominately associated with virtual laser-driven transitions between the states of the cation. Finally, the last group of channels represent strongly coupled cross-channels CB and BC: laser-driven transitions

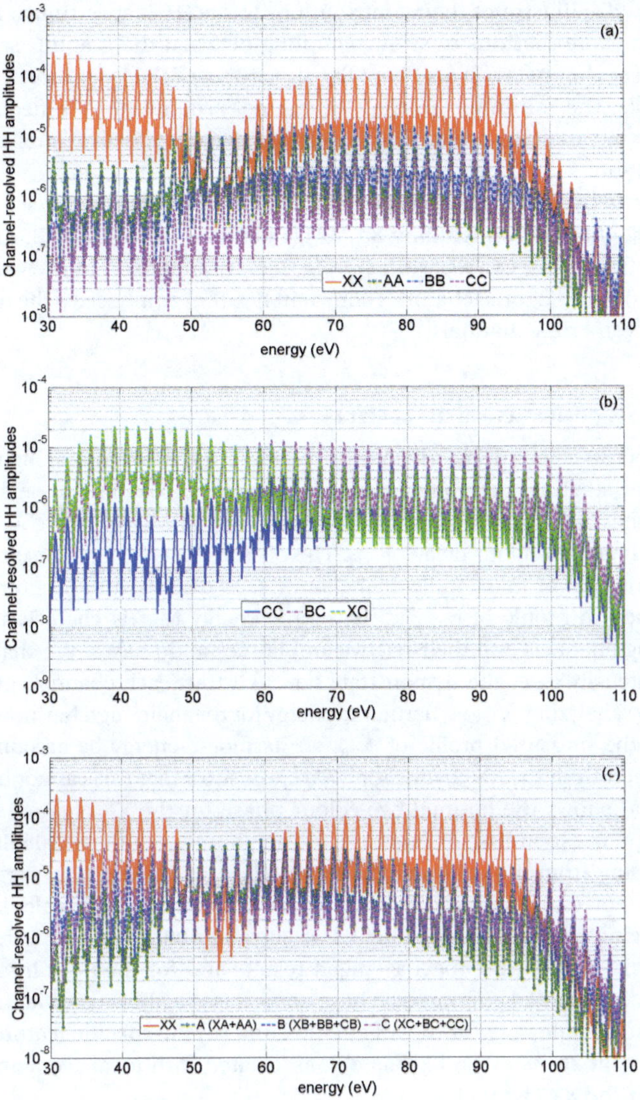

Fig. 9 Channel-resolved contributions to the high harmonic spectrum. Theoretical results for high intensity and parallel alignment. The peak field strength $F = 0.057$ a.u. The second harmonic field is set to $F_2 = 0.14F$, the fundamental wavelength is 1480 nm. The results are averaged over a distribution of laser intensities in the focus. The two-colour delay is 1.98 rad. (a) Amplitudes of direct HHG channels, XX (red, solid), AA (green, dashed, stars), BB (blue, dash-dot), CC (magenta, dashed). (b) Amplitudes of HHG channels recombining to state C: CC (blue, solid), BC (magenta, dash-dotted), XC (green, dashed). (c) Amplitudes of HHG channels grouped by recombination: XX (red, solid), A (includes AA and XA) (green, dashed, stars), B (includes BB, CB, XB) (blue, dashed), C (includes XC, CC, BC) (magenta, dash-dotted).

between the cationic states B and C that lead to real population transfer. Our following analysis shows that cross-channels leave a mark in the two-dimensional HHG spectrum. The channels recombining to the C-state become particularly visible. Fig. 9(b) illustrates their contributions to the HHG spectrum at the level of amplitudes. Note that due to strong coupling between the B and C states of the cation, direct channels can also bear the signatures of the laser-driven dynamics. For example, the minimum in the CC channel around 46 eV reflects a loss of population in this state due to the laser-driven dynamics between ionization and recombination.

It is instructive to group all HHG channels into four groups by recombination: channel XX, channel A (including XA, AA), channel B (including BB, XB, CB) and channel C (including CC, BC, XC), see Fig. 9(c). Comparison of their amplitudes with those of the diagonal channels shown in Fig. 9(a) emphasizes the importance of laser-driven cross-channels.

5.5 Analysis of theoretical 2D spectra

The value of two-colour delay, which corresponds to the maximum of high harmonic intensity for a given harmonic, plays an important role in our analysis. This observable has been introduced and analysed in our earlier work.[7,42–44,56] We refer to it as the "optimal two-colour delay" ϕ_{max}. The observed HHG signal results from the coherent addition of many individual channels, as discussed in Section 2. This in turn produces a profile of ϕ_{max} *vs.* harmonic energy that is characteristic of the contributing channels and their relative amplitudes and phases. We shall focus on the high intensity case, and demonstrate how to extract each channel contribution. We begin by analyzing ϕ_{max} *vs.* harmonic energy for the main high harmonic channel XX. Could the measured profile of ϕ_{max} *vs.* harmonic energy be accounted for by a contribution from the XX channel only? We will show that this cannot be the case.

Fig. 10(a) shows the optimal two-colour phase for the XX channel at parallel alignment (corresponding in our case to the coherent sum of calculated HHG signals arising from molecules aligned at $\pm 35°$, $\pm 40°$, and $\pm 45°$), perpendicular alignment (corresponding in our case to the coherent sum of calculated HHG signals arising from molecules aligned at 80° and 100°), for high intensity. We also compare ϕ_{max} for the focal averaged results and for the peak intensity ($F = 0.057$ a.u.), for parallel alignment. These comparisons allow us to identify (i) the characteristic shape ϕ_{max} for the XX channel and (ii) the specific features (jumps) below 60 eV (at 30–35 eV and 50–55 eV), associated with focal averaging and the structure of the XX channel.

The 0.6 rad jump in ϕ_{max} around 55 eV is associated with the structural minimum in recombination to the X-state, but has a purely interference nature. The interference is between the contributions of two adjacent half-cycles. The signal for each half-cycle, as a function of the two-colour delay, is slightly different due to the difference in the recombination angles. Thus, the modulation of the harmonic signal *vs.* two-colour delay is slightly sheared for the two successive half-cycles. As a result, their interference leads to a differential signal when the half-cycle contribution amplitudes change sign. This can occur in the vicinity of any spectral minimum, either structural, or caused by depletion and a sign change in the laser-driven population amplitude for a particular channel. Better angular averaging involving many more angles (with a smaller step than used here) is likely to smooth out the feature around 55 eV.

While the structural minimum around 55 eV is observed in the 1D spectra for high intensity and parallel alignment (see Fig. 3) it does not manifest itself in the

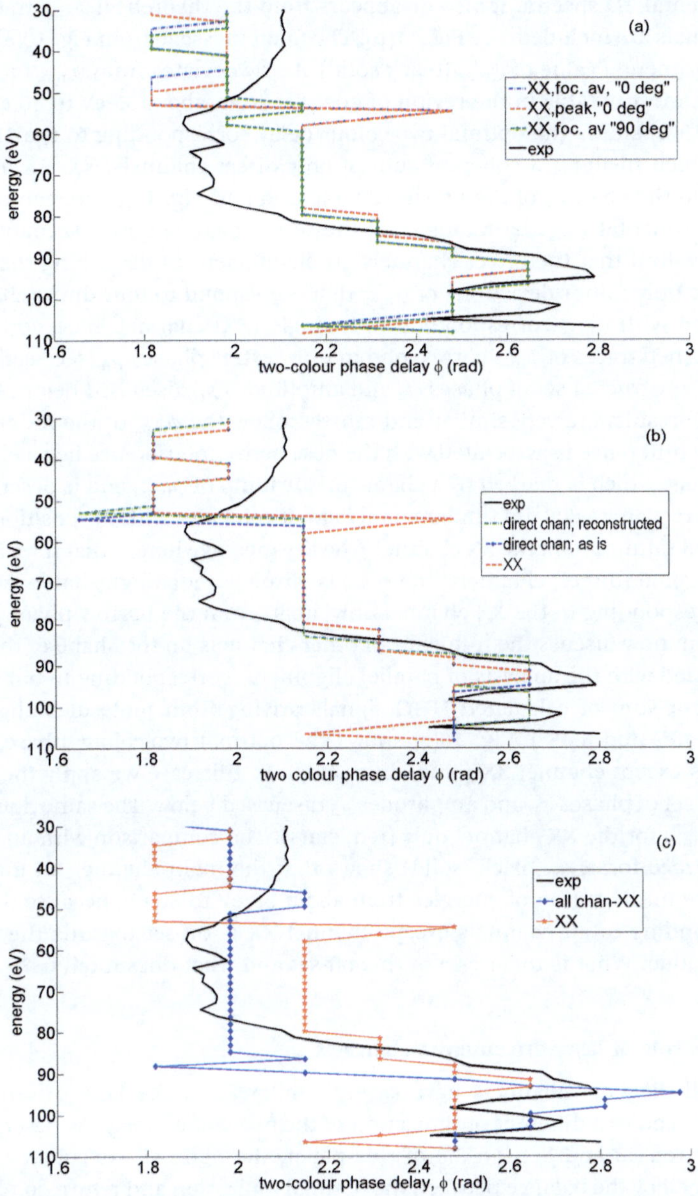

Fig. 10 Analysis of channel contributions to the shape of the optimal two-colour delay ϕ_{max}: theoretical results for high intensity and parallel alignment. The peak field strength $F = 0.057$ a.u. The second harmonic field is $F_2 = 0.14F$, the fundamental wavelength is 1480 nm. The results are averaged over the distribution of laser intensities in the focus. The experimental value of ϕ_{max} (solid, black) is given for reference in all panels. (a) ϕ_{max} resulting from the XX channel only, parallel alignment, focal averaged (green, dotted), perpendicular alignment, focal averaged (blue, dash-dotted), ϕ_{max} for a single field strength corresponding to the peak value of $F = 0.057$ a.u. (red, dashed). (b) ϕ_{max} resulting from direct channels only, parallel alignment, focal averaged: as is, *i.e.* with zero extra phases $\varphi_n = 0$ (blue, dashed-dotted), with reconstructed phases φ_n and weights ρ_n (green, dotted), ϕ_{max} resulting from the XX channel only, parallel alignment, focal averaged (red, dashed). (c) ϕ_{max} for all channels without the XX channel (blue, diamonds), ϕ_{max} for the XX channel only (red, stars), experimental ϕ_{max} (solid, black).

experimental 2D spectra. It also disappears from the theoretical ϕ_{max} trace once all channels are included, see Fig. 11(b). The main message from Fig. 10(a) is that the experimental value of ϕ_{max} (black solid) clearly deviates from ϕ_{max} for the XX channel, most notably in the region of energies from about 55 eV to 80 eV.

Fig. 10(b) shows the optimal two-colour delay corresponding to the 2D spectrum, which includes a coherent sum of only direct channels: XX, AA, BB, CC. Similar to the analysis of the previous paragraph and Fig. 10(a), we can try to fit the experimental ϕ_{max} by taking only the direct channels into account. Once again, we find that the direct channels are insufficient in describing the data – there are only minor deviations of ϕ_{max} that correspond to four direct channels, compared with ϕ_{max} corresponding to a single (XX) channel. Blue dotted and green dashed traces of ϕ_{max} correspond to zero "extra" phases φ_n (see Section 5.2) and a reconstructed set of phases φ_n and amplitudes ρ_n discussed below. In both cases the results are very similar and mostly follow the ϕ_{max} of the XX channel. The only difference is associated with the destructive interference between X and B channels, which is marked by a characteristic jump of ϕ_{max}, and is described in detail in ref. 7 for a 800 nm fundamental field. It coincides with the position of the structural minimum in the XX channel. The key message here is that if we include only diagonal (direct) channels, the ϕ_{max} is virtually indistinguishable from the one corresponding to the XX channel only, apart from the narrow phase jump.

We can now discuss the influence of other channels on the shape of the ϕ_{max}. We proceed with the analysis of parallel alignment (corresponding in our case to a coherent sum of calculated HHG signals arising from molecules aligned at $\pm 35°$, $\pm 40°$, and $\pm 45°$). Fig. 10(c) shows the optimal two-colour phase for all channels except channel XX (blue, diamonds). In this case we apply the reconstructed set of phases φ_n and amplitudes ρ_n discussed below. The same figure also shows ϕ_{max} for the XX channel only (red, stars). The comparison with an experimental trace for ϕ_{max} (black, solid) shows that the main channel XX must not dominate in the region of energies from about 55 eV to 80 eV, because the ϕ_{max} corresponding to all channels except channel XX is off-set towards the experimental value. What is the origin of this off-set and what does it tell us?

5.6 The role of laser-driven cross-channels

While all diagonal channels have a very similar ϕ_{max}, the laser-driven cross-channels lead to a different optimization of the two-colour delay. As described in detail in ref. 43 and Section 3.3 of this paper, the optimal two-colour delay is determined by the balance between the optimal ionization and return conditions, which are, in turn, determined by ionization and recombination times. For cross-channels, such as *e.g.* XB, XC and XA, the ionization and recombination times correspond to different ionization potentials I_p. For direct (diagonal) channels, such as XX they correspond to the same I_p. Therefore, the two-colour delay $\phi = \phi_{max}$ that maximises the HHG signal for direct channels, will not be optimal for cross-channels.

Fundamentally, this difference is related to the entanglement between the electron and the ion. For a given harmonic order, if a laser-driven "real" or "virtual" transition occurs in the ion, the continuum electron trajectory leading to this harmonic must adjust accordingly. For example, channels XX and XC will have the same ionization time but different recombination times for the same

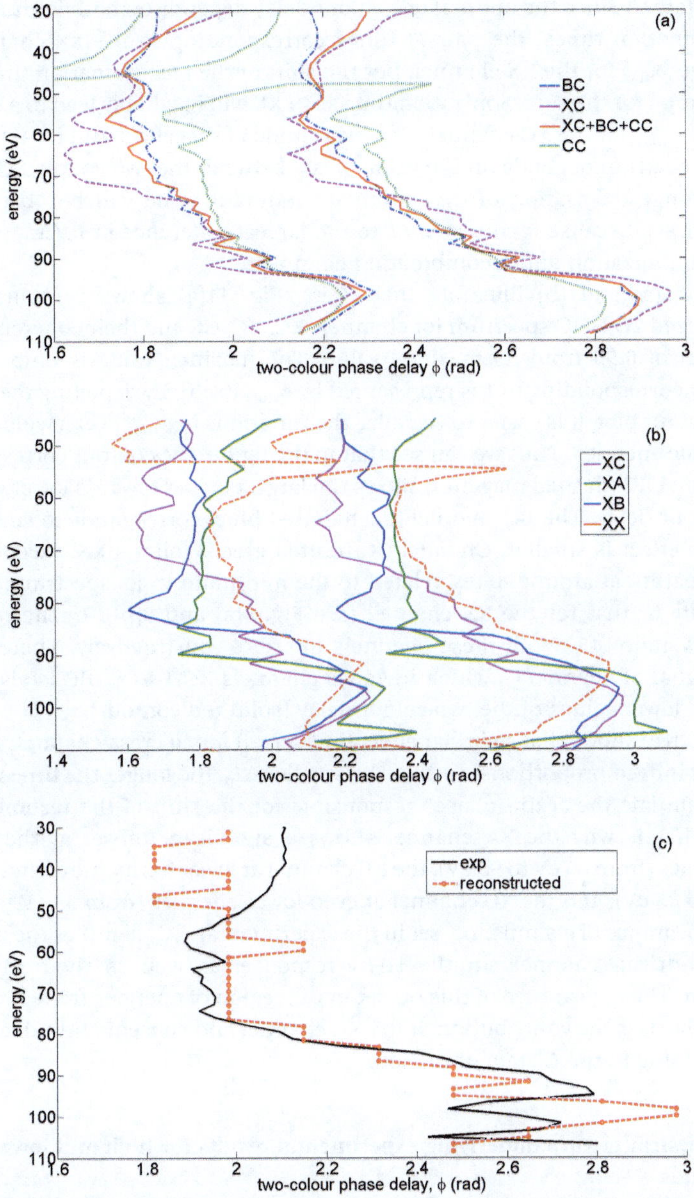

Fig. 11 Analysis of the optimal two-colour phase: theoretical results for high intensity and parallel alignment. The peak field strength $F = 0.057$ a.u., the second harmonic field is $F_2 = 0.14F$, the fundamental wavelength is 1480 nm. The results are averaged over the distribution of laser intensities in the focus. (a) Channel-resolved contours of 2D HHG spectra at 0.95 of the normalized signal. Channel XC (magenta, dashed), BC (blue, dash-dotted), C (green, dotted), coherent sum of XC, BC, CC (red, solid) (b) channel-resolved contours of 2D HHG spectra at 0.95 of the normalized signal. Channel XC (magenta), XA (green), XB (blue), XX (red, dashed). (c) Optimal two-colour delay ϕ_{max} corresponding to all channels (XX, A, B, C) summed with reconstructed phases φ_n and weights ρ_n *vs.* experimental ϕ_{max} (solid black).

harmonic, due to the extra phase accumulated in the XC channel upon transition to the state C. Since the optimal two-colour delay depends on both ionization and recombination times, the pair of times corresponding to the XX channel will optimize ϕ_{max} for the XX channel, but the shift of the recombination time in the XC channel for the same ionization time as in XX will inevitably lead to a different value of ϕ_{max} for the XC channel. The magnitude of this effect and its appearance in the spectrum depends on the value of ΔI_p between the two coupled channels. The strongest deviation of ϕ_{max} from its standard value will be observed for channel XC, because it corresponds to the largest difference in I_p ($\Delta I_p = 5.6$ eV) between ionization and recombination channels.

Fig. 11(a) and (b) illustrate this effect. Fig. 11(a) shows contours of the normalized 2D HHG spectrum for channels XC, BC, CC and their coherent sum at the level of 0.95 from its maximal value. The maximal value is unity (*i.e.* the contour corresponding to 1 is represented by ϕ_{max} itself). By depicting the 2D data as a contour plot, it is easier to visualize the variations in ϕ_{max} over a wide range of harmonic energies. Thus we can see that in the figure, the contour corresponding to channel XC (dashed magenta) shows the largest off-set towards lower values of two-colour delay. The BC channel (dash-dotted blue) corresponds to smaller ΔI_p and the effect is smaller. Channel CC (dotted green) follows XX, except for the sharp feature at around 45 eV related to the minimum in its spectrum, in close similarity to that for the XX channel (see Fig. 9(a) and 9(b)). In the region of energies above 45 eV all these channels interfere constructively. Therefore the total signal of channel C, which includes channels XC + CC + BC is also off-set towards lower values of the two-colour delay (solid red contour).

Fig. 11(b) shows that a similar effect is observed for all cross-channels and this effect is indeed proportional to ΔI_p. The smaller ΔI_p, the longer the time required to accumulate the "extra phase" responsible for the shift of the recombination time. That is why the XA channel shows a significant off-set at the highest harmonics (from 65 eV to 95 eV), the XB channel at lower harmonic energies (from 60 eV to 85 eV), and the XC channel at even lower energies (from 55 eV to 80 eV). The appearance of a similar off-set in the experimental ϕ_{max} signifies the presence of laser-driven channels in the HHG response, as well as their significant strength. The appearance of this off-set in the region of energies from 55 eV to 80 eV emphasises the contribution of the XC channel and consequently all channels recombining to the C state of the ion.

5.7 Physical picture underlying experimental results for high and low intensities

In this subsection we will make direct comparison between the theoretical and experimental 2D spectra to find the set of "extra" phases φ_n and amplitudes ρ_n between channels XX, A, B, and C, if they are required.

Channels A, B, and C arise once we group all essential channels by recombination (Fig. 9(c)). The channel A includes XA, AA, the channel B includes XB, BB, CB and the channel C includes CC, BC, XC. What is the strategy for finding the possibly missing relative phases between the four different channels XX, A, B and C?

We will follow the variation of the optimal two-colour delay ϕ_{max} to uncover the underlying multielectron dynamics. First of all, as it was pointed out in ref. 7, if ϕ_{max} deviates sharply from the trace corresponding to the XX-channel, it means that there is a destructive interference between XX and another channel. The position of the destructive interference in the spectrum is determined by the evolution of the relative phase between these channels *vs.* harmonic number, while the magnitude and the "direction" of the deviation of ϕ_{max} depends on the relative weights of the two destructively interfering channels (see Supplementary Information to ref. 7 for details). Thus, in principle, for each harmonic experiencing the phase jump in ϕ_{max} due to destructive interference the relative phase (*i.e.* $\approx\pi$) and relative weight of the two contributing channels could be approximately reconstructed. Such features are definitely observed in the spectra for both high and low intensities (and parallel alignment) and correspond to energies at about 95 eV (high intensity, see also minimum in 1D spectrum in Fig. 3) and 70 eV to 90 eV (low intensity). These phase jumps scale with intensity, as discussed in ref. 7.

In addition to these jumps, for high intensity and parallel alignment the experimental ϕ_{max} smoothly deviates from its XX value towards lower values of the two-colour delay in a broad region of energies from about 55 eV to 80 eV, as extensively discussed above. Fig. 10(c) unambiguously shows that this effect is related to switching between the main channel XX and all other channels: A + B + C. Specifically, the off-set is mainly due to channel C, as discussed in the previous subsection and shown in Fig. 11(a) and (b). The switching can occur in the region of energies where all channels interfere constructively (or very mildly destructively), see Fig. 12. The horizontal double arrow shows the region of energies where A + B + C channels interfere constructively or mildly destructively. The vertical double arrow shows a specific energy, for which the destructive interference between B + A and C channels occurs. For this energy the XX channel has already become unimportant, since it reaches its cut-off, but the C channel with the highest cut-off dominates. The dominance of A + B + C channels before the cut-off may be achieved either due to their constructive interference (this condition defines the relative phases between the channels φ_n), or, if the constructive interference is still not enough to observe the switching from XX to A + B + C, it requires an additional increase of the weights of the A, B, C channels (this condition approximately determines the amplitudes ρ_n).

The condition of constructive interference in the region of energies from about 55 eV to 80 eV and destructive interference at about 95 eV allowed us to establish the missing phases between the four recombination channels: $\varphi_X = 0$, $\varphi_A \approx 0.4\pi$, $\varphi_B \approx 0$, $\varphi_C \approx -0.12\pi$. The error in this reconstruction is about 0.2π. The condition that the total spectrum must switch from channel XX to A + B + C channels at 55 eV and back at about 80 eV have determined the high harmonic order N-dependent "extra" weights $\rho_n(N)$ that we have introduced:

$$\sigma = (N - N_0)/\Delta N, \tag{7}$$

$$\rho_n(N) = \left[0.8 + 1.7/\left(1 + \exp^{-2\sigma}\right)\right]\frac{1}{\sqrt{\sum(N)}}, \tag{8}$$

where $N_0 = 75$, $\Delta N = 10$, n = A, B, C and $\sum(N)$ ensures that the norm of the vector $\mathbf{d}(t)$ is preserved upon multiplication by matrix Φ. N-dependent "extra" weights $\rho_n(N)$ introduce a gradual increase of channels A, B, C after a structural minimum of channel XX, saturating at the increase factor 2.5 at the cut-off region. These missing relative phases and "extra" relative weights were included in the results shown in Fig. 10(c) and 12, yielding the 2D spectrum shown in Fig. 13(a) and respective optimal phase shown in Fig. 11(b). The extra relative phases are very small for all channels, except channel A. However, the A channel gives a small contribution to the off-set of the optimal two-colour phase from the XX-channel everywhere, except for the region of the spectrum around 95 eV. We have checked it by varying the weight $\rho_A(N)$ and the phase φ_A, and established that the result is weakly sensitive to the choice of phase φ_A and virtually insensitive to the choice of the "extra" weight $\rho_A(N)$ everywhere except for the region of the spectrum around 95 eV. Thus, we can not attempt reconstructing the sub-cycle ionization dynamics for channel A.

In Fig. 13(a) we show the calculated normalized 2D spectrum for high intensity. The results capture all essential features of the corresponding experimental spectrum in Fig. 6(a), including the deviation from the single-channel picture. A similar comparison is presented for the low intensity case in Fig. 13(b) and (c), showing the same qualitative difference in the 2D spectra between high and low intensities as in the experiment. For low intensity we used the same phases φ_n and amplitude corrections $\rho_n(N)$ that we have established from the analysis of theoretical and experimental results for high intensity.

In the low intensity case, the detailed mechanism underlying the deviation of ϕ_{max} from its value corresponding to XX is different from the high intensity case. It is associated with a broad region of destructive interference between XX and all

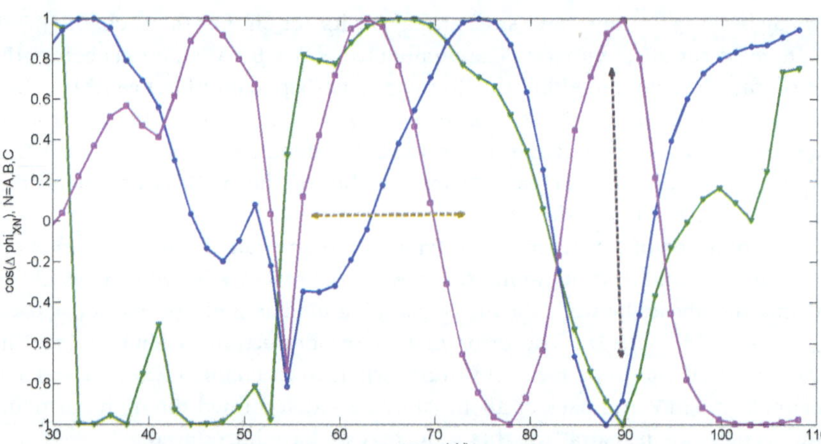

Fig. 12 Cosine of relative phases between channels, grouped by recombination: theoretical results for high intensity and parallel alignment. The peak field strength $F = 0.057$ a.u. The second harmonic field is $F_2 = 0.14F$, the fundamental wavelength is 1480 nm. The results are averaged over the distribution of laser intensities in the focus. The curves show the cosine of the relative phase between XX and B (includes XB, BB, CB) channels (blue, circles), XX and C (includes CC, XC, BC) channels (magenta, squares) and XX and A (includes XA, AA) channels (green, triangles). Horizontal double arrows show the region of energies where A, B, C channels interfere constructively or mildly destructively. Vertical double arrows show the energy, for which the destructive interference between C + A and B channels occurs. For this energy the XX channel is already unimportant, since it has reached its cut-off.

other channels (A + B + C) ranging from 70 eV to 93 eV (see Fig. 14(b) to gauge the cosine of the relative phase between XX and all other channels). This destructive interference is due to a "conveyor" built from the channels A, B, and C "passing" the destructive interference with XX from channel to channel as it "moves" from lower to higher energies in the region from 70 eV to 93 eV, see Fig. 14(c) illustrating this effect. In contrast, in the high intensity experiment the constructive interference is more dominant.

To summarize, the constructive interference between the A, B, and C channels, and the constructive interference between the A + B + C and XX channels together with a markedly different behaviour of ϕ_{max} for laser-driven cross channels allowed us to identify and detect the switching between the XX and A + B + C channels, highlighting the dominance of the C and B channels in the region of energies from 55 eV to 80 eV. These observations allowed us to reconstruct the missing phases φ_n and weights of the recombination channels ρ_n.

Overall, our analysis provides clear evidence of the coupled multichannel light–matter interaction that manifests itself in different ways in the low and high intensity experiments.

The resulting 2D HHG spectra for high and low intensities are shown in Fig. 13(a) and (b). These spectra are in good agreement with the experiment, within experimental error bars. Note that for low intensity and parallel alignment the non-normalized experimental HHG signal is extremely weak in the energy region from 55 eV to 75 eV, thus reliable assignment of ϕ_{max} in this region becomes very challenging. This spectral region is indicated by horizontal lines in Fig. 6(b) and 13(b). Outside of this energy region, the theoretical and experimental ϕ_{max} are in excellent agreement.

5.8 Pilot reconstruction of sub-cycle ionization dynamics

We are now in a position to find the deviation from the original prediction of the sub-cycle ionization amplitudes and phases. However, we note that a full reconstruction of sub-cycle ionization dynamics from the experimental data requires a detailed account of the molecular alignment distribution; here, we have only made illustrative calculations with a limited range of alignment angles.

The results of our analysis are shown in Fig. 15, presented in the basis of the field free cationic states. Back-propagation is performed for each harmonic, starting with the real-valued recombination time $t = t(N)$ corresponding to this harmonic, and finishing at the real part of the complex ionization time $t_i(N)$ corresponding to the same harmonic. The ionization and recombination times are found for the average ionization potential between the four participating channels. This averaging results in the error in phase reconstruction, contributing to the $\sim 0.2\pi$ error associated with the reconstructed recombination phases. Note that the results are obtained for the so-called short trajectories, relevant for the experiment. Hence, the plateau trajectories correspond to electrons that are ionized well after the peak of the laser field oscillation.

Fig. 15(a) shows the reconstructed ionization amplitude *vs.* ionization time, corresponding to channels X (red, solid), B (blue, solid), C (magenta, solid) along with the original (initial) sub-cycle amplitudes (dashed), which served as an input for our calculations. Fig. 15(b) shows the respective ionization phases for channels X (red), B (blue), C (magenta).

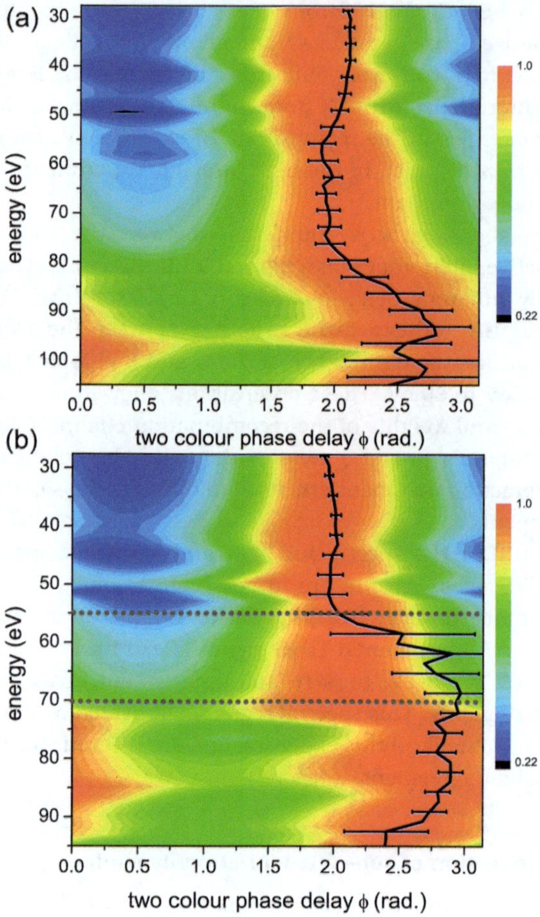

Fig. 13 Normalized two-dimensional HHG spectra of CO_2 for parallel alignment. The peak field strength $F = 0.057$ a.u. (a) and $F = 0.051$ a.u. (b). The second harmonic field is $F_2 = 0.14F$, the fundamental wavelength is 1480 nm. The results are averaged over the distribution of laser intensities in the focus. Black lines: experimental ϕ_{max} including estimated error in extracting ϕ_{max} for the respective laser intensities, adapted from Fig. 6(a) and (b). For low intensity the non-normalized experimental HHG signal is extremely weak in the energy region from 55 eV to 70 eV, thus reliable assignment of ϕ_{max} in this region becomes very challenging. This spectral region is indicated by grey dotted horizontal lines in Fig. 6(b) and also in panel (b). Outside of this energy region, the theoretical and experimental ϕ_{max} are in excellent agreement.

The reconstructed ionization phases show that channels B and C are generally in phase. States B and C are strongly coupled by the laser field. It makes the ionization process highly non-trivial and puts it outside of the applicability region of standard Keldysh-type strong field ionization theories. In the limit of strongly coupled two-level system, one could think of the two quasi energy states, formed by the "+" and the "−" superposition of B and C states. The results of our reconstruction indicate that the quasi energy state B + C as opposed to B − C was initially populated in our experiment. The B + C quasi energy state has a higher I_p than the state B − C, but its Dyson orbital has asymptotic behaviour that is

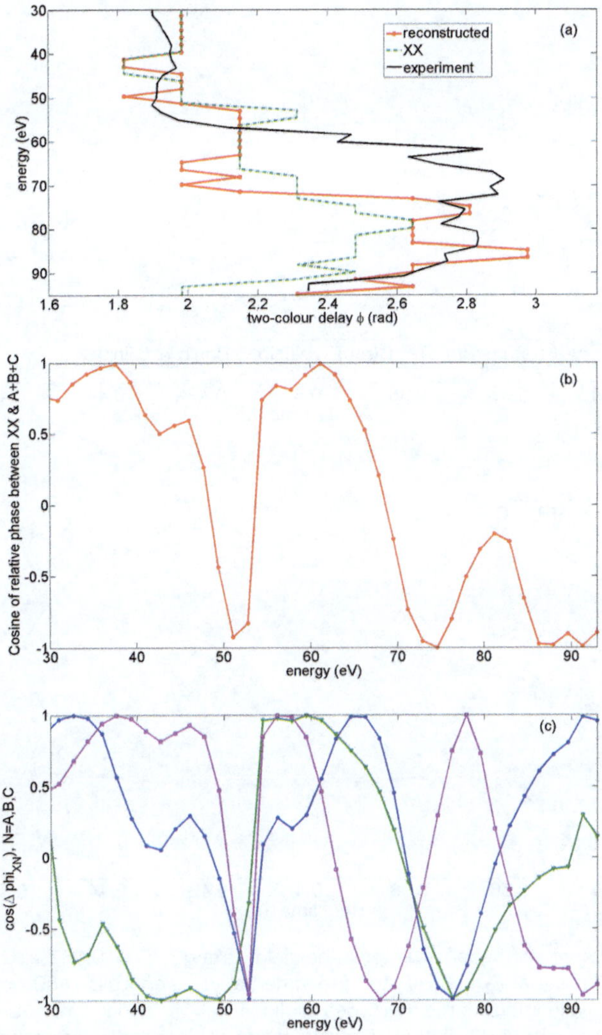

Fig. 14 Analysis of the optimal two-colour phase: theoretical results for low intensity and parallel alignment. The peak field strength $F = 0.051$ a.u., the second harmonic field is $F_2 = 0.14F$, the fundamental wavelength is 1480 nm. The results are averaged over the distribution of laser intensities in the focus. (a) Optimal two-colour delay ϕ_{max} corresponding to all channels (XX, A, B, C) summed with reconstructed phases φ_n and weights ρ_n (red, dashed) *vs.* experimental ϕ_{max} (black solid) and ϕ_{max} corresponding to the XX channel (green, dash-dotted). The deviation between experimental and theoretical results (from 55 eV to 70 eV) coincides with the region where the non-normalized experimental HHG signal is extremely weak and the reliable assignment of ϕ_{max} is very challenging. (b) Cosine of the relative phase between the main channel XX and the rest of the channels (A + B + C). Destructive interference in the wide region of energies from 70 eV to 90 eV defines the deviation of ϕ_{max} (red, dashed, in panel (a)) from the trace corresponding to the XX channel only (green, dash-dotted, panel (a)). (c) "Conveyor" of destructive interference. Cosine of the relative phases between XX-channel and A (green, triangles), B (blue, circles), C (magenta, squares) show how the destructive interference between XX and all other channels can exist in a wide region of energies from 70 eV to 93 eV (see panel (b)).

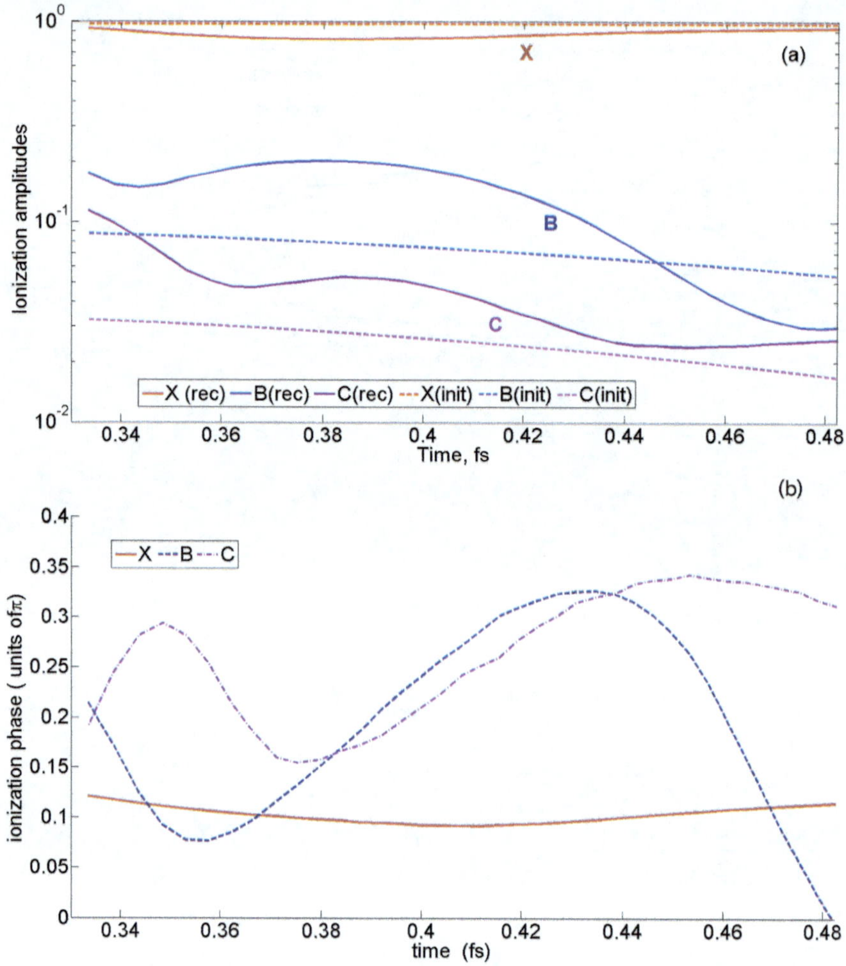

Fig. 15 Results of the reconstruction. The field strength $F = 0.057$ a.u. The second harmonic field is set at $F_2 = 0.14F$, the fundamental wavelength is 1480 nm. (a) Reconstructed sub-cycle ionization amplitudes (solid lines) *vs.* initial ionization amplitudes (dashed), that were originally included in calculation corresponding to channels X (red), B (blue), C (magenta). (b) Reconstructed sub-cycle ionization phases, corresponding to channels X (red), B (blue), C (magenta).

advantageous for tunneling. Specifically, it extends towards the tunneling barrier, but not in the opposite direction, as would be in the case of the B − C state.

According to our reconstruction, the relative phase between the B and X channels is also small, especially in the high photon energy region, corresponding to early ionization times. This conclusion is in agreement with our earlier observation reported in ref. 16. There, we have followed the evolution of destructive interference in the high harmonic spectrum with laser intensity near the cut-off harmonic and have found that the phase between the X and B channels is close to zero. Our current investigation extends these earlier conclusions to the long-wavelength regime.

In contrast to very minor relative inter-channel ionization phases reconstructed from our experiment, the reconstructed ionization amplitudes (solid curves in Fig. 15(a)) are significantly higher than those expected from single channel strong field ionization theory (dashed curves). This result is not surprising from the point of view of our analysis, since we have increased the contribution of the B and C recombination channels to ensure switching between the dominant channel XX at low harmonic energies to dominant channels A + B + C in the region from 55 eV to 80 eV. The deviation from a single channel model is expected due to strong laser coupling between B and C states, as discussed above. Thus, the experiment suggests that the single-channel model of strong field ionization we have used in our calculations substantially underestimates the relative population of the B and C states.

6 Conclusions

In contrast to standard assumptions, our findings reveal the important role of sub-cycle laser-driven dynamics between cationic states, even for mid-IR driving fields. These dynamics leave a clear fingerprint in the 2D HHG spectra. In general, the optimal phase delay ϕ_{max} is very sensitive to the presence of multiple channels – information that may be hidden in the 1D HHG spectrum. In particular, the dependence of ϕ_{max} on the harmonic number is sensitive to the presence of the laser-driven hole dynamics between ionization and recombination. This property allows one to reconstruct the relative phases between different ionization channels and their relative amplitudes at the moment of ionization, on the sub-cycle time scale. These phases and amplitudes encode the dynamics of multi-electron rearrangement upon strong field ionization. We have found substantial deviations of reconstructed sub-cycle ionization amplitudes from the values predicted by single-channel tunneling theory, which signifies the non-adiabatic nature of the ionization process and may be related to the coupling between the ionization channels, both due to the laser field and also due to correlation-driven interactions during tunnelling. Similar information is present and can be extracted at different intensities and wavelengths of the fundamental field. Additional scans of the wavelength and the intensity of the fundamental field can then be used to reveal the intensity and wavelength-dependent aspects of the dynamics.

Our results on the unexpectedly strong contributions of deeper-lying channels to strong-field ionization are pertinent to understanding a wide range of phenomena initiated by ionization in strong IR and mid-IR fields, from laser filamentation and air lasing to multiple ionization and plasma dynamics.

Fundamentally, our study allows us to identify the initial conditions for the hole dynamics – the coherent electronic wavepacket that can be created on multiple potential energy surfaces of a molecule. The role of this coherence in systems with more complex coupled electronic and nuclear dynamics, especially at conical intersections, and the possibility of laser control of these dynamics on the sub-cycle time scale[69] are the future challenges of this field.

Authors contributions

B. D. B., M. N., M. D., D. F., C. V., and S. S. performed the experiments. B. D. B., H. S., Y. M., C. V. and S. S. analyzed experimental data. A. G. H. developed the R-matrix code for photo-ionization and photo-recombination and the model for CO_2 calculations. Z. M., D. B., and A. G. H. performed R-matrix calculations. Z. M. extended R-matrix codes to high energy region and implemented a phase-consistent approach to photoionization. F. M. and O. S. developed the HHG codes and performed HHG calculations. V. S. contributed at the initial stages of the project. S. P. performed quantum chemical calculations. M. I. and O. S. analysed the experimental and theoretical results and performed the reconstruction. B. D. B., M. I., N. D., and O. S. wrote the paper. All authors discussed the results and provided comments on the manuscript. N. D. and O. S. supervised the project.

Acknowledgements

Z. M., F. M. and O. S. gratefully acknowledge the support of Deutsche Forschungsgemeinschaft, project Sm292/5-1, A. G. H. and M. I. gratefully acknowledge the support of Deutsche Forschungsgemeinschaft, project IV152/7-1. The MBI (Z. M., F. M., A. G. H., O. S.) and Polimi (C. V., S. S.) nodes gratefully acknowledge the support of their collaboration *via* EU ITN MEDEA – AMD-641789-17 project. M. I. gratefully acknowledges the support of EPSRC/DSTL MURI grant EP/N018680/1. The authors gratefully acknowledge the support of their collaboration trough European Cooperation in Science and Technology (COST), program CM1204 XLIC. Prof. Nirit Dudovich is the incumbent of the Robin Chemers Neustein Professorial Chair. N. D. gratefully acknowledges the Minerva Foundation, the Israeli Science Foundation, the European Research Council Starting Research Grant MIDAS, the Crown Photonics Center, and the I-Core Center for financial support. This project has received funding from the European Union's Horizon 2020 research and innovation programme under the Marie Sklodowska-Curie grant agreement No. 641272.

References

1 F. Krausz and M. Y. Ivanov, *Rev. Mod. Phys.*, 2009, **81**, 163–234.

2 M. Schultze, M. Fieß, N. Karpowicz, J. Gagnon, M. Korbman, M. Hofstetter, S. Neppl, A. Cavalieri, Y. Komninos, T. Mercouris, C. Nicolaides, R. Pazourek, S. Nagele, J. J. Feist, J. Burgdörfer, A. M. Azzeer, R. Ernstorfer, R. Kienberger, U. Kleineberg, E. Goulielmakis, F. Krausz and V. S. Yakovlev, *Science*, 2010, **328**, 1658–1662.

3 K. Klünder, J. M. Dahlström, M. Gisselbrecht, T. Fordell, M. Swoboda, D. Guénot, P. Johnsson, J. Caillat, J. Mauritsson, A. Maquet, R. Taïeb and A. L'Huillier, *Phys. Rev. Lett.*, 2011, **106**, 143002.

4 R. Pazourek, S. Nagele and J. Burgdörfer, *Rev. Mod. Phys.*, 2015, **87**, 765–802.

5 P. Eckle, A. N. Pfeiffer, C. Cirelli, A. Staudte, R. Dörner, H. G. Muller, M. Büttiker and U. Keller, *Science*, 2008, **322**, 1525–1529.

6 L. Torlina, F. Morales, J. Kaushal, I. Ivanov, A. Kheifets, A. Zielinski, A. Scrinzi, H. G. Muller, S. Sukiasyan, M. Y. Ivanov and O. Smirnova, *Nat. Phys.*, 2015, **11**, 503–508.

7 D. Shafir, H. Soifer, B. D. Bruner, M. Dagan, Y. Mairesse, S. Patchkovskii, M. Y. Ivanov, O. Smirnova and N. Dudovich, *Nature*, 2012, **485**, 343–346.

8 O. Pedatzur, G. Orenstein, V. Serbinenko, H. Soifer, B. D. Bruner, A. J. Uzan, D. S. Brambila, A. G. Harvey, L. Torlina, F. Morales, O. Smirnova and N. Dudovich, *Nat. Phys.*, 2015, **11**, 815–820.
9 E. Goulielmakis, Z. H. Loh, A. Wirth, R. Santra, N. Rohringer, V. S. Yakovlev, S. Zherebtsov, T. Pfeifer, A. M. Azzeer, M. F. Kling, S. R. Leone and F. Krausz, *Nature*, 2010, **466**, 739–743.
10 A. Fleischer, H. J. Wörner, L. Arissian, L. R. Liu, M. Meckel, A. Rippert, R. Dörner, D. M. Villeneuve, P. B. Corkum and A. Staudte, *Phys. Rev. Lett.*, 2011, **107**, 113003.
11 Y. Mairesse, J. Higuet, N. Dudovich, D. Shafir, B. Fabre, E. Mével, E. Constant, S. Patchkovskii, Z. Walters, M. Y. Ivanov and O. Smirnova, *Phys. Rev. Lett.*, 2010, **104**, 213601.
12 P. M. Kraus, B. Mignolet, D. Baykusheva, A. Rupenyan, L. Horný, E. F. Penka, G. Grassi, O. I. Tolstikhin, J. Schneider, F. Jensen, L. B. Madsen, A. D. Bandrauk, F. Remacle and H. J. Wörner, *Science*, 2015, **350**, 790–795.
13 A. I. Kuleff and L. S. Cederbaum, *J. Phys. B: At., Mol. Opt. Phys.*, 2014, **47**, 124002.
14 F. Calegari, D. Ayuso, A. Trabattoni, L. Belshaw, S. de Camillis, S. Anumula, L. Frassetto, L. Poletto, A. Palacios, P. Decleva, J. B. Greenwood, F. Martín and M. Nisoli, *Science*, 2014, **346**, 336–339.
15 J. Leeuwenburgh, B. Cooper, V. Averbukh, J. P. Marangos and M. Ivanov, *Phys. Rev. Lett.*, 2013, **111**, 123002.
16 O. Smirnova, Y. Mairesse, S. Patchkovskii, N. Dudovich, D. Villeneuve, P. Corkum and M. Y. Ivanov, *Nature*, 2009, **460**, 972–977.
17 O. Smirnova, S. Patchkovskii, Y. Mairesse, N. Dudovich and M. Y. Ivanov, *Proc. Natl. Acad. Sci. U. S. A.*, 2009, **106**, 16556–16561.
18 O. Smirnova, S. Patchkovskii, Y. Mairesse, N. Dudovich, D. Villeneuve, P. B. Corkum and M. Y. Ivanov, *Phys. Rev. Lett.*, 2009, **102**, 063601.
19 S. Haessler, J. Caillat, W. Boutu, C. Giovanetti-Teixeira, T. Ruchon, T. Auguste, Z. Diveki, P. Breger, A. Maquet, B. Carré, R. Taïeb and P. Salières, *Nat. Phys.*, 2010, **6**, 200–206.
20 P. B. Corkum, *Phys. Today*, 2011, **64**, 36–41.
21 O. Smirnova and M. Ivanov, *How to for Beginners – Multi-electron Dynamics in HHG*, Max Born Institut, 2016, https://www.youtube.com/watch?v=fA2pdLYVkQQ.
22 P. B. Corkum, *Phys. Rev. Lett.*, 1993, **71**, 1994–1997.
23 M. Lein, N. Hay, R. Velotta, J. P. Marangos and P. L. Knight, *Phys. Rev. Lett.*, 2002, **88**, 183903.
24 M. Lein, *J. Phys. B: At., Mol. Opt. Phys.*, 2007, **40**, R135–R173.
25 A. Zaïr, T. Siegel, S. Sukiasyan, F. Risoud, L. Brugnera, C. Hutchison, Z. Diveki, T. Auguste, J. W. Tisch, P. Salières, M. Y. Ivanov and J. P. Marangos, *Chem. Phys.*, 2013, **414**, 184–191.
26 R. Cireasa, A. E. Boguslavskiy, B. Pons, M. C. H. Wong, D. Descamps, S. Petit, H. Ruf, N. Thiré, A. Ferré, J. Suarez, J. Higuet, B. E. Schmidt, A. F. Alharbi, F. Légaré, V. Blanchet, B. Fabre, S. Patchkovskii, O. Smirnova, Y. Mairesse and V. R. Bhardwaj, *Nat. Phys.*, 2015, **11**, 654–658.
27 S. Baker, J. S. Robinson, C. A. Haworth, H. Teng, R. A. Smith, C. C. Chirila, M. Lein, J. W. G. Tisch and J. P. Marangos, *Science*, 2006, **312**, 424–427.

28 N. L. Wagner, A. Wüest, I. P. Christov, T. Popmintchev, X. Zhou, M. M. Murnane and H. C. Kapteyn, *Proc. Natl. Acad. Sci. U. S. A.*, 2006, **103**, 13279–13285.

29 W. Li, X. Zhou, R. Lock, S. Patchkovskii, A. Stolow, H. C. Kapteyn and M. M. Murnane, *Science*, 2008, **322**, 1207–1211.

30 A. Ferré, D. Staedter, F. Burgy, M. Dagan, D. Descamps, N. Dudovich, S. Petit, H. Soifer, V. Blanchet and Y. Mairesse, *J. Phys. B: At., Mol. Opt. Phys.*, 2014, **47**, 124023.

31 A. Ferré, A. E. Boguslavskiy, M. Dagan, V. Blanchet, B. D. Bruner, F. Burgy, A. Camper, D. Descamps, B. Fabre, N. Fedorov, J. Gaudin, G. Geoffroy, J. Mikosch, S. Patchkovskii, S. Petit, T. Ruchon, H. Soifer, D. Staedter, I. Wilkinson, A. Stolow, N. Dudovich and Y. Mairesse, *Nat. Commun.*, 2015, **6**, 5952.

32 A. Ferré, H. Soifer, O. Pedatzur, C. Bourassin-Bouchet, B. D. Bruner, R. Canonge, F. Catoire, D. Descamps, B. Fabre, E. Mével, S. Petit, N. Dudovich and Y. Mairesse, *Phys. Rev. Lett.*, 2016, **116**, 053002.

33 H. J. Wörner, J. B. Bertrand, D. V. Kartashov, P. B. Corkum and D. M. Villeneuve, *Nature*, 2010, **466**, 604–607.

34 H. J. Wörner, J. B. Bertrand, P. B. Corkum and D. M. Villeneuve, *Phys. Rev. Lett.*, 2010, **105**, 103002.

35 J. Itatani, J. Levesque, D. Zeidler, H. Niikura, H. Pépin, J. C. Kieffer, P. B. Corkum and D. M. Villeneuve, *Nature*, 2004, **432**, 867–871.

36 C. Vozzi, F. Calegari, E. Benedetti, J.-P. Caumes, G. Sansone, S. Stagira, M. Nisoli, R. Torres, E. Heesel, N. Kajumba, J. P. Marangos, C. Altucci and R. Velotta, *Phys. Rev. Lett.*, 2005, **95**, 153902.

37 C. Vozzi, M. Negro, F. Calegari, G. Sansone, M. Nisoli, S. D. Silvestri and S. Stagira, *Nat. Phys.*, 2011, **7**, 822–826.

38 T. Kanai, S. Minemoto and H. Sakai, *Nature*, 2005, **435**, 470–474.

39 P. Hamm and M. Zanni, *Concepts and Methods of 2D Infrared Spectroscopy*, Cambridge University Press, 1st edn, 2011.

40 M. H. Cho, *Chem. Rev.*, 2008, **108**, 1331–1418.

41 F. D. Fuller and J. P. Ogilvie, *Annu. Rev. Phys. Chem.*, 2015, **66**, 667–690.

42 H. Soifer, M. Dagan, D. Shafir, B. D. Bruner, M. Y. Ivanov, V. Serbinenko, I. Barth, O. Smirnova and N. Dudovich, *Chem. Phys.*, 2013, **414**, 176–183.

43 V. Serbinenko and O. Smirnova, *J. Phys. B: At., Mol. Opt. Phys.*, 2013, **46**, 171001.

44 H. Soifer, B. D. Bruner, M. Negro, M. Devetta, D. Faccialà, C. Vozzi, S. de Silvestri, S. Stagira and N. Dudovich, *J. Phys.B: At., Mol. Opt. Phys.*, 2014, **47**, 204029.

45 L. Torlina, M. Ivanov, Z. B. Walters and O. Smirnova, *Phys. Rev. A*, 2012, **86**, 043409.

46 Z. B. Walters and O. Smirnova, *J. Phys. B: At., Mol. Opt. Phys.*, 2010, **43**, 161002.

47 V. P. Majety and A. Scrinzi, *Phys. Rev. Lett.*, 2015, **115**, 103002.

48 J. Muth-Böhm, A. Becker and F. H. M. Faisal, *Phys. Rev. Lett.*, 2000, **85**, 2280–2283.

49 R. Murray, M. Spanner, S. Patchkovskii and M. Y. Ivanov, *Phys. Rev. Lett.*, 2011, **106**, 173001.

50 P. Salières, A. Maquet, S. Haessler, J. Caillat and R. Taïeb, *Rep. Prog. Phys.*, 2012, **75**, 062401.

51 Z. Diveki, R. Guichard, J. Caillat, A. Camper, S. Haessler, T. Auguste, T. Ruchon, B. Carré, A. Maquet, R. Taïeb and P. Salières, *Chem. Phys.*, 2013, **414**, 121–129.
52 A. Rupenyan, P. M. Kraus, J. Schneider and H. J. Wörner, *Phys. Rev. A*, 2013, **87**, 013401.
53 A. Rupenyan, P. M. Kraus, J. Schneider and H. J. Wörner, *Phys. Rev. A*, 2013, **87**, 033409.
54 R. Torres, T. Siegel, L. Brugnera, I. Procino, J. G. Underwood, C. Altucci, R. Velotta, E. Springate, C. Froud, I. C. E. Turcu, S. Patchkovskii, M. Y. Ivanov, O. Smirnova and J. P. Marangos, *Phys. Rev. A*, 2010, **81**, 051802.
55 Z. Diveki, A. Camper, S. Haessler, T. Auguste, T. Ruchon, B. Carré, P. Salières, R. Guichard, J. Caillat, A. Maquet and R. Taïeb, *New J. Phys.*, 2012, **14**, 023062.
56 B. D. Bruner, H. Soifer, D. Shafir, V. Serbinenko, O. Smirnova and N. Dudovich, *J. Phys. B: At., Mol. Opt. Phys.*, 2015, **48**, 174006.
57 C. Vozzi, F. Calegari, E. Benedetti, S. Gasilov, G. Sansone, G. Cerullo, M. Nisoli, S. de Silvestri and S. Stagira, *Opt. Lett.*, 2007, **32**, 2957–2959.
58 R. Torres, T. Siegel, L. Brugnera, I. Procino, J. G. Underwood, C. Altucci, R. Velotta, E. Springate, C. Froud, I. C. E. Turcu, M. Y. Ivanov, O. Smirnova and J. P. Marangos, *Opt. Express*, 2010, **18**, 3174–3180.
59 C. Vozzi, M. Negro and S. Stagira, *J. Mod. Opt.*, 2012, **59**, 1283–1302.
60 K. D. Schultz, C. I. Blaga, R. Chirla, P. Colosimo, J. Cryan, A. M. March, C. Roedig, E. Sistrunk, J. Tate, J. Wheeler, P. Agostini and L. F. DiMauro, *J. Mod. Opt.*, 2007, **54**, 1075–1085.
61 *Atomic and Molecular Beam Methods*, ed. G. Scoles, Oxford University Press, Inc., 1988, vol. 1.
62 H. Stapelfeldt and T. Seideman, *Rev. Mod. Phys.*, 2003, **75**, 543–557.
63 O. Smirnova and M. Ivanov, *Attosecond and XUV Physics: Ultrafast Dynamics and Spectroscopy*, Wiley-VCH Verlag GmbH and Co. KGaA, 2014, pp. 201–256.
64 O. Smirnova and M. Ivanov, Multielectron High Harmonic Generation: simple man on a complex plane, arXiv:1304.2413, 2013.
65 A. G. Harvey, D. S. Brambila, F. Morales and O. Smirnova, *J. Phys. B: At., Mol. Opt. Phys.*, 2014, **47**, 215005.
66 D. S. Brambila, A. G. Harvey, Z. Mašín, J. D. Gorfinkiel and O. Smirnova, *J. Phys. B: At., Mol. Opt. Phys.*, 2015, **48**, 245101.
67 F. Morales, I. Barth, V. Serbinenko, S. Patchkovskii and O. Smirnova, *J. Mod. Opt.*, 2012, **59**, 1303–1311.
68 P. Salières, B. Carré, L. L. Déroff, F. Grasbon, G. G. Paulus, H. Walther, R. Kopold, W. Becker, D. B. Milošecvić, A. Sanpera and M. Lewenstein, *Science*, 2001, **292**, 902–905.
69 M. Richter, F. Bouakline, J. González-Vázquez, L. Martínez-Fernández, I. Corral, S. Patchkovskii, F. Morales, M. Ivanov, F. Martín and O. Smirnova, *New J. Phys.*, 2015, **17**, 113023.

Faraday Discussions

PAPER

Using high harmonic radiation to reveal the ultrafast dynamics of radiosensitiser molecules

Pierre Çarçabal,[a] Dominique Descamps,[b] Stéphane Petit,[b] Yann Mairesse,[b] Valérie Blanchet[b] and Raluca Cireasa*[a]

Received 11th May 2016, Accepted 11th July 2016
DOI: 10.1039/c6fd00129g

5-Fluorouracil (5FU) is a radiosensitiser molecule routinely used in combined chemo- and radio-therapies to enhance and localize cancer treatments. We have employed ultra-short XUV pulses produced by high harmonic generation (HHG) as a pump pulse to study the dynamics underlying the photo-stability and the radiation damage of this molecule. This work shows that it is possible to resolve individual dynamics even when using unselected HH. By comparing the results with those obtained in the multiphoton absorption at 400 nm, we were able to identify the frequencies of the HH comb relevant to the recorded dynamics: HH5 and HH3. The latter excites a high-lying Rydberg state interacting with a valence state and its dynamics is revealed by a 30 fs decay signal in the parent ion transient. Our results suggest that the same photoprotection mechanisms as those conferring photostability to the neutral nucleobases and to the DNA appear to be activated: HH5 excites the molecule to a state around 10.5 eV that undergoes an ultrafast relaxation on a timescale of 30 fs due to nonadiabatic interactions. This is followed sequentially by a 2.3 ps internal conversion as revealed by the dynamics observed for another fragment ion. These dynamics are extracted from the fragment ion signals. Proton or hydrogen transfer processes are required for the formation of three fragments and we speculate that the time scale of one of the processes is revealed by a H^+ transient signal.

1 Introduction

In the last decade, the high harmonic generation (HHG) has emerged as a new light source providing radiation in the XUV and soft X-ray domains with femtosecond (attosecond pulse trains) and attosecond pulse duration.[1-3] Furthermore, it became an extremely powerful spectroscopic tool capable of providing structural and dynamical information, including (sub)-femtosecond arrangements of

[a]*Institut des Sciences Moléculaires d'Orsay, CNRS, Université Paris Sud, Orsay, France. E-mail: raluca.cireasa@u-psud.fr; raluca.cireasa@gmail.com*
[b]*Centre Lasers Intenses et Applications, CNRS, Université de Bordeaux, Talence, France*

nuclei and electrons.[4–9] Most of the studies concern small molecules, but recently, the HH radiation has been used for investigating more complex molecular systems of interest for chemistry and biology.[10–15] In 2003, the seminal work of Breidbach and Cederbaum and later, in 2006, that of Remacle and Levine, has shown theoretically that sudden ionisation by attosecond pulses could lead to ultrafast charge migration along the molecular backbone before the nuclear dynamics set in.[16,17] Since then, the possibility to study electron correlation processes, such as charge migration, which are ubiquitous in biological molecules have inspired experimentalists and strongly driven the improvement of the temporal and spectral characteristics of the HH radiation. Two years ago, a stark question was raised: 'Attosecond molecular dynamics: fact or fiction?'.[18] With the first observation of charge migration in amino-acids, attosecond molecular dynamics became a fact and also an experimental field of research.[13]

Another important advance that HH radiation and Free Electron Lasers (FEL) have brought to molecular science is the extension of the femtosecond time-resolved experiments to XUV and X-ray domains. This opens up the possibility to study in real time the underlying mechanisms of the radiation damage and it has a particular impact on the understanding of the biological processes of damage and repair, as those occurring in the DNA.

In this paper, we will discuss the interaction between the ionising XUV radiation and some biologically relevant molecules: 5-halouracils (5-HaloU). They belong to a particular class of biomolecules called radiosensitisers that enable the use of the radiation damage for therapeutic purposes. When utilised in combined chemo- and radio-therapy treatments to replace the thymine in the DNA of tumour cells, they enhance the lethal effects of the UV, X-ray, proton and γ radiation on these cells.[19–21] The radiation damage processes are at the heart of the therapeutic effect of 5-HaloU and investigating this process may shed light on the mechanisms of their radiosensitivity.

Owing to the same pyrimidinic structure and because it also pairs with adenine in DNA, 5-HaloU can be used to replace thymine in DNA. The only difference concerns the substituent attached to C5, which in the case of thymine is a CH_3 group and for 5HaloU is a halogen atom. At room temperature, they exist as the di-keto tautomers. The enol form can be obtained by hydrogen or proton transfer. This form mispairs in the DNA with guanine. It was found that ionisation increases the percentage of the mispairing for 5FU and BrU.[22] Thereby, proton and hydrogen transfers are important for the mispairing processes and are considered to be involved in mutagenesis.

To study the dynamics underlying the photo-stability and the radiation damage of 5-HaloU, we have employed ultra-short XUV pulses produced by HHG and several pump-probe schemes, involving different spectra and wavelengths. In addition to providing the first measurements of ultrafast dynamics involved in the photo-stability and the photo-damage of 5-HaloU, this work also shows that it is possible to resolve individual dynamics even when they are investigated with unselected HH radiation. The methods used to select a single harmonic reduce the photon flux/harmonic/pulse by a factor of $\sim$3. But, employing such a broad spectrum, can lead to many excitation channels producing many fragments and to an averaging of the many overlapping dynamics. Our results obtained for 5FU demonstrate that this is not necessarily the case if the HH are generated at 400 nm to produce a spectrum in which the HH frequencies are largely separated

(6.2 eV). Actually, the multitude of the fragmentation channels provides the selectivity that it lacks: the different dynamics are revealed through different fragmentation channels. Together with the absorption cross section dependence on wavelength and the Franck–Condon factors, they enable us to select and disentangle dynamics started in the same time by different energy components of the broad XUV spectrum. We discuss these dynamics in comparison with those induced by multiphoton excitation with 400 nm to identify which of the frequencies of the comb are relevant to the recorded dynamics.

2 Experimental

The experiments presented herein were performed using the AURORE laser system and the experimental set-up from CELIA. In Fig. 1, we show a schematic view of the set-up which consists of four parts: the high harmonic generation (HHG) chamber, the high harmonic spectral characterisation chamber, a chamber containing focalisation optics and filters and finally, the molecular beam source chamber and the velocity map imaging spectrometer (VMIS). All these chambers, as well as the VMIS are pumped down to achieve secondary vacuum conditions. The 1 kHz Ti:sapphire laser system produces 22 mJ pulses with a central wavelength of 795 nm. Part of the beam (~10 mJ/pulse) is delivered to the gas-phase beamline and recompressed to 23 fs. 7 mJ/pulse of this energy is further split into two parts to be used for the generation of the pump and of the probe beams. About 4 mJ/pulse was frequency doubled and subsequently, 600 μJ/pulse of spatially cleaned 400 nm radiation was focused using a 500 mm focal lens into a 4 mm cell placed in the high harmonic generation chamber and filled with 20–30 mbar of argon or krypton. The periodic HH emission over several optical cycles gives rise to a periodic emission in frequency, thereby creating a spectrum of odd harmonics, separated by twice the driving field frequency (6.2 eV) and an attosecond pulse train with a total duration of about 10 fs. The emitted harmonic spectrum was recorded using a flat field XUV spectrometer equipped with a 1200 grooves/mm spherical grating and an imaging detector consisting of two micro-channel plates (MCP), a phosphor screen and a CCD camera. For the conditions used in this experiment, it was possible to produce XUV radiation in Ar containing a comb of harmonics, from HH3 (~9.3 eV) to HH9 (~27.9 eV), with the HH5 being the most intense one. In the spectrum generated in Kr, HH9 is not present and

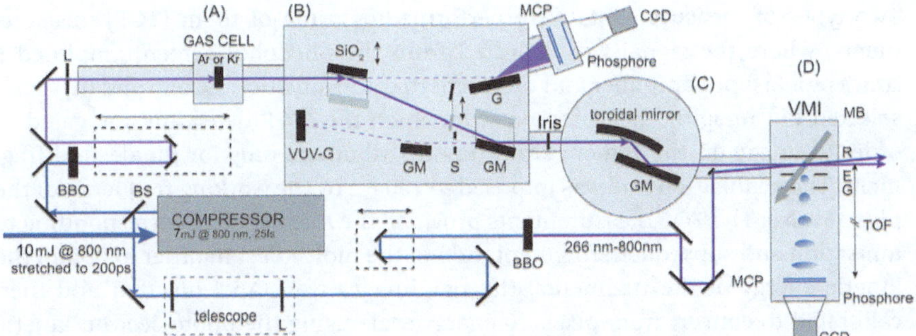

Fig. 1 Experimental set-up.

HH7 is 10 times weaker than HH5, while for the spectrum generated in Ar, it is only 1.5 times weaker. All the pump–probe experiments were performed using the entire XUV spectrum. Due to the low photon flux/harmonic/pulse, no attempt was made to select individual harmonics as that would make the experiments very challenging indeed. In order to remove the fundamental 400 nm beam, the HH radiation was produced by employing 400 nm radiation with an annular spot profile. A motorised diaphragm placed before the toroidal mirror is then able to block this 400 nm beam and to transmit the HH beam having a lower divergence. The XUV beam was employed either as a pump beam to ionise and induce dynamics in the parent cation or as a probe beam to investigate the relaxation dynamics following the excitation by 266 nm pulses of the $S_2(\pi\pi^*)$ state of the neutral molecules. In some experiments, we have used directly the 400 nm (9 nm full width half maximum, FWHM, bandwidth) radiation instead of the XUV. The remaining part of the fundamental beam was sent on a different optical line and used as such (795 nm, 60 nm FWHM bandwidth) or tripled (266 nm, 4 nm FWHM bandwidth) to generate the probe beam. The probe beam was focused using a fused silica lens ($f = 600$ mm) and then reflected using a D-shaped mirror to recombine at a small angle with the pump onto the molecular beam. We estimate that the intensities of the 800 nm, 400 nm and 266 nm radiation in the interaction region were of the order: 10^{13} W cm^{-2}, 10^{12} W cm^{-2} and 10^{11} to 10^{12} W cm^{-2}, respectively.

The molecular beam of gas-phase molecules was produced by using a pulsed General valve and an oven mounted on the face plate of the valve. Solid samples of 5-Halouracils: 5-Fluorouracil (5FU), 5-Chlorouracil (5ClU) and 5-Bromouracil (5BrU) were placed in the oven and then sublimated at 190 °C, 200 °C and 240 °C, respectively. The General valve was run at 100 Hz and it was equipped with an adjustment mechanism in order to optimise the vapour densities directly on the signal. The vapours were entrained from the oven by a pulsed beam of Ne produced by the General valve. For specific working conditions of the pulsed valve, it was possible to obtain very short 5-halouracil pulses of 20 μs duration and thus larger molecular densities in the interaction region. The so-formed molecular beam was intersected by the laser beam(s) in the region between the repeller and the extractor of the VMIS. The ions were extracted perpendicularly to the plane defined by the molecular and laser beam directions and it was velocity mapped onto the imaging detector consisting of a pair of 80 mm chevron MCP and a P43 phosphor screen mounted on a fibre optic window. The image on the phosphor screen was captured by a charge-coupled device (CCD) PCO camera. Two types of measurements were performed: (i) time of flight (TOF) measurements, where the signal was collected from the phosphor screen, amplified 5 times by a fast pre-amplifier and then sent to the acquisition system and (ii) mass-selected ion imaging measurements, in which the MCP detection was gated in order to image on the camera scattering distributions only for the desired fragment. The acquisition rate was imposed to 100 Hz by the working frequency of the pulsed valve. The TOF measurements provided the mass spectra corresponding to ionisation and subsequent fragmentation of the molecules. In order to obtain the kinetic energy of the fragments, the raw images were Abel inverted and then calibrated to convert from pixels to energy scale using the photoelectron kinetic energy distributions for the Ar photoionisation by the HH frequency comb. Each image corresponds to a sum of 2000 frames each integrated over 200 ms.

In the pump–probe experiments, we have only recorded mass spectra as a function of the time delay. As mentioned above, we have employed several pump–probe wavelength combinations that allow investigation of both the dynamics of the neutral and cation. The short dynamics were recorded using 5 fs steps, while the longer dynamics (time delays > 150 fs) were recorded with time steps of 20, 50 and 100 fs. A full scan consisted of 5 to 10 series and for each delay an average over 200–400 mass spectra was performed. The time resolution as obtained from the full width half maximum (FWHM) cross correlation signals of Ne was 45 ± 1 fs (for XUV + 800 nm) and of H_2O was 50 ± 1 fs (for 400 + 800 nm). The peak of the cross correlation signal defines the 0 fs time delay.

In the following of the paper, we only discuss the 5FU results and those obtained for 5ClU and 5BrU will be presented elsewhere.[23]

3 Results and discussions

A. Mass spectra and fragmentation

Absorption of XUV radiation leads to ionisation and subsequent dissociation of the parent ion. In Fig. 2b we present the mass spectrum recorded using the entire XUV spectrum (HH3–HH9). For comparison, we also show the mass spectra resulting from multiphoton ionisation (MPI) with 800 nm, 400 nm and 266 nm. The fragmentation pathways depicted schematically in Fig. 2a, indicate that, as expected due to the cyclic form of the molecule, at least two bonds have to break in order to produce any of the fragments, except perhaps for H^+. The formation of some fragments requires a hydrogen or a proton transfer, while for others, there

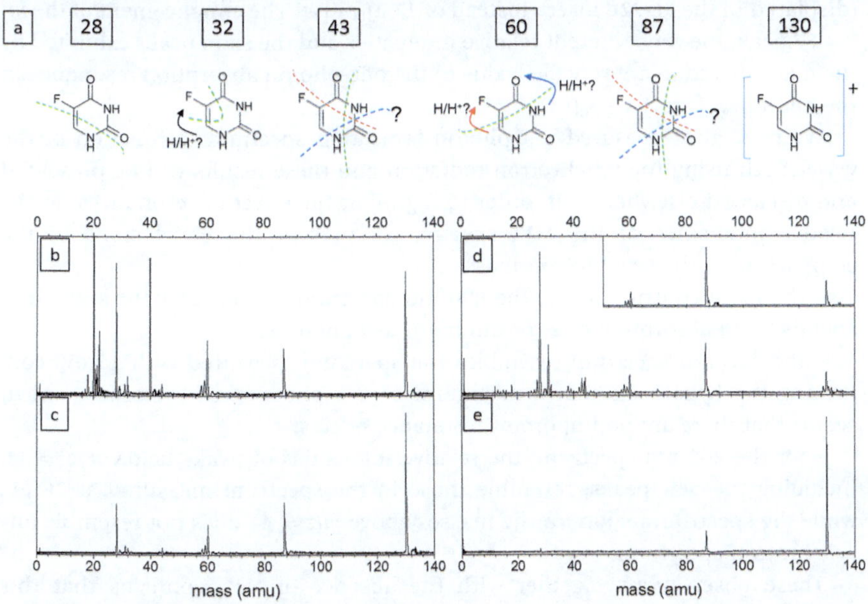

Fig. 2 Fragmentation pathways (a) corresponding to some of the fragments appearing in the mass spectra recorded using XUV radiation (b), 400 nm (c), 266 nm (d) and 800 nm (e). (d) also contains an inset with a spectrum recorded using a pulse energy 5 times lower: 3 μJ/pulse.

are several fragmentation possibilities. So, the mass spectra already indicates that the parent ionisation is followed by complex dynamics involving significant nuclear rearrangements, such as ring opening reactions, H/H^+ transfer and multichannel parallel and/or sequential dissociations.

The main peaks in the mass spectrum recorded with XUV radiation are m/z = 130 (parent ion), 87, 60 and 28. The peaks m/z = 40 and m/z = 20 and 22 are due to the ionisation of some Ar pollution and of the Ne carrier gas, respectively. Additionally, there are much weaker peaks: m/z = 59, 58, 44, 43, 32, 31, 18 and 1. The main peaks are present in all MPI mass spectra, while this is not the case for the weak peaks. Their relative intensities vary greatly from one excitation scheme to another because of the differences in the total excitation energies and the intermediate resonance encountered. As expected, the spectrum that most resembles the XUV (Fig. 2b) is that recorded using 400 nm (Fig. 2c), which is the wavelength employed to generate the XUV HH. All the peaks except m/z = 18 are present, though their relative intensities are somewhat different than in the XUV mass spectrum.

In the mass spectrum recorded at 800 nm (Fig. 2e) the parent cation peak is by far the strongest and only the m/z = 87 peak has some reasonable intensity, while the other two peaks m/z = 60 and 28 are barely visible.

The mass spectrum recorded at 266 nm (Fig. 2d) contains a larger number of weak peaks and has a very different pattern of peak relative intensities when compared to the other three spectra. The absence of the Ne peak (IP = 21.57 eV) suggests that the maximum number of non-resonantly absorbed photons is 4. The most salient feature of this spectrum is the large enhancement of the m/z = 87 peak which becomes 2–3 times more intense than m/z = 130 and m/z = 60. It is even more striking in the spectrum taken with a lower energy of 10 μJ/pulse (displayed in the Fig. 2d inset), instead of 17 μJ/pulse. The enhancement of the m/z = 87 peak, the very different relative intensities and the extra peaks exhibited by the mass spectrum are very likely due to the one-photon absorption resonance in the neutral: $S_2(\pi\pi^*) \leftarrow S_0(n\pi)$.[24–27]

We have also measured one-photon ionisation spectra as a function of the wavelength using the synchrotron radiation and these results will be presented and discussed elsewhere.[28] In order to highlight the effect of resonances on the relative intensities of the MPI spectra, we present the conclusions of the comparison to the synchrotron mass spectra:

- the main contribution to the 800 nm spectrum is non-resonant and corresponds to an absorption of a maximum of 8–9 photons,

- the 400 nm spectrum resembles the spectrum measured with 15 eV, confirming the hypothesis that a maximum of 4–5 photons are absorbed, and it seems that there are no important resonance effects,

- for the 266 nm spectrum, the relative intensities of peaks below m/z = 44, including the new peaks, resemble those in the spectrum measured at 26 eV, while the spectrum region for the masses above m/z = 44 does not resemble any synchrotron spectrum.

These observations together with the absence of Ne^+, confirms that this spectrum is strongly influenced by the resonance in the neutral and possibly by some resonance in the lower cation states.

A comparison of the XUV mass spectrum to the synchrotron spectra reveals as in the case of 266 nm spectra that the relative intensities of the group of heavier peaks and that of lighter peaks resemble spectra taken at different photon energies: at 15.5 eV (HH5) and >21.5 eV (HH7), respectively. The relative intensities of lighter peaks are very similar to that of an XUV spectrum recorded using an Al filter in order to select HH7 and HH9 (not shown).

We discuss below the different fragmentation pathways:

***m*/*z* = 87, 43 and 44.** The ion $m/z = 87$ $[C_3H_2OF]^+$ is produced by a loss of the HNCO neutral moiety, while $[HNCO]^+$ is produced by the loss of $m/z = 87$ neutral. The formation of the fragment $m/z = 44$ requires prior nuclear rearrangements involving tautomerisation *via* H/H^+ transfer. This fragment is more intense than its partner, $m/z = 43$, in all the mass spectra, but the 266 nm spectra, where they have similar intensities.

These three fragments can be produced in at least three different ways, as shown in Fig. 2a, that can all contribute to the intensity of the peaks. Nevertheless, for $m/z = 87$, each fragmentation pathway results in the formation of an ion with a different chemical structure: $[NHCOCFCH]^+$, $[NHCHCFCO]^+$ and $[CFCHNHCO]^+$. The mass detection that is not sensitive to the chemical structure, the bandwidth of the lasers and the lack of information on the spectroscopy of these isomers prevented their selective formation, detection and identification. For sure, additive contributions can account for the intensity of the $m/z = 87$ peak, but are at variance with the weak peak $m/z = 43$. $[HNCO]^+$ fragment can also be produced by secondary dissociation of the ion fragment $m/z = 87$. In addition, $m/z = 43$ can also be assigned to $[C_2F]^+$ and thus produced *via* a fifth pathway.

***m*/*z* = 60, 59 and 58.** The first two peaks correspond to $[C_2HOF]^+$ and $[C_2OF]^+$. While the former requires isomerisation or tautomerisation *via* H/H^+ transfer as shown in Fig. 2a, the latter can be formed either directly or by de-hydrogenation of the former. The fragment $m/z = 58$ can be assigned to $[CHN_2O]^+$ (NHCONH) the largest fragment not containing F. In the 266 nm spectrum this fragment has a comparable intensity or stronger $m/z = 59$.

***m*/*z* = 32 and 31.** These two fragments correspond to $[CHF]^+$ and $[CF]^+$, respectively. As shown in Fig. 2a, the formation of $[CHF]^+$ requires prior isomerisation. Their relative intensities depend on the excitation: XUV or MPI. $m/z = 31$ becomes quite intense in the 266 nm spectrum.

***m*/*z* = 28.** This mass can be assigned to $[CH_2N]^+$, $[CO]^+$ or $[N_2]^+$. The formation of the former two requires the breaking of two bonds and the latter a more complicated formation mechanism. It has the strongest peak in all the mass spectra except for that recorded at 800 nm and this is also the case in the ionisation of uracil by MPI.[29] The satellite peaks appearing in the 266 nm mass spectrum ($m/z = 27$ and 26) can be explained by sequential de-hydrogenations of $[HCNH]^+$, lending support to this assignment.

***m*/*z* = 40, 39, 38, 19, 18, 1.** The first four fragments are extremely weak, very probably because their formation requires the breaking of a least three bonds. $m/z = 40$, 39 and 38 $[C_2H_2N]^+$ followed by sequential de-hydrogenations) can be observed only in the spectrum recorded using 266 nm and $m/z = 19$ (F^+) can barely be observed in the XUV spectrum. $m/z = 18$ appears in the XUV and 266 nm mass spectra, while $m/z = 1$ is present also in the 400 nm mass spectrum. The latter could be produced by a simple cleavage of the H^+ from one of the NH or CH bonds as it occurs with H in the neutral uracil.[30] Nevertheless, as in the case of $[H_2O]^+$, it could also be the result of complicated dynamics involving nuclear rearrangements leading to isomerisation and tautomerisation.

B. Kinetic energy distributions of the fragment IONS

In Fig. 3 we present the kinetic energy distributions together with the associated images recorded for the main fragments ($m/z = 87$, $m/z = 60$ and $m/z = 28$) produced in the dissociative ionization of 5FU by the XUV HH frequency comb produced in Ar.

From Fig. 3, one can observe that the maximum kinetic energy for any of the three fragments is low, typically less than 0.5 eV. Likewise, the maximum kinetic energy release (KER) calculated assuming a two-body dissociation, does not exceed 1.3 eV, indicating that very little energy is transferred into fragment translation. Therefore, the remaining excess energy should be shared between the electron kinetic energy and the internal energy of the fragments. In general, the kinetic energy increases as the fragment mass decreases. This is indeed the case for fragments $m/z = 87$ and $m/z = 60$, where the maximum kinetic doubles (0.3 eV and 0.6 eV, respectively). Surprisingly, the light fragment, $m/z = 28$, has an important component at low kinetic energies up to 0.1 eV. But, in addition it also has a 'long tail' contribution. Generally, the low kinetic energy distributions suggest a slow statistical dissociation over a potential surface (PES) with a loose transition state (TS). But dynamical effects in the exit channel can affect the kinetic distribution width. Another striking feature characterising the kinetic energy distributions of all fragments obtained with either XUV or 400 nm radiation, is due to the small 'modulations' in the low energy part of the distributions (0–0.15 eV). We also observed this for the distributions recorded with the synchrotron radiation of the same energy as HH5 or HH7.[28] This may be explained by multiple contributions due to excitation at several wavelengths (energies) of the HH comb and with different numbers of photons for the MPI. In the case of

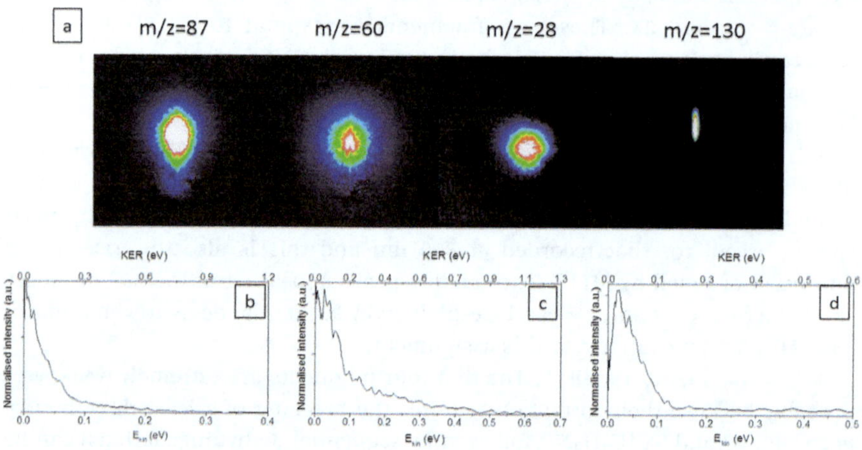

Fig. 3 Velocity map images recorded with XUV radiation for some fragments and the parent (a). The kinetic energy distributions of the fragments: $m/z = 87$ (b), $m/z = 60$ (c) and $m/z = 28$ (d). An image with the parent ion is given to show that even though the kinetic energies are small, they are significantly larger than the translational distribution of the molecular beam. The KER energies were calculated from the $E_{kin}[m_1^+]$ using the formula $\frac{M}{m_2} \times E_{kin}[m_1^+]$ (m_1^+, m_2 and M are the ion fragment, the neutral co-fragment and the parent masses).

m/*z* = 87, this observation also supports the idea that there is more than one pathway to produce it as described in the previous sub-section.

C. Temporal dependence of the parent and fragment ions

In Fig. 4 and 5 we present the temporal dependence of the main (*m*/*z* = 130, 87, 60, 28) and weak peaks (*m*/*z* = 44, 43, 32, 31, 1) for different combinations of the pump and probe wavelengths. The positive delays for Fig. 4a, b, d and e and 5a-b, correspond to the situation when the XUV is the pump and for Fig. 4c and f and 5c correspond to the situation when 400 nm is the pump. The probe beam is either 800 nm or 266 nm. For negative delays, they become the pump beam and XUV or 400 nm become the probe beam. All the spectra shown in these figures were normalised to [0, 1], meaning that the minimum value was subtracted from all the data points and then the maximum values were used to divide all the values. One has to bear in mind that in this representation, a zero signal, does not necessarily mean a background-free signal. We have specifically chosen this representation to facilitate the direct comparison of the relative signal intensities among different pump–probe excitation schemes.

A quick inspection of Fig. 4 and 5 reveals that all *m*/*z* exhibit an asymmetric strong peak located around 0 fs time delay. The peak asymmetry is due to some ultrafast dynamics. The peak shift from 0 fs is not the same for all *m*/*z* and all excitation conditions. The largest ones are observed for the transients obtained with XUV generated in Kr.

Besides this main feature, the signals of all *m*/*z* have additional contributions due to some slower dynamics. For the time dependence recorded with 800 nm (Fig. 4a–d), the signals are typically due to dynamics induced by XUV radiation or by 400 nm, although some dynamics initiated by 800 nm is also visible for some *m*/*z* transients. In general, for these excitation schemes (XUV + 266, 400 + 266) the

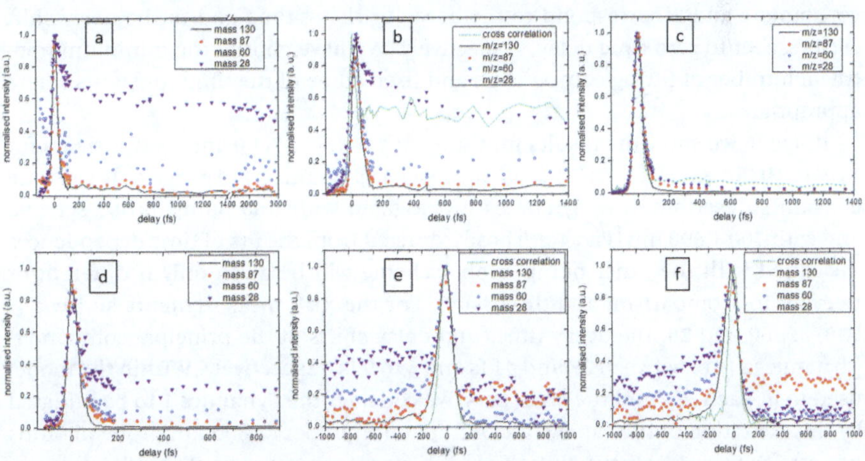

Fig. 4 Dynamics of the parent ion and main fragments (*m*/*z* = 130, 87, 60, 28) recorded using different pump–probe schemes (a) XUV (HHG in Ar, reduced spectrum) and 800 nm, (b) XUV (HHG in Kr, full spectrum) and 800 nm, (c) 400 nm and 800 nm, (d) XUV (HHG in Ar, full spectrum) and 800 nm, (e) XUV (HHG in Ar, full spectrum) and 266 nm, (f) 400 nm and 266 nm. For HHG in Ar, reduce spectrum means that HH9 is absent.

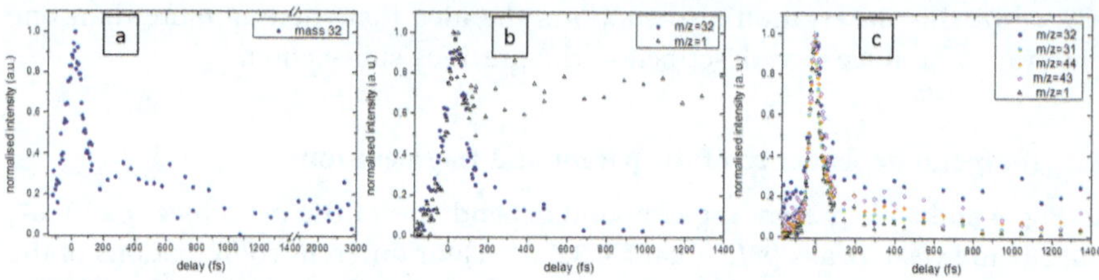

Fig. 5 Dynamics of the weak fragments ($m/z = 44, 43, 32, 31, 1$) recorded using different pump–probe schemes. (a) XUV (HHG in Ar, reduced spectrum) and 800 nm, (b) XUV (HHG in Kr, full spectrum) and 800 nm, (c) 400 nm and 800 nm.

time dependences are quite similar. In turn, in all m/z time dependences recorded with 266 nm (Fig. 4e and f), the strong signals correspond to the relaxation dynamics initiated by 266 nm excitation of the $S_2(\pi\pi^*)$ state of the neutral molecule. They are shown herein only for comparison and discussion, whereas their associated dynamics will be discussed elsewhere.[23]

The time dependences shown in Figs. 4a–d and 5 were fitted using the functional:

$$S(t) = A_0 + \mathrm{cc} + \sum_{i>0} A_i \cdot P_i(t) \otimes G(t), \tag{1}$$

where A_0 is the constant signal (background), cc is the cross correlation, P_i is the time dependence and A_i the corresponding amplitude. P_i was a decay or a step function:

$$P_i(t) = \left(\mathrm{e}^{-\frac{t-T}{\tau_{\mathrm{decay}}}}\right) \text{ or } \left(1 - \mathrm{e}^{-\frac{t-T}{\tau_{\mathrm{rising}}}}\right), \tag{2}$$

which was convoluted with the experimentally determined Gaussian cross correlation function, $G(t)$, to account for the finite time resolution. The time constants τ and T correspond to the decay or rising time and to the delay shift with respect to zero time delay, respectively. We have chosen the minimum non-trivial number of fitting components and fixed some of the time constants where appropriate.

In Fig. 6 we show the results for fitting $m/z = 130$, 87, 60 and 28 transients for XUV (HHG in Kr: HH3–HH7) + 800 nm (Fig. 4b). In Table 1, we give only the time constants describing the dynamics investigated with this pump–probe scheme and with 400 + 800 nm (Fig. 4c). Those extracted from the fits of time dependences measured with the other pump–probe scheme will be given only if different or relevant for comparison and discussion. For the XUV measurements and $m/z =$ 130, 87, 60 and 28, the decay times and delay shifts of the principal component (dynamics 1) are between 27 and 41 fs and 6 to 9 fs, respectively. Within the model used to fit each m/z and the error bars, we consider the dynamics 1 to be initiated by the excitation of the same state. A different number of 800 nm probe photons projects the wavepacket evolution onto different templates provided by the fragmentation channels into the ion continuum. The dynamics 2 revealed by $m/z = 87$ corresponds to a step function with an ultrafast rising time and delay from 0 fs, both of about 9 fs. Likewise, the $m/z = 60$ signal contains a step contribution. For XUV experiments, its amplitude is too weak to be fitted, but it certainly affects the

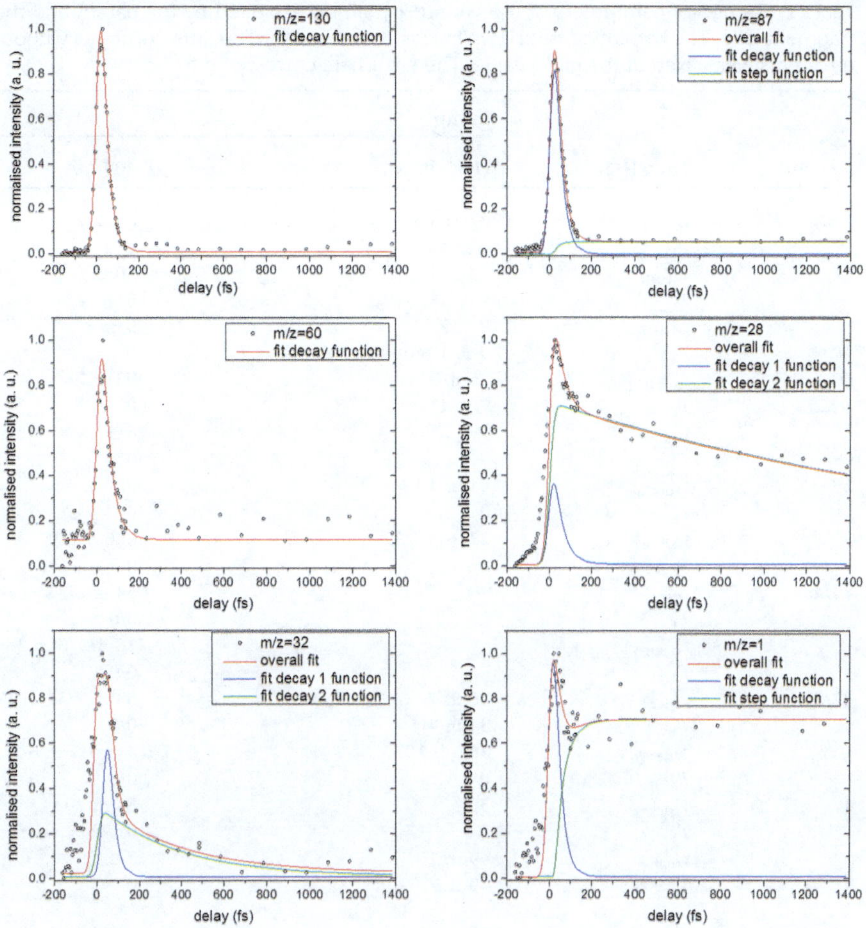

Fig. 6 The results of the fits for transients m/z = 130, 87, 60 and 28 detected in the measurements XUV (HHG in Kr) + 800 nm.

transient resulting in an increase of the dynamics 1 decay time to 36 fs. In the case of $m/z = 28$, the dynamics 2 corresponds to a very slow decay of about 2.3 ps. Its amplitude relative to the other m/z dynamics is quite significant for all pump–probe schemes, but it is largest for the XUV spectra generated in Kr and Ar (reduced spectrum).

So, there are three main conclusions which can be drawn from the fits:

- for all excitation schemes, most of the time constants obtained for different m/z are very similar,

- the most important contribution for all m/z comes from the cross correlation signal for all the pump–probe schemes except for that involving XUV (HHG in Kr) where no cross correlation signal was apparent,

- the dynamics induced by XUV (HHG in Kr) are delayed from 0. Allowing for such delay shifts was not required for improving the fits of the 400 nm data.

The possible origins and the physical interpretation of these time constants will be discussed in the next section.

Table 1 Temporal parameters of the various dynamics exhibited by the parent and the fragment ions. The dynamics are shown in Figs. 4 and 5, b–c. The time constants without error bars were fixed at the given value. The error bars correspond to 1σ

		Pump	
m/z (au)	Time (fs)	HHG in Kr	400 nm
130	τ_{decay}	27 ± 1	34 ± 2
	T_1	6 ± 1	0
87	τ_{decay}	29 ± 1	50 ± 5
	T_1	9 ± 1	0
	τ_{rising}	9 ± 7	9 ± 8
	T_2	9 ± 1 (constrained: $T_2 = T_1$)	0
60	τ_{decay}	36 ± 3	53 ± 7
	T_1	7 ± 1	0
	τ_{rising}	—	6 ± 3
	T_2	—	0
28	τ_{1decay}	41 ± 10	50
	T_1	0	0
	τ_{2decay}	2326 ± 732	606 ± 526
	T_2	0	0
44/43	τ_{decay}	—	54 ± 8/52 ± 14
	T_1		0/0
	τ_{rising}		10 ± 7/5 ± 2
	T_2		0/0
32/31	τ_{1decay}	26 ± 7	53 ± 7/51 ± 19
	T_1	34 ± 4	0/0
	$\tau_{2decay}/\tau_{rising}$	387 ± 107	4 ± 1/7 ± 6
	T_2	0	0/0
1	τ_{1decay}	30	51 ± 20
	T_1	0	0
	τ_{rising}	49 ± 10	7 ± 5
	T_2	19 ± 7	0

D. Disentangling and interpreting dynamics using HH radiation and multiphoton excitation

Fig. 7 illustrates schematically the excitations that can be promoted in $5FU^+$ by the frequency comb of the full XUV spectrum (HH3–HH9) and by MPI reaching similar energies. Given the range of the energetically accessible calculated states[31,32] encompassed by the XUV spectrum (9.3–27.9 eV), it would not be surprising to expect numerous open channels. Each component of the comb will lead to the ionisation of many states as well as to potential excitations of the same state by all frequency components. In principle, the latter should result, only in additive contributions to the same dynamics weighted by the cross sections. However, such a broad band multi-frequency excitation will result in launching parallel dynamics and vibrational wavepackets on many potential energy surfaces (PESs). Moreover, the relaxation mechanisms could imply dynamics of a sequential nature, where the initial populated state relaxes onto a state that further relaxes, due to nonadiabatic interactions, as for example, in the photostability of the DNA bases.[24–27] All these concur to suggest that the ionisation and subsequent fragmentation of a polyatomic molecule by an XUV HH comb could produce a very complicated and incomprehensible picture: many excitation channels open

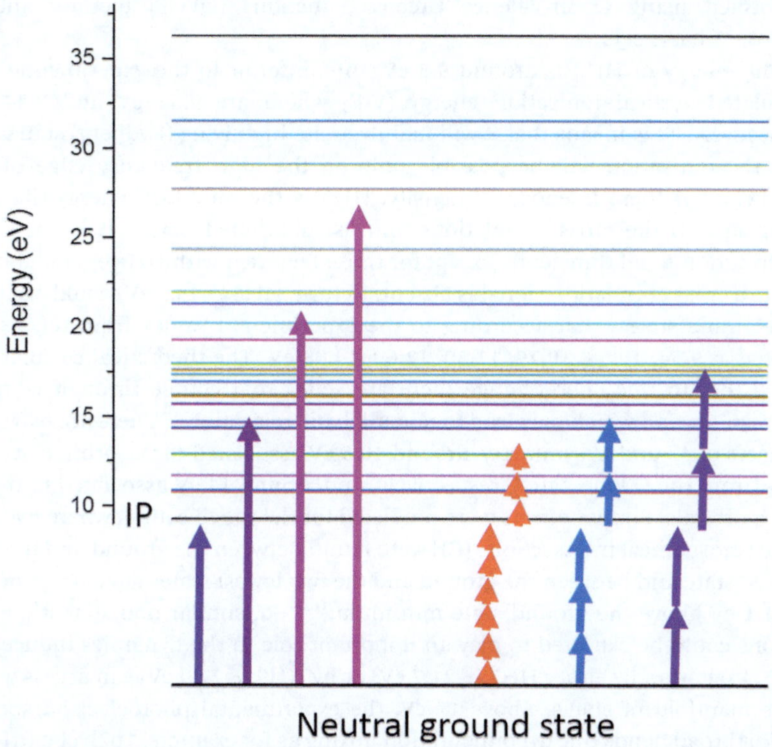

Fig. 7 Excitation by the HH frequency comb (HH3–HH9) (arrows in shades of violet) and by multiphoton absorption of 800 nm (red arrows), 400 nm (blue arrows) and 266 nm (violet arrows) radiation. The values used to depict the energy levels of the cation states in the diagram were taken from ref. 31 and 32.

leading to the production of many fragments and to an averaging of the many overlapping dynamics.

Indeed, as we discussed in Section 3A, the mass spectra confirm that there are many fragmentation channels open. The slight 'modulations' in the kinetic energy distributions of fragments $m/z = 87$, 60 and 28 reveal that they formed *via* several fragmentation pathways, either due to excitation by different HH and/or to the opening of different fragmentation channels. When taking into account all these considerations and observations, the time dependences shown in Fig. 4 and 5 look unexpectedly simple. This is the result of the selectivity conferred by the absorption cross section dependence on wavelength and the Franck–Condon factors.

To understand the origins of the dynamics, it is essential to separate their contributions in the measured time dependences and to associate them with excitation energies and electronic states. While the former can be obtained from the fitting procedure, the latter requires additional experimental and/or theoretical information. For this, we will use the experimental and calculated values given in ref. 31. The experimental values were determined by measuring photoelectron spectra with 40 eV synchrotron radiation and the calculations were performed using Outer-Valence Green's Functions (OVGF) method and 6-311++G** basis sets.

The energy of HH3 is around 9.3 eV, thus inferior to the experimental and calculated vertical ionisation energy (VIE) which are 9.54 eV and 9.24 eV, respectively. This means that it will mainly excite high-lying (Rydberg) states and that the ionisation will be possible only on the high frequency edge of the bandwidth if there is enough intensity. HH3 is the only XUV energy that can contribute to the cross correlations and as mentioned before it is the main contribution to all transients, except for those recorded with HH generated in Kr where a cross correlation signal is absent. In turn, HH5 (~15.5 eV) could excite 6–7 electronic states, that according to the experimental values from ref. 28 are located at 9.54, 10.58, 10.79, 11.30, 12.69, 14.08 eV. The theoretical calculations reveal that in this energy range there are states that couple through vibronic interactions and which may lead to nonadiabatic relaxation.[28] There are two pairs of states of A$''$ and A$'$ symmetry, around 10.5 eV associated to the orbitals π_4 and σ_{OLP} (from the O lone pairs), respectively, and around 14 eV associated to the π_2 and σ orbitals. Similar effects were predicted for the uracil cation, where two and three state conical intersections (CI) were found between the ground and the first excited state and between the ground and the two lowest states at energies of less than 1 eV above the ground state minimum.[31,33] So, similar nonadiabatic interactions could be expected to play an important role in the dynamics induced by HH5. Excitation by either HH7 (~21.7 eV) or by HH9 (~27.9 eV) can access a very large manifold of states. Above 17 eV, the experimental photoelectron spectra exhibit broad bands due to configuration mixing as for example, 1p2h (1 particle 2 holes) configurations, and multi-electronic dynamics related to shake-up processes can be initiated in this energy domain.

The relevant contributions of the HHs to the dynamics can be inferred from the comparison of measurements taken with different XUV spectra (generation in Kr or in Ar) and with 400 nm and 266 nm. The 400 nm MPI mass spectra give an upper limit to the excitation energy at 15.5 eV. According to this, for single-photon ionisation, HH5 should make the most significant contribution to the dynamics displayed by the transients of these fragments.

The time dependences look very similar, although some time constants are somewhat longer for all m/z signals recorded with 400 nm. This suggests that the dynamics are very likely initiated around the same excitation energies. The fact that the same components are present in all time dependences recorded with XUV radiation, implies that HH9, which is only present in the full HH spectrum generated in Ar, does not make any distinguishable contribution. HH7 is ten times weaker than HH5 in the XUV spectrum generated in Kr, therefore it should contribute less for these measurements. No conspicuous difference in the time dependences recorded with different XUV spectra can be unambiguously related to HH7 excitation. Thereby, the comparison of the time dependences also supports the idea that the excitation and the subsequent dynamics is very likely due to HH5 or possibly HH3.

As mentioned in the previous section, all the transient signals contain two time dependences. To determine the possible origins of the dynamics 1: the fast decay (~30 fs; observed for all m/z) and the dynamics 2: the fast rising step (~9 fs; observed for $m/z = 87$ and 60) or the slow decay (2.3 ps; observed for $m/z = 28$), we will discuss them in the context of two mechanisms that can be initiated:

Parallel dynamics. The same wavelength or different wavelengths simultaneously prepare two different cation states that decay independently.

Sequential dynamics. The same wavelength prepare a state that decays to a lower state which will further decay to a third state. However, since the delay shifts for $m/z = 87$ and $m/z = 60$ transients are the same (~9 fs) for both dynamics 1 and 2, a sequential mechanism cannot be envisaged.

Note that here wavelength designates an energy that can be attained either by multiphoton absorption of 400 nm or by a HH.

Dynamics initiated at 9.3 eV. We consider first the case where HH3 initiates the dynamics in the neutral molecule. When using HH generated in Kr, there is no cross-correlation contribution to the transients and the main contribution comes from the dynamics 1. For the generation in Kr, we expect a more intense HH3 and thus an enhancement of contributions from the resonances with high-lying Rydberg states located around 9.3 eV. Then the dynamics 1 (~30 fs) should very likely reflect a Rydberg–valence interaction, as for example, the π3sRydberg–ππ* in ethylene or a fast Rydberg–Rydberg interaction, as detected recently in CH_2IBr.[34–36] For the other XUV measurements (HHG in Ar) these resonances might be reached only by an edge of the harmonic bandwidth and HH3 will principally contribute non-resonantly to the cross correlation by the central part of the bandwidth. However, considering the HH3 bandwidth and the high density of states close to the IP, it is rather surprising that the signal produced by the dynamics initiated by HH3 does not overcome the cross correlation signal for all excitation schemes (HHG in Ar, 400 nm). The 6–9 fs delay shift observed for the dynamics 1 when using HH generated in Kr is of the order of the NH stretch (3500 cm^{-1} for the neutral ground state[37]). This may suggest that nuclear rearrangements involving this mode are necessary to reach the geometries favouring photoionisation.

A second dynamics (2) could be fitted for the $m/z = 87$ signal (step function ~ 9 fs) and $m/z = 28$ signal (slow decay ~ 2.3 ps). Note that the $m/z = 60$ transient also exhibits a step contribution, but it is too weak to be fitted. Similar step dynamics are also present in the 400 nm MPI measurements and within the error bar, it has a similar value (7–9 fs). Despite the 45 fs time resolution, the fitting procedure revealed that delay shifts of 5–9 fs are required to reproduce the transients from Fig. 4b (XUV-HHG in Kr), but no delay shifts are required for those from Fig. 4c (400 m). This an important result of our experiments and can be rationalised as the effect of the possible intermediate resonances reached by two 400 nm photons at 6.2 eV. We know that the dynamics initiated at 4.5 eV are as fast as ~50 fs.[23] If a similar situation occurs at 6.2 eV, a different Franck–Condon window may be available after the absorption of three 400 nm photons, such that dynamics 2 (and also dynamics 1) takes place immediately as a result of these intermediate resonances.

If HH3 excites in parallel the dynamics 1 and 2 of $m/z = 87$ and 60, then the fast rising dynamics 2 could also correspond to the direct excitation of a high-lying Rydberg state. In this case, the Rydberg state has to be long lived because the signal does not decay within the 1 ps measurement duration. The 800 nm probe photons will ionise this state towards two different dissociative ionisation states. As observed experimentally, its contribution would be lower for $m/z = 60$ because more probe photons are required to reach the dissociation ionisation limit. The rising time would correspond to additional nuclear rearrangements or to ultrafast electron dynamics. If one considers that the Franck–Condon windows opened by one photon and multiphoton excitation should have different widths, obtaining very similar values is somewhat unexpected.

The dynamics 2 exhibited by $m/z = 28$ correspond to slower decays: 2.3 ps and 606 fs for XUV and 400 nm excitation, respectively. The large difference between these values indicates that these dynamics are probably initiated at different energies where different processes are at play. If this dynamics 2 is due to the dissociation of a valence state, then one should also see a rising step function on the same timescale in another transient signal produced by the probe photons by either dissociative ionisation of the valence state or ionisation of the neutral co-fragment. This behaviour should be expected whether the mechanism is parallel or sequential. However, it was not observed on any transient signal, indicating against the possibility that $m/z = 28$ slow dynamics could be initiated by HH3.

Dynamics initiated at 15.5 eV. From the 3D coincidence imaging experiments recorded with synchrotron radiation, we know that photoionisation at 15.5 eV typically leaves $5FU^+$ with an internal energy of 10.5–11 eV and that the states up to 12 eV are non-dissociative.[28] As discussed above, in this region there is a couple of states (A″, A′) that interact strongly *via* vibronic coupling. The 800 nm probe photons then project this vibronic dynamics onto the fragmentation channels producing $m/z = 87$, 60, 28. In order to be sensitive to the dynamics, one state should have a lower dissociative ionisation probability than the other. In this case, the decay dynamics (1) of 30 fs reveal their nonadiabatic interaction. However this excitation scheme cannot explain the similar dynamics observed on the parent transient, which are very likely initiated by HH3.

Considering the large number of states encompassed by HH5, one can easily imagine that HH5 may excite a $5FU^+$ state other than those located at 10.5 eV and may initiate the dynamics 2. The 800 nm probe photons could project this nascent dynamics onto the dissociating templates: $m/z = 87$ and 60. In this case, the rising time (~9 fs) of dynamics 2 may be associated with nuclear rearrangements necessary to reach the relevant Franck–Condon region to obtain efficient dissociative ionisation by the probe pulse. This scenario is possible only if the state excited by HH5 is not dissociative on the time scale relevant for our experiments. With regard to the slow dynamics observed for the $m/z = 28$ signal, the same arguments as in the case of the HH3 excitation, forbid a parallel mechanism. In a sequential relaxation following HH5 excitation of the couple A″ and A′ states located around 10.5–11 eV, the 2.3 ps dynamics recorded for $m/z = 28$ could be the signature of an internal conversion to the ground cation state after the nonadiabatic interaction.

E. Proton transfer

Finally, we will briefly discuss the time constants of the $m/z = 1$ and $m/z = 32$ transient signals observed in the XUV + 800 nm measurements (Fig. 5b).

In order to be formed, the fragment $m/z = 32$ requires an H/H^+ transfer reaction leading to the isomerisation of $5FU^+$. If the H/H^+ transfer occurs after the XUV excitation, then it should be possible to record it with the probe photons.

We suppose that the isomerisation is initiated by HH5 through ionisation of 5FU to a non-dissociative state and that $m/z = 32$ is produced by dissociative ionisation in the probe step. This hypothesis is supported by the fact that the intensity of $m/z = 32$ does not seem to be affected when using an Al filter to remove the contributions from harmonics below HH7. The time constants we have extracted for $m/z = 32$ are quite different from those obtained for the other masses and correspond to a relaxation time (26 fs) similar to that of $m/z = 130$ and to relaxation on a slower timescale, 400 fs. The first dynamics also involved a significant delay shift of 34 fs. Based solely on this information, it is difficult to interpret these dynamics and relate them to the H/H^+ transfer.

During the isomerisation, H/H^+ migrates from the initial site to the final site of transfer and during this time it becomes very loosely bound. The absorption of 800 nm probe photons may lead to complete breaking of the bond and the detachment of H/H^+. One may then speculate, that a part of or the whole dynamics exhibited by H^+ may reflect an H/H^+ transfer taking place during the isomerisation process required to yield $m/z = 32$. The time constants extracted for $m/z = 1$ correspond to a relaxation time of 30 fs and a rising time of 49 fs delayed by 19 fs. The rising step could be interpreted as being the isomerisation time. In order to observe a rising H^+ signal it is necessary that the bond at the initial site be strong and that the bond at the final site can be easily broken by the probe photons. Of course this is a very simplistic picture and additional experiments with higher temporal resolution or Deuterium labelled 5FU are needed in order to confirm it.

4 Conclusions

The main point we would like to highlight through the work presented herein concerns the fact that HH radiation is a powerful tool for investigating ultrafast dynamics even when using the whole spectrum. Using a comb composed of frequencies ranging from 9.3 to 27. 9 eV, separated by 6.2 eV and with FWHM bandwidth of about 0.15 eV, we could measure and resolve the dynamics induced following ionization of 5FU. At these photon energies a large manifold of neutral and cation states can be accessed and many channels (dissociative or not) are open. Actually, the absorption cross sections and discrimination of the open dissociation channels provide our experiments with the selectivity that is lacking spectroscopically. The multitude of the fragmentation channels reveals the different dynamics through different fragmentation channels.

These experiments also show that for such molecular systems, where the kinetic energy of the fragments is too small and the photoelectron spectra too complicated, time-resolved mass spectra are sensitive and capable of resolving multiscale dynamics occurring in the cation, even when detected on singly charged ions and not on the doubly charged ions, as commonly done.

Relying on comparisons with measurements performed in a multiphoton regime (400 nm) and changing the XUV spectrum (Ar or Kr for the generation gas, different driving laser intensity) enabled us to disentangle and assign the components of the transient signals. These comparisons reveal that the time dependences are very similar. But, few key differences, such as the lack of delay shift in the dynamics induced by 400 nm and the large contributions to the cross correlation in all the measurement schemes but one, helped us in assigning an excitation energy to each of the dynamics. HH3 excites a high-lying Rydberg state interacting with a valence state and its dynamics is revealed by a 30 fs decay signal in the parent ion transient. HH5 excites a couple of cation states located at 10.5 eV and their interaction is visible on the ion fragment transients. According to the theoretical calculations of Holland *et al.*[28] these states exhibit vibronic interactions that lead to nonadiabatic dynamics. The transients we have measured for $m/z = 87$ and 60 provide the time constant for the nonadiabatic interaction: 30 fs (dynamics 1). Nuclear rearrangements seem to be required before this relaxation can take place. Also, they are associated with an ultrafast dynamics 2 (~9 fs for $m/z = 87$) initiated in parallel with dynamics 1. Further sequential relaxation occurs through slow decay (2.3 ps) and it is observed on the $m/z = 28$ signal. These explanations reveal that the same photoprotection mechanisms, as those conferring photostability to the neutral nucleobases and DNA, appear to be activated for the $5FU^+$. It is very interesting to note here that the time scale is much faster than in the deactivation of the $S_2(\pi\pi^*)$ state of the neutral. For this state, we have obtained a primary relaxation time of ~50 fs,[23] which is also a factor of 2–2.5 times faster than the relaxation time reported for the same state of the neutral thymine and uracil.[24]

Acknowledgements

R. C. and P. Ç. wish to acknowledge the beam time granted at CELIA *via* the LOA-IRAMIS-CELIA femtosecond lasers facilities and to the GDR EMIE (Edifices Moléculaires Isolés et Environés) for financial support.

References

1 D. Popmintchev, C. Hernández-García, F. Dollar, C. Mancuso, J. Pérez-Hernández, M. C. Chen, A. Hankla, X. Gao, B. Shim, A. Gaeta, M. Tarazkar, D. Romanov, R. Levis, J. Gaffney, M. Foord, S. Libby, A. Jaron-Becker, A. Becker, L. Plaja, M. Murnane, H. Kapteyn and T. Popmintchev, *Science*, 2015, **350**, 1225.
2 M. Chini, K. Zhao and Z. Chang, *Nat. Photonics*, 2014, **8**, 178.
3 F. Calegari, G. Sansone, S. Stagira, C. Vozzi and M. Nisoli, *J. Phys. B: At., Mol. Opt. Phys.*, 2016, **49**, 062001.
4 S. Zamith, V. Blanchet, B. Girard, J. Andersson, S. L. Sorensen, I. Hjelte, O. Björneholm, D. Gauyacq, J. Norin, J. Mauritsson and A. L'Huillier, *J. Chem. Phys.*, 2003, **119**, 3763.
5 F. Lépine, G. Sansone and M. J. J. Vrakking, *Chem. Phys. Lett.*, 2013, **578**, 1.
6 A. Camper, A. Ferré, V. Blanchet, F. Burgy, D. Descamps, S. Petit, T. Ruchon and Y. Mairesse, *Opt. Lett.*, 2015, **40**, 5387.
7 A. Ferré, A. E. Boguslavskiy, M. Dagan, V. Blanchet, B. D. Bruner, F. Burgy, A. Camper, *et al.*, *Nat. Commun.*, 2015, **6**, 5952.
8 L. Medišauskas, S. Patchkovskii, A. Harvey, D. S. Brambila, C. Neidel, J. Klei, A. Rouzée, M. J. J. Vrakking and M. Yu. Ivanov, *Phys. Rev. A*, 2015, **92**, 053403.
9 A. Tehlar and H. J. Worner, *Mol. Phys.*, 2013, **111**, 2057.
10 J. van Tilborg, T. K. Allison, T. W. Wright, M. P. Hertelein, R. W. Falcone, Y. Liu, H. Merdji and A. Belkacem, *J. Phys. B: At., Mol. Opt. Phys.*, 2009, **42**, 081002.
11 M. P. Ziemkiewicz, C. Bacellar, K. R. Siefermann, S. R. Leone, D. M. Neumark and O. Gessner, *J. Chem. Phys.*, 2014, **141**, 174306.

12 L. Belshaw, F. Calegari, M. J. Duffy, A. Trabattoni, L. Poletto, M. Nisoli and J. B. Greenwood, *J. Phys. Chem. Lett.*, 2012, **3**, 3751.

13 F. Calegari, D. Ayuso, A. Trabattoni, L. Belshaw, S. De Camillis, S. Anumula, F. Frassetto, L. Poletto, A. Palacios, P. Decleva, J. B. Greenwood, F. Martín and M. Nisoli, *Science*, 2014, **346**, 336.

14 R. Cireasa, A. E. Boguslavskiy, B. Pons, M. C. H. Wong, D. Descamps, S. Petit, H. Ruf, N. Thiré, A. Ferré, J. Suarez, J. Higuet, B. E. Schmidt, A. F. Alharbi, F. Légaré, V. Blanchet, B. Fabre, S. Patchkovskii, O. Smirnova, Y. Mairesse and V. R. Bhardwaj, *Nat. Phys.*, 2015, **11**, 654.

15 A. Marciniak, V. Despré, T. Barrilot, A. Rouzée, M. C. E. Galbraith, J. Klei, C.-H. Yang, C. T. L. Smeenk, V. Loriot, S. Nagaprasad Reddy, A. G. G. M. Tielens, S. Mahapatra, A. I. Kuleff, M. J. J. Vrakking and F. Lépine, *Nat. Commun.*, 2015, **6**, 7909.

16 J. Breidbach and L. S. Cederbaum, *J. Chem. Phys.*, 2003, **118**, 3983.

17 F. Remacle and R. D. Levine, *Proc. Natl. Acad. Sci. U. S. A.*, 2006, **103**, 6793.

18 F. Lépine, M. Y. Ivanov and M. J. J. Vrakking, *Nat. Photonics*, 2014, **8**, 195.

19 P. Wardman, *Clin. Oncol.*, 2007, **19**, 397.

20 M. D. Prados, C. Scott, H. Sandler, J. C. Buckner, T. Phillips, C. Schultz, R. Urtasun, R. Davis, P. Gutin, T. L. Cascino, H. S. Greenberg and W. J. Curran Jr, *Int. J. Radiat. Oncol., Biol., Phys.*, 1999, **45**, 1109.

21 T. Ito, K. Tanabe, H. Yamada, H. Hatta and S. Nishimoto, *Molecules*, 2008, **13**, 2370.

22 H. Yu, R. Eritja, L. B. Bloom and M. F. Goodman, *J. Biol. Chem.*, 1993, **268**, 1593.

23 R. Cireasa, *et al.*, work in progress.

24 C. Canuel, M. Mons, F. Piuzzi, B. Tardivel, I. Dimicoli and M. Elhanine, *J. Chem. Phys.*, 2005, **122**, 074316.

25 S. Ullrich, T. Schultz, M. Z. Zgierski and A. Stolow, *Phys. Chem. Chem. Phys.*, 2004, **6**, 2796.

26 H. R. Hudock, B. G. Levine, A. L. Thomson, H. Satzger, D. Townsend, N. Gador, S. Ullrich, A. Stolow and T. J. Todd-Martinez, *J. Phys. Chem. A*, 2007, **111**, 8500.

27 S. Matsika, M. Spanner, M. Kotur and T. C. Weinacht, *J. Phys. Chem. A*, 2013, **117**, 12796.

28 P. Çarçabal, J.-P. Champeaux, D. Dowek and R. Cireasa, in preparation.

29 B. Barc, M. Ryszka, J. Spurell, M. Dampc, P. Limão-Vieira, R. Parajuli, N. J. Mason and S. Eden, *J. Chem. Phys.*, 2013, **139**, 244311.

30 M. Schneider, C. Schon, I. Fischer, L. Rubio-Lago and T. Kitsopoulos, *Phys. Chem. Chem. Phys.*, 2007, **9**, 6021.

31 D. M. P. Holland, A. W. Potts, L. Karlsson, I. L. Zaytseva, A. B. Trofimov and J. Schirmer, *Chem. Phys.*, 2008, **352**, 205.

32 D. Ayuso and F. Martin, private communication.

33 S. Matsika, *Chem. Phys.*, 2008, **349**, 356.

34 O. Schalk, A. E. Boguslavskiy, A. Stolow and M. S. Schuurman, *J. Am. Chem. Soc.*, 2011, **133**, 16451.

35 G. Wu, A. E. Boguslavskiy, O. Schalk, M. S. Schuurman and A. Stolow, *J. Chem. Phys.*, 2011, **135**, 164309.

36 P. Sándor, V. Tagliamonti, A. Zhao, T. Rozgonyi, M. Ruckenbauer, P. Marquetand and T. Weinacht, *Phys. Rev. Lett.*, 2016, **116**, 063002.

37 V. K. Rastogi and M. Alcolea Palafox, *Spectrochim. Acta, Part A*, 2011, **79**, 970.

Faraday Discussions

DISCUSSIONS

Attosecond processes and X-ray spectroscopy: general discussion

Chris J. Milne, Peter M. Weber, Markus Kowalewski, Jon P. Marangos, Allan S. Johnson, Ruaridh Forbes, Hans Jakob Wörner, Daniel Rolles, Dave Townsend, Oliver Schalk, Sebastian Mai, Morgane Vacher, R. J. Dwayne Miller, Martin Centurion, Ágnes Vibók, Wolfgang Domcke, Raluca Cireasa, Kiyoshi Ueda, Filippo Bencivenga, Daniel M. Neumark, Albert Stolow, Artem Rudenko, Adam Kirrander, Danielle Dowek, Fernando Martín, Misha Ivanov, Jan Marcus Dahlström, Nirit Dudovich, Shaul Mukamel, Alvaro Sanchez-Gonzalez, Michael P. Minitti, Dane R. Austin, Victor Kimberg and Zdenek Masin

DOI: 10.1039/c6fd90071b

Martin Centurion opened a discussion of the paper by Shaul Mukamel: There seems to be an issue of two different time scales. The resonant cavity will have a long lifetime, while the excited molecules will evolve on a much faster time, which can change their absorption spectrum and thus the strength of interaction with the cavity. Will this be a problem for the method that you are proposing?

Shaul Mukamel replied: This is not a problem in the applications I presented. The cavity field is stationary and its only role is static: It dresses the bare molecular states to create joint molecule/field (polariton) states. Time dependent polaritons in multimode cavities is an interesting issue which was not addressed in the present studies.

Jon Marangos said: You include no decay of the photon field, will this be a good approximation for the dynamics you model?

Shaul Mukamel replied: Decay of the photon field like loss in mirrors may limit the spectral resolution and may limit the observation time. Chemical reactions and nonadiabatic dynamics in cavities are typically affected by other broadening effects so that the photon decay may be neglected.

Martin Centurion remarked: If the cavity has a high Q, and the absorption of the molecules changes, what will be the effect on the cavity fields?

Shaul Mukamel replied: In our applications the cavity field is given, and is unaffected by the molecule. This assumption is valid in many cases. It is made in all classical treatments of spectroscopy. It is an open interesting question when should a self consistent treatment of the cavity field and molecule be required. This could lead to interesting new effects.

Misha Ivanov commented: Coupling a molecule to a cavity field modifies the potential energy surfaces of a molecule in a way that is controlled by the strength of the interaction and the frequency of the cavity mode. However, it is often assumed that a cavity supports a spectrally narrow range of modes. Does it affect or limit the control, especially if we are interested in ultrafast approaches to control?

Shaul Mukamel responded: The cavity polariton effect is static. It affects the zero-order Hamiltonian by dressing the molecular states with photons. This is a very powerful but limited degree of control. Dynamical control using shaped pulses is different and may not be achieved in cavities. It is a different control regime.

Misha Ivanov said: What is the advantage of using a cavity field, as opposed to a laser pulse, for modifying the potential energy surfaces of the molecule? And is there an aspect of the molecule–cavity field interaction where the quantum nature of the cavity field is important, as opposed to the classical treatment of the laser field?

Shaul Mukamel responded: These are two excellent questions.

Placing a molecule in a cavity subjects it to strong coupling with a single field mode. This is formally identical to subjecting the molecule to a strong field (without a cavity). Under certain conditions the two techniques may yield the same information. However strong fields can cause many other processes such as multiphoton absorption and photodissociation. These are interesting in their own right but compete with and complicate the analysis of spectroscopy. Cavity strong coupling can be realized even by the vacuum field and does not suffer from these complications.

Some aspects of polaritions can be fully understood with classical fields. The celebrated work of Hopfield (who coined the term polariton) used an equation of motion for coupled field and exciton modes and only involves a classical field. The linear response of polaritons is thus purely classical. However, multipolariton states that are accessible by nonlinear spectroscopy may require taking the quantum nature of the field into account.

Jon Marangos remarked: In the molecular dynamics calculation – what defines t 0 and is it some initial excitation of the molecule and/or the cavity field included?

Shaul Mukamel answered: This is correct. The process is initiated by a preparation(actinic) pulse which defines the zero time. We assumed that the cavity plays no role in this step. It only affects the subsequent nonadiabatic dynamics

Ágnes Vibók addressed Markus Kowalewski and Shaul Mukamel: Would it be possible to take into account the rotation in the cavity (without any external field) so as to describe properly the LICI (for diatomics)?

Markus Kowalewski answered: The coupling of a dipole to a cavity mode is formally very similar to the coupling to a laser field. Both cases can be described with a molecular dressed states (Jaynes–Cummings) model.[1]

The angle θ, which serves as a second coordinate for the LICI, is then defined by the angle between the molecular axis and the cavity axis. A subtle difference here is that, depending on the cavity geometry, the polarization of the vacuum state might be undefined.

1 M. Kowalewski, K. Bennett, S. Mukamel, *J. Chem. Phys.*, 2016, **144**, 054309.

Shaul Mukamel answered: yes, this is a very interesting application that could potentially lead to the experimental observation of the Berry phase. Such an observation has attracted considerable attention but no clear evidence has been achieved so far.

Ágnes Vibók asked: Do you include a 2nd degree of freedom in the dynamical calculations? It would be necessary to establish a branching space in order to obtain the proper conditions for a light induced conical intersection situation. What is your opinion: Should there be any difference between the dynamical results of the numerical simulations obtained by classically or by QED description of the light?

Shaul Mukamel replied: yes, conical intersections require at least two nuclear coordinates; our simulation includes both a one dimensional model that can represent an avoided crossing and a two dimensional model of conical intersection.

The presented applications are equivalent to placing the molecule in a strong classical field. The quantum nature of the field does not play a role. However, they may be required in other measurements say photon coincidence statistics.

Jon Marangos addressed Shaul Mukamel and Markus Kowalewski: Is this a step towards a model for nanophotonic catalysis – or in general why would one want to do this? Nanostructured materials are after all used already in catalysis but won't in practice there be limits due to the cavity field resonance (width, control over and variability)?

Markus Kowalewski replied: We agree that this is a step towards nanophotonic catalysis. In contrast to a conventional catalyst it acts *via* vacuum field modifications rather than chemical interactions. This opens up a whole new field of possibilities. The intended use would be to modify the outcome of a photochemical reaction under the influence of sun light. Here, the catalyst would have to be tailor made to match a substance specific resonance frequency. The question about its limits, like collective enhancement will be part of future work.

Shaul Mukamel responded: That is correct. One can view the cavity role as a novel nanocatalyst. It offers a different control space that is not available through conventional catalysts. Its control over the potential landscape is unique and can be easily varied by cavity parameters.

Misha Ivanov said: It would be interesting to extend these schemes to several atoms or molecules inside a cavity, where one could possibly think of controlling the interaction of these molecules *via* the cavity-induced dipoles. The quantum nature of the cavity field, with just a single photon in the cavity, could make this molecule–molecule interaction rather interesting.

Markus Kowalewski said: A subtle difference between a classical and a quantised electromagnetic field mode is the back action of the field mode on the molecule. Assuming the field mode is initially in the vacuum state, then a change in photon number introduces a large relative change in the intra-cavity intensity (or cavity coupling). The field intensity scales like $0.5I_0$, $1.5I_0$, $2.5I_0$, *etc.* In a classical laser field with a very large photon number this relative change of intensity becomes irrelevant.

Shaul Mukamel responded: Note that quantum optical fields offer numerous new opportunities for spectroscopy. Parameters of the photon field wavefunction may be varied to reveal matter information that is not accessible by classical light. This includes high temporal and spectral resolution and kinetics information imprinted in photon statistics. Entangled photons may be used to create entanglement in matter and control energy transfer in aggregates.

R. J. Dwayne Miller opened a general discussion of the paper by Fillipo Bencivenga: These are really remarkable experiments in that you were able to get such stable amplitude and spatial mode quality with seeding your XFEL source. The use of soft X-ray excitation in 4-wave mixing geometries is new territory. However, you are interpreting your signals in terms of carrier dynamics in SiN. How do you explain the nonlinearity in the dynamics while you show the excitation is within linear response? (I understand you are considering Auger processes. Do the estimated carrier density and observed decays make sense relative to previous studies which have extensively examined Auger?)

Filippo Bencivenga replied: The estimated carrier density is in the order of 10^{20} cm^{-3}, which is enough high to have a significant Auger contribution on sub-ps timescales (see, *e.g.*, the case of Si, extensively discussed in ref. 1 or 2, Fig. 3). We carried out some calculations, following the approach outlined in these papers and considering the range of parameters of the present experiment. We found a qualitative agreement [the larger is the fluence and the faster is the decay: see Fig. 1 (A. Calvi *et al.*, unpublished)] and we found that the decay is essentially dominated by the Auger. However, our calculations are very preliminary and don't consider other effects (*e.g.* surface fields, see question N. 215) nor the spatial modulation of the transient grating. The fact that we are in the weak field regime (signal intensity proportional to the product of the intensities of the three input beams) is reasonable in light of the relatively low input field strength, not larger than in many optical TG experiments.

1 A. Rämer, O. Osmani and B. Rethfeld, *J. Appl. Phys.*, 2014, **116**, 053508.
2 Yong Gan and J. K. Chen, *Computer Phys. Commun.*, 2012, **183**, 278.

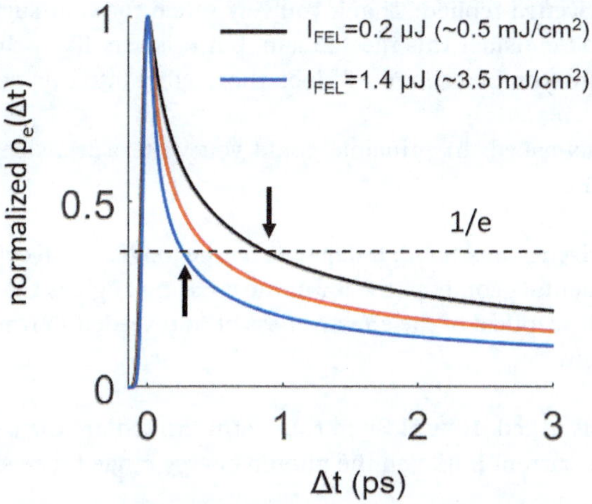

Fig. 1 Time dependence of the free electron density for different values of I_{FEL}, as estimated following the approach reported in [A. Ramer, O. Osmani and B. Rethfeld, *J. Appl. Phys.*, 2014, **116**, 053508]. Data sets are scaled by the respective peak values, the horizontal dashed line indicates the 1/*e* value.

Jon Marangos asked: As in essence you're creating an electron density grating – is the fact the recombination rate due to two body interactions the reason that the decay rate is proportional to electron density?

Filippo Bencivenga responded: We found that the proportionality between the decay rate and the (average) electron density can be ascribed to a term (accounting for Auger decay) proportional to the cube of the electron density in the rate equations describing the energy transfer from photo-excited electrons and the lattice (see, *e.g.* ref. 1). However, other effects could play a role. We are working on that.

1 A. Rämer, O. Osmani and B. Rethfeld, *J. Appl. Phys.*, 2014, **116**, 053508.

R. J. Dwayne Miller asked: I understand that the carrier dynamics can be nonlinear if you have carrier densities in the Auger regime ($>10^{19}$ cm^{-3}). However, the electrons will undergo photoemission at the surface region at your excitation wavelength. Your penetration depth is on the order of a few 10s of nm and the electron mean free path is on the few nm to 10 nm range. The photoemission creates a surface field, which for mj cm^{-2} excitation conditions could translate into a very large surface field that would affect your carrier dynamics. This near surface region of free electrons becomes trapped by the resulting charged surface potential with the electrons recombining with the surface on a time scale very similar to your observations (see work of C. Hebeisen *et al.*[1] in imaging surface emitted photoelectrons using fs electron imaging of the charge distribution). This is just a suggestion that other processes may be involved that are unique to the soft X-ray excitation conditions employed.

1 C. Hebeisen *et al.*, *Phys. Rev. B*, **78**, 081403(R).

Filippo Bencivenga replied: Thank you very much for your suggestion! To be honest, we didn't consider this mechanism, but it is very likely that it is of relevance for data interpretation. We will look thoroughly into this point.

Jon Marangos asked: In principle could you write a transient grating with a non-seeded FEL?

Filippo Bencivenga answered: It depends on the longitudinal coherence length and the experimental geometry (the lower the crossing angle is the better it is). In general I'm a bit sceptical of the effectiveness of non-seeded FEL and nobody has tried it up to now.

Jon Marangos asked: How close to transform limited are the seeded pulses? What are the current limits on the photon energy range for seeded operation?

Filippo Bencivenga responded: The time-bandwidth product is ~1.5 and essentially reflects the degradation of the time-bandwith product of the seed laser pulse when it travels through the optical transport line (several 10s of meters from the seed laser room to the undulator hall). It is very likely that the FEL pulse is as transform limited as the optical seeding laser. The larger photon energy achievable at FERMI (with pulse energy >1 μJ) is 310 eV, the facility is evaluating plans to extend this limit to ~600 eV. Larger photon energies are unlikely with optical seeding

Daniel Neumark asked: Can you explain the different decay dynamics in the transient gratings formed in the two materials, silicon nitride and silicon oxide

Filippo Bencivenga answered: We think that we generate an electron density grating in both materials, but in silicon oxide the decay rate is faster than our time resolution (~100 fs). This may be connected to the different bandgap of the two materials (~9 eV in silicon oxide *vs.* ~4 eV in silicon nitride)

Chris Milne enquired: What are your plans for extending the transient-grating set-up beyond semiconductor samples and into perhaps thin films of molecular samples or even liquids?

Filippo Bencivenga answered: We recently tested a set-up for the detection of the transient grating signal in reflection geometry. In this configuration we are not limited by the sample absorption (that is very strong at XUV/soft X-ray wavelengths). Using this set-up it is possible to probe thin samples deposited onto a substrate as well as bulk opaque samples. Liquid samples are already possible (in principle) using injectors and we are going to design a container for liquid samples with a thin Si_3N_4 window (a single window is enough by exploiting the TG detection in reflection geometry)

Jon Marangos asked: A technical question, you have a low rep rate 10 Hz – how long do you have to integrate to get your signals?

Filippo Bencivenga answered: An appreciable signal can be seen in a single shot (at the time coincidence), typical integration time is 600 shot/delay point, then we have some death times (~30%) in the acquisition system. Let's say that a typical transient-grating measurement can take 2–6 hours.

Jon Marangos asked: Another technical question on the effect of seeding on the FEL field. Does it reduce the energy fluctuations?

Filippo Bencivenga answered: Yes, dramatically!

Hans Jakob Wörner asked: What is the diffraction efficiency of the XUV-induced grating? Can you conclude from this whether it is a pure amplitude grating or whether a phase grating is also involved?

Filippo Bencivenga replied: It is in the 10^{-5} to 10^{-8} range, depending on the sample. Up to now we worked with the field polarization perpendicular to the plane of the crossed FEL beams, so we have a pure amplitude grating. One of our planned next steps is to play with the FEL polarization to generate a mixed amplitude-phase grating (in plane polarization) and a pure polarization grating (crossed FEL polarizations); the latter may be possible with a special configuration of the FEL

Jon Marangos asked: What is the minimum pulse duration you can go down to with the seeded FEL? Can it get to a few fs?

Filippo Bencivenga replied: In principle, yes. In the present configuration of FERMI the FEL pulse duration ($\Delta t_{\rm FEL}$) is related to that of the seed laser ($\Delta t_{\rm seed}$) by the relation: $\Delta t_{\rm FEL} = (7/6) \times \Delta t_{\rm seed}/N^{(1/3)}$, where N is the harmonic number. At present $\Delta t_{\rm seed}$ ~100 fs and N is in the 4–60 range, hence $\Delta t_{\rm FEL}$ ~60–25 fs. However, by reducing $\Delta t_{\rm seed}$ down to ~20 fs $\Delta t_{\rm FEL}$ can be reduced by a factor 5 (shorter seed pulses will be not effective in light of the FEL hardware of FERMI). More effective schemes are in principle possible, *e.g.*: (*i*) superradiance (FEL pulses ~10 fs with the present hardware but a large pedestral in the spectrum); (*ii*) coherent undulator emission (in this case $\Delta t_{\rm FEL} = \Delta t_{\rm seed}/\sqrt{N} \rightarrow \Delta t_{\rm FEL} = 50$–13 fs with the present hardware but <1 μJ pulse^{-1}); (*iii*) chirped pulse amplification (in this case $\Delta t_{\rm FEL} = \Delta t_{\rm seed}/N^{(2/3)} \rightarrow \Delta t_{\rm FEL}$ down to ~1 fs using $\Delta t_{\rm seed} = 20$ fs, but an XUV/soft X-ray compressor downstream of the FEL is required).

Shaul Mukamel opened a general discussion of the paper by Victor Kimberg: Could you elaborate more on the merits and the sensitivity of your covariance technique? Which aspects of the stochastic nature of the field does it overcome?

Victor Kimberg answered: The covariance technique allows, in general, to take full advantage of the stochastic SASE radiation, widely available from present XFEL facilities. In a recent theoretical study[1] we showed the possibility for ultrahigh (vibrational) resolution using two-color SASE pulses by means of statistical analysis in a stimulated X-ray Raman scheme. Indeed, the broad band spectrum of the SASE pulse (~5 eV) allows us to cover a wide vibrational excitation range for all electronic states involved in the X-ray Raman scattering. On the other hand, narrow spectral width of an individual spike (~0.05 eV) of the SASE pulse is smaller than the typical vibrational spacing. Applying the covariance technique allows us to overcome the stochastic nature of the SASE radiation and to obtain vibrationally resolved RIXS maps for a broad frequency range.

However, the experimental situation is more complicated: The stochasticity of SASE radiation is often "contaminated" by some XFEL-related correlations, which should be excluded from the statistical analysis. For example, due to the instability of the XFEL source, the central energy of the X-ray beam varied in a rather broad range (+/− 3 eV) from shot to shot, while the energy spacing between the two X-ray frequency components is rather stable. This results in large XFEL-related features in the RIXS covariance map which mask completely the low-intensity fine vibrational structure related to the studied system. One can overcome this obstacle by recording the pulse energy for each XFEL shot and performing a partial covariance analysis over a sub-ensemble of XFEL shots with a small deviation of the SASE central frequencies and intensities (see also my reply to Jon Marangos regarding the ideal pulse to do simulated Raman). More sophisticated experimental schemes, such as self-seeding pump radiation considered in the present paper, brings about more complex spectral features of XFEL-related shot-to-shot covariance.

1 *Struct. Dyn.*, 2016, **3**, 34101.

Alvaro Sanchez-Gonzalez remarked: In relation to the use of covariance analysis in the case of a self-seeded pump SASE dump scheme, the main factor limiting covariance analysis to detect SRIXS was a strong correlation between the intensity of the seeded pump pulse and the depth of a hole appearing in the spectrum of the dump pulse for the particular FEL settings used. As a result of this, the covariance maps show a strong negative covariance peak exactly in the region where the positive SRIXS peak should appear. As a consequence, the background signal in the region of interest is really large, making it very difficult to observe a small SRIXS signal.

Peter Weber remarked: In laser spectroscopy, stimulated Raman experiments benefit greatly from heterodyne detection, where one looks at radiofrequency side bands. Is it possible to devise a similar method for X-ray stimulated Raman experiments?

Victor Kimberg replied: Indeed, heterodyne detection looks in general applicable for the case of nonlinear X-ray Raman processes, by using optical and X-ray frequencies. Recently, four-wave mixing of X-ray and optical frequencies in a crystal was reported,[1] opening the gateway to new types of nonlinear optical-X-ray spectroscopy. However, the heterodyne detection method is rather difficult to implement at the present stage of the nonlinear X-ray Raman experiments due to limited longitudinal coherence of the currently available soft X-ray pulses and the complexity to design soft X-ray optics that would allow for splitting the X-ray beam in several beams and combining them at the sample with different angles of incidence. A discussion of "dream experiments" of stimulated Raman scattering can be found in several papers by the group of Shaul Mukamel. Although at the moment not quite implementable, I however believe that stimulated coherent

X-ray Raman spectroscopy will have a strong impact and will soon be accessible with development of X-ray sources in terms of their stability and coherence.

1 Steven Chu and Majumdar, *Nature*, 2012, **488**, 603.

Daniel Neumark asked: What is the lifetime of the final CO state produced by the stimulated X-ray Raman process? Is it long enough for the product to be directly detected *via* ionization, absorption, impact on a microchannel plate, or some other "flop-in" experiment that might be more sensitive than measuring gain in the X-ray emission?

Victor Kimberg answered: The final states of the stimulated X-ray Raman process are the valence excited states of the neural molecule (CO). The lifetimes of these states are rather large (varying from nanoseconds to microseconds) allowing in principle their direct measurement. At the present experimental settings of stimulated RIXS within a gas cell the direct measurement of the products is impossible. Indeed, we need to use rather high gas pressures (100–700 mbar pressure was used in our experiments) and a long gas-cell (up to 1 cm) in order to obtain high gain and high amplification for the nonlinear X-ray scattering. Moreover, as we have shown in our paper with the help of X-ray absorption analysis, strong XFEL radiation opens up many additional non-resonant absorption and ionization channels forming highly core- and valence-excited cations. These unstable products decay *via* various cascades with pico- to femtosecond timescales and thus can be hardly registered directly. In my opinion, the fingerprints of these intermediate products can be traced effectively from the advanced analysis of the stimulated RIXS measurements. Development of the theoretical framework for such analysis is one of the important future tasks.

Jon Marangos remarked: In response to the suggestion of detecting molecular products (rather than the stimulated field). I commented that in the latter case a high gas density is needed – but wouldn't you need to go to a low gas density (*i.e.* a different experimental configuration) to collect the products with high fidelity?

Victor Kimberg answered: I agree, that this will be the main problem for direct measurements of the products, along with the non-resonant multi-channel excitation and decay processes in strong XFEL radiation.

Filippo Bencivenga asked: A possibility to implement heterodyne detection in the EUV/soft X-ray is to exploit the directionality of the signal in four-wave-mixing schemes. For instance by recording the interference between the (time-independent) diffraction from a physical grating and the (time-dependent) diffraction from a transient grating, as performed by E. Sistrunk *et al.*[1]

1 E. Sistrunk *et al.*, *Opt. Lett.*, 2015, **23**, 4340.

Jon Marangos said: What would be the ideal pulse to do simulated Raman? Comment: Given the stability, possibility to reach the carbon edge, and two 2 colour operation being developed maybe the Trieste seeded FEL is more like the ideal source.

Victor Kimberg replied: In fact, we would be rather satisfied with narrow-band (self-seeded) pump and broad-band (SASE) dump pulses, as it was used in our last experiment (July 2015), if the pump flux is strong and stable enough to launch the SRIXS. Use of the broad-band SASE pulse on the dump channel allows us to avoid a dump-energy scan, which would be necessary in the case of narrow-band dump radiation. We have shown theoretically[1] that with the help of covariance analysis, applied to a large statistical ensemble of individual XFEL shots, high resolution (~0.1 eV) spectra can be obtained even in the case of broad-band SASE (~5 eV) pulses used in the SRIXS scheme.

In reality, however, the covariance analysis can be complicated due to strong XFEL-related correlations observed in the spectra. For example, in the two-color XFEL regime, the energy difference between the two X-ray components is kept constant at each XFEL shot, while the shot-to-shot variation of the central frequency is rather large (+/− 3 eV around 530 eV). As a result, XFEL-related features in the covariance map completely mask the fine low-scale structure of the SRIXS-related correlations. This problem can be solved by increasing the XFEL stability, or simply by using a partial covariance analysis for a sub-ensemble of XFEL shots with small deviation of the SASE central frequency and intensity. Partial covariance analysis however requires a much larger data set. (See also my reply to Daniel Neumark's regarding the lifetime of the final CO state.) The suggested use of the two narrow band X-ray pulses would require a scan of both the frequency components in order to obtain a complete SRIXS map (2D function of the pump and dump frequencies), which is rather difficult to realize experimentally.

On the other hand, the use of a coherent X-ray radiation source is absolutely necessary for the study of nonlinear optical effects that require a stable phase relation, such as for example for four-wave mixing. We therefore would be looking forward to seeing a wavelength extension of seeded XFELs, for example an extension up to the carbon K-edge at the FERMI FEL. (See also my reply to Jon Marangos regarding the requirements of the synchronization of pump and probe pulses).

1 *Struct. Dyn.*, 2016, **3**, 34101.

Jon Marangos asked: What are the advantages of your scheme with respect to the impulsive Raman excitation – as you are in the "transient limit", neither adiabatic or impulsive limit – what are the pros and cons of this?

Victor Kimberg replied: The main advantage of our "transient limit", is its possible applications for nonlinear X-ray spectroscopic methods to study the nuclear dynamics in molecular systems. We apply our scheme to the soft X-ray range excitation, which has potential applications for the study of larger organic molecules, for which C, N, O K-edges, as well as many metal L-edges (vibrational energy transfer in bio-molecules, in metal-to-ligand transfer systems, *etc.*) can be addressed. Let us also note, that our scheme is fully adapted to presently available realistic parameters of XFELs operating in the soft X-ray range. Presently available X-ray pulses are still far from the ultrashort and coherent realization required for the impulsive Raman scheme. The adiabatic regime, on the other hand, would not allow for stimulated emission and other X-ray nonlinear processes, due to pulse durations much longer than the typical lifetime of core–hole states. This would not allow the creation of a sufficient population of the core-excited state due to the strong non-radiative Auger decay channel in the soft X-ray energy range.

Jon Marangos asked: If you had a larger coherent bandwidth you could work in the impulsive regime. Not only would this give access to faster processes but it would ensure the phase coherence of the different spectral components of the field driving the Raman transition.

Victor Kimberg responded: The impulsive X-ray Raman scheme, developed theoretically in more details by the group of Prof. Shaul Mukamel, uses ultra-short (attosecond range) X-ray pulses with a rather broad bandwidth of a few eVs. This scheme addresses the dynamics of the electronic wave packet, since several electronic states can be core-excited coherently thanks to a broad-band transform-limited attosecond X-ray pulse. The nuclear dynamics is usually ignored in this scheme, since the scattering duration and the pulse duration are typically much shorter than the typical time scale for nuclear motion.

The main goal of our study was, however, to address the vibrational dynamics and nuclear wave packets by nonlinear X-ray processes. In this case, using the broadband spectrum of a short (below 1 fs) coherent pulse would not allow us to resolve the vibrational structure. In that sense, the use of a SASE pulse on the dump transition is even advantageous: the broadband SASE spectrum covers a broad range of the emission energies at each single shot, while the spectral coherence (average width of an individual spectral SASE spike) can be smaller than the vibrational level spacing. For this, we need to use rather long SASE pulses (~100 fs), since the spectral coherence is inversely proportional to the pulse duration. The high spectral resolution can then be achieved with the help of statistical analysis of a large ensemble of SASE shots, as explained in the reply to the question regarding the ideal pulse.

Shaul Mukamel remarked: The combination of a narrowband (picosecond) and a broadband (femtosecond) optical pulse is a powerful tool in stimulated Raman spectroscopy of molecular vibrations which yields a good optimal combination of spectral and temporal resolutions. This idea can be imported to the X-ray pump/probe technique by using a combination of a femtosecond and an attosecond pulse to study electron motion in a similar fashion. One possibility is to use an FEL femtosecond pulse and a HHG attosecond pulse.[1]

1 J. Biggs, D. Healion, Y. Zhang, and S. Mukamel, Multidimensional Attosecond Resonant X-ray Spectroscopy of Molecules; Lessons from the Optical Regime, *Ann. Rev. Phys. Chem.*, 2013, **64**, 101–127.

Markus Kowalewski presented a comment:

DETECTION OF CONICAL INTERSECTIONS THROUGH ELECTRONIC COHERENCES

We want to point out three theoretically proposed experimental techniques, which allow for monitoring the nonadiabatic dynamics in the vicinity of conical intersections (CIs) (see Fig. 2). The first method is the Transient Redistribution

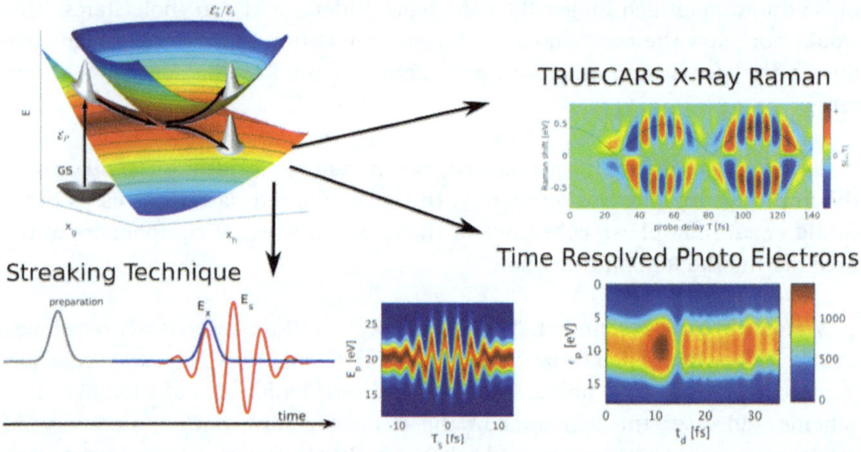

Fig. 2 A molecule is initially excited by a UV/VIS pump pulse, launching a nuclear wave packet in the excited state. When the wave packet reaches a CI or the seam of a CI the wave packet branches and creates an electronic coherence. The created coherence can be detected with attosecond pulses by different methods: The TRUECARS technique[1] is background free and allows for the direct determination of the energy gap. Time resolved photo electron spectroscopy allows its detection through rapid coherent temporal oscillations of the top of the normal TRPES signal.[2] The streaking technique is an extension of TPRES adding a strong IR field as an additional control knob, which allows for the use of longer ionization pulses.[3]

of Ultrafast Electronic Coherences by Attosecond Raman Spectroscopy (TRUECARS).[1]

A nuclear wavepacket is photo-excited to a higher electronic potential surface where it then evolves towards a CI. At the CI, the wavepacket splits and a coherent oscillation is created. We then use a linear, hybrid narrowband–broadband (femtosecond–attosecond) Raman process to connect the two wavepackets (*i.e.*, the upper (lower) wavepacket can be de-excited (excited) *via* a single off-resonant Raman interaction). In the absence of electronic coherences (or if the nuclear wavepackets no longer overlap) this linear Raman signal vanishes:

$$S(\omega, T) = 2\Im \int_{-\infty}^{+\infty} \mathrm{d}t e^{i\omega(t-T)} \varepsilon_0^*(\omega)\varepsilon_1(t-T)\langle\psi(t)|\widehat{\alpha}|\psi(t)\rangle \tag{1}$$

where $\Im =$ is the imaginary part, ε^0 (ε_1) is the broad (narrow) band field, T is the pump–probe time delay, and α is the molecular polarizability. Through its frequency dispersed, linear detection TRUECARS provides a direct and background free measurement of the electronic energy gap between the electronic states. The second method is time resolved photo electron spectroscopy (TRPES). It can be shown[2] that with sufficiently short ionization pulses the created electronic coherence can be detected in the form of coherent oscillations on top of the population dominated TRPES signal. The third method is the streaking of photo electrons,[3] which can be viewed as an extension of TRPES. Here, a strong infrared laser field is superimposed on the XUV ionization field, allowing to "time stamp" the released photo electrons. With the streaking technique adding the streaking delay to the detection scheme, the Fourier limitation of the TPRES method is lifted and the time resolution is enhanced.

1 M. Kowalewski, K. Bennett, K. E. Dorfman, and S. Mukamel, *Phys. Rev. Lett.*, 2015, **115**, 193003.
2 K. Bennett, M. Kowalewski, and S. Mukamel, *J. Chem. Theor. Comput.*, 2016, **12**, 740.
3 M. Kowalewski, K. Bennett, J. R. Rouxel, and S. Mukamel, *Phys. Rev. Lett.*, 2016, **117**, 043201.

Daniel Rolles asked: Please give us a few more details about the results of the time-resolved photoelectron spectroscopy. What are the advantages and disadvantages of doing time-resolved photoelectron spectroscopy *vs.* Raman?

Markus Kowalewski answered: The major difference between time resolved photoelectrons (TRPES) and Raman is the fact that the TRUECARS scheme is background free, *i.e.* the signal contains only contributions from the electronic coherences. The TRPES signal is mainly composed from the population contributions with the oscillating coherence contribution on top of it. TRPES is thus not a background free detection method.

From an experimental point of view a TRPES measurement might be easier to realize nowadays with technology, possibly with an HHG based light source. The off-resonant Raman scheme requires intense attosecond laser pulses.

Jon Marangos asked: What are the requirements on the synchronization of pump and probe pulses?

Comment: A high degree of synchronization is achievable with a HHG source, but there the fields are weak. The alternative is a seeded FEL scheme.

Markus Kowalewski responded: In the TRUECARS detection scheme there is no requirement for phase synchronisation between UV pump pulse and the pulses in the probe sequence. However, the broadband and narrowband Raman pulses in the probe need to be phase stable with respect to each other.

Filippo Bencivenga said: The possibility to scale down the "ps-narrowband and fs-broadband" approach, by coupling (fs) FEL and (as) HHG sources is a very exciting idea! The synchronization issue is very demanding, in this respect laser-seeded FEL's offer the advantage that the same optical laser may drive both FEL and HHG sources.

Daniel Neumark addressed Shaul Mukamel and Markus Kowalewski :Would it be possible to observe similar effects to those predicted in your calculations using transient X-ray absorption to monitor signals associated with Br and Br*?

Markus Kowalewski answered: If you prepare an electronic coherence between the Br and Br* state, the Raman detection would show up, given there is a non vanishing transition polarization matrix element between the two states.

Shaul Mukamel responded: In principle, transient X ray absorption carries information about specific oxidation states of various species. However, stimulated Raman signals are much better resolved spectrally since they are not broadened by the core lifetime, and the broad bandwidth gives a snapshot of many valence states in the vicinity of the selected atom. The technique is thus much more selective and sensitive than transient absorption.

Jon Marangos addressed Shaul Mukamel and Markus Kowalewski: In the schemes for Raman experiments with a broad band probe and a narrow band pump do the broad and narrow band fields have to be coincident in time?

Markus Kowalewski answered: The TRUECARS Raman detection technique we showed is supposed to be an off-resonant detection. The two components of the hybrid pulse must be temporally coincident. It is noteworthy that the purpose of utilizing a hybrid pulse is to exploit the dependence of the TRUECARS signal on the relative phase of the two components of the pulse (*e.g.*, to distinguish it from the transient absorption signal). In practice, there is no spectro-temporal resolution gain achieved by employing a hybrid pulse specifically composed of broadband and narrowband components. This is due to the signal being defined as the frequency-dispersed transmission of the broadband component.

Shaul Mukamel responded: This is an interesting and subtle issue. If the molecule is stationary, the Raman process is instantaneous and the two photons are time coincident (within the electronic dephasing time). Time resolution is not an issue in that case since no dynamical process takes place. However when the molecule undergoes a process such as nonadiabatic dynamics, the signal does not simply depend on the temporal overlap of the two pulses since the two photon process is convoluted with the dynamics. This is clearly seen from the expressions for the signal

Jan Marcus Dahlström asked: I have two questions about the use of short and long pulses to obtain simultaneous resolution in both time and frequency. Recently, I was involved in a theoretical study of photoionization by one short pump pulse and two "monochromatic" probe fields.[1] In our work, we found that the effective two-photon interaction was limited to the short duration of overlap between the pump pulse and the probe fields (in a resonance free region). In practice this means that only a small fraction of the probe field energy is used efficiently. Are the processes your study also suffering from this overlap issue? How high would the probe pulse energies need to be to reach a meaningful signal without complete ionization of the target during pump ramp up?

1 Jhih-An You, Nina Rohringer and Jan Marcus Dahlström, *Phys. Rev. A*, 2016, **93**, 033413.

Shaul Mukamel responded: The temporal and spectral resolutions in nonlinear optical signals are not fully controlled by experimental knobs and depend on the molecular response as well. The molecule decides which spectral components of the pulses are relevant in a given measurement which affects the resolutions. This is discussed in ref. 1 and 2.

Stimulated Raman signals have been measured in atoms (Ne) using the LCLS. This demonstrates their ability to compete with Auger and other ionization processes.

1 J. Biggs and S. Mukamel, Comment on the Effective Temporal and Spectral Resolution of Impulsive Stimulated Raman Signals, *J. Chem. Phys.*, 2011, **134**, 161101.
2 K. E. Dorfman, B. P. Fingerhut, and S. Mukamel, Time-Resolved Broadband Raman Spectroscopies; A Unified Six-wave mixing Representation, *J. Chem. Phys.*, 2013, **139**, 124113.

Morgane Vacher commented: In your Raman simulations, the nuclear wave packet initially on one electronic state approaches the conical intersection and then splits on the two electronic states, leading to coherent oscillations in the simulated spectra. As time goes by, the nuclear wave packet moves back and forth, crosses several times the conical intersection and the coherent oscillations seems to survive for a long time.

However, the harmonic potentials used look very similar, just slightly shifted in position. The energies of the two electronic states may be more realistically described by potentials of different frequencies and shifted both in energy (along the vertical axis, *i.e.* minima at different energies) and in position (along the horizontal axis, *i.e.* minima at different nuclear geometries). I would then expect the nuclear motion on the two potentials to be quite different. As the overlap between the nuclear wave packets on the two electronic states decreases with time, don't you anticipate that the coherences should disappear with time in this more realistic case?

Markus Kowalewski replied: The harmonic system is suitable to clearly demonstrate the features of the signal. As you already pointed out this is not a realistic system, because the coherences, *i.e.* the wave packet overlap is too long lived. In ref. 1 we also show a simulation for a more realistic 2D system with non-harmonic potential energy surfaces and a conical intersection. The decay of the signal and the electronic coherence depend on the gradient difference between the surfaces. The TRUECARS signal thus contains not only information about the electronic energy gap but indirectly also information about the gradient difference.

1 M. Kowalewski, K. Bennett, K. E. Dorfman, and S. Mukamel, *Phys. Rev. Lett.*, 2015, **115**, 193003.

Hans Jakob Wörner opened a general discussion of the paper by Dane Austin: In your article, you are discussing the importance of the asymptotic part of the Dyson orbitals in defining the properties of the continuum wave packets in high-harmonic spectroscopy. In a recent study of high-harmonic spectroscopy of the methyl halides,[1] we have found that the existence of a permanent dipole moment can qualitatively change the shape of Dyson orbitals, when the Stark shift becomes larger than the separation of the electronic states (which were degenerate in our case). Can you explain which role this effect plays in your work?

1 P. M. Kraus *et al.*, *Nat. Commun.*, 2015, **6**, 7039.

Dane Austin answered: In principle, laser-induced coupling of the cation states is included in our calculation for through equation (4) in the paper, which describes the coupled evolution of the phases of the cation states in the field-free basis. One could derive the laser-induced shape change of the Dyson orbitals by diagonalizing the matrix $[\bar{I}_{\mathrm{p}} - \boldsymbol{E}(t)\cdot\Delta\bar{\mu}]$that appears in (4), and forming appropriate linear combinations of the orbitals. Figure 3 shows the result of such a calculation for benzene and *o*-xylene with a static field corresponding to 0.3×10^{14} W cm^{-2} along the *x*-axis.

The Dyson orbitals of the field-dressed states are indeed significantly distorted with this particular field direction. This occurs even in benzene, which has no permanent dipole. However, we find that the overall effect on the angular dependence of the harmonic emission is small. Figure 4 (which is the same as Fig. 2 of the paper) shows the orientation-resolved emission from toluene. Figure 5 shows the same simulation with the laser-induced coupling switched off. The change in the relative amplitude of the emission from the different orientations is of order 10%, and the phase flip between the upper and lower half-planes is still present.

Albert Stolow stated: The dynamic polarizability of the ion core can have a big effect on strong field ionization processes. As an example, it is remarkable that, in the same laser field, transition metal atoms (I.P. around 7 eV) can have ionization rates comparable to Xenon atoms (I.P> around 12 eV).[1] This is due to dynamic screening in the core which reduces the field applied to the departing electron. Have you considered the polarizability of the ion core in your modelling?

1 M. Smits, C. A. de Lange, A. Stolow and D. M. Rayner, Absolute ionization rates of multielectron transition metal atoms in strong infrared laser fields, *Phys. Rev. Lett.*, 2004, **93**, 213003.

Dane Austin replied: The polarizability is not included. Including such effects in HHG molecular calculations is beyond the capability of current models. We agree it is likely to significantly alter the orientation-independent part of the ionization rate. Indeed, the saturation intensities of benzene and toluene were shown in 2000,[1] to be much higher than that expected from ADK. Regarding the

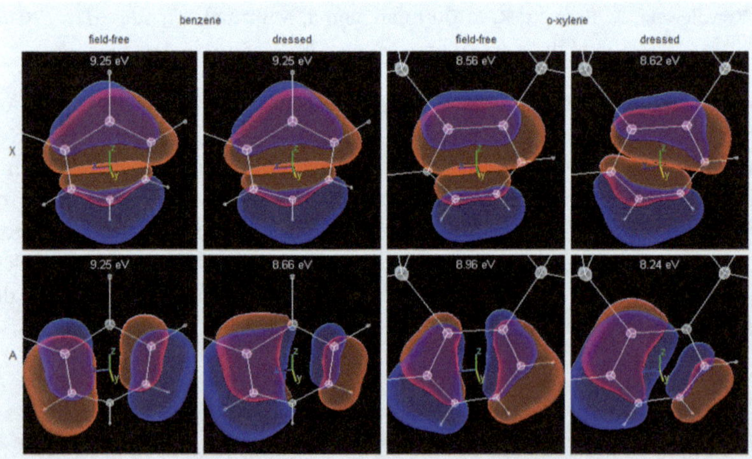

Fig. 3 Distortion of the X (top row) and A (bottom row) Dyson orbitals of benzene (leftmost two columns) and *ortho*-xylene (rightmost two columns) caused by a static field corresponding to an intensity of 0.3×10^{14} W cm^{-2} along the *x*-axis. The first and third columns show field-free orbitals, and the second and fourth show the distorted orbitals.

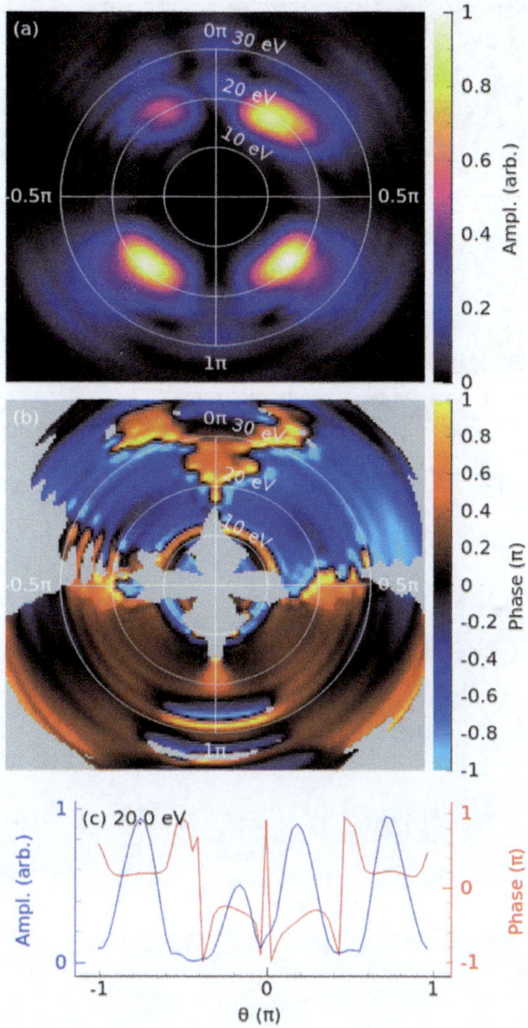

Fig. 4 Spectral amplitude (a) and phase (b) of the attosecond burst from a single laser half-cycle in toluene with the laser field in the HOMO symmetry plane; the radial direction corresponds to photon energy and the angle θ is between the ring carbon–methyl carbon bond and the direction of electron motion at ionization. The phase is plotted relative to the average over all angles (removing the angle-independent attochirp), and its color is set to grey where the amplitude is less than 1% of its maximum value. (c) Amplitude (blue, left y-axis) and phase (red, right y-axis) at 20 eV. Both y-axes have the same scale as the respective color scales in (a) and (b).

orientation dependence, some indication of the anisotropy is given by the measured polarizabilities along the principal axes of the molecule.[2] Of the molecules considered in this study, the most anisotropic is p-xylene, which is 50% more polarizable along its major axis than perpendicular to the molecular plane. Connecting this bulk effective medium polarizability to the dynamic screening relevant to tunnel ionization requires more theory however. We note that the

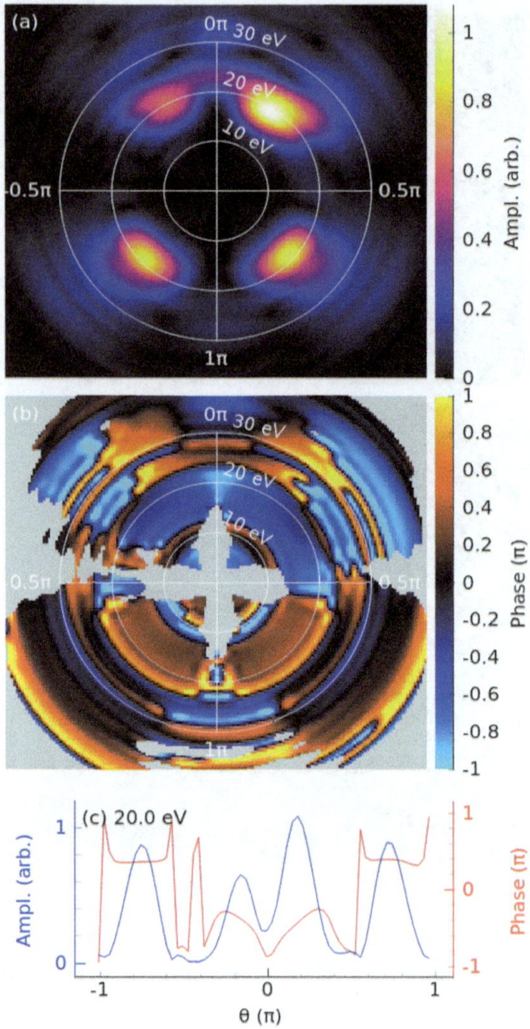

Fig. 5 The same simulation as Fig. 4 but with the laser-induced coupling switched off.

effect of a linear polarizability possesses inversion symmetry, and hence will not affect the asymmetries caused by the substituent groups.

1 S. M. Hankin, D. M. Villeneuve, P. B. Corkum and D. M. Rayner, Intense-field laser ionization rates in atoms and molecules, *Phys. Rev. A*, 2001, **64**, 013405.

2 M. J. Aroney and S. J. Pratten, Electric birefringences and molecular anisotropic polarisabilities of benzene and of nine methylbenzenes and *t*-butylbenzenes, *J. Chem. Soc., Faraday Trans. 1*, 1984, **80**, 1201–1204.

Jon Marangos asked: Dane looked at a number of different cation channels, he found that the different channels had different dependencies on alignment. To what extent in a given experiment should we consider an individual molecule in a coherent superposition state and when should we regard the system as a statistical mixture?

Dane Austin replied: The different channels do indeed have completely different dependencies on alignment. The plot (Fig. 6) shows the intensities (left column) and phases (right column) of the emission at 20 eV, resolved by orbital (rows) and molecular orientation (horizontal and vertical axes of each subplot).

The intensity distributions of the different orbitals are nearly orthogonal. Therefore, to a good approximation the system is a statistical mixture, with the participating orbital determined by the molecular orientation. That said, there are particular angles where the X and A states contribute similarly to the emission, but these are a small fraction of the total.

Morgane Vacher addressed Dane Austin: In atoms or small molecules, orbitals may have well-defined symmetry and potentially (perpendicular) nodal planes. Therefore, different electric field polarisations (or rather different orientations of the molecule with respect to the electric field polarisation) would open different ionisation channels *i.e.* lead to ionisation from different orbitals. An important consequence seems indeed to be the population of a single, but different, electronic state in each molecule across an ensemble of unaligned molecules, rather than the population of a coherent electronic wave packet in each molecule of the sample. The magnitude of this effect depends on the molecular system and especially on the structure of the molecular orbitals. As we go to larger and less symmetric molecules, this effect may be less important which suggests we would "recover" the population of coherent superpositions in each molecule.

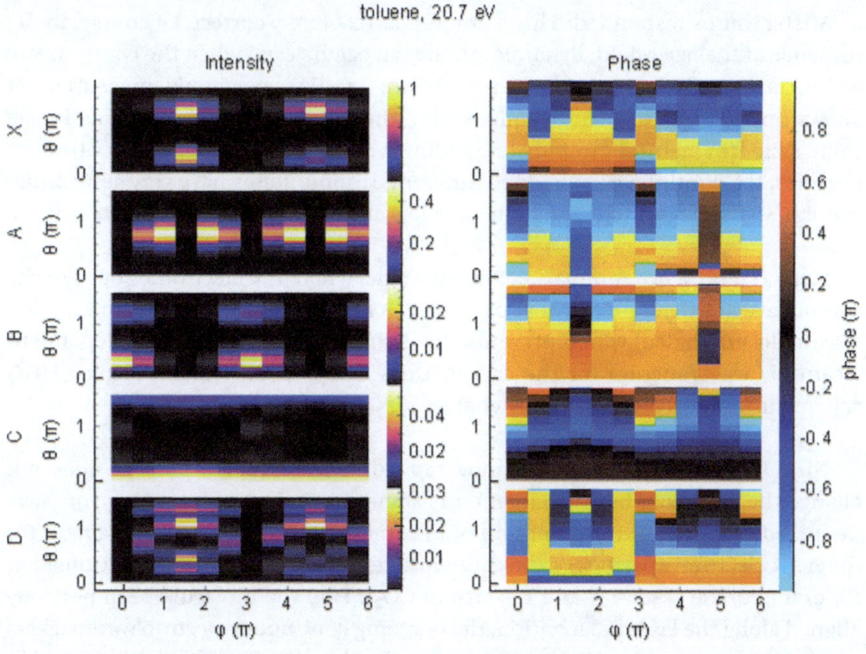

Fig. 6 Plot showing the intensities (left column) and phases (right column) of the emission at 20 eV, resolved by orbital (rows) and molecular orientation (horizontal and vertical axes of each subplot).

Dane Austin answered: I agree, and Figure 6 in this discussion in my earlier response illustrates this.

Hans Jakob Wörner addressed Dane Austin and Misha Ivanov: I would like to comment about the role of electron-ion entanglement in reconstructing ionic dynamics from high-harmonic spectroscopy. Let us consider two electronic states of the cation (ψ_X and ψ_A) and their associated electronic continua ψ_X and ψ_A. When these two continua are orthogonal to each other and laser-induced transitions in the ion (real or virtual) are neglected, the ion-electron wavefunction is fully entangled, *i.e.* it must be written as $\psi = c_{X*}\psi_{X*}\psi_X + c_{A*}\psi_{A*}\psi_A$, where c_X and c_A are complex coefficients and cannot be factorized into a product of the ionic and continuum wavefunctions. In this case, nothing can be said about the electronic state of the cation and no dynamics can be reconstructed. If, in contrast the continua ψ_X and ψ_A have partial overlap, then partial information about the ionic state can – in principle – be extracted from a high-harmonic measurement.

Dane Austin responded: I basically agree, but would express it slightly differently. Regardless of the number of participating electronic states, HHG is only sensitive to the component of the ion-electron wavefunction in which the continuum electron has appropriate momentum for recollision (*i.e.* zero transverse momentum in a linearly polarized laser). Whenever one speaks of reconstructing dynamics with HHG, one is always – implicitly or explicitly – referring to this component of the wavefunction. Regardless of the orthogonality of the wavefunctions or the degree of entanglement, this component is at least well-defined, although as you point out it may not be the full picture.

Misha Ivanov responded: This comment is absolutely correct. Of course, in the presence of the laser field, dynamics in the ion occurs even when the continua are orthogonal, and the interferometric nature of high harmonic measurement allows one to capture such dynamics: the whole process starts and ends in the same state (the neutral ground state). Moreover, for sufficiently high continuum energies, the continua correlated to different cationic states have very substantial overlap, so the degree of entanglement is low and the degree of coherence is high.

Hans Hakob Wörner remarked: In situations where the electronic states of the cation have a different electronic parity (*e.g.* X *vs.* B in CO_2^+), electron and ion are maximally entangled, which prevents the definition of a unique time-dependent electronic wavefunction for the cation. Does the perpendicular two-color HHG scheme used here fundamentally change this situation?

Nirit Dudovich and Misha Ivanov replied: The two-color scheme does not change the situation, but the situation is somewhat different from how you have described it. In fact, the assumption of maximum electron-core entanglement for different electronic parities of the cationic states is not straightforward. Consider, for example, the case of X and B states of CO_2^+. Had the molecule been perfectly aligned along the laser polarization, the outgoing continuum electron wavepackets would have been, indeed, orthogonal to each other. The wavepacket originating from the X state would have had a node along the laser polarization, while the wavepacket emerging from the B state would have had no node. However, there are no perfectly aligned molecules. In our experiments, the characteristic alignment angle is about 30°. In this case the nodal structure of the continuum wavepacket correlated to the X state of the cation is gone, and the wavepacket is very similar to the continuum wavepacket correlated to the cation in the B state. Of course, none of this changes the analysis of the high harmonic emission itself, since in this process the initial and the final states of the system are the same and the interference of different pathways proceeding *via* different ionic states is not affected by entanglement. Specifically, the reconstruction procedure for the amplitudes of different ionization channels correlated to different ionic states is not affected by the degree of electron–core entanglement.

Kiyoshi Ueda addressed Dane Austin and Misha Ivanov: Let us consider the case when HOMO is doubly degenerate, like benzene (or CO_2). I suppose field ionization may be regarded to occur from one of the HOMO orbitals, or a linear combination of them from which the ionization rate is maximal. Then, the returning electron wavepacket sees the polarized ion (the ion part would not have time evolution in this case) and so the recombination takes place for the polarized ion. Does theory take account of such a polarization effect of the ion or can the effect be negligible?

Dane Austin replied: The theory takes into account ionization occurring from all of the included orbitals (5 in this study). Linear combinations are certainly possible, depending on the relative ionization potentials of the orbitals and the orientation of the molecule. My previous answer to Jon Marangos on coherent superposition state illustrates this – there are some angles at which emission from only one orbital dominates, and others were multiple orbitals participate equally. In the model, the electron can only recombine into the orbital with which it is correlated. Transitions of the ion between different orbital vacancy states driven by the laser are possible, but do not appear to play a large role.

Misha Ivanov answered: Yes, our theory takes this into account fully. This is one of the key aspects of the paper: we include both real and virtual excitations of the ion between ionization and recombination, which leads to coherent addition of multiple recombination pathways *via* different ionic channels.

Kiyoshi Ueda addressed Dane Austin and Misha Ivanov: If two orbitals participating in the tunnel ionization are not degenerate, does the coupling between the two hole states *via* the IR field play any role? Are such effects included in theory?

Dane Austin responded: The orbitals are degenerate in all the molecules except benzene. The coupling is included in the theory, but does not play a large role in our study because (*i*) for the most part (all results except Fig. 5 and Fig. 6 in the paper) the electric field is restricted to a nodal plane of the HOMO-1, so that it does not participate, and (*ii*) the coupling is fairly small.

Misha Ivanov responded: Yes, this coupling is included, both due to the IR field and due to the returning electron. The IR-induced coupling is incorporated by solving the Schrödinger equation for the ion in the IR field. The electron-induced coupling of the ionic states is naturally included *via* multi-channel *R*-matrix calculations of the recombination dipole.

Hans Jakob Wörner addressed Misha Ivanov: I understand from your article that you have calculated the dynamics of the cation during the "propagation step" by solving the TDSE in the basis of the lowest four electronic states of the cation. Is such a calculation really converged for CO_2^+ with only four states and without including the ionization continuum of the cation?

Misha Ivanov replied: Given the intensity and the mid-IR wavelength of the driving field, yes. Most of the dynamics in the mid-IR field is due to laser-induced polarization of the core. We have checked that the four states were sufficient to represent the relative Stark shifts of the cationic states. In fact, the important role of the C state was a bit of a surprise, given how far it lies above the ground state of the cation – the reason turned out to be the strong coupling between the C and B states. Of course, a lot more states are needed to converge the recombination step due to the high energy of the returning electron – up to 300 states where used, including representation of the double continuum.

Albert Stolow opened a general discussion of the paper by Nirit Dudovich: Field-driven non-adiabatic interchannel coupling, also termed Nonadiabatic Multi-Electron (NME) ionization, can induce sub-cycle transitions in the field, particularly when the field is changing most rapidly.[1,2] This is expected to be near the zero crossing of the field, leading to the expectation of a quarter-cycle phase shift between adiabatic multi-channel strong field ionization and these field-induced non-adiabatic processes. Have you seen evidence of such processes in your results or analyses.

1 M. Lezius, V. Blanchet, D. M. Rayner, D. M. Villeneuve, A. Stolow and M. Yu. Ivanov, Non-adiabatic multi-electron dynamics in strong field molecular ionization, *Phys. Rev. Lett.*, 2001, **86**, 51.
2 M. Lezius, V. Blanchet, M. Yu. Ivanov and A. Stolow, Polyatomic molecules in strong laser fields: Nonadiabatic multielectron dynamics, *J. Chem. Phys.*, 2002, **117**, 1575.

Nirit Dudovich answered: In our previous study[1] using a perturbative field in parallel polarization we observed a signature of nonadiabatic dynamics in tunneling. Indeed as you suggest, nonadiabatic dynamics leads to a shift in the two color phase, compared with the adiabatic picture.

1 Oren Pedatzur *et al., Nat. Phys.*, 2015, **11**, 815.

Danielle Dowek remarked: The method presented in the paper "Multidimensional HHS of polyatomic molecules" provides a means to characterize the multi-electron dynamics induced by strong field ionization in molecules (first step of the HHG process), featured by the contribution of different ionization channels and their coupling during ionization. The question is about the ability, with the support of the AZUR method presented, to disentangle the coupling between different ionic states (here of the CO_2^+ molecular ion) induced by the IR laser field from that driven by electron correlation dynamics, also evidenced in weak field photoionization, and to determine the relative weighting due to both types of interaction.

Misha Ivanov responded: Disentangling the two contributions is a very important point. We believe this becomes possible when mid-IR drivers are used, so that the electron excursion is much larger than the characteristic size of the "inner region", where ion-electron correlation cannot be ignored. In this regime, one cannot disentangle the role of the tunneling electron and the laser field in the inner region, and this complex dynamics creates the target of reconstruction: the unknown amplitudes and phases of the cationic states created when the electron leaves the vicinity of the molecule. In the AZUR method, we treat these as unknowns and aim at reconstructing them from the best possible calculation of the remaining two steps of the three-step HHG process: electron propagation far away from the laser-driven core, and electron recombination with the laser polarized core characterized by IR-induced amplitudes in the different cationic states.

Danielle Dowek asked: Would the polarization analysis of the HHG radiation constitute an interesting additional dimension to the multidimensional high harmonic spectroscopy study presented in this work?

Nirit Dudovich answered: Polarization analysis holds the potential of adding valuable information. When the molecule is aligned at 0 or 90° such an analysis does not require a polarizer. Due to symmetry odd harmonics are polarized along the fundamental field's polarization while even harmonics are polarized along the SH field's polarization.[1] The polarization vector is dictated by the recollision angle and the polarization of the dipole moment. Extracting the modification of the recollision angle provides an additional signature of strong field's trajectories that participate the interaction.

1 D. Shafir *et al.*, *Nat. Phys.*, 2009, **5**, 412.

Allan Johnson asked: Given that the two-colour method functions by disturbing the phase of the recolliding electron, this must also change the phase of the HHG emission, which should change the phase matching conditions as well as the single emitter response. What effect do you expect this to have upon the macroscopic signal, and how do you take this into account for comparisons with single emitter theory calculations?

Nirit Dudovich answered: Phase matching conditions are modified when the propagation length is sufficiently long. Our setup minimizes modifications of the phase matching conditions, by generating HHG in a thin jet and low gas pressure. We verify that phase matching does not play an important role by performing a systematic study. We perform the two color scan for different pressure values and focusing geometry. Our experiment is invariant to these modifications, demonstrating that they reflect the single molecule response.

Allan Johnson remarked: When would you expect your phase matching controls to fail? For instance, I can imagine in a system with many absorption resonances the phase matching conditions becoming too complicated to simply isolate the single emitter response.

Nirit Dudovich answered: The single molecule response approximation can fail in the case of long propagation distances, or in case of multiple absorption lines. In this case the 1D HHG spectrum will reflect the macroscopic response. This is a general limitation of HHG spectroscopy which is not unique to the two color measurement.

Hans Jakob Wörner said: In his article, Fernando Martín has shown that in single-photon ionization, the leaving photoelectron can perturb the dynamics over the first 0.5 fs. Whereas the exact length of the relevant time span will depend on the velocity of the electron, this raises a question about the possible influence of the continuum electron on the ionic dynamics during the "electron propagation step" of high-harmonic spectroscopy. I understand that the influence of the continuum electron is accounted for in properly correlated multi-channel calculations for recombination and it is accounted for in the ionization step by adjusting the relevant populations and phases of the ionic-state coefficients, as also performed in our recent work.[1]

1 P. M. Kraus *et al.*, *Science*, 2015, **350**, 790.

Misha Ivanov replied: The complex dynamics associated with the interaction between the outgoing electron and the core is one of the key reasons for nontrivial ionization phases and amplitudes. Reconstructing these phases and amplitudes is the key aspect of the paper. In fact, excitation of the ionic core driven by the outgoing electron on its way out are the source of the surprisingly large probability of ionic excitation during tunneling.[1,2] The real question here is whether one can split the whole high harmonic generation process into three steps (ionization, propagation, recombination) and treat the "propagation" step separately. The short answer is "yes", for sufficiently long driving wavelengths. Indeed, the influence of the outgoing electron on the core is determined not by the time (say, 0.5 fs), but by the distance between the electron and the core. The characteristic distance we should worry about in this context is about 10 Bohr or so – the inner region in the *R*-matrix approach. As long as the electron excursion in the intense, mid-IR laser field is large compared to this distance, the "simple" propagation step can still be factored out, relegating all complexity to the "ionization" step.[2]

1 L. Torlina, M. Ivanov, Z. B. Walters and O. Smirnova, Time-dependent analytical *R*-matrix approach for strong-field dynamics. II. Many-electron systems, *Phys. Rev. A*, **86**(*4*), 043409.
2 Z. B. Walters and O. Smirnova, Attosecond correlation dynamics during electron tunnelling from molecules, *J. Phys. B: At., Mol. Opt. Phys.*, 2010, **43**, 161002.

Artem Rudenko addressed Misha Ivanov: The pulse duration used in the experiment is comparable to the vibrational period. Does the vibrational motion disturb the reported reconstruction of the phases for different ionization channels?

Misha Ivanov responded: Firstly, there is some (rather limited) effect of vibrational dynamics in the cation, between ionization and recombination. It is easily accounted for *via* a complex nuclear auto-correlation function between the nuclear wavepacket created at the moment of ionization and the nuclear wavepacket at the moment of recombination, see ref. 1. Secondly, there might be some vibrational excitation of the neutral molecule, which could change the harmonic emission at the beginning and at the end of the pulse. This has not been included in the analysis.

1 Serguei Patchkovskii, Nuclear dynamics in polyatomic molecules and high-order harmonic generation, *Phys. Rev. Lett.*, 2009, **102**, 25.

Raluca Cireasa asked: In photoionisation, shape resonances lead to changes in the phase sign of the outgoing electron wavefunction. The CO_2 molecule has a shape resonance in the C state. This may induce a flip of the phase sign for the harmonics emitted at energies in its vicinity. Did you include in the calculations the shape resonance and if so, how does it contribute to the phase?

Zdenek Masin replied: The formation of the shape resonance in the C-channel has been included in the *R*-matrix modeling of the dipole matrix elements. For details of this model see ref. 1. For the purpose of this work, where we aim at describing HHG driven by mid-IR pulses, which results in extremely high energies (up to approx. 100eV) of the returning electron capable of exciting the ion, this model has been extended to include a large number of 300 cationic states and a more accurate description of the continuum wavefunction. However, these changes did not result in significant differences in the description of the shape resonance which remains in an excellent agreement with experimental data. The shape resonance appears in the partial photoionization cross section as a broad peak between approx. 30 eV and 50 eV of photon energy with a peak around 40 eV. Both the amplitude and the phase of the resonance are included in our calculations. For example, its effect on the HHG signal can be seen from Fig. 9 of the paper being discussed which shows the channel-resolved contributions to the total signal. The direct contribution of the state C, see panel (b), shows an enhancement with a peak around 40 eV whose width approximately corresponds to the width of the resonance in the partial photoionization cross section. This enhancement of the C–C contribution probably comes from the shape resonance.

Figures 12 and 14c of the paper show the relative phases between different HHG channels. Due to laser-driven dynamics in the ion between ionization and recombination, electron recombination with the C state of the ion imprints the recombination phase onto several HHG channels, such as BC, XC, CC. The phase of all channels recombining to C (BC, XC, CC) with respect to channel XX does not show specific structure-related features in the vicinity of the resonance due to the following reasons: (*i*) the resonant feature is quite broad in energy (*ii*) other contributions to harmonic phases evolve faster as a function of harmonic number.

1 Harvey *et al.*, *J. Phys. B: At. Mol. Opt. Phys.*, 2014, **47**, 215005.

R. J. Dwayne Miller asked: The use of 2 colour HHG as the first entry into multidimensional methods seems very promising with respect to extension to more complex systems. The additional information as you show helps distinguish mechanisms. In terms of chemical dynamics, the relevant time scales are nuclear (femtosecond domain). The changes in electron distribution with excitation create forces (the excited state potential energy surface is displaced from the ground state) that lead to nuclear motions and subsequent relaxation to some minima. The attosecond methods for following the electron dynamics hold great potential for giving us information on the electron distribution during this process. The electron dynamics do, however, become rate limiting with respect to the dynamics at CI. There have been a few small molecule systems studied in which it has been possible to access this information (N_2 orbitals for example). Do you think multidimensional HHG methods are important to advancing the imaging electron distributions/dynamics at CI's? I recognize that you are currently using VUV transitions related to core transitions and hole redistribution. I am referring to optical preparation with HHG probes.

Nirit Dudovich replied: The use of HHG to follow electronic dynamics near CI's is indeed an exciting direction for the field. We are very interested in pursuing this direction using multidimensional HHG methods. We have already used such methods to provide full reconstructions of the electronic dipole moments in vibrating molecular dimers, see Ferre *et al.*[1] The concepts presented in this work could be generalized to provide a picture of the nuclear and electronic dynamics near CI's. This would entail optical preparation of the excited state, as the questioner implies, followed by a time delayed two colour HHG probe. Since the probing takes place on attosecond time scales, the measurement can be made on sufficiently fast time scales to resolve the electronic distributions and dynamics near the CI. This measurement will be an exciting extension of the current study – we will be able to resolve how multielectron dynamics evolves while we change the potential surfaces.

1 Ferre *et al., Phys. Rev. Lett.*, 2016, **116**, 053002.

Albert Stolow asked: A non-zero polarizability in the ion core means that there could be an "image charge" or "hole" in the ionic charge distribution during strong field ionization[1] which is itself driven in the field but presumably has a different effective mass. How would the non-zero polarizability of the ion core affect the interpretation and modelling of your results?

1 M. Smits, C. A. de Lange, A. Stolow and D. M. Rayner, Dynamic polarization in the strong field ionization of small metal clusters, *Phys. Rev. Lett.*, 2004, **93**, 203402.

Nirit Dudovich and Misha Ivanov answered: The target of our analysis is the reconstruction of ionization amplitudes. The complexity of the ionization process does affect the reconstruction procedure. However, it will affect the outcome – the amplitudes we reconstruct.

We have found that channels B and C are a lot stronger than standard, single-channel tunneling theory would predict. This has to be related to the interaction between different cationic states, and between the cation and the outgoing electron, during the ionization process. The interaction can be induced by the laser field and by the electron–electron correlation, the latter changing the state of the cation during the departure of the tunneling electron, a process we have described previously,[1,2] which can lead to substantial enhancement of the ionization amplitudes from inner orbitals. Regarding the polarization of the core described by M. Smits *et al.*,[3] it can be viewed as the adiabatic limit of the non-adiabatic electron–core interaction during tunneling,[1,2] which we think is likely to be responsible for the enhanced ionization amplitudes from the deeper orbitals, relative to the ionization amplitude from the highest orbital.

1 Lisa Torlina, Misha Ivanov, Zachary B. Walters and Olga Smirnova, Time-dependent analytical *R*-matrix approach for strong-field dynamics. II. Many-electron systems, *Phys. Rev. A*, 2012, **86**, 043409.
2 Zachary B. Walters and Olga Smirnova, Attosecond correlation dynamics during electron tunnelling from molecules, *J. Phys. B: At., Mol. Opt. Phys.*, 2010, **43**, 161002.
3 M. Smits, C. A. de Lange, A. Stolow and D. M. Rayner, Dynamic polarization in the strong field ionization of small metal clusters, *Phys. Rev. Lett.*, 2004, **93**, 203402.

Jon Marangos asked: You demonstrate the importance of good molecular alignment, how then can we apply these techniques when we have not got the controls?

Nirit Dudovich responded: Indeed, if the alignment angle cannot be directly controlled the measurement will reflect an averaged response only. However, in the case of a poor alignment contrast, applying a more advanced measurement scheme can be applied. Adding an XUV polarizer will isolate the weak contribution of the aligned molecules from the large signal obtained from the unaligned molecules.

Albert Stolow asked: Would it be possible to observe field-induced non-adiabatic interchannel coupling, which leads to a “memory” (non-periodicity) effect between one field cycle and the next, in the form of the envelope of the total high harmonic spectrum?

Nirit Dudovich replied: Field induced non-adiabatic interchannel coupling that leads to “memory” will leave a clear signature on the spectral response within each harmonic order. Such a memory will break the periodicity between the different cycles, leading to spectral variation of the two color phase on a sub-harmonic scale. In the experiment described in the paper such a modification has not been observed.

Wolfgang Domcke opened a general discussion of the paper by Raluca Cireasa: Could you please explain the connection between your measurements and radiotherapy. Which kind of radiation is employed in radiotherapy, what are the chemical mechanisms and how is radiotherapy related to your results?

Raluca Cireasa answered: The most common radiotherapies employ either photon-based sources (X-, γ-radiation) or particle-based sources (proton, heavy ion beams). It is generally agreed that the mechanisms responsible for the destruction of DNA (single and double strand breaks) by the ionising radiation occur following indirect processes: interactions of the DNA with primary and secondary low energy electrons and radicals (OH, OH*,...), ions, pre- and hydrated electrons.[1–3] The radiation produces all these species as it travels through matter (water and other biomolecules) to reach the tumour site thereby leading also to the destruction of healthy cells. In the case of the photon-based radiotherapies, the energy is distributed and attenuated along the irradiation path, while in the case of particle therapies, the energy can be delivered in a more controlled and localised manner to the tumour site to reduce the damage of the healthy cells. In combination with radiosensitiser molecules (including 5FU), they permit to devise more efficient treatments, while reducing the radiation dosage. In the particular case of the halogenated radiosensitisers, it was suggested that the interaction with hydrated electrons or low energy electrons, generated in the radiotherapies, leads to anion formation and their subsequent dissociative electron detachment produces reactive species: halogen atom/anion and uracilyl anion/uracilyl neutral radical, respectively.[4,5] Although most of the chemical studies focused on the indirect processes induced by the ionising radiation in the DNA, nevertheless, about half of the processes initiated in the DNA result from the direct interaction with the radiation.[3] With the development of new therapies that target better the tumour site, perhaps these direct processes may become prevalent and their studies could reveal new mechanisms underlying the DNA damage and radiosensitivity. In our project, we investigate the direct interaction between the 5FU radiosensitiser molecules and ionising radiation. We have chosen XUV radiation to study the ionisation in the valence region where the dissociation processes occur. We have performed both time- and energy-resolved measurements. The fragmentation spectra recorded for 5FU (see the paper) and 5 BrU in our experiments have very similar relative intensities of the peaks as those obtained in 100 keV proton beam experiments.[6] At such high projectile velocities only a small fraction of a few tens of eV is deposited in the molecules and the proton induced ionisation bears similarity with the photoionisation processes. Therefore, our studies have direct relevance for these experiments and our time-resolved results may be used to validate mechanisms proposed in ref. 6 to explain dissociation processes induced by proton beams. At higher XUV energies (not presented in the paper), we measured timescales for tautomerisation and isomerisation processes which were also observed through the fragments detected in the proton mass spectra. The information we obtained for the electronic states and the dynamics timescales helps us to build a picture of the photo-damage and photo-stability processes occurring in the gas-phase 5FU. Our studies relate to the studies of the interaction of gas-phase radiosensitisers with proton beams, which is the radiation employed in radiotherapies. The complementary information gained from both these studies about the intrinsic properties of 5FU radiosensitiser molecule is extremely valuable for modelling the interaction between radiosensitisers and ionising radiation, and such experiments are the first step in a bottom-up approach aimed at the investigation of the mechanisms underlying radiosensitivity (from isolated molecules to *in-vitro* and *in-vivo*).

1 B. Boudaiffa, P. Cloutier, D. Hunting, M. A. Huels and L. Sanche, *Science*, 2000, **287**, 1658.
2 C.-R. Wang, J. Nguyen and Q.-B. Lu, *J. Am. Chem. Soc.*, 2009, 131, 11320
3 S. Lehnert, Biomolecular Action of Ionizing Radiation, Taylor and Francis, New York, 2008, ch. 6.
4 K. Bhatia and R. H. Schuler, *J. Phys. Chem.*, 1973, 77, 1888.
5 H. Abdoul-Carime, M. A. Huels, E. Illenberger and L. Sanche, *J. Am. Chem. Soc.*, 2000, **123**, 5354.
6 J.-P. Champeaux, P. Çarçabal, J. Rabier, P. Cafarelli, M. Sence and P. Moretto-Capelle, *Phys. Chem. Chem. Phys.*, 2010, **12**, 5454.

Jon Marangos asked for a clarification of the pump–probe pulse sequence used in the experiments.

Raluca Cireasa answered: We have employed several pump–probe excitation schemes to investigate 5FU cation dynamics: XUV + 800 nm, XUV + 266 nm, 400 nm + 800 nm and 400 nm + 266 nm. The molecules are ionised and excited either by the absorption of a single XUV photon, or alternatively, of several 400 nm photons. These dynamics initiated by single photon ionisation (SPI) or multi-photon ionization (MPI), respectively, are probed either by 800 nm photons or by 266 nm photons. The corresponding transient signals appear for positive delays in Fig. 4 of the paper. For photon flux requirements, we have used all the high-order harmonics (HHs) contained in the XUV spectrum (HH3–HH9). In order to disentangle the various channels that could be excited simultaneously by different or the same HH, we employed excitation schemes based on MPI and also different probe wavelengths. The 400 nm was the driving wavelength in the high harmonic generation process and MPI provides excitation energies commensurate with individual HHs. Based on the comparison of the dynamics initiated by SPI and by MPI, we were able to identify the HH responsible for initiating particular dynamics.

Adam Kirrander asked: Looking at the transients in the XUV + 800 nm measurements (Fig. 6 in your paper), it appears that one may distinguish some periodic oscillations in the signals? Is this correct, and if so, have you attempted to fit them?

Raluca Cireasa answered: There are visible fluctuations in some of the signals that lend to associations with oscillatory behaviors. The signals for $m/z = 60$, 32 and 1 are low and the fluctuations reflect low S/N ratios. For $m/z = 28$, they seem to be on a longer time scale and have more an appearance of proper oscillations, however, there is no regular pattern and we have not attempted to fit them.

Ruaridh Forbes remarked: In Table 1 of the presented paper time constants are listed which were obtained from fitting the time-resolved ion yield data. In several of the mass channels, for example m/z 87, numerous time constants are extracted which have values significantly less than the temporal resolution of your experiment. The reported time resolution in your experiment, obtained from cross correlation signals in Ne, was 45 fs (for XUV + 800 nm). Could you comment on the reliability of fitting and the absolute values obtained from the fit?

Raluca Cireasa answered: Indeed, several time constants, corresponding to time delays and a rising time, have values in the range of 6–9 fs, which represent 15–20% of the time resolution value of 45 fs. They are non-trivial parameters that were included in the model to accurately reproduce the time dependences. The transient signals from Fig. 6 in the paper obtained with an excitation scheme: XUV (HHG in Kr) + 800 nm, exhibit significant shifts of the peak maxima from 0 fs time delay, thus indicating the necessity to account for time delays in the dynamics and in the fitting procedure. These parameters are unambiguously determined with a confidence level of 95% ($1\ \sigma = 1$ fs: corresponding to 15–10% of their value of 6–9 fs). The $m/z = 87$ ion signal exhibits a visible step contribution. The fit yielded a time constant associated with an ultrafast rising time of 9 ± 7 fs, which is less accurately determined, but still significant at the same confidence level of 95%. Without accounting for this step contribution in the model, the data could not be reproduced and the time constant and the error extracted for the other dynamics would be affected. Consequently, all these parameters are physically meaningful for the fitting models employed. As for the reliability of the extracted values with regard to the experimental resolution, if the 0 fs time delay and the time resolution are well determined through an independent cross correlation measurement, as it was in our experiments, a fitting analysis giving parameter values down to about 20% of the resolution is trustworthy.[1] Taking this into account, as well as, the error bars, we believe that the absolute value for the rising step parameter is rather the upper limit. This was confirmed by an experiment in which the temporal resolution was about 7 fs.[2]

1 S. Ullrich, T. Schultz, M. Z. Zgierski and A. Stolow, *Phys. Chem. Chem. Phys.*, 2004, **6**, 2796.
2 R. Cireasa, work in progress.

Sebastian Mai asked: Which of your results relates to the neutral dynamics of 5-fluorouracil? What can be learned from these results?

Raluca Cireasa answered: We interpret the dynamics observed at positive delays on the parent cation transient signal ($m/z = 130$) as being the result of an interaction between a Rydberg state and a valence or another Rydberg state, where the former is excited by HH3. In addition, the transient signals appearing for some m/z at negative delays in Fig. 4 and 5 (of our paper) also correspond to dynamics initiated in the neutral 5FU molecules, by 266 nm or 800 nm photons and probed through either SPI using XUV radiation or MPI using 400 nm photons. Absorption of one or three pump photons, respectively, leads to the excitation of $S_2(\pi\pi^*)$ state as in the case of the DNA/RNA bases. A preliminary analysis indicates that a primary relaxation occurs in about 50 fs, faster than for thymine and uracil.[1] The full analysis of the data obtained for the three halogenated uracils (5FU, 5ClU and 5BrU) using both SPI and MPI to probe the dynamics, should help in solving the long-standing debate about the occurrence of population trapping on the $S_2(\pi\pi^*)$ state prior to the relaxation mechanisms responsible for the photostability of the other DNA/RNA pyrimidines.

1 R. Cireasa, work in progress.

Daniel Rolles asked: In the paper, you also show results for 266 and 400 nm as pump wavelengths. How are those results different from those using 800 nm as the pump? What was the pulse length of the 266 nm pulses?

Raluca Cireasa replied: The results recorded using 400 nm pump photons correspond to the excitations by three 400 nm of some Rydberg states and by four or five 400 nm photons of some cation states. The ensuing dynamics are similar with those discussed in the paper for the XUV pump excitation scheme (see Table 1 of the paper). In turn, the absorption of one 266 nm photon or alternatively, three 800 nm photons results in the excitation of the $S_2(\pi\pi^*)$ state of the neutral 5FU. This will subsequently launch the dynamics leading to relaxation back to the ground state. It is difficult to compare the results obtained with 266 nm photons with those obtained with 800 nm photons as the latter were recorded only in a very narrow temporal range of about 150 fs and thus, cannot be properly fitted. The use of 800 nm pulses was intended only as an alternative probe, not also as a pump. We have not measured the duration of the 266 nm pulses, only their cross correlation with 400 nm pulses (in Xe), which gave a FWHM of ~ 80 fs.

Dave Townsend remarked: When using 267 nm, your measurements potentially provide some information on the excited state dynamics taking place in the neutral 5-fluorouracil molecule – specifically, this is shown in Fig. 4(e) and (f) of your paper. There is an extensive body of literature relating to the analogous dynamics in the bare uracil system and, in particular, the origin of a ~2 ps decay lifetime seen in this species is a matter of some contention [being assigned to either non-adiabatic crossing from the $S_2(\pi\pi^*)$ state to the $S_1(n\pi^*)$ state or, alternatively, population trapping on the $S_2(\pi\pi^*)$ state]. Are your findings in 5-fluorouracil able to say anything more about this issue? Additionally, the fragment ion transients you present are different in appearance to that of the parent ion, displaying much greater (and seemingly longer-lived) signals at extended delay times. Once again focusing specifically on the 267 nm data, what is the reason for this and does it offer any additional insight into the neutral excited state dynamics?

Raluca Cireasa responded: The time dependencies recorded for the $S_2(\pi\pi^*)$ state dynamics initiated by 266 nm photons in the neutral 5FU are shown in Fig. 4 e–f (of our paper) only for comparison purposes. Their analysis and interpretation are in progress and will be published elsewhere. In order to tackle the long-standing debate about the existence of a barrier on the $S_2(\pi\pi^*)$ state PES leading to population trapping dynamics in uracil and thymine, we have investigated the effect of the substituent bonded to C5, by performing measurements for three halogenated uracils (5XU, with X = F, Cl, Br). Moreover, we have employed two probe schemes: XUV SPI and 400 nm MPI, to enable some selectivity in the ionisation or dissociative ionisation of the population evolving on different regions of the PESs (S_2, S_1 and S_0 states) involved in the relaxation mechanisms responsible for the photostability. We expect that the differences observed for the transient ion signals and the associated dynamics, as a function of the substituent and probe excitation scheme, should contribute to clarifying this issue for 5-Halouracils and the other DNA/RNA pyrimidines.

Oliver Schalk commented: In Figure 6 of your paper, you show a transient for the *m*/*z* 28 channel for the HHG pump, 800 nm probe. Your transient shows a signal at negative delays. Does that mean that you have a signal for the 800 nm pump? If yes, that means that you can have a multiphoton probe process which causes you to have an undefined number of probe photons. Does this cause problems with the interpretation of your data (*e.g.* because you do not know which ionization channels are open, *etc.*)?

Raluca Cireasa answered: Indeed, some of the *m*/*z* transients exhibit dynamics launched following the excitation by three 800 nm photons of the $S_2(\pi\pi^*)$ state. Moreover, the mass spectra recorded with 800 nm photons (Fig. 2e of the paper) indicate that as many as 8–9 photons can be absorbed leading to ionisation and subsequent fragmentation. These observations confirm that multiphoton absorption is involved in the excitation and ionisation of the neutral molecules. That is also the case when using the 800 nm photons to probe the dynamics initiated in the cation. Note though that we did not observe any di-cation signal. The intensity of the 800 nm radiation was adjusted to minimise the one-colour MPI signal, while still obtaining a decent XUV pump – 800 nm probe signal. The ability to absorb several photons enables access to different dissociation limits of the cation and hence, probe the dynamics *via* the different fragmentation templates. As concerning the ionisation channels accessed in the pump step, we identify which HH of the XUV comb contributes to them by comparing the dynamics induced by XUV radiation with those induced by MPI with 400 nm photons. From additional spectroscopic experiments (3D coincident photoelectron–photoion momentum imaging measurements of photoionisation induced by synchrotron radiation) we gain knowledge about ionisation and dissociation energies, and thus we find out which channels are open by HH and which by 800 nm probe photons. On the signals recorded for $m/z = 87$, 60, 28 and 1 with a different number of probe photons, we have identified contributions occurring on a very similar timescale (within the error bars) that we have related to the same dynamics (dynamics 1). In turn, the ps dynamics was observed only on the $m/z = 28$ transient. For a particular m/z fragment, the absorption cross section dependence on the number of photons and the Franck–Condon factors, leads to weighting or selecting contributions of different dynamics to the total m/z signal. Actually, the multitude of fragmentation channels provides the selectivity that lacks spectroscopically: the dynamics can be revealed through different fragmentation channels. Of course, the experiments, where the pump ionises the neutral molecules and the probe reveals the dynamics occurring in the cation through its dissociation, *i.e.* (same) ions are produced by both pump and probe, are more difficult to interpret than those where the dynamics are recorded *via* chemical species that can only be accessed by the probe: dynamics initiated in the neutral and probed *via* ionisation of the neutral parent or fragments, and dynamics initiated in the cation and probed through double ionisation. However, our results show that as for the other two, it is possible to disentangle and interpret such experiments relying on additional experiments and/or calculations. In all cases where IR photons are used for probing, one should be careful about the possible intensity effects on the extracted dynamics and on the interpretation.

Fernando Martín asked: You showed fragmentation spectra obtained by using different radiation sources but also referred to experiments in which fragmentation was induced by ion impact. How do spectra obtained with ions compare with those obtained with synchrotron radiation? Since the energy deposited in both kinds of experiments is in principle quite different, I would expect that the fragmentation spectra are also different. Could you comment on this?

Raluca Cireasa responded: In addition to the mass spectra presented in the paper, in Fig. 7 we show here spectra obtained with 21.5 eV synchrotron radiation (panel a) and with HH7 (21.5 eV) and HH9 (27.7 eV) (panel b).[1] We have chosen a photon energy of 21.5 eV because HH9 is only 10% of the HH7. When visually comparing the signals of mass spectra recorded with synchrotron and with HH, one should take into account the shape of the peaks, *i.e.* the widths and the resolutions due to the detection methods. The relative ratios of the signals are very similar as expected from the photoexcitation rules. A comparison of the synchrotron and HH spectra with proton impact spectra published in ref. 2, Fig.1, shows that the ratios of the main ion fragment signals: 87, 60 and 28 to the parent ion signal are very similar and that some differences can be seen for $m/z = 44$, 43, 32 and 31. The only stark differences appear for the weaker fragments: $m/z = 40$, 39 and 38 which are barely visible in the synchrotron spectra and for $m/z = 15$–12 which are absent. The energy of the proton beam is 100 keV, but only a small fraction of a few tens of eV is deposited in the molecules.[2] In principle, ionisation by proton impact in this (high) collision energy regime bears similarities with the photoionisation processes. From ref. 2, we learn that the kinetic energies of the halogen cations can be as large as 10 eV and that the double ionisation amounts about 10% of the total events. In our experiments, no double ionisation was detected and the halogen cation (Br) exhibit extremely low kinetic energy ~ 100–200 meV. Therefore, we expect that the differences present in the mass spectra produced by proton impact would mainly arise from the excitation of electronic states lying above 21.5 eV, which are accessible in those experiments, but not in ours. This is consistent with the above observations indicating that the differences essentially concern the light fragments with higher appearance energies. The energy region above 21.5 eV corresponds to deeper inner valence states. On the one hand, these states dissociate into smaller fragments. On the other hand, for such high-lying electronic states and molecules of this size, the electronic correlation and relaxation effects are significant and may substantially

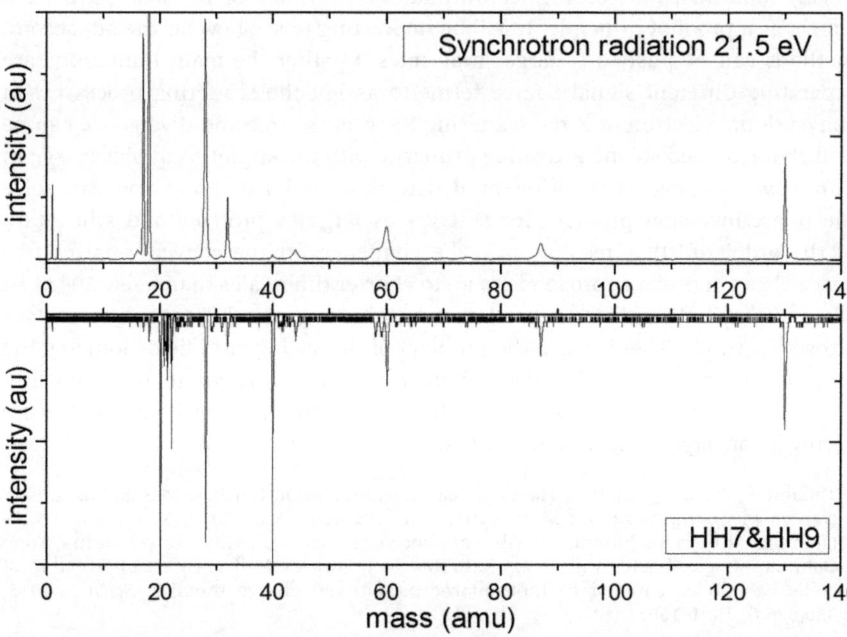

Fig. 7 Mass spectra recorded employing: (a) synchrotron radiation at 21.5 eV and 3D coincident photoelectron–photoion momentum imaging detection[1] and (b) HH7 and HH9 generated in Ar and Velocity Map Imaging detection; a 100 μm Aluminium filter was used to remove the low harmonics: HH3 and HH5 and to keep only HH7 and HH9. The Y axis ticks were chosen to guide the eye in the comparisons. Note: in panel b $m/z = 20$ and 22 are due to Ne carrier gas, and $m/z = 40$ is due to pollution with Ar, while in panel a, $m/z = 18$, 17 and 16 are largely due to H_2O "pollution".

influence the nuclear dynamics and the ensuing fragmentation. The electron dynamics occurring in inner valence states may be responsible for the differences between the mass spectra obtained by proton impact and by XUV ionisation (synchrotron or HH).

1 R. Cireasa, in preparation.
2 J.-P. Champeaux, P. Çarçabal, J. Rabier, P. Cafarelli, M. Sence and P. Moretto-Capelle, *Phys. Chem. Chem. Phys.*, 2010, **12**, 5454.

Misha Ivanov opened a general discussion of the papers by Dane Austin, Nirit Dudovich and Raluca Ciresa: It is fair to say that there are successful techniques for imaging subfemtosecond dynamics with both spatial and temporal resolution: these are based on electron recollision with the parent ion, driven by the laser field. High harmonic generation spectroscopy, laser-induced electron diffraction and electron holography are theory-intensive but have been quite successful in unravelling electron dynamics inside the core on sub-femtosecond time scale.

R. J. Dwayne Miller answered: I agree. These methods are most important for following core electron dynamics on subfs timescales. The problem is connecting the change in the spectroscopic or electron scattered signatures to the nuclear dynamics directing the chemistry on the 10–100 fs timescale of sampling different nuclear configurations. To date only small molecules have been demonstrated, mostly diatomic, but recently conformational motions of benzene have been shown as a proof of principle. It will be interesting to see how far the attosecond methods can be pushed to larger molecules. I gather the main limitations are separating different signal source terms to assign the scattering processes. In conventional electron or X-ray scattering the wave source and divergence can be well characterized so one is dealing primarily with uncertainties in phase, *e.g.*, in diffractive imaging. With differential detection we have shown you can solve the phase/inversion problem for systems as large as proteins and still attain of the order of .01 Å resolution.[1]. The challenge for the conventional or free space electron probe approach is to go to shorter timescales than a few 10s of fs; whereas the challenge is going to larger systems to probe different chemistry for attosec methods. There is also the problem of the strong laser fields ionizing the molecule or perturbing the electron distribution that needs to be taken into account in connecting attosecond methods to 1-photon photochemistry. We are seeing a convergence from both fronts.

1 Tadahiko Ishikawa, Stuart A. Hayes, Sercan Keskin, Gastón Corthey, Masaki Hada, Kostyantyn Pichugin, Alexander Marx, Julian Hirscht, Kenta Shionuma, Ken Onda, Yoichi Okimoto, Shin-ya Koshihara, Takashi Yamamoto, Hengbo Cui, Mitsushiro Nomura, Yugo Oshima, Majed Abdel-Jawad, Reizo Kato and R. J. Dwayne Miller, Direct observation of collective modes coupled to molecular orbital-driven charge transfer, *Science*, 2015, **350**(*6267*), 1501–1505.

Misha Ivanov addressed Raluca Cireasa: I was wondering if using only one harmonic, say H5, in combination with measuring photoelectrons in coincidence with fragments would give any additional insight into the underlying dynamics. What limits the possibility of such a measurement?

Raluca Cireasa answered: Selecting one harmonic and using 3D momentum imaging detection of coincident photoions and photoelectrons allow us to define the excitation energy and to identify the dissociative channels open in the cation. For time-resolved measurements, this approach enables disentangling energetically the temporal contributions from various orbitals/ionisation channels in order to extract their associated nuclear and electron dynamics and electronic correlations. Because of the unique association of the photoion kinetic energies with those of the photoelectrons, the energy selected photoion time dependences may reveal the effect of the electron motion and correlations on the nuclear dynamics and the electronic origin of non-adiabatic dynamics. This type of detection is particularly important for electron dynamics studies of polyatomic molecules for which the photoelectron spectra are complicated by overlapping competing contributions from many orbitals/ionisation channels, making their time dependences difficult to interpret based on the photoelectron detection only. Due to the low counting rates, time-resolved coincidence experiments are very demanding in terms of acquisition time and stability of molecular and laser sources, and of the pump–probe spatial overlap and temporal definition. Advances in optical technologies show that (time-resolved) coincidence measurements can be performed at 50 Hz for atoms and volatile molecules.[1] Currently, these experiments are not feasible for non-volatile molecules as in our case, where the vaporisation source stability is the key limiting factor. The HH are commonly generated using lasers running at 1kHz and such a low repetition rate will imply many hours of acquisition, making thus impossible to fulfil this stability requirement. An alternative solution would be the use of covariance methods.[2] Until now, these methods were generally applied for other types of experiments where ion–ion and electron–electron energy correlation were measured.[3] Using covariance methods instead of the coincidence ones, could enable us to extract the information for the correlated electrons and photoions on acquisition timescales compatible with the stability requirements.[3,4]

1 F. P. Sturm, T. W. Wright, D. Ray, I. Zalyubovskaya, N. Shivaram, D. S. Slaughter, P. Ranitovic, A. Belkacem and Th. Weber, *Rev. Sci. Instrum.*, 2016, **87**, 063110.
2 L. J. Frasinski, K. Codling and P. A. Hatherly, *Science*, 1989, **246**, 1029
3 L. J. Frasinski, *J. Phys. B: At. Mol. Opt. Phys.*, 2016, **49**, 152004 and references therein.
4 A. E. Boguslavskiy, J. Mikosch, A. Gijsbertsen, M. Spanner, S. Patchkovskii, N. Gador, M. J. J. Vrakking and A. Stolow, *Science*, 2012, **335**, 1336.

Nirit Dudovich asked: Can we extract additional information by resolving the dynamics for the different fragments?

Raluca Cireasa replied: In the paper, we show that on the signals recorded for different fragments ($m/z = 87$, 60, 28 and 1) with a different number of IR probe photons, we identified contributions occurring on a very similar timescale (within the error bars) that we have related to the same nonadiabatic dynamics (dynamics 1). We have not observed any coherent motion due to vibrational wavepackets excited by the XUV pulse, but, in principle, it could be possible to observe such dynamics. If the probe pulse projects the vibrational wavepacket on higher excited states that further dissociates, the vibrational dynamics is recorded on the ion fragment signals. If these states correlate to different dissociation limits, then the dynamics can be revealed on the time-dependence of different fragments. In order to achieve dissociation, the vibrations of the wavepacket should correspond or couple to the vibrations of the dissociation coordinates. Information on both the nature of the vibrational wavepacket (vibration frequency and type) and the nature of the potential surface (chemical bond) excited by the pump can be extracted from the signal of different fragments.[1] Because we have not selected a single high-order harmonic (HH), but employ the whole XUV spectrum, it is possible to excite coherently and selectively different electronic states by two (consecutive) HH of the attosecond pulse train. As in the experiments performed for D_2, by adding an IR pump pulse it may be possible to create interferences of electronic wavepackets to control the excitation and the ionisation on the attosecond timescale.[2] Another IR pulse (probe) delayed from the XUV+IR pump pulse will lead to dissociative ionisation. Selective bond-breaking will be achieved by controlling the excitation wavelength as well as the delay time between the pump and probe pulses. Nevertheless, for a molecule of this size, it is challenging to devise and understand such control schemes due to the high density of states, multielectronic effects and numerous nonadiabatic interactions.

1 C. Jouvet, S. Martrenchard, D. Solgadi, and C. Dedonder-Lardeux, M. Mons, G. Grégoire, I. Dimicoli, F. Piuzzi, J. P. Visticot, J. M. Mestdagh, P. D'Oliveira, P. Meynadier and M. Perdrix, *J. Phys. Chem. A*, 1997, **101**, 2555.
2 P. Ranitovic, C. W. Hogle, P. Rivière, A. Palacios, X.-M. Tong, N. Toshima, A. González-Castrillo, L. Martin, F. Martín, M. M. Murnane and H. Kapteyn, *Proc. Nat. Am. Soc.*, 2014, **111**, 912.

Faraday Discussions

PAPER

Strong-field-induced wave packet dynamics in carbon dioxide molecule

Artem Rudenko,*[ab] Varun Makhija,[ac] Aram Vajdi,[ad] Thorsten Ergler,[b] Markus Schürholz,[b] Rajesh K. Kushawaha,[a] Joachim Ullrich,[be] Robert Moshammer[b] and Vinod Kumarappan[a]

Received 23rd June 2016, Accepted 29th June 2016
DOI: 10.1039/c6fd00152a

Temporal evolution of electronic and nuclear wave packets created in strong-field excitation of the carbon dioxide molecule is studied employing momentum-resolved ion spectroscopy and channel-selective Fourier analysis. Combining the data obtained with two different pump-probe set-ups, we observed signatures of vibrational dynamics in both, ionic and neutral states of the molecule. We consider far-off-resonance two-photon Raman scattering to be the most likely mechanism of vibrational excitation in the electronic ground state of the neutral CO_2. Using the measured phase relation between the time-dependent yields of different fragmentation channels, which is consistent with the proposed mechanism, we suggest an intuitive picture of the underlying vibrational dynamics. For ionic states, we found signatures of both, electronic and vibrational excitations, which involve the ground and the first excited electronic states, depending on the particular final state of the fragmentation. While our results for ionic states are consistent with the recent observations by Erattupuzha *et al.* [*J. Chem. Phys.* **144**, 024306 (2016)], the neutral state contribution was not observed there, which we attribute to a larger bandwidth of the 8 fs pulses we used for this experiment. In a complementary measurement employing longer, 35 fs pulses in a 30 ps delay range, we study the influence of rotational excitation on our observables, and demonstrate how the coherent electronic wave packet created in the ground electronic state of the ion completely decays within 10 ps due to the coupling to rotational motion.

1. Introduction

Molecular dynamics in strong laser fields continues to attract increasing attention, driven by the desire to understand and by the vision to control ultrafast

[a]J.R Macdonald Laboratory, Department of Physics, Kansas State University, Manhattan, Kansas 66506, USA. E-mail: rudenko@phys.ksu.edu
[b]Max-Planck-Institut für Kernphysik, 69117 Heidelberg, Germany
[c]Department of Physics, University of Ottawa, Ottawa, ON K1N 6N5, Canada
[d]Department of Electrical and Computer Engineering, Kansas State University, Manhattan, Kansas 66506, USA
[e]Physikalisch-Technische Bundesanstalt, 38116 Braunschweig, Germany

light-driven chemical reactions. Whereas the time scales of chemical transformations are determined by the kinematics of molecular nuclei, it is the change of the electronic structure which typically drives the reaction. Such an interplay between the motion of electrons and nuclei offers possibilities to steer even the fastest reactions *via* control over electron dynamics. However, this requires a detailed understanding of the behavior of the complete molecular wave packet, including nuclear (vibrational and rotational) dynamics in the bound and dissociating states, as well as the evolution of the electronic structure and the couplings between different degrees of freedom.[1]

Another important aspect of studying a strong-field-induced molecular response is the role of dynamics in the neutral and ionic molecular states. Most of the "conventional" gas-phase photochemistry occurs in neutral species. However, when a molecule interacts with a strong laser field, it is likely to be ionized. Understanding strong-field ionization is an essential step for attosecond pulse generation and development of strong-field coherent control schemes.[2] It also provides a basis for imaging molecular structure and dynamics *via* high-harmonic generation (HHG) spectroscopy.[2–4] At the same time, vibrational[4–6] and electronic[7,8] excitations of the neutral molecule were shown to play an important role for strong-field ionization and HHG. Therefore, distinguishing the effects due to the propagation on neutral or ionic potential surfaces of the intermediate states is one of the key issues in dealing with strong-field-induced reactions.

Here we report on a pump-probe experiment that visualizes the signatures of vibrational, rotational and electronic components of strong-field-induced molecular wave packets, and involves both, neutral and ionic excitations. We chose a carbon dioxide molecule as a prototype system for this study. CO_2 represents one of the simplest polyatomic molecules with a well-known structure and a comprehensive set of spectroscopic data for both, neutral and ionic states (see, *e.g.*, ref. 9–17). It has been recently employed as a target for many ultrafast and strong-field experiments, including the demonstration of the importance of different electronic excited states for HHG,[3,18–20] mapping of coherent electron–hole dynamics near the conical intersection[21] and studies of vibrational and electronic wave packet motion in ground and excited ionic states.[22] Combining the data obtained with two different pump-probe set-ups and employing channel-selective Fourier spectroscopy, we (i) demonstrate that a broadband, 8 fs pump pulse may result in an efficient vibrational excitation in the ground state of a neutral polyatomic molecule; (ii) visualize the dynamics in the ground ionic state, including features reflecting both, vibrational and electronic coherence; and (iii) show how the electronic component of this wave packet decays within $\sim$10 ps due to coupling to rotational motion. We compare our findings to recent pump-probe results obtained with two 25 fs pulses,[22] which were interpreted in terms of dynamics occurring solely in ionic states, and discuss likely reasons for differences between two sets of experimental observations.

2. Experimental

In order to address both, ultrafast motion of strong-field-induced molecular wave packets and their long-term evolution, the experiments reported here were carried out in two distinctly different modes. As sketched in Fig. 1, we employed a COLTRIMS (COLd Target Recoil-Ion Momentum Spectroscopy)[23–25] set-up in combination

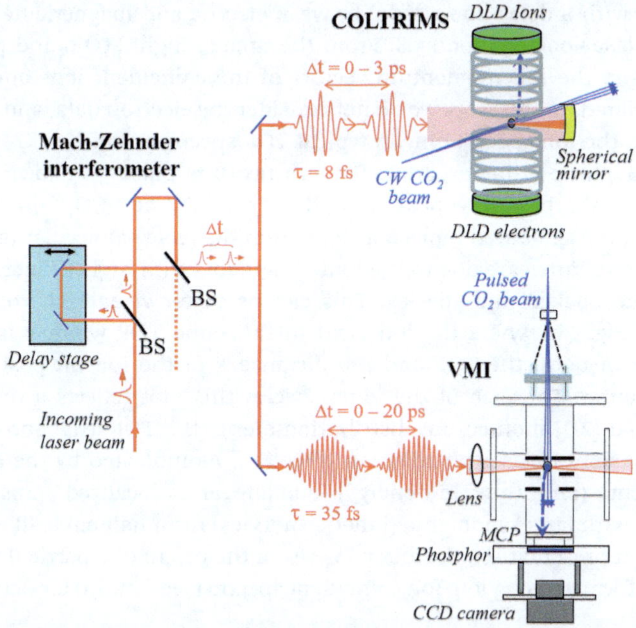

Fig. 1 Schematics of the experimental set-up. Left: Sketch of the pump-probe setup used. BS: beam splitter. Right, top: Sketch of the COLTRIMS apparatus. DLD: delay line detector. Only the ion data are discussed in this article. Right, bottom: Sketch of the VMI apparatus.

with a pair of 8 fs pump-probe pulses to perform channel-resolved measurements focused on ultrafast dynamics, and used a Velocity Map Imaging (VMI) apparatus combined with two longer, 35 fs pulses to study the wave packet propagation in a 20 ps time window, thus, addressing rotational time scales.

2.1 Experiments with COLTRIMS and ultrashort pulses

Measurements involving the COLTRIMS set-up were performed using linearly polarized radiation from a Kerr-lens mode locked Ti:sapphire laser at a 795 nm central wavelength with 25 fs pulse width (FWHM). To generate sub-10 fs few-cycle pulses, they were spectrally broadened in a gas-filled hollow fiber and then compressed to ~8 fs (FWHM) by a chirped-mirror-based compressor. The laser beam was fed through a Mach–Zehnder-type interferometer with one arm variable in length, yielding two pulses separated by an adjustable time delay which can be scanned from 0 to 3000 fs with a resolution of better than 1 fs. Both laser beams were focused to a spot size of ~10 μm (FWHM) onto the collimated supersonic CO_2 gas beam in the center of an ultra-high vacuum reaction chamber (background pressure < 10^{-10} mbar). The molecular beam was operated in a continuous-wave (CW) mode with a CO_2 background pressure of 3 bar without additional carrier gas. An upper limit for the temperature of the molecular beam (estimated from the momentum resolution achieved in ion-electron coincidence measurement) was ~15–20 K, resulting in a vibrationally cold target.

Charged reaction fragments created during laser–molecule interaction were guided onto two time- and position-sensitive microchannel plate detectors (equipped with a delay-line anode) by weak electric and magnetic fields applied along the laser polarization axis. From the time-of-flight (TOF) and position on the detectors the full momentum vectors of the coincident ions and electrons were calculated. In this work we do not consider the electron data, and restrict the analyses to the ion spectra only. A typical TOF spectrum for the CO_2 fragmentation in our pump-probe experiments with two 8 fs pulses is shown in Fig. 2a. Whereas the sharp, narrow peaks (like those for CO_2^+ and CO_2^{2+} ions) originate from the non-fragmented molecular ions from the (cold) molecular target beam, the broad structures are due to the ionic fragments with higher kinetic energies, or from the background species. This can be better visualized from the data presented Fig. 2b, where the ion yield in the same TOF window is shown as a function of both, the TOF and the distance *R* of the ion hit position to the detector center. For each of the ionic species this plot reflects a rescaled two-dimensional (2D) momentum distribution along (the TOF axis) and perpendicular to (*R*-axis) the laser polarization direction. As indicated by the arrows, the low-momenta (and, thus, low-energy) ions appear as localized spots, while the fragments with larger momenta (kinetic energies) form half-circle like structures in this 2D representation. Further analyses of the origin of a particular structure can be performed using ion–ion coincident spectra (see (ref. 25) for details of such analyses).

In this work we will focus on the considerations involving the delay-dependent yields of CO_2^{2+}, CO^+ and O^+ ions. For the latter two species, two clearly distinct contributions can be identified in the data shown in Fig. 2b. In both cases, the inner structures (marked as "O^+ or CO^+ from diss.") result from the dissociative ionization with a neutral partner (CO and O, respectively), while the outer features originate from the double ionization with subsequent Coulomb Explosion (CE). For the latter case, both O^+ and CO^+ fragments are detected in coincidence with each other.

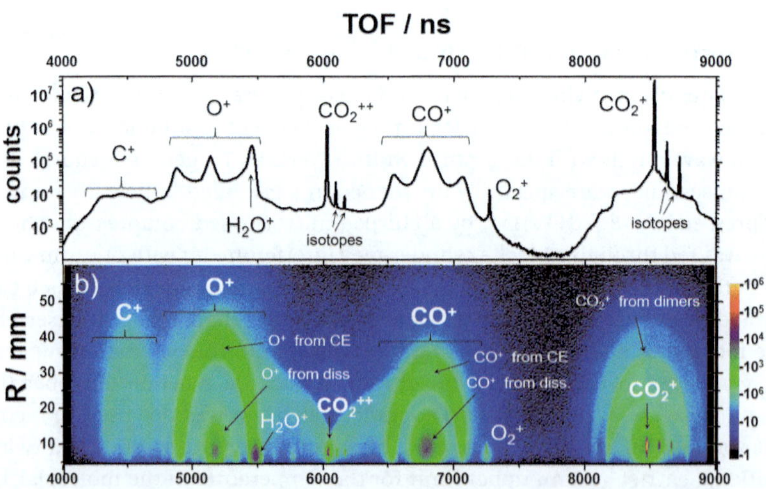

Fig. 2 (a) Ion time-of-flight (TOF) spectrum obtained in the pump-probe experiment on CO_2 fragmentation (integrated over all pump-probe delays). (b) Here the same ion yield as shown in (a) is plotted as a function of the TOF and the distance *R* to the detector center.

2.2 Experiments with 35 fs pulses in a long delay window

Since the measurements using the coincident momentum imaging set-up described in the previous section need rather low target density (limited with the requirement to have, on average, less than one fragmentation event per laser shot), the range for the accessible delay scans was limited practically by the acquisition time, and by the long-term stability of the pair of ultrashort, sub-10 fs pulses. In order to extend these measurements to the delays of order of tens of picoseconds, where we expect the rotational dynamics of the CO_2 molecule to unfold, we performed a complementary set of experiments using considerably longer pulses (35 fs FWHM) and a VMI-based set-up with considerably higher target density, as sketched in Fig. 1. The experimental set-up is essentially similar to the one described in ref. 26. In brief, the output of a Ti:sapphire laser (central wavelength 790 nm, pulse duration 35 fs) was sent through a Mach–Zehnder-type interferometer to produce a pair of pump-probe pulses with the delay adjustable in a range of 0 to 50 ps. The laser pulses were focused onto a pulsed, doubly skimmed molecular beam based on the Even-Lavie valve. The beam contained a 5% CO_2–95% helium mixture with a total pressure of 70 bar, and the estimated target temperature was ~5 K. The created ionic fragments were projected onto a microchannel plate detector by the VMI spectrometer operated in time-of-flight mode. It should be noted that we also applied this detection set-up in combination with two ~7 fs pulses to repeat an ultrashort-pulse measurement in a few ps time-window described above, and were able to reproduce most of the essential findings.

3. Results obtained with two 8 fs pulses

3.1 Channel-selective Fourier spectroscopy and identification of wave packet constituents

From the variety of different ionic fragments visible in Fig. 2 for the delay-dependent analysis we consider four major fragmentation channels: dissociative ionization into (1) $CO^+ + O$ and (2) $CO + O^+$ final states; (3) double ionization producing a bound parent ion CO_2^{2+}; and (4) double ionization resulting in a Coulomb explosion $CO^+ + O^+$. Fig. 3a displays the measured yields of these four channels as a function of time delay between two 8 fs pump and probe pulses (intensities 1.5×10^{14} and 3.5×10^{14} W cm^{-2}, respectively). Fig. 3b shows the delay-dependent yields of the same four channels measured under identical experimental conditions in a much shorter time window (marked with the shaded area in Fig. 3a). Besides a strong enhancement of the signal close to zero delay, which results from the overlap of both pulses and, thus, considerably higher light intensity, three major features can be observed in the data. First, a pronounced oscillation with ~25 fs period is clearly visible, in particular, in Fig. 3b. Whereas the CO^+ yield from dissociative ionization (black) and the yield of Coulomb explosion channel (green) oscillate in phase, the CO_2^{2+} and O^+ signals are shifted by half of the period. Second, an oscillatory structure on ~200 fs time scale can be seen in the CO_2^{2+} yield in Fig. 3a (red curve). Finally, the delay-dependent yields of $CO^+ + O$ dissociation and Coulomb explosion (black and green curves, respectively) in Fig. 3a manifest a broad maximum centered around 500–600 fs delays. The latter feature results from the excitation of the rotational wave packet

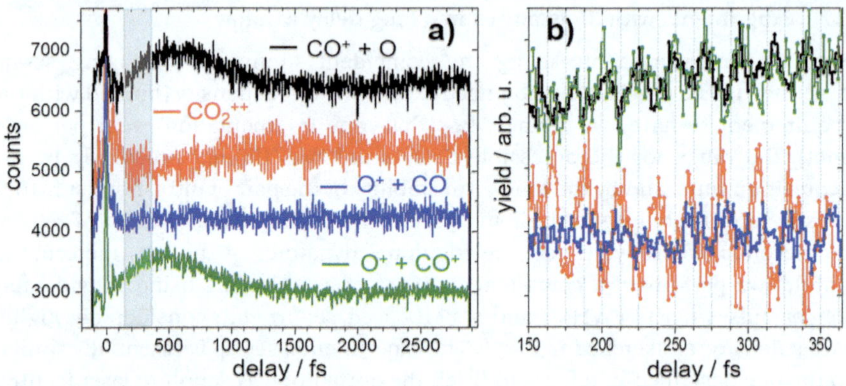

Fig. 3 (a) Yields of different fragmentation pathways measured as a function of pump-probe delay between two 8 fs pulses. Intensities are 1.5×10^{14} and 3.5×10^{14} W cm^{-2} for the pump and the probe pulses, respectively. (b) Yields of the four channels shown in (a) measured in a short delay window (marked by the shaded area in (a)) under identical experimental conditions. The yield of each channel is shown in the same color as in (a). The data are shifted in the vertical direction for visual convenience. In the text the O^+ + CO^+ channel is referred to as "Coulomb explosion", and the CO^+ + O/O^+ + CO channels are called "dissociative ionization".

(*i.e.*, weak impulsive alignment of the molecule) by the pump pulse. This can be verified by determining the molecular alignment parameter $\langle\cos^2\theta\rangle$ (where θ is the angle between the molecular axis and the laser polarization) from the emission angles of the detected Coulomb explosion fragments. For the latter the measured delay-dependence of the $\langle\cos^2\theta\rangle$ closely resembles the observed structure. The same effect also causes a dip in the CO_2^{2+} signal for the delay range of ~300–500 fs, owing to the reduction of the ionization probability for molecules aligned parallel to the laser polarization direction.[22,27–30]

To understand the origin of the other two delay-dependent features, we carry out a Fourier analysis of the yields presented in Fig. 3a. Fig. 4 displays the obtained fast Fourier transform (FFT) spectra for all four channels discussed above. The Fourier spectra of the channel-selective pump-probe data are expected to reflect frequency differences between different intermediate molecular states populated by the pump pulse. Depending on the binning of the data, the energy resolution for these data is ~2–3 meV. All four spectra in Fig. 4 contain significant components centered at 159 meV (1283 cm^{-1}) and 172 meV (1387 cm^{-1}). These correspond to the well-known Fermi resonance that involves the symmetric stretch (ν_1^0) and the first overtone of the bending mode ($2\nu_2^0$) of the ground state ($X^1\Sigma_g^+$) of the neutral molecule.[9–11] As such, both frequencies correspond to a mixture of symmetric stretching and bending of the molecule. These modes are Raman active, and the 8 fs pump pulse has enough bandwidth to excite these modes through far-off-resonance Raman transitions. The lower of these two frequencies was also observed in a recent measurement by Erattupuzha *et al.*[22] (Vienna group), where the authors employed two 25 fs pulses to study the wave packet dynamics in CO_2, and was attributed to the symmetric stretching vibrations in the ground state of the CO_2^+ ion ($X^2\Pi_g$). The peak at 172 meV was not observed in the previous 25 fs experiment[22] and does not match any frequencies reported for three lowest electronic states of CO_2^+.[12–15]

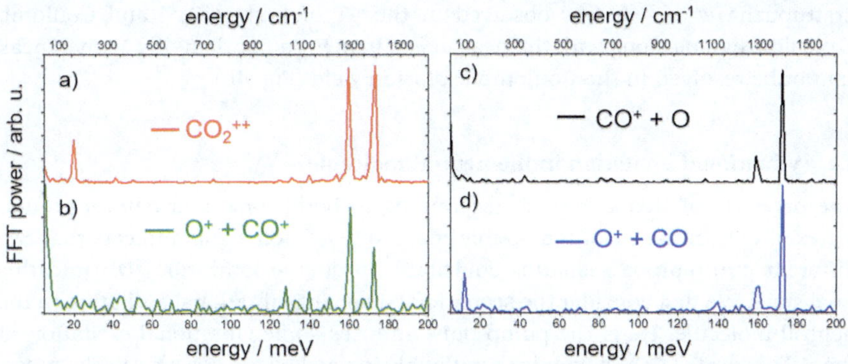

Fig. 4 The power spectra obtained from the Fourier transforms of the ion yields shown in Fig. 3a for CO_2^{2+} (a), $O^+ + CO^+$ (b), $CO^+ + O$ (c) and $O^+ + CO$ (d) channels.

The appearance of the 172 meV peak is a clear signature of vibrational dynamics excited in the neutral molecule. To the best of our knowledge, the only vibrational frequency of the CO_2^+ ion that might overlap with this feature within our resolution is the symmetric stretching mode ν_1^+ of the $C^2\Sigma_g^+$ state.[15] However, this state is predissociative on a time scale of ~1 ps (ref. 21 and 31) and, thus, the corresponding frequency would disappear in experimental data after a couple of picoseconds, which does not fit our observations. Moreover, at the intensity of our pump pulse, the population of this rather high-lying state is expected to be small compared to the ionization out of the highest occupied molecular orbital (HOMO), which leaves the ion in its ground state ($X^2\Pi_g$). The absence of the corresponding peak in the data of the Vienna group[22] can be explained by comparably narrow photon spectrum of the 25 fs pulses used in that work. The most likely mechanism of vibrational excitation in the ground electronic state of a neutral molecule with our experimental conditions is a far-off-resonance two-photon Raman scattering (ref. 4, 32 and 33, see also ref. 5 and 6 for discussion). At 800 nm such vibrational excitation would require the wavelength difference of the two photons exceeding 90 nm, which is rather unlikely with 25 fs pulses but can be readily fulfilled with the broadband 8 fs pulses we use here.

While the peak at 172 meV can be uniquely assigned to the electronic ground state of the neutral molecule, the second prominent vibrational feature (centered around 159 meV) might originate from several different states. As mentioned above, within the experimental resolution it coincides with the lower component $2\nu_2^0$ of the Fermi dyad of the ground state of neutral CO_2.[9–11] However, it also lies close to several transitions in the $X^2\Pi_g$ ground state of the CO_2^+ ion,[12,14] and was assigned to its symmetric stretching vibrational mode in ref. 22. Moreover, the symmetric stretching vibrational frequency of the second excited ionic state $B^2\Sigma_u^+$ was also reported to be 158–159 meV (see ref. 15 and references therein). Thus, this feature cannot be unambiguously associated with a particular fragmentation pathway. However, a third peak at ~20 meV, which can be observed in the CO_2^{2+} Fourier spectrum in Fig. 4a represents a unique signature of the dynamics unfolding in the ionic ground state. This feature can be attributed to the beating frequency between the two electronic states ($X^2\Pi_{1/2g}$ and $X^2\Pi_{3/2g}$) that result from spin–orbit splitting of the doubly degenerate X-state of CO_2^+. In the experiment of

Erattupuzha *et al.*[22] it was observed in the FT of both, CO_2^{2+} and Coulomb explosion signals along with the peak at ~159 meV. In our data this 20 meV peak cannot be resolved in the Coulomb explosion yield (Fig. 4b).

3.2 Vibrational excitation in the neutral molecule

The presence of frequencies corresponding to both, ionic and neutral ground states of CO_2 in the FT of the doubly charged CO_2^{2+} ion signal indicate that two different pump-probe scenarios contribute to double ionization (DI) into this final state. We first consider the scenario(s) involving vibrational excitation in the neutral molecule. Here, the pump pulse induces some vibrational excitation in the $X^1\Sigma_g^+$ state of CO_2, which is then doubly ionized by the stronger probe pulse. The observed modulation means that the DI probability is sensitive to changes of molecular geometry, in particular, symmetric stretch. According to the study of the intensity dependence of CO_2 DI,[34,35] the intensity of our probe pulse (~3.5×10^{14} W cm^{-2}) lies in the transition region between recollision-induced non-sequential (NS) and sequential mechanisms, even though it should be noted that for few-cycle pulses like those used in this work, this transition was shown to occur at somewhat higher intensities than for longer pulses.[36,37] Since very short pulse duration essentially excludes any significant nuclear motion on the intermediate potential curves during the pulse, for sequential DI one would expect somewhat higher probability at the inner turning point (with respect to the C–O distance), where the total ionization potential for DI is somewhat lower.[16,22] For NSDI, which was shown to include contributions of different pathways involving both, HOMO and HOMO−1, the expected dependence on the C–O distance is less intuitive.

As can be seen from the Fourier spectra in Fig. 4c and d, which are dominated by the peak at 172 meV, vibrational excitation in the neutral molecule also plays an essential role for dissociative ionization into singly charged final states. Since in our experiment the probe pulse is considerably stronger than the pump, the more likely pump-probe scenario implies that, both ionization and dissociation occur during one (probe) pulse. This significantly reduces the contributions of intermediate dynamics in the ionic states, which is the likely reason for the suppression of both, 159 meV and 20 meV peaks in the CO^+ and O^+ data of Fig. 4c and d. The mechanisms of CO_2 dissociative ionization are discussed in detail in ref. 30. The alignment-dependent enhancement of the CO^+ yield observed in Fig. 3a (black curve) is consistent with the findings of that work. The authors also reported some alignment dependence for the dissociation yielding O^+ fragments, which does not show any alignment-related behavior in our experiment (Fig. 3a, blue curve). However, since the O^+ alignment dependence observed in ref. 30 was considerably weaker than for CO^+, and our degree of alignment is very low, there is no contradiction between the two observations. More importantly, the origin of the sensitivity of both, CO^+ and O^+ dissociation to the symmetric stretching – which is involved in both of the Fermi dyad modes – of the neutral molecule can be readily understood. As discussed in ref. 21 and 30, essentially all dissociation channels in the CO_2^+ ion proceed *via* the predissociative $C^2\Sigma_g^+$ state. While the population of the low-lying vibrational levels of the $C^2\Sigma_g^+$ state leads to O^+ production in the lowest dissociation limit, the excitation of the higher vibrational levels strongly favors dissociation into CO^+ + O final state.[30] Since our 8 fs probe pulse essentially induces vertical transition from the neutral ground electronic state, the predissociative vibrational distribution depends on the initial position with respect to C–O distance. For the $C^2\Sigma_g^+$ ionic state, which is shifted inwards compared to the ground state of the neutral, the higher vibrational levels will be reached at the outer turning point, which should thus favor CO^+ ion production, while at the inner turning point the O^+ yield is expected to reach its maximum value. This behavior can be clearly observed in Fig. 3b, in which both, CO^+ and O^+ signals (black and blue, respectively) oscillate out of phase. Noticeably, the double ionization yield (red curve) oscillates in phase with the O^+ signal, in agreement with the previous assumption that it is enhanced at the inner turning point.

In earlier studies of vibrational wave packets in the electronic ground states of neutral diatomics[5,6,33] it was shown that the initial direction of the wave packet motion can be deduced from the phase of observed oscillations. Along with the above considerations, phase analysis of the data shown in Fig. 3b indicates that in our experiment the wave packet in the neutral CO_2 initially moves towards larger C–O distances. This is consistent with the assumption that vibrational motion is excited by stimulated off-resonance two-photon Raman scattering, in contrast to distance-dependent ionization observed in ref. 5 and 6.

3.3 Vibrational and electronic dynamics in the ionic states

Besides the vibrational dynamics in the neutral molecule discussed above, the wave packets in the ionic states clearly contribute to the data presented in Fig. 3 and 4. A prominent signature of the ionic state dynamics can be traced in the FT of the CO_2^{2+} signal in Fig. 4a. There, the 20 meV feature results from the electronic wave packet propagation in the $X^2\Pi_{1/2g}$ and $X^2\Pi_{3/2g}$ states of the ion, while the 160 meV peak very likely contains contributions from the symmetric stretching vibration in the X-state.[14] As shown in ref. 38 for atoms, tunneling ionization from the ground state of the ion is sensitive to the instantaneous configuration of the electronic wave packet resulting from spin–orbit splitting.[39–41] In our experiment this causes slow, ~200 fs modulation of the CO_2^{2+} signal in Fig. 3a. The contribution of the ionic X-state to this channel is similar to the results obtained by the Vienna group,[22] in contrast to the Coulomb explosion channel. For the latter, the absence of the 20 meV peak as well as alignment dependence not observed in ref. 22 indicate a different fragmentation mechanism at work here. One possibility might be a dominance of the dissociative DI of the vibrationally excited neutral molecule by the probe pulse. However, it should be noted that due to the limited statistics, the FT of this channel is rather noisy, and a weak feature due to the electronic wave packet might be just buried in the low-contrast data.

Both, the CO_2^{2+} and CE yield show only weak signatures of vibrational modes corresponding to the $A^2\Pi_u$ state of CO_2^+ (around 127–140 meV). This is consistent with the findings of ref. 22, where the authors saw that the DI pathways involving this state are strongly suppressed in the "weak pump–strong probe" configuration similar to the one employed here. The main reason for this is simply that the weak pump pulse does not efficiently ionize the HOMO−1, which results in the population of the $A^2\Pi_u$ state of the ion. However, a clear signature of this state can be observed in the O^+ yield in Fig. 4d. There, beyond a rather weak vibrational feature at ~140 meV, a pronounced peak at 12 meV can be seen in the spectrum. This frequency corresponds to the energy difference between the components of the electronic wave packet due to spin–orbit splitting in the A-state ($A^2\Pi_{1/2u}$ and $A^2\Pi_{3/2u}$).[12,13] Most likely it appears in the FT of the O^+ signal due to a small population of the $A^2\Pi_u$ state in the pump pulse, with subsequent probe-induced dissociation. Interestingly, this spectral feature was not observed by Erattupuzha *et al.*[22] even in the configuration where the vibrational dynamics in the $A^2\Pi_u$ state totally dominated the DI and CE signals.

3.4 Results for a weaker probe pulse

The contributions of intermediate ionic states appear to be considerably suppressed by a rather asymmetric choice of pump and probe pulse intensities, especially for dissociative ionization. In order to complement the above analysis, in Fig. 5 we consider the results of the pump-probe experiment with identical pump and probe pulses. The measurement was performed in a delay range of 0 to 1000 fs, and both pulses had the same intensity as the pump pulse in the experiments discussed above (1.5×10^{14} W cm^{-2}). Fig. 5a displays the yield of the doubly charged ions (red) as well as the yields of CO^+ and O^+ ions resulting from dissociative ionization (black and blue curves, respectively) as a function of pump-probe delay. Fig. 5b shows the FT spectra of these delay-dependent ion signals.

Because of its rather low yield, the CE channel is not analyzed here. Whereas the CO^+ yield as well as its Fourier transform is in general similar to the data with a more intense probe presented in Fig. 3 and 4, the results for CO_2^{2+} and O^+ ions are significantly different. In the yield curves in Fig. 5a a pronounced fast (~25 fs) oscillation appears on top of a slow modulation, which clearly differs in its time scale for both traces. This can be understood analyzing the corresponding Fourier transforms in Fig. 5b. Both, the CO_2^{2+} and O^+ spectra show a clear low-frequency peak reflecting the motion of the electronic wave packet due to the spin–orbit

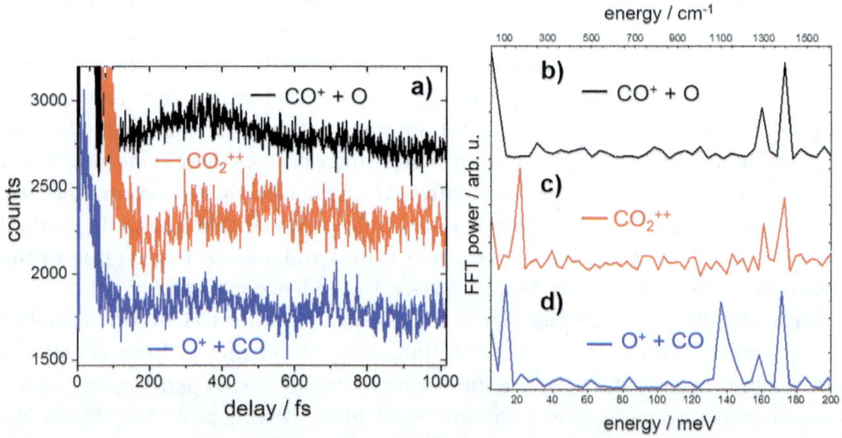

Fig. 5 (a) Yields of three different fragmentation pathways measured as a function of pump-probe delay between two identical 8 fs pulses at the intensity of 1.5×10^{14} W cm^{-2}. (b)–(d) The power spectra obtained from the Fourier transforms of the ion yields shown in (a) for CO^+ + O (b), CO_2^{2+} (c), and O^+ + CO (d) channels. The shorter delay range makes these FFT spectra noisier compared to those shown in Fig. 4.

splitting in the ion. However, the frequency of this feature is clearly different for CO_2^{2+} and O^+, corresponding to the splitting between $X^2\Pi_{1/2g}$ and $X^2\Pi_{3/2g}$ components (~20 meV) in the ground and $A^2\Pi_{1/2u}$ and $A^2\Pi_{3/2u}$ (~12 meV) in the first excited electronic states of CO_2^+. In the O^+ Fourier spectra the enhancement of the 12 meV line due to electronic coherence in the $A^2\Pi_u$ state is accompanied by the appearance of a strong peak at 140 meV, reflecting the symmetric stretching vibrational motion in the same electronic state (compare Fig. 4d and 5d). Thus, the reduction of the probe pulse intensity resulted in the enhanced contribution of the pathways, where ionization and dissociation occur not in one pulse, but rather in a pump and a probe step, respectively. It should be noted that this is the change of the contribution of different pathways, and not a difference in the initial excitation, which causes the modification of the observed dynamics, since the pump pulse, and, thus the whole electronic and nuclear wave packet it produces, remains the same for both Fig. 3–5. Correspondingly, all three Fourier spectra in Fig. 5 carry the signatures of vibrational dynamics in the neutral molecule discussed above.

Besides the different relative contributions of several competing pathways, another reason for the observed change of the spectra might originate from the larger sensitivity of the weaker probe to the changes of molecular structure. A detailed analysis of how the dissociation or ionization probability by the probe pulse depends on evolving nuclei geometry and electronic structure requires serious theoretical modelling and is definitely beyond the scope of this paper. However, qualitatively it appears plausible that a stronger probe might at least partly wash out such dependence, eliminating it completely in the limiting case of the saturation regime. Correspondingly, a more pronounced signature of the electronic wave packet motion in data with a weaker pulse is understandable and consistent with earlier observations of similar dependence in atoms.[38]

4. Experiments with two 35 fs pulses: the role of rotational motion

Up to now we mainly considered electronic and vibrational components of a strong-field induced molecular wave packet. Rotational excitation resulted in some yield modulations in Fig. 3 and 5, but was not considered in detail. With a rotational constant of 0.39 cm^{-1} for the neutral CO_2 molecule,[9,42] the influence of the created rotational wave packet on the observed signals within the time range below 3 ps considered above is expected to be restricted to the "prompt" alignment feature. The alignment by a very short, 8 fs pulse is also typically rather weak, which is consistent with our experimental results. Rotational excitation was recently shown to significantly alter the evolution of the coherent electronic wave packet in a neutral NO molecule, in which the signatures of the electronic coherence were traced in the time-dependent HHG signal.[7,8] In order to analyze the influence of rotational excitation on the electronic wave packet produced in the ground ionic state of the CO_2^+ ion observed in our experiment, we carried out a similar measurement in a much larger delay window (up to ~50 ps). As discussed in Section 2.2, to perform such a long scan with sufficient statistical significance, we used a different experimental set-up and longer, 35 fs pulses. The results of this experiment are shown in Fig. 6. In Fig. 6a the measured yield of the doubly charged CO_2^{2+} ions in a 30 ps delay window is shown as a function of the pump-probe delay. The intensities of the pump and the probe pulses are 7×10^{13} and 9×10^{13} W cm^{-2}, respectively. The signal shows pronounced peaks and dips at the positions of the fractional revivals of the CO_2 rotational wave packet.[18]

Fig. 6b displays the same data in a short delay range (from 2.2 to 3.3 ps). This zoomed-in view reveals a pronounced oscillation with a period slightly above 200 fs. This oscillation results in a 20 meV peak in a Fourier transform of the delay-dependent CO_2^{2+} signal shown in Fig. 6c. Similar to the CO_2^{2+} data obtained with ultrashort pulses discussed in the previous section, this feature can be uniquely identified as the beat frequency between the spin–orbit split states $X^2\Pi_{1/2g}$ and $X^2\Pi_{3/2g}$ of CO_2^+. Besides this peak, a manifold of lines below 15 meV can be seen in the spectra, reflecting a broad distribution of rotational levels excited by the pump pulse. Since the ground state geometries of the ion and the neutral molecule are essentially the same, it is difficult to disentangle the states reflecting rotational dynamics in the ion and neutral. Two different scenarios might contribute to the formation of the delay-dependent pump-probe signal in Fig. 6. First, the pump pulse can ionize the CO_2 molecule, preparing the ro-vibrational and electronic wave packet in the ion. At the intensity of our pump pulse the dominant contribution within this scenario is expected to come from the ionization of the HOMO, leaving the CO_2^+ ion in its ground state. It is then tunnel-ionized by the probe pulse, resulting in the production of the doubly charged ion.

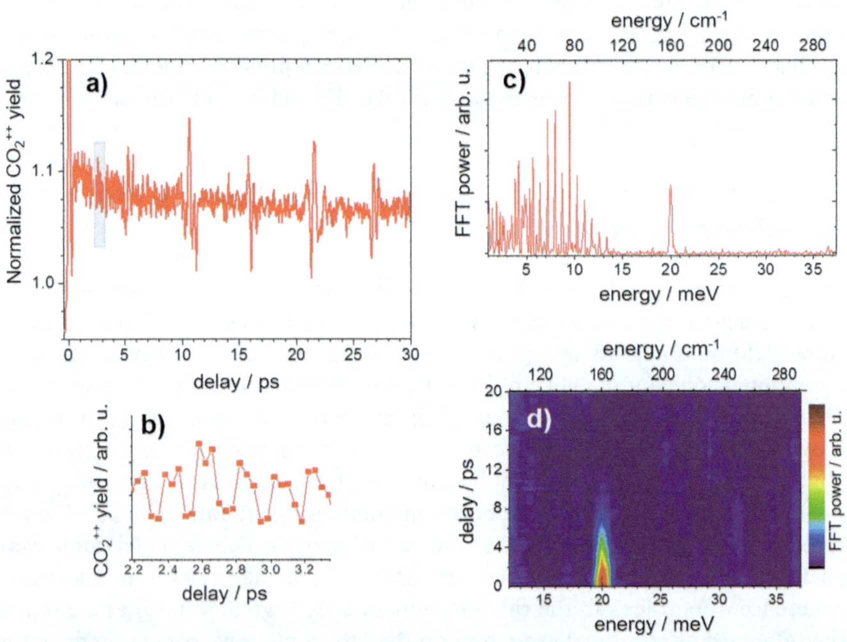

Fig. 6 (a) The measured CO_2^{2+} ion yield as a function of pump-probe delay between two 35 fs pulses. Intensities are 7×10^{13} and 9×10^{13} W cm^{-2} for the pump and the probe pulses, respectively. (b) CO_2^{2+} yield from (a) shown in a short delay range (marked by the shaded area in (a)). (c) The power spectra obtained from the Fourier transform of the ion yield shown in (a). (d) A sliding window Fourier transform of the delay-dependent CO_2^{2+} signal shown in (a). The vertical axis indicates the center of the 2 ps delay window.

As discussed in the previous section, this process is sensitive to the instantaneous structure of the electronic wave packet produced by spin–orbit splitting, resulting in the modulation of the delay-dependent CO_2^{2+} signal with the corresponding frequency. The contribution from HOMO−1 (resulting in the first excited ionic state $A^2\Pi_u$) is expected to be essentially negligible.[22,30] However, it should be noted that the analogous peak due to the spin–orbit splitting in the ionic $A^2\Pi_u$ state (observed at 12 meV in Fig. 4d and 5d) would lie below the onset of rotational transitions in Fig. 6c, such that it could be resolved only if its strength would be comparable with that of the ground state. Vibrational dynamics in either of the ionic states is too fast to be resolved with 35 fs pulses used here.

The second pump-probe scenario implies rotational excitation of the neutral molecule by the pump, with subsequent double ionization by the probe. At the probe intensity of 9×10^{13} W cm^{-2} the latter step is dominated by the recollision-induced NSDI,[29,34,35] most likely involving a recollision-excitation scenario due to the rather low energy of the recolliding electron at this intensity.[43] The detailed structure of the CO_2^{2+} signal in Fig. 6a closely resembles the alignment-dependent HHG signal (yield of the 23rd harmonic) reported in ref. 18 for similar pump intensity, and is quite different from the CO_2^+ trace presented in the same article. However, in ref. 29 a similar angular dependence for single and double ionization of the CO_2 molecule has been reported. In view of the above-mentioned similarity of the ionic and neutral ground state geometries, a contribution from this pump-probe scenario cannot be uniquely identified in the data.

The oscillations due to the coherent electronic wave packet motion in the ionic ground state appear to be more pronounced in the region of small delays in Fig. 6a. In order to quantify this, in Fig. 6d we present the results of the sliding-window Fourier transform of the measured delay-dependent CO_2^{2+} yield. Since the chosen length of the window (2 ps) is considerably shorter than the characteristic time scales of rotational dynamics, we did not include the region of rotational frequencies in this representation. The decay of the electronic wave packet for the increasing window position is evident in the spectrum. The peak at 20 meV completely disappears within less than 10 ps. Given the time scale of this decay, the most likely reason for this is coupling to rotational motion. The latter was recently predicted to reduce the amplitude of similar electronic wave packet oscillations in the neutral NO molecule,[7] and corresponding complicated evolution of the wave packet was observed in the experimental data.[7,8] In our case, the electronic coherence seems to be completely lost and no signatures of its reappearance can be traced in the delay range up to 50 ps.

5. Conclusions and outlook

In this work we have studied temporal evolution of electronic and nuclear wave packets created in strong field interaction with the carbon dioxide molecule employing momentum-resolved ion spectroscopy and channel-selective Fourier analysis. Using broadband, few-cycle pulse excitation, we found signatures of vibrational dynamics in both, ionic and neutral states of the molecule. Whereas our observations for the ionic ground state look rather similar to the recent results obtained by the Vienna group,[22] the frequencies of the neutral molecule excitation were absent in their data. We attribute this striking difference between the two otherwise similar experiments to the difference in the bandwidth of the pulses used, since a broad photon spectrum is required for vibrational excitation *via* far-off-resonance two-photon stimulated Raman scattering, which we consider to be the most likely mechanism for the dynamics in the ground electronic state of the neutral CO_2. Using the data from two different dissociative ionization channels, we suggest a simple, intuitive explanation of the vibrational dynamics of the neutral, which is consistent with the proposed mechanism and with the observed phase relations between the time-dependent yields of different fragmentation channels. For ionic states, we found signatures of both, electronic and vibrational excitations. The frequencies corresponding to the electronic ground and first excited states were found in the Fourier transforms of the double ionization signal and of the yield of dissociative ionization leading to the $O^+ + CO$ final state, respectively. Varying the intensity of the probe pulse, we demonstrated how the relative contribution of the pathways proceeding *via* different intermediate states can be changed. Finally, in the experiment with longer pulses we demonstrated how the coherent electronic wave packet created in the ground electronic state of the ion might completely decay within $\sim$10 ps due to coupling to rotational motion.

Whereas the experimental data presented here are rather sophisticated and allow for deep multidimensional analysis, the interpretation of the results is mostly based on rather simple, intuitive considerations and boundary conditions deduced from the data. The next level of understanding of the obtained experimental results would require a more rigorous analysis involving theoretical modelling. Particularly important here is the quantitative description of the probe step, and the reasons for its (often different) sensitivity to instantaneous electronic structure and nuclear geometry of the molecule for different final states, which were not addressed in detail in the present work. Quantitative analyses of (time-dependent) experimental and theoretical branching ratios for different ionization and fragmentation channels could serve as a reliable gauge of the adequacy of the conclusions made.

Channel-selective Fourier spectroscopy employed here can be efficiently extended to the differential yields of various final states (*i.e.*, momentum-, energy- or/and angle-resolved analysis). An example of such an extension can be found in our earlier work on the D_2 molecule.[44,45] Whereas technically this possibility exists already for the data sets presented here, it is hampered by the use of the same wavelength of the pump and the probe, which often results in overlapping contributions from both pulses. This limitation can be lifted in two-color pump probe experiments, where the use of a different probe wavelength would also reduce the difficulty in disentangling the dynamics within the probe pulse. Particularly promising for such extensions are the pump-probe schemes employing high-frequency XUV or X-ray pulses to map light-induced molecular wave packet dynamics.[46]

In combination with the analysis of differential observables and coincident few-particle detection, in particular, channel-resolved photoelectron spectroscopy,[47] the experimental approach used here can be very efficient for studies of multimode wave packets in small polyatomic molecules. As demonstrated here for the case of coherent electronic wave packet in CO_2^+, in certain cases it can directly yield information on the couplings between different degrees of freedom. A promising application would be a time domain study of coupled multimode (stretching and bending) vibrational motions (*e.g.*, in halomethanes), with a possibility to visualize and quantify the effects of nonadiabatic couplings of vibrational and electronic degrees of freedom.

Acknowledgements

This work was supported by the U.S. Department of Energy, Office of Science, Office of Basic Energy Sciences, Division of Chemical, Geological, and Biological Sciences, under Contract No. DE-FG02-86ER1349. The authors are grateful to O. Smirnova and S. Patchkovskii for fruitful discussions.

References

1 A. Stolow, *Annu. Rev. Phys. Chem.*, 2003, **54**, 89.
2 F. Krausz and M. Ivanov, *Rev. Mod. Phys.*, 2009, **81**, 163.
3 X. Zhou, *et al.*, *Phys. Rev. Lett.*, 2008, **100**, 073902.
4 W. Li, *et al.*, *Science*, 2008, **322**, 1207.
5 T. Ergler, B. Feuerstein, A. Rudenko, K. Zrost, C. D. Schröter, R. Moshammer and J. Ullrich, *Phys. Rev. Lett.*, 2006, **97**, 103004.
6 L. Fang and G. N. Gibson, *Phys. Rev. Lett.*, 2008, **100**, 103003.
7 P. M. Kraus, *et al.*, *Phys. Rev. Lett.*, 2013, **111**, 243005.
8 D. Baykusheva, *et al.*, *Faraday Discuss.*, 2014, **171**, 113.
9 G. Herzberg, *Molecular Spectra and Molecular Structure: Electronic Spectra and Electronic Structure of Polyatomic Molecules*, Van Nostrand Reinhold, New York, 1966, vol. 3.
10 P. Linstrom and W. Mallard, *NIST Chemistry WebBook*, NIST Standard Reference Database Number 69, http://webbook.nist.gov/chemistry/.
11 C. Roche, G. Millot, R. Chaux and R. Saint-Loup, *J. Chem. Phys.*, 1994, **101**, 2863.
12 P. Baltzer, *et al.*, *J. Chem. Phys.*, 1996, **104**, 8922.
13 J. Liu, M. Hochlaf and C. Y. Ng, *J. Chem. Phys.*, 2000, **113**, 7988.
14 J. Liu, W. Chen, C.-W. Hsu, M. Hochlaf, M. Evans, S. Stimson and C. Y. Ng, *J. Chem. Phys.*, 2000, **112**, 10767.
15 D. A. Shaw, *et al.*, *Chem. Phys.*, 2006, **324**, 515.
16 M. Hochlaf, F. R. Bennett, G. Chambaud and P. Rosmus, *J. Phys. B: At., Mol. Opt. Phys.*, 1998, **31**, 2163.
17 Th. Zimmermann, H. Koppel and L. S. Cederbaum, *J. Chem. Phys.*, 1985, **83**, 4697.
18 T. Kanai, S. Minemoto and H. Sakai, *Nature*, 2005, **435**, 470–474.
19 C. Vozzi, *et al.*, *Phys. Rev. Lett.*, 2005, **95**, 153902.
20 O. Smirnova, *et al.*, *Nature*, 2009, **460**, 972–977.
21 H. Timmers, *et al.*, *Phys. Rev. Lett.*, 2014, **113**, 113003.
22 S. Erattupuzha, *et al.*, *J. Chem. Phys.*, 2016, **144**, 024306.
23 R. Dorner, V. Mergel, O. Jagutzki, L. Spielberger, J. Ullrich, R. Moshammer and H. Schmidt-Bocking, *Phys. Rep.*, 2000, **330**, 95.
24 J. Ullrich, R. Moshammer, A. Dorn, R. Dorner, L. P. H. Schmidt and H. Schmidt-Bocking, *Rep. Prog. Phys.*, 2003, **66**, 1463.
25 V. L. B. de Jesus, *et al.*, *J. Electron Spectrosc. Relat. Phenom.*, 2004, **141**, 127.
26 X. Ren, V. Makhija, F. Li, M. F. Kling and V. Kumarappan, *Phys. Rev. A*, 2014, **90**, 013419.

27 D. Pavicic, K. Lee, D. Rayner, P. Corkum and D. Villeneuve, *Phys. Rev. Lett.*, 2007, **98**, 243001.
28 A. S. Alnaser, *et al.*, *Phys. Rev. A*, 2005, **71**, 031403.
29 M. Opperman, S. J. Weber, L. J. Frazinski, M. Y. Ivanov and J. P. MArangos, *Phys. Rev. A*, 2013, **88**, 043432.
30 M. Opperman, *et al.*, *J. Phys. B: At., Mol. Opt. Phys.*, 2014, **47**, 124025.
31 J. Liu, W. Chen, M. Hochlaf, X. Qian, C. Chang and C. Y. Ng, *J. Chem. Phys.*, 2003, **118**, 149.
32 S. Chelkowski and A. D. Bandrauk, *J. Raman Spectrosc.*, 1997, **28**, 459.
33 E. Goll, G. Wunner and A. Saenz, *Phys. Rev. Lett.*, 2006, **97**, 103003.
34 L. Pei and C. Guo, *Phys. Rev. A*, 2010, **82**, 021401.
35 Z. Jun-Feng, *et al.*, *Chin. Phys. B*, 2015, **24**, 033302.
36 V. R. Bhardwaj, *et al.*, *Phys. Rev. Lett.*, 2001, **86**, 3522.
37 A. Rudenko, *et al.*, *IOP Conf. Ser.: Mater. Sci. Eng.*, 2005, **183**, 93.
38 L. Fechner, *et al.*, *Phys. Rev. Lett.*, 2014, **112**, 213001.
39 Z.-H. Loh, *et al.*, *Phys. Rev. Lett.*, 2007, **98**, 143601.
40 E. Goulielmakis, *et al.*, *Nature*, 2010, **466**, 739.
41 N. Rohringer and R. Santra, *Phys. Rev. A*, 2009, **79**, 053402.
42 B. J. Sussman, *et al.*, *Phys. Rev. A*, 2006, **73**, 053403.
43 Y. Liu, *et al.*, *Phys. Rev. Lett.*, 2008, **101**, 053001.
44 T. Ergler, *et al.*, *Phys. Rev. Lett.*, 2006, **97**, 193001.
45 B. Feuerstein, *et al.*, *Phys. Rev. Lett.*, 2007, **99**, 153002.
46 J. Ullrich, A. Rudenko and R. Moshammer, *Annu. Rev. Phys. Chem.*, 2012, **63**, 635.
47 A. E. Boguslavskiy, *et al.*, *Science*, 2012, **335**, 1336.

Faraday Discussions

PAPER

Towards controlling the dissociation probability by light-induced conical intersections

András Csehi,*[a] Gábor J. Halász,[b] Lorenz S. Cederbaum[c] and Ágnes Vibók*[ad]

Received 26th May 2016, Accepted 7th July 2016

DOI: 10.1039/c6fd00139d

Light-induced conical intersections (LICIs) can be formed both by standing or by running laser waves. The position of a LICI is determined by the laser frequency while the laser intensity controls the strength of the nonadiabatic coupling. Recently, it was shown within the LICI framework that linearly chirped laser pulses have an impact on the dissociation dynamics of the D_2^+ molecule (*J. Chem. Phys.*, **143**, 014305, (2015); *J. Chem. Phys.*, **144**, 074309, (2016)). In this work we exploit this finding and perform calculations using chirped laser pulses in which the time dependence of the laser frequency is designed so as to force the LICI to move together with the field-free vibrational wave packet as much as possible. Since nonadiabaticity is strongest in the vicinity of the conical intersection, this is the first step towards controlling the dissociation process *via* the LICI. Our showcase example is again the D_2^+ molecular ion. To demonstrate the impact of the LICIs on the dynamical properties of diatomics, the total dissociation probabilities and the population of the different vibrational levels after the dissociation process are studied and discussed.

I. Introduction

Conical intersections (CIs) provide a particularly striking and important cause for nonadiabatic dynamics in polyatomic molecules not available in diatomics. At a CI the nonadiabatic couplings become singular and this gives rise to dramatic nonadiabatic effects widely studied in the literature.[1,2] Clearly, the position of a CI appearing in nature and the strength of the nonadiabatic effects are inherent properties of the electronic states of a molecule and are hard to manipulate. For brevity, we rename the naturally occurring CIs "natural CIs". While exposed to

[a]*Department of Theoretical Physics, University of Debrecen, PO Box 5, H-4010 Debrecen, Hungary. E-mail: csehi.andras@science.unideb.hu; vibok@phys.unieb.hu*

[b]*Department of Information Technology, University of Debrecen, PO Box 12, H-4010 Debrecen, Hungary*

[c]*Theoretische Chemie, Physikalish-Chemisches Institut, Universität Heidelberg, H-69120 Heidelberg, Germany*

[d]*ELI-ALPS, ELI-HU Non-Profit Ltd, Dugonics tér 13, H-6720 Szeged, Hungary. E-mail: Agnes.Vibok@eli-alps.hu*

resonant laser light, a new feature emerges. This feature is a CI induced by the light which cannot be avoided even in the case of diatomic molecules.[3,4] The angle θ between the laser polarization and the molecular axis now becomes the missing dynamical variable that together with the stretching coordinate constitute the space in which the induced CI can live. The phenomenon is rather general and not restricted to propagating waves. Light-induced CIs (LICI) automatically also emerge in standing waves which form optical lattices widely used in cold-atom physics.[3] Actually, this is the case where they have been first predicted. In optical lattices not only the electronic, the stretching and rotational motion couple strongly *via* the induced CI, but also the translational motion enters the scene and becomes a dynamical variable. This gives cooling mechanisms new aspects. It has been demonstrated that the LICI in diatomics gives rise to a variety of nonadiabatic phenomena. Of course, they exhibit a topological phase and provide singular nonadiabatic couplings like the natural CIs do.[5,6] Already in traveling waves of rather weak intensity the LICI has a substantial impact on the spectrum of the molecule.[4] Moreover, even in short laser pulses one finds remarkable consequences of the LICI. The degree of alignment of a diatomic molecule as a function of time differs substantially from the rigid rotor predictions.[7] A wealth of nuclear-wave-packet quantum interferences show up in intense laser dissociation due to the nonadiabatic involvement of the rotation.[8,9] A robust effect in the angular distribution of the photofragments has been found that serves as a direct signature of the LICI and should be amenable to observation.[10] If the resonant laser excitation is to an electronically metastable state which decays by emitting an electron, the LICI is in the continuum and changes its topology like all natural CIs which are embedded in the continuum. It takes on the appearance of doubly-intersecting complex surfaces (DICES) which also exhibit singular nonadiabatic couplings.[11] Examples where light-induced DICES appear are resonant Auger decay in diatomics and interatomic coulombic decay by two-photon transitions in intense pulses. LICIs are ubiquitous in polyatomic molecules both fixed in space or freely rotating. Because of the presence of several vibrational degrees of freedom, LICIs exist also without rotations and become multidimensional in nuclear coordinate space.[12] This opens the door for manipulating and controlling nonadiabatic effects by light.[12–17]

Currently, we are investigating the impact of linearly chirped laser pulses on the dissociation of diatomic molecules (D_2^+).[18,19] As the frequency of the pulse changes in time, the position of the LICI changes analogously. It was found that using laser pulses which are long enough ($T_p \geq 10$ fs) the amplitude of the periodic change in the total dissociation probability is significantly squeezed by the chirped pulse compared to the transform limited situation.[18] By chirping the frequency the effective pulse duration is increased and the effective intensity is decreased by the same factor resulting in suppression of the dissociation yields.[19] Several works in the literature demonstrate the controllability of different dynamical properties of molecular systems by using chirped laser pulses.[20–43]

To be consistent with our former chirp type investigations, the showcase example will again be the D_2^+ molecule. In the present work we plan to go further and perform calculations using chirped laser pulses in which the time dependence of the laser frequency is designed so as to constrain the motion of the LICI's position to follow, as much as possible, the time evolution of the nuclear wave packet. To achieve this goal, the time dependence of the laser frequency is approximated by the energy difference between the potential energy curves of the participating electronic states of the D_2^+ ion. We note, that the ideal case would be to let the LICI follow the true wave packet and not the field-free one. This can be achieved by a self-consistent procedure (optimal control theory). But, since this is the first study of its kind, we would like to first see whether there are gains to be expected from the LICI following the wave packet, and this is the purpose of the present investigation.

The photodissociation probabilities and the population of the different vibrational levels after the dynamics are studied. Two dimensional (2D) simulations are applied to the problem in which the rotational angle is taken into account as a dynamic variable and, therefore, the LICI is explicitly included.

II. The Hamiltonian

Fig. 1 shows the potential energy curves for D_2^+ in the Floquet representation.[44] $V_1(R_1) = 1s\sigma_g$, $V_2(R) = 2p\sigma_u$ and $(V_2(R) - \hbar\omega_L)$ denote the energies of the electronic ground, the first excited and the dressed Eigen states, which are considered in the calculations. Assuming the scenario where the neutral D_2 molecule is ionized at $t = 0$, its vibrational ground state is transferred vertically to the potential energy curve of the ground electronic state ($V_1 = 1s\sigma_g$) of the ion. This defines the initial wave packet for the nuclear dynamics, which can be considered as the Franck–Condon (FC) distribution of all the vibrational states of the ion D_2^+. Exciting the

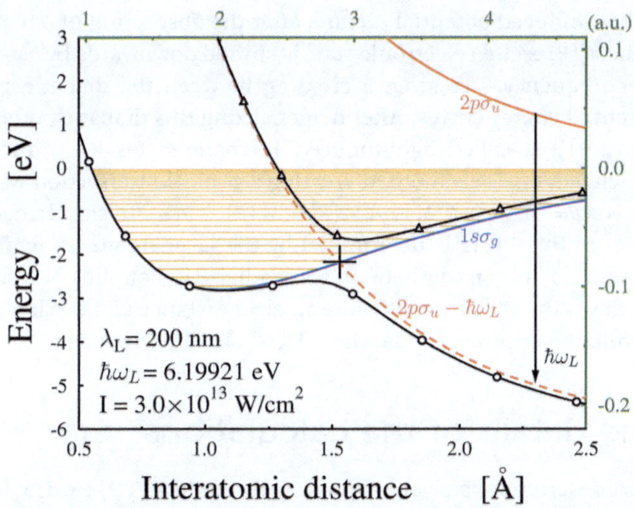

Fig. 1 Potential energy curves of the D_2^+ ion. Diabatic energies of the ground $1s\sigma_g$ and the first excited $2p\sigma_u$ states are displayed with solid blue and red lines, respectively. The field dressed excited state ($2p\sigma_u - \omega_L$) is represented with a dashed red line. The field dressed excited state ($2p\sigma_u - \omega_L$) forms a light-induced conical intersection (LICIs) with the ground state. A cut through the adiabatic surfaces at $\theta = 0$ (parallel to the field) is also shown for a particular frequency ($\omega_L = 6.19921$ eV) and field intensity (3×10^{13} W cm^{-2}) depicted by solid black lines marked with circles (V_1) and triangles (V_u). Alternative x–y axes (top and right hand side) with a.u.'s are provided. The horizontal orange lines denote the field-free vibrational levels of the $1s\sigma_g$ state. We denote with a cross the position of the LICI ($R_{LICI} = 1.53$ Å $= 2.891$ a.u. and $E_{LICI} = -2.16611$ eV).

$1s\sigma_g$ ground electronic state of the D_2^+ molecule by a resonant laser pulse to the repulsive $2p\sigma_u$ state (see also Fig. 1) these two electronic states are radiatively coupled by the electric field. An electronic transition occurs due to the nonvanishing transition dipole moment and light-induced states are formed. In the space of the $V_1(R)$ and $V_2(R)$ electronic states the following time-dependent Hamiltonian can be written for the rovibronic nuclear dynamics of the D_2^+:

$$H = \begin{pmatrix} -\frac{1}{2\mu}\frac{\partial^2}{\partial R^2} + \frac{L_{\theta\varphi}{}^2}{2\mu R^2} & 0 \\ 0 & -\frac{1}{2\mu}\frac{\partial^2}{\partial R^2} + \frac{L_{\theta\varphi}{}^2}{2\mu R^2} \end{pmatrix} + \begin{pmatrix} V_1(R) & -E(t)d(R)\cos\theta \\ -E(t)d(R)\cos\theta & V_2(R) \end{pmatrix}. \tag{2.1}$$

The off-diagonal elements of eqn (2.1) represent the radiative couplings, where the electric field $E(t)$ characterizes the laser field. $d(R)\left(= -\left\langle \psi_1{}^e \middle| \sum_j r_j \middle| \psi_2{}^e \right\rangle \right)$ is the transition dipole matrix element and θ is the angle between the polarization direction of the light and the direction of the molecular axes. Here, R and (θ,φ) are the molecular vibrational and rotational coordinates, respectively, μ is the reduced mass, and $L_{\theta\varphi}$ is the angular momentum operator of the nuclei ($e = m_e = \hbar = 1$; atomic units are used throughout the article). The potential energies $V_1(R)$ and $V_2(R)$ and the transition dipole moment were taken from ref. 45 and 46.

In the light-induced potential picture, after the absorption of one photon, the energy of the $V_2(R)$ excited electronic state is shifted downwards by $\hbar\omega_L$ – where ω_L is the laser frequency – creating a crossing between the diabatic ground and excited potential energy curves. After diagonalizing this diabatic potential energy matrix (eqn (2.1)) so-called light-induced adiabatic states V_{lower} and V_{upper} are formed which can cross each other, creating a conical intersection whenever the conditions $\cos\theta = 0$, $(\theta = \pi/2)$ and $V_1(R) = V_2(R) - \hbar\omega_L$ are simultaneously met.[3] The position of this LICIs is determined by the laser frequency while the laser intensity controls the strength of the nonadiabatic coupling.[3] Increasing the frequency moves the CI to a smaller internuclear distance and to a lower energetic position while the opposite holds when decreasing the frequency.

III. The details of the calculations

Here a short description is given of the time-dependent (TD) electric fields of the pulses and of the wave packet propagation method used to solve the TD nuclear Schrödinger equation employing the Hamiltonian of eqn (2.1).

A. The applied electric fields

At time $t = 0$ the system is ionized and the vibrational wave packet on the ground ionic state starts to oscillate on the $V_1 = 1s\sigma_g$ potential curve. The initial nuclear wave packet is built up as the FC distribution of the ionic vibrational states and the molecule was initially in its rotational ground state ($J = 0$). Then a laser pulse is applied after a time delay of t_{delay}. A linear polarized electric field is considered which is a product of a Gaussian envelope and a $\cos(\phi(t))$ function

$$E(t) = \varepsilon_0 f(t) \cos(\phi(t)). \tag{3.1}$$

here ε_0 is the maximum amplitude of the electric field and $f(t) = e^{-\frac{1}{2\sigma^2}(t - t_{delay})^2}$, $\left(\sigma = T_p/\sqrt{4 \log 2}\right)$ is the Gaussian shape of the pulse. T_p and t_{delay} are the pulse duration (at FWHM) and the temporal maximum of the pulse, respectively. $\omega(t) = \frac{d}{dt}\phi(t)$ is the laser frequency and its time dependence is the control knob. The $\omega(t) = V_2(R) - V_1(R)$ function ($\hbar = 1$) controls the position of the nuclear coordinate R at which the resonance condition for the appearance of the LICI is fulfilled. The initial nuclear wave packet is built up as the FC distribution of the ionic vibrational states and its field-free time evolution can be seen in Fig. 2A. The time period of the field-free oscillation is about 24 fs. As the wave packet is a superposition of several different vibrational Eigen states, it spreads during the time evolution relatively quickly, but exhibits recurrences. In order to force the LICI to move together with the field-free vibrational wave packet, we estimate the time-dependent frequency $\omega(t)$ from the motion of the field-free wave packet shown in Fig. 2A.

In our calculations the pulse duration was chosen to be $T_p = 6$ fs in all examples studied, t_{delay} was varied in the range between 0 and 100 fs, and the phase of the electric field at time t was derived according to the following formula:

$$\phi(t) = \int_0^t \omega\left(t'\right)dt'. \tag{3.2}$$

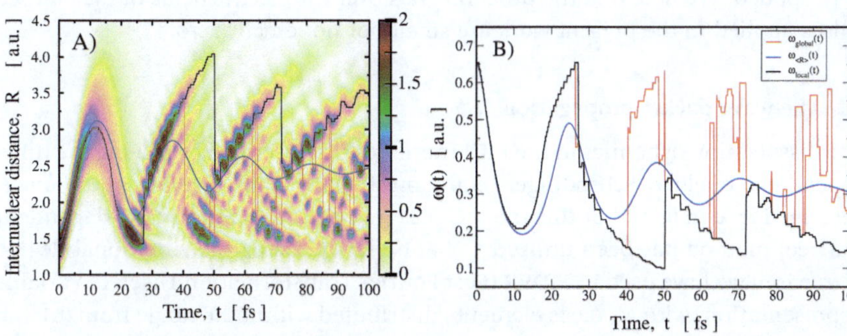

Fig. 2 The field-free vibration of the D_2^+ system as a function of time t after ionization of the neutral at time $t = 0$ and the deduced chirped pulse frequency. In panel (A) the square of the vibrational wave packet is represented by a color code: the darker the color the larger the amplitude of the wave packet. Additionally, three curves are shown which are used to determine the chirped frequency of the pulses employed in this investigation: the time-dependent average of the internuclear distance (blue curve), a curve following the global (red curve) and one following the local (black curve) maximum positions of the density of the field-free wave packet. For more details see the main text. Panel (B) shows the resulting three different forms of the $\omega(t)$ frequency function (for definition see in Section IIIA.) of the laser pulse used in the calculations of the nuclear dynamics. The respective frequency functions are denoted by $\omega_{\langle R\rangle}(t)$, $\omega_{global}(t)$ and $\omega_{local}(t)$.

In the present numerical study three kinds of chirped $\omega(t)$ functions were applied. They were obtained as explained in the following and are shown in Fig. 2B. The so-called $\omega_{global}(t)$ function (red line on Fig. 2B) is obtained by following the global maximum of the density of the field-free wave packet ($\omega_{global}(t) = V_2(R_{global}(t)) - V_1(R_{global}(t))$) shown in Fig. 2A. This function is seen to exhibit abrupt strong oscillations at certain times. In order to eliminate these undesirable oscillations we applied a technique described below and obtained the so-called $\omega_{local}(t)$ function (black line in Fig. 2B). Instead of following the global maximum of the density of the nuclear wave packet, this function follows the maximum value of the local density of the wave packet ($\omega_{local}(t) = V_2(R_{local}(t)) - V_1(R_{local}(t))$). For deriving the $R_{local}(t)$ at time t, the maximum of the product of the density $|\psi(R,t)|^2$ and an exponential weight factor was considered rather than the maxima of the density themselves. This was performed by maximizing the following expression at each value of t:

$$|\psi(R,t)|^2 e^{-\frac{1}{2\sigma_R{}^2}(R - R_{local}(t-\Delta t))^2}. \tag{3.3}$$

here $R_{local}(t - \Delta t)$ is the position of maximum in the preceding time step ($\Delta t = 0.1$ fs) and σ_R has a value of 0.77 a.u. for $t < 80$ fs and 0.28 a.u. for later times.

Finally, the third kind of chirped $\omega(t)$ function was simply obtained by following the average value $\langle R \rangle$ of the variable R as a function of time using as usual $\langle R(t) \rangle = \langle \psi(R,t)|R|\psi(R,t) \rangle$. This chirped frequency function is named $\omega_{\langle R \rangle}(t)$.

Since the pulses used in the present calculations are relatively short, we also compare the results with those of pulses of fixed frequencies, where for each time delay this constant frequency is chosen to correspond to $\omega_{local}(t)$ at that time delay. We will address the respective calculations as frequency varying transform limited (vTL) calculations. In many experiments a particular pulse is used for all time delays, and we have, therefore, also performed a reference transform limited (TL) calculation where the central frequency is the same ($\omega = 0.23518$ a.u.) at all time delays.

In the simulations the intensities $I = 1 \times 10^{12}$ W cm^{-2} and $I = 1 \times 10^{13}$ W cm^{-2} were applied. We note that the time integrals over the electric fields of the chirped pulses applied in the present work are small but not exactly zero.[56]

B. The wave packet propagation

The dissociation dynamics in the LICI framework is described by solving the time-dependent nuclear Schrödinger equation (TDSE) with the Hamiltonian H described by eqn (2.1). For this the MCTDH (multi configuration time-dependent Hartree) method has been utilized.[47–51] For characterizing the vibrational degree of freedom we have used FFT-DVR (Fast Fourier Transformation-Discrete Variable Representation) with N_R basis elements distributed within the range from 0.1 a.u. to 80 a.u. for the internuclear separation. The rotational degree of freedom was described by Legendre polynomials $\{P_j(\cos\theta)\}_{j=0,1,2,\cdots,N_\theta}$. These so-called primitive basis sets (χ) were used to represent the single particle functions (ϕ), which in turn were used to represent the wave function:

$$\begin{aligned}\phi_{j_q}{}^{(q)}(q,t) &= \sum_{l=1}^{N_q} c_{j_q l}{}^{(q)}(t)\chi_l{}^{(q)}(q) \quad q = R,\theta \\ \psi(R,\theta,t) &= \sum_{j_R=1}^{n_R}\sum_{j_\theta=1}^{n_\theta} A_{j_R j_\theta}(t)\phi_{j_R}{}^{(R)}(R,t)\phi_{j_\theta}{}^{(\theta)}(\theta,t).\end{aligned} \tag{3.4}$$

The actual number of primitive basis functions in the numerical simulations were chosen to be $N_R = 2048$ and $N_\theta = 61$ for the vibrational and rotational degrees of freedom, respectively. On both diabatic surfaces and for both degrees of freedom a set of $n_R = n_\theta = 4,\cdots, 20$ single particle functions were applied to form the nuclear wave packet of the system. (The actual value of N_θ and $n_R = n_\theta$ was chosen depending on the peak field intensity I_0.) Attention has been paid to the proper choice of basis so that convergence has been reached in each propagation.

The TD nuclear wave packet is used to calculate the total dissociation probability $P_{\rm diss}$ and the populations. The probability reads

$$P_{\rm diss} = \int_0^\infty {\rm d}t\langle\psi(t)|W|\psi(t)\rangle. \tag{3.5}$$

where $-{\rm i}W$ is the complex absorbing potential (CAP) applied at the last 10 a.u. of the vibrational (R) grid.

To obtain the population of the individual vibrational levels we have used two steps of the simulation. The first one is initiated from the FC state averaged nuclear wave function and contains the interaction to the external electric field and the dissociation process. It is technically finished when the CAP eliminates the flying away portion of the wave packet (t = 350 fs). The second step of the simulation uses the final wave function of the first one as an initial wave function. From this second simulation we calculate the autocorrelation function as:

$$C(t) = \langle\psi(R,\theta,0)|\psi(R,\theta,T)\rangle \tag{3.6}$$

where T is the time period of the second simulation. Next, we calculate the Fourier transform of this function to obtain the spectra of the bounded wave function:[52–54]

$$\sigma(\omega) = \frac{1}{\pi}\int_0^T \mathscr{R}\left[C(t){\rm e}^{{\rm i}\omega t}\right]{\rm e}^{-t/\tau}\cos^2\frac{\pi t}{2T}{\rm d}t. \tag{3.7}$$

The τ damping parameter was chosen to be 400 fs. Using a long enough second simulation (T = 350 fs) we can resolve the individual lines of the vibrational levels in the spectra. Finally, we integrate these individual peaks to derive the population of the corresponding vibrational level:

$$P_\nu = \int_{(E_{\nu-1}+E_\nu)/2}^{(E_\nu+E_{\nu+1})/2} \sigma(\omega){\rm d}\omega. \tag{3.8}$$

IV. Results and discussion

In our previous work linearly chirped laser pulses were used, which can be easily produced experimentally.[55] To be specific, the pulses preserved the spectrum of their transform limited (TL) equivalents and their time integral was exactly equal to zero. In this work, our attention is focused on constructing pulses in which the time dependence of the frequency is such that the light-induced conical intersection moves together with the field-free vibrational wave packet. This could be a first step towards developing a control protocol by conical intersection, and it is meaningful, because in the case of short laser pulses the field-free wave packet does not change dramatically during the dynamical process. We are interested in finding out how much this can enhance the dissociation yield of the D_2^+ ion and also studying how such a chirp influences the behavior of the photodissociation probabilities as a function of delay time and the populations on the different vibrational levels. Throughout the paper $T_p = 6$ fs pulse length is used.

A. Dissociation probability

1. Chirped pulses. Fig. 3 (panel A) shows the total dissociation probability (eqn (3.5)) as a function of time delay for the TL and for the chirped simulations employing a 1×10^{12} W cm^{-2} intensity. The central frequency of the TL pulse is the same for all particular time delays and its value ($\omega = 0.23518$ a.u.) corresponds to the value of the second minimum of the $\omega_{\langle R\rangle}(t)$ function in Fig. 2B. Three different chirped $\omega(t)$ functions ($\omega_{\text{global}}(t)$, $\omega_{\text{local}}(t)$ and $\omega_{\langle R\rangle}(t)$) were applied in the simulations.

The most striking phenomenon is that each dissociation probability curve shows a periodic oscillatory behavior. One can easily understand it for the case of TL calculation, as in this situation the wave packet is periodically located at a distance where the resonance condition is satisfied, *i.e.* the LICI is positioned at the high density region of the wave packet. One might expect more balanced results over time for the other curves, since in these cases the resonance condition is almost always fulfilled in such places, where the density of the wave packet possesses great value. Instead, we obtained curves which are more or less in accord with the TL one, possessing well defined maxima and minima and the ratios between these extreme values are significant. The latter is true both for the $\omega_{\text{global}}(t)$ and $\omega_{\text{local}}(t)$ curves, but perhaps it is more relevant for the case of $\omega_{\text{global}}(t)$. This is probably due to the finding that at larger nuclear distances the resonance condition induces dissociation more efficiently than at shorter distances. Indeed, starting, for instance, from $t_{\text{delay}} = 35$ fs, when the position of the actual selected local maximum of the nuclear density differs significantly

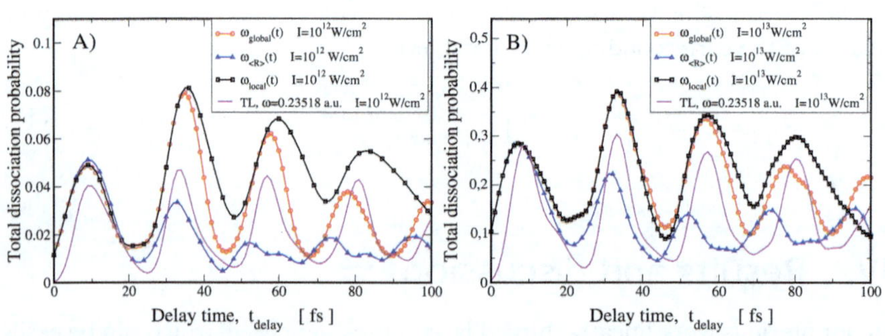

Fig. 3 Dissociation probabilities of D_2^+ as a function of the delay time (t_{delay}) of the pulse. In panel (A) the black (marked with squares), the red (marked with circles) and the blue curves (marked with triangles) denote the dissociation probabilities corresponding to the $\omega_{\text{local}}(t)$, $\omega_{\text{global}}(t)$ and $\omega_{\langle R\rangle}(t)$ frequency functions, respectively, at 1×10^{12} W cm^{-2} intensity. The fourth curve (magenta solid) shows the reference transform limited (TL) result calculated with $\omega = 0.23518$ a.u. Panel (B) shows the same curves as in panel (A), but for 1×10^{13} W cm^{-2} intensity. In all calculations the pulse length is $T_p = 6$ fs.

from the position of the global time maximum, the $\omega_{local}(t)$ curve gives the higher dissociation yield. Comparing Fig. 2A and 3A we see that the large values of the dissociation rate correspond to large internuclear distances. That is, if the resonance condition is satisfied at larger internuclear distances a more efficient dissociation will result. Therefore, by determining the $\omega_{local}(t)$ function we have chosen that particular maximum value of the wave packet density from all the possible maximum values which belongs to a larger internuclear distance. It can be seen that the dissociation rates obtained by the $\omega_{global}(t)$ and $\omega_{local}(t)$ functions show good agreement up to 32–34 fs because in this case the positions of the global and the chosen local maximum density values coincide with each other. After $t_{delay} = 34$ fs, however, with a few exceptions, the $\omega_{local}(t)$ function always provides better results, *i.e.*, higher dissociation probabilities. Although the first strong deviation between the $\omega_{global}(t)$ and $\omega_{local}(t)$ functions occurs at around $t_{delay} = 39$ fs, the impact of this deviation on the dissociation rate has already been felt at the end of the pulses centered at previous delay times. This manifests itself in the dissociation yield from $t_{delay} = 34$ fs onwards.

Let us now discuss the results obtained with the $\omega_{\langle R\rangle}(t)$ frequency function which has been determined by following the time-dependent average of the internuclear distance. Although only marginally, the dissociation probability obtained by the $\omega_{\langle R\rangle}(t)$ frequency function is the largest one up to three quarters of the first period of the oscillation of the field-free nuclear wave packet ($t_{delay} = 18$ fs). Accordingly, in this interval – when the wave packet is relatively well localized – it is more beneficial to adjust the $\omega(t)$ function to the time-dependent average of the internuclear distance $\langle R\rangle$ than to the maximum of the nuclear wave packet density. For $t_{delay} > 18$ fs however, this dissociation yield (with $\omega_{\langle R\rangle}(t)$) is basically always the lowest and by far at large time delays. The reason for this behavior can be found in Fig. 2A. It is clearly seen that for $t_{delay} > 18$ fs the $\omega_{\langle R\rangle}(t)$ function avoids the internuclear distances which correspond to the largest global or any local nuclear density values.

The dissociation probabilities for the pulses with 1×10^{13} W cm^{-2} peak intensity are shown in Fig. 3B. The situation is roughly similar to that discussed above for the previous case of 1×10^{12} W cm^{-2} peak intensity, but there are differences as well. It is eye-catching that while the intensity has increased by a factor of 10, the maximum yield of the dissociation increased only about 5-fold. This implies that the dissociation probability has begun to saturate at this intensity. This is also related to the observation that while at the lower intensity $\omega_{local}(t)$ provides the highest dissociation yields between the $t_{delay} = 35$–50 fs, now at higher intensity, the yield curves obtained with $\omega_{local}(t)$ and $\omega_{global}(t)$ are in good agreement up to $t_{delay} = 40$ fs. This saturation also affects several other features of the curves. For example, while at lower intensity in the $t_{delay} = 40$–50 fs interval the $\omega_{local}(t)$ dissociation curve goes above the $\omega_{global}(t)$ one, the opposite is true at the higher intensity. This is in spite of the finding that the local nuclear density maxima are located at large internuclear distances and the dissociation process is more efficient here. Due to saturation, the large nuclear density present at small internuclear distances together with the 10-fold higher intensity altogether produces more dissociation. Saturation is also responsible for finding that the differences between the dissociation yields obtained with various frequency functions are slightly smaller than those found at low intensity. Altogether, however, the use of $\omega_{local}(t)$ clearly provides the best choice for obtaining the largest dissociation yield.

In our previous studies[18,19] the performance of TL and linearly chirped laser pulses in obtaining the dissociation probability has been investigated. There, the pulses used have fulfilled several conditions. We stress again that in the present work the time-dependence of the frequency functions of the pulses is designed in a completely different way following a different objective. As a consequence, the conditions put forward in ref. 18 and 19 do not hold anymore. Due to the different choice of the chirped pulses used in this work, a considerable growth of the total dissociation yield is obtained compared to the former results (see Fig. 3B in ref. 19).

2. Varying transform limited pulses. In the varying transform limited (vTL) calculations, the frequency of the laser pulse was kept constant in every individual calculation at each time delay, corresponding to the ω value given by the $\omega_{\mathrm{local}}(t)$ function at the respective time delay. The results for the dissociation rates obtained by the vTL and $\omega_{\mathrm{local}}(t)$ pulses as well as by the TL reference pulse are displayed in Fig. 4.

Inspecting the curves, we see that across the whole time interval the dissociation probabilities calculated by vTL pulses are smaller than those obtained by $\omega_{\mathrm{local}}(t)$, except for the tiny $t_{\mathrm{delay}} = 8$–12 fs interval. In this time period the LICI corresponding to the constant frequencies of the vTL pulses is located at large internuclear distances during the pulse duration providing a more efficient dissociation, while in the case of the chirped pulse this applies only for the center of the pulse. However, because the nuclear wave packet is well localized in space during this time and its average momentum is close to 0, there is no significant difference in the nuclear densities corresponding to the two different positions of the LICI. This results in a higher dissociation yield in this short time interval.

Up to $t_{\mathrm{delay}} = 18$ fs the TL and vTL pulses provide very similar results and this applies at many other time delays too. Such time periods are *e.g.* $t_{\mathrm{delay}} = 28$–30 fs and $t_{\mathrm{delay}} = 50$–55 fs, at which the dissociation yield shows a growing trend. This

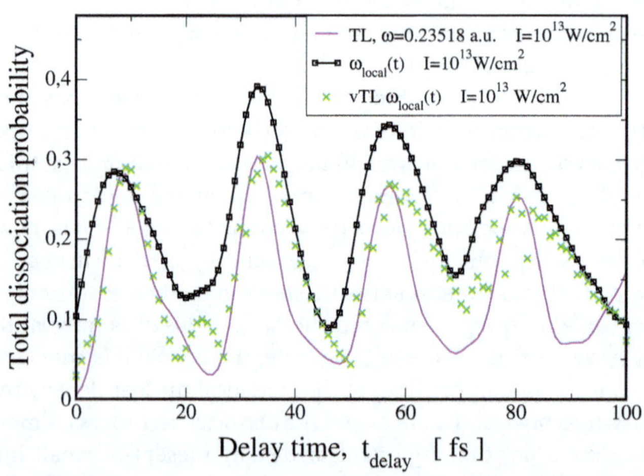

Fig. 4 Dissociation probabilities of D_2^+ as a function of the delay time (t_{delay}) of the pulse. The black (marked with square), the green crosses and the magenta (solid) curves show the dissociation probabilities corresponding to the $\omega_{\mathrm{local}}(t)$, vTL $\omega_{\mathrm{local}}(t)$ and the TL $\omega = 0.23518$ a.u. frequencies, respectively. In the calculations the pulse length is $T_p = 6$ fs and the intensity is 1×10^{13} W cm^{-2}.

Table 1 The changes of population of the individual vibrational levels during the light matter interaction for three different simulations. In all three cases the intensity of the field was 1×10^{13} W cm^{-2} and the frequency of the electric field was applied according to the $\omega_{local}(t)$ function. In the first three columns the vibrational levels (ν), the corresponding energies (E_ν) and the initial populations (P_ν^{FC}) of the vibrational Eigen states are displayed. The next three main columns each containing two sub-columns contain the absolute and relative changes of the populations for the chosen $t_{delay} = 8$, 30 and 59 fs. In the bottom of the table the last two rows contain the sum of the absolute changes (the dissociation probability) and the energetic position of the LICI at time $t = t_{delay}$

			$t_{delay} = 8$ fs		$t_{delay} = 30$ fs		$t_{delay} = 59$ fs	
ν	E_ν	P_ν^{FC}	ΔP_ν	$\Delta P_\nu/P_\nu^{FC}$	ΔP_ν	$\Delta P_\nu/P_\nu^{FC}$	ΔP_ν	$\Delta P_\nu/P_\nu^{FC}$
0	−2.69	0.03655	−0.00009	−0.23%	−0.00485	−13.27%	0.00252	6.90%
1	−2.49	0.09619	−0.00489	−5.08%	−0.02095	−21.78%	0.00170	1.77%
2	−2.30	0.14372	−0.02038	−14.18%	−0.04211	−29.30%	−0.03123	−21.73%
3	−2.11	0.16054	−0.04411	−27.48%	−0.05510	−34.32%	−0.06932	−43.18%
4	−1.94	0.14976	−0.06280	−41.94%	−0.05652	−37.74%	−0.07537	−50.33%
5	−1.77	0.12374	−0.06475	−52.32%	−0.05176	−41.83%	−0.05902	−47.69%
6	−1.61	0.09385	−0.05115	−54.50%	−0.04408	−46.97%	−0.04995	−53.23%
7	−1.46	0.06693	−0.03216	−48.05%	−0.03289	−49.14%	−0.04012	−59.94%
8	−1.31	0.04564	−0.01569	−34.37%	−0.01633	−35.78%	−0.01018	−22.30%
9	−1.17	0.03013	−0.00512	−16.98%	−0.00154	−5.12%	0.01627	54.01%
10	−1.04	0.01945	0.00066	3.37%	0.00350	18.00%	0.00371	19.10%
11	−0.92	0.01237	0.00274	22.17%	0.00071	5.76%	−0.00420	−33.99%
12	−0.80	0.00779	0.00309	39.71%	−0.00055	−7.12%	0.00352	45.20%
13	−0.70	0.00489	0.00274	56.00%	−0.00050	−10.31%	−0.00033	−6.85%
14	−0.60	0.00307	0.00209	68.17%	−0.00027	−8.64%	−0.00081	−26.24%
15	−0.50	0.00193	0.00156	80.74%	0.00012	6.33%	0.00057	29.37%
16	−0.42	0.00123	0.00111	90.11%	−0.00017	−13.86%	0.00045	36.61%
17	−0.34	0.00079	0.00077	98.13%	0.00003	4.36%	0.00019	24.25%
18	−0.28	0.00051	0.00053	103.34%	0.00002	4.32%	0.00005	10.12%
19	−0.21	0.00033	0.00036	106.16%	−0.00009	−26.25%	−0.00007	−22.66%
20	−0.16	0.00022	0.00024	109.22%	0.00005	23.44%	−0.00001	−4.16%
21	−0.11	0.00015	0.00016	108.50%	0.00003	17.23%	0.00005	33.79%
22	−0.08	0.00010	0.00011	106.53%	−0.00003	−28.77%	−0.00002	−21.05%
23	−0.05	0.00006	0.00007	104.77%	−0.00002	−33.46%	0.00001	10.44%
24	−0.02	0.00004	0.00004	100.79%	0.00000	−7.41%	0.00002	59.29%
25	−0.01	0.00003	0.00003	97.74%	0.00000	14.39%	0.00000	13.71%
$\sum_\nu \Delta P_\nu$			−0.28483		−0.32329		−0.31155	
E_{CI} $(t = t_{delay})$ [eV]			−2.26379		−2.48587		−2.18610	

is surprising, because the vTL and TL pulses work on different frequencies. Mostly, the dissociation yield obtained by the TL pulses exhibits a smooth behavior. In particular beyond $t_{delay} = 50$ fs, the vTL results do not lie on a smooth curve. This is related to the fact that several islands of large density regions of the nuclear wave packet exist. Since the vTL dissociation yield is calculated by varying the time delay by 1 fs steps, the frequency at two consecutive calculations can refer to two different islands and thus be rather different leading to the non-smooth behavior seen in Fig. 4.

As time proceeds, the peaks in the dissociation yield as a function of time delay become broader in the vTL calculations than in the reference TL curve. This implies that it is beneficial to use a varying frequency in order to increase the total dissociation probability. Although the pulse duration is rather short ($T_p = 6$ fs), it is even more beneficial to use the $\omega_{local}(t)$ frequency function where the frequency changes during the pulse. As seen in Fig. 4, the gain is then substantial.

B. Populations of the individual vibrational levels after the pulse is over

We now examine the effect of photodissociation dynamics initiated by the $\omega_{global}(t)$ chirped laser pulses for the population of the different individual vibrational levels of the residual D_2^+ ion which has not undergone dissociation as the pulse terminated. Comparison has been made between the initial FC population of the D_2^+ molecule and the population of the different vibrational levels after the dissociation has taken place. The results are collected in Table 1.

At the bottom of the table the actual energy positions of the LICI corresponding to the relevant delay times are indicated. At $t_{delay} = 8$ fs and $t_{delay} = 59$ fs the LICI lies energetically between the $\nu = 2$ and $\nu = 3$ vibrational levels, while at $t_{delay} = 30$ fs it lies very close to the $\nu = 1$ level. It is striking that in the close vicinity of the LICI the population drastically decreases and the depletion is continuously held up to the $\nu = 8$ vibrational level. This is the interval from where the ion can easily dissociate. Then, above the $\nu = 8$ vibrational level a sudden change occurs resulting in a different behavior of the populations for the studied delay times. While for $t_{delay} = 8$ fs the population is continuously increasing above the $\nu = 9$ vibrational level, for the other two delay times an alternating increase or decrease of the populations occur. Because of the strong nonadiabatic effects present it is difficult to explain this interesting reshuffling of the populations. Probably, the presence of the nonadiabatic effects can, in addition to the direct dissociation process, induce vibrational excitations resulting in a newly mixed population distribution among the various vibrational levels. This process can be dominant above $\nu = 8$ level because in this region practically no dissociation occurs. The populations of the vibrational levels should be amenable to measurements.

V. Conclusions

Applying the light-induced conical intersection framework, numerical simulations have been performed for studying the photodissociation dynamics of the D_2^+ molecular ion in order to understand better the relation between the evolving nuclear vibrational wave packet and the frequency chirping of the laser pulse. We present a strategy to directly follow the time evolution of the nuclear wave packet meanwhile forcing it to move together with the LICI as much as possible. Based on this idea three different types of chirped frequencies were designed. The effects of the laser pulses constructed by using these $\omega_{global}(t)$, $\omega_{local}(t)$ and $\omega_{\langle R \rangle}(t)$ functions have been explored in terms of the total dissociation rate as a function of the time delay between the ionization of D_2 and the pulse, as well as in terms of the population of the different vibrational levels after the pulse has terminated. We have found that of all the ω functions applied so far, the $\omega_{local}(t)$ provides the overall highest yield for the dissociation rate, which is probably close to the largest possible outcome or dissociation limit for the given pulse length and intensity values.

It was found that employing a frequency which adapted to the time delay in question was very beneficial for the dissociation rate. Moreover, although the pulses were rather short ($T_p = 6$ fs), it was clearly demonstrated that varying the frequency during the pulse, *i.e.*, using a chirped pulse, gave rise to a further large gain in the dissociation rate. The reshuffling of the populations of the individual vibrational levels of the residual pulse after the pulse has terminated has been found to depend markedly on the time delay employed. These populations contain much information on nonadiabatic effects due to the LICI and should be amenable to experiments.

In our work the motion of the LICI was adjusted to the time evolution of the nuclear field-free wave packet. As we applied only relatively short laser pulses ($T_p = 6$ fs) it seems meaningful as a first approximation to use the field-free wave packet because the pulses have not caused any dramatic effect on the structure of the wave packet. The presented results are very encouraging and call for including the effect of the pulse on the nuclear wave packet in determining a more refined chirp function useful in particular for longer pulses and also for high intensity pulses. This will include new studies using laser pulses in which the time dependence is designed to follow the dynamical evolution of the realistic wave packet. Namely, by means of the feedback-optimized frequency chirped laser pulses one could try to guide the time evolution of a molecular system together with the LICI towards a desired target state. In any case, the LICI opens the door to investigate light-induced nonadiabatic phenomena.

Acknowledgements

The authors acknowledge the financial support by the Deutsche Forschungsgemeinschaft (Project ID CE10/50-3). The ELI-ALPS project (GOP-1.1.1-12/B-2012-000, GINOP-2.3.6-15-2015-00001) is supported by the European Union and co-financed by the European Regional Development Fund. For this work, the supercomputing service of NIIF has been used. The authors thank H.-D. Meyer for very helpful discussions about the MCTDH calculations.

References

1 G. A. Worth and L. S. Cederbaum, *Annu. Rev. Phys. Chem.*, 2004, **55**, 127.
2 M. Baer, *Beyond Born Oppenheimer: Electronic Non-Adiabatic Coupling Terms and Conical Intersections*, Wiley, New York, 2006.
3 N. Moiseyev, M. Sindelka and L. S. Cederbaum, *J. Phys. B: At., Mol. Opt. Phys.*, 2008, **41**, 221001.
4 M. Sindelka, N. Moiseyev and L. S. Cederbaum, *J. Phys. B: At., Mol. Opt. Phys.*, 2011, **44**, 045603.
5 G. J. Halász, Á. Vibók, M. Sindelka, N. Moiseyev and L. S. Cederbaum, *J. Phys. B: At., Mol. Opt. Phys.*, 2011, **44**, 175102.
6 G. J. Halász, M. Sindelka, N. Moiseyev, L. S. Cederbaum and Á. Vibók, *J. Phys. Chem. A*, 2012, **116**, 2636.
7 G. J. Halász, Á. Vibók, M. Sindelka, L. S. Cederbaum and N. Moiseyev, *Chem. Phys.*, 2012, **399**, 146.
8 G. J. Halász, Á. Vibók, N. Moiseyev and L. S. Cederbaum, *Phys. Rev. A*, 2013, **88**, 043413.
9 A. Natan, M. R. Ware and P. H. Bucksbaum, *Book of Ultrafast Phenomena XIX Springer Proceedings in Physics Volume*, 2015, vol. 162, p. 122.

10 G. J. Halász, Á. Vibók and L. S. Cederbaum, *J. Phys. Chem. Lett.*, 2015, **6**, 348.
11 L. S. Cederbaum, Y. C. Chiang, P. V. Demekhin and N. Moiseyev, *Phys. Rev. Lett.*, 2011, **106**, 123001.
12 P. V. Demekhin and L. S. Cederbaum, *J. Chem. Phys.*, 2013, **46**, 164008.
13 J. Kim, H. Tao, J. L. White, V. S. Petrovi, T. J. Martinez and P. H. Bucksbaum, *J. Phys. Chem. A*, 2012, **116**, 2758.
14 M. E. Corrales, J. González-Vázquez, G. Balerdi, I. R. Solá, R. de Nalda and L. Bañares, *Nat. Chem.*, 2014, **6**, 785.
15 I. R. Solá, J. González-Vázquez, R. de Nalda and L. Bañares, Strong field laser control of photochemistry, *Phys. Chem. Chem. Phys.*, 2015, **17**, 13183.
16 A. Stolow, *Nat. Chem.*, 2014, **6**, 759.
17 J. Kim, H. Tao, T. J. Martinez and P. H. Bucksbaum, *J. Phys. B: At., Mol. Opt. Phys.*, 2015, **48**, 164003.
18 A. Csehi, G. J. Halász, L. S. Cederbaum and Á. Vibók, *J. Chem. Phys.*, 2015, **143**, 014305.
19 A. Csehi, G. J. Halász, L. S. Cederbaum and Á. Vibók, *J. Chem. Phys.*, 2016, **144**, 074309.
20 J. Cao, C. J. Bardeen and K. R. Wilson, *Phys. Rev. Lett.*, 1998, **80**, 1406.
21 J. Cao, C. J. Bardeen and K. R. Wilson, *J. Chem. Phys.*, 2000, **113**, 1898.
22 V. S. Malinovsky and J. L. Krause, *Eur. Phys. J. D*, 2001, **14**, 147.
23 A. Datta, S. S. Bhattacharyya and B. Kim, *Phys. Rev. A*, 2002, **65**, 043404.
24 E. Heesel, B. M. Garraway and J. P. Marangos, *J. Chem. Phys.*, 2006, **124**, 024320.
25 A. Kirrander, H. H. Fielding and C. Jungen, *J. Chem. Phys.*, 2010, **132**, 024313.
26 A. Kirrander, C. Jungen and H. H. Fielding, *J. Phys. B: At., Mol. Opt. Phys.*, 2008, **41**, 074022.
27 A. Kirrander and H. H. Fielding, *J. Phys. B: At., Mol. Opt. Phys.*, 2007, **40**, 897.
28 G. Katz, M. A. Ratner and R. Kosloff, *New J. Phys.*, 2010, **12**, 015003.
29 J. Vala and R. Kosloff, *Opt. Express*, 2001, **8**, 238.
30 L. Rybak, Z. Amitay, S. Amaran, R. Kosloff, M. Tomza, R. Moszynski and C. P. Koch, *Faraday Discuss.*, 2011, **153**, 383.
31 L. Rybak, S. Amaran, L. Levin, M. Tomza, R. Moszynski, Z. R. Kosloff, C. P. Koch and Z. Amitay, *Phys. Rev. Lett.*, 2011, **107**, 273001.
32 L. Levin, W. Skomorowski, L. Rybak, R. Kosloff, C. P. Koch and Z. Amitay, *Phys. Rev. Lett.*, 2015, **114**, 233003.
33 V. S. Prabhudesai, U. Lev, A. Natan, B. D. Bruner, A. Diner, O. Heber, D. Strasser, D. Schwalm, I. Ben-Itzhak, J. J. Hua, B. D. Esry, Y. Silberberg and D. Zajfman, *Phys. Rev. A*, 2010, **81**, 023401.
34 A. Natan, U. Lev, V. S. Prabhudesai, B. D. Bruner, D. Strasser, D. Schwalm, I. Ben-Itzhak, O. Heber, D. Zajfman and Y. Silberberg, *Phys. Rev. A*, 2012, **86**, 043418.
35 V. S. Prabhudesai, A. Natan, B. D. Bruner, Y. Silberberg, U. Lev, O. Heber, D. Strasser, D. Schwalm, D. Zajfman and I. Ben-Itzhak, *J. Korean Phys. Soc.*, 2011, **59**, 2890.
36 B. Y. Chang, S. Shin, J. Santamaria and I. R. Sola, *J. Phys. Chem. A*, 2012, **116**, 2691.
37 S. Askeland and M. Førre, *Phys. Rev. A*, 2013, **88**, 043411.
38 C. P. Zhang and X. Y. Miao, *Spectrosc. Lett.*, 2014, **47**, 267.

39 U. Lev, L. Graham, C. B. Madsen, I. Ben-Itzhak, B. D. Bruner, B. D. Esry, H. Frosting, O. Heber, A. Natan and V. S. Prabhudesai, *J. Phys. B: At., Mol. Opt. Phys.*, 2015, **48**, 201001.
40 D. V. Novitsky, *Opt. Commun.*, 2016, **358**, 202.
41 C. C. Shu and N. E. Henriksen, *J. Chem. Phys.*, 2011, **134**, 164308.
42 C. C. Shu and N. E. Henriksen, *J. Chem. Phys.*, 2012, **136**, 044303.
43 A. K. Tiwari, D. Dey and N. E. Henriksen, *Phys. Rev. A*, 2014, **89**, 023417.
44 S. I. Chu, *J. Chem. Phys.*, 1981, **75**, 2215.
45 S. I. Chu, C. Laughlin and K. Datta, *Chem. Phys. Lett.*, 1983, **98**, 476.
46 F. V. Bunkin and I. I. Tugov, *Phys. Rev. A*, 1973, **8**, 601.
47 H. D. Meyer, U. Manthe and L. S. Cederbaum, *Chem. Phys. Lett.*, 1990, **165**, 73.
48 U. Manthe, H. D. Meyer and L. S. Cederbaum, *J. Chem. Phys.*, 1992, **97**, 3199.
49 M. H. Beck, A. Jäckle, G. A. Worth and H. D. Meyer, *Phys. Rep.*, 2000, **324**, 1.
50 G. A. Worth, *et al.*, *The MCTDH Package, Version 8.2*, University of Heidelberg, Heidelberg, Germany, 2000; H. D. Meyer, *et al.*, *The MCTDH Package, Versions 8.3 and 8.4*, University of Heidelberg, Germany, 2002 and 2007, http://mctdh.uni-hd.de/.
51 Multidimensional Quantum Dynamics, *MCTDH Theory and Applications*, ed. H. D. Meyer, F. Gatti and G. A. Worth, Wiley-VCH, Weinheim, 2009.
52 R. Schinke, *Photodissociation Dynamics*, Cambridge University Press, Cambridge, 1993.
53 G. G. Balint-Kurti, R. N. Dixon and C. C. Marston, *J. Chem. Soc., Faraday Trans.*, 1990, **86**, 1741.
54 O. Vendrell, F. Gatti and H.-D. Meyer, *J. Chem. Phys.*, 2007, **127**, 184303.
55 D. B. Milosevic, G. G. Paulus, D. Bauer and W. Becker, *J. Phys. B: At., Mol. Opt. Phys.*, 2006, **39**, R203.
56 For TL cases the ratio of the integrals relative to the Gaussian envelope are below 10^{-8}, while for the other frequency functions ($\omega_{\rm local}(t)$ and $\omega_{\rm global}(t)$) are below 1.5%.

Faraday Discussions

PAPER

Visualization of bond rearrangements in acetylene using near single-cycle laser pulses

Christian Burger,*[ab] Nora G. Kling,[a] Robert Siemering,[c] Ali S. Alnaser,[bd] Boris Bergues,[b] Abdallah M. Azzeer,[e] Robert Moshammer,[f] Regina de Vivie-Riedle,[c] Matthias Kübel[a] and Matthias F. Kling[ab]

Received 15th April 2016, Accepted 10th May 2016

DOI: 10.1039/c6fd00082g

The migration of hydrogen atoms resulting in the isomerization of hydrocarbons is an important process which can occur on ultrafast timescales. Here, we visualize the light-induced hydrogen migration of acetylene to vinylidene in an ionic state using two synchronized 4 fs intense laser pulses. The first pulse induces hydrogen migration, and the second is used for monitoring transient structural changes *via* Coulomb explosion imaging. Varying the time delay between the pulses reveals the migration dynamics with a time constant of 54 ± 4 fs as observed in the $H^+ + H^+ + CC^+$ channel. Due to the high temporal resolution, vibrational wave-packet motions along the CC- and CH-bonds are observed. Even though a maximum in isomerization yield for kinetic energy releases above 16 eV is measured, we find no indication for a backwards isomerization – in contrast to previous measurements. Here, we propose an alternative explanation for the maximum in isomerization yield, namely the surpassing of the transition state to the vinylidene configuration within the excited dication state.

1 Introduction

The photo-excitation of hydrocarbons and the reactions of their nuclei have been studied extensively in order to reveal the underlying physical processes, such as vibrations,[1,2] rotations,[3] bond rearrangements[4] and molecular fragmentation.[5] Each process has a corresponding natural timescale, with the fastest nuclear motions occurring in less than 10 fs, and therefore ultrafast laser pulses are ideally suited for their study. Furthermore, bond rearrangement and molecular

[a]Department of Physics, LMU Munich, D-85748 Garching, Germany. E-mail: christian.burger@mpq.mpg.de
[b]Max Planck Institute of Quantum Optics, D-85748 Garching, Germany
[c]Department of Chemistry and Biochemistry, LMU Munich, D-81377 Munich, Germany
[d]Physics Department, American University of Sharjah, POB2666, Sharjah, UAE
[e]Department of Physics and Astronomy, King-Saud University, Riyadh 11451, Saudi Arabia
[f]Max Planck Institute of Nuclear Physics, D-69117 Heidelberg, Germany

fragmentation of hydrocarbons can be influenced by tuning the laser parameters. For example, Xie *et al.* showed that they can control the fragmentation pathways in ethylene by varying the laser pulse duration,[6] and Alnaser *et al.* showed that they could steer the directional emission of the deprotonation from acetylene with the carrier-envelope phase (CEP) of few-cycle laser pulses.[7] The second example was interpreted with a vibrational state coupling model. Recently, CEP-control has been extended to the isomerization of acetylene and allene, demonstrating the general importance of the vibrational state coupling mechanism.[8]

In fact, the process of hydrogen migration in acetylene has been the subject of numerous past studies, both theoretically[9] and experimentally.[2,10,11] More recent experiments have taken advantage of the Coulomb explosion imaging technique using a reaction microscope (REMI) in order to visualize the temporal evolution of the isomerization process in "real time".[12] Specifically, the isomerization of acetylene into vinylidene has been visualized for the dissociation channel: $C_2H_2^{3+} \rightarrow H^+ + C^+ + CH^+$.[12–15] The isomerization time was consistently found to be around 50 fs, independent from the chosen excitation wavelength and intensity, as well as independent of whether the dynamics occurred on the cation or dication intermediate states.

Here, we show how to exploit the additional channel $C_2H_2^{3+} \rightarrow H^+ + H^+ + CC^+$ to obtain a more complete understanding of the proton migration as suggested in ref. 16. We use ultrashort laser pulses as pump and probe pulses to visualize the migration process. After a short introduction into our experimental and theoretical methods we present our results on the triple coincidence channels, focusing on the comparison between the channels $H^+ + C^+ + CH^+$ and $H^+ + H^+ + CC^+$. Additionally, we determine the isomerization time for the latter channel, and by means of Newton diagrams we present snapshots of the hydrogen migration for various time delays between the pump and probe pulses.

2 Experimental and theoretical methods

2.1 Experimental methods

In order to investigate the hydrogen migration from acetylene to vinylidene, the experimental setup depicted in Fig. 1 was used. The titanium:sapphire system ("FEMTOPOWER™ HR 10 kHz CEP4", Spectra-Physics) produces laser pulses at a 10 kHz repetition rate with a pulse duration of 22 fs, a pulse energy of 800 μJ and a central wavelength of 790 nm. The pulses are spectrally broadened by a 1 m long hollow-core-fiber with a 275 μm inner core diameter, filled with 500 mbar of argon with the final spectrum covering a range from 400 nm to 1000 nm with a new central wavelength of 690 nm. Temporal compression is attained in a double-angle chirped mirror compressor,[17] while the fine tuning to 4 fs full-width half-maximum (FWHM) is done with fused silica wedges. A variable neutral density filter allows us to change the intensity without changing the beam diameter or the dispersion and thus preserves the pulse parameters. The pulse duration is determined by transient-grating frequency-resolved optical gating (TG-FROG).[18] The carrier-envelope-phase (CEP) is recorded by an f–2f interferometer throughout the measurement.[19] The laser is split into two identical pump- and probe-pulses by a 50 : 50 Mach–Zehnder interferometer prior to the entrance window of a REMI (base pressure: 4×10^{-10} mbar). The pulses are back-focused using a 12.5 cm focal length mirror onto a cold acetylene jet. The thickness of the

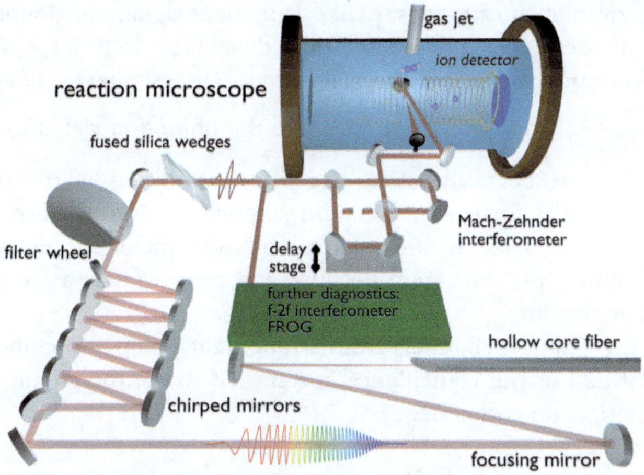

Fig. 1 Experimental setup including all major optical components after the amplifier and a schematic drawing of the experimental chamber.

jet can be varied using automated skimmers to maintain a constant overall count rate. Each pulse has an intensity around 5×10^{14} W cm^{-2}. The time delay between the pulses is continuously swept from −267 fs to 267 fs in steps of 33 as. One period takes around 5 minutes, while the phase of both pulses changes by 2π within 7 s. This difference in timing ensures that no artificial CEP dependencies on the delay measurement are introduced, which is further verified by a flat CEP distribution for the complete delay range.

The data acquisition for the results shown here was carried out over three days and consists of four individual measurements, in between which the pulse duration and spatial overlap were verified to ensure similar parameters throughout the entire duration of the experiment.

In Table 1, the accumulated amount of events is shown for each run and each coincidence channel. The statistics indicate that the $H^+ + C^+ + CH^+$ channel has around 80% more counts than the $H^+ + H^+ + CC^+$ channel. The total statistics are about 5 times higher than previously reported,[14] attributed to the high repetition rate laser and the long-term stability of the overall system. Such high statistics, combined with the high temporal resolution, allows us to investigate the vibrational motion within the intermediate state in addition to the isomerization process.

To combine the CEP-dependent data from each measurement, the relative phase for each scan is shifted to match an independent reference data set. This

Table 1 Presented data are combined from the following measurements

	Coincidences per channel			
Measurement run	1	2	3	4
$H^+ + C^+ + CH^+$	245 552	89 274	32 796	146 586
$H^+ + H^+ + CC^+$	129 131	55 396	18 770	80 105
$H^+ + H^+ + C^+ + C^+$	53 695	8156	5944	44 955

reference is given by the strongest phase dependent signal, the double ionization asymmetry of acetylene, at a specific time delay (100 ± 10 fs) — far from the temporal overlap to avoid interference effects. The asymmetry is calculated by $A = \frac{N_{\text{left}} - N_{\text{right}}}{N_{\text{left}} + N_{\text{right}}}$, where N_{left} and N_{right} are the photo-ion yields to the left and right directions, respectively, along the polarization axis in the time-of-flight (TOF) spectrum. For further information see ref. 20. Furthermore, we checked that the measured data are independent of the sign of the delay. With this condition fulfilled, the data from negative and positive delays are combined to achieve better statistics.

Various dissociation channels from strong field pump–probe measurements can be identified in the coincidence spectra. Of these, the channels involving a clear isomerization component are:

$$C_2H_2^{2+} \rightarrow C^+ + CH_2^+ \quad \text{(i)}$$

$$C_2H_2^{2+} \rightarrow C_2^+ + H_2^+ \quad \text{(ii)}$$

$$C_2H_2^{3+} \rightarrow H_2^+ + C^+ + C^+ \quad \text{(iii)}$$

$$C_2H_2^{3+} \rightarrow H^+ + C^+ + CH^+ \quad \text{(iv)}$$

$$C_2H_2^{3+} \rightarrow H^+ + H^+ + CC^+ \quad \text{(v)}$$

$$C_2H_2^{4+} \rightarrow H^+ + H^+ + C^+ + C^+ \quad \text{(vi)}$$

Channel (i) permits us to measure the fragmentation yield of vinylidene and therefore to determine the average isomerization time to 52 ± 15 fs, as investigated by Jiang *et al.*[21] Channels (ii) and (iii) are relatively weak, most likely because they require both hydrogen migration and bond formation between the hydrogens. Despite being interesting channels, the low statistics prevented us from being able to determine much about them and therefore, they are left out of the main discussion here. Channels (iv) and (v) have excellent statistics, and while channel (iv) has been extensively studied (see ref. 12–15), the investigation of channel (v) has been proposed[16] but reports have been lacking so far. Here, we compare both channels and highlight the advantages of channel H^+ + H^+ + CC^+. For channel (vi), preliminary analysis also shows signs of isomerization. However, due to the additional complexity of the quadruply charged parent ion, and the data analysis of the four-body coincidence, it will not be discussed in detail here. Instead, our main focus is on channels (iv) and (v). These two channels enable us to compare our measurements with previous experiments on the H^+ + C^+ + CH^+ channel and gain further information by analyzing the H^+ + H^+ + CC^+ channel. However, we would also like to point out that a considerable advantage of channel (v) lies in the large difference in mass of the C_2^+ and H^+ fragments, avoiding any overlap of their TOF's. This is especially important in contrast to the H^+ + C^+ + CH^+ channel, where the C^+ and CH^+ fragment have overlapping TOF's, which make the molecular reconstruction more difficult.

2.2 Theoretical methods

We used a Complete Active Space Self-Consistent Field model (CASSCF) to calculate the potential energy surfaces for the ground states of the neutral, cation, dication, and trication states, as well as for different excited states. The active space provided for the electron correlation was adapted to the number of the valence electrons and consists of the same 10 molecular orbitals and all electrons except the 1s-core electrons of the carbon atoms. The calculations were performed using the program package MOLPRO with a 6-311++G** basis set.[22] Using this model we utilize the precise description of excited states to calculate the molecular configurations. These results enable us to understand the pathway of the migrating hydrogen atom within the dication state.

3 Results and discussion

In Fig. 2, the kinetic energy release (KER) of the three-fold coincidence channels is shown as a function of the time delay between the pump and probe pulses. The color scale in (a) and (b) indicates the yield, which is normalized to one for each time step. The integrated counts for each time step without normalization are plotted below in (c) and (d). Clearly visible in Fig. 2(a) and (b) is the strong signal at 16–17 eV that is largely delay independent and the contributions at lower KER values which are time dependent. Both areas can be separated at a KER of 13 eV, indicated by the black dashed line, which is later used for energy filtering. The low KER features are the result of bond stretching induced by the pump pulse, as seen in many other pump–probe studies on dissociating molecules.[23–25]

While the $H^+ + C^+ + CH^+$ channel is sensitive to both the CC- and CH-bonds, the $H^+ + H^+ + CC^+$ channel exclusively probes the CH-bonds. The slower temporal evolution of the low KER contribution in the $H^+ + C^+ + CH^+$ channel indicates a slower bond stretching of the CC-bond in comparison to the CH-bond, as expected. Furthermore, the data shown in Fig. 2 contains two independent indications for a wave-packet motion in the intermediate state. First, we present in Fig. 2(c) and (d) the yield as a function of the pulse delay which shows

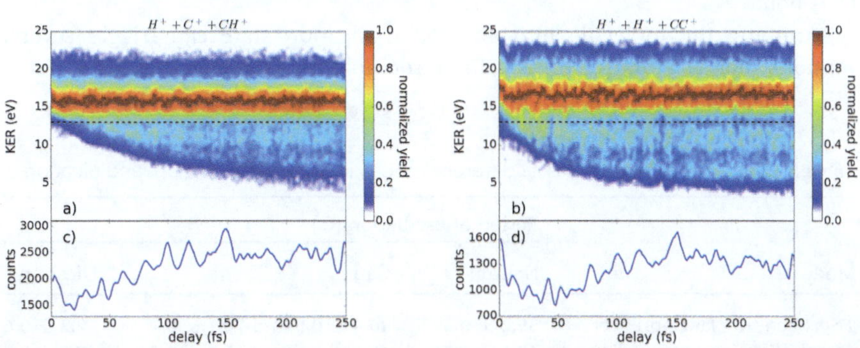

Fig. 2 The yield of (a) $H^+ + C^+ + CH^+$ and (b) $H^+ + H^+ + CC^+$ as a function of the kinetic energy release (KER) and pulse delay. The yield for each time bin is normalized to its peak, revealing a small oscillation of the KER with the delay. The black dashed line indicates the separation of the regions investigated in Fig. 4. The total yield corresponding to (a) and (b) is plotted in (c) and (d), respectively.

a fine-structured oscillation. Secondly, we observe a periodical oscillation of the maximum position of the KER distribution around 17 eV which becomes observable only by using ultrashort pulses.[26,27] Applying a Fourier analysis to the oscillations of the highest KER values, as well as to the oscillations of the total yield dependent of the pulse delay, reveals three major contributions: 11.5 fs $\pm$ 1.7 fs, 20.4 fs $\pm$ 1.5 fs, and 33.4 fs $\pm$ 1.2 fs. To attribute these experimentally determined periods to the corresponding wave-packet motion we calculated the periods of various modes within the neutral, cation, and dication, presented in Table 2.

Comparing our experimental results to the calculated modes in Table 2 we identify the 11.5 fs $\pm$ 1.7 fs oscillation with a CH-bond stretching mode and the 20.4 fs $\pm$ 1.5 fs vibration with a CC-bond stretching mode within the dication. Thus we can directly observe the CH- and CC-vibrations in acetylene *via* strong-field pump–probe measurements of the fragment ions. The slowest experimentally observed oscillation of 33.4 fs $\pm$ 1.2 fs is not identified by our theoretical calculations as the various bending modes are slower than 40 fs, while the stretching modes are too fast to explain our results. Even though we cannot link the observed oscillation to one distinct mode, it is likely that we observe a combination of the stretching modes resulting in slower oscillations. To investigate the populated state further, we present in Fig. 3 the different potential energy curves of acetylene.

Selected potential energy curves along the CH-bond internuclear distance are depicted in Fig. 3 along with a sketch of the various configurations of acetylene in the intermediate state. First, the pump pulse is ionizing the molecule, bringing it into an intermediate state, either the cation or dication state, where hydrogen migration can take place. The probe pulse ionizes the molecule further until it dissociates from a triply or even higher charged state. Depending on the excitation wavelength and intensity, different populated intermediate states have been reported so far. While using XUV or UV light for excitation results in a populated cation state,[14,15,28] IR strong field excitation populates the dication state.[12,13] In our experimental setup, a dication state is the most probable to be populated, as we use identical pump and probe pulses (see Fig. 3). Reachable are the ground state $X^3\Sigma_g^-$ and excited states $X^1\Delta_g$, and $A^3\Pi_u$. The latter is purely repulsive along the C–H bond.

Pumping the molecule into an excited dication state can trigger several motions such as vibration, dissociation, rotation and isomerization, as depicted

Table 2 Calculated frequencies of different modes in the neutral, cation, and dication

	Period of oscillation (fs)		
Mode	Neutral	Cation	Dication
Trans-bending (*x*–*z* and *y*–*z*)	59.2 $\pm$ 0.6	40.4 $\pm$ 0.5 and 70.1 $\pm$ 0.7	52.2 $\pm$ 0.6
Cis-bending (*x*–*z* and *y*–*z*)	48.1 $\pm$ 0.5	40.5 $\pm$ 0.5 and 45.4 $\pm$ 0.5	47.0 $\pm$ 0.5
CC-stretching	16.3 $\pm$ 0.5	23.0 $\pm$ 0.5	21.6 $\pm$ 0.5
Antisymmetric CH-stretching	10.0 $\pm$ 0.4	11.2 $\pm$ 0.4	11.4 $\pm$ 0.4
Symmetric CH-stretching	9.7 $\pm$ 0.4	10.9 $\pm$ 0.4	11.0 $\pm$ 0.4

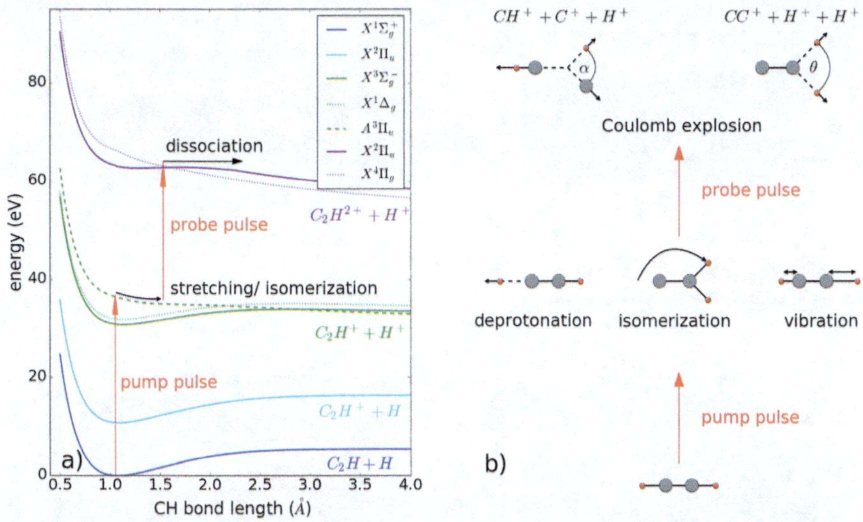

Fig. 3 (a) Potential energy curves for the neutral (blue), cation (cyan), dication (green), and trication (purple) ground states and select excited states as a function of the internuclear distance of one CH-bond. (b) Sketch of the molecular configurations at different stages of the experiment. A pump pulse doubly ionizes the molecule and excites it to a reactive state that permits hydrogen migration. The probe pulse ionizes the molecule further bringing it into a triply charged dissociative state. The angles α and θ are used to distinguish between the acetylene and vinylidene configurations.

schematically in Fig. 3(b). Experimentally, we observe the 3D momenta of each ionic fragment, enabling us to reconstruct the molecular arrangement at the time of Coulomb explosion. Together with the KER, this information allows us to disentangle the various molecular motions.

To describe isomerization, it is useful to define the angles α and θ of the migrating hydrogen (see Fig. 3(b)). In order to compare our results to previous measurements, α is defined as in ref. 12 and 13 and describes the angle between the proton and the carbon ion with respect to the center of gravity of the molecule (see Fig. 3(b)). The angle θ corresponds to the angle in between both protons seen from the C_2^+ ion. Using the C_2^+ ion as a center enables us to directly compare our results with theoretical calculations performed on the dication of acetylene. Both angles are calculated *via*: $\alpha, \theta = \cos^{-1}\left(\frac{\vec{p}_1 \cdot \vec{p}_2}{|\vec{p}_1||\vec{p}_2|}\right)$, where $\vec{p}_1$ and $\vec{p}_2$ are the momenta of the hydrogen and carbon ions for channel (iv) and of both protons for channel (v).

By this definition of angles, we assign α at around 0° and θ at 180° to the acetylene configuration and α above 90° and θ below 90° to vinylidene. In the following part we will extend our analysis to the reconstruction of the molecular configuration — the angles α and θ provide us with the information about the angular position of the migrating proton.

In Fig. 4, the fragmentation yield as a function of the angles α and θ and the pulse delay is shown for both coincidence channels with two different energy filters. The energy filter separates the time independent KER region (above 13 eV)

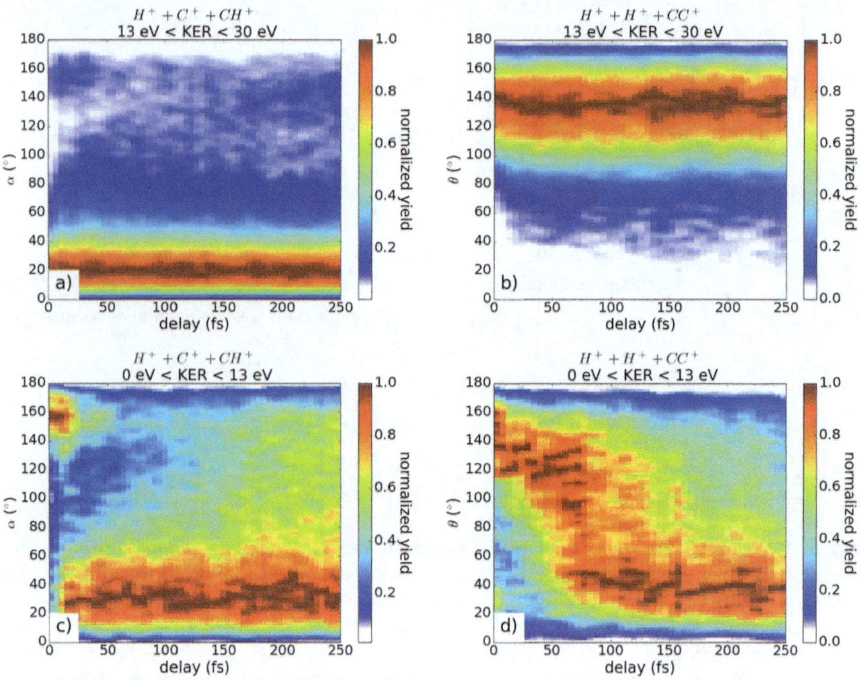

Fig. 4 The fragmentation yield is shown as a function of pump–probe delay and the angle α for the channel $H^+ + C^+ + CH^+$ in panels (a) and (c), and θ for channel $H^+ + H^+ + CC^+$ in panels (b) and (d). The applied KER filters (indicated above each plot) are used to separate the bound (13–30 eV, (a) and (b)) from the dissociating contribution (0–13 eV, (c) and (d)). The contributions in (a) and (c) at large angles and small delays are attributed to an artifact (see text).

from the dissociating contributions at lower KER values (see Fig. 2). For the high KER region in both channels, the angles stay relatively constant with broad angular distributions around 20° and 140°, respectively, indicating that the molecules tend to stay in the acetylene configuration. These results fit very well to the calculated configuration of excited acetylene molecules within the dication, where the hydrogen atoms are bent with respect to the CC axis. Likely, the excitation energy is thereby stored in a wave-packet motion in the acetylene configuration, which results in a widespread momenta distribution around the acetylene configuration, visible here by the broad angular distribution.

For the low KER contributions plotted in Fig. 4(c) and (d), a clear time dependence is observed showing the movement to smaller angles. This indicates the isomerization to the vinylidene configuration. In contrast to the $H^+ + C^+ + CH^+$ channel, the majority of the molecules change their molecular configuration in the $H^+ + H^+ + CC^+$ channel (see Fig. 4(d)). After approximately 100 fs, more than 60% of all molecules in the $H^+ + C^+ + CH^+$ channel have reached the vinylidene configuration. We note that in contrast to ref. 12, the migration back to the acetylene configuration is not observed, but rather, the molecules tend to favor the vinylidene configuration for the remaining duration of the scanned time delay. We note that this discrepancy could, at least in part, be due to the fact that

Hishikawa *et al.* used a background subtraction, which accounted for three-fold coincidences from only one pulse.[12]

For the coincidence channel $H^+ + C^+ + CH^+$, it is visible that we have a considerable amount of molecules in the vinylidene configuration at small time delays. This contribution results from the overlap in the TOF spectrum for the slow C^+ fragments and the fast CH^+ fragments. Inevitably, the identification of the two fragments in the data analysis is convoluted, potentially leading to a wrong assessment of the molecular configuration. For the $H^+ + H^+ + CC^+$ channel, the TOF's of each fragment are clearly separated, and the assessment of the molecular configuration is thus much more reliable.

Using the angular information as a filter, the molecules which ultimately dissociated from the vinylidene configuration are separated from those dissociated from the acetylene configuration. For channel $H^+ + C^+ + CH^+$, all angles α above 90° (see Fig. 3(b)) are referred to as vinylidene while all angles below 90° are attributed to the acetylene configuration. For channel $H^+ + H^+ + CC^+$, the vinylidene configuration is reached for all angles θ below 90° (see Fig. 3(b)) and all angles above 90° correspond to the acetylene configuration. The resulting KER distribution is plotted *versus* the pulse delay in Fig. 5.

In Fig. 5, the yield is plotted as a function of the KER and time delay for molecules in the vinylidene configuration. Compared to Fig. 2, the contribution at 16–17 eV is considerably weaker, while the delay dependent contribution below 16 eV becomes clearly visible. For the investigated pump–probe delays, this indicates that the isomerized molecules in the dication dissociate too fast to allow for a back migration of the proton into an acetylene-like configuration.

Taking a closer look at Fig. 5(a), one can see that the KER extends to higher values at delays between 0 fs and 100 fs, marked by an arrow in the plot. This increase in KER values in channel $H^+ + C^+ + CH^+$, as well as the delay dependence in both KER plots in Fig. 5, can be well explained within the $A^3\Pi_u$ excited dication state, where an isomerization can take place as shown by theoretical calculations, see Fig. 6.[29]

According to our calculations in full dimension, acetylene in the $A^3\Pi_u$ state relaxes into the two local minima t1′ and t1″ with bent configurations having

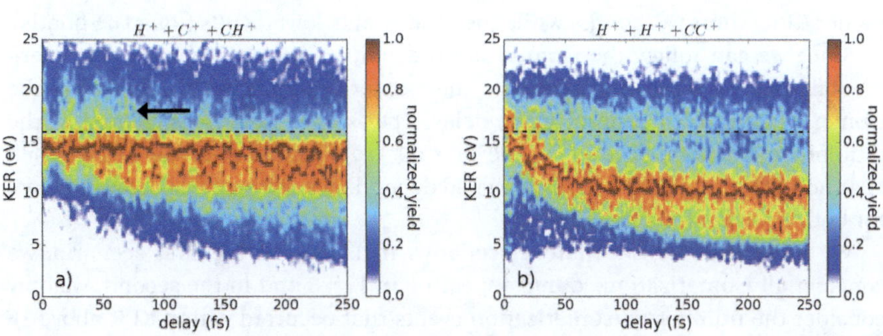

Fig. 5 Kinetic energy release from (a) $H^+ + C^+ + CH^+$ and (b) $H^+ + H^+ + CC^+$ *versus* the pulse delay, fulfilling the condition: $\alpha > 90°$ and $\theta < 90°$, indicating the molecule dissociated from the vinylidene configuration. The dashed black lines indicates the new energy filter used for Fig. 7, while the arrow points to an area of interest used to identify the intermediate state, as explained in the text.

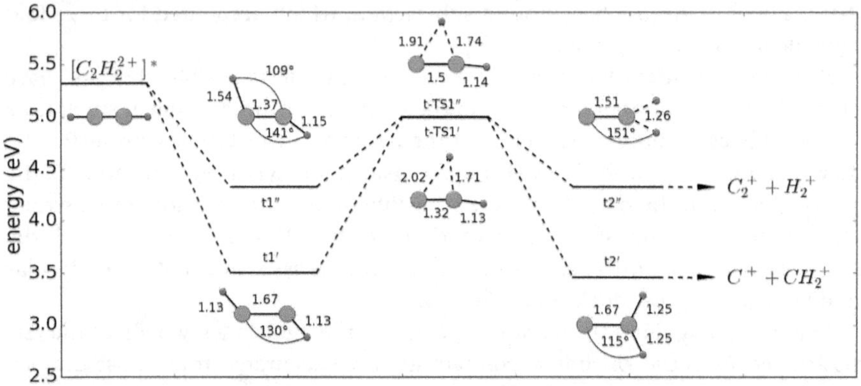

Fig. 6 Sketch of the energies and corresponding configurations of an acetylene molecule in the excited dication state. All configurations were calculated using the CASSCF[8,10] model with a 6-311++G** basis set, where the bond lengths are given in units of 10^{-10} m. The idea for this figure was taken from ref. 29.

angles around 140° between the CH-and CC-bonds. From these local minima, the energetically higher lying transition states t-TS1′ and t-TS1″ have to be overcome. The transition states are close in energy, the angle between both protons is around 90°, and the CH-bond of the migrating proton is elongated while the CC-bond is slightly shortened. In a last step, the molecule finishes its migration into vinylidene from where it can dissociate in either of two channels: $[C_2H_2^{2+}]^* \rightarrow C_2^+ + H_2^+$ or $\rightarrow C^+ + CH_2^+$. The probe pulse in our experiment is used to further ionize and dissociate the H_2^+ or CH_2^+, respectively. These dissociations are very well observed within both cases, visible in the time dependent low KER regions, whereas the intermediate transition states t-TS1′ and t-TS1″ are observed in the following two ways. Within the $H^+ + H^+ + CC^+$ channel, the stretched CH-bond of the migrating proton is probed, resulting in the rapidly decreasing KER within the first 100 fs. In the $H^+ + C^+ + CH^+$ channel, the CC-bond can be probed resulting in a higher KER, seen at around 40 fs (see arrow in Fig. 5). This effect cannot be observed as easily as the CH-bond stretching because within this channel both effects occur simultaneously, washing out the signal as one contribution delivers higher KER (short CC-bonds) while the other yields lower KERs (long CH-bonds).

Since we can follow the isomerization dynamics in our pump–probe experimental scheme, we can also extract the isomerization time by measuring the isomer yield as a function of pulse delay. The isomer yield is calculated as the ratio of vinylidene molecules ($\alpha > 90°$, $\theta < 90°$) to all triple coincidences for each channel. In Fig. 7, the isomerization yield dependence on the pulse delay is shown for both investigated channels.

We investigated two different scenarios in Fig. 7. In the first scenario, we consider all isomerization counts appearing in Fig. 5 and in the second, we only consider the number of isomerization events that occurred with a KER above 16 eV, indicated by the dashed line in Fig. 5. For the first scenario, we see an increase of the isomer yield with pulse delay. In the $H^+ + H^+ + CC^+$ channel an asymptotic behavior is observed for long delays indicating that the final isomerization yield is around 30%. Note that these 30% result from the investigation of all measured

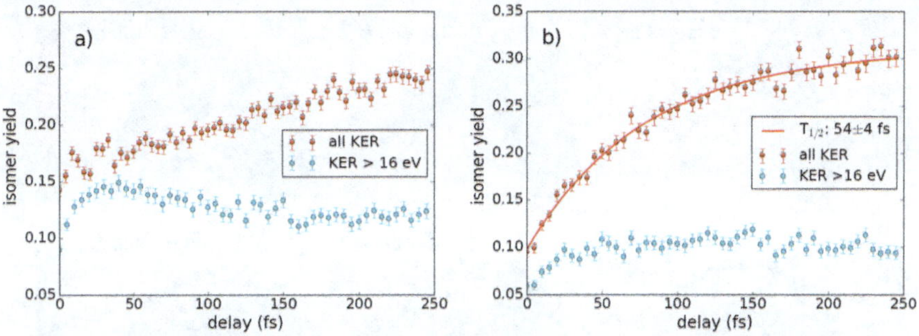

Fig. 7 Isomerization yield of (a) $H^+ + C^+ + CH^+$ and (b) $H^+ + H^+ + CC^+$ as a function of the pulse delay. The overall vinylidene yield divided by the total yield is shown in red, while an additional high-energy filter (>16 eV) is applied to obtain the cyan colored data (error bars denote statistical errors). The solid red line indicates an exponential fit of the experimental data. Note that the isomer yield at zero pulse delay in (a) is overestimated due to the artifact discussed above.

data without applying any filter. Fitting this data with an exponential function allows us to determine the isomerization time to be 54 ± 4 fs, which is consistent with previously reported isomerization times from both measurements and calculations.[14,21,30,31]

The $H^+ + C^+ + CH^+$ channel, in contrast, shows more of a linear trend for the pulse delays considered, which makes a determination of the isomerization time difficult. After 250 fs, the isomerization yield is at 25%, which is considerably lower than for $H^+ + H^+ + CC^+$. As the dynamics within the $H^+ + C^+ + CH^+$ channel seems to occur on longer timescales, it is expected that the isomerization yield would also saturate for longer pulse delays, from which we can only speculate that the isomerization time for this channel is longer than for the $H^+ + H^+ + CC^+$ channel.

For the second scenario, applying a high energy filter (KER > 16 eV), we reproduce previous measurements in the $H^+ + C^+ + CH^+$ channel showing a peak at 40 fs, which was interpreted as an isomerization maximum, where afterwards the proton may migrate back into the acetylene configuration.[14] However, as we do not observe a similar behavior in the second coincidence channel, we propose an alternative explanation for the peak around 40 fs. In the intermediate state t-TS1″ (see Fig. 6), the CC-bond has a reduced equilibrium bond length, which may result in high KER values when dissociation is induced. For higher time delays the bond length increases again and the molecule stays within the vinylidene configuration without backwards migration. In other words, the peak of isomerization yield for the applied high KER filter is not due to a maximal isomerization yield, but rather to a CC vibrational motion. One would maybe think that a similar feature should appear in the $H^+ + H^+ + CC^+$ channel, but this channel is less sensitive to the CC-bond length, as this bond remains intact. Indeed, for the channel $H^+ + H^+ + CC^+$, the isomerization yield fluctuates always around 10% without any other clear trend (see Fig. 7(b)).

Furthermore, we observe in Fig. 7 that the isomer yield at zero delay is always above 5% — independent of the investigated channel. This is likely due to a population of the trication where the isomerization takes place followed by Coulomb explosion. The direct population becomes possible by the high

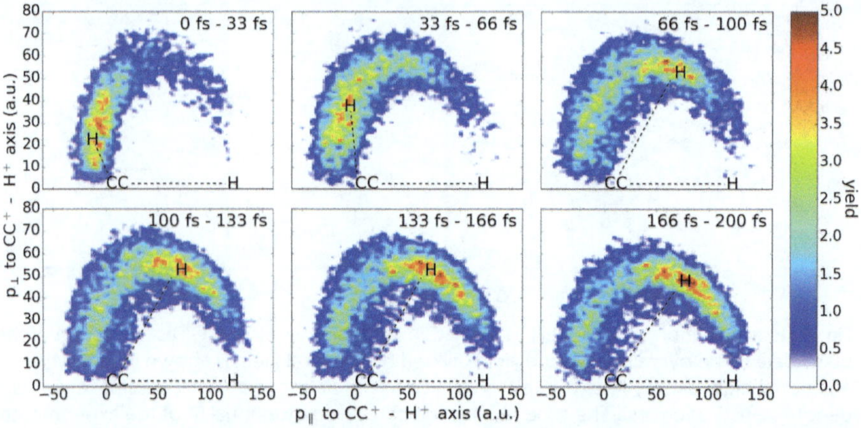

Fig. 8 Newton diagrams of the hydrogen migration for different pulse delays (indicated at the top right in each plot). As after 200 fs the Newton plots do not change anymore, only the first 200 fs are presented here. The position of each atom is schematically shown as a guide to the eye.

intensities in the temporal overlap of both pulses. For higher pulse delays a direct ionization into the trication becomes unlikely as the intensity of one pulse alone is too low. The even higher isomer yields for Fig. 7(a) result from the artifact discussed above (see Fig. 4).

Finally, to visualize the hydrogen migration process, we plot a series of Newton diagrams for the $H^+ + H^+ + CC^+$ channel in Fig. 8. Filtering only the events of one dissociating contribution (with a KER between 8 eV and 13 eV) highlights the isomerization process. Here, the Newton diagrams depict the 3D momenta in the following way: first, the momentum of the CC^+ ion is subtracted from the proton momenta in order to use the heaviest fragment as the center. Secondly, the momentum of the first proton is used as a reference to turn the coordinate system such that the $CC^+ - H_1^+$ axis is fixed along the positive horizontal axis. Finally, the momentum components of the second hydrogen along this $CC^+ - H_1^+$ axis ($p_{\parallel}$) and the momentum perpendicular ($p_{\perp}$) are plotted for different time delays. It is well visible that the isomerization takes place in the first 100 fs until the majority of molecules are in the vinylidene configuration. For longer pulse delays, a maximal isomerization is reached and the momentum distribution stays mostly unchanged showing no sign of a backwards isomerization.

As mentioned in the Experimental methods section, the CEP was also recorded such that CEP effects can be investigated. We have not observed clear CEP-dependent signatures in the pump–probe data, but it should be mentioned that the CEP effects in single pulse data were relatively small (asymmetries of around 7% have been detected).[8] With the data at hand we cannot rule out that time-dependent CEP-effects may be present, but if so, they are very small.

4 Conclusion

In summary, we have presented the molecular response in acetylene at various times after excitation by a near single-cycle pulse using the Coulomb explosion imaging technique. Investigating both major triple coincidence channels, we have observed the evolution of the hydrogen migration process for the isomerization to the vinylidene configuration. However, under our experimental conditions, no backwards isomerization to acetylene was observed. Despite this discrepancy, the isomerization time of the acetylene ion of 54 ± 4 fs, which we have clearly determined *via* the $H^+ + H^+ + CC^+$ channel, is in excellent agreement with previous experimental studies[21,30,31] and theoretical predictions.[29] The maximum in isomerization yield is explained within the structural evolution of the excited dication, namely the surpassing of the energetically higher lying transition states t-TS1′ and t-TS1″ to the vinylidene configuration. The unprecedented temporal resolution in our experiment allowed us to identify vibrational wave-packet motions along the CH- and CC-bonds with periods of 11.5 fs ± 1.7 fs and 20.4 fs ± 1.5 fs, respectively.

Acknowledgements

We are grateful to F. Krausz for his support and fruitful discussions. We are thankful for support by the Max Planck Society and by the DFG *via* LMUexcellent and *via* the Cluster of Excellence: Munich Center for Advanced Photonics (MAP). We are also grateful for support from the King-Saud University in the framework of the MPQ-KSU-LMU collaboration.

References

1 P. Jungner and L. Halonen, *J. Chem. Phys.*, 1997, **55**, 1680–1682.
2 R. Thissen, J. Delwiche, J. M. Robbe, D. Duflot, J. P. Flament and J. H. D. Eland, *J. Chem. Phys.*, 1993, **99**, 6590.
3 N. Ochi and S. Tsuchiya, *Chem. Phys.*, 1991, **152**, 319–336.
4 W. F. Manders and J. M. Bellama, *Polymer*, 1985, **23**, 351–357.
5 C. Cornaggia, M. Schmidt and D. Normand, *Phys. Rev. A*, 1995, **51**, 1431–1437.
6 X. Xie, S. Roither, M. Schöffler, E. Lötstedt, D. Kartashov, L. Zhang, G. G. Paulus, A. Iwasaki, A. Baltuška, K. Yamanouchi and M. Kitzler, *Phys. Rev. X*, 2014, **4**, 1–10.
7 A. S. Alnaser, M. Kübel, R. Siemering, B. Bergues, N. G. Kling, K. J. Betsch, Y. Deng, J. Schmidt, Z. A. Alahmed, A. M. Azzeer, J. Ullrich, I. Ben-Itzhak, R. Moshammer, U. Kleineberg, F. Krausz, R. de Vivie-Riedle and M. F. Kling, *Nat. Commun.*, 2014, **5**, 3800.
8 M. Kübel, R. Siemering, C. Burger, N. G. Kling, H. Li, A. S. Alnaser, B. Bergues, S. Zherebtsov, A. M. Azzeer, I. Ben-Itzhak, R. Moshammer, R. de Vivie-Riedle and M. F. Kling, 2015, arxiv, 1508.04018.
9 D. Duflot, J.-M. Robbe and J.-P. Flament, *J. Chem. Phys.*, 1995, **102**, 355.
10 K. M. Ervin, J. Ho and W. C. Lineberger, *J. Chem. Phys.*, 1989, **91**, 5974.
11 M. P. Jacobson and R. W. Field, *J. Phys. Chem. A*, 2000, **104**, 3073–3086.
12 A. Hishikawa, A. Matsuda, M. Fushitani and E. J. Takahashi, *Phys. Rev. Lett.*, 2007, **99**, 258302.
13 A. Hishikawa, A. Matsuda, E. J. Takahashi and M. Fushitani, *J. Chem. Phys.*, 2008, **128**, 084302.
14 H. Ibrahim, B. Wales, S. Beaulieu, B. E. Schmidt, N. Thiré, E. P. Fowe, É. Bisson, C. T. Hebeisen, V. Wanie, M. Giguére, J.-C. Kieffer, M. Spanner, A. D. Bandrauk, J. Sanderson, M. S. Schuurman and F. Légaré, *Nat. Commun.*, 2014, **5**, 4422.

15 A. Matsuda, M. Fushitani, E. J. Takahashi and A. Hishikawa, *Springer Proc. Phys.*, 2012, **125**, 317–322.

16 A. S. Alnaser, I. Litvinyuk, T. Osipov, B. Ulrich, A. Landers, E. Wells, C. M. Maharjan, P. Ranitovic, I. Bochareva, D. Ray and C. L. Cocke, *J. Phys. B: At., Mol. Opt. Phys.*, 2006, **39**, 485–492.

17 V. Pervak, I. Ahmad, M. K. Trubetskov, a. V. Tikhonravov and F. Krausz, *Opt. Express*, 2009, **17**, 7943–7951.

18 D. J. Kane and R. Trebino, *IEEE J. Quantum Electron.*, 1993, **29**, 571–579.

19 M. Schultze, A. Wirth, I. Grguras, M. Uiberacker, T. Uphues, A. Verhoef, J. Gagnon, M. Hofstetter, U. Kleineberg, E. Goulielmakis and F. Krausz, *J. Electron Spectrosc. Relat. Phenom.*, 2011, **184**, 68–77.

20 M. Kübel, K. J. Betsch, N. G. Johnson, U. Kleineberg, R. Moshammer, J. Ullrich, G. G. Paulus, M. F. Kling and B. Bergues, *New J. Phys.*, 2012, **14**, 093027.

21 Y. H. Jiang, A. Rudenko, O. Herrwerth, L. Foucar, M. Kurka, K. U. Kühnel, M. Lezius, M. F. Kling, J. Van Tilborg, A. Belkacem, K. Ueda, S. Düsterer, R. Treusch, C. D. Schröter, R. Moshammer and J. Ullrich, *Phys. Rev. Lett.*, 2010, **105**, 263002.

22 H.-J. Werner, P. J. Knowles, G. Knizia, F. R. Manby, M. Schütz, P. Celani, W. Györffy, D. Kats, T. Korona, R. Lindh, A. Mitrushenkov, G. Rauhut, K. R. Shamasundar, T. B. Adler, R. D. Amos, A. Bernhardsson, A. Berning, D. L. Cooper, M. J. O. Deegan, A. J. Dobbyn, F. Eckert, E. Goll, C. Hampel, A. Hesselmann, G. Hetzer, T. Hrenar, G. Jansen, C. Köppl, Y. Liu, A. W. Lloyd, R. A. Mata, A. J. May, S. J. McNicholas, W. Meyer, M. E. Mura, A. Nicklass, D. P. O'Neill, P. Palmieri, D. Peng, K. Pflüger, R. Pitzer, M. Reiher, T. Shiozaki, H. Stoll, A. J. Stone, R. Tarroni, T. Thorsteinsson and M. Wang, *MOLPRO, Version 2006.1, a Package of Ab Initio Programs*, 2006.

23 H. Stapelfeldt, E. Constant, H. Sakai and P. Corkum, *Phys. Rev. A*, 1998, **58**, 426–433.

24 I. A. Bocharova, A. S. Alnaser, U. Thumm, T. Niederhausen, D. Ray, C. L. Cocke and I. V. Litvinyuk, *Phys. Rev. A: At., Mol., Opt. Phys.*, 2011, **83**, 1–17.

25 T. Ergler, A. Rudenko, B. Feuerstein, K. Zrost, C. D. Schröter, R. Moshammer and J. Ullrich, *Phys. Rev. Lett.*, 2005, **95**, 14–17.

26 M. Magrakvelidze, O. Herrwerth, Y. H. Jiang, A. Rudenko, M. Kurka, L. Foucar, K. U. Kühnel, M. Kübel, N. G. Johnson, C. D. Schröter, S. Düsterer, R. Treusch, M. Lezius, I. Ben-Itzhak, R. Moshammer, J. Ullrich, M. F. Kling and U. Thumm, *Phys. Rev. A: At., Mol., Opt. Phys.*, 2012, **86**, 1–11.

27 B. Fischer, M. Kremer, T. Pfeifer, B. Feuerstein, V. Sharma, U. Thumm, C. D. Schröter, R. Moshammer and J. Ullrich, *Phys. Rev. Lett.*, 2010, **105**, 1–4.

28 Y. H. Jiang, A. Senftleben, M. Kurka, A. Rudenko, L. Foucar, O. Herrwerth, M. F. Kling, M. Lezius, J. V. Tilborg, A. Belkacem, K. Ueda, D. Rolles, R. Treusch, Y. Z. Zhang, Y. F. Liu, C. D. Schröter, J. Ullrich and R. Moshammer, *J. Phys. B: At., Mol. Opt. Phys.*, 2013, **46**, 164027.

29 T. S. Zyubina, Y. A. Dyakov, S. H. Lin, A. D. Bandrauk and A. M. Mebel, *J. Chem. Phys.*, 2005, **123**, 134320.

30 Y. H. Jiang, A. Senftleben, A. Rudenko, M. E. Madjet, O. Vendrell, M. Kurka and K. Schnorr, 2010, arxiv, 1402.4874.

31 T. Osipov, C. L. Cocke, M. H. Prior, A. Landers, T. Weber, O. Jagutzki, L. Schmidt, H. Schmidt-Böcking and R. Dörner, *Phys. Rev. Lett.*, 2003, **90**, 233002.

Faraday Discussions

PAPER

Interference stabilization of autoionizing states in molecular N_2 studied by time- and angular-resolved photoelectron spectroscopy

Martin Eckstein,[a] Nicola Mayer,[a] Chung-Hsin Yang,[a] Giuseppe Sansone,[b] Marc J. J. Vrakking,[a] Misha Ivanov[a] and Oleg Kornilov*[a]

Received 25th April 2016, Accepted 9th June 2016
DOI: 10.1039/c6fd00093b

An autoionizing resonance in molecular N_2 is excited by an ultrashort XUV pulse and probed by a subsequent weak IR pulse, which ionizes the contributing Rydberg states. Time- and angular-resolved photoelectron spectra recorded with a velocity map imaging spectrometer reveal two electronic contributions with different angular distributions. One of them has an exponential decay rate of 20 ± 5 fs, while the other one is shorter than 10 fs. This observation is interpreted as a manifestation of interference stabilization involving the two overlapping discrete Rydberg states. A formalism of interference stabilization for molecular ionization is developed and applied to describe the autoionizing resonance. The results of calculations suggest, that the effect of the interference stabilization is facilitated by rotationally-induced couplings of electronic states with different symmetry.

1 Introduction

One of the main challenges in quantum chemistry is the description of electron–electron correlations, which are at the root of electron dynamics. The recent development of experimental methods with time resolution comparable to the characteristic time of electron motion opens the possibility to study electron–electron correlations directly in the time domain. These methods have recently been applied to autoionization in helium,[1,2] neon,[3] and argon atoms,[4] paving the way to studies of multielectron dynamics from the time domain perspective. All these studies are concerned with isolated autoionizing resonances, which are represented in the spectral domain as Fano line profiles.[5] In our recent work we used time-, energy-, and angular-resolved photoelectron spectroscopy to investigate for the first time dynamics of autoionization at a complex resonance in

[a]*Max Born Institute, Max-Born-Straße 2A, 12489 Berlin, Germany. E-mail: kornilov@mbi-berlin.de*
[b]*Dipartimento di Fisica, Politecnico, Piazza Leonardo da Vinci 32, 20133 Milano, Italy*

nitrogen molecules.[6] In this article we expand on the previous report providing a detailed analysis of the angular distributions and developing a theoretical framework for the description of coupled molecular resonances based on the theory of interference stabilization.[7–10] Our experimental and theoretical results demonstrate for the first time the presence of this so-far elusive mechanism in molecular autoionization.

Autoionizing resonances in nitrogen molecules were first discovered in 1936 by Hopfield, who observed two sequences of sharp spectral lines converging to a common threshold.[11] It was soon realized that the lines are caused by the excitation of electronic Rydberg states attached to the nitrogen ion in the $B^2\Sigma_u^+$ electronic state, which decays *via* electron rearrangement resulting in autoionization. However, it took until 1961 to explain the asymmetric line profiles, later called Fano profiles, which are caused by an interference between the direct ionization and the autoionization.[5] Even today the details of the line profiles of the two series observed by Hopfield (the absorption and the "emission" series) have not been fully described theoretically. The most detailed calculations of Raoult *et al.*[12] provided correct assignments of the resonance lines to the responsible Rydberg series, but could not reproduce the line shapes. Later experiments of Huber *et al.*[13] with room temperature nitrogen and rotationally-cold supersonic jets demonstrated, that rotational couplings of electronic states with different angular momenta are important. These couplings were not included in the calculations of ref. 12 obtained in the space-fixed molecular frame.

In this work the resonance at 17.33 eV is investigated. The results of Raoult *et al.*[12] suggest that at the position of this member of the emission series two Rydberg states overlap, namely the $3d\pi_g$ and $4's'\sigma_g$ states. This prediction is confirmed in our time-resolved experiments,[6] which reveal time-dependent asymmetry parameters in the transient photoelectron signals pointing to contributions of at least two electronic states. These results further show that the two participating electronic states have different lifetimes contrary to the predictions of ref. 12. This discrepancy suggests that the so-called phenomenon of interference stabilization may play a role.

The phenomenon of interference stabilization, also called interference narrowing, was previously observed and described for excited atoms in static electric fields and laser fields.[7–9,14] It was described in detail analytically by Fedorov and Movsesian[10] for a set of overlapping resonances in atomic ionization by a strong laser field. In their model a strong laser field ionizes two or more Rydberg states, *i.e.* couples them to the ionization continuum. As the laser field strength increases, the lifetimes of individual Rydberg states decrease and hence their corresponding energy widths increase. When the widths of the states become comparable to the energy spacing between them, the continuum couplings (*i.e.* the interactions between the resonances *via* the continuum) start to play a significant role and lead to a non-intuitive effect: the increase of their lifetimes. In the case of two resonances, the energies of the two states approach each other and the width of one of the resonances continues to grow with the laser field strength, while the width of the other state starts to shrink. That is, one of the resonances becomes progressively more and more stabilized, even though the coupling constants for both discrete states continue to increase. This effect was termed "interference stabilization" by Fedorov *et al.*[10] because it originates from quantum interference of the contributions to ionization resulting from the two states. Its origin is in general similar to that of the electromagnetically induced transparency (EIT).

The origin of the stabilization effect is generic and its theory is applicable to all types of discrete–continuum couplings, not necessarily induced by static electric fields or laser fields. It is therefore reasonable to expect it to influence the case of two overlapping Fano resonances, where electron–electron interactions are responsible for coupling autoionizing Rydberg states to the continuum. In fact, such an effect was numerically predicted by Mies in 1968,[15] although he did not analyze it analytically. In this article we generalize the theory of Fedorov *et al.*[16] to the realistic case of autoionization of N_2, combining the treatment of Fano resonances with the interference stabilization effect, and analyze the results of time-resolved photoelectron spectroscopy experiments in terms of the theory of interference stabilization.

The article is structured as follows: the next section briefly describes the experiment and presents results on time-dependent angular distributions of the N_2 autoionizing resonance at 17.33 eV. The third section presents a theoretical description of the interference stabilization effect for the case of N_2. The fourth section provides a comparison of the theoretical and experimental results.

2 Experimental results

The experiment has already been described in detail in our previous report.[6] Here, we briefly summarize the experimental set-up, the main findings and analyze in detail the time-dependent photoelectron angular distributions corresponding to the 17.33 eV resonance investigated in the present work. Time-resolved study of the Fano resonance in N_2 was performed at a recently constructed XUV time delay compensating monochromator beamline connected to a velocity map imaging (VMI) spectrometer.[17] In the experiment 2 mJ, 20 fs near-infrared (IR) pulses from a commercial Ti:sapphire laser system were used to drive a high-order harmonic generation process in Ar gas[18] to produce linearly polarized XUV pulses with photon energies ranging from 10 to 70 eV. The time delay compensating monochromator[17] was used to select the 11th harmonic of the fundamental wavelength and slightly tune its spectrum to predominantly excite the resonance at 17.33 eV. For this purpose the XUV excitation spectrum was centered at 17.5 eV and had a full width at half maximum (FWHM) of 0.5 eV.

Autoionizing Rydberg states excited by the XUV pulse were probed by a weak IR pulse with a pulse energy of 50 μJ (intensity of 2 TW cm^{-2}) and a linear polarization parallel to that of the XUV pulse. The pulse arrives with a variable time delay with respect to the XUV pulse. The resulting photoelectron spectra were recorded using the VMI spectrometer,[19] which provides both energy- and angular-resolved photoelectron spectra. The changes induced by the IR pulse are modest and therefore the results are best represented in the form of difference VMI images, *i.e.* images resulting from subtraction of an XUV-only image from an image recorded when both the XUV and IR pulses are present. Such a difference image recorded at zero time delay between the XUV and IR pulses is presented in Fig. 1a. The figure shows a slice of the 3D momentum distribution reconstructed from the raw experimental data using the BASEX reconstruction algorithm.[20] The image contains several features having both positive (red, additional signal due to

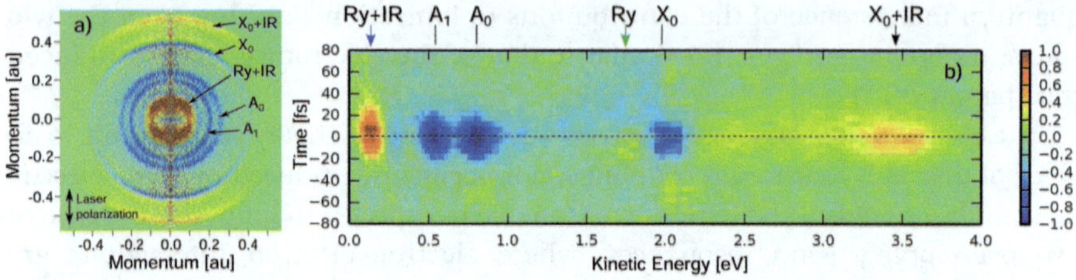

Fig. 1 (a) A difference VMI image at zero time delay between the XUV and the IR pulses. The image is obtained by subtracting an XUV-only signal from an XUV + IR one. The double arrow in the bottom left corner indicates the XUV and IR laser polarization. Features corresponding to direct ionization to the $X^2\Sigma_g^+$ and $A^2\Pi_u$ channels are labeled as X_0, $A_{0,1,2}$, where the subscript denotes the vibrational state. The most intense IR-assisted sideband signal is labeled X_0 + IR. The feature corresponding to IR ionization of the Rydberg state attached to the $B^2\Sigma_u^+$ ionization threshold is labeled as Ry + IR. (b) False color map of photoelectron kinetic energy spectra plotted *vs.* kinetic energy and time delay between the XUV and IR pulses. The dashed line indicates the position of time "zero" determined from the X_0 + IR sideband signal. The map clearly shows that only the Ry + IR feature extends towards positive time delays. Other intense features are symmetric with respect to $t = 0$ fs, *i.e.* show no dynamics. (Adopted from ref. 6).

the IR pulse) and negative (blue, signal reduced by the IR pulse) intensities. The outermost feature, labeled X_0 + IR, corresponds to the generation of a $n = +1$ sideband[21,22] of the photoelectron feature in the $X^2\Sigma_g^+$ ($\nu = 0$) ionization channel, *i.e.* the channel in which the nitrogen molecular ion N_2^+ is in the ground electronic and vibrational state. The negative (blue) feature labeled X_0 is a depletion of the signal in the $X^2\Sigma_g^+$ ($\nu = 0$) channel due to formation of the sideband. Similar signals are observed for the $A^2\Pi_u$ ($\nu = 0, 1, 2$) ionization channels, but the corresponding sideband signals are partially masked by the dominant X_0 feature. The signal corresponding to ionization of the Rydberg electron by one IR photon appears as an intense positive (red) ring close to the center of the image. It is labeled Ry + IR to reflect its nature: ionization of a Rydberg state with one IR photon leaving the core in the excited $B^2\Sigma_u$ state.

The temporal evolution of the features observed in the photoelectron spectra was studied by scanning the time delay between the XUV and IR pulses. The false color map of the photoelectron kinetic energy spectra (integrated over all angles) as a function of the XUV-IR delay is shown in Fig. 1b). As expected, all signals corresponding to a formation of sidebands and depletion of the original photoelectron bands have temporal profiles symmetric with respect to the zero time delay (dashed line). On the contrary, feature Ry + IR extends towards positive time delays (where the XUV pulse arrives before the IR pulse), which reflects the finite lifetime of the Rydberg state.

The short lifetime of the resonance in Fig. 1b precludes an analysis of its composition based on the transient intensity data alone. However, following and extending the approach of ref. 6 the time-dependent angular distributions of this feature can be used to analyze its origin. Fig. 2a shows the angular distributions of feature Ry + IR integrated over a range of energies from 0.1 to 0.3 eV as a function of time delay. The distributions are normalized to the maximum signal intensity at each pump–probe delay. This normalization procedure removes the time dependence of the signal intensity and thus reveals pure variations of the angular

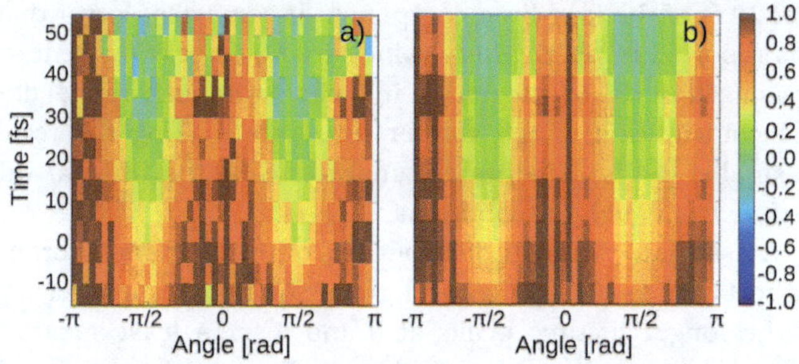

Fig. 2 (a) False color map of the photoelectron angular distributions of feature Ry + IR normalized to the intensity at zero emission angle (relative to the light polarization). This plot emphasizes the time-dependent changes of the angular distributions as the signal becomes more peaked along the 0 and $\pm\pi$ directions for positive time delays. The data are plotted in the range of time delays, where the signal intensity exceeds the noise. (b) False color map of the photoelectron angular distributions composed of the two dominant components extracted from singular value decomposition of the experimental data matrix plotted in (a). This procedure removes experimental noise and allows for extraction of decay-associated angular distributions (see text for details).

distributions of the photoemission as a function of time delay. One can immediately see that close to zero time delay the emission is relatively isotropic (the red color is uniformly distributed over all emission angles), while at later time delays the signal is peaked at 0 and $\pm\pi$, corresponding to emission along the laser polarization.

Thus, the angular distribution of photoemission significantly changes as a function of the time delay between the XUV pulse exciting the Rydberg state and the IR pulse ionizing it. This observation can only be explained by the presence of two overlapping electronic states with different lifetimes, since isolated electronic states (and hence IR-induced photoemission) cannot change with time. To confirm this observation and extract parameters of the two contributing states we decompose the data matrix in Fig. 2a using the method of singular value decomposition.[23,24] This method leads to decomposition of the matrix into principal components. For the case of two electronic states with different angular emission characteristics and different lifetimes we expect to observe a decomposition of the data into two dominant contributions. This is indeed the case as can be seen in Fig. 2b, which shows the data matrix composed of these two dominant components with all other contributions removed. Clearly this map reproduces all features of the experimental data in part a, while simultaneously strongly suppressing the noise.

The singular value decomposition thus allows us to extract angular distributions and decay times of the two electronic states contributing to the resonance at 17.33 eV.[23,24] Exponential decay fits convoluted with a Gaussian cross correlation function (FWHM of 36 fs) yield the lifetimes of 20 ± 5 fs and 9 ± 3 fs for the two contributions. The errors represent the statistical uncertainties of the fit. Systematic uncertainties arising from the delay step size and the exact shape of the cross correlation function are relatively small for the longer timescale, but are significant for the short one substantially extending the lower boundary of uncertainty. We therefore assume it to be less than 10 fs in the following text. The timescale of 20 fs of the long-lived component is somewhat larger than the value of 14 ± 1 fs reported in ref. 6. These values are however consistent, because the data analysis used in ref. 6 (integration in a narrow angular range) does not fully remove the contribution of the shorter timescale making the resulting value somewhat smaller (14 fs). The present analysis using the singular value decomposition is free from such contributions.

The decay-associated angular distributions of the long and short components are plotted in Fig. 3 as blue circles and green squares, respectively. The component with the longer lifetime peaks at 0 and $\pm\pi$. A least squares fit to this distribution, shown as a blue line, yields an asymmetry parameter of 1.3 ± 0.2. The short-lived component has a relatively uniform angular distribution (with an asymmetry parameter $\beta \approx 0$). We thus extract the parameters of the two electronic states contributing to the resonance at 17.33 eV, which will be discussed later in the text in the context of the interference stabilization theory developed in the following section.

3 Theory of interference stabilization in autoionization

The observation of two overlapping resonances with substantially different lifetimes immediately reminds of the interference stabilization theory (or interference narrowing) developed[7–10,16] for atoms in static electric fields and intense laser fields. In the description of ref. 16 two Rydberg states of an atom are laser-coupled to the ionization continuum. When the coupling strength, expressed in units of energy, exceeds the energy gap between the states, the mixing of the two states *via* the continuum leads to stabilization (increased lifetime) of one of them at the expense of de-stabilization (decreased lifetime) of the other. This scenario is very general and can be applied to any type of discrete–continuum couplings, *i.e.* also

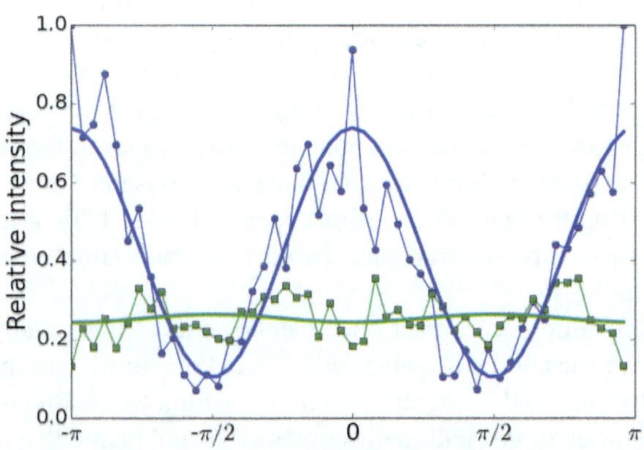

Fig. 3 Decay-associated angular distributions of the two dominant components of the singular value decomposition, presented in Fig. 2. Symbols are data points extracted in the SVD analysis. The two distributions correspond to 20 ± 5 fs decay (blue circles) and a decay shorter than 10 fs (green squares). Thick lines are fits of the asymmetry parameters yielding β values of 1.3 ± 0.2 and 0.05 ± 0.06, respectively.

in the case of two overlapping autoionizing states of an atom or molecule interacting with the same ionization continuum. In the following we derive the theory of interference stabilization for the case of two autoionizing states coupled to two or more continua, which reflects the realistic case of the N_2 molecule investigated in our experiment, where autoionization can proceed to the $X^2\Sigma_g^+$ and $A^2\Pi_u$ ionization channels.

3.1 Stabilization effect

We express the wavefunction of the molecule upon absorption of an XUV photon in terms of two discrete state wavefunctions ϕ_1 and ϕ_2 and two continuum wavefunctions ψ_A and ψ_X, with the total energies E_1, E_2, E_A and E_X, respectively:

$$\Psi(t) = c_1(t)\phi_1 e^{-iE_1 t} + c_2(t)\phi_2 e^{-iE_2 t} + \int dE_A c_A(t)\psi_A e^{-iE_A t} + \int dE_X c_X(t)\psi_X e^{-iE_X t} \quad (1)$$

The labels A and X for the two continua are chosen to facilitate the subsequent comparison with the case of N_2. The matrix elements of the system Hamiltonian H are given by the following set of equations:

$$\langle\phi_n|H|\phi_m\rangle = E_n\delta_{nm} \quad (2)$$

$$\langle\psi_A|H|\psi'_A\rangle = E_A\delta(E_A - E'_A) \quad (3)$$

$$\langle\psi_X|H|\psi'_X\rangle = E_X\delta(E_X - E'_X) \quad (4)$$

$$\langle\psi_A|H|\psi_X\rangle = 0 \quad (5)$$

$$\langle\phi_n|H|\psi_A\rangle = V_{An} \quad (6)$$

$$\langle\phi_n|H|\psi_X\rangle = V_{Xn} \quad (7)$$

here δ denotes the Dirac delta-function and V_{An}, V_{Xn} are the terms describing the electron–electron interaction, which leads to autoionization.[5] Substituting the wavefunction Ψ into the time-dependent Schrödinger equation and taking the projection of the total wavefunction Ψ onto the wavefunctions of the individual states one arrives at the following system of coupled differential equations for state population coefficients $c(t)$:

$$i\dot{c}_1(t) = \int dE_A c_A(t) e^{-i(E_A - E_1)t} V_{A1} + \int dE_X c_X(t) e^{-i(E_X - E_1)t} V_{X1} \quad (8)$$

$$i\dot{c}_2(t) = \int dE_A c_A(t) e^{-i(E_A - E_2)t} V_{A2} + \int dE_X c_X(t) e^{-i(E_X - E_2)t} V_{X2} \quad (9)$$

$$i\dot{c}_A(t) = c_1(t) e^{-i(E_1 - E_A)t} V_{A1} + c_2(t) e^{-i(E_2 - E_A)t} V_{A2} \quad (10)$$

$$i\dot{c}_X(t) = c_1(t) e^{-i(E_1 - E_X)t} V_{X1} + c_2(t) e^{-i(E_2 - E_X)t} V_{X2} \quad (11)$$

As one can see from these equations, the coupling arises because both ϕ_1 and ϕ_2 couple to the same continuum states ψ_A and ψ_X. Following the treatment of

Fedorov[16] one can formally integrate equations for the continuum coefficients and substitute them into the expressions for discrete states. Assuming that the continuum coefficients c_A, c_X and transition matrix elements V_{An}, V_{Xn} are sufficiently slow functions of the energy in the continuum channel (E_A and E_X), the following system of differential equations is obtained:

$$i\dot{c}_1(t) = -2\pi i[c_1(t)(|V_{A1}|^2 + |V_{X1}|^2) + c_2(t)e^{i(E_1-E_2)t}(V_{A1}V_{A2} + V_{X1}V_{X2})] \quad (12)$$

$$i\dot{c}_2(t) = -2\pi i[c_2(t)(|V_{A2}|^2 + |V_{X2}|^2) + c_1(t)e^{i(E_2-E_1)t}(V_{A1}V_{A2} + V_{X1}V_{X2})] \quad (13)$$

To compare with ref. 16 we may choose equal coupling strengths for both discrete states: $V_{A1} \simeq V_{A2}$ ($V_{X1} \simeq V_{X2}$) and define the decay rates $\Gamma_A \equiv 2\pi|V_{An}|^2$ and $\Gamma_X \equiv 2\pi|V_{Xn}|^2$. Looking for a non-trivial solution of the form $c_n(t) = f_n e^{i(E_n-\gamma)t}$ a quadratic equation for the quasi-energy γ is obtained, which has two solutions:

$$\gamma_\pm = \frac{1}{2}\left[(E_1 + E_2) - i(\Gamma_A + \Gamma_X) \pm \sqrt{(E_1 - E_2)^2 - (\Gamma_A + \Gamma_X)^2}\right] \quad (14)$$

This is equivalent to the eqn (8.1.21) of ref. 16 with the total decay rate of each discrete state equal to the sum $\Gamma = \Gamma_A + \Gamma_X$. The resonance widths of the two interacting states are given by twice the imaginary part of the two quasi-energies $2\mathrm{Im}(\gamma_\pm)$ and the energy positions of the resonances are given by the real part $\mathrm{Re}(\gamma_\pm)$. Fig. 4a shows both the positions and the widths of the resonances as a function of the total rate Γ. The positions are shown as black lines and the widths are represented by shaded areas enclosed by effective resonance boundaries given by $\mathrm{Re}(\gamma_\pm) \pm \mathrm{Im}(\gamma_\pm)$. The energies and widths are given in reduced units chosen to yield the energy spacing of the states at $\Gamma = 0$ equal to 1. One can see that at a critical value of $\Gamma = E_2 - E_1$ the resonance positions merge and from there on the width of one of the resonances continues to grow, while the width of the other, remarkably, shrinks, indicating an increase of the lifetime with increasing coupling strength. Therefore one of the states becomes more and more stabilized. The critical rate Γ corresponds to a change of the sign of the expression under the square root in eqn (14). For the case of molecular autoionization considered here the continuum couplings are given by electron–electron interactions and cannot be changed experimentally. Therefore the case of the molecule would correspond to a particular value of Γ given by the electron–electron interactions.

In general, the couplings of two discrete states to the two continua are not identical. Introducing decay rates $\Gamma_{An} = 2\pi|V_{An}|^2$, $\Gamma_{Xn} = 2\pi|V_{Xn}|^2$ and carrying out the same procedure as outlined above we arrive at the following expression for the quasi-energy:

$$\gamma_\pm = \frac{1}{2}[(E_1 + E_2) - i(\Gamma_1 + \Gamma_2) \pm \sqrt{\Delta}], \quad (15)$$

where $\Gamma_1 = \Gamma_{A1} + \Gamma_{X1}$, $\Gamma_2 = \Gamma_{A2} + \Gamma_{X2}$ and Δ is given by:

$$\Delta = (E_1 - E_2)^2 + i(\Gamma_2 - \Gamma_1) - 4\left(\sqrt{\Gamma_{A1}\Gamma_{A2}} + \sqrt{\Gamma_{X1}\Gamma_{X2}}\right)^2 \quad (16)$$

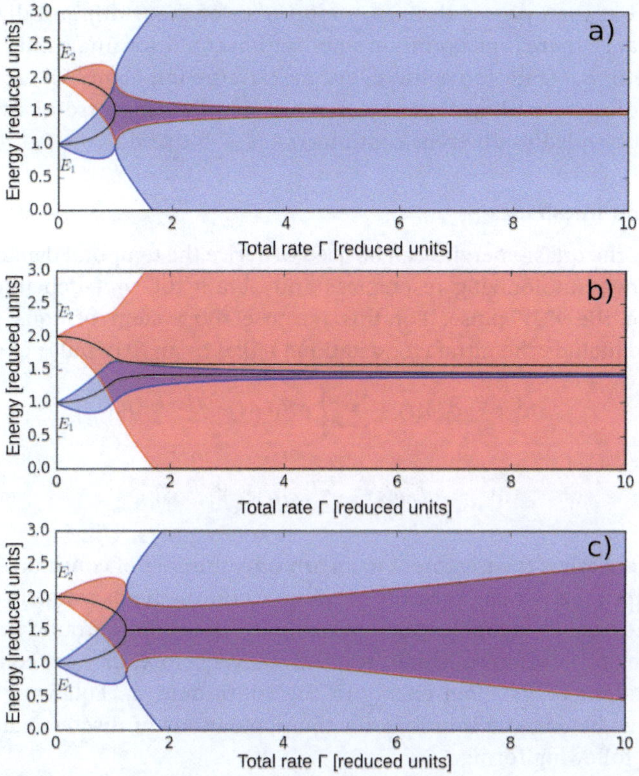

Fig. 4 Resonance positions $\mathrm{Re}(\gamma_{\pm})$ (black) and effective resonance boundaries $\mathrm{Re}(\gamma_{\pm}) \pm \mathrm{Im}(\gamma_{\pm})$ (red and blue) for three sets of parameters describing the discrete–continuum interactions, plotted against the total rate Γ given in reduced units. (a) Balanced case: the two discrete states have identical coupling strengths to the two continua, *i.e.* $V_{A1} = V_{A2}$ and $V_{X1} = V_{X2}$. (b) Partially balanced case: the coupling strengths of both discrete states to the two continua are balanced ($V_{A1} = V_{X1}$ and $V_{A2} = V_{X2}$), but the resonance 1 is coupled to the continua weaker than the resonance 2 ($\Gamma_1 = 0.75\Gamma_2$). (c) Imbalanced case: the coupling constants of two discrete states are different for each continuum: $\Gamma_{A1} = 0.7\Gamma$, $\Gamma_{X1} = 0.3\Gamma$, $\Gamma_{A2} = 0.3\Gamma$, $\Gamma_{X2} = 0.7\Gamma$.

One can immediately see that for the case $\Gamma_1 = \Gamma_2$ the value of $\sqrt{\Delta}$ is always real and the system behaves similar to that of ref. 16 when all continuum couplings are uniformly increased leading to stabilization of one of the resonances. In the general case, however, the couplings of the two resonances to the two continua are different and this leads to a modification of the system behavior. If $\Gamma_1 \neq \Gamma_2$, but the coupling strengths of the discrete states to the two continua remain balanced, *i.e.* $\Gamma_{A1} = \Gamma_{X1}$ and $\Gamma_{A2} = \Gamma_{X2}$, the resonance positions do not merge anymore, but the stabilization effect is still observed, as can be seen in Fig. 4b. The positions of the two resonances do move closer together and the resonance with the initially weaker coupling to the continua (*i.e.* resonance 1 for $\Gamma_2 > \Gamma_1$) becomes stabilized.

Finally, an interesting effect is observed for the general case of imbalanced coupling strengths, *i.e.* when either $\Gamma_{A1} \neq \Gamma_{X1}$ or $\Gamma_{A2} \neq \Gamma_{X2}$ or both. In this case, for a certain degree of imbalance the weaker coupled state first becomes partially stabilized, but then passes through a minimum of the width and diverges (see Fig. 4c). That is, there is an optimum stabilization condition and a minimum non-infinite lifetime a state can achieve. For an extreme imbalance the stabilization effect can disappear altogether. An example is a limiting case of two discrete resonances coupled to different continua, *i.e.* $\Gamma_{A2} = 0$ and $\Gamma_{X1} = 0$.

3.2 Spectral lineshapes

Eqn (15) for the quasi-energies can be used to derive the temporal dynamics of the system of two autoionizing resonances and obtain the time-dependent dipole response to the XUV pulse. For this purpose the system of eqns (8)–(12) is modified to include the ultrafast excitation (δ-like) by an XUV pulse at time $t = 0$:

$$i\dot{c}_i = -d_{gi}\delta(t) + \sum_{J}\int \mathrm{d}E_J c_J(t)\mathrm{e}^{-i(E_J-E_i)t}V_{iJ} \tag{17}$$

$$i\dot{c}_J = -d_{gJ}\delta(t) + \sum_{l} c_l(t)\mathrm{e}^{-i(E_l-E_J)t}V_{lJ} \tag{18}$$

where we generalize the treatment to an arbitrary number of continua denoted by the subscript J (*i.e.* J = {A, X} in the previous section). The parameters d_{gi} and d_{gJ} describe the transition dipole moments coupling the neutral ground state $|g\rangle$ and the discrete and continuum states, respectively. As before, they are considered to be independent of the final energy in the continuum E_J. Following the same procedure as before, the solutions for the populations of discrete states can be cast in the following form:

$$c_i(t) = \sum_{\alpha=\pm} C_\alpha A_i^\alpha \mathrm{e}^{-i\gamma_\alpha t}\mathrm{e}^{iE_i t}, \tag{19}$$

where the coefficients $A_i^{\pm}$ describe the mixing of the discrete states *via* interference stabilization:

$$A_1^{\pm} = 1 \tag{20}$$

$$A_2^{\pm} = \frac{i}{2}\left(\frac{(E_2 - E_1) - i(\Gamma_2 - \Gamma_1) \pm \sqrt{\Delta}}{\sum_J \sqrt{\Gamma_{J1}\Gamma_{J2}}}\right) \tag{21}$$

and the coefficients $C_{\pm}$ are set to fulfil the initial conditions given by the XUV pulse:

$$C_+ = \frac{2\pi i}{A_2^- - A_2^+}\sum_J d_{gJ}\left[V_{1J}(q_1 - i)A_2^- - V_{2J}(q_2 - i)A_1^-\right] \tag{22}$$

$$C_- = \frac{2\pi i}{A_2^- - A_2^+}\sum_J d_{gJ}\left[V_{2J}(q_2 - i)A_1^+ - V_{1J}(q_1 - i)A_2^+\right] \tag{23}$$

where we set the Fano indices of the re-normalized discrete states as:

$$q_i = \frac{d_{gi}}{2\pi\sum_J d_{gJ}V_{iJ}}. \tag{24}$$

The time-dependent dipole response is then given by the following expression:

$$d(t) \equiv \langle g | \hat{H}_{\text{XUV}} | \Psi(t) \rangle = \sum_{\text{J}} \left[\sum_{i,\alpha} C_\alpha A_i^\alpha e^{-i\gamma_\alpha t} 2\pi d_{g\text{J}} V_{\text{J}i}(q_i - i) + 2i\pi |d_{g\text{J}}|^2 \delta(t) \right] \Theta(t) \tag{25}$$

The energy-domain absorption profile can be obtained by carrying out the Fourier transform of the time-dependent dipole response $d(t)$ and taking the imaginary part of the resulting energy-domain complex response function Im $[d(E)]$.[1] This procedure is implemented in the next section to compare the formalism developed here with the experimental data.

4 Discussion

The theoretical treatment of interference stabilization in the presence of several continua given above can now be used to analyze the Fano resonance in N_2 at 17.33 eV, using the current experimental observations as well as the most detailed theoretical calculations available of Raoult *et al.*[12] According to the calculations of Raoult, two Rydberg states may contribute to this resonance: the $3d\pi_g$ and the $4\text{'s'}\sigma_g$ states. Based on calculated absorption strengths the authors assign the dominant contribution to the $3d\pi_g$ state, which is coupled to two continua with the $N_2{}^+$ ion in the $X^2\Sigma_g^+$ and $A^2\Pi_u$ states. Since the calculations of ref. 12 are performed in a space-fixed molecular frame, the full symmetry dictated by the electronic wave function of the excited Rydberg state should be preserved during autoionization and therefore the $3d\pi_g$ state is coupled to XΠ and AΠ continua, where Π describes the product symmetry of the ion core and the outgoing electron. Similarly, the $4\text{'s'}\sigma_g$ Rydberg state in their calculations is coupled to the XΣ and AΣ continua. The coupling constants and relative populations of the discrete and continuum states can be extracted from Fig. 5 of ref. 12 by fitting Fano profiles to the theoretical curves. The parameters of the Fano profiles obtained in this way are given in Table 1.

The treatment of ref. 12 does not include the effect of interference stabilization, because in the fixed molecular frame the two contributing discrete states are coupled to ionization continua of different symmetry, as can be seen in Table 1. The resonance widths Γ given in the table are all rather similar, ranging from 63 to 88 meV. This, however, contradicts our experimental observations which clearly show that the two states have substantially different lifetimes and,

Table 1 Parameters of Fano profiles extracted from Fig. 5 of ref. 12 for two discrete states coupled to four possible ionization continua

$3d\pi_g$	XΠ	AΠ
E_0 [eV]	17.35	17.35
Γ [meV]	63	74
q	1.18	0.37
$4\text{'s'}\sigma_g$	XΣ	AΣ
E_0 [eV]	17.405	17.405
Γ [meV]	63	88
q	−1.08	4.87

therefore, must have different spectral widths. This controversy can be resolved if we recall that the experiments are not carried out in a fixed molecular frame. In the gas phase, nitrogen molecules can freely rotate and the autoionizing transitions may involve changes in the rotational state of the molecule. Such rotational transitions lead to a coupling of states with different symmetries, because the angular momentum of the total wave function includes the angular momentum of the changing rotational state. Indeed in 1993 Huber *et al.*[13] re-investigated the Hopfield series both at room temperature and in rotationally cold supersonic jets and found significant distortions of the line profiles induced by rotation of the molecules. They pointed out that for high rotational temperatures the effect of "l-uncoupling" should be taken into account, that is the coupling of the electronic orbital angular momentum to the molecular axis becomes weaker than the rotational couplings. In molecular spectroscopy this corresponds to transition from the Hund's case (b) (fixed frame) to the Hund's case (d) (rotational couplings). In the latter case the projection Λ of the electronic orbital angular momentum on the molecular axis is not a good quantum number, the Rydberg states acquire mixed Σ/Π character and can interact with each other *via* ionization continuum.

We therefore propose that rotational couplings relax the symmetry constraints in autoionizing transitions and allow for electron–electron interactions between discrete states and continua of different symmetry. This in turn leads to interactions between the two resonances affected by the effect of interference stabilization. To test this assumption we implement the model described in the previous section for two discrete states (state 1, $3d\pi_g$, and state 2, $4's'\sigma_g$) coupled to four continua ($X\Pi$, $A\Pi$, $X\Sigma$ and $A\Sigma$). We set the coupling strengths for the symmetry-allowed transitions based on the values of the resonance widths Γ given in Table 1. The Fano indices q given in the table are used to set the relative populations of the discrete states and the continua induced by absorption of the XUV. The only remaining parameter, *i.e.* the coupling strengths of the two discrete states to the neutral ground state d_{g1} and d_{g2}, are set to $d_{g2}/d_{g1} = 3$ to match the XUV absorption spectrum calculated by Raoult *et al.* (Fig. 4 of ref. 12). The results of the calculation corresponding to the case of ref. 12, *i.e.* without symmetry-changing transitions and thus, without interference stabilization, are shown in Fig. 5 together with the calculated absorption spectrum of ref. 12. The general

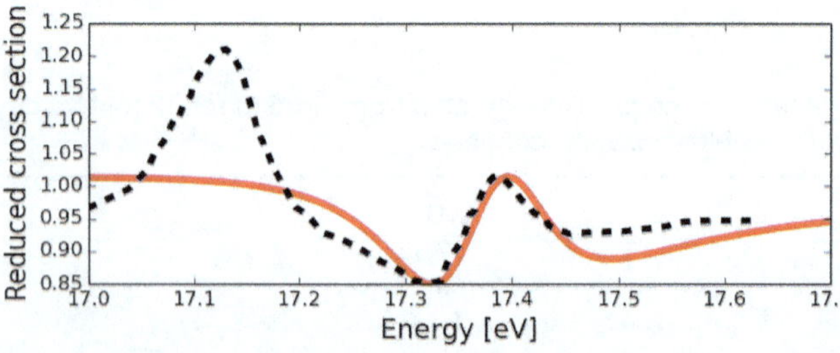

Fig. 5 Calculated XUV absorption profile of the resonance at 17.33 eV taken from ref. 12 (black dashed line) and calculated using the theory developed in the present work not including the stabilization effect. The resonance at 17.14 eV is not included in our treatment.

shape of the resonance at 17.33 eV is well reproduced with all differences attributable to the resonance at 17.14 eV, which is omitted in our calculations.

We can now model the effects induced by rotations. To the best of our knowledge there is no information in the literature on rotational coupling strengths for these short-lived Rydberg states. To test our assumption we vary the values of the corresponding symmetry-changing matrix elements, describing coupling between the $3d\pi_g$ Rydberg state and the $X\Sigma$ and $A\Sigma$ continuum states, and between the $4\text{'s'}\sigma_g$ Rydberg state and the $X\Pi$ and $A\Pi$ continuum states. For simplicity we take all couplings, initially forbidden by symmetry, to be equal and set them to values ranging from 0 to 25 meV, which corresponds to the maximum total rotational coupling of 100 meV. For a comparison, remember that the symmetry-allowed couplings in Table 1 are all in the range of 63 to 88 meV. The quasi-energy diagram for this range of total rotational couplings Γ_{rot} is shown in Fig. 6a. Note that only rotational couplings are varied in the plot and therefore at the $\Gamma_{rot} = 0$ the widths of the states are given by sums of the values in Table 1. One can already see that for the total rotational coupling on the order of 5 meV the

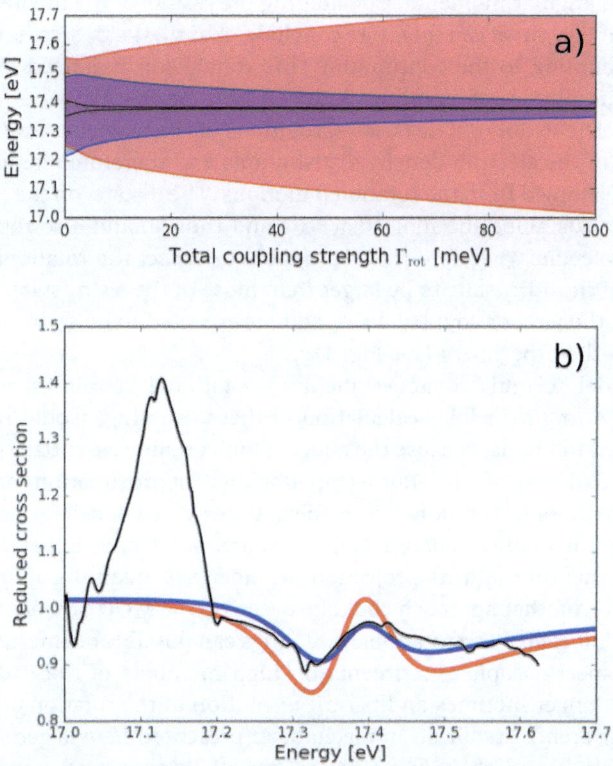

Fig. 6 (a) Quasi-energy diagram similar to those of Fig. 4 demonstrating the effect of interference stabilization induced by the symmetry-changing rotational couplings Γ_{rot}. The case of $\Gamma_{rot} = 0$ corresponds to the calculations of ref. 12. (b) XUV absorption profile calculated with (thick blue line) and without (thick red line) the effect of interference stabilization compared to the experimental spectrum of Morin *et al.*[25] (black line). This plot demonstrates how the inclusion of interference stabilization can largely remove the discrepancies observed in the fixed molecular frame calculations of Raoult *et al.*[12]

positions of the two discrete states (shown by black lines in the plot) approach each other, leading to substantial modification of the absorption profile. The stabilization effect also sets in at very small values of the total coupling strength and for $\Gamma_{\text{rot}} = 20$ meV the ratio of the lifetimes (or spectral widths) exceeds 2. For much larger values of Γ_{rot} (not shown in the plot) the resonance is destabilized as predicted in the general imbalanced case of Fig. 4c. In Fig. 6b XUV absorption profiles calculated with the total rotational couplings of 0 meV and 1 meV are compared with the experimental data of Morin *et al.*[25] The plot shows that even very small couplings can explain the discrepancies observed in the calculations of Raoult *et al.*[12]

The relative contributions of the resonance positions and resonance width to the combined XUV absorption profile sensitively depend on the specific values of the couplings to different continua. An analysis of the photoelectron angular distributions in Fig. 3 shows that the long-lived component has a relatively large positive asymmetry parameter $\beta = 1.3 \pm 0.2$. For linearly polarized light such a distribution corresponds to one-photon ionization of a state with a dominant s-type contribution. Therefore the angular distributions observed in the experiment are consistent with stabilization of the 4's'σ_g Rydberg state induced by rotational couplings of angular momenta. Considering the results of the theoretical section presented in Fig. 4b we can therefore conclude, that the 4's'σ_g state is the one with a weaker coupling to the continuum. This conclusion is indeed logical if one considers the physics of couplings induced by rotations. For a perfectly spherically symmetric Rydberg state (s-state) rotations of the molecular axis do not lead to changes in the electron density distributions and therefore one may consider this state decoupled from the rotational motions. The d-state, on the contrary, has a symmetry axis along the molecular axis and thus should also rotate together with the molecule. Therefore it is reasonable to expect the rotationally-induced couplings of the 3dπ_g state to be larger than those of the 4's'σ_g state. This in turn means that the weaker coupled 4's'σ_g state is expected to be stabilized, which is observed in the experiment (see Fig. 3).

The model formulated above includes rotational couplings as a variable parameter. A first principles calculation of these couplings would require quite sophisticated methods, because the effect of l-uncoupling essentially corresponds to the breakdown of the Born–Oppenheimer approximation and requires a special treatment. Additionally, Rydberg states discussed here have lifetimes shorter than the rotational period of the ion core, which means that the rotational angular momentum and its projection are not good quantum numbers in the excited state. Another approach could be based on the MQDT theory, such as that used by Raoult *et al.*[12] The rotational couplings can possibly be inferred from high resolution spectroscopic experiments on high members of the Rydberg series, which have longer lifetimes and permit resolution of the rotational structure.

The interference stabilization mechanism presented here is general and may play a role in many phenomena where several discrete states are sufficiently strongly coupled to the same continuum. Such situations can arise, for example, in autoionization or Auger decays of atoms and molecules and lead to very complex resonance structures, which cannot be easily assigned based on static spectroscopic methods. In these situations time-dependent techniques prove to be very useful, as demonstrated here, and could help to resolve many challenging problems in spectroscopy.

5 Conclusions

This article presents results of the time-, energy- and angular-resolved photoelectron spectroscopy of a single autoionizing resonance of N_2 at 17.33 eV. The experimental photoelectron spectra reveal a transient feature corresponding to ionization of autoionizing Rydberg states excited by a short XUV pulse. Analysis of angular distributions reveals two dominant contributions to the photoelectron spectra corresponding to two transient electronic states. One state has a relatively long lifetime of 20 $\pm$ 5 fs and a peaked angular distribution of s-like character. The other state has a very short lifetime (<10 fs) and a uniform angular distribution. Such a combination of two overlapping resonances with long and short lifetimes has been previously predicted and observed for the case of laser-induced interaction of Rydberg states with an ionization continuum and is referred to as the effect of interference stabilization. Here we develop a formalism of interference stabilization for the case of autoionizing molecular resonances and apply it to describe the window resonance investigated in our experiments. We propose, that rotations of the molecule induce the otherwise symmetry-forbidden coupling between the two contributing Rydberg states, the $3d\pi_g$ and 4's'σ_g states attached to the $B^2\Sigma_u^+$ ionic core, and lead to a stabilization of one at the expense of the other. Using published values for the coupling strengths of the two contributing Rydberg states, and adjusting the strength of the unknown rotational couplings we are able to remove previously unexplained discrepancies between the calculated and experimental XUV absorption spectra. This observation suggests that even very weak couplings of electronic states can be amplified by the effect of interference stabilization and significantly modify congested XUV spectra. Our results demonstrate both the advantages of time-resolved methods for the analysis of complex overlapping autoionizing resonances and provide a general theoretical treatment of interference stabilization in molecular autoionization, which can be applied to a large variety of problems in autoionization, Auger decay and similar phenomena in atoms and molecules.

Acknowledgements

We thank Iason Katechis for help with the SVD analysis.

References

1 C. Ott, A. Kaldun, P. Raith, K. Meyer, M. Laux, J. Evers, C. H. Keitel, C. H. Greene and T. Pfeifer, *Science*, 2013, **340**, 716–720.

2 C. Ott, A. Kaldun, L. Argenti, P. Raith, K. Meyer, M. Laux, Y. Zhang, A. Blattermann, S. Hagstotz, T. Ding, R. Heck, J. Madronero, F. Martin and T. Pfeifer, *Nature*, 2014, **516**, 374–378.

3 H. Mashiko, T. Yamaguchi, K. Oguri, A. Suda and H. Gotoh, *Nat. Commun.*, 2014, **5**, 5599.

4 M. Kotur, D. Guenot, A. Jimenez-Galan, D. Kroon, E. W. Larsen, M. Louisy, S. Bengtsson, M. Miranda, J. Mauritsson, C. L. Arnold, S. E. Canton, M. Gisselbrecht, T. Carette, J. M. Dahlstrom, E. Lindroth, A. Maquet, L. Argenti, F. Martin and A. L'Huillier, *Nat. Commun.*, 2016, **7**, 10566.

5 U. Fano, *Phys. Rev.*, 1961, **124**, 1866–1878.

6 M. Eckstein, C.-H. Yang, F. Frassetto, L. Poletto, G. Sansone, M. J. J. Vrakking and O. Kornilov, *Phys. Rev. Lett.*, 2016, **116**, 163003.
7 Y. Heller and A. Popov, *Phys. Lett. A*, 1976, **56**, 453–454.
8 S. Feneuille, S. Liberman, E. Luc-Koenig, J. Pinard and A. Taleb, *J. Phys. B: At., Mol. Opt. Phys.*, 1982, **15**, 1205–1216.
9 J. Y. Liu, P. McNicholl, D. A. Harmin, J. Ivri, T. Bergeman and H. J. Metcalf, *Phys. Rev. Lett.*, 1985, **55**, 189–192.
10 M. V. Fedorov and A. Movsesian, *J. Phys. B: At., Mol. Opt. Phys.*, 1988, **21**, L155–L158.
11 J. J. Hopfield, *Phys. Rev.*, 1930, **35**, 1133–1134.
12 M. Raoult, H. L. Rouzo, G. Raseev and H. Lefebvre-Brion, *J. Phys. B: At. Mol. Phys.*, 1983, **16**, 4601–4617.
13 K. P. Huber, G. Stark and K. Ito, *J. Chem. Phys.*, 1993, **98**, 4471–4477.
14 A. S. Stodolna, F. Lépine, T. Bergeman, F. Robicheaux, A. Gijsbertsen, J. H. Jungmann, C. Bordas and M. J. J. Vrakking, *Phys. Rev. Lett.*, 2014, **113**, 103002.
15 F. H. Mies, *Phys. Rev.*, 1968, **175**, 164–175.
16 M. V. Fedorov, *Atomic and Free Electrons in a Strong Light Field*, World Scientific, 1998.
17 M. Eckstein, C.-H. Yang, M. Kubin, F. Frassetto, L. Poletto, H.-H. Ritze, M. J. J. Vrakking and O. Kornilov, *J. Phys. Chem. Lett.*, 2015, **6**, 419–425.
18 T. Pfeifer, C. Spielmann and G. Gerber, *Rep. Prog. Phys.*, 2006, **69**, 443–505.
19 O. Ghafur, W. Siu, P. Johnsson, M. F. Kling, M. Drescher and M. J. J. Vrakking, *Rev. Sci. Instrum.*, 2009, **80**, 033110.
20 V. Dribinski, A. Ossadtchi, V. A. Mandelshtam and H. Reisler, *Rev. Sci. Instrum.*, 2002, **73**, 2634–2642.
21 N. M. Kroll and K. M. Watson, *Phys. Rev. A*, 1973, **8**, 804–809.
22 A. Maquet and R. Taïeb, *J. Mod. Opt.*, 2007, **54**, 1847–1857.
23 G. H. Golub and C. Reinsch, *Numer. Math.*, 1970, **14**, 403–420.
24 S. Yamaguchi and H.-o. Hamaguchi, *J. Chem. Phys.*, 1998, **109**, 1397–1408.
25 P. Morin, M. Adam, I. Nenner, J. Delwiche, M. Hubin-Franskin and P. Lablanquie, *Nucl. Instrum. Methods Phys. Res.*, 1983, **208**, 761–766.

Faraday Discussions

PAPER

Femtosecond photodissociation dynamics of 1,4-diiodobenzene by gas-phase X-ray scattering and photoelectron spectroscopy

Brian Stankus,[a] James M. Budarz,[ab] Adam Kirrander,[c] David Rogers,[c] Joseph Robinson,[b] Thomas J. Lane,[b] Daniel Ratner,[b] Jerome Hastings,[b] Michael P. Minitti*[b] and Peter M. Weber*[a]

Received 13th May 2016, Accepted 2nd June 2016
DOI: 10.1039/c6fd00135a

We present a multifaceted investigation into the initial photodissociation dynamics of 1,4-diiodobenzene (DIB) following absorption of 267 nm radiation. We combine ultrafast time-resolved photoelectron spectroscopy and X-ray scattering experiments performed at the Linac Coherent Light Source (LCLS) to study the initial electronic excitation and subsequent rotational alignment, and interpret the experiments in light of Complete Active Space Self-Consistent Field (CASSCF) calculations of the excited electronic landscape. The initially excited state is found to be a bound $^1B_1$ surface, which undergoes ultrafast population transfer to a nearby state in 35 ± 10 fs. The internal conversion most likely leads to one or more singlet repulsive surfaces that initiate the dissociation. This initial study is an essential and prerequisite component of a comprehensive study of the complete photodissociation pathway(s) of DIB at 267 nm. Assignment of the initially excited electronic state as a bound state identifies the mechanism as predissociative, and measurement of its lifetime establishes the time between excitation and initiation of dissociation, which is crucial for direct comparison of photoelectron and scattering experiments.

Introduction

Photodissociation reactions are ideal prototypical phenomena to explore the dynamics of unimolecular systems and the conversion of photon energy into chemical energy. Photochemical bond cleavage can occur *via* a variety of mechanisms, including direct promotion to a repulsive potential energy surface or predissociation after excitation to a bound state coupled to a dissociation

[a]*Dept. of Chemistry, Brown University, Providence, RI 02912, USA. E-mail: peter_weber@brown.edu*
[b]*SLAC National Accelerator Laboratory, Menlo Park, CA 94025, USA. E-mail: minitti@slac.stanford.edu*
[c]*EaStCHEM, School of Chemistry, University of Edinburgh, David Brewster Road, Edinburgh EH9 3FJ, UK*

continuum. Because they often proceed on very fast time scales, photodissociation reactions are ideally investigated using ultrafast techniques.

The photolysis of halobenzenes has been the subject of extensive experimental study,[1–24] typically *via* analysis of photofragment energy distributions. Because of the strong coupling of the π system to a series of repulsive σ* states, their photodissociation dynamics can be quite complex. 1,4-Dichlorobenzene, for example, has been shown to undergo predissociation,[19] yet iodobenzene dissociates *via* multiple competing pathways.[12,16,23] The photodissociation of 1,4-diiodobenzene (DIB) has been observed for many years, but the mechanism, or set of mechanisms, remains largely unknown. The early photofragment studies by Bersohn *et al.* indicated that the translational energy distribution of the iodophenyl radical did not depend on the excitation wavelength, leading those researchers to propose a predissociation mechanism.[6] No time-resolved studies of DIB have yet been reported to support that mechanism explanation or to provide time scales for the predissociation reaction.

The present investigation aims to explore the time-resolved reaction dynamics of DIB, illustrated in Fig. 1, including the initial excitation and subsequent wavepacket dynamics. By combining femtosecond spectroscopic and X-ray scattering experiments with high-level calculations of potential energy surfaces, we aim to map both the electronic and structural dynamics as a function of time. In this manuscript chronicling our initial results, we focus on the nature and lifetime of the initially excited state. Understanding this initial excitation establishes the starting point for the reaction, which is a prerequisite to understanding the mechanism of the subsequent electronic and structural dynamics. In addition, the character of the initially excited state will reveal the temporal relationship between the electronic excitation and the initiation of structural dynamics, as is

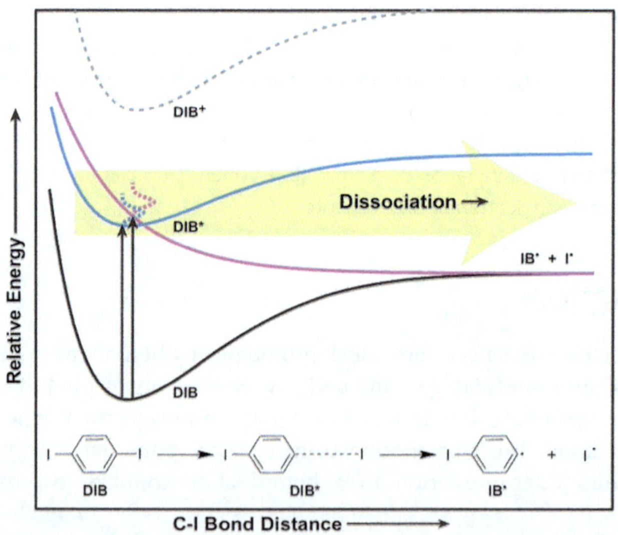

Fig. 1 Illustrative schematic of the possible photodissociation pathways of DIB. The graph shows the ground and two excited state potentials corresponding to direct and indirect dissociation. The ionization potential, corresponding to the lowest DIB$^+$ energy, is also included (dashed line).

required to allow for a direct comparison of spectroscopic and scattering experiments.

Experimental and computational approaches

The ultrafast photodissociation of 1,4-diiodobenzene (DIB) is investigated experimentally using pump–probe methodologies. In the first step, excitation at 267 nm launches a wavepacket on the excited-state surface. The nature of the excited-state surface and the related wavepacket dynamics govern the timescale of the photodissociation reaction. In the second step, the progress of the reaction is probed at variable time delays. We use two complementary approaches to illuminate the reaction from very different vantage points.

In the time-resolved spectroscopy experiment, the molecule is probed *via* photoionization, and the kinetic energies of ejected electrons are analyzed using a time-of-flight photoelectron spectrometer. A Ti:Sapphire oscillator (Coherent, Mantis) generates ~35 fs pulses at 800 nm, seeding a regenerative amplifier (Coherent Legend Elite Duo) operating at 5 kHz. This output is split, with 10% of the light being upconverted to the third harmonic at 267 nm using nonlinear optical crystals to serve as the pump pulse. The remaining 90% of the amplifier output seeds an optical parametric amplifier (Coherent OPerA Solo) that is tuned to 300 nm, which serves as the ionizing probe pulse. The laser pulses perpendicularly intersect a molecular beam generated by expanding a gas mixture of ~7 torr of DIB in ~850 torr of helium through a 100 μm diameter nozzle and a 210 μm diameter skimmer. By delaying the probe pulse relative to the pump pulse, we measure the kinetic energy of the ejected electrons as a function of time. The instrument response function was determined by measuring the two-color photoionization signal of acetone, which has a response time shorter than the temporal resolution of the experiment. Further details of the apparatus have been described in detail elsewhere.[25,26]

In the scattering experiment, the time-evolving molecule is probed *via* X-ray scattering using ultrafast X-ray pulses generated at the Linac Coherent Light Source (LCLS).[27,28] This apparatus has also been described in detail elsewhere.[29] The 267 nm pump pulse has a duration of 65 fs and is synchronized to the 120 Hz repetition rate pulse stream of the LCLS. Each X-ray probe pulse, with a photon energy of 8.3 keV, contains about 10^{12} photons and has a 30 fs pulse duration. For each delay time setting, the temporal jitter between the UV and X-ray pulses is monitored with $<$10 fs accuracy using a specialized timing tool that has been described previously.[30] The optical and X-ray pulses are focused and combined collinearly into the diffraction chamber, which was heated to 90 °C and contained less than 0.5 torr of DIB. The scattering pattern in the range of 1.0 to 4.2 $Å^{-1}$ is measured using a CSPAD.[31] The ultrafast optical pump, X-ray probe technique was previously implemented in the study of the ring-opening of 1,3-cyclohexadiene.[32,33] Given the complexity of the DIB system, to date we have been able to investigate only the angular dependence of the scattering signal after the dissociation is complete. This is a first and necessary step for a more comprehensive study that will rest on the results presented here.

It is important to note that the scattering experiments are performed on a warm, low-pressure static gas, while the spectroscopic experiments probe the dynamics of molecules that have been cooled by a free-jet expansion. This implies that while the molecules in the scattering experiments are in thermally populated vibrational states, almost all the molecules in the spectroscopy experiments are in their vibrational ground state. But since the experiments by Bersohn *et al.* showed no dependence on the excitation wavelength,[6] we postulate that this difference does not invalidate the comparison of the results.

As for the computations, Complete Active Space Self-Consistent Field (CASSCF) calculations were performed using a correlation-consistent Dunning basis set of triple-zeta plus polarization quality (cc-pVTZ) for carbon and hydrogen atoms.[34] Iodine atoms were described by an effective core potential (ECP) to describe a 46-electron core, *i.e.* large-core [Kr]$5d^{10}$, with the larger cc-pVTZ-pp basis set of Peterson *et al.*[35] to describe the seven valence electrons. The ECP for iodine contains terms for the evaluation of spin–orbit coupling for the valence electrons.

To account for the effect of dynamical electron correlation on the predicted vertical transition properties, complete active space with second order perturbation theory (CASPT2) calculations were performed, using the implementation where the doubly external configurations are internally contracted and in which also subspaces of the singly external and internal configuration spaces are internally contracted.[36] An imaginary level shift of 0.3 hartree was employed to avoid intruder state problems and the carbon 1s orbitals were frozen in the CASPT2 calculations.

All calculations employed C_{2v} symmetry with the DIB molecule lying in the *yz*-plane. Following Ajitha *et al.*'s work on iodobenzene,[37] two active spaces were employed for the CASSCF calculations on DIB: a fourteen electron in twelve orbital active space, denoted (14,12), and a (18,14) active space. The (14,12) space comprises the six benzene-like π orbitals and electrons and, for each iodine, the two C–I σ bond orbitals and electrons and the iodine p_x (out-of-plane) lone pair of electrons. The (18,14) space comprises all orbitals and electrons from the (14,12) space plus the iodine p_y (in-plane) lone pair for each iodine atom. The smaller (14,12) space is necessary due to an orbital rotation between a C–C σ core orbital (irreducible representation, IR, b_2) with the iodine p_y (in-plane) lone pair orbital (IR b_2) during unconstrained CASSCF optimization of the (18,14) active space's orbitals. Hence for the (18,14) active space, core orbitals were frozen, following an initial CASSCF(14,12) calculation, which places correctly the iodine p_y orbitals in the larger active space.

The geometry of the $^1A_1$ ground state of DIB was optimized at the CASSCF(14,12)/cc-pVTZ-pp level of theory. A 16 state-averaged (SA) CASSCF(18,14) wave function was used to study the ground and excited states: two roots in each IR of C_{2v} symmetry (A_1, B_1, B_2, A_2) of both singlet and triplet spin multiplicity. The CASPT2 calculation used the 16 SA-CASSCF(18,14) wave function as the reference wave function. Calculations on DIB^+ employed an analogous (17,14) active space in an 8 SA-CASSCF wave function: two roots of doublet spin multiplicity for each IR. All calculations were performed using the Molpro *ab initio* electronic structure program.[38]

Potential energy surfaces of DIB

To understand the structural dynamics and to aid the analysis of the experimental results we performed high-level computations to determine the potential energy surfaces for the relevant excited electronic states. Because of its large number of electrons, DIB is an exceedingly challenging system for sophisticated calculations.

The calculated potential energy surfaces of DIB are shown as a function of the dissociating C–I bond distance in Fig. 2. In addition, the CASPT2 calculated vertical excitation energies to the same excited states are compared with the CASSCF results in Table 1, which also includes transition dipole moments and oscillator strengths. Although the CASSCF curves are a reasonably accurate representation of the shapes of the excited states, the CASPT2 excitation energies serve as more accurate relative energies. Considering these results in combination, there are three energetically accessible singlet states. The transition dipole moments from the ground state to the $^1B_1(1)$, $^1B_1(2)$, and $^1B_2(1)$ states are computed as 0.19, 0.31, and 0.12, respectively. This indicates that the dominant transition is from the ground $^1A_1(1)$ surface to the $^1B_1(2)$ state, which has a transition dipole moment perpendicular to the molecular plane. At the excitation energy considered in this study, the $^1B_1(2)$ state is bound and dissociation can only proceed *via* crossing to a nearby repulsive surface. The computational results show that absorptions to the $^1B_1(1)$ and $^1B_2(1)$ states are also possible. Those excitations have their transition dipoles oriented perpendicular to the molecular plane, and in-plane but perpendicular to the I–I axis, respectively. Excitation to the $^1B_1(1)$ state would initiate barrierless dissociation. Excitation to the $^1B_2(1)$ state would launch a wavepacket that can immediately clear a barrier, but might incur some reflection since the pump energy of 267 nm places the system only slightly above the barrier.

Energetically, there are two repulsive singlet states and three repulsive triplet states that could potentially initiate dissociation at our experimental excitation wavelength of 267 nm. Although it has not been explicitly accounted for in this calculation, it is clear that the presence of two heavy iodine atoms introduces a large amount of spin–orbit coupling, which could facilitate intersystem crossing. There are six energetically accessible triplet states in the excitation region. However, the two accessible repulsive singlet states, $^1B_1(1)$ and/or the $^1B_2(1)$, are closer in energy than any repulsive triplet to the dominantly excited $^1B_1(2)$ state, and are therefore likely most strongly coupled. The $^3B_2(2)$ state is also

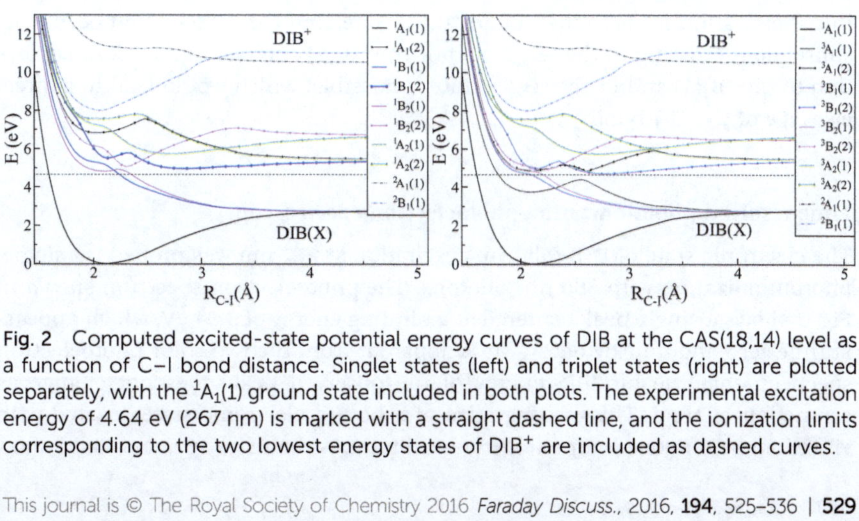

Fig. 2 Computed excited-state potential energy curves of DIB at the CAS(18,14) level as a function of C–I bond distance. Singlet states (left) and triplet states (right) are plotted separately, with the $^1A_1(1)$ ground state included in both plots. The experimental excitation energy of 4.64 eV (267 nm) is marked with a straight dashed line, and the ionization limits corresponding to the two lowest energy states of DIB$^+$ are included as dashed curves.

Table 1 Vertical excitation energies, dipole transition moments (TDM), and oscillator strengths (*f*) for DIB at the CAS(14,12)/cc-pVTZ-pp optimized geometry with $R_{C-I} = 2.10$ Å. The CASSCF calculations are performed using 16-state-averaged CAS(18,14)-SCF/cc-pVTZ-pp with ECP46, with the CASPT2 based on the CASSCF. The ionized DIB^+ ground state energy is taken from a 8-state-averaged CAS(17,14) and corresponding CASPT2 calculation. All calculations use C_{2v} symmetry. The I–I molecular axis is aligned with the *z*-coordinate axis, with the molecule in the *yz*-plane. The electronic character of each ground state to singlet excited state transition is indicated in the column 'Transitions'. Only singlet–singlet TDM's and *f*'s are given (singlet–triplet transitions are strictly zero in the absence of spin–orbit coupling)

	SA-CAS(18,14)-SCF			CASPT2		
State	*E* (eV)	TDM (au)	*f* (au)	*E* (eV)	*f* (au)	Transition
$^1A_1(1)$	0.00	0.000	0.000	0.00	0.000	—
$^1A_1(2)$	6.85	2.327	1.818	5.14	1.363	$\pi \rightarrow \pi^*(z)$
$^1B_1(1)$	5.22	0.190	0.009	4.31	0.008	$\pi \rightarrow \sigma^*(x)$
$^1B_1(2)$	5.57	0.312	0.027	4.41	0.021	$\pi \rightarrow \sigma^*(x)$
$^1B_2(1)$	4.87	0.122	0.004	4.40	0.003	$\pi \rightarrow \pi^*(y)$
$^1B_2(2)$	6.04	0.351	0.036	4.99	0.030	$n \rightarrow \sigma^*(y)$
$^1A_2(1)$	7.18	0.000	0.000	5.89	0.000	$\pi \rightarrow \pi^*$
$^1A_2(2)$	7.59	0.000	0.000	6.11	0.000	$n \rightarrow \pi^*$
$^3A_1(1)$	3.74	—	—	3.64	—	—
$^3A_1(2)$	4.88	—	—	4.27	—	—
$^3B_1(1)$	4.69	—	—	3.93	—	—
$^3B_1(2)$	5.03	—	—	4.02	—	—
$^3B_2(1)$	4.89	—	—	4.20	—	—
$^3B_2(2)$	5.11	—	—	4.33	—	—
$^3A_2(1)$	7.19	—	—	5.90	—	—
$^3A_2(2)$	7.58	—	—	6.10	—	—
$^2B_1(1)$ [DIB^+]	8.05	—	—	8.43	—	—

energetically close to the $^1B_1(2)$ state, but it is bound at this energy and would not directly initiate dissociation. It is also worth noting that the spin–orbit coupling will be critical to evaluate the asymptotic dissociation limits of the system beyond the initial early time dynamics.

The calculated potential energy surfaces suggest that the dominant decay mechanism following absorption of 267 nm radiation is predissociation. The initial excitation leads to the bound $^1B_1(2)$ state, and the dissociation occurs *via* subsequent crossing to the $^1B_1(1)$ and/or the $^1B_2(1)$ surface. In addition, absorption to the $^1B_1(1)$ and/or the $^1B_2(1)$ state is possible, which would lead to a direct cleavage of the C–I bond.

Time-resolved photoionization-photoelectron spectroscopy

The electronic state of DIB following excitation at 267 nm was probed *via* single-photon ionization with 300 nm photons. The photoelectron spectrum shown in Fig. 3 shows a single peak centered at a binding energy of 3.94 eV, which appears at time zero and quickly decays. It is apparent from the two-color photoelectron spectrum that the binding energy of the observed peak does not change as a function of time. This is suggestive of a bound electronic state, excited with vibrational energy near an energetic minimum. A repulsive surface would show

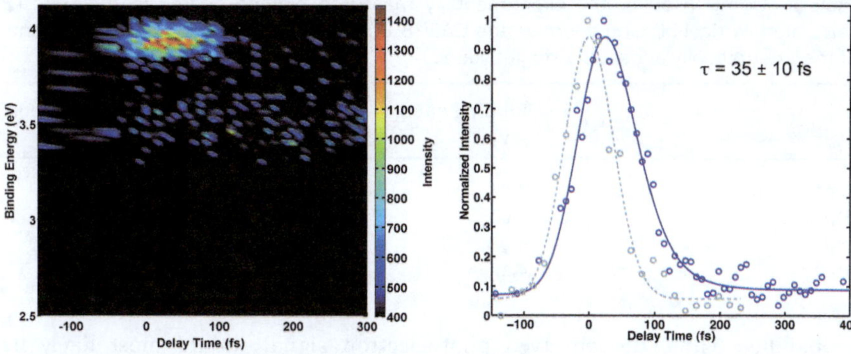

Fig. 3 (Left) Two-color photoelectron spectrum of DIB, with 267 nm pump and 300 nm probe with one-color photoelectron signal subtracted. (Right) Temporal evolution of the two-color photoelectron signal centered at a binding energy of 3.94 eV (blue), and the instrument response function measured *via* the photoionization signal of acetone (grey). Circles represent the experimental normalized intensities, and the solid lines represent the least-squares fits.

a decreasing binding energy as a function of delay time. Thus, the decay of the photoelectron signal is interpreted as arising from population transfer to a nearby electronic state, likely the repulsive surface(s) that leads to dissociation.

The temporal evolution of the observed peak was fitted to a single exponential decay, with time constant τ, plus a slight baseline shift at $t = 0$. The exponential decay represents population transfer from the initially excited state (observed) to another coupled surface (unobserved). The lifetime of the initially excited state was found to be 35 $\pm$ 10 fs (all errors reported to 3σ), indicating ultrafast population transfer to a strongly coupled surface. The wavepacket is unobserved after crossing either because of a significant decrease in ionization cross-section, or because the second state quickly decreases in energy. In the present experiment, our probe energy allows for single-photon ionization of states with binding energies up to 4.13 eV.

To assign the initially excited bound state, we compare our measured photoelectron spectrum to the computed potential energy surfaces. As suggested by the computational results, initial absorption to the $^1B_1(1)$, $^1B_1(2)$, and/or $^1B_2(1)$ states is possible, with excitation to the $^1B_1(2)$ state having the highest probability. The experimentally observed signal indicates a bound state, which is consistent with absorption to $^1B_1(2)$. The $^1B_2(1)$ state also has a slight barrier and could cause partial reflection, but a significant portion of the wavepacket would likely proceed immediately over the barrier. We conclude that the observed time dependence of the spectral signature is most consistent with single-photon ionization out of the $^1B_1(2)$ state.

Vertical CASPT2 binding energies were obtained by subtracting the electronic excitation energies listed in Table 1 from the ionization energy. Table 2 provides a comparison of the experimentally observed binding energy with the computed binding energies of the relevant states. The experimentally measured binding energy best agrees with the computed binding energy of the $^1B_1(2)$ state. As previously discussed, this state also has the highest computed excitation

Table 2 Comparison of the experimentally measured binding energy to the CASPT2 calculated vertical binding energies and CAS(18,14) calculated transition dipole moments of the energetically accessible singlet states

	Binding energy (eV)	Transition dipole moment
Experiment	3.94	—
$^1B_1(2)$	4.02	0.31
$^1B_2(1)$	4.03	0.12
$^1B_1(1)$	4.12	0.19

probability. Thus, the observed photoelectron signal should most likely be attributed to single-photon ionization out of the $^1B_1(2)$ state.

Additional spectroscopic signatures that could be attributed to transitions to the $^1B_1(1)$ and/or the $^1B_2(1)$ state are not observed. No significant two-color photoelectron signal was observed at the binding energy predicted for the $^1B_1(1)$ state, and no signal with a sloping binding energy (which would be consistent with excitation to the repulsive $^1B_2(1)$ state) was observed. In addition, a single exponential decay fully captures the time dependence of the observed signal, indicating that population transfer occurs from only a single electronic state. If other processes occur, they either overlap the dominant process energetically and have very similar temporal profiles, or they give significantly lower photoelectron intensity and can therefore not be observed.

X-ray scattering results

The percent difference in pump–probe X-ray scattering signal is defined as $\Delta I = 100 \times \frac{I_{\text{pumped}}(\phi) - I_{\text{unpumped}}(\phi)}{I_{\text{unpumped}}(\phi)}$ where $I(\phi)$ is the sum of $I(q, \phi)$ over all

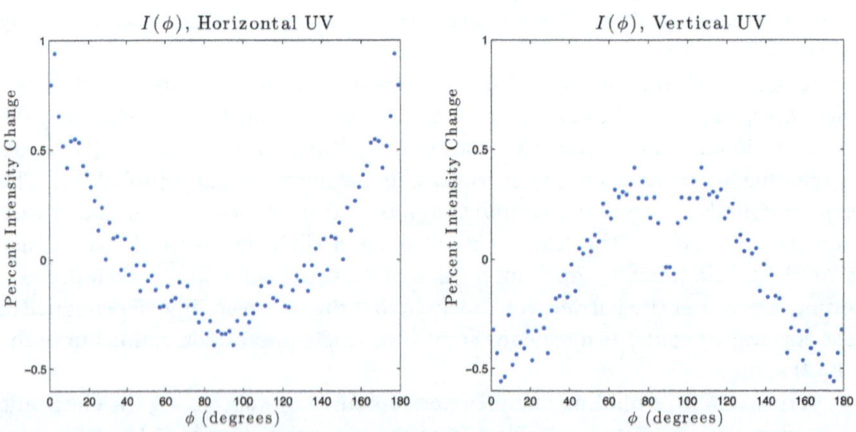

Fig. 4 Overall intensity change of the X-ray scattering signal, as a function of azimuthal angle, at a time of 1 ps after excitation. The X-rays are polarized horizontally, and an azimuthal angle of 0° is defined to be the horizontal plane. Based on the symmetry of the system the signal from each quadrant must be identical and their signals have been added to reduce the noise. For visual display, the data points are plotted from 0° to 180°, so that each data point is plotted twice.

observed q. This difference signal at a time of 1 ps after the photodissociation of DIB is shown as a function of the azimuthal detection angle ϕ in Fig. 4. Measurements were taken with the UV pump laser polarized either horizontally or vertically, while the X-ray probe is horizontally polarized in all cases. Ejection of an iodine atom results in an oscillatory difference signal which, in an experiment with limited q range, leads to an apparent change in the scattering signal as shown in Fig. 4. The dip at 90° in the plot for vertical UV excitation is not yet understood but might be caused by the gaps in the CSPAD detector in the vertical direction. For either horizontal or vertical UV polarization, the percent difference in the scattering signal is most negative perpendicular to the direction of the respective laser polarization. This indicates that for the laser-excited molecules the I–I axis is preferentially oriented in the vertical (90°) direction for a horizontal pump and in the horizontal (0°) direction for a vertical pump, respectively. Although the anisotropy will dephase on a picosecond timescale, the preferential alignment immediately after excitation provides additional support for the nature of the initially excited state. The depletion of molecules with an I–I axis perpendicular to the UV pump pulse indicates that the transition dipole moment of the excitation must be oriented perpendicular to the I–I axis. This agrees with our assignment of the initial excitation as an absorption to the $^1B_1(2)$ state, which has a transition dipole moment oriented perpendicular to the molecular plane.

Conclusions

We present a combined experimental and computational study of the photodissociation of 1,4-diiodobenzene following excitation at 267 nm. The experimental investigations include both ultrafast time-resolved photoionization-photoelectron spectroscopy and X-ray scattering. Combined with high-level calculations, we establish the nature of the initially excited electronic state. This lays the foundation for future explorations of the electronic landscape and reaction pathways by time-resolved photoelectron spectroscopy and X-ray scattering.

Our computational results indicate that the dominant process is the initial excitation to the $^1B_1(2)$ bound electronic state and that the dissociation must be initiated by internal conversion to one or more nearby repulsive surfaces. This finding is consistent with the photoelectron results, which indicate absorption to a bound state with a lifetime of 35 ± 10 fs. Although two other nearby singlet states were calculated to have sizeable transition dipole moments, only one excited electronic state was observed. This state was assigned as the $^1B_1(2)$ bound state because its absorption has the largest calculated transition dipole moment, and its bound character at our experimental excitation energy agrees qualitatively with the observed photoelectron spectrum. In addition, the measured binding energy of the initially excited state agrees best with the computed binding energy of the $^1B_1(2)$ state. Furthermore, our X-ray scattering results show a substantial rotational alignment of excited molecules that is consistent with the orientation of the transition dipole moment of the initial excitation.

Although several time-resolved studies of related halobenzenes have been reported,[12,16,17,19,24] these experiments recorded the appearance of photofragments, leaving the lifetimes of the involved electronic states undetermined except for a qualitative comparison to rotational time scales. Dissociations faster than rotations (<1 ps) are often ascribed to repulsive surfaces, while slower photofragment appearances are attributed to inherent lifetimes of an initially excited state. For example, the photodissociation of 1,4-dichlorobenzene proceeds in 122 ps and follows a predissociation mechanism[19] with the initially excited π, π^* state having a lifetime of ~122 ps. Similarly, 1,4-dibromobenzene also undergoes predissociation, dissociating in 18.2 ps.[17] In contrast, our findings on DIB indicate a predissociation with a curve crossing, but on a time scale that is typically associated with a direct dissociation. This continues the trend of increased curve crossing rates with heavier halogen substitution. A similar trend has been observed for singly substituted halobenzenes.

Because the rate of curve crossing increases for heavier halogens, the effect has been attributed to intersystem crossing facilitated by spin–orbit coupling. However, our computational results indicate that the singlet repulsive surfaces are closest in energy to the initially excited state so that the mechanism in DIB is likely an internal conversion. Predissociation *via* intersystem crossing has not been directly observed in the other halobenzenes, but it is known that the dominant absorption leading to the photodissociation of methyl iodide is a triplet.[39] An analogous state has been assumed to initiate dissociation in halobenzenes. Instead, we find that while it is possible that a minor pathway might involve intersystem crossing, it is likely that the dominant pathway involves internal conversion.

Looking forward, we plan to investigate further the wavepacket dynamics and structural evolution in DIB upon optical excitation. Our attention will turn to a full CASPT2 calculation of the excited-state potential energy surfaces wherein we will attempt to incorporate state mixing *via* empirical spin–orbit coupling constants. Although *ab initio* electronic structure calculations are very costly for this system, one should consider all-atom quantum molecular dynamics simulations of the photodissociation dynamics to interpret the data as fully as possible.[40] Experimentally, we will study the reaction dynamics using both time-resolved ultrafast pump–probe X-ray scattering and photoionization photoelectron spectroscopy. We expect that combined ultrafast experimental studies, such as in this article and in a prior publication,[26] will become increasingly influential in interpreting complex photochemical processes, as they provide a more comprehensive signal and important cross-checks of the models used to interpret the results.

Acknowledgements

Portions of this research were carried out on the XPP Instrument at the Linac Coherent Light Source (LCLS), SLAC National Accelerator Laboratory, and were supported by the U.S. Department of Energy, Office of Science, Office of Basic Energy Sciences under Contract No. DE-AC02-76SF00515. PWM acknowledges funding by the Defense Threat Reduction Agency (DTRA), Grant No. HDTRA1-14-1-0008. BMS acknowledges funding by the NASA Rhode Island Space Grant Consortium. We also thank Jeremy Harvey for sharing the ECP46 basis set with us. The computational work used the ARCHER UK National Supercomputing Service (http://www.archer.ac.uk).

References

1 W. Wolf and N. Kharasch, *J. Org. Chem.*, 1961, **26**, 283.

2 J. A. Kampmeier and E. Hoffmeister, *J. Am. Chem. Soc.*, 1962, **84**, 3787.
3 A. P. Marchetti and D. R. Kearns, *J. Am. Chem. Soc.*, 1967, **89**, 5335.
4 A. P. Marchetti and D. R. Kearns, *J. Am. Chem. Soc.*, 1967, **89**, 768.
5 M. Dzvonik, S. Yang and R. Bersohn, *J. Chem. Phys.*, 1974, **61**, 4408.
6 M. Kawasaki, S. J. Lee and R. Bersohn, *J. Chem. Phys.*, 1977, **66**, 2647.
7 A. Freedman, S. C. Yang, M. Kawasaki and R. Bersohn, *J. Chem. Phys.*, 1980, **72**, 1028.
8 T. Ichimura, Y. Mori, H. Shinohara and N. Nishi, *Chem. Phys. Lett.*, 1985, **122**, 55.
9 T. Ichimura, Y. Mori, H. Shinohara and N. Nishi, *J. Spectrosc. Soc. Jpn.*, 1986, **35**, 299.
10 H. J. Hwang and M. a. El-Sayed, *J. Chem. Phys.*, 1992, **96**, 856.
11 J. E. Freitas, H. J. Hwang and M. A. El-Sayed, *J. Phys. Chem.*, 1993, **97**, 12481.
12 P. Y. Cheng, D. Zhong and A. H. Zewail, *Chem. Phys. Lett.*, 1995, **237**, 399.
13 H. J. Hwang and M. A. El-Sayed, *J. Photochem. Photobiol., A*, 1996, **102**, 13.
14 G.-J. Wang, R.-S. Zhu, H. Zhang, K.-L. Han, G.-Z. He and N.-Q. Lou, *Chem. Phys. Lett.*, 1998, **288**, 429.
15 R.-S. Zhu, H. Zhang, G.-J. Wang, X.-B. Gu, K.-L. Han, G.-Z. He and N.-Q. Lou, *Chem. Phys.*, 1999, **248**, 285.
16 M. Kadi, J. Davidsson, A. N. Tarnovsky, M. Rasmusson and E. Åkesson, *Chem. Phys. Lett.*, 2001, **350**, 93.
17 M. Kadi and J. Davidsson, *Chem. Phys. Lett.*, 2003, **378**, 172.
18 D. Senapati, S. Maity and P. K. Das, *J. Phys. Chem. A*, 2004, **108**, 7949.
19 L.-W. Yuan, J.-Y. Zhu, Y.-Q. Wang, L. Wang, J.-L. Bai and G.-Z. He, *Chem. Phys. Lett.*, 2005, **410**, 352.
20 M. D. Poulsen, E. Skovsen and H. Stapelfeldt, *J. Chem. Phys.*, 2002, **117**, 2097.
21 K.-L. Han and G.-Z. He, *J. Photochem. Photobiol., C*, 2007, **8**, 55.
22 X.-P. Zhang, Z.-R. Wei, Y. Tang, T.-J. Chao, B. Zhang and K.-C. Lin, *ChemPhysChem*, 2008, **9**, 1130.
23 A. G. Sage, T. A. A. Oliver, D. Murdock, M. B. Crow, G. A. D. Ritchie, J. N. Harvey and M. N. R. Ashfold, *Phys. Chem. Chem. Phys.*, 2011, **13**, 8075.
24 O. A. Borg, Y. J. Liu, P. Persson, S. Lunell, D. Karlsson, M. Kadi and J. Davidsson, *J. Phys. Chem. A*, 2006, **110**, 7045.
25 B. Kim, N. Thantu and P. M. Weber, *J. Chem. Phys.*, 1992, **97**, 5384.
26 C. C. Pemberton, Y. Zhang, K. Saita, A. Kirrander and P. M. Weber, *J. Phys. Chem. A*, 2015, **119**, 8832.
27 C. Bostedt, D. M. Fritz, Z. Huang, H. J. Lee, H. Lemke, W. F. Schlotter, J. J. Turner and G. J. Williams, *Rev. Mod. Phys.*, 2016, **88**, 015007.
28 W. E. White, A. Robert and M. Dunne, *J. Synchrotron Radiat.*, 2015, **22**, 472.
29 J. M. Budarz, M. P. Minitti, B. Stankus and A. Kirrander, *J. Phys. B: At., Mol. Opt. Phys.*, 2016, **49**, 34001.
30 M. Harmand, R. Coffee, M. R. Bionta, M. Chollet, D. French, D. Zhu, D. M. Fritz, H. T. Lemke, N. Medvedev, B. Ziaja, S. Toleikis and M. Cammarata, *Nat. Photonics*, 2013, **7**, 215.
31 H. T. Philipp, M. Hromalik, M. Tate, L. Koerner and S. M. Gruner, *Nucl. Instrum. Methods Phys. Res., Sect. A*, 2011, **649**, 67.
32 M. P. Minitti, J. M. Budarz, A. Kirrander, J. Robinson, T. J. Lane, D. Ratner, K. Saita, T. Northey, B. Stankus, V. Cofer-Shabica, J. Hastings and P. M. Weber, *Faraday Discuss.*, 2014, **171**, 81.

33 M. P. Minitti, J. M. Budarz, A. Kirrander, J. S. Robinson, D. Ratner, T. J. Lane, D. Zhu, J. M. Glownia, M. Kozina, H. T. Lemke, M. Sikorski, Y. Feng, S. Nelson, K. Saita, B. Stankus, T. Northey, J. B. Hastings and P. M. Weber, *Phys. Rev. Lett.*, 2015, **114**, 255501.
34 T. H. Dunning Jr, *J. Chem. Phys.*, 1989, **90**, 1007.
35 K. a. Peterson, B. C. Shepler, D. Figgen and H. Stoll, *J. Phys. Chem. A*, 2006, **110**, 13877.
36 P. Celani and H.-J. Werner, *J. Chem. Phys.*, 2000, **112**, 5546.
37 D. Ajitha, D. G. Fedorov, J. P. Finley and K. Hirao, *J. Chem. Phys.*, 2002, **117**, 7068.
38 (*a*) H. J. Werner, P. J. Knowles, G. Knizia, F. R. Manby and M. Schütz, *Wiley Interdiscip. Rev.: Comput. Mol. Sci.*, 2012, **2**, 242; (*b*) H.-J. Werner, P. J. Knowles, G. Knizia, F. R. Manby, M. Schütz, P. Celani, T. Korona, R. Lindh, A. Mitrushenkov, G. Rauhut, K. R. Shamasundar, T. B. Adler, R. D. Amos, A. Bernhardsson, A. Berning, D. L. Cooper, M. J. O. Deegan, A. J. Dobbyn, F. Eckert, E. Goll, C. Hampel, A. Hesselmann, G. Hetzer, T. Hrenar, G. Jansen, C. Köppl, Y. Liu, A. W. Lloyd, R. A. Mata, A. J. May, S. J. McNicholas, W. Meyer, M. E. Mura, A. Nicklass, D. P. O'Neill, P. Palmieri, D. Peng, K. Pflüger, R. Pitzer, M. Reiher, T. Shiozaki, H. Stoll, A. J. Stone, R. Tarroni, T. Thorsteinsson and M. Wang, *MOLPRO, version 2012.1, a package of ab initio programs*, see http://www.molpro.net.
39 S. H. Gardiner, M. L. Lipciuc, T. N. V. Karsili, M. N. R. Ashfold and C. Vallance, *Phys. Chem. Chem. Phys.*, 2015, **17**, 4096.
40 A. Kirrander, K. Saita and D. V Shalashilin, *J. Chem. Theory Comput.*, 2016, **12**, 957.

Faraday Discussions

PAPER

Femtosecond charge and molecular dynamics of I-containing organic molecules induced by intense X-ray free-electron laser pulses†

K. Nagaya,[ab] K. Motomura,[c] E. Kukk,[cd] Y. Takahashi,[e] K. Yamazaki,[ef] S. Ohmura,[ag] H. Fukuzawa,[bc] S. Wada,[bh] S. Mondal,[c] T. Tachibana,[c] Y. Ito,[c] R. Koga,[h] T. Sakai,[a] K. Matsunami,[a] K. Nakamura,[e] M. Kanno,[e] A. Rudenko,[i] C. Nicolas,[j] X.-J. Liu,[jk] C. Miron,[jq] Y. Zhang,[l] Y. Jiang,[l] J. Chen,[m] M. Anand,[no] D. E. Kim,[no] K. Tono,[p] M. Yabashi,[bp] M. Yao,[a] H. Kono[e] and K. Ueda*[bc]

Received 19th April 2016, Accepted 27th May 2016
DOI: 10.1039/c6fd00085a

We studied the electronic and nuclear dynamics of I-containing organic molecules induced by intense hard X-ray pulses at the XFEL facility SACLA in Japan. The interaction with the intense XFEL pulse causes absorption of multiple X-ray photons by the iodine atom, which results in the creation of many electronic vacancies (positive

[a]*Department of Physics, Graduate School of Science, Kyoto University, Kyoto 606-8502, Japan*
[b]*RIKEN SPring-8 Center, Sayo, Hyogo 679-5148, Japan. E-mail: ueda@tagen.tohoku.ac.jp*
[c]*Institute of Multidisciplinary Research for Advanced Materials, Tohoku University, Sendai 980-8577, Japan*
[d]*Department of Physics and Astronomy, University of Turku, FI-20014, Finland*
[e]*Department of Chemistry, Graduate School of Science, Tohoku University, Sendai 980-8578, Japan*
[f]*Department of Chemistry, Graduate School of Science, Hokkaido University, Sapporo 060-0810, Japan*
[g]*Research Center for Condensed Matter Physics, Hiroshima Institute of Technology, Hiroshima 731-5193, Japan*
[h]*Department of Physical Science, Hiroshima University, Higashi-Hiroshima 739-8526, Japan*
[i]*J.R. Macdonald Laboratory, Department of Physics, Kansas State University, 66506 Manhattan, KS, USA*
[j]*Synchrotron SOLEIL, L'Orme des Merisiers, Saint-Aubin, BP 48, FR-91192 Gif-sur-Yvette Cedex, France*
[k]*School of Physics and Nuclear Energy Engineering, Beihang University, Beijing 100191, China*
[l]*Shanghai Advanced Research Institute, Chinese Academy of Sciences, 201210 Shanghai, China*
[m]*Shanghai Institute of Applied Physics, Chinese Academy of Sciences, 201800 Shanghai, China*
[n]*Department of Physics, Center for Attosecond Science and Technology, Pohang University of Science and Technology, Pohang, 37673, South Korea*
[o]*Max Planck Center for Attosecond Science, Max Planck POSTECH/KOREA Res. Init., Pohang, 37673, South Korea*
[p]*Japan Synchrotron Radiation Research Institute (JASRI), Sayo, Hyogo 679-5198, Japan*
[q]*Extreme Light Infrastructure – Nuclear Physics (ELI-NP), "Horia Hulubei" National Institute for Physics and Nuclear Engineering, 30 Reactorului Street, RO-077125 Măgurele, Jud. Ilfov, Romania*

† Electronic supplementary information (ESI) available. See DOI: 10.1039/c6fd00085a

charges) *via* the sequential electronic relaxation in the iodine, followed by intramolecular charge redistribution. In a previous study we investigated the subsequent fragmentation by Coulomb explosion of the simplest I-substituted hydrocarbon, iodomethane (CH_3I). We carried out three-dimensional momentum correlation measurements of the atomic ions created *via* Coulomb explosion of the molecule and found that a classical Coulomb explosion model including charge evolution (CCE-CE model), which accounts for the concerted dynamics of nuclear motion and charge creation/charge redistribution, reproduces well the observed momentum correlation maps of fragment ions emitted after XFEL irradiation. Then we extended the study to 5-iodouracil ($C_4H_3IN_2O_2$, 5-IU), which is a more complex molecule of biological relevance, and confirmed that, in both CH_3I and 5-IU, the charge build-up takes about 10 fs, while the charge is redistributed among atoms within only a few fs. We also adopted a self-consistent charge density-functional based tight-binding (SCC-DFTB) method to treat the fragmentations of highly charged 5-IU ions created by XFEL pulses. Our SCC-DFTB modeling reproduces well the experimental and CCE-CE results. We have also investigated the influence of the nuclear dynamics on the charge redistribution (charge transfer) using nonadiabatic quantum-mechanical molecular dynamics (NAQMD) simulation. The time scale of the charge transfer from the iodine atomic site to the uracil ring induced by nuclear motion turned out to be only ~5 fs, indicating that, besides the molecular Auger decay in which molecular orbitals delocalized over the iodine site and the uracil ring are involved, the nuclear dynamics also play a role for ultrafast charge redistribution. The present study illustrates that the CCE-CE model as well as the SCC-DFTB method can be used for reconstructing the positions of atoms in motion, in combination with the momentum correlation measurement of the atomic ions created *via* XFEL-induced Coulomb explosion of molecules.

1. Introduction

X-ray free-electron lasers (XFELs)[1–3] provide extremely intense femtosecond X-ray pulses that promise direct observations of atoms in action during various types of reactions.[4–8] Creating such "molecular movies" of atomic motions is indeed a grand challenge of femtochemistry.[9–11] X-ray scattering experiments on gas-phase reactions with XFELs are, however, still very challenging, though not impossible.[12,13] An alternative approach, photoelectron diffraction combined with the XFEL pulses, was suggested for visualizing the atomic motion in gas-phase intramolecular reactions[14,15] and proof-of principle experiments were carried out.[16–18]

Another promising approach for probing atoms in action is the three-dimensional momentum correlation measurement for the atomic ions created *via* Coulomb explosion of the molecule,[19–21] called Coulomb Explosion Momentum Imaging (CEMI) throughout the present paper. In CEMI, the ultra-fast creation of highly charged ions is the key for monitoring the dynamics of intermediate states. Intense XFEL pulses create extremely highly charged ions on the femtosecond time scale by multiple inner-shell ionization[22–24] and are thus also expected to realize femtosecond time-resolved CEMI. For small molecules containing a heavy atom, CEMI by inner-shell X-ray absorption provides insight into the interplay of electron and ion/atom dynamics relevant to radiation damage at the atomic scale.[25–29] As an example, Erk *et al.* studied the ionization and fragmentation of methylselenol (CH_3SeH) molecules by intense ($>10^{17}$ W cm^{-2}), 5 fs X-ray pulses (~2 keV) with CEMI.[25] They found signatures of ultrafast charge redistribution from the inner-shell ionized Se atom to other atoms and noted significant displacements of atoms in the course of multiple ionization. Later, they extended the CEMI technique to probe laser-induced photodissociation of CH_3I[26] and I_2.[27]

Recent calculations of the double hole state of glycine (NH_2CH_2COOH), created after K-shell ionization and subsequent Auger decay,[30] revealed that the double hole is transferred within 3 to 4 fs between the two functional end groups of the glycine molecule, driven solely by specific nuclear displacements and non-Born–Oppenheimer effects. The calculations also showed that the concurrent valence electron redistribution caused by nuclear dynamics is much faster than previously anticipated. It was noted that the non-Born–Oppenheimer effects affect the apparent electron densities and are thus relevant for X-ray imaging experiments, in which the ionization of inner atomic shells is unavoidable.

We have ourselves studied, by CEMI, the charge and molecular dynamics in I-containing organic molecules induced by intense XFEL pulses delivered from the hard X-ray FEL facility SACLA in Japan. In a first study we investigated the charge redistribution and fragmentation in the simplest I-substituted hydrocarbon, iodomethane (CH_3I),[28] using very short (~10 fs) and intense XFEL pulses. We also simulated the momentum correlations of fragment ions emitted after XFEL irradiation by adopting a Coulomb explosion model, where only the Coulomb repulsions between charged atoms (ions) are the forces in a classical molecular dynamics (MD) simulation and there are no forces to form chemical bonds. The charge of a molecule builds up in time owing to the multiple XFEL ionization and charge redistribution occurs among atoms. These two effects are introduced into the classical Coulomb explosion model by their characteristic time constants. The present model, which accounts for the concerted dynamics between nuclear motion and charge build-up/charge redistribution, may be called the classical Coulomb explosion model with charge evolution (CCE-CE model). Fitting the simulation results to the experimental data, we were able to estimate the specific time scales of charge build-up in the molecule (~9 fs) and of charge redistribution over the molecule (~3 fs).

To deepen the understanding of the charge build-up, charge redistribution and fragmentation processes induced by XFEL pulses in more complex systems, we have extended our CEMI study to 5-iodouracil $C_4H_3IN_2O_2$ (5-IU).[29] The kinetic energy distributions as well as the momentum correlations of fragments are derived from the observed momentum vectors of each fragment ion. Comparing the simulation results obtained by the CCE-CE model with the experimental data, we discussed the electronic and nuclear dynamics of 5-IU under XFEL irradiation and confirmed that the charge build-up takes ~10 fs, while the charge redistributes from the I atom to the other atoms and localizes in a couple of fs. We found that within the 10 fs XFEL pulse duration, the C–H bonds are significantly elongated, say, by a factor of two or more, but the other bonds are elongated only up to 5 to 10%.

In the present paper we review the above experimental and CCE-CE model studies,[28,29] but also present new numerical simulations for 5-IU obtained by a higher-level approach and discuss what is behind the validity of the CCE-CE model. The employed approach is a semi-empirical one in which the forces on atoms are calculated by the self-consistent charge density-functional based tight-binding (SCC-DFTB) method.[31] Chemical bonds are explicitly taken into account in the SCC-DFTB approach, unlike in the case of the CCE-CE model. The SCC-DFTB approach has been applied to simulate the fragmentation of C_{60} (ref. 32) and of its highly charged cations such as $C_{60}{}^{60+}$.[33] This approach has been also used to simulate the dissociation processes of doubly-ionized thiophene (C_4H_4S) molecules triggered by X-ray photoionization from the 2p core level of the S atom.[34] In this paper, we present the results of SCC-DFTB simulations for 5-IU. Good agreement with the experimental results was obtained by incorporating the effects of vibrational and electronic excitation upon XFEL-induced multiple ionization into the SCC-DFTB approach. Another approach we employed to evaluate the contribution of molecular dynamics to the charge dynamics is a molecular dynamics (MD) simulation based on a first-principles electronically nonadiabatic quantum-mechanical (NAQ) method, abbreviated as the NAQMD approach. We carried out NAQMD simulations for a high charge state of 5-IU to investigate the charge transfer dynamics in this molecule exposed to an XFEL pulse. The CCE-CE model deals with the charge redistribution in a molecule phenomenologically, assuming that it occurs, for example, as a result of molecular Auger decay in the later stage of the Auger cascades. The NAQMD approach, on the other hand, can simulate the electronic dynamics induced by the nuclear motion that occurs after the Auger decay, in real-time and real-space. The time scale of intramolecular charge transfer for 5-IU, estimated from the NAQMD simulation, was ~5 fs, elucidating that nuclear dynamics play also a significant role in the ultrafast charge redistribution. These two approaches at the higher levels show that the CCE-CE model captures well the essential features of Coulomb explosion in a pragmatic way. The consistency between the experiment and all these computational approaches validates our understanding of the electronic and nuclear dynamics in Coulomb explosion under XFEL irradiation of sample molecules. The present study thus demonstrates that the CCE-CE model as well as the SCC-DFTB approach, combined with the experimental CEMI technique, can be used for the characterization of ultrafast dynamics of the molecular samples and provide a solid basis for time-resolved CEMI with XFEL.

In the following section, the experimental methods, as well as the CCE-CE, SCC-DFTB and NAQMD simulations are described. The results for all these methods are presented in Section 3 and the motions of individual atoms extracted from CCE-CE model simulations as well as the advantages and disadvantages of each theoretical approach are discussed in Section 4. Section 5 is the Conclusion.

2. Methods

2.1. Experimental

The experiments for CH_3I and 5-IU have been carried out at experimental hutch 3 of beam line 3 of SACLA,[35] where the XFEL beam is focused by a Kirkpatrick–Baez (KB) mirror system to a focal spot of ~1 μm FWHM in diameter.[36] The photon energy was set at 5.5 keV, corresponding to the energy above the L-absorption edge of the iodine atom. The repetition rate of the XFEL pulses was 10 Hz. The temporal FWHM of the pulse was estimated to be ~10 fs.[37] The pulse energies were measured by the beam-position monitor[38] located upstream of the beam line. The relative X-ray pulse energy passing through the interaction point was also measured shot-to-shot by a p–intrinsic–n photodiode. The shot-to-shot pulse energy fluctuation was ∼7% (14% FWHM), and the peak fluence was 26 μJ μm^{-2} on average. The absolute value of the peak fluence was determined just after the experiment *via* the calibration using Ar.[24]

The CH_3I (>99.5%) reagent was purchased from Wako Pure Chemical Industries, Ltd. and used without further purification. The CH_3I sample was provided as a liquid and it was introduced to the focal point of the XFEL beam as a pulsed supersonic gas jet seeded in helium gas. 5-IU (>99.0%) was purchased from Tokyo Chemical Industry Co., Ltd. and used without further purification. The 5-IU sample was heated up to 190 °C in the reservoir with a modified electromagnetic solenoid valve,[39] and the resulting sample vapor was seeded in helium gas.

The molecular beam was crossed with the focused XFEL beam at the reaction point, and the emitted ions were projected onto a microchannel plate (MCP) in front of a delay-line anode. The three dimensional momentum vectors of the emitted ions were determined by using both the arrival time and the position of each ion. A three-layer type delay-line anode (Roentdek HEX80) was employed to minimize the detection dead time.[40] The design of the ion spectrometer used was basically the same as the one described in ref. 41 but the electrodes were modified to extract high energy ions from the reaction point.[29] The maximum acceptable kinetic energy was ∼$80q$ eV for a multiply charged ion with charge q.

In the 5-IU experiments, the CEMI data have been recorded for 6.3×10^6 XFEL shots, keeping the particle hit rate at less than 0.1 events per XFEL shot so that the false coincidences become negligible in the coincidence data analysis.

2.2. CCE-CE model simulations

We carried out CCE-CE simulations to translate the experimental results to spatiotemporal motion. The charge build-up induced by an XFEL pulse as well as the charge redistribution from the iodine atom to its counterparts (CH_3 for iodomethane and $C_4H_3N_2O_2$ for 5-IU, respectively) have been taken into account in the CCE-CE model.[28] Briefly, several photons were absorbed within the XFEL pulse duration (∼10 fs) by the deep atomic inner shells of iodine in the present experiments. In parallel and partly also subsequently to the XFEL pulse, Auger cascades should be a dominant mechanism of charge multiplication which will generate charges at an exponentially decreasing rate. The multiplication process is modelled statistically by assuming that the total charge of a molecule (Q^{tot}) increases with time t as

$$Q^{\text{tot}} = Q^{\max}\left(1 - \exp\left(-\frac{t}{\tau}\right)\right), \tag{1}$$

where τ is a charge build-up time constant and $Q^{\max}$ is the final charge. The charge redistribution from iodine to the rest of the molecule is represented by the rate constant R as

$$\frac{\mathrm{d}Q^{\text{Body}}[\mathrm{t}]}{\mathrm{d}t} = R\mathrm{d}Q^{\mathrm{I}}[\mathrm{t}]. \tag{2}$$

here $Q^{\mathrm{I}}[\mathrm{t}]$ and $Q^{\text{Body}}[\mathrm{t}]$ represent the charge of the iodine atom and of the rest of the molecule, respectively. Although the model itself is empirical and does not concern the physical mechanisms behind the charge redistribution, we note that it typically happens during the later stages of the Auger cascades that involve molecular orbitals, not localized on the iodine atom. Nuclear dynamics after the Auger decay may further proceed the charge redistribution (charge transfer) as well, as will be discussed later. In the CCE-CE model, all these contributions are taken into account by a single parameter R.

For trajectory simulations of 5-IU by the CCE-CE model, we prepared ensembles of initial positions and velocities of the atoms in the parent molecule. The initial positions and velocities were sampled from MD trajectories at an effective temperature T to provide an initial distribution that mimics the vibrational wavefunctions. The MD trajectories were obtained by performing quantum chemical calculations at the B3LYP/3-21G(d) level of theory using the GAMESS package.[42] The classical Coulomb explosion model such as the CCE-CE model counts on the idea that the Coulomb explosion process may be modelled through simple classical trajectory calculations of point charges; *i.e.*, the effect of chemical bonds in a molecule (*e.g.*, 5-IU) may not practically play a significant role in the Coulomb explosion processes. We evaluated the Coulombic repulsive forces between ions for the trajectory calculations. We constructed kinetic energy distributions (KED) of ions, two-body angular correlations and three-body angular correlations among the momenta of fragment ions from the results of the CCE-CE model simulations averaged over the distribution of final charges $Q^{\max}$ estimated from CH_3I experiments. The charge distribution peaked at $Q^{\max} = 10$, ranging from +4 to +14. We scanned the parameter space of charge build-up time τ, charge redistribution rate R and temperature T so that the results of the CCE-CE model simulations reproduce the experimental observations.

2.3. SCC-DFTB simulations

Another theoretical approach we developed is an adaptation of the self-consistent charge density-functional based tight-binding (SCC-DFTB) theory for electronic structure calculations.[31] We applied it to on-the-fly classical trajectory calculations for the fragmentation dynamics of 5-IU ions. The density-functional based tight-binding (DFTB) method used to describe molecular structures or chemical bonds is computationally extremely fast and semiquantitatively reproduces the results of density functional theory (DFT). The DFTB method is based on the zeroth- or second-order expansion of the Kohn–Sham total energy with respect to electron density fluctuations. The second-order expansion is the SCC-DFTB method, where the charge distribution in a molecule is obtained in an iterative self-consistent manner by taking into account charge interactions between atoms. This method is applicable to systems where the charge balance between atoms is crucial, as for instance in biomolecules and other heteroatomic molecules.[43,44] We carried out all DFTB calculations by using the DFTB+ program package (ver. 1.2.2).[45] The adopted Slater–Koster files were parameter sets halorg-0-1 (ref. 46) and mio-1-1.[31] The forces on atoms for the use in the MD simulations are calculated by SCC-DFTB; this MD simulation is abbreviated as SCC-DFTB/MD. The initial positions and velocities of atoms in a neutral molecule just before the interaction with an XFEL pulse are taken from an equilibrium ensemble sampled from SCC-DFTB/MD trajectories at temperature T.

Core ionization takes place in molecules exposed to X-ray radiation, followed by internal electronic relaxations such as Auger processes, which may occur in cascades in iodine-containing molecules. During core ionization and subsequent electronic relaxation processes, the molecule experiences different potential surfaces and is consequently heated. Part of the electronic energy is converted to vibrational energy. For instance, in the case of C_{60}, a vibrational energy of 10–20 eV is deposited for each core ionization accompanied by an Auger process, which amounts to 5–10 eV on average per unit charge increase.[47] In reference to these values, we infer the vibrational energy deposited in 5-IU that reproduces the experimental results. To incorporate the increasing process of vibrational energy into the SCC-DFTB approach, we employed a sequential ionization model[32,48] where the molecule is vertically ionized from the $q-1$ charge state to the q charge state when Q^{tot} in eqn (1) reaches a positive integer q.

In the sequential ionization model, the momentum vector of atom i is instantaneously changed from $\vec{p}_i$ to $\vec{p}'_i$ when the charge of the molecule switches to the higher one:

$$\vec{p}'_i = \vec{p}_i + \Delta\vec{p}_i \tag{3}$$

where $\Delta\vec{p}_i$ is a momentum added according to the acquired vibrational energy upon each ionization. We assume that the momentum $\Delta\vec{p}_i$ is constructed from momenta parallel to the bond axes between atom i and its surrounding atoms, in line with the experimental observation that only bond stretching is induced by core ionization;[49] the momentum $\Delta\vec{p}_i$ that atom i receives is the composition of momentum vectors parallel to its bond axes. Various modifications can be made. We examined if the results of simulations depend on the types of energy injection significantly or not. For 5-IU, the vertical component of $\Delta\vec{p}_i$ to the molecular plane, $\Delta p_{\mathrm{i}\perp}$, vanishes on average over all trajectories. We also designed a scheme in which nonzero $\Delta p_{i\perp}$ are replaced by $\Delta p_{i\perp} = 0$. We carried out simulations for the two cases of $\Delta p_{i\perp} = 0$ and $\Delta p_{i\perp} \neq 0$ and found that the choice of $\Delta p_{i\perp} = 0$ leads to better results, as discussed later.

We prepare the momenta along bond axes following a one-dimensional Maxwell–Boltzmann distribution. The magnitudes of Δp_i are then scaled by a factor so that the total kinetic energy of atoms increases by an average vibrational energy that the molecule gains per unit charge increase. Denoting this quantity by ε, we can express the total kinetic energy added up to the final charge state of Q^{max} as

$$\Delta E = Q^{\mathrm{max}}\varepsilon. \tag{4}$$

In other words, we assume that the vibrational energy deposited in the molecule is more or less proportional to the final charge of the molecule. If the acquired vibrational energy ΔE is larger than the atomization energy ΔH_{a} of 5-IU (~80 eV), the molecule is immediately decomposed into individual atoms. As ΔE approaches ΔH_{a}, all atoms are separated so that vibrational energy is no longer accumulated upon further ionization. We therefore assume that the upper limit of ΔE is around ΔH_{a}.

In the SCC-DFTB approach, charge redistribution in real time is ignored and it is assumed that the electronic state instantaneously relaxes to the ground electronic state or an electronic state distribution characterized by the electronic temperature T_{e} which represents the width of the Fermi–Dirac distribution. The charge distribution is that in the ground electronic state or the relaxed electronic state distribution. We set T_e here to be in the range from a few eV to several eV to include the effects of electronic excitation on the charge distribution. The value of T_e should be as high as the energies of excited states. The energy of the first absorption band of 5-IU was experimentally determined to be 4.4 eV (280 nm).[50] According to our TD-DFT calculation at the B3LYP/CEP-121G level of theory implemented in the Gaussian 09 program package,[51] the transition energy to the absorption band was 4.5 eV. The optimal value of T_e was slightly higher than these values, *i.e.*, ~6 eV. A more detailed discussion about the range of T_e will be given in Section 3.5 and 4. In the present simulations, T_e was set to be constant for different charge states, once T_e was determined as a parameter. The final output of the simulation results is the average over the distribution of Q^{max} estimated from CH_3I experiments, as in the case of the CCE-CE model.

2.4. NAQMD simulations

In order to examine the charge redistribution dynamics or charge transfer induced by the nuclear dynamics in a highly charged parent ion, which is generated through Auger cascades initiated by XFEL irradiation, we performed NAQMD simulations that include the effects of nonadiabatic electronic transitions with a surface-hopping approach.[52–56] The nuclei are treated classically in the adiabatic representation, *i.e.*, the atomic forces at the current nuclear positions are calculated from the potential surfaces of ground and excited electronic eigenstates. The switching probability from the current adiabatic state to another is computed from the density-matrix elements obtained by solving the time-dependent Kohn–Sham (TDKS) equations, and the occurrence of nonadiabatic transitions between adiabatic states is determined stochastically, according to the switching probability. In the present treatment, the electron configuration representing the current electronic state does not switch with another one unless the molecular structure changes toward a region where surface hopping can occur. By using the above NAQMD, we can investigate the charge dynamics solely driven by nuclear motion. We have previously applied it to investigate charge transfer in several photo-excited organic molecules.[56–59] In the present study, we extend the method to the charge dynamics in 5-IU molecular ions with many holes.

3. Results

3.1. Experimental results for CH_3I

Fig. 1a depicts the partial ion yields released from CH_3I recorded at a photon energy of 5.5 keV and peak fluence of 26 μJ μm^{-2}. We observed only atomic ions H^+ (not shown), C^{n+} with $n \leq 4$ and I^{m+} with $m \leq 15$.

We extracted momentum-correlated iodine and carbon ion-pairs released from single molecules assuming three hydrogen atoms are released as three protons and the sum of momenta of three protons is parallel to C–I bond axis. Then we can apply the momentum conservation law as

$$\vec{p}(I^{m+}) \times k_{mn} + \vec{p}(C^{n+}) = 0, \quad (5)$$

where $\vec{p}(I^{m+})$ and $\vec{p}(C^{n+})$ are momenta of I^{m+} and C^{n+}, respectively, and k_{mn} is the momentum correlation coefficient that results from the missing momentum of

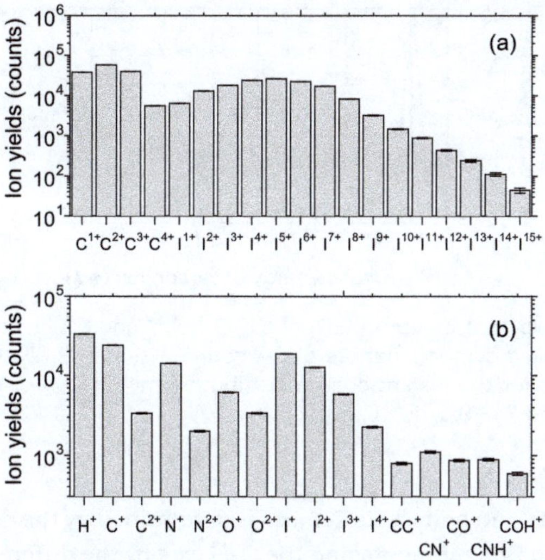

Fig. 1 Partial ion yields released from (a) CH_3I and (b) 5-iodouracil.

the unobserved protons. The coefficients k_{mn} for all observed combinations of (m,n) are plotted in Fig. 2. Assuming three protons are released, we can estimate the total charges of ions released from a single molecule as $m + n + 3$. We then deduce that the maximum total charge reaches +22 as a result of multiphoton absorption and Auger cascade cycles.[24]

Fig. 3 depicts the kinetic energy distributions (KEDs) of C^+, C^{2+}, C^{3+} and C^{4+} in coincidence with I^{4+}, I^{6+}, I^{8+} and I^{9+} from CH_3I. Fig. 4 depicts proton emission angular distributions with respect to the C–I axis in coincidence with C^{2+}–I^{6+},

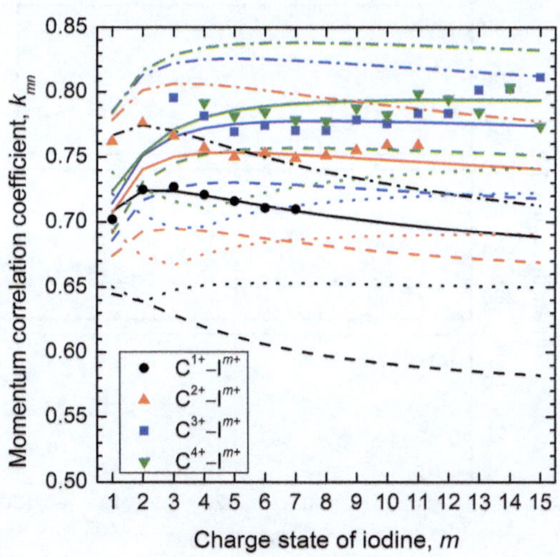

Fig. 2 Momentum correlation coefficients k_{mn}. Filled symbols denote the experimentally determined values. Dashed lines denote the results of the instant charge distribution model ($\tau = 0$). Dotted lines, solid lines and chain lines denote the results of the CCE-CE model with a rate constant R of 0.1, 0.37 and 1 fs^{-1}, respectively ($\tau = 9$ fs).

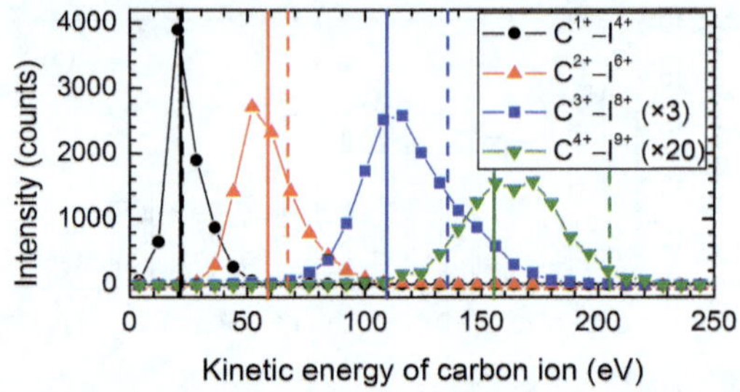

Fig. 3 Kinetic energy distributions (KED) of C^+, C^{2+}, C^{3+} and C^{4+} in coincidence with I^{4+}, I^{6+}, I^{8+} and I^{9+}. Filled symbols denote the experimental results. Dashed and solid lines denote the results for the instant charge distribution model ($\tau = 0$ fs) and the CCE-CE model ($\tau = 9$ fs and $R = 0.37$ fs^{-1}), respectively.

C^{3+}–I^{8+} and C^{4+}–I^{9+} ion pairs. The C–I axis is determined by the difference between the carbon and iodine momenta and the C–H axis by the difference of the carbon and hydrogen momenta. Thus the angle I–C–H formed by the three fragment ions is also affected by the motion of the carbon ions during the Coulomb explosion.

3.2. Experimental results for 5-IU

Fig. 1b depicts the partial ion yields released from 5-IU recorded at the same photon energy and the peak fluence as we employed for CH_3I. As we can see in the figure, atomic ions are the dominant fragments from 5-IU. We also observed singly charged polyatomic fragments, CC^+, CN^+, CO^+, CNH^+, and COH^+ but their fractions are much smaller than those of atomic fragments, indicating strong fragmentation of 5-IU by intense XFEL pulses. Multiply charged ions such as

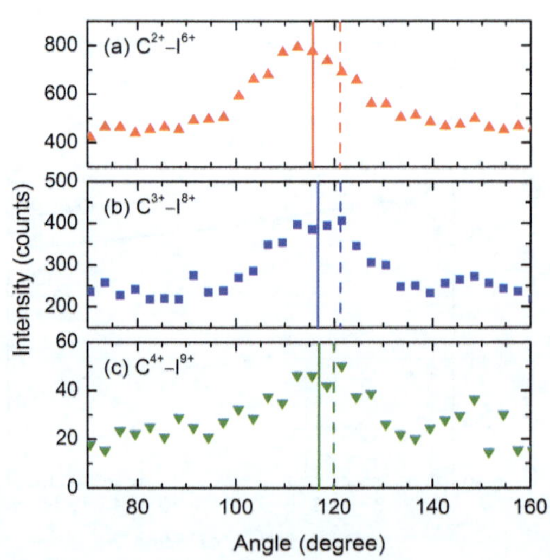

Fig. 4 Proton emission angular distributions with respect to C–I axis in coincidence with C^{2+}–I^{6+}, C^{3+}–I^{8+} and C^{4+}–I^{9+} ion pairs. Filled symbols denote the experimental results. Dashed and solid lines denote the results for the instant charge distribution model and the CCE-CE model, respectively.

doubly charged ions C^{2+}, O^{2+}, and N^{2+}, as well as the iodine ions I^{q+} with $q \leq 4$, have also been observed. In contrast to CH_3I, the yields of the highly charged iodine ions are significantly suppressed for 5-IU. The predominance of the lower charge states of iodine ions suggests an efficient charge redistribution from iodine to the uracil ring before the break-up of 5-IU molecule.

Fig. 5 shows the KEDs of atomic fragment ions, C^+, H^+, N^+, O^+ and I^{q+} emitted from 5-IU. For the KED of fragment ions, we included all iodine ions I^{q+} from $q = 1$ to 4 detected in coincidence with fragment ions. We show the KEDs for the individual charge states of the iodine ion in the ESI.†

The Coulomb explosion of 5-IU almost always create more than two ions, where the correlations among the momenta of emitted fragments are expected to provide us with rich information about fragmentation dynamics. Let us consider the angles formed between the momenta of atomic fragments in the Coulomb explosion of 5-IU. The cosine of angle θ between the momenta of two fragments A and B is defined by the scalar product of momentum vectors as follows:

$$\cos(\theta) = \frac{\vec{P}_A \cdot \vec{P}_B}{|\vec{P}_A||\vec{P}_B|}. \tag{6}$$

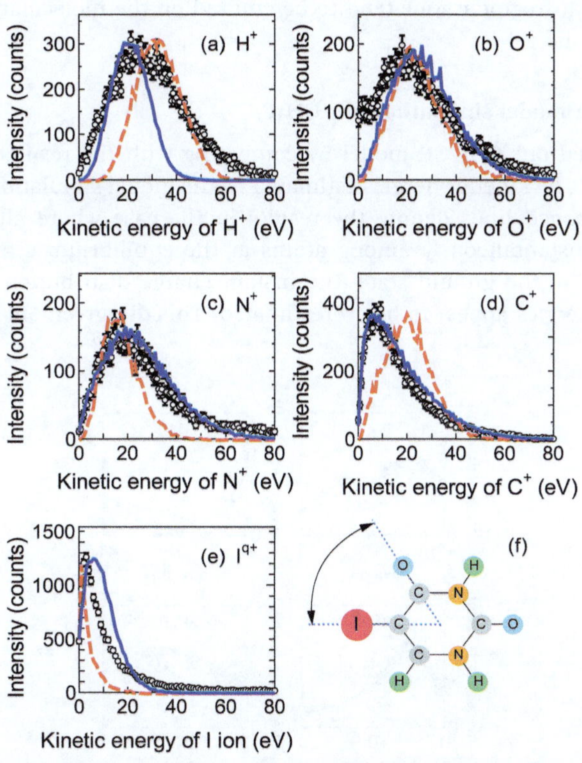

Fig. 5 Kinetic energy distributions (KED) of fragment ions emitted from XFEL-irradiated 5-iodouracil for (a) H^+, (b) O^+, (c) N^+, (d) C^+ and (e) I^{q+}. KED experimental data are summed over I^{q+} of $q = 1$ to 4. The open circles, red dashed lines and blue lines are the experimental data, CCE-CE results and SCC-DFTB results, respectively. (f) Molecular structure of the neutral 5-IU molecule and the definition of the equilibrium angle between two atoms (the I atom and its nearest O atom).

The distributions of $\cos(\theta)$ for the pair of A and B ions are denoted by $SP_2(A,B)$ and are shown in Fig. 6a–d. Peaks observed in $SP_2(I^{q+},H^+)$, $SP_2(I^{q+},O^+)$ and $SP_2(I^{q+},N^+)$ correspond to the angles in the equilibrium structure of the neutral 5-IU molecule. In order to see whether the momenta of three ions A, B and C lie on a plane or not, we calculated the following triple product from the observed momenta:

$$\cos(\varphi) = \frac{\left(\vec{P}_A \times \vec{P}_B\right) \cdot \vec{P}_C}{\left|\vec{P}_A \times \vec{P}_B\right|\left|\vec{P}_C\right|}. \tag{7}$$

The experimental distribution of $\cos(\varphi)$, $SP_3(A,B,C)$, are displayed in Fig. 7a–d. Note that the distribution function SP_3 for three randomly oriented and uncorrelated vectors is flat against $\cos(\varphi)$, so that the presented curves reflect directly the deviation from the random orientation of the three vectors. The Coulomb explosion that occurs preferentially in the plane then results in the distributions of $SP_3(A,B,C)$ peaked around $\cos(\varphi) = 0$. In Fig. 7a–d, we chose four combinations of ions including an iodine ion and a hydrogen ion. Experimental results show evident peaks at $\cos(\varphi) = 0$ for $SP_3(I^{q+},H^+,H^+)$, $SP_3(I^{q+},H^+,O^+)$ and $SP_3(I^{q+},H^+,N^+)$, indicating that fragment ions tend to be emitted on the molecular plane of the parent molecule.

3.3. CCE-CE model simulations for CH_3I

We first tested our CCE-CE model by comparing with the results of the CH_3I experiments. The kinetic energies estimated by the model simulations are shown in Fig. 3. Dashed lines denote the results in the case where all charges are distributed instantaneously among atoms at the equilibrium structure of the neutral CH_3I in the ground state (the instant charge distribution model). The calculated energies appear to be overestimated. This disagreement indicates the

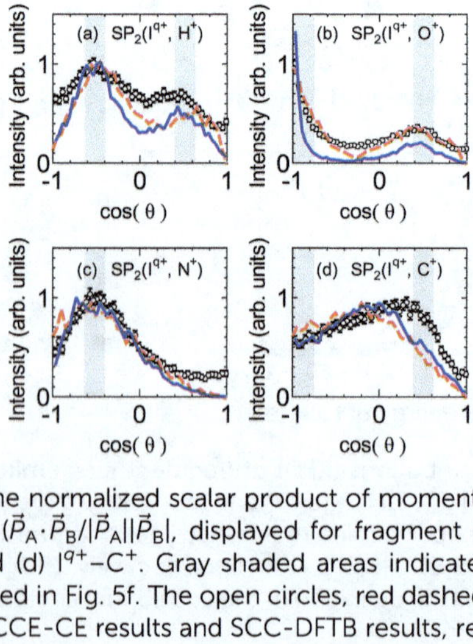

Fig. 6 Distribution of the normalized scalar product of momentum vectors, $SP_2(A,B)$, as a function of $\cos(\theta) = (\vec{P}_A \cdot \vec{P}_B / |\vec{P}_A||\vec{P}_B|$, displayed for fragment ion pairs (a) $I^{q+}-H^+$, (b) $I^{q+}-O^+$, (c) $I^{q+}-N^+$, and (d) $I^{q+}-C^+$. Gray shaded areas indicate the equilibrium angles between atoms as defined in Fig. 5f. The open circles, red dashed lines and blue lines are the experimental data, CCE-CE results and SCC-DFTB results, respectively.

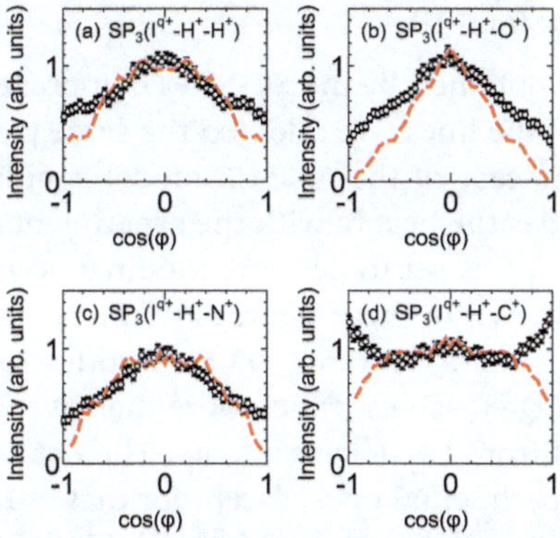

Fig. 7 Distribution of the normalized scalar triple products of momentum vectors $SP_3(A,B,C)$ plotted as a function of $\cos(\varphi) = (\vec{P}_A \times \vec{P}_B)\cdot\vec{P}_C/|\vec{P}_A \times \vec{P}_B||\vec{P}_C|$ for (a) $I^{q+}-H^+-H^+$, (b) $I^{q+}-H^+-O^+$, (c) $I^{q+}-H^+-N^+$, and (d) $I^{q+}-H^+-C^+$. The open circles and red dashed lines are the experimental data and CCE-CE results, respectively.

fact that the interatomic distances have considerably increased before the atoms reached their final charge states. It is thus essential to include the charge build-up time τ in order to obtain reasonable agreement of the model simulation with the measured kinetic energies of the ions. It also appeared that the charge redistribution rate R introduced in our CCE-CE model is very sensitive to the momentum correlation coefficient k_{mn} in Fig. 2. We scanned the combination of parameters of τ and R and obtained a good agreement of the model simulation with both measured kinetic energies and momentum correlation coefficients k_{mn} when we chose $\tau = 9$ fs and $R = 0.37$ fs^{-1}, as seen in Fig. 2 and 3. To illustrate the sensitivity of R to the coefficient k_{mn}, k_{mn} values estimated with R of 0.1 and 1 fs^{-1} are also shown in Fig. 2. Comparisons between experiment and simulation for proton emission angular distributions are also shown in Fig. 4. The CCE-CE model also shows better agreement with the experimental results in the angular distributions. From the model simulation, we learned that charge build-up time and charge redistribution time must be considered in order to describe the molecular dynamics appropriately.

3.4. CCE-CE model simulations for 5-IU

We extended the CCE-CE model for 5-IU by incorporating the thermal vibration effects in addition to the charge build-up time and charge redistribution rate. We scanned the combination of parameters of τ, R and T, and found good agreement between experiment and classical MD when we adopted the parameter set of $T = 300$ K, $\tau = 10$ fs and $R = 0.5$ fs^{-1}. Calculated KED, $SP_2(A,B)$ and $SP_3(A,B,C)$ are displayed by red dashed lines in the Fig. 5–7, respectively. The values of τ and R deduced from the 5-IU experiments are almost the same as those we obtained from CH_3I (τ = 9 fs and R = 0.37 fs^{-1}). We emphasize that both the thermal vibrations (represented by T) and charge dynamics (represented by τ and R) in the molecules are important to reproduce momentum correlations for 5-IU. There are still discrepancies in the KEDs of iodine ions and carbon ions; they may be attributed to the neglect of chemical bonds in the classical MD.

3.5. SCC-DFTB results for 5-IU

The KED and SP_2(A,B) obtained by the SCC-DFTB approach are shown in Fig. 5 and 6, respectively, by blue lines. We adopted the same parameters of $T = 300$ K and $\tau = 10$ fs as in the case of the CCE-CE model simulations. We chose that $T_e = 6$ eV, which provides the best fit with the experimental results. The injected energy per ionization, ε, was set to be 6 eV; the final accumulated energy ΔE is 60 eV for $Q^{max} = 10$. Shown in these figures are the results for the zero vertical momentum case ($\Delta \vec{p}_{i\perp} = 0$). We ran 1000 trajectories to obtain the KED and SP_2(A,B) for individual Q^{max} values, which were then averaged over the distribution of Q^{max} estimated from CH_3I experiments. The obtained KEDs are in good agreement with the experimental ones, except for that of H^+. This approach also well reproduces the observed peaks in $SP_2(I^{q+},H^+)$, $SP_2(I^{q+},O^+)$, and $SP_2(I^{q+},N^+)$; it holds the feature of the broad peak in $SP_2(I^{q+},C^+)$.

The general applicability of the SCC-DFTB approach depends on how the results are sensitive to ε and T_e. Almost the same results were obtained for the range of $\varepsilon = 3$–9 eV ($\Delta E = 30$–90 eV for $Q^{max} = 10$), though the KED and SP_2(A,B) gradually became broader with increasing ε. Overall, the results are insensitive to the choice of ε (or ΔE). The dependence on T_e is not critical either around $T_e = 6$ eV. While the peaks in the KED of N^+ and O^+ were shifted toward the higher energy side at extremely high T_e (>9 eV), the KED for H^+ and C^+ were insensitive to T_e. We would like to point out that 9 eV is equal to the ionization potential of 5-IU.[60] All SP_2(A,B) remained unchanged for $T_e > 3$ eV and the peaks in SP_2(A,B) disappear below $T_e = 3$ eV. We also investigated the case of $\Delta \vec{p}_{i\perp} \neq 0$. The results for $\Delta \vec{p}_{i\perp} \neq 0$ were almost the same as those for $\Delta \vec{p}_{i\perp} = 0$. The main differences that we noticed are as follows (see Fig. S5 and S6 of the ESI†): the KEDs of O^+ and I^{q+} for $\Delta \vec{p}_{i\perp} \neq 0$ were broadened toward higher energy; the peak at $\cos(\theta) \approx 0$ in $SP_2(I^{q+},C^+)$ is shifted to $\cos(\theta) \approx 1/4$ and a peak appears around $\cos(\theta) = -1$.

3.6. NAQMD results for 5-IU

Finally, we carried out NAQMD simulations to estimate the time scale of the charge transfer induced by the nuclear motion in a molecular parent ion with a 6+ charge, as an example. As the initial state of 5-IU^{6+}, we assumed an electron configuration where six electrons are removed from three molecular orbitals (MOs) of a neutral 5-IU: HOMO, HOMO−1, and HOMO−4. As shown in Fig. S7,† these MOs are mainly localized around the iodine atom. In contrast, HOMO−2 and HOMO−3 do not have iodine atomic populations and thus their holes will not be created *via* Auger cascades of the I atom. To investigate the charge dynamics in 5-IU^{6+}, we first optimized the MOs so that the energy of the ground state electron configuration of 5-IU^{6+} is minimized. We then transferred two electrons from HOMO−1 to LUMO of 5-IU^{6+}. These two MOs correspond to HOMO−4 and HOMO−2 of 5-IU; that is, the resultant electron configuration corresponds to the hole state in which six electrons are removed from HOMO, HOMO−1, and HOMO−4 of 5-IU. Fig. 8 shows the time evolution of the existence probability $R_\alpha(t)$ (α = iodine or the molecular counterpart) of holes obtained from the ensemble average over 15 different trajectories. The initial positions and velocities of the nuclei are taken from a classical trajectory at $T = 300$ K. The solid and dashed lines indicate $R_\alpha(t)$ for the iodine and the molecular counterpart, respectively. The holes transfer from the iodine to the molecular counterpart in

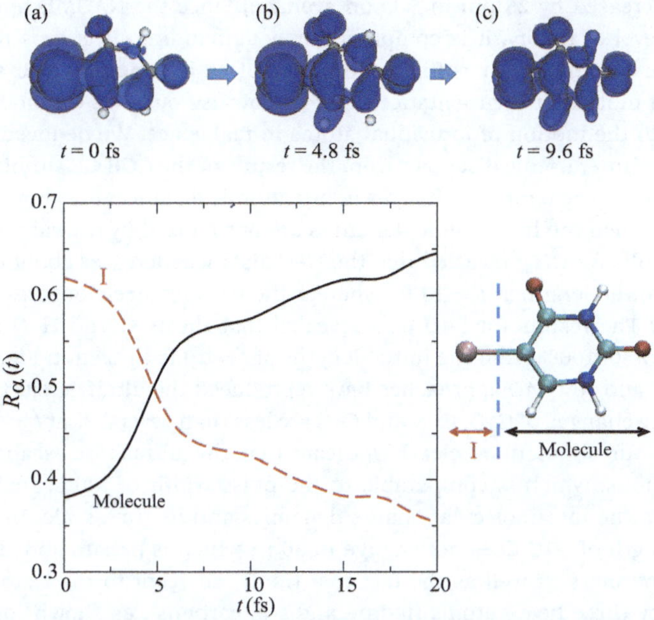

Fig. 8 Time evolution of the hole probability $R_\alpha(t)$ of 5-UI^{6+} in NAQMD simulation. The red dashed line and black solid line denote $R_\alpha(t)$ for iodine (I) and the molecular counterpart (molecule) region, respectively. The spatial distribution of the hole (one of the 15 trajectories) is also shown at time t = (a) 0 fs, (b) 4.8 fs, and (c) 9.6 fs. The hole density is displayed at a contour level of $0.03e/a_0^3$, where a_0 is the Bohr radius.

~5 fs. The threshold time of the charge transfer in each trajectory is almost the same within ±1 fs. The value is in the same order of the charge redistribution time scale estimated from experiments (~3 fs for CH_3I and ~2 fs for 5-IU), indicating that ultrafast nuclear dynamics also play a significant role in charge redistribution.

4. Discussion

The experimental results of the CEMI presented in the above section were analyzed in detail using three theoretical approaches ranging from a classical model, namely, the CCE-CE model to a first-principles approach (the NAQMD approach). We confirmed that the CCE-CE model reproduces well the experimental observables such as the kinetic energy distributions of fragment ions both for CH_3I and 5-IU. The results obtained by the SCC-DFTB semi-empirical approach for fragmentation of 5-IU are in reasonable agreement with the results of our CCE-CE model simulations wherever comparisons are possible. The intramolecular charge redistribution in 5-IU induced by nuclear motion was investigated by using the first principles approach NAQMD. The results of the first principles approach NAQMD illustrates that besides Auger cascades, nuclear dynamics plays a significant role in the ultrafast intramolecular charge redistribution in 5-IU. (The charge transfer from the iodine to the molecular counterpart more or less follows the fast C–H stretching motion. The C–H and N–H bond lengths increased by 25 pm in 5 fs on average.) Since the NAQMD approach is a higher level of theory, it is computationally demanding. It is thus practically impossible to generate a sufficient number of trajectories for the statistical analysis of molecular fragmentations. Let us now use our CCE-CE model to give insight into the motion of individual atoms in real space. We deduced the time evolution of interatomic distances from the results of the CCE-CE simulations for CH_3I and 5-IU. The temporal changes in interatomic distances are shown in Fig. 9 and 10 in which the interatomic distances are normalized by the values at $t = 0$.

The results for CH_3I revealed that the H–C distance increases about two times from the initial length at $t = 10$ fs, whereas the I–C distance elongates less than several %. The results for 5-IU also revealed that the H–C and H–N distances elongate by 50–100% from the initial lengths at $t = 10$ fs as seen in Fig. 10a. The SCC-DFTB and NAQMD approaches have reproduced the ultrafast ejection of H^+ as well. The changes of C–O, C–N and C–C are less than several % at $t = 10$ fs. The results of our simulations clearly indicate that the ultrashort escape time of hydrogen ions, which is comparable to the pulse width of XFEL, makes them sensitive to the intramolecular charge dynamics and its time scale. In contrast, the I–C length of 5-IU does not change in 30 fs, which is presumably due to the large iodine mass as well as the fact that the C_1 adjacent to the I atom is surrounded by three heavy atoms (iodine and two carbons), as shown in Fig. 10c. This effect was examined by evaluating the site dependence of $SP_2(I^{q+},C^+)$ on C^+ sites as well as the off-planar displacement of each carbon ion. Site-resolved $SP_2(I^{q+},C^+)$ is shown in the Fig. 10f. The angles between the momenta of I^+ and that of its neighbor C_1 has a broad peak at $\cos(\theta) = 0$, indicating that C_1 tends to be emitted at right angles to the direction of I^+ ejection. The off-planar motion of emitted ions is quantified by $\alpha(t)$ which is defined by the angle between $\vec{n_0}$ and $\vec{r}(t)$, where $\vec{n_0}$ is the normal vector of the molecular plane defined at $t = 0$ and $\vec{r}(t)$ is the position vector of an ion at t, respectively (see the definition in Fig. 10d). An increase of $\alpha(t)$ corresponds to the off-planar displacement of ions. The time evolution of $\alpha(t)$ shown in Fig. 10e indicates that the off-planar displacement of C_1 is apparently larger than those of other carbons. The results suggest a strong frustration of the Coulombic repulsive forces exerted on the C_1 site of 5-IU.

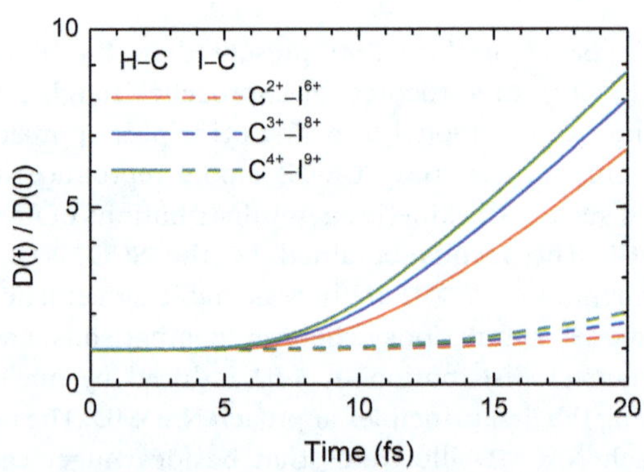

Fig. 9 Simulated time evolution of H–C (solid lines) and I–C (broken lines) distances in C^{2+}–I^{6+} (red), C^{3+}–I^{8+} (blue) and C^{4+}–I^{9+} (green) ion-pair formations. The distances are normalized by those at time zero.

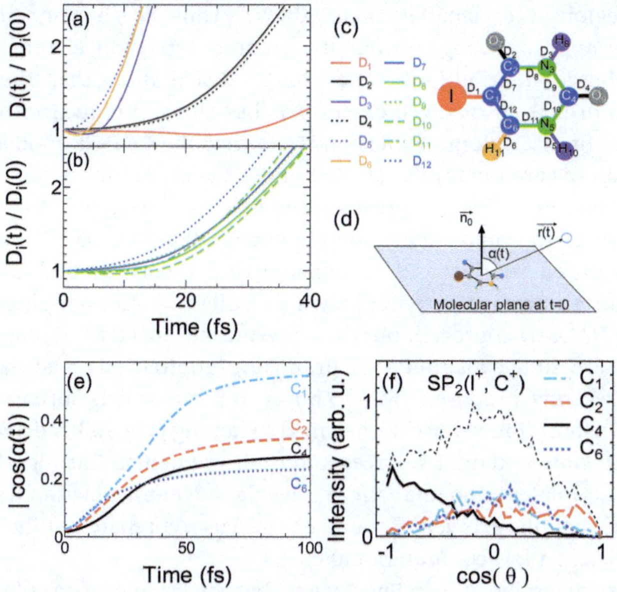

Fig. 10 (a) and (b) Time evolution of interatomic distances of the 5-iodouracil molecule averaged over the CCE-CE trajectories. Interatomic distances are normalized by the values at $t = 0$. (c) Definition of bonds. (d) Definition of $\alpha(t)$ to quantify the degree of off-planar ejection. (e) Time evolution of $|\cos(\alpha(t))|$. (f) Site-resolved $SP_2(I^+,C^+)$.

The success of the SCC-DFTB approach is attributed to the introduction of T_e. While the charge of the C atom adjacent to I^{q+} tends to be negative for 5-IU, it becomes positive for appropriate values of T_e; all the atoms are positively charged. As a result, the bond between I^{q+} and its adjacent C atom is weakened by Coulomb repulsion and I^{q+} are emitted almost parallel to the initial direction of the I–C bond axis of the molecule (or on the molecular plane), which is essential to reproduce, especially the two-body momentum correlations with I^{q+} observed in the experiments. In the range of $T_e > 3$ eV, all the atoms are positively charged if the total charge of the molecule is positive. This type of charge distribution is the key to reproduce the experimental results and reveals why the CCE-CE model is so successful. Another point to be noted is that the chemical bonds (covalent bonds in the case of 5-IU) in a molecule are almost broken, *i.e.*, ΔE is as high as the atomization energy of the molecule when the Coulomb explosion is about to proceed. These two points are the reasons why the CCE-CE model and the SCC-DFTB approach provide nearly the same results, though differences exist.

As we have shown, the performance of the SCC-DFTB approach is overall satisfactory. There is however a serious discrepancy with the experiment. As shown in Fig. 5, the high energy components (>40 eV) in the KED of H^+ are not reproduced by the SCC-DFTB approach. We performed SCC-DFTB simulations at high temperatures of $T = 2000$–3000 K, which is as high as the zero point energy of the C–H or N–H stretching vibration, but the KED sharply dropped in the high energy region as in the lower temperature cases. This denies the possibility that the high energy components of H^+ originate from different molecular structures of short C–H or N–H distances. (The Coulomb explosion energies of H^+ in the explored region were smaller than 50 eV.) Those high-energy components presumably appear through different Coulomb explosion channels that have different charge distributions. The charge distribution that the SCC-DFTB approach affords is an averaged one over T_e but almost unique for the structures at the moment of Coulomb explosion. Note that the CCE-CE model, where the charges of atoms are randomly selected under the constraint of the conservation of total charge, gives better agreement with the experimental KED of H^+. The reproduction of the high-energy components in the KED of H^+ is challenging. What is required to this end is nonadiabatic theory beyond the Ehrenfest dynamics on an average potential, such as Tully's surface-hopping algorithm[52] used in the NAQMD approach, but the conventional surface hopping methods or algorithms are time-consuming. A promising approach is a surface hopping algorithm recently proposed by C. Zhu *et al.*,[61] in which surface hopping is executed efficiently following the analytical switching probability determined only from excited state potential surfaces and their gradients. This algorithm, which efficiently simulates nonadiabatic molecular dynamics without calculating nonadiabatic coupling, is very fast and enables us to prepare a sufficient number of trajectories at a low computational cost.

The experimental results in Fig. 5 show that the fragment ions of peripheral H and O atoms have high kinetic energies, of which peaks are higher than or around 20 eV. This due to the fact that H^+ and O^+ immediately escape from the parent molecule (see Fig. 10a). One may thus expect that the fragment ions of ring atoms have lower kinetic energies. In fact, the kinetic energy distribution of N^+ is peaked around 20 eV, whereas the peak in the kinetic energy distribution of C^+ is much lower than 20 eV. The ring atoms C and N are almost considered the same type of atom without chemical bonds (the difference in mass is small): in the CCE-CE model, *i.e.*, a free atom model, the kinetic energy distribution for N^+ is similar to that for C^+. While the experimental peak in the kinetic energy distribution of C^+ is lower than the peak in the CCE-CE simulation, the experimental peak of N^+ is higher than the CCE-CE peak. This suggests that N^+ tends to be released earlier than C^+ in the Coulomb explosion of 5-IU. The chemical environment of N is different from that of C. We attribute the difference between C^+ and N^+ to the existence of chemical bonds. This conclusion is reinforced by the fact that the SCC-DFTB approach reproduces the experimental KEDs for N^+ and C^+.

In short, the CCE-CE model and the SCC-DFTB approach attained a level to reproduce the CEMI experiment semiquantitatively. Both can therefore serve as a Coulomb explosion simulator, though there is still room for improvement. Their advantages and disadvantages can be summarized as follows. The SCC-DFTB approach deals with chemical bonds appropriately, whereas the CCE-CE model ignores them. The SCC-DFTB approach hence predicts more accurately when fragments ions are ejected from the parent molecule and how much kinetic energies the fragment ions gain, except for H^+. The advantage of the CCE-CE model over the SCC-DFTB approach is the flexibility to allow various charge distributions for a molecule. In the CCE-CE model, intramolecular charge redistribution from the I atom to its molecular partner in 5-IU is well characterized by a rate equation with the charge redistribution rate *R*. In the SCC-DFTB approach, intramolecular charge redistribution is not handled in real time; instead, the charge redistributions that result from intramolecular charge redistribution are imitated by those obtained by the introduction of T_e. The best choice at present is to do a cross-check on the results of simulations by making the most of the advantages of the CCE-CE model and the SCC-DFTB approach. The CCE-CE model is computationally cheaper than the SCC-DFTB approach, but the differences are no longer significant considering current computer resources.

5. Conclusion

We carried out CEMI measurements for halogenated molecules, CH_3I and 5-iodouracul (5-IU), by using intense XFEL pulses at SACLA. The CCE-CE model, which includes the charge dynamics and nuclear dynamics in the classical Coulomb explosion model, has been applied to the analysis of CEMI, *i.e.*, 3D momentum correlations between fragment ions. We confirmed for CH_3I that the CCE-CE model reproduces the main features of the experimentally observed two-body momentum correlations and kinetic energies of fragment ions; we succeeded in determining the charge build-up time of $\tau = 9$ fs and the charge redistribution rate of $R = 0.37$ fs^{-1}. We then applied the model to 5-IU and reproduced the experimental observations by using the parameters of $T = 300$ K, $\tau = 10$ fs and $R = 0.5$ fs^{-1}. The effective temperature T is introduced to prepare a proper initial distribution of the positions and velocities of the atoms. The initial distribution, sampled from trajectories of the thermal molecular vibration, is essential to reproduce the experimentally observed two-body momentum correlations. The range of optimal values of τ and R was almost the same between CH_3I and 5-IU. The present results indicate that both the charge build-up and charge redistribution take place within the XFEL pulse duration of 10 fs, at the early stages of the Coulombic explosion. The charge transfer dynamics for 5-IU, induced by nuclear dynamics, was simulated by the first-principles NAQMD approach. The ultrafast charge transfer from the iodine atom to its molecular counterpart follows the ultrafast nuclear dynamics such as hydrogen motion. The estimated time scale of charge transfer is in the same order of the value of the charge redistribution rate R used in the CCE-CE model.

We also developed a novel approach based on the SCC-DFTB method, where the effects of sequential ionization and concurrent electronic excitation are described in the framework of the sequential ionization model. The SCC-DFTB approach is more realistic than the CCE-CE model in the sense that the chemical bonds in molecules are accurately taken into account in the SCC-DFTB approach, though semiempirically. In the SCC-DFTB approach, bond dissociation processes are triggered by the excess vibrational energy injected to molecules during the Auger cascade, as well as Coulomb repulsion. For 5-IU, the SCC-DFTB approach employed the values of T and τ used in the CCE-CE model, whose parameter set consistently reproduced the experimental observations of ion kinetic energy and two-body momentum correlation. The dynamics of charge redistribution, represented by R, cannot be explicitly treated in our SCC-DFTB approach. We had to introduce an electronic temperature T_e in order to model the charge distribution that reflects electronic excitation and charge redistribution, *i.e.*, the contribution from various excited electronic states.

The SCC-DFTB approach is robust. It reproduces the experimental results for reasonable choices of ε and T_e. The analysis of the SCC-DFTB results clarified why two apparently contradictory approaches, namely the CCE-CE model and the SCC-DFTB approach, function well. Our understanding of the mechanism of

XFEL-induced Coulomb explosion is founded on the consistency among the three theoretical approaches which are adapted or developed for this study. The present study demonstrates that both approaches can be used for extracting the physics of Coulomb explosion from the experimental data and therefore for reconstructing molecular structures. The successful extension from CH_3I to 5-IU suggests that the applicability is more general and that photo-induced ultrafast dynamics of many molecules can be adequately characterized by the laser–pump–XFEL–probe CEMI combined with CCE-CE and/or SCC-DFTB simulations. Such studies are in progress at both SACLA and LCLS.

Author contributions

K. U. conceived the research; K. Nag., K. Mo., E. Ku., S. W., H. F., T. T., Y. I., S. M., R. K., T. S., K. Ma., A. R., C. N., X.-J. L., C. M., Y. Z., J. C., Y. J., D. E. Ki., M. A., K. T., M. Yab. and K. U. performed the experiments; K. Nag., K. Mo. and E. Ku. analyzed the data and performed the CCE-CE model simulations; K. Y. developed the major parts of the protocols and Python scripts for the SCC-DFTB simulations and Y. T. performed SCC-DFTB simulations consulting with K. Y., K. Nak., M. K. and H. K.; S. O. performed NAQMD simulations consulting with M. Yao and H. K.; all the authors participated in the discussion of the results; and K. Nag., H. F., S. O., H. K. and K. U. wrote the paper consulting with all other authors.

Conflict of interest

The authors declare no conflict of interest.

Acknowledgements

The experiments were performed at SACLA with the approval of JASRI and the program review committee (No. 2012B8045). This study was supported by the X-ray Free Electron Laser Utilization Research Project and the X-ray Free Electron Laser Priority Strategy Program of the Ministry of Education, Culture, Sports, Science and Technology of Japan (MEXT), by the Japan Society for the Promotion of Science (JSPS), by the Proposal Program of SACLA Experimental Instruments of RIKEN and by the IMRAM project. H. K. acknowledges Grant-in-Aid for Scientific Research by MEXT (No. 24245001 and 16H04091). Part of the calculations was carried out by using supercomputing resources at the Cyberscience Center, Tohoku University and at the Research Center for Computational Science, Okazaki, Japan. K. Nak. is supported by Research Fellowships of Institute for Quantum Chemical Exploration for Young Scientists. S. O. acknowledges Grant-in-Aid for Scientific Research by MEXT (No. 15K17722) and also thanks the Supercomputer Center, Institute for Solid State Physics, the University of Tokyo for the use of facilities. Part of calculations was also carried out using the computer facilities at the Research Institute for information Technology, Kyushu University. K. Nag. and S. O. acknowledge support by the Cooperative Research Program of "Network Joint Research Center for Materials and Devices" of Japan. E. Ku. acknowledges support by the Academy of Finland, and Y. H. J. by the National Basic Research Program of China (2013CB922200), the Natural Science Foundation of China (11420101003, 11274232 and 61308068). A. R. was supported by the Chemical Sciences, Geosciences, and Biosciences Division of the Office of Basic Energy Sciences, Office of Science, U.S. Department of Energy under contract No. DE-FG02-86ER1349. M. A. and D. E. Ki., acknowledge the Global Research Laboratory Program (No. 2009-00439) and the Max Planck POSTECH/KOREA Research Initiative Program (No. 2011-0031558) through the National Research Foundation of Korea (NRF) funded by the Ministry of Science ICT & Future Planning. J. H. C. acknowledges support by the Shanghai Natural Science Foundation (No. 13ZR1464700) and the Knowledge Innovation Project of the Chinese Academy of Sciences (No. 255015061).

References

1 W. Ackermann, G. Asova, V. Ayvazyan, A. Azima, N. Baboi, J. Bähr, V. Balandin, B. Beutner, A. Brandt, A. Bolzmann, R. Brinkmann, O. I. Brovko, M. Castellano, P. Castro, L. Catani, E. Chiadroni, S. Choroba, A. Cianchi, J. T. Costello, D. Cubaynes, J. Dardis, W. Decking, H. Delsim-Hashemi, A. Delserieys, G. Di Pirro, M. Dohlus, S. Düsterer, A. Eckhardt, H. T. Edwards, B. Faatz, J. Feldhaus, K. Flöttmann, J. Frisch, L. Fröhlich, T. Garvey, U. Gensch, C. Gerth, M. Görler, N. Golubeva, H.-J. Grabosch, M. Grecki, O. Grimm, K. Hacker, U. Hahn, J. H. Han, K. Honkavaara, T. Hott, M. Hüning, Y. Ivanisenko, E. Jaeschke, W. Jalmuzna, T. Jezynski, R. Kammering, V. Katalev, K. Kavanagh, E. T. Kennedy, S. Khodyachykh, K. Klose, V. Kocharyan, M. Körfer, M. Kollewe, W. Koprek, S. Korepanov, D. Kostin, M. Krassilnikov, G. Kube, M. Kuhlmann, C. L. S. Lewis, L. Lilje, T. Limberg, D. Lipka, F. Löhl, H. Luna, M. Luong, M. Martins, M. Meyer, P. Michelato, V. Miltchev, W. D. Möller, L. Monaco, W. F. O. Müller, O. Napieralski, O. Napoly, P. Nicolosi, D. Nölle, T. Nuñez, A. Oppelt, C. Pagani, R. Paparella, N. Pchalek, J. Pedregosa-Gutierrez, B. Petersen, B. Petrosyan, G. Petrosyan, L. Petrosyan, J. Pflüger, E. Plönjes, L. Poletto, K. Pozniak, E. Prat, D. Proch, P. Pucyk, P. Radcliffe, H. Redlin, K. Rehlich, M. Richter, M. Roehrs, J. Roensch, R. Romaniuk, M. Ross, J. Rossbach, V. Rybnikov, M. Sachwitz, E. L. Saldin, W. Sandner, H. Schlarb, B. Schmidt, M. Schmitz, P. Schmüser, J. R. Schneider, E. A. Schneidmiller, S. Schnepp, S. Schreiber, M. Seidel, D. Sertore, A. V. Shabunov, C. Simon, S. Simrock, E. Sombrowski, A. A. Sorokin, P. Spanknebel, R. Spesyvtsev, L. Staykov, B. Steffen, F. Stephan, F. Stulle, H. Thom, K. Tiedtke, M. Tischer, S. Toleikis, R. Treusch, D. Trines, I. Tsakov, E. Vogel, T. Weiland, H. Weise, M. Wellhöfer, M. Wendt, I. Will, A. Winter, K. Wittenburg, W. Wurth, P. Yeates, M. V. Yurkov, I. Zagorodnov and K. Zapfe, *Nat. Photonics*, 2007, **1**(6), 336–342.

2 P. Emma, R. Akre, J. Arthur, R. Bionta, C. Bostedt, J. Bozek, A. Brachmann, P. Bucksbaum, R. Coffee, F.-J. Decker, Y. Ding, D. Dowell, S. Edstrom, A. Fisher, J. Frisch, S. Gilevich, J. Hastings, G. Hays, P. Hering, Z. Huang, R. Iverson, H. Loos, M. Messerschmidt, A. Miahnahri, S. Moeller, H.-D. Nuhn, G. Pile, D. Ratner, J. Rzepiela, D. Schultz, T. Smith, P. Stefan, H. Tompkins, J. Turner, J. Welch, W. White, J. Wu, G. Yocky and J. Galayda, *Nat. Photonics*, 2010, **4**(9), 641–647.

3 T. Ishikawa, H. Aoyagi, T. Asaka, Y. Asano, N. Azumi, T. Bizen, H. Ego, K. Fukami, T. Fukui, Y. Furukawa, S. Goto, H. Hanaki, T. Hara, T. Hasegawa,

T. Hatsui, A. Higashiya, T. Hirono, N. Hosoda, M. Ishii, T. Inagaki, Y. Inubushi, T. Itoga, Y. Joti, M. Kago, T. Kameshima, H. Kimura, Y. Kirihara, A. Kiyomichi, T. Kobayashi, C. Kondo, T. Kudo, H. Maesaka, X. M. Maréchal, T. Masuda, S. Matsubara, T. Matsumoto, T. Matsushita, S. Matsui, M. Nagasono, N. Nariyama, H. Ohashi, T. Ohata, T. Ohshima, S. Ono, Y. Otake, C. Saji, T. Sakurai, T. Sato, K. Sawada, T. Seike, K. Shirasawa, T. Sugimoto, S. Suzuki, S. Takahashi, H. Takebe, K. Takeshita, K. Tamasaku, H. Tanaka, R. Tanaka, T. Tanaka, T. Togashi, K. Togawa, A. Tokuhisa, H. Tomizawa, K. Tono, S. Wu, M. Yabashi, M. Yamaga, A. Yamashita, K. Yanagida, C. Zhang, T. Shintake, H. Kitamura and N. Kumagai, *Nat. Photonics*, 2012, **6**(8), 540–544.

4 R. Mankowsky, A. Subedi, M. Först, S. O. Mariager, M. Chollet, H. T. Lemke, J. S. Robinson, J. M. Glownia, M. P. Minitti, A. Frano, M. Fechner, N. A. Spaldin, T. Loew, B. Keimer, A. Georges and A. Cavalleri, *Nature*, 2014, **516**(7529), 71–73.

5 K. H. Kim, J. G. Kim, S. Nozawa, T. Sato, K. Y. Oang, T. W. Kim, H. Ki, J. Jo, S. Park, C. Song, T. Sato, K. Ogawa, T. Togashi, K. Tono, M. Yabashi, T. Ishikawa, J. Kim, R. Ryoo, J. Kim, H. Ihee and S. Adachi, *Nature*, 2015, **518**(7539), 385–389.

6 T. R. M. Barends, L. Foucar, A. Ardevol, K. Nass, A. Aquila, S. Botha, R. B. Doak, K. Falahati, E. Hartmann, M. Hilpert, M. Heinz, M. C. Hoffmann, J. Köfinger, J. E. Koglin, G. Kovacsova, M. Liang, D. Milathianaki, H. T. Lemke, J. Reinstein, C. M. Roome, R. L. Shoeman, G. J. Williams, I. Burghardt, G. Hummer, S. Boutet and I. Schlichting, *Science*, 2015, **350**(6259), 445–450.

7 P. Wernet, K. Kunnus, I. Josefsson, I. Rajkovic, W. Quevedo, M. Beye, S. Schreck, S. Grübel, M. Scholz, D. Nordlund, W. Zhang, R. W. Hartsock, W. F. Schlotter, J. J. Turner, B. Kennedy, F. Hennies, F. M. F. de Groot, K. J. Gaffney, S. Techert, M. Odelius and A. Föhlisch, *Nature*, 2015, **520**(7545), 78–81.

8 K. R. Ferguson, M. Bucher, T. Gorkhover, S. Boutet, H. Fukuzawa, J. E. Koglin, Y. Kumagai, A. Lutman, A. Marinelli, M. Messerschmidt, K. Nagaya, J. Turner, K. Ueda, G. J. Williams, P. H. Bucksbaum and C. Bostedt, *Sci. Adv.*, 2016, **2**(1), e1500837.

9 A. H. Zewail, *Science*, 2010, **328**, 187–193.

10 J. A. Weinstein and N. T. Hunt, *Nat. Chem.*, 2012, **4**(3), 157–5.

11 R. J. D. Miller, *Science*, 2014, **343**(6175), 1108–1116.

12 J. Kupper, S. Stern, L. Holmegaard, F. Filsinger, A. Rouzée, A. Rudenko, P. Johnsson, A. V. Martin, M. Adolph, A. Aquila, S. Bajt, A. Barty, C. Bostedt, J. Bozek, C. Caleman, R. Coffee, N. Coppola, T. Delmas, S. Epp, B. Erk, L. Foucar, T. Gorkhover, L. Gumprecht, A. Hartmann, R. Hartmann, G. Hauser, P. Holl, A. Hömke, N. Kimmel, F. Krasniqi, K.-U. Kühnel, J. Maurer, M. Messerschmidt, R. Moshammer, C. Reich, B. Rudek, R. Santra, I. Schlichting, C. Schmidt, S. Schorb, J. Schulz, H. Soltau, J. C. H. Spence, D. Starodub, L. Strüder, J. Thøgersen, M. J. J. Vrakking, G. Weidenspointner, T. A. White, C. Wunderer, G. Meijer, J. Ullrich, H. Stapelfeldt, D. Rolles and H. N. Chapman, *Phys. Rev. Lett.*, 2014, **112**, 083002.

13 M. P. Minitti, J. M. Budarz, A. Kirrander, J. S. Robinson, D. Ratner, T. J. Lane, D. Zhu, J. M. Glownia, M. Kozina, H. T. Lemke, M. Sikorski, Y. Feng, S. Nelson,

K. Saita, B. Stankus, T. Northey, J. B. Hastings and P. M. Weber, *Phys. Rev. Lett.*, 2015, **114**, 255501.

14 F. Krasniqi, B. Najjari, L. Strüder, D. Rolles, A. Voitkiv and J. Ullrich, *Phys. Rev. A*, 2010, **81**, 033411.

15 M. Kazama, T. Fujikawa, N. Kishimoto, T. Mizuno, J. Adachi and A. Yagishita, *Phys. Rev. A*, 2013, **87**, 063417.

16 A. Rouzée, P. Johnsson, L. Rading, A. Hundertmark, W. Siu, Y. Huismans, S. Düsterer, H. Redlin, F. Tavella, N. Stojanovic, A. Al-Shemmary, F. Lépine, D. M. P. Holland, T. Schlatholter, R. Hoekstra, H. Fukuzawa, K. Ueda and M. J. J. Vrakking, *J. Phys. B: At., Mol. Opt. Phys.*, 2013, **46**, 164029.

17 R. Boll, D. Anielski, C. Bostedt, J. D. Bozek, L. Christensen, R. Coffee, S. De, P. Decleva, S. W. Epp, B. Erk, L. Foucar, F. Krasniqi, J. Küpper, A. Rouzée, B. Rudek, A. Rudenko, S. Schorb, H. Stapelfeldt, M. Stener, S. Stern, S. Techert, S. Trippel, M. J. J. Vrakking, J. Ullrich and D. Rolles, *Phys. Rev. A*, 2013, **88**, 061402.

18 K. Nakajima, T. Teramoto, H. Akagi, T. Fujikawa, T. Majima, S. Minemoto, K. Ogawa, H. Sakai, T. Togashi, K. Tono, S. Tsuru, K. Wada, M. Yabashi and A. Yagishita, *Sci. Rep.*, 2015, **5**, 14065.

19 F. Légaré, K. F. Lee, I. V. Litvinyuk, P. W. Dooley, A. D. Bandrauk, D. M. Villeneuve and P. B. Corkum, *Phys. Rev. A*, 2005, **72**, 1–4.

20 I. Bocharova, R. Karimi, E. F. Penka, J.-P. Brichta, P. Lassonde, X. Fu, J.-C. Kieffer, A. D. Bandrauk, I. Litvinyuk, J. Sanderson and F. Légaré, *Phys. Rev. Lett.*, 2011, **107**, 63201.

21 A. Matsuda, M. Fushitani, E. J. Takahashi and A. Hishikawa, *Phys. Chem. Chem. Phys.*, 2011, **13**, 8697–8704.

22 L. Young, E. P. Kanter, B. Krässig, Y. Li, A. M. March, S. T. Pratt, R. Santra, S. H. Southworth, N. Rohringer, L. F. DiMauro, G. Doumy, C. A. Roedig, N. Berrah, L. Fang, M. Hoener, P. H. Bucksbaum, J. P. Cryan, S. Ghimire, J. M. Glownia, D. A. Reis, J. D. Bozek, C. Bostedt and M. Messerschmidt, *Nature*, 2010, **466**(7302), 56–61.

23 B. Rudek, S.-K. Son, L. Foucar, S. W. Epp, B. Erk, R. Hartmann, M. Adolph, R. Andritschke, A. Aquila, N. Berrah, C. Bostedt, J. Bozek, N. Coppola, F. Filsinger, H. Gorke, T. Gorkhover, H. Graafsma, L. Gumprecht, A. Hartmann, G. Hauser, S. Herrmann, H. Hirsemann, P. Holl, A. Hömke, L. Journel, C. Kaiser, N. Kimmel, F. Krasniqi, K.-U. Kühnel, M. Matysek, M. Messerschmidt, D. Miesner, T. Möller, R. Moshammer, K. Nagaya, B. Nilsson, G. Potdevin, D. Pietschner, C. Reich, D. Rupp, G. Schaller, I. Schlichting, C. Schmidt, F. Schopper, S. Schorb, C.-D. Schröter, J. Schulz, M. Simon, H. Soltau, L. Strüder, K. Ueda, G. Weidenspointner, R. Santra, J. Ullrich, A. Rudenko and D. Rolles, *Nat. Photonics*, 2012, **6**, 858–865.

24 H. Fukuzawa, S.-K. Son, K. Motomura, S. Mondal, K. Nagaya, S. Wada, X.-J. Liu, R. Feifel, T. Tachibana, Y. Ito, M. Kimura, T. Sakai, K. Matsunami, H. Hayashita, J. Kajikawa, P. Johnsson, M. Siano, E. Kukk, B. Rudek, B. Erk, L. Foucar, E. Robert, C. Miron, K. Tono, Y. Inubushi, T. Hatsui, M. Yabashi, M. Yao, R. Santra and K. Ueda, *Phys. Rev. Lett.*, 2013, **110**(17), 173005.

25 B. Erk, D. Rolles, L. Foucar, B. Rudek, S. W. Epp, M. Cryle, C. Bostedt, S. Schorb, J. Bozek, A. Rouzee, A. Hundertmark, T. Marchenko, M. Simon, F. Filsinger, L. Christensen, S. De, S. Trippel, J. Küpper, H. Stapelfeldt, S. Wada, K. Ueda, M. Swiggers, M. Messerschmidt, C. D. Schröter,

R. Moshammer, I. Schlichting, J. Ullrich and A. Rudenko, *Phys. Rev. Lett.*, 2013, **110**(5), 053003.

26 B. Erk, R. Boll, S. Trippel, D. Anielski, L. Foucar, B. Rudek, S. W. Epp, R. Coffee, S. Carron, S. Schorb, K. R. Ferguson, M. Swiggers, J. D. Bozek, M. Simon, T. Marchenko, J. Küpper, I. Schlichting, J. Ullrich, C. Bostedt, D. Rolles and A. Rudenko, *Science*, 2014, **345**(6194), 288–291.

27 K. Schnorr, A. Senftleben, M. Kurka, A. Rudenko, G. Schmid, T. Pfeifer, K. Meyer, M. Kübel, M. F. Kling, Y. H. Jiang, R. Treusch, S. Düsterer, B. Siemer, M. Wöstmann, H. Zacharias, R. Mitzner, T. J. M. Zouros, J. Ullrich, C. D. Schröter and R. Moshammer, *Phys. Rev. Lett.*, 2014, **113**, 073001.

28 K. Motomura, E. Kukk, H. Fukuzawa, S. Wada, K. Nagaya, S. Ohmura, S. Mondal, T. Tachibana, Y. Ito, R. Koga, T. Sakai, K. Matsunami, A. Rudenko, C. Nicolas, X.-J. Liu, C. Miron, Y. Zhang, Y. Jiang, J. Chen, M. Anand, D. E. Kim, K. Tono, M. Yabashi, M. Yao and K. Ueda, *J. Phys. Chem. Lett.*, 2015, **6**(15), 2944–2949.

29 K. Nagaya, K. Motomura, E. Kukk, H. Fukuzawa, S. Wada, T. Tachibana, Y. Ito, S. Mondal, T. Sakai, K. Matsunami, R. Koga, S. Ohmura, Y. Takahashi, M. Kanno, A. Rudenko, C. Nicolas, X.-J. Liu, Y. Zhang, J. Chen, M. Anand, Y. H. Jiang, D.-E. Kim, K. Tono, M. Yabashi, H. Kono, C. Miron, M. Yao and K. Ueda, *Phys. Rev. X*, 2016, **6**, 021035.

30 Z. Li, O. Vendrell and R. Santra, *Phys. Rev. Lett.*, 2015, **115**, 143002.

31 M. Elstner, D. Porezag, G. Jungnickel, J. Elsner, M. Haugk, Th. Frauenheim, S. Suhai and G. Seifert, *Phys. Rev. B: Condens. Matter Mater. Phys.*, 1998, **58**, 7260.

32 N. Niitsu, M. Kikuchi, H. Ikeda, K. Yamazaki, M. Kanno, H. Kono, K. Mitsuke, M. Toda and K. Nakai, *J. Chem. Phys.*, 2012, **136**, 164304.

33 K. Yamazaki, T. Nakamura, N. Niitsu, M. Kanno, K. Ueda and H. Kono, *J. Chem. Phys.*, 2014, **141**, 121105.

34 E. Kukk, D. T. Ha, Y. Wang, D. G. Piekarski, S. Diaz-Tendero, K. Kooser, E. Itälä, H. Levola, M. Alcamí, E. Rachlew and F. Martín, *Phys. Rev. A*, 2015, **91**, 043417.

35 K. Tono, T. Togashi, Y. Inubushi, T. Sato, T. Katayama, K. Ogawa, H. Ohashi, H. Kimura, S. Takahashi, K. Takeshita, H. Tomizawa, S. Goto, T. Ishikawa and M. Yabashi, *New J. Phys.*, 2013, **15**(8), 083035.

36 H. Yumoto, H. Mimura, T. Koyama, S. Matsuyama, K. Tono, T. Togashi, Y. Inubushi, T. Sato, T. Tanaka, T. Kimura, H. Yokoyama, J. Kim, Y. Sano, Y. Hachisu, M. Yabashi, H. Ohashi, H. Ohmori, T. Ishikawa and K. Yamauchi, *Nat. Photonics*, 2013, **7**(1), 43–47.

37 Y. Inubushi, K. Tono, T. Togashi, T. Sato, T. Hatsui, T. Kameshima, K. Togawa, T. Hara, T. Tanaka, H. Tanaka, T. Ishikawa and M. Yabashi, *Phys. Rev. Lett.*, 2012, **109**(14), 144801.

38 K. Tono, T. Kudo, M. Yabashi, T. Tachibana, Y. Feng, D. Fritz, J. Hastings and T. Ishikawa, *Rev. Sci. Instrum.*, 2011, **82**(2), 023108.

39 T. Ebata, *Bull. Chem. Soc. Jpn.*, 2009, **82**(2), 127–151.

40 O. Jagutzki, A. Cerezo, A. Czasch, R. Dörner, M. Hattaß, M. Huang, V. Mergel, U. Spillmann, K. Ullmann-Pfleger, T. Weber, H. Schmidt-Böcking and G. D. W. Smith, *IEEE Trans. Nucl. Sci.*, 2002, **49**(5), 2477–2483.

41 H. Fukuzawa, T. Tachibana, K. Motomura, W. Q. Xu, K. Nagaya, S. Wada, P. Johnsson, M. Siano, S. Mondal, Y. Ito, M. Kimura, T. Sakai, K. Matsunami, H. Hayashita, J. Kajikawa, X.-J. Liu, E. Robert, C. Miron, R. Feifel, J. P. Marangos, K. Tono, Y. Inubushi, M. Yabashi, M. Yao and K. Ueda, *J. Phys. B: At., Mol. Opt. Phys.*, 2016, **49**, 034004.
42 M. W. Schmidt, K. K. Baldridge, J. A. Boatz, S. T. Elbert, M. S. Gordon, J. H. Jensen, S. Koseki, N. Matsunaga, K. A. Nguyen, S. Su, T. L. Windus, M. Dupuis and J. A. Montgomery Jr, *J. Comput. Chem.*, 1993, **14**(11), 1347–1363.
43 D. K. Chakravorty and K. M. Merz Jr, *Acc. Chem. Res.*, 2015, **48**(2), 439–448.
44 Y. Guo, F. E. Beyle, B. M. Bold, H. C. Watanabe, A. Koslowski, W. Thiel, P. Hegemann, M. Marazzi and M. Elstner, *Chem. Sci.*, 2016, **7**, 3879.
45 B. Aradi, B. Hourahine and T. Frauenheim, *J. Phys. Chem. A*, 2007, **111**, 5678–5684.
46 T. Kubař, Z. Bodrog, M. Gaus, C. Köhler, B. Aradi, T. Frauenheim and M. Elstner, *J. Chem. Theory Comput.*, 2013, **9**, 2939–2949.
47 K. Endo, S. Koizumi, T. Otsuka, T. Ida, T. Morohashi, J. Onoe, A. Nakao, E. Z. Kurmaev, A. Moewes and D. P. Chong, *J. Phys. Chem. A*, 2003, **107**, 9403.
48 K. Yamazaki, PhD thesis, Tohoku University, 2015.
49 M. Hoshino, K. Nakagawa, T. Tanaka, M. Kitajima, H. Tanaka, A. De Fanis, K. Wang, B. Zimmermann, V. McKoy and K. Ueda, *J. Phys. B: At., Mol. Opt. Phys.*, 2006, **39**, 3047.
50 H. Görner, *J. Photochem. Photobiol., A*, 1993, **72**, 197.
51 M. J. Frisch, G. W. Trucks, H. B. Schlegel, G. E. Scuseria, M. A. Robb, J. R. Cheeseman, G. Scalmani, V. Barone, B. Mennucci, G. A. Petersson, H. Nakatsuji, M. Caricato, X. Li, H. P. Hratchian, A. F. Izmaylov, J. Bloino, G. Zheng, J. L. Sonnenberg, M. Hada, M. Ehara, K. Toyota, R. Fukuda, J. Hasegawa, M. Ishida, T. Nakajima, Y. Honda, O. Kitao, H. Nakai, T. Vreven, J. A. Montgomery Jr, J. E. Peralta, F. Ogliaro, M. Bearpark, J. J. Heyd, E. Brothers, K. N. Kudin, V. N. Staroverov, R. Kobayashi, J. Normand, K. Raghavachari, A. Rendell, J. C. Burant, S. S. Iyengar, J. Tomasi, M. Cossi, N. Rega, J. M. Millam, M. Klene, J. E. Knox, J. B. Cross, V. Bakken, C. Adamo, J. Jaramillo, R. Gomperts, R. E. Stratmann, O. Yazyev, A. J. Austin, R. Cammi, C. Pomelli, J. W. Ochterski, R. L. Martin, K. Morokuma, V. G. Zakrzewski, G. A. Voth, P. Salvador, J. J. Dannenberg, S. Dapprich, A. D. Daniels, O. Farkas, J. B. Foresman, J. V. Ortiz, J. Cioslowski and D. J. Fox, *Gaussian 09, Revision A.02*, Gaussian, Inc., Wallingford CT, 2009.
52 J. C. Tully, *J. Chem. Phys.*, 1990, **93**, 1061.
53 W. R. Duncan, C. F. Craig and O. V. Prezhdo, *J. Am. Chem. Soc.*, 2007, **129**, 8528.
54 M. E. Casida, in *Recent Advances in Density Functional Methods (Part I)*, ed. D. P. Chong, World Scientific, Singapore, 1995, p. 155.
55 F. Shimojo, S. Ohmura, W. Mou, R. K. Kalia, A. Nakano and P. Vashishta, *Comput. Phys. Commun.*, 2013, **184**, 1.
56 S. Ohmura, S. Koga, I. Akai, F. Shimojo, R. K. Kalia, A. Nakano and P. Vashishta, *Appl. Phys. Lett.*, 2011, **98**, 113302.
57 W. Mou, S. Ohmura, F. Shimojo and A. Nakano, *Appl. Phys. Lett.*, 2012, **100**, 203306.
58 W. Mou, S. Ohmura, S. Hattori, K. Nomura, F. Shimojo and A. Nakano, *J. Chem. Phys.*, 2012, **136**, 184705.

59 S. Ohmura, K. Tsuruta, F. Shimojo and A. Nakano, *AIP Adv.*, 2016, **6**, 015305.
60 D. M. P. Holland, A. W. Potts, L. Karlsson, I. L. Zaytseva, A. B. Trofimov and J. Schirmer, *Chem. Phys.*, 2008, **352**, 205.
61 L. Yu, C. Xu, Y. Lei, C. Zhu and Z. Wen, *Phys. Chem. Chem. Phys.*, 2014, **16**(47), 25883–25895.

Faraday Discussions

PAPER

Femtosecond gas phase electron diffraction with MeV electrons

Jie Yang,[a] Markus Guehr,[bc] Theodore Vecchione,[b] Matthew S. Robinson,[a] Renkai Li,[b] Nick Hartmann,[b] Xiaozhe Shen,[b] Ryan Coffee,[b] Jeff Corbett,[b] Alan Fry,[b] Kelly Gaffney,[b] Tais Gorkhover,[b] Carsten Hast,[b] Keith Jobe,[b] Igor Makasyuk,[b] Alexander Reid,[b] Joseph Robinson,[b] Sharon Vetter,[b] Fenglin Wang,[b] Stephen Weathersby,[b] Charles Yoneda,[b] Xijie Wang[b] and Martin Centurion*[a]

Received 10th April 2016, Accepted 2nd June 2016

DOI: 10.1039/c6fd00071a

We present results on ultrafast gas electron diffraction (UGED) experiments with femtosecond resolution using the MeV electron gun at SLAC National Accelerator Laboratory. UGED is a promising method to investigate molecular dynamics in the gas phase because electron pulses can probe the structure with a high spatial resolution. Until recently, however, it was not possible for UGED to reach the relevant timescale for the motion of the nuclei during a molecular reaction. Using MeV electron pulses has allowed us to overcome the main challenges in reaching femtosecond resolution, namely delivering short electron pulses on a gas target, overcoming the effect of velocity mismatch between pump laser pulses and the probe electron pulses, and maintaining a low timing jitter. At electron kinetic energies above 3 MeV, the velocity mismatch between laser and electron pulses becomes negligible. The relativistic electrons are also less susceptible to temporal broadening due to the Coulomb force. One of the challenges of diffraction with relativistic electrons is that the small de Broglie wavelength results in very small diffraction angles. In this paper we describe the new setup and its characterization, including capturing static diffraction patterns of molecules in the gas phase, finding time-zero with sub-picosecond accuracy and first time-resolved diffraction experiments. The new device can achieve a temporal resolution of 100 fs root-mean-square, and sub-angstrom spatial resolution. The collimation of the beam is sufficient to measure the diffraction pattern, and the transverse coherence is on the order of 2 nm. Currently, the temporal resolution is limited both by the pulse duration of the electron pulse on target and by the timing jitter, while the spatial resolution is limited by the average electron beam current and

[a]*University of Nebraska-Lincoln, 855 N 16th Street, Lincoln, Nebraska 68588, USA. E-mail: martin.centurion@unl.edu*

[b]*SLAC National Accelerator Laboratory, Menlo Park, CA 94025, USA*

[c]*Physics and Astronomy, Potsdam University, Potsdam, 14476, Germany*

the signal-to-noise ratio of the detection system. We also discuss plans for improving both the temporal resolution and the spatial resolution.

1. Introduction

Ultrafast gas electron diffraction (UGED) can in principle provide sufficient temporal and spatial resolution to observe bond length changes in isolated molecules during a photo-induced reaction. In order to capture the trajectory of individual atoms during a reaction, femtosecond resolution is required. Recent experiments using relativistic electron pulses with MeV energy have succeeded in reaching a temporal resolution of 230 fs full-width at half maximum (FWHM) or 100 fs root-mean-squared (RMS).[1] In this manuscript we will expand on the previously reported results. Additional information is provided about the operation of the MeV gun during the gas phase experiments, and the detection system is described in detail. A crucial step towards performing pump–probe experiments is to find temporal overlap between laser and electron pulses. Here we describe how we found the temporal overlap by producing a laser plasma that deflects the electron beam. We also provide additional data on the noise level of the measurements and the temporal fitting that was used to extract the temporal resolution from the rotational dynamics.

1.1. Background

Time resolved electron diffraction has been applied to gas phase samples over the years with resolutions from microseconds to nanoseconds.[2,3] The promise of UGED was demonstrated by the Zewail group when they were able to retrieve the structure of transient states with few-picosecond resolution.[4–6] Over the last 15 years, though, the achievable resolution in UGED was limited to the picosecond scale.[7–9]

While capturing the motion of atoms in isolated molecules has remained elusive, this goal has been achieved for crystallized molecules by the Miller group.[10,11] In condensed matter ultrafast electron diffraction (UED), femtosecond dynamics have been observed using compact electron guns[12,13] where the charge per pulse is limited and the source-to-target distance is kept very short to reduce the Coulomb broadening of the electron pulses. An alternative approach has been the use of radio-frequency (RF) cavities to deliver temporally compressed electron pulses at the sample position.[14–16] These methods have so far not been successfully implemented on UGED, although they could be applied if some technical challenges are overcome. For example, compact guns require a high vacuum on the photocathode to avoid discharges, while the pressure near the gas target is typically much higher. For RF compression of keV electrons, the synchronization between the RF compression fields and the laser pulses has limited the available temporal resolution to $\sim$300 fs.

Several other methods are available to probe molecules in the gas phase with high spatio-temporal resolution using X-ray and laser sources. The first is hard X-ray diffraction using femtosecond pulses from an X-ray free-electron laser (X-FEL). X-ray diffraction experiments on the ring-opening reaction of CHD were recently carried out with temporal resolution of 80 fs, where the transient structures were retrieved by comparing the experimental results with simulations.[17]

Molecular structures can also be probed using femtosecond X-ray photoelectron diffraction[18] and laser-induced electron diffraction.[19–21] In these methods the molecule is ionized either by an X-ray pulse or an intense laser pulse, and the emitted electron diffracts from the parent molecules. The advantage of UGED compared with these methods is that the probing electrons are not generated on the molecule itself, so knowledge of the ionization process is not required. In addition, UGED does not expose the molecule to strong laser fields. Here we show how the challenges associated with achieving high temporal resolution in UGED can be overcome using electron pulses accelerated to MeV energy.

1.2. Spatial and temporal resolution in UGED

The spatial resolution δ is the smallest feature which can be resolved in an object. It is determined by the de Broglie wavelength of the electrons and the range of diffraction angles that are captured. For small diffraction angles, the resolution can be written as

$$\delta = \frac{\lambda}{\theta_{\text{max}}},$$

where λ is the de Broglie wavelength of the electron beam, and θ_{max} is the maximum scattering angle captured in the diffraction pattern. For a relativistic electron beam with a kinetic energy of 3.7 MeV, λ is 0.30 pm. In UGED, the interatomic distances in a molecule can often be determined with an accuracy significantly higher than the spatial resolution, provided that distances do not overlap within the resolution, or that there is some prior knowledge of the structure, such as the number of atoms in the molecule. The resolution determines the width of each peak corresponding to an atomic position (or interatomic distance) in the retrieved object, while the accuracy determines how accurately the center of each peak can be found. The spatial resolution of a retrieved object is mostly limited by the signal-to-noise ratio (SNR) in the experiment. The differential scattering cross section drops rapidly with scattering angle, thus detecting electrons at higher scattering angles requires higher signal levels. Spatial resolution can be improved by increasing the beam current, the integration time, or the sample density, or by reducing the detector noise.

The maximum size of an object that can be measured is determined by the transverse coherence length of the electron beam $L_c = \lambda/2\pi\Delta\phi$, where $\Delta\phi$ is the angular divergence of the electron beam. The transverse coherence of the MeV beam used in our experiments is 1.7 nm. The transverse coherence of a beam can be increased by expanding the beam or by using apertures, at the expense of beam intensity. The ratio of the transverse coherence length to the beam size depends on the emittance of the beam.

The temporal resolution of a UED experiment, where a sample is excited by a laser pulse and probed by an electron pulse, can be approximated as:

$$\tau_{\text{EXP}} = \sqrt{{\tau_{\text{laser}}}^2 + {\tau_{\text{electron}}}^2 + {\tau_{\text{TOA}}}^2 + {\tau_{\text{VM}}}^2} \tag{1}$$

where τ_{laser} is the duration of the laser pulse, τ_{electron} is the duration of the electron pulse, τ_{TOA} is a measure of the jitter and drift in the relative time of arrival (TOA) between laser and electron pulses at the sample, and τ_{VM} accounts for the effect of the velocity mismatch (VM) of the laser and electron pulses as they traverse the sample. The contribution of the laser pulse duration to the overall resolution is minimal, since laser pulses with a duration of 30 fs or less can be made routinely. The electron pulses are generated by photoemission, and can be as short as the laser pulse at the point of emission, but the pulse duration rapidly increases as the pulse is accelerated and propagates to the sample. This broadening is caused by the initial energy spread of the photoelectrons and by the Coulomb force (space charge effect). Several methods are available for delivering short pulses on a target: placing the sample very close to the source, limiting the number of electrons per pulse, and using time-dependent fields to recompress the pulses. The jitter in the TOA arises from temporal variations in the acceleration electric field in the gun and/or the RF compression field. The VM results from the fact that laser and electron pulses traverse the sample at different velocities.[22] The VM effect can be minimized by using thin samples, by tilting the laser pulse front,[23,24] or by using relativistic electron pulses.[25–29] For condensed matter experiments in transmission, the sample is usually thinner than a micrometer, resulting in negligible blurring due to velocity mismatch.

We identify three unique challenges towards achieving high spatio-temporal resolution in UGED: (i) the target density is very low. Usually, gas jets are used to deliver the sample molecules, resulting in a low column density on the order of 10^{15} molecules per cm^3, and a high electron beam current is needed to have a sufficient number of scattering events. This often results in a trade-off between spatial and temporal resolution, since the duration of the pulse will elongate with increasing charge.

(ii) The sample molecules in the gas beam are randomly oriented, which reduces the amount of information that is encoded in the diffraction pattern due to the angular averaging. The result is that typically only 1D information (corresponding to the interatomic distances) can be accessed.[30] It has been shown that diffraction from aligned molecules can be used to retrieve both the distances and angles in a molecule.[7] The effects of alignment have been described elsewhere[31,32] and in this manuscript we will focus on the effects of the temporal resolution.

(iii) The gas target is typically at least 100 μm thick, which results in significant VM when non-relativistic electrons are used. For example, a 100 keV electron pulse travels with a speed of 0.55c, where c is the speed of light in vacuum. The 100 keV electron pulse would take 1.2 ps to traverse a 200 μm sample, while the laser would take only 0.67 ps, resulting in a significant change in the relative delay between the laser and electrons throughout the sample.

2. Experimental setup

The main components of the setup are the photoelectron RF gun, the target chamber and the detection system. Here we describe each in detail.

2.1. The electron gun

Fig. 1 shows a diagram of the main parts of the setup. The experiment comprises the radiofrequency (RF) electron gun, beam transport and electron optics, the target chamber, a second beam transport line and the detector. The RF electron gun is identical to the one that drives the Linac Coherent Light Source (LCLS) at SLAC National Accelerator Laboratory. The gun is a 1.6 cell structure with

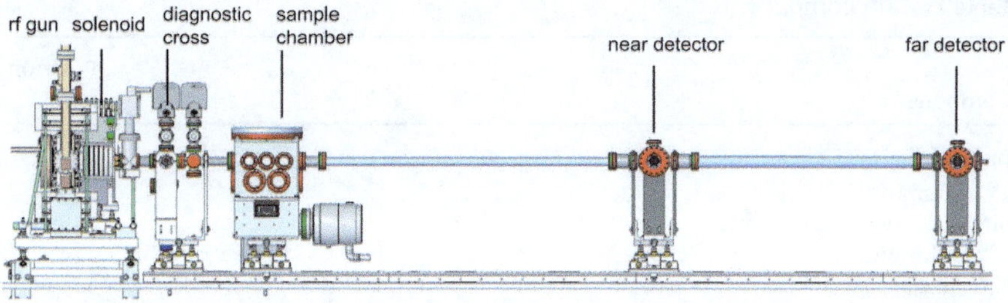

Fig. 1 Experimental layout for gas phase diffraction experiments. Electron pulses with 3.7 MeV energy are produced at the photoelectron RF gun. A series of collimators and solenoids are used to collimate the beam. The far detector position was used for gas phase experiments. Figure reprinted from ref. 25 with the permission of AIP Publishing.

2.856 GHz resonant frequency and a maximum accelerating electric field of 120 MV m^{-1}. For the UGED experiments, we operated the gun at 80 MV m^{-1} gradient and the electron beams were accelerated to 3.7 MeV kinetic energy. This beam energy was chosen to give the desired *s*-range of 1.6–15 $Å^{-1}$ in combination with the locations and dimensions of the collimators and diffraction detector.

The high acceleration gradient and high beam energy of photocathode RF guns are very effective in suppressing space-charge induced bunch lengthening. However, there are technical challenges for the control of the time or arrival jitter τ_{TOA} in RF guns. This jitter is caused by fluctuations of the RF fields amplitude and phase as well as the synchronization between the RF phase and the photocathode drive laser. In the last decade, SLAC has developed relevant key technologies mostly motivated by more precise timing control of the LCLS injector. These technologies naturally benefited the MeV UGED application as well. A top-off circuit was added to a conventional pulse-forming-network-based modulator and improved the voltage stability to <50 ppm. The modulator drives a klystron, which is the high power RF source for the gun. The amplitude and phase stabilities of the RF field are 2×10^{-4} RMS and 25 fs RMS, respectively – a factor 2–5 better than typical performances. A high-precision femtosecond timing system developed for LCLS was implemented and locks the laser and the low-level RF signal at the level of 30 fs RMS. Based on the RF and RF-laser timing performance, the estimated τ_{TOA} is 50 fs RMS for the gas-phase UED experiment.

The vacuum levels inside the UGED beamline are strongly varying. The electron gun requires $<1 \times 10^{-9}$ torr to reliably operate at high gradient with low field emission and no structure damage, while the vacuum in the sample chamber can be as high as mid-10^{-4} torr. Moreover, some gas sample species are corrosive for copper – the material the entire electron gun is made of. We separated the gun vacuum using a 50 nm thick, 1.5 mm^2 area silicon nitride (SiN) window. We experimentally verified that 75% of the electrons in the beam penetrate the SiN window with a negligible change of divergence. These electrons will contribute to the diffraction signal. The location of the SiN window, as well as other main beamline components are summarized in Table 1. A first collimator of 200 μm diameter defines the beam size at the SiN window and reduced the field emission electrons from the gun which otherwise increase the background in the diffraction pattern. A second collimator of 400 μm diameter is used to block the large-

Table 1 Gun components

Component	Position (cm)
Photocathode	0
1st collimator	55.8
SiN window	74.0
2nd collimator	144
Gas nozzle	156
Diffraction detector	458

angle scattering from the SiN window, which is much stronger than the diffraction signal from the gas molecules. The size of the 2nd collimator matched the beam block (actually a hole in the phosphor screen used for detection) of the diffraction detector. Four sets of steering magnets are used to guide the electron beam throughout the apertures along the beam line. They are located at: (i), the position of the solenoid; (ii), right after the diagnostic cross; (iii), right before the sample chamber; (iv), right after the sample chamber.

2.2. The target chamber

Fig. 2 shows a schematic view of the interaction region in the experiment. The sample molecules are introduced into the chamber using a pulsed nozzle with an exit hole of 100 μm. The nozzle operates at 120 Hz, the same repetition rate as the gun. Two differential pumping stages and the 50 nm SiN window are used to separate the gun and target chambers.

A Ti:sapphire laser with 40 fs pulses and 2 mJ energy is available to excite the sample and to trigger photoemission in the cathode. A frequency tripler is used to convert the 800 nm light into a wavelength of 267 nm for the photoemission. On the pump side, a range of wavelengths can be delivered using a dedicated optical

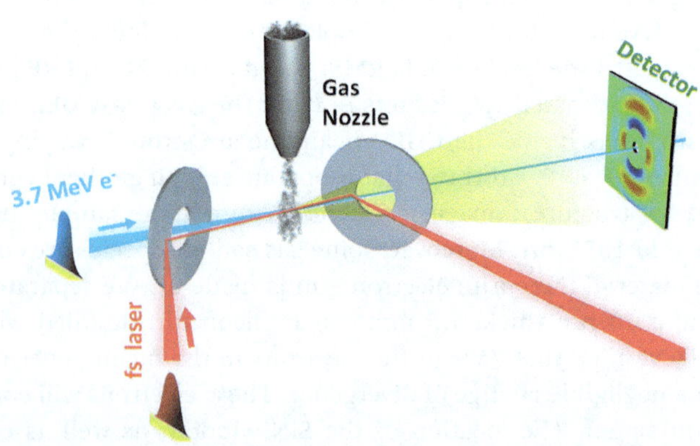

Fig. 2 Interaction region in the experimental chamber. The sample gas molecules are introduced using a pulsed nozzle. The laser beam (red) is deflected towards the target and out of the chamber using a pair of mirrors with center holes. The electron beam passes through the holes in the mirrors. The laser and electrons are nearly parallel as they traverse the sample. Figure is reproduced from ref. 1.

parametric amplifier, frequency doubler and frequency tripler. The laser is introduced into the setup using two mirrors with a hole in the center. The laser is reflected off the side of the mirrors while the electron beam passes through a hole in the center. The laser and electron pulses traverse the sample with a small angle of 5 degrees between them. The diameter of the laser beam, electron beams and gas jet are all approximately 200 μm in the interaction region. The scattered electrons exit through the hole in the second mirror and are captured by the detector.

2.3. The detector

The diffracted electrons are captured with a phosphor screen that has a hole in the center to transmit the unscattered electron beam. The size of the screen is 40 mm × 40 mm, and the center hole has a diameter of 6.3 mm to prevent saturation of the detector due to the much brighter central beam. The original phosphor screen was later replaced by a screen with a 2.9 mm hole to capture diffraction at smaller angles. The light emitted by the phosphor screen is optically coupled into an EMCCD camera. A mirror with a 6.5 mm × 9.0 mm elliptical hole, oriented at an angle of 45° to the incoming electron beam, reflects the light from the phosphor screen towards the camera. An *f*/0.85 lens is used to image the screen onto the camera.

Fig. 3a shows a schematic of the detector showing the holey phosphor screen, mirror, vacuum port, lens and EMCCD camera positions. Fig. 3b and c show diffraction patterns captured with the phosphor screens with the 6.3 mm and with the 2.9 mm diameter holes, respectively. The larger hole was originally used to avoid saturation due to the main beam, but it was later found that the smaller hole was sufficient to pass the main beam, and it also allowed for capturing more

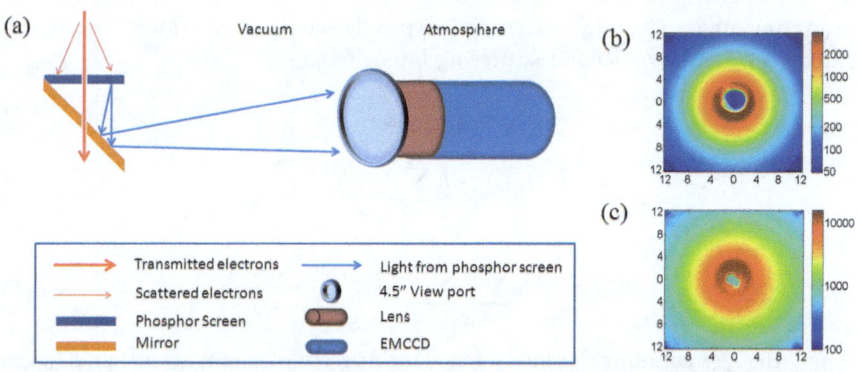

Fig. 3 (a) A sketch of the detection system. The diffraction pattern is captured by a phosphor screen, and then imaged by a 45 degree mirror and a lens on to an EMCCD camera. The main (unscattered) electron beam is transmitted through two holes in the center of the phosphor screen and mirror. A 4.5″ view port separates the components under vacuum (mirror, phosphor screen) and the components under atmosphere (lens, EMCCD). The phosphor screen size is 40 mm × 40 mm with a hole in the center. The mirror size is 40 mm × 56 mm with an elliptical hole of 6.5 mm × 9 mm. The distance between the center of the mirror to the lens entrance is 12 cm. The lens has a focal length of 40 mm and an *f*-number of 0.85. (b) and (c) are raw static diffraction patterns for nitrogen (b) and iodine (c), shown in logarithmic scale.

of the diffraction pattern at small angles. The pattern in Fig. 3b was captured with nitrogen, and is the configuration used for the time-resolved experiments described later in this manuscript. The pattern in Fig. 3c was captured with iodine molecules using the screen with the smaller hole. In the current detection system, the hole in the 45 degree mirror transmits a fraction of photons from the phosphor screen and thus creates non-uniformity in the low-s range of the diffraction patterns. This effect can be compensated by calibrating the image transfer function of the optical system, for example by using a static diffraction pattern from a known molecule.

3. Static gas electron diffraction with MeV electrons

Diffraction from molecules in the gas phase can be modelled using the independent atom approximation, and assuming the molecules are randomly oriented.[33] The transverse coherence of the electron beam is assumed to be larger than the size of the molecules but smaller than the distance between molecules. Thus, the scattering from different atoms within a molecule are added coherently, while the scattering from different molecules is added incoherently. Under these assumptions the scattering intensity $I(s)$ on the detector can be described as:

$$I(s) = I_0 \sum_{i=1}^{N} \sum_{j=1}^{N} f_i(s) f_j^*(s) \frac{\sin(r_{ij}s)}{r_{ij}s} \tag{2}$$

here $f_i(s)$ is the complex elastic scattering amplitude of the ith atom in the molecule, r_{ij} is the distance between atoms i and j, I_0 is a constant and N is the total number of atoms in a molecule. The scattering is expressed in terms of the momentum change of the scattered electrons $s = \frac{4\pi}{\lambda}\sin(\theta/2)$, where θ is the scattering angle. The total scattering is typically separated into atomic scattering intensity I_{at} and a molecular scattering intensity I_{mol}:

$$I = I_{\text{at}} + I_{\text{mol}}, \tag{3}$$

$$I_{\text{at}}(s) = I_0 \sum_{i=1}^{N} |f_i(s)|^2, \tag{4}$$

$$I_{\text{mol}}(s) = I_0 \sum_{i=1}^{N} \sum_{j=1, j\neq i}^{N} f_i(s) f_j^*(s) \frac{\sin(sr_{ij})}{sr_{ij}}, \tag{5}$$

where the I_{mol} contains the interference terms from different atoms, and I_{at} acts like a background scattering that is modulated by I_{mol}. The scattering intensity decreases rapidly with s, so I_{mol} is rescaled to form the modified scattering intensity.

$$\text{sM}(s) = s\frac{I_{\text{mol}}}{I_{\text{at}}}. \tag{6}$$

Each pair of atoms will give rise to a sinusoidal modulation in sM(s), with the amplitude proportional to the product of the scattering amplitudes and inversely proportional to the interatomic distance. For a diatomic molecule with interatomic distance r, the modified scattering becomes

$$sM(s) = \frac{\sin(sr)}{r} \tag{7}$$

In this case a single sinusoidal modulation is present. In this formulation only elastic scattering is included. The inelastic scattering contributes to the experimental background but does not contain any structural information. The interatomic distance can be extracted either by a Fourier transform of sM or by fitting a sine function to the measured signal. Vibrational excitations within the electronic ground state are not taken into account in this formalism. The inclusion of vibration was found to be helpful for structural refinement in static gas electron diffraction,[30] but for UGED experiments the effect of vibrations is typically much smaller than the effect of structural rearrangements.

Experimentally, the sM must be extracted from a background of atomic scattering, inelastic scattering and background noise. The background noise is produced by electrons scattered towards the detector by collimating apertures or by background gas, by any background photons that reach the detector and by camera noise. Fig. 4a shows the different levels of background observed experimentally with nitrogen as the target molecule. The figure is obtained by azimuthally averaging a diffraction pattern, such as the one shown in Fig. 3b. Data below $s = 3$ Å^{-1} is not available due to the hole in the phosphor screen. The blue line in Fig. 4a shows the total counts on the detector, on a logarithmic scale. The image was acquired using 38k electrons per pulse at a repetition rate of 120 Hz and a total integration time of 30 minutes. The backing pressure on the nozzle was 1 bar, which results in an estimated sample density of 2×10^{17} molecules per cm^3. The green line in Fig. 4a shows the count levels when the experiment is run under the same conditions but with the gas jet off. The detected electrons are produced mostly by scattering from apertures and from the silicon nitride window. Scattering from the target gas is more than an order of magnitude stronger than the background. Finally, the magenta line shows the detector counts when both the gas jet and electron beam are turned off. This reflects

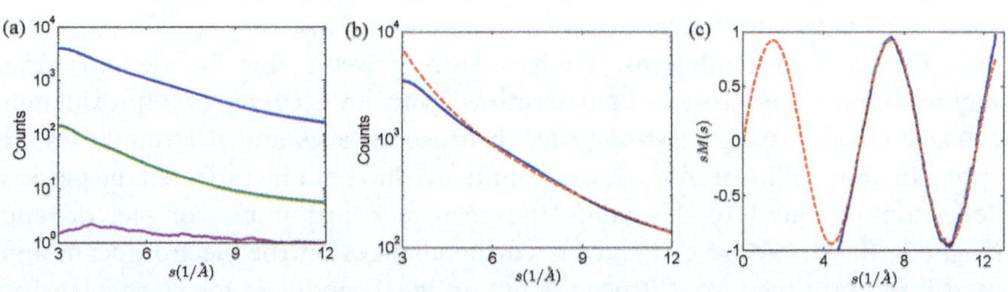

Fig. 4 Diffraction intensity of static N_2 molecules. (a) Azimuthally averaged diffraction intensity (blue), background taken without gas (green) and background taken without electron beam (magenta). The vertical axis is the raw camera counts per pixel per one minute integration time. (b) Azimuthally averaged diffraction intensity of static N_2 molecules (blue solid) and the scattering background obtained by a standard fitting routine (red dashed). (c) Experimental (blue solid) and simulated (red dashed) modified diffraction intensity sM(s). (c) is adapted from ref. 1

mostly the detector noise and is about three orders of magnitude below the scattering signal level.

The total experimental background needs to be removed from the data to extract the sM and the interatomic distance. We follow a standard iterative fitting procedure to remove the experimental background.[34] First, an initial guess of the bond length is used to simulate sM. The position of the zero crossings is extracted from the simulated sM. Then the background is determined by fitting an exponentially decaying function at these zero points, and the experimental sM is obtained by subtracting the fitted background. The error is calculated as the RMS difference between the experimental and simulated sM. The process is then repeated for a number of values of the bond length, until the best fit is found. Fig. 4b shows the total counts (blue line) and the fitted background (dashed red line), in logarithmic scale. Fig. 4c shows the experimental (blue line) and simulated (dashed red line) sM for the best fit parameter. The data below $s = 4$ Å^{-1} is missing in Fig. 4c because the zero crossing of sM is used to remove the background.

The bond length extracted from the sM in Fig. 4b is 1.073 ± 0.027 Å, in agreement with the previously measured bond length of nitrogen of 1.098 Å.[35] There is an uncertainty in the measurement of 2.5% due to the calibration of the sample to detector distance. The distance was calibrated using diffraction from a gold sample, and the uncertainty is caused by the scattering from the SiN window which makes the electron beam emittance larger. For time resolved experiments, if the ground state structure is known, a static gas diffraction pattern can be used as a calibration, reducing the uncertainties, as is shown later in this manuscript.

4. Finding the temporal overlap of electron and laser pulses

The temporal overlap, or time-zero, of the laser and electron pulses in UGED with keV electrons can be determined to an accuracy of 1 ps using plasma lensing.[22] The laser ionizes the sample molecules and the electron beam is deflected by the plasma fields as the charges separate. The deflection typically grows on a scale of picoseconds.[36,37] It was not clear whether this method could also be applied to relativistic electrons with MeV energies because they need higher fields to be deflected on the detector. We have seen, however, that the method works very well and time-zero can be determined with an accuracy of approximately 200 fs. We believe we see a strong effect because the laser and electron pulses are approximately collinear through the sample. We have not investigated the plasma deflection in detail to determine the relative contributions of electric and magnetic fields. We have characterized the changes on the electron beam and used it to find time-zero. Nitrogen is not an ideal candidate for plasma-lensing because it has a high ionization potential of 15.58 eV, nevertheless we are able to generate a sufficiently strong plasma by increasing the 800 nm Ti:sapphire fundamental laser intensity to 8×10^{14} W cm^{-2}. The laser focus spot size is 50 μm FWHM, and the electron beam is 200 μm FWHM at the interaction region. We expect that for molecules with a lower ionization potential a significantly lower laser intensity will be needed.

After traversing the laser plasma, the electron beam is deflected along the direction of the laser polarization. The effect is quantified by measuring the number of deflected electrons, *i.e.* electrons that are detected outside of the main beam. Fig. 5a shows the time trace obtained from a plasma lensing scan where the relative time delay between the electrons and laser is varied. Significant changes can be seen on a timescale of 200 fs. Fig. 5b shows an image of the electron beam without the plasma lensing effect, at a time of 100 fs before time-zero. Fig. 5c and d show the corresponding images of the beams for time delays of 400 fs and 900 fs after time zero, respectively. A significant number of electrons are deflected along the direction of laser polarization (vertical in the images). The white rectangle in Fig. 5c shows the region over which the deflected electrons are integrated.

5. Determination of temporal resolution and spatial resolution

We have used the rotational dynamics in nitrogen molecules to characterize the temporal resolution of the instrument. A femtosecond laser pulse with a wavelength of 800 nm is used to create a rotational wavepacket and impulsively align the molecules.[38] The laser pulse energy is 800 μJ, its duration 34 fs, and the laser focal size is 200 μm in diameter, resulting in a laser intensity of 5×10^{13} W cm^{-2}. When the molecules are aligned, the diffraction pattern becomes anisotropic,[7,39,40] and this anisotropy can be used to follow the temporal evolution of the angular distribution. We have chosen nitrogen for this experiment because the temporal evolution is very fast and dynamics of the alignment can be simulated

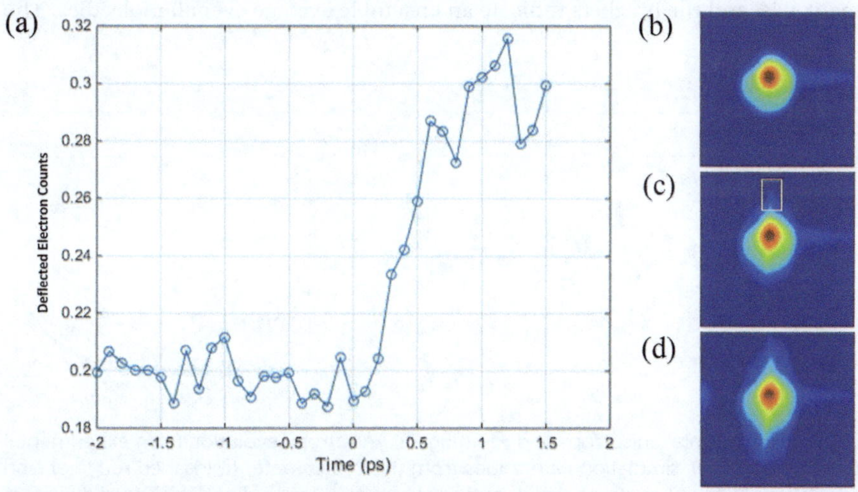

Fig. 5 Time-zero determined by plasma-induced lensing effect. (a) The deflected electron counts (normalized) as a function of delay time. Time-zero can be determined with an accuracy of approximately 200 fs. (b)–(d) Images of the electron beam on detector, at delay times of (b) −100 fs, (c) 400 fs and (d), 900 fs. The deflected electron counts shown in part (a) are calculated by summing over all the pixel counts in the white rectangular shape displayed in part (c). The plasma is generated by focusing an intense IR beam (8×10^{14} W cm^{-2}) on the N_2 gas jet.

accurately.[41,42] For example, the distribution changes from aligned (prolate) to anti-aligned (oblate) in 300 fs, and the fast changes can be used to accurately characterize the temporal resolution by comparing simulations and experiment.

Diffraction patterns were recorded as a function of delay between the alignment laser pulse and the probing electrons. The step size was 100 fs and the delay was varied from −1 ps to 10 ps, just over the time for the full revival of the alignment at 8.35 ps. The anisotropy of the diffraction pattern was determined for each time delay. Prior to this, each diffraction pattern was pre-processed by removing a dark background, finding the center of the diffraction pattern and normalizing to the total counts. The anisotropy was calculated by dividing the total counts in a horizontal cone with a half-angle of 35° by the total counts in a vertical cone with a half angle of 55°. A region of the pattern that shows high anisotropy, between $s = 3$ Å^{-1} and $s = 4.5$ Å^{-1}, was used for the calculation.

Fig. 6a shows the temporal evolution of the anisotropy in the diffraction pattern. The blue dots show the anisotropy extracted from the measured diffraction patterns. For each data point, the diffraction pattern was captured continuously for two minutes at a repetition rate of 120 Hz. In order to quantify the uncertainty in the anisotropy measurement, additional measurements were taken at a few points before time zero. The standard deviation for these points is displayed in Fig. 6a as error bars. The solid green and dashed red lines in Fig. 6a show the simulated anisotropy for two different sets of parameters. The angular distribution as a function of time was simulated using a rigid-rotor model.[42] A diffraction pattern was calculated from for each time delay, using the calculated angular distribution.[43] Finally, the anisotropy of the simulated pattern was calculated using the same procedure applied to the experimental data. The right hand axis of Fig. 6a shows the degree of alignment extracted from the simulation in terms of $\langle \cos^2 \alpha \rangle$, where α is the angle between each molecule and the alignment axis, and the brackets indicate an ensemble average over all molecules. The

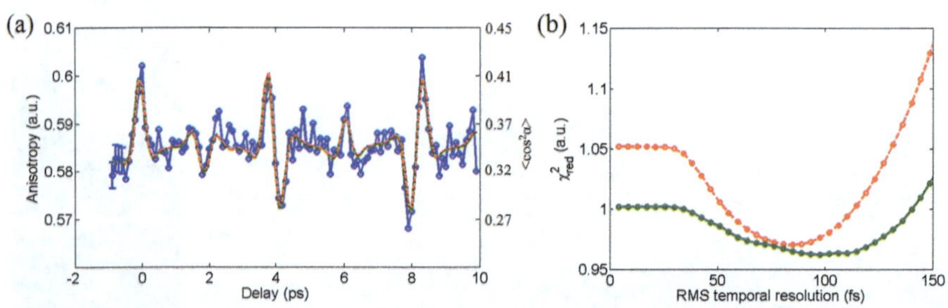

Fig. 6 Experimental anisotropy and χ^2 fitting. (a) Anisotropy evolution from experimental data (blue circles), simulation with values from the 2-parameter fit (dashed red line), and simulations with values from the 4-parameter fit (solid green line). (b) The reduced χ^2 metric for the 4-parameter fit (circles and solid green line) and 2-parameter fit (circles and dashed red line). The 4-parameter fit involves laser fluence, initial temperature, temporal resolution and a re-scaling factor that accounts for the excitation percentage of the target molecules. The 2-parameter fit uses the measured laser fluence and estimated initial temperature, while varying the temporal resolution and the re-scaling factor. The optimal temporal resolution from the 4-parameter fit and 2-parameter fit are 100 fs RMS and 85 fs RMS, respectively.

angular distribution changes rapidly with time, oscillating between alignment (increased anisotropy) and anti-alignment (decreased anisotropy).

The sharp feature around 4 ps corresponds to a half revival, while the feature around 8 ps corresponds to a full revival. The distribution changes very rapidly between the two extremes of maximum alignment (prolate distribution) and maximum anti-alignment (oblate distribution). The temporal evolution of the measured anisotropy signal is determined by three parameters: the initial temperature of the molecules, the fluence of the alignment laser pulse and the overall temporal resolution of the experimental set-up. The laser fluence was measured to be 2.0 J cm^{-2}. We estimated that the rotational temperature of the molecules in the jet before laser excitation was 65 K, based on the backing pressure, the electron-nozzle exit distance and the nozzle geometry.[44]

We have used a χ^2 fitting routine to extract the temporal resolution of the experiment, by comparing the data with the simulation results. The whole temporal scan shown in Fig. 6a is used for the fit. We performed the fit in two different ways, first fixing the values of the laser fluence and rotational temperature cited above, and then letting them also be free parameters in the fit. In the first case we run a two-parameter fit in which only the temporal resolution and a scaling factor were allowed to vary. The temporal resolution is introduced by convolving the simulated anisotropy with a Gaussian envelope. As the temporal resolution becomes longer in time, the fast-changing features in the anisotropy will start to disappear. The two-parameter fit returned a resolution of 85 fs RMS (200 fs FWHM) and scaling factor of 0.28 with a reduced χ^2 of 0.97. The rescaling factor of 0.28 means that 28% of the molecules were excited by the laser due to imperfect spatial overlap. The results of this fit are shown with the red circles in Fig. 6b. We also run a four-parameter fit in which the temporal resolution, scaling factor, laser fluence and the initial rotational temperature were allowed to vary. The fit returned a temporal resolution of 100 fs RMS (230 fs FWHM), laser fluence of 1.8 J cm^{-2}, rotational temperature of 54 K, scaling factor of 0.42 and reduced χ^2 of 0.96. The results are consistent with the two-parameter fit.

The changing angular distribution can be observed directly in the diffraction patterns. Fig. 7 shows the difference diffraction patterns corresponding to the peak alignment and peak anti-alignment. The difference diffraction patterns are generated by taking the difference between a diffraction pattern at a positive delay time (the electrons reach the sample after the laser) and a diffraction pattern at a negative time (the electrons reach the sample before the laser). For capturing high quality diffraction patterns, the integration time was increased to a total of 60 minutes for each pattern. Fig. 7a and b show the experimental and simulated difference patterns at the maximum alignment at the half revival (around 4 ps after laser excitation). There is good agreement between the experiment and simulation. The missing data in the experimental pattern (s < 3.5 Å) is filled in by letting the pattern smoothly go to zero at the center. The spatial resolution can be calculated from the maximum value of s captured in the experiment, $s_{max} = 8.3$ Å^{-1}, which gives a spatial resolution $\delta = 0.76$ Å.

Fig. 7c and d show the experimental and simulated 2D Fourier transform (FT) of the difference pattern. The FT shows the changes in the distribution of molecules in real space. For the prolate distribution, this means increased population along the direction of alignment and decreased population in the region perpendicular to the alignment. Since the FT gives a representation of the

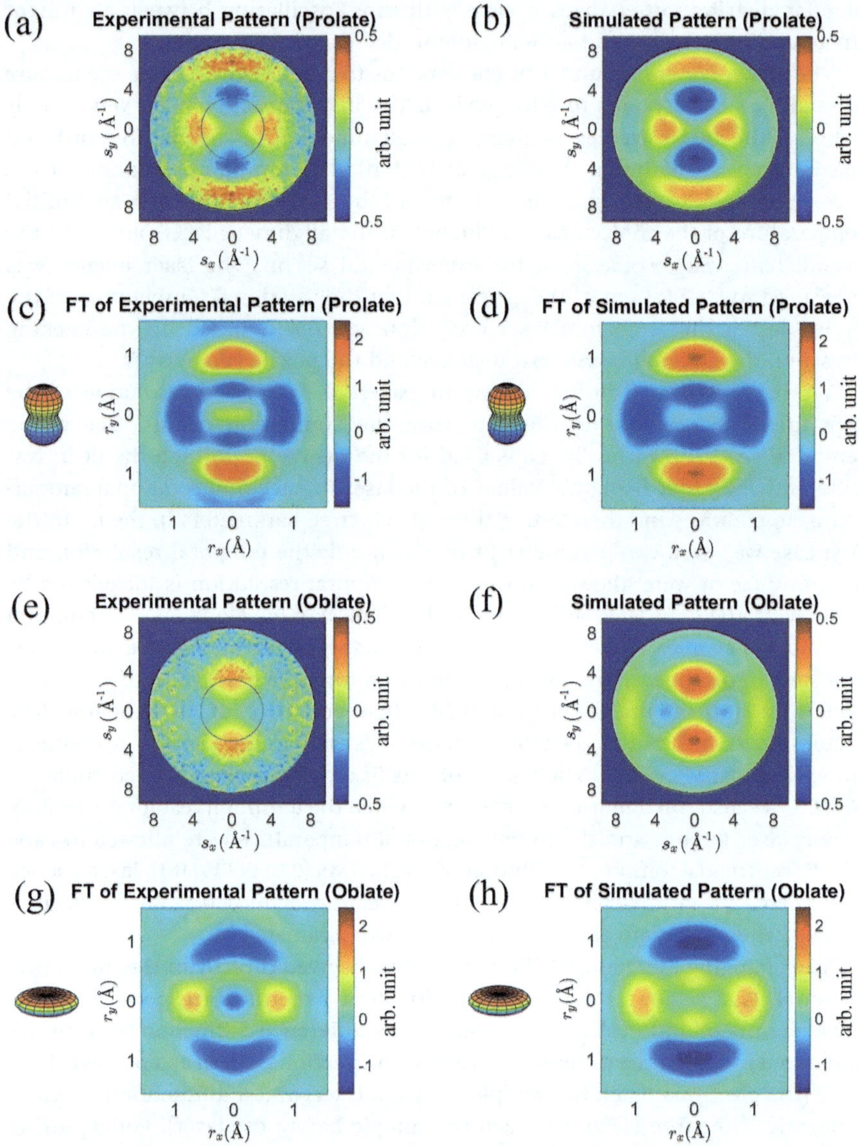

Fig. 7 2D N_2 diffraction patterns at half revival. (a) Experimentally measured and (b) simulated diffraction-difference patterns of the prolate distribution. (c) and (d) are Fourier transforms of (a) and (b), respectively. (e) Experimentally measured and (f) simulated diffraction-difference pattern of the oblate distribution. (g) and (h) are Fourier transforms of (e) and (f), respectively. In patterns (a) and (e), the data inside the black circles are missing due to the beam stop. They are obtained by extrapolating the pattern and letting the counts smoothly go to zero towards the center. For illustrative purpose, angular distributions are shown on the side of panel (c), (d), (g) and (h) for visual guidance. Figure is reproduced from ref. 1.

molecule in real space, the interatomic distance can be extracted from the diffraction patterns. The measured object is a 2D projection of a 3D distribution of molecules, so the interatomic distance is not equal to the distance of the bright spots from the origin, as would be the case for perfectly aligned molecules. A least squares fitting method was used to extract the bond length form the real space object in Fig. 7c, using the static diffraction pattern as a calibration. The fitting was done by simulating the diffraction pattern for different bond lengths of nitrogen, and then finding the RMS difference between the simulated and measured objects. The angular distribution for the simulated patterns was extracted from the alignment simulation shown in Fig. 6a. The extracted bond length was 1.091 ± 0.011 Å, in good agreement with the ground state bond length of nitrogen of 1.098 Å.

Fig. 7e and f show the experimental and simulated difference diffraction patterns at the maximum anti-alignment during the half revival, where the distribution is oblate. In this case, there is increased population in the direction perpendicular to the alignment axis, as can be seen in the FTs in Fig. 7g and h. A similar fit for the bond length was performed, which returned a value of 1.096 ± 0.018 Å, in good agreement with the previous values. The fit was performed for ten different data sets (6 minutes of integration each), and the mean value is reported as the best estimate with the standard deviation of the mean reported as the uncertainty. Note that we had previously reported the uncertainty as the standard deviation,[1] but the standard deviation of the mean is a more accurate measure of the uncertainty. The experimental uncertainty in the bond length is significantly less than the spatial resolution of 0.76 Å. The resolution determines the width of the distribution along the radial direction (see Fig. 7c and d), while determining the bond length can be thought of as finding the center of the distribution (accounting for the fact that the alignment is not perfect). The resolution becomes important when there are multiple closely spaced distances to be measured simultaneously.

6. Conclusion and outlook

In summary, we have demonstrated that relativistic electrons can be used to break a longstanding barrier in the temporal resolution of gas phase electron diffraction. We have achieved a temporal resolution of 100 fs RMS (230 fs FWHM), and spatial resolution of 0.76 Å. Our results also show that for simple molecules the bond lengths can be determined with an accuracy significantly lower than the spatial resolution. These results open the door to imaging photo-induced chemical reactions in isolated molecules with a combined spatio-temporal resolution that will allow for resolving the motion of the nuclei on the sub-Å scale.

To further improve the spatial and temporal resolution, a number of system upgrades can be implemented in the near and far future. A direct detection sensor[45] will significantly reduce the detector noise, which will increase the SNR and thus the spatial resolution. The direct detection camera will also obviate the need for a 45 degree mirror with a hole, thus resolving the issue of a non-uniform transfer function and the missing data at small scattering angles. The signal levels could be further improved by replacing the circular orifice gas nozzle with a slit to generate a longer gas target. An RF-based compression cavity has been proposed to compress the pulse duration of MeV electron pulses down to the order of 10 fs.[46] The RF compression would also increase the number of electrons per pulse, thus to improving both the temporal and spatial resolution. Further improvements in temporal resolution require not only compressing the electron pulses but also compensating the timing jitter. These could be achieved by measuring the TOA of each pulse and resorting the data on a shot-by-shot basis using time-stamping techniques.[47,48] Ultimately, a temporal resolution of 30 fs would be sufficient to capture most photochemical dynamics processes, but these improvements must be accompanied by an improvement of the spatial resolution such that small structural changes can be observed. With high spatio-temporal resolution, we can envision that this method will be able to capture not only structures, but also spatial information on the moving nuclear wavepackets. For example, as a reaction proceeds, the probability density of a nuclear wavepacket can broaden spatially, which can in principle be detected by UGED. It will be interesting to observe transition states where, on the relevant spatial and temporal scales, parts of a molecule may be delocalized.

There are several types of photo-reactions that could be studied with this setup, such as ultrafast relaxation, *cis–trans* isomerization, dissociation, and charge transfer. These reactions take place on femtosecond timescales, and in most cases the structure of the intermediate states has not been measured. Many of the molecules of interest have a low vapor pressure, so will only be available at densities well below the atmosphere. One of the major challenges will be to have a sufficiently high signal-to-noise ratio for structure retrieval.

We can make some estimates based on the nitrogen diffraction patterns presented in this manuscript. Here the sample was introduced into the nozzle with a backing pressure of approximately 1 bar and then entered the vacuum as an expanding gas jet. For many samples, the vapor pressure that can be expected (with some heating) to be on the order of 10 mBar, so the signal level (measured in scattered electrons per second) will decrease by a factor of a hundred. This decrease can be compensated by straightforward improvements in multiple components of the setup. One option is to increase the integration time of the diffraction patterns, but this is the least desirable route as the duration of the experiments will become very long and more susceptible to long term instabilities. In the current setup, the backing pressure that could be used was limited by the speed with which the gas could be pumped out of the system. With lower pressure in the nozzle (or more efficient pumping), the length of the gas jet could be increased by a factor of 2 to 5 without significantly degrading the temporal resolution. This would directly translate into 2–5 times more scattering. The repetition rate of the experiment can be increased from 120 Hz to 1 kHz with the current laser system, and it could be increased even further with an upgrade to the electron gun. This would result in an additional factor of 8 or more in average beam current. The use of RF compression, as mentioned above, can increase the charge of each pulse by a factor of 10 or more, also increasing the beam current and thus the total number of scattered electrons. Combining these improvements it is reasonable to expect a two order of magnitude increase in the signal level, opening the door to investigating a large number of possible target molecules.

Acknowledgements

The authors would like to thank SLAC management for the strong support. The technical support by the SLAC Accelerator Directorate, Technology Innovation Directorate, LCLS Laser Science & Technology division and Test Facilities

Department is gratefully acknowledged. This work was supported in part by the U.S. Department of Energy (DOE) Contract No. DE-AC02-76SF00515, DOE Office of Basic Energy Sciences Scientific User Facilities Division, the SLAC UED/UEM Initiative Program Development Fund, and by the AMOS program within the Chemical Sciences, Geosciences, and Biosciences Division of the Office of Basic Energy Sciences, Office of Science, U.S. Department of Energy. J. Yang, and M. Centurion were supported by the U.S. Department of Energy Office of Science, Office of Basic Energy Sciences under Award Number DE-SC0014170. M. S. Robinson was supported by the National Science Foundation EPSCoR RII Track-2 CA Award No. IIA-1430519. M. Guehr thanks the Volkswagen Foundation for a Lichtenberg Professorship.

References

1 J. Yang, *et al.*, Diffractive imaging of a rotational wavepacket in nitrogen molecules with femtosecond Megaelectronvolt electron pulses, *Nat. Commun.*, 2016, **7**, 11232.

2 A. A. Ischenko, J. D. Ewbank and S. Lothar, Structural kinetics by stroboscopic gas electron diffraction Part 1. Time-dependent molecular intensities of dissociative states, *J. Mol. Struct.*, 1994, **320**, 147–158.

3 J. D. Ewbank, *et al.*, Instrumentation for gas electron diffraction employing a pulsed electron beam synchronous with photoexcitation, *Rev. Sci. Instrum.*, 1992, **63**, 3352–3358.

4 C. Y. Ruan, *et al.*, Ultrafast diffraction and structural dynamics: the nature of complex molecules far from equilibrium, *Proc. Natl. Acad. Sci. U. S. A.*, 2001, **98**, 7117–7122.

5 H. Ihee, *et al.*, Direct imaging of transient molecular structures with ultrafast diffraction, *Science*, 2001, **291**, 458–462.

6 R. Srinivasan, J. S. Feenstra, S. T. Park, S. Xu and A. H. Zewail, Dark structures in molecular radiationless transitions determined by ultrafast diffraction, *Science*, 2005, **307**, 558–563.

7 C. J. Hensley, J. Yang and M. Centurion, Imaging of isolated molecules with ultrafast electron pulses, *Phys. Rev. Lett.*, 2012, **109**, 133202.

8 J. Yang, J. Beck, C. J. Uiterwaal and M. Centurion, Imaging of alignment and structural changes of carbon disulfide molecules using ultrafast electron diffraction, *Nat. Commun.*, 2015, **6**, 8172.

9 M. Centurion, Ultrafast imaging of isolated molecules with electron diffraction, *J. Phys. B: At., Mol. Opt. Phys.*, 2016, **49**, 062002.

10 M. Gao, *et al.*, Mapping molecular motions leading to charge delocalization with ultrabright electrons, *Nature*, 2013, **496**, 343–346.

11 T. Ishikawa, *et al.*, Direct observation of collective modes coupled to molecular orbital-driven charge transfer, *Science*, 2015, **350**, 1501–1505.

12 L. Waldecker, R. Bertoni and R. Ernstorfer, Compact femtosecond electron diffractometer with 100 keV electron bunches approaching the single-electron pulse duration limit, *J. Appl. Phys.*, 2015, **117**, 044903.

13 C. Gerbig, A. Senftleben, S. Morgenstern, C. Sarpe and T. Baumert, Spatio-temporal resolution studies on a highly compact ultrafast electron diffractometer, *New J. Phys.*, 2015, **17**, 043050.

14 T. Van Oudheusden, *et al.*, Compression of subrelativistic space-charge-dominated electron bunches for single-shot femtosecond electron diffraction, *Phys. Rev. Lett.*, 2010, **105**, 264801.
15 V. R. Morrison, *et al.*, A photoinduced metal-like phase of monoclinic vanadium dioxide revealed by ultrafast electron diffraction, *Science*, 2014, **346**, 445–448.
16 M. Gao, *et al.*, Full characterization of RF compressed femtosecond electron pulses using ponderomotive scattering, *Opt. Express*, 2012, **20**, 12048.
17 M. P. Minitti, *et al.*, Imaging Molecular Motion: Femtosecond X-Ray Scattering of an Electrocyclic Chemical Reaction, *Phys. Rev. Lett.*, 2015, **114**, 255501.
18 D. Rolles, *et al.*, Femtosecond x-ray photoelectron diffraction on gas-phase dibromobenzene molecules, *J. Phys. B: At., Mol. Opt. Phys.*, 2014, **47**, 124035.
19 C. I. Blaga, *et al.*, Imaging ultrafast molecular dynamics with laser-induced electron diffraction, *Nature*, 2012, **483**, 194–197.
20 M. G. Pullen, *et al.*, Imaging an aligned polyatomic molecule with laser-induced electron diffraction, *Nat. Commun.*, 2015, **6**, 7262.
21 M. Meckel, *et al.*, Laser-induced electron tunneling and diffraction, *Science*, 2008, **320**, 1478–1482.
22 M. Dantus, S. B. Kim, J. C. Williamson and A. H. Zewail, Ultrafast electron diffraction. 5. Experimental time resolution and applications, *J. Phys. Chem.*, 1994, **98**, 2782–2796.
23 P. Baum and A. H. Zewail, Breaking resolution limits in ultrafast electron diffraction and microscopy, *Proc. Natl. Acad. Sci. U. S. A.*, 2006, **103**, 16105–16110.
24 P. Zhang, J. Yang and M. Centurion, Tilted femtosecond pulses for velocity matching in gas-phase ultrafast electron diffraction, *New J. Phys.*, 2014, **16**, 083008.
25 S. P. Weathersby, *et al.*, Mega-electron-volt ultrafast electron diffraction at SLAC National Accelerator Laboratory, *Rev. Sci. Instrum.*, 2015, **86**, 073702.
26 S. Manz, *et al.*, Mapping atomic motions with ultrabright electrons: towards fundamental limits in space-time resolution, *Faraday Discuss.*, 2015, **177**, 467–491.
27 F. Fu, *et al.*, High quality single shot ultrafast MeV electron diffraction from a photocathode radio-frequency gun, *Rev. Sci. Instrum.*, 2014, **85**, 083701.
28 J. B. Hastings, *et al.*, Ultrafast time-resolved electron diffraction with megavolt electron beams, *Appl. Phys. Lett.*, 2006, **89**, 184109.
29 X. J. Wang, D. Xiang, T. K. Kim and H. Ihee, Potential of Femtosecond Electron Diffraction Using Near-Relativistic Electrons from a Photocathode RF Electron Gun, *J. Korean Phys. Soc.*, 2006, **48**, 390–396.
30 I. Hargittai and M. Hargittai, *Stereochemical Applications of Gas-Phase Electron Diffraction, Part A: Methods in Stereochemical Analysis*, Wiley-VCH, 1988.
31 J. Yang and M. Centurion, Gas-phase electron diffraction from laser-aligned molecules, *Struct. Chem.*, 2015, **26**, 1513–1520.
32 J. C. H. Spence, *et al.*, Diffraction and imaging from a beam of laser-aligned proteins: Resolution limits, *Acta Crystallogr., Sect. A: Found. Crystallogr.*, 2005, **61**, 237–245.
33 L. O. Brockway, Electron diffraction by gas molecules, *Rev. Mod. Phys.*, 1936, **8**, 231–266.

34 R. Srinivasan, V. A. Lobastov, C.-Y. Ruan and A. H. Zewail, Ultrafast electron diffraction (UED): A new development for the 4D determination of transient molecular structures, *Helv. Chim. Acta*, 2003, **86**, 1763–1838.

35 K. P. Huber and G. Herzberg, Molecular Spectra and Molecular Structure, *Constants of Diatomic Molecules*, Van Nostrand Reinhold Inc., 1979, vol. 4.

36 M. Centurion, P. Reckenthaeler, S. A. Trushin, F. Krausz and E. E. Fill, Picosecond electron deflectometry of optical-field ionized plasmas, *Nat. Photonics*, 2008, **2**, 315–318.

37 M. Centurion, P. Reckenthaeler, F. Krausz and E. E. Fill, Picosecond imaging of low-density plasmas by electron deflectometry, *Opt. Lett.*, 2009, **34**, 539–541.

38 H. Stapelfeldt and T. Seideman, Colloquium: Aligning molecules with strong laser pulses, *Rev. Mod. Phys.*, 2003, **75**, 543–557.

39 J. S. Baskin and A. H. Zewail, Oriented ensembles in ultrafast electron diffraction, *ChemPhysChem*, 2006, **7**, 1562–1574.

40 P. Reckenthaeler, *et al.*, Time-resolved electron diffraction from selectively aligned molecules, *Phys. Rev. Lett.*, 2009, **102**, 213001.

41 I. V. Litvinyuk, *et al.*, Alignment-dependent strong field ionization of molecules, *Phys. Rev. Lett.*, 2003, **90**, 233003.

42 J. Ortigoso, M. Rodriguez, M. Gupta and B. Friedrich, Time evolution of pendular states created by the interaction of molecular polarizability with a pulsed nonresonant laser field, *J. Chem. Phys.*, 1999, **110**, 3870–3875.

43 J. Yang, V. Makhija, V. Kumarappan and M. Centurion, Reconstruction of three-dimensional molecular structure from diffraction of laser-aligned molecules, *Struct. Dyn.*, 2014, **1**, 044101.

44 O. F. Hagena, Nucleation and growth of clusters in expanding nozzle flows, *Surf. Sci. Lett.*, 1981, **106**, 101–116.

45 M. Battaglia, *et al.*, Characterisation of a CMOS active pixel sensor for use in the TEAM microscope, *Nucl. Instrum. Methods Phys. Res., Sect. A*, 2010, **622**, 669–677.

46 R. K. Li, P. Musumeci, H. A. Bender, N. S. Wilcox and M. Wu, Imaging single electrons to enable the generation of ultrashort beams for single-shot femtosecond relativistic electron diffraction, *J. Appl. Phys.*, 2011, **110**, 074512.

47 M. Gao, Y. Jiang, G. H. Kassier and R. J. D. Miller, Single shot time stamping of ultrabright radio frequency compressed electron pulses, *Appl. Phys. Lett.*, 2013, **103**, 033503.

48 F. O. Kirchner, A. Gliserin, F. Krausz and P. Baum, Laser streaking of free electrons at 25 keV, *Nat. Photonics*, 2013, **8**, 52–57.

Faraday Discussions

DISCUSSIONS

Structural dynamics: general discussion†

Piero Decleva, Andrew J. Orr-Ewing, Markus Kowalewski, Oleg Kornilov, Jon P. Marangos, Hans Jakob Wörner, Allan S. Johnson, Ruaridh Forbes, Daniel Rolles, Dave Townsend, Oliver Schalk, Sebastian Mai, Tom J. Penfold, R. J. Dwayne Miller, Martin Centurion, Kiyoshi Ueda, Wolfgang Domcke, Peter M. Weber, Kyoung Koo Baeck, Oksana Travnikova, Chelsea Liekhus-Schmaltz, João Pedro Figueira Nunes, Daniel M. Neumark, Oliver Gessner, Albert Stolow, Artem Rudenko, Pankaj Kumar Mishra, Adam Kirrander, Danielle Dowek, Fernando Martín, Ágnes Vibók, Michael P. Minitti, Brian Stankus and Christian Burger

DOI: 10.1039/c6fd90072k

Hans Jakob Wörner opened a general discussion of the paper by Artem Rudenko: The CO_2^+ yield seems to be exclusively dominated by the spin–orbit frequency of the ground state of CO_2^+, whereas the CO + O^+ yield exclusively the spin–orbit frequency of the first excited state. Can you explain why this is the case?

Artem Rudenko responded: It indicates that the dissociation to the CO + O^+ channel does not proceed *via* the ground state of the CO_2^+ ion, in contrast to sequential double ionization. This is consistent with the conclusions made in M. Opperman, *et al.*[1] The appearance of the spin–orbit frequency of the excited A-state indicates that this dissociation channel at least partly proceeds *via* the excitation to the A-state.

1 M. Opperman, *et al.*, *J. Phys. B: At., Mol. Opt. Phys.*, 2014, **47**,124025.

Oleg Kornilov commented: The coherent beating detected at 20 meV and corresponding to the spin–orbit splitting decays on the timescale of a few picoseconds. The authors suggest this is a dephasing due to coupling to rotational degrees of freedom. What would be the rephasing time of this feature given the known rotational constants of CO_2?

Artem Rudenko replied: If there would be any rephasing time, we would expect it to be way longer than our window of observation of ~ 30 ps.

† Yann Mairesse was unable to attend the Faraday Discussion meeting to present his article and there was no discussion of this article: Y. Mairesse *et al.*, Probing ultrafast dynamics of chiral molecules using time-resolved photoelectron circular dichroism (DOI: 10.1039/C6FD00113K).

Hans Jakob Wörner said: Is there any evidence for the population of electronically excited states of the singly-charged ion, *i.e.* the A or B states which are known to be populated from high-harmonic spectroscopy?

Artem Rudenko replied: At our experimental conditions we do not populate these states very efficiently because of a rather weak pump pulse. In a recent work of the TU Vienna group[1] it was shown that the signature of the symmetric stretching mode of the A-state becomes very pronounced if one increases the intensity of the pump pulse. In our data we also see a signature of the dynamics in the A state, but mainly in the dissociation channel with an O^+ + CO final state, and its relative contribution becomes stronger with a weaker probe pulse. As can be seen in a Fig. 5d of our manuscript (here shown as Fig. 1), the peak at ~139 meV reflecting the symmetric stretching in the A-state becomes very clear. In addition, the peak at ~13 meV due to the electronic wave packet because of the spin–orbit splitting in the A-state of the CO_2^+ ion also becomes rather strong. As for the B state, the difficulty there is that its symmetric stretching mode lies very close to the corresponding frequencies of the ground states of both, the ion and the neutral molecule (a peak at ~158–160 meV in our data), so it is hard to identify its contribution, especially since we expect it to be rather weak under our conditions (weak pump pulse).

1 Sonia Erattupuzha, Seyedreza Larimian, Andrius Baltuška, Xinhua Xie and Markus Kitzler Erattupuzha, *J. Chem. Phys.*, 2016, **144**, 024306.

Danielle Dowek commented: In the reported results for strong field ionization of the CO_2 molecule studied by pump–probe momentum-resolved ion spectroscopy, you do not comment on the angular distribution of the ion fragments: would the measured distributions provide any insight into the studied ionization processes? Are differences or evolutions observed for different channels and pump–probe delays?

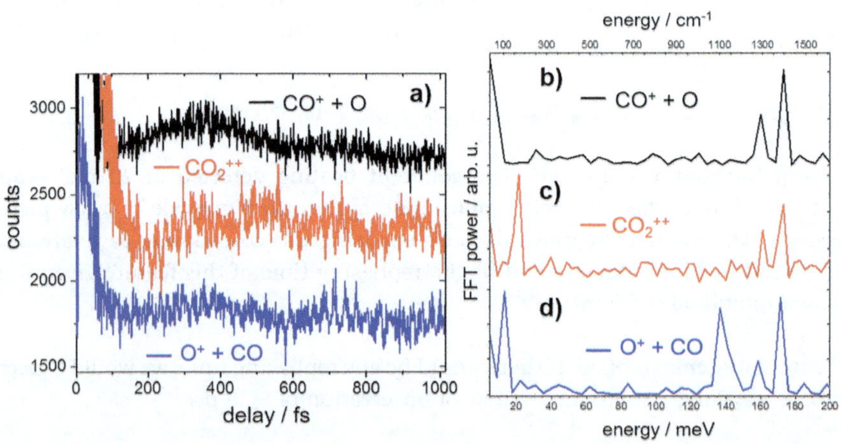

Fig. 1 (a) Yields of three different fragmentation pathways measured as a function of pump-probe delay between two identical 8 fs pulses at the intensity of 1.5×10^{14}–3.5×10^{14} W cm^{-2}. (b), (c), (d): The power spectra obtained from the Fourier transforms of the ion yields shown in Fig. 1(a) for CO^+ + O (a), CO_2^{++} (b), and O^++ CO (c) channels.

Artem Rudenko answered: First of all, the measured angular distribution of the Coulomb explosion fragments provides a good estimate of the time-dependent evolution of molecular alignment. The 8 fs pump pulse we used does induce noticeable (even though rather weak) alignment of the molecule, and we can trace a signature of this in the delay-dependent angular distributions for Coulomb explosion. However, since we used a linearly polarized probe pulse, this angular distribution is also convolved with the angular-dependent ionization probability for the probe step. The latter is even more important for angular distributions of other fragments (*e.g.*, those from the dissociation in the singly ionized states). In this work we focused on the delay-dependent yields, which carry the signature of the initial rotational excitation (alignment). However, the angular distribution of a particular ionic channel, especially at a fixed pump–probe delay, can help us to uniquely identify the pathway, which led to this final state.

Hans Jakob Wörner asked: Have you tried to measure electrons in coincidence with the ions?

Artem Rudenko responded: We did. However, the quality of the electron data was not very good, and we did not learn much there. In general, it is often pretty challenging to interpret electron spectra in strong-field molecular ionization at 800 nm.

Hans Jakob Wörner remarked: Could you use measurements at different relative polarizations to distinguish the electronic states of the cation that play a role in your measurements?

Artem Rudenko responded: This is a good point. We probably could, but we did not vary the polarization during the experiment.

Hans Jakob Wörner asked: What are the relative roles of impulsive stimulated Raman excitation and ionization depletion in creating rotational excitation in the ground state of CO_2?

Artem Rudenko answered: I guess, the question addresses vibrational excitation in the ground state of CO_2. We did not see any signatures of ionization depletion ("Lochfrass") mechanisms in our CO_2 data. The phases of the vibrational wave packet we observe match those expected for impulsive stimulated Raman excitation.

Markus Kowalewski opened a general discussion of the paper by Ágnes Vibók: For the outcome of the control experiment, does it make a difference if the molecule is modeled with a LICI or could the dissociation probability also be steered through a light induced avoided crossing. Is it necessary to include the rotation to steer the dissociation probability?

Ágnes Vibók replied: Of course it is not necessary to include the rotation (as a dynamical variable) to steer the dissociation probability (that is what we do in the 1D calculations). But these results then often strongly differ from the realistic 2D calculations (where the LICI is properly taken into account) which includes the rotational degree of freedom as a dynamical variable. Since the laser rotates the molecule the 2D model should provide more accurate results generally.

Hans Jakob Wörner said: Have you observed the expected signatures of geometric phase in your calculations concerning dynamics at light-induced conical intersections?

Ágnes Vibók responded: Yes, we could show the quantum interference effect of the LICI by numerical simulations.[1] This effect was also measured experimentally.[2] By analyzing the photodissociation process of the D_2^+ molecule carefully, we found a robust effect in the angular distribution of the photofragments that serves as a direct signature of the LICI, providing undoubted evidence of its existence.[3]

1 G. J. Halász *et al.*, *Phys. Rev. A*, 2013, **88**, 043413.
2 I. Kruse, K. Lange, J. Peise, B. Lücke, L. Pezzè, J. Arlt, W. Ertmer, C. Lisdat, L. Santos, A. Smerzi and C. Klempt, *Phys. Rev. Lett.*, 2016, **116**, 143004.
3 G. J. Halász, Á. Vibók, and L. S. Cederbaum, *J. Phys. Chem. Lett.*, 2015, **6**, 348.

Albert Stolow remarked: For Hamiltonians with band structure (*e.g.* molecules, solids), the non-resonant, non-perturbative dynamic Stark effect introduces two source terms which drive the coupled equations of motion of the state coefficients.[1,2] One of these is a dipole-type coupling (analogous to a first order Stark effect) and the other is a polarizability-type coupling (analogous to a second order Stark effect). Which term dominates the equations of motion depends greatly on the Hamiltonian: specifically, it depends on whether or not the states of interest are directly dipole-coupled to each other. If the polarizability term dominates, as for neutral homonuclear diatomics, then the superposition in the field follows the envelope of the laser pulse, a method we have termed Dynamic Stark Control.[3] However, if the dipole term dominates, then the superposition in the field oscillates at the carrier frequency (*i.e.* every couple of femtoseconds). These two regimes will lead to very different quantum dynamics. Have you considered these two regimes in your simulations?

1 B. J. Sussman, J. G. Underwood, R. Lausten, M. Yu. Ivanov and A. Stolow, Quantum control *via* the dynamic Stark effect: Application to switched rotational wave packets and molecular axis alignment, *Phys. Rev. A*, 2006, **73**, 053403.
2 B. J. Sussman, M. Yu. Ivanov and A. Stolow, Non-perturbative quantum control *via* the non-resonant dynamic Stark effect, *Phys. Rev. A*, 2005, **71**, 051401R.
3 B. J. Sussman, D. Townsend, M. Yu. Ivanov and A. Stolow, Dynamic Stark control of molecular photodissociation, *Science*, 2006, **314**, 278.

Ágnes Vibók replied: The papers mentioned in the question are interesting, but the situation in our studies is different. The Stark effect is obtained non-resonantly by an infrared field which is then used to modify the curve-crossing barrier at a specific time for a pair of molecular states which exhibit an avoided crossing in the absence of the field. In our scenario we wish to have a resonant excitation which then leads to the appearance of one or more LICI. Let us briefly address the example studied in the references 1–3 addressed in the question. There, the diatomic molecule (IBr) possesses 3 electronic states 1(X), 2(B) and 3(Y). The latter two are dissociative and exhibit an avoided crossing.

The dissociation is initiated by absorption of a visible photon bringing population from 1(X) to 2(B). An applied infrared field modifies the crossing region by inducing a non-resonant Stark effect. Manipulation of this field can be used to control the relative output of the two dissociation channels. In our context, we would use a pulse to excite 1(X) to 2(B) and in this way create a LICI between these two adiabatic states. This LICI produces it's own dynamics which can compete with that resulting from the avoided crossing between 2(B) and 3(Y). A totally different situation. Of course, one could combine such an exciting pulse with an infrared field which induces Stark effects.

Hans Jakob Wörner addressed Ágnes Vibók and Albert Stolow: This is also a comment on a previous question from Albert Stolow. It seems to me that the light-induced conical intersections emerge within a dressed-state formalism. Therefore the rapid oscillation of the carrier wave of the coupling laser pulse is already included in the nature of the dressed state and does not need to be considered further in describing the dynamics. Do you agree?

Albert Stolow responded: Formally, of course, the carrier wave is always there. In their quantum dynamics, all states are coupled to all (allowed) others by dipole matrix elements, which contain the carrier frequency. The dynamics of this set of coupled states is the formal solution: anything else is an approximation of one sort or another. This is true even for the well-known case of laser-induced molecular alignment, where the carrier field is typically (adiabatically) eliminated in the derivation of an effective Hamiltonian.[1] One starts by assuming a Hamiltonian having band structure (*e.g.* electronic states, each with vibrational structure) and a strong non-resonant laser field interaction, leading to nearby 'states of interest' and a set of 'far away' states. The question is: are the strongly coupled 'states of interest' (*e.g.* those forming the LICI) directly dipole-coupled together or are they rather indirectly coupled by the set of all 'far away' states. In either case, the laser electric field is always present and strongly interacting – this is not perturbation theory. It is in the limit where the strongly coupled 'states of interest' only interact through the set of 'far away' states that the carrier wave plays a minor role in their quantum dynamics (*e.g.* alignment of a homonuclear diatomic, dynamic Stark control of IBr photodissociation). If, in contrast, the strongly coupled 'states of interest' directly interact through dipole matrix elements, the carrier wave then plays a major role. Which situation is attained will depend, therefore, on the specific Hamiltonian under study. The quantum dynamics of the 'states of interest' will be very different for these two limiting cases. Of course, real molecular systems will generally be somewhere in between these limits, therefore having aspects of each.

1 Benjamin J. Sussman, Jonathan G. Underwood, R. Lausten, Misha Yu. Ivanov and Albert Stolow, *Phys. Rev. A*, 2006, **73**, 053403.

Ágnes Vibók replied: The LICIs were first demonstrated by applying the Floquet representation for the nuclear Hamiltonian.[1,2] This presentation provides a very illustrative picture and is often used to explain various phenomena in the area of strong field physics. The appearance of the LICI in diatomics can also be demonstrated impressively in this picture. The question arises: what is the performance of the 2 × 2 Floquet approach? Namely to what extent the results produced by the static 2 × 2 Floquet picture can be compared to those obtained from the experimental measurements using real laser pulses with switching times comparable or even longer than the period of the laser. It has been discussed in ref. 3, where the performance of the 2 × 2 Floquet representation, the rotating wave approximation and the "exact" method based on the time-dependent exact Hamiltonian have been compared in the presence of a LICI. Avoiding the numerical inaccuracy that originates from the Floquet approximation, the time-dependent exact Hamiltonians are used in our dynamical simulations.

1 Nimrod Moiseyev, Milan Šindelka and Lorenz S. Cederbaum, *J. Phys. B.*, 2008, **41**, 221001.
2 Milan Šindelka, Nimrod Moiseyev and Lorenz S. Cederbaum, *J. Phys. B.*, 2011, **44**, 045603.
3 Gábor J. Halász, Ágnes Vibók, Nimrod Moiseyev and Lorenz S. Cederbaum, *J. Phys. B.*, 2012, **45**, 135101.

Oleg Kornilov said: The wavepacket dynamics are treated using the MCTDH method and the potential energy surfaces (PES) in the light-induced potential picture. However, for the chirped pulses these PES change as quickly as the wavepacket propagates. Do equations of motion of the MCTDH method still hold in this case (the PES changing with the speed of nuclear propagation)? Is the explicit time dependence of the Hamiltonian treated exactly in MCTHD?

Ágnes Vibók responded: We use 2 × 2 Floquet representation only to demonstrate the concept of the LICI. The validity of this description is strongly dependent upon the intensity of the applied laser field. During the numerical simulations we did not use this picture. In our working Hamiltonian we use the true time dependent electric field and so the PES itself does not depend on the external electric field. On the other hand, in this case the coupling between the surfaces is changing even more rapidly. Using a proper propagation scheme MCTDH can handle correctly even these fast time dependent couplings in the Hamiltonian.

Hans Jakob Wörner asked: What would in your opinion be the ideal time-resolved measurement on the manifestation of light-induced conical intersections?

Ágnes Vibók answered: Probably the experimental investigation of the dissociation probability of the D_2^+ would be appropriate. First to ionize the D_2 molecule with a pump pulse and then applying a probe laser pulse with time delay of *t*_delay to measure the angular distribution of the total dissociation rate. For certain situations one should obtain enhanced dissociation yield approaching the $\theta = 90°$ position. However, this effect would be more prominent starting with the initial nuclear wave packet from one of the vibrational eigenstates, which of course experimentally is a more difficult task.

Hans Jakob Wörner asked: Could you comment on the importance of light-induced conical intersections in the dynamics of polyatomic molecules?

Ágnes Vibók responded: In polyatomics there are, of course, also laser-induced CIs, but the situation is more complicated than the one described for diatomics.

In polyatomics the interplay of LICIs with the CIs given by nature will lead to a wealth of new phenomena.

However, the investigation of polyatomic systems in intense laser fields is more involved than in the case of diatomics, the nuclear dynamics has more dimensions and is expected to be more complex and the consequences more diverse.

Artem Rudenko asked: You present the calculation for the D_2^+ ionic target. Would experimental observation of the light-induced conical intersection become possible with angle-resolved and vibrational state-resolved measurement of the fragmentation of such an ionic target?

Ágnes Vibók responded: Yes. To our present knowledge one possible direct observation of the effect of the light induced conical intersection would be for measuring of an enhanced dissociation yield of the D_2^+ photofragments approaching the $\theta = 90$ position by applying a vibrational state-resolved experimental technique.

Wolfgang Domcke opened a general discussion of the paper by Christian Burger: You are starting the dynamics by double ionization of acetylene. There have been earlier experiments considering the dynamics triggered by the excited state of the mono-cation. Could you comment on the differences of these experiments and their results. What is the motivation to prepare the di-cation?

Christian Burger answered: While the timescale for isomerization in the cation was determined as (41 ± 10) fs by Ibrahim *et al.* (reference 14 in the paper) and (52 ± 15) fs by Jiang *et al.* (reference 21 in the paper), precise measurements of the isomerization time in the dication are scarce. So far, only upper limits, *e.g.* 100 fs (reference 12 in the paper) and 60 fs (see references 13 and 31 in the paper), were known, which motivated us to investigate the reactions in the dication. For a detailed discussion concerning the differences and similarities to previous experiments, please also refer to our response to Oliver Gessner.

Wolfgang Domcke asked: It could be that an H-atom isomerizes or a proton. In the first case, the system remains essentially nonpolar. In the second case, it would acquire a substantial dipole moment. Can your experiment distinguish between these two cases?

Christian Burger replied: Within this experiment we were not able to distinguish between both scenarios. According to the theory isomerization takes place *via* hydrogen migration rather than proton migration.

Oliver Gessner asked: While your experiment is slightly different from those described in Hishikawa *et al.*,[1] and Ibrahim *et al.*,[2] I wonder why you do not observe the reverse isomerization dynamics described in these articles. Can you discuss the differences in the experiments and possible reasons for the different observations in more detail?

1 Akiyoshi Hishikawa, Akitaka Matsuda, Mizuho Fushitani and Eiji J. Takahashi, *Phys. Rev. Lett.*, 2007, **99**, 258302.
2 Heide Ibrahim, Benji Wales, Samuel Beaulieu, Bruno E. Schmidt, Nicolas Thiré, Emmanuel P. Fowe, Éric Bisson, Christoph T. Hebeisen, Vincent Wanie, Mathieu Giguére, Jean-Claude Kieffer, Michael Spanner, André D. Bandrauk, Joseph Sanderson, Michael S. Schuurman and François Légaré, *Nat. Commun.*, 2014, **5**, 4422.

Christian Burger replied: Thank you for this question, as this is likely the most debatable aspect in this experiment. An interpretation of the different experimental results is not straightforward as the parameters of earlier experiments and ours are very similar and our statistics are considerably higher. The main difference to Hishikawa *et al.*,[1] appears to be that we did not record sufficient quality single pulse data to allow for the background subtraction applied in their publication. We therefore might not have been as sensitive to the back-isomerization. When we, however, take the new analysis of the 4-fold coincidence channel into account and apply a high-energy filter, indications for such a back-isomerization are seen (see response to Oliver Schalk regarding the presence of 4-fold coincidences), resolving this discrepancy.

When compared to Ibrahim *et al.*,[2] we obtain similar results after the application of a high energy filter. In addition, however, we investigated another dissociation channel. We see that both channels yield different behavior indicating an alternative explanation for the observed maximum in the isomerization yield (Figure 7a). We cannot exclude, however, that we do not populate the state from which isomerization takes place as efficiently as done in previous experiments, which could be due to the broader and slightly shifted spectrum of the pulses used in our experiment. A low population probability would result in low signal strength and the high energy filter alone might not provide the same contrast as in earlier work for the visualization of the backwards isomerization.

1 Akiyoshi Hishikawa, Akitaka Matsuda, Mizuho Fushitani and Eiji J. Takahashi, *Phys. Rev. Lett.*, 2007, **99**, 258302.
2 Heide Ibrahim, Benji Wales, Samuel Beaulieu, Bruno E. Schmidt, Nicolas Thiré, Emmanuel P. Fowe, Éric Bisson, Christoph T. Hebeisen, Vincent Wanie, Mathieu Giguére, Jean-Claude Kieffer, Michael Spanner, André D. Bandrauk, Joseph Sanderson, Michael S. Schuurman and François Légaré, *Nat. Commun.*, 2014, **5**, 4422.

Daniel Neumark said: Your plot of the angle θ *vs.* time in Fig. 4 shows strong evidence for isomerization. What do you learn from the time-dependent signal for the angle a?

Christian Burger answered: From angle a we can learn that the major part of molecules stay in the acetylene configuration independent of the chosen energy filter. This is a very important difference of the H–C–CH channel to the H–H–CC channel which shows the isomerization very nicely. Additionally, the angle a indicates a small amount of isomerization at short time delays which disappears for larger delay times (compare Figure 4a). This could be similar to the isomerization and back-isomerization seen by Hishikawa *et al.*[1] but with smaller isomerization probability.

1 Akiyoshi Hishikawa, Akitaka Matsuda, Mizuho Fushitani and Eiji J. Takahashi, *Phys. Rev. Lett.*, 2007, **99**, 258302.

Daniel Neumark asked: In your experiment, you don't have control over the total energy deposited in the vinylidene dication since it is prepared by strong-field ionization. It would seem that the isomerization dynamics would depend sensitively on this. Do you have any means of estimating the total available energy?

Christian Burger replied: The maximal recollision electron energy is given by 3.17 U_p. Considering the cycle-averaged laser intensity of 5×10^{14} W cm^{-2} and a central wavelength of 690 nm we estimate the ponderomotive energy to be 22.2 eV and the maximal recollision energy to be 70.4 eV:

$$U_p = 9.33 \times 10^{-14}\, I\,[\text{W cm}^{-2}]\lambda\,[\mu\text{m}]\,\text{eV} = 9.33 \times 10^{-14} \times 5 \times 10^{14} \times 0.69 \times 0.69\ \text{eV} = 22.2\ \text{eV} \rightarrow 3.17\ U_p = 70.4\ \text{eV}.$$

As the energy difference between the $A^3\Pi_u$ state and the cation ground state is 25.6 eV a maximum excess energy of 44.8 eV is available. While most of this energy is carried away by the recolliding electron we can certainly efficiently populate the $A^3\Pi_u$ state.

Kiyoshi Ueda said: In the case of Hishikawa *et al.*,[1] they employed a 9 fs, 800 nm pulse and observed that the hydrogen migration in deuterated acetylene dication proceeds in a recurrent manner. In case of Ibrahim *et al.*,[2] they employed a 266 nm pulse and observed that hydrogen migration in the acetylene cation proceeds in a recurrent manner. You employed a 4 fs, 690 nm pulse and observed hydrogen migration in the acetylene dication with no trace of recurrence. It seems the states you probed were different from the above two previous observations. Could you identify the states you probed? What is the reason for the no recurrence?

1 Akiyoshi Hishikawa, Akitaka Matsuda, Mizuho Fushitani and Eiji J. Takahashi, *Phys. Rev. Lett.*, 2007, **99**, 258302.
2 Heide Ibrahim, Benji Wales, Samuel Beaulieu, Bruno E. Schmidt, Nicolas Thiré, Emmanuel P. Fowe, Éric Bisson, Christoph T. Hebeisen, Vincent Wanie, Mathieu Giguére, Jean-Claude Kieffer, Michael Spanner, André D. Bandrauk, Joseph Sanderson, Michael S. Schuurman and François Légaré, *Nat. Commun.*, 2014, **5**, 4422.

Christian Burger responded: We were able to identify the dication as the intermediate state by studying the vibrational wave packets and therefore, we conclude that the isomerizing molecules are in the $A^3\Pi_u$ state, which is the lowest energy state of the dication from which isomerization can take place.

For a detailed discussion concerning the differences and similarities to previous experiments, please also refer to our response to Oliver Gessner.

Chelsea Liekhus-Schmaltz commented: What electronic state do you believe is involved in the isomerization, and is there a barrier to isomerization on that state?

Christian Burger replied: As shown in Figure 6 of the paper, the theory suggests that the isomerization starts from the $A^3\Pi_u$ state of the dication. Beginning from the Franck–Condon point the molecules can either relax into the local minima ($t1'$

and $t1''$) or surpass the energy barrier (t-TS1$'$ and t-TS1$''$) allowing the molecule to isomerize.

Chelsea Liekhus-Schmaltz commented: One issue that can arise in Coulomb explosion experiments is that the molecule is not stationary when it explodes. As a result, it is not necessarily easy to connect the momentum image that you have collected to the original geometry of the molecule at the time of Coulomb explosion. This problem could be exasperated because you additionally have a carbon–carbon ion fragment that would have it's own angular momentum. Have you found any way to resolve these kinds of issues in your analysis of the data?

Christian Burger answered: This problem is indeed difficult to approach, but previous experiments on small molecules have also relied on using the assumption that the extracted momenta correspond to the configuration at the time of Coulomb-explosion. Not only the molecular configuration of acetylene (Hishikawa *et al.*[1] and Ibrahim *et al.*[2]) but also the chirality of molecules can be extracted utilizing this Coulomb-explosion imaging technique.[3]

Likely, the mentioned problem will broaden the retrieved angular distribution but still the average angle will stay similar enabling the reconstruction of the molecular configuration at the time of Coulomb-explosion. Furthermore, in our previous single pulse experiments under similar conditions we see a directional emission dependence of the fragments on the carrier envelope phase (CEP), see reference 8 of the paper. If the relation between retrieved momenta and original geometry is not valid we would not have been able to observe such CEP-effects as all directionalities would wash out. In summary, we agree that it is important to consider the validity of the recoil approximation for larger molecules in Coulomb-explosion imaging, but in our case it appears reasonable to rely on the approximation.

1 Akiyoshi Hishikawa, Akitaka Matsuda, Mizuho Fushitani and Eiji J. Takahashi, *Phys. Rev. Lett.*, 2007, **99**, 258302.
2 Heide Ibrahim, Benji Wales, Samuel Beaulieu, Bruno E. Schmidt, Nicolas Thiré, Emmanuel P. Fowe, Éric Bisson, Christoph T. Hebeisen, Vincent Wanie, Mathieu Giguére, Jean-Claude Kieffer, Michael Spanner, André D. Bandrauk, Joseph Sanderson, Michael S. Schuurman and François Légaré, *Nat. Commun.*, 2014, **5**, 4422.
3 M. Pitzer *et al.*, *Science*, 2013, **341**, 1096.

Oliver Schalk asked: In your paper, you mentioned the presence of 4-fold coincidences. Were you able to obtain relevant information from these coincidences? The data should not be prone to rotational momentum stored into, *e.g.* the C–C bond. In addition, they should provide a sanity check of the observed dynamics.

Christian Burger replied: Thank you for this nice question. You are right, that the 4-fold coincidence enables us to extract even further information compared to 3-fold coincidences. It is difficult, however, to directly compare this channel to the other channels presented in the paper as it is not completely known which intermediate channels we populate.

In the following, we show the evolution of the angle between the protons obtained from the 4-fold coincidence channel (see Fig. 2). The momentum sum of both carbon ions acts as center of gravity.

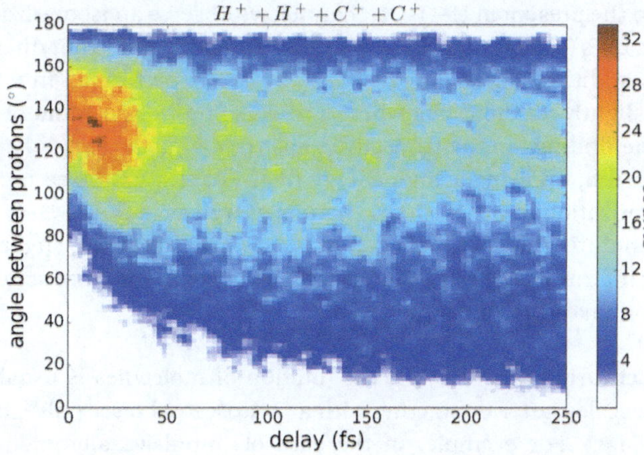

Fig. 2 Angle between the protons *vs.* the delay between the few-cycle pulses.

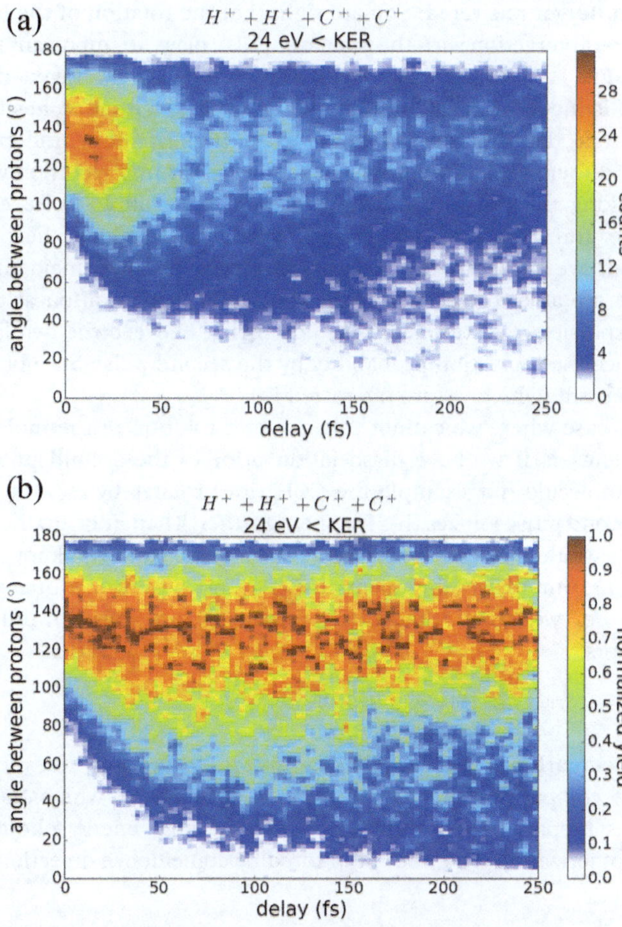

Fig. 3 (a) Same as Fig. 2, but with an applied filter of KER > 24 eV. (b) Same as (a), but with the yield normalized to facilitate easier comparison at different pulse delays.

Similar to the presented H–H–CC channel, we observe an isomerization within the first tens of fs (see Fig. 2), whereas a back-isomerization is not directly visible. Applying an additional high energy filter (>24 eV, as performed in reference 15) results in a slightly different plot where a back-isomerization could be identified, similar to the findings in the respective paper (see Fig. 3).

In conclusion, applying a high-energy filter enables us to see indications of a back-isomerization in the H–C–CH channel as well as in the H–H–C–C channel. As elaborated during the discussion, it would be nice to excite a larger part of the isomerizing intermediate states to facilitate their separation from states where no isomerization takes place.

Martin Centurion asked: While the rotation of molecules is usually on picosecond time scales, after interacting with a femtosecond laser pulse, the rotation can be very fast. For example, in the case of impulsive alignment for small molecules, the angular distribution can change significantly on a time scale of 100 fs. How do you account for this in your experiments?

Christian Burger answered: You are right that the rotation of the isomerizing acetylene after interaction with the first laser pulse plays an important role. As you can see in Fig. 1(a) of reference 8 (of the paper), the averaged momentum distribution is indeed almost isotropic indicating that a rotation has taken place (assuming that the ionization probability is highest along the polarization direction and therefore a Coulomb explosion without prior rotation would yield momenta along the polarization direction). In our double pulse experiment, however, we are not only investigating fragmentation along the polarization direction but we measure the full 3D momenta of all coincident particles. Thereby, we are able to reconstruct the molecular configuration at the time of Coulomb explosion. Rotational changes occurring between the two laser pulses may influence the ionization probability by the second pulse but not the molecular arrangement (which we are interested in).

The only case where we cannot rule out that rotation of the molecule introduces problems is if we have dissociation prior to the second pulse and the remaining molecule (for example the CCH^+ ion) rotates by *e.g.* 180°, at which point the second pulse ionizes this fragment further. Then, indeed, the rotation of the molecule would obscure our experimental results. However, knowing that Xie *et al.*,[1] applied impulsive alignment to C_2H_2 and observed a quarter rotation within 400 fs, we think that this process is too slow to influence our measurements.

1 Xie *et al.*, *Phys. Rev. Lett.*, 2014, **112**, 163003.

Daniel Neumark remarked: Following up on my previous question, how do your results compare to the synchrotron experiments in which the acetylene dication was prepared by absorption of a single high energy photon and the resulting isomerization dynamics monitored by coincidence detection of the ion fragments?

Christian Burger answered: Compared to the synchrotron measurements (see reference 31 of our paper and Osipov *et al.*[1]) we see that similar isomerization times can be obtained indicating similar dynamics. In strong-field ionization the characterization of populated intermediate states is challenging, whereas in the case of synchrotron measurements this issue was solved by investigating the Auger electron energy in coincidence with ionic fragments (Osipov *et al.*[1]). In our study we utilize the observation of vibrational wave packets for an assignment of the intermediate state.

Another important difference between both experiments is the extraction of isomerization times. In synchrotron measurements only an upper limit of the isomerization times is obtained. We can directly measure the temporal evolution of the migration process.

1 T. Osipov, T. N. Rescigno, T. Weber, S. Miyabe, T. Jahnke, A. S. Alnaser, M. P. Hertlein, O. Jagutzki, L. Ph. H. Schmidt, M. Schöffler, L. Foucar, S. Schössler, T. Havermeier, M. Odenweller, S. Voss, B. Feinberg, A. L. Landers, M. H. Prior, R. Dörner, C. L. Cocke and A. Belkacem, *J. Phys. B: At. Mol. Opt. Phys.*, 2008, **41**, 091001.

Hans Jakob Wörner opened a general discussion of the paper by Oleg Kornilov: You extracted anisotropy parameters for the two states involved in the measured autoionization dynamics. What are the error bars on these parameters?

Oleg Kornilov replied: Fig. 3 of the paper shows fits to the angular distributions of the two contributions. The asymmetry parameters are $\beta_2 = 1.3(2)$ and $\beta_2 = 0.05(6)$, where the number in the parentheses is the 1σ uncertainty.

Albert Stolow commented: A time-evolving photoelectron angular distribution (PAD) can also be due to the nuclear coordinate dependence of the ionization transition dipole? In other words, the failure of the Franck–Condon approximation. For example, as was discussed for excited state dynamics in CS_2, time-resolved Molecular Frame PADS can give a direct view of the evolution of excited state electron character upon nuclear motion.[1] Have you considered this as a possible contribution to the fitting of the PADS shown in your data?

1 P. Hockett, C. Z. Bisgaard, O. J. Clarkin and A. Stolow, Time-resolved imaging of purely valence-electron dynamics during a chemical reaction, *Nat. Phys.*, 2011, 7, 612.

Oleg Kornilov replied: This is an absolutely valid point. We have carefully considered the possibility, that the time-dependence of PADs comes from nuclear dynamics and to the best of our knowledge this is not the case. High kinetic energy resolution of our photoelectron images allows us to establish with certainty, that the IR-induced feature (labeled Ry+IR) corresponds to the IR ionization of the $3d\pi_g$ member of the Hopfield emission series. This series is attached to the $B_2\Sigma_u(\nu = 0)$ ionic threshold, where $\nu = 0$ labels the lowest vibrational state of the ion. According to the literature excitations of higher vibrational states are negligible. Therefore no nuclear (vibrational) dynamics could be observed in the intermediate state contrary to the experiment of Hockett *et al.* (see ref. in the question), where a group of vibrational (quasibound) states forms a nuclear wavepacket, which evolves on the electronic potential energy surface and leads to the time dependence of PADs.

Hans Jakob Wörner remarked: Can you attribute the two zero-order states of your model to Rydberg series converging to specific ionic states?

Oleg Kornilov replied: We assume in our model, that the two states are attributed to the members of the two Rydberg series predicted by the calculations of Raoult *et al.* (ref. 12 of the paper), namely the 4"s"σ_g and the 3dπ_g states. These series both converge to the same ionic threshold, the $B^2\Sigma_u$ state of the N_2^+ molecular ion.

Oliver Gessner asked: The experiment is essentially designed such that the critical dynamics take place while the XUV-pump pulse and the IR-probe pulse overlap in time. Have you considered possible contributions to the observed dynamics from processes other than the continuum coupling of the two Rydberg states discussed in the manuscript, such as coupling between a greater manifold of Rydberg states?

Oleg Kornilov responded: We have considered a range of other possible explanations for the observed time dependence of the photoelectron angular distributions (PADs): (*i*) nuclear dynamics are excluded based on the considerations described in the answer to Albert Stolow; (*ii*) attribution of the signal to a $n = -1$ sideband of one of the states is excluded, because the sideband signals peak at directions parallel to the IR light polarization and vanish perpendicular to it, whereas in our experiment at $t = 0$ photoelectrons are emitted perpendicular to the laser polarization with high probability; (*iii*) two other possible explanations include an IR-enabled XUV excitation of other Rydberg states or a coupling of the excited Ry state to the continuum by the IR field. Both options require multiphoton interactions by the IR field. We exclude these possibilities considering the fact that the second order ($n = 2$) sidebands are not observed in our experiments indicating that two-photon processes (IR) are too weak. Of course, contributions of such higher order process can only be fully excluded if the particular matrix elements are evaluated, which goes beyond the scope of the present paper; (*iv*) it is possible to reconstruct the observed signals by assuming saturation in the second ionization step by IR, but this is excluded based on the conditions of our experiment and the successful fits of the exponential decays to the transients extracted from the SVD analysis; (*v*) one more possible reason is discussed in the answer to the question posed by Kiyoshi Ueda regarding the change of β_2 as a function of delay.

Daniel Neumark commented: If you look at other peaks of this Rydberg series, do you see similar dynamical effects attributable to interference between overlapped autoionizing features?

Oleg Kornilov replied: In this work we limited the XUV spectrum bandwidth such that it only excites one resonance of the Hopfield series. This is confirmed by photoelectron spectra showing only one peak corresponding to an autoionizing state. We expect a similar behavior for other members of the "emission" series up to some high *n* numbers, where rotations might completely decouple the resonances (see Huber *et al.*, ref. 13 of our paper). The absorption series, on the other hand, consists of isolated Rydberg resonances and should not show these dynamics effects.

Hans Jakob Wörner commented: The XUV pulses generated from your monochromator appear to be quite far from the Fourier limit. How much temporal broadening is caused by the monochromator?

Oleg Kornilov responded: A properly aligned monochromator induces 2–4 fs pulse stretch in the photon energy range of 10 to 50 eV caused by geometric aberrations of the toroidal optics. Bear in mind, that XUV pulses produced *via* HHG already carry a substantial chirp induced by the generation process (*e.g.* K. Varjull *et al.*[1]), which can actually be reduced for each photon energy by a slight realignment of the monochromator optics. More information can be found in M. Eckstein *et al.*,[2] and M. Eckstein, PhD thesis.[3]

1 K. Varjull *et al.*, *J. Mod. Optics*, 2005, **52**, 379.
2 M. Eckstein *et al.*, arXiv:1604.02650 [physics.ins-det]
3 M. Eckstein, PhD thesis, Berlin, 2015.

Artem Rudenko asked: What is the bandwidth used in your experiment?

Oleg Kornilov replied: The bandwidth of the monochromatized XUV pulse is about 0.5 eV (FWHM).

Hans Jakob Wörner commented: Does predissociation play a role for your experiments?

Oleg Kornilov responded: It is well established in the literature, that photodissociation plays a negligible role for these resonances. See, for example, ref. 13 in our paper and the references therein.

Kiyoshi Ueda said: Your observation of the change of β_2 as a function of delay might be interpreted by additional phase accumulations *via* non-resonant contribution when the two pulses are overlapped (rather than autoionization dynamics). Please have a look at our theoretical[1] and experimental[2] investigations on this issue (on atomic non-autoionizing Rydberg states) and see if it is relevant to your observations. At least this effect should also be in your observations.

1 Kenichi L. Ishikawa, A. K. Kazansky, N. M. Kabachnik and Kiyoshi Ueda, *Phys. Rev. A*, **90**, 023408.
2 S. Mondal, H. Fukuzawa, K. Motomura, T. Tachibana, K. Nagaya, T. Sakai, K. Matsunami, S. Yase, M. Yao, S. Wada, H. Hayashita, N. Saito, C. Callegari, K. C. Prince, C. Miron, M. Nagasono, T. Togashi, M. Yabashi, K. L. Ishikawa, A. K. Kazansky, N. M. Kabachnik and K. Ueda, *Phys. Rev. A*, **89**, 013415.

Oleg Kornilov responded: We are very grateful to Kiyoshi Ueda for this point. It is indeed important to understand what role this effect may play in our experiments. In our work we neglected all non-resonant effects based on the argument, that two photon processes (XUV + IR) induced by pulses with parallel polarizations should lead to enhanced signals in the direction of the polarization and result in no enhancement or depletion in the perpendicular direction. Analyzing the theoretical results of Ishikawa *et al.*[1] we indeed see that for all cases of strong asymmetry variation (H atom, Ne and Ar) the intensity in the direction perpendicular to the polarization decreases at time zero. This is contrary to our experimental observation, where at time zero the perpendicular photoemission is large and decreases down to zero for longer delays between the pulses (see Fig. 2 of the paper). Also, no strong variation of the β_4 is observed in the experiment, contrary to the calculations of Ishikawa *et al.*[1] We thus think that non-resonant contributions cannot explain our observations, although the interpretation of the asymmetry parameter for the short-lived state (the one with the almost symmetric PAD) may indeed be influenced by this effect.

1 Kenichi L. Ishikawa, A. K. Kazansky, N. M. Kabachnik and Kiyoshi Ueda, *Phys. Rev. A*, **90**, 023408.

Kyoung Koo Baeck opened a general discussion of the papers by Artem Rudenko, Ágnes Vı́bok, Christian Burger: The noticeable non-adiabatic effect near '*natural*' conical intersections (NCIs) corresponds to the mixing or coupling between nearby adiabatic electronic states by the '*natural*' movement of nuclei. On the other hand, the light induced conical intersections (LICIs) may be thought as an artificial or intentional mixing between the adiabatic states due to the external fields. In either case of NCIs/LICIs, the mixing between different adiabatic electronic states could make a noticeable effect on the electron dynamics within the molecular system. It is probably too early to ask this question, but I wonder if you (Ágnes Vibók) or someone else has ever tried to investigate the noticeable change of the electron dynamics around LICI theoretically or experimentally. Concerning to this, as I already talked to Gabor J. Halász (one of the coworkers of Ágnes Vibók) in person, I wonder what will be the results if the current study of LICI in the D_2^+ system is extended to alkali-halide systems, like LiH and NaH, because the potential energy curves of their ground and first excited electronic states make the avoided-crossing, correspond to a 'natural' CI, very similar to that of LICI in the D_2^+ system. The alkali-halide systems correspond to the simplest prototype of the Harpoon reaction involving very fast temporal and wide spacial movement of electrons.

Ágnes Vibók answered: Yes, the LICI can be used intentionally to mix the adiabatic states. Without the external field such a mixing is not possible. However, as is the case in the presence of NCIs, the motion of the nuclei is dictating the mixing dynamics. For diatomics we need two nuclear degrees of freedom to have a LICI, and these are the internuclear coordinate and the rotational angle. In polyatomics, all internal nuclear coordinates may participate as well as the rotations, but there is no need for the rotations to form LICIs as there are sufficiently many other nuclear degrees of freedom. Once the external field "dresses" the potential surfaces, the nuclei see these surfaces and move accordingly. This behavior can, indeed, be intentionally exploited to control a quantity of interest, *e.g.*, dissociation rate.

If I correctly understand the question about the mixing of the adiabatic electronic states, this is exactly what we study in our examples, namely, we compute fully the non-adiabatic dynamics for different field strengths and also by freezing the rotation. Comparing the results of the two different simulations we could show the impact of the LICI on different measurable quantities.

The extension to alkali-halide systems is indeed an interesting issue. In principle, one can extend the theory and computations to different scenarios including that which is addressed in the question. Let's assume that the system has three relevant adiabatic states in the energy range of interest, for example the ground state X and two excited states A and B, and that the states A and B exhibit a natural CI, or as is the case for diatomics, a natural avoided crossing. By the way, from theory and calculations we know that a natural CI and a natural avoided crossing are not similar and behave qualitatively different,[1] but this is not of relevance to our reply. We can now have two scenarios. In the first, the pulse excites the state X to the manifold A+B. Then, according to the theory [polyatomics], there will be LICIs between X and this manifold and the natural CI will persist between A and B. We will have a 3×3 time-dependent Hamiltonian which governs the dynamics. In the second scenario, the pulse is between A and B (may be after a weak pump pulse populating A and/or B initially). Then, the situation can be quite intricate and we will have competition between the LICI and the natural CI which is subject to a shift. The operative Hamiltonian will be 2×2 in this case. Both situations are interesting and show that there is great potential for manipulating the non-adiabatic dynamics by LICIs.

1 H. Köppel, W. Domcke and L. S. Cederbaum, *Adv. Chem. Phys.*, 1984, **57**, 59–246.

Daniel Neumark addressed Christian Burger and Artem Rudenko: It seems that the strong-field pump–probe experiments on CO_2 and C_2H_2 that both of you are carrying out are poorly defined, since neither the number of pump nor probe photons is known. Now that intense high harmonic sources are becoming available, would it be more desirable to carry out experiments in which at least one of the pump–probe steps involved absorption of a single high energy photon? Or is there intrinsic value in experiments in which only strong-field ionization/excitation is used?

Artem Rudenko answered: The goal of our particular experiment on CO_2 was to characterize the wave packet produced by strong field excitation. Therefore, we have to use the intense IR (*e.g.*, 800 nm) as a pump pulse. It would be beneficial though to use a single photon probe, if the corresponding pulses of sub-10 fs duration and enough flux to efficiently probe the wave packet were available.

Christian Burger replied: We agree that it would be beneficial to have a light source providing high energy photons, where a defined state from which isomerization occurs can be populated by the pump pulse. Additionally, the probe step could be replaced by a high-energy photon, but it is less crucial in this case. Since high-flux high-photon energy, ultrashort light sources are not yet as stable as few-cycle near-infrared sources, it poses a challenge to carry out experiments over long acquisition times. In our experiments using strong-field ionization allowed for a combination of very good temporal resolution and stability over several days of data acquisition for the price of less control over the population step. Extreme ultraviolet sources with high repetition rate and stability over several days of operation may, however, now be within reach, making their application in pump–probe experiments indeed very desirable.

Hans Jakob Wörner addressed Artem Rudenko and Christian Burger: The three experimental papers in this session made use of some of the most advanced detection techniques or light sources. What would you consider the ideal combination of light sources and detectors to be for probing the dynamics of small molecules in the gas phase?

Artem Rudenko responded: For reactions, which do not require non-linear XUV/X-ray interactions, the ideal combination in my opinion would be a high repetition rate lab-based HHG source (with photon energies from ~ 10 eV up to the water window) and a coincident ion/electron momentum imaging setup (COLTRIMS/"reaction microscope" or double-sided VMI with delay line detectors). For the reactions involving non-linear XUV/X-ray interactions we currently need FELs. With their current repetition rates VMI with phosphor screen detectors (ideally, double-sided) seems to be an optimal detection system. At the future high rep. rate FEL (European XFEL, LCLS-2), the coincident fragment detection would become a routine tool.

Oleg Kornilov responded: The papers of Rudenko *et al.* and Burger *et al.* use a reaction microscope combined with multiphoton ionization using ultrashort (sub 10fs) infrared pulses. In our experiment the pulse duration is more conventional (20–25 fs), but we employ an XUV pump pulse with tunable wavelength using the recently constructed time delay compensating monochromator coupled to an HHG source. For an ideal experiment one would certainly envision a combination of these techniques, *i.e.* a pump–probe experiment with the wavelength-tunable XUV sub-10fs or even attosecond pulse, an IR or visible pulse and a momentum-resolved coincident detection of both ions and electrons. However one should be careful in choosing the experimental arrangement. In one class of experiments the XUV pulse can be used in the excitation step (usually *via* ionization and shake-up), which is then probed by a single or few-photon ionization with IR. Here the combination of the well-defined XUV photon energy and the coincident electron-ion detection will allow us to single out the electronic states of interest. Even more interesting could be an experiment, in which the molecules are pumped by a tunable (but still short) IR/vis/UV pulse and probed by XUV ionization. The molecular frame angular distributions and multiply-charged states could be fully characterized. In either ideal case the detection of the photoelectrons in coincidence with the photoions is the key to accessing the full information on the molecular dynamics.

Christian Burger replied: Depending on the experiment, a combination of the following parameters would be ideal:

(1) A light source providing a specific wavelength to excite only a limited amount of states – this would simplify the data analysis/data interpretation.

(2) Short pulse durations for pump and probe pulses to allow for good temporal resolution.

(3) A high repetition rate (50–100 kHz) to achieve sufficient statistics and meanwhile being able to separate events from shot to shot.

(4) A reaction microscope as detection apparatus as it delivers an almost complete picture of the processes involved (including coincidence measurements).

(5) Velocity-map imaging (VMI) measurements to allow for good single-shot signal, *e.g.* for recording CEP-dependencies in background-free channels or processes with very low cross sections.

Hans Jakob Wörner addressed Oleg Kornilov and Artem Rudenko: What would you consider to be the ideal experimental technique if you had simultaneous access to all experimental tools discussed in this session?

Artem Rudenko and Oleg Kornilov responded: We think the answer to this question depends on the type of experiment. Two broad categories of experiments can be suggested in our field: the experiments aiming at understanding the photoinduced (*e.g.*, molecular) dynamics and the experiments focused on the light–matter interaction itself. In the first category the light–matter interaction should be as simple as possible to allow one to extract the intrinsic properties of the reaction of interest, while in the second category optical setup and detection scheme should be optimized to reveal the mechanisms of the interaction. Consequently, for an ideal experiment of the first type we would imagine XUV/IR/vis/UV one photon pump, specifically triggering the process under consideration, one photon probe at high rep. rate, (most likely, from HHG sources) combined with the 3D momentum-resolved coincident detection of the fragments (generalized "reaction microscope" type of experiment). All this can be nowadays realized in a lab-based, table-top setup. For studying mechanisms of light–matter interactions, which are often non-linear, we might need a very particular, often intense light source. For example, if we want to study non-linear interactions in the XUV or X-ray regime, we need free-electron lasers. There, a simpler detection scheme, (*e.g.* VMI or TOF) is often a benefit, because of a limited rep. rate and large number of fragments produced. However, with the development of high rep. rate FELs like European XFEL or LCLS2, we expect a multi-coincident fragment detection and 3D momentum imaging, possibly combined with simultaneous detection of X-ray fluorescent photons, to become a standard and most efficient technique there as well.

Hans Jakob Wörner addressed Oliver Gessner: What is the state-of-the-art time resolution from timing tools at free-electron lasers?

Oliver Gessner responded: Few-femtosecond (Harmand *et al.*[1]) and even sub-femtosecond (Hartmann *et al.*[2]) resolution have been demonstrated with timing tools at the LCLS. Note, however, that the performance of such tools may vary depending on the particular experimental parameters.

1 M. Harmand, R. Coffee, M. R. Bionta, M. Chollet, D. French, D. Zhu, D. M. Fritz, H. T. Lemke, N. Medvedev, B. Ziaja, S. Toleikis and M. Cammarata, *Nat. Photon.*, 2013, 7, 215.
2 N. Hartmann, W. Helml, A. Galler, M. R. Bionta, J. Grünert, S. L. Molodtsov, K. R. Ferguson, S. Schorb, M. L. Swiggers, S. Carron, C. Bostedt, J.-C. Castagna, J. Bozek, J. M. Glownia, D. J. Kane, A. R. Fry, W. E. White, C. P. Hauri, T. Feurerand and R. N. Coffee, *Nat. Photon.*, 2014, 8, 706.

Jon Marangos opened a general discussion of the paper by Michael Minitti: Comment: molecules aren't aligned so it is like powder diffraction.

This is in the "diffract and destroy" regime – in these small molecules is the lack of inertia an issue *i.e.* does the molecule significantly distort through Coulomb explosion during the diffraction process? Could it distort the structural information?

Note: This point was discussed later in questions by several other participants.

Michael Minitti responded: On the comment of non-aligned molecules: The absorption of the optical pump laser gives the molecule a mildly aligned character, but this is far from being properly "aligned". Techniques such as impulsive or adiabatic alignment with ps to ns laser pulses is the way to go. Attempts will be made to employ these methods in future XFEL experiments if we are so fortunate to receive beamtime. The diffraction images obtained from the mildly aligned gas-phase targets are rotationally averaged, resulting in diffuse patterns.

Considering the experimental X-ray parameters we used (8.2 keV, 30 fs, 30 μm FWHM focal spot), the dilute gas-phase targets would not absorb a tremendous amount of energy from the X-ray probe in order to Coulomb explode. Moreover if it did, we'd never observe it as we'd remove it as part of our data analysis.

During the experiment, we take a series of optical laser (pump) off as well as X-ray (probe) off frames to help us in handling X-ray only background as well as reference X-ray dark shots for the optical/X-ray cross-correlator (the time-tool). In the end, we take all background data and subtract them from each pumped and un-pumped X-ray frame. Then we take the difference between those images and report them as a percentage change. If damage from the probe happens, one would argue that the damage is always happening, no matter whether the molecule is excited with the pump or not. Therefore since we subtract this out through the procedure described above, we'd never see it. Lastly, the diffraction is instantaneous in this single scattering event, multiple scattering or rescattering are tremendously low probability events.

Wolfgang Domcke addressed Michael Minitti and Brian Stankus: You are showing the potential energy curves of singlet and triplet states. However, the SO coupling in iodine is of the order of one electron volt and singlets and triplets are strongly mixed. What are the "high-level" *ab initio* calculations mentioned in the paper. Are these fully relativistic calculations?

Michael Minitti answered: Great point. We did not include the SO coupling in these CASSCF calculations. We are aware that by excluding this, the relative energies and oscillator strengths of the excited states could be effected. Since publication, higher level calculations at the CASPT2 level of theory have been done, which give significantly more accurate relative energies, but still don't account for SO coupling. We are working to incorporate it, but those calculations are expensive and are not yet complete.

Brian Stankus replied: The CAS(18,14) calculated potential energy surfaces do not account for spin–orbit coupling, so the oscillator strengths for triplet absorptions are strictly zero. We are currently working on two strategies for incorporating the spin–orbit coupling. The first is an approximate scheme that uses the empirical atomic spin–orbit coupling constant for iodine, which is the dominant contribution of spin–orbit coupling in this molecule. Second, we pursue relativistic calculations of spin-adiabatic states, using the same effective core potential as the current spin–orbit free calculations but a smaller basis set. It is worth noting that these calculations are computationally very expensive. As shown in iodobenzene by Sage *et al.*[1], the spin–orbit coupling will have some effect on the oscillator strengths and vertical excitation energies, and thus could impact our present discussion, but arguably the most important effect of spin–orbit coupling is in terms of the asymptotic limits for dissociation.

1 Alan G. Sage, Thomas A. A. Oliver, Daniel Murdock, Martin B. Crow, Grant A. D. Ritchie, Jeremy N. Harvey and Michael N. R. Ashfold, *Phys. Chem. Chem. Phys.*, 2011, **13**, 8075–8093.

Andrew Orr-Ewing asked: The spin–orbit coupling will contribute oscillator strength to the states nominally assigned as triplets in your Table 1 of excited states. These states therefore need to be considered as candidates for the photon excitation step. A velocity map imaging study of the photofragment recoil anisotropy (*i.e.* a determination of the anisotropy parameter, β), ideally as a function of excitation wavelength, might help to unravel the contributions to the absorption from the various accessible excited states.

Michael Minitti responded: Agreed. This would be a great addition to the experiment. Now we just need to get our hands on a time-resolved VMI setup! Thankfully, while at the Discussion, we kickstarted this idea with Theis Sølling and his group in Denmark to make such a measurement.

Dave Townsend addressed Michael Minitti and Brian Stankus: Your time-resolved photoelectron spectroscopy measurements on 1,4-diiodobenzene were done with a 267 nm pump and a 300 nm probe. This gives a total photon energy of approximately 8.8 eV. The ionization potential of 1,4-diiodobenzene is approximately 8.5 eV. As such, your experiments don't project very deeply into the ionization continuum and so potentially have a very restricted view of the reaction coordinate(s) along which the dynamics evolve. Is there a chance that you may be missing additional information that could be picked up by using a shorter wavelength probe?

Michael Minitti replied: This is a fantastic point, speaking to an experimental parameter we did not choose well. While we are confident of the dynamics and lifetimes we observe and report in our manuscript, revisiting this measurement whilst employing a shorter wavelength probe is definitely worth while.

Brian Stankus answered: The choice of probe wavelength was a delicate balance – the wavelength had to be short enough to allow for single-photon ionization out of the initially excited valence state, but long enough to avoid exciting the same absorption band that is pumped with 267 nm. Of course, more probe energy would allow access to a larger energy range, but we were limited by this valence absorption band. Our current assignment of the observed band as a bound state is based on the fact that the energy of the band does not change as a function of time. Even though, as pointed out, the probe wavelength does not project very deeply into the ionization continuum, the available energy range should suffice to reveal a direct dissociation if that were taking place. A larger range of valence band energies were probed *via* resonant two-photon ionization with 405 nm, but this data is still being interpreted.

Artem Rudenko remarked: You did not specify how tight you focus the LCLS beam. Does the (possibly non-linear) photoabsorption disturb the structure obtained from your scattering data?

Michael Minitti answered: Like in our 1,3-cyclohexadiene studies, the X-ray focus we used in the 1,4-diiodobenzene was a modest 30 microns FWHM. Typically, the X-ray parameters were at 8.2 keV, 30 micron FWHM focus and about 2.5 mJ per pulse. We have no reason to believe that there is sufficient enough photon density in this regime to induce any non-linear effects in these single scattering events.

Daniel Rolles addressed R. J. Dwayne Miller and Michael Minitti: Regarding the radiation damage in X-ray scattering: Based on the cross section ratio between elastic and inelastic X-ray scattering, one will always get more ionization (=damage) than elastic scattering. This is why photoelectron diffraction seems like a good technique because there, we use the ionization for imaging.

R. J. Dwayne Miller replied: I agree with this statement. I would only add that one could use HHG sources as 100 eV photons which would give short enough de Broglie wavelengths to give atomic structure information, which I guess was one of your previous points. Surface studies have a major advantage as the photoemission process can avoid multiple scattering with a large number of emission centres – all from a single plane in space. This approach is used for surface studies at synchrotrons as shown by Adam Hitchcock and colleagues.

Michael Minitti answered: This may be true, and photoelectron diffractive imaging has potential to be a strong alternative technique to traditional gas-phase X-ray scattering, but one concern I have about X-ray induced, photoelectron diffraction at current XFELs are the very low signal count rates obtained. These extremely low signal levels force the requirement of having extremely low beamline noise, *e.g.*, fractions of background X-ray counts per second. This is experimentally very demanding to get right. While worth developing, techniques like X-ray induced photoelectron diffraction will truly benefit from high-repetiton rate XFELs like the European XFEL and LCLS-II when they come online in the near future.

Daniel Rolles addressed R. J. Dwayne Miller and Michael Minitti: This is a comment regarding the question of how much radiation damage affects the X-ray scattering. If more than one photon interacts with the molecule during the duration of the X-ray pulse, radiation damage may play a role since absorption of one X-ray photon results in ejection of 5 of more electrons. Given the pulse parameters given in the paper, this effect is not negligible for iodine at 8 keV.

Michael Minitti responded: This question is very similar in nature to Jon Marangos' question regarding the "diffract and destroy" regime, so I'll respond with the same answer. Considering the experimental X-ray parameters we used (8.2 keV, 30 fs, 30 μm FWHM focal spot), the dilute gas-phase targets would not absorb a tremendous amount of energy from the X-ray probe in order to Coulomb explode. Moreover if it did, we'd never observe it as we'd remove it as part of our data analysis. During the experiment, we take a series of optical laser (pump) off as well as X-ray (probe) off frames to help us in handling X-ray only background as well as reference X-ray dark shots for the optical/X-ray cross-correlator (the time-tool). In the end, we take all background data and subtract them from each pumped and un-pumped X-ray frame. Then we take the difference between those images and report them as a percentage change. If damage from the probe happens, one would argue that the damage is always happening, no matter whether the molecule is excited with the pump or not. Therefore since we subtract this out through the procedure described above, we'd never see it.

Lastly, the diffraction is instantaneous in this single scattering event, multiple scattering or rescattering are tremendously low probability events.

R. J. Dwayne Miller answered: This mechanism requires there to be multiple X-ray scattering events from the same molecule in which case I agree. The ionization from the first X-ray photon will modify the molecular structure on timescales within the X-ray pulse width and/or induce a time dependent structure factor. It is relatively straightforward to expand the focusing to avoid this problem in diffraction. It is only for imaging single molecules (without crystals) that 100 nm focusing conditions are used to ensure one elastically scattered X-ray photon per atom for which there are 10 inelastic. In this observable, it is impossible to avoid this problem plus the issue of dramatically changing the electron distribution (per previous discussion). In our XFEL studies, we typically have 10^{14} molecules within the sampled volume with 10^{11} X-ray photon incident such that the probability for multiple X-ray scattering from the same atom is exceedingly low (10^{-6} without consideration of elemental differences in scattering cross section). The scattering efficiency for these conditions is typically on the order of a few percent as another condition. The gas phase numbers should be similar. Your point is well taken when one is using very tight focusing conditions.

Pankaj Kumar Mishra commented: Using pump–probe methodologies, the authors have presented the ultrafast photodissociation of 1,4-diiodobenzene (DIB). The authors have also used an efficient computational technique to compare the experimental results, which are fascinating. My question is related to the polarisation direction of the X-ray used during the experiment. Why was the X-ray probe horizontally polarized in all cases for the X-ray scattering probe in the experiment? What do you expect if X-rays are polarised along the vertical direction?

Michael Minitti replied: Knowing how the polarization of the optical laser is oriented with respect to the location of the iodine atoms in 1,4-diiodobenzene is important. By varying the polarization of the optical pump laser, we can change the observed I(*f*). This helps us verify the relationship between the iodine atoms and the dipole moment of the molecule. When the large I atoms are in plane with the horizontally polarized XFEL pulse, the observed I(*f*) signal will have intensities at the poles of the X-ray CCD camera. Conversely, when rotated 90° by flipping the optical laser polarization, the I atoms will be aligned normal to the X-ray polarization, and thus preferentially scatter along the equator of the detector. As for the X-ray polarization, we had no choice but horizontal. This is simply what the LCLS produces.

Sebastian Mai said: As the paper explains, all quantum-chemical calculations were performed within *C2v* symmetry. Did you consider that there might be bound-state, non-symmetric minima in diiodobenzene? If they existed, such minima could be located at energies too low to be detected by the experimental setup, where the pump and probe lasers provide just enough energy to reach the ionization potential at the Franck–Condon point.

In this case, might it be possible that the observed 35fs time constant does not relate to dissociation, but simply to relaxation to a bound-state minimum?

Michael Minitti replied: Yes, this could be a possibility. Since we did not employ a sufficiently short ionization pulse to probe the state, we could be missing dynamics to a lower bound minimum. Although we are confident in our state assignment and observed lifetime, it is possible we are missing some lower state dynamics after the initial 35 fs decay time. Revisiting the experiment with a shorter probe pulse would be good, and plans are in the works to remeasure these dynamics.

Hans Jakob Wörner addressed Michael Minitti and Kiyoshi Ueda: Could you compare X-ray diffraction and angle-resolved X-ray photoelectron measurements for imaging molecular dynamics, such as the dissociation of diiodobutane, with respect to feasibility, signal levels and information content?

Kiyoshi Ueda responded: Angle-resolved X-ray photoelectron measurements for randomly oriented molecules result in a single parameter β_2, which has indeed the information about the electronic structure. But I assume here the comparison should be between molecular-frame core-level photoelectron angular distributions or photoelectron diffractions (PED) and X-ray diffractions (XRD). Historically the structural information of gaseous molecules have been determined by electron diffraction (ED) with the external electron source, not XRD, due to low X-ray scattering cross sections. Knowing it, and knowing the experimental difficulty of fs time-resolved ED, originally I myself as well as some of our colleagues in the AMO community thought that PED would be the way to capture the fs structural change of isolated molecules in photoreactions. However, only proof of principle experiments of PED for the stable molecules with known structures have been reported so far, using synchrotron radiations and FELs. The difficulties may stem from how to record the 3D diffraction pattern with high energy photoelectrons, ideally of keV, from the molecules fixed in space, ideally in field free conditions. On the other hand, as we learned today, other colleagues started to see photoreaction of isolated molecules using XRD and even fs time-resolved ED with the external pulsed electron source (instead of internally produced photoelectrons). I cannot say PED is better in terms of feasibility any more. I would say that all techniques have some difficulties and some advantages and so we should try all. Also, as we saw on other days, and as seen in our work, for relatively small molecules, Coulomb explosion momentum imaging will also catch the structural change.

Michael Minitti responded: I agree with your comments. Combining these techniques to fully resolve PEDs, electronic energy level structure and molecular structural dynamics would be a fascinating *tour de force.*

R. J. Dwayne Miller remarked: I was extremely impressed when your first work came out showing X-ray diffraction from gas phase molecules (CHD in this case). Given the extremely small X-ray scattering cross section, this work is a real testimony to the value of ultrabright sources. How does the X-ray case compare to electron diffraction, which has been the standard for structure determination of gas phase molecules, in terms of structural resolution? The X-ray wavelength being 1.5 Å and the weak scattering of X-rays limit the number of resolvable diffraction orders. I understand the inversion process using theory for refinement is more straightforward than electron diffraction. However, both sources can use the same level of theory for refinement. What do you see as the limit in spatial resolution? The clear advantage of X-rays at present is the much higher time resolution.

Michael Minitti answered: I agree with all you say here. Ultimately we need harder and harder X-ray sources to remotely approach the spatial resolution of electron diffraction. Unfortunately, the limit now at most XFELs is about 12.8 keV, translating into roughly just over 6 $Å^{-1}$ spatial resolution. Until harder, ultra-bright, ultrafast XFELs come online, UED will win this battle. Sources like LCLS-II will be capable of generating hard X-rays up to 25 keV. Impressive upgrades will make it possible to extend to nearly 10 $Å^{-1}$ spatial resolution. This advancement will really assist theory in generating better and better structural refinement calculations from XRD results. So who will the real winner be in the spatio-temporal resolved race? It boils down to two things. Can UED make sufficiently short electron bunches, overcoming space charge? Or, will future XFEL technology outpace and produce 100s of keV X-rays, extending their overall resolution? Only time will tell. It should be a good show.

Kiyoshi Ueda communicated after the meeting as a clarification of the paper: In our paper, we mentioned that for the SCC-DFTB approach to simulate XFEL-induced Coulomb explosion, "various modification can be made" on the way of injecting kinetic energies into atoms. We have extensively explored different energy injection models and found that the best way is to treat the molecule as a sum of diatomic molecules upon kinetic energy injection, which is a version slightly modified from the one described in the paper. In the present sequential ionization model, the momentum vector of atom i is instantaneously changed from $\vec{p}_i$ to $\vec{p}_i{}'$ when the charge of the molecule switches to the higher one:

$$\vec{p}_i{}' = \vec{p}_i + \Delta\vec{p}_i \quad (1)$$

where $\Delta\vec{p}_i$ is a momentum added according to the acquired vibrational energy upon each ionization. We assumed that the momentum $\Delta\vec{p}_A$ that the atom A receives is constructed from the momenta $\Delta\vec{p}_A$ (A–B) generated parallel to the bond axes between atom A and its surrounding atoms B. We prepared random values for the momenta along bond axes following a one-dimensional Maxwell–Boltzmann distribution. In the new model, we treat two atoms connected by a chemical bond like a diatomic molecule; *i.e.*, the momenta $\Delta\vec{p}_A$ (A–B) and $\Delta\vec{p}_B$ (A–B) of atoms A and B are generated under the additional constraint of momentum conservation, $\Delta\vec{p}_A$ (A–B) + $\Delta\vec{p}_B$ (A–B) = 0.

The kinetic energy distributions (KEDs) for 5-IU obtained by following the above momentum constraint are shown in Fig. 4. All the parameters such as the electronic temperature T_e are the same as those in Fig. 5 in the paper. Comparing the new figure (Fig. 4) with Fig. 5 in the paper, we notice that the SCC-DFTB results based on the new model agree better with experiments and are indeed more satisfactory. The discrepancies in the KED of H^+ in the previous models were significantly reduced. The KED of H^+ is well reproduced by the new model as shown in Fig. 4, though the peak energy of H^+ was much lower in the SCC-DFTB approach without the momentum constraint than in the experiment. The KED of H^+ is shifted to the higher energy side and that of I^{q+} is shifted to the lower energy side by consequences of the imposed momentum constraint introduced in the SCC-DFTB approach.

Daniel Rolles asked: In order to achieve sub-Ångström resolution with photoelectron diffraction, a few hundred eV photoelectron kinetic energy is sufficient. The question is at what kinetic energy do multiple scattering effects become small enough that we can interpret the scattering images in a single-scattering picture. My feeling is that this will be the case above 300 eV, but this is something we are currently investigating.

Kiyoshi Ueda responded: I agree. A well established Muffin-Tin potential approach (including multiple scattering) will work down to the range 100–200 eV.[1] It is also possible to employ DFT in the least square fitting in a similar energy range.[2] Using a more elaborated approach (*e.g.* multichannel Schwinger configuration interaction MCSCI method), we can handle ~ 50 eV photoelectrons and extract the bond length from PED, at least for diatomic molecules and CO_2. All these studies have so far been based on SR. High rep rate FELs will be a big advantage to extend such studies to time-resolved structural changes of photoexcited molecules.

1 Vladislav V. Serov, Vladimir L. Derbov and Tatyana A. Sergeeva, *Phys. Rev. A*, 2013, **87**, 063414.
2 E. Kukk, D. Ayuso, T. D. Thomas, P. Decleva, M. Patanen, L. Argenti, E. Plésiat, A. Palacios, K. Kooser, O. Travnikova, S. Mondal, M. Kimura, K. Sakai, C. Miron, F. Martín, and K. Ueda, *Phys. Rev. A*, **88**, 033412.

Allan Johnson remarked: We've done some work on using CEI to determine the structure of chiral molecules in Frankfurt and found a similar problem with hydrogen motion when using strong laser fields. We've found better results since using synchrotron radiation to trigger ultrafast Auger processes. For larger molecules it would be necessary to trigger Auger at multiple sites simultaneously, and an FEL is an obvious choice for the light source. How badly would the hydrogen motion you've observed affect a CEI experiment at SACLA? Are there any parameters for an XFEL you could use to suppress the hydrogen motion, making it more appropriate for CEI?

Kiyoshi Ueda replied: In the case of our SACLA experiments, the X-ray photon is absorbed by a single I atom in a molecule. *Via* Auger cascades the charges are built up and redistributed to the entire molecule. If the X-ray pulse is 10fs or below, the time scale of charge build up and redistribution is determined *via* the nature of the Auger cascades. Our empirical model confirmed that the charge buildup and redistribution time is about 10 fs for similar experiments at both SACLA (10 fs pulse width) and LCLS (5 fs pulse width). Hydrogen atoms significantly move within such time scales. However, what we found is that the direction of the H^+ ejections still reflects the original conformation of the molecules (see Fig. 6 and 7 of the paper, though the distributions were broadened) while the bond lengths are significantly elongated (see Fig. 8 of the paper). To use FEL as a source of Coulomb explosion momentum imaging, it is not absolutely necessary to use hard X-ray FELs; if one uses hard X-rays the time scale of charge buildup is determined by the Auger cascade and it is about 10 fs, it cannot be shorter. Use of ultra-intense (>10^{16} W cm^{-2}), ultra-brief (<5 fs), EUV or soft X-ray pulse may be better to simultaneously remove electrons as much as possible, within the pulse width <5 fs.

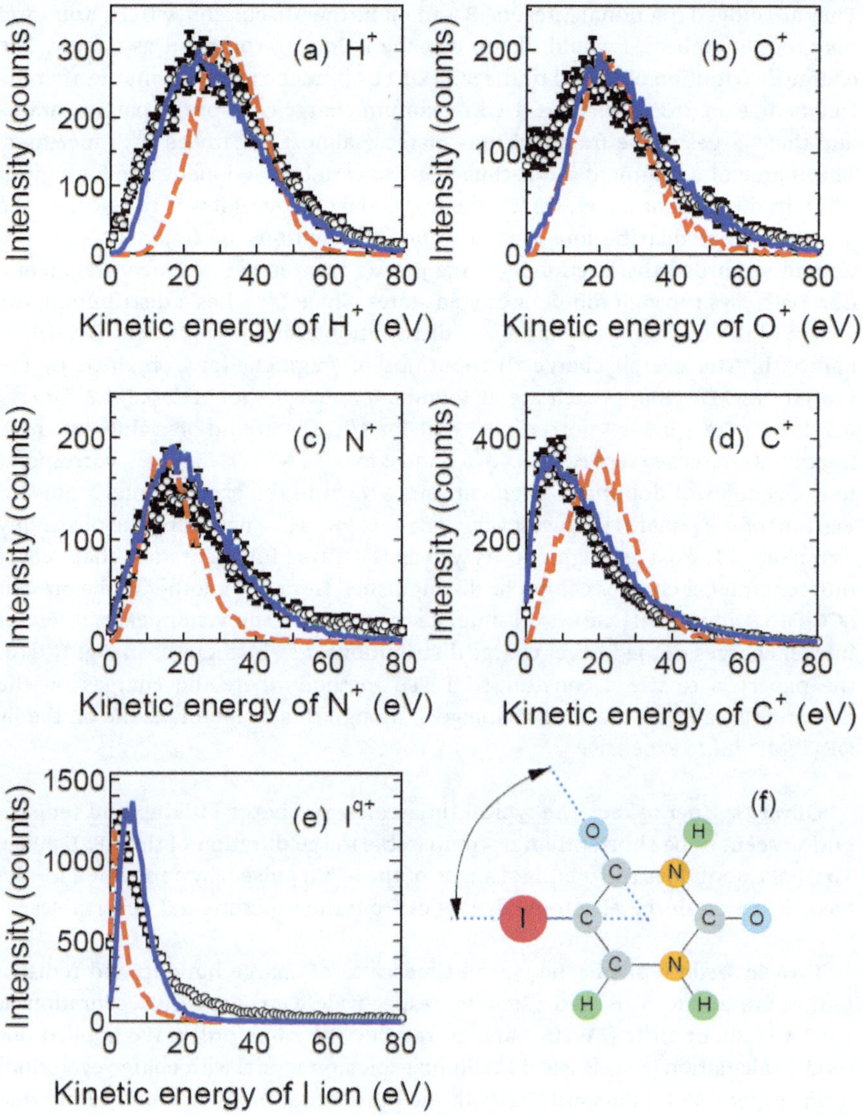

Fig. 4 Kinetic energy distributions (KED) of fragment ions emitted from XFEL-irradiated 5-iodouracil for (a) H^+, (b) O^+, (c) N^+, (d) e I^{q+}. KED data are summed up over I^{q+} of $q = 1$ to 4. The open circles, red dashed lines and blue lines are the experimental data, CCE-CE results and SCC-DFTB results, respectively. (f) Molecular structure of neutral 5-IU molecule. In the SCC-DFTB approach.

Fernando Martín commented: To understand the results of your experiments, in particular, how the charge distributes in the molecule, you showed the results of molecular dynamics simulations. Molecular dynamics simulations are based on DFT approaches, but it is known that these methods can lead to unphysical fractional charges in the fragments – how do you deal with this?

Kiyoshi Ueda replied: Fernando Martín asked a question about fractional integers of charges for the fragment ions obtained from the SCC-DFTB approach.

They are indeed fractional integers. Based on further discussion with H. Kono and his group members, I would like to add the following comment as a reply. The charge distribution obtained by the SCC-DFTB approach is nearly unique after the fragmentation processes for a fixed maximum charge Q^{max} of the parent cations and the charges of the fragment ions increase almost linearly as Q^{max} increases. The charge of a fragment ion is thus nearly a certain fractional value for a given Q^{max}. In the present paper, we left the fractional charge as it was to calculate the kinetic energy distributions and angular distributions of fragment ions; we virtually regarded the fractional charge just as an averaged one over fragmentation pathways through different excited states. Since Q^{max} has a distribution, the charge of the fragment ion has also a distribution even in the present SCC-DFTB approach. The overall charge distributions of fragment ions obtained by the present SCC-DFTB approach are as follows: q = 0.3–0.8 for H^{q+}; 0.5–1.5 for C^{q+}; 0.5–1.5 for N^{q+}; 0.5–1.5 for O^{q+}; 0.5–2.0 for I^{q+}. The round-off values of these fractional charges except H^{q+}, *i.e.*, $q = 1$ and 2 for C^{q+}, N^{q+}, O^{q+}, and I^{q+}, correspond to the charges of dominant fragment ions shown in the experimental results of Fig.1 (b) of the paper. The lowest value of $q = 0.3$ for H^{q+} suggesting the possibility that neutral H exist as fragments as well as H^+. These fragmentation channels of different integer charges cannot be distinguished from one another in the present SCC-DFTB approach. One way to simulate the fragmentation channels of different integer charges or the integer charge distributions, *i.e.*, histograms in Fig. 1(b) (in the paper) is to use a constrained DFTB method where the charges on the departing fragments are fixed as integers, though this would make the on-the-fly DFTB/MD more expensive.

Oliver Gessner asked: The typical timescales for charge buildup and redistribution seem to be shorter than or comparable to the duration of the XFEL pulse. To what extent does the temporal shape of the XFEL pulse play a role then for the reconstruction of the electronic dynamics from the experimental observables?

Kiyoshi Ueda replied: The typical timescales of charge buildup and redistribution are about 10 fs and a few fs, respectively. Our X-ray pulse duration at SACLA is about 10 fs (FWHM) and so roughly the same order. We applied our model calculation (the classical Coulomb explosion model with charge evolution) to the experimental data at LCLS with the pulse duration of 5 fs[1] and found that charge buildup time is about the same as the SACLA experiments. In this case Auger cascades seem to dominate the charge build up time scale. If the pulse duration is much larger than 10 fs, then the charge build up time is expected to be determined by the pulse duration. For example, for the N_2 Coulomb explosion experiment of FLASH with the pulse duration of 100 fs, the FEL power density and the pulse duration determined the full Coulomb explosion dynamics.[2] From such studies (fitting the classical model calculations to the experimental data) we cannot learn much about the influence of the temporal shape of the XFEL pulse beyond the temporal duration of the pulse. However, according to the latest *ab initio* molecular dynamics calculations by Robin Santra's group, it seems the temporal shape also plays a role to fully reproduce the experimental observables.

1 B. Erk *et al.*, *Phys. Rev. Lett.*, 2013, **110**, 053003.
2 O. Kornilov *et al.*, *J. Phys. B*, 2013, **46**, 164028.

Artem Rudenko addressed Oliver Gessner: We have recently performed similar experiments with longer pulses and much higher intensities at LCLS. There, the comparison between the experiment and theory (by R. Santra's group) indicates that the exact pulse shape (pulse envelope) is important.

Oliver Gessner answered: It would be interesting to learn more about these experiments.

Hans Jakob Wörner remarked: From your experiments and calculations, is it possible to extract a general picture about the sequence in which chemical bonds are breaking, *e.g.* according to bond energies or the distribution of positive charges?

Kiyoshi Ueda responded: In principle yes. Comparing SCC-DFTB calculations that take account of the chemical bonds and CCE-CE calculations that neglect the chemical bonds with the experimental results, we can extract such information, but only to some extent. To fully answer the question, I discussed it with Hirohiko Kono (the second to last author on the paper). Let me now explain further details. Let us focus on the experimental and theoretical KEDs in Fig. 5 of the paper to discuss the role of chemical bonds in the sequence of bond breaking. In the case the CCE-CE results where chemical bonds are not taken into account, the kinetic energies of fragment ions are more or less governed by the mass ratio; the heavier the mass of a fragment under consideration is, the lower the kinetic energy. For the fragments of moderate weight, namely, for C^+, N^+, and O^+, the CCE-CE results show that the peak energies in KEDs are around 20 eV, which is significantly lower than that of a light fragment H^+ and higher than that of a heavy fragment I^{q+}. Without chemical bonding, the lighter fragment simply takes away more Coulomb repulsion energy.

This picture of Coulomb explosion is not always applicable to interpreting experimental results. In the experimental or SCC-DFTB results, the peak energy in the KED of C^+ is smaller than that of O^+ by 10–15 eV, though C^+ is even lighter than O^+. According to our SCC-DFTB simulation, among C^+, N^+, and O^+, the O^+ fragments leave the gathering of atoms early (after H^+ ejection), and then the C^+ and N^+ that are capped by a H atom follow. The remaining three C^+ capped by the heavier atoms O or I are finally ejected. The fragments that are ejected earlier take away more Coulomb repulsion energy than those ejected later if the difference in mass is small. This "rule" is in accord with the experimental fact that the peak energy decreases in the order of O^+, N^+, and C^+, not in the order of mass, C^+, N^+, and O^+. Appropriate theoretical treatment of chemical bonds is inevitable to reproduce the accurate temporal order of fragment ejections. In the SCC-DFTB simulation, the bond distances between adjacent atoms in the uracil ring stretch almost at the same speed on statistical average. The mode of bond stretching between ring atoms is insensitive to the chemical nature of the bond, *i.e.*, whether the bond is a single or double bond in the neutral equilibrium structure. The energy that a fragment ion gains is rather sensitive to when it is ejected from the gathering of atoms.

Oksana Travnikova said: In your simulations the nuclear dynamics only in the final valence ionised states, which are formed after the charge buildup step, are considered. However, during the charge buildup step core–hole states are generated in the course of Auger cascades with potential energy surfaces, which can be rather different from that of valence ionised states. Can nuclear dynamics during Auger cascades have an effect on charge redistribution and fragmentation?

Kiyoshi Ueda answered: In both the classical Coulomb explosion model with charge evolution (CCE-CE) and the self-consistent charge density-functional based tight-binding (SCC-DFTB) method, we took account of Auger cascades by introducing one parameter, a charge build-up time, which describes the exponentially increasing charge in the molecule. We took account of the nuclear dynamics during this charge build up time of about 10 fs. To be more precise, we employed this charge build up time as a fitting parameter to reproduce our multiple ion momentum correlation data. So we know that nuclear dynamics during Auger cascades certainly has an effect on fragmentation. As you pointed out, however, the potential surfaces of the inner shell hole states produced in the early stage of Auger cascades may be rather different from valence ionized states, even with the same total charge. Such differences were not explicitly taken into account in the current models but implicitly partly taken into account by another fitting parameter, *i.e*, the charge redistribution rate in the CCE-CE model, which introduces a slight delay of the charge build up of other atoms than iodine (X-ray absorber). In the CCE-CE model, we empirically introduced the charge redistribution rate and used it as a fitting parameter. The resulting time scale is only a few fs. We consider the main mechanism of the charge redistribution is the later stages of Auger cascades where the electrons in the delocalized molecular orbitals participate in the decay. We checked the influence of the nuclear motion to the charge redistribution using the non-adiabatic quantum molecular dynamics (NAQMD) approach, where we assumed a 6+ charge state of the parent 5-UI and saw how the charge was redistributed (transferred) *via* nuclear motion. It turned out that the charge of ionidine changes by one, following the H vibrational motion, in 5 fs and the charges of most of the atoms stabilize in 10–20 fs (*via* crossing conical intersections). This just gives us the idea of the time scale when the nuclear dynamics sets in and influences the charge dynamics. As you noted, however, the assumption here was that we have six valence holes as the initial state. From our current study, we cannot say much about competition between the nuclear dynamics and the charge redistribution during the Auger cascades.

Tom Penfold opened a general discussion of the paper by Martin Centurion: You stated that you have been able to extract not only the atomic positions but also the wavepacket itself from the experimental data. Have you compared this to quantum dynamics calculations? Also how far, in terms of number of nuclear degrees of freedom, do you think it will be possible to extend this in future experiments?

Martin Centurion responded: We have been able to extract the amplitude of the wavepacket of iodine molecules that were excited into a bound vibrational wavepacket. We did compare with simulations, with good agreement, and the results are published.[1] We don't yet have a general method to expand this to many degrees of freedom, but we expect that it should be possible to do it for a few degrees of freedom. Many photochemical reactions can be described by a few degrees of freedom, so we expect that even with a few degrees of freedom the method will provide valuable information. How far this method can be pushed in terms of large molecules is still unknown.

1 J. Yang *et al.*, Diffractive Imaging of Coherent Nuclear Motion in Isolated Molecules, *Phys. Rev. Lett.*, 2016, **117**, 153002.

Piero Decleva asked: I would like to clarify the possibility of observing the vibrational wavepacket. I imagine that it is the probability distribution of nuclear positions, *i.e.* the modulus squared of the nuclear wavepacket that is imaged. Still I feel confused in the case of a split wavepacket moving onto two different electronic surfaces, what would be observed?

Martin Centurion replied: Yes, that is correct. What can be measured with this method is the absolute value squared of the nuclear wavepacket. In the case that the wavepacket splits onto two parts, what will be measured is still the absolute value squared of the wavepacket, so both parts will be detected. This is not too different from detecting the spreading of the wavepacket. As the wavepacket spreads, a broader wavepacket is observed. If it were to spread and split in two parts, two parts would be observed, assuming they are far enough apart to be resolved. This method is directly sensitive to the spatial distribution of charge, so it does not matter on which electronic surface a wavepacket is moving.

R. J. Dwayne Miller said: It is interesting to note that Ischenko *et al.*[1,2] were able to retrieve information on nuclear distributions, which your refer to as probability distributions. They treated theoretically dissociation of IBr, a reaction in this case, as well as CS product channels in the photodissociation of CS_2. Similarly, it was suggested that with sufficient SNR one could even use their quantum tomography approach to determine electron distributions. In your case you are looking at the bound state vibration of I_2. There is only 1 mode possible. Do you think you can disentangle this information from more complicated systems such as cyclohexadiene where there are multiple near resonant modes (C=C, C–C stretch)? If so, then it might be possible to observe the specific details of the mode coupling leading to photodissociation in this system as a prototypical example.

1 A. A. Ischenko, L. Schäfer, J. Y. Luo and J. D. Ewbank, *J. Phys. Chem.*, 1994, **98**, 8673.
2 A. A. Ischenko, J. D Ewbank and L. Schäfer, *J. Phys. Chem.*, 1995, **99**, 15790.

Martin Centurion answered: In Ischenko *et al.*[1,2] they retrieved the nuclear probability density from simulated diffraction patterns of linear diatomic and triatomic molecules. In our case, because we looked at a diatomic molecule, iodine, we were able to use an algorithm similar to the standard algorithms used in gas electron diffraction to retrieve not only the interatomic distance but also the shape of the moving wavepacket.[3] Our method can in principle be extended to more complex molecules if there is sufficient alignment to retrieve the structure. In some cases, the alignment due to photoselection may be sufficient for this. What we would reconstruct is nuclear positions that are more spread out in space, where the spreading is due to the spread of the nuclear wavepacket. In practice, the limited spatial and temporal resolution also contribute to broadening of the moving wavepacket so these will have to be accounted for to retrieve wavepackets accurately. Extending the method to molecules such as cyclohexadiene is in principle possible, but still a work in progress.

1 A. A. Ischenko, L. Schäfer, J. Y. Luo and J. D. Ewbank, *J. Phys. Chem.*, 1994, **98**, 8673.
2 A. A. Ischenko, J. D Ewbank and L. Schäfer, *J. Phys. Chem.*, 1995, **99**, 15790.
3 J. Yang *et al.*, Diffractive Imaging of Coherent Nuclear Motion in Isolated Molecules, *Phys. Rev. Lett.*, 2016, **117**, 153002.

Peter M. Weber asked: In your paper you write that the electron beam has a diameter of 200 μm. In the N_2 plasma experiments, the laser has a diameter of 50 μm, and in the alignment experiment it is 200 μm. This implies that the plasma experiment only captures 1/16th of the molecules. Is this a problem? And, in the alignment experiment, one would expect a distribution of laser intensities across the nitrogen molecules. How does this distribution affect the interpretation of the experimental data?

Martin Centurion responded: For the plasma experiments, the only goal is to determine the temporal overlap of the laser and electron pulses. The plasma is used to deflect some of the electrons in the beam, and the deflected electrons are used to find the temporal overlap. While a larger laser spot size would increase the overlap with the electron, a higher intensity is more advantageous because the ionization rate depends nonlinearly on the intensity. Therefore, given a limited pulse energy, the number of deflected electrons was maximized with a smaller spot size. For the alignment experiments, the diameter of the laser and electron beams on the sample was the same. Ideally, the laser spot would be larger so that all of the molecules probed by the electron would experience the same laser intensity. We could not operate under these conditions because with the current setup it was not possible to reduce the electron beam size on the sample further without increasing the beam divergence, which would reduce the quality of the diffraction patterns. At the same time, making the laser spot size larger would have decreased the intensity. The way we accounted for this in the data analysis was that we used a scaling factor and considered an average intensity experienced by the molecules. This gave very good agreement between the simulations and the data. Effectively, the molecules under the more intense part of the field would be more aligned, but what we measure is the overall distribution, so the alignment is slightly lower than would be expected for uniform illumination.

João Pedro Figueira Nunes said: In diffraction patterns obtained from laser aligned molecules, do you observe any contribution from unaligned species? and if so, do you account for their isotropic contribution to the diffraction data?

Martin Centurion responded: When we analyze the diffraction patterns, we look at differences between patterns with and without the laser alignment. By taking a difference, we are removing the contribution of the molecules that are not excited by the laser, because their diffraction pattern does not change.

Ruaridh Forbes remarked: In experiments related to alignment or strong field processes a concern is intensity averaging over the laser focus. Intensity averaging typically results in a series of different radial intensities within the laser focus.

How did you deal with this in the data processing as well as minimise these effects during the data collection?

Martin Centurion replied: In order to minimize this effect, we made the laser spot size as large as possible, given the constraints of the experimental setup. This resulted in laser and electron spots of approximately the same diameter. This was taken into account in the data analysis by considering the average intensity experienced by the molecules. The molecules within the more intense part of the laser beam will be more strongly aligned, while the others will be less aligned. The angular distribution probed by the electrons will be the average over all the molecules, which can be well approximated by considering alignment with an "average" or effective intensity.

R. J. Dwayne Miller commented: Relativistic electrons have advantages over nonrelativistic sources in terms of being able to go to higher bunch densities before the onset of space charge effects. The problem is not so much in the longitudinal direction due the relativistic nature of the source but rather in the transverse direction. The difference in charge densities before the onset of undesirable emittance growth is however not too dissimilar to the same situation for nonrelativistic and leads very rapidly to a loss in spatial resolution (loss of transverse coherence). The most unique feature for gas phase studies is the near elimination of velocity mismatch between the laser excitation and electron pulses with respect to time resolution. However, you may be limited in the size of the system you can study because of transverse coherence issues. How far do you feel you can push the size of molecular systems you can study, assuming conventional laser excitation conditions (mj laser systems)? What is the transverse coherence of your source at the sample position?

Martin Centurion replied: With the current setup, the transverse coherence length is 1.7 nm. The transverse coherence can be increased at the expense of beam charge. In order to do this effectively, one would need to increase the repetition rate of the system to increase the average beam current. Currently, the electron gun operates at 120 Hz, which is significantly less than what lasers can currently do. The technology to run electron guns at high repetition rates already exists,[1] and could be implemented in gas phase UED. Using a radio-frequency cavity to compress the electron pulses can also increase the beam current by two orders of magnitude [see for example ref. 2 and 3]. If a high repetition rate gun operating at around 100 kHz could be combined with pulse compression, the beam current could be increased by five or six orders of magnitude, which would allow to tune the beam to a transverse coherence of 100 nm or more, while maintaining a sufficient beam current for diffraction. Ultimately, photocathodes that are capable of generating brighter electron beams would be needed to investigate more complex systems and increase both the beam current and transverse coherence.

1 D. Filippetto and H. Qian, *J. Phys B.*, 2015, **49**(*10*), 104003.
2 X. H. Lu, CX Tang, RK Li, H To, G Andonian and P Musumeci, *Phys. Rev. ST Accel. Beams*, 2015, **18**, 032802.
3 T. van Oudheusden, P. L. E. M. Pasmans, S. B. van der Geer, M. J. de Loos, M. J. van der Wiel and O. J. Luiten, *Phys. Rev. Lett.*, 2010, **105**, 264801.

Peter M. Weber said: In our X-ray scattering we found that after accounting for an overall scale factor for the intensity of the X-ray beam, the X-ray scattering signal is in quantitative agreement with the computed signal. As your paper describes very carefully, the analysis of the electron scattering signals is much more convoluted, requiring multiple steps and analysis. This can probably be traced to the rapid $1/s^4$ dependence in electron diffraction, which gives rise to the underlying shape of the scattering signal. Could you comment on how you see the advantages and limitations of electron scattering in this light? Does the convoluted nature of the analysis complicate the extraction of pump–probe data?

Martin Centurion replied: So far, in the pump–probe experiments, the structure as a function of time was retrieved from the diffraction pattern for each time step. For this, we used well established data analysis methods of gas phase electron diffraction. While there are several steps in the process, this method has been tried and tested many times over the last few decades in static patterns, and so far this has not been a limitation. For pump–probe data with more complex molecules, one could take advantage that the ground state structure is known, and use it as a starting point. In addition, one could add additional constraints, for example that the structure changes are continuous in time. I think what will be most interesting, and challenging, is to capture the dynamics when the position of the nuclei becomes delocalized, for example when there are multiple moving nuclear wavepackets spread in space such that they become very broad. This will present a challenge and an opportunity for both X-ray and electron diffraction.

Adam Kirrander said: In light of your previous work on reconstruction of molecular structure from diffraction data[1] and your recent measurements of vibrational wave packets in the iodine molecule,[2] how do you see the future of reconstructing structural dynamics from ultrafast scattering experiments? How feasible will it be to reconstruct the time-evolving molecular wave packet in molecules larger than diatomics? Laser-alignment and possibly holographic techniques would help, but imaging a wave packet that spreads across many degrees of freedom is hard. How do you see the state-of-the-art evolving, and what are the main limitations in your view?

I should perhaps add that our own take on this problem has been to calculate the scattering signal from a molecular wave packet,[3] and then bias this wave packet to reproduce the experimental scattering data as best possible, as a way of including pre-existing knowledge about molecular geometry, electronic structure, and dynamics. This was essentially the procedure followed in our ultrafast X-ray scattering study of the ring-opening reaction of 1,3-cyclohexadiene,[4] an analysis which was supported by time-resolved photoelectron spectroscopy.[5]

1 J. Yang *et al.*, *Struct. Dyn.*, 2014, **1**, 044101.
2 J. Yang *et al.*, *Phys. Rev. Lett.*, 2016, **117**, 153002.
3 A. Kirrander, K. Saita and D. Shalashilin, *J. Chem. Theor. Comput.*, 2016, **12**, 957–967.
4 M. Minitti *et al.*, *Phys. Rev. Lett.*, 2015, **114**, 255501.
5 C. Pemberton *et al.*, *J. Phys. Chem. A*, 2015, **119**, 8832–8845.

Martin Centurion replied: In the ideal case, the dynamics of the molecule would be completely measured by experiment, simulated using state-of-the-art theory (without interaction between the two), and then theory and experiment can be compared. We are, of course, far from reaching that state with diffraction measurements, except for the simplest molecules. It is clear that for more complex dynamics an interaction between theory and experiment is needed to retrieve the dynamics. However, it is important to move towards minimizing the theoretical input in the interpretation of the experimental data. This will allow for more stringent tests on the theory, and more confidence in the dynamics retrieved from the experiment. As mentioned in the question, there are several approaches to extracting structures from experimental data. Currently, the main limitations in the experiment are the lack of a general algorithm for structure retrieval, and the limited spatial and temporal resolution that tend to blur out the details of the dynamics. Lastly, most diffraction experiments still suffer from low signal-to-noise ratios, so brighter sources will be important for the continued development of this research area.

Oliver Gessner opened a general discussion of the papers by Michael Minetti, Kiyoshi Ueda and Martin Centurion: As you have demonstrated, both ultrafast X-ray scattering and ultrafast electron diffraction provide access to structural dynamics in small molecules. However, there are important differences between the two methods. X-rays scatter predominantly off bound electrons, electrons off the Coulomb potential. Have you considered studying an example for unimolecular dynamics, such as the photoinduced ring-opening in CHD, with both methods and to compare the results? What differences would you expect and where do you see the particular strengths/weaknesses of each method?

Michael Minitti replied: Boy have we ever. These two techniques are extremely complimentary and need to be combined in some way. Now that MeV electron sources are becoming more and more prevalent, the task is to perform side by side measurements of each technique on a common target. 1,3-cyclohexadiene would be an ideal candidate.

A collaboration between Martin Centurion, SLAC, and LCLS will begin an ultrafast electron diffraction (UED) campaign at the ASTA electron source at SLAC, studying various gas phase targets, and 1,3-CHD will be one of them. It will be interesting.

As for strengths and weaknesses of the techniques, clearly UED offers exquisite spatial resolution when compared to X-ray diffraction. However, in the current state of the art, UED suffers from a lack of temporal resolution when compared to currently offered capabilities at XFELs. The trade-off with UED pulses in order to make them short enough, is to reduce the number of total electrons per pulse, due to the space–charge effect. This in turn will make the s/n ratio lower, thus needing longer and longer acquisitions times. Additionally, a challenge with UED is the ability to precisely time stamp the electron arrival time with that of any optical laser pump. Like the X-ray time-tools (cross-correlators, arrival time monitors... *etc.*), UED will greatly benefit from shot-to-shot time correction.

Martin Centurion responded: In most cases, X-rays and electrons will provide equivalent structural information because most of the charge is indeed concentrated near the nucleus. One interesting example where they will provide different information is in proton transfer reactions. If only a proton is moving during the reaction, the change will be captured by electron diffraction but not by X-ray diffraction. One can envision a combination of X-ray and electron diffraction to determine whether a proton or a hydrogen atom is moving during the reaction, and to determine the path of proton and electron. In comparing the current state of X-ray and electron diffraction, X-ray diffraction provides higher temporal resolution because photons do not experience the repulsive Coulomb force that broadens electron pulses, while electron diffraction provides higher spatial resolution because of the smaller de Broglie wavelength and higher scattering cross section. The spatial resolution of X-ray diffraction is expected to increase as the repetition rate and photon energy of the FELs increases, while for the case of MeV UED there is a lot of room for improvement in both spatial and temporal resolution. For example, with RF compression, the pulse duration can be compressed to 20 fs, while also significantly increasing the charge per pulse, and thus improving the spatial resolution [see for example ref. 1] . The joint X-ray and electron experiments will be most interesting once both sources can reach temporal and spatial resolutions on the order of 50 fs and 0.1 Å. In the meantime, the main advantage of combining X-ray and electron diffraction data will be to take advantage of the high temporal resolution of X-rays and high spatial resolution of electrons.

1 X. H. Lu, C. X. Tang, R. K. Li, H. To, G. Andonian and P. Musumeci, *Phys. Rev. ST Accel. Beams*, **18**(*3*), 032802.

Adam Kirrander commented in response to an informal remark by Dwayne Miller (not recorded): Indirect observations of electronic state *via* changes in molecular geometry have been around for a while (see *e.g.*, ref. 1) but I agree with Dwayne that it would be incredibly interesting to directly observe excited electronic states using electron or X-ray scattering. In fact, we recently developed a code to predict elastic scattering from molecules in their ground and excited states based on their *ab initio* electronic structure wavefunctions[2,3] in anticipation of such experiments. However, in fairness one should remember that powerful techniques already exist to selectively study valence electronic structure, *e.g.* photoelectron spectroscopy and photoelectron imaging (including MF-PAD). In contrast, the scattering will contain contributions from all electrons, including core electrons.

1 M. R. Pressprich, M. A. White and P. Coppens, *J. Am. Chem. Soc.*, 1993, **115**, 6444–6445.
2 T. Northey, N. Zotev and A. Kirrander, *J. Chem. Theor. Comput.*, 2014, **10**, 4911–4920.
3 Thomas Northey, Andrés Moreno Carrascosa, Steffen Schäfer and Adam Kirrander, *J. Chem. Phys.*, 2016, **145**, 154304.

R. J. Dwayne Miller answered: I share your enthusiasm. Given the success of electron diffraction to enable quantum tomography on ns timescales to follow nuclear distribution of reaction products,[1,2] it is clear it should be possible to do the same on shorter timescales and with the theoretical advances you describe to be able to follow changes in electron distribution. The prospect of directly imaging electronic distributions is exciting as fs electron diffraction methods would enable a complete mapping of the forces at play in reaction dynamics and defining excited state potentials. As you point out, the problem will be the required signal-to-noise ratio to pull out these small changes. The changes at any atomic site relative will be relatively small compared to the total electron density and Coulomb potential to give excitation dependent structure factors. From an experimental standpoint, the explicit use of differential detection and known initial structures should provide the needed contrast. Other improvements in source technology (stability), sample delivery methods, and new analysis methods should give us the tools to attain the required SNR to pull out these details.

1 A. A. Ischenko, J. D. Ewbank and L. J. Schäfer, *Phys. Chem.*, 1995, **99**, 15790.
2 A. A. Ischenko, L. Schäfer and J. D. Ewbank, *Proc. SPIE*, 1998, **3516**, 580.

R. J. Dwayne Miller addressed Michael Minitti and Martin Centurion: It is not clear to me that alignment is needed for enhancing structural resolution for the time resolved differential changes in structure involving known initial and final structures in the gas phase. This approach does provide additional information that would improve spatial resolution of unknown structures, however, one is always dealing with known structures in time resolved studies of structural dynamics. It is imperative that the excited state dynamics for the system are known as well as the electronic states excited in the preparation of the excited state. Therefore, only systems with known structures (and ideally excited state dynamics) are used in this class of experiments. There is also the additional complication of ionization and perturbation of the initial state preparation under the high field conditions needed for alignment. If there are high field effects for systems such as CS_2 with one of the largest Raman/differential polarizations for alignment, it seems relatively unlikely this approach will help in resolving structural dynamics. In the case of the use of alignment for unknown structures, even with adiabatic fields, the field ionization limits the utility of alignment. There are other ways to achieve alignment in this case without high laser fields.

Michael Minitti answered: I agree with you on some level, but an experiment to test the validity of alignment and whether or not it enhances the observation of time-resolved structural changes is to try it in a weak field case, such as adiabatically aligned targets. This should be done. These alignment techniques employ longer IR pulses, 100s of ps to ns at 1064 nm, to weak field align rotationally cold targets. The longer pulse widths tend to avoid imparting significant ionization effects or perturbation to the system. However, in order to achieve any decent alignment, these pulses require the target to be rotationally cold (a few K at best). This means lower target density, which in the case of gas phase X-ray scattering, is a bad thing given X-ray cross-sections. It is a definite trade off scenario, one that I don't have a good grasp of which parameter would win out and improve the overall resolution of the experiment. Only one way to find out, and that is to attempt it!

Martin Centurion answered: If the initial structure is known, this will indeed provide an advantage in determining the structure of the excited states, in particular if the structural changes are small, and it may be possible to determine the structure without alignment. In most experiments, a one-photon excitation is used, which results in selective alignment (even without active alignment) because of the angular dependence of the excitation. This already produces a significant alignment that can be used for structure retrieval. What we have seen in our experiments and simulations, is that in order to determine structures directly form the data it is very beneficial to have alignment, in particular if two or more patterns are recorded where the angular distribution is different (and known). Regarding the complication of ionization and initial state perturbation, with impulsive alignment the alignment pulse can be applied before the excitation pulse, so that when the molecules are excited the alignment pulse is no longer present. Typically the alignment peaks at one or a few picoseconds after the interaction with the alignment pulse. Our previous experiments with CS_2 have shown that there is an intensity regime where significant alignment can be achieved without ionization and without multiphoton excitation.[1] If the laser intensity is increased beyond a certain threshold, the alignment no longer improves, and additional effects such as multiphoton excitation and ionization become significant. We are confident that impulsive alignment can be used [see also ref. 2] at least for the small molecules tested so far. For large molecules this will become more challenging, as the alignment pulse may also excite vibrations in the molecule. The regime of impulsive alignment of large molecules still needs to be explored.

1 J. Yang, J. Beck, C. J. Uiterwaal and M. Centurion, *Nat. Commun.*, 2015, **6**, 8172.
2 C. J. Hensley, J. Yang and M. Centurion, *Phys. Rev. Lett.*, 2012, **109**, 133202.

Kiyoshi Ueda commented: I am familiar only with the fluence of SACLA. At SACLA, a typical "peak" fluence of so called 1μm focus at 5.5 keV is ~6 $\times$ 10^{10} photons $\mu m-^{2}$. The photoionization cross section of neutral xenon atoms at 5.5 keV is ~0.166 Mb. So the "peak" fluence nearly saturates the single photon absorption. Due to volume focusing effects, however, still single photon absorption events dominate in the ion yield mass spectra. In the mild focusing conditions at LCLS, the single photon absorption of iodine atoms might not fully saturate. This might be a reason that the significant electronic damage in the hard X-ray scattering experiments at LCLS for the I-contained gaseous molecules was not observed in the X-ray diffraction.

Faraday Discussions

PAPER

Towards characterization of photo-excited electron transfer and catalysis in natural and artificial systems using XFELs

R. Alonso-Mori,[a] K. Asa,[b] U. Bergmann,[c] A. S. Brewster,[d] R. Chatterjee,[d] J. K. Cooper,[e] H. M. Frei,[d] F. D. Fuller,[d] E. Goggins,[f] S. Gul,[d] H. Fukuzawa,[gh] D. Iablonskyi,[g] M. Ibrahim,[i] T. Katayama,[j] T. Kroll,[k] Y. Kumagai,[g] B. A. McClure,[d] J. Messinger,[l] K. Motomura,[gh] K. Nagaya,[bh] T. Nishiyama,[b] C. Saracini,†[d] Y. Sato,[b] N. K. Sauter,[d] D. Sokaras,[k] T. Takanashi,[g] T. Togashi,[j] K. Ueda,[gh] W. W. Weare,[f] T.-C. Weng,[m] M. Yabashi,[j] V. K. Yachandra,[d] I. D. Young,[d] A. Zouni,[i] J. F. Kern*[ad] and J. Yano*[deg]

Received 18th April 2016, Accepted 2nd June 2016
DOI: 10.1039/c6fd00084c

The ultra-bright femtosecond X-ray pulses provided by X-ray Free Electron Lasers (XFELs) open capabilities for studying the structure and dynamics of a wide variety of biological and inorganic systems beyond what is possible at synchrotron sources. Although the structure and chemistry at the catalytic sites have been studied intensively in both biological and inorganic systems, a full understanding of the atomic-scale chemistry requires new approaches beyond the steady state X-ray crystallography and X-ray spectroscopy at cryogenic temperatures. Following the dynamic changes in the

[a]*Linac Coherent Light Source, SLAC National Accelerator Laboratory, Menlo Park, CA 94025, USA*
[b]*Department of Physics, Graduate School of Science, Kyoto U., Kyoto, 606-8502, Japan*
[c]*Stanford PULSE Institute, SLAC National Accelerator Laboratory, Menlo Park, CA 94025, USA*
[d]*Molecular Biophysics and Integrated Bioimaging Division, Lawrence Berkeley National Laboratory, 1 Cyclotron Rd., Berkeley, CA 94720, USA. E-mail: JFKern@lbl.gov; JYano@lbl.gov*
[e]*Joint Center for Artificial Photosynthesis (JCAP), Lawrence Berkeley National Laboratory, 1 Cyclotron Rd., Berkeley, CA 94720, USA*
[f]*Dept. of Chemistry, North Carolina State University, 2620 Yarborough Rd., Raleigh, NC 27695-8204, USA*
[g]*IMRAM, Tohoku U., Sendai 980-8577, Japan*
[h]*RIKEN SPring-8 Center, Kouto, Sayo, Hyogo 679-5148, Japan*
[i]*Institut für Biologie, Humboldt-Universität zu Berlin, D-10099 Berlin, Germany*
[j]*Japan Synchrotron Radiation Research Institute (JASRI), SPring-8/SACLA, Kouto, Sayo-cho, Sayo-gun, Hyogo 679-5198, Japan*
[k]*Stanford Synchrotron Radiation Lightsource (SSRL), SLAC National Accelerator Laboratory, Menlo Park, CA 94025, USA*
[l]*Institutionen för Kemi, Kemiskt Biologiskt Centrum, Umeå Universitet, Umeå, Sweden*
[m]*Center for High Pressure Science & Technology Advanced Research, Shanghai, China*
† Current address: Center for Biomimetic Systems (CBS), Ewha Womans University, Department of Bioinspired Science, Seoul, 120-750, Korea.

geometric and electronic structure at ambient conditions, while overcoming X-ray damage to the redox active catalytic center, is key for deriving reaction mechanisms. Such studies become possible by using the intense and ultra-short femtosecond X-ray pulses from an XFEL, where sample is probed before it is damaged. We have developed methodology for simultaneously collecting X-ray diffraction data and X-ray emission spectra, using an energy dispersive spectrometer, at ambient conditions, and used this approach to study the room temperature structure and intermediate states of the photosynthetic water oxidizing metallo-protein, photosystem II. Moreover, we have also used this setup to simultaneously collect the X-ray emission spectra from multiple metals to follow the ultrafast dynamics of light-induced charge transfer between multiple metal sites. A Mn–Ti containing system was studied at an XFEL to demonstrate the efficacy and potential of this method.

1. Introduction

Many of the catalytic reactions in inorganic systems and natural enzymes involve multiple electrons, and proceed through several intermediate steps. For example, photosynthetic water oxidation in nature is catalyzed by an oxo-bridged Mn_4Ca metal center located in a multi-subunit membrane protein, photosystem II (PS II) (Fig. 1). An understanding of the mechanism of light-harvesting, charge separation and catalysis is well-connected to the function of this molecular machine. A PS II monomer consists of 17 or 18 membrane integral subunits composed of 35–36 *trans*-membrane helices and 3 peripheral subunits. In addition, there are many cofactors, 35 chlorophyll a (Chl), 11–12 all-*trans* β-carotene molecules, 1 OEC (Oxygen Evolving Complex, Mn_4CaO_5 cluster), 1 heme b, one heme c, 2 or 3 plastoquinones, 2 pheophytins, and 1 nonheme Fe, that are important for light-harvesting, charge separation and stabilization, and electron transfer. The high-

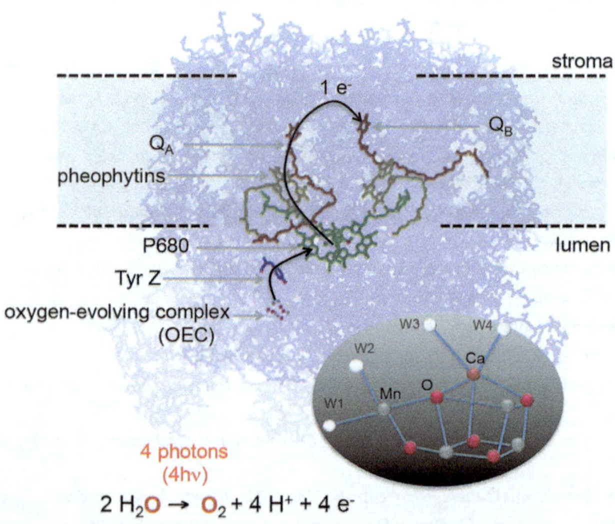

Fig. 1 The electron transfer chain of photosystem II. The inset shows the structure of the water oxidizing Mn_4CaO_5 catalyst.[1]

resolution X-ray crystal structures of PS II from a thermophilic cyanobacterium has been reported to a resolution of 1.9 Å, revealing the detailed architecture of this enzyme.[1,2]

In parallel to the basic understanding of natural enzymes, learning from natural systems on how to control electron flow between multiple sites that can undergo an ultrafast photo-excited charge separation is a key issue for developing materials such as artificial photosynthetic devices and magnetic materials. For example, a rapid charge separation in inorganic chromophores is often accomplished *via* a metal-to-metal charge transfer event (MMCT). To be useful for driving a multi-electron catalyst, as required for processes such as CO_2 reduction or water oxidation, the excited state that is formed by the initial ultrafast processes in the binuclear charge transfer unit has to be sufficiently long lived to allow for electron transport to/from the catalyst. To understand such sequential chemistry, it is important to probe signals from multiple metal sites simultaneously in a time-resolved manner.

The introduction of X-ray Free Electron Lasers (XFELs) has made it possible to follow photochemical reactions in biological systems at room temperature, using time-resolved detection methods, before the onset of radiation-induced changes propagates, by taking advantage of the ultrashort femtosecond X-ray pulses. Similarly, detecting early picosecond to sub picosecond time-scale phenomena using X-ray techniques has become possible, expanding the impact of such photochemical studies to many inorganic systems with ultrafast excited state dynamics.

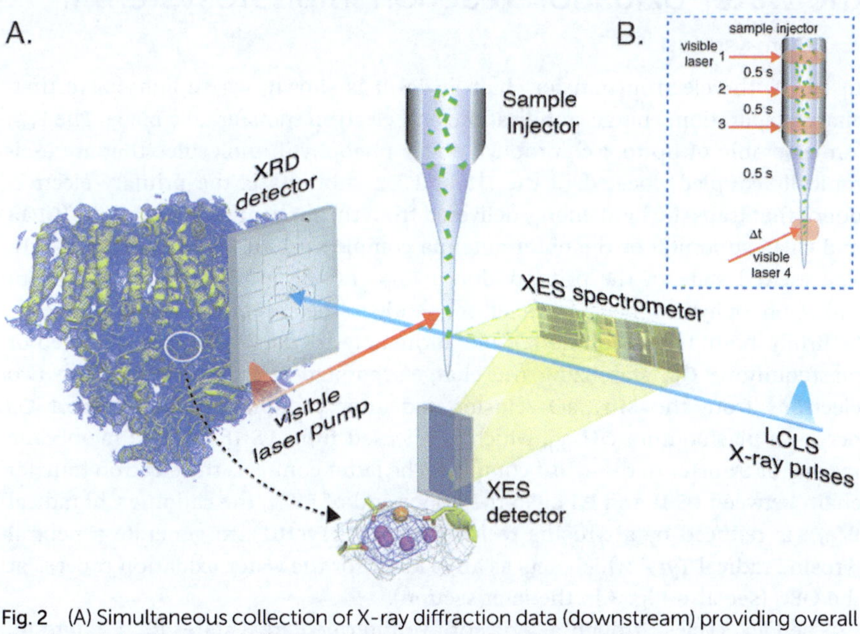

Fig. 2 (A) Simultaneous collection of X-ray diffraction data (downstream) providing overall structural information and X-ray emission spectra (at 90 degrees) using the wavelength-dispersive spectrometer providing electronic structure information of the metal site.[3,4] Adapted from Kern *et al.*[5] (B) Multiple illumination protocol used for triggering photochemical reactions that can be probed using the femtosecond X-ray pulses from the XFEL.

We have developed several X-ray-based techniques suitable for studying photochemical reactions in biological and inorganic systems at XFEL facilities (XFELs).[3–8] Fig. 2 shows a wavelength-dispersive X-ray emission spectrometer that has been developed and used for collecting X-ray emission spectra (XES) at the LCLS (Linac Coherent Light Source, SLAC, USA) and SACLA (Spring 8, Japan). This setup is suitable for shot-by-shot data collection at XFELs, as it can collect the entire emission spectrum at once. It also has several advantages for dealing with the inherent properties of the XFELs, *e.g.* it does not require a beamline monochromator and it is compatible with the XFEL SASE (Self-Amplified Spontaneous Emission) spectral profile in which the energy bandwidth is 0.2–0.5% (FWHM) of the total SASE energy for hard X-rays. In addition, the shot-by-shot variation of the intensity and the energy distribution of the SASE beam is not an issue. Wavelength-dispersive XES can also be combined with other techniques like crystallography in which the same incoming X-rays can be used for simultaneous collection of the diffraction (XRD) signal to obtain complementary structural information. Moreover, this setup allows detecting emission signals from multiple elements simultaneously, which is useful to follow the time-course of sequential reactions in multimetallic sites in biological and inorganic systems.

In this work, we present the XES and related methodology we have developed at XFELs and its application to study photochemical reactions in both biological and inorganic systems.

2. Photochemically driven electron transfer and the water oxidation reaction in photosystem II

A. Photosystem II

In Fig. 1, the electron transfer chain in PS II is shown, where light-harvesting, charge separation, charge stabilization, and electron transfer take place. The P_{680} (an ensemble of up to 4 chlorophyll and 2 pheophytin molecules that are excitonically coupled) located in the D1 and D2 subunits is the primary electron donor that traps the light energy delivered from the inner antenna subunits (CP43 and CP47 subunits) or the outer antenna complexes (LHC1 and LHC2) of PS II. The excited state of the primary donor P_{680}^{*} rapidly transfers the electron to Chl_{D1}, pheophytin ($Pheo_{D1}$) and subsequently to the acceptor, plastoquinone Q_A (a firmly bound plastoquinone) and ultimately to the final electron acceptor plastoquinone Q_B, stabilizing the charge-separated state. After accepting two electrons from the Mn_4CaO_5 cluster and after protonation the acceptor Q_B becomes plastoquinol $QH_{2(B)}$, which is released from PS II into the membrane matrix for transfer to the cytb$_6$f complex. The latter connects the electron transfer chain between PS II and PS I. On the donor end of PS II, the cationic Chl radical $P_{680}^{\bullet +}$ is reduced by a tyrosine residue, Tyr_Z (D1Tyr161), to generate a neutral tyrosine radical $Tyr_Z^{\bullet}$ which acts as an oxidant for the water oxidation process at the OEC (see also Fig. 3 in the later section).

The OEC cycles through a series of five intermediate S-states (S_i, $i = 0$ to 4), representing the number of oxidizing equivalents stored in the OEC driven by the energy of the four successive photons absorbed by the PS II reaction center (see Fig. 4 in the later section).[9] When PS II is dark-adapted, it relaxes to the S_1 state (note that although S_0 is the most reduced state of the OEC, the S_0 state is oxidized by tyrosine D(Y_D^{*+}) during the dark adaptation and therefore the S_1 state becomes the dark-stable state). Illumination of dark-adapted PS II (the S_1 state) with saturating flashes of visible light leads to a maximum O_2 yield after the 3rd flash, and then after every 4th successive flash. Each flash advances the oxidation state of the OEC by removing one electron, and the OEC acts like a redox capacitor for the water oxidation reaction until the four oxidizing equivalents are accumulated (S_4-state). Once four oxidizing equivalents are accumulated in the OEC, a spontaneous reaction occurs that results in the oxidation of water, release of O_2 and the formation of the S_0-state. Thus, the Mn_4CaO_5 complex in the OEC couples the four-electron four-proton oxidation of water with the one-electron photochemistry occurring at the PS II reaction center by acting as the locus of charge accumulation.[10]

During the reaction, the Mn cluster provides a high degree of redox and chemical flexibility, while protein residues are critical for mediating the reaction by modulating redox potentials and providing pathways for electrons, protons, substrate H_2O, and product O_2.[11,12] Thus, PS II orchestrates a well-controlled catalytic reaction at close to the thermodynamic potential, while avoiding the release of chemical intermediate species, such as superoxide or peroxide, during the water oxidation reaction that can be detrimental to the protein matrix and to the chemistry occurring at the OEC.

To follow catalysis under physiological conditions in PS II, X-ray crystallography and X-ray spectroscopy methods have been used at XFELs.[5] In particular, the simultaneous collection of crystallography and X-ray spectroscopy data has been proven to be a powerful method for studying metalloenzymes like PS II (Fig. 2), providing important insights from both the atomic structure of the protein crystals and the emission signals from the metal catalytic centers. This setup allows accessing time resolved data during photochemical reactions triggered using visible light.

B. Room temperature crystallography and spectroscopy of PS II

XFEL serial femtosecond X-ray crystallography has been proven to be a new way of doing protein crystallography.[13] In general, biological crystallography at a synchrotron facility is carried out at cryogenic temperatures to minimize the radiation-induced changes during data collection, summarized as radiation damage.[14] Nevertheless, in the case of metalloenzymes, often damage to the active site is observed even under cryogenic conditions and at a dose well below the dose threshold leading to an observable loss of diffractivity.[15–17] It is accepted that the cooling process, typically to 100–150 K, does not perturb the functional structure of enzymes. However, it is also known that the temperature could shift the intrinsic population of conformers in many proteins.[18–20] Protein dynamics play a critical role in enzymatic functions, and therefore many enzymes do not function at freezing temperatures. For some systems, the conformational landscape that is important for enzyme function may not be captured under cryo-cooled conditions. In addition, data collection at cryogenic temperatures can prevent capturing short-lived reactive intermediates unless such species can be cryo-trapped within the time-scale of state-of the art freeze-quenching methods. Thus, data collection under physiological conditions is required in such cases.

We have collected room temperature X-ray diffraction data from PS II at the CXI instrument[21,22] at the LCLS using 9.5 keV incident X-rays. The rmsd (root-mean-square deviation) of the room temperature dark state crystal structure of PS II in comparison with the cryogenic structure of the same protein reported by Suga *et al.*[1] showed that the changes are mostly within 2 Å, and they are in the solvent-exposed region of the protein. This implies that the cofactor distances in the electron transfer chain remain similar to those obtained at cryogenic temperature, as the cofactors are present in the hydrophobic interior of the PS II membrane complex.

In terms of the chemical states of the Mn_4CaO_5 catalytic center in PS II, a similar XES spectrum in the dark state was observed in both XFEL room temperature and synchrotron cryogenic temperature data.[5] This evidence demonstrates that the catalytic center at room temperature remains intact under the XFEL data collection conditions used in this study. The XES data also serves as a diagnostic tool for evaluating the chemical state of the samples.[23]

C. S-State advancement and acceptor side chemistry

The dark S_1 state can be advanced to higher S-states with a laser flash(es) at room temperature. By cryo-trapping the sample after the laser flash, each S-state (S_0, S_2, S_3, in addition to the dark S_1 state) has been characterized at synchrotron X-ray facilities. As the S-states do not decay rapidly, freeze-quenching of the illuminated samples to liquid N_2 temperature within a couple of seconds is sufficient for trapping the stable intermediates like S_2, S_3, and S_0 states. The yield is never 100% due to the intrinsic limitations in PS II S-state advancement because of misses and double-hits by visible light photons. In the best case, a S-state distribution after each flash is similar to what is shown in the inset table in Fig. 3. The $T_{1/2}$ of each S-state has been obtained from the literature,[24–26] and also summarized in this figure.

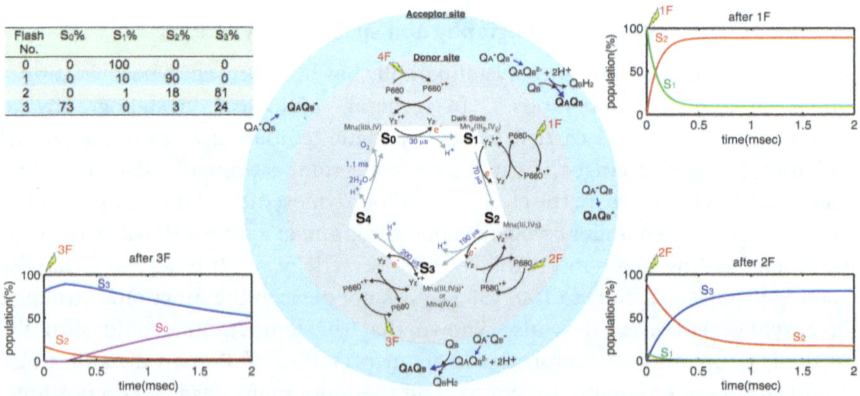

Flash No.	S_0%	S_1%	S_2%	S_3%
0	0	100	0	0
1	0	10	90	0
2	0	1	18	81
3	73	0	3	24

Fig. 3 Center: The Kok S-state cycle for water oxidation by PS II. Gray – donor side: each of the four photons absorbed by the PS II P_{680} reaction center oxidizes the tyrosine Y_Z intermediate, which in turn oxidizes the Mn_4CaO_5 complex. Blue – acceptor side: Q_A and Q_B are reduced on the acceptor side. Left top: A best case steady state S-state population as a function of laser-flash illuminations. 10% miss hit is assumed. Insets: The evolution of the populations of the S-states as a function of time are shown after 1flash (1F), 2 flashes (2F) and 3 flashes (3F).[24–26]

Stable charge separation is the prerequisite for advancing the Mn_4CaO_5 cluster of the oxygen-evolving complex (OEC) from one oxidation state to the next higher one. It is well known that stable charge separation can only be achieved if the acceptor quinone, Q_A, is present in the oxidized state (Fig. 3).[27] It is thus impossible to uncouple the S state advancement from the often slower acceptor side kinetics, *i.e.* the electron transfer from Q_A^- to Q_B or Q_B^-, and the formation and exchange of Q_BH_2 with a plastoquinone molecule from the thylakoid membrane (or a pre-bound Q_C molecule).[28,29] Thus, an important parameter governing the yield of the S_3 state after two flashes is the amount of open centers (Q_A) at the time point of the second flash. A stable charge separation can only happen in centers where Q_A is present in the oxidized form (see *e.g.* ref. 27), whereas in closed centers only a very small fraction (<10%) can proceed to the next higher S-state. If the second light excitation flash is given in the earlier stage after the first flash, a large fraction of the centers still have reduced Q_A (Q_A^-) and are thus closed, meaning they cannot produce a stable $P_{680}{}^{\bullet+}$ to advance the Mn_4CaO_5 cluster from the S_2 into the S_3 state. To establish the illumination conditions at XFELs, a replica of the XFEL illumination set up was built to study the advancement of the sample using isotope labeling and membrane inlet mass spectrometry (MIMS) under practically identical conditions as at the beamline. Optimization of all these parameters is absolutely crucial for achieving optimal sample turnover with a low miss factor.

The electron transfer rate between Q_A^- and Q_B or Q_B^-, and the exchange rate of Q_BH_2 with PQ, are species- and sample-dependent. We thus measured the time scale of such electron transfer events in the preparations used for our experiments. The fastest rates reported so far is by de Wijn & van Gorkom[30] for spinach membrane fragments. We note that measurements on intact cells show that wait-times between flashes in the order of 10's to 100's of ms are necessary to ensure turnover of the OEC and the acceptor side.[31] The S-state advancement in crystals should not be fundamentally different from all the literature values available.

We have collected the 2F XRD and XES data at CXI at LCLS using conditions similar to those described in the previous section. The dark-adapted PS II sample was illuminated *in situ* with 532 nm visible laser light (2 flashes) with the setup shown in Fig. 2, by turning-on fiber-coupled lasers 2 and 3. Incident X-rays of 9.5 keV were used for the experiment. Fig. 4A shows the Mn XES $K\beta_{1,3}$ peak shift observed between the dark (0F) and 2F data. The $K\beta_{1,3}$ transition of Mn corresponds to a fluorescence decay of 3p to 1s orbitals, that occurs after the excitation of the 1s core electron to the continuum. Similar difference spectra were observed for the synchrotron (cryogenic) and the XFEL (room temperature) data, indicating that the S-state advancement under our experimental conditions was comparable to that of samples prepared by the freeze-quenching method for synchrotron experiments.[3] The same setup was used to collect the XRD data of the 2F sample, and the data has been published.[23]

As mentioned above, the simultaneous collection of crystallography and X-ray spectroscopy data using crystalline samples is a valuable approach for studying the overall protein structure together with the chemistry that occurs at the metal catalytic centers. XES can also be used for diagnostic purposes during XRD data collection to ensure the intactness of the metal catalytic centers. However, some caution is necessary when crystallography and spectroscopic data are collected under the same conditions. Fig. 4B shows a single-shot diffraction pattern from

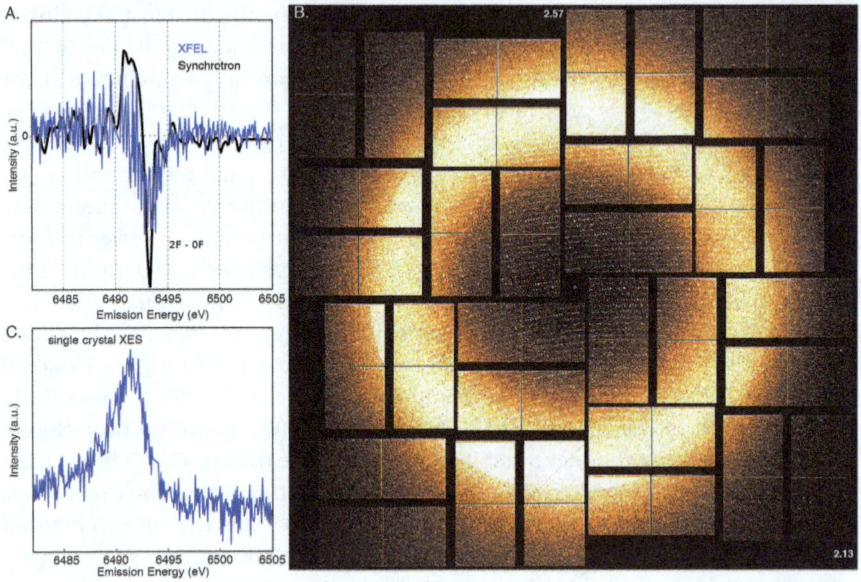

Fig. 4 (A) Overlay of the 2F-dark Mn $K\beta_{1,3}$ XES difference spectra using an XFEL (blue) and a synchrotron (black). (B) fs diffraction pattern of a PS II crystal in the 2 flash state recorded at room temperature, resolution in the corners is 2.13 Å and at the edges 2.57 Å, diffraction spots are visible beyond 2.3 Å resolution. C. Mn $K\beta_{1,3}$ XES from PS II crystals in the 2F state. The spectrum was collected from an experimental run in which 1466 crystal hits were identified in the diffraction detector.

a PS II crystal obtained from injecting a suspension of crystals, with a size distribution of 10–20 micrometers, using an electrofocusing liquid injector with a liquid shield flow (coMESH).[32] An XES spectrum obtained from more than 1400 crystal hits is shown in Fig. 4C, thus providing a diagnostic use of spectroscopy for the chemical state and the intactness of the sample, while collecting XRD data. For XES, each crystal provides only a weak XES signal and accumulation over many individual shots is necessary to obtain an interpretable spectrum for samples like metalloenzymes in which the metal concentration is very low. The characteristics of the individual crystal samples determine the maximum achievable concentration of crystals, and these are often limited by factors like aggregation and sedimentation of the crystals leading to clogging of the injection setup. The necessary reduction of the particle concentration leads then to a lower hit rate, *e.g.* often only a few percent in liquid injector-based crystallography. Therefore the average concentration in the sample volume probed by the X-ray beam is lower, although the intrinsic metal concentration in a protein crystal is generally higher compared to a solution of the same protein. In the case of PS II, the concentration difference between crystals and solution samples is about six times but the hit rate difference is a factor of 20–50. In addition, the jet volume probed by the X-ray beam is not always filled completely by a crystal, giving rise to another reduction in signal level from crystals compared to solution samples. This leads to an average improvement in signal rates for PS II solution samples over crystals by a factor of ~10. In PS II solution samples, with a Mn concentration of ~1 mM and a path length of 10 micrometers, a total of 600 Mn Kβ fluorescence photons are emitted per shot (40 fs XFEL pulses, 9.5 keV excitation, with 2 mJ per pulse) out of which 8 photons can be theoretically collected by our spectrometer (solid angle of 1.3%).[4] Therefore, the spectroscopic signal needs to be averaged over many shots. In practice, a collection time of 20–30 minutes at 120 Hz X-ray repetition rate for a $K\beta_{1,3}$ spectrum from solutions was necessary to achieve sufficient S/N. To obtain a spectrum of similar quality from a crystal suspension, collection times of 4–5 hours would be necessary. To study the time-evolution of chemical states of the metal cluster, based on small changes in the spectral shape and energy position of the Mn emission, high-quality $K\beta_{1,3}$ XES spectra are required. This is only achievable in a reasonable amount of measurement time with solution samples as they provide a more efficient way of collecting data (100% sample hit rate).

Another point that needs to be taken into account for the simultaneous data collection of diffraction and spectroscopic signal is that both methods are sensitive to different degrees of sample damage. We have shown that XES data, free of electronic damage, can be collected at XFELs together with the XRD data[3,5] under the conditions that were used for these studies (<50 fs pulse duration, spot size of $\sim 1.5 \times 1.5\ \mu m^2$, $3–6 \times 10^{11}$ photons per pulse at energies between 7 and 9.5 keV, equivalent to an X-ray dose of 50–300 MGy). However, X-ray induced changes to the electronic structure will become visible when the X-ray dose increases (*i.e. via* the increase of incoming X-ray photons/shot, the use of a lower X-ray energy leading to a higher cross section, a better focused beam, *etc.*), and the effect will become noticeable in the XES more easily than in the XRD. Such effects could for example be caused by multiple excitation processes at the same site (either at the same atom or in the same electronically coupled moiety, for example at two neighboring Mn in the Mn_4CaO_5 cluster in PS II). Such effects may not be readily apparent in the diffraction data until the atomic positions are affected, especially as the time scales for electronic and atomic motions are very different. Recent studies, both in the soft- and hard X-ray regime, indicated the onset of such effects under certain conditions, for example in the oxygen K-edge of water,[33] in the Fe L-edge RIXS spectrum of $Fe(CO)_5$ in solution,[34] and in the XRD data of ferredoxin obtained using a very high dose (30 GGy) at the 100 nm focus instrument at CXI, LCLS.[35]

3. Towards studying inorganic electron transfer chromophores

The functionality of metalloenzymes arises as a consequence of the spin state, electronic structure, and ligand environment of the metal catalytic center, cofactor arrangement for electron and proton transfer in both space and time, as well as the pigment–pigment interactions at interfaces. To be useful for driving a multi-electron catalyst, the excited state has to be sufficiently long-lived to allow for electron transport from one pigment/cofactor to another. There are some promising synthetic candidates for this type of photodynamics based on a multimetallic unit anchored on a silica nanoporous surface.[36–40] While the back electron transfer rate and slower changes following the initial MMCT excitation have been studied by various techniques, very little is known about the ultrafast steps that immediately follow the light excitation. Besides the MMCT there are excited state dynamics, structural rearrangements, and, possibly, spin crossover processes that are critical for the function of these and related systems.

In the case of the TiOMn system ($Ti^{IV}OMn^{II} \rightarrow Ti^{III}OMn^{III}$) (Fig. 5A and B), an excited state lifetime of 1.8 microseconds was established at room temperature (monitored by the dynamics of the ground state MMCT depletion and the Ti^{III} excited state absorption). The unusually slow back electron transfer (k_{BET2}) enables the photocatalytic activity, yet the photochemical yield is determined by competition between two ultrafast processes shown in Fig. 5C, namely ultrafast intersystem crossing (k_{ISC}) *versus* ultrafast back electron transfer in the initially excited S = 5/2 MMCT state (k_{BET1}) following photo excitation. It is the S = 3/2 state with the long lifetime of 1.8 microseconds that gives rise to the observed reactivity. In order to improve and optimize the photocatalytic yield, an understanding of the factors that determine the branching between ultrafast (spin allowed) back electron transfer in the S = 5/2 state *versus* intersystem crossing to the S = 3/2 excited state is essential. Although some of the optical methods can be applied in the femtosecond time regime, the element-specific methods based on X-ray techniques are useful for understanding the local geometric and electronic structure of the metals. At synchrotron radiation sources, X-ray studies have been generally limited to the ~100 ps temporal resolution[41] and therefore XFEL X-ray pulses are required for capturing faster phenomena.

Using wavelength-dispersive XES, we have developed a method to probe multiple elements based on an ultrafast optical pump and X-ray probe to simultaneously detect signals from both metal centers in a time-resolved manner (Fig. 6A and B).[42] The setup, shown in Fig. 6A, allows the time-points to

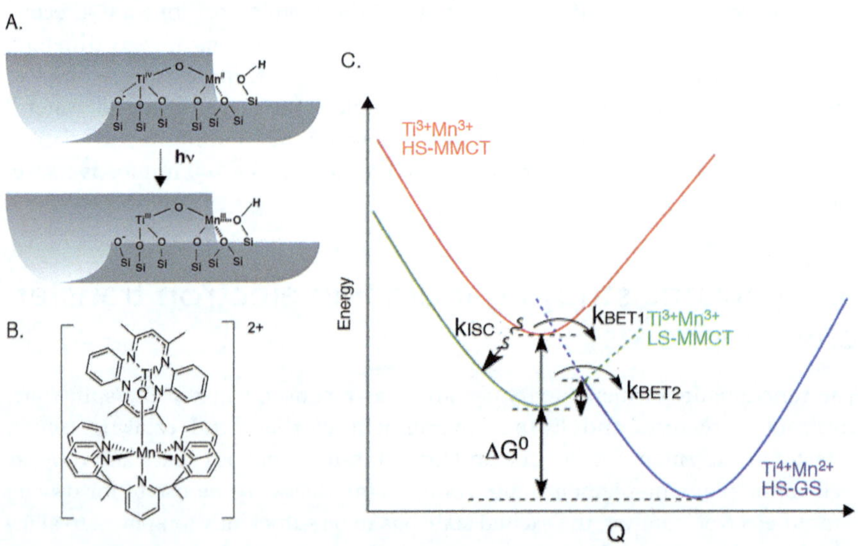

Fig. 5 (A) Schematic of the Ti–O–Mn complex on a mesoporous silica substrate. (B) Structure of a Ti–O–Mn bridged complex. (C) Proposed energy level diagram for the photo-induced metal-to-metal charge transfer (MMCT) processes that determine the photocatalytic yields for the $TiOMn^{II}$ charge transfer unit. k_{BET1} and k_{ISC} are ultrafast, k_{BET2} is the 1.8 μs process.

be accessed within the accuracy of the X-ray jitter and the laser timing. This setup allows the measurement of time-points starting from 300 femtoseconds, which is limited by the accuracy of the timing between the X-ray probe and laser pump.

A multi-crystal von Hamos spectrometer combined with two position-sensitive detectors was used for the data collection of the photo-excitation of

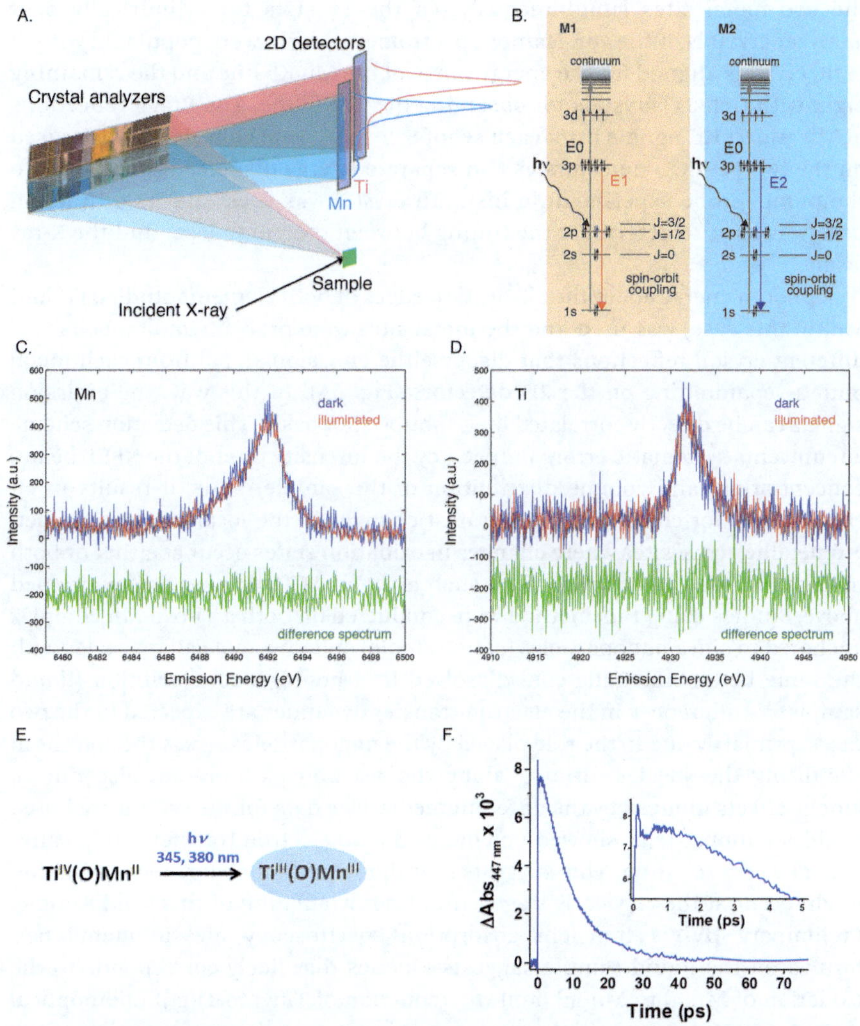

Fig. 6 (A) The schematic shows the method used for the simultaneous collection of the emission spectra from two elements, Mn and Ti, using an energy dispersive X-ray spectrometer and a 2D detector. (B) Energy level diagram showing the excitation, E_1 and E_2, and Kβ emission processes for Mn and Ti. (C) Mn $K\beta_{1,3}$ spectra of the Mn/Ti molecular compound in solution in the dark (blue) and illuminated (1.2 ps) (red). (D) Ti $K\beta_{1,3}$ spectra of the Mn/Ti molecular compound in solution in the dark (blue) and illuminated (1.2 ps) (red). The difference spectra are shown in green. (E) Expected changes of the metal oxidation states during the MMCT. (F) Transient absorption spectroscopy of Mn/Ti molecular compound in solution.

the MnTi system. The data was collected at Hutch 3 in SACLA (SPring-8, Japan), using 7 keV incident X-rays with a pulse duration of ~10 femtoseconds and a beamsize of 50 micrometers. A $Ti^{IV}OMn^{II}$ molecular complex in acetonitrile solution (5 mM Mn and Ti) and a $Ti^{IV}OMn^{II}$ complex anchored to silica nanoparticles (solid) was used for the data collection. The solid sample was mounted on a solid-target stage and was moved to a fresh sample spot shot-by-shot. The liquid sample was injected with a liquid injector (100 μm I.D.). Mn and Ti Kβ emission lines are used to detect the electronic structural changes of the two metal sites simultaneously. Of the 16 slots for cylindrically bent analyzer crystals of the von Hamos spectrometer, eight were populated with Si (440) crystals aligned for the energy range of the Mn Kβ line and the remaining eight with Ge(331) crystals for observing the Ti Kβ line. The Bragg reflections of Mn and Ti Kβ signals from each set of 8 crystals were collected, and focused on the two MPCCD detectors as two separate energy dispersed lines. Prior to the pump–probe experiment, a bismuth crystal was used and its diffraction peak was used to determine the timing between the pump laser and the X-ray laser.

A photon energy above the absorption edges of both elements studied (Ti and Mn, in this case) was used and the metal sites were probed simultaneously by different crystal reflections that disperse the emission signal from each metal onto a separate line on the 2D detectors (Fig. 6A). In this way, the emission signals can be directly correlated on a shot-by-shot basis. This detection scheme circumvents systematic errors induced by the intensity jitter of the XFEL beam, concentration, and volume distribution of the sample. Thus, it results in an efficient way for characterizing ultrafast changes in the local structure of heterometallic active sites where changes in oxidation states occur at either or both metal sites upon light irradiation, such as in the Ti(O)Mn system mentioned above. Studies using this method were conducted on both a Ti-oxo-Mn complex anchored to silica nanoparticles (solid sample) and on a molecular species with the same heterobimetallic core dissolved in a homogeneous solution (liquid sample).[43] Differences in the electron transfer dynamics are expected in the two cases, primarily due to the role played by the nanoparticles *versus* the solvent in stabilizing the species forming along the reaction pathway but also due to kinetic effects induced by the different media. The data for the system anchored to silica nanoparticles showed no change due to electron transfer ~10 ps after optical laser excitation. This suggests that the important charge recombination might occur at the very early stages after laser irradiation of this solid sample. Preliminary UV-Vis transient absorption spectroscopy measurements performed on the liquid sample suggests kinetics that likely corresponds to the oxidation of Mn(II) to Mn(III) (and the reduction of Ti(IV) to Ti(III)) upon optical laser excitation of the initial sample, showing the maximum formation of the product around 1 ps after laser irradiation (Fig. 6E and F). However, data from the pump–probe experiment performed at the SACLA XFEL collected after 1.2 ps does not show clear changes of the metal oxidation states (Fig. 6C and D). Likely reasons for this are the current data quality or the population of the excited state being under the detection limit. More detailed data analysis including shorter delay times is necessary for both the silica-anchored and the Ti(O)Mn liquid samples, and such a study is underway.

4. Conclusion and future direction

We have developed XFEL-based methods for studying the structure of complex biological metalloenzymes and for studying the electronic structural dynamics of both biological and inorganic systems. The simultaneous detection of XRD and XES from metalloenzymes allows the investigation of both the geometric and electronic structure of enzymatic reactions in real time at physiological temperatures, and to ensure the integrity and intermediate state characterization of short-lived intermediate states. The method of simultaneous detection of the X-ray emission spectra from multiple elements after optical excitation has great potential for probing the dynamics of MMCT or charge-separated states where charge transfer occurs on ultrafast time scales. This method will help with understanding the fundamentals that govern multidimensionally (time and space) controlled chemistry, by detecting competing pathways of energy loss and observing energy transfer processes. Femtosecond X-ray spectroscopy is a unique tool that provides access to these insights and can guide design improvements for maximizing branching in favor of the photochemically active charge-separated or MMCT states. The knowledge gained from these studies will not only help the understanding of currently known biochemical processes, but may also lead to the improvement/design of novel catalytic systems.

Besides the various advantages that the use of XFELs has brought to the field of biology and chemistry by using the fs X-ray probes to interrogate ps and sub ps timescales and outrun damage processes that otherwise often hinder X-ray studies at conventional synchrotron sources, there are some limitations to this approach. The most important point is that further increasing the X-ray dose may not help to get the intact form of the XRD and spectroscopic data, due to the limitation that comes from electronic damage to the samples in these high-flux regimes. Instead, increasing the X-ray repetition rate will be a huge advantage, in particular, for spectroscopic data collection at time delays that are large compared to the X-ray and laser timing jitter, as in these cases often it is not necessary to collect shot-by-shot information and the signal can be averaged over many shots without the need of a high repetition rate detection system. Nevertheless, at very high repetition rates, one has to consider the effect of the shock wave created by an X-ray pulse hitting the sample on the neighboring sample volume both in a continuous jet or a stream of droplets (see ref. 44) and high speed sample delivery schemes have to be designed accordingly. Another approach will be to use stimulated X-ray emission processes to enhance the signal intensity. The possibility of a non linear X-ray spectroscopy approach has been postulated theoretically (*e.g.* ref. 45–47) and recently the induction of stimulated emission processes was demonstrated experimentally for neon gas in the soft X-ray regime[48,49] and a Cu foil in the hard X-ray regime.[50] When established, this approach will be truly beneficial for collecting data from dilute samples or less-probable transitions. Whether the stimulated process is applicable to more dilute systems, in which the shot-by-shot spectroscopic information is currently not accessible due to the limited S/N ratio, is still an open question.

Author contributions

Data collection at XFELs LCLS/SACLA – R. A.-M., K. A., U. B., R. C., F. D. F., H. F., D. I., T. Ka., J. F. K., T. Kr., Y. K., K. M., K. N., T. N., C. S., Y. S., T. T., K. U., T.-C. W., V. K. Y., J. Y. Sample injector – J. F. K., C. S., F. D. F., S. G., R. G. S. Emission spectrometer – R.-A. M., U. B., D. S., T.-C. W. SACLA scientific support – T. K., T. T., M. Y. Samples Mn–Ti – W. W. W., E. G., B. A. M., H. M. F. Samples PS II – M. I., A. Z., R. C. Optical spectroscopy – C. S., J. K. C. X-ray emission spectra analysis – T. Kr., F. D. F., K. M. X-ray diffraction data analysis – N. K. S., A. S. B., I. D. Y. Experiment design and writing – T. K., J. F. K., R.-A. M., U. B., H. M. F., W. W. W., K. N., J. M., K. U., M. Y., A. Z., V. K. Y., J. Y.

Acknowledgements

The authors acknowledge research support from the Director, Office of Science, Office of Basic Energy Sciences, Division of Chemical Sciences, Geosciences, and Biosciences of the Department of Energy (DOE) under contract DE-AC02-05CH11231 (J. Y., V. K. Y., and H. F.), the NIH grants GM116423 (F. D. F.), GM55302 (V. K. Y.), GM095887 (N. K. S.), GM102520 (N. K. S.), and GM110501 (J. Y.), and the Human Frontier Research grant RGP0063/2013 (J. Y., U. B., A. Z.). Use of the Linac Coherent Light Source (LCLS), SLAC National Accelerator Laboratory, is supported by the U.S. Department of Energy, Office of Science, Office of Basic Energy Sciences under contract No. DE-AC02-76SF00515. H. F., K. M., K. N. and K. U. acknowledge research support by the X-ray Free Electron Laser Utilization Research Project and the X-ray Free Electron Laser Priority Strategy Program of the Ministry of Education, Culture, Sports, Science and Technology of Japan (MEXT). H. F., K. N. and K. U. are supported by the Japan Society for the Promotion of Science (JSPS), H. F. and K. U. by the IMRAM project, and K. N. by the Cooperative Research Program of the "Network Joint Research Center for Materials and Devices" of Japan. The DFG-Cluster of Excellence "UniCat" coordinated by the Technische Universität Berlin and Sfb1078, TP A5 (A. Z.); the Solar Fuels Strong Research Environment (Umeå University), the Artificial Leaf Project (K&A Wallenberg Foundation 2011.0055) and Energimyndigheten (36648-1) (J. M.) are acknowledged for supporting this project. The Mn/Ti data collection was done at SACLA (Japan), under proposal numbers 2014A8013 and 2015A8007. The transient absorption spectra were taken at the Joint Center for Artificial Photosynthesis (JCAP), DOE Energy Innovation Hub under award no. DE-SC0004993. We are grateful to our present and former group members and all our collaborators, who contributed to the research from our group presented in this review. We especially like to thank the CXI beamline scientists at the LCLS, Sébastien Boutet, Garth Williams, and Mengning Liang and all the CXI support staff. We thank Raymond Sierra, Hartawan Laksmono, and Claudiu Stan for help with the MESH injector. We thank Gabriella Carini, Sven Herrmann, and Jack Pines from LCLS for support with data collection at SACLA. We also thank the excellent support staff at LCLS and SACLA, where the XFEL experiments were conducted and at SSRL (BL 6-2) and ALS (BL 5.0.2), where the synchrotron experiments were conducted.

Notes and references

1 M. Suga, F. Akita, K. Hirata, G. Ueno, H. Murakami, Y. Nakajima, T. Shimizu, K. Yamashita, M. Yamamoto, H. Ago and J. R. Shen, *Nature*, 2015, **517**, 99–103.
2 Y. Umena, K. Kawakami, J. R. Shen and N. Kamiya, *Nature*, 2011, **473**, 55–60.
3 R. Alonso-Mori, J. Kern, R. J. Gildea, D. Sokaras, T. C. Weng, B. Lassalle-Kaiser, R. Tran, J. Hattne, H. Laksmono, J. Hellmich, C. Glockner, N. Echols, R. G. Sierra, D. W. Schafer, J. Sellberg, C. Kenney, R. Herbst, J. Pines, P. Hart, S. Herrmann, R. W. Grosse-Kunstleve, M. J. Latimer, A. R. Fry, M. M. Messerschmidt, A. Miahnahri, M. M. Seibert, P. H. Zwart, W. E. White, P. D. Adams, M. J. Bogan, S. Boutet, G. J. Williams, A. Zouni, J. Messinger, P. Glatzel, N. K. Sauter, V. K. Yachandra, J. Yano and U. Bergmann, *Proc. Natl. Acad. Sci. U. S. A.*, 2012, **109**, 19103–19107.
4 R. Alonso-Mori, J. Kern, D. Sokaras, T. C. Weng, D. Nordlund, R. Tran, P. Montanez, J. Delor, V. K. Yachandra, J. Yano and U. Bergmann, *Rev. Sci. Instrum.*, 2012, **83**, 073114.
5 J. Kern, R. Alonso-Mori, R. Tran, J. Hattne, R. J. Gildea, N. Echols, C. Glockner, J. Hellmich, H. Laksmono, R. G. Sierra, B. Lassalle-Kaiser, S. Koroidov, A. Lampe, G. Y. Han, S. Gul, D. DiFiore, D. Milathianaki, A. R. Fry, A. Miahnahri, D. W. Schafer, M. Messerschmidt, M. M. Seibert, J. E. Koglin, D. Sokaras, T. C. Weng, J. Sellberg, M. J. Latimer, R. W. Grosse-Kunstleve, P. H. Zwart, W. E. White, P. Glatzel, P. D. Adams, M. J. Bogan, G. J. Williams, S. Boutet, J. Messinger, A. Zouni, N. K. Sauter, V. K. Yachandra, U. Bergmann and J. Yano, *Science*, 2013, **340**, 491–495.
6 J. Hattne, N. Echols, R. Tran, J. Kern, R. J. Gildea, A. S. Brewster, R. Alonso-Mori, C. Glockner, J. Hellmich, H. Laksmono, R. G. Sierra, B. Lassalle-Kaiser, A. Lampe, G. Han, S. Gul, D. DiFiore, D. Milathianaki, A. R. Fry, A. Miahnahri, W. E. White, D. W. Schafer, M. M. Seibert, J. E. Koglin, D. Sokaras, T. C. Weng, J. Sellberg, M. J. Latimers, P. Glatzel, P. H. Zwart, R. W. Grosse-Kunstleve, M. J. Bogan, M. Messerschmidt, G. J. Williams, S. Boutet, J. Messinger, A. Zouni, J. Yano, U. Bergmann, V. K. Yachandra, P. D. Adams and N. K. Sauter, *Nat. Methods*, 2014, **11**, 545–548.
7 R. Mitzner, J. Rehanek, J. Kern, S. Gul, J. Hattne, T. Taguchi, R. Alonso-Mori, R. Tran, C. Weniger, H. Schroder, W. Quevedo, H. Laksmono, R. G. Sierra, G. Y. Han, B. Lassalle-Kaiser, S. Koroidov, K. Kubicek, S. Schreck, K. Kunnus, M. Brzhezinskaya, A. Firsov, M. P. Minitti, J. J. Turner, S. Moeller, N. K. Sauter, M. J. Bogan, D. Nordlund, W. F. Schlotter, J. Messinger, A. Borovik, S. Techert, F. M. F. de Groot, A. Fohlisch, A. Erko, U. Bergmann, V. K. Yachandra, P. Wernet and J. Yano, *J. Phys. Chem. Lett.*, 2013, **4**, 3641–3647.
8 R. Alonso-Mori, D. Sokaras, D. L. Zhu, T. Kroll, M. Chollet, Y. P. Feng, J. M. Glownia, J. Kern, H. T. Lemke, D. Nordlund, A. Robert, M. Sikorski, S. Song, T. C. Weng and U. Bergmann, *J. Synchrotron Radiat.*, 2015, **22**, 612–620.
9 B. Kok, B. Forbush and M. Mcgloin, *Photochem. Photobiol.*, 1970, **11**, 457–475.
10 *Photosystem II: The Light-Driven Water: Plastoquinone Oxidoreductase*, ed. T. Wydrzynski and S. Satoh, Springer, Dordrecht, 2005.
11 R. J. Debus, *Coord. Chem. Rev.*, 2008, **252**, 244–258.

12 J. Messinger, T. Noguchi and J. Yano, Photosynthetic O_2 evolution, in *Molecular solar fuels*, ed. T. J. Wydrzynski and W. Hillier, RSC, London, 2012, ch. 7, pp. 163–207.

13 H. N. Chapman, P. Fromme, A. Barty, T. A. White, R. A. Kirian, A. Aquila, M. S. Hunter, J. Schulz, D. P. DePonte, U. Weierstall, R. B. Doak, F. R. N. C. Maia, A. V. Martin, I. Schlichting, L. Lomb, N. Coppola, R. L. Shoeman, S. W. Epp, R. Hartmann, D. Rolles, A. Rudenko, L. Foucar, N. Kimmel, G. Weidenspointner, P. Holl, M. N. Liang, M. Barthelmess, C. Caleman, S. Boutet, M. J. Bogan, J. Krzywinski, C. Bostedt, S. Bajt, L. Gumprecht, B. Rudek, B. Erk, C. Schmidt, A. Homke, C. Reich, D. Pietschner, L. Struder, G. Hauser, H. Gorke, J. Ullrich, S. Herrmann, G. Schaller, F. Schopper, H. Soltau, K. U. Kuhnel, M. Messerschmidt, J. D. Bozek, S. P. Hau-Riege, M. Frank, C. Y. Hampton, R. G. Sierra, D. Starodub, G. J. Williams, J. Hajdu, N. Timneanu, M. M. Seibert, J. Andreasson, A. Rocker, O. Jonsson, M. Svenda, S. Stern, K. Nass, R. Andritschke, C. D. Schroter, F. Krasniqi, M. Bott, K. E. Schmidt, X. Y. Wang, I. Grotjohann, J. M. Holton, T. R. M. Barends, R. Neutze, S. Marchesini, R. Fromme, S. Schorb, D. Rupp, M. Adolph, T. Gorkhover, I. Andersson, H. Hirsemann, G. Potdevin, H. Graafsma, B. Nilsson and J. C. H. Spence, *Nature*, 2011, **470**, 73–U81.

14 E. F. Garman and M. Weik, *J. Synchrotron Radiat.*, 2015, **22**, 195–200.

15 J. Yano, J. Kern, K. D. Irrgang, M. J. Latimer, U. Bergmann, P. Glatzel, Y. Pushkar, J. Biesiadka, B. Loll, K. Sauer, J. Messinger, A. Zouni and V. K. Yachandra, *Proc. Natl. Acad. Sci. U. S. A.*, 2005, **102**, 12047–12052.

16 K. D. Daughtry, Y. Xiao, D. Stoner-Ma, E. Cho, A. M. Orville, P. Liu and K. N. Allen, *J. Am. Chem. Soc.*, 2012, **134**, 2823–2834.

17 K. G. Sigfridsson, P. Chernev, N. Leidel, A. Popovic-Bijelic, A. Graslund and M. Haumann, *J. Biol. Chem.*, 2013, **288**, 9648–9661.

18 J. S. Fraser, H. van den Bedem, A. J. Samelson, P. T. Lang, J. M. Holton, N. Echols and T. Alber, *Proc. Natl. Acad. Sci. U. S. A.*, 2011, **108**, 16247–16252.

19 D. H. Juers and B. W. Matthews, *J. Mol. Biol.*, 2001, **311**, 851–862.

20 R. F. Tilton, J. C. Dewan and G. A. Petsko, *Biochemistry*, 1992, **31**, 2469–2481.

21 S. Boutet and G. J. Williams, *New J. Phys.*, 2010, **12**, 035024.

22 M. N. Liang, G. J. Williams, M. Messerschmidt, M. M. Seibert, P. A. Montanez, M. Hayes, D. Milathianaki, A. Aquila, M. S. Hunter, J. E. Koglin, D. W. Schafer, S. Guillet, A. Busse, R. Bergan, W. Olson, K. Fox, N. Stewart, R. Curtis, A. A. Miahnahri and S. Boutet, *J. Synchrotron Radiat.*, 2015, **22**, 514–519.

23 J. Kern, R. Tran, R. Alonso-Mori, S. Koroidov, N. Echols, J. Hattne, M. Ibrahim, S. Gul, H. Laksmono, R. G. Sierra, R. J. Gildea, G. Han, J. Hellmich, B. Lassalle-Kaiser, R. Chatterjee, A. S. Brewster, C. A. Stan, C. Glockner, A. Lampe, D. DiFiore, D. Milathianaki, A. R. Fry, M. M. Seibert, J. E. Koglin, E. Gallo, J. Uhlig, D. Sokaras, T. C. Weng, P. H. Zwart, D. E. Skinner, M. J. Bogan, M. Messerschmidt, P. Glatzel, G. J. Williams, S. Boutet, P. D. Adams, A. Zouni, J. Messinger, N. K. Sauter, U. Bergmann, J. Yano and V. K. Yachandra, *Nat. Commun.*, 2014, **5**, 4371.

24 G. Renger, *Physiol. Plant.*, 1997, **100**, 828–841.

25 M. R. Razeghifard and R. J. Pace, *Biochim. Biophys. Acta, Bioenerg.*, 1997, **1322**, 141–150.

26 M. Haumann, P. Liebisch, C. Muller, M. Barra, M. Grabolle and H. Dau, *Science*, 2005, **310**, 1019–1021.

27 H. J. Eckert and G. Renger, *Photochem. Photobiol.*, 1980, **31**, 501–511.

28 R. Krivanek, J. Kern, A. Zouni, H. Dau and M. Haumann, *Biochim. Biophys. Acta, Bioenerg.*, 2007, **1767**, 520–527.

29 A. Guskov, A. Gabdulkhakov, M. Broser, J. Kern, A. Zouni and W. Saenger, *J. Biomol. Struct. Dyn.*, 2009, **26**, 865.

30 R. de Wijn and H. J. van Gorkom, *Biochemistry*, 2001, **40**, 11912–11922.

31 G. Ananyev and G. C. Dismukes, *Photosynth. Res.*, 2005, **84**, 355–365.

32 R. G. Sierra, C. Gati, H. Laksmono, E. H. Dao, S. Gul, F. Fuller, J. Kern, R. Chatterjee, M. Ibrahim, A. S. Brewster, I. D. Young, T. Michels-Clark, A. Aquila, M. Liang, M. S. Hunter, J. E. Koglin, S. Boutet, E. A. Junco, B. Hayes, M. J. Bogan, C. Y. Hampton, E. V. Puglisi, N. K. Sauter, C. A. Stan, A. Zouni, J. Yano, V. K. Yachandra, S. M. Soltis, J. D. Puglisi and H. DeMirci, *Nat. Methods*, 2016, **13**, 59–62.

33 S. Schreck, M. Beye, J. A. Sellberg, T. McQueen, H. Laksmono, B. Kennedy, S. Eckert, D. Schlesinger, D. Nordlund, H. Ogasawara, R. G. Sierra, V. H. Segtnan, K. Kubicek, W. F. Schlotter, G. L. Dakovski, S. P. Moeller, U. Bergmann, S. Techert, L. G. M. Pettersson, P. Wernet, M. J. Bogan, Y. Harada, A. Nilsson and A. Föhlisch, *Phys. Rev. Lett.*, 2014, **113**, 153002.

34 P. Wernet, K. Kunnus, I. Josefsson, I. Rajkovic, W. Quevedo, M. Beye, S. Schreck, S. Grubel, M. Scholz, D. Nordlund, W. Zhang, R. W. Hartsock, W. F. Schlotter, J. J. Turner, B. Kennedy, F. Hennies, F. M. de Groot, K. J. Gaffney, S. Techert, M. Odelius and A. Fohlisch, *Nature*, 2015, **520**, 78–81.

35 K. Nass, L. Foucar, T. R. Barends, E. Hartmann, S. Botha, R. L. Shoeman, R. B. Doak, R. Alonso-Mori, A. Aquila, S. Bajt, A. Barty, R. Bean, K. R. Beyerlein, M. Bublitz, N. Drachmann, J. Gregersen, H. O. Jonsson, W. Kabsch, S. Kassemeyer, J. E. Koglin, M. Krumrey, D. Mattle, M. Messerschmidt, P. Nissen, L. Reinhard, O. Sitsel, D. Sokaras, G. J. Williams, S. Hau-Riege, N. Timneanu, C. Caleman, H. N. Chapman, S. Boutet and I. Schlichting, *J. Synchrotron Radiat.*, 2015, **22**, 225–238.

36 H. Frei, *Chimia*, 2009, **63**, 721–730.

37 H. S. Soo, M. L. Macnaughtan, W. W. Weare, J. Yano and H. Frei, *J. Phys. Chem. C*, 2011, **115**, 24893–24905.

38 X. N. Wu, W. W. Weare and H. Frei, *Dalton Trans.*, 2009, 10114.

39 B. A. McClure and H. Frei, *J. Phys. Chem. C*, 2014, **118**, 11601–11611.

40 W. Kim, B. A. McClure, E. Edri and H. Frei, *Chem. Soc. Rev.*, 2016, **45**, 3221–3243.

41 G. Vanko, P. Glatzel, V. T. Pham, R. Abela, D. Grolimund, C. N. Borca, S. L. Johnson, C. J. Milne and C. Bressler, *Angew. Chem., Int. Ed.*, 2010, **49**, 5910–5912.

42 S. Gul, J. W. D. Ng, R. Alonso-Mori, J. Kern, D. Sokaras, E. Anzenberg, B. Lassalle-Kaiser, Y. Gorlin, T.-C. Weng, P. H. Zwart, J. Z. Zhang, U. Bergmann, V. K. Yachandra, T. F. Jaramillo and J. Yano, *Phys. Chem. Chem. Phys.*, 2015, **17**, 8901–8912.

43 E. M. Goggins, T. T. Lekich, W. W. Weare, R. D. Sommer, M. A. Ribeiro and C. B. Pinheiro, *Eur. J. Inorg. Chem.*, 2016, **2016**, 1054–1059.

44 C. A. Stan, D. Milathianaki, H. Laksmono, R. G. Sierra, T. A. McQueen, M. Messerschmidt, G. J. Williams, J. E. Koglin, T. J. Lane, M. J. Hayes,

S. A. H. Guillet, M. Liang, A. L. Aquila, P. R. Willmott, J. S. Robinson, K. L. Gumerlock, S. Botha, K. Nass, I. Schlichting, R. L. Shoeman, H. A. Stone and S. Boutet, *Nat. Phys.*, 2016, DOI: 10.1038/nphys3779.
45 S. Tanaka and S. Mukamel, *Phys. Rev. Lett.*, 2002, **89**, 043001.
46 V. Kimberg and N. Rohringer, *Phys. Rev. Lett.*, 2013, **110**, 043901.
47 Y. Zhang, J. D. Biggs and S. Mukamel, *ChemPhysChem*, 2015, **16**, 2006–2014.
48 N. Rohringer, D. Ryan, R. A. London, M. Purvis, F. Albert, J. Dunn, J. D. Bozek, C. Bostedt, A. Graf, R. Hill, S. P. Hau-Riege and J. J. Rocca, *Nature*, 2012, **481**, 488–491.
49 C. Weninger, M. Purvis, D. Ryan, R. A. London, J. D. Bozek, C. Bostedt, A. Graf, G. Brown, J. J. Rocca and N. Rohringer, *Phys. Rev. Lett.*, 2013, **111**, 233902.
50 H. Yoneda, Y. Inubushi, K. Nagamine, Y. Michine, H. Ohashi, H. Yumoto, K. Yamauchi, H. Mimura, H. Kitamura, T. Katayama, T. Ishikawa and M. Yabashi, *Nature*, 2015, **524**, 446–449.

Faraday Discussions

PAPER

Imaging ultrafast excited state pathways in transition metal complexes by X-ray transient absorption and scattering using X-ray free electron laser source

Lin X. Chen,*[ab] Megan L. Shelby,[ab] Patrick J. Lestrange,[c] Nicholas E. Jackson,[ab] Kristoffer Haldrup,[d] Michael W. Mara,[b] Andrew B. Stickrath,[a] Diling Zhu,[e] Henrik Lemke,[e] Matthieu Chollet,[e] Brian M. Hoffman[b] and Xiaosong Li[c]

Received 16th April 2016, Accepted 3rd June 2016
DOI: 10.1039/c6fd00083e

This report will describe our recent studies of transition metal complex structural dynamics on the fs and ps time scales using an X-ray free electron laser source, Linac Coherent Light Source (LCLS). Ultrafast XANES spectra at the Ni K-edge of nickel(II) tetramesitylporphyrin (NiTMP) were measured for optically excited states at a timescale from 100 fs to 50 ps, providing insight into its sub-ps electronic and structural relaxation processes. Importantly, a transient reduced state Ni(I) (π, $3d_{x^2-y^2}$) electronic state is captured through the interpretation of a short-lived excited state absorption on the low-energy shoulder of the edge, which is aided by the computation of X-ray transitions for postulated excited electronic states. The observed and computed inner shell to valence orbital transition energies demonstrate and quantify the influence of the electronic configuration on specific metal orbital energies. A strong influence of the valence orbital occupation on the inner shell orbital energies indicates that one should not use the transition energy from 1s to other orbitals to draw conclusions about the d-orbital energies. For photocatalysis, a transient electronic configuration could influence d-orbital energies up to a few eV and any attempt to steer the reaction pathway should account for this to ensure that external energies can be used optimally in driving desirable processes. NiTMP structural evolution and the influence of the porphyrin macrocycle conformation on relaxation kinetics can be likewise inferred from this study.

[a]*Chemical Sciences and Engineering Division, Argonne National Laboratory, Lemont, Illinois, USA. E-mail: lchen@anl.gov*
[b]*Department of Chemistry, Northwestern University, Evanston, Illinois, USA. E-mail: l-chen@northwestern.edu*
[c]*Department of Chemistry, University of Washington, Seattle, WA 98195, USA*
[d]*Physics Department, Technical University of Denmark, 2800 Kgs. Lyngby, Denmark*
[e]*LCLS, SLAC National Laboratory, Menlo Park, CA 94025, USA*

Introduction

Transition metal complexes (TMCs) have been used broadly in photochemistry, especially in solar energy conversion processes[1–3] based on their versatility in light absorption, excited state redox properties as well as chemical tunability. One of the important criteria to determine if a TMC is suitable for solar energy conversion applications is the excited state lifetime, often the triplet state lifetime, while the excited states before thermalization with short lifetimes are not fully explored. Historically, quantum mechanical calculations on TMCs were limited which also hindered our understanding of the unthermalized and relatively short-lived excited state TMCs as well as their applications. An active frontier in solar energy research is in search of low cost TMCs from the first row transition metals to replace those of high cost and relatively rare noble metal elements in many applications.[4–6] Recent studies have shown that some singlet excited state TMCs with picosecond lifetimes could be applied to useful solar energy conversion processes if the charge separation/injection processes are fast enough to compete with other excited state decay pathways.[6–9] X-ray transient absorption (XTA) spectroscopy using synchrotron sources with approximately 100 ps time resolution has enhanced our understanding of TMS excited state structural dynamics significantly in the past two decades,[10–18] but many TMC excited state lifetimes are even shorter than 100 ps, which requires shorter pulsed X-ray sources, such as X-ray free electron lasers (XFELs) to study in addition to commonly used ultrafast laser spectroscopic methods. We have seen so far several examples of XTA from the new sources[9,19–28] and expect many exciting results will appear from these X-ray sources as well as table top X-ray sources with high harmonic generation with femtosecond time resolution in the coming years.

Here we report our recent studies of TMS structural dynamics using an XFEL source, the Linac Coherent Light Source (LCLS). Compared to synchrotron sources, X-ray absorption spectroscopic measurements face new challenges due to the limited spectral range accessible to cover the extended X-ray fine structure (EXAFS) region. Currently, X-ray absorption measurements rely on the spectral fluctuation of self-amplified spontaneous emission (SASE) process of the source, which, depending on the energy, can cover the 40–50 eV range near a central energy at the LCLS, sufficient for obtaining X-ray absorption near edge structure (XANES) spectra and X-ray emission (XES) spectra as shown in the literature. Therefore, the current femtosecond X-ray spectroscopy is focused on resolving excited state electronic structure while the nuclear geometry information is largely obtained by wide angle solution X-ray scattering.

Using this state-of-the-art source, we have obtained a number of XANES spectra at the Ni K-edge of nickel(II) tetramesitylporphyrin (NiTMP) and at Cu K-edge of copper(I) bis-dimethyl-phenanthroline $[Cu(dmp)_2]^+$ in dilute solutions on a timescale from 100 fs to 50 ps, providing insight into their sub-picosecond electronic and structural relaxation processes. We mainly present here the results on NiTMP and highlight the discovery of a transient reduced state Ni(I) (π, $3d_{x^2-y^2}$) electronic state that was only captured through femtosecond time resolution at the LCLS. Moreover, quantum mechanical calculations were carried out for both structures of the valence excited state and inner shell transitions which can be used to interpret our observed experimental results and gain insightful information on the electronic structures of the excited state as well as the interplay between energy levels of molecular orbitals with nuclear geometry changes. The observed and computed inner shell to valence orbital transition energies demonstrate and quantify the influence of electronic configuration on specific metal orbital energies. A strong influence of the valence orbital occupation on the inner shell orbital energies indicates that one should not use the transition energy from 1s to other orbitals to draw conclusions about the d-orbital energies. These studies have brought new insights into electron and nuclear motions as the result of the light excitation. The information obtained here will help in designing new TMC based artificial photosynthetic systems, such as photocatalysts for solar fuel generation.

Results and discussion

The results from experimental measurements and calculations described below will here establish the NiTMP excited state kinetic scheme of Fig. 1 and provide insight into how the evolving nuclear structure and electronic structure are interrelated as this decay occurs. This study allows us to capture the electronic configuration of the short-lived Ni(I) T′ state on the sub-ps timescale. The combined experimental and computational results that describe electronic transitions involving the inner shell electrons enable us to characterize properties that are not directly measurable using other ultrafast methods.

1. Excited state pathways on ultrafast time scales

NiTMP in solution has been chosen because it represents metalloporphyrins with open-shell transition metals whose excited states are short-lived and poorly understood due to the coupling between the molecular orbitals (MOs) involved in the initial (π, π*) excited state with metal d electron dominant MOs involved in light-activated redox reactions.[29] In our previous studies as well as others[30–36]

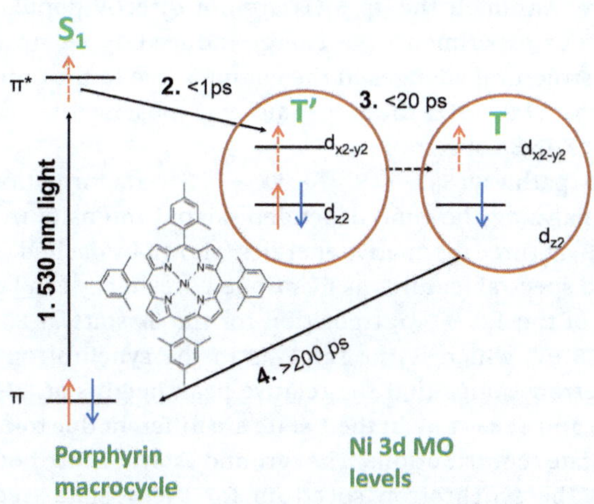

Fig. 1 NiTMP excited state pathways and decay time constants obtained from optical transient absorption spectroscopy.

using both optical and X-ray transient absorption spectroscopies, the excited state decay kinetics of NiTMP can be outlined by Fig. 1.

Briefly, the electronic configuration of Ni(II) ($3d^8$) in a nearly square-planar ground state S_0 has a vacant $3d_{x^2-y^2}$ MO and a fully occupied $3d_{z^2}$ MO.[37] The $S_0 \rightarrow S_1$ transitions in NiTMP can be induced by $\pi \rightarrow \pi^*$ transition at the Q-band centered at 530 nm. Within 300 fs of the excitation, S_1 is believed to convert, through energy transfer to an intermediate state T′ that then undergoes vibrational relaxation in less than 20 ps to a relaxed singlet or triplet state, T with a presumed[1,3] ($3d_{x^2-y^2}$, $3d_{z^2}$) configuration,[38] where $3d_{x^2-y^2}$ and $3d_{z^2}$ MOs are each singly occupied with an unsolved spin state.[39–42] The $T \rightarrow S_0$ transition takes place with a time constant of ~200 ps. The electronic configurations of these excited states have been assigned largely according to computational results that describe charge-transfer states of transition metalloporphyrins with unfilled d orbitals.[43] Using the Advanced Photon Source (APS), a synchrotron source at Argonne National Laboratory, we previously carried out XTA measurements for NiTMP where the Ni–N bond elongation and porphyrin planarization at the T state were determined. The identity of the T′ state, however, remained unknown then. The previous study also identified the electronic configuration at the T state with single occupation in the two highest energy 3d MOs, $3d_{x^2-y^2}$, $3d_{z^2}$.

From the LCLS results, Fig. 2 displays the time evolution of the NiTMP XANES spectra as a function of the delay time between the laser pump pulse and X-ray probe pulse, ranging from −5 ps to 100 ps covering the time window that the synchrotron studies could not access. As shown in Fig. 2, the ground state XANES spectrum collected at −5 ps time delay and the T state XANES at 100 ps time delay are largely the same as those obtained previously shown in the inset of Fig. 2A.[10,44] These peaks are assigned to the $1s \rightarrow 4p_z$ transition.[45] Spectral features at 8351 eV are contributions from multiple scattering resonances and transitions from Ni 1s to $4p_x$ and $4p_y$ orbitals hybridized with porphyrin N 2s orbitals. In the pre-edge region from 8330 to 8336 eV are the weak quadrupole-allowed $1s \rightarrow 3d$ transitions, which directly probe the vacant the 3d electronic occupations,[46] according to the Ni coordination geometry and electronic state. Although the 4p MOs are not directly populated by valence transitions in our experiment, the energy induced $1s \rightarrow 4p$ transitions are indicators for structural changes in the complex due to the correlations of the sensitivity of the TMC MOs to the geometry as observed in many complexes from the static XANES spectra.

The excited pathway $S_0 \rightarrow S_1 \rightarrow T' \rightarrow T$ transformation kinetics are extracted by analyzing the time dependent signal intensity in the difference spectra (Fig. 2B) at three distinctive energies (shown by dashed vertical lines) at which assigned spectral features as mentioned previously can be followed. The peak position of the $1s \rightarrow 4p_z$ transition for the S_0 state at 8339.2 eV clearly shifts to 8340.8 eV, which is the same as in the synchrotron results within experimental error. Notice that the relative peak heights at 8339 eV in the S_0 state spectrum and at 8341 eV in the T state are different due to compositions of the ground state contribution. The ground state contribution has been removed from the synchrotron spectrum for the T state (red curve in the inset of Fig. 2A), while it remains in the LCLS spectra due to the difficulty in resolving sequential kinetics within a few picoseconds after excitation and

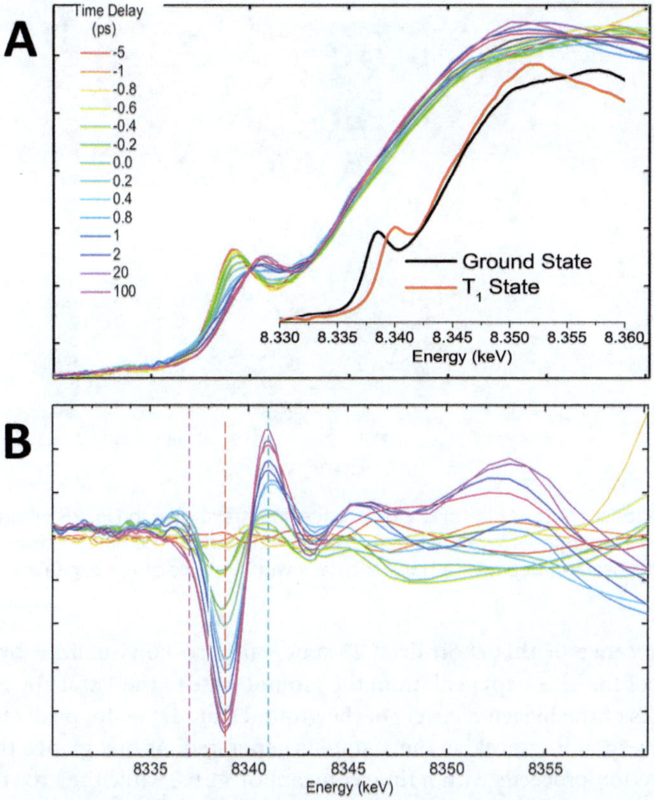

Fig. 2 (A). Ni K-edge XANES spectra of NiTMP obtained at the LCLS at different time delays and at APS (inset) for the ground and T state (~100 ps delay limited by the APS pulse duration); (B) difference XANES spectra with three dashed lines aligned with three X-ray photon energies at which the kinetic curves were collected.

distinguishing the spectral difference from more than two states. In comparison, the synchrotron data with ~100 ps time resolution only involves two states, S_0 and T with contributions from S_1 and T′ neglected because they cannot be resolved and have minimal contribution beyond a few picoseconds delay. In the white line region, a peak feature shifts from 8359 eV to 8353 eV. While the $\sigma^*_{x,y}$ orbitals should shift to lower energy in the excited state as the Ni–N distances expand and the hybridization between the Ni $4p_{x,y}$ and Ni 2s orbitals is weakened,[47] the $4p_z$ transition blue shift cannot be explained initially because this MO should have minimal influence for the valence transitions considered here induced by the laser excitation because main structural changes as observed in previous synchrotron results are Ni–N bond expansion in the x–y plane.

A different view from the difference spectra in Fig. 2B is presented in a contour plot in Fig. 3. A transient feature at 8337 eV rises within ~100 fs of the pump pulse, peaks to its maximum amplitude in 400–800 fs, and disappears within 2 ps (see the orange patch). From then on, the XANES difference reflects an increasing population of T state that rises to its maximum within ~10 ps after excitation.

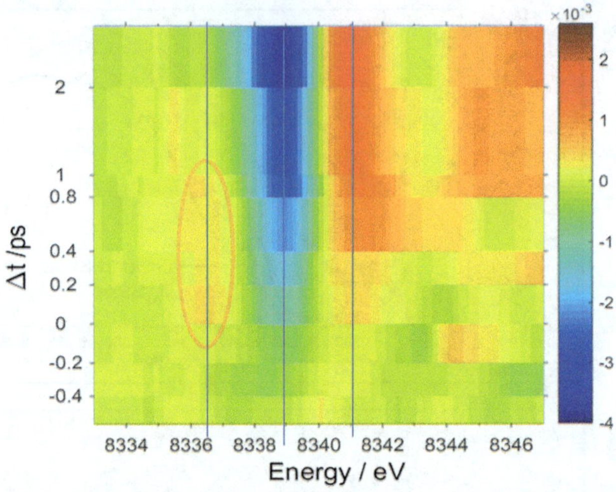

Fig. 3 The local contour plot of the difference spectra shown in Fig. 2B where the three lines mark the energies where the kinetics data are obtained. The orange patch at 8337 eV suggests the rise and decay of a transient state within ~2 ps of the excitation.

The presence of this short-lived T′ state can also be visualized by the time evolution of the 1s → $4p_z$ peak from the ground state to the T state by comparing the kinetics of the bleach recovery of the ground state 1s → $4p_z$ peak and the rise of the corresponding peak as the T state is generated. At first glance the ground state depletion proceeds with a time constant of <1 ps, while the growth of the T state seems to have a rapid rise within ~200 fs and then a slower rise to the maximum peak height. The apparent kinetics seem to be contrary to common sense that the ground state depletion must be faster than other processes. Other dynamic spectral features on the same timescale include an apparent delay between the rise of the white line peak feature at 8353 eV, which is an indication of the Ni–N bond elongation as seen in the T state, and the rise of the 1s → $4p_z$ the T state associated peak. Although at the high energy boundary of the XANES spectrum the X-ray photon flux drops, resulting in noise difference spectra, a growth of this feature in about 2 ps is clearly observed, suggesting a nuclear geometry change, most likely the Ni–N bond expansion due to the electronic configuration changes. The complications of sequential kinetics and a limited accessible spectral region prompted us to examine the kinetics beyond those that appear in the time dependent signals at the three energies mentioned above.

2. The sequential excited state decay kinetics

Averaged traces at the three characteristic energies were globally fit to a sequential kinetics model below:

$$S_1 \xrightarrow{k_1} T'_1 \xrightarrow{k_2} T_1 \xrightarrow{k_3} S_0 \tag{1}$$

where k_i ($i = 1$–3) is the rate constant and its inverse $1/k_i$ is τ_i, the time constant for the i-th step of the reaction. The sequential kinetics for each species therefore can be expressed by eqn (2) below:

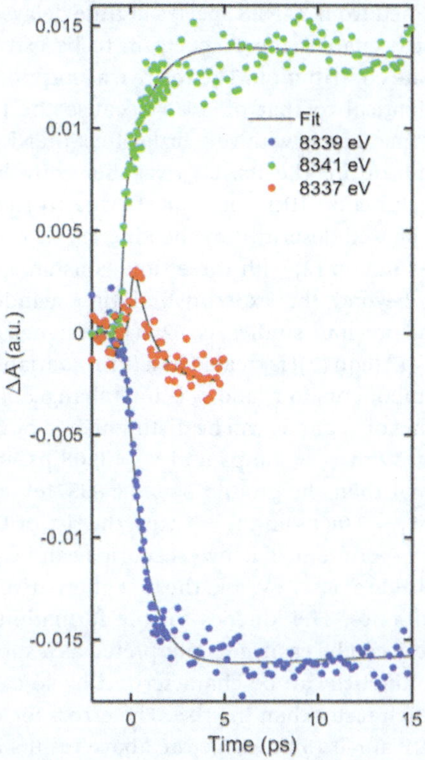

Fig. 4 Kinetics curves taken at the three X-ray photon energies as labelled by the vertical lines in Fig. 2 and 3.

$$
\begin{aligned}
[\mathrm{S}_1(t)] &= [\mathrm{S}_1(0)]\mathrm{e}^{-k_1 t} \\
[\mathrm{T'}_1(t)] &= \frac{k_1[\mathrm{S}_1(0)]}{k_2 - k_1}\left[\mathrm{e}^{-k_1 t} - \mathrm{e}^{-k_2 t}\right] \\
[\mathrm{T}_1(t)] &= \left[\frac{k_1 k_2[\mathrm{S}_1(0)]}{(k_2 - k_1)(k_3 - k_1)}\right]\mathrm{e}^{-k_1 t} + \left[\frac{k_1 k_2[\mathrm{S}_1(0)]}{(k_1 - k_2)(k_3 - k_2)}\right]\mathrm{e}^{-k_2 t} \\
&\quad + \left[\frac{k_1 k_2[\mathrm{S}_1(0)]}{(k_3 - k_1)(k_2 - k_3)}\right]\mathrm{e}^{-k_3 t} \\
[\mathrm{S}_0(t)] &= [\mathrm{S}_1(0)]\left[1 - \frac{k_2 k_3 \mathrm{e}^{-k_1 t}}{(k_2 - k_1)(k_3 - k_1)} - \frac{k_1 k_3 \mathrm{e}^{-k_2 t}}{(k_1 - k_2)(k_3 - k_2)} - \frac{k_1 k_2 \mathrm{e}^{-k_3 t}}{(k_3 - k_1)(k_3 - k_2)}\right]
\end{aligned}
\tag{2}
$$

The signals at different energies can therefore be described as eqn (3):

$$A_{\mathrm{total}}(E,t) = \sum_{i=0}^{3} A_i(E)P_i(t) \tag{3}$$

where i is the index of the states, $i = 0$–3, corresponding to S_0, S_1, T′, and T states, $A_i(E)$ is the relative absorption of the ith state at E, and $P_i(t)$ is the population of ith state at t. The relative absorption of the ground state S_0, $A_0(E)$ and that of the T state, $A_3(E)$ were obtained from XANES spectra at time delays with only the S_0 and T states present, which enables the T spectrum to be extracted by subtracting a fraction of the S_0 state contribution. The relative absorption $A_1(E)$ of the S_1 state is assumed to be identical to that of $A_0(E)$ because the (π, π^*) excitation is confined only on the macrocycle with very little effect on Ni. $A_2(E)$ for the T′ state was treated as a variable during the fits. A_{total} was then convoluted with a Gaussian instrument response function (IRF) of ~300 fs due to the time drift of X-ray pulses. The data can be well-described by the kinetics scheme in eqn (1) and the spectral contributions in eqn (2) with three time constants, 1.0 ps, 0.08 ps, and a much longer one beyond the experimental time window. This long time constant, based on the previous studies, is ~200 ps assigned as the lifetime of the T state.[44,48] Because $A_2(E)$ (eqn (2)) is treated as a fitting variable, the assignment of 1.0 ps and 0.08 ps components to τ_1 and τ_2 is interchangeable in eqn (2). The two alternative assignments of τ_1 and τ_2 can be distinguished by taking account of the T′ absorption signals. With $\tau_1 = 1.0$ ps and $\tau_2 = 0.08$ ps, it requires T′ to have a much stronger signal than the ground state at 8337 eV and little population accumulation. When $\tau_1 = 0.08$ ps and $\tau_2 = 1.0$ ps, the rise of T will be much slower than observed in the experiment. The two scenarios can be distinguished based on the T′ signal amplitude at 8337 eV, and the first alternative is preferred in order to fit the observed kinetics. The kinetics for the formation and decay of T′ by apparent rise and decay can be easily misinterpreted as a sub-ps rise and a few ps decay while the true kinetics can be characterized by so called reverse kinetics where the decay is much faster than the rise. The errors for τ_1 and τ_2 are defined by 0.5 fwhm of the IRF at ~0.15 ps, while the above results are best fitted in the global analysis. The IRF has been improved at the XPP station recently to be sub-100 fs, but it was not completely implemented at the time of our experiment.

Although the global fitting of kinetic traces at multiple X-ray photon energies can assist us to obtain rate constants for the sequential kinetics in eqn (2), with the relative contributions from the species the XANES signals could only be confidently determined with the assignment of the difference XANES spectral feature at the 1s → $4p_z$ transition peak of the T′ state. The global fitting under a kinetic scheme obtained from ultrafast optical transient absorption as well as the spectral assignments in terms of the inner shell to valent orbital transition energies from the calculations (shown below) are required to extract the true kinetics and capture the unknown species. The global fitting at three X-ray photon energies shows that while T′ has a small contribution to the total signal, its formation and decay lifetimes have a significant impact on the XAS signal kinetics. Although the kinetics of the T′ state absorption initially appear to be the result of an ultrafast generation rate $(\tau_1)^{-1}$ and a relatively slow decay rate $(\tau_2)^{-1}$ of the T′ species, the T′ population evolution can be described by the integrated rate law in eqn (2).

Relaxation through a charge transfer (CT) state that produces a transient Ni(I) intermediate has been previously hypothesized based on double-excitation optical methods.[49] Such a CT state has also been implicated as the route of excited porphyrin deactivation[50,51] by pump-probe photoelectron spectroscopy measurements which report fast time constants for the evolution of the NiOEP excited state absorption very similar to this work (100 fs and 1.2 ps). A short-lived T′ state is also compatible with previous work confirming the presence of the T state on the picosecond time scale,[52,53] which depicted the S_1 relaxation process as a simultaneous transfer of the excitation from S_1 to T, and observed slower decay processes as vibrational cooling of a hot T state. During the first 10 ps after excitation, vibrational relaxation processes may also influence the XANES in subtle ways before the T state is fully thermalized. Nuclear movement during vibrational cooling will affect the 3d orbital energies, especially those involved in coordination to the porphyrin ring, which would be reflected in the dynamics of the 1s → 3d transition region as general broadening of the 3d transitions in T. Such an influence of the coordination geometry to the linewidth of the 1s → 3d transition was also observed in the ground state with coexisting multiple conformers.[2] This may provide some explanation of the shape of the 3d transitions at 10 ps, although a full description of the relaxation processes in terms of nuclear movement requires additional measurement with more direct structurally sensitive techniques, such as extended X-ray absorption fine structure (EXAFS) or wide angle X-ray scattering (WAXS). In the past synchrotron XTA experiments, we normally dealt with only two species at two delay times, the ground state at negative delay before the laser pulse, and the excited state at a positive delay with the laser excitation. Because the spectral feature of the ground state is known, its contribution can be subtracted from the total signal to extract the excited state spectrum if the fraction of the remaining ground state is known. At the LCLS, as shown by eqn (3), the relative contributions of each species to the total signal vary as a function of the delay time, and a very good kinetic scheme as well as the spectral assignments will be required to extract the kinetics of each transient species in the excited state trajectory.

3. Ni 3d electron configuration dynamics

The electronic configurations of Ni transient states can be captured by time evolution of the pre-edge features from quadrupole-allowed transitions from 1s to vacant 3d which provide directly the electronic configuration information of the vacant 3d orbitals. We have determined the pre-edge features for S_0, a single peak assigned to the 1s → $3d_{x^2-y^2}$ transition arising from the Ni(II) $3d^8$ electronic configuration of $(3d_{z^2})^2(3d_{x^2-y^2})^0$.[44] Also determined were the pre-edge features for the T_1 state where the two peaks were assigned to the 1s → $3d_{x^2-y^2}$ and 1s → $3d_{z^2}$ transitions, respectively, for the $(3d_{z^2})^1(3d_{x^2-y^2})^1$ electronic configuration. In the LCLS experiment, the time evolution of the pre-edge features were captured in a much higher time resolution as shown in Fig. 5. Before the laser excitation, the

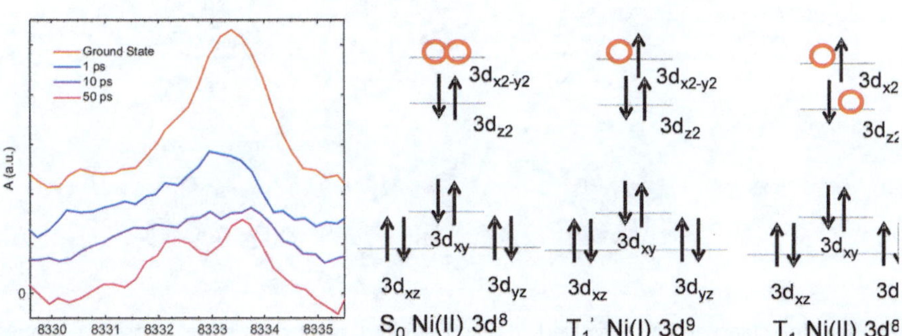

Fig. 5 The time evolution of the pre-edge 1s to 3d transition features.

pre-edge of S_0 again shows a single peak as we observed before. At $t = 1$ ps, the magnitude of this peak is diminished and slightly red shifted, while by $t = 10$ ps the pre-edge features are significantly broadened with some intensity at lower energy. At $t = 50$ ps, the double peak features for T_1 are observed again. Due to the weak signals from the pre-edge, we could only take a few discrete time points, which may have missed the rise of the double peak feature, while we can still suggest that the double single occupancy at $3d_{x^2-y^2}$ and $3d_{z^2}$ could have appeared between $t = 10$–50 ps. The absence of sharp split peaks by 1 ps is consistent with the presence of the intermediate T'_1 at this delay time. The interesting part of the result is the visualization of the transient Ni(I) species in T'_1 an intramolecular charge transfer state where the electronic occupation is $(3d_{z^2})^2(3d_{x^2-y^2})^1$ from the relaxation of the $S_1(\pi, \pi^*)$ state *via* charge transfer to Ni(II). The partial occupation of the $3d_{x^2-y^2}$ orbital diminishes the probability of the $1s \rightarrow 3d_{x^2-y^2}$ transition. The T' decay to T_1 is a second charge transfer event, filling the hole in the π orbital and resulting in a Ni(II) d-centered excitation.

4. Excited state electronic structures of NiTMP from XANES and calculations

Although one can obtain XANES spectral features for different transient states of NiTMP, the precise assignments of these features can benefit from electronic structural calculations from which the XANES transition features can be obtained to compare with the experimental results. Although calculations for the ground state TMCs have become relatively routine processes, the corresponding calculations for the valence excited state as well as the corresponding inner shell transitions are scarce. DFT and TDDFT calculations identified possible electronic states participating in the relaxation pathway. Ground and excited state geometry optimizations of these potential intermediate electronic states provided their relative energies and structural characteristics as shown in Fig. 6.

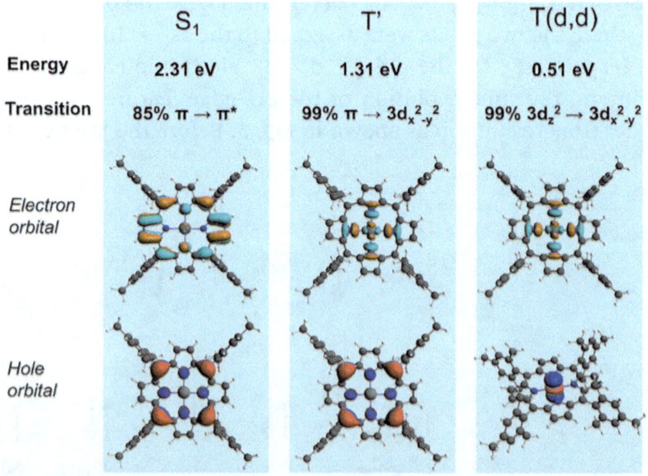

Fig. 6 Molecular orbitals involved in each excited electronic state transition. Many features of the excited state XANES can be explained based on the movement of electron density on or off the metal center.

Using the optimized ground state geometry, the S_1 state Frank–Condon excitation energy is computed which is 2.31 eV above the ground state. Any intermediate states in the singlet manifold should be rapidly populated. The TDDFT excited state geometry optimizations performed on the lowest lying singlet excited states found the lowest A state at 1.31 eV and the lowest B state at 1.81 eV. The lowest A singlet state, which possesses (π, $3d_{x^2-y^2}$) orbital character, is chosen as a point of probable intersystem crossing and triplet states computed using the geometry optimized in this lowest A state provides a probable candidate for the identity of T′ because it resides in the singlet manifold of the lowest energy state and its energy proximity of the singlet and triplet of (π, $3d_{x^2-y^2}$) character. The T′ state with a (π, d) configuration is identified with a $(3d_{z^2})^2(3d_{x^2-y^2})^1$ configuration due to the observed XANES pre-edges.

Relaxed geometries for the ground, S_1, T′, and T states show significant differences in their structures, with the calculated Ni–N bond lengths of 1.94, 2.02 and 2.04 Å, respectively, and the macrocycle structure from a significant out-of-plane ruffling distortion to largely planar conformation. The Ni–N bond lengthening for the T state is very similar to the synchrotron results.[44,54] Interestingly, the calculated Ni–N bond lengths for the T′ state are almost the same as the T state, suggesting a structural similarity between the geometry of the relaxed T′ and T states around the Ni centre.

In order to compute the XANES features based on the molecular structures and valence electronic configurations, a series of inner shell bound transitions were calculated using the solutions to the Self Consistent Field (SCF) equations that converge to at a higher energy than ground state solutions that have long been recognized as useful approximations to excited state wave functions.[55–57] In order to obtain these higher-energy SCF solutions, a set of natural transition orbitals (NTOs)[58] for the state of interest was first generated at the excited state geometry. The initial guess for the SCF density of this higher-energy solution was then formed by a HOMO–LUMO swap of the dominant NTOs. A second-order optimization scheme was then used to converge to the state of interest.[59] The XANES features for the intermediate states were calculated with ES-TDDFT. Due to the neglect of scalar and spin–orbit relativistic effects in this simulation, the calculated transitions are much lower in energy than the experimental results. However, it has been shown that these operators have little effect on relative transition energies[60] and uniform shifts are routinely applied to calculated XANES spectra to better compare transition energies with experimental data.[60–62] A uniform shift of 172 eV (~2% of the Ni K-edge energy) is applied to all calculated spectra and all transitions are convoluted with Gaussian functions to match experimental lineshapes; 2.2 and 1.8 eV fwhm for the $4p_z$ and 3d transitions, respectively.

The electronic environment around the Ni remains essentially the same from S_0 to S_1, so little change is seen in the dominating 1s → $4p_z$ transition from the calculation. The observed blue-shift of the T 1s → $4p_z$ transition feature compared to S_0 is also reproduced in the calculation. More interestingly, a red shift for the 1s → $4p_z$ transition is produced from the calculation also as observed for the T′ state. The calculated results clearly identified the origin of these shifts. For the T′ state, the red-shift of its 1s → $4p_z$ transition peak, based on the calculated results (Table 1), is due to the energy rise of the 1s core level energy by about 2 eV, reducing the energy gap between 1s and $4p_z$ orbitals, while the $4p_z$

Table 1 Orbital energies for the electronic ground state and changes in those energies in various excited states

	Inputs for excited-state XAS modeling					
Wave function[1]	S_0	$S_1(\pi, \pi^*)$	T′	T(d, d)	S_0	S_0
Geometry[2]	S_0	$S_1(\pi, \pi^*)$	T′	T(d, d)	T′	T(d, d)
	Calculated Ni orbital energies					
	eV	ΔeV				
1s α	−8167.47	−0.05	2.02	−1.17	−0.37	−0.34
1s β	−8167.47	−0.05	2.03	−1.17	−0.37	−0.34
$4p_z$ α	1.39	−0.07	0.18	−0.23	−0.06	−0.03
$4p_z$ β	1.39	−0.06	0.05	−0.44	−0.06	−0.03

orbital energy only changes on the order of 0.1 eV. This could be due to the transient reduction of the Ni(II) to Ni(I) which gives rise of the red-shift at the edge. For the T state, the blue shift of the 1s → $4p_z$ peak could also originate from simultaneous lowering of the 1s core level energy by ~1.2 eV and the $4p_z$ energy level by <0.2 eV to give rise to the blue shift by 1 eV compared to the experimental value of 1.5 eV, with the difference comparable to the energy resolution of the experiments. In addition, the calculations also successfully reproduced the trend of the 1s → 3d transitions with the expected 3d electronic configuration and nuclear geometry (not shown).

We can therefore conclude that the $4p_z$ transition energy is largely determined by the electronic state of the Ni, which is in line with the fact that the shift in the 1s energy dominates the observed shifts and that $4p_z$ has less interaction with ligand orbitals than other valence Ni orbitals.

5. What have we learned about interplay between electronic and nuclear structural changes in the excited state NiTMP?

The time scale in structural dynamics studies enabled by the LCLS allows us to directly probe transient metal orbital energies and occupancy as well as the interplay of photoinduced electronic configuration changes and molecular orbital energies before excited state thermalization. While ultrafast optical transient absorption measurements can obtain kinetics of the S_1 state decay, optically dark metal centered electronic state, or metal centered orbital energy changes in the subsequent excited states are often elusive by these methods. This study has directly obtained energies of the transitions 1s → 3d and 1s → $4p_z$ in Ni for different electronic states from the experimental and computational studies which appear to be key to extracting information regarding the interplay between electronic and nuclear movement in the excited state since the transient EXAFS is not currently available in the X-ray free electron source.

Moreover, only the comparison between the 1s to 3d and 4p transition energies obtained from XANES and calculated energies of these orbitals can provide information about the origin of the transition energy changes and quantify the influence of the valence excitation and the energy level of the core orbital. The experimental results only could not identify the red-shift of $1s \rightarrow 4p_z$ transition as an evidence of the T′ state formation and the evidence could be misinterpreted or missed entirely without the computation. With the assignment of T′ to Ni(I), T′ state formation can be understood as the intramolecular electron movement from the macrocycle to the metal center $3d_{x^2-y^2}$ orbital, causing greater nuclear shielding and therefore a lower electron binding energy. The rise of the 1s orbital energy in turn reduces the $1s \rightarrow 4p_z$ transition energy in the T′ state, analogous to the core-level shift seen in X-ray photoelectron spectroscopy (XPS)[63] and the effect of oxidation transition edge energies in previous XAS experiments.[64,65] The 2 eV rise of the Ni 1s orbital energy due to the formation of Ni(I) and associated nuclear structural change reminds us that the energy of the core orbital can be affected by the valence excitation although only less than 0.1% with respect to the core transition energy, it is sufficiently large to affect our interpretation of the XANES results.

The results reported here also provide direct information on the mutual influence between molecular orbitals associated with the metal center which has been difficult to obtain by other means. This red shift resulted from the formation of the transient Ni(I) is the first direct observation of this intermediate state in this system. A strong influence of the valence orbital occupation on the inner shell orbital energies indicates that one should not use the transition energy from 1s to other orbitals to draw conclusions about the d-orbital energies. As calculated, only ~0.2 eV of the ~2 eV shift in the $1s \rightarrow 4p_z$ transition energy change is from the changes in the $4p_z$ orbital from S_0 to T′. The blue shift in the $1s \rightarrow 4p_z$ transition energy of the T state can again be attributed to a change in the repulsive potential felt by the core electrons due to more antibonding electrons with weaker interactions with the core. Consequently, this change reduces the repulsive potential felt by the Ni 1s electrons, lowering the 1s orbital energy, and increasing the energy gap for the $1s \rightarrow 4p_z$ transition by 1.5 eV relative to that of the ground state. The identification by TDDFT of T′ as the lowest energy singlet as well as a state with a triplet state lying close in energy lends credence to the identification of T′ as a (π, d) state. The optimized geometries of the involved electronic states (Fig. 6) provide insight into the behavior of the NiTMP structure in response to electronic changes and the subsequent energetic rearrangement of Ni orbitals.

Macrocycle expansion and flattening in the T state is explained based on the net movement of electron density from $3d_{z^2}$ to $3d_{x^2-y^2}$ in the porphyrin plane destabilizing the Ni–N bonds, which leads to the bond elongation. Such an electronic movement makes the effective radius of Ni(II) larger than in the S_0 state and significantly removes the need for the non-planar distortion of the porphin to accommodate the unusually short Ni–N bond length. Interestingly, the calculated Ni–N bonds in the Ni(I) of the T′ state already have the majority of the lengthening of the Ni–N bond, suggesting that even with only one additional electron on the $3d_{x^2-y^2}$ orbital could cause substantial nuclear geometry change. This is direct evidence for the interplay between the electronic and nuclear structures of a TMC.

In the ground state, Ni porphyrins have a low barrier to interconversion between various distorted conformers of the macrocycle,[66,67] relative broadness of the $1s \rightarrow 3d_{x^2-y^2}$ transition peak.[54] The lowest in energy of these conformers based on our calculations is a ruffled geometry. Out-of-plane distortions are structural factors that influence electronic state energies mainly through the destabilization of the π and $3d_{x^2-y^2}$ orbitals, an effect that when taken to a greater extreme in substituted Ni phthalocyanines (Pc) leads to longer CT state lifetimes. The octa-butoxy substituted Ni(II)Pc have been described as having long-lived LMCT states with lifetimes in the hundreds of picoseconds.[68,69] Based on our studies on Ni(II) TMP with a preferred planar structure to accommodate the Ni(I) species, one can relate the results from Ni(II)Pc to suggest using planar Ni(II) species as potential reducing catalysts, even though such correlation remains to be investigated in the future.

Experimental

1. Ultrafast XANES spectroscopy

Ultrafast XANES spectra were collected for an 8 mM solution of NiTMP in toluene at the X-ray Pump–Probe (XPP) station of the LCLS[70] using a similar experimental configuration as previous XANES measurements at XPP.[19] The sample solution was delivered as a flat 100 μm thick liquid jet where the laser and X-ray overlap nearly collinear in a nitrogen filled chamber. The "pump–probe" cycle was at a 120 Hz repetition rate. The pump laser pulses are from the output of an optical parametric amplifier (OPA) at 527 nm with a pulse duration of 50 fs (fwhm). The laser pulse energy at the sample was between 15 and 18 μJ with a spot size of ~0.1 mm diameter. Ni K-edge (8333 eV) XANES spectra were collected by scanning a Si(111) double crystal X-ray monochromator across the XFEL SASE bandwidth. The ~50 fs pulse duration X-ray probe pulses have significant bunch-to-bunch spectral and temporal fluctuations, resulting in a time-integrated energy bandwidth of ~50 eV at the Ni K-edge. The monochromatic X-ray beam was focused to a ~0.1 mm diameter spot at the sample. To obtain the kinetics at a specific energy, scans of the pump–probe delay were performed at the fixed photon energies as described in the results. For both scan types, the XANES signal was collected with two seconds integration time and a total of 240 single pulses were used at each energy step.

XANES spectra *via* Ni K_α X-ray fluorescence were collected using two solid state passivated implanted planar silicon (PIPS) point detectors (Canberra, Inc.) each equipped with a set of Z-1 filter and Soller slits to minimize the elastic scattering signals. The raw data from XANES scans were smoothed with the locally weighted regression method.

To account for timing jitter in the X-ray pulses, individual shots were re-binned in time for time delay scans according to up-stream diagnostic RF cavities ("phase cavities") that record the average electron bunch arrival time. This signal is used to account for the long-time drift in the average pulse arrival time and replaces and enhances the overall time resolution of the experiments. After phase cavity re-binning, the time resolution of the experiment was reduced from an estimated 400 fs to 300 fs based on the fwhm of the Gaussian instrument response function (IRF) obtained as a fitted variable in the fits of the time delay scans.

To obtain kinetics for the NiTMP XANES evolution, signals at a chosen X-ray photon energy were collected as a function of the pump–probe delay time as shown in Fig. 4. These kinetic traces were globally fitted to a sequential kinetic scheme in eqn (1). Based on this scheme, the relative populations of each species as a function of the delay time were simulated by numerical integration of a set of differential equations shown in eqn (2). These are the differential rate expressions for each species included in the kinetic scheme. τ_1, τ_2, and τ_3 are assigned to the first second and third steps in eqn (1), respectively. These components were weighted by their relative absorption at each energy and the total simulated kinetic traces were fitted to the experimental traces using a non-linear least squares method to obtain the time constants for each step as shown in eqn (3) as described earlier.

2. Electronic structure calculations

To understand the overall electronic and structural evolution of the NiTMP excited states and to ascertain how changes in the electronic and nuclear structure as the molecule relaxes are reflected in the experimentally observed XTA signals, each electronic state in the proposed mechanism (Fig. 1) was modeled separately. Initial DFT and TDDFT calculations using the BP86 functional were employed to probe the basic orbital structure of NiTMP and its excited states using the ADF package (ADF2013.01 (ref. 71 and 72)). A double-ζ polarized (DZP) basis set was used for the description of C, N, and H atoms, and a triple-ζ polarized (TZP) basis set was used to describe the atomic orbitals of Ni. This combination of functional and basis set has previously described the orbital structure of a related Ni compound with high accuracy.[73] Subsequently, the geometries of the ground and each intermediate excited state were optimized with the BP86 functional[74–76] and the 6-31G(d) basis set[77,78] using a development version of the Gaussian software package.[76] The X-ray absorption was calculated with energy-specific TDDFT (ES-TDDFT)[79,80] using the PBE1PBE and Ahlrichs' def2-TZVP basis set[81] with diffuse functions on the nickel atom.[82–84] For all calculations of X-ray absorption spectra, the mesityl groups on the porphyrin have been replaced with methyl groups to reduce computational cost.

Conclusions

Ultrafast excited state pathways for an open shell metalloporphyrin Ni(II)TMP are obtained based on XANES spectra at the Ni K-edge with ~100 fs time resolution at the LCLS, providing insight into its ultrafast electronic and structural relaxation processes. Importantly, a transient Ni(I) (π, d) electronic state (T′) is captured as an intermediate from analyzing the spectral data and quantum mechanical calculations.

The observed and computed inner shell-to-valence orbital transition energies demonstrate and quantify the influence of electronic configuration on specific metal orbital energies. The strong influence of the valence orbital occupation on the inner shell orbital energies indicates that one should not use the transition energy from 1s alone to draw conclusions about the d-orbital energies of different states. A transient electronic configuration could influence d-orbital energies up to a few eV and any potential photocatalytic application should account for this to ensure that energy levels are well matched and that the photoexcitation can be used optimally in driving desirable processes.

NiTMP structural dynamics have been deduced from DFT optimized geometries and structurally sensitive features in the Ni K-edge XANES. The influence of the porphyrin macrocycle conformation on relaxation kinetics may be significant enough to affect the kinetics that govern the limited population accumulation of

Ni(I). The short lifetime of the T′ state and the delay of the nuclear rearrangement to a longer Ni–N bond length and flattened macrocycle with respect to its formation suggests the still-ruffled macrocycle expedites T′ decay.

Acknowledgements

We acknowledge support for this work from the Solar Energy Photochemistry program (experimental work) and Ultrafast Initiative (theoretical work) of the U. S. Department of Energy, Office of Science, Office of Basic Energy Sciences, through Argonne National Laboratory under Contract No. DE-AC02-06CH11357 and MLS is supported by the National Institute of Health, under Contract No. R01-GM115761 (LXC) and R01-HL63203 (BMH). Use of the Linac Coherent Light Source (LCLS), SLAC National Accelerator Laboratory, is supported by the U.S. Department of Energy, Office of Science, Office of Basic Energy Sciences under Contract No. DE-AC02-76SF00515. Computations on modeled spectra were facilitated through the use of advanced computational, storage, and networking infrastructure provided by the Hyak supercomputer system at the University of Washington, funded by the Student Technology Fee. PJL is also grateful for support by the State of Washington through the University of Washington Clean Energy Institute. MLS also thanks the National Institute of General Medical Sciences of NIH for support through the Molecular Biophysics training grant administered by Northwestern University (5T32 GM008382). KH gratefully acknowledges support from DANSCATT and from the Villum and Carlsberg Foundations. The authors would like to thank Tim Brandt Van Driel for invaluable assistance with the phase cavity timing correction by providing a means to calibrate the phase cavity data.

References

1 V. Balzani, A. Credi and M. Venturi, *Coord. Chem. Rev.*, 1998, **171**, 3–16.
2 T. J. Meyer, *Acc. Chem. Res.*, 1989, **22**, 163–170.
3 H. B. Gray, *Nat. Chem.*, 2009, **1**, 7.
4 A. L. Smeigh and J. K. McCusker, *Ultrafast Phenomena*, 2006, pp. 273–275.
5 F. N. Castellano, M. Ruthkosky and G. J. Meyer, *Inorg. Chem.*, 1995, **34**, 3–4.
6 J. Huang, O. Buyukcakir, M. W. Mara, A. Coskun, N. M. Dimitrijevic, G. Barin, O. Kokhan, A. B. Stickrath, R. Ruppert, D. M. Tiede, J. F. Stoddart, J. P. Sauvage and L. X. Chen, *Angew. Chem., Int. Ed.*, 2012, **51**, 12711–12715.
7 S. Ferrere, A. Zaban and B. A. Gregg, *J. Phys. Chem. B*, 1997, **101**, 4490–4493.
8 L. L. Jamula, A. M. Brown, D. Guo and J. K. McCusker, *Inorg. Chem.*, 2014, **53**, 15–17.
9 M. W. Mara, D. N. Bowman, O. Buyukcakir, M. L. Shelby, K. Haldrup, J. Huang, M. R. Harpham, A. B. Stickrath, X. Zhang, J. F. Stoddart, A. Coskun, E. Jakubikova and L. X. Chen, *J. Am. Chem. Soc.*, 2015, **137**, 9670–9684.
10 L. X. Chen, W. J. H. Jäger, G. Jennings, D. J. Gosztola, A. Munkholm and J. P. Hessler, *Science*, 2001, **292**, 262–264.
11 L. X. Chen, *Angew. Chem., Int. Ed.*, 2004, **43**, 2886–2905.
12 L. X. Chen, *Annu. Rev. Phys. Chem.*, 2005, **56**, 221–254.
13 L. X. Chen and X. Zhang, *J. Phys. Chem. Lett.*, 2013, **4**, 4000–4013.
14 L. X. Chen, X. Zhang and M. L. Shelby, *Chem. Sci.*, 2014, **5**, 4136–4152.

15 M. Saes, C. Bressler, R. Abela, D. Grolimund, S. L. Johnson, P. A. Heimann and M. Chergui, *Phys. Rev. Lett.*, 2003, **90**, 047403/047401–047403/047404.

16 C. Bressler and M. Chergui, *Chem. Rev.*, 2004, **104**, 1781–1812.

17 C. Bressler, C. Milne, V. T. Pham, A. ElNahhas, R. M. van der Veen, W. Gawelda, S. Johnson, P. Beaud, D. Grolimund, M. Kaiser, C. N. Borca, G. Ingold, R. Abela and M. Chergui, *Science*, 2009, **323**, 489–492.

18 M. Chergui, *Acta Crystallogr., Sect. A: Found. Crystallogr.*, 2010, **66**, 229–239.

19 H. T. Lemke, C. Bressler, L. X. Chen, D. M. Fritz, K. J. Gaffney, A. Galler, W. Gawelda, K. Haldrup, R. W. Hartsock, H. Ihee, J. Kim, K. H. Kim, J. H. Lee, M. M. Nielsen, A. B. Stickrath, W. K. Zhang, D. L. Zhu and M. Cammarata, *J. Phys. Chem. A*, 2013, **117**, 735–740.

20 M. Chollet, R. Alonso-Mori, M. Cammarata, D. Damiani, J. Defever, J. T. Delor, Y. Feng, J. M. Glownia, J. B. Langton, S. Nelson, K. Ramsey, A. Robert, M. Sikorski, S. Song, D. Stefanescu, V. Srinivasan, D. Zhu, H. T. Lemke and D. M. Fritz, *J. Synchrotron Radiat.*, 2015, **22**, 503–507.

21 J.-P. Colletier, M. Sliwa, F.-X. Gallat, M. Sugahara, V. Guillon, G. Schiro, N. Coquelle, J. Woodhouse, L. Roux, G. Gotthard, A. Royant, L. M. Uriarte, C. Ruckebusch, Y. Joti, M. Byrdin, E. Mizohata, E. Nango, T. Tanaka, K. Tono, M. Yabashi, V. Adam, M. Cammarata, I. Schlichting, D. Bourgeois and M. Weik, *J. Phys. Chem. Lett.*, 2016, **7**, 882–887.

22 D. J. Higley, K. Hirsch, G. L. Dakovski, E. Jal, E. Yuan, L. Tianmin, A. A. Lutman, J. P. MacArthur, E. Arenholz, C. Zhao, G. Coslovich, P. Denes, P. W. Granitzka, P. Hart, M. C. Hoffmann, J. Joseph, L. Le Guyader, A. Mitra, S. Moeller, H. Ohldag, M. Seaberg, P. Shafer, J. Stohr, A. Tsukamoto, H. D. Nuhn, A. H. Reid, H. A. Durr and W. F. Schlotter, *Rev. Sci. Instrum.*, 2016, **87**, 033110–033115.

23 T. Katayama, Y. Inubushi, Y. Obara, T. Sato, T. Togashi, K. Tono, T. Hatsui, T. Kameshima, A. Bhattacharya, Y. Ogi, N. Kurahashi, K. Misawa, T. Suzuki and M. Yabashi, *Appl. Phys. Lett.*, 2013, **103**, 131105.

24 J. Kern, R. Tran, R. Alonso-Mori, S. Koroidov, N. Echols, J. Hattne, M. Ibrahim, S. Gul, H. Laksmono, R. G. Sierra, R. J. Gildea, G. Han, J. Hellmich, B. Lassalle-Kaiser, R. Chatterjee, A. S. Brewster, C. A. Stan, C. Gloeckner, A. Lampe, D. DiFiore, D. Milathianaki, A. R. Fry, M. M. Seibert, J. E. Koglin, E. Gallo, J. Uhlig, D. Sokaras, T.-C. Weng, P. H. Zwart, D. E. Skinner, M. J. Bogan, M. Messerschmidt, P. Glatzel, G. J. Williams, S. Boutet, P. D. Adams, A. Zouni, J. Messinger, N. K. Sauter, U. Bergmann, J. Yano and V. K. Yachandra, *Nat. Commun.*, 2014, **5**, 4371.

25 J. Kim, K. H. Kim, K. Y. Oang, J. H. Lee, K. Hong, H. Cho, N. Huse, R. W. Schoenlein, T. K. Kim and H. Ihee, *Chem. Commun.*, 2016, **52**, 3734–3749.

26 Y. Obara, T. Katayama, Y. Ogi, T. Suzuki, N. Kurahashi, S. Karashima, Y. Chiba, Y. Isokawa, T. Togashi, Y. Inubushi, M. Yabashi, T. Suzuki and K. Misawa, *Opt. Express*, 2014, **22**, 1105–1113.

27 Y. Ogi, Y. Obara, T. Katayama, Y. I. Suzuki, S. Y. Liu, N. C. M. Bartlett, N. Kurahashi, S. Karashima, T. Togashi, Y. Inubushi, K. Ogawa, S. Owada, M. Rubesova, M. Yabashi, K. Misawa, P. Slavicek and T. Suzuki, *Struct. Dyn.*, 2015, **2**, 034901.

28 Y. Uemura, D. Kido, Y. Wakisaka, H. Uehara, T. Ohba, Y. Niwa, S. Nozawa, T. Sato, K. Ichiyanagi, R. Fukaya, S.-i. Adachi, T. Katayama, T. Togashi,

S. Owada, K. Ogawa, M. Yabashi, K. Hatada, S. Takakusagi, T. Yokoyama, B. Ohtani and K. Asakura, *Angew. Chem., Int. Ed.*, 2016, **55**, 1364–1367.
29 K. Kalyanasundaram, *Photochemistry of Polypyridine and Porphyrin Complexes*, Academic Press, London, 1992.
30 L. X. Chen, X. Zhang, E. C. Wasinger, K. Attenkofer, G. Jennings, A. Z. Muresan and J. S. Lindsey, *J. Am. Chem. Soc.*, 2007, **129**, 9616–9618.
31 L. X. Chen, X. Zhang, E. C. Wasinger, J. V. Lockard, A. B. Stickrath, M. W. Mara, K. Attenkofer, G. Jennings, G. Smolentsev and A. Soldatov, *Chem. Sci.*, 2010, **1**, 642.
32 D. Kim and D. Holten, *Chem. Phys. Lett.*, 1983, **98**, 584–589.
33 J. Rodriguez and D. Holten, *J. Chem. Phys.*, 1990, **92**, 5944–5950.
34 A. V. Zamyatin, A. V. Gusev and M. A. J. Rodgers, *J. Am. Chem. Soc.*, 2004, **126**, 15934–15935.
35 W. Jentzen, E. Unger, G. Karvounis, J. A. Shelnutt, W. Dreybrodt and R. Schweitzer-Stenner, *J. Phys. Chem.*, 1996, **100**, 14184–14191.
36 C. M. Drain, S. Gentemann, J. A. Roberts, N. Y. Nelson, C. J. Medforth, S. Jia, M. C. Simpson, K. M. Smith, J. Fajer, J. A. Shelnutt and D. Holten, *J. Am. Chem. Soc.*, 1998, **120**, 3781–3791.
37 C. J. Ballhausen, *Introduction to Ligand Field Theory*, McGraw-Hill, New York, 1962.
38 X. Zhang, E. C. Wasinger, A. Z. Muresan, K. Attenkofer, G. Jennings, J. S. Lindsey and L. X. Chen, *J. Phys. Chem. A*, 2007, **111**, 11736–11742.
39 D. Kim, C. Kirmaier and D. Holten, *Chem. Phys.*, 1983, **75**, 305–322.
40 J. Rodriguez and D. Holten, *J. Chem. Phys.*, 1989, **91**, 3525–3531.
41 S. Gentemann, N. Y. Nelson, L. Jaquinod, D. J. Nurco, S. H. Leung, C. J. Medforth, K. M. Smith, J. Fajer and D. Holten, *J. Phys. Chem. B*, 1997, **101**, 1247–1254.
42 H. S. Eom, S. C. Jeoung, D. Kim, J.-H. Ha and Y.-R. Kim, *J. Phys. Chem. A*, 1997, **101**, 3661–3669.
43 A. V. Zamyatin, A. V. Gusev and M. A. J. Rodgers, *J. Am. Chem. Soc.*, 2004, **126**, 15934–15935.
44 L. X. Chen, X. Zhang, E. C. Wasinger, K. Attenkofer, G. Jennings, A. Muresan and S. Lindsey Jonathan, *J. Am. Chem. Soc.*, 2007, **129**, 9616–9618.
45 L.-S. Kau, D. J. Spira-Solomon, J. E. Penner-Hahn, K. O. Hodgson and E. I. Solomon, *J. Am. Chem. Soc.*, 1987, **109**, 6433–6442.
46 T. E. Westre, P. Kennepohl, J. G. DeWitt, B. Hedman, K. O. Hodgson and E. I. Solomon, *J. Am. Chem. Soc.*, 1997, **119**, 6297–6314.
47 L. Campbell, S. Tanaka and S. Mukamel, *Chem. Phys.*, 2004, **299**, 225–231.
48 X. Y. Zhang, E. C. Wasinger, A. Z. Muresan, K. Attenkofer, G. Jennings, J. S. Lindsey and L. X. Chen, *J. Phys. Chem. A*, 2007, **111**, 11736–11742.
49 A. V. Zamyatin, A. V. Gusev and M. A. J. Rodgers, *J. Am. Chem. Soc.*, 2004, **126**, 15934–15935.
50 M.-H. Ha-Thi, N. Shafizadeh, L. Poisson and B. Soep, *Phys. Chem. Chem. Phys.*, 2010, **12**, 14985.
51 S. Sorgues, L. Poisson, K. Raffael, L. Krim, B. Soep and N. Shafizadeh, *J. Chem. Phys.*, 2006, **124**, 114302–114310.
52 Y. Mizutani, Y. Uesugi and T. Kitagawa, *J. Chem. Phys.*, 1999, **111**, 8950–8962.
53 J. Rodriguez and D. Holten, *J. Chem. Phys.*, 1989, **91**, 3525–3531.

54 L. X. Chen, X. Zhang, E. C. Wasinger, J. V. Lockard, A. B. Stickrath, M. W. Mara, K. Attenkofer, G. Jennings, G. Smolentsev and A. Soldatov, *Chem. Sci.*, 2010, **1**, 642–650.
55 A. T. B. Gilbert, N. A. Besley and P. M. W. Gill, *J. Phys. Chem. A*, 2008, **112**, 13164–13171.
56 E. R. Davidson, *J. Chem. Phys.*, 1964, **41**, 656.
57 B. Peng, B. E. Van Kuiken, F. Ding and X. Li, *J. Chem. Theory Comput.*, 2013, **9**, 3933–3938.
58 R. L. Martin, *J. Chem. Phys.*, 2003, **118**, 4775.
59 G. B. Bacskay, *Chem. Phys.*, 1981, **61**, 385–404.
60 S. DeBeer George, T. Petrenko and F. Neese, *Inorg. Chim. Acta*, 2008, **361**, 965–972.
61 N. A. Besley and F. A. Asmuruf, *Phys. Chem. Chem. Phys.*, 2010, **12**, 12024.
62 G. Fronzoni, R. De Francesco and M. Stener, *J. Phys. Chem. B*, 2005, **109**, 10332–10340.
63 C. S. Fadley, S. B. M. Hagstrom, M. P. Klein and D. A. Shirley, *J. Chem. Phys.*, 1968, **48**, 3779–3794.
64 R. Sarangi, S. DeBeer George, D. J. Rudd, R. K. Szilagyi, X. Ribas, C. Rovira, M. Almeida, K. O. Hodgson, B. Hedman and E. I. Solomon, *J. Am. Chem. Soc.*, 2007, **129**, 2316–2326.
65 W. Gu, H. Wang and K. Wang, *Dalton Trans.*, 2014, **43**, 6406.
66 R. G. Alden, B. A. Crawford, R. Doolen, M. R. Ondrias and J. A. Shelnutt, *J. Am. Chem. Soc.*, 1989, **111**, 2070–2072.
67 R. G. Alden, M. R. Ondrias and J. A. Shelnutt, *J. Am. Chem. Soc.*, 1990, **112**, 691–697.
68 G. Balakrishnan, A. V. Soldatova, P. J. Reid and T. G. Spiro, *J. Am. Chem. Soc.*, 2014, **136**, 8746–8754.
69 C. M. Drain, S. Gentemann, J. A. Roberts, N. Y. Nelson, C. J. Medforth, S. Jia, M. C. Simpson, K. M. Smith, J. Fajer, J. A. Shelnutt and D. Holten, *J. Am. Chem. Soc.*, 1998, **120**, 3781–3791.
70 M. Chollet, R. Alonso-Mori, M. Cammarata, D. Damiani, J. Defever, J. T. Delor, Y. Feng, J. M. Glownia, J. B. Langton, S. Nelson, K. Ramsey, A. Robert, M. Sikorski, S. Song, D. Stefanescu, V. Srinivasan, D. Zhu, H. T. Lemke and D. M. Fritz, *J. Synchrotron Radiat.*, 2015, **22**, 503–507.
71 C. Fonseca Guerra, J. G. Snijders, G. te Velde and E. J. Baerends, *Theor. Chem. Acc.*, 1998, **99**, 391–403.
72 G. te Velde, F. M. Bickelhaupt, E. J. Baerends, C. Fonseca Guerra, S. J. A. van Gisbergen, J. G. Snijders and T. Ziegler, *J. Comput. Chem.*, 2001, **22**, 931–967.
73 S. Patchkovskii, P. M. Kozlowski and M. Z. Zgierski, *J. Chem. Phys.*, 2004, **121**, 1317–1324.
74 A. D. Becke, *J. Chem. Phys.*, 1986, **84**, 4524.
75 J. Perdew, *Phys. Rev. B: Condens. Matter Mater. Phys.*, 1986, **33**, 8822–8824.
76 M. J. Frisch, G. W. Trucks, H. B. Schlegel, G. E. Scuseria, M. A. Robb, J. R. Cheeseman, G. Scalmani, V. Barone, B. Mennucci, G. A. Petersson, H. Nakatsuji, M. Caricato, X. Li, H. P. Hratchian, A. F. Izmaylov, J. Bloino, G. Zheng, J. L. Sonnenberg, M. Hada, M. Ehara, K. Toyota, R. Fukuda, J. Hasegawa, M. Ishida, T. Nakajima, Y. Honda, O. Kitao, H. Nakai, T. Vreven, J. A. Montgomery, Jr, J. E. Peralta, F. Ogliaro, M. Bearpark, J. J. Heyd, E. Brothers, K. N. Kudin, V. N. Staroverov, T. Keith, R. Kobayashi,

J. Normand, K. Raghavachari, A. Rendell, J. C. Burant, S. S. Iyengar, J. Tomasi, M. Cossi, N. Rega, J. M. Millam, M. Klene, J. E. Knox, J. B. Cross, V. Bakken, C. Adamo, J. Jaramillo, R. Gomperts, R. E. Stratmann, O. Yazyev, A. J. Austin, R. Cammi, C. Pomelli, J. W. Ochterski, R. L. Martin, K. Morokuma, V. G. Zakrzewski, G. A. Voth, P. Salvador, J. J. Dannenberg, S. Dapprich, A. D. Daniels, O. Farkas, J. B. Foresman, J. V. Ortiz, J. Cioslowski and D. J. Fox, *Gaussian 9*, 2010.
77 V. A. Rassolov, J. A. Pople, M. A. Ratner and T. L. Windus, *J. Chem. Phys.*, 1998, **109**, 1223–1229.
78 P. C. Hariharan and J. A. Pople, *Theor. Chim. Acta*, 1973, **28**, 213–222.
79 W. Liang, S. A. Fischer, M. J. Frisch and X. Li, *J. Chem. Theory Comput.*, 2011, 7, 3540–3547.
80 P. J. Lestrange, P. D. Nguyen and X. Li, *J. Chem. Theory Comput.*, 2015, **11**, 2994–2999.
81 F. Weigend and R. Ahlrichs, *Phys. Chem. Chem. Phys.*, 2005, **7**, 3297.
82 D. Rappoport and F. Furche, *J. Chem. Phys.*, 2010, **133**, 134105.
83 K. L. Schuchardt, B. T. Didier, T. Elsethagen, L. Sun, V. Gurumoorthi, J. Chase, J. Li and T. L. Windus, *J. Chem. Inf. Model.*, 2007, **47**, 1045–1052.
84 D. Feller, *J. Comput. Chem.*, 1996, **17**, 1571–1586.

Faraday Discussions

PAPER

Towards efficient time-resolved X-ray absorption studies of electron dynamics at photocatalytic interfaces

Stefan Neppl,*[a] Johannes Mahl,[a] Anton S. Tremsin,[b] Bruce Rude,[c] Ruimin Qiao,[c] Wanli Yang,[c] Jinghua Guo[c] and Oliver Gessner*[a]

Received 10th May 2016, Accepted 18th May 2016
DOI: 10.1039/c6fd00125d

We present a picosecond time-resolved X-ray absorption spectroscopy (tr-XAS) setup designed for synchrotron-based studies of interfacial photochemical dynamics. The apparatus combines a high power, variable repetition rate picosecond laser system with a time-resolved X-ray fluorescence yield detection technique. Time-tagging of the detected fluorescence signals enables the parallel acquisition of X-ray absorption spectra at a variety of pump–probe delays employing the well-defined time structure of the X-ray pulse trains. The viability of the setup is demonstrated by resolving dynamic changes in the fine structure near the O1s X-ray absorption edge of cuprous oxide (Cu_2O) after photo-excitation with a 355 nm laser pulse. Two distinct responses are detected. A pronounced, quasi-static, reversible change of the Cu_2O O1s X-ray absorption spectrum by up to ~30% compared to its static line shape corresponds to a redshift of the absorption edge by ~1 eV. This value is small compared to the 2.2 eV band gap of Cu_2O but in agreement with previously published results. The lifetime of this effect exceeds the laser pulse-to-pulse period of 8 μs, resulting in a quasi-static spectral change that persists as long as the sample is exposed to the laser light, and completely vanishes once the laser is blocked. Additionally, a short-lived response corresponding to a laser-induced shift of the main absorption line by ~2 eV to lower energies appears within <200 ps and decays with a characteristic timescale of 43 ± 5 ns. Both the picosecond rise and nanosecond decay of this X-ray response are simultaneously captured by making use of a time-tagging approach – highlighting the prospects of the experimental setup for efficient probing of the electronic and structural dynamics in photocatalytic systems on multiple timescales.

[a]*Chemical Sciences Division, Lawrence Berkeley National Laboratory, Berkeley, California, USA. E-mail: ogessner@lbl.gov; sneppl@lbl.gov*
[b]*Space Sciences Laboratory, University of California, Berkeley, California, USA*
[c]*Advanced Light Source, Lawrence Berkeley National Laboratory, Berkeley, California, USA*

Introduction

At the heart of many emerging sunlight-to-fuel and sunlight-to-electricity concepts are interfacial processes that require an optimized, concerted flow of charge and energy on the molecular level. Microscopic processes evolving on spatial and temporal scales spanning orders of magnitudes have to be connected in order to gain comprehensive insight into the mechanisms that govern the efficiency of macroscopic charge and energy transport. Soft X-ray spectroscopy techniques have the potential to probe electronic and chemical states of matter with atomic site-specificity and chemical sensitivity, which is required to improve our fundamental understanding of interfacial chemistry and photophysics in complex systems. Over the last decade, *in situ* and *in operando* X-ray absorption and X-ray photoemission techniques have been developed at several synchrotron radiation facilities and are now routinely being used to probe the chemical and structural properties of photo-catalytic interfaces including oxidation state changes, local redox activity, elemental composition, as well as the electronic characters of chemically active states under realistic working conditions.[1–8]

While such (quasi-) static experiments are valuable for shedding light on fundamental chemical transformations, they are only sensitive to chemical species and electronic configurations with sufficiently long lifetimes such that their time-averaged concentration is above the detection limit. The identification of short-lived intermediates states, however, which frequently play a decisive role for the overall outcome and efficiency of photochemical reactions, is often impossible. This limitation has been overcome in the optical regime by the development of ultrafast NIR/Vis and UV time-resolved spectroscopy techniques, which nowadays form the experimental backbone of the field of femtochemistry.[9] Despite their remarkable sensitivity and superb time-resolution, however, gaining comprehensive microscopic insight into electronic and nuclear dynamics in large, heterogeneous systems, such as complex interfaces using solely ultrafast all-optical spectroscopy, is extremely challenging: in particular, spectroscopic fingerprints of different system components may overlap due to the delocalized nature of optically driven valence electron transitions, which denies an atomic-scale perspective on the ensuing chemical dynamics.

In contrast, the simultaneous element specificity and chemical sensitivity of X-ray transitions can potentially offer a significantly more distinct probe of local chemical transformations. Consequently, tremendous efforts have been devoted over the past decade to extend the power of ultrafast spectroscopy for real-time probing of chemical dynamics into the X-ray regime. Laboratory,[10–12] synchrotron[13–19] and free-electron laser based[20–25] methods have been explored in order to combine the sensitivity of inner-shell transitions to the local valence electronic structure with the capability of time-domain techniques to reveal transient charge configurations.

Here, we report on the progress toward developing a picosecond time-resolved transient X-ray absorption spectroscopy (tr-XAS) capability for time-domain *in situ* studies of interfacial photochemistry, which is currently being implemented at the Advanced Light Source (ALS). Absorption spectra are recorded in total X-ray fluorescence yield (TFY) mode, *i.e.*, in a photon-in/photon-out configuration rather than in transmission geometry, in order to be compatible with a wide variety of materials and existing *in situ/operando* soft X-ray spectroscopy techniques.[1,2,26] The setup is based on the combination of a high repetition rate picosecond laser system with a microchannel plate (MCP) based time- and position-sensitive delay-line detector.[27] The detector is used to simultaneously record and discriminate the TFY signals induced by individual X-ray pulses of the ALS bunch train by adopting a time-tagging data acquisition approach.[28] The mobile system can be deployed at various beamlines and is compatible with all synchrotron operating modes and electron bunch filling patterns.

The viability of the picosecond tr-XAS apparatus is demonstrated by monitoring photo-induced dynamics in the near-edge X-ray absorption fine structure of cuprous oxide (Cu_2O) in the vicinity of the O1s absorption edge. This metal oxide semiconductor is a promising electrode material for a number of photochemical applications, *e.g.* photocatalytic water splitting, where a detailed understanding of the photo-induced charge carrier dynamics in the bulk and at the metal oxide–liquid interface is crucial for the design of successful device architectures.[29–31]

In the experiment presented here, a Cu_2O powder sample is excited with 10 ps, 355 nm (3.5 eV) pulses, *i.e.*, significantly above the ~2.2 eV optical band gap,[32–34] using laser pulse fluences of ~6–10 mJ cm^{-2}. A pronounced, reversible spectral change of the O1s XAS line shape is observed upon laser illumination that remains constant as long as the sample is exposed to laser light and completely vanishes when the laser is turned off. Considering the laser repetition rate of 127 kHz, the lifetime of this effect must be much longer than 8 μs. The quasi-static spectral response may be described by an effective shift of a Gaussian shaped near-edge spectral component to ~1 eV lower photon energies. The magnitude of this shift is in agreement with the findings by Gaffney and co-workers, who probed the dynamics in a thin polycrystalline film of Cu_2O by tr-XAS in transmission mode at a similar laser excitation fluence.[35]

In addition to this quasi-static response, a short-lived laser-induced modification of the O1s XAS spectrum develops within a rise time of less than 200 ps after photo-excitation and may be described by an effective spectral shift of ~2 eV to lower photon energies. The temporal evolution of this transient XAS response follows a mono-exponential decay with a characteristic time constant of 43 ± 5 ns. Notably, the spectral shift associated with this short-lived response is comparable to the band gap energy of Cu_2O (2.2 eV), while the quasi-static response manifests itself in a significantly smaller shift.[32–34] The results are discussed in the light of previous static and time-resolved XAS experiments on Cu_2O and with respect to possible electronic and structural dynamics that may underlie the photo-induced changes in the O1s XAS spectrum.

Experimental details

The feasibility of picosecond tr-XAS has been demonstrated at several synchrotron radiation facilities,[14,28,36–38] with the main focus of research being the investigation of photochemical dynamics in liquid jets.[13,14,16] A temporal resolution limited only by the length of the electron bunches inside the storage ring (30–100 ps) can be routinely achieved. The efficiency of these experiments has recently been significantly improved by using high-power, high-repetition rate picosecond lasers as excitation sources instead of femtosecond kHz Ti:sapphire systems.[37,38]

In the commonly adapted pump–probe tr-XAS approach, gated detectors (typically avalanche photodiodes) are used to record the X-ray fluorescence or transmission signal related to a single X-ray pulse at one specific pump–probe delay at a time. Evidently, such techniques can only use a tiny fraction of the MHz X-ray pulse trains available at third-generation synchrotrons to probe the laser-induced dynamics in the target, which results in a significantly reduced data collection efficiency compared to static X-ray measurements. Recently, Stebel *et al.* demonstrated that replacing the gated (slow) diode with a fast time-resolving detector enables laser-pump/multiple X-ray probe data acquisition schemes that can efficiently sample transient changes in XAS spectra over hundreds of nanoseconds simultaneously.[28]

Fig. 1 illustrates the tr-XAS setup implemented at the Advanced Light Source (ALS). It combines a flexible picosecond laser system[37–39] with a fast MCP/delay line detector[27] with the goal to enable studies of photochemical reactions in a wide range of solids and solid–molecular interfaces on pico- to nanosecond

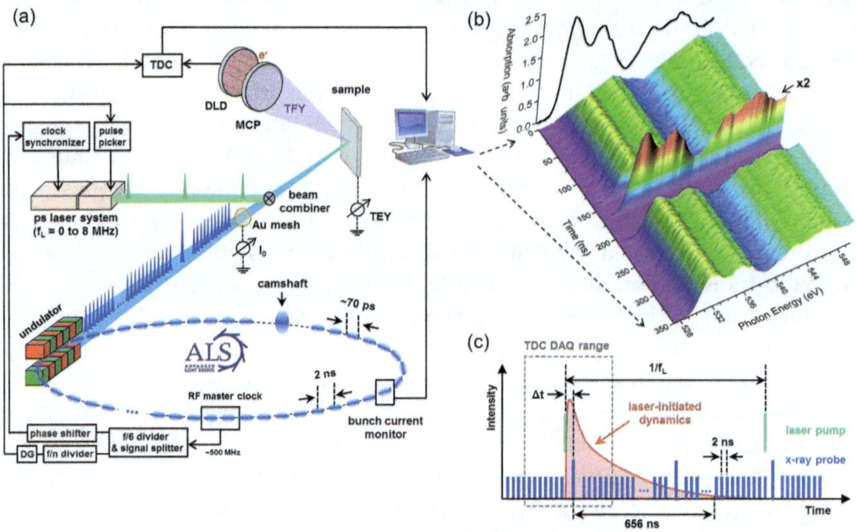

Fig. 1 Overview of the picosecond time-resolved X-ray absorption (tr-XAS) setup currently operated at the Advanced Light Source (ALS). (a) Schematic view of key components of the experiment including the electronic timing scheme applied to achieve and control the synchronisation between the laser pulses and the ALS X-ray pulse train. See text for details. (b) Example of time- and photon-energy-dependent total X-ray fluorescence yield (TFY) spectra obtained from a TiO_2 nanoparticle film at the O1s edge. The fast time-resolved detector and associated electronics enable the unambiguous assignment of TFY signals to individual X-ray pulses within the 500 MHz X-ray pulse train. Note that the apparent merging of the multi-bunch signals is the result of the image rendering routine. A more detailed view of the individual bunch signals is presented in Fig. 2. All tr-XAS traces are individually normalized to the X-ray pulse fluence based on the corresponding electron bunch charges measured inside the storage ring. The data is further corrected for the photon-energy-dependent X-ray transmission of the beamline by using the photocurrent of a gold mesh inserted into the X-ray beam as a reference. The time-averaged total electron yield (TEY) spectrum of the sample (black solid line) is recorded simultaneously with the time-resolved TFY data. (c) Principle of laser-pump/multiple X-ray probe spectroscopy during ALS multi-bunch operating mode.

timescales. Both the laser system and the high-vacuum endstation (base pressure $<5 \times 10^{-9}$ mbar) that accommodates the detector, beam steering optics and diagnostic tools, as well as the motorized sample manipulator, are mobile, flexible units, enabling tr-XAS experiments at a variety of ALS beamlines.

The measurements presented here have been performed at the soft X-ray undulator beamline 8.0.1. The sample is excited with 355 nm laser pulses at a repetition rate of 127 kHz. The corresponding photon energy of 3.5 eV enables efficient generation of electron–hole pairs by excitation above the ~2.2 eV direct band gap of Cu_2O.[32–34] The ensuing charge dynamics due to, *e.g.* carrier trapping and recombination, are probed by monitoring transient spectral changes near the O1s X-ray absorption edge. Since both the valence band (VB) and the bottom of the conduction band (CB) in Cu_2O are derived predominantly from O2p and Cu3d states,[40–42] the dipole allowed O1s → O2p X-ray transitions provide – in principle – simultaneous sensitivity to the generation of VB vacancies and the population of the CB states upon laser excitation.

Laser-X-ray synchronization

A critical issue in synchrotron-based laser-pump/X-ray probe experiments is the precise and long-term stable synchronization of the two independent light sources. The electronic layout to control the phase relation between the laser and the X-ray pulse train in the tr-XAS experiment presented here is shown Fig. 1(a). It is based on a previously described synchronisation scheme that has been successfully applied in time-resolved X-ray photoemission experiments at the ALS.[39,43] Briefly, the ~500 MHz master clock signal of the ALS storage ring is down-converted to ~83 MHz and split into two signal branches. A clock synchronizer is used to phase-lock the laser oscillator with a typical jitter of <1 ps (rms) to the reference signal supplied by the first branch. The frequency of the second signal output can be further adjusted and delayed using a programmable frequency divider and an electronic delay generator (DG). This signal is used to trigger both the pulse picker inside the laser amplifier, and to start the data acquisitions of the time-to-digital converter (TDC) that counts all photons registered by the detector and determines their arrival time with respect to the start signal within a time window of ~350 ns. The time delay between the laser pulses and the X-ray pulse train can be adjusted in steps of the 12 ns oscillator roundtrip time using the DG in the pulse picker synchronization branch. Additionally, time delays with a minimum step size of 5 ps can be set electronically using a phase shifter unit with a total delay range 5 ns in the oscillator feedback loop. This robust and flexible timing scheme has proven to provide reliable synchronisation between the X-ray and laser pulse with sub-picosecond accuracy for several days.[43]

The employed laser system is a commercial high-power picosecond laser (DUETTO™, Time-Bandwidth Products) delivering 10 ps pulses at user-selectable repetition rates between the single-shot limit and 8 MHz. In addition to the fundamental wavelength of 1064 nm, internal frequency doubling and tripling stages provide 532 nm and 355 nm laser pulses.

Data acquisition and normalisation scheme

Typical time-resolved total fluorescence yield XAS data obtained at the O1s edge of a TiO_2 test sample is shown in Fig. 1(b). A cut along the detection time axis at a fixed photon energy yields a time histogram as shown in Fig. 2(a and b). Evidently, the time-response of the detector is sufficiently fast to resolve all the characteristic features imprinted on the TFY by the standard ALS electron bunch filling pattern, *i.e.*, the 2 ns spaced multi-bunch pulses and the more intense "camshaft" pulse that resides within a 60 ns wide gap between the remaining bunches. Compared to conventional pump–probe methods that record data at one specific pump–probe delay at a time, the time-tagging technique enabled by the fast detector/electronics eliminates the need to explore extended delay ranges with a large number of separate measurements. Instead, the long-term dynamics are automatically sampled within the TDC time window (Fig. 1(c)).

In order to gain accurate spectral information, the measured fluorescence yields have to be normalized with respect to the X-ray flux incident on the sample at each photon energy. For static XAS spectra, it is sufficient to determine the time/bunch-averaged incident X-ray flux by measuring, for example, the photocurrent from a gold mesh inserted into the X-ray beam. This procedure is mainly

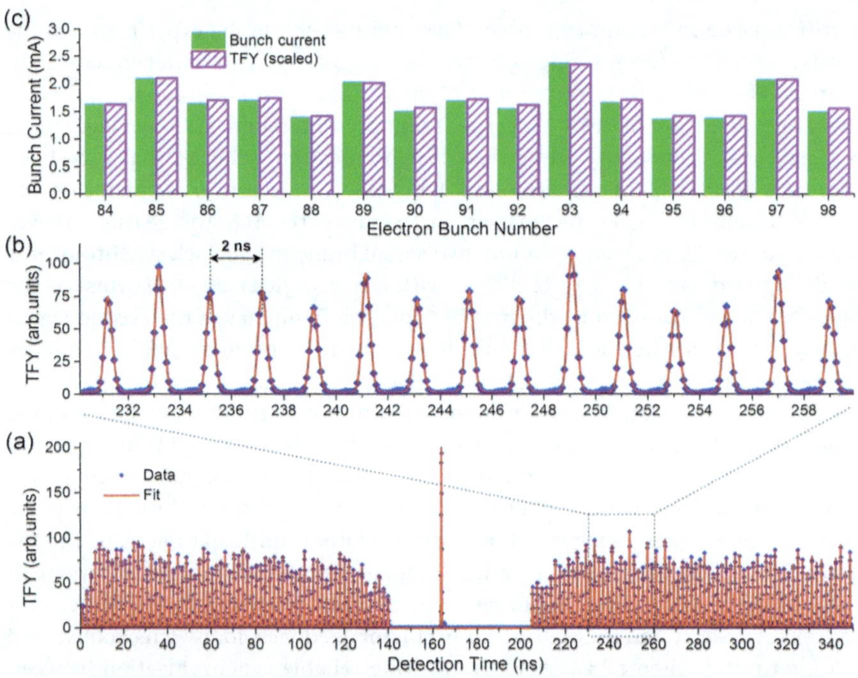

Fig. 2 Correlation between individual bunch currents and TFY signals. (a) Time-resolved TFY (blue markers) from a Cu_2O sample at a fixed photon energy of 532 eV accumulated over 2 min (typical data acquisition time per energy step for high-quality tr-XAS spectra). A global fit of the time histogram to a series of Gaussians with identical FWHMs (red solid line) is used to quantify the TFY intensities (peak areas) induced by the individual X-ray pulses. (b) Enlarged section of the data shown in (a). The overall time resolution of the employed detector/electronics enables an unambiguous separation of the TFY originating from the camshaft pulse and from the 2 ns spaced X-ray pulses within the multi-bunch section of the ALS electron filling pattern. (c) Comparison of the time-resolved TFY to the respective electron bunch currents measured inside the storage ring. A single, global scaling factor has been applied to the TFY intensities to match the TFY of bunch #84 with its bunch current value.

used to correct for the photon-energy dependent transmission of the beamline. In contrast, the tr-XAS traces induced by individual X-ray pulses have to be separately normalized with respect to the X-ray flux associated with specific X-ray pulses. For the data presented here, this bunch-specific flux is estimated by an electron-bunch charge measurement inside the ALS storage ring, which is recorded simultaneously with the time-resolved TFY data.

As illustrated in Fig. 2(c), the bunch charge and the corresponding TFY may vary significantly from bunch to bunch. Moreover, both quantities vary as a function of time due to scattering-induced electron beam losses and frequent bunch refilling events during ALS top-off operating mode. Fig. 3(a) compares the bunch current associated with a specific bunch (#23, red) to the average current of a series of bunches (#1–72, blue) while scanning the photon energy across the O1s

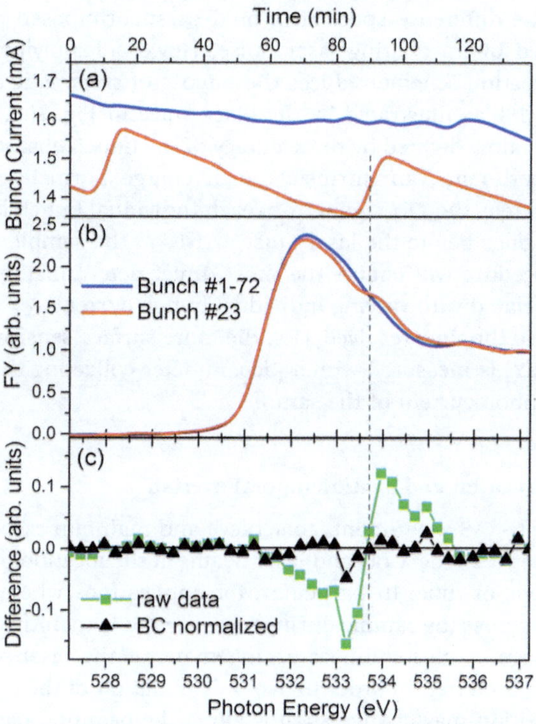

Fig. 3 Importance of individual bunch charge normalization in the tr-XAS experiments. All data were recorded during a single photon energy scan across the entire range starting from the lowest value. Therefore, the photon energy axis represents a time axis with a 2 min point-to-point spacing. (a) The time dependence of individual electron bunch currents inside the ALS storage ring is exemplified by the current of bunch #23 (red solid line). The spurious effects of the characteristic slow decay and step-like increase of the bunch current due to electron scattering losses and electron bunch refilling events, respectively, are clearly visible. They cannot be completely eliminated by averaging the signal of a large number of bunches, *e.g.* bunches #1–72 (blue solid line). (b) Raw O1s tr-XAS spectra of Cu_2O (without any laser excitation) recorded with the indicated X-ray pulses over the same time span as in (a). A significant distortion of the XAS line shape, mainly because of the refilling of bunch #23 during the photon energy scan, is evident. (c) Difference of the two XAS spectra before (green squares) and after (black triangles) normalization to the individual bunch current values shown in (a).

edge region of a Cu_2O sample. In contrast to the relatively stable average current, the individual bunch current undergoes large (~10%) amplitude variations during the time required to record a high-quality XAS spectrum. The importance of an individual bunch charge calibration is highlighted in Fig. 3(b and c). Fig. 3(b) compares the O1s tr-XAS spectra of Cu_2O associated with the same specific bunch (#23, red) and with the series of bunches (#1–72, blue) as used in Fig. 3(a). No laser was employed in this measurement so all X-ray bunches should produce the same spectrum. However, as can be seen in Fig. 3(b), the variations in bunch current and, in particular, the abrupt increase of the bunch charge during refilling leads to significant distortions of the tr-XAS spectral line shape. These intensity variations translate into corresponding amplitude modulations on a similar order of magnitude in the tr-XAS difference spectra. This effect is illustrated by the green curve in Fig. 3(c), which shows the difference between the red and the blue curves in Fig. 3(b). Without further normalization, the significant structures in the difference spectrum would be superimposed on any effect of interest induced by an exciting laser pulse. However, applying the individual bunch normalization scheme reduces these types of systematic artifacts to typically less than 1% as illustrated by the black trace in Fig. 3(c). The remaining variations are mainly defined by the accuracy of the bunch charge measurement. For future experiments, an intrinsic bunch charge normalization scheme is envisioned, whereby the TFY signal from each individual bunch is sampled twice per laser shot, once before the laser pulse arrives at the sample and once thereafter. This procedure will enable the most direct determination of TFY signal variations associated with varying individual bunch currents.

In addition to the time-resolved TFY, the more surface sensitive total electron yield signal (TEY) is measured with a picoammeter collecting the time-averaged X-ray induced photocurrent of the sample.

Beam characterisation and spatiotemporal overlap

It is crucial for tr-XAS experiments to achieve and maintain excellent spatiotemporal overlap between the X-ray and laser beams at the location of the sample for extended periods of time. In particular, for applications where the beams are raster-scanned across the sample during measurements to mitigate the impact of radiation damage, a near-collinear arrangement of the laser-pump and X-ray probe beams is preferred in order to avoid degradation of the spatial overlap as the sample position may change slightly along the beam propagation direction due to mechanical and/or alignment imperfections. The geometric arrangement of the light beams, sample, and detector in the tr-XAS apparatus discussed here is illustrated in Fig. 4(a): a drilled-through laser mirror is positioned (in vacuum) at a 45° angle with respect to the X-rays, which pass through the *ca.* Ø 1 mm central hole of the mirror and hit the sample under normal incidence geometry. The laser beam is collinearly combined with the X-rays by reflecting it off the mirror surface, which forms the upper part of a periscope. The TFY is measured in grazing emission geometry, in order to minimize saturation/self-absorption effects in the XAS spectra.[44] The MCP detector may be optionally covered by a 500 nm thick Al or Ti filter to eliminate background signals due to stray laser light. Fine adjustment of the laser pointing and spot size at the sample position is accomplished with a combination of a picomotor-controlled steering mirror and

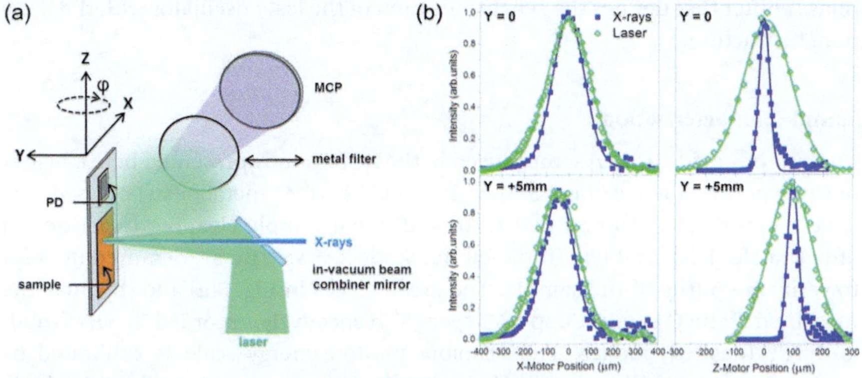

Fig. 4 Details of the pump–probe tr-XAS measurement geometry. (a) Schematic view of the laser and X-ray beam paths, and the arrangement of the sample holder and detector. A drilled-through 45° mirror acts as a beam-combiner that enables a collinear laser/X-ray beam geometry. A 500 nm thick aluminium or titanium metal filter can be translated in front of the detector to block laser stray light. The detector is mounted at a grazing fluorescence detection geometry to minimize saturation/self-absorption effects. (b) Knife-edge scans of the X-ray (blue) and laser (green) beams at different Y positions (along the beam propagation direction) of the sample manipulator. Left and right columns correspond to scans along the horizontal and vertical dimension, respectively. Upper and lower rows correspond to different sample positions along the beam propagation direction. The relative positions of the two beams remain unaffected by moving the sample along the beam propagation direction, demonstrating the robustness of the spatial overlap achieved with the in-vacuum collinear beam-combiner. Minor changes in the absolute beam positions are the result of a small misalignment of the *Y*-axis of the sample manipulator with respect to the beam propagation direction.

a focusing lens mounted on a translation stage outside the vacuum chamber. The sample holder is mounted on a fully motorized, computer-controlled XYZ manipulator, with a total surface area of $5 \times 30\ cm^2$ available for raster scanning. The large scan area will enable future experiments, *e.g.*, on radiation-sensitive semiconductor–molecule interfaces.[43]

The spatial profiles of the laser and the X-ray beams at the sample position are obtained by performing knife-edge scans with a photodiode mounted directly on the sample holder. Typical results are shown in Fig. 4(b) with the X-ray and laser beam profiles marked by blue and green traces, respectively. The left and right columns of Fig. 4(b) correspond to scans along the horizontal and vertical directions, respectively. A significant astigmatism in the X-ray focus, which is unavoidable at the roll-up port of beamline 8.0.1 of the ALS, is readily apparent. The upper and lower rows of Fig. 4(b) correspond to two sample positions that are 5 mm apart along the beam propagation direction. As a result of the collinear beam geometry, the spatial overlap between the two beams is virtually unaffected by this shift.

The relative timing between the laser pulses and the ALS X-ray pulse train can be adjusted with an accuracy of $\pm$50 ps by monitoring the response of a fast metal–semiconductor–metal photodiode (rise/fall time < 30 ps) to both beams on a GHz bandwidth oscilloscope. Fine-tuning of the zero pump–probe delay is achieved by using the photoresponse of a suitable sample while scanning the phase shifter that defines the synchronization of the laser oscillator with the X-ray bunch structure.

Sample characterisation

The cuprous oxide (Cu_2O) sample used in the tr-XAS study presented here consists of Cu_2O powder (566282 Sigma-Aldrich, ≥99.99% trace metal basis) pressed into a *ca.* 2 mm thick Ø 10 mm pellet. No additional sample cleaning is performed after transfer from ambient into vacuum. Static XAS spectra are acquired in order to verify the purity of the sample. The green traces in Fig. 5(a) and (b) show the measured static O1s and Cu2p XAS spectra, respectively, recorded in TEY (solid) and TFY (dashed) modes. The absolute photon energy scale is calibrated by setting the t_{2g} transition energy in the O1s XAS spectrum of a rutile TiO_2 single crystal reference to $h\nu = 530.7$ eV.[45] Except for an overall spectral broadening, the shape of the XAS spectra recorded in TEY and TFY modes agree well with each other, suggesting a very similar chemical composition near the surface and in the bulk of the sample. The broadening in the TFY spectra is likely the result of slightly more pronounced saturation/self-absorption effects compared to the TEY measurements in spite of the grazing detection geometry shown in Fig. 4.[44]

A comparison of the measured spectra with XAS reference data from single crystalline Cu_2O (blue) and CuO (red)[40] confirms the chemical integrity of the Cu_2O powder sample. The main pre-edge features in the O1s and Cu2p XAS spectra of the powder sample at $h\nu = 532.6$ eV and $h\nu = 933.8$ eV, respectively, are

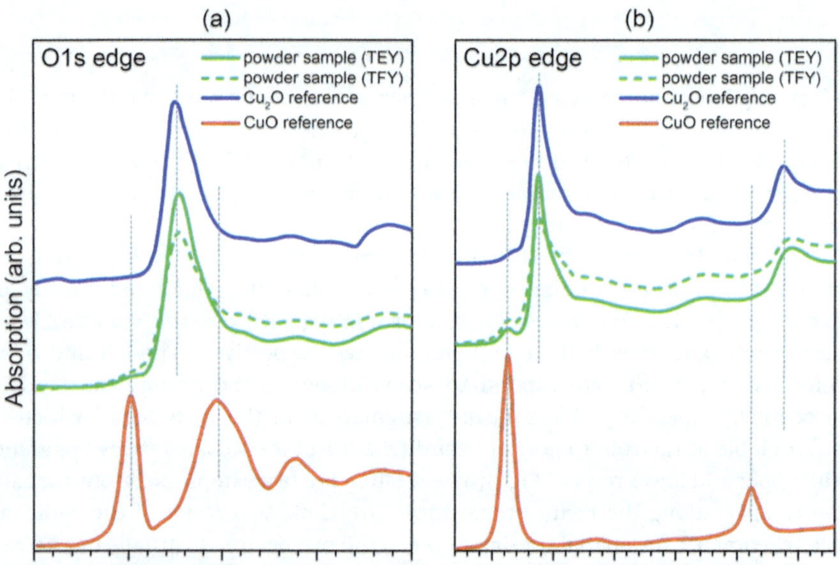

Fig. 5 Characterization of the Cu_2O powder sample. Static O1s (a) and Cu2p (b) XAS spectra recorded in TEY (green solid line) and TFY (green dashed line) are compared to the corresponding reference TEY spectra of single crystalline Cu_2O (blue line) and CuO (red line).[40] Dashed vertical lines indicate transition energies of the most intense absorption features in the reference spectra.

perfectly aligned with the corresponding XAS peaks of the Cu_2O reference. These transitions are due to the excitation of O1s and $Cu2p_{3/2}$ electrons into empty conduction band states with strongly hybridized O2p/Cu3d character.[40,41] The corresponding transitions in CuO occur at *ca.* 2.5 eV lower photon energies. Indeed, a closer inspection of the XAS data in Fig. 5 reveals a small (<5%) but detectable spectral contribution of these transitions to the Cu_2O powder XAS traces.

Results and analysis

For the tr-XAS experiments discussed in the following, the X-ray/laser timing was adjusted such that the laser pulse arrives at the sample $\Delta t = 250$ ps before the camshaft X-ray pulse (Fig. 1(c)). The start trigger for the TDC is adjusted to centre the camshaft signal in the *ca.* 350 ns wide time-recording window of the TDC (Fig. 2(a)). In this way, the tr-XAS traces induced by the ~70 X-ray multi-bunch pulses preceding the laser can be averaged to obtain a high signal-to-noise "pre-laser" ground-state reference, while the X-ray pulses following the camshaft pulse monitor the dynamic evolution of the photo-induced X-ray response at pump–probe delays up to ~180 ns within the same measurement. The laser fluence is varied between 6 mJ cm^{-2} and 10 mJ cm^{-2}, the laser pulse repetition rate is 127 kHz.

The most notable response of the sample to laser excitation is a pronounced, quasi-static broadening and redshift of the main absorption feature near $h\nu = 532$ eV compared to the static O1s XAS spectrum as illustrated in Fig. 6(a). As long as the laser is illuminating the sample, the steady-state "laser-on" XAS spectrum is represented by the red curve, rather than the blue "laser-off" curve that has been measured before laser exposure with 10 mJ cm^{-2}. The dashed green laser-off curve has been measured after laser illumination of the sample. The two laser-off curves are almost identical, demonstrating that the dramatic photo-induced spectral changes are entirely reversible. The quasi-static nature of the effect indicates that the 355 nm laser pulse modulates the X-ray response of the Cu_2O sample on a timescale that significantly exceeds the 8 μs pulse-to-pulse spacing of the 127 kHz laser excitation. The excellent agreement between the blue and green traces in Fig. 6(a) verifies that the effect is not related to an irreversible chemical or structural modification of the Cu_2O sample. Moreover, the corresponding TEY spectra exhibit a very similar behaviour (not shown), suggesting that the mechanism provoking this phenomenon involves both the bulk and the surface of the Cu_2O sample.

The blue trace in Fig. 6(b) shows the difference between the laser-on and the laser-off spectra. Positive features indicate laser-induced enhancement of X-ray absorption, negative features indicate reduced absorption. Applying the method proposed by Hillyard *et al.*,[35] the predominant response between ~529 eV and ~535 eV is analysed in terms of an XAS edge shift that is represented by two Gaussian-shaped spectral contributions: a negative Gaussian at higher energies representing suppressed absorption and a positive Gaussian at lower energies representing enhanced absorption. Using a nonlinear least-squares fit procedure, the positions, widths, and intensities of both model peaks are derived from the measured difference curve, leading to an estimated redshift of the absorption edge by ~0.8 eV.

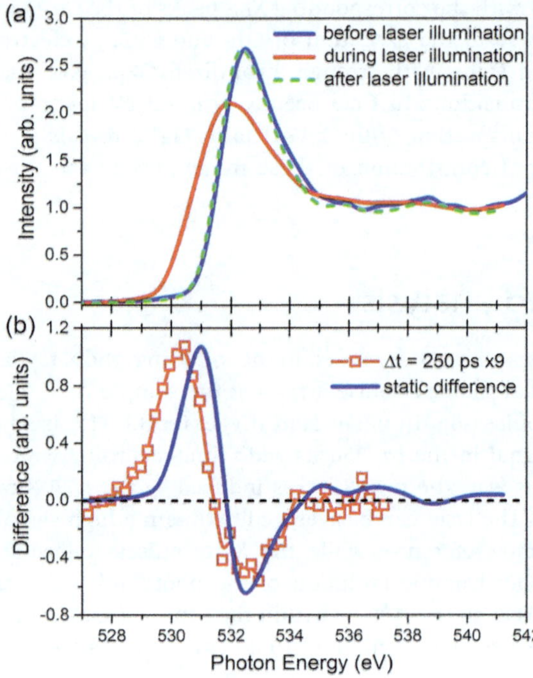

Fig. 6 Evidence for a transient X-ray response with a micro-to-millisecond lifetime. (a) Static O1s XAS spectra recorded before (blue solid line), during (red solid line) and after (green dashed line) illumination of the of Cu_2O sample with the 355 nm laser (10 mJ cm^{-2}; 127 kHz; 0.9 W). (b) The difference of the static laser-on and laser-off XAS spectrum (blue line) is compared to the scaled tr-XAS difference spectrum at $\Delta t = 250$ ps (red squares).

In addition to the quasi-static response, a fast decaying response is observed as illustrated in Fig. 6 and 7. Fig. 7(a) compares the camshaft-related tr-XAS spectrum at $\Delta t = 250$ ps to the pre-laser reference (note that "pre-laser" refers to a spectrum recorded just before the next laser pulse arrives at the sample, which is different from the "laser-off" spectra recorded with the laser output being completely blocked). A distinct laser-induced redshift of the main absorption feature is clearly discernible, which manifests itself in a characteristic bimodal difference spectrum exhibiting a positive and negative amplitude as shown in Fig. 7(b) for a variety of pump–probe delays. The amplitude of the positive peak is a strong function of the pump–probe delay, decaying by more than 70% over a 180 ns time span, whereas the amplitude of the negative peak remains almost constant within the accuracy of the measurement. Moreover, Fig. 7(c) reveals that the spectral shape of the positive peak in the difference spectrum is also affected by the laser fluence, featuring a significant broadening towards lower photon energies when the laser fluence is increased from 6 mJ cm^{-2} to 10 mJ cm^{-2}.

Fig. 8 illustrates the temporal evolution of the positive difference peak amplitude, measured as the peak height ($\Delta t < 500$ ps) or the peak area ($\Delta t > 500$ ps) for different delays. The onset of the response is measured by varying the pump–probe delay (using the phase-shifter unit) between $\Delta t = -500$ and +500 ps. As shown in the inset in Fig. 8, the transient X-ray response is already fully developed within 200 ps. We note that a time resolution of < 100 ps has been demonstrated in time-

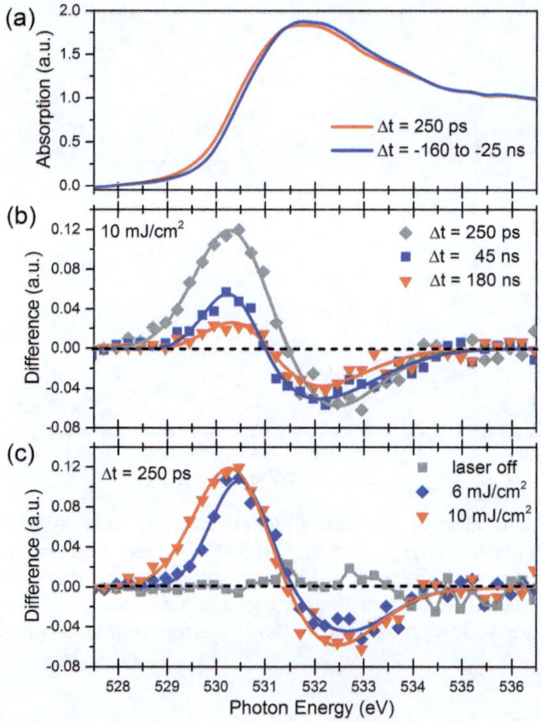

Fig. 7 Laser-induced transient XAS response at the O1s edge of Cu_2O. (a) The tr-XAS spectrum at a pump–probe delay $\Delta t = 250$ ps after 355 nm laser excitation (red solid line) is compared to the tr-XAS spectrum obtained by averaging the XAS signals induced by the 72 X-ray pulses preceding the laser pulse (blue solid line). Post-laser–pre-laser tr-XAS difference spectra recorded with the same laser fluence at different delays Δt, and at a fixed Δt but different laser fluences are shown in (b) and (c), respectively. Solid lines are fits to the tr-XAS difference traces using the positive/negative Gaussian fit model (see Table 1).

resolved X-ray photoelectron experiments relying on the same X-ray/laser synchronization scheme.[39,43] Future experiments are planned to disentangle the characteristic rise time of the transient Cu_2O X-ray response and the instrument-limited time resolution. By analysing the tr-XAS spectra stemming from the multi-bunches after the camshaft, we find a mono-exponential decay with a characteristic time constant of 43 ± 5 ns as shown in Fig. 8.

The magnitude of the edge shift associated with the fast decaying response is analysed by applying the same fit procedure as for the quasi-static response.[35] Within this particular modelling, a strong dependence of the differential energy shift with pump–probe delay is found, decreasing from ~2.2 eV at $\Delta t = 250$ ps down to ~1.2 eV at $\Delta t = 180$ ns (see Table 1). However, we would like to emphasize that this analysis is associated with large uncertainties concerning the Gaussian peak line shapes. We therefore additionally report the energetic difference of the two extrema in the difference spectra, which do not depend on a specific fit model. This procedure yields spectral shifts of ~−2 eV for all pump–probe delays, as noted in Table 2. Interestingly, this differential shift is comparable to the band gap energy of Cu_2O.[32–34] The corresponding model-free shift of the quasi-static

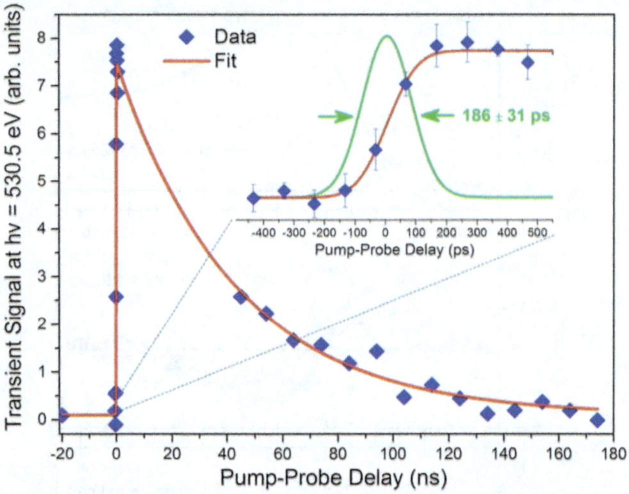

Fig. 8 Temporal evolution of the laser-induced transient change of the O1s XAS line shape probed at a photon energy of $h\nu = 530.5$ eV. The red solid line is a fit to a sum of a step function and a mono-exponential decay model (time constant 43 ± 5 ns) convoluted with a Gaussian. The inset highlights the sub-nanosecond rise time of the laser-induced X-ray response. Error bars denote standard errors of an average of 4 pump–probe delay scans.

response is 1.6 eV. The qualitative differences between the fast decaying and the quasi-static responses are readily apparent from the comparison of the differential tr-XAS curves in Fig. 6(b). While the fast decaying response corresponds to a larger shift in energy, its amplitude is about a factor of 9 smaller than that of the quasi-static response.

Discussion

Cuprous oxide is among the class of metal oxides under consideration for a variety of photo- and electrochemical applications.[29,30,32–35,46–50] Understanding the material's electronic and chemical properties under non-equilibrium conditions is important for the evaluation of its potential and fundamental limitations.

Table 1 Parameters extracted from a fit of the O1s tr-XAS difference spectra to a sum of positive- and negative-amplitude Gaussians (A = peak area; FWHM = full width at half maximum; E_{fit} = peak-to-peak separation of the two Gaussians)

Data	$A(+)$	$A(-)$	FWHM(+) [eV]	FWHM(−) [eV]	E_{fit} [eV]
Hillyard *et al.*[35]	0.19	0.28	0.99	2.15	1.14
Static diff. 10 mJ cm^{-2}	3.09	2.52	1.60	2.01	0.78
$\Delta t = 250$ ps@6 mJ cm^{-2}	0.16	0.10	1.40	2.07	2.17
$\Delta t = 250$ ps@10 mJ cm^{-2}	0.25	0.15	1.81	2.36	1.94
$\Delta t = 45$ ns@10 mJ cm^{-2}	0.11	0.16	1.37	2.94	1.65
$\Delta t = 180$ ns@10 mJ cm^{-2}	0.09	0.13	1.67	2.76	1.15

Table 2 Model-independent key parameters of the tr-XAS O1s difference spectra of Cu_2O: integrals over the positive (Int(+)) and negative (Int(−)) parts of the difference spectra and the energetic separation between the maximum and minimum Δmax/min

Data	Int(+)	Int(−)	Δmax/min [eV]
Hillyard *et al.*[35]	0.097	0.185	1.3
Static diff. 10 mJ cm^{-2}	1.471	0.910	1.6
$\Delta t = 250$ ps@6 mJ cm^{-2}	0.145	0.081	2.2
$\Delta t = 250$ ps@10 mJ cm^{-2}	0.197	0.104	2.2
$\Delta t = 45$ ns@10 mJ cm^{-2}	0.032	0.070	2.0
$\Delta t = 180$ ns@10 mJ cm^{-2}	0.035	0.064	1.9

The ground state electronic structure and the electronic excitation spectrum of Cu_2O have long been a matter of continued discussions.[32–35,40–42,50–55] Whereas the excitonic spectrum of Cu_2O is well understood and often serves as a textbook example for bound electron–hole excitations in condensed matter systems,[33,53] other fundamental properties such as the extent of the band gap and the relative orbital contributions to different portions of the VB and CB are subject to ongoing debate.[35,40,42,52,53,56–59] X-ray spectroscopy studies are particularly well suited to probe the density and character of occupied and unoccupied states, employing element specific transitions with well-defined selection rules.[40,57,59–61] An extension of this class of experiments to optical-pump/X-ray-probe studies, whereby inner-shell transitions are employed to monitor *transient* valence electronic configurations, may provide not only complementary information on the character of VB and CB states, but also their dynamic evolution under non-equilibrium conditions that are relevant for many envisioned applications of the material.[16,35,62–65]

The picosecond tr-XAS experiments described here reveal two types of laser-induced dynamic responses in the O1s XAS spectrum of Cu_2O. Excitation with 355 nm laser pulses at fluences of ~6–10 mJ cm^{-2} leads to both a fast (~ns) and a slowly ($\gg$8 μs) decaying, reversible line shape change of the dominant XAS feature at $h\nu = 532$ eV. Without state-of-the-art theory support, it is challenging to provide details on the nature of the electronic configurations that induce the observed tr-XAS signals. Nevertheless, some qualitative results are readily apparent and may be discussed in the context of previous studies and general considerations concerning transient XAS experiments on condensed phase systems.

The slow response manifests itself in a quasi-static redshift of the absorption edge by approximately 1 eV. Interestingly, this value reproduces the results of Gaffney and co-workers who studied a 50 nm thick film of Cu_2O by tr-XAS in transmission geometry using 400 nm excitation at essentially the same pulse fluences as used in our experiment but at a laser repetition rate of only 1 kHz.[35] In their study, a ~1 eV redshift was the only observed response within the ±200 ps pump–probe delay range accessible in their experiment, and it was observed as a transient response, *i.e.*, it fully relaxed on the 1 ms timescale between laser pulses. Given the similarities between the observed shifts, we speculate that the quasi-static response observed in our experiment may be linked to the same underlying physics as the transient response detected by Hillyard *et al.*, and that the different temporal evolution (quasi-static *vs.* transient) are caused by the significantly different laser repetition rates. Within this picture, the lifetime of the configurations leading to the ~1 eV O1s XAS shift would have to be on the order of 8 μs ≪ Δ*t* ≪ 1 ms. In fact, among direct-band gap semiconductors, Cu_2O is known to exhibit unusually long excitonic lifetimes due to dipole-forbidden recombination channels.[66,67] Previous studies reported charge carrier recombination timescales ranging from sub-nanosecond to milliseconds depending on the quality of the sample and its temperature.[67–69] It is, therefore, possible that the combination of high repetition rate laser excitation and long-lived excited electronic configurations leads to a build-up of electron–hole states that appear as a significant, quasi-static differential tr-XAS signal as shown in Fig. 6.

As noted by Hillyard *et al.*,[35] the observed XAS edge shift corresponds to only about half of the 2.2 eV optical band gap energy of Cu_2O.[32–34] At first glance this seems somewhat surprising. On the timescales accessible with the experiment, holes and photoexcited electrons are expected to have relaxed to the top of the VB and the bottom of the CB, respectively.[35,66] As holes in previously occupied VB states may lead to enhanced inner-shell absorption below the ground state XAS edge and the filling of usually unoccupied CB states with photoexcited electrons may suppress transitions in the XAS spectrum at higher energies, one may expect an effective redshift in the tr-XAS spectrum where the depletion and enhancement regions are separated by the band gap energy. While the direction of the observed shift conforms to this basic model, its magnitude clearly does not. We note that even more advanced multi-body models predict an X-ray edge shift after super-band gap excitation comparable to the band gap provided that only VB and CB states are involved in the photo-induced dynamics. The discrepancies between the experimental results and general expectations for tr-XAS spectra can, therefore, not readily be explained and we resort to a discussion of possible contributing factors.

One tempting explanation for an apparently reduced energy gap may be the involvement of mid-gap states. For example, impurity and/or defect induced trap states may be abundant in the pellet sample used in this study and could be located in the nominal band gap of single crystalline Cu_2O. Such states may also be associated with very long lifetimes, which would explain the observed saturation effect leading to the quasi-static response. Even then, however, the amplitude of the quasi-static photo-induced effect, which amounts to ~30% of the maximum static XAS signal, seems unusually large to be associated with mid-gap states. More information on the possible involvement of such states will be gained from a complementary optical characterization of the samples.

More generally, the simple model outlined above for expected photo-induced XAS edge shifts does not take the localized nature of inner-shell transitions and final state effects in core-excited configurations into account.[41,70,71] Previous time-resolved inner-shell absorption studies have demonstrated that, depending on the character of the involved valence configurations, transient XAS shifts in condensed phase systems are sometimes better described in a localized picture of atomic oxidation state changes rather than within a delocalized band model.[16,62,63] This trend is expected to be particularly pronounced for materials with strong ionic bonding character, such as transition metal oxides, where the valence orbitals are significantly localized.[53,63] Several studies have pointed out that, despite a nominally filled Cu3d shell, which dominates both the occupied and unoccupied electronic states near the VB and CB edges, respectively, bonding in Cu_2O is not purely ionic but also exhibits notable contributions of hybridization and covalent bonding involving Cu3d, Cu4s, and O2p states.[40–42,57] X-ray transitions in this material may therefore bear signatures of both localized and delocalized valence orbital configurations and the correlation between optical excitations and corresponding changes in X-ray absorption is non-trivial, even without taking core-hole relaxation into account. We note, however, that a strong charge-transfer between Cu and O atomic sites is not expected for electron/hole configurations that have relaxed to the CB/VB edges. This is supported by the tr-XAS results of Hillyard *et al.* who find almost identical redshifts for the Cu2p and O1s edges and theoretical calculations that predict a mixed Cu3d/O2p orbital character near both the VB and the CB edges.[41,42,57]

Self-energy (also termed relaxation, screening, polarization, or reorganization) corrections in the core-excited states may further complicate a straightforward interpretation of the transient difference spectrum.[41,61,70,71] These corrections are the result of the response of all electrons in a system to the creation of a core-hole. The electron/core-hole interaction may lower the energy of the final state by several tens of eV compared to the first-order picture of non-interacting electrons (for atoms with $Z \leq 30$).[71] However, in view of the results presented here, the corrections would have to be *different* for excitations of O1s electrons into VB and CB vacancies in order to affect the magnitude of the observed transient XAS shift, which would imply different electron/core-hole interactions and, therefore, different orbital configurations at the VB and CB edges. In this case, however, one may also expect a difference in the cross sections for core-excitation into VB and CB vacancies, which would result in different magnitudes of the tr-XAS suppression and enhancement features. Interestingly, such an asymmetry is indeed observed (see Fig. 7(b and c)). Integration over the positive and negative slopes of the differential absorption spectra at 250 ps delay reveals that the transient signal enhancement exceeds the signal suppression by a factor of 1.9 and 2.0 for pump laser fluences of 6 mJ cm^{-2} and 10 mJ cm^{-2}, respectively. While the difference between these two numbers may not be significant, their absolute values indicate an enhanced X-ray absorption involving VB compared to CB vacancies. This finding would be consistent with a difference in electronic character of the occupied (VB) and unoccupied (CB) band edge states and, therefore, a difference in core-hole relaxation energies seems possible.

We note that Hillyard *et al.* found no definitive evidence for an asymmetry in their tr-XAS difference spectra and concluded that the contributions of Cu3d and O2p orbitals to the VB and CB edges must be very similar.[35] We are unable to provide a conclusive explanation for the differences in the observed asymmetries but we point out that the two studies were not performed under identical conditions. The sample for the transmission-based experiment was a 50 nm polycrystalline film of Cu_2O deposited on a silicon nitride membrane, whereas an ~2 mm thick pellet of Cu_2O powder was used for the TFY based experiment reported here. Perhaps the most significant difference is an increase in the average laser exposure by about two orders of magnitude in the current experiment (127 kHz compared to 1 kHz), which may lead to saturation effects as indicated by the quasi-static response illustrated in Fig. 6. Future studies using thin films of Cu_2O, lower pulse repetition rates, and, potentially, lower peak fluences, will provide more information on the interplay between these parameters and the observed spectral X-ray responses.

Principally, structural effects (in addition to purely electronic dynamics) need to be taken into account. Reversible solid–liquid phase changes in germanium (Ge), *e.g.*, have previously been shown to produce tr-XAS changes on similar orders of magnitude as observed for the quasi-static Cu_2O response.[28] Inducing these changes, however, required more than an order of magnitude higher laser fluences ($\sim$120 mJ cm^{-2}) compared to the experiment presented here. Given the very similar melting points of Cu_2O and Ge, it is unlikely that the comparably small fluence of the current experiment could induce a similar phase change. The solid phase of Ge was completely restored within $\sim$200 ns after laser excitation, which is very small compared to the $\gg$8 μs lifetime of the quasi-static response observed here.

For completeness, we mention that, in principle, a completely different effect could also cause a deviation of tr-XAS redshift from the optical band gap energy of Cu_2O. Based on first principles, self-consistent orthogonalized linear combination of atomic orbitals (OLCAO) calculations, Ching and co-workers predicted that the intrinsic band gap of Cu_2O is on the order of only 0.8 eV and that the difference between this value and the optically measured band gap of 2.2 eV is due to symmetry-forbidden transitions.[55] Given the similarity of this intrinsic band gap estimate with the spectral shifts observed in both tr-XAS studies on Cu_2O, it might be tempting to draw parallels between the theoretical and experimental results. However, the local density approximation (LDA) approach associated with OLCAO is known to underestimate band gap values by significant margins,[57,72] and neither current state-of-the-art methods that go beyond LDA[53,72] nor alternative experimental methods to measure the band gap[60] seem to reproduce the OLCAO result. We, therefore, conclude that the most likely reasons for the relatively small photo-induced tr-XAS shift compared to the optical band gap may be related to mid-gap states, core-hole relaxation effects and, potentially, the impact of localized valence electron configurations. Clearly, more high-level theory efforts are needed to provide a more conclusive explanation of this phenomenon.

Concerning the fast-decaying response, a closer inspection of the tr-XAS spectra reveals that the maximum relative changes in the absorption spectrum associated with this response are on the order of $\approx$4–12% ($\approx$18–56 mOD). These difference amplitudes are essentially the same for both laser pulse fluences of 6 mJ cm^{-2} and 10 mJ cm^{-2}. However, the absorption enhancement feature recorded with the higher fluence extends to lower photon energies by $\approx$0.5 eV while the depletion feature remains almost unchanged (Fig. 7(c)). Note that the peak of the enhancement feature appears at the same energy for both fluences (530.5 eV), only the shape changes. This effect already demonstrates a shortcoming of the simple Gaussian-based modelling of the tr-XAS features, which does not capture such subtle changes in a physically insightful way. Without further experimental and theoretical investigation, it is challenging to provide a better description and deeper interpretation of the observed spectral changes. Nevertheless, the observation provides additional evidence that, at least in the study presented here, transient depletion and enhancement features in the O1s tr-XAS spectrum of photoexcited Cu_2O are not mirror images of each other. This effect may be related to differences in electronic characters at the VB and CB edges, different relaxation dynamics of photogenerated holes and electrons, and/or the impact of saturation effects.

The $\sim$2 eV shift associated with the short-lived response has not been reported previously. Given the discussion on expected photo-induced XAS edge shifts outlined above, it may be tempting to draw connections between the short-lived response and the 2.2 eV optical band gap of Cu_2O. While we cannot exclude a deeper physical connection, we emphasize that the saturation conditions that are evident by the quasi-static response may induce a non-trivial connection between the two observed effects. In the case of a saturated density of hole states at the VB edge, for example, additionally created holes may not be able to relax all the way to the VB edge within the expected picosecond time frame and, instead, lead to inner-shell transitions at lower energies on the nanosecond timescales observed in the experiment (Fig. 8(b)). The same concept would provide a qualitative explanation for the observed extension of the fast decaying response to lower energies upon increasing the pump fluence from 6 mJ cm^{-2} to 10 mJ cm^{-2} as illustrated in Fig. 7(c). Note that such an effect would not necessarily have to be symmetric for photo-induced holes and electrons (*i.e.*, tr-XAS enhancement and suppression) since the densities of states on both sides of the band gap are not identical.[60]

Conclusion and outlook

A newly developed tr-XAS experimental setup is presented that employs a fast time-tagging technique to simultaneously monitor transient changes in the XAS spectra of condensed phase samples across temporal scales from tens of picoseconds to hundreds of nanoseconds. A proof-of-concept experiment shows that photo-induced dynamics in a microporous Cu_2O powder sample after 355 nm excitation at fluences of 6–10 mJ cm^{-2} lead to two distinct responses in the O1s tr-XAS spectrum. A quasi-static shift of the ground state O1s XAS spectrum by ~1 eV to lower energies indicates a lifetime of the corresponding transient configurations well beyond the 8 μs delay between laser pulses. An additional ~2 eV redshift of the absorption edge appears within less than 200 ps and decays on a timescale of 43 ± 5 ns. While the amplitude of this effect is insensitive to a doubling of the laser fluence, the shape of the photo-induced enhanced absorption feature does vary and extends to lower energies for the higher fluence. Absorption enhancement and depletion features in both the fast decaying and the quasi-static response are not symmetric, with the photo-induced enhancement exceeding the depletion by a factor of ~1.7–2.0.

All observed differential tr-XAS effects are tentatively assigned to transient, photo-induced electron–hole configurations governed by different lifetimes. The dominant quasi-static shift reproduces the transient shift observed by Gaffney and co-workers. We speculate that both observations are connected to similar underlying electronic dynamics and that the different appearances (quasi-static *vs.* transient) are linked to saturation effects in the present experiment due to the high laser repetition rate. The quasi-static shift cannot be explained within a picture that would suggest a shift of the X-ray absorption edge by the extent of the optical band gap energy of 2.2 eV. The qualitative difference between the expected and measured shifts may be connected to mid-gap states, core-hole relaxation effects, and/or the localized nature of core-hole transitions in combination with the at least partially localized nature of the Cu_2O valence orbitals. The XAS edge shift associated with the fast decaying response approximately matches the extent of the band gap. However, in view of the by far dominant quasi-static response, a direct connection might not be appropriate. A simple picture of saturated hole/electron densities at the VB/CB edges associated with the quasi-static response provides a possible qualitative explanation for the difference between the two shifts and the observed trend of an increased redshift in the short-lived tr-XAS response with increasing laser fluence.

The observed asymmetries in all differential tr-XAS spectra may be an indication for different orbital mixtures in VB and CB states. However, we cannot exclude that excitation saturation effects may also play a role. A much larger body of experimental and, in particular, theoretical work is needed to give a conclusive interpretation of the observed effects. Nevertheless, the discussion demonstrates that the newly developed technique provides a wealth of information on the electronic structure and dynamics of photoactive metal oxide semiconductors from the viewpoint of well-defined atomic sites.

Extending the realm of time-domain X-ray spectroscopy techniques to non-replenishing condensed phase systems and, ultimately, operating photochemical devices holds great promise to capture transient electronic and chemical configurations that may be crucial for the overall function of a material or device but that are too short-lived to be tracked by steady-state spectroscopy techniques. X-ray absorption measurements using photon-in/photon-out techniques are already playing an important role in the static domain where *in situ* and *in operando* methods for the study of ongoing photochemical processes have significantly advanced over the past decade. Time-resolved XAS techniques – such as presented here – may provide a route toward monitoring ongoing chemical transformations with the atomic site specificity and chemical sensitivity provided by inner-shell transitions, in particular, when combined with state-of-the-art *in operando* X-ray spectroscopy components.

Acknowledgements

This work was supported by the U.S. Department of Energy, Office of Basic Energy Sciences, Chemical Sciences, Geosciences and Biosciences Division, through Contract No. DE-AC02-05CH11231. O. G. was supported by the Department of Energy Office of Science Early Career Research Program. S. N. acknowledges support by the Alexander von Humboldt foundation. We would like to thank Dr C. D. Pemmaraju for stimulating discussions and the staff of the Advanced Light Source for help and assistance. The Advanced Light Source is supported by the Director, Office of Science, Office of Basic Energy Sciences, of the U.S. Department of Energy under Contract No. DE-AC02-05CH11231.

References

1 O. Fuchs, F. Maier, L. Weinhardt, M. Weigand, M. Blum, M. Zharnikov, J. Denlinger, M. Grunze, C. Heske and E. Umbach, *Nucl. Instrum. Methods Phys. Res., Sect. A*, 2008, **585**, 172–177.

2 A. Braun, K. Sivula, D. K. Bora, J. Zhu, L. Zhang, M. Grätzel, J. Guo and E. C. Constable, *J. Phys. Chem. C*, 2012, **116**, 16870–16875.

3 N. Becknell, Y. Kang, C. Chen, J. Resasco, N. Kornienko, J. Guo, N. M. Markovic, G. A. Somorjai, V. R. Stamenkovic and P. Yang, *J. Am. Chem. Soc.*, 2015, **137**, 15817–15824.

4 S. Nemšák, A. Shavorskiy, O. Karslioglu, I. Zegkinoglou, A. Rattanachata, C. S. Conlon, A. Keqi, P. K. Greene, E. C. Burks, F. Salmassi, E. M. Gullikson, S.-H. Yang, K. Liu, H. Bluhm and C. S. Fadley, *Nat. Commun.*, 2014, **5**, 5441.

5 E. Gross, X.-Z. Shu, S. Alayoglu, H. A. Bechtel, M. C. Martin, F. D. Toste and G. A. Somorjai, *J. Am. Chem. Soc.*, 2014, **136**, 3624–3629.

6 D. N. Mueller, M. L. Machala, H. Bluhm and W. C. Chueh, *Nat. Commun.*, 2015, **6**, 6097.

7 E. J. Crumlin, Z. Liu, H. Bluhm, W. Yang, J. Guo and Z. Hussain, *J. Electron. Spectros. Relat. Phenom.*, 2015, **200**, 264–273.

8 H. Ali-Löytty, M. W. Louie, M. R. Singh, L. Li, H. G. Sanchez Casalongue, H. Ogasawara, E. J. Crumlin, Z. Liu, A. T. Bell, A. Nilsson and D. Friebel, *J. Phys. Chem. C*, 2016, **120**, 2247–2253.

9 A. H. Zewail, *J. Phys. Chem. A*, 2000, **104**, 5660–5694.

10 Z.-H. Loh and S. R. Leone, *J. Phys. Chem. Lett.*, 2013, **4**, 292–302.

11 A. R. Attar, A. Bhattacherjee and S. R. Leone, *J. Phys. Chem. Lett.*, 2015, **6**, 5072–5077.

12 H. Dachraoui, M. Michelswirth, P. Siffalovic, P. Bartz, C. Schäfer, B. Schnatwinkel, J. Mattay, W. Pfeiffer, M. Drescher and U. Heinzmann, *Phys. Rev. Lett.*, 2011, **106**, 107401.

13 C. Bressler and M. Chergui, *Annu. Rev. Phys. Chem.*, 2010, **61**, 263–282.

14 M. Saes, C. Bressler, R. Abela, D. Grolimund, S. Johnson, P. Heimann and M. Chergui, *Phys. Rev. Lett.*, 2003, **90**, 047403.

15 W. Gawelda, M. Johnson, F. M. F. de Groot, R. Abela, C. Bressler and M. Chergui, *J. Am. Chem. Soc.*, 2006, **128**, 5001–5009.

16 M. H. Rittmann-Frank, C. J. Milne, J. Rittmann, M. Reinhard, T. J. Penfold and M. Chergui, *Angew. Chem., Int. Ed.*, 2014, **53**, 5858–5862.

17 L. X. Chen, *Annu. Rev. Phys. Chem.*, 2005, **56**, 221–254.

18 B. E. Van Kuiken, N. Huse, H. Cho, M. L. Strader, M. S. Lynch, R. W. Schoenlein and M. Khalil, *J. Phys. Chem. Lett.*, 2012, **3**, 1695–1700.

19 N. Huse, H. Cho, K. Hong, L. Jamula, F. M. F. de Groot, T. K. Kim, J. K. McCusker and R. W. Schoenlein, *J. Phys. Chem. Lett.*, 2011, **2**, 880–884.

20 P. Wernet, K. Kunnus, I. Josefsson, I. Rajkovic, W. Quevedo, M. Beye, S. Schreck, S. Grubel, M. Scholz, D. Nordlund, W. Zhang, R. W. Hartsock, W. F. Schlotter, J. J. Turner, B. Kennedy, F. Hennies, F. M. F. de Groot, K. J. Gaffney, S. Techert, M. Odelius and A. Fohlisch, *Nature*, 2015, **520**, 78–81.

21 H. Öström, H. Öberg, H. Xin, J. LaRue, M. Beye, M. Dell'Angela, J. Gladh, M. L. Ng, J. A. Sellberg, S. Kaya, G. Mercurio, D. Nordlund, M. Hantschmann, F. Hieke, D. Kühn, W. F. Schlotter, G. L. Dakovski, J. J. Turner, M. P. Minitti, A. Mitra, S. P. Moeller, A. Föhlisch, M. Wolf, W. Wurth, M. Persson, J. K. Nørskov, F. Abild-Pedersen, H. Ogasawara, L. G. M. Pettersson and A. Nilsson, *Science*, 2015, **347**, 978–982.

22 M. Dell'Angela, T. Anniyev, M. Beye, R. Coffee, A. Föhlisch, J. Gladh, T. Katayama, S. Kaya, O. Krupin, J. LaRue, A. Møgelhøj, D. Nordlund, J. K. Nørskov, H. Öberg, H. Ogasawara, H. Öström, L. G. M. Pettersson, W. F. Schlotter, J. A. Sellberg, F. Sorgenfrei, J. J. Turner, M. Wolf, W. Wurth and A. Nilsson, *Science*, 2013, **339**, 1302–1305.

23 K. R. Siefermann, C. D. Pemmaraju, S. Neppl, A. Shavorskiy, A. A. Cordones, J. Vura-Weis, D. S. Slaughter, F. P. Sturm, F. Weise, H. Bluhm, M. L. Strader, H. Cho, M.-F. Lin, C. Bacellar, C. Khurmi, J. Guo, G. Coslovich,

J. S. Robinson, R. A. Kaindl, R. W. Schoenlein, A. Belkacem, D. M. Neumark, S. R. Leone, D. Nordlund, H. Ogasawara, O. Krupin, J. J. Turner, W. F. Schlotter, M. R. Holmes, M. Messerschmidt, M. P. Minitti, S. Gul, J. Z. Zhang, N. Huse, D. Prendergast and O. Gessner, *J. Phys. Chem. Lett.*, 2014, **5**, 2753–2759.

24 W. K. Zhang, R. Alonso-Mori, U. Bergmann, C. Bressler, M. Chollet, A. Galler, W. Gawelda, R. G. Hadt, R. W. Hartsock, T. Kroll, K. S. Kjaer, K. Kubicek, H. T. Lemke, H. Y. W. Liang, D. A. Meyer, M. M. Nielsen, C. Purser, J. S. Robinson, E. I. Solomon, Z. Sun, D. Sokaras, T. B. van Driel, G. Vanko, T. C. Weng, D. L. Zhu and K. J. Gaffney, *Nature*, 2014, **509**, 345.

25 S. E. Canton, K. S. Kjær, G. Vankó, T. B. van Driel, S.-i. Adachi, A. Bordage, C. Bressler, P. Chabera, M. Christensen, A. O. Dohn, A. Galler, W. Gawelda, D. Gosztola, K. Haldrup, T. Harlang, Y. Liu, K. B. Møller, Z. Németh, S. Nozawa, M. Pápai, T. Sato, T. Sato, K. Suarez-Alcantara, T. Togashi, K. Tono, J. Uhlig, D. A. Vithanage, K. Wärnmark, M. Yabashi, J. Zhang, V. Sundström and M. M. Nielsen, *Nat. Commun.*, 2015, **6**, 6359.

26 S. Neppl, Y.-S. Liu, C.-H. Wu, A. Shavorskiy, I. Zegkinoglou, T. Troy, D. S. Slaughter, M. Ahmed, A. S. Tremsin, J.-H. Guo, P.-A. Glans, M. Salmeron, H. Bluhm and O. Gessner, *Ultrafast Phenomena XIX*, 2014, vol. 162, pp. 325–327.

27 A. S. Tremsin, O. H. W. Siegmund, J. S. Hull, J. V. Vallerga, J. B. McPhate, J. Soderstrom, J. W. Chiou, J. Guo and Z. Hussain, *IEEE Nucl. Sci. Symp. Conf. Rec.*, 2006, **2**, 735–739.

28 L. Stebel, M. Malvestuto, V. Capogrosso, P. Sigalotti, B. Ressel, F. Bondino, E. Magnano, G. Cautero and F. Parmigiani, *Rev. Sci. Instrum.*, 2011, **82**, 123109.

29 M. Hara, T. Kondo, M. Komoda, S. Ikeda, J. N. Kondo, K. Domen, M. Hara, K. Shinohara and A. Tanaka, *Chem. Commun.*, 1998, 357–358, DOI: 10.1039/A707440I.

30 V. S. K. Yadav and M. K. Purkait, *Energy Fuels*, 2015, **29**, 6670–6677.

31 J. Y. Xiang, X. L. Wang, X. H. Xia, L. Zhang, Y. Zhou, S. J. Shi and J. P. Tu, *Electrochim. Acta*, 2010, **55**, 4921–4925.

32 M. A. Rafea and N. Roushdy, *J. Phys. D: Appl. Phys.*, 2009, **42**, 015413.

33 P. W. Baumeister, *Phys. Rev.*, 1961, **121**, 359–362.

34 S. Nikitine, J. B. Grun and M. Sieskind, *J. Phys. Chem. Solids*, 1961, **17**, 292–300.

35 P. W. Hillyard, S. V. N. T. Kuchibhatla, T. E. Glover, M. P. Hertlein, N. Huse, P. Nachimuthu, L. V. Saraf, S. Thevuthasan and K. J. Gaffney, *Phys. Rev. B: Condens. Matter Mater. Phys.*, 2009, **80**, 125210.

36 M. Saes, F. van Mourik, W. Gawelda, M. Kaiser, M. Chergui, C. Bressler, D. Grolimund, R. Abela, T. E. Glover, P. A. Heimann, R. W. Schoenlein, S. L. Johnson, A. M. Lindenberg and R. W. Falcone, *Rev. Sci. Instrum.*, 2004, **75**, 24–30.

37 F. A. Lima, C. J. Milne, D. C. V. Amarasinghe, M. H. Rittmann-Frank, R. M. v. d. Veen, M. Reinhard, V.-T. Pham, S. Karlsson, S. L. Johnson, D. Grolimund, C. Borca, T. Huthwelker, M. Janousch, F. van Mourik, R. Abela and M. Chergui, *Rev. Sci. Instrum.*, 2011, **82**, 063111.

38 A. M. March, A. Stickrath, G. Doumy, E. P. Kanter, B. Krässig, S. H. Southworth, K. Attenkofer, C. A. Kurtz, L. X. Chen and L. Young, *Rev. Sci. Instrum.*, 2011, **82**, 073110.

39 S. Neppl, A. Shavorskiy, I. Zegkinoglou, M. Fraund, D. S. Slaughter, T. Troy, M. P. Ziemkiewicz, M. Ahmed, S. Gul, B. Rude, J. Z. Zhang, A. S. Tremsin, P.-A. Glans, Y.-S. Liu, C. H. Wu, J. Guo, M. Salmeron, H. Bluhm and O. Gessner, *Faraday Discuss.*, 2014, **171**, 219–241.
40 P. Jiang, D. Prendergast, F. Borondics, S. Porsgaard, L. Giovanetti, E. Pach, J. Newberg, H. Bluhm, F. Besenbacher and M. Salmeron, *J. Chem. Phys.*, 2013, **138**, 024704.
41 M. Grioni, J. F. van Acker, M. T. Czyžyk and J. C. Fuggle, *Phys. Rev. B: Condens. Matter Mater. Phys.*, 1992, **45**, 3309–3318.
42 P. Marksteiner, P. Blaha and K. Schwarz, *Z. Phys. B: Condens. Matter*, 1986, **64**, 119–127.
43 S. Neppl and O. Gessner, *J. Electron Spectrosc. Relat. Phenom.*, 2015, **200**, 64–77.
44 R. Nakajima, J. Stöhr and Y. U. Idzerda, *Phys. Rev. B: Condens. Matter*, 1999, **59**, 6421–6429.
45 F. M. F. de Groot, M. Grioni, J. C. Fuggle, J. Ghijsen, G. A. Sawatzky and H. Petersen, *Phys. Rev. B: Condens. Matter Mater. Phys.*, 1989, **40**, 5715–5723.
46 B. K. Meyer, A. Polity, D. Reppin, M. Becker, P. Hering, P. J. Klar, T. Sander, C. Reindl, J. Benz, M. Eickhoff, C. Heiliger, M. Heinemann, J. Bläsing, A. Krost, S. Shokovets, C. Müller and C. Ronning, *Phys. Status Solidi B*, 2012, **249**, 1487–1509.
47 A. S. Zoolfakar, R. A. Rani, A. J. Morfa, A. P. O'Mullane and K. Kalantar-zadeh, *J. Mater. Chem. C*, 2014, **2**, 5247–5270.
48 P. E. de Jongh, D. Vanmaekelbergh and J. J. Kelly, *J. Electrochem. Soc.*, 2000, **147**, 486–489.
49 S. Somasundaram, C. Raman Nair Chenthamarakshan, N. R. de Tacconi and K. Rajeshwar, *Int. J. Hydrogen Energy*, 2007, **32**, 4661–4669.
50 A. Paracchino, J. C. Brauer, J.-E. Moser, E. Thimsen and M. Graetzel, *J. Phys. Chem. C*, 2012, **116**, 7341–7350.
51 S. Brahms, S. Nikitine and J. P. Dahl, *Phys. Lett.*, 1966, **22**, 31–33.
52 A. Filippetti and V. Fiorentini, *Phys. Rev. B: Condens.Matter Mater. Phys.*, 2005, **72**, 035128.
53 F. Bruneval, N. Vast, L. Reining, M. Izquierdo, F. Sirotti and N. Barrett, *Phys. Rev. Lett.*, 2006, **97**, 267601.
54 M. Nolan and S. D. Elliott, *Phys. Chem. Chem. Phys.*, 2006, **8**, 5350–5358.
55 W. Y. Ching, Y.-N. Xu and K. W. Wong, *Phys. Rev. B: Condens. Matter Mater. Phys.*, 1989, **40**, 7684–7695.
56 A. Önsten, M. Månsson, T. Claesson, T. Muro, T. Matsushita, T. Nakamura, T. Kinoshita, U. O. Karlsson and O. Tjernberg, *Phys. Rev. B: Condens. Matter Mater. Phys.*, 2007, **76**, 115127.
57 J. P. Hu, D. J. Payne, R. G. Egdell, P. A. Glans, T. Learmonth, K. E. Smith, J. Guo and N. M. Harrison, *Phys. Rev. B: Condens. Matter Mater. Phys.*, 2008, **77**, 155115.
58 M. M. Islam, B. Diawara, V. Maurice and P. Marcus, *J. Mol. Struct.*, 2009, **903**, 41–48.
59 Y.-J. Kim, J. P. Hill, H. Yamaguchi, T. Gog and D. Casa, *Phys. Rev. B: Condens. Matter Mater. Phys.*, 2010, **81**, 195202.
60 J. Ghijsen, L. H. Tjeng, J. van Elp, H. Eskes, J. Westerink, G. A. Sawatzky and M. T. Czyzyk, *Phys. Rev. B: Condens. Matter Mater. Phys.*, 1988, **38**, 11322–11330.
61 L. G. Parratt, *Rev. Mod. Phys.*, 1959, **31**, 616–645.

62 J. Vura-Weis, C.-M. Jiang, C. Liu, H. Gao, J. M. Lucas, F. M. F. de Groot, P. Yang, A. P. Alivisatos and S. R. Leone, *J. Phys. Chem. Lett.*, 2013, **4**, 3667–3671.

63 C.-M. Jiang, L. R. Baker, J. M. Lucas, J. Vura-Weis, A. P. Alivisatos and S. R. Leone, *J. Phys. Chem. C*, 2014, **118**, 22774–22784.

64 A. Cavalleri, M. Rini, H. H. W. Chong, S. Fourmaux, T. E. Glover, P. A. Heimann, J. C. Kieffer and R. W. Schoenlein, *Phys. Rev. Lett.*, 2005, **95**, 067405.

65 S. L. Johnson, P. A. Heimann, A. G. MacPhee, A. M. Lindenberg, O. R. Monteiro, Z. Chang, R. W. Lee and R. W. Falcone, *Phys. Rev. Lett.*, 2005, **94**, 057407.

66 D. W. Snoke, D. Braun and M. Cardona, *Phys. Rev. B: Condens. Matter Mater. Phys.*, 1991, **44**, 2991–3000.

67 A. Mysyrowicz, D. Hulin and A. Antonetti, *Phys. Rev. Lett.*, 1979, **43**, 1123.

68 R. Habiger and A. Compaan, *Solid State Commun.*, 1976, **18**, 1531–1534.

69 A. Jolk, M. Jörger and C. Klingshirn, *Phys. Rev. B: Condens. Matter Mater. Phys.*, 2002, **65**, 245209.

70 J. Stöhr, *NEXAFS spectroscopy*, Springer Science & Business Media, 2013.

71 F. Bechstedt, *Phys. Status Solidi B*, 1982, **112**, 9–49.

72 M. van Schilfgaarde, T. Kotani and S. Faleev, *Phys. Rev. Lett.*, 2006, **96**, 226402.

Faraday Discussions

PAPER

Probing the excited state relaxation dynamics of pyrimidine nucleosides in chloroform solution†

Katharina Röttger,[ab] Hugo J. B. Marroux,[a] Hendrik Böhnke,[b] David T. J. Morris,[a] Angus T. Voice,[a] Friedrich Temps,[b] Gareth M. Roberts*[a] and Andrew J. Orr-Ewing*[a]

Received 8th April 2016, Accepted 3rd May 2016
DOI: 10.1039/c6fd00068a

Ultrafast transient electronic and vibrational absorption spectroscopy (TEAS and TVAS) of 2′-deoxy-cytidine (dC) and 2′-deoxy-thymidine (dT) dissolved in chloroform examines their excited-state dynamics and the recovery of ground electronic state molecules following absorption of ultraviolet light. The chloroform serves as a weakly interacting solvent, allowing comparisons to be drawn with prior experimental studies of the photodynamics of these nucleosides in the gas phase and in polar solvents such as water. The pyrimidine base nucleosides have some propensity to dimerize in aprotic solvents, but the monomer photochemistry can be resolved clearly and is the focus of this study. UV absorption at a wavelength of 260 nm excites a $^1\pi\pi^* \leftarrow S_0$ transition, but prompt crossing of a significant fraction (50% in dC, 17% in dT) of the $^1\pi\pi^*$ population into a nearby $^1n\pi^*$ state is too fast for the experiments to resolve. The remaining flux on the $^1\pi\pi^*$ state leaves the vertical Franck–Condon region and encounters a conical intersection with the ground electronic state of ethylenic twist character. In dC, the $^1\pi\pi^*$ state decays to the ground state with a time constant of 1.1 ± 0.1 ps. The lifetime of the $^1n\pi^*$ state is much longer in the canonical forms of both molecules: recovery of the ground state population from these states occurs with time constants of 18.6 ± 1.1 ps in amino-oxo dC and ~114 ps in dT, indicating potential energy barriers to the $^1n\pi^*/S_0$ conical intersections. The small fraction of the imino-oxo tautomer of dC present in solution has a longer-lived $^1n\pi^*$ state with a lifetime for ground state recovery of 193 ± 55 ps. No evidence is found for photo-induced tautomerization of amino-oxo dC to the imino-oxo form, or for population of low lying triplet states of this nucleoside. In contrast, ~8% of the UV-excited dT molecules access the long-lived T_1 ($^3\pi\pi^*$) state through the $^1n\pi^*$ state. The primary influence of the solvent appears to be

[a]*School of Chemistry, University of Bristol, Cantock's Close, Bristol, BS8 1TS, UK. E-mail: a.orr-ewing@bristol.ac.uk; g.m.roberts@bristol.ac.uk*

[b]*Institut für Physikalische Chemie, Christian-Albrechts-Universität zu Kiel, Olshausenstr. 40, D-24098 Kiel, Germany*

† All experimental data are archived in the University of Bristol's Research Data Storage Facility (DOI: 10.5523/bris.re6ei611x5in1quskmq7dft3m).

the degree to which it destabilizes the states of $^1n\pi^*$ character, with consequences for the lifetimes of these states as well as the triplet state yields.

1. Introduction

The photochemistry of the four canonical DNA nucleobases has undergone intense study within the chemical physics community,[1,2] largely through a drive to better understand their avoidance of photochemical damage following the absorption of solar ultraviolet (UV) radiation (*i.e.*, their photostability). The advent of ultrafast spectroscopy[3] and advances in theoretical methods for treating electronically excited states[4] have revealed that this photostability is a direct consequence of the ultrashort lifetimes of optically bright electronically excited $^1\pi\pi^*$ states,[1] brought about by (often barrierless) access to an array of conical intersections between the ground (S_0) and excited states.[5,6] However, even with their intrinsic resistance to UV-induced degradation, photodamage can still occur when the nucleobases are embedded within double- or single-stranded DNA. Unlike the purine derived bases (adenine and guanine), the pyrimidine bases, cytosine and thymine (Fig. 1, R = H), are particularly susceptible to mutagenic photodimerization pathways, such as (2 + 2) cycloaddition and (6 − 4) adduct formation, which must then undergo energetically costly enzymatic photo-repair.[7–9] Various schools of thought have emerged to rationalize the susceptibility of pyrimidines to photo-reactions, which invoke both the trapping of excited population in reactive triplet states[10–12] and, more generally, the role of dynamic conformational fluctuations in the DNA duplex.[13] In this work, we specifically revisit the photochemistry of the pyrimidine nucleosides (Fig. 1). We examine the ultrafast photodynamics of chloroform solutions, and compare with prior studies in water, acetonitrile and methanol to explore what impact (if any) this more weakly interacting solvent environment has on the excited state relaxation dynamics of the pyrimidine nucleosides. Chloroform also provides a dielectric

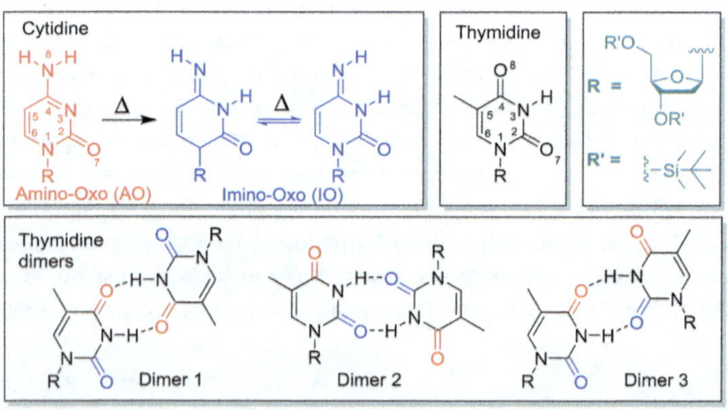

Fig. 1 Molecular structures of the amino-oxo (AO, red) and imino-oxo (IO, blue) tautomers of the protected cytidine (dC) and protected thymidine (dT, black) nucleosides. Possible hydrogen-bonded dT·dT dimer structures and the structures of the silyl protected 2′-deoxy-ribose groups (grey) are also shown. The scheme for dC shows thermally induced tautomerization between AO and IO.

medium for the nucleobases similar to that experienced within the core of a DNA double helix.[14]

The relaxation dynamics of the cytidine nucleoside (C) have been studied in various solution phase environments[15–21] as well as in isolated species in the gas phase.[22] Following excitation to the bright $^1\pi\pi^*$ state at wavelengths from 260–270 nm in aqueous solution, a range of transient absorption methods reveal that excited state population decays bi-exponentially with time constants of 720 fs and ~30 ps.[15–18,21] Similar dynamics have also been found for C in chloroform through ultrafast fluorescence up-conversion.[19] Numerous theoretical studies[23–35] on the relaxation pathways available to the simpler cytosine nucleobase do not yield a single unified picture of the dynamics, returning (at least) three calculated conical intersections with their relative importance to excited state deactivation depending on the level of theoretical treatment, particularly in dynamics simulations.[31,36–38] However, experimentalists have broadly elected to ascribe the sub-picosecond time constant to direct relaxation to S_0, most likely *via* an 'ethylenic-twist' type $^1\pi\pi^*/S_0$ conical intersection, while the slower 30 ps component has been associated with the decay of population trapped in a $^1n\pi^*$ state, formed following initial ultrafast $^1\pi\pi^* \rightarrow {}^1n\pi^*$ bifurcation in the vertical Franck–Condon (vFC) region.[17] This slower $^1n\pi^*$ channel was only revealed through bleach recovery kinetics[17,18] and transient absorption decay at $\lambda \leq 340$ nm,[17] indicating that this state is optically dark to excited state absorption (ESA) in the visible wavelength region.[16,17] Most importantly though, the lifetime of the $^1n\pi^*$ decay channel shows a clear dependence on both the chemical substituent (R) present in the N1 position (Fig. 1) and the solvent environment; in aqueous cytosine (R = H) it is reduced to 12 ps (*cf.* ~30 ps in aqueous C).[17,18,21] This substituent effect is also found in the gas phase,[39] although the exact origins of this behaviour are not yet fully understood.

Ultrafast spectroscopy studies on the thymidine nucleoside (T) at wavelengths spanning 260–270 nm return a qualitatively similar picture to the relaxation dynamics observed for C;[11,15–17,20,22,40–43] in aqueous solution, bi-exponential decay of the excited state population to S_0 occurs *via* the bright $^1\pi\pi^*$ (540 fs) and dark $^1n\pi^*$ (127 ps) states.[15–17] A host of theoretical calculations[28,37,44–52] on the isolated thymine nucleobase led experimentalists to propose the above interpretation for the extracted time constants. However, transient vibrational absorption spectroscopy (TVAS) measurements by Kohler and co-workers on T in acetonitrile-d_3 indicate a significant fraction of excited state flux also undergoes intersystem crossing (ISC) into a long-lived T_1 ($^3\pi\pi^*$) triplet state (with lifetime $\tau \approx 0.6$ μs).[11] Note that although the quantum yield for ISC is $\phi_{ISC} \approx 0.2$ for dT in acetonitrile-d_3, ISC is an order of magnitude less probable for aqueous T ($\phi_{ISC} \leq 0.02$).[53] In agreement with theory,[46,51] these authors suggested that ISC most likely occurs from the dark $^1n\pi^*$ surface, rather than $^1\pi\pi^*$, although they were only able to determine a lower-limit of ≤10 ps for $^1n\pi^* \rightarrow {}^3\pi\pi^*$ transfer. To the best of our knowledge, no analogous TVAS experiments have studied triplet state formation in C, although, in principle, ISC may be possible, given a small but measurable phosphorescence quantum yield for C,[54] and bleach recovery kinetics suggesting $\phi_{ISC} \leq 0.07$ in water.[17]

In this discussion paper, we report the use of ultrafast optical spectroscopy, in both the UV/visible and mid-infrared (IR) spectral regions, supported by spectral assignments based on density functional theory (DFT) calculations, to probe the excited state relaxation dynamics of chemically modified cytidine and thymidine 2′-deoxy-nucleosides (dC and dT, with d denoting the 2′-deoxyribose form). All experiments were performed on low concentrations (≤50 mM) of silyl-protected 2′-deoxy-nucleosides in chloroform. For dC, the excited state relaxation dynamics, following excitation at 260 nm, were tracked by both transient electronic (UV/visible) absorption spectroscopy (TEAS) and TVAS. Unlike in water, dC in chloroform at 298 K is present in two different tautomers: the canonical amino-oxo (AO) and minor imino-oxo (IO) forms (see Fig. 1). Although both tautomers display bi-phasic excited state relaxation dynamics, the $^1n\pi^*$ channel is ~9 times slower in the IO form, relative to AO. In the case of dT, only TVAS is used to probe the relaxation dynamics. Unlike dC, we see the presence of only a single dT tautomer and clear evidence for triplet state formation on a picosecond timescale with a quantum yield of $\phi_{ISC} \approx 0.08$. Triplet state decay for dT in chloroform occurs over more than 1 ns. Although triplet states are discussed to be the precursor for thymidine dimerization in solution,[55–60] we do not see evidence for the diffusion-limited formation of dT dimers within 1.3 ns. Minor contributions from hydrogen-bonded dC·dC and dT·dT homodimers are also observed in both cases, although a detailed analysis of their dynamics will be the subject of a future study.

2. Methods

2.1 Synthesis

To enable sufficient solubility of dC and dT in aprotic solvents, such as chloroform, the ribose OH groups of the 2′-deoxy-nucleosides were protected by bulky, apolar *tert*-butyldimethylsilyl (TBDMS) groups (see Fig. 1). The synthesis in Kiel followed a modified protocol of Ogilvie.[61] The synthesis for protected dC and dT at the University of Bristol was described in detail in a recent publication.[62]

2.2 Transient electronic absorption spectroscopy

TEAS experiments were performed at the University of Kiel, and the details of the apparatus have been presented elsewhere.[63,64] The transient absorption experiment was driven by the output of a Ti:sapphire laser (ClarkMXR CPA 2001) which delivers pulses with pulse lengths of 150 fs (FWHM) at 775 nm. Half of the overall output of a 1000 μJ per pulse was used for the transient absorption experiment. The excitation pulses were generated in a non-collinear optical parametric amplifier (NOPA) with subsequent temporal compression and frequency doubling. The resulting pulses had a centre wavelength of 260 nm and a bandwidth of ~4 nm. For the generation of the broadband probe pulses, ~70 μJ per pulse of the laser fundamental was used. The laser pulses passed a delay stage (M-531.DG, Physik Instrumente) equipped with a retroreflector (CVI Melles Griot). They were attenuated to ~2 μJ by an absorptive neutral density filter and a combination of a $\lambda/2$ waveplate and a polarizer. These pulses were focused into a CaF_2 plate ($d = 2$ mm, Korth Kristalle) to produce a white-light continuum. The CaF_2 plate was rastered vertically and horizontally so that consecutive laser pulses hit different spots, thus ensuring that the low damage threshold of CaF_2 was not exceeded. The white light pulses were split into probe and reference beams using the front and back reflections from a planar glass plate. Pump and probe pulses were recollimated using reflective optics to reduce the chirp and obtain an optimal time resolution of the experiment. The pump pulses were set to magic angle polarization with respect to the probe pulses using a $\lambda/2$ waveplate. An optical chopper (MC2000, Thorlabs) equipped with a 10 shot blade (MC1F10, Thorlabs) was used to cut out every second laser pulse, thus enabling the measurement of background signals without excitation. The transmitted broadband light pulses were dispersed in a prism spectrograph and detected by two full-frame transfer back-thinned CCD cameras (Entwicklungsbüro Stresing, Berlin).

The sample solutions were circulated through a home-built flow cell equipped with two quartz windows ($d = 0.2$ mm, diameter 15 mm, Korth Kristalle) and a PTFE spacer of 100 μm thickness. A gear pump (Reglo-Z analog, Ismatec) equipped with an organic solvent resistant pumphead (Z-186 with PPS gears, Ismatech) and PTFE tubes was employed to flow the sample solution continuously through the cell. Transient absorption measurements for each sample and the neat solvent were measured in one experimental run to avoid a change of coherent solvent signals. The cross-correlation of pump and probe pulses and the instrument response function were both ~50 fs. Each sample and solvent measurement was repeated three times to ensure reproducibility, and all measurements were repeated at least two times on different days with fresh sample solutions. The purity and integrity of the samples were checked before and after each measurement *via* UV/vis absorption spectroscopy. The water content was reduced by preparing all sample solutions with anhydrous $CHCl_3$ in a flow box purged with dry air and checked before and after the transient absorption measurements *via* IR spectroscopy. For the TEAS experiments, 2′,3′,5′-tri-*O*-(*t*-butyldimethylsilyl)-cytidine (C) was used instead of the protected 2′-deoxynucleoside (dC). As was shown previously,[65] the dynamics are virtually the same for the protected C and dC molecules. All TEAS measurements were performed with sample concentrations of $c_0 = 10$ mM.

2.3 Transient vibrational absorption spectroscopy

TVAS experiments were performed at the University of Bristol, using an apparatus that has been described in detail previously.[66] Briefly, the system consists of a Coherent Vitara-S oscillator and Coherent Legend Elite HE+ regenerative amplifier, operating at 1 kHz and configured to produce 40 fs duration pulses at 800 nm with a total output power of 5 W. This fundamental beam was split into three parts using a series of beam splitters. Only two of these beams were used for the current measurements, each with an energy of 2.45 mJ per pulse. The two beams seeded two Coherent OPerA Solo optical parametric amplifiers (OPAs). One of these OPAs produced spectrally tunable light spanning the UV to IR range (220–20 000 nm) and was used to generate the 260 nm (~100 fs) pump pulses for all the TVAS experiments reported here. The remaining OPA generated broadband (~300 cm^{-1}) tunable IR pulses for use as a probe in TVAS experiments, with energies of ~1 μJ per pulse at the sample.

For the TVAS measurements on dC and dT, the 260 nm pump pulse was attenuated to between ~0.6 and 1 μJ per pulse by cross-polarization using a $\lambda/2$ waveplate and wire-grid polarizer, and then focused ~2 cm behind the sample by a $f = 250$ mm CaF_2 lens. A beam profiler was used to determine a ~250 μm beam diameter for the Gaussian profile (full-width at half maximum) of the pump beam at the sample, returning pump fluences in the range 1.2–2 mJ cm^{-2}. The polarization of the UV pump was maintained at the magic angle (54.7°), relative to the polarization of the IR probe pulse, by using the $\lambda/2$ waveplate in the UV pump beam line.

Broadband IR probe pulses were generated by difference frequency generation of the signal and idler beams from the IR OPA. In the experiments described here, the IR light was centred at ~1680 cm^{-1} (carbonyl stretching region) or ~3400 cm^{-1} (N–H stretching region). The entire IR probe beam line was enclosed by sealed plastic tubes and continuously purged by dry N_2 to avoid undesired absorption by atmospheric water vapour and CO_2. The IR probe pulses were reflectively focused into the sample to a tight ~50 μm beam diameter, so that the probed region of the sample was uniformly excited by the more loosely focused UV pump beam. The cross correlation of pump and probe pulses was 120 fs, but the instrument response function was longer (200–400 fs) because of thermal lensing effects.

The temporal delay (t) between the UV pump and IR probe pulses was controlled by changing the path length of the pump beam with an aluminium retro-reflector mounted on a motorized delay stage, providing a maximum possible delay of $t = 1.3$ ns. The pump and probe beams then intersected the sample with a small crossing angle of ~5°. Prior to interaction with the sample, the UV pump beam was modulated at 500 Hz (blocking every other pulse) with an optical chopper wheel to obtain pump on/off spectral pairs at each t, which were then used to generate individual transient absorption spectra. After passing through the sample, the transmitted IR probe light was detected by a 128 pixel, liquid N_2 cooled Mercury Cadmium Telluride array (Infrared Associates Inc., MCT-10-128) coupled to a spectrometer (HORIBA Scientific, iHR320), providing a spectral resolution of ~2 cm^{-1}.

Sample solutions of protected dC ($c_0 = 5$ mM) and dT ($c_0 = 50$ mM) were prepared in anhydrous $CDCl_3$ or $CHCl_3$ (Sigma-Aldrich, 99.99%). These sample solutions were then delivered through a stainless steel flow cell, containing two 1.5 mm thick CaF_2 windows separated by either a 200 or 380 μm thick Teflon spacer, which defined the absorption path length. The sample solution was flowed continuously through the cell by a peristaltic pump with PTFE tubing throughout.

3. Results and discussion

Time-resolved vibrational and electronic absorption spectroscopies provide information on both the UV-photoexcited states of dC and dT and their relaxation pathways back to the ground state. We first present TEAS and TVAS measurements for dC solutions in chloroform and then TVAS results obtained for the corresponding dT solutions. The amino-oxo (AO) and the minor imino-oxo (IO) tautomeric structures adopted by the monomeric dC nucleosides in chloroform are shown in Fig. 1, together with the single structure for monomeric dT. Possible hydrogen-bonded dT·dT dimer structures are also shown. In the gas phase, the cytosine nucleobase also adopts an enol tautomer of the AO form,[67] but connection of a ribose group to the N1 atom in the ring to form the nucleoside removes this possibility in our solutions. The minority IO tautomer of dC has two conformers distinguished by the *trans* or *cis* arrangement of the imino group. It can be identified by a characteristic IR absorption band at 1725 cm^{-1} in chloroform solution.[62,68] Computational studies of the photochemistry of cytosine and its 1-methyl derivative identified conical intersections between the $^1\pi\pi^*$ (S_1) and S_0 states, passage through which leads to bifurcation on the ground S_0 state to produce both the AO and IO tautomers.[32,33] These computations therefore suggest a photochemical route for AO to IO conversion, as observed in argon matrix studies of 1-methylcytosine.[69]

The propensity for the silyl-derivatized nucleosides to dimerize in aprotic solvents can be quantified by a solvent-dependent equilibrium constant, and in the case of dC in chloroform the association constant $K_{CC} = [dC{\cdot}dC]/[dC]^2 = 42.8 \pm 2.1\ M^{-1}$ was previously determined by Schwalb *et al.*[19] The degree of association is given by the parameter $\beta_{CC} = 2[dC{\cdot}dC]/c_0$, where $c_0 = [dC] + 2[dC{\cdot}dC]$ is the prepared concentration of the dC solution. The degree of association and the fractions of dC and dC·dC can be evaluated from c_0 using the known value of K_{CC}. We also note that this previous study by Schwalb *et al.* identified only a single symmetric dC·dC dimer structure in chloroform (see Fig. 1 in ref. 19). We undertook a similar analysis of concentration dependent steady state FTIR spectra of dT in chloroform solutions, examples of which are shown in panels (a) and (b) of Fig. 2, to evaluate the corresponding association constant K_{TT} and degree of association β_{TT} for dT solutions (Fig. 2(c)). The analysis used spectra obtained in the N–H stretching region (Fig. 2(b), 3100–3450 cm^{-1}); in the carbonyl stretching region (Fig. 2(a), 1600–1800 cm^{-1}), the monomer and dimer bands in the corresponding spectra overlap too much to be analysed. We obtained $K_{TT} = 2.1 \pm 0.7\ M^{-1}$, which indicates a lower propensity to dimerize than dC. Table 1 summarizes the degrees of association for different solutions with c_0 values in the range used in our current study. Fig. 1 shows three possible isomers of the dT·dT dimer, and our FTIR spectra in the carbonyl stretching region suggest dimer 1 is

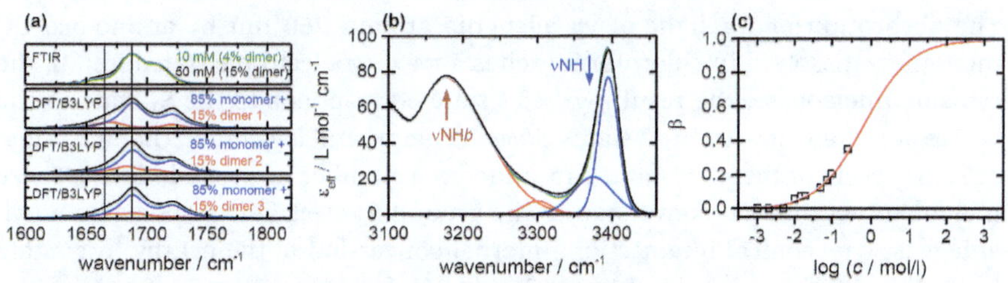

Fig. 2 (a) Top panel: concentration dependent FTIR spectra of dT solutions in $CHCl_3$ recorded at $c_0 = 10$ mM (green) and $c_0 = 50$ mM (black) in the carbonyl stretch region. dT·dT dimer fractions are also shown. Lower panels: simulated FTIR spectra of 50 mM dT solutions in chloroform, calculated at the PCM-B3LYP/6-311++G** level of theory. Calculated harmonic vibrational frequencies are scaled by 0.993 for monomers and dimers. The three simulated spectra (black) are constructed assuming an 85% dT monomer contribution (blue) together with a 15% dT·dT fraction (red) from either dimer 1, dimer 2 or dimer 3 (see Fig. 1). (b) Example FTIR spectrum for a 50 mM dT solution in $CHCl_3$ in the N–H stretching region (black) together with a fit (green) consisting of monomer (νNH, blue) and dimer (νNHb, red) contributions. (c) Degree of association (β_{TT}) obtained from the concentration dependent FTIR band fitting analysis in the N–H stretching region (b) for dT·dT dimerization in $CHCl_3$ as a function of log(c_0).

Table 1 Degrees of association (β) into dC·dC or dT·dT dimers for different concentrations of cytidine and thymidine in chloroform

c_0/mM	β_{CC}	β_{TT}
5	0.24	0.02
10	0.35	0.04
25	0.51	0.09
50	0.62	0.15

the form most commonly adopted in chloroform. Comparison of the measured FTIR spectra for 10 mM and 50 mM solutions of dT (Fig. 2(a)) reveals a dimer contribution at the low wavenumber side of the carbonyl stretching band of the monomer. Predicted IR spectra for all possible dimer structures (with R = CH_3, Fig. 1) were obtained using Gaussian 09 (ref. 70) calculations at the B3LYP/6-311++G** level of theory, including a polarizable continuum model (PCM) for the chloroform solvent, and confirm that dimer 1 is the only candidate that shows a similar band shift to lower wavenumber.

We concentrate on TVAS data obtained in the carbonyl and N–H stretching regions of the nucleosides, but also considered the possibility that UV photoexcitation might cleave bonds in the heterocyclic rings, as seen in related systems.[68,71] Homolytic ring-opening photochemistry should produce photoproducts with characteristically strong isocyanate bands around 2100–2200 cm^{-1}.[68] While no such features were observed for dT, TVAS measurements in this spectral region for dC revealed very weak (μΔOD changes in optical density) bands that might be assigned to isocyanate species. However, comparisons with band intensities in the carbonyl region under comparable conditions, demonstrates that the ring-opening pathways make negligible contributions to the pyrimidine nucleoside photochemistry.

3.1 Transient spectroscopy of dC solutions in chloroform

The absorption of UV light of wavelengths around 260 nm by amino-oxo dC monomers dissolved in chloroform excites a $\pi^* \leftarrow \pi$ electronic transition. In the cytosine nucleobase, the resulting $^1\pi\pi^*$ state corresponds to the S_1 state in the vFC region[30] (albeit with $^1n\pi^*$ states close in energy, and lower than the $^1\pi\pi^*$ state at some levels of theory[6,38]) and can undergo a number of possible fates. These include direct internal conversion to the ground S_0 electronic state *via* an ethylenic-twist type conical intersection, internal conversion to the nearby $^1n\pi^*$ state (S_2 in the vFC region), or perhaps ISC to the T_2 ($^3n\pi^*/^3\pi\pi^*$) or T_1 ($^3\pi\pi^*$) triplet states.[6] The $S_1 \rightarrow T_1$ conversion is forbidden by the El Sayed rules[72] and ISC has therefore been proposed to occur following internal conversion to the $^1n\pi^*$ state,[17] or *via* $S_1 \rightarrow T_2$ conversion.[26,34,35]

Fig. 3(a) shows examples of TEA spectra obtained following 260 nm excitation of a 10 mM C solution in $CHCl_3$. At this concentration, approximately 65% of the C in solution is present as monomers and the remainder is dimers. The time-dependent TEA spectra are displayed as a 2D map, with colours indicating the intensities of transient features. These features are weak, but two components dominate the spectra: a broad, positive band at visible wavelengths is attributed to excited state absorption (ESA), while a negative feature in the near-UV

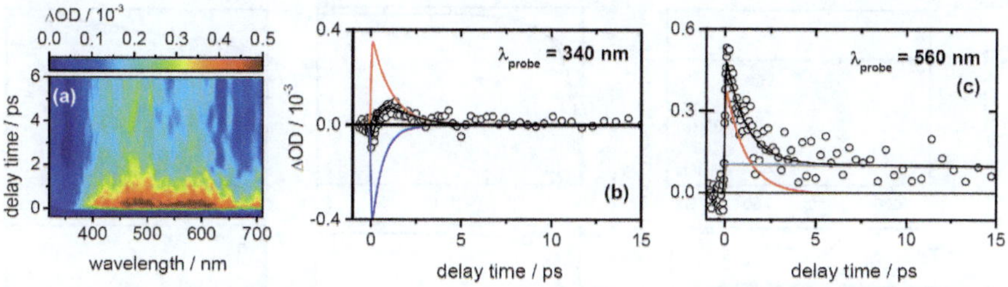

Fig. 3 (a) Two-dimensional false colour intensity map of the TEA spectrum obtained from C (c_0 = 10 mM) in $CHCl_3$ following excitation at 260 nm (C = 65%, C·C = 35%). Absorption change profiles as a function of delay time obtained from (a) at probe wavelengths of (b) λ_{probe} = 340 nm and (c) λ_{probe} = 560 nm. Open circles = experimental data, black line = total fit. Coloured lines are decay components with time constants of τ_1 = 0.6 ps (blue line), τ_2 = 1.1 ps (red line) and τ_3 = 193 ps (grey line). See Table 2 for details of τ_n fit amplitudes (A_n).

corresponds to stimulated emission from the excited state. Neither band shows significant wavelength shifts over time. Decay of the majority of the signal intensity in both bands within a few picoseconds is consistent with the short excited state lifetimes measured previously for cytidine samples in other solvents or the gas phase.[15–17,19–22]

The TEA spectra were analysed by selecting time-dependent intensity profiles at 8 wavelengths in the range 340–600 nm, two examples of which are shown in panels (b) and (c) of Fig. 3 for probe wavelengths of 340 nm and 560 nm, and simultaneously fitting to exponential decays. Three time constants emerge from these fits: $\tau_1 = 0.6 \pm 0.5$ ps for the decay of the stimulated emission signal, and $\tau_2 = 1.1 \pm 0.1$ ps and $\tau_3 = 193 \pm 55$ ps for the decay of the ESA. The large relative uncertainty in the value of τ_1 is a consequence of the overlapping stimulated emission and ESA signals at early times. Because the three time constants are deduced from only a single sample concentration, we cannot decompose them further into contributions from C monomers and C·C dimers. Depopulation of the vFC region on the excited state potential energy surface accounts for the loss of the stimulated emission signal, and is commensurate with the shortest time-scale reported in fluorescence up-conversion experiments.[19] The τ_2 value would appear to reflect the lifetime of molecules in the $^1\pi\pi^*$ state. It is therefore tempting to assign τ_3 to molecules that transfer from the $^1\pi\pi^*$ to the $^1n\pi^*$ state, but fluorescence up-conversion measurements suggested a lifetime of 21 ± 2 ps for the $^1n\pi^*$ state population,[19] which is hard to reconcile with the longer lifetime ESA component reported here. We return to this point below after consideration of the information obtained from TVAS.

TVA spectra in the carbonyl and N–H stretching regions provide further detailed insights, and example spectra are displayed in panels (a) and (d) of Fig. 4, respectively. These spectra were obtained using 5 mM solutions of dC in $CDCl_3$, for which we expect 76% dC monomers and 24% dC·dC dimers. In panels 4(a) and (d), negative changes in optical density (ΔOD) correspond to depletion of the ground state molecules by the UV laser pulse. We refer to these as ground state bleach features. Positive ΔOD values indicate intermediate or product species. The dominant bleach feature centred close to 1650 cm^{-1} in Fig. 4(a) has two

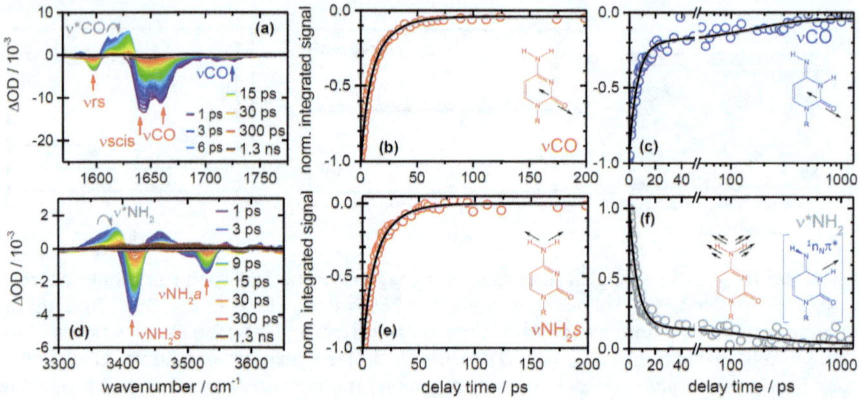

Fig. 4 TVA spectra recorded for dC (c_0 = 5 mM) in $CDCl_3$ following excitation at 260 nm and probing in (a) the carbonyl stretch region (1550–1775 cm^{-1}) and (d) the NH_2 stretching region (3300–3650 cm^{-1}) (dC = 76%, dC·dC = 24%). Normalized integrated signal traces (open circles) and kinetic fits (black lines) as a function of delay time for (b) the carbonyl bleach (νCO, 1660 cm^{-1}) of the AO tautomer, (c) the carbonyl bleach (νCO, 1720 cm^{-1}) of the IO tautomer, (e) the symmetric NH_2 stretch bleach (νNH_2s, 3410 cm^{-1}) of the AO tautomer, and (f) vibrationally hot NH_2 ground state (ν^*NH_2, 3380 cm^{-1}). ν^*NH_2 also contains contributions from the $^1n_N\pi^*$ state of the IO tautomer – see main text for details. See Table 2 for details of τ_n fit amplitudes (A_n).

overlapping components assigned to a carbonyl stretch (νCO) and a scissoring motion of the NH_2 group (νscis). A further weak bleach to lower wavenumber is attributed to a ring-stretching mode (νrs). The assignment of all these features to the AO tautomer of dC is supported by observation of corresponding absorption bands in the steady state FTIR spectrum.[19,62] The weak bleach feature at 1725 cm^{-1} was previously assigned to a νCO mode of the IO tautomer of dC.[62,68] Positive going features to the low wavenumber side of the bleaches are characteristic of absorptions by vibrationally hot, but electronically S_0 molecules formed by internal conversion from higher lying electronically excited states. These features grow and then decay over time as the excess vibrational energy of the molecules dissipates into the surrounding solvent.

In the N–H stretching region shown in Fig. 4(d), the two main bleach features are assigned to the symmetric (νNH_2s) and antisymmetric (νNH_2a) stretches of the NH_2 group of dC.[19] Again, signatures of vibrationally hot S_0 state molecules can be seen as transient absorption features to the low wavenumber sides of the bleaches. In both IR regions, the bleach features reduce in depth over time, and return to the baseline absorbance level. The absence of significant residual bleaches at times approaching the upper limit of 1.3 ns for our measurements demonstrates almost complete recovery of ground state dC molecules following absorption of 260 nm UV light, and we conclude that there is negligible formation of photoproducts (including long-lived triplet states). A suggestion of a small remaining depletion of the νCO mode at 1660 cm^{-1} and the νscis mode at 1645 cm^{-1} cannot be indicative of photoproduct formation because it is only observed for this pair of overlapping bands; instead, we attribute this residual bleach to an overlapping band of dC·dC dimers (see below).

Panels (b), (c), (e) and (f) of Fig. 4 show the time dependences of selected bleach and transient absorption features in the TVA spectra. These time-dependent intensities were obtained by integrating the band intensities over 10 cm^{-1} wide intervals about the band centres. Fits of the time-dependent band intensities to bi- or tri-exponential functions give the time constants (τ_n) and relative amplitudes (A_n) of each exponential component reported in Table 2. The recoveries of the bleaches of the carbonyl stretching band of the AO tautomer (νCO, 1660 cm^{-1}, panel (b)) and the NH_2 symmetric stretching band (νNH_2s, 3410 cm^{-1}, panel (e)) show similar time dependences. The same fast initial decays in amplitude are also seen in the carbonyl band bleach corresponding to the IO tautomer (νCO, 1720 cm^{-1}, panel (c)) and the transient absorption in the NH_2 symmetric stretching region by vibrationally hot S_0 molecules (ν^*NH_2, 3380 cm^{-1}, panel (f)). We therefore globally fitted all these kinetic decays to the same set of exponential functions to obtain a single set of time constants representing all the band intensity changes.

The recoveries of the bleach features of the AO tautomer are described well by biexponential functions with the two components having similar amplitudes, and time constants of $\tau_4 = 6.2 \pm 0.3$ and $\tau_5 = 18.6 \pm 1.1$ ps. The combination of these two decays returns the bleach depths to within a few percent of their initial values. The majority of the decay of the ν^*NH_2 feature at 3380 cm^{-1} also occurs with a 6.2 $\pm$ 0.3 ps time constant, indicating that this timescale corresponds to vibrational cooling of hot S_0 molecules formed by rapid internal conversion from the $^1\pi\pi^*$ (S_1) state. The prompt rise of the ν^*NH_2 feature confirms that this internal conversion is comparable to, or faster than, our experimental time resolution of 0.5 ps for TVA, consistent with the 1.1 ps lifetime of the $^1\pi\pi^*$ state deduced from our TEA spectra. As the molecules vibrationally cool, the ground electronic and vibrational level populations are recovered and the transient bleach of this ground state steadily disappears. The 6.2 $\pm$ 0.3 ps value is likely to be indicative of the timescales for the last few steps down the ladder of vibrational levels, or perhaps just the final cooling step from $v = 1 \rightarrow v = 0$ of a particular mode, which are expected to be rate determining.[66,71] The same time constant describes the

Table 2 Extracted time constant values (τ_n), their respective fit amplitudes (A_n) and photophysical assignments, from a global fitting analysis of the kinetic traces obtained from TEAS (Fig. 3) and TVAS (Fig. 4) measurements on dC in chloroform at 260 nm

		TEAS		TVAS				
Time const. (τ_n)	Value/ps	A_n (λ_{pr} = 340 nm)	A_n (λ_{pr} = 560 nm)	A_n (νCO_{AO})	A_n (νCO_{IO})	A_n (νNH_2s)	A_n (ν^*NH_2)	Assignment
τ_1	0.6 ± 0.5	−0.55	—	—	—	—	—	vFC SE
τ_2	1.1 ± 0.1	0.45	0.80	—	—	—	—	$^1\pi\pi^*$ decay
τ_3	193 ± 55^a	—	0.20	—	0.17	—	0.12	$^1n_N\pi^*$ IO decay
τ_4	6.2 ± 0.3	—	—	0.49	0.78	0.44	0.85	S_0 vib. cooling
τ_5	18.6 ± 1.1	—	—	0.48	—	0.56	—	$^1n_O\pi^*$ AO decay
τ_6	∞	—	—	0.03	0.05	—	0.03	dC·dC dimer

^{a} A modified value of $\tau'_3 = 162 \pm 48$ ps is required to fit the TVAS data.

initial vibrational cooling of the IO tautomer, suggesting similar coupling of the internal modes of both tautomers to the solvent bath. We also observed similar time constants for vibrational cooling in our recent study of G·C Watson–Crick base pair photochemistry in chloroform.[62]

A significant component of the intensity (~50%) decays with a time constant of 18.6 ± 1.1 ps demonstrating a second, slower pathway to repopulation of the S_0 state. This lifetime agrees with the 21 ± 2 ps time constant reported from fluorescence up-conversion studies of dC in chloroform solution.[19] The evidence from electronic structure calculations and prior experimental studies of dC photochemistry points to the $^1n\pi^*$ state as an intermediate in this relaxation pathway. The 18.6 ± 1.1 ps time constant mostly reflects the lifetime of this state, because after the $^1n\pi^* \rightarrow S_0$ internal conversion, the vibrationally hot S_0 molecules will relax with the aforementioned 6.2 ± 0.3 ps time constant. The population of the $^1n\pi^*$ state is seen indirectly in our measurements through recovery of the ground state bleach, but not directly by TEAS. Previous studies of dC in water concluded that the $^1n\pi^*$ state is optically dark to further absorption in the near-UV and visible regions.[16,17] Observation of this state by fluorescence up-conversion at wavelengths around 340 nm indicates a non-zero transition dipole moment to S_0,[19,21] yet we do not see persistence of a stimulated emission feature over this timescale in our TEA spectra. Likely reasons are that the transition dipole moment to S_0 is weaker than for the $^1\pi\pi^* \rightarrow S_0$ emission, and not detected at our low ΔOD levels, and that the emission lies at the short-wavelength edge of our TEAS detection window.

A residual offset of 3% of the maximum amplitude is seen for the bleach feature of the AO tautomer νCO band, and in the IO tautomer the corresponding residual offset is ~5%. This offset persists for the duration of our experiments (up to 1.3 ns) and is therefore described in Table 2 as an exponential decay component with 'infinite' time constant (τ_6). The most plausible explanation for this feature is UV photo-depletion of dC·dC dimers in the chloroform solution, with recovery of an equilibrium concentration of the dimers taking much longer than 1 ns. This assignment is supported by calculations of vibrational frequencies of the dimer which indicate bands in the 1625–1650 cm^{-1} region, and by concentration-dependent FTIR spectra of the dC solutions in chloroform. We discount formation of a long-lived triplet state of dC because the TVA spectra obtained in the N–H stretching region above 3300 cm^{-1}, exclusively feature monomer bands (the broad hydrogen-bonded dC·dC dimer bands are observed below 3300 cm^{-1}, *cf.* Fig. 2(b)),[19] showing complete recovery of the ground state population. The νNH_2s and νNH_2a bands of the T_1 state of the AO form of dC are calculated to shift by ~50 cm^{-1} to higher wavenumber from the ground state bleaches, and we do not observe product bands at the corresponding locations in our TVA spectra. The fast $S_1 \rightarrow T_2$ ISC with ~13% branching predicted by Richter *et al.*[34] in dynamical calculations for isolated AO cytosine is therefore not apparent in our study of the corresponding nucleoside in chloroform.

The bleach feature assigned to the IO tautomer (νCO at 1720 cm^{-1}) does not show a decay component with 18.6 ± 1.1 ps time constant. Instead, a much slower component with τ'_3 = 162 ± 48 ps accounts for ~17% of the recovery of this bleach (Fig. 4(c)). The timescale for this decay appears to agree with the τ_3 = 193 ± 55 ps process observed in TEAS experiments. The likely explanation for this slower component is again diversion of some of the ensemble of relaxing molecules into the $^1n\pi^*$ state, but in the case of the IO tautomer this fraction is significantly smaller than for the AO tautomer, and the $^1n\pi^*$ state lifetime is almost an order of magnitude longer. The extended lifetime suggests that access to the conical intersection connecting the IO $^1n\pi^*$ state to the S_0 state is more restricted, perhaps lying at higher energy above the minimum of the $^1n\pi^*$ state than is the case for the AO tautomer.[27] Unlike the case of the AO tautomer, the $^1n\pi^*$ state of the IO tautomer does not appear to be optically dark to ESA in the visible region. The fraction of IO is small in dC samples in chloroform, so the $^1n\pi^*$ state emission was not resolved from AO tautomer signals in fluorescence up-conversion experiments.[19]

A slow component fitted by the 162 ± 48 ps time constant also contributes ~12% of the amplitude of the relaxation of the ν^*NH_2 feature at 3380 cm^{-1} (Fig. 4(f)). This observation is consistent with an overlapping contribution to this spectral feature from the IO tautomer, with slow relaxation from the longer lived IO $^1n\pi^*$ state feeding into the later time intensity changes of this transient band.

We see no evidence for growth of the IO band corresponding to the previously proposed bifurcation of flux at the $^1\pi\pi^*/S_0$ conical intersection in theoretical calculations;[32,33,69] instead, the weak νCO bleach of the IO tautomer band returns to the baseline because of (nearly) complete ground state recovery of the initially photo-excited IO molecules. Nor do we observe any significant long-time bleaching of the ground state bands of the AO tautomers that might indicate photochemical conversion to the IO form, or to triplet states of the AO form of dC.

The various deductions from our TEAS and TVAS measurements of the AO and IO tautomers of dC in chloroform solution are summarized in Fig. 5. This figure shows schematic potential energy cuts along coordinates corresponding to the relaxation of the initially photo-excited molecules on the $^1\pi\pi^*$ state and *via* the $^1n\pi^*$ state. In the latter case, we consider distinct $^1n\pi^*$ potential energy landscapes for the AO and IO tautomers. Excitation in the vFC region by UV photon

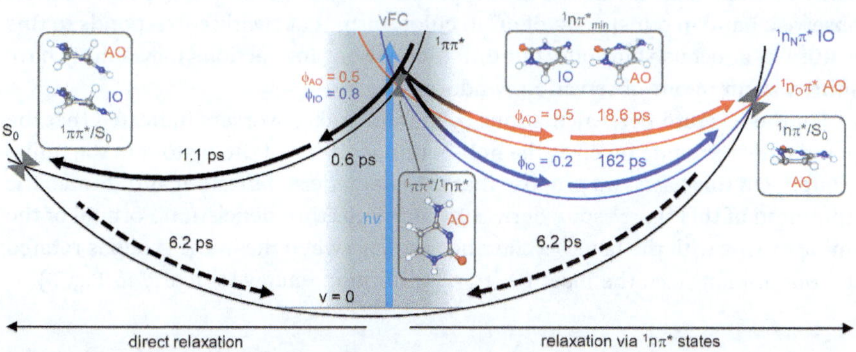

Fig. 5 Schematic potential energy cuts through the S_0, $^1\pi\pi^*$, $^1n_O\pi^*$ and $^1n_N\pi^*$ electronic states for the AO and IO tautomers of dC, depicting a plausible kinetic scheme for the excited state relaxation of dC in chloroform based on TVAS and TEAS measurements with 260 nm excitation. Black = IO and AO, red = AO only, blue = IO only. Possible structures involved in the relaxation dynamics are obtained from theoretical calculations on the AO and IO tautomers of the simpler cytosine nucleobase in the gas-phase; $^1\pi\pi^*/S_0$ AO and $^1\pi\pi^*/^1n\pi^*$ AO from ref. 37, $^1n\pi^*/S_0$ AO and $^1n\pi^*_{min}$ AO from ref. 38, $^1\pi\pi^*/S_0$ IO and $^1n\pi^*_{min}$ IO from ref. 27. The values of ϕ denote branching between competing channels.

absorption draws its transition strength from the $^1\pi\pi^* \leftarrow S_0$ electronic character, but bifurcation along the two competing pathways in almost equal measure (for the AO tautomer) suggests strong coupling between the S_1 ($^1\pi\pi^*$) and S_2 ($^1n\pi^*$) states in the vFC region. The vFC region depopulates in ~0.6 ps, with subsequent relaxation on the $^1\pi\pi^*$ state accessing a conical intersection with the S_0 state. The total lifetime of the $^1\pi\pi^*$ state is ~1.1 ps under our experimental conditions. These timescales are controlled by the gradient of the path from the vFC region to the $^1\pi\pi^*/S_0$ conical intersection and a small barrier to this conical intersection from the minimum of the $^1\pi\pi^*$ state.[6] Passage through the conical intersection returns flux to the S_0 state where vibrational cooling occurs with a 6.2 ps time constant. Approximately half the photo-excited AO tautomers instead branch into the $^1n\pi^*$ state where they are trapped with excited state lifetimes of ~18 ps before returning to the S_0 state through (or in the vicinity of) a conical intersection, and vibrationally cooling with a characteristic 6.2 ps time constant. The relaxation pathway on the $^1\pi\pi^*$ state is also observed in the IO tautomer, and is now the dominant (~80%) route back to the ground state. The remaining ~20% of photo-excited IO tautomers cross onto the $^1n\pi^*$ state where they remain with lifetimes of 162 ps before reaching the $^1n\pi^*/S_0$ conical intersection. Fig. 5 includes approximate molecular structures at the conical intersections, which are based on calculations for cytosine in the gas phase.[27,37,38]

The character of the $^1n\pi^*$ states merits some further consideration because the unpaired electron remaining in a non-bonding orbital might reside preferentially on an O or an N atom (denoted $^1n_O\pi^*$ or $^1n_N\pi^*$, respectively). Most prior computational and experimental studies of the excited states of the AO tautomer consider the $^1n\pi^*$ state to involve the non-bonding orbital on the carbonyl group (hence $^1n_O\pi^*$).[1,6,17] We have adopted this assignment in Fig. 5, but note that Quinn *et al.* presented arguments favouring a state involving excitation of an electron from the N-atom lone pair ($^1n_N\pi^*$).[18] They reported a band at 1574 cm^{-1} seen for aqueous solutions of dC and 2′-deoxy-cytidine 5′-monophosphate (dCMP) nucleotide, but not for cytosine, which they assigned to a vibration of the $^1n_N\pi^*$ state. This spectral feature indicated an excited state lifetime of 37 ps. We do not observe a band in our spectra of dC in chloroform that clearly corresponds to this feature in aqueous solutions, but different solvent interactions may simply have shifted it out of our observation window.

For the IO tautomer, calculations by Marian and co-workers indicated that the populated $^1n\pi^*$ state involves the non-bonding orbital of the N atom in the imine group.[27] On this basis, we propose that the energetic stabilization at the adiabatic minimum of this $^1n_N\pi^*$ state derives from the electron-deficient n_N-orbital of the imine mixing with the heterocyclic ring π-system when the imine group is rotated 90° out-of-plane (see the inset IO $^1n_N\pi^*$ minimum energy structure in Fig. 5).

3.2 Transient spectroscopy of dT solutions in chloroform

The smaller association constant for dT than dC in chloroform means that the time resolved spectra presented here for dT solutions are dominated by monomer photochemistry. Comparisons of TVA spectra obtained for dT concentrations of 10 mM and 50 mM (4% and 15% association to dT·dT dimers, Table 1) show no differences, confirming the negligible contribution from dT·dT dimers (or dynamics of the dimer that are indistinguishable from those of the monomer).

Fig. 6 displays TVA spectra of dT solutions in chloroform in the carbonyl (Fig. 6(a)) and N–H (Fig. 6(d)) stretching regions. These spectra were analysed by fitting to sets of Gaussian functions to obtain time-dependent band intensities. The fits were conducted using Origin 9.0. In the carbonyl stretching region, separate Gaussian functions described the νCO bleach (requiring 3 Gaussian functions with tied amplitudes), two hot ν*CO bands centred at 1657 and 1689 cm^{-1} and a νCO_{T_1} product band (attributed to the T_1 ($^3\pi\pi^*$) triplet state – see below) centred at 1629 cm^{-1}. In the N–H stretching region, single Gaussian functions were used to fit the νNH bleach (3400 cm^{-1}), a ν*NH vibrational hot band (3380 cm^{-1}) and a νNH_{T_1} product feature (3412 cm^{-1}). The plots in Fig. 6(b) and (d) show the time-dependences of the bands in these two spectral regions. As with the analysis of dC, the time-dependent band intensities are once again fitted to bi- or tri-exponential functions, returning the τ_n and A_n values collated in Table 3. The νCO bleach feature recovers ~80% of its amplitude with an exponential time constant of $\tau_1 = 5.3 \pm 0.6$ ps, with two slower components with ill-determined time constants estimated to be $\tau_2 \approx 114$ ps and $\tau_3 > 1$ ns (referred to as infinity within our experimental time window) accounting for the remaining recovery. When the νCO bleach recovery is fitted with the corresponding signals in the N–H stretching region, the average value obtained for the fastest resolved

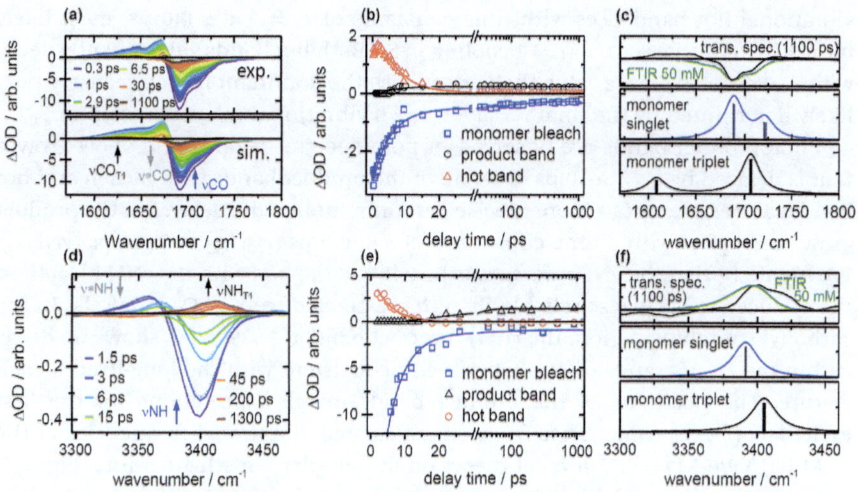

Fig. 6 TVA spectra recorded for dT ($c_0 = 50$ mM) in $CHCl_3$ following excitation at 260 nm and probing in (a) the carbonyl stretch region (1550–1800 cm^{-1}) and (d) the N–H stretching region (3300–3475 cm^{-1}) (dT = 85%, dT·dT = 15%). Experimental (upper panel) and simulated fitted spectra (lower panel) are shown in (a), while in (d) solid and dashed lines represent experiment and fits, respectively. Integrated fitted band signals (symbols) and kinetic fits (solid lines) as a function of delay time for: (b) the carbonyl bleach (νCO, blue), the vibrationally hot CO ground state (ν*CO, red) and T_1 triplet state product band (νCO_{T_1}, black); and (e) the NH stretch bleach (νNH, blue), the vibrationally hot N–H stretch ground state (ν*NH, red) and T_1 triplet state product band (νNH_{T_1}, black). (c) and (f) compare (upper panels) TVA spectra recorded at $t = 1.1$ ns (black) and FTIR spectra (green), together with (middle panels) calculated vibrational spectra for the S_0 ground state of the dT monomer and (lower panels) T_1 ($^3\pi\pi^*$) state of the dT monomer at the PCM-B3LYP/6-311++G** level of theory. Calculated harmonic vibrational frequencies are scaled by 0.993 in the carbonyl stretching region and 0.945 in the N–H stretching region.

Table 3 Extracted time constant values (τ_n), their respective fit amplitudes (A_n) and photophysical assignments, from a global fitting analysis of the kinetic traces obtained from TVAS (Fig. 6) measurements on dT in chloroform at 260 nm excitation

Time const. (τ_n)	Value/ ps	A_n (νCO)	A_n (ν*CO)	A_n (νCO_{T_1})	A_n (νNH)	A_n (ν*NH)	A_n (νNH_{T_1})	Assignment
τ_1	7 ± 2[a]	0.83	Decay time	Rise time	0.96	Decay time	Rise time	$\nu = 1 \rightarrow \nu = 0$ vib. cooling
τ_2	~114	0.09	—	—	0.01	—	—	$^1n\pi^*$ decay
τ_3	∞	0.08	—	Decay time	0.03	—	Decay time	$^3\pi\pi^*$ decay
τ_4	2.4 ± 0.6[b]	—	Rise time	—	—	[b]	—	$\nu > 1 \rightarrow \nu = 1$ S_0 vib. cooling
τ_5	5–10	—	—	Onset	—	—	Onset	$^1n\pi^* \rightarrow {}^3\pi\pi^*$ ISC

[a] Averaged value for the N–H stretch and carbonyl regions. [b] The rise time of ν*NH is <1 ps and masked by experimental complications and/or our time resolution.

component of the S_0 recovery is $\tau_1 = 7 \pm 2$ ps, and this value is reported in Table 3. Approximately 8% of the initial νCO bleach amplitude remains at a time delay of 1 ns. Comparable observations are made for the νNH bleach recovery. The ν*CO vibrational hot band rises with time constants of $\tau_4 = 2.4 \pm 0.6$ ps, most likely indicative of initial $\nu > 1 \rightarrow \nu = 1$ cooling in S_0. Both hot bands subsequently decay with $\tau_1 = 5.3 \pm 0.6$ ps, a value that agrees with the dominant bleach recovery rate, likely determined by the final $\nu = 1 \rightarrow \nu = 0$ vibrational relaxation step in S_0.

Photoproduct bands are observed in both spectral regions and show growth that is delayed by $\tau_5 = 5$–10 ps. Overlap of the product bands with bleach and hot band features prevents a more precise determination of the delay, but the product grows thereafter with a time constant of 5.3 ± 0.6 ps (νCO_{T_1}) or 7–14 ps (νNH_{T_1}). We highlight that the νNH_{T_1} feature is heavily overlapped with the νNH bleach, so the deduced delay is less reliable than that extracted from νCO_{T_1} analysis. In the carbonyl stretching region, the νCO_{T_1} product band at 1629 cm^{-1} shows no decay within the 1.1 ns range of our experiment, consistent with the long-lived bleach feature. The positions of the product bands match expectations for the first excited T_1 ($^3\pi\pi^*$) triplet state from our calculated vibrational frequencies at the PCM-B3LYP/6-311++G** level of theory on the simpler 1-methylthymine (Fig. 1, R = CH_3) (Fig. 6(c) and (f)). This assignment accords with a previous report of thymine and dT photochemistry in acetonitrile-d_3, in which bands at 1603 and ~1700 cm^{-1} were assigned to the T_1 ($^3\pi\pi^*$) state.[11] This prior study observed complete growth of the T_1 ($^3\pi\pi^*$) triplet state population within 10 ps of UV photoexcitation. Kohler and co-workers also examined the photochemistry of thymine and thymidine 5′-monophosphate (dTMP) in aqueous solution by TEAS,[17] and deduced bifurcation of the excited state flux between ultrafast relaxation on the $^1\pi\pi^*$ state (accessing a conical intersection with the S_0 state), and slower picosecond relaxation *via* the $^1n\pi^*$ state. These deductions are broadly supported by subsequent calculations of excited state pathways and dynamics in the thymine nucleobase and various derivatives using a variety of theoretical methods, although the precise outcomes differ between calculations.[6,44,45,47,48,50,52]

The $^1n\pi^*$ state lifetimes were deduced to be 30 ± 13 ps and 127 ± 15 ps for thymine and dTMP, respectively, in aqueous solution.[17]

Fig. 7 summarizes the photochemical pathways we observe, together with relevant calculated structures on the simpler thymine nucleobase.[37,46] Guided by the earlier work on thymine and its derivatives, we assign the rapid rise of the vibrational hot bands of the S_0 state of dT in chloroform to ultrafast $^1\pi\pi^* \rightarrow S_0$ internal conversion *via* a conical intersection between these two electronic states (most likely involving an ethylenic-twist), and the ~114 ps time constant to recovery of the S_0 ground state *via* the longer-lived $^1n\pi^*$ state. Our measurements suggest ~17% of the excited state flux undergoes $^1\pi\pi^* \rightarrow {}^1n\pi^*$ internal conversion around the vFC region. The >1 ns persistence of the ground state bleaches is indicative of long-lived population in the T_1 ($^3\pi\pi^*$) state, formed with a quantum yield of $\phi_{ISC} \approx 0.08$.

The T_1 ($^3\pi\pi^*$) state of dT is unlikely to form directly from the photoexcited $^1\pi\pi^*$ state, both on orbital symmetry grounds[72] and because the lifetime of the UV populated $^1\pi\pi^*$ state is so short.[6,49] Instead, it will be populated from the $^1n\pi^*$ state. Hare *et al.* previously argued that this $^1n\pi^* \rightarrow {}^3\pi\pi^*$ ISC process must occur before vibrational relaxation to the minimum of the $^1n\pi^*$ state to account for their TVAS observations in acetonitrile-d_3 solution.[11,17] Our results in chloroform solution are not so clear-cut because the bands assigned to the T_1 state show an apparently delayed onset of 5–10 ps, whereas the T_1 state population growth was complete on this timescale according to the measurements in acetonitrile-d_3. Nevertheless, the picosecond time frame on which we observe evolution of the T_1 bands in chloroform is consistent with population *via* the longer lived $^1n\pi^*$ state and not the initially excited $^1\pi\pi^*$ state, and the timescale for growth of the T_1 band intensities is comparable with our expectations for vibrational cooling (*cf.* τ_5 *vs.* τ_1, see Table 3). This cooling may occur in the $^1n\pi^*$ state or the $^3\pi\pi^*$ state, or both. Two candidates for the delayed onset of the triplet bands are plausible: (i) ISC occurs only after the $^1n\pi^*$ state has vibrationally cooled sufficiently to access

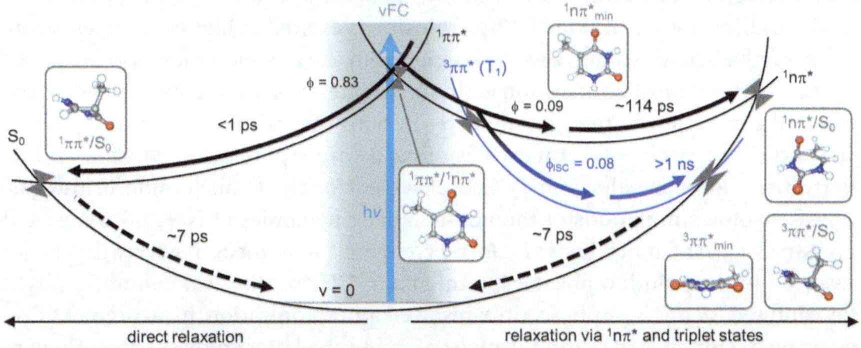

Fig. 7 Schematic potential energy cuts through the S_0, $^1\pi\pi^*$, $^1n\pi^*$ and $^3\pi\pi^*$ electronic states of dT, depicting a plausible kinetic scheme for the excited state relaxation of dT in chloroform based on TVAS measurements at 260 nm. Black = singlet states, blue = triplet states. Possible structures involved in the relaxation dynamics are obtained from theoretical calculations on the simpler thymine nucleobase in the gas-phase; $^1\pi\pi^*/S_0$, $^1\pi\pi^*/^1n\pi^*$, $^1n\pi^*_{min}$ and $^1n\pi^*/S_0$ from ref. 37, $^3\pi\pi^*/S_0$ and $^3\pi\pi^*_{min}$ from ref. 46. The values of ϕ denote branching between competing channels.

the region of ISC, which must lie above the minimum of the $^1n\pi^*$ state otherwise population would continue to feed into T_1 over the ~114 ps lifetime of the excited $^1n\pi^*$ state; (ii) $^1n\pi^* \rightarrow {}^3\pi\pi^*$ ISC is prompt, but our TVAS measurements fail to observe molecules in the initially hot $^3\pi\pi^*$ state until they have relaxed to low vibrational levels. We observe one weak band at around 1600 cm^{-1} which appears to form promptly, and might be a signature of hot triplet state molecules, but its time dependence cannot be isolated in spectral fits. Similar difficulties in resolving absorption by vibrationally excited product molecules arose in our previous TVAS study of G·C Watson–Crick base pairs in chloroform.[62] Possible causes include spectral overlap (*e.g.* with strong ground state bleaches) and anharmonic spreading of the hot band intensities.

3.3 Solvent effects on the photodynamics of cytosine and thymine derivatives

The nuclear dynamics that follow UV excitation to the vFC region of the $^1\pi\pi^*$ states of pyrimidine bases, and their nucleoside and nucleotides, are governed by the energetic proximity of $^1n\pi^*$ states and downhill pathways to conical intersections with the ground electronic state. In the case of thymine and its derivatives, interaction with the low lying triplet states is also significant. The relative energies of the $^1\pi\pi^*$ and $^1n\pi^*$ states, and their triplet counterparts, will be sensitive to the properties of the surrounding medium, as will the locations of conical intersections. The dielectric constant of the medium is a measure of electrostatic interactions, but the molecularity of a solvent, and the ultrafast dynamics of solvent–solute couplings may also play a role. Table 4 draws together the time constants and ISC quantum yields (ϕ_{ISC}) reported in previous and the current ultrafast laser studies for the various photochemical pathways of the pyrimidine bases and their derivatives in different solvents.

The gas-phase spectroscopy studies of Leutwyler and co-workers demonstrate that low-lying vibrational levels of the S_1 ($^1\pi\pi^*$) state of the AO form of cytosine, accessed at wavelengths of ~310 nm, have lifetimes of tens of picoseconds,[73] with the S_1 $\nu = 0$ level surviving for 730 ps.[39] Ultrafast fluorescence up-conversion measurements at 296 nm did not show this behaviour for the related dC nucleoside in chloroform solution.[19] When cytosine is excited to higher energies, using wavelengths below 290 nm, the low barrier from the zero-point level ($\nu = 0$) of the S_1 state to the conical intersection with the S_0 state does not substantially slow the internal conversion to the ground state; in both the gas-phase[67,73–75] and several solvents,[21,76–79] $^1\pi\pi^* \rightarrow S_0$ timescales of <2 ps are reported for cytosine and its derivatives. Similar behaviour is also observed for the C nucleoside in the gas-phase.[22] Some simulations of the non-adiabatic dynamics of isolated UV-excited cytosine suggest a much faster (<100 fs) contribution to the S_0 recovery that is not resolved in the solution-phase experiments.[31,38] These ultrafast dynamics might be captured in the gas-phase time-resolved photoionization measurements on cytosine by Ullrich *et al.*[80] and Kosma *et al.*,[67] but the latter study ascribed them to wavepacket motion out of the vFC region, more consistent with the simulations of Hudock & Martinez[36] and Lan *et al.*,[37] with internal conversion to S_0 on the 1.1 ps timescale. The 0.69 ps lifetime observed in the non-adiabatic dynamics calculations of Barbatti *et al.* was instead attributed to decay of the $^1n_O\pi^*$ state, but the 1.2 ps limit of the simulations precluded consideration of processes occurring over longer timescales.[38] Conversely, Hudock & Martinez[36] suggested that

Table 4 Comparison of dynamical process time constants and intersystem crossing quantum yields (ϕ_{ISC}) obtained for the photodynamics of cytosine and thymine derivatives in different solvents following excitation at wavelengths between 260 and 270 nm (unless stated otherwise). C = cytidine, CMP = cytidine 5′-monophosphate, T = thymidine, TMP = thymidine 5′-monophosphate, d = 2′-deoxy-ribose form

	Dynamical process time constant/ps				
Derivative & solvent	$^1\pi\pi^* \rightarrow S_0$	$^1n\pi^* \rightarrow S_0$	$^1n\pi^* \rightarrow T_1$	ϕ_{ISC} (T_1)	References
Cytosine in water	0.7	12	—	≤0.04	17
Cytosine in methanol	1.1	52	—	—	21
CMP in water	0.7	34	—	≤0.07	17
dCMP in water	1.3	33	—	—	18 and 21
C in water	1.8	35	—	—	21
C in methanol	1.3	144	—	—	21
dC in water	0.7	37	—	—	18,21 and 33
dC in methanol	0.8	186	—	—	21
dC in chloroform	1.1[e]	18 (193[a])	—	—	This work
Cytosine in gas-phase[c]					
λ_{exc} > 310 nm	30–730	—	—	<0.05	39,73 and 81
λ_{exc} = 290 nm	1.1	(≥150[a])	—	—	67
C in gas-phase	0.74	—	—	—	22
Thymine in water	0.72	30	—	≤0.02	17
Thymine in acetonitrile	<1.1	—	≤10[d]	~0.2[d]	11 and 42
dTMP in water	0.41	127	—	≤0.013	17
dT in water	0.54	127[b]	—	—	15–17
dT in acetonitrile	—	—	≤10[d]	~0.2[d]	11
dT in chloroform	<1	~114	5–10	0.08	This work
Thymine in gas-phase	0.46	6.4	—	—	36 and 80
dT in gas-phase	2.9	—	—	—	22

[a] $^1n_N\pi^*$ state lifetime obtained for the minor imino-oxo tautomer. [b] Hare *et al.* stated equal $^1n\pi^*$ lifetimes for dT and dTMP. [c] Values are for the amino-oxo form in the gas-phase. [d] Recorded in acetonitrile-d_3. [e] Obtained from TEAS on C.

population on the $^1n_O\pi^*$ state is trapped for >1 ps, in-line with experimental interpretations.

Estimated triplet state yields for cytosine are at most a few percent, either in the gas-phase[39,81] or aqueous solution.[17] Our measurements suggest that for the nucleoside in chloroform, triplet state formation is also negligible. The most significant effect of the choice of solvent appears to be on the lifetimes of the $^1n\pi^*$ states,[21] which are also sensitive to whether cytosine has a ribose group/chemical derivative attached at the N1 site.[17,18,39] In the case of the 2′-deoxy-ribose nucleoside, the $^1n\pi^*$ state lifetime is reduced in chloroform when compared with water and methanol,[18,21] an effect that we consider unlikely to be caused by the peripheral TBDMS protecting groups present in our nucleoside derivative. The polar, protic solvents are expected to destabilize the $^1n\pi^*$ state with respect to the conical intersection with the S_0 state to a greater degree than chloroform does, which should extend the $^1n\pi^*$ lifetime. Alternatively, protic solvents might promote excited state isomerization to the imino-oxo tautomer which has a longer lived $^1n\pi$ state; the extended $^1n\pi^*$ lifetime is particularly noticeable in methanol solution,[21] with a lifetime similar to the minor IO form $^1n_N\pi^*$ state in chloroform reported here. Both values are also similar to the $^1n\pi^*$ lifetime deduced from gas-phase studies on IO cytosine.[67]

More generally, the foregoing discussion goes some way to highlighting that, despite extensive research, a unified picture has not been reached for the relaxation dynamics of cytosine and its derivatives, as recently acknowledged in a review by Improta *et al.*[6] The outcomes of computational studies depend sensitively on the chosen electronic structure calculation method, reflecting the importance of energy orderings of close lying excited states, energy gradients and the locations of conical intersections. The lack of consensus aptly demonstrates that the task of completely disentangling DNA photodynamics (even in the simpler nucleobases) is still a complex undertaking.

The UV photo-induced dynamics of thymine and its derivatives involve sub-picosecond relaxation from the $^1\pi\pi^*$ state to the S_0 state in water,[16,17,20,40,41,43] acetonitrile[42] and chloroform. Fluorescence up-conversion studies by Gustavsson *et al.* implied that there is some degree of solvent dependence on the direct $^1\pi\pi^* \rightarrow S_0$ process, being (on average) faster in acetonitrile (235 fs) than in water (388 fs).[42] A sub-200 fs decay component was also reported in fluorescence up-conversion experiments on aqueous thymine, dT and dTMP and assigned to relaxation of the $^1\pi\pi^*$ state.[41] This faster $^1\pi\pi^*$ loss could reflect a more direct route to the ethylenic-twist conical intersection with the S_0 ground state,[50] but may also be associated with wavepacket motion out of the fluorescence observation window. Kwok *et al.* observed similarly fast processes by both time-resolved fluorescence and TEAS of an aqueous solution of dT (with $\tau \sim 150$ fs and 760 fs components) and instead argued for sub-ps ISC to the T_1 state.[43] In their time-resolved photoelectron spectroscopy studies of 267 nm excited gas-phase thymine, Ullrich *et al.* reported unassigned time constants of 490 fs and 6.4 ps that bear some resemblance to the solution-phase results, but also observed a <50 fs decay that has not been resolved in solution.[80] Qualitatively similar dynamics were found for the isolated dT nucleoside.[22] A later study by Hudock *et al.*[45] revisited the analysis of the isolated thymine data with the aid of dynamical calculations, and suggested that the sub-500 fs time constants correspond to wavepacket evolution to a local minimum of the $^1\pi\pi^*$ state, rather than any internal conversion process. Meanwhile, the longer picosecond component likely reflects population in the $^1n\pi^*$ state. Other dynamics calculations present a conflicting picture of $^1\pi\pi^* \rightarrow S_0$ internal conversion through an ethylenic-twist conical intersection within ~200 fs,[37] and competing $^1\pi\pi^* \rightarrow {}^1n\pi^*$ internal conversion within ~50 fs.[37,52] Time-resolved ionization experiments on photoexcited gaseous thymine by Gonzalez *et al.*[82] observed transient intermediates with lifetimes of <100 fs, 7 ps and >200 ps, with the two longer timescale components either much weaker or missing in clusters of thymine with one or two water molecules. These observations are consistent with sub-ps relaxation on the $^1\pi\pi^*$ state, and longer-time dynamics corresponding to relaxation and then decay of the $^1n\pi^*$ state or involvement of a triplet state. The suppression of these latter channels by water clustering may signal destabilization of the $^1n\pi^*$ state by (even a few) protic solvent molecules.

Once populated from the $^1\pi\pi^*$ state, the $^1n\pi^*$ state of thymine and its derivatives is comparatively long lived,[17] but (as with dC) the lifetime of this excited state is notably sensitive to both solvent and the derivatization at the N1 site. The lifetime variation reflects modifications of the $^1n\pi^*$ potential energy surface and the energy barrier to the conical intersection with the S_0 state. Unlike cytosine derivatives, the quantum yield for ISC to the T_1 ($^3\pi\pi^*$) state is notably enhanced in a number of solvents, and its population is mediated by a vibrationally hot $^1n\pi^*$ state.[17] The growth of T_1 ($^3\pi\pi^*$) absorption bands occurs on comparable ($\leq$10 ps) timescales in both acetonitrile-d_3 and chloroform.[11] A similar timescale emerges from ISC calculations by Etinski *et al.* on isolated thymine.[49] The ISC quantum yields differ by a factor of ~2 in these two solvents, but are an order of magnitude higher than for thymine derivatives in aqueous solution,[11] an effect attributed to destabilization of the intermediate $^1n\pi^*$ state by water solvation.[83]

As was recently summarized by Improta *et al.*,[6] a more qualitatively consistent picture of the relaxation dynamics in thymine derivatives emerges from the recent literature than for cytosine and its derivatives. While there is a degree of quantitative discrepancy, in general, $^1\pi\pi^* \rightarrow S_0$ relaxation is found to occur on a sub-ps timeframe *via* a dominant ethylenic-twist conical intersection. In competition with this direct relaxation, ultrafast $^1\pi\pi^* \rightarrow {}^1n\pi^*$ internal conversion populates a vibrationally hot $^1n\pi^*$ state, which can then either undergo slower relaxation to the S_0 ground state over many picoseconds or act as a doorway for ISC to a long-lived T_1 ($^3\pi\pi^*$) state. Once vibrational cooling has occurred on the $^1n\pi^*$ surface, the ISC channel closes. Consultation of the broad literature shows that the relative importance of these three main relaxation pathways in thymine derivatives depends on the solvation environment.

4. Conclusions

Ultraviolet (λ = 260 nm) absorption by silyl-protected 2′-deoxy-cytidine and 2′-deoxy-thymidine in chloroform solutions initiates ultrafast non-adiabatic dynamics on the photoexcited $^1\pi\pi^*$ state, and on the $^1n\pi^*$ states lying close in energy. These dynamics are revealed by a combination of ultrafast transient electronic and vibrational absorption spectroscopies. Nuclear motion on the excited states encounters conical intersections with the ground electronic state, and in the case of the amino-oxo canonical form of dC, recovery of the ground state population is essentially complete within 100 ps. Similarly fast repopulation of the ground state occurs in dT, but ~8% of the photoexcited molecules are instead trapped in a long-lived triplet state. The pathways for these various processes involve prompt bifurcation between the $^1\pi\pi^*$ and $^1n\pi^*$ states in the vertical Franck–Condon region, rapid ($\lesssim$ 1 ps) $^1\pi\pi^* \rightarrow S_0$ internal conversion through a conical intersection with ethylenic-twist character, and slower $^1n\pi^* \rightarrow S_0$ internal conversion. The general features of these dynamics appear to be common to both the chosen pyrimidine bases, as well as their nucleosides and nucleotides, but the details of time constants and branching between pathways are sensitive to the functionality at the N1 site of the pyrimidine ring and the properties of the surrounding environment. Recovery of thermalized molecules in their electronic ground state *via* the $^1n\pi^*$ state takes 18.6 $\pm$ 1.1 ps in the amino-oxo form of dC in chloroform, ~114 ps in dT, and 193 $\pm$ 55 ps in the minor imino-oxo form of dC. These timescales are mostly controlled by the excited state lifetimes because vibrational cooling of hot ground state molecules requires ~6 ps.

A region of singlet–triplet coupling encourages intersystem crossing from the $^1n\pi^*$ state of dT to the T_1 ($^3\pi\pi^*$) state. The $^1n\pi^*$ state lifetime of amino-oxo dC in chloroform is distinctly shorter than those previously reported in methanol and aqueous solutions, indicating either a lower energy $^1n\pi^*/S_0$ conical intersection in this more weakly interacting solvent, or promotion of AO $^1n\pi^* \rightarrow$ IO $^1n\pi^*$ tautomerization by the protic solvents.

The above interpretation of the dynamics revealed by our TEAS and TVAS measurements is qualitatively in agreement with some, but not all computational studies of the non-adiabatic dynamics of UV-excited cytosine and thymine nucleobases and their derivatives. The outcomes of these computational simulations are highly sensitive to the level of electronic structure theory employed, because of the close proximities of several electronically excited states and the variations in predicted locations of conical intersections which govern the non-adiabatic pathways. Moreover, the dynamics simulations concentrate on processes no longer than a few picoseconds because of the computational expense of propagating the nuclear dynamics over longer timescales relevant to the $^1n\pi^*$ and triplet state pathways. Conversely, the theoretical simulations predict some <100 fs non-adiabatic dynamics that cannot be distinguished with our current experimental time resolution. We find no evidence to support computational predictions of triplet state formation in photo-excited dC, or bifurcation at the $^1\pi\pi^*/S_0$ conical intersection which has been proposed as a mechanism for AO $\rightarrow$ IO photo-isomerization.

Unravelling the precise photochemical mechanisms of the monomeric cytosine and thymine nucleobases, and their nucleosides and nucleotides, continues to present a challenge to both experiment and theory. This challenge is still greater when these species are incorporated into Watson–Crick base pairs or nucleic acid strands, where the electronic character of the photoexcitation and the available dynamical and chemical pathways are affected by spatial proximity to other purine or pyrimidine bases.

Acknowledgements

The Bristol group gratefully acknowledges funding from the ERC Advanced Grant 290966 CAPRI. GMR thanks the Ramsay Memorial Trust for the award of a Fellowship. The Kiel group thanks the German Science Foundation (DFG) for financial support through grant SFB 677 and KR acknowledges the DFG for the award of a Research Fellowship. We are grateful to A. S. Henderson, and M. C. Galan (University of Bristol) for help with the synthesis of the derivatized nucleosides, and to M. P. Grubb and P. M. Coulter for assistance with the Bristol transient absorption spectroscopy measurements.

References

1 C. E. Crespo-Hernandez, B. Cohen, P. M. Hare and B. Kohler, *Chem. Rev.*, 2004, **104**, 1977–2019.

2 C. T. Middleton, K. de La Harpe, C. Su, Y. K. Law, C. E. Crespo-Hernandez and B. Kohler, *Annu. Rev. Phys. Chem.*, 2009, **60**, 217–239.

3 A. H. Zewail, *J. Phys. Chem. A*, 2000, **104**, 5660–5694.

4 *Conical Intersections: Theory, Computation and Experiment*, ed. W. Domcke, D. R. Yarkony and H. Koppel, World Scientific, 2011.
5 M. Barbatti, A. J. A. Aquino, J. J. Szymczak, D. Nachtigallova, P. Hobza and H. Lischka, *Proc. Natl. Acad. Sci. U. S. A.*, 2010, **107**, 21453–21458.
6 R. Improta, F. Santoro and L. Blancafort, *Chem. Rev.*, 2016, **116**, 3540–3593.
7 Z. Y. Liu, C. Tan, X. M. Guo, Y. T. Kao, J. Li, L. J. Wang, A. Sancar and D. P. Zhong, *Proc. Natl. Acad. Sci. U. S. A.*, 2011, **108**, 14831–14836.
8 J. Li, Z. Y. Liu, C. Tan, X. M. Guo, L. J. Wang, A. Sancar and D. P. Zhong, *Nature*, 2010, **466**, 887.
9 Z. Y. Liu, L. J. Wang and D. P. Zhong, *Phys. Chem. Chem. Phys.*, 2015, **17**, 11933–11949.
10 D. W. Whillans and H. E. Johns, *J. Am. Chem. Soc.*, 1971, **93**, 1358.
11 P. M. Hare, C. T. Middleton, K. I. Mertel, J. M. Herbert and B. Kohler, *Chem. Phys.*, 2008, **347**, 383–392.
12 A. A. Lamola and J. P. Mittal, *Science*, 1966, **154**, 1560.
13 W. J. Schreier, T. E. Schrader, F. O. Koller, P. Gilch, C. E. Crespo-Hernandez, V. N. Swaminathan, T. Carell, W. Zinth and B. Kohler, *Science*, 2007, **315**, 625–629.
14 K. Siriwong, A. A. Voityuk, M. D. Newton and N. Rosch, *J. Phys. Chem. B*, 2003, **107**, 2595–2601.
15 J. M. L. Pecourt, J. Peon and B. Kohler, *J. Am. Chem. Soc.*, 2000, **122**, 9348–9349.
16 J. M. L. Pecourt, J. Peon and B. Kohler, *J. Am. Chem. Soc.*, 2001, **123**, 10370–10378.
17 P. M. Hare, C. E. Crespo-Hernandez and B. Kohler, *Proc. Natl. Acad. Sci. U. S. A.*, 2007, **104**, 435–440.
18 S. Quinn, G. W. Doorley, G. W. Watson, A. J. Cowan, M. W. George, A. W. Parker, K. L. Ronayne, M. Towrie and J. M. Kelly, *Chem. Commun.*, 2007, 2130–2132.
19 N. K. Schwalb, T. Michalak and F. Temps, *J. Phys. Chem. B*, 2009, **113**, 16365–16376.
20 J. Peon and A. H. Zewail, *Chem. Phys. Lett.*, 2001, **348**, 255–262.
21 C. S. Ma, C. C. W. Cheng, C. T. L. Chan, R. C. T. Chan and W. M. Kwok, *Phys. Chem. Chem. Phys.*, 2015, **17**, 19045–19057.
22 S. De Camillis, J. Miles, G. Alexander, O. Ghafur, I. D. Williams, D. Townsend and J. B. Greenwood, *Phys. Chem. Chem. Phys.*, 2015, **17**, 23643–23650.
23 N. Ismail, L. Blancafort, M. Olivucci, B. Kohler and M. A. Robb, *J. Am. Chem. Soc.*, 2002, **124**, 6818–6819.
24 M. Merchán and L. Serrano-Andres, *J. Am. Chem. Soc.*, 2003, **125**, 8108–8109.
25 L. Blancafort and M. A. Robb, *J. Phys. Chem. A*, 2004, **108**, 10609–10614.
26 M. Merchán, L. Serrano-Andres, M. A. Robb and L. Blancafort, *J. Am. Chem. Soc.*, 2005, **127**, 1820–1825.
27 K. Tomic, J. Tatchen and C. M. Marian, *J. Phys. Chem. A*, 2005, **109**, 8410–8418.
28 M. Merchán, R. Gonzalez-Luque, T. Climent, L. Serrano-Andres, E. Rodríguez, M. Reguero and D. Pelaez, *J. Phys. Chem. B*, 2006, **110**, 26471–26476.
29 L. Blancafort, *Photochem. Photobiol.*, 2007, **83**, 603–610.
30 K. A. Kistler and S. Matsika, *J. Phys. Chem. A*, 2007, **111**, 2650–2661.
31 J. González-Vázquez and L. Gonzalez, *ChemPhysChem*, 2010, **11**, 3617–3624.
32 Q. S. Li and L. Blancafort, *Photochem. Photobiol. Sci.*, 2013, **12**, 1401–1408.
33 C. G. Triandafillou and S. Matsika, *J. Phys. Chem. A*, 2013, **117**, 12165–12174.

34 M. Richter, P. Marquetand, J. González-Vázquez, I. Sola and L. Gonzalez, *J. Phys. Chem. Lett.*, 2012, **3**, 3090–3095.
35 S. Mai, P. Marquetand, M. Richter, J. González-Vázquez and L. Gonzalez, *ChemPhysChem*, 2013, **14**, 2920–2931.
36 H. R. Hudock and T. J. Martinez, *ChemPhysChem*, 2008, **9**, 2486–2490.
37 Z. G. Lan, E. Fabiano and W. Thiel, *J. Phys. Chem. B*, 2009, **113**, 3548–3555.
38 M. Barbatti, A. J. A. Aquino, J. J. Szymczak, D. Nachtigallova and H. Lischka, *Phys. Chem. Chem. Phys.*, 2011, **13**, 6145–6155.
39 S. Blaser, M. A. Trachsel, S. Lobsiger, T. Wiedmer, H. Frey and S. Leutwyler, *J. Phys. Chem. Lett.*, 2016, **7**, 752–757.
40 F. Buchner, A. Nakayama, S. Yamazaki, H. H. Ritze and A. Lübcke, *J. Am. Chem. Soc.*, 2015, **137**, 2931–2938.
41 T. Gustavsson, A. Sharonov and D. Markovitsi, *Chem. Phys. Lett.*, 2002, **351**, 195–200.
42 T. Gustavsson, N. Sarkar, E. Lazzarotto, D. Markovitsi and R. Improta, *Chem. Phys. Lett.*, 2006, **429**, 551–557.
43 W. M. Kwok, C. Ma and D. L. Phillips, *J. Am. Chem. Soc.*, 2008, **130**, 5131–5139.
44 S. Perun, A. L. Sobolewski and W. Domcke, *J. Phys. Chem. A*, 2006, **110**, 13238–13244.
45 H. R. Hudock, B. G. Levine, A. L. Thompson, H. Satzger, D. Townsend, N. Gador, S. Ullrich, A. Stolow and T. J. Martinez, *J. Phys. Chem. A*, 2007, **111**, 8500–8508.
46 J. J. Serrano-Pérez, R. Gonzalez-Luque, M. Merchán and L. Serrano-Andres, *J. Phys. Chem. B*, 2007, **111**, 11880–11883.
47 G. Zechmann and M. Barbatti, *J. Phys. Chem. A*, 2008, **112**, 8273–8279.
48 J. J. Szymczak, M. Barbatti, J. T. S. Hoo, J. A. Adkins, T. L. Windus, D. Nachtigallova and H. Lischka, *J. Phys. Chem. A*, 2009, **113**, 12686–12693.
49 M. Etinski, T. Fleig and C. A. Marian, *J. Phys. Chem. A*, 2009, **113**, 11809–11816.
50 D. Asturiol, B. Lasorne, M. A. Robb and L. Blancafort, *J. Phys. Chem. A*, 2009, **113**, 10211–10218.
51 R. Gonzalez-Luque, T. Climent, I. Gonzalez-Ramirez, M. Merchán and L. Serrano-Andres, *J. Chem. Theory Comput.*, 2010, **6**, 2103–2114.
52 D. Picconi, V. Barone, A. Lami, F. Santoro and R. Improta, *ChemPhysChem*, 2011, **12**, 1957–1968.
53 C. Salet, R. Bensasson and R. S. Becker, *Photochem. Photobiol.*, 1979, **30**, 325–329.
54 H. Görner, *J. Photochem. Photobiol., B*, 1990, **5**, 359–377.
55 T. Climent, I. Gonzalez-Ramirez, R. Gonzalez-Luque, M. Merchán and L. Serrano-Andres, *J. Phys. Chem. Lett.*, 2010, **1**, 2072–2076.
56 M. M. Brister and C. E. Crespo-Hernandez, *J. Phys. Chem. Lett.*, 2015, **6**, 4404–4409.
57 H. E. Johns, M. Delbruck and S. A. Rapaport, *J. Mol. Biol.*, 1962, **4**, 104.
58 H. E. Johns, M. L. Pearson, C. W. Helleiner and J. C. Leblanc, *J. Mol. Biol.*, 1964, **9**, 503–524.
59 C. F. Yang, Y. Q. Yu, K. H. Liu, D. Song, L. D. Wu and H. M. Su, *J. Phys. Chem. A*, 2011, **115**, 5335–5345.
60 M. C. Cuquerella, V. Lhiaubet-Vallet, F. Bosca and M. A. Miranda, *Chem. Sci.*, 2011, **2**, 1219–1232.
61 K. K. Ogilvie, *Can. J. Chem.*, 1973, **51**, 3799–3807.

62 K. Röttger, H. J. B. Marroux, M. P. Grubb, P. M. Coulter, H. Böhnke, A. S. Henderson, M. C. Galan, F. Temps, A. J. Orr-Ewing and G. M. Roberts, *Angew. Chem., Int. Ed.*, 2015, **54**, 14719–14722.
63 K. Röttger, R. Siewertsen and F. Temps, *Chem. Phys. Lett.*, 2012, **536**, 140–146.
64 K. Röttger, S. Wang, F. Renth, J. Bahrenburg and F. Temps, *Appl. Phys. B*, 2015, **118**, 185–193.
65 L. Biemann, S. A. Kovalenko, K. Kleinermanns, R. Mahrwald, M. Markert and R. Improta, *J. Am. Chem. Soc.*, 2011, **133**, 19664–19667.
66 G. M. Roberts, H. J. B. Marroux, M. P. Grubb, M. N. R. Ashfold and A. J. Orr-Ewing, *J. Phys. Chem. A*, 2014, **118**, 11211–11225.
67 K. Kosma, C. Schröter, E. Samoylova, I. V. Hertel and T. Schultz, *J. Am. Chem. Soc.*, 2009, **131**, 16939–16943.
68 L. Buschhaus, J. Rolf and K. Kleinermanns, *Phys. Chem. Chem. Phys.*, 2013, **15**, 18371–18377.
69 I. Reva, M. J. Nowak, L. Lapinski and R. Fausto, *J. Phys. Chem. B*, 2012, **116**, 5703–5710.
70 M. J. Frisch, G. W. Trucks, H. B. Schlegel, G. E. Scuseria, M. A. Robb, J. R. Cheeseman, G. Scalmani, V. Barone, B. Mennucci, G. A. Petersson, H. Nakatsuji, M. Caricato, X. Li, H. P. Hratchian, A. F. Izmaylov, J. Bloino, G. Zheng, J. L. Sonnenberg, M. Hada, M. Ehara, K. Toyota, R. Fukuda, J. Hasegawa, M. Ishida, T. Nakajima, Y. Honda, O. Kitao, H. Nakai, T. Vreven, J. A. Montgomery, J. E. Peralta, F. Ogliaro, M. J. Bearpark, J. J. Heyd, E. Brothers, K. N. Kudin, V. N. Staroverov, R. Kobayashi, J. Normand, K. Raghavachari, A. Rendell, J. C. Burant, S. S. Iyengar, J. Tomasi, M. Cossi, N. Rega, J. M. Millam, M. Klene, J. E. Knox, J. B. Cross, V. Bakken, C. Adamo, J. Jaramillo, R. Gomperts, R. E. Stratmann, O. Yazyev, A. J. Austin, R. Cammi, C. Pomelli, J. W. Ochterski, R. L. Martin, K. Morokuma, V. G. Zakrzewski, G. A. Voth, P. Salvador, J. J. Dannenberg, S. Dapprich, A. D. Daniels, O. Farkas, J. B. Foresman, J. V. Ortiz, J. Cioslowski and D. J. Fox, *Gaussian 09*, Gaussian Inc, Wallingford, CT, 2009.
71 D. Murdock, S. J. Harris, J. Luke, M. P. Grubb, A. J. Orr-Ewing and M. N. R. Ashfold, *Phys. Chem. Chem. Phys.*, 2014, **16**, 21271–21279.
72 M. A. El-Sayed, *Acc. Chem. Res.*, 1968, **1**, 8–16.
73 S. Lobsiger, M. A. T'rachsel, H. M. Frey and S. Leutwyler, *J. Phys. Chem. B*, 2013, **117**, 6106–6115.
74 C. Canuel, M. Mons, F. Piuzzi, B. Tardivel, I. Dimicoli and M. Elhanine, *J. Chem. Phys.*, 2005, **122**, 074316.
75 J. W. Ho, H. C. Yen, W. K. Chou, C. N. Weng, L. H. Cheng, H. Q. Shi, S. H. Lai and P. Y. Cheng, *J. Phys. Chem. A*, 2011, **115**, 8406–8418.
76 R. J. Malone, A. M. Miller and B. Kohler, *Photochem. Photobiol.*, 2003, **77**, 158–164.
77 A. Sharonov, T. Gustavsson, V. Carre, E. Renault and D. Markovitsi, *Chem. Phys. Lett.*, 2003, **380**, 173–180.
78 A. Sharonov, T. Gustavsson, S. Marguet and D. Markovitsi, *Photochem. Photobiol. Sci.*, 2003, **2**, 362–364.
79 L. Blancafort, B. Cohen, P. M. Hare, B. Kohler and M. A. Robb, *J. Phys. Chem. A*, 2005, **109**, 4431–4436.
80 S. Ullrich, T. Schultz, M. Z. Zgierski and A. Stolow, *Phys. Chem. Chem. Phys.*, 2004, **6**, 2796–2801.

81 S. Lobsiger, M. Etinski, S. Blaser, H. M. Frey, C. Marian and S. Leutwyler, *J. Chem. Phys.*, 2015, **143**, 234301.
82 J. González-Vázquez, L. Gonzalez, E. Samoylova and T. Schultz, *Phys. Chem. Chem. Phys.*, 2009, **11**, 3927–3934.
83 M. Etinski and C. M. Marian, *Phys. Chem. Chem. Phys.*, 2010, **12**, 4915–4923.

Faraday Discussions

PAPER

Towards elucidating the photochemistry of the sunscreen filter ethyl ferulate using time-resolved gas-phase spectroscopy†

N. D. N. Rodrigues,[a] M. Staniforth,[a] J. D. Young,[a] Y. Peperstraete,[ab] N. C. Cole-Filipiak,[a] J. R. Gord,[c] P. S. Walsh,[c] D. M. Hewett,[c] T. S. Zwier[c] and V. G. Stavros*[a]

Received 13th April 2016, Accepted 28th April 2016

DOI: 10.1039/c6fd00079g

Ultrafast time-resolved ion yield (TR-IY) and velocity map imaging spectroscopies are employed to reveal the relaxation dynamics after photoexcitation in ethyl 4-hydroxy-3-methoxycinnamate (ethyl ferulate, EF), an active ingredient in commercially available sunscreens. In keeping with a bottom-up strategy, the building blocks of EF, 2-methoxy-4-vinylphenol (MVP) and 4-hydroxy-3-methoxycinnamyl alcohol (coniferyl alcohol, ConA), were also studied to assist in our understanding of the dynamics of EF as we build up in molecular complexity. In contrast to the excited state dynamics of MVP and ConA, which are described by a single time constant (>900 ps), the dynamics of EF are described by three time constants (15 ± 4 ps, 148 ± 47 ps, and >900 ps). A mechanism is proposed involving internal conversion (IC) between the initially excited $S_1(1^1\pi\pi^*)$ and $S_2(1^1n\pi^*)$ states followed by intramolecular vibrational redistribution (IVR) on both states, in competition with intersystem crossing onto neighbouring triplet states (15 ± 4 ps). IVR and IC within the triplet manifold then ensues (148 ± 47 ps) to populate a low-lying triplet state (>900 ps). Importantly, the fluorescence spectrum of EF at the S_1 origin, along with the associated lifetime (6.9 ± 0.1 ns), suggests that population is trapped, during initial IVR, on the $S_1(1^1\pi\pi^*)$ state. This serves to demonstrate the complex, competing dynamics in this sunscreen filter molecule.

Introduction

Many living organisms have an intrinsic need for protection against ultraviolet (UV) radiation emitted by the Sun.[1] While the solar spectrum peaks in the visible

[a]University of Warwick, Department of Chemistry, CV4 7AL, Coventry, UK. E-mail: v.stavros@warwick.ac.uk
[b]ENS de Cachan, 61 Avenue du Président Wilson, 94230, Cachan, France
[c]Department of Chemistry, Purdue University, West Lafayette, Indiana 47907-2084, USA

† Electronic supplementary information (ESI) available: UV/Vis spectrum of EF; details of kinetic fits; comparative kinetic fits of EF TR-IY data; H^+ TKER spectra for MVP, ConA, and EF; TOF MS of EF; calculated triplet state energies in EF; EF fluorescence lifetimes; and further TR-IY data of EF. See DOI: 10.1039/c6fd00079g

range (700–400 nm), a significant amount of UV-A and UV-B radiation (400–315 nm and 315–280 nm, respectively) reaches the surface of the Earth; on a sunny day, approximately 220 μW cm^{-2} globally.[2,3] UV initiated reactions resulting in DNA damage within skin cells are well-established phenomena, with such reactions leading to skin aging and carcinogenic effects, including melanoma and carcinoma.[4–9] Both plant and animal species are naturally equipped with defence mechanisms against potentially toxic radiation,[10] either within their DNA itself[11,12] or through radiation filtering, "sunscreen", molecules, *e.g.* (eu) melanin in humans.[13–15] Such filters absorb efficiently in the UV, yet their physical and chemical properties allow for the safe dissipation of this potentially harmful energy, returning the molecule to its original ground state.

The increasing incidence of UV-A and UV-B radiation on the Earth's surface[16,17] and the emerging trend for more Sun exposure over recent years has meant that this natural protection is often insufficient, with skin cancer cases increasing steadily between 1975 and 2011.[18–21] There is, therefore, a strong impetus for understanding the photoprotective mechanisms of both natural and artificial sunscreen molecules.

The interest in extra chemical protection from the Sun's radiation dates back to 1936, when the first sunscreen product was made available by L'Oréal's founder, E. Schueller.[22] Since then, while the sunscreen industry has expanded greatly, little progress has been made in understanding the photophysical mechanisms through which sunscreens provide photoprotection. For instance, it is not clear whether the components of the current commercially available sunscreens release reactive oxygen species upon absorption of UV radiation. The creation of such reactive species would imply sunscreens can potentially increase the risk of melanoma, rather than reduce it.[23] Efforts have recently begun towards spectroscopic studies of natural (plant and human) and synthetic sunscreen molecules in order to garner information on their excited state photodynamics and consequently contribute to the development of more efficient, safer sunscreens.[24–29]

Ethyl 4-hydroxy-3-methoxycinnamate (ethyl ferulate, EF), shown in Fig. 1(c), was chosen for the present studies due to its use in commercial sunscreens as a UV filter and antioxidant.[30–32] Similar to all active sunscreen ingredients, EF is used in these products as part of a combination of radiation filtering species in order to achieve optimum filtering efficiency across the entire UV-A and UV-B range (usually measured in terms of the Sun protection factor, SPF).[33] The broad absorption of EF in the UV region (see Fig. S1 in ESI†), justifies the presence of EF in sunscreen formulas.

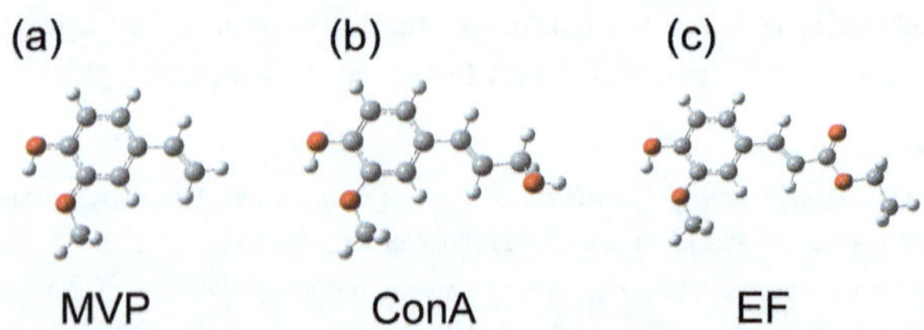

Fig. 1 (a) 2-Methoxy-4-vinylphenol (MVP), (b) 4-hydroxy-3-methoxycinnamyl alcohol (coniferyl alcohol, ConA), and (c) ethyl 4-hydroxy-3-methoxycinnamate (ethyl ferulate, EF).

The research presented here explores the excited state photodynamics of EF, building on previous studies of sunscreen molecules published in recent years.[24–29] EF presents itself as an extremely relevant system to be studied, not only due to its practical relevance as a sunscreen filter molecule, but also due to the insight provided into how increasing molecular complexity, in particular the addition of a carbonyl moiety (through the ester group functionality), influences the photochemistry of EF and other structurally related sunscreen molecules. To achieve this, we also study the subunits of EF, 2-methoxy-4-vinylphenol (MVP) and 4-hydroxy-3-methoxycinnamyl alcohol (ConA) shown in Fig. 1(a) and (b) respectively.

Experimental methods

a. Time-resolved ion yield (TR-IY)

The experimental setup has been described previously,[34,35] and therefore is only summarised here, being expanded as necessary with regards to the present experimental conditions. A commercial femtosecond (fs) Ti:sapphire oscillator (Spectra-Physics Tsunami) and regenerative amplifier (Spectra-Physics Spitfire XP) are used to produce ~40 fs laser pulses centred at 800 nm at ~3 mJ per pulse. The fundamental 800 nm output is subsequently split into three beams, each ~1 mJ per pulse, with one of these beams used to pump an optical parametric amplifier (Light Conversion, TOPAS-C) in order to generate the pump ($h\nu_{pu}$) pulses. $h\nu_{pu}$ is centred on the S_1 origin band (S_1 ($\nu = 0$)) of each molecule studied (305 nm, 306 nm and 317.5 nm for MVP, ConA and EF, respectively; *vide infra*). A second beam is up-converted *via* frequency doubling the 800 nm fundamental output in a type-I β-barium borate (BBO) crystal to generate 400 nm, followed by successive frequency mixing with the residual 800 nm pulses to generate the third harmonic (267 nm) and finally the desired 200 nm fourth harmonic (using type-II and type-I BBO crystals, respectively) which is used as the probe ($h\nu_{pr}$) pulse for these experiments. Alternatively, a second TOPAS-C is used to generate 243 nm probe pulses. The pump and probe pulses are temporally delayed with respect to each other by way of a hollow gold retroreflector mounted on a motorised delay stage, allowing a maximum temporal delay (Δt) of 900 ps ($h\nu_{pr}$ = 200 nm) or 1.2 ns ($h\nu_{pr}$ = 243 nm). The relative polarisation between the pump and probe beams is held at the magic angle (54.7°).

The two laser beams intercept a molecular beam which is produced by seeding the target molecules (EF, Sigma-Aldrich, 98%; ConA, Sigma-Aldrich, 98%; or MVP, VWR International Ltd., 97%), heated to their melting points (55–95 °C), into helium (~3 bar). This is then expanded into a vacuum (~10^{-7} mbar) using an Even-Lavie pulsed solenoid valve.[36] At the point of intersection, $h\nu_{pu}$ excites the species in the molecular beam and $h\nu_{pr}$ ionises any excited (or photodissociated; *vide infra*) species. The resulting ions are focused onto a detector, consisting of two microchannel plates (MCPs), by three ion optics similar to the set up described by Eppink and Parker.[37] The current output from the MCP, gated in ion flight time over the mass channel of each parent ion ($[MVP]^+$, $[ConA]^+$, and $[EF]^+$), is measured on a digital oscilloscope (LeCroy LT372 Waverunner) and integrated as a function of Δt in order to produce time-resolved ion yield (TR-IY) transients.

b. Velocity map imaging (VMI) of H atoms

Velocity map imaging (VMI) is used to monitor any H atom elimination from the molecules of interest, following excitation. Any photodissociated H atoms are resonantly ionised to form H^+ with a 243 nm probe pulse set at $\Delta t = 1.2$ ns relative to the pump pulse. H^+ ions are then accelerated by an electrostatic potential onto a position-sensitive detector such that ions with the same initial velocity are mapped onto a single pixel on the detector.[37] The original 3D H^+ distribution is reconstructed from the measured 2D projection using a polar onion peeling algorithm[38] from which the desired 1D total kinetic energy release (TKER) spectrum can be derived using an appropriate Jacobian, energy calibration factor, and consideration of the co-fragment's mass. The spectrometer is calibrated using the TKER spectrum of photolysed HBr.[39]

c. Resonant two-photon ionisation (R2PI) and ionisation potential (IP) determination of EF

Mass-resolved resonant two-photon ionisation (R2PI) studies were carried out for EF using a molecular beam time-of-flight mass spectrometer (TOF-MS) described in detail elsewhere.[40] The solid sample is wrapped in glass wool and inserted into a glass sample holder, which is placed inside a stainless-steel reservoir behind a general valve (Parker, Series 9) operating at 20 Hz. The sample and valve assembly are heated (90–95 °C) in order to achieve sufficient vapour pressure of the sample. The sample is then entrained in a Ne backing gas (2–4 bar) and expanded through a pin-hole nozzle (800 μm) forming a supersonic expansion. The supersonic expansion is skimmed downstream forming a molecular beam which travels to the extraction region of a TOF-MS where the cold, isolated molecules interact with the laser beams.

The frequency doubled output of a Nd:YAG (Continuum SureLite II) pumped dye laser is used to generate a UV beam (310–319 nm, DCM dye, <1 mJ per pulse) which is loosely focused into the interaction region. R2PI spectra are recorded by collecting the molecular ion yield as a function of the laser wavelength.

The ionisation potential (IP) of each of the EF conformers was determined by employing two-color R2PI (2C-R2PI), which has been previously described in detail.[41–43] In order to determine the IP, the two-colour signal enhancement is maximised; the second colour is then scanned towards increasingly longer wavelength until the two-colour contribution to the ion signal goes to zero at the threshold. If the geometry change between the excited state and the ion is sufficiently small, a sharp onset is observed at the threshold, providing an accurate measure of the IP. The energy of the two photons at the threshold of ionisation corresponds to the IP of the molecule.

d. UV–UV hole-burning (UVHB)

Conformer specific UV spectra were also recorded for EF, using a setup described previously.[40] The doubled output of a Continuum SureLite II Nd:YAG pumped dye laser (Lambda-Physik ScanMate), operating at 10 Hz and <2 mJ per pulse, is fixed on the origin band (S_1 ($\nu = 0$)) of each of the two conformers of EF present (determined from analysis of the R2PI spectra obtained as described above) thus serving as the hole-burn laser. The burn laser is spatially overlapped with a UV probe laser (Radiant Dyes NarrowScan, 0.5–1 mJ per pulse, 20 Hz) and precedes it temporally by ~200 ns. The probe laser is scanned over the frequency range of interest (310–319 nm). The difference between the signal with and without the presence of the hole-burn laser is recorded using active baseline subtraction in a gated integrator (Stanford Research SRS250).

e. Dispersed fluorescence and total fluorescence lifetimes

The laser induced fluorescence apparatus is described elsewhere.[40] EF was heated to approximately 90–95 °C, entrained in ~3 bar of helium, and pulsed into a vacuum through an 800 μm orifice (Parker General Valve Series 9) to produce a supersonic jet expansion. The sample was excited downstream in the collision-free regime using the doubled output of a Lambda-Physik ScanMate dye laser and the total resulting fluorescence was detected with a photomultiplier tube as a function of time. A pair of filters (340/345 nm) were used to cut out scattered light from the excitation laser when measuring fluorescence lifetimes; filters transmitting >400 nm were used to search for phosphorescence. Dispersed fluorescence measurements were achieved by dispersing the fluorescence in a 0.75 m monochromator (Horiba Jobin Yvon, 1200 grooves per mm) outfitted with an iCCD camera (Andor, iStar 720, 25 mm). Dispersed fluorescence measurements were background subtracted with scans with no sample to remove any laser scatter present in the spectra.

Theoretical methods

Quantum mechanical calculations were performed to obtain ground and excited state geometries for MVP, ConA, and EF. Geometry optimised structures were calculated using Gaussian09 (ref. 44) with density functional theory (DFT) and the Coulomb attenuated model Becke-3rd parameter Lee–Yang–Parr (CAM-B3LYP) functional[45] combined with a 6-31G* basis set.[46]

The S_0/S_3 (*E*/*Z* isomerisation) conical intersection (CI) structure in EF was optimised at the complete active space self-consistent field (CASSCF)[47,48]/6-31G* level of theory using Gaussian09 and a 6 electron, 6 orbital (6,6) active space consisting of three occupied π and three unoccupied π* orbitals. The CI structure between the S_1 and S_2 states (ππ* and nπ* states, respectively) was optimised at the same level of theory as above, however, employing a (10,9) active space, adding to the previous active space one occupied σ orbital, one unoccupied σ* orbital (localised on the C=O bond), and one occupied n orbital (localised on the double bonded O atom).

Potential energy cuts (PECs) were produced by calculating the energies of the first four singlet electronic states (S_0, $S_1(1^1\pi\pi^*)$, $S_2(1^1n\pi^*)$, and $S_3(2^1\pi\pi^*)$) of EF at various points along each of the linearly interpolated internal coordinates (LIIC) from the various minimum energy geometries to their respective CIs. This was done using Molpro 2010.1 (ref. 49) and employing fully state averaged CASSCF along with its second order perturbation theory extension (CASPT2)[50] using a cc-pVDZ basis set[51] and a (12,10) active space, adding an additional occupied π orbital to the previously described (10,9) active space.

Results

a. Excited state dynamics in MVP

We start this section with the simplest molecule of the series studied, MVP, the structure of which is shown in Fig. 1(a). MVP has been shown to exist as two stable conformers which form a *syn*/*anti* pair.[25] The molecule was photoexcited to the S_1 origin band ($\nu = 0$) of the *syn*-conformer, which lies at 305 nm. The S_1 origin band of the *anti*-conformer is 805 cm^{-1} higher in energy.[25] As the bandwidth of the laser pulses used in these TR-IY experiments is ~500 cm^{-1}, it is not possible to access the origin of the *syn*- and *anti*-conformers simultaneously, and as such the dynamics inferred herein can be considered to be conformer specific.

The parent TR-IY transient resulting from exciting MVP at 305 nm and probing with 200 nm is shown in Fig. 2(a), along with the kinetic fit for these data (red line). This transient was fitted with two exponential decay functions (see ESI† for details pertaining to the kinetic fits) convolved with the Gaussian instrument response function (IRF ~ 160 fs). The time constants returned, as summarised in Table 1, are (i) an ultrafast component of <50 fs, which, by comparison with previous work,[52] we attribute to multiphoton absorption and ionisation, and (ii) a slow decay, with a time constant outside of the temporal window of these experiments (>900 ps). This latter time constant is reminiscent of that seen in previous studies on the MVP subunit, guaiacol.[53] The addition of a vinyl group to guaiacol, therefore, does not result in a dramatic difference between the dynamics of these two molecules. While there are notable reverse (probe-pump) dynamics observable in MVP compared to guaiacol, the probe wavelength of 200 nm puts these dynamics well outside the UV-A and UV-B regions and so beyond the scope of this work. As such here, and for our ensuing discussions of ConA and EF, we do not discuss these in any great detail but note their presence where appropriate.

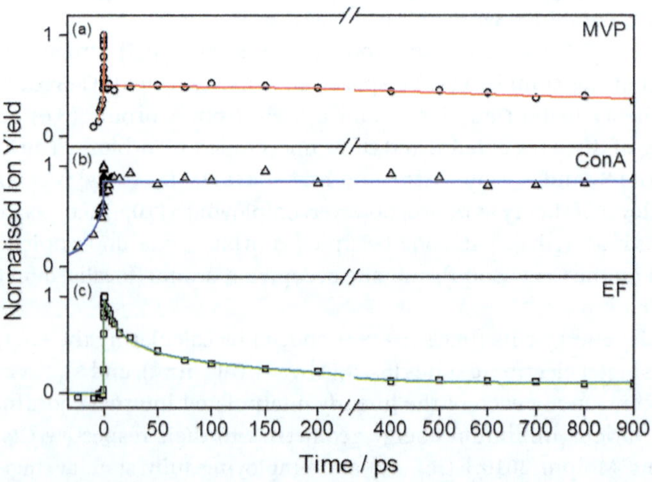

Fig. 2 TR-IY transients of the (a) MVP, (b) ConA, and (c) EF parent ions following excitation at 305 nm, 306 nm, and 317.5 nm, respectively, corresponding to their S_1 origin bands. Also shown are the overall kinetic fits for MVP (red line), ConA (blue line), and EF (green line).

Table 1 Summary of the time constants extracted for each of the molecules studied

Molecule	Time constants
MVP	$\tau_1 < 50$ fs, $\tau_2 > 900$ ps
ConA	$\tau_1 > 900$ ps
EF	$\tau_1 = 15 \pm 4$ ps, $\tau_2 = 148 \pm 47$ ps, $\tau_3 > 900$ ps

b. Excited state dynamics in ConA

We now increase the molecular complexity of our target molecule through the study of ConA. Once again, the molecule is photoexcited to the S_1 origin ($\nu = 0$) of its lowest energy conformer which, for the *syn*-conformer, corresponds to 306 nm.[25] The *syn*-conformer of ConA lies 723 cm^{-1} lower in energy relative to the *anti*-conformer,[25] which, as previously discussed for MVP, allows for conformer specific results to be obtained (*vide supra*).

The TR-IY parent transient of ConA is shown in Fig. 2(b) along with the overall kinetic fit (blue line). The TR-IY transient of ConA can be fitted with a single exponential decay function returning a time constant outside our temporal window (>900 ps). Unlike MVP, no multiphoton absorption/ionisation artifact in ConA is observed, which serves as the only main difference between the excited state dynamics of ConA and MVP. We once again note for completeness that there are reverse dynamics following photoexcitation with 200 nm in ConA.

c. Excited state dynamics in EF

i. R2PI and UVHB spectroscopy. The final molecule studied is EF. The very limited literature on the spectroscopy of EF necessitated the acquisition of high energy resolution spectroscopy data to elucidate both the origin band(s) and the conformational make-up of this molecule. To this end, R2PI and UVHB spectra for EF are presented in Fig. 3. UVHB spectroscopy reveals the existence of two conformers of EF, lying very close in energy (~15 cm^{-1}), with origin bands lying at 317.39 nm and 317.55 nm. Given the small spectral shift between the proposed *syn*- and *anti*-conformers of EF, and their close energetic proximity, we have not attempted to assign which UVHB spectrum in Fig. 3 is due to which conformer, nor did we undertake assignment of the vibronic bands, as we used the spectra in Fig. 3 primarily to provide a guide to the spectral location of the origin bands of the conformers present in the expansion.

ii. Laser induced and dispersed fluorescence. The total fluorescence lifetime of EF when excited at the origin of conformer A was measured to be 6.9 ± 0.1 ns (see ESI†). Attempts to observe any phosphorescence at longer wavelengths (>400 nm) and/or time delays (100 s of ns) produced no measurable signal. Dispersed fluorescence measurements (Fig. 4) following excitation of the electronic origin of the two conformers shows Franck–Condon activity consistent with emission almost exclusively from the S_1 origin in the 315–345 nm region, although there is likely a contribution to this signal from scatter of the excitation laser.

iii. Time-resolved ion yield. EF was photoexcited at 317.5 nm (S_1 origin) and probed at 200 nm in order to ensure photoionisation from any lower lying states accessed upon relaxation of the initially photoexcited S_1 state (see Discussion).

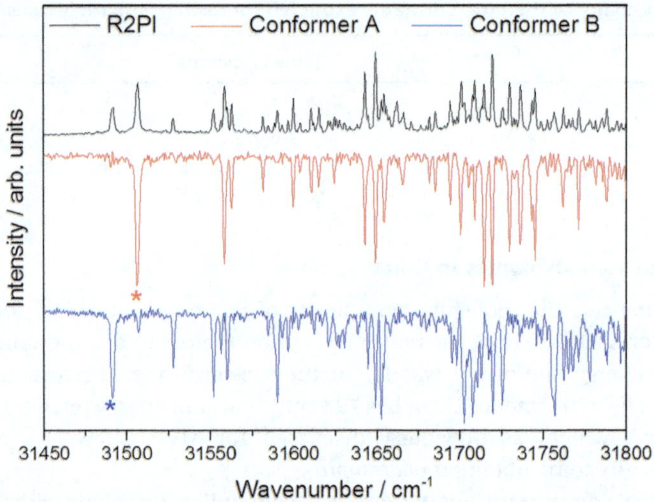

Fig. 3 R2PI spectrum (black) and UV–UV hole burn spectra of conformers A (red) and B (blue) for EF. The origin bands (labelled with *) are 317.39 nm and 317.55 nm for conformers A and B, respectively. The IP for both conformers was determined to be 7.73 eV. We note here that the signal from the R2PI spectrum has likely saturated the detector at the highest peaks. This was not avoided as, for the purposes of the current work, we were concerned only with the number of conformers present and the energies of their respective origin bands.

The ~500 cm^{-1} bandwidth of the pump accesses both conformers simultaneously and thus conformer specific dynamics are not possible (contrary to the case for MVP and ConA). The resulting TR-IY parent transient is shown in Fig. 2(c), along with the overall kinetic fit (green line). The TR-IY parent transient requires

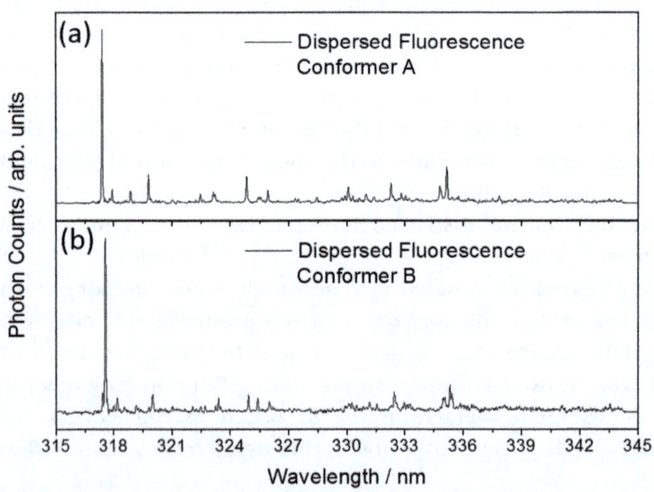

Fig. 4 Dispersed fluorescence from the origin transitions of (a) conformer A and (b) conformer B of EF. We note that scatter is likely to contribute to the large relative intensity of the peak at the origin of both conformers.

a minimum of three exponential decays (see ESI† for attempted fits with one and two decays), as summarised in Table 1. The time constants extracted are 15 ± 4 ps, 148 ± 47 ps and a long lived component which persists beyond the temporal limit of our experiment (>900 ps). These dynamics are clearly different from those observed for MVP and ConA and these differences are explored in the Discussion section below. For completeness, we note that in the case of EF, no reverse dynamics are observed.

TR-IY studies at lower probe energies (322 nm *cf.* 200 nm, Fig. 5) yield, within error, similar time constants, but the data can be reasonably fitted by exclusion of the long lived component, suggesting that if it is present at all, the amplitude of the long lived component (>900 ps) is now greatly reduced. This implies that the excited state population responsible for this component is trapped in a low energy potential well which cannot be ionised by this longer probe wavelength.

d. H atom elimination

There has been considerable interest in the H atom elimination dynamics along heteroatom-hydrogen bonds (X–H, where X is usually O or N) mediated by dissociative $\pi\sigma^*$ states localised along this coordinate. The existence of such states, localised specifically here along O–H, is usually identified by the production of high kinetic energy (KE) H atoms[52,55,56] in accordance with the original postulate of Sobolewski and Domcke.[57] The H^+ TKER spectrum for EF at $\Delta t = 1.2$ ns is shown in Fig. 6 (equivalent spectra for MVP and ConA are shown in the ESI†). The TKER spectrum of EF is evidently noisy, due to poor signal level, although one can discern two features; the first centred at low KE ($\sim$2500 cm^{-1}) and a second centred at higher KE ($\sim$12 500 cm^{-1}). Using a 20 point moving average (red line, Fig. 6), these two features are more clearly distinguishable. In keeping with our previous studies, the low KE feature is very likely due to multiphoton processes and statistical unimolecular decay on S_0.[52,53] The second, high KE feature is also attributed to multiphoton processes, based on energy considerations; this peak appears at much higher KEs than one would anticipate

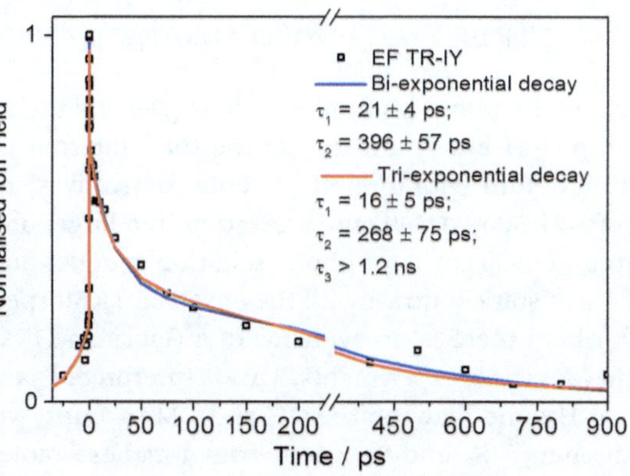

Fig. 5 TR-IY data for EF with 317.5 nm pump and 322 nm probe with two different kinetic fits. Kinetic fit 1 (blue line) yields two time constants: 21 ± 4 ps and 396 ± 57 ps. No long lived step was included in this fit. Kinetic fit 2 yields a 16 ± 5 ps and a 268 ± 75 ps time constant, and also a > 1.2 ns step.

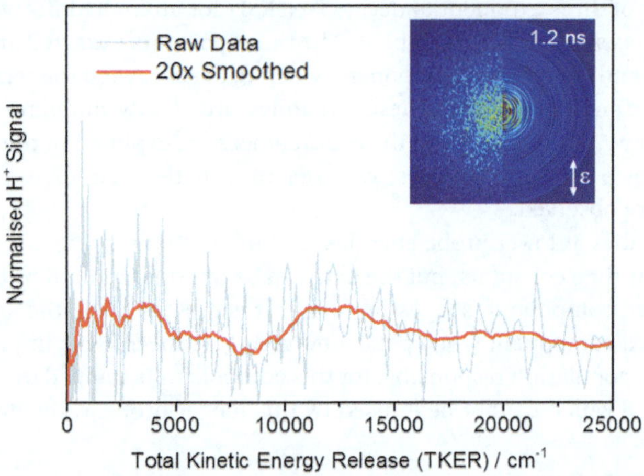

Fig. 6 Normalised TKER spectrum obtained for EF at 317.5 nm with a pump-probe time delay of $\Delta t = 1.2$ ns. The red line corresponds to a 20 point moving average through the raw data (grey). Inset: H^+ velocity map image from which the TKER spectrum of EF was derived: raw image (left half) together with a reconstructed slice through the centre of the original 3D ion distribution (right half). The vertical white arrow indicates the electric field polarisation vector of the pump pulse, ε.

for one photon absorption, $\pi\sigma^*$ state mediated O–H dissociation (see Discussion). MVP and ConA produce essentially featureless TKER spectra. We return to the ramifications of these findings below.

Discussion

We begin our discussion by focusing on the H atom elimination results presented above. We first note that the expected TKER of the photofragments after O–H dissociation in MVP/ConA/EF can be determined (assuming the molecules begin in their ground vibronic state) according to the relationship:

$$\text{TKER} = h\nu_{\text{pu}} - D_0(\text{O–H}) - E_{\text{int}} \qquad (1)$$

In eqn (1), $h\nu_{\text{pu}}$ is the pump photon energy, D_0(O–H) is the adiabatic dissociation energy of the O–H bond, and E_{int} is the total internal energy of the co-fragment (electronic and vibrational). Phenol derivatives typically possess D_0(O–H) values of ~27 500–31 500 cm^{-1} resulting in a large, broad Gaussian H^+ signal centred around 6500 cm^{-1} for photoexcitation energies around their origin bands.[53,58–63] This is in stark contrast with the observed TKER spectra for MVP and ConA (see ESI†), where there is no evidence of a Gaussian H^+ signal within the 1.2 ns time window for these experiments. This is interpreted as a consequence of the intramolecular H bond between the OH and OMe groups, which can be seen in the minimum energy (S_0 and S_1) geometries for these molecules, shown in Fig. 7. This intramolecular H bond is likely to introduce a large barrier to O–H dissociation, hindering the fast (within our time window) formation of RO + H photoproducts (where R represents MVP, ConA, or EF excluding the H on the OH group).[62] It is worth noting, however, that H atom elimination might still occur

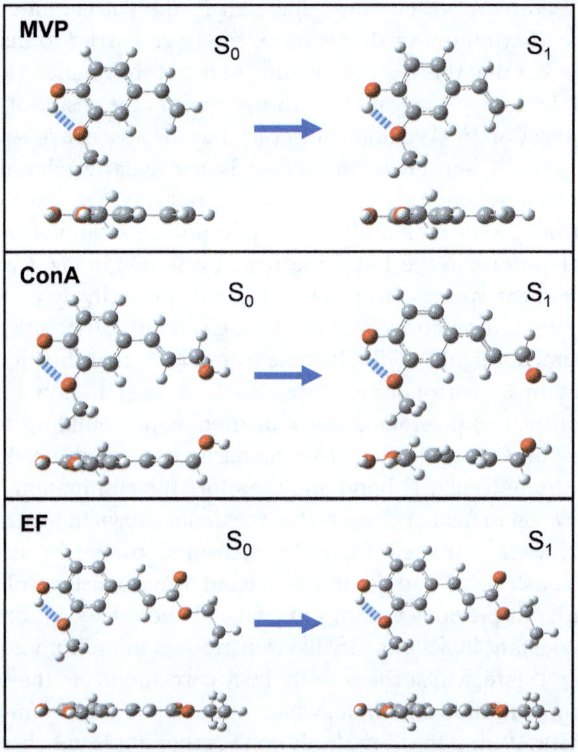

Fig. 7 Optimised S_0 and S_1 geometries of MVP, ConA, and EF. Geometry optimised structures were calculated using Gaussian09 (ref. 44) with density functional theory (DFT) and the Coulomb attenuated model Becke-3rd parameter Lee–Yang–Parr (CAM-B3LYP) functional[45] combined with a 6-31G* basis set.[46] Dashed blue lines indicate intramolecular hydrogen bonds.

beyond the 1.2 ns timescale that can be probed in these experiments, although, given the fluorescence lifetime of ~7 ns, it will not likely be a major competitive relaxation pathway.

In the case of EF, energy considerations dictate that at $h\nu_{pu} = 317.5$ nm, O–H dissociation should produce H atoms (and therefore H^+) at energies of 0–4000 cm^{-1} – definitive assignment is not possible as there is no literature (to the best of our knowledge) on the experimental value of D_0(O–H) for EF. The small Gaussian feature at ~12 500 cm^{-1} is therefore likely too high in energy for single photon, $\pi\sigma^*$ mediated photodissociation. Indeed, with regards to eqn (1), this feature would necessitate a D_0(O–H) value of approximately 19 000 cm^{-1} which is very low (*cf.* typical values ~27 500–31 500 cm^{-1}).[53,58–63] This feature is not present when detuning the probe wavelength from the H atom 2s ← 1s resonance (see ESI†). This is usually indicative of a one-photon dissociation process. However, absorption of two photons of $h\nu_{pu} = 317.5$ nm may promote to higher-lying dissociative states that yield H atoms with much higher KEs, as evidenced in the TKER spectrum of Fig. 6. Alternatively, H atoms could result from dissociative ionisation where the charge remains on the co-fragment.

While the feature at ~2500 cm^{-1} does fall in the correct energy range, the Boltzmann-like distribution of this feature, the large barrier to dissociation expected, and the fact that this feature remains in the off-resonance TKER spectrum of EF (see ESI†) strongly suggests that this feature is the result of multiphoton dissociative ionisation.[52,53] We note, however, that the absence of a significant H^+ signal in the present sunscreen molecules is particularly relevant, since this process would be a potential source of toxic free radicals. The low incidence of H atom release from EF (or MVP and ConA) upon photoexcitation (evidenced from the poor signal-to-noise in the TKER spectrum presented in Fig. 6 and ESI†) may indicate that radical species (from O–H fission) are unlikely to form *via* this mechanism under illumination from the Sun, given the comparatively low power density of natural sunlight. We finally note from Fig. 7, that the high planarity of the geometries in S_1 corroborate the absence of O–H fission in the present systems, in contrast to previous studies in monolignol building blocks, which showed how significant distortion from planarity in S_1 could lead to the weakening of the intramolecular H bond and therefore the elimination of H atoms.[54]

We now move on to further discuss the transients shown in Fig. 2. In MVP, two time constants were extracted from the measured transient; an instrument-limited time constant of <50 fs and a second time constant of >900 ps. In accordance with our previous findings in guaiacol and related chromophores,[52,54] the first time constant in MVP is very likely attributed to multiphoton ionisation, *via* the $S_1(1^1\pi\pi^*)$ state, whilst the second likely corresponds to the persistence of excited state population in this state, which outlasts the maximum time window of the experiment. With ConA, a multiphoton component is not observed (as with EF). In these two molecules, multiphoton-induced fragmentation of the ester/alcohol group side chain (see ESI† for EF TOF-MS) required TR-IY measurements to be taken at lower laser powers. As such, observation of a feature from multiphoton absorption is significantly less likely in ConA and EF than in MVP (which possesses the smallest side chain of the three systems).

In EF, the clear presence of two additional time constants of 15 ± 4 ps and 148 ± 47 ps, in contrast with MVP and ConA, suggests that a different (or competing) relaxation mechanism is in operation. It would perhaps be reasonable to expect at least one of these time constants to correspond to *E*/*Z* isomerisation, a process found to be operative for similar molecules in previous studies.[26,64–66] To explore this possibility further, we have carried out calculations to map out the PECs along the *E*/*Z* isomerisation coordinate of EF akin to those in ref. 26. This was achieved by calculating the energies of the ground and first three excited electronic states of EF at various points along the LIIC from the ground state minimum energy geometries of each isomer towards the S_0/S_3 CI, resulting in the PECs in Fig. 8(a). These calculations suggest that, while the *E*/*Z* isomers' energy minima are similar, there is a substantial energy barrier (~1.1 eV) that must be overcome upon photoexcitation to the S_1 origin in order for *E*/*Z* isomerisation, mediated by the $S_1(1^1\pi\pi^*)/S_3(3^1\pi\pi^*)$ CI, to occur. This finding implies that this relaxation pathway in EF can be ruled out *in vacuo*.

In light of previous work[24,27,67] which has speculated low energy $^1n\pi^*$ states to be involved in the photodynamics of similar systems to those studied herein, calculations were performed to search for such states. The PECs of the S_0, $S_1(1^1\pi\pi^*)$, $S_2(1^1n\pi^*)$, and S_3 $(2^1\pi\pi^*)$ states were produced along the LIIC from the DFT optimised geometries of (i) the S_0 to the $S_1(1^1\pi\pi^*)$ state and (ii) the

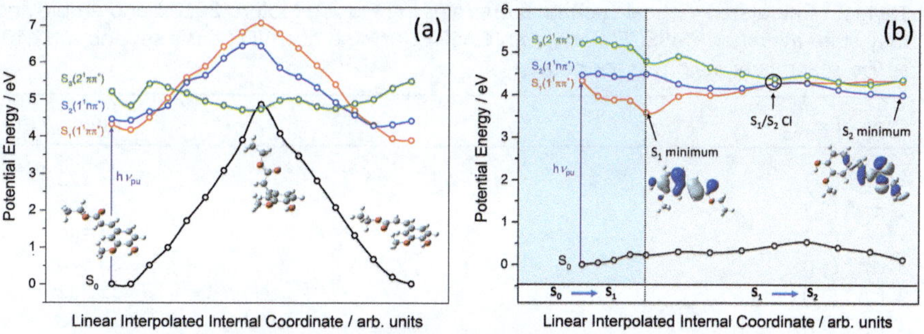

Fig. 8 (a) PECs of the S_0 (black line), $S_1(1^1\pi\pi^*)$ (red line), $S_2(1^1n\pi^*)$ (blue line), and $S_3(2^1\pi\pi^*)$ (green line) states of EF along the *E*/*Z* isomerisation coordinate. Optimised ground state geometries for each isomer and the S_0/S_3 conical intersection (CI) are also presented. (b) PECs of the S_0 (black line), $S_1(1^1\pi\pi^*)$ (red line), $S_2(1^1n\pi^*)$ (blue line), and $S_3(2^1\pi\pi^*)$ (green line) states of EF. The blue and grey shaded areas represent the S_0 to S_1 and S_1 to S_2 (*via* the S_1/S_2 CI) LIICs, respectively. Optimised geometries for the S_1 and S_2 states are also presented.

$S_1(1^1\pi\pi^*)$ to the $S_2(1^1n\pi^*)$ state *via* the $S_1(1^1\pi\pi^*)/S_2(1^1n\pi^*)$ CI, in the *E* isomer, as shown in Fig. 8(b). There appears to be a barrierless pathway through which the minimum of the $S_2(1^1n\pi^*)$ state can be accessed *via* a CI with the $S_1(1^1\pi\pi^*)$ state. Therefore it is possible for an $S_1(1^1\pi\pi^*) \rightarrow S_2(1^1n\pi^*)$ transition to occur, with the $S_2(1^1n\pi^*)$ state then being long lived, as observed in previous studies.[24] The path to the $S_2(1^1n\pi^*)$ minimum passes through a deep minimum in $S_1(1^1\pi\pi^*)$, where population may become trapped, which could then be responsible for the fluorescence observed. However, neither of these scenarios accounts for all the dynamics observed in EF since they would only yield two time constants; one to account for the $S_1(1^1\pi\pi^*) \rightarrow S_2(1^1n\pi^*)$ transition and/or motion to the $S_1(1^1\pi\pi^*)$ minimum, and the second to account for the lifetime of the final $S_1(1^1\pi\pi^*)/S_2(1^1n\pi^*)$ state. We therefore need to consider alternative mechanisms.

The photophysics of aromatic carbonyl compounds have been extensively reported in the literature to involve low-lying $n\pi^*$ states of both singlet and triplet character,[24,27,68–75] with intersystem crossing (ISC) being a kinetically competitive photophysical process.[76,77] For example, Spighi *et al.*[77] suggested a kinetic model for benzophenone, in which a $^1\pi\pi^*$ state decays into a $^3n\pi^*$ state within 5 ps, followed by repopulation to the ground state within 400 ns. While there are structural differences between benzophenone and EF, the presence of the carbonyl moiety, common to both, suggests that these results may be relevant to the interpretation of the present data. In order, therefore, to evaluate the role of triplet states in the photodynamics of EF, further *ab initio* calculations were performed using CASSCF along with CASPT2 with a cc-pVDZ basis set and a (12,10) active space (*vide supra*). Four triplet states were identified below the $S_1(1^1\pi\pi^*)$ state energy in the vertical Franck–Condon region and their energies are presented in Table 2. The $T_2(2^3\pi\pi^*)$, $T_3(3^3\pi\pi^*)$, and $T_4(1^3n\pi^*)$ states are near isoenergetic and lie on average ~2000 cm^{-1} lower in energy in relation to the vertical $S_1(1^1\pi\pi^*)$ state excitation energy; the $T_1(1^3\pi\pi^*)$ state lies ~10 500 cm^{-1} below these. These calculations match well with triplet state energies calculated using CAM-B3LYP and a cc-pVDZ basis set (see ESI†). Apart from being close in

Table 2 Calculated vertical excitation energies in EF using Molpro 2010.1 and employing fully state averaged CASSCF along with CASPT2 using a cc-pVDZ basis set and a (12,10) active space (see main text for details)

Type of state	Energy/cm^{-1}
$1^1n\pi^*$ (S_2)	36 122
$1^1\pi\pi^*$ (S_1)	35 148
$1^3n\pi^*$ (T_4)	33 782
$3^3\pi\pi^*$ (T_3)	33 620
$2^3\pi\pi^*$ (T_2)	32 752
$1^3\pi\pi^*$ (T_1)	23 258

energy to both the $S_1(1^1\pi\pi^*)$ and $S_2(1^1n\pi^*)$ states, these triplet states are also of the correct orbital type to allow for ISC to occur from either accessible singlet state in accord with El-Sayed's rule,[78] which states that the rate of ISC will be higher for a transition where there is a change in orbital type (for example, n to π or *vice versa*). Once within the triplet manifold, the population could then undergo internal conversion (IC) to the lowest triplet state, and from there return to the ground state (or undergo photoreaction).

In light of the information presented above, we suggest the following mechanisms are involved in the photophysics of EF, presented graphically in Fig. 9.

Firstly, following photoexcitation of EF to the $S_1(1^1\pi\pi^*)$ state, wavepacket bifurcation[79,80] results in population of both the $S_1(1^1\pi\pi^*)$ and $S_2(1^1n\pi^*)$ states, mediated *via* an energetically accessible CI between these states. This is evidenced by the relatively flat potential energy landscape in regions removed from the vertical Franck–Condon region (Fig. 8(b)). Intramolecular vibrational redistribution (IVR) then likely occurs on both the $S_1(1^1\pi\pi^*)$ and $S_2(1^1n\pi^*)$ states, with population moving into the deep energy well of the $S_1(1^1\pi\pi^*)$ state and the shallower well of $S_2(1^1n\pi^*)$. Energy is redistributed into coordinates orthogonal to the reaction coordinate in Fig. 8(b), so that population gets trapped in the wells of the $S_1(1^1\pi\pi^*)$ and $S_2(1^1n\pi^*)$ states along this reaction coordinate, as represented in Fig. 9. In competition with IVR, ISC onto several predicted high lying triplet states may also occur from either the $S_1(1^1\pi\pi^*)$ or $S_2(1^1n\pi^*)$, in accordance with

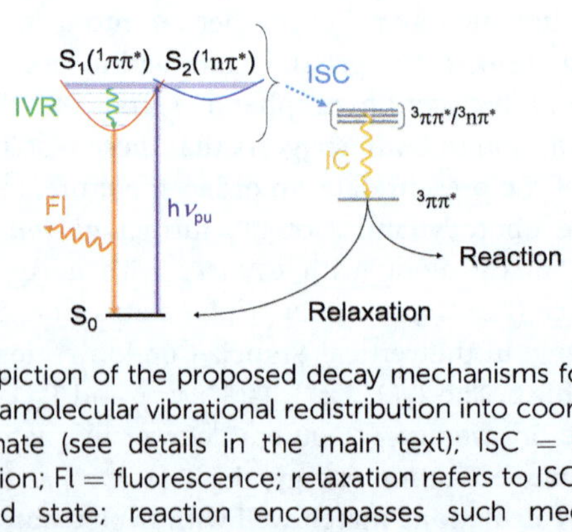

Fig. 9 Schematic depiction of the proposed decay mechanisms for excited state relaxation in EF. IVR = intramolecular vibrational redistribution into coordinates orthogonal to the reaction coordinate (see details in the main text); ISC = intersystem crossing; IC = internal conversion; Fl = fluorescence; relaxation refers to ISC or phosphorescence back to the ground state; reaction encompasses such mechanisms as photofragmentation/photocyclisation. Note that 'Reaction' has been omitted from S_1 for clarity, although it is still a valid pathway.

El-Sayed's rule.[78] Importantly, the potential well of $S_1(1^1\pi\pi^*)$ lies lower in energy than the vertical excitation energies of the higher lying triplet states. As such, it is conceivable that, as the wavepacket progresses towards the minimum of $S_1(1^1\pi\pi^*)$, it encounters regions of strong spin–orbit (SO) coupling between the $S_1(1^1\pi\pi^*)$ state and isoenergetic triplet state(s), facilitating efficient ISC.[81–84] This SO coupling would allow ISC to occur while still retaining population trapped in the $S_1(1^1\pi\pi^*)$ well. We suggest that all these processes will be acting in competition with each other and are overall described by τ_1 (15 ± 4 ps), which is consistent with timescales seen in similar systems.[77,85–87]

Following ISC onto the triplet state manifold, we suggest that τ_2 (148 ± 47 ps) corresponds to IVR and IC occurring down the ladder of triplet states, to the lowest triplet state $T_1(1^3\pi\pi^*)$.[87] The longer lifetime for IC compared to ISC could then be explained by the relatively large energy gap between the $T_1(1^3\pi\pi^*)$ and $T_2(2^3\pi\pi^*)$ states. The remaining long lived step would then be attributed to population trapped in $S_1(1^1\pi\pi^*)$ – which decays through fluorescence and/or photoreaction (on the order of ns; as is solely the case in MVP and ConA) – in conjunction with population of the lowest lying triplet state (which again decays on the order of ns). While one might expect the triplet state to phosphoresce, the emission wavelengths observed in EF, coupled with the nanosecond lifetime of this emission, suggest that there is little or no phosphorescence from, or fluorescence between, triplet states in EF. This may be due to photoreaction occurring on the excited triplet states or, due to the low energy of the $T_1(1^3\pi\pi^*)$ state, ISC back onto S_0 may be a kinetically competitive pathway.

We note that the TR-IY of the EF parent ion taken at long probe wavelengths ($h\nu_{pr}$ = 322 nm) suggest that the long lifetime (>900 ps) represents a state lower in energy than those represented by τ_1 and τ_2 (*vide supra*). This is in agreement with the proposed mechanisms, in which either IVR on the $S_1(1^1\pi\pi^*)$, or ISC followed by IC to $T_1(1^3\pi\pi^*)$, would move excited state population outside of the ionisation probe window of the 322 nm laser. An elongation of τ_2 was derived when using two exponential decay functions to fit the 322 nm probe data. Given that both a two- or three-exponential decay function fits the data equally well (*vide supra*), it is possible that this elongation is merely an artefact of the absence of the third time constant. Since the pump wavelength is unchanged, the elongation of τ_2 is likely an effect of the final (ionic) state accessed in the probe step, rather than a change in the dynamics of EF. However, since this change in τ_2 has no qualitative effect on the interpretation of our results, such exploration is considered beyond the scope of this work.

Two caveats to the mechanisms proposed are in order here. Firstly, while there is precedent for ISC occurring in systems such as this on picosecond timescales, it has been shown to occur in other related systems in hundreds of picoseconds.[88] As such, one cannot rule out τ_2 corresponding to ISC from the singlet to the triplet states, with IVR/IC then occurring between the triplet states on much faster timescales.[77] τ_1 would then correspond to IC between, and IVR on, the $S_1(1^1\pi\pi^*)$ and $S_2(1^1n\pi^*)$ states. Population trapped on these two states would then undergo ISC with time constant τ_2. Secondly, as we cannot selectively probe the individual conformers of EF identified in the UVHB experiments described above, it is possible that τ_1 and τ_2 may correspond to the dynamics of the two separate conformers. This is considered unlikely however, as the structural difference, as well as the energy separation, of the two conformers is minimal. As such, one would not expect any significant difference in the excited state landscapes between conformer A and B, and hence the dynamics, in contrast to the order of magnitude difference observed between τ_1 and τ_2.

From the discussion above, it is clear that the photochemistry and photophysics of EF is highly complex, very likely involving a number of competing deactivation pathways. While we have proposed scenarios that may account for the present experimental results, we do not discount the possibility that there may be others. For example, photofragmentation is a viable pathway in EF. In fact, several fragments are observed in the TOF trace (see ESI†) but attempts to obtain TR-IY transients for these were not successful due to very weak signal intensities. However, it would certainly be worth evaluating the extent of photofragmentation processes in EF, especially given the impact such fragmentation would have on the photostability and photoreactivity of this sunscreen molecule. Further work, such as time-resolved photoelectron spectroscopy, would be highly beneficial, in enabling us to directly track excited state population as it evolves along the various excited states. Likewise, more extensive calculations involving PECs for both singlet and triplet states will inevitably assist in ascertaining regions of strong SO coupling, putative in facilitating efficient ISC.

Conclusions

The excited state dynamics of EF, a commercial sunscreen filter molecule, have been studied using TR-IY spectroscopy and H^+ TR-VMI. MVP and ConA were also studied so that the effects of increasing molecular complexity on the excited state dynamics could be evaluated.

Upon photoexcitation to their respective $S_1(1^1\pi\pi^*)$ origin bands, MVP and ConA show similar excited state dynamics, in contrast to EF, which shows markedly different dynamics. The excited state dynamics of MVP and ConA, both resemble the case of guaiacol;[53,54] population of a long lived $S_1(1^1\pi\pi^*)$ state which persists beyond the temporal window of our measurements (>900 ps), returning to the ground state *via* fluorescence. The resemblance of MVP and ConA to guaiacol extends to H atom elimination, where there is no clear high KE feature in the TKER spectra, suggesting O–H bond fission mediated by $\pi\sigma^*$ states is not a viable relaxation pathway in these two systems. This is unsurprising, given the presence of the intramolecular H bond, which likely provides a substantial barrier to dissociation. We believe this decay pathway is also absent in EF for the same reasons; the presence of two weak KE energy features centred at $\sim$2500 and $\sim$12 500 cm^{-1} are assigned to multiphoton processes and/or statistical unimolecular decay. By way of contrast, the excited state dynamics of EF are notably different to MVP and ConA, as they now contain three components: one component which decays with a time constant of 15 $\pm$ 4 ps, a second which decays with a time constant of 148 $\pm$ 47 ps, and finally a much longer lived component which, as with MVP and ConA, decays with a time constant >900 ps. In addition, LIF studies show that EF fluoresces (similarly to MVP and ConA) with a lifetime of 6.9 $\pm$ 0.1 ns.

Upon photoexcitation of EF, IC to the $S_2(1^1n\pi^*)$ state, IVR on both the $S_1(1^1\pi\pi^*)$ and $S_2(1^1n\pi^*)$ states, and ISC from these states onto nearby triplet states are proposed to account for the $\sim$15 ps time constant. IC then occurs between the triplet states down to the lowest triplet state, $T_1(1^3\pi\pi^*)$, in $\sim$150 ps.

We note a caveat that this second lifetime may also be due to ISC from the accessible singlet states. The third time constant then accounts for the excited state lifetime of both the $S_1(1^1\pi\pi^*)$ and $T_1(1^3\pi\pi^*)$ states, the former of which fluoresces whilst the latter might undergo photoreactions or ISC to S_0.

One may surmise from the present measurements that, if the long lived state of EF is particularly reactive, be it through emission of radiation or photofragmentation/photocyclisation, it may interact with other sunscreen components or even DNA possibly leading to detrimental consequences. These measurements also suggest that further pump-probe studies on EF prepared in a more realistic environment, such as microsolvation and/or solution phase studies, are imperative in assessing how external perturbations influence the electronic structure which, in turn, may impact the dynamics. Importantly however, we believe that the present studies serve to highlight the significance of garnering a molecular level understanding of the underlying mechanisms of photoprotection in a commercial sunscreen constituent in the absence of any external perturbations. We hope that such a study will lay further foundations for future research, with potentially far reaching repercussions in developing safer and more efficient UV radiation blocking products.

Acknowledgements

N. D. N. R. and M. S. thank the EPSRC for doctoral and postdoctoral funding, respectively; M. S. would also like to thank Prof. Martin Paterson for the use of computational facilities. J. D. Y. thanks the University of Warwick for a doctoral training award; N. C. C. F. thanks the Leverhulme Trust for postdoctoral funding; Y. P. thanks the ENS-Cachan for funding. J. R. G., P. S. W., D. M. H., and T. S. Z. thank the National Science Foundation (NSF CHE 1465028) for funding. V. G. S. thanks the EPSRC for an equipment grant (EP/J007153) and the Royal Society for a University Research Fellowship.

References

1 N. D. Paul and D. Gwynn-Jones, *Trends Ecol. Evol.*, 2003, **18**, 48–55.
2 A. R. Webb and M. D. Steven, *Q. J. R. Meteorol. Soc.*, 2007, **113**, 393–400.
3 K. Coulson, *Solar and Terrestrial Radiation: Methods and Measurements*, Academic Press Inc., New York, 1975.
4 B. A. Gilchrest, M. S. Eller, A. C. Geller and M. Yaar, *N. Engl. J. Med.*, 1999, **340**, 1341–1348.
5 L. Marrot, *J. Am. Acad. Dermatol.*, 2008, **58**, S139–S148.
6 F. R. de Gruijl, *Eur. J. Cancer*, 1999, **35**, 2003–2009.
7 A. Ziegler, A. S. Jonason, D. J. Leffell, J. A. Simon, H. W. Sharma, J. Kimmelman, L. Remington, T. Jacks and D. E. Brash, *Nature*, 1994, **372**, 773–776.
8 J. A. Lo and D. E. Fisher, *Science*, 2014, **346**, 945–949.
9 T. L. Diepgen and V. Mahler, *Br. J. Dermatol.*, 2002, **146**, 1–6.
10 H. C. Wulf, J. Sandby-Moller, T. Kobayasi and R. Gniadecki, *Micron*, 2004, **35**, 185–191.
11 C. T. Middleton, K. de La Harpe, C. Su, Y. K. Law, C. E. Crespo-Hernandez and B. Kohler, *Annu. Rev. Phys. Chem.*, 2009, **60**, 217–239.

12 H. Kang, K. T. Lee, B. Jung, Y. J. Ko and S. K. Kim, *J. Am. Chem. Soc.*, 2002, **124**, 12958–12959.
13 J.-L. Ravanat, T. Douki and J. Cadet, *J. Photochem. Photobiol., B*, 2001, **63**, 88–102.
14 N. Agar and A. R. Young, *Mutat. Res.*, 2005, **571**, 121–132.
15 J. P. Ortonne, *Br. J. Dermatol.*, 2002, **146**, 7–10.
16 R. L. McKenzie, P. J. Aucamp, A. F. Bais, L. O. Bjorn, M. Ilyas and S. Madronich, *Photochem. Photobiol. Sci.*, 2011, **10**, 182–198.
17 M. M. Caldwell, L. O. Björn, J. F. Bornman, S. D. Flint, G. Kulandaivelu, A. H. Teramura and M. Tevini, *J. Photochem. Photobiol., B*, 1998, **46**, 40–52.
18 D. M. Parkin, F. Bray, J. Ferlay and P. Pisani, *Ca-Cancer J. Clin.*, 2005, **55**, 74–108.
19 A. Baldi, E. P. Spugnini and P. Pasquali, *Skin cancer: a practical approach*, 2013.
20 R. Dummer, M. R. Pittlekow, K. Iwatsuki, A. Green and N. M. Elwan, *Skin Cancer – A World-Wide Perspective*, Springer, Berlin Heidelberg, 2011.
21 R. L. Siegel, K. D. Miller and A. Jemal, *Ca-Cancer J. Clin.*, 2015, **65**, 5–29.
22 D. S. Rigel, *Photoaging*, Marcel Dekker, Inc., New York, 2005.
23 D. L. Maslin, *Int. J. Dermatol.*, 2014, **53**, 1319–1323.
24 E. M. M. Tan, M. Hilbers and W. J. Buma, *J. Phys. Chem. Lett.*, 2014, **5**, 2464–2468.
25 C. P. Rodrigo, W. H. James III and T. S. Zwier, *J. Am. Chem. Soc.*, 2011, **133**, 2632–2641.
26 T. N. V. Karsili, B. Marchetti, M. N. R. Ashfold and W. Domcke, *J. Phys. Chem. A*, 2014, **118**, 11999–12010.
27 J. C. Dean, R. Kusaka, P. S. Walsh, F. Allais and T. S. Zwier, *J. Am. Chem. Soc.*, 2014, **136**, 14780–14795.
28 N. Serpone, A. Salinaro, A. V. Emeline, S. Horikoshi, H. Hidaka and J. Zhao, *Photochem. Photobiol. Sci.*, 2002, **1**, 970–981.
29 N. J. Lowe, *Sunscreens: Development: Evaluation, and Regulatory Aspects*, CRC Press, 1996.
30 D. L. Compton and J. A. Laszlo, Sunscreens from vegetable oil and plant phenols, *US pat.*, US6346236, 2002.
31 B. Choquenet, C. Couteau, E. Paparis and L. J. Coiffard, *Nat. Prod. Res.*, 2008, **22**, 1467–1471.
32 H. Kikuzaki, M. Hisamoto, K. Hirose, K. Akiyama and H. Taniguchi, *J. Agric. Food Chem.*, 2002, **50**, 2161–2168.
33 S. Fuehring, *Standardisation Mandate Assigned to CEN Concerning Methods for Testing Efficacy of Sunscreen Products*, European Commission, 2006.
34 A. Iqbal, M. S. Cheung, M. G. D. Nix and V. G. Stavros, *J. Phys. Chem. A*, 2009, **113**, 8157–8163.
35 K. L. Wells, G. Perriam and V. G. Stavros, *J. Chem. Phys.*, 2009, **130**, 074308.
36 U. Even, J. Jortner, D. Noy, N. Lavie and C. Cossart-Magos, *J. Chem. Phys.*, 2000, **112**, 8068–8071.
37 A. T. J. B. Eppink and D. H. Parker, *Rev. Sci. Instrum.*, 1997, **68**, 3477–3484.
38 G. M. Roberts, J. L. Nixon, J. Lecointre, E. Wrede and J. R. Verlet, *Rev. Sci. Instrum.*, 2009, **80**, 053104.
39 P. M. Regan, S. R. Langford, A. J. Orr-Ewing and M. N. R. Ashfold, *J. Chem. Phys.*, 1999, **110**, 281–288.
40 T. S. Zwier, *J. Phys. Chem. A*, 2001, **105**, 8827–8839.

41 P. S. Walsh, E. G. Buchanan, J. R. Gord and T. S. Zwier, *J. Chem. Phys.*, 2015, **142**, 154304.

42 P. S. Walsh, E. G. Buchanan, J. R. Gord and T. S. Zwier, *J. Chem. Phys.*, 2015, **142**, 154303.

43 J. C. Dean, E. G. Buchanan, W. H. James III, A. Gutberlet, B. Biswas, P. V. Ramachandran and T. S. Zwier, *J. Phys. Chem. A*, 2011, **115**, 8464–8478.

44 M. J. Frisch, G. W. Trucks, H. B. Schlegel, G. E. Scuseria, M. A. Robb, J. R. Cheeseman, G. Scalmani, V. Barone, B. Mennucci, G. A. Petersson, H. Nakatsuji, M. Caricato, X. Li, H. P. Hratchian, A. F. Izmaylov, J. Bloino, G. Zheng, J. L. Sonnenberg, M. Hada, M. Ehara, K. Toyota, R. Fukuda, J. Hasegawa, M. Ishida, T. Nakajima, Y. Honda, O. Kitao, H. Nakai, T. Vreven, J. A. Montgomery Jr, J. E. Peralta, F. Ogliaro, M. J. Bearpark, J. Heyd, E. N. Brothers, K. N. Kudin, V. N. Staroverov, R. Kobayashi, J. Normand, K. Raghavachari, A. P. Rendell, J. C. Burant, S. S. Iyengar, J. Tomasi, M. Cossi, N. Rega, N. J. Millam, M. Klene, J. E. Knox, J. B. Cross, V. Bakken, C. Adamo, J. Jaramillo, R. Gomperts, R. E. Stratmann, O. Yazyev, A. J. Austin, R. Cammi, C. Pomelli, J. W. Ochterski, R. L. Martin, K. Morokuma, V. G. Zakrzewski, G. A. Voth, P. Salvador, J. J. Dannenberg, S. Dapprich, A. D. Daniels, Ö. Farkas, J. B. Foresman, J. V. Ortiz, J. Cioslowski and D. J. Fox, *Gaussian 09, Revision E.01*, Gaussian, Inc., Wallingford CT, 2009.

45 T. Yanai, D. P. Tew and N. C. Handy, *Chem. Phys. Lett.*, 2004, **393**, 51–57.

46 G. A. Petersson, A. Bennett, T. G. Tensfeldt, M. A. Al-Laham, W. A. Shirley and J. Mantzaris, *J. Chem. Phys.*, 1988, **89**, 2193–2218.

47 M. Klene, M. A. Robb, M. J. Frisch and P. Celani, *J. Chem. Phys.*, 2000, **113**, 5653–5665.

48 P. E. M. Siegbahn, *Chem. Phys. Lett.*, 1984, **109**, 417–423.

49 H.-J. Werner, P. J. Knowles, G. Knizia, F. R. Manby, M. Schutz, P. Celani, W. Gyorffy, D. Kats, T. T. Korona, R. Lindh, A. Mitrushenkov, G. Rauhut, K. R. Shamasundar, T. B. Adler, R. D. Amos, A. Bernhardsson, A. Berning, D. L. Cooper, M. J. O. Deegan, A. J. Dobbyn, F. Eckert, E. Goll, C. Hampel, A. Hesselmann, G. Hetzer, T. Hrenar, G. Jansen, C. Koppl, Y. Liu, A. W. Lloyd, R. A. Mata, A. J. May, S. J. McNicholas, W. Meyer, M. E. Mura, A. Nicklass, D. P. O'Neill, P. Palmieri, D. Peng, K. Pfluger, R. Pitzer, M. Reiher, T. Shiozaki, H. Stoll, A. J. Stone, R. Tarroni, T. Thorsteinsson and M. Wang, *MOLPRO, Version 2012.1, a Package of Ab Initio Programs*, see http://www.molpro.net.

50 K. Andersson, P.-Å. Malmqvist and B. O. Roos, *J. Chem. Phys.*, 1992, **96**, 1218–1226.

51 T. H. Dunning, *J. Chem. Phys.*, 1989, **90**, 1007–1023.

52 G. M. Roberts and V. G. Stavros, *Chem. Sci.*, 2014, **5**, 1698–1722.

53 A. S. Chatterley, J. D. Young, D. Townsend, J. M. Zurek, M. J. Paterson, G. M. Roberts and V. G. Stavros, *Phys. Chem. Chem. Phys.*, 2013, **15**, 6879–6892.

54 J. D. Young, M. Staniforth, J. C. Dean, G. M. Roberts, F. Mazzoni, T. N. V. Karsili, M. N. R. Ashfold, T. S. Zwier and V. G. Stavros, *J. Phys. Chem. Lett.*, 2014, **5**, 2138–2143.

55 R. N. Dixon, T. A. A. Oliver and M. N. R. Ashfold, *J. Chem. Phys.*, 2011, **134**, 194303.

56 M. N. R. Ashfold, A. L. Devine, R. N. Dixon, G. A. King, M. G. D. Nix and T. A. A. Oliver, *Proc. Natl. Acad. Sci. U. S. A.*, 2008, **105**, 12701–12706.
57 A. L. Sobolewski, W. Domcke, C. Dedonder-Lardeux and C. Jouvet, *Phys. Chem. Chem. Phys.*, 2002, **4**, 1093–1100.
58 M. G. D. Nix, A. L. Devine, B. Cronin, R. N. Dixon and M. N. R. Ashfold, *J. Chem. Phys.*, 2006, **125**, 133318.
59 J. D. Young, M. Staniforth, A. S. Chatterley, M. J. Paterson, G. M. Roberts and V. G. Stavros, *Phys. Chem. Chem. Phys.*, 2014, **16**, 550–562.
60 M. Staniforth, A. S. Chatterley, G. M. Roberts and V. G. Stavros, *Biomed. Spectrosc. Imaging*, 2014, **3**, 271–279.
61 M. Lucarini, P. Pedrielli and G. F. Pedulli, *J. Org. Chem.*, 1996, **61**, 9259–9263.
62 S. E. Greenough, M. D. Horbury, J. O. F. Thompson, G. M. Roberts, T. N. V. Karsili, B. Marchetti, D. Townsend and V. G. Stavros, *Phys. Chem. Chem. Phys.*, 2014, **16**, 16187–16195.
63 G. A. King, T. A. A. Oliver, R. N. Dixon and M. N. R. Ashfold, *Phys. Chem. Chem. Phys.*, 2012, **14**, 3338–3345.
64 T. Fujino, S. Y. Arzhantsev and T. Tahara, *J. Phys. Chem. A*, 2001, **105**, 8123–8129.
65 B. I. Greene and R. C. Farrow, *J. Chem. Phys.*, 1983, **78**, 3336–3338.
66 W. L. Ryan, D. J. Gordon and D. H. Levy, *J. Am. Chem. Soc.*, 2002, **124**, 6194–6201.
67 X.-P. Chang, C.-X. Li, B.-B. Xie and G. Cui, *J. Phys. Chem. A*, 2015, **119**, 11488–11497.
68 S. Dym, *J. Chem. Phys.*, 1968, **48**, 646–652.
69 M. B. Ledger and G. Porter, *J. Chem. Soc., Faraday Trans. 1*, 1972, **68**, 539–553.
70 H. Satzger, B. Schmidt, C. Root, W. Zinth, B. Fierz, F. Krieger, T. Kiefhaber and P. Gilch, *J. Phys. Chem. A*, 2004, **108**, 10072–10079.
71 K. S. Peters and J. Lee, *J. Phys. Chem.*, 1993, **97**, 3761–3764.
72 V. R. Nicholas, J. Turro and J. C. Scaiano, *Modern Molecular Photochemistry of Organic Molecules*, University Science Books, USA, 2010.
73 S. Yabumoto, S. Shigeto, Y.-P. Lee and H. O. Hamaguchi, *Angew. Chem.*, 2010, **49**, 9201–9205.
74 F. Tomas, A. Olba, P. Medina and I. Zabala, *J. Mol. Struct.*, 1986, **142**, 143–146.
75 J. J. Cavaleri, K. Prater and R. M. Bowman, *Chem. Phys. Lett.*, 1996, **259**, 495–502.
76 J. D. Coyle and H. A. J. Carless, *Chem. Soc. Rev.*, 1972, **1**, 465–480.
77 G. Spighi, M.-A. Gaveau, J.-M. Mestdagh, L. Poisson and B. Soep, *Phys. Chem. Chem. Phys.*, 2014, **16**, 9610–9618.
78 M. A. El-Sayed, *Acc. Chem. Res.*, 1968, **1**, 8–16.
79 K. Takatsuka, Y. Arasaki, K. Wang and V. McKoy, *Faraday Discuss.*, 2000, **115**, 1–15.
80 Y. Arasaki, K. Takatsuka, K. Wang and V. McKoy, *Phys. Rev. Lett.*, 2003, **90**, 248303.
81 Q. Ou and J. E. Subotnik, *J. Phys. Chem. C*, 2013, **117**, 19839–19849.
82 M. A. El-Sayed, *J. Chem. Phys.*, 1963, **38**, 2834–2838.
83 C. M. Marian, *WIREs Comput Mol Sci.*, 2012, **2**, 187–203.
84 D. Beljonne, Z. Shuai, G. Portois and J. L. Bredas, *J. Phys. Chem. A*, 2001, **105**, 3899–3907.

85 O. Schalk, M. S. Schuurman, G. Wu, P. Lang, M. Mucke, R. Feifel and A. Stolow, *J. Phys. Chem. A*, 2014, **118**, 2279–2287.
86 R. Moore, F. E. Doany, E. J. Heilweil and R. M. Hochstrasser, *Faraday Discuss. Chem. Soc.*, 1983, **75**, 331–340.
87 S. K. Lower and M. A. El-Sayed, *Chem. Rev.*, 1966, **66**, 199–241.
88 S. Lobsiger, M. Etinski, S. Blaser, H.-M. Fray, C. Marian and D. Leutwyler, *J. Chem. Phys.*, 2015, **143**, 234301.

Faraday Discussions

PAPER

Probing spin–vibronic dynamics using femtosecond X-ray spectroscopy

T. J. Penfold,*[a] M. Pápai,[bc] T. Rozgonyi,[d] K. B. Møller[b] and G. Vankó[c]

Received 10th April 2016, Accepted 17th May 2016
DOI: 10.1039/c6fd00070c

Ultrafast pump-probe spectroscopy within the X-ray regime is now possible owing to the development of X-ray Free Electrons Lasers (X-FELs) and is opening new opportunities for the direct probing of femtosecond evolution of the nuclei, the electronic and spin degrees of freedom. In this contribution we use wavepacket dynamics of the photoexcited decay of a new Fe(II) complex, $[Fe(bmip)_2]^{2+}$ (bmip = 2,6-bis(3-methyl-imidazole-1-ylidine) pyridine), to simulate the experimental observables associated with femtosecond Fe K-edge X-ray Absorption Near-Edge Structure (XANES) and X-ray emission (XES) spectroscopy. We show how the evolution of the nuclear wavepacket is translated into the spectroscopic signal and the sensitivity of these approaches for following excited state dynamics.

1 Introduction

To observe, understand and exploit the simultaneous evolution of nuclear, electronic and spin degrees of freedom on the atomic scale of time, *i.e.* femtoseconds, is a great challenge with far reaching implications. This is especially pertinent for our understanding of processes occurring within electronically excited states. Metal organic complexes are excellent model systems to study the correlation between these fundamental degrees of freedom,[1] but are also of significant interest in their own right due to often remarkable photophysical behaviour and versatility.[2,3] Indeed after photoexcitation, these molecular complexes may undergo any of a plethora of important phenomena including radiative (fluorescence and phosphorescence) decay, non-radiative intramolecular relaxation processes, such as internal conversion (IC), intersystem crossings (ISC, *i.e.* a spin change) and intramolecular vibrational redistribution (IVR). Most of these processes are accompanied or driven by structural changes occurring as the nuclei adapt to the new electronic structure. Consequently, the relative fraction of

[a]*School of Chemistry, Newcastle University, Newcastle upon Tyne, NE1 7RU, UK. E-mail: tom.penfold@ncl.ac.uk*
[b]*Department of Chemistry, Technical University of Denmark, DK-2800, Kongens Lyngby, Denmark*
[c]*Wigner Research Centre for Physics, Hungarian Academy of Sciences, P.O. Box 49, H-1525 Budapest, Hungary*
[d]*Institute of Materials and Environmental Chemistry, Research Centre for Natural Sciences, Hungarian Academy of Sciences, P.O. Box 286, H-1519 Budapest, Hungary*

these pathways can often be manipulated by modifying the chemical composition[4] or by an external perturbation.[5]

Besides the study of fundamental photophysical processes, transition metal complexes have also gathered significant attention owing to their potential applications, such as photosensitisers in photovoltaics or photocatalysts.[6,7] At the heart of such applications are the photoactive metal-to-ligand charge transfer (MLCT) states. The importance of these states has meant that Fe(II) complexes are generally considered inappropriate as their MLCT states are typically very short lived[8,9] owing to the presence of low-lying metal centered (MC) states.[10] However, Liu *et al.*[11] recently demonstrated that by exploiting the strong field effects of NHC ligands to create $[Fe(bmip)_2]^{2+}$ (bmip = 2,6-bis(3-methyl-imidazole-1-ylidine)pyridine), as shown in Fig. 1a, it was possible to destabilise the MC states and significantly increase MLCT lifetime. Using ultrafast transient absorption they reported a ^{1}MLCT → ^{3}MLCT conversion of ∼100 fs, no population of a high spin $^5T_2$ state and a ^{3}MLCT lifetime of ∼9 ps.[11]

Our understanding of these ultrafast processes within metal organic complexes has increased enormously with the emergence of femtosecond pump-probe spectroscopies, and the rapid development within this area has made it possible to tackle increasingly complicated dynamical processes. This complexity provides a strong driving force for combining complementary spectroscopic techniques to resolve a full understanding of excited state dynamics. However, until recently this spectroscopic toolkit has not included techniques within the short wavelength regime.[12] In particular ultrafast X-ray experiments, made possible by the advent of X-ray free electron lasers[13,14] which are beginning to provide complementary information about a wide range of interesting systems.[8,15–18]

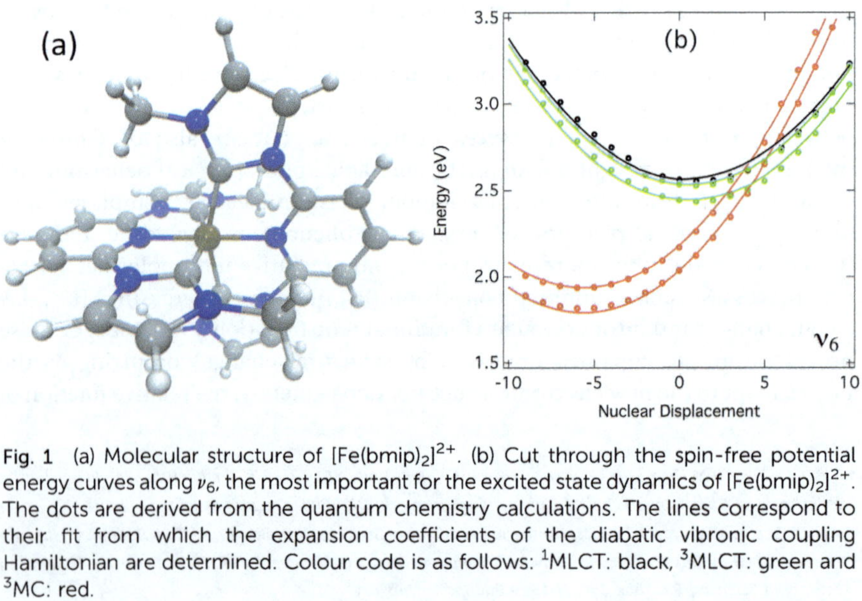

Fig. 1 (a) Molecular structure of $[Fe(bmip)_2]^{2+}$. (b) Cut through the spin-free potential energy curves along ν_6, the most important for the excited state dynamics of $[Fe(bmip)_2]^{2+}$. The dots are derived from the quantum chemistry calculations. The lines correspond to their fit from which the expansion coefficients of the diabatic vibronic coupling Hamiltonian are determined. Colour code is as follows: ^{1}MLCT: black, ^{3}MLCT: green and ^{3}MC: red.

Advancements in experimental techniques also call for the development of theoretical methodologies. In terms of ultrafast X-ray spectroscopy, this means going beyond simulations of meta-stable states[19] and requires accurately including excited states dynamics with nonadiabatic effects. Recently, Capano *et al.* applied quantum dynamics simulations of a prototypical Cu(I)-phenanthroline complex, $[Cu(dmp)_2]^+$ (dmp = 2,9-dimethyl-1,10-phenanthroline)[20,21] to simulate femtosecond X-ray spectroscopic signals.[22] Using realistic experimental parameters they also derived the anticipated signal strengths. This showed that it would be possible to resolve the fine spectral details associated with wavepacket dynamics within realistic experimental acquisition times. Indeed, this was recently confirmed experimentally by Lemke *et al.*[23] for $[Fe(bipy)_3]^{2+}$. In addition, the time-resolved Extended X-ray Absorption Fine Structure (EXAFS) signals were shown to be very effective for probing the width as well as the position of the wavepacket and therefore will provide strong insight into the dynamics associated with vibrational cooling. In this contribution we extend our recent quantum dynamics simulations of $[Fe(bmip)_2]^{2+}$ (ref. 24) to predict femtosecond X-ray spectroscopy, and in particular the experimental observables associated with X-ray Absorption Near Edge Spectroscopy (XANES) and X-ray emission (XES). This combination provides insight into structural, electronic and spin dynamics of the ^{1,3}MLCT and ^{3}MC states. The implications of these simulations on future studies of molecular photophysics at X-FELs is also discussed.

2 Methods and computations

2.1 Vibronic coupling Hamiltonian

The potential used in the present work is based upon the vibronic coupling Hamiltonian model[25] and has been described in detail in ref. 24. Briefly, the diabatic basis is fixed by choosing a point at which it is equal to the adiabatic basis (*i.e.* the coupling is zero), in this case the Franck–Condon geometry. The Hamiltonian is expanded as a Taylor series around this point $\mathbf{Q}_0$, using dimensionless (mass-frequency scaled) normal mode coordinates:

$$\mathbf{H} = \mathbf{H}^{(0)} + \mathbf{W}^{(0)} + \mathbf{W}^{(1)} + \ldots \quad (1)$$

The zeroth order term ($\mathbf{H}^{(0)}$) is the ground state harmonic oscillator approximation. The zeroth order coupling matrix ($\mathbf{W}^{(0)}$) contains the adiabatic state energies at $\mathbf{Q}_0$. The first order linear coupling matrix elements are written:

$$W_{ij}^{(1)} = \sum_{\alpha} \left\langle \Phi_i(\mathbf{Q}_0) \left| \frac{\partial H_{\mathrm{el}}}{\partial Q_\alpha} \right| \Phi_j(\mathbf{Q}_0) \right\rangle Q_\alpha \quad (2)$$

here, the on-diagonal elements are the forces acting within an electronic surface and the off-diagonal elements are the nonadiabatic couplings. The expansion coefficients for these terms, and those at higher order are obtained from a fit to the potential energy curves along the most important normal modes (α). Similar approaches have recently been used in the case of Cu(I)[20,21] and Re(I)[26] transition metal complexes.

In ref. 24 we used symmetry rules of the D_{2d} point group to simplify the determination of nonadiabatic coupling elements. We identified four normal modes, which are most important for describing the excited state dynamics.

These are a totally symmetric breathing mode (ν_6), two antisymmetric stretching modes (ν_{11} and ν_{25}), predominately acting on the Fe–C bonds with symmetry b_2, and another totally symmetric breathing (ν_{36}) mode with symmetry a_1, predominantly acting on the Fe–N bonds. ν_6, whose potential energy curves are shown in Fig. 1b, is the dominant vibrational mode of $[Fe(bmip)_2]^{2+}$ and is used during the calculations of the excited state core-hole spectra.

Although a model including only four nuclear degrees of freedom clearly represents a significant reduction in the dimensionality of the potential, this is consistent with the excited state dynamics of many Fe(II) complexes, whose dynamics are usually well represented schematically using one vibrational coordination. Consequently, although the present Hamiltonian is unable to capture longer time effects, such as vibrational cooling, this component has little influence on the ultrafast dynamics.

The potential energy surfaces of the singlet and triplet states were calculated and fitted separately yielding two sets of spin-free potential energy surfaces. The coupling of the two manifolds to create the spin–vibronic potential energy surfaces was treated as a perturbation to the spin-free potential energy curves and was achieved by including the spin–orbit coupling matrix elements between all of the states considered. This mixing between the singlet and triplet states promoted by SOC makes it impossible to rigorously use spin state labeling. However, throughout this paper, labeling which follows the convention of the spin free states is retained for simplicity. The potential energy surfaces were calculated using TDDFT(B3LYP*)[27–30] within the Tamm–Dancoff approximation (TDA)[31] and a TZVP basis set as implemented within the ORCA quantum chemistry package.[32] The SOC matrix elements were computed at the ground state equilibrium geometry using the perturbative approach developed by Wang and Ziegler[33] as implemented within ADF.[34–36] A TZP basis set was used for all atoms and scalar relativistic effects were included using the ZORA[37,38] approximation. All the parameters for the potentials, nonadiabatic and spin–orbit couplings for the model used below are available in ref. 24.

2.2 Quantum dynamics

The excited state wavepacket simulations were performed using the Heidelberg Multi Configuration Time Dependent Hartree (MCTDH) package.[39,40] In this approach the wavefunction ansatz is written as a linear combination of Hartree products:

$$\Psi(Q_1, \cdots, Q_f, t) = \sum_{j_1=1}^{n_1} \cdots \sum_{j_f=1}^{n_f} A_{j_1 \ldots j_f}(t) \prod_{k=1}^{f} \varphi_{j_k}^{(k)}(Q_k, t) \tag{3}$$

where Q_1, …, Q_f are the nuclear coordinates, $A_{j_1 \cdots j_f}$ are the time-dependent expansion coefficients and $\varphi_{j_k}^{(k)}$ are the time dependent basis functions for each k (degree of freedom), known as single particle functions (SPFs). The SPFs used in MCTDH have two advantages: (1) fewer are required as they are variationally determined and (2) the functions can be multi-dimensional particles containing more than one degree of freedom, thus reducing the effective number of degrees of freedom.

The computational details for the quantum dynamics simulations are shown in Table 1. These ensured convergence of the dynamics for 4 ps. To initiate the excited state dynamics, the wavefunction in the ground state built using

Table 1 Computational details for the MCTDH simulations of the 26 state spin–vibronic Hamiltonian. N_i is the number of primitive harmonic oscillator discrete variable representation (DVR) basis functions used to describe each mode. n_i are the number of single-particle functions used to describe the wavepacket on each state. Note that each triplet state has three components ($m_s = -1, 0, 1$) which are treated with the same number of basis functions

Modes	N_i	$n_{S_0}, n_{S_1}, n_{S_2}, n_{S_3}, n_{S_4}, n_{T_1}, n_{T_2}, n_{T_3}, n_{T_4}, n_{T_5}, n_{T_6}, n_{T_7}$
ν_6	191	1, 12, 12, 5, 5, 25, 26, 24, 24, 17, 17, 17
ν_{11}	61	1, 12, 12, 5, 5, 25, 26, 24, 24, 17, 17, 17
ν_{25}	61	1, 12, 12, 5, 5, 25, 26, 24, 24, 17, 17, 17
ν_{36}	61	1, 12, 12, 5, 5, 25, 26, 24, 24, 17, 17, 17

one-dimensional harmonic oscillator functions with zero initial momentum was impulsively excited into the lowest ^{1}MLCT state at the Franck–Condon geometry.

2.3 Simulating the X-ray spectra

The X-ray spectra of the non-stationary wavepacket was computed using a weighted sum of the spectra calculated at each grid point used in the quantum dynamics simulations. The weighting corresponds to the magnitude of the nuclear wavepacket at that grid point. As the nuclear motion in the dynamics is dominated by the motion along ν_6, all of the spectroscopic observables were calculated by projecting the wavepacket along this mode. For each spectra, a sum over all of the electronic states was performed, thus achieving a description of the full nuclear wavepacket.

The XANES simulations were performed using the finite difference method as implemented within the FDMNES package.[41,42] This uses a free form potential of radius 7.0 Å around the absorbing atom and includes scalar relativistic effects. Broadening due to the finite mean-free path of the photoelectron and the core-hole lifetime were accounted for using an arctangent convolution.[43] The XES spectra were computed within the one-electron approach[44,45] as implemented in the ORCA quantum chemistry package.[32] Computations used the B3LYP* functional[27–30] and the TZVP basis set.[46,47] All calculations included spin–orbit coupling (SOC), for which the SOC operator was approximated by the spin–orbit mean field method (SOMF).[48] A Lorentzian lifetime broadening with full-width at half maximum (FWHM) of 2.5 eV was applied after the calculation to account for the initial and final state broadening.

For both the XANES and XES spectra, the excited ^{1,3}MLCT states were simulated by increasing the overall charge on the complex, *i.e.* using $[Fe(bmip)_2]^{3+}$. This, as discussed in more detail below, reflects the sensitivity of X-ray spectra to the local electronic structure of the iron, rather than the electronic structure of the whole complex. Finally, the relative energies of the calculated transitions are generally well reproduced compared to experiment. However, it is well documented that the absolute transition energies are usually in poor agreement with experiment.[49] This failure stems from the approximate exchange description within the exchange-correlation functionals and is associated with the self interaction error (SIE).[50] This is usually corrected by applying a constant shift to the spectrum *a posteriori*.[51,52] In this case, as the spectra presented herein are not directly compared to experimental data, these shifts have not been included.

3 Results

3.1 Photoexcited decay of $[Fe(bmip)_2]^{2+}$

Fig. 2a shows the population kinetics of the ^{1,3}MLCT and ^{3}MC states following excitation into the lowest ^{1}MLCT states as obtained using the vibronic Hamiltonian outlined above and detailed in ref. 24. The dynamics are dominated by two main processes. The first is an ultrafast ISC from ^{1}MLCT → ^{3}MLCT, which has a time constant of ~100 fs (ref. 24) in excellent agreement with previous experimental observations.[11] The rapid nature of this step is a result of the close energetic proximity of the ^{1}MLCT and ^{3}MLCT states, which promotes strong mixing and leads to efficient population transfer. Fig. 2b shows the population kinetics for which the two MLCT states have been summed together. As mentioned above and discussed below this is important in the context of the sensitivity of the X-ray spectroscopic approaches used herein.

The dominant dynamics in the context of the present work corresponds to the decay of the ^{3}MLCT into the ^{3}MC states. Our simulations show two components, a faster one, which dominates the timescale of the present study, especially at early times (<2 ps), and which has a time constant of ~1 ps. This is followed by a slower component consistent with a time constant ≥4 ps, consistent with ref. 11. This population transfer (^{3}MLCT → ^{3}MC) step is between two triplet states and therefore can occur through two mechanisms, nonadiabatic or spin–orbit coupling. During the first 1–2 ps, both mechanisms contribute and this gives rise to the initial faster component of the decay dynamics. However at later times the contribution of nonadiabatic coupling is reduced due to a redistribution of population within the ^{3}MLCT manifold, and this gives rise to the lower population kinetics.[24] This population transfer is slower than the ^{1}MLCT → ^{3}MLCT transition because the population transfer predominantly occurs away from the crossing point between the ^{3}MLCT–^{3}MC states, at or near the energy minimum geometry of the ^{3}MLCT states. Crucially, the energetic separation of the ^{3}MLCT–^{3}MC states is >0.1 eV and therefore, regardless of the coupling strength, efficient population transfer is difficult.

Finally, previous experimental data[11] exhibits a weak oscillatory period close to 300 fs. This oscillatory component is very clearly observed in the wavepacket along ν_6 (Fig. 2c). These oscillations are stronger than those observed experimentally due to the absence of the redistribution of vibrational energy within our model Hamiltonian. However, this does reveal that the principal nuclear dynamics during the excited state decay are dominated by the nuclear motion along one nuclear degree of freedom, in this present case ν_6 with a period of 300 fs.

3.2 Femtosecond X-ray spectroscopy

In this section we translate our quantum dynamics simulations[24] into ultrafast X-ray spectroscopic signals. As previously mentioned, the X-ray techniques addressed herein are unable to distinguish between the ^{1,3}MLCT states even in the case where the coupling between the two states is weak. XANES spectra are sensitive to the formal oxidation state and local geometry around the absorbing atom. Therefore the ^{1,3}MLCT states are differentiated from the ground state and ^{3}MC states due to a shift in the edge position to higher energies arising from the increased formal charge of the Fe following transfer of an electron from the metal

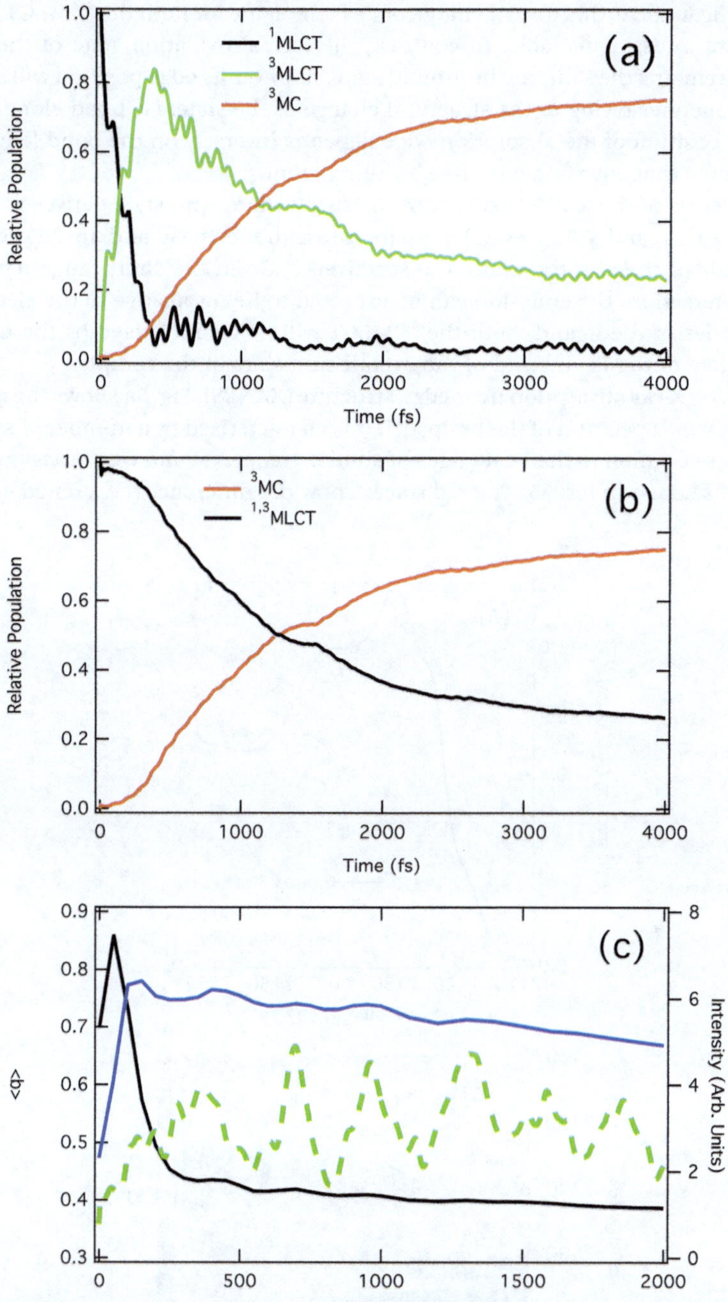

Fig. 2 (a) Relative diabatic state populations of the ^{1}MLCT (black), ^{3}MLCT (green) and ^{3}MC (red) states for 4 ps following photoexcitation. (b) Relative diabatic state populations of the ^{1,3}MLCT (black) and ^{3}MC (red) states for 4 ps following photoexcitation. (c) Expectation value of the position, $\langle q \rangle$ of the wavepacket in the ^{3}MLCT state along ν_6 (green) plotted with the transient absorption data in ref. 11 integrated between 600–680 nm (black) and 510–538 nm (blue) corresponding to the ^{1}MLCT and ^{3}MLCT states, respectively.

to the ligands. As this formal charge of Fe is the same for both the ^{1,3}MLCT states, they are indistinguishable. In contrast, the formal oxidation state of the latter (^{3}MC) remains the same as the ground state, however its edge position will shift to lower energies owing to the structural change of the state, *i.e.* bond elongations, as the position of the absorption edge depends inversely on the bond lengths of the atoms that coordinate to the absorbing atom.[53]

In terms of the XES, sensitivity of the spectrum to spin state derives from the Kα(2p → 1s) and Kβ(3p → 1s) emission through the 2p–3d and 3p–3d exchange integrals of the absorbing atom, respectively.[54] Given the short range nature of this interaction, the emission can be expected to be insensitive to the electronic properties of the ligand, both the ^{1,3}MLCT will be characterised by the doublet spin state of the Fe, instead of the overall spin state of the complex.

3.2.1 X-ray absorption near-edge structure (XANES). Fig. 3a shows the ground state XANES spectrum of $[Fe(bmip)_2]^{2+}$. It is characterised by a number of spectral features common to the Fe K-edges of similar complexes and whose assignments were discussed in ref. 55. The red traces show the difference (*i.e.* excited–ground

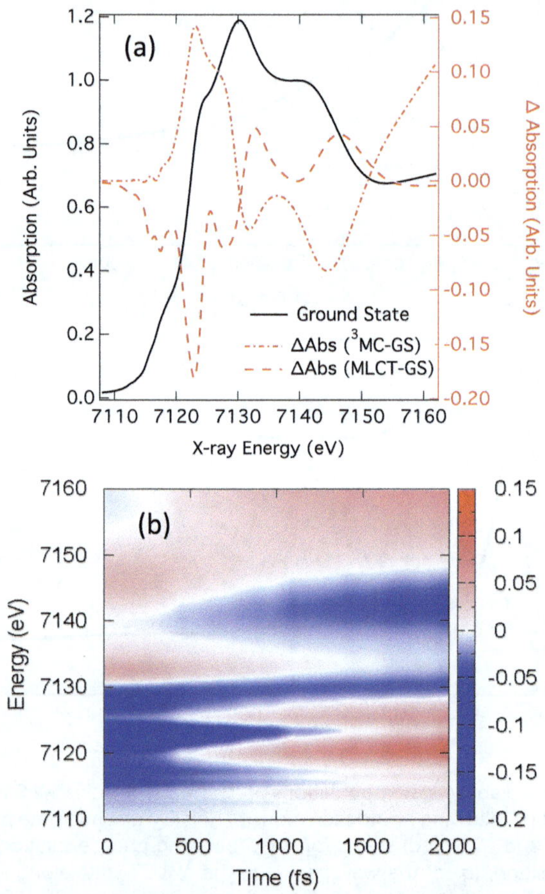

Fig. 3 (a) The simulated ground state Fe K-edge XANES spectra and transient spectra corresponding to ^{1,3}MLCT-GS (dashed line) and ^{3}MC-GS (dotted and dashed line). (b) The transient XANES spectrum as a function of time, during the first 2 ps after photoexcitation.

state) spectra calculated for the MLCT states at the Franck–Condon geometry and the ^{3}MC state at its optimised geometry. The former (dashed line) bears a strong resemblance to the difference spectrum of the MLCT states of $[Cu(dmp)_2]^+$ (ref. 56). In this case the spectrum is dominated by the edge shift and can, to a large extent, be reproduced using a shifted difference spectrum, *i.e.* energy-shifted ground-state spectrum minus the original ground-state spectrum. A similar situation is expected here, as the structure of the ground and ^{1,3}MLCT states of $[Fe(bmip)_2]^{2+}$ are very similar and therefore the main change is expected to be an edge shift associated with the change of oxidation state of the Fe. The transient spectrum of the ^{3}MC states is characterised by a shift of the absorption edge to lower energy, owing to the expansion of the first coordination sphere. This spectrum exhibits, as expected, a similar profile to the transient spectrum of the metal centred states of $[Fe(bipy)_3]^{2+}$ reported in ref. 57.

Fig. 3b shows the temporal evolution of the transient XANES spectra during the first 2 ps after photoexcitation. At early times, the spectrum, as expected from the population kinetics in Fig. 2a, closely resembles the transient ^{1,3}MLCT spectrum in Fig. 3a. As the dynamics progresses the spectrum begins to exhibit increasing characteristics of the ^{3}MC spectrum. There are also clear 300 fs oscillations in the transient signal, consistent with those observed in the transient absorption study.[11,24] As assigned above, this corresponds to the wavepacket motion along the ν_6 breathing mode[24] and consequently it is similar to the widely reported wavepacket dynamics of $[Fe(bipy)_3]^{2+}$.[9]

Fig. 4 shows kinetic traces integrated over four main regions, namely 7119–7121 eV, 7122–7124 eV, 7131–7135 eV and 7136–7143 eV. The lowest two, which probe the pre- and rising edge of the Fe K-edge spectrum show a monotonic rise largely associated with the population transfer between the ^{1,3}MLCT and ^{3}MC states. This is not the case of the latter two (7131–7135 eV and 7136–7143 eV), which are slightly non-linear. Indeed, these actually better reflect the population kinetics shown in Fig. 2b. This difference derives from the varying sensitivity of the two regions. The former is most sensitive to the electronic structure, the latter the geometric structure. Indeed, as recently shown by Lemke *et al.*[23] the higher lying resonances in the XANES regions are not only sensitive to the position of the wavepacket, but also to higher moments of the wavepacket distribution, *e.g.* width. This derives from the $\exp(-k)$ scaling of the disorder (Debye–Waller) effects. This is consistent with the vibrationally very hot wavepacket in the ^{3}MC state.

The kinetics of each of the 4 spectral regions are imprinted, to a greater or lesser extent, with the 300 fs oscillations associated with the wavepacket dynamics along ν_6. This is least clear in latter two (7131–7135 eV and 7136–7143 eV), which is because of the sensitivity of the spectra features to both the position and width of the wavepacket, which somewhat convolute these transient features. These spectral changes associated with these dynamics, once a realistic photolysis yield of $\sim$10% is included, are $\sim$0.01% of the spectral features in the ground state and are therefore similar to those reported in ref. 22. These are about an order of magnitude smaller than the coherent wavepacket motion reported for $[Fe(bipy)_3]^{2+}$ in ref. 23, which is consistent with the transient absorption study of ref. 11 showing only weak wavepacket oscillations.

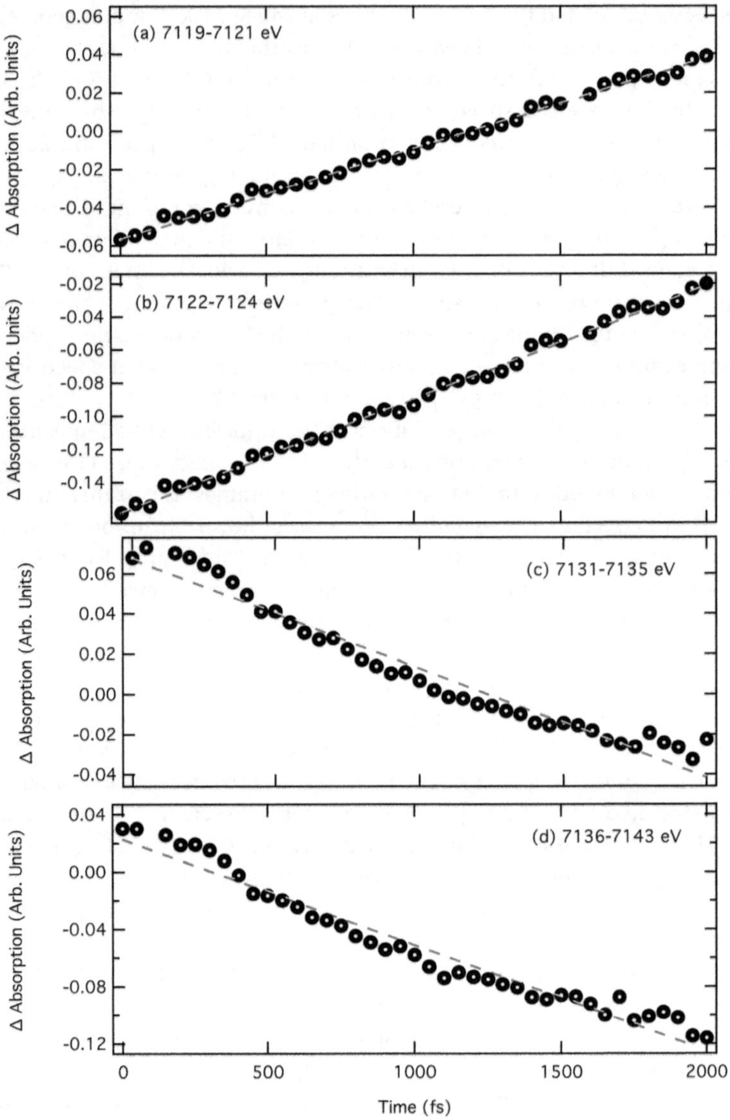

Fig. 4 Time traces of the transient spectrum integrated between 7119 and 7121 eV (a), 7122–7124 eV (b), 7131–7135 eV (c) and 7136–7143 eV (d). The red dashed line corresponds to a linear fit of the kinetic traces.

3.3 X-ray emission (XES)

Complementary to the XANES spectra presented in the previous section is XES, which provides a direct probe of the electronic structure of the occupied density of states and the spin state of the absorbing atom. The sensitivity of the latter derives from the electron exchange integrals, which is a short range interaction around the absorbing atom. This is highlighted in Fig. 5 showing the K$\alpha_{1,2}$, K$\beta_{1,3}$ and K$\beta_{2,5}$ emission spectra for $[Fe(bmip)_2]^{2+}$ simulated at a distorted geometry along ν_6 where the ^{3}MLCT is the lowest triplet state. The red trace is the ground state

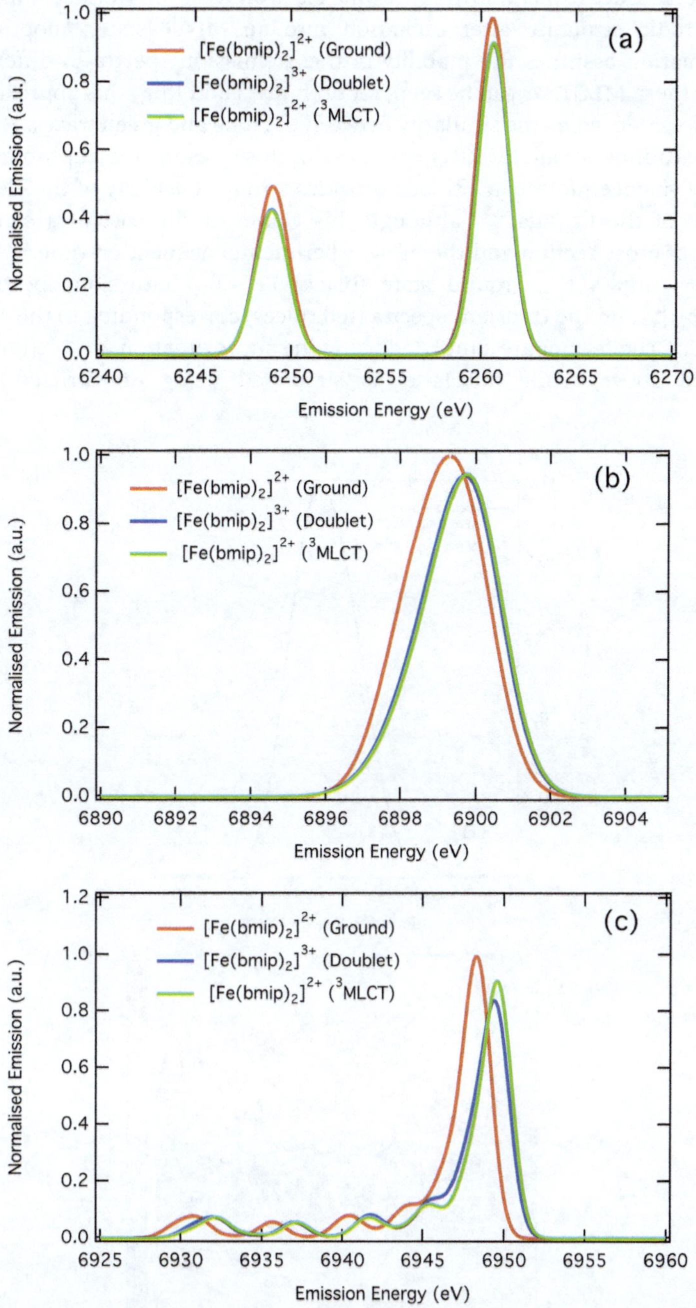

Fig. 5 The Kα (a), $K\beta_{1,3}$ (b) and $K\beta_{2,5}$ (c) XES spectra of $[Fe(bmip)_2]^{2+}$ at a slightly distorted geometry for which the ^{3}MLCT is the lowest triplet excited state. The ground state (*i.e.* singlet) spectrum is shown in red, the ground state triplet (*i.e.* lowest ^{3}MLCT) is shown in green and the spectrum for $[Fe(bmip)_2]^{3+}$, *i.e.* a doublet Fe(III) used to simulate MLCT excited states, is shown in blue.

spectrum of $[Fe(bmip)_2]^{2+}$, the green trace is the lowest triplet state, *i.e.* a ^{3}MLCT, and the blue trace is $[Fe(bmip)_2]^{3+}$, *i.e.* one electron removed from Fe, mimicking its electronic structure after excitation into an MLCT state. Adopting this approximation assumes the inability of these emission spectra to differentiate between the ^{1,3}MLCT. As can be seen, for both $K\alpha_{1,2}$ and $K\beta_{1,3}$ this approximation holds, as evidenced by the similarity between the blue and green traces. However, the same cannot be said for $K\beta_{2,5}$ emission. In this case, the direct probing of the occupied valence molecular orbitals provides strong sensitivity of the electronic structure of the ligands,[58–60] although this comes at the cost of a significant sacrifice of cross section and therefore experimental acquisition time.[22,64]

Fig. 6a shows the ground state (black) Fe $K\beta_{1,3}$ emission spectrum of $[Fe(bmip)_2]^{2+}$ and the transient spectra (red traces) corresponding to the ^{1,3}MLCT and ^{3}MC.[61] The former are simulated using the approximation (*i.e.* $[Fe(bmip)_2]^{3+}$) described above, while the latter is simulated using unrestricted density

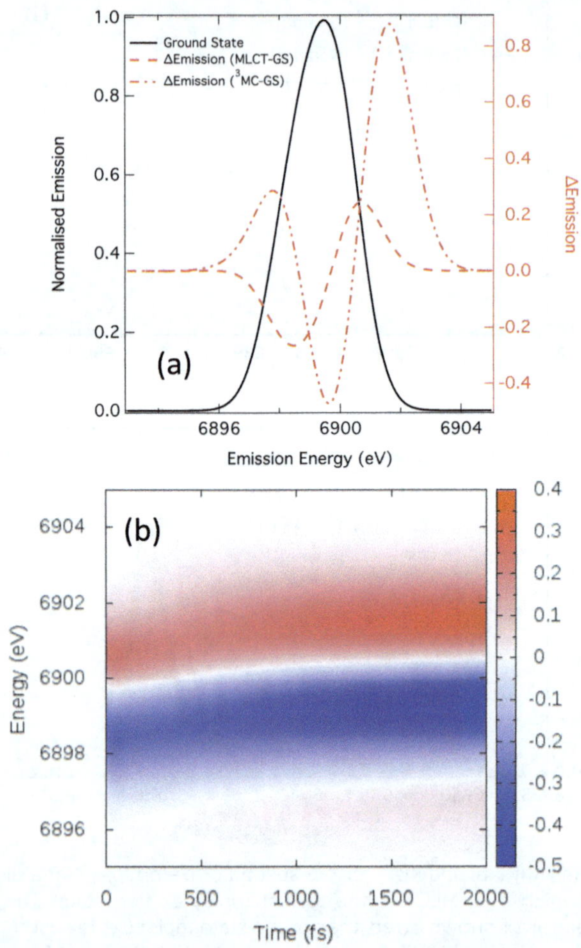

Fig. 6 (a) The simulated ground state Fe K-edge XES spectra and transient spectra corresponding to ^{1,3}MLCT-GS (dashed line) and ^{3}MC-GS (dotted and dashed line). (b) The transient XANES spectrum as a function of time, during the first 2 ps after photoexcitation.

functional simulations in the triplet state.[61] The transient spectra for the ^{1,3}MLCT states are dominated by a shift of the emission energy associated with the larger nuclear charge on the Fe, in contrast the transient spectrum of ^{3}MC indicates a broader emission as well as a shift to higher energies as observed for $[Fe(bipy)_3]^{2+}$.[8]

Fig. 6b plots the spectral evolution as a function of time. This shows an increase in the intensity of the main features associated with an increasing population of the ^{3}MC states, as well as a shift of the main positive and negative features to large emission energies, as would be expected from Fig. 6a. As reported previously in ref. 22 and expected for a core–core transition, these spectra exhibit no signature of the coherent wavepacket dynamics.

4 Discussion and conclusions

Building on our recent quantum dynamics simulations studying the ultrafast photoexcited dynamics of $[Fe(bmip)_2]^{2+}$,[24] in this present work we reinforce the ability of X-ray spectroscopy to reveal fine details about excited state dynamics, including coherent wavepacket motion around the absorbing atom. In addition, the spectral signatures in the XANES region for the two sets of excited states (^{1,3}MLCT and ^{3}MC) are shown to be largely out of phase with each other. Consequently, a change in sign of the transient signal provides a clear signature of the switching between the two manifolds. Indeed, this was recently used for $[Fe(bipy)_3]^{2+}$, albeit for a significantly shorter time.[23] We also confirm that X-ray emission, with the exception of $K\beta_{2,5}$, is not an effective approach to retrieve coherent wavepacket motion, and consequently is best suited to probing the character of the electronic structure and spin state of the absorbing atom. However, it is emphasised that the local nature of the electron exchange interaction means that the changes must be associated with those on the absorbing atom and not necessarily the molecule as a whole.

Importantly, the details of the excited state dynamics that would emerge from an experimental study on the present system suggests that in comparison to other time-resolved spectroscopies and especially in comparison to the transient optical absorption study on $[Fe(bmip)_2]^{2+}$ reported in ref. 11, femtosecond X-ray spectroscopic studies are unlikely to shed significantly new insight on the photoexcited decay mechanisms. Here, one should bear in mind that the strength of X-ray spectroscopy is providing a local interpretation around the absorbing atom. Consequently, its use will be most valuable in disentangling the dynamics of complex systems which may exhibit important optically dark states[62] and which can give a large number of broad overlapping bands, obvious examples in this direction are multi-nuclear metal complexes[15] or nanoparticles.[63]

Moving forward, an increasing emphasis should be placed upon obtaining sensitivity to the electronic structure of the ligands *via* $K\beta_{2,5}$ XES. The biggest challenge for these experiments is the low cross section, which means that recording a transient spectrum with a reasonable signal to noise ratio will require $\sim 10^{17}$ incident photons on the sample.[22,64] This equates to $\sim 10^{5}$ X-FEL pulses. Such a high number of pulses will only be realistic with the development of high-repetition X-FELs, such as the European X-FEL. Nevertheless, this would yield the most detailed insight into excited state dynamics. Importantly, as recently demonstrated by Wernet *et al.*,[16] if implemented within a resonant inelastic X-ray scattering (RIXS) setup, it would open the opportunity to probe transitions corresponding to negative energy transfers. These transitions are those whose incident photon energy is smaller than the scattered photon energy, and therefore correspond exclusively to electronically excited states. The characterisation and dynamical evolution of these transitions, without a strong background arising from molecules in their ground state, could provide crucial insights into excited state dynamics.

Acknowledgements

TJP acknowledges support from the Leverhulme trust grant RPG-2016-103. This work is also supported by the European Research Council *via* contract ERC-StG-259709 (X-cited!), the 'Lendület' (Momentum) Program of the Hungarian Academy of Sciences (LP2013-59), and the Hungarian Scientific Research Fund (OTKA) under contract K29724 and the People Programme (Marie Curie Actions) of the European Union's Seventh Framework Programme (FP7/2007-2013) under REA grant agreement number 609405 (COFUNDPostdocDTU).

References

1 M. Chergui, *Dalton Trans.*, 2012, **41**, 13022–13029.
2 V. W.-W. Yam and K. M.-C. Wong, *Chem. Commun.*, 2011, **47**, 11579–11592.
3 A. Barbieri, G. Accorsi and N. Armaroli, *Chem. Commun.*, 2008, 2185–2193.
4 Z. R. Grabowski, K. Rotkiewicz and W. Rettig, *Chem. Rev.*, 2003, **103**, 3899–4032.
5 M. Delor, P. A. Scattergood, I. V. Sazanovich, A. W. Parker, G. M. Greetham, A. J. Meijer, M. Towrie and J. A. Weinstein, *Science*, 2014, **346**, 1492–1495.
6 A. Hagfeldt and M. Graetzel, *Chem. Rev.*, 1995, **95**, 49–68.
7 W. J. Youngblood, S.-H. A. Lee, K. Maeda and T. E. Mallouk, *Acc. Chem. Res.*, 2009, **42**, 1966–1973.
8 W. Zhang, R. Alonso-Mori, U. Bergmann, C. Bressler, M. Chollet, A. Galler, W. Gawelda, R. G. Hadt, R. W. Hartsock, T. Kroll, K. S. Kjaer, K. Kubiček, H. T. Lemke, H. W. Liang, D. A. Meyer, M. M. Nielsen, C. Purser, J. S. Robinson, E. Solomon, Z. Sun, D. Sokaras, T. B. van Driel, G. Vanko, T.-C. Weng, D. Zhu and K. J. Gaffney, *Nature*, 2014, **509**, 345–348.
9 G. Auböck and M. Chergui, *Nat. Chem.*, 2015, **7**, 629–633.
10 A. Hauser, *Spin Crossover in Transition Metal Compounds II*, Springer, 2004, pp. 155–198.
11 Y. Liu, T. Harlang, S. Canton, P. Chabera, K. Suarez-Alcantara, A. Fleckhaus, D. A. Vithanage, E. Göransson, A. Corani, R. Lomoth, V. Sundström and K. Wärnmark, *Chem. Commun.*, 2013, **49**, 6412.
12 R. M. van der Veen, T. J. Penfold and A. H. Zewail, *Struct. Dyn.*, 2015, **2**, 024302.
13 C. Milne, T. Penfold and M. Chergui, *Coord. Chem. Rev.*, 2014, **277**, 44–68.
14 C. Bostedt, S. Boutet, D. M. Fritz, Z. Huang, H. J. Lee, H. T. Lemke, A. Robert, W. F. Schlotter, J. J. Turner and G. J. Williams, *Rev. Mod. Phys.*, 2016, **88**, 015007.
15 S. E. Canton, K. S. Kjær, G. Vankó, T. B. van Driel, S.-i. Adachi, A. Bordage, C. Bressler, P. Chabera, M. Christensen, A. O. Dohn, *et al.*, *Nat. Commun.*, 2015, **6**, 6359.

16 P. Wernet, K. Kunnus, I. Josefsson, I. Rajkovic, W. Quevedo, M. Beye, S. Schreck, S. Grubel, M. Scholz, D. Nordlund, W. Zhang, R. W. Hartsock, W. F. Schlotter, J. J. Turner, B. Kennedy, F. Hennies, F. M. F. de Groot, K. J. Gaffney, S. Techert, M. Odelius and A. Föhlisch, *Nature*, 2015, **520**, 78–81.
17 M. Minitti, J. Budarz, A. Kirrander, J. Robinson, D. Ratner, T. Lane, D. Zhu, J. Glownia, M. Kozina, H. Lemke, *et al.*, *Phys. Rev. Lett.*, 2015, **114**, 255501.
18 K. H. Kim, J. G. Kim, S. Nozawa, T. Sato, K. Y. Oang, T. W. Kim, H. Ki, J. Jo, S. Park, C. Song, *et al.*, *Nature*, 2015, **518**, 385–389.
19 T. Penfold, C. Milne and M. Chergui, *Adv. Chem. Phys.*, 2013, **153**, 1–41.
20 G. Capano, T. J. Penfold, U. Rothlisberger and I. Tavernelli, *Chimia*, 2014, **68**, 227–230.
21 G. Capano, M. Chergui, U. Rothlisberger, I. Tavernelli and T. J. Penfold, *J. Phys. Chem. A*, 2014, **118**, 9861–9869.
22 G. Capano, C. J. Milne, M. Chergui, U. Rothlisberger, I. Tavernelli and T. J. Penfold, *J. Phys. B: At., Mol. Opt. Phys.*, 2015, **48**, 1–11.
23 H. T. Lemke, K. S. Kjær, R. Hartsock, T. B. van Driel, M. Chollet, J. Glownia, S. Song, D. Zhu, E. Pace and M. M. Nielsen, *et al.*, 2016, arXiv preprint arXiv:1511.01294.
24 M. Pápai, T. Rozgonyi, G. Vanko and T. Penfold, *J. Phys. Chem. Lett.*, 2016, **7**, 2009–2014.
25 H. Köppel, W. Domcke and L. S. Cederbaum, *Adv. Chem. Phys.*, 1984, **57**, 59–246.
26 J. Eng, C. Gourlaouen, E. Gindensperger and C. Daniel, *Acc. Chem. Res.*, 2015, **48**, 809–817.
27 A. D. Becke, *J. Chem. Phys.*, 1993, **98**, 5648–5652.
28 C. Lee, W. Yang and R. G. Parr, *Phys. Rev. B: Condens. Matter Mater. Phys.*, 1988, **37**, 785–789.
29 S. Vosko, L. Wilk and M. Nusair, *Can. J. Phys.*, 1980, **58**, 1200–1211.
30 M. Reiher, O. Salomon and B. Artur Hess, *Theor. Chem. Acc.*, 2001, **107**, 48–55.
31 S. Hirata and M. Head-Gordon, *Chem. Phys. Lett.*, 1999, **314**, 291–299.
32 F. Neese, *Wiley Interdiscip. Rev.: Comput. Mol. Sci.*, 2012, **2**, 73–78.
33 F. Wang and T. Ziegler, *J. Chem. Phys.*, 2005, **123**, 154102.
34 C. Fonseca Guerra, J. G. Snijders, G. Te Velde and E. J. Baerends, *Theor. Chem. Acc.*, 1998, **99**, 391–403.
35 S. van Gisbergen, J. Snijders and E. Baerends, *Comput. Phys. Commun.*, 1999, **118**, 119–138.
36 *ADF2014.07, SCM, Theoretical Chemistry*, Vrije Universiteit, Amsterdam, The Netherlands, Scientific Computation and Modelling, 2014.
37 E. v. Lenthe, E. J. Baerends and J. G. Snijders, *J. Chem. Phys.*, 1993, **99**, 4597–4610.
38 E. van Lenthe, E. J. Baerends and J. G. Snijders, *J. Chem. Phys.*, 1994, **101**, 9783–9792.
39 H.-D. Meyer, U. Manthe and L. S. Cederbaum, *Chem. Phys. Lett.*, 1990, **165**, 73–78.
40 M. H. Beck, A. Jäckle, G. A. Worth and H.-D. Meyer, *Phys. Rep.*, 2000, **324**, 1–105.
41 Y. Joly, *Phys. Rev. B: Condens. Matter Mater. Phys.*, 2001, **63**, 125120.
42 S. A. Guda, A. A. Guda, M. A. Soldatov, K. A. Lomachenko, A. L. Bugaev, C. Lamberti, W. Gawelda, C. Bressler, G. Smolentsev, A. V. Soldatov and Y. Joly, *J. Chem. Theory Comput.*, 2015, **11**, 4512–4521.

43 O. Bunău and Y. Joly, *J. Phys.: Condens. Matter*, 2009, **21**, 345501.

44 M. A. Beckwith, M. Roemelt, M.-N. Collomb, C. DuBoc, T.-C. Weng, U. Bergmann, P. Glatzel, F. Neese and S. DeBeer, *Inorg. Chem.*, 2011, **50**, 8397–8409.

45 N. Lee, T. Petrenko, U. Bergmann, F. Neese and S. DeBeer, *J. Am. Chem. Soc.*, 2010, **132**, 9715–9727.

46 A. Schäfer, H. Horn and R. Ahlrichs, *J. Chem. Phys.*, 1992, **97**, 2571–2577.

47 F. Weigend and R. Ahlrichs, *Phys. Chem. Chem. Phys.*, 2005, **7**, 3297–3305.

48 B. A. Hess, C. M. Marian, U. Wahlgren and O. Gropen, *Chem. Phys. Lett.*, 1996, **251**, 365–371.

49 M. Stener, *Chem. Phys. Lett.*, 2003, **373**, 115–123.

50 N. Besley, M. Peach and D. Tozer, *Phys. Chem. Chem. Phys.*, 2009, **11**, 10350.

51 S. DeBeer-George, T. Petrenko and F. Neese, *J. Phys. Chem. A*, 2008, **112**, 12936–12943.

52 G. Capano, T. J. Penfold, N. A. Besley, C. J. Milne, M. Reinhard, H. Rittmann-Frank, P. Glatzel, R. Abela, U. Rothlisberger, M. Chergui and I. Tavernelli, *Chem. Phys. Lett.*, 2013, **580**, 179–184.

53 G. Bunker, *Introduction to XAFS*, Cambridge University Press, 2010.

54 F. de Groot, *Chem. Rev.*, 2001, **101**, 1779–1808.

55 V. Briois, P. Sainctavit, G. Long and F. Grandjean, *Inorg. Chem.*, 2001, **40**, 912–918.

56 T. J. Penfold, S. Karlsson, G. Capano, F. A. Lima, J. Rittmann, M. Reinhard, M. H. Rittmann-Frank, O. Braem, E. Baranoff, R. Abela, I. Tavernelli, U. Rothlisberger, C. J. Milne and M. Chergui, *J. Phys. Chem. A*, 2013, **117**, 4591–4601.

57 W. Gawelda, V. Pham, M. Benfatto, Y. Zaushitsyn, M. Kaiser, D. Grolimund, S. Johnson, R. Abela, A. Hauser, C. Bressler and M. Chergui, *Phys. Rev. Lett.*, 2007, **98**, 057401.

58 C. J. Pollock and S. DeBeer, *Acc. Chem. Res.*, 2015, **48**, 2967–2975.

59 G. Vankó, T. Neisius, G. Molnár, F. Renz, S. Kárpáti, A. Shukla and F. M. F. de Groot, *J. Phys. Chem. B*, 2006, **110**, 11647–11653.

60 G. Vankó, A. Bordage, P. Glatzel, E. Gallo, M. Rovezzi, W. Gawelda, A. Galler, C. Bressler, G. Doumy, A. M. March, E. P. Kanter, L. Young, S. H. Southworth, S. E. Canton, J. Uhlig, G. Smolentsev, V. Sundstrom, K. Haldrup, T. B. van Driel, M. M. Nielsen, K. S. Kjaer and H. T. Lemk, *J. Electron Spectrosc. Relat. Phenom.*, 2013, **188**, 166–171.

61 It is noted that at distorted geometries along mode 6, the lowest triplet state is no longer the metal centred state, as exploited in Fig. 5. This means a uDFT simulation for the metal centred state is strictly no longer valid. However, the fraction of wavepacket that visits this region of the potential is very small and therefore this has no influence on the transient spectra reported.

62 M. Pápai, T. J. Penfold and K. B. Møller, *J. Phys. Chem. C*, 2016, **120**, 17234–17241.

63 M. Rittmann-Frank, C. Milne, J. Rittmann, M. Reinhard, T. Penfold and M. Chergui, *Angew. Chem., Int. Ed.*, 2013, **53**, 5858–5862.

64 A. M. March, T. A. Assefa, C. Bressler, G. Doumy, A. Galler, W. Gawelda, E. P. Kanter, Z. Németh, M. Pápai, S. H. Southworth, *et al.*, *J. Phys. Chem. C*, 2015, **119**, 14571–14578.

Faraday Discussions

DISCUSSIONS

Vibrational and condensed phase dynamics: general discussion†

Andrew J. Orr-Ewing, Oleg Kornilov, Theis I. Sølling, Theo Keane, Michael P. Minitti, Hans Jakob Wörner, Oliver Schalk, Gareth M. Roberts, Russell S. Minns, Chris J. Milne, Lukas Miseikis, Tom J. Penfold, R. J. Dwayne Miller, Wolfgang Domcke, Martin Centurion, Kiyoshi Ueda, Peter M. Weber, Oliver Gessner, Daniel M. Neumark, Albert Stolow, Junko Yano, Shaul Mukamel and Vasilios G. Stavros

DOI: 10.1039/c6fd90073a

R. J. Dwayne Miller opened a general discussion of the paper by Junko Yano: Assuming one can obtain the structure of the Mn/Ca water splitting site, is this sufficient information to understand the water splitting mechanism? I would think one would want to catch the actual water splitting event in the barrier crossing region. The Mn/Ca positions could be strongly modulated by the protein environment to drive this process. Synthetically, it would be relatively easy to do this but the dynamic coupling to the surrounding protein and the charged environment could be the difference in the low efficiency so far of synthetic photocatalysts. Put another way, what is the key question that will be answered by getting the structure of the water splitting complex in the fully primed S_4 state?

Junko Yano responded: We don't know which two oxygens come together to form the O–O bond. I think getting the structural information of the S_3, S_3' and S_0 states will help determine the site of the reaction. We indeed also aim to catch the S_4 state (the transient state where the O–O bond is formed), but it may be present at very low concentrations only, if the rate of its decay is faster than the rate of its formation. Nevertheless, together with various spectroscopic and mass spectrometric techniques, plus theoretical calculations, we should be able to understand the mechanism. There have been huge efforts to synthesize catalysts that are akin to the OEC, while structural mimics have become very similar to the Mn_4CaO_5 cluster, making a close functional mimic has not been that successful so far.

Kiyoshi Ueda asked: Could you please explain more about how the structure and oxidization stage change in the Mn–Ca cluster in PS II when one goes through

† Lin Chen was unable to attend the Faraday Discussion meeting to present her article and there was no discussion of this article: L. X. Chen *et al.*, Imaging ultrafast excited state pathways in transition metal complexes by X-ray transient absorption and scattering using X-ray free electron laser source (DOI: 10.1039/C6FD00083E).

$S_1 \rightarrow S_2 \rightarrow S_3 \rightarrow S_4 \rightarrow S_0$ *via* photon absorption at each step? How much do you know about the structures for these states? Also, what is the time scale of the reaction of each step?

Junko Yano answered: Regarding the structure, the resting S_1 structure is known from crystallography. For other stable S-states, there are many spectroscopic studies, where these S-states are generated with laser flash excitation and then captured using a cryo-trap method, with the data collection being at cryogenic temperature. There are no published high-resolution crystal structure data on the other S-states yet. This is why we and some other groups work on improving the XFEL approach, with the focus on getting such structures at room temperature. The transitions from one S state to the next occur in the 20–2000 microsecond timescales. Once formed, the S_0–S_3 states are stable within the time scales of the XFEL experiment. In contrast, the S_4 state decays rapidly into S_0 with the release of O_2. It is not certain yet if it will be possible to obtain a spectroscopic or crystallographic signature of it. It is presently assumed that the S_0 state has three Mn(III) ions and one Mn(IV) ion, and that on each S state transition one Mn(III) is oxidized to Mn(IV). The S_4 state formation may be coupled to oxygen radical or Mn(V) formation.

Wolfgang Domcke commented: In the natural photosystem II, the photons are absorbed by a pair of chlorophylls. In your experiment, are the photons absorbed by the manganese cluster or by chlorophyll? What happens between the photon absorption and the structural changes observed in the manganese cluster?

Junko Yano answered: The light is absorbed primarily by the chlorophyll molecules in the antenna of photosystem II. This creates an electronically excited state. By excitation energy transfer, this excitation energy can be transferred into the reaction center, where it leads to the primary charge separation between P_{680} and Pheophytin. $P_{680}{}^{+}$ then oxidizes the side chain tyrosine D, which in turn oxidizes the Mn_4CaO_5 cluster. After four such events, the Mn_4CaO_5 cluster oxidizes two water molecules and thus returns to its lowest oxidation state (S_0). The time between the photoabsorption and the oxidation of the Mn cluster varies in each S-state transition. For example, the half time of the S_2 state formation after the illumination of the S_1 state is about ~90 microseconds. We note that the Mn cluster can absorb IR and UV light, but that the absorption coefficients for these are rather low. Such direct light absorption may contribute in part to photodamage of the OEC. In contrast, it does not induce S state transitions.

Hans Jakob Wörner remarked: What are typical numbers for the collection efficiency and the energy resolution in your X-ray emission measurements? Would it be feasible and/or beneficial to perform X-ray photoelectron spectroscopy on the samples that you have studied?

Junko Yano replied: For example, the data collection of the $K\beta_{1,3}$ XES of 1 mM Mn in solution with a 150 micrometer sample path length will take approximately 10 min at the LCLS with ~1 x 10^{12} incident photons per pulse. The resolution depends on the element studied, since it depends on the type of crystals used and the geometry, for 3d transition metals it is between 0.5 and 1.5 eV. XPS has not been done so far on these samples. As far as I know, XPS has been attempted on biological samples only at metal L-edges, but with serious damage to the systems. Part of the reason is the concentration of metals in the sample, the penetration depth of the X-rays and the path length of the photoelectrons. Even if high-energy X-rays are used for XPS, it will be still difficult to probe metals in aqueous solution with concentrations of ~1 mM.

Michael Minitti asked: In your von Hamos spectrometer, I believe you use a 16 crystal element. What would be the effect on your spectra if you used a spectrometer with more elements? Would the resolution get better? What are the trade offs?

Junko Yano replied: The efficiency of the spectrometer scales with the number of crystals since we are covering a more or less solid angle. The energy resolution is similar as it is mainly determined by the type of crystals used and the geometry. In fact, the resolution would start degrading significantly due to geometric effects if we were to use more crystals in the vertical direction (*i.e.* more than 4 rows of crystals).

Michael Minitti remarked: There is a definite race to achieve a certain resolution in the X-ray diffraction images. What is the goal? And as a follow up, what is the resolution limit in the current state-of-the-art?

Junko Yano replied: Ideally, we would like to achieve better than 2 Å resolution in order to define oxygen positions. Currently, 1.9 Å is achieved in the dark resting state structure at cryogenic temperature. At room temperature, the published data is around 4.5 Å for photosystem II but it is getting better than 2.5 Å.

Shaul Mukamel commented: An exciting goal of X-ray spectroscopy will be to resolve the different manganeses and their evolving oxidation states. Can you do that at this point? This could clarify the mechanism of the water splitting complex in PS II. Stimulated resonant X-ray Raman spectroscopy is a promising candidate.

Junko Yano responded: It is not possible to do stimulated resonant X-ray Raman spectroscopy right now for dilute samples like PS II, but there are some experiments that could pave the road to do such types of experiments at XFELs.

Chris Milne asked: Have you thought about doing the experiment in solution? X-ray absorption would potentially give you more information – would you be interested in exploring this at SACLA where you can easily do long-range X-ray energy scans?

Junko Yano responded: We are doing XES data collection on solution samples as well, and it is far more efficient than collecting XES from crystals due to the high hit rate. Collecting X-ray absorption spectra is certainly a good idea, to get more detailed electronic structural information of the metal cluster, and we are also working in this direction.

R. J. Dwayne Miller remarked: Don't you need to observe the point of barrier crossing (dynamics) and not just the 'static' structure of this water splitting step? Admittedly, this structural information will help enormously into thinking about the rearrangements involved to prime the water splitting step.

Junko Yano answered: Yes, that is what we are aiming for, in order to understand the structural changes. Getting the 'static' structures of the S_3, S_3' and S_0 states are however excellent starting points.

Chris Milne commented: To look at the valence to core XES, would it make sense to use a Johann-geometry spectrometer for better efficiency over a narrow energy range? Would the experiment be possible using this approach (are there enough photons)?

Junko Yano answered: The Rowland approach would be indeed more efficient over that particular narrow energy range (all of the crystal solid angle is used for one energy), but then you would be collecting only one point of your spectrum and you would need to scan the spectrometer and detector. This could be problematic for pump–probe experiments at the LCLS due to the intrinsic shot-by-shot fluctuations and the need for good normalization. The flux would be enough and this might be the way to go to avoid the large background effects associated with the von Hamos approach, and we are considering that option.

Tom Penfold said: How close are you to achieving a valence-to-core XES measurement on these low concentration proteins? Besides the low cross section, what are the obstacles in the way of this?

Junko Yano responded: The obstacle is the background that comes from the von Hamos spectrometer. While the data collection of the valence-to-core emissions of dilute samples will be possible in the current setup of XFELs, the data collection time is too long and our estimation will be about 4 to 5 h. At synchrotron facilities, we have collected the valence-to-core XES of PS II from frozen solution samples using Rowland circle geometry, and that is possible. However, we need to change samples quite often in order to work within the threshold of a certain X-ray dose to avoid radiation-induced changes. Collecting one spectrum with a reasonable S/N will take about half a day.

R. J. Dwayne Miller commented: In trying to populate 100% of the S_4 state, you need a 4-photon absorption process. Since you are not dealing with a strictly 3-level system with a bottleneck state to accumulate population, it is not possible to get 100% population of a given excited state. To be specific, the transition to a third level must be off-resonant from that of the second bottleneck state. The fact that you use 532 nm excitation and have rapid vibrational relaxation to the exciton manifold helps act as the donor state but the excited states are still resonant for excited state absorption processes. The excited state cross section is not so different from the ground state or is at least non-negligible. Each photon cycle needed to push the population to as high an excited fraction of S_1 to S_n as possible requires multiple photon absorption per PS II site. Much of this energy goes in as heat. There should be a significant temperature rise as the crystals have very low thermal conductivity and are on the 10 micron scale. I expect thermal diffusion times on the order of 1 ms. Has anyone measured the temperature rise in the crystals and how does it compare to barriers and rates for potential conformational changes? If not, can you walk us through the estimate of the ΔT for the excitation conditions used.

Junko Yano replied: There seems to be a misunderstanding because of the confusing nature of the nomenclature for the intermediate states of the water-oxidizing Mn_4CaO_5 catalyst (S_n, $n = 0$ to 4, states). These S_n states do not refer to a multi-photon excitation in the PS II reaction center, which is not necessary to populate higher S-states. In PS II, the system is comprised of an ultrafast light capture and charge separation process taking place in the reaction center that is coupled to a slow 'charge capacitor', the water oxidizing catalytic complex. The reaction center resets itself after electron extraction from the cluster and the oxidation equivalents are stored in the Mn_4CaO_5 cluster after each light-induced charge separation leading to the formation of the stable S-states (*e.g.* starting from the dark stable S_1 state, one charge separation leads to the formation of the next highest oxidized state, the S_2 state). In this respect the nomenclature of the S-states, as explained above, of the water oxidizing complex in PS II should not be confused with the S state nomenclature used in photochemistry to denote different photoexcited states of a pigment. One might expect to see a small change in the electron transfer rate due to electrostatic effects of the cluster, but the oxidation of the cluster is accompanied by proton transfer reactions, compensating the charge accumulation, and therefore no highly charged species are formed. The S-states of the water oxidizing complex are very long lived (on the time scale of seconds) and can be transformed to the next state in the photochemical cycle by a subsequent photon absorption and charge separation. Our illumination conditions are chosen to drive one charge separation event per flash and per center to obtain the maximum yield of holes on the cluster in each turnover pulse and the overall S-state populations achieved under these conditions are shown in Fig. 3 in our paper. Using, on average, a power of 100–130 μJ pulse^{-1}, a laser spot size of 500 micron in diameter and taking into account the size and volume of the irradiated sample (~250 micron in diameter) we deposited 8–10 J ml^{-1} of the crystal suspension. Assuming the worst case of 100% conversion of all these photons into heat, this would lead to a temperature change of ~2 K. When we instead look at a single crystal of 10 micron thickness and take into account the irradiation at the surface of the sample volume and the absorption coefficient of PS II of ~2 mM^{-1} cm^{-1} at 530 nm, a full conversion of all absorbed photons in the crystal into heat would lead to a temperature change of 3.8 K. This number does not take into account that in reality the largest fraction of the excess absorbed light is converted to fluorescence and not to heat. The sample is fully thermally equilibrated in the timescale of our delay time (0.5 s) between subsequent pulses so we do not expect any additional heating due to us applying multiple light excitation pulses to the same sample volume.

R. J. Dwayne Miller said: In the pulse cycling of PS II used for the preparation of the S_3 and S_4 states, it is not possible to excite 100% of each step in the overall photocycle for water splitting. These are not examples of 3-level systems with well separated resonances where fast relaxation can lead to a population bottleneck to give inversion and, in principle, 100% excitation. As the excited population increases with increasing excitation intensities, there will be an equal probability for either stimulated emission back to the ground state or stimulated absorption to upper excited states. The use of 532 nm helps but the excited state absorption, as determined from 2D spectroscopy and fs studies, has nearly the same absorption cross section as the ground state absorption. Even for multiple-excited chlorophyll in the light harvesting units, there will be a significant fraction of excited molecules experiencing excited state absorption. The other problem, as you state in the paper, is that if the time interval is too short, the photocycle does not have time to fully develop. The same issue would occur within your pulse duration. I understand that indirect methods have been used to estimate the degree of S_n state preparation. However, I don't see how you can prepare 100% of any of the S_n levels and, as such, each distribution of S_1–S_4 will contribute to a distribution with fractional populations among all the S_1–S_4 states. This point is known from 2D photon echo studies of related systems in which it is possible to directly measure distributions and remove inhomogenous broadening effects from the population dynamics. How accurate is the use of XES/XAS and ESR for population distributions to state that one has 100% conversion to S_4? This question is prefaced by the above comment. It seems to me you would need a baseline by which one could prepare 100% to know the absolute maximum signal possible for normalization with these methods. As I argue above, it is not possible. There will always be a distribution. Here I think 2D spectroscopic methods employing the same excitation pulse sequence would be essential to get the fraction S_n needed for the diffraction analysis. This comment is to point out that other methods that might prove useful to solve this problem.

Junko Yano responded: There is no way for making 100% of each S-state (except for the dark-stable S_1 state) due to the miss-hit and double hit connected to each light-induced S state transition. However, it is not necessary to make 100% pure S-states, as long as we have a way to estimate the population distribution of each S-state after a given number of flashes. The best case scenario of the S-state distribution is shown in Fig. 3 in our manuscript. These numbers are backed up by extensive oxygen evolution measurements under various experimental conditions (laser frequency, sample conditions, *etc.*) that were obtained under conditions highly similar to those during the XFEL experiments. We also record XES data at the beamline that confirm the turnover of the OEC, but they are not accurate enough yet to determine the exact S state populations. We note that one needs to be careful to not confuse the S_n states of the Kok cycle that describes the oxidation states of the Mn_4CaO_5 cluster with the S_n state nomenclature of singlet excited states of *e.g.* pigment molecules in the antenna system.

Chris Milne opened a general discussion of the paper by Oliver Gessner: from the shift, it looks like there is a shift to red from the O edge – so there is a shift from Cu to O? Is this a charge transfer?

Oliver Gessner replied: Picosecond time-resolved XANES experiments on thin films of Cu_2O by the Gaffney group indicate that both the O 1s and the Cu $2p_{3/2}$ edges exhibit a red-shift as a result of photoexcitation with 400 nm pulses at peak fluences of ~5-10 mJ cm^{-2}.[1] This result would be inconsistent with a charge transfer between the oxygen and copper atoms, provided the X-ray spectroscopic responses are dominated by electronic dynamics. Furthermore, the fact that both the top of the valence band (VB) and the bottom of the conduction band (CB) are largely comprised of states with strongly hybridized O 2p/Cu 3d character (see, *e.g.*, Hu *et al.*)[2] does not favor a strong charge transfer character of the super-band gap excitations detected in picosecond time-resolved experiments. However, the relative weight of oxygen- and copper-centered orbitals may differ for VB and CB states, which may enable some charge transfer contributions. Femtosecond time-resolved experiments that access states before carrier relaxation to the band edges may also provide a different picture compared to measurements on picosecond timescales.

1 Hillyard *et al.*, *Phys. Rev. B*, 2009, **80**, 125210.
2 Hu *et al.*, *Phys. Rev. B*, 2008, 77, 155115.

Chris Milne remarked: Can you look at the Cu edge? Is that possible? This might help clarify the nature of the local charge around the Cu as compared to the O.

Oliver Gessner responded: We can measure at any edge accessible with the beamline where the mobile transient XAS setup is used, including the Cu L-edge on BL 8.0.1 of the Advanced Light Source (ALS). Fig. 5b of our paper already shows static Cu $L_{2,3}$-edge XANES spectra recorded with the apparatus. Corresponding time-resolved spectra are foreseen for future beam times.

Oleg Kornilov asked: A quasi-static shift of the O 1s absorption edge of Cu_2O is observed in the experiments. It can potentially be attributed to the overall heating of the sample induced by the pump pulse. Did the authors look at the static temperature-dependent absorption or do they know of published temperature-dependent results for Cu_2O?

Oliver Gessner replied: Unfortunately, we have not done this yet but plan to do so in the near future.

Oleg Kornilov asked: The transient signal in Fig. 8 has the following sampling pattern: a point at −20 ns, 10 points spaced by 0.1 ns around t=0 and 14 points spaced by about 10 ns starting from approximately t=43 ns (accidentally coincides with the extracted decay time). How is this pattern generated? Why have the authors chosen to have a gap in the first 40 ns?

Oliver Gessner responded: The distribution of sampling points is the result of the bunch fill pattern of the ALS in combination with the desire to map out the early, picosecond range dynamics with the more intense camshaft pulse. The camshaft bunch of the ALS resides in a ~60 ns wide gap within the bunch fill pattern, as can be seen in Fig. 2a. If only nanosecond dynamics are of interest, one could choose to move the laser pulse to a different point in the ALS fill pattern, which would remove the large gap in the temporal sampling.

Daniel Neumark asked: Does the shift in your X-ray spectrum represent excitation into the conduction band or the valence band of Cu_2O? Is it possible that the UV pump pulse moves enough of the population from the valence to conduction band to result in significant band gap narrowing, lowering the minimum energy of the conduction band and thereby causing the shift that you observe?

Oliver Gessner replied: The extent of the observed inner shell absorption energy shifts, on the order of 1–2 eV, does not seem to be consistent with band narrowing being the sole source of these shifts. Band gap narrowing or renormalization due to significant excited carrier generation usually results from symmetric energy shifts of both valence and conduction bands in opposite directions. The transition energies from core levels to conduction band states would therefore be expected to shift by approximately half the amount of the total gap narrowing (when neglecting additional core level shifts). Within this picture, observed XANES shifts of 1–2 eV would correspond to a gap narrowing on the order of 2–4 eV, a range of values that mostly exceed the band gap of Cu_2O of 2.3 eV. Note that this does not preclude some contribution of band renormalization effects to the observed shifts. While it is beyond the scope of this work to derive quantitative estimates for band narrowing in Cu_2O, it is instructive to compare the shifts reported here to high-level estimates for gap renormalization energies in other semiconductors. Using a GW approach, Pemmaraju recently estimated that the gap narrowing in silicon for electron doping intensities of 2.3×10^{21} cm^{-3} amounts to ~180 meV.[1] For the experiments on Cu_2O we estimate that the upper limit for excited carrier densities is ~7×10^{21} cm^{-3} at a pump laser fluence of 6 mJ cm^{-2}. Thus, a notable contribution of band gap renormalization to the observed XANES shifts cannot be excluded, it is however not expected to be the dominant effect.

1 Schultze *et al.*, *Science*, **2014**, 346, 1348.

Hans Jakob Wörner asked: You mentioned that the heating of the sample may be difficult to distinguish from phase transitions in the XANES spectra. Would it be feasible to use EXAFS and what are the limitations?

Oliver Gessner replied: Heating, phase transitions, and electronic dynamics may indeed contribute simultaneously and in a non-trivial fashion to time-resolved XANES spectra. Time-resolved EXAFS would be suitable to gain further insight into the nature of the dynamics underlying the near edge spectra. Note, however, that this would also require additional data acquisition time as, at least in the energy-scanning approach used in our experiment, the time required to record the spectra scales linearly with the number of energy points required to sample the XANES/EXAFS regions of interest.

Peter M. Weber remarked: Is it possible for you to continuously move the sample so as to minimize the local heating by the laser beam?

Oliver Gessner replied: Yes, we use relatively large (up to >10 cm^2) samples that we continuously scan throughout the entire measurement by means of a long travel range manipulator.

R. J. Dwayne Miller asked: I am impressed that you can bin the data at such a high repetition rate to get the quality of data you have. The fact that each X-ray pulse only contains 1000 photons per pulse is interesting as this level of X-rays/pulse could be generated with a table top electron-Cu target source. Would this work for such studies or is the bandwidth of the emitted X-rays (from perhaps multiple different targets) too narrow? It would be interesting to have a table top source for *in situ* studies of catalysis.

Oliver Gessner replied: Near the oxygen K-edge, the estimated photon fluence in the experiment is more on the order of ~6000 up to ~30 000 per pulse depending on the chosen energy resolution (E/ΔE ~6000–1200) and can reach ~100 000 for the camshaft pulse. The experiment was performed at a laser repetition rate of 127 kHz, resulting in an effective photon fluence per standard bunch (*i.e.* per time delay) of ~8×10^8–4×10^9 photons s^{-1} and up to ~10^{10} photons s^{-1} for the camshaft bunch. I am not sure if such fluences can be reached with laboratory-based sources within a comparably narrow photon energy window and a comparable tunability across 10s or even 100s of eV.

Lukas Miseikis said: Do you think that the 1 eV shift in the O absorption edge can be due to surface or defect states, which result from a non-uniform sample made by compressing powdered Cu oxide? What would be your comment on the possibility to study polaron dynamics if you believe such species exist in this material, provided the electron–hole pair is tightly bound?

Oliver Gessner responded: As discussed in the manuscript, the existence of defect and/or surface states is very likely in the powder sample used in this study and it may be tempting to link the ~1 eV shift to states located within the nominal Cu_2O band gap. However, the amplitude of the XANES spectra change associated with this shift (up to ~30% of the ground state spectrum) seems unexpectedly large to be caused entirely by gap states.

Kiyoshi Ueda opened a general discussion of the papers by Junko Yano and Oliver Gessner: Sometimes synchrotron radiation data of protein crystallography suffers from radiation damage. This was also the case of PS II when the structure of the Mn–Ca cluster was first determined.[1] They re-determined the data using SACLA under damage-free conditions.[2] In the general discussion of the session on structural dynamics, however, R. J. Dwayne Miller commented on electronic damage, where the electron densities of the individual atoms, especially heavy atoms, may be strongly reduced *via* ionization and cascading Auger decay and it may affect the X-ray scattering, even though nuclei do not move, if the FEL pulse is intense enough. Have you ever seen such electronic damage for metal atoms at the reaction center? If not, what was the probability for, *e.g.* Mn atoms to absorb X-rays under the employed experimental condition?

1 Umena *et al.*, *Nature* 2011, **473**, 55.
2 Suga *et al.*, *Nature* 2015, **517**, 99.

Junko Yano answered: In our XFEL experiment, we always try to stay below the ionization or Auger cascade becoming noticeable in the data. Thus, we have not seen such effects in our experimental data. This is based on our observation of whether we see an XES spectrum that is comparable to that of the low dose synchrotron data collected at cryogenic temperature (to avoid radiation damage, but in this case we are avoiding the secondary radiation damage caused by the diffusion of hydroxl radicals induced by the interaction of X-rays and water molecules in the sample). We also argue the fact that we observe the expected spectral shift of the $K\beta_{1,3}$ spectrum between each S-state in the XFEL data that tells us the electronic damage is not visible in the data. Even if there is such damage, under our experimental conditions it is small (we calculated that in a single XFEL pulse, one Mn will only see 0.1 photons), and this will not prevent us from extracting the structural and spectroscopic information that we would like to get from the XFEL data.

Kiyoshi Ueda addressed Junko Yano and Oliver Gessner: In other sessions, we heard about charge migration where electronic wavepackets move or oscillate in a molecule without nuclear motion, at least in the timescale less than 10 fs; after that the nuclear motion starts to influence the charge migration. Do you think such kind of ultrafast charge migration might also play a role in your systems? In your case (or photochemical reactions in general) does charge move only with nuclei? How do you think charge transfer takes place in large systems?

Junko Yano replied: Charge transfer between the co-factors in photosystem II is well-described by Marcus theory and the Moser-Dutton approximation. The rate of intermolecular electron transfer is determined by the distance of the co-factors, the medium in between, and the reorganization energies. So this is different to the above described case of electron movement to an excited state within a molecule.

Oliver Gessner answered: Unfortunately, our current experiments cannot resolve dynamics proceeding on femtosecond timescales. These are of course of great interest and several groups are pursuing either high-harmonics or free-electron laser-based routes to probe electronic and nuclear dynamics in condensed phase materials with few femtosecond or even sub-femtosecond temporal resolution. Generally, I would expect that nuclear motion may play an important role for charge migration in these complex systems in order to enable an effective net transfer of charge as opposed to pure oscillations. Indeed, some theoretical models predict that vibronic coupling plays an essential role during charge injection, for example, from a molecular dye into a semiconductor substrate[1] and a recent experimental study of photoinduced charge transfer in an organic photovoltaic blend has identified vibronic coupling as an enabling factor.[2]

1 Jiao *et al.*, *Physical Chemistry Chemical Physics*, 2011, **13**, 13196.
2 Falke *et al.*, *Science*, 2014, **344**, 1001.

Chris Milne remarked: Kiyoshi Ueda raised the question about coherent electronic energy transfer in PS II. I commented that the experiment Junko Yano performed won't see this since it used a long laser pulse, so no impulsive excitation and wave packet creation. But this doesn't address whether this is possible in PS II. Note that my original intended comment was on Kiyoshi Ueda's initial comment about X-ray damage in these kinds of measurements (SFX on PS II). We have measured both nanocrystals and molecules in solution at XPP (LCLS) and looked at the Kα and Kβ XES signals and we see significant distortions in the XES signals at almost all fluences. The focus used was on the order of 3 microns at the smallest and the mode was pink beam (non-monochromatic, ~3 mJ). So there is significant electronic damage around the transition metals in our samples at fluences comparable to the SFX experiments performed at CXI.

Wolfgang Domcke addressed Chris Milne and Kiyoshi Ueda: Kiyoshi Ueda asked whether electron migration processes may play a role in natural photosynthesis. In my opinion, it is unlikely that electrons and protons are moving around independently in photosynthesis. Protons and electrons attract each other too strongly by Coulomb forces in nonpolar environments such as proteins. It is more likely that electrons and protons are moving together as neutral radicals.

Peter M. Weber remarked: The paper points out that protein conformeric structures may depend on temperature. Do you think the lasers in your experiment have a heating effect on the proteins that should be considered?

Junko Yano responded: No. We do not see a heating effect from the visible (532 nm) laser flashes that we employ for S state advancement. We estimated that the maximum temperature rise by the visible laser (if the entire energy results in heat, which is not the case in reality) will be only a couple of degrees. If a substantial heating effect existed, it would make the diffraction quality worse as we increase the number of flashes (we go up to 3 flashes when we make S_0-rich sample). We do not see any differences in the crystal quality between the dark state and the 3F state. In contrast, the XFEL pulse will heat up the sample significantly, but the data collection is outrunning the sample damage.

Peter M. Weber said: Following on the previous question, do you think that the heating effect is larger in your samples?

Oliver Gessner responded: While we are trying to minimize heating effects, *e.g.* by fast sample scanning and small laser/X-ray spot sizes, we are only beginning to investigate heating-related effects in greater detail. A quantitative statement is currently not possible. It is a very important issue, though, and other groups such as Lin Chen's at Northwestern University have already performed more systematic studies that will hopefully soon lead to a better insight into thermally induced effects.

Daniel Neumark commented: I appreciate that you are concerned about thermal effects in your experiment. However, my understanding is that your samples are quite thick, since you are carrying out X-ray emission as opposed to absorption experiments. Do the thick samples mitigate thermal effects owing to their heat capacity?

Oliver Gessner replied: Generally, the X-ray emission-based technique gives us greater freedom in the design of the samples compared to transmission-based absorption measurements and, therefore, also the mediation of thermal effects. In this particular case, however, the sample was a pellet consisting of compressed Cu_2O powder, which we don't expect to have good thermal conductivity. So even if the total heat capacity of the sample may be large compared to a thin film, a limited thermal conductivity may be an issue.

Hans Jakob Wörner said: You showed that X-ray absorption measurements on complex systems are often not feasible in transmission mode, which is however the most convenient approach for broad-band sources, such as high-harmonic generation. Could you propose solutions to this problem?

Oliver Gessner replied: For some samples and particular energy ranges, one may consider experiments in reflection geometry. However, this is not a universal solution and comes with its own challenges. Among them are the requirement for grazing incidence geometry and the low X-ray reflectivity of many materials even under grazing incidence. A notable surface roughness of many interesting materials may further complicate a reflection approach.

Hans Jakob Wörner opened a general discussion of the paper by Andrew Orr-Ewing: I was impressed by the level of detail that you can extract from your measurements. Could you comment on the possibility of obtaining additional information by performing time-resolved photoelectron spectroscopy on these solvated systems?

Andrew Orr-Ewing answered: Transient infra-red absorption spectroscopy with broadband, ultrafast laser pulses is indeed rich in detail because of the chemical information contained within vibrational spectra. Nevertheless, alternative spectroscopic techniques such as time-resolved photoelectron spectroscopy (PES) will provide valuable and complementary information on non-adiabatic photochemical dynamics. In the gas phase, time-resolved PES has been shown to be a powerful method to observe changes in electronic state (*via* internal conversion and intersystem crossing) on ultrafast timescales,[1] and is perhaps better at capturing the dynamics at very short times after initiation. For example, in our time-resolved vibrational absorption spectroscopy (TVAS) measurements, our instrument response function is limited to 200–300 fs. Implementing PES in solution presents experimental challenges because of the need for high-vacuum to measure the kinetic energies of the emitted electrons. However, advances with liquid jet technologies have made these types of measurement possible.[2] Observation of ground electronic state dynamics remains a further challenge for PES because vacuum ultraviolet photons are required to ionize most ground-state molecules. Again, the complementary nature of different ultrafast techniques becomes evident because TVAS can report on the recovery of ground state molecules, or the formation of new photoproducts in their ground states, as illustrated in our paper and recent publications.[3,4]

1 T. Suzuki, *Annu. Rev. Phys. Chem.*, 2006, **57**, 555–592.
2 M. Faubel, B. Steiner and J. P. Toennies, *J. Chem. Phys.*, 1997, **106**, 9013–9031.
3 K. Röttger, H. J. B. Marroux, M. P. Grubb, P. M. Coulter, H. Böhnke, A. S. Henderson, M. C. Galan, F. Temps, A. J. Orr-Ewing and G. M. Roberts, *Angew. Chem. Int. Ed.*, 2015, **54**, 14719–14722.

4 D. Murdock, S. J. Harris, J. Luke, M. P. Grubb, A. J. Orr-Ewing and M. N. R. Ashfold, *Phys. Chem. Chem. Phys.*, 2014, **16**, 21271–21279.

Wolfgang Domcke commented: Could you explain which specific new information you obtain by the investigation of these nucleosides in a condensed phase environment compared to the wealth of information that is available from experiments in the gas phase or in molecular beams.

Andrew Orr-Ewing replied: The dynamical pathways followed by photoexcited molecules depend sensitively on the topography and character of electronically excited states, their energetic orderings, and their crossings with other excited states and the ground state. The locations and shapes of conical intersections are key to controlling non-adiabatic photochemical dynamics, and the pyrimidine bases present a particularly challenging class of compounds for experiment and theory to examine.[1] All these excited-state properties are sensitive to the environment in which the molecule is found, and hence are modified by the presence of a solvent. To some extent, solvent effects can be explored in molecular clusters under molecular beam conditions, but a bulk solvent at ambient temperatures is a rapidly fluctuating environment, and couplings between the solute molecules and solvent bath dissipate energy on timescales comparable to those for the photodynamics. Hence, there is much to be learned by studying the photodynamics in liquid solutions and contrasting with the gas phase studies. Moreover, studies in a range of solvent environments allow us to explore the effects of protic *vs.* non-protic and polar *vs.* non-polar intermolecular interactions.

The gas phase studies of nucleobases (and other molecules), together with high-level electronic structure calculations and dynamical simulations for isolated molecules, provide a very important starting point for the interpretation of our data from solution phase measurements. As the question notes, a wealth of detailed information is available from state-of-the-art experiments in molecular beams, whether performed with high time (*e.g.* ultrafast photoelectron spectroscopy) or energy (*e.g.* velocity map imaging, or H-atom photofragment translational spectroscopy) resolution, particularly when combined with complementary theoretical studies. We are now in the exciting position of being able to make in-depth comparisons between the gas and condensed phase studies, which provide new insights and quantitative information about the multitude of ways in which solvents modify photochemical dynamics. In the specific case of the nucleoside studies reported in our paper, we observe how the $^1n\pi^*$ states are shifted in energy by interaction with a solvent, and thus the branchings between dynamics on the $^1\pi\pi^*$ and $^1n\pi^*$ states are modified, with consequences for the triplet-state population in the case of thymidine. The efficient quenching of excess vibrational energy by the solvent bath also allows us to determine timescales for the recovery of ground-state molecules *via* different routes, and to quantify the quantum yields for competing photochemical pathways.

1 R. Improta, F. Santoro and L. Blancafort, *Chem. Rev.*, 2016, **116**, 3540–3593.

Oliver Schalk said: In your paper, you described the dynamics of pyrimidine nucleosides in chloroform. The dynamics have been previously studied in methanol, water and the gas phase. What new information do you expect by using yet another environment and why is it worth doing such an investigation?

Andrew Orr-Ewing responded: The photo-induced dynamics of the pyrimidine nucleosides involve excited states of different electronic character, and internal conversion and intersystem crossing, mediated by conical intersections, on femtosecond to picosecond timescales. Comparisons between measurements made in different solvents show that the timescales for these processes are sensitive to the properties of the solvent such as its dielectric constant and hydrogen bonding (as shown in Table 4 of our paper, where we collect the available experimental data). We can only develop a quantitative understanding of the influence of the solvent if we make measurements of excited-state lifetimes, and quantum yields for competing pathways such as triplet-state population and ground-state recovery, in a range of different environments. These types of comparative studies will test and improve theoretical and computational descriptions of the excited-state solute interactions with the surrounding medium. Ultimately, we would like to understand how the environment at the core of double-stranded DNA affects the photochemistry of the pyrimidine and purine nucleobases. This environment is not aqueous, and is also not well-represented by isolated-molecule studies. Instead, the dielectric constant in the DNA core is similar to that in chloroform. However, the detail that we are able to extract from ultrafast spectroscopy studies of solutions of nucleosides is difficult to reproduce in DNA strands because the closely stacked nucleobases do not act like isolated chromophores. Therefore, we need to produce quantitative and accurate theoretical models of how the local environment modifies the energies and crossings of excited states to be able to understand what happens in UV-excited DNA. Our work seeks to provide data that will facilitate this objective by using simpler, but well-chosen, model systems.

Oliver Schalk asked: In your study, you addressed the differences between the dynamics of pyrimidine nucleosides in chloroform and the protic solvents water and methanol as well as the gas phase. If the dynamical changes between different solvents are important, why did you chose a target with dynamics as complicated as those in the nucleosides?

Andrew Orr-Ewing responded: There is already a rich literature of computational and experimental studies of the photochemistry of heterocylic organic molecules representative of some of the structural features of the nucleobases. We are therefore now well placed to push the limits of our studies to more complicated molecules such as the nucleosides, with good reason to expect that we can extract quantitative dynamical insights because of the remarkable capabilities of modern theory and experiment. Indeed, I do not think we should be satisfied with restricting ourselves to ever-more-detailed studies of simpler molecules: the electronic structures, nuclear dynamics and solute–solvent interactions that determine the photochemical behaviour of the nucleobases are, after all, largely specific to these molecules. We must therefore be prepared to tackle these complex molecular systems directly if we are to advance our understanding of their photochemistry.

Chris Milne said: Are you interested in doing more complicated systems, such as those with a couple of nucleobases?

Andrew Orr-Ewing answered: Yes, the study of canonical Watson–Crick dimers is a major part of our research activity in nucleobase photochemistry. For example, we recently reported a detailed study of the electron-driven proton transfer (EDPT) between guanine and cytosine nucleobases in their Watson–Crick dimers, using ultrafast transient electronic and vibrational absorption spectroscopy.[1] These studies necessitated the use of a non-protic solvent to allow the hydrogen-bonded Watson–Crick dimers to form. We were able to resolve both an ultrafast EDPT step across the central hydrogen bond, and a slower EDPT process across a second hydrogen bond that produced a tautomeric form of the guanine–cytosine dimer. We have subsequently extended this work to studies of the photodynamics in guanine–guanine dimers, and in adenine–thymine dimers adopting the Watson–Crick structure.

1 K. Röttger, H. J. B. Marroux, M. P. Grubb, P. M. Coulter, H. Böhnke, A. S. Henderson, M. C. Galan, F. Temps, A. J. Orr-Ewing and G. M. Roberts, *Angew. Chem. Int. Ed.*, 2015, **54**, 14719–14722.

Daniel Neumark remarked: One of the appealing features of your experiment is that you are able to monitor relaxation dynamics on the ground state surfaces of the nucleic acid constituents, in contrast to methods such as time-resolved photoemission where only the excited state dynamics are energetically accessible. However, are you able to see any transient infrared signals associated with excited states? The $^1n\pi^*$ state in particular is quite long-lived (Fig. 5 and 7). Do you see any signature of that state in your experiments?

Andrew Orr-Ewing responded: There is no reason in principle why we should not see transient infra-red absorptions from molecules in the $^1n\pi^*$ state, given their lifetimes. The IR probe is sensitive to vibrational motions of the molecule, regardless of the electronic state, and sufficient population should accumulate transiently in the $^1n\pi^*$ states to be observable with our signal-to-noise levels. However, we were only able to report one such band (Fig. 4 of our paper), which we assigned to an N–H stretch of the $^1n\pi^*$ state of the imino-oxo form of 2′-deoxy-cytidine. We did not see corresponding transient absorptions in the carbonyl stretching region in either the 2′-deoxy-thymidine or 2′-deoxy-cytidine measurements. The most likely reason is that the spectra are congested, and transient absorption bands from $^1n\pi^*$-state molecules cannot be resolved if they sit on top of the stronger ground state bleach features. The timescales for depopulation of the $^1n\pi^*$ state and recovery of ground-state molecules are similar, which means the spectroscopic signatures cannot be distinguished kinetically. It is also possible that certain transient absorption features associated with the $^1n\pi$-state populations are shifted to vibrational frequencies that lie outside our probe wavenumber ranges, and we did not shift the probe window to look for such bands.

Albert Stolow commented: In the low energy electron attachment experiments by Sanche,[1] damage to DNA bases such as adenine seems to occur at the N9 position. Theoretical considerations by Simons *et al.* indicate that electron attachment to the nucleobase can lead to rupture of the C–O bond between the sugar and the phosphate.[2] The electron-withdrawing nature of the phosphate is thought to play a key role. Although the excess electron will have significant effects on the dynamics (*i.e.* a transient anion *versus* a photoexcited neutral), we have speculated that an electron attached to the π orbital of adenine couples rapidly and efficiently to the σ* orbital at the N9 position in a manner similar to that discussed in our time-resolved photoelectron spectroscopy measurements of isolated adenine.[3] Do you have any evidence of electron withdrawing groups having such effects in your results?

1 L. Sanche,*Eur. Phys. J. D*, 2015, **35**, 367–390.
2 I. Anusiewicz, M. Sobczyk, J. Berdys-Kochanska, P. Skurski and J. Simons,*J. Phys. Chem. A*, 2005, **109**, 484–492.
3 H. Satzger, D. Townsend, M. Z. Zgierski, S. Patchkovskii, S. Ullrich and A. Stolow, *Proc. Natl. Acad. Sci. U. S. A.*, 2006, **103**, 10196.

Gareth Roberts answered: Albert Stolow's comment is not without merit, although in some ways UV- and low-energy-electron (LEE)-induced damage in DNA are separate sides of the same coin, in so much that they can both give rise to damage, the dynamical mechanisms initiated by these two scenarios could be very different.

In our measurements there is no obvious evidence that UV-initiated cleavage of the N–C glycosidic bond occurs at 260 nm in the pyrimidine nucleosides, analogous to theoretically predicted outcomes of LEE attachment into a π* hole on the nucleoside.[1] It should be said that in the gas phase, N9–H bond cleavage was observed for the adenine nucleobase at shorter excitation wavelengths (<233 nm)[2,3] and calculated for other nucleobases;[4] our studies to date do not find evidence for this same channel in aqueous adenine at >220 nm.[5] Second, our experiments are conducted in chloroform solutions, so the formation of solvated LEEs does not occur, shutting off the possibility of any diffusion-limited reactions between LEEs and nucleosides.

Finally, for nucleotides (rather than nucleosides) much theory has suggested that LEE-attachment will predominantly cleave the C–O sugar–phosphate bond (and not the glycosidic bond), leading to the ejection of free phosphate anions.[1] This behaviour is a model for LEE-induced strand breaks in double helix DNA, the ultrafast dynamics of which are currently poorly understood from an experimental perspective. In water (or more specifically D_2O), one could envisage a TVAS experiment which uses UV light to generate aqueous LEEs (either from the water solvent or I^{-1} (aq)) and subsequently monitor the build-up of vibrational bands for any free phosphate as a signature for LEE-induced C–O cleavage in the nucleotide. Such an experiment is not without its challenges though, which could include prohibitively low signal levels due to the bimolecular nature of the LEE + nucleotide reaction (*cf.* Ref. 6) and disentangling the LEE-induced nucleotide dynamics from the UV-induced nucleotide dynamics.

1 J. Gu, J. Leszczynski and H. F. Schaefer III, *Chem. Rev.*, 2012, **112**, 5603–5640.
2 M. G. D. Nix, A. L. Devine, B. Cronin and M. N. R. Ashfold, *J. Chem. Phys.*, 2007, **126**, 124312.
3 K. L. Wells, D. J. Hadden, M. G. D. Nix and V. G. Stavros,*J. Phys. Chem. Lett.*, 2010, **1**, 993–996.
4 S. Yamazaki, W. Domcke and A. L. Sobolewski,*J. Phys. Chem. A*, 2008, **112**, 11965–11968.

5 G. M. Roberts, H. J. B. Marroux, M. P. Grubb, M. N. R. Ashfold and A. J. Orr-Ewing, *J. Phys. Chem. A*, 2014, **118**, 11211–11225.
6 A. J. Orr-Ewing, *Annu. Rev. Phys. Chem.*, 2015, **66**, 119–141.

Wolfgang Domcke asked: Saigusa and coworkers[1] showed a few years ago in a molecular beam experiment that the excited-state lifetime of adenosine is shorter than that of adenine. We could explain this result by demonstrating the existence of an additional relaxation channel in adenosine. This process is an electron-driven proton transfer from the ribose to adenine along an existing hydrogen bond. This relaxation mechanism may also be present in the nucleosides considered here. It would therefore be interesting to perform a detailed comparison of the excited-state decay constants of the bases and the corresponding nucleosides in solution.

1 H. Asami, K. Yagi, M. Ohba, S. Urashima and H. Saigusa, *Chem. Phys.*, 2013, 419, 84–89.

Andrew Orr-Ewing answered: Unfortunately, we are unable to make this type of direct comparison between the nucleobases (cytosine and thymine) and their nucleosides (2′-deoxy-cytidine and 2′-deoxy-thymidine) because the former are not sufficiently soluble in our solvent of choice (chloroform). Alternative studies in aqueous solution are challenging because of the strong IR-absorption of water. To engineer sufficient solubility for the nucleosides in chloroform, we prepared them with silyl-group substitutions of the H-atoms attached to the hydroxyl sites in the 2′-deoxyribose moiety (see Fig. 1 in our paper). This chemical modification will switch off the electron-driven proton transfer (EDPT) mechanism that you previously identified.[1,2]

1 D. Tuna, A. L. Sobolewski and W. Domcke, *J. Phys. Chem. A*, 2014, **118**, 122–127.
2 B. Marchetti, T. N. V. Karsili, M. N. R. Ashfold and W. Domcke, *Phys. Chem. Chem. Phys.*, 2016, **18**, 20007–20027.

Albert Stolow asked: In principle, chemical substituents such as a methyl group or a halogen atom can lead to both inertial and potential (*i.e.* electronic structure) effects on the dynamics at conical intersections. Do you have any general comments on how substituents may affect dynamics at conical intersections?

Andrew Orr-Ewing replied: This is a complicated question, and one that has been extensively explored both computationally and experimentally by judicious substitution studies. It is now recognized, for example through the work of Robb and coworkers,[1–3] that the locations and the shapes of conical intersections between states have a profound effect on adiabatic *vs.* non-adiabatic branching, as do inertial effects. Our current study does not directly shed any further light on this question. However, there may be some inertial effects of the 2′-deoxyribose group (and indeed the bulky silyl groups attached to it) on the dynamics at conical intersections between the $^1\pi\pi^*$, $^1n\pi^*$ and S_0 states because, as Fig. 5 and 7 show, molecular deformation is necessary to access these conical intersections. There may also be subtle effects of solvent friction on the dynamics, compared to isolated molecules in the gas phase. These are interesting ideas that must await further investigation to resolve.

1 A Computational Strategy for Organic Photochemistry, M. A. Robb, M. Garavelli, M. Olivucci, and F. Bernardi, in *Reviews in Computational Chemistry, Volume 15*, ed. K. B. Lipkowitz and D. B. Boyd, Wiley–VCH Publishers, New York, 2000, 87-146.
2 Conical Intersections in Organic Photochemistry, M. A. Robb, in *Conical Intersections, Theory, Computation and Experiment*, ed. W. Domcke, D. R. Yarkony and H. Köppel, World Scientific Publishing, 2011, 3–50).
3 M. A. Robb, *Adv. Phys. Org. Chem.*, 2014, 48, 190–239.

Martin Centurion opened a general discussion of the paper by Vasilios Stavros: In Fig. 2a of your manuscript you mentioned a 900 ps lifetime, and there is also an ultrafast femtosecond component, which you attribute to multiphoton absorption. This fast peak does not appear as prominently on the experiments with the other two molecules shown in parts b and c. Can you explain why?

Vasilios Stavros responded: Both ConA and EF showed significant fragmentation and depletion of signal (where the action of the sum of the pump and probe pulses results in a decrease in the parent ion signal). In order to avoid this, lower pump and probe powers were used when recording TR-IY transients. The lower pump and probe powers used meant that multiphoton effects around time zero were reduced significantly in these transients, which is why ConA and EF do not display this very large signal spike around time-zero.

Russell Minns commented: Your study concentrates on the behavior of the parent ion and I wonder if you see any fragmentation in your spectrum and if this provides any extra information on the dynamics?

Vasilios Stavros answered: EF did indeed undergo photofragmentation, and several fragment ions were detectable, as we show in Fig. S5 of the ESI in our paper. However, given that we ran the TR-IY measurements in EF with a reduced pump and probe power, to reduce extensive fragmentation of the parent, we obtained quite modest signal intensities of these fragments. As a result of this, trying to carry out TR-IY measurements of these fragments resulted in very poor transients owing to the poor signal-to-noise levels. As such, no useful information could be drawn from the time-dependent behaviour of these fragment mass channels.

Andrew Orr-Ewing asked: Can you estimate the yield of triplet states in your ethyl ferulate study? What might the consequences be of a sunscreen component that has a significant propensity for triplet state formation?

Vasilios Stavros answered: Unfortunately, our experiments do not allow the measurement of quantum yields for each relaxation pathway. Such measurements of quantum yield are more amenable to transient absorption spectroscopy, such as transient vibrational absorption spectroscopy, where the extent of the bleach recovery may be used to extract quantum yields. As regards to significant triplet state formation in sunscreens, this is likely not ideal, given that there will be a long lived, and potentially reactive state. This excited-state molecule may, in turn, facilitate potentially harmful chemistry with other components within the multicomponent sunscreen, or, providing there is enough skin penetration, with skin cells. For this reason, off-the-shelf sunscreens also contain photostabilisers that can quench potentially reactive states.

Daniel Neumark asked: When molecules such as phenol are excited in the ultraviolet, dissociation *via* H atom loss is an important channel. This channel appears to be largely suppressed in the more complex systems that you have studied here. Is this because the H atom in question is hydrogen-bonded to a nearby heteroatom, or is it because the energy initially deposited into the aromatic ring *via* $\pi\pi^*$ excitation leaks out quickly into the side chains?

Vasilios Stavros replied: As you correctly point out, we believe that the dissociation *via* H atom loss is not a viable pathway in any of these molecules (MVP, ConA and EF) due to the intramolecular H bond between the OH and OMe groups, which results in too high (and wide) a barrier for H atom dissociation. This is analogous to previous observations in similar molecules, specifically the most recent work we carried out on the building blocks of lignin.[1] Here we demonstrated that adding an OCH_3 group *ortho* to the OH group promoted the intramolecular hydrogen bond, leading to a significant decrease in the H^+ signal measured in our total kinetic energy release spectra (TKER, for both guaiacol and syringol, *cf.* phenol). I note here that these observations led us to conclude that the intramolecular hydrogen bond weakens as the S_1 state deviates from planarity; in the present work, these more complex systems are planar in the S_1 state (see Fig. 7 in the manuscript). This is in agreement with our present findings of the low H^+ signal (from one-photon mediated O–H bond fission) in our measured TKER spectra.

1 J. D. Young, M. Staniforth, J. C. Dean, G. M. Roberts, F. Mazzoni, T. N. V. Karsili, M. N. R. Ashfold, T. S. Zwier, and V. G. Stavros, *J. Phys. Chem. Lett.*, 2014, **5**, 2138–2143.

Daniel Neumark remarked: You mentioned at the end of your talk that isomerization of EF might be one of the main relaxation pathways based on a growing spectroscopic signal at long times. Is that *trans*-to-*cis* isomerization in the side-chain, and does the rise time for this signal match any of the decay times reported in your paper?

Vasilios Stavros answered: Indeed, the suggested *trans*/*cis* isomerisation I briefly alluded to for EF in my short presentation would occur with respect to the C=C double bond of the side-chain. This pathway seems to be operative in the most recent solution phase measurements we have carried out for an EF/cyclohexane mixture. Whether we can link the 2.5 picosecond time-constant we propose for *trans*/*cis* isomerisation in the solution phase measurements to the 15 picosecond time-constant (τ_1=15 ps) in the gas phase can, in our opinion, only be tentative. This is largely based around our calculations (see Fig. 8), although we once again acknowledge that we cannot completely rule out this mechanism in the gas-phase. I'd also like to point out that for the solution phase measurements, the decay in the excited state absorption of the *trans* isomer does indeed match the rise in the appearance of *cis* isomer, as you would anticipate.

Andrew Orr-Ewing remarked: The attribution of the ~150 ps time constant to relaxation within the triplet manifold seems slow for internal conversion, and perhaps inconsistent with Kasha's rule. Is there a reason why the internal conversion might plausibly be so slow?

Vasilios Stavros replied: We recognise that this time constant is perhaps longer than what would be expected for internal conversion. However, we attribute this to the large energy separation between the first (lower in energy) and second triplet states. That being said, and in light of the most recent solution phase measurements we have carried out which imply that a *cis* photoproduct may be formed following photoexcitation around the S_1 origin in EF/cyclohexane, the first time constant may have a component of the *trans/cis* isomerisation pathway (although our calculations do point towards this pathway being inactive), whilst the second, 150 ps time constant, may be an overall reflection of the ISC/IC process, the latter of which is in line with one of the caveats we proposed in the manuscript.

R. J. Dwayne Miller opened a general discussion of the paper by Tom Penfold: The dynamics of going from the metal–ligand charge transfer band into the high spin quintet state for $Fe(bp)_3$ has been reported to be less than 150 fs and now updated to be less than 50 fs.[1] This time scale is extremely fast and it seems to involve very fast relaxation to the fully relaxed quintet state to explain the relatively narrow transient absorption in the UV region assigned to the quintet state. The absorption lineshape in which 2 eV of excess energy has been dissipated should be quite broad but it seems to be fully relaxed. I don't know what we learn from these time scales. The transition probability for intersystem crossing should depend on the spin–orbit coupling weighted by the Franck–Condon overlap connecting the singlet–triplet–quintet manifolds. From a single number, say 50 fs, how can we connect this measurement to the microscopic processes? Also the dynamics seem extremely fast. The assignments are based on optical spectra and the actual spin transition could be unrelated to the putative transitions. To date there has been no direct measurement of the spin transition (*e.g.* XES or MOKE probes) under excitation conditions low enough to avoid multiphoton ionization (which would create unpaired spins). I realize that you have studied a different Fe system, however, can you explain these extremely fast timescales based on your theoretical studies? Is a $<$50 fs timescale consistent with the spin–orbit coupling for such systems as found from your theoretical studies? Also, in the limit, the dynamics are pushed to $<$20 fs, at what point can we call the initial state a singlet? It is a heavily mixed state. Can you comment on this point?

1 G. Auböck and M. Chergui, *Nat. Chem.*, 2015, 7, 629–633.

Tom Penfold answered: The single timescale, say 50 fs, tells us that kinetics leading to the population of the quintet state is extremely fast. But this does not directly give us any insight into the microscopic processes responsible for driving this transfer. The timescale is so rapid that I would argue that to unravel the microscopic processes, quantum dynamics simulations such as those presented here, are possible and would very informative.

The transition between the states labelled ^{1}MLCT and ^{3}MLCT from their spin–orbit coupling free character in the model presented in our paper and in Ref. 1 is ~100 fs. This arises from the sizeable spin–orbit coupling and the small energy gap between the states. As spin–orbit coupling in these complexes will derive mostly from the iron 3d orbitals, I would not expect much difference in the size of spin–orbit coupling between iron complexes. However, the density of excited states in the low lying MLCT will increase significantly from $[Fe(CAB)_2]^{2+}$ to $[Fe(bpy)_3]^{2+}$. Consequently, I find a <50 fs time scale entirely consistent with our model.

In terms of describing the initial state as a singlet, my feeling is that even before the dynamics becomes less than 20 fs it is not rigorous to label the initial state a singlet. For both the present case, $[Fe(CAB)_2]^{2+}$, and $[Fe(bpy)_3]^{2+}$ there is a high density of ^{1}MLCT and ^{3}MLCT states, which are close in energy, have very good Franck–Condon overlap and sizeable spin–orbit coupling. In the case of $[Fe(bpy)_3]^{2+}$, the triplet and quintet metal centred states also intersect at or near the Franck–Condon geometry.[2] The size of the spin–orbit coupling and the small energy gap between these states means that a full description has to go beyond perturbation theory, *i.e.* Fermi's golden rule. It would be better to describe it as a MLCT manifold rather than attempting to distinguish between the two spin manifolds.

1 M. Pápai *et al.*, *J. Phys. Chem. Lett.*, 2016, 7, 2009–2014.
2 C. Sousa *et al.*, *Chem. - Eur. J.*, 2013, **19**, 17541–17551.

R. J. Dwayne Miller commented: The other question that is keeping me up nights is the connection of the high spin transition of Fe spin crossover materials and $Fe(bp)_3$ involving a specific bond displacement. Typically it involves an Fe–N bond elongation and various probes claim excitation of this mode is involved and critical to the spin transition. The Fe–N bond elongation is treated as a 'smoking gun' for the spin transition to high spin states. On what physics is this connection made (other than the high spin state is calculated to have an elongated Fe–N bond and the antibonding nature of this excited state). If the motion of the Fe–N mode couples to the electron density to modulate the orbital angular momentum then this would provide a fluctuating magnetic dipole moment on the molecular frame that would in turn provide the force to couple to the electron spin (much like a change in charge distribution gives rise to a fluctuating dipole moment that can couple to the electric field in absorption of a photon). Does this mode contribute to the spin–orbit coupling significantly? If not, there are a lot of modes that would act to dissipatively decouple the low and high spin states if the spin–orbit coupling was sufficient without this consideration.

Tom Penfold answered: The Fe-N bond elongation is totally symmetric and therefore should not have a significant impact on the magnitude of the spin–orbit coupling matrix elements. In our present work we have treated the spin–orbit coupling matrix elements as a constant coupling, using their values at the Franck–Condon geometry. While spin–orbit coupling enhanced by vibrational motion (spin–vibronic) can be important in many systems, I do not think it contributes significantly in this case.

Oliver Schalk commented: The model Hamiltonian in your study uses four different vibrational modes. Which changes in the dynamics do you expect if you were to consider more vibrational modes?

Tom Penfold responded: In this model we used excited state geometry optimisations and the symmetry of the molecule to identify the vibrational degrees of freedom, which would be expected to dominate following excitation.[1] The fact that we identified 4 normal modes is generally consistent with the excited-state dynamics of other Fe(II) complexes which are usually schematically represented with the vibrational motion reduced to one vibrational coordinate. Given the restricted nuclear configuration space in our model, we would obviously expect the accuracy of the simulations to reduce with time, as this model is unable to capture longer time effects, such as vibrational redistribution. However, this is not expected to change the mechanistic description of the dynamics. A similar model has been successfully used to describe the ultrafast dynamics of other transition metal complexes.[2–4]

1 M. Pápai *et al.*, *J. Phys. Chem. Lett.*, 2016, 7, 2009–2014.
2 G. Capano *et al.*, *J. Phys. Chem. A*, 2014, **118**, 9861–9869.
3 Y. Harabuchi *et al.*, *J. Chem. Theory Comput.*, 2016, **12**, 2335–2345.
4 M. Pápai *et al.*, *J. Phys. Chem. C*, 2016, **120**, 17234–17241.

Oliver Schalk commented: In smaller molecules (*e.g.* butadiene), several studies with a reduced number of vibrational modes have been performed and can now be compared to more advanced studies like "on-the-fly" studies which intrinsically include all vibrational modes. The reduced dimensionality models all seem to get part of the dynamics wrong and while some of the initial dynamics might be represented correctly, failures are observed on a 50–100 fs timescale, particularly due to the energy flux not being able to distribute over the molecule correctly. In that sense, what are the limitations implied by the reduced Hamiltonian in this study?

Tom Penfold replied: The limitations of any reduced coordinate space Hamiltonian is that one is restricting the nuclear motion/vibrational energy to a subset of modes, rather than letting it redistribute throughout the whole molecule as it should. In the case of small organic molecules, such as butadiene, this can be a significant problem as they can often be quite floppy in the excited state. Consequently, describing the dynamics within a small subset of normal modes could represent a significant approximation. This is obviously not the case for all systems and model Hamiltonians have been successfully used, most famously for pyrazine. The complex studied here, $[Fe(bmip)_2]^{2+}$ does not exhibit very large structural rearrangements and therefore lends itself nicely to a reduced space representation. The accuracy of any reduced space representation clearly hinges upon an suitable description of the important vibrational degrees of freedom. In the case presented here, due to the molecular symmetry and the relatively small structural changes, this choice is relatively straightforward. However, this is not always be the case. In such cases, an "on-the-fly" approach which includes all the nuclear degrees of freedom is clearly advantageous. The problem for these simulations, is that the computational expense associated with converging such dynamics for larger systems (>50 atoms) is very significant. I suspect that in such cases the best way to proceed is to use the initial "on-the-fly" simulations to inform the coordinate choice for model dynamics.

Albert Stolow asked: Do you have any general comments about the relative utility of pre-edge *versus* post-edge inner shell transitions in time-resolved X-ray absorption spectroscopy probes of molecular dynamics?

Tom Penfold replied: In terms of the post-edge, our simulations indicate that one will be able to extract important information about the femtosecond structural change, energy redistribution and vibrational cooling *i.e.* the position and width of the excited nuclear wavepacket.

We did not calculate the transient pre-edge peaks as we are unable, at present, to simulate the core–hole spectra of electronically excited states in a sufficiently computationally efficient manner to complete such calculations in a reasonable timeframe. However, the paper from this meeting by Lin Chen (DOI: 10.1039/C6FD00083E) illustrates that it is possible to extract some electronic information from the pre-edge, although it is nontrivial and, in comparison to your paper (DOI: 10.1039/C6FD00117C) hampered by the large core–hole lifetime broadening. This issue could be reduced and the information content significantly enhanced by using resonant inelastic X-ray scattering[1] and becomes increasingly feasible using high repetition rate hard X-ray FELs.

1 P. Wernet, *et al.*, *Nature*, 2015, **520**, 78–81.

Daniel Neumark said: Your calculated X-ray emission spectra appear to be much less sensitive to the time-evolving nuclear dynamics than the X-ray absorption spectra. Why is that the case?

Tom Penfold answered: This is because the $K\alpha$ and $K\beta_{1,3}$ X-ray emission that we calculated in the paper are core-to-core and inner valence-to-core transitions, respectively. The transition density is therefore quite localised upon the iron and therefore less sensitive to structural changes. This is not the case for $K\beta_{2,5}$ (valence-to-core) X-ray emission as we have shown previously.[1] This is very sensitive to the evolving nuclear dynamics. The challenge in this case is the small cross sections of these transitions, a limitation that can certainly be overcome with high-repetition rate X-FEL sources.

1 G. Capano *et al.*, *J. Phys. B*, 2015, **48**, 1–11 (2015).

Junko Yano remarked: Regarding the chemical sensitivity of the $K\beta_{1,3}$ spectra, the calculation shows the insensitivity of this transition to the MLCT in comparison to $K\beta_{2,5}$. However, the $K\beta_{1,3}$ does shows small changes, and thus there are some experimental advantages of using this transition to follow the changes. You mentioned that 10^{17} photons are required to get $K\beta_{2,5}$ with a reasonable signal to noise ratio. What is the number of photons that is required to get $K\beta_{1,3}$?

Tom Penfold replied: Yes there is a very slight change. It is approximately ~0.1% of the total signal, assuming 100% excitation yield. Using this signal strength, and the assumptions outlined in our previous work,[1] I would estimate one would require at least 10^{16} photons.

1 G. Capano *et al.*, *J. Phys. B*, 2015, **48**, 1–11 (2015).

Wolfgang Domcke addressed R. J. Dwayne Miller: The photon can only couple to the excited singlet states, even if the singlet and triplet states are strongly mixed by spin–orbit coupling. Without coupling to vibrations, we have an electronic few-level system and the photoinduced dynamics can only consist of periodic oscillations of the electronic population between the singlet, triplet and quintet states. Coupling to the vibrations is necessary for irreversible transitions from the singlet states to the triplet and quintet states to occur.

R. J. Dwayne Miller responded: I agree with your comment. I think the issue here is whether or not the dynamics intrinsically reflect the spin–orbit coupling and/or the Franck–Condon factor overlap that weights the transition probability. The problem is that the dynamics have been interpreted as solely propagated by one vibrational mode, the Fe–N mode typically for $[Fe(bpy)_3]^{2+}$ or other Fe-based spin crossover materials. The assignment is based on shifts or inferred absorption bands in which the spin transition was reported to be <150 fs and now it is less than 50 fs. Since the spin transition is not fully resolved as yet, at some point one needs to ask what is the mechanism for such fast spin transitions. If the Fe–N motion is involved in driving the electron spin state into the high spin manifold, which is along your statement, then the dynamics are too fast for this mode. Issues of the initial state preparation come into play much like the case of charge transfer bands *vs.* excited state electron transfer in interpreting photoinduced electron transfer dynamics. I should add that on a preliminary basis, we have now completed fs electron diffraction studies on two Fe-based systems and do not see the purported Fe–N bond elongation until the ps timescale.

In this respect, I rather like your comment in that it would mean that the very high density of vibrational modes, if the spin–orbit coupling is sufficient, would serve this relaxation role. The spin transition would then occur on the IVR timescale, consistent with the observed dynamics. It would involve many modes not just the Fe–N displacement. I would still like to see a direct measurement of the spin transition – THz/ESR or at least magneto-optical measurements to really pin down the dynamics of the spin transition. We need the spin–orbit coupling and information on the nuclear Franck–Condon overlap between spin manifolds. A single number does not give the operating physics. Thanks for this insightful comment.

Theo Keane asked: How mixed are the spin states in $[Fe(bmip)_2]^{2+}$ along the lowest energy pathway from the Franck–Condon region of the ^{1}MLCT state down to the final ^{3}MC state and how does this compare to the equivalent pathway in $[Fe(bipy)_3]^{2+}$?

Tom Penfold replied: The lowest singlet metal–ligand charge transfer (^{1}MLCT) excited states included in the model are heavily mixed, approaching 50%, with the close lying ^{3}MLCT states, but less so with the ^{3}MC states. This contributes to the slower population decay in the case of the latter. The mixing between the ^{1}MLCT and ^{3}MLCT states is comparable to that reported for $[Fe(bipy)_3]^{2+}$.

Gareth Roberts presented a comment: In the conclusions of our discussion paper (DOI: 10.1039/C6FD00068A), we allude to the fact that upon being embedded in double helix DNA, the nucleobases are subjected to additional chemical interactions which open new photochemical relaxation channels. Intrastrand base stacking interactions can open relaxation *via* long-lived excimer/exciplex states and photo-damage channels, such as (2+2) cycloaddition and (6-4) adduct production. Both have been studied in detailed over recent years.[1] Our research group has recently focused on the role of interstand dynamics through Watson–Crick (WC) base pairing, which opens the possibility of UV-induced hydrogen(H)/proton transfer across intermolecular H-bonds. The role of inter-strand proton or H atom transfer in double helix DNA has been debated for over five decades.[2] In its most recent guise, the debate has predominately stemmed from theoretical studies which postulated that ultrafast interstrand electron-driven proton transfer (EDPT) contributes to the prevention of mutagenic pho-tolesions in DNA excited by absorption of solar UV radiation.[3] Despite extensive prior study of DNA photodynamics,[1] the question of whether UV-induced EDPT is active in WC base pairs has until very recently remained contentious and contradictory experimental results have been published.[4–7]

In collaboration with the research groups of Professor Friedrich Temps (Kiel) and Professor M. Carmen Galan (Bristol), ultrafast transient electronic absorption spectroscopy (TEAS) and transient vibrational absorption spectroscopy (TVAS) have been utilised to track the decay dynamics of UV-excited Guanine-Cytosine (G•C) WC base pairs in chloroform solution.[7] As summarized in the top of Scheme 1, the results show direct evidence for the involvement of EDPT in the deactivation dynamics of the G•C WC pair. The mechanism that emerges from these measurements is a fully kinetically resolvable step-wise double H-atom transfer process. After excitation of G•C at 260 nm, TEAS monitored ultrafast internal conversion from the locally excited $^1\pi_G\pi_G{}^*$ state to a charge transfer (CT) $^1\pi_G\pi_C{}^*$ state and subsequent proton transfer, all of which takes place in <40 fs. This first EDPT process forms an intermediate G[−H]•C[+H] biradical species with a quantum yield of $\geq 0.6 \pm 0.1$, and is trapped in a local minimum on the CT state for 2.9 ps. Alternatively, the initially excited $^1\pi_G\pi_G{}^*$ state can be depopulated via monomer-like deactivation pathways with a quantum yield $\leq 0.4 \pm 0.1$. The majority of G[−H]•C[+H] biradicals decay via a $^1\pi_G\pi_C{}^*/S_0$ conical intersection back to the original WC structure. However, TVAS measurements in the carbonyl

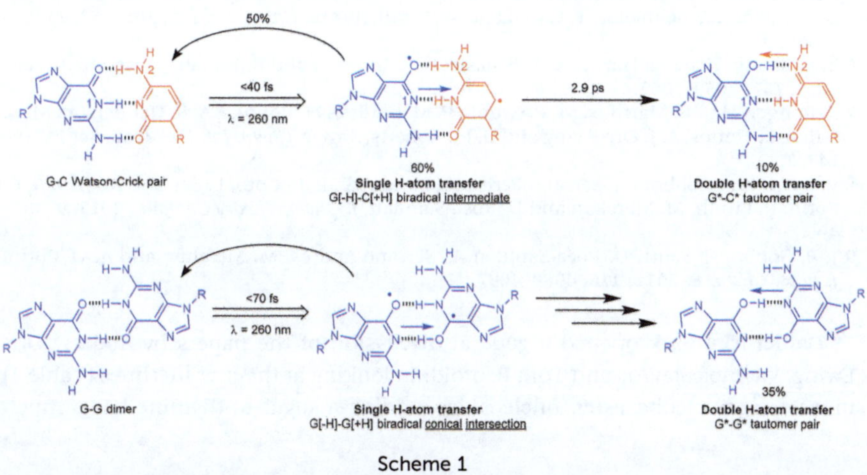

Scheme 1

stretching region reveal that up to 10% of the overall excited state population instead undergoes a second H atom transfer along the $N_{2,C}$–H••O_G H-bond to form the long-lived (>1 ns) double H atom transferred photoproduct (G*•C*).

The EDPT dynamics observed for G•C show clear evidence for a step-wise double H atom transfer pathway, which is kinetically resolvable with our chosen experimental probes. However, theoretical studies of double H atom transfer in WC base pairs have also considered the possibility of a concerted rather than sequential mechanism.[8,9] Although a step-wise process is clearly resolvable through our TEAS and TVAS for G•C, it raises the question whether (with currently available technology) two sequential ultrafast (<40 fs) EDPT processes could be distinguished from a truly concerted pathway? Our current studies of the H-bonded G•G homodimer in chloroform solution serve as a test-bed to illustrate this issue. The EDPT mechanism after 260 nm excitation of G•G is summarized at the bottom of Scheme 1. Provisional theoretical calculations indicate that a concerted double H-atom transfer mechanism is unfeasible in G•G, although a sequential G•G → G[−H]•G[+H] → G*•G* path is open at 260 nm. In contrast to the G•C pair, where the G[−H]•C[+H] biradical could be transiently trapped in an excited state minimum, the G[−H]•G[+H] structure instead lies at a point of conical intersection with S_0, with no local minima en route. After photoexcitation of G•G to a $^1\pi\pi^*$ state, we observe population decay out of the vertical Franck–Condon window within 70 fs through TEAS, with no observable spectral signature(s) for a G[−H]•G[+H] species. However, TVAS in the N–H stretching region clearly tracks ultrafast formation of a G*•G* double H-atom transferred photoproduct, with a lifetime >1.4 ns and a quantum yield of ~0.35. Although theory indicates double H atom transfer should be sequential in G•G, our current experiments are unable to resolve the ultrafast step-wise dynamics. Instead, it is experimentally observed as a quasi-concerted process, due to the nature of the excited state potential energy landscape along the sequential H atom transfer coordinate.

1 C. T. Middleton, K. de La Harpe, C. Su, Y. K. Law, C. E. Crespo-Hernandez and B. Kohler, *Annu. Rev. Phys. Chem.*, 2009, **60**, 217–239.
2 P. O. Löwdin, *Rev. Mod. Phys.*, 1963, **35**, 724–732.
3 A. L. Sobolewski and W. Domcke, *Phys. Chem. Chem. Phys.*, 2004, **6**, 2763–2771.
4 L. Biemann, S. A. Kovalenko, K. Kleinermanns, R. Mahrwald, M. Markert and R. Improta, *J. Am. Chem. Soc.*, 2011, **133**, 19664–19667.
5 D. B. Bucher, A. Schlueter, T. Carell and W. Zinth, *Angew. Chem. Int. Ed.*, 2014, **53**, 11366–11369.
6 Y. Y. Zhang, K. de La Harpe, A. A. Beckstead, R. Improta and B. Kohler, *J. Am. Chem. Soc.*, 2015, **137**, 7059–7062.
7 K. Röttger, H. J. B. Marroux, M. P. Grubb, P. M. Coulter, H. Böhnke, A. S. Henderson, M. C. Galan, F. Temps, A. J. Orr-Ewing and G. M. Roberts, *Angew. Chem. Int. Ed.*, 2015, 54, 14719–14722.
8 V. Sauri, J. P. Gobbo, J. J. Serrano-Pérez, M. Lundberg, P. B. Coto, L. Serrano-Andrés, A. C. Borin, R. Lindh, M. Merchán and D. Roca-Sanjuán, *J. Chem. Theory Comput.*, 2013, **9**, 481–496.
9 J. P. Gobbo, C. Sauri, D. Roca-Sanjuán, L. Serrano-Andrés, M. Merchán and A. C. Borin, *J. Phys. Chem. B*, 2012, **116**, 4089–4097.

Daniel Neumark opened a general discussion of the papers by Andrew Orr-Ewing, Vasilios Stavros and Tom Penfold: In looking at the nπ^* lifetimes (Table 4) in water of nucleobases *vs.* nucleosides, adding a sugar to thymine has a much larger effect than to cytosine (127 ps *vs.* 37 ps). What is the origin of such a large effect?

Andrew Orr-Ewing responded: The lengthening of the lifetimes of the $^1n\pi^*$ states is likely to be a consequence of their destabilization, and hence a raising of the energies of conical intersections with the electronic ground state to which they decay. This effect is evident in different solvents: the protic solvents can hydrogen bond with the non-bonding electrons in the carbonyl groups of the pyrimidine nucleosides, and excitation of one of these electrons to a π^* state (giving the $^1n\pi^*$ configuration) weakens this stabilizing interaction. Destabilization of the $^1n\pi^*$ state may also arise when an alkyl or ribose group replaces the N_1 hydrogen atom in either cytosine or thymine, although the effects apparently differ as the question notes. A quantitative understanding will be best achieved through high-level *ab initio* electronic structure calculations that map the energies and locations of conical intersections between the $^1\pi\pi^*$, $^1n\pi^*$ and ground states, and examine the effects of the substituent on the N_1 atom. Some progress towards this goal was reported in Ref. 39 of our paper.[1]

1 S. Blaser, M. A. Trachsel, S. Lobsiger, T. Wiedmer, H. Frey and S. Leutwyler, *J. Phys. Chem. Lett.*, 2016, 7, 752–757.

Oleg Kornilov remarked: The authors conclude that the primary role of the solvent is to destabilize the $n\pi^*$ and this way influence the triplet state yields. Can this conclusion be connected to the dynamics observed in DNA? Do $n\pi^*$ states of nucleobases in DNA or base pairs play a similar role as in the isolated nucleobases?

Andrew Orr-Ewing responded: We need to be cautious in translating the results of our studies on solvated nucleosides to the photodynamics of double-stranded DNA for two reasons. One is that the environment in the core of the DNA strand differs markedly from chloroform or aqueous solutions. Consequently, the degree of destabilization of the $^1n\pi^*$ state will differ, with consequences for $^1n\pi^*$ state lifetimes and triplet state yields. The second reason is that the chromophore excited in a DNA molecule may differ significantly in character from that in an isolated nucleobase (for which we assign a localized $\pi^* \leftarrow \pi$ transition). Within a DNA strand, the nucleobases are stacked closely, and electronic excitation is likely to be delocalized over multiple bases (*i.e.* the UV absorption prepares an excitonic state, with character that will depend on the precise base sequence). The subsequent ultrafast photodynamics may therefore differ, and are an active field of investigation. Nevertheless, we believe that the exploration of the photodynamics in isolated, solvated nucleobases is interesting because it offers deep insights into available deactivation pathways, helps to explain the stability of the nucleobases to UV radiation exposure in the pre-biotic Earth, and provides excellent candidates for fundamental studies of non-adiabatic photochemical dynamics.

Russell Minns commented: For the hydrogen transfer measurements, are the measurements sensitive to the ground and excited electronic states? If they are sensitive to both, can you differentiate which parts of the dynamics occur on the excited and ground electronic states?

Andrew Orr-Ewing responded: The TVAS measurements are, in general, sensitive to molecules in both the ground and electronically excited states. Indeed, this is one of the strengths of the technique because vibrational frequencies of functional groups such as the carbonyl and N–H stretches used in our study change with the electronic state of the molecule (because vibrational force constants are determined by the bonding character). The vibrational frequencies are also modified by intermolecular interactions such as solvation or hydrogen bonding. We can therefore use TVAS both to resolve dynamics on ground and excited states, and in principle to explore how solvation modifies the properties and dynamics of these states.

Theis Sølling asked: Did you see any indication that the triplet yield increased in the $CHCl_3$ because of a solvent heavy atom effect?

Andrew Orr-Ewing replied: The triplet yield in 2′-deoxy-thymidine is modified by the chloroform solvent (see Table 4), but we do not think it is a result of an intermolecular heavy-atom effect on the singlet–triplet spin–orbit coupling interaction. Instead, we attribute this change to modifications to the intersections between the $^1n\pi^*$ and T_1 ($^3\pi\pi^*$) state caused by solvent-induced shifts of the $^1n\pi^*$ state, and perhaps also to the rate of vibrational relaxation within the $^1n\pi^*$ state.

Gareth Roberts commented to Theis Sølling: In extension of Andrew Orr-Ewing's response, one additional way in which triplet state features could be confirmed in our TVAS measurements on thymidine in chloroform, is to perform comparative quenching experiments in both Ar purged and O_2 saturated solutions. Although we have not presented these in our paper, previous TVAS studies of both thymine and thymidine in acetonitrile-d_3 by Kohler and co-workers (Ref. 11 in our paper)[1] have taken this approach and indeed confirm the quenching of triplet marker bands at 1603 and ~1700 cm^{-1} in the carbonyl stretching region. These triplet bands are at similar wavenumbers to the bands we assign to a triplet state in chloroform (see Fig. 6 in our paper), which together with our frequency calculations gives us confidence in our assignment.

1 P. M. Hare, C. T. Middleton, K. I. Mertel, J. M. Herbert and B. Kohler, *Chem. Phys.*, 2008, **347**, 383–392.

Theis Sølling addressed Andrew Orr-Ewing and Vasilios Stavros: It seems like the common denominator of the three presentations was the multiplicity of the states involved. How was the multiplicity determined?

Vasilios Stavros responded: For these types of measurements, the presence of triplet states could be determined by measuring, for example, phosphorescence. We attempted to measure phosphorescence in our experiment by cutting out the fluorescence in our signal (the addition of filters that absorb radiation).

Andrew Orr-Ewing responded: We use both kinetic and spectroscopic information to identify IR bands of triplet-state molecules in our transient vibrational absorption spectroscopy measurements. We first identify IR bands of the ground electronic singlet state (S_0) by steady-state FTIR spectroscopy of solutions of the molecule of interest. The T_1 state commonly has a much longer lifetime than excited singlet state molecules, so we then use the time constants for decay of transient bands (*i.e.* those that are not associated with the S_0 state) to guide identification of triplet state features. In the case of 2′-deoxy-thymidine, a >1 ns lifetime for certain bands is indicative of T_1 state population. We also use high-level electronic structure calculations to assist in the assignment of transient IR spectral bands associated with excited singlet and triplet states. The calculated harmonic band frequencies generally need to be multiplied by a scaling factor to account for anharmonicity and the effects of a solvent. We determine the linear scaling factors by comparison of calculated frequencies with those measured by steady state FTIR of S_0 molecules, and apply the same scaling factors to computed frequencies for excited-state absorption bands. This procedure, in combination with the kinetic information, is usually sufficient to identify vibrational bands of excited singlet and triplet states.

Faraday Discussions

PAPER

Ultrafast imaging of photochemical dynamics: roadmap to a new conceptual basis for chemistry†

R. J. Dwayne Miller[ab]

Received 25th November 2016, Accepted 28th November 2016
DOI: 10.1039/c6fd00241b

1. Introduction – from static structures to a dynamical view of chemistry

One of the long standing objectives in chemistry has now been realized. We can watch directly atomic motions during the defining moments of chemistry.[1,2] Given the rare nature of barrier crossings, we currently rely on optical methods to trigger the chemistry of interest on the 100 femtosecond time scale, *i.e.* current observations are limited to photochemical processes ergo the use of photochemical as a qualifier for this Faraday Discussion. This meeting was dedicated to exploring the latest advances in atomically resolved reaction dynamics to expose current issues and to chart a course for both theory and experiment to fully exploit the information content from this class of experiments. We are now at the fundamental space–time resolution effectively needed to image chemistry.

What new can be learned from the direct observation of chemistry? It is not about reconfirming what we already know about chemistry. As with all things, once you observe a phenomenon at higher and higher spatial and temporal resolution, you start to see things you did not see before and gain a deeper understanding of the phenomenon. There is always a surprise. Observing chemistry on the primary atomic timescales is no different. There is great

[a]*The Max Planck Institute for the Structure and Dynamics of Matter, The Hamburg Centre for Ultrafast Imaging, Luruper Chausse 149, Hamburg 22607, Germany. E-mail: dwayne.miller@mpsd.mpg.de*

[b]*Departments of Chemistry and Physics, University of Toronto, 80 St. George Street, Toronto, Ontario L5G 3J6, Canada*

† Electronic supplementary information (ESI) available: Movie 1 From R. J. D. Miller, *Annu. Rev. Phys. Chem.*, 2014, **65**, 583–604. Copyright © 2014, Annual Reviews. Movie 2 From M. Gao, C. Lu, H. Jean-Ruel, L. C. Liu, A. Marx, K. Onda, S. Koshihara, Y. Nakano, X. Shao, T. Hiramatsu, G. Saito, H. Yamochi, R. R. Cooney, G. Moriena, G. Sciaini and R. J. D. Miller, *Nature*, 2013, **496**, 343–346. Copyright © 2013, Rights Managed by Nature Publishing Group. Movie 3 From T. Ishikawa, S. A. Hayes, S. Keskin, G. Corthey, M. Hada, K. Pichugin, A. Marx, J. Hirscht, K. Shionuma, K. Onda, Y. Okimoto, S. Koshihara, T. Yamamoto, H. Cui, M. Nomura, Y. Oshima, M. Abdel-Jawad, R. Kato and R. J. D. Miller, *Science*, 2015, **350**, 1501–1505. Used with permission from AAAS. See DOI: 10.1039/c6fd00241b

potential for research in this area to provide a new conceptual basis for chemistry – one in which the fundamental connection between structure and dynamics is considered in a unifying manner.

In this respect, I have taken the opportunity in writing this summary article to give my perspective on this question and personal motivation for entering this field with the goal of achieving the quantum limit in space–time resolution for imaging chemistry. This effort now spans nearly 30 years for my group, with the ultimate goal to resolve the structure–function relationship of biological systems. This issue is related to how these systems so efficiently transduce stored chemical potential into doing work on the surroundings, *i.e.* breathe life into otherwise inanimate matter.[3–5] Our original observations indicated that the protein response function couples to collective modes as the most efficient means of coupling chemistry into biological functions. After many different experiments exploring different time length scales, energetics, and probes, we demonstrated this is indeed the dominant mechanism.[3,4,6,7] However, we could only guess at what the collective modes were and how they were encoded in the protein structure. These systems are such complex high dimensional problems that the only way to come to a first principles understanding of the connection between chemistry and biology would be to directly observe proteins in action at the atomic level. It was this rather simple observation, in retrospect, that motivated a nearly 30-year effort to image chemical and biological processes at the atomic level. We now have our first full atom resolved chemical reaction and can now directly see how chemistry scales with increasing complexity up to the level of proteins.

There is a principle of reduced dimensions that naturally occurs through the inherent space–time coupling of molecular motions. It is through these observations I believe we are on the cusp of a new dynamical structural basis for chemistry that explicitly includes the many body nature of the problem and recasts chemistry into a reduced set of key reaction modes for different classes of reactions. Nature has exploited this effect on an even grander scale. This new conceptual basis could change the way we think about chemistry in terms of controlling barrier heights, as well as providing more efficient theoretical sampling of reaction space to enable interactive guidance to synthetic strategies. To put it simply, chemistry is defined by the atomic motions occurring during barrier crossing events. We can now see these processes directly. From the time dependent velocities and known effective masses for the motion, we can directly determine potential energy surfaces in a single measurement, *i.e.*, directly observe reaction forces – the very essence of chemistry. This information will necessarily lead to higher levels of optimization for harnessing these forces.

Historical overview: space–time conundrum in understanding chemistry

In way of background, it is useful to trace the steps that led us to the present status in imaging chemistry to better appreciate what we should expect in terms of providing new insights into chemistry. There has always been an understanding that structure and dynamics are important but there has been a tension in terms of which observable provides more useful information for guiding our thinking. Since the very foundation of the atomic theory for chemistry was laid, we have been fascinated with the atomic details of matter. It is even difficult to imagine that there was a time when the idea of an atomic basis for matter was considered to be theoretical conjecture. We have to go back to the time of Dalton in which matter was established to be conserved (for chemical processes) and the early work of Max Laue as recounted by Eckert[8] and W. Lawrence Bragg[9] that definitely showed the atomic structure of matter using the newly discovered X-ray sources.[10] There is no better example of the connection of atomic structure to chemical processes than the work of Lonsdale that showed that all 6 carbon atoms of benzene lie in a plane;[11] hardly a chemistry student does not know the story of Kekule's proposed structure of benzene and the origin of the extra stability or aromaticity of benzene (based on Kekule's account of a dream he had of a serpent slithering to form a ring within the benzene structure).[12] The structure of benzene led to the concept of electron delocalization and sharply focused discussion on the relationship between time scales and structure. The debate centered on whether the structure of benzene consisted of 3 double bonds and 3 single bonds in which rapid interconversion between two equivalent forms occurred. The determination of the structure of benzene and many more molecular structures contributed greatly to Pauling's formulation of the hybridization of atom centered orbitals[13] and the Valence Shell Electron Pair Repulsion Theory of Gillespie to specifically take into account the effect of neighboring nuclei on electron correlation energies and corresponding electron probability distribution.[14,15] The celebrated argument between Olah and Brown on radical cations is yet another example of questioning whether the chemistry involves rapidly interconverting equivalent intermediate structures or new minima involving electron delocalization.[16] We see a constant struggle in terms of trying to rationalize nuclear arrangements with its effect on electron distributions and the time needed for creating local stable minima along a reaction pathway. The debate revolves around the connection of structure to the underlying dynamics. This time–spatial relationship for understanding chemistry goes back to the earliest days of formulating an atomic basis for chemistry. Structures were known to us. The dynamics were not.

In parallel, to formulating a structural basis for chemistry, there were very important developments in trying to understand the time scales of chemistry with respect to reaction kinetics. The work of van't Hoff and Arrhenius revealed the barriers to chemical processes and showed the exponential dependence of reaction rates in terms of the statistics of barrier crossing.[17] Since this seminal work, it has been appreciated that chemistry is really a race against time. It has long been recognized that it is the manipulation of barrier heights that give chemists such exquisite (exponential) control over different types of chemistry in our quest to make virtually any imaginable molecule. We obtain the highest yield for the process that occurs fastest, and not necessarily the one leading to the most thermodynamically stable state. The work of Eyring and M. Polanyi towards absolute rate theories ultimately led to the formulation of transition state theory and the general concept that there are particular nuclear arrangements that define the barrier height, or mathematically the minimum nuclear flux point in nuclear configuration space.[17,18] Since then, the one unifying conceptual basis across all fields in chemistry is this idea of a transition state through which one can gain control over the chemistry. We try to imagine what this rare moment in space–time looks like and to introduce changes in structure to lower or raise the barrier to arrive at the desired chemistry.

The interesting conundrum we have is that we base our present view of chemistry, and the dynamical exploration of nuclear configurations to achieve the near mythical transition state region, on static structures. Chemistry is by definition dynamic; yet we use static structures to project to the barrier region. This present situation has arisen from the great success in correlating material properties with static structures and our previous inability to determine the structural changes occurring during the fleeting moments involved in barrier crossings. How is it that we can get away with describing what is purely a dynamical phenomenon (aka chemistry) with the longest lived structures within a constant blur of motion? Or do we? There is no reason to believe it is even possible to project from any given minima or static structure to the narrow region in nuclear configuration space defining the barrier region that governs the chemistry of interest. The simple rules and general reaction mechanisms found empirically by synthetic chemists seemingly defy statistics.

This issue has perplexed me my entire research career. To put this problem in context, imagine a box with a large number of marbles (N) of different size, shape and colour in which there is one spatial location in which you would like to localize one particularly coloured, say blue, marble to represent the sampling of the transition state for molecular systems. (To make this analogy more closely related to the chemical problem, the spatial location for trapping the blue marble would change depending on the specific arrangement of the other marbles by some hidden rules.) You will find by shaking the box to model the thermal sampling that there are a very large number of permutations. We can formalize this as either a dilute or condensed system with respect to cross section of the particles and the area available for motion to give the precise probability for sampling the designated position. The chances of getting the blue coloured marble at this one site is exceedingly small for $N > 100$ and would be short lived. This gives some idea of these rare events in chemistry. Now pick one random starting point for the coloured marble. My main point should be made clear by trying to predict the trajectory by which the coloured marble would arrive at the defined transition state region. There is no clear pathway nor arrangement of the other marbles at this position. You would not expect to be able to predict the pathway or arrangement of particles at this point from any given initial starting point. In this problem statement, all the motions are uncorrelated. One way to solve the problem is to impose correlations on the sampling of different configurations to reduce the number of possible arrangements. Clearly the formation of molecular bonds strongly correlates and maintains fixed spatial relationships. We can repeat the thought experiment with a coloured marble connected to some subset of marbles as beads on a string. Again is this a sufficient condition to allow static structures to be used to predict the structural relationship at specific points in configuration space? If you repeat the exercise with spatially correlated particles, you will have reduced the problem to a smaller effective N but the pathway to a given position (in nuclear configuration with respect to particle configurations within this analogy) is still quite complex. Nevertheless, we rely heavily on the static structure of reactants to project to critical points. How can we use static structures as inputs to thinking about nuclear configurations in barrier crossing events?

The problem becomes more complex as we increase N. It is clear that molecular properties depend on the specific 3D structure of the molecule. (This is the entire basis of the pharmaceutical industry.) Within transition state theory, there are a number of key assumptions that merit scrutiny in this regard. (1) The transition state is defined mathematically to be the minimum flux point. (2) The energy is assumed to be equipartitioned within all degrees of freedom, and (3) the passage from reactant to product surface is assumed to be unity at the reaction saddle point or transition state.[17] The problem with the assumption of unit probability for reaction is that it does not take into account friction and recurrences at the critical point so that TS theory always overestimates reaction rates.[17] One can argue about how to define the transition state region and statistical distribution. The one point that stands out is the assumption of equipartitioning of energy in all modes in terms of the statistical sampling this region. If all modes were equally coupled to the reaction coordinate, this condition would be met for all points in nuclear configuration. If this condition was rigorously true, then every molecule would show different chemistry. We would not be able to find such generalized reaction mechanisms as each molecule would have a unique many body potential, unique properties, and by extension would have significantly different barriers to desired pathways and more importantly other competing pathways steering the system away from a given chemical pathway. In this respect, there is always a caveat on the condition of equipartitioning of the sampling process to all modes of the system. If there are some motions that are more strongly coupled to the reaction coordinate (as is the case – *vide infra*), statistical fluctuations from this condition will be amplified. This coupling would represent the hidden rules in the above analogy. How indeed have we, over the course of time, found chemical reactions to be transferable from one molecule to another for a given class of compounds? Chemists have discovered generalized reactions that hold true for a wide range of systems much like physics has found generalized equations to describe the operating physics within a range of boundary conditions. The two must meet. We are missing what must be some simple physics or unifying principle that has allowed us to find such generalized reaction mechanisms.

The scaling to more complex systems is yet another question. The number of independent degrees of freedom to arrange the atoms, or dimensions to chemical problems, scale as $3N$ where N now refers to the number of atoms. Current objectives in chemistry involve targets often involving hundreds of atoms, verging on the mesoscale to emulate biological systems. The increased complexity with requirement for unique properties for such targeted synthesis necessitates that the details of the many body potential matter. Again, this statement begs the question in terms of the underlying physics that makes chemistry transferrable to arbitrarily complex systems.

In scaling to higher levels of complexity, there are lessons to be learned from biological systems. For example, let us scale up the problem to the size of a protein ($N > 1000$). Mother Nature has found solutions by which these large molecular systems are able to transduce relatively small amounts of stored chemical potential into particular functions. There must be an enormous reduction in dimensionality to couple chemistry to function on this scale. The number of possible permutations, especially within the context of a highly associated or glass-like system,[19] are astronomical and with any reasonable sampling time seem to contradict the time scales by which biological systems are observed to execute their functions. In this context, it is important to note that one of the definitions of a living system is that it must be catalytic.[20] The evolutionarily optimised structure leads to control over barrier heights by which processes occur faster than without the biological construct. This observation would appear to be at odds with the requirement that protein structures must be robust to have reproducible properties (extension of the required property of self-replication for living systems). Such highly associated constructs should behave as glassy systems, which by the very nature of glasses would greatly retard motions and timescale for sampling barriers and directing chemistry into functions.

This problem in understanding how such large molecules can direct chemistry can be recast following the same logic as Levinthal's paradox for trying to rationalise protein structure and the millions of different structures connected to different biological functions.[21] The number of possible nuclear configurations is so enormous that it would seemingly preclude the possibility of a protein finding its active structure, never mind executing its functionalized chemistry. For example, consider the structure of a protein with N amino acids. Each amino acid can rotate along the protein backbone for which there are 3 near degenerate orientations (within kT). The total number of permutations is 3^N, just considering this one rotational coordinate. For even a relatively small protein of 100 amino acids, this corresponds to 3^{100} or 10^{47} different possible permutations along a single rotational coordinate for each amino acid of which there are many other possible types of motion. If each possible configuration was sampled at the fastest possible sampling rate (10^{13} s^{-1}, less than half period for the rotation), it would take longer than the age of the universe for the system to find its minimum energy structure. This paradox, as with all paradoxes, is used to point out the problem with a common assumption. The system clearly does not have equal weighting or resident times in sampling possible nuclear configurations. This obvious fact is not taken into account in the way the problem is posed. One could ask very similar questions about forming an ice cube, with even more atoms. Once the potential energy minimum is formed to create a nucleation site for the formation of ice, there is a strong driving force for additional water to bind and conserve the lattice symmetry. There are high barriers to explore other configurations so the water spends an exponentially longer time in the ice configuration, which only grows in time ($T < 273$ K). Here a clear distinction is needed. If protein structures were so rigid, as in the counter example of ice, we would be dead. The real issue with Levinthal's paradox is that protein systems must be mutable in order to respond to changing environmental conditions. Thus, they must have an active structure with only small barriers separating different conformations such that the system is only marginally bound within kT. This distinction is what makes Levinthal's paradox so profound. There must be a driving force sufficiently strong to reproduce the protein's active structure, yet provide thermally accessible minima to enable function. The two conditions would seem to be mutually exclusive. However, if there were spatial correlations of the protein fluctuations on some comparable length scale of the protein itself, it would reduce the number of minima from a myriad of possible structures to a smaller subset, evidently to the limit of just a few biologically active forms. These spatial correlations represent again an enormous reduction in dimensionality of the problem, from 1000s of degrees of freedom for fully uncorrelated motion of the atoms (as in the above marble problem) to a few strongly correlated motions. The basic physics of these inherent spatial correlations has been discussed within the context of the collective mode coupling model for protein function.[3,4,6,7]

Returning to our coloured marble analogy involving fully uncorrelated motions, we have now basically deduced that there must be inherent spatial correlations or colored noise on the sampling process for barrier crossing for systems as large as proteins. However, the degree of correlation and specific subset of motions out of the enormous number of degrees of freedom are not known. This point is made to show that we cannot take static structures and predict the nuclear configuration(s) defining the barrier crossing region. There are too many possible configurations and we are missing information on how strongly coupled the different types of nuclear motion are to the electronic degrees of freedom, the convolution of which determines a dynamic potential energy surface. It is impossible to imagine all the possible trajectories and the effective potential. To give yet another analogy, it is like hiking from one valley to another in the mountains where you represent the effective mass of some heavy reduced coordinate depicting motion along the potential energy surface of the mountainous topology. The difference is that your every step, changes the landscape. It would be very difficult to predict the lowest barrier route to go from one valley to another as the landscape continually changes with your very motion.

Given the complexity of the problem, how then have our colleagues in synthetic chemistry managed to make nearly any imaginable molecular structure up to the nm scale of molecular architecture? In the classic exercise of retrosynthesis, various reaction mechanisms are explored to create the required molecular stepping stones to the final product. Over the last 100 years, hundreds of reaction mechanisms have been found empirically using synthetic or molecular probes of the process.[22] With the understanding that changes in molecular structure affect the barrier height to the desired process, it is possible to make remarkable progress in understanding qualitatively reaction mechanisms to the point of establishing complex synthetic routes—with remarkable overall yields. Within the spirit of the above discussion of the myriad of possible nuclear configurations leading to the barrier region controlling the chemistry, how is this possible? Consider the typical strategy involved. The desired product structure is the objective. The input reactant structures at each step are known or assumed. From the difference in structure, one can readily discern the minimum atomic displacements needed to create the product structure. In chemistry, we are trained to imagine the structure of the transition state, as the half way point along the reaction path assumed in the above construct. This exercise is central to chemical means to control barrier heights (see Fig. 1 for a tongue-in-cheek presentation of this classic thought experiment). The intellectual challenge is to use various known reaction mechanisms to direct these motions. The reaction forces are controlled by introducing various groups and rearranging atomic positions to manipulate barrier heights. The qualitative effect on barrier heights is typically envisioned through arrow pushing exercises to describe how the electrons will rearrange in the presence of electron donating or withdrawing groups and relative stability of intermediate states along the overall reaction pathway. Reactive complexes are inferred as lowest energy barrier configurations based on similar ideas. This reasoning is based only on very qualitative input; yet it works!

Presently, quantum chemistry calculations are conducted to better assist in planning synthetic routes but the majority of chemistry has historically been developed through the above thought process. There is an important lesson here.

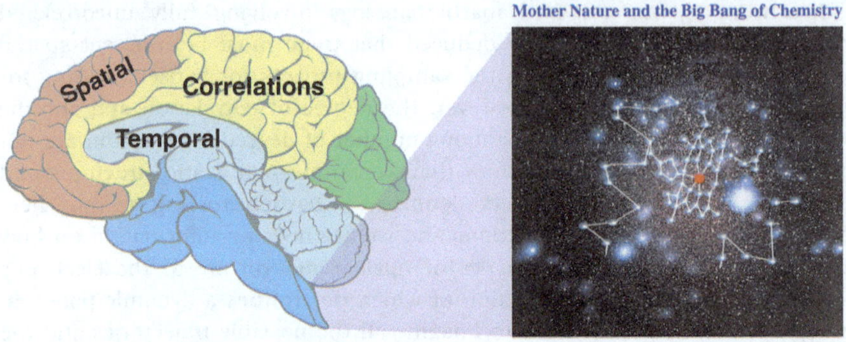

Fig. 1 Dissecting a Chemist's brain. This figure shows the classic thought experiment all chemists are trained to execute in imagining the critical moment in space time when a chemical event occurs. Here, the particular case chosen is oxygen dissociation or release from myoglobin (my favourite part of the universe where I spend a lot of time thinking about these motions). The star-like objects are used to represent heavy atoms with the giant red star as the central Fe atom of the heme binding site. The bond breaking event is modeled as a super nova-like event with the dumbbell shaped diatomic O_2 whisking out of the picture. The motions in this figure are derived using conventional transition state concepts. From the difference in static structure of the reactant and product states, it appears the motion is all along the normal to the heme-porphyrin ring in which highly sequential motions are envisaged transmitted along the proximal histidine in response to this force. The actual process involves coupling to collective modes of the protein and porphyrin ring as well as heme doming coordinates. The specific protein motions are still an open question. (Reproduced with permission from https://www.vecteezy.com © pauldizonr; inset modified from ref. 4; R. D. Miller, *Can. J. Chem.*, 2002, **80**, 1–24. © 2008 Canadian Science Publishing or its licensors. Reproduced with permission.

The methods used to conceptualize the chemistry focus on the largest net atomic motions needed to make the new structure and the forces considered in directing the atomic motions are viewed on a local basis. There is no consideration of spatially delocalized motions involving the entire many body potential of the molecule or other competing trajectories – only the principle motions needed. This is a highly localized view of chemistry. Some spatial delocalization of the reaction forces are taken into account in terms of the change in electron distribution and associated degree of bonding but this is really a secondary consideration. The high degree of success of this approach cannot be denied. Synthetic probes of chemistry have empirically discovered that the process of barrier crossing must distil down to a small subset of modes that lead to localized motions. There are undoubtedly other important motions and nuclear configurations involved that affect the barrier crossing but evidently not to the point to change the balance of dynamic flux across barriers of other competing pathways. It is this enormous reduction in dimensionality in the barrier crossing region that enables chemistry to be transferrable from one molecule to another. Further, it needs to be emphasized that the success of this approach demonstrates that it is not essential to have a high level accuracy on the details of the potential energy surface or reaction coordinate. The relative relationship of the desired process relative to other possible pathways and barrier heights needs to be preserved to direct chemistry along the desired pathway.

The main point of the above discussion is to illustrate that there is both considerable empirical evidence and straightforward deductive reasoning using simple analogies that chemistry reduces to a few dominant types of motions or modes. Even within this reduced dimensionality, the actual modes involved are not known. The number of possible permutations of all possible motions that could direct the system from the reactant to product surface are too many to realistically project the initial structures to transition state structures to truly understand what needs to be done to optimally direct the chemistry of interest.

With this in mind, it can be readily appreciated that one of the dream experiments in chemistry would be to directly observe atomic motions during the barrier crossing event to see the specific motions involved (Fig. 1). This information would provide new insight to optimally directing chemistry within the myriad of possibilities. For this reason, there has been a long fascination in going to higher and higher spatial and time resolution to follow chemical processes. Thanks to the perturbative approaches introduced by Porter[23] and Eigen,[24] we have been able to push the time aspect in the observation of chemistry from milliseconds to nanoseconds and beyond. However, the corresponding spatial resolution always has been the most imposing hurdle. Until the last 10–15 years, I think it is fair to say that it was largely believed that it would be impossible to attain – simultaneously – sufficient spatial and temporal resolution to directly observe atomic motions. The primary processes of chemistry were viewed to be immeasurably fast and one would always have to infer the underlying dynamics from indirect measurements. The above exercises in conceptualizing atomic motion would remain the purest form of a Gedanken experiment.

From Gendanken experiment to realization of atom resolved chemistry

To appreciate the technical challenge to directly resolving the atomic motions involved in chemistry, consider the challenge to build a molecular movie camera to capture chemistry on-the-fly. First, consider the shutter speed required for the simple act of bond breaking. How long does a bond take to break? To answer this question, you have to define when the bond is broken. For typical potentials, the displacement of atoms one additional bond length past the equilibrium position places the atoms within kT of the continuum. We can use this as the point of bond breaking. Using a bond length of 1 Å and for atoms moving along the dissociation coordinate at the speed of sound to approximate thermal fluctuations, we arrive at a required shutter speed of 10^{-13} s or 100 fs ($t_{\text{diss}} = 10^{-8}$ cm/10^5 cm s^{-1}). The required time resolution is on the order of 100 fs.[5] This time scale should be familiar as it corresponds to the typical prefactor used in the Arrhenius rate constant expression for unimolecular reactions. There are faster processes involving repulsive potentials. For example the photoinduced bond breaking of Fe–CO in myoglobin and in the ring opening of cyclohexadiene occur within 25 fs.[6,25] There are faster nuclear motions such as the O–H vibration of water, which defines the fastest nuclear motions. The period of the OH vibration is 10 fs but the rms motions are less than 0.05 Å, *i.e.*, within harmonic limits of a bound potential, which are not chemically relevant. As in the bond breaking example, chemistry is determined by far from equilibrium motions with net displacements typically larger than 0.1 Å, *i.e.*, beyond harmonic approximations to the local potential. (This point is crucial to understanding the extremely strong anharmonic coupling in this region.) In order for a chemical process to occur, nuclear fluctuations need to sample the reaction saddle point, or be optically prepared at this point, and the rate of passage depends on coupling to dissipative modes that collapse the wavefunction on the product state. There will be a distribution of time scales depending on the particular reaction. However, the overall dynamics are dictated by rearrangements of both intramolecular and intermolecular coordinates, which are dominated by dissipation involving low frequency modes in the highest density of states region. These modes undergo the largest amplitude motion and have the highest degree of anharmonic coupling in relation to the high density of near degenerate states to conserve energy and momentum in the relaxation process to the new configuration.[26,27] Whether one is talking about solvent reorganization involved in electron transfer, isomerization, ring closing *etc.*, the most highly damped modes undergoing the largest amplitude motion are the low frequency intramolecular and intermolecular modes on the order of 100 cm^{-1}.[1] The half period motion of such modes is again on the time scale of 100 fs. The point being made here is extremely important in terms of defining what constitutes chemistry and the relevant time scales for the far from equilibrium motions which govern the process.

The quantum mechanics of the problem also need to be considered. At any instant, the molecular wavefunction of the system can be described by a time dependent superposition state using any convenient basis to describe the coupling between the reactant and product surfaces. The phase relationships of this superposition state define the temporal and spatial coherence of the wavefunction. At the saddle point, the subsequent relaxation and dissipation collapses this coherence onto the product surface. This point needs to be emphasized. In order to have dynamics and completely irreversible collapse of the wavefunction requires coupling to a continuum of states. Otherwise, at some point there will be recurrence or recrossing to reform the initial state.[28] The time scale of chemistry is the time scale by which the system samples the nuclear continuum of states. It is this process which makes the passage through the transition state region irreversible. The relevant time scale for sampling the nuclear continuum of states is as given above. In the case of dissociation, the product states are coupled to the continuum of translational states, whereas in the case of isomerization, bond formation, or electron/proton transfer, it is the coupling of the reaction modes to the nuclear continuum leading to redistribution of energy among the system and bath degrees of freedom. In all cases, the relevant time scale is defined by the nuclear motion along the reaction coordinate convolved to the sampling time of this continuum of states or dissipation dynamics.

The only exception to this statement is for purely electronic derived chemistry involving electron (hole) transfer to a conduction (valence) band state. In this case, the relevant electronic band forms a continuum of states and provides the necessary continuum condition for irreversible collapse of the wavefunction on the product state.[29] In this case, the relevant time scale is the electronic coupling between the donor molecule and band states. The lifetime of excited molecular states resonant with band states gives a direct measurement of this coupling.[29,30] It is only for this class of chemistry that the time scale of chemistry can be faster than nuclear sampling of states with electron transfer times reported as fast as a few fs.[28–31] No nuclear motion is needed to collapse the wavefunction or to define the localization of the electron in the product channel. Dye sensitized solar cells explicitly exploit this process to achieve high quantum yields for photoinduced electron transfer.[30]

Apart from the electron/hole transfer process to band states, which can be purely electronic, the time scale for the primary processes responsible for barrier passage is determined by the time scale for nuclear fluctuations along the reaction coordinate convolved to the dissipative coupling to the nuclear continuum of states. This timescale will vary depending on the particular chemistry. For most chemical processes, the relevant time scale is related to the effective period of the key modes along the reaction coordinate, which in turn are anharmonically coupled to system-bath modes leading to energy redistribution and relaxation at the reaction saddle point to give the new structure. The upper limit for the rate of passage then is effectively the same as intramolecular vibrational relaxation (IVR)/energy redistribution and intermolecular vibrational energy transfer that traps the system on the product surface.[27,32] This tends to be on 100 fs time scales for molecular systems with $N > 10$ atoms. Thus, the approximate shutter speed of 100 fs can be taken as the canonical shutter speed to capture the atomic motions involved in chemistry.

The lighting source for the molecular movie camera now needs to be considered. The experimental protocol is based on femtosecond spectroscopy methods in which one uses a femtosecond laser pulse to create the excited state under barrierless conditions to play out the chemistry of interest. The ensuing dynamics are generally followed with a probe pulse with a given time delay to follow the dynamics in a stroboscopic fashion. Each time delay between the excitation or preparation pulse and the probe pulse constitutes one time frame in what can be considered a movie of the system response. The observable depends on the probe pulse. If optical or IR probe pulses are used, the observable is related to excited electronic state dynamics. For photochemical processes, this form of transient absorption spectroscopy tracks the population dynamics from the initially excited reactant state through any intermediates to the final photoproduct state. Using IR probes, it is possible to get structural information through the changes in vibrational spectra, however, just as in IR spectroscopy of ground state species, this information is insufficient to fully determine molecular structure. To directly observe atomic motions, the probe pulse must have sufficiently short wavelengths, on the order of interatomic distances, or Å to sub-Å, as in conventional imaging based on the Rayleigh criterion to resolve two points in space. This condition requires either hard X-rays or high energy electrons to achieve sufficiently short carrier wavelengths for direct imaging.

In principle, it would have been possible to achieve these structural probes immediately with the advent of femtosecond lasers using either laser driven X-ray plasma sources or photoemission for electrons. The major obstacle was in fact not the attainment of simultaneous femtosecond time resolution with atomic spatial resolution but rather source brightness, getting enough X-ray photons or electrons to image the atomic motions above background. In all imaging problems, this detail is related to source brightness to illuminate sufficiently the features of interest with appropriate contrast above noise.

Keeping within this movie camera analogy, as one moves to faster and faster shutter speeds, the camera needs a brighter and brighter source in order not to lose image contrast within the noise. Flashbulbs were invented for precisely this reason. In the case of photography, the objective is to have a single frame with enough contrast to see the image. In the case of imaging chemistry, the very act of photoinducing the reaction in general leads to irreversible change in the sample. The sample in this case is the film, a very expensive film, in limited supply. Ideally, we also need single frame imaging but now at the atomic level. It is essential to extract the most information as possible from either amplitude or phase contrast to capture the image and use the least amount of sample as possible. The most sensitive means for imaging is based on diffraction using primarily phase contrast (phase shifts in the scattering process) in which the repeating structure of N identical objects leads to an N^2 amplification of the image contrast as the amplitude of the scattered light or electron waves interfere in quadrature. The film in this case requires single crystals for the highest spatial resolution and to fully exploit the N^2 amplification factor. The very act of exciting the sample, irreversibly changes the sample, and packing order leading to sample damage. Problems of X-ray or electron induced sample damage are often discussed for static structure determination.[33] In the case of time resolved structural dynamics, it is the very act of preparing the system on a reactive excited state surface that damages the sample. Given the limitation in sample, it is not good enough to have imaging sources capable of femtosecond time resolution and atomic spatial resolution. The source brightness must be sufficient to achieve this space–time resolution under or near single shot conditions. In the case of gas phase or solution phase systems, there is the possibility for rapid sample exchange to average more frames with low brightness sources to build up the image. However, the enormous background for solution phase studies or limited density for gas phase studies greatly reduces the SNR (signal-to-noise ratio) and thereby the spatial and time resolution. The higher the brightness the better the image contrast in all cases. The main point is that the brightness criterion has to be fully appreciated. The machine physics for imaging chemistry at the atomic level has been solved with high brightness electron and X-ray sources (*vide infra*). It is the samples now that are the primary bottleneck – as it should be.

In keeping with the above historical account, the first experiments to combine both the structural and dynamical basis of chemistry was the work of Ischenko *et al.*[34] This work used an elegant beam deflection method to study gas phase reaction dynamics to achieve microsecond time resolution, sufficiently short to capture the structure of reactive intermediates. It was the first resolved structural study of a molecular system, clearly illustrating the importance of time resolved structural dynamics, and laid the foundation for the work to follow. Follow up studies by Ischenko and Ewbank pointed out that it is possible to extract nuclear probability distributions[35,36] and introduced a form of quantum tomography using certain constraints on the initial and intermediate states. This approach pointed out the higher information content possible from electron diffraction when the full quantum scattering problem is treated. The subsequent seminal work of Mourou *et al.* introduced laser excitation of a photocathode to generate short electron pulses *via* photoemission for the structural probe.[37] This work achieved 20 ps time resolution to following the heating of Al, which was originally attributed to melting but the process was reversible with no sample damage. It appears they did not include the heat of fusion to pick their excitation conditions to sufficiently drive the structural phase transition. Nevertheless, this work, and other independent work at the same time (see ref. 38 and 39), introduced the basic source technology that has only recently been supplanted with rf pulse compression methods and compact *e* gun designs (*vide infra*). The Zewail group followed this lead, introducing a molecular beam for sample delivery, which is pioneering work on resolving structural intermediates in gas phase photochemistry.[40–42] This work would still be considered to be a major accomplishment by today's standard given the very small signal levels in gas phase electron diffraction. Weber and his group introduced the idea of high repetition sampling to improve the time resolution and SNR to the few ps domain to follow the ring opening reaction of cyclohexadiene (CHD) at approximately the same time.[43,44] The time resolution for all gas electron diffraction (GED) studies of reaction dynamics to date is limited to the few ps domain by the difference in the electron pulse velocity and the speed of light for the excitation pulse.[45] This time resolution is sufficient to resolve intermediates but not to directly observe the key reaction modes, which requires sub-ps time resolution as discussed previously. The velocity mismatch problem in electron diffraction can be solved by using tilted phase fronts for the excitation[46,47] to match the electron velocity. The implementation of this has recently been solved for studies of surface dynamics involving reversible phase transitions.[48] The alternative is to use relativistic electron sources as will be discussed below.

The first work to achieve atomic resolution on sub-ps timescales was the work of Siwick *et al.*,[49] which was based on the use of a new concept in ultrabright electron sources to provide the structural probe. The use of high bunch charge fs electron pulses, or ultrabright electrons in this context, was the key enabling technology that made real time viewing of atomic motions possible. Prior to this work, it was largely believed that the inherent electron–electron repulsion between electrons, or space charge effects, would destroy any chance of achieving fs time resolution with sufficient electrons for single shot diffraction. The advance in electron source brightness for time resolved studies was based on an effectively exact solution to the coupled equations of motion of some 10^4 electrons within an approximately 100 fs temporal profile.[50] This number of electrons is sufficient for single shot structure determination for simple unit cells (assuming 100% efficient detectors). The exact numerical simulation of the electron propagation showed that there is indeed strong Coulombic repulsion or space charge effects at the required bunch charge densities. However, one does not lose space–time correlation in the longitudinal or temporal profile related to the electron pulse duration. This finding was completely missed in previous treatments of electron pulse propagation dynamics. Most important, the transverse coherence of the electrons was found to be sufficiently conserved to enable atomic imaging of even complex systems. This discussion is relevant to our above discussion of source brightness for imaging. Two solutions to producing ultrabright electron sources were immediately apparent from this work. One solution is to use a dispersive electron optic to provide the opposite linear chirp to temporally refocus the electron pulse at the sample position.[50,51] The work of the Luiten group in the development of rf cavities for this purpose is particularly notable.[51,52] The most robust solution is to use a compact electron gun design and not allow the pulse to broaden after the high field extraction zone.[49,53–55] To give some idea of the source brightness, it is now possible with the newest generations of compact electron guns to achieve electron numbers of 10^5 to 10^6 electrons per pulse (100 micron beam diameter) with pulses as short as ~200 fs and coherences of several nm, sufficient for systems as large as proteins.[56,57] These systems reduce the technical difficulty of the experiment to the same level as conventional all optical pump–probe time resolved studies. More technically involved rf pulse compression systems can achieve 30 fs time resolution with time stamping to remove rf timing jitter[58] and approximately 10× more electrons per pulse. Both source concepts are capable of single shot atomic resolution. For comparison, XFELs (X-ray Free Electron Lasers) are major international facilities with fully coherent beams and fluxes of 10^{12} X-ray photons per pulse. Given the 10^6 higher scattering cross section of electrons relative to hard X-rays for this purpose, the diffracted signal intensity (detected electrons or photons) for typical samples are comparable. These are very bright sources for atomic imaging.

The term atomic movie needs to be defined in relation to imaging chemistry, the central theme of this discussion. The definition must embrace the ability to observe a correlation of events in time to understand causal connections of force and action. Here, I define an atomic movie to mean a measurement of atomic structure with sufficient space–time resolution to observe atomic motions faster than collisional processes blur out the details, *i.e.* resolve how atoms move relative to one another during changes in atomic structure. As discussed above, the first atomic movie meeting this criterion focused on the simplest possible structural transition, melting, albeit under strongly driven conditions to capture the primary motions.[49] High source brightness was essential as each laser shot destroyed the sample, which was in limited supply. This work is distinctive from similar work that followed the ps acoustic modes of Al, a fully reversible process with forces on the 10 nm scale (not atomic scale),[59] and from earlier reports of Al melting that could not have resolved this process and were near the melting threshold.[37] The first atomic movie resolved a long standing issue on the speed limit for driving phase transitions.[49,60] It was discovered that at sufficient degrees of laser induced superheating the system could melt faster than conventional heterogeneous nucleation by over an order of magnitude through the long predicted mechanism of homogeneous nucleation. The time resolution achieved was 600 fs, which was required to clearly distinguish a controversial point on thermal *vs.* electronic or nonthermal driving forces for the phase transition.[60] This work showed that the strongly driven phase transition of Al was a purely thermal process and occurred through homogeneous nucleation (now confirmed by all optical studies[61]). The remarkable finding was that for sufficient superheating, the Al lattice underwent spinoidal decomposition within 1 ps. All the bonds in the lattice broke within 1 ps, with complete collapse of the transverse barrier to allow atomic motions to acquire the radial distribution of the disordered liquid state. The time scale for the collective bond breaking is faster than collisions could wash out the details. Inspection of the time dependent reduced density (atom pair correlation function) shows the major motions involve shear type motions to the interstitial spaces in the lattice. We teach in physical chemistry that it is thermal motions exceeding the shear barrier to transverse motions that leads to the liquid state. We can now see this at the atomic level. These results also demonstrated that the nucleation sites had to form homogeneously throughout the sample and could only be on the order of 10 atoms. The insight gained from the first atom resolved structural transition helped solve the problem in laser driven ablation and shock wave formation that leads to collateral damage in laser surgery.[62] The long held promise of the laser to achieve the fundamental (single cell) limit to minimally invasive surgery has now been achieved and for the first time without scar tissue formation.[63,64] This application is noted here as it is the perfect example of how basic research can lead in unexpected ways to important real life applications. This outcome was derived from an atomic level perspective of the simplest structural transition, *i.e.*, melting. One can well imagine what the outcome might be for a complete atomic level understanding of catalysis, or the chemistry of structure–function relationships in biological systems, as just two other examples.

In the years that followed, most the work using high brightness electron sources focused on photoinduced phase transitions from collective many body effects in the formation of charge density waves and bond hardening or changes in nuclear screening with the onset of warm dense matter at electronic temperatures of several hundred thousand K.[65–71] There has been a continual increase in source brightness, but equally important developments in source characterization, and detector development. The evolution in source brightness and improvements in space–time resolution of structural transitions are shown in Fig. 2 to give an impression of the rapid progress made towards imaging chemistry once source brightness was realized as the major impediment.

The first molecular movie of a chemical reaction was that of electrocyclization with conserved stereochemistry.[74] The specific system was the ring closing reaction of diarylethene, which is a photochromic material pioneered by Irie and co-workers.[75] They deliberately designed the electrocyclization process by including electron withdrawing groups on the aryl rings to specifically weaken the ethene double bond in the bridge position to allow torsional motions to close the ring. The ideas behind the synthesis of this molecular platform and its photoactive functionalization are not too dissimilar to the previous discussion in thinking about reaction routes to strategic molecular targets. The photoactive process is centered on the bond formation process in electrocyclization and as such represents the formation of collisional complexes in otherwise two body or 2nd order reaction mechanisms. The spatial relationship of the interacting carbon atoms for bond formation had to be fixed by the molecular moiety to reduce the nuclear configuration space so that the dynamics are determined primarily by the very motions involved in bond formation. It was critical to maintain the stereochemistry to ensure a single photoproduct state and some degree of reversibility. The system is set up as if to watch the formation of a collision complex upon photoexcitation. In this respect, it is an ideal system to test the ideas used for maintaining stereochemistry in synthetic routes in general. What are the key motions involved in electrocyclization? Given the enormous phase space for random rotation about multiple points, how is stereochemistry conserved? This latter question is related to the prospect of any long lived intermediate states and the relative dynamics by which the reaction forces (potential energy gradients) in the barrier crossing region drive the system from the open to the closed ring form. Ultrafast time scales for barrier crossing are implied just from consideration of the quantum yield and competing timescales for random rotational/torsional motions sampling other configurations than the desired pathway.

The ultrafast aspect of the reaction dynamics was fully characterized *in situ* with femtosecond absorption spectroscopy on 100 nm thin single crystals.[76] There is an initial fast 200–300 fs inertial like component to the excited state dynamics followed by a 4–5 ps component leading to the formation of the ring closed form. This full characterization of the excited state dynamics is essential background

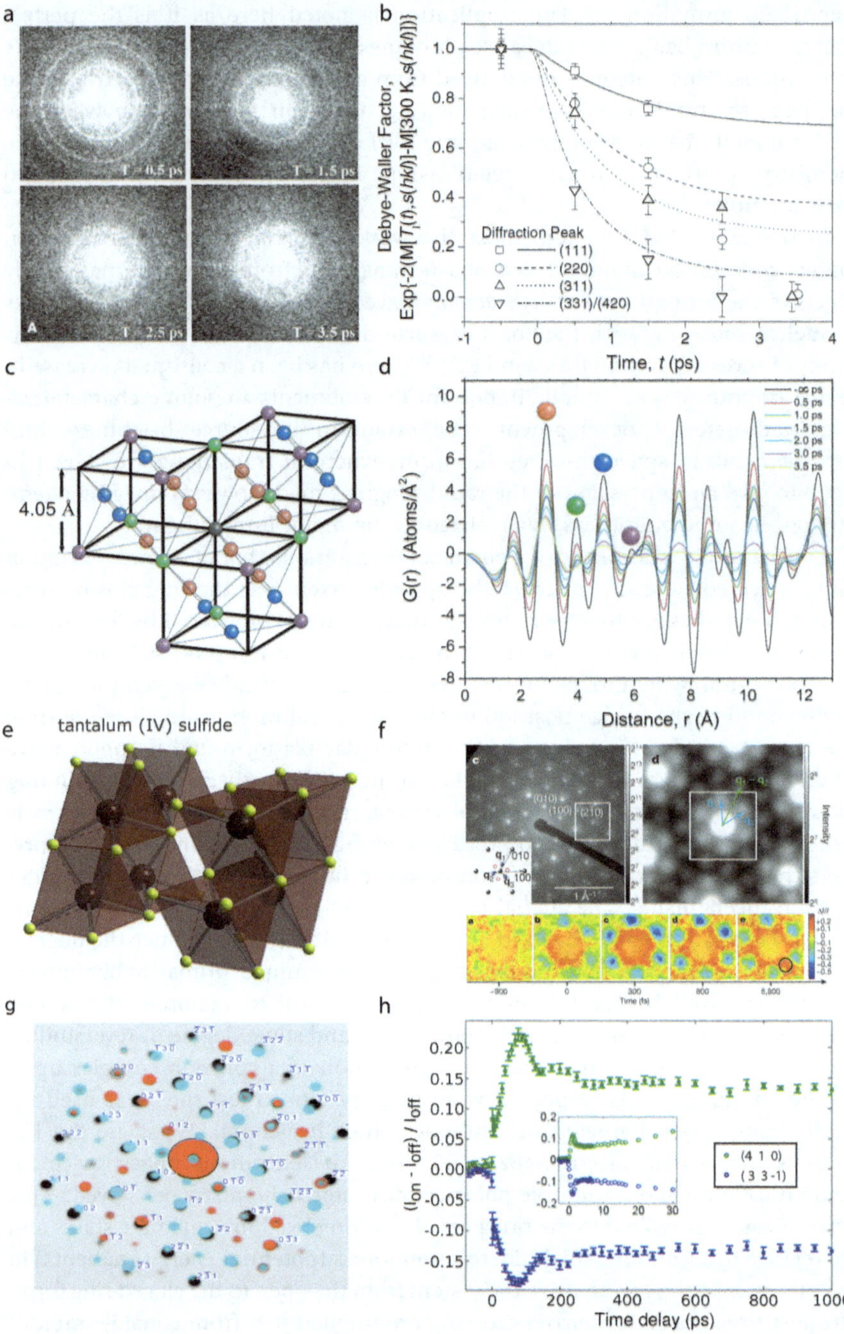

Fig. 2 Evolution in atom resolved structural dynamics. (a) shows the very grainy images from the first atomic movie, which appear much like the first Duguerre type images in the early days of photography.[49] From B. J. Siwick, J. R. Dwyer, R. E. Jordan and R. J. D. Miller, *Science*, 2003, 302, 1382–1385. Reprinted with permission from AAAS. One can still see by eye that the onset of melting occurs within 1 ps between $t = 2.5$ and 3.5 ps. The time resolution and ability to clearly distinguish nonthermal from thermal structural changes is shown in (b).[72] The corresponding real space transform is shown to the right and second

work to this class of studies to correlate the electronic and nuclear degrees of freedom in undergoing photochemical processes. At the excitation level needed for the structural studies, the photochromic nature of the open and closed ring forms was exploited to optically cycle the closed ring photoproduct back to the open ring form.[74] Even with the increased sampling per crystal location, the repetition rate was limited to the 0.1 Hz range due to the slow photorecovery and low thermal conductivity of organic crystals. Again, the ultrabright nature of the electron source used, in this case an rf pulse compression scheme, was essential to obtaining sufficient SNR to capture the atomic motions.

The main findings are shown in Fig. 3. The initial fast motion is due to a flexing of the backbone and slight ring torsional motion to bring the two labile carbon atoms into wavefunction overlap. This specific mode involved is a spatially delocalized mode at approximately 55 cm^{-1} assigned from DFT calculations of the full molecule ground state vibrational mode, which matches well the initial projections based on an *ab initio* calculation at the CASSCF level for a truncated model system of the diarylethene core. There was however no well defined minimum but rather continuous motion involving primarily the backbone flexing and 3 localized torsions at opposing sides of the closed ring which close the ring in a conrotary fashion. We now see directly at the atomic level how stereochemistry is conserved. The steric groups instinctively put in place by synthetic chemists do create forces that are sufficiently large to steer the system into bond formation. The ensuing vibrational relaxation to the product minimum occurs faster than sampling other coordinate space. The process is dynamically controlled. The transient state in which the labile carbon atoms are within wavefunction overlap of each other could have involved a flat potential with multiple barriers to other orientations that would otherwise cascade into other possible photoproducts with torsional motions leading to loss of stereochemistry. However, this is not the case. The system is strongly driven directly to the closed ring form with conserved stereochemistry. The other main point is that this system has hundreds of degrees of freedom including the surrounding lattice. We see directly now for the first time this distillation of chemistry down to a few key

row panels, giving the (c) colour coded unit cell and corresponding (d) time dependent motions of the relative atomic positions. Reprinted from B. J. Siwick, J. R. Dwyer, R. E. Jordan and R. J. D. Miller, *Chem. Phys.*, 2004, **299**, 285–305. Copyright (2003), with permission from Elsevier. The 3rd row shows dramatically improved diffraction quality in the study of the dynamics of collective phenomena associated with charge density waves (TaS_2, left panel).[68] (e) Source: WebElements [http://www.webelements.com/]. Adapted from M. Eichberger, H. Schäfer, M. Krumova, M. Beyer, J. Demsar, H. Berger, G. Moriena, G. Sciaini and R. J. D. Miller, *Nature*, 2010, 468, 799–802. Copyright © 2010, Rights Managed by Nature Publishing Group. The final row shows the extremely high spatial imaging quality for even poorly diffracting organic systems with diffraction well beyond 0.1 Å (reciprocal space).[73] The SNR of just 2 diffraction orders are shown in the bottom right panel, comparable to a high quality all optical pump probe signal, to give an impression of the outstanding SNR now possible in the image contrast. There are hundreds of such diffraction orders of this quality to help constrain the inversion to real space atom resolved movies of chemistry. (g) Adapted from M. Gao, C. Lu, H. Jean-Ruel, L. C. Liu, A. Marx, K. Onda, S. Koshihara, Y. Nakano, X. Shao, T. Hiramatsu, G. Saito, H. Yamochi, R. R. Cooney, G. Moriena, G. Sciaini and R. J. D. Miller, *Nature*, 2013, 496, 343–346. Copyright © 2013, Rights Managed by Nature Publishing Group. (h) Adapted from H. Jean-Ruel, *PhD Thesis*, University of Toronto, 2014.

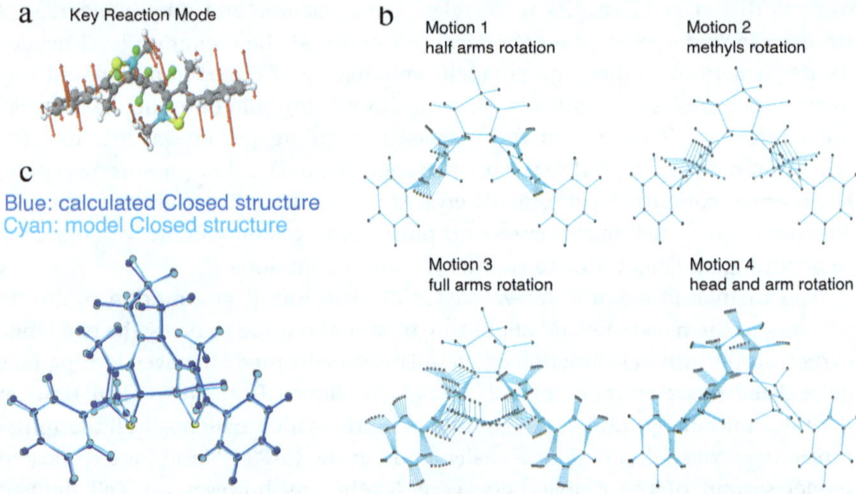

Fig. 3 Modal analysis of the key reaction modes involved in electrocyclization with conserved stereochemistry for diarylethene. The main mode is shown in the upper left panel, which corresponds to motion 3 in the second row. These theoretically generated classes of motions were used in structure refinement to provide a converged fit to the data with hundreds of diffraction orders for constraints. These are the key reaction modes and the time ordering leading to ring closing. There are no long lived intermediates that would allow torsional sampling and loss of stereochemistry, but rather a continual progression from a very fast inertial motion to bring the labile carbon atoms involved in bond formation to be within wavefunction overlap and torsions to click the bond formation into place to close the ring. (a) Adapted with permission from H. Jean-Ruel, R. R. Cooney, M. Gao, C. Lu, M. A. Kochman, C. A. Morrison and R. J. D. Miller, *J. Phys. Chem. A*, 2011, **115**, 13158–13168. Copyright (2011) American Chemical Society. (b), (c) Adapted with permission from H. Jean-Ruel, M. Gao, M. A. Kochman, C. Lu, L. C. Liu, R. R. Cooney, C. A. Morrison and R. J. D. Miller, *J. Phys. Chem. B*, 2013, **117**, 15894–15902. Copyright (2013) American Chemical Society.

modes. This point will be further reinforced below. For an appreciation of the specific time ordering of the motions involved in conserved stereochemistry see the movie (molecular movie 1 (ESI†)).

The next system to receive atomic scrutiny was the intermolecular electron transfer process exhibited in the charge ordered organic crystal ethylenedioxy tetrathiafulvalene (EDO-TTF).[73] This system undergoes a thermally driven phase transition from an insulator to a metal at 280 K. This phase transition involves a net transfer of an electron from one EDO-TTF molecule to another within the unit cell to create a delocalized electron distribution. In the low temperature insulating phase, the alternating EDO-TTF cations have a deformed ring as part of the intramolecular reorganization to minimize the energy of the cation state. There is a structural change between the insulating to conducting phases in which the rings all flatten and shift slightly for better π–π wavefunction overlap as part of creating the delocalized band structure commensurate with the onset of metallic properties. The electron transfer process can also be photoinduced under barrierless conditions by exciting a charge transfer band. The ensuing structural dynamics cast out the motions most strongly coupled to the electron transfer coordinate. These motions are expected to include both intramolecular bond length and angle changes as well as intermolecular motions as part of the intra- and intermolecular reorganizational energy defining the electron transfer reaction coordinate. No bonds are made or broken. Electron transfer processes are determined by the intra and intermolecular reorganization or medium repolarization to both bring electronic states into resonance and to stabilize the change in charge distribution. The change in structure of the EDO-TTF lattice with T is known from X-ray diffraction of the insulating and metallic phases. Using the traditional thinking in conceptualizing the transition state, one would expect the key reaction modes would be the flattening of the thio rings to increase wavefunction overlap as part of the process of forming an electronic band. In keeping with the previous discussion, it is not possible to project static structural end points to the motions involved in barrier crossing. There are always surprises from conventional expectations when chemistry is observed at the atomic level.

This experimental study also exploited rf pulse compression to provide the necessary number of electrons and source brightness to overcome the very low thermal conductivity and thermal accumulation artifacts. The experiment had to be conducted at sampling repetition rates of 10 Hz. It would not have been possible to obtain sufficient SNR before the onset of crystal degradation with low brightness sources. The diffraction quality was extraordinary with rms amplitude variation in diffraction intensities on the order of a few percent reflecting the high stability of the electron source. This crystal diffracted well beyond 0.1 Å (reciprocal space parlance) to enable very high spatial resolution (the camera was cut off at 0.2 Å). This is an order of magnitude higher than typical X-ray studies where 2 Å is the cut off in reciprocal space for atomic resolution for proteins (large molecules). For small molecules the resolution is better (< 0.5 A). There were hundreds of diffraction peaks to constrain the diffraction analysis in the to real space.

The first major surprise that strikes you immediately is the enormous reduction in dimensionality. Of the myriad of possible nuclear configurations, the motions collapsed to just a few key motions. There are over 280 degrees of freedom in this problem including all possible molecular motions within a unit cell as part of the intermolecular reorganization. Using a new method for differential diffraction analyses to remove background, and a Pearson correlation analysis of an exhaustive basis of possible atomic motions connecting the initially known structure to the final photoproduct state, it was found that the hundreds of diffraction orders over multiple time points could be described by just 3 key reduced reaction modes. The second major surprise was the specific mode most strongly modulating the electron transfer coordinate or medium reorganization. The dominant mode was not the expected ring flattening but rather the motion of the PF_6^- counter ion. This observation could not have been guessed as we don't have enough information on how the counterion is coupled to the electronic states. The coupling of the PF_6^- anion to the electron transfer is substantial from the magnitudes of the relative motions. Most importantly, the three dominant reaction modes were fully correlated. The trajectories appear to be shadow projections of each other. This observation means that the slowest mode of the three is dominating the overall medium repolarization in response to the change in electron distribution. A slow mode would not be able to respond in lock step with a higher frequency mode and would necessarily appear to execute relaxation on separated time scales. This was not the case. The slowest mode is related to the motion of the PF_6^- counterion. In retrospect, the role of the counterion is obvious. The photoinduced electron transfer from one EDO-TTF to another creates a local electric field, which would electrostatically create the largest force on the PF_6^-. This counterion is so large that its displacement driven by the reaction field leads to renormalization of the potential energy surface local to the EDO-TTF to assist ring flattening, lowering further the energy for the new charge state, and also the intermolecular potential leading to the change in the centre of mass locations of the EDO-TTF rings. See the atomic movie of these correlated motions for an impression of the significant effect of the counterion in stabilizing the electron transfer (molecular movie 2 (ESI†))

This observation is what inspired the analogy of the hike in the Alps where your every step changes the landscape opening secret passages to hidden valleys. At the hazard of overdoing analogies, the very motion of the PF_6^- is like an elephant going into a crowded room. Everyone would move out of the way, and if the floor was weak enough, it would bend under the weight to create new minima depending on the location of the elephant in the room, to which this many body/person potential would involve movement to accommodate (the heavy slow mode correlates the lighter mass, faster, motions). This direct observation of the importance of the counterion helps rationalize previously inexplicable counterion dependence of material properties for this class of organic materials. The counterion in organic salts is not screened by the motions of other ions as it would be in a liquid. It is the nearest neighbor and dominant source term to the local electric field. Its motion strongly modulates the energetics, or barrier heights, of the electron transfer reaction coordinate and modifies the potential energy surface to slave other motions that lead to increased wavefunction overlap. The system is poised to interconvert from insulating to metallic through these motions over a collective coordinate. We now see the effect of the bath on electron transfer processes directly at the atomic level.

The most significant advance in ultrafast imaging of chemistry was the recent work on the electron transfer process in the organometallic system $Me_4P[Pt(dmit)_2]_2$ (Me_4P = tetramethylphosphonium, dmit = 1,3-dithiol-2-thione-4,5-dithiolate) or $Pt(dmit)_2$ for short.[57] This system, similar to EDO-TTF is a charge separated ionic crystal that undergoes a structural phase transition in which formally an electron is transferred from one $[Pt(dmit)_2]_2$ dimer to another but this is triggered by excitation of an electron between frontier molecular orbitals. The local dimer structure therefore acts as a switch mediating the charge transfer process. The motions most strongly coupled to the photoinduced excited state would be the same motions involved in intramolecular and intermolecular reorganization for the thermally accessed electron transfer process. This reaction mechanism is specifically exploited in biological processes involving metal centers for mediating charge transfer and contains all the physics involved in classic metal-to-metal electron transfer in solution phase that we teach in first year chemistry courses. The difference is that we have a well-defined solid state lattice from which we can pull out the details of the reaction forces on the atomic motions, which would not be discernable in the ensemble average of disordered structures in solution phase. We effectively have well defined initial atomic coordinates for the bath to directly image medium repolarization effects. It is also important to note that $Pt(dmit)_2$ has a crystal structure with unit cell dimensions of 3.7 nm, which puts it on a par with protein crystals for complexity. A compact electron gun was used for these studies. This simple robust design concept has now achieved the level of brightness needed to capture the atomic motions for such a high dimensional system. The results were stunning.

Fig. 4 shows the femtosecond electron diffraction (FED) results for $Pt(dmit)_2$ using a new mathematical inversion method. To appreciate the space–time resolution achieved, some comments are warranted. In conventional diffraction analysis, it is not possible to uniquely invert the diffraction pattern to real space as the measurement only contains information on the intensity of the different diffraction orders or structure factors. The scattered wave amplitude also has a phase and interference effects between scattering centres that lead to ambiguity in the reconstruction. Simply put there is not enough information in the intensities to uniquely invert from reciprocal space to real space as there are two spatially dependent variables for which you only have the intensity. This problem cannot be solved for imaging unknown structures. The published atomic structures for molecules based on diffraction data are refined using constraints from either theory or molecular models based on other absolute structures for which some known scattering center was exploited to phase the data. In time resolved diffraction experiments, we are dealing with known initial structures and final structures. We are not trying to solve unknown structures. Here differential detection in combination with this additional information is sufficient to solve the phase problem. The work of Ishikawa, Hayes *et al.* introduced a new method in which the known initial structure, conserved bond lengths, bond angles and the causal constraint of continuous progression to the product state (no discontinuous jumps) were sufficient constraints to give unique convergence to structures (see supporting online material (SOM) of ref. 57) In effect, the known conserved features act mathematically as phasing centers to uniquely solve the phase ambiguity problem. The atomic motions you see in Fig. 4 are not the result of a fit to a simplified model or computer simulations used for refinement. The results shown in Fig. 4 constitute the first full atom resolved chemical reaction.

As you zoom in and focus on the details in this figure a number of features become apparent. In this respect, we are very good at discerning space–time correlations (as attempted to convey in Fig. 1), *i.e.*, how one event leads to another and over what length and time scales. We now can directly observe what we tried to imagine previously (Fig. 1). For this reason, movies of causally connected events make a particularly powerful means to convey the operating physics of a phenomenon. This distinguishes the time series of connected actions from

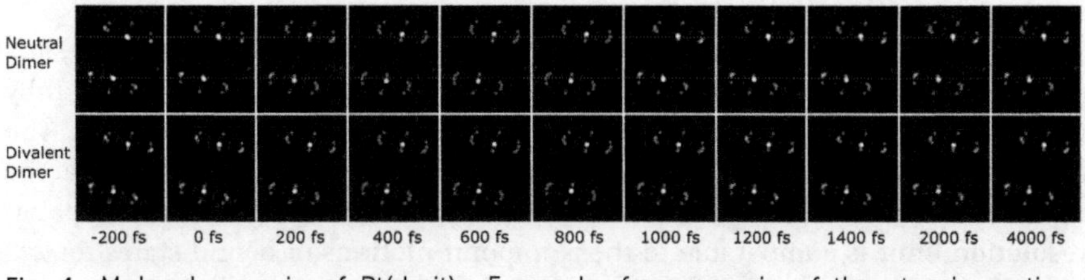

Fig. 4 Molecular movie of $Pt(dmit)_2$. Frame by frame movie of the atomic motions involved in metal-to-metal electron transfer. The atomic positions are indicated as white circles, the largest brightest circles as the Pt atoms, and the other atoms are the sulphur atoms in the rings. The dotted yellow line is given as a reference to enable observation of the motions. The change in Pt motions can be discerned by eye and the tilting of the rings can be seen by the change in the projections of the sulphur atoms along this viewing axis.

separate intermediates, as observed in time lapse photography, for which you see the "before" and "after" structures but not the pathway. To not bias your eye with balls and sticks or van der Waals' surfaces, Fig. 4 is shown with the atoms depicted as stars against a night sky. The brightest "star" is the Pt atom, which is the largest scattering source term. You can see immediately that not all possible atomic motions are involved or there would be a blur and no well resolved motion. Instead you can make out by eye that there are at least 2 dominant motions involving primarily the Pt–Pt separation and direction of displacement. This system has on the order of 1000 degrees of freedom including all intra- and intermolecular motions that could be involved in the reorganization to stabilize the photoinduced electron transfer. This effect is seen more dramatically than stills if you observe the time correlated motions as a movie to fully exploit our ability to see space time correlations (compare Fig. 4 to molecular movie 3 (ESI†)). From the movie, there is much greater clarity of the causally connected events. You can now see directly the collapse of innumerable possible nuclear configurations to just a few key modes. The timescale is slower than that for IVR in this large molecule such that the condition of equipartition of energy (*vide supra*) is approximately correct, at least for individual excited sites. Even in this case, not all degrees of freedom are involved. The reaction forces are highly anisotropic and drive specific modes. This enormous reduction in dimensionality during structural transitions is in keeping with the study of the reaction dynamics in diarylethenes, EDO-TTF, and other systems we are exploring. This reduction in dimensionality during structural transitions is a general phenomenon.

The degree of space–time resolution achieved in these experiments to light up the specific reaction modes can only be fully appreciated by examining the relevant motions quantitatively. The time resolved structural changes involving the full set of diffraction orders for constraints could be mapped on to 6 dominant reduced reaction modes, as shown in Fig. 5. The change in electron density upon the photoinduced electron transfer affects largely the Pt metal centres. For this reason, the largest motions involve the Pt atoms. This aspect of the structural changes could have been expected, however, neither the magnitudes nor the anisotropic coupling to the other displacements could be *a priori* inferred. Also, this system is too large and involves high Z atoms that make it too difficult to properly treat with *ab initio* methods. The level of detail from these experiments should provide rigorous benchmarks to explore different approximations to theoretically treat reactions on this scale. This step towards treating larger systems is essential if we wish this level of understanding to be relevant to current synthetic objectives.

From the inspection of the time dependent changes in the Pt–Pt distance, it is possible to observe the oscillations and damping of this mode on to the fully relaxed position corresponding to the high temperature charge distribution. The amplitude of the motion is only 0.1 Å but is clearly resolved. The rms noise can be used to estimate the resolution limit to be on the order of 0.01–0.02 Å. This spatial resolution limit is comparable to the zero point motions of bound states for 2nd row atoms in the periodic table. The resolution only depends on the diffraction efficiency such that the same resolution can be achieved for all systems. With high brightness electron sources, we have effectively achieved the fundamental limit for space–time imaging of chemical reactions. We can now directly observe the far

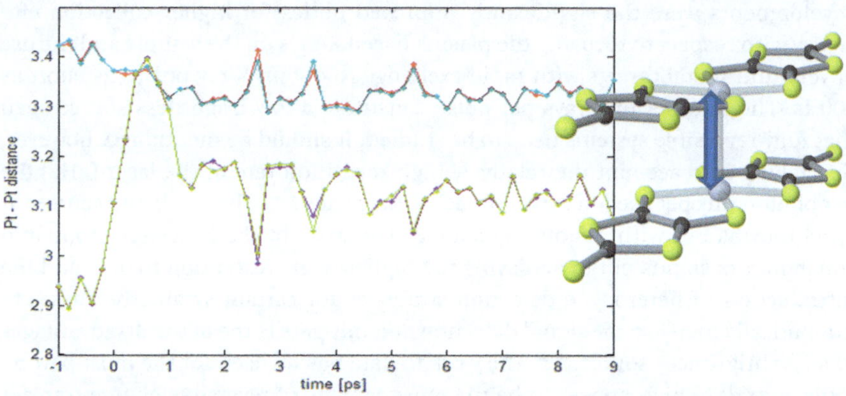

Fig. 5 Quantifying the atomic motions: analysis of the Pt–Pt separation with photoinduced electron transfer. There are two different Pt–Pt dimers in the crystal depending on the spatially ordered charge state. The motion is schematically depicted for the ball-and-stick structure to the right showing the Pt atoms (large grey spheres), sulphurs in the ring (yellow spheres) and carbon atoms (black spheres). The upper curve is for the −2 dimer and the lower curve is the neutral dimer. The spatial resolution is sufficient to resolve minute changes in energetics as the system undergoes lattice reorganization, with the Pt distances being one of the dominant terms. There are 5 other key modes involved (see SOM ref. 57; Adapted from T. Ishikawa, S. A. Hayes, S. Keskin, G. Corthey, M. Hada, K. Pichugin, A. Marx, J. Hirscht, K. Shionuma, K. Onda, Y. Okimoto, S. Koshihara, T. Yamamoto, H. Cui, M. Nomura, Y. Oshima, M. Abdel-Jawad, R. Kato and R. J. D. Miller, *Science*, 2015, **350**, 1501–1505. Reprinted with permission from AAAS).

from equilibrium motions that direct chemistry with sufficient resolution to provide new experimental benchmarks for theory.

The overall message that jumps out of this data is that the simple bit of missing physics discussed above that makes chemical reaction mechanisms transferable is the enormous reduction in dimensionality that occurs in barrier crossing regions. The high degree of anharmonicity in this region of nuclear configuration space couples high frequency modes with more spatially delocalized modes that are most strongly damped. It is the quantum aspect of this coupled state that leads to more localized motions through interference terms. One can readily see from the atomic motions for the photoinduced electron transfer for $Pt(dmit_2)_2$ that there are spatially delocalized motions involving the entire $Pt(dmit_2)$ moiety and these cascade to larger, more localized motions of the Pt atoms. The proposal here is that different reaction mechanisms involve a certain class of reaction modes that direct the chemistry and it is these conserved features that make chemical reaction mechanisms transferable from one molecule to another. It is the reduction of a high dimensional many body problem of order $3N$ to just a few key modes that is the "magic" of chemistry. This is the key physics that we need to determine in a systematic way by exploring different classes of reaction mechanisms.

In parallel, there has been tremendous progress in X-ray source development to enable imaging chemical reaction dynamics. The first work to achieve sufficient time resolution to follow chemical reaction dynamics with X-ray probes was based on a table top laser driven X-ray plasma source.[77–79] The major new developments were the significantly improved optics for higher collection efficiency with respect to focusing the plasma based X-rays on the sample and the use of very thin metal targets with rapid exchange to obtain X-ray pulses as short as 100 fs. The number of X-rays per pulse constitute a low brightness source such that fully reversible systems need to be studied. It should be mentioned, however, that taking into account the relatively high repetition rate of the laser (kHz) the net photon flux per second can approach X-ray fluxes from beam slicing sources at synchrotrons but with a table top source. The work of the Elsaesser group and collaborators stands out in achieving the highest time resolution to date and the introduction of heterodyne detection strategies for certain symmetry classes to dramatically increase the signal detection and mitigate some of the disadvantages of a low brightness source.[77–80] The work to date has focused for the most part on ionic crystals, which appear to be the most capable of reversible charge transfer processes. The most interesting finding is the discovery of a concerted electron–proton transfer and evidence for a highly collective reorganization of the lattice in the formation of a high spin state of the classic iron bispyridine system. The low photon numbers of the source require the use of relatively thick crystals to obtain the maximum diffraction efficiency for a given excitation volume. Given the large mismatch between resonant excitation of the sample and the X-ray probed volume, these experiments have had to use 2-photon excitation to obtain sufficient penetration into the required crystals thicknesses for optimal diffraction. In order to excite at least 10% of the lattice to observe the small signal changes, there will necessarily be higher order multiphoton excitation processes. Further excitation levels above 2-photon are all resonant ionization channels that would strongly compete with pure non-resonant 2-photon state preparation. The observed structural dynamics could involve ionization processes as opposed to photochemistry from well-defined excited state surfaces. It is not clear what excited states or ionization processes may be contributing to the observed structural changes. This is the same limitation for all X-ray studies as there is a large mismatch between the needed X-ray probed volume for sufficient diffraction and resonant 1-photon preparation of well-defined excited states. The very recent development of extremely high power femtosecond IR lasers promises to push the X-ray flux several of orders of magnitude higher, which should be sufficient to enable resonant 1-photon excitation to well defined excited states. The experiments to date illustrate the structural capabilities. This approach will provide a new tool comparable to the high brightness electron sources for the study of systems that are not readily amenable to electron probes such as solution phase and protein systems.

In the past year, X-ray free electron lasers have also achieved real time atomic movies of chemical reaction dynamics. These sources represent an increase of several orders of magnitude in X-ray source brightness over previous generations of major X-ray facilities.[81] The source is spatially coherent in the transverse plane like a conventional laser operating at visible wavelengths so it can be focused down to near transform limits of nm dimensions for imaging applications. The enormous X-ray flux is generated by a Self Amplified Spontaneous Emission (SASE) process that is subject to noise fluctuations in initiating the amplification cascade. As such, these sources undergo large stochastic variations in intensity. The accompanying reduced SNR for differential detection of structural dynamics and the longer wavelength intrinsically decrease the effective space–time resolution relative to electron sources, all other factors being constant. However, there are other variables at play that determine the overall imaging efficacy. Samples are a big issue. Source brightness can also play a defining parameter in what systems can be studied. The importance of source brightness and XFELs in particular for X-ray diffraction studies has been amply demonstrated through the work of Minitti *et al.* in the gas phase study of the ring opening reaction of the classic cyclohexadiene problem.[25] This system as in the diarylethene study discussed above serves as a model system for electrocyclization processes. Given the extremely low X-ray scattering cross section and gas density, this experiment would have been impossible without an XFEL source. The diffracted signal went out to a couple of diffraction orders that allowed resolving the bond dissociation dynamics, which correlated well with optical studies. By using theoretical constraints, it was possible to project the resulting dynamics from bond breaking to the ring open structure and onset to the different conformational states of the photoproduct. With a higher number of diffraction orders, it will be possible to directly map these motions on to specific modes. In the gas phase studies, the sample limitations are not an issue and rapid sample exchange is a major advantage over solid state studies. Here the excitation conditions match well those of nonperturbative all optical studies such that the nature of the excited state leading to the observed chemistry is well defined.

As emphasized, the highest spatial resolution is obtained with crystalline samples *via* the N^2 dependence in the diffraction process and reduced diffuse background scatter from disordered systems, *e.g.* gas and solution phase. In this regard, the major advantage of X-ray sources over electrons is the ability to use conventional crystals or solution phase pathlengths by the very virtue of the much weaker scattering cross section for X-rays. The use of X-ray probes has, however, its own sample limitations. Conventional crystals giving high quality diffraction patterns are typically on the order of 10 microns or thicker. In order to excite 10% of the unit cells to be above background, the peak power typically used to obtain sufficient changes in diffraction intensities for this crystal thickness is on the order of TW cm^{-2} for nearly every experiment performed to date. Peak powers of the order of 100 GW cm^{-2} have led to significant non-resonant 3-photon ionization of water for similar pathlengths.[82] The phototriggered optical processes for time resolved X-ray diffraction are fully resonant processes, which are greatly amplified above the aforementioned problem with non-resonant 2-photon excitation. The resonance condition creates amplification for all n-photon excitation pathways. Multiphoton processes could be the dominant source for the induced structural changes. This excitation mechanism changes significantly the chemistry from well-defined excited state surfaces undergoing isomerization, cyclization *etc.* to ionization driven structural changes from ill-defined states. The effect of multiphoton ionization has not been taken into account in terms of contributions to the observed structural changes.

With this caveat, XFELs are still the only source currently capable of studying biological systems, simply due to sample issues. The main limitation for electron sources is the difficulty in obtaining sufficient protein crystals on the 100 nm scale for time resolved studies. This problem of obtaining large numbers of protein crystals has been solved for serial crystallography applications. In this respect, XFELs enable probing the issue of reduced dimensionality for systems of increasing complexity. The ultimate limit in scaling to larger systems is to observe the transduction of chemistry into biological functions. As discussed above in the case of proteins, one is now dealing with highly associated glass-like systems (on long time scales) with >10 000 atoms; yet the chemistry driving functions and biochemical pathways is highly robust. The system does not get trapped in local minima along the chemically relevant reaction coordinate. In this context, the reduction of the structure–function relationship of biological systems to a few key modes is a truly profound observation in relation to how highly optimized nature has become. In this regard, the most significant results have come from the study of the structural changes involved in ligand dissociation within carboxymyoglobin as a model system for the tertiary level structural changes involved in molecular co-operativity in hemoglobin.[83] This response forms the cornerstone of our understanding in general of molecular cooperativity in biological systems as one form of feedback in biochemical pathways of living systems. As discussed in the Introduction in the context of Levinthal's paradox, this system is highly associated and involves chemistry at the extreme end of the complexity scale. The Collective Mode Coupling Model was proposed as the mechanism by which such highly associated systems beat being trapped in local minima in executing their biological function.[3,4,6,7,84,85] This model was based initially on an observation of acoustic-like collective motions of the protein driven by the ligand dissociation and exhaustive all optical studies of different length scales of motion and energetics to show that this is the dominant mechanism. It is the inherent spatial correlations in the fluctuations of biological systems that enables the effective reduction in dimensionality and escape from glass-like dynamics in dictating functional responses. This statement returns to the initial discussion of correlated motions and reaction phase space. The issue in terms of making the connection between the evolutionarily optimized structure and function can only be solved by a direct observation of protein function at the atomic level. The latest XFEL experiments have fully confirmed that the protein motions involved collective modes and these correlated motions are close to those based on molecular dynamics calculations.[83] The issue of ionization effects is still a problem. The motions involved may not be biologically relevant. However, it is clear from this work that there are correlated motions, which can be discussed in terms of collective modes bearing out the collective mode coupling model. These studies need to be redone at excitation levels for which the forces driving the protein motion clearly originate from ligand dissociation. The other key experiment is the XFEL study of the light induced structural changes involved in the isomerization of the photoactive yellow protein (PYP).[86] Again, this work shows beautifully that there are highly localized changes directing the isomerization process. The excitation process in the sampled volume is $\sim$1 TW cm^{-2}, which again raises issues of ionization in this interpretation. Power dependences are needed to sort out functionally relevant motions from possible ionization effects.

The main point in light of the above discussion is that even for systems as large as proteins, the chemistry involved reduces to a few key modes. This reduction in dimensionality holds in general for chemical reactions and has been highly exploited by nature in scaling to ever more complex systems. At the extreme end of complexity as defined by proteins, the reaction forces of this reduced chemically active space are further coupled even to mesoscale dimensions by the imposed structural constraints of the highly optimized protein structure. The net effect is to further reduce the phase space (as in the beads on a string analogy) and direct reaction forces to execute biological functions, often at the maximum possible efficiency.[3] It is an extraordinary feat in directing chemistry into action.

This discussion has focused on direct atom resolved chemical reaction dynamics – direct imaging of chemistry. As with any observation, as one looks closer and closer with greater acuity, one finds surprises in how things work. Here we are talking about observing chemistry at the fundamental space–time limit, watching chemistry unfold at the atomic level of detail. Our previous conceptual models for chemical reaction mechanisms have been hard earned through the extensive use of molecular probes as in the case of synthetic chemistry or equally extensive spectroscopic and molecular beam studies. We have historically focused on details that were particular to a given molecule. In the process, we have failed to see the generalities of chemical processes. For each system, the concept of reaction modes has long been discussed. However, it is only now that we see directly the innumerable nuclear possibilities collapse to a few key reaction modes that the significance of the anharmonic coupling in the barrier crossing region becomes fully apparent. We can now witness these far from equilibrium motions reduce the dimensionality of the possible nuclear rearrangements to a few key reaction modes for systems as small as a few atoms up to systems as complex as proteins. The first full atom resolved chemical reactions have experimentally established that a general physical principle is indeed at work.

The many body potential is so complex and the degree of nonadiabatic coupling between electronic and nuclear degrees of freedom is not known with the required accuracy to predict reaction pathways. The major advance made is that we can now observe these motions directly. What can we do with this information and new insight into reaction dynamics? This question will now be addressed in relation to developing a new conceptual basis for chemistry.

2. Overview of the discussion

This Faraday Discussion was organized to discuss the latest developments in imaging chemistry, as part of coming to a better understanding of the forces at play that ultimately determine the chemistry, *i.e.*, to basically address the questions posed above. In this regard, I will not review all the papers presented as the work is published in this issue with sufficient detail to appreciate the main findings. I will instead aim to give a flavor of the discussion and where I see the field going.

The meeting had a nice balance between new and established methods in imaging photochemical dynamics, as well as recent theoretical advances. Since photochemistry by definition involves excited states, we need to know the details of the excited state surface or initial state preparation. In other words, we need information on both electronic and nuclear degrees of freedom to understand the chemistry. The nuclear motion and electronic state energies are intrinsically coupled. It is essential to correlate the reaction dynamics to the excited state dynamics. The use of time resolved X-ray spectroscopies and electron spectroscopies, coherent multidimensional spectroscopies, as well as conventional time resolved transient spectroscopy discussed at the meeting provides this critical link. This point was made abundantly clear by the work presented by Orr-Ewing *et al.* (DOI: 10.1039/C6FD00068A) discussing the classic case of the ultrafast relaxation processes in DNA bases and the effect of the surroundings. In addition, there are other probes of reaction dynamics that provide a different perspective of the nuclear dynamics. The Discussion emphasized the importance of theory in fully understanding reaction dynamics, particularly the aforementioned interplay between nuclear coordinates and electronic states with respect to shaping the potential energy surface. In this regard, there are equally dramatic advances in theory that are perfectly timed to take full advantage of the detailed atomically resolved reaction dynamics that are now coming from experiment. The papers in this *Faraday Discussions* allow the reader to see the full spectrum of experimental and theoretical approaches taken to understand photochemical reaction dynamics.

The alternative approaches to direct imaging of atomic motions (as discussed above) were treated in a number of papers. They provide important additional information on the interplay between electronic and nuclear degrees of freedom in defining the reaction coordinate. One such approach is to take advantage of high field recollision physics of electrons that form the heart of attosecond spectroscopy.[87] The interaction of a molecule with extremely short femtosecond pulses in the visible to IR leads to the generation of high harmonics through nonlinear induced polarizability. The major advance of this methodology is that it provides more correlations than spectroscopic methods that typically probe a single quantum state regime (*e.g.* valence electronic states or vibrations). It is the constraints of higher order correlations in the harmonic generation that gives the additional information. The induced nonlinear polarizability explicitly depends on the electron distribution within the molecule. The specific correlation of the high harmonic intensities can be related to the molecular structure to determine the electronic distribution on the attosecond time scale. These attosecond spectroscopic methods have generally been applied to small molecules. In scaling to larger systems, there is a complication in connecting the high harmonics to electron distribution as the number of channels for generating harmonics and role of ion states become more complex. The field strength driving the harmonics is above the atomic Coulomb potential such that ionization processes occur in concert with high harmonic generation. The ionization processes are expected for the wavelength range of the high harmonics, which is well into the VUV to soft X-ray range where 1-photon resonant processes are significantly above the ionization potential of molecular systems. For non-resonant excitation fields, the field strength is sufficient to lead to tunnel ionization effects. Thus, there are ionization effects, dynamics of the resultant cation state, and recombination processes to consider in the overall harmonic generation. To understand these processes and connection to valence molecular orbitals, a detailed theoretical treatment was presented for benzene and substituted benzenes as a system with a high degree of symmetry and relative ease with substitution to change the electron distribution in known ways (Austin *et al.*, DOI: 10.1039/C6FD00116E). This work was motivated by previous experimental work and found that it is the tunnel ionization interaction that makes the observable more sensitive to the details of the electron distribution. The number of contributing factors to the high harmonic generation is quite complex as further discussed in the work of (Dudovich *et al.*, DOI: 10.1039/C6FD00130K) (see Fig. 2 of this paper). One of the goals in attosecond spectroscopy is to separately determine all the different channels contributing to the high harmonic generation. To separate the different contributing terms, the use of multidimensional spectroscopy was employed to spread the interactions out along a separate time axis (phase) with polarization control between the two excitation fields. Apart from tunnel ionization, the laser field drives electron redistribution. By using multidimensional (2D) correlations of the high harmonic generation, it was possible to obtain information on the laser driven hole dynamics of CO_2 upon high IR field ionization. This problem is similar to the inversion problem in converting diffraction data to real space. This work provides additional means to invert high harmonic profiles to imaging of electron dynamics. During the discussion, the prospect of using this information to map out changes in electron distribution during structural changes was pointed out. This information becomes particularly important at a conical intersection (CI) where the Born–Oppenheimer approximation breaks down and the ensuing electron redistribution becomes slower than the nuclear fluctuations and ultimately defines the transition probability between the two coupled surfaces. This is an exciting prospect in terms of obtaining a new spectroscopic observable with sufficiently high order correlations to give information of the electron distribution during a chemical event. The open issue will be to connect the observed molecular dynamics under strong field conditions undergoing simultaneous ionization and laser driven electron redistribution with the weak field conditions of 1-photon photochemistry.

In addition to observing the high harmonics, it is possible to exploit the laser driven electrons to take a snap shot of the molecule in the process of ionization.[87] The spatial resolution depends on the electron energy. By pushing the laser field to longer wavelengths into the IR, there is more time for electrons to interact with the field and to pick up more energy before undergoing the recollision process and scattering. The general idea is similar to Rydberg spectroscopy[88] in terms of using internal electrons to scatter off nuclear positions to report on structure, with the exception the energy of the electrons is much higher, approaching the 100 eV range. The action of one half cycle of the applied field is to drive the electrons away from the molecular core and then the attractive Coulomb potential adds to the electric field of the subsequent half cycle field of opposite sign to accelerate the electrons to the point of coherently scattering from the molecules. The effect has been referred to as Laser Induced Electron Diffraction (LIED).[89] The diffraction comes directly from the molecular frame of reference. The energy range of the electrons involved in LIED is in the same range as Low Energy Electron Diffraction (LEED) used typically for surface studies. The difference is that the source coherence cannot be independently determined such that again theory is essential to interpret the observed electron photoemission patterns. Recently, this approach was used in a *tour de force* experiment to observe proton motion in dissociation from the second ionization state of acetylene.[90] The scattering cross section of the proton is extremely small and it is not clear how convolved the observable is to all the other scattering processes as delineated in the aforementioned papers.

The energy of the photoemitted electrons can be used to determine the energetics of the excited state dynamics. In this way, photoelectron spectroscopy in conjunction with structural dynamics gives the coupling between the electronic and nuclear degrees of freedom, *i.e.* the energetics. There were a number of papers that exploited some form of photoelectron spectroscopy to probe the reaction dynamics. For example the work of (Iikubo *et al.*, DOI: 10.1039/C6FD00063K) used

VUV probes to track the ring opening reaction of cyclohexadiene and could track the changes in energy of the molecular orbitals upon passage through the conical intersection in the process of ring opening. This study complements previous work using fs diffraction to follow the ring opening[25] and previous optical studies.[44,91] The ring opening was found to occur an order of magnitude slower than that found in the diffraction studies, which was ascribed to the difference in the excited state preparation. This work excited the 3^1A and 2^1B levels as opposed to the lower energy 1^1B electronic surface in the aforementioned X-ray diffraction studies. This excitation creates an excited state wavepacket nearly 2 eV higher energy than excitation directly to the 1^1B state of the previous work. The dramatically different time scales observed was ascribed to the time for IVR to partition the excess energy into the reactive modes and sample the CI. This difference highlights the importance of the excited state preparation, which will be discussed later. The most important point is that the details of the molecular orbitals and energetics of the photochemical reaction can be directly observed in this manner. The information content can be enhanced by spatially imaging the photoelectron distribution as shown by (Zawadzki *et al.*, DOI: 10.1039/C6FD00092D) in the systematic study of the nonadiabatic relaxation dynamics of aniline derivatives using UV excitation mixing Rydberg states with valence states. With time dependent *ab initio* cluster methods, there is enough information to connect the spatial distribution of photoelectrons to specific molecular processes involving bond elongation along key coordinates as well as projection to two competing channels of nonradiative relaxation to longer lived S_1 states or direct dissociation along N–CH_3 channels. The lone pair of the N of the aniline moiety is implicated to play a central role in how the Rydberg-valence mixed states evolve. Again, the nature of the details of the excited state preparation come into play.

One of the limitations in Rydberg spectroscopy to resolving structural dynamics is the lack of sensitivity to vibrational modes. This feature arises from the fact that the Rydberg orbitals are very similar to the ionic states as the electron is effectively only weakly bound by this condition in terms of its spatial distribution about the molecule. This means that the Franck–Condon factors for the Rydberg transitions are highly focused near the origin and other vibrational modes that may play a role in the photochemistry are largely invisible. The work of Beaulieu *et al.* (DOI: 10.1039/C6FD00113K) have explored the use of chiral probes for the photoelectron process using a combination of circularly polarized, and linearly polarized excitation conditions for 1-photon and multiphoton/tunnel ionization regimes to follow dynamics in chiral molecules using fenchone as a model system. This work is able to distinguish chirality in the angle resolved photoelectron spectra, and from modulation in the dynamics which affect the chirality, it is possible to tease out information on vibrational modes generally hidden in Rydberg spectroscopy. The number of observable modes is still very sparse compared to those observed in fs transient absorption spectroscopy where time dependent FC factors help to observe the excited state wavepacket dynamics. The spreading of wavepackets in due course of relaxation along the excited state potential may wash out the contrast in the photoelectron dichroism spectra. The ability to observe vibronic states with photoelectron spectroscopy is in all cases an important development.

In light of the above work, the importance of the nature of the excited state was further amplified in the use of attosecond pulse trains and angle resolved photoemission during the process of high harmonic generation. All attosecond studies involve strong laser field interactions in which tunnel ionization occurs with multiple scattering source terms (*vide supra*). The work of Veyrinas *et al.* (DOI: 10.1039/C6FD00137H) illustrated how angle resolved photoemission gives the molecular reference frame photoelectron angular distributions, to give a spatial picture of the different scattering processes leading to photoemission. The states probed and the ensuing nuclear motion of the resultant ion states can be followed in this manner but it needs to be understood that these processes are high field ionization physics and cannot be connected to 1-photon photochemistry, as can be well appreciated in the sensitivity of the initial state conditions observed for cyclohexadiene or the chirality problem in terms of 1-photon or multiphoton effects, discussed above.

Attosecond spectroscopic methods in general involve strong fields and as such inherently involve ionization processes and ensuing molecular dynamics of highly excited ion states. The other way to probe the same physics is to use Coulomb Explosion Imaging methods made possible with COLTRIMS and Velocity Map Imaging detection schemes to follow the partitioning of the multiphoton excitation into different molecular degrees of freedom and different ionization channels. The work by Rudenko *et al.* (DOI: 10.1039/C6FD00152A) on the CO_2 ionization was able to map out vibrational energy redistribution pattern for the different ion fragmentation states. Similarly, the studies of Burger *et al.* (DOI: 10.1039/C6FD00082G) were able to follow proton isomerization in the formation of vinylidene. Here Coulomb Explosion Imaging was exploited to post sample the excited state dynamics of H migration on the doubly ionized excited state surface of acetylene. The high time resolution of this experiment was able to identify specific vibrational wavepacket motions along C–C and C–H bonds in picturing the proton rearrangement. There was very interesting related work using XFELs to excite core level transitions in I (iodine) substituted molecules (Ueda *et al.*, DOI: 10.1039/C6FD00085A). Here rather than involving multiphoton electronic excitations, the intense XFEL pulse can selectively cause multiple resonant absorption of the I atom. The subsequent redistribution of charge (created vacancies) and electronic relaxation leads to the Coulomb explosion process. This mechanism gives a much more localized view to the ensuing ionization processes. The highly localized nature of the initial forces made the observed dynamics particularly amenable to Classic Coulomb Explosion with Charge Exchange (CCE-CE) models and self-consistent charge density functional based tight binding (SCC-DFTB) theory to depict the charge transfer from the I to the adjacent moiety and atomic motions coupled to this change in charge density. This level of theory is less than that typically needed to understand 1-photon excited state processes, which seems counterintuitive in terms of the highly excited nature of the XFEL driven process.

The important role XFELs will play in imaging photochemistry was well represented by the studies of the photodissociation dynamics of diiodobenzene (DIB) (Stankus *et al.*, DOI: 10.1039/C6FD00135A), using laser based time resolved photoelectron spectroscopy and XFEL based X-ray scattering (8.3 keV). The photoelectron spectroscopy work showed that the excited state of DIB accessed with 267 nm is a nominally bound predissociative state with a lifetime of approximately 35 fs and could be understood in light of the supportive *ab initio* calculations. The very fast relaxation to another strongly coupled surface follows the trends found for halogen substitutions but other resonant conditions need to be considered in a systematic way. The X-ray scattering studies of DIB were limited in the dynamic range and can be viewed as preliminary work towards resolving the structural details. The very fact that X-ray scattering can be observed for typical gas phase sources illustrates the brightness of XFELs. In this regard, one of the main advantages of ultrabright electron sources as discussed in the Introduction at length is the 6 orders higher scattering cross section of electrons over X-rays. Electrons are ideally suited for gas phase diffraction studies. The problem in the past has been the time resolution due to velocity mismatch between the laser excitation and the nonrelativistic electron probe. The development of relativistic electron sources solves the velocity mismatch problem. In addition, with the introduction of rebunching cavities it is possible to generate high bunch charge, pulses as short as 10 fs with sufficient transverse coherence for all classes of samples,[92] especially gas phase systems. The work of Yang *et al.* (DOI: 10.1039/C6FD00071A) have demonstrated the ability to observe the vibrational wavepacket of excited I_2, as a first step in resolving the primary motions involved in gas phase reaction dynamics. The relativistic source technology as well as nonrelativistic electron sources using tilted phase fronts for 100 fs time resolution will fill this experimental niche in imaging photochemistry.

The other major application of XFELs was presented in the form of X-ray absorption spectroscopy (XAS) of excited state dynamics. One of the most important applications will be in elucidation of the structure of the active site for photocatalytic water splitting in PSII. We currently have no viable photocatalysts for water splitting and nothing on par with the solution found by nature in the form of the PSII photocycle to harness 4 photons to split water. The paper by Yano *et al.* (DOI: 10.1039/C6FD00084C) was informative in setting the stage for the studies that are forthcoming. One of the most innovative applications of time resolved XAS was presented by Gessner *et al.* (DOI: 10.1039/C6FD00125D) in which a synchrotron X-ray source was used as a probe to follow the interfacial dynamics of cuprous oxide as a test case.The dynamics could be understood in the form of carrier dynamics and interfacial states in line with previous work. XAS probes for interfacial processes with sufficient sensitivity are now in place to follow photocatalysis *in situ* and even *in operando* of catalytic reactors, as discussed at the meeting. In this context, the theoretical work of Neville *et al.* (DOI: 10.1039/C6FD00117C) showed the use of various levels of theory to fit time dependent XAS spectra to the nuclear motions involved in the isomerization of acetylene in a highly satisfactory way. It will be informative to see how sensitive time resolved XAS spectra are to more complex systems and how unique such effective lineshape analysis is. This aspect of relating XFEL based XAS studies to structural dynamics is needed, especially at the high excitation typically used where the excited state is not well defined.

All of the above requires some level of theory to fully understand the connection between the electronic and nuclear degrees of freedom. The recent advances in theory were well represented by the work of Martín *et al.* (DOI: 10.1039/C6FD00074F) in treating the ionization and various scattering, hole dynamics, and ionizing electrons on the nuclear dynamics as studied in attosecond pump–probe measurements. They resorted to treating the strong IR probe field used in these experiments as a weak XUV probe pulse to isolate the hole dynamics using various levels of time dependent *ab initio* theory based on DFT approaches with time dependent perturbation theory and Ehrenfest MD to study the ionization and nuclear dynamics, respectively. The hole dynamics are superpositions of cationic states, which depend on the spectrum especially near the threshold. This work illustrated that with this insight it should be possible to use spectrally shaped attosecond pulses to control hole or charge dynamics in molecules at least as large as their glycine model system. Similar concepts were explored by Vacher *et al.* (DOI: 10.1039/C6FD00067C) in the valence ionization of bismethylene-adamantane. This work used higher levels of theory at the CASSCF level to more accurately treat the electron dynamics in which the electronic states may have gaps of less than 0.3 eV for which the redistribution of electron density slows down to nuclear time scales. The ionization process was found to involve symmetric and antisymmetric C=C bond elongations that could potentially be controlled with shaped pulses to give what they called "charge directed reactivity". With respect to treating 1-photon resonant photochemistry involving well defined excited states, the work of Chen *et al.* (DOI: 10.1039/C6FD00083E) and Makhov *et al.* (DOI: 10.1039/C6FD00073H) treated the effect of the bath dissipation on the overall reaction yield and new means of more efficiently sampling MD trajectories for a full quantum description of photochemistry, respectively. The effect of the bath on the passage through a conical intersection is critical to understand in the context of open quantum systems. This work used the hierarchical equation of motions approach developed by Tanimura and Kubo to treat the system-bath coupling as described in ref. 93. This formalism is exact for certain types of system-bath coupling and allows the exploration of the effect of dissipation on reaction dynamics, in a very similar vein to Kraemer's classic work on friction in transition state crossing.[17] This work found that dissipation had a pronounced effect depending on the details of the coupling or tuning modes in the formation of the conical intersection, which should inspire future experiments in testing these findings. Finally, the work discussed by Makhov *et al.* (DOI: 10.1039/C6FD00073H) gave some appreciation of the enormous gains made recently to fully exploit trajectory guided quantum dynamics. A new means of introducing multiple quantum states to sample the same initial trajectories using so-called cloning and spawning methods enables a much more computationally efficient means to calculate on-the-fly quantum state dynamics with sufficient statistics to compare to experiment. This paper included tunneling effects to calculate the total kinetic energy release spectrum of hydrogen atoms for the photodissociation dynamics of pyrrole, which appears to be essential to get the physics correct. The larger implication is that this new methodology allows the calculation of reaction dynamics along excited state surfaces for relatively large systems.

The main message from this Discussion is that theory and experiment are reaching the same level of capability in terms of "imaging" photochemical dynamics for increasingly complex systems.

3. Roadmap to a reaction mode basis for chemistry

The above introduction was intended to point out one of the truly remarkable features of chemistry that makes this discipline unique. Fundamentally, chemical principles involve nonlinear many body problems of extremely high dimension.

Each different molecule has a distinct many body potential and properties. The smallest changes in molecular composition or structural arrangements in space change their properties, *i.e.*, the details of the many body potential matter. The very fact that chemical reaction mechanisms have been discovered that can be applied to a near infinite number of molecular systems indicates there is some reduction principle at play. It can be deduced from the simple observations made in the introduction that the reaction forces do not sense the entire molecule but must be localized to some self-similar features in the dynamic sampling of the molecular framework.

We have now achieved the fundamental space–time limit needed to image chemistry to directly observe this reduction in dimensionality. "Seeing is believing". This new ability will change how we think about chemistry. This assertion is based on the fact that we cannot, save for a few small molecule systems, predict the nuclear configuration(s) of the system in the barrier crossing region. The information that will be forthcoming will help to create a new reaction mode basis to better predict reaction pathways and categorize chemical reactions.

There are a number of challenges on both the experimental and theoretical fronts that need to be discussed to achieve this vision of developing a reaction mode basis for chemistry. Firstly, barrier crossings are rare events such that some means must be used to trigger the chemistry of interest. This rather obvious point may be the Achilles Heel of this overall methodology. There are very few molecular systems that can be optically prepared along barrierless, reactive, excited state surfaces. New means need to be found to initiate chemistry. Secondly, there are significant constraints on the sample preparation both to avoid peak power limits in the initial state preparation and with respect to developing means for a systematic study of different reaction mechanisms. Ideally, we would like to isolate different aspects of the bath with respect to coupling to the nuclear continuum of states that is ultimately responsible for the formation of the product state. Thirdly, it is highly desirable to push the space–time resolution to the highest possible limit. The current limit to spatially resolved reaction dynamics is on the order of 0.01 Å. These very motions can be connected to the energetics of the reaction coordinate. Finally, new theoretical frameworks are needed to cast the problem within tractable computation limits and explicitly exploit the experimental information to predict the key reaction modes involved in barrier crossing. The experimental results will provide the proving ground for theoretical advances aimed to reduce computational costs and focus on what matters most, the key modes driving the chemistry. This coordinated comparative approach will enable a limited number of experimentally testable systems to be generalized. In this limit, we will have arrived at a formulation of chemistry that can be readily transported to guide synthetic strategies.

Phototrigger challenge

There have been thousands of photochemical processes studied over the past several decades. The classic systems involve photoisomerization, electron/proton transfer, keto–enol tautomerization, dissociation processes, cyclization reactions as a short list. Of these, only a small subset can be photoinduced along barrierless reaction coordinates. As discussed above, in the limit of a barrier in the excited state surface, the observed dynamics involve uncorrelated events and only the intermediate structures can be resolved – not the pathway. Moreover, there is the problem of connecting the photoinduced processes to thermally driven reactions along ground state surfaces. This latter issue was brought up in the discussions (see comments by Dave Townsend). Excited state processes often involve conical intersections that naturally reduce the problem to a few key types of motion. It is these motions that create degeneracies in the electronic surfaces, and thereby create curvatures in the potential energy surface that direct the chemistry. I would argue that if this pathway intersects the ground state reaction coordinate, typically in the barrier region, then these modes are also involved in the chemistry along the ground state surface. The point can be understood by simply playing the atomic movie of the photoinduced reaction dynamics backwards in time and the barrier region is reconstructed. Even if this is only one subset of possibilities of the thermally sampled barrier, it is precisely the needed information to recreate a saddle point in the barrier region. It is the back projection that gives the critical nuclear arrangements that open a seam to the product surface and define the barrier region. It is these key reaction modes we wish to characterize to develop a new dynamic reaction mode basis for conceptualizing chemistry. It is not completely inclusive of all possible pathways but it gives enough information on the critical arrangement of atoms at the barrier crossing region to more accurately think about manipulating the barrier height and better controlling the chemistry.

The problem is that there are precious few systems that can be triggered under barrierless conditions and it is highly unlikely that a systematic means of discovering new photochemical systems can be found at this point. The ultimate solution is to drive chemistry along the ground state surface and explicitly take advantage of what has been learned from atomically resolved reaction dynamics to date, namely there are only a few key modes in the barrier crossing process. Again, one needs 100 fs time resolution which dictates that any experimental protocol must involve femtosecond laser approaches to initiate the chemistry. There are at least two possible avenues to explore. The most obvious is to use femtosecond IR pulses to drive the purported reaction modes. This idea dates back to the early discussion of laser selective chemistry, which has largely been abandoned due the extremely fast intramolecular vibrational redistribution processes that quickly dissipates the vibrational energy entropically as heat into the system modes. In this case, laser selective chemistry is no more selective than a Bunsen burner. The problem with past attempts is that generally only a single mode was adiabatically excited in a ladder climbing fashion up the anharmonic potential. At some point, well below the barrier crossing region, the energy is anharmonically coupled into modes orthogonal to the reaction coordinate. Another way to look at the wavepacket dynamics is that below the barrier crossing region, the potential is necessarily repulsive and keeps the system in the reactant half space. Also, as stated, previous studies have focused on a single vibrational mode. As shown above, chemistry undergoes this remarkable reduction in dimensionality, however, reactions requires at least 2–3 modes at the bare minimum to direct the chemistry. Previous failures in pursuing laser selectivity chemistry are due to this neglect. The challenge is to develop new laser sources capable of driving at least 2–3 different modes, with addressable phase relations, to give the last proper kick into the product channel on the canonical 100 fs timescale of barrier crossing. New high power femtosecond IR laser systems with IR pulse shaping now puts this prospect at least within reach.[94]

The other prospect discussed at the meeting is to use attosecond methods with shaped VUV pulses to mix electronic states and direct electron density into labile coordinates (Martín *et al.*, DOI: 10.1039/C6FD00074F, Vacher, Robb *et al.*, DOI: 10.1039/C6FD00067C).[87] The fields involved must be on the order of the atom Coulomb potential such that this process will occur from highly ionized excited states. It will not be possible to connect the reaction dynamics of these highly excited ion states to ground state chemistry, however, as pointed out in the Discussion this process does provide an interesting new proving ground for theory.

Other means for controlling the potential energy surface or perturbative means needs to be found. The work discussed on the use of optical cavities to shape excited state surfaces by Mukamel *et al.* (DOI: 10.1039/C6FD00095A) holds promise for controlling chemistry with modest input fields. The local field is still a high field problem. This approach necessarily involves extremely low number densities of molecules to single molecule regimes. Here, the velocity map imaging and other charged particle detection schemes are nearly perfect platforms for this class of studies. With increased electron source brightness, it is conceivable that even imaging of single molecule reaction dynamics within optical cavities is possible with repetitive sampling of identically prepared systems. The significant advances in electron source brightness would also enable imaging of cross molecular beam reactions to open up the study of bimolecular, second order, reaction mechanisms.

Other perturbations need to be explored as well. Most molecules are unstable in the anion state and undergo electron induced reactions. The use of either hot electrons to drive chemistry as recently shown for initiating phase transitions[95] or the use of femtosecond electron pulses *in lieu* of optical excitation can be used to induce chemistry and this process would be along ground state anion surfaces. Other perturbative variables not involving photoexcitation such as extreme pressure and temperature conditions can be used to induce chemical processes very near the primary timescales. This prospect is possible if the perturbation is of sufficient magnitude to emulate the instantaneous conditions corresponding to the rare statistical events where the system has sufficient energy to overcome the barrier.

If the phototrigger challenge can be solved and in particular if multimode IR driven chemistry can be developed, it would enable atomically resolved reaction dynamics for all classes of chemistry. The single most important outcome would be to open up the systematic study of catalysts. We presently have a very poor understanding of why certain metals and substituents make better catalysts than others. The single most glaring example to illustrate this lack of understanding is our inability to develop a water splitting photocatalyst that operates even remotely close to the quantum efficiency of PSII (see discussion of Yano *et al.*, DOI: 10.1039/C6FD00084C).

Sample challenge

There are different problems for sample preparation that depend on the structural probe to image the chemistry of interest. The method of inducing the chemistry also needs to be considered. For photochemistry, the photoexcited and probed volumes must match well with a sufficient fraction excited above background. The needed fraction excited is typically on the order of 10%. This absorbed photon requirement needs to be scaled with respect to absorption cross sections to ensure the excitation peak power stays below thresholds leading to multiphoton ionization effects. This threshold is on the order of 100 GW cm^{-2}. This statement is reinforced by the reported optical damage thresholds for high power laser optics that are generally below 10 GW cm^{-2}. For condensed phase samples (solid or liquid state), the pathlength must be on the order of microns, and must employ very strong absorbing states to decrease the number of incident photons (peak power) to stay within safe peak power limits. This statement is predicated on typical densities for condensed phase systems and does not pertain to gas phase samples. In the gas phase case, it only depends on the number density needed to observe signal and velocity mismatch issues. The maximum incident peak power in all cases should not exceed 100–200 GW cm^{-2} as a general guideline. For studies of condensed phase systems using electron probes, the samples have to be on the order of 100 nm thin (depending on electron energy) irrespective of the laser excitation conditions but the point of the peak power limitations needs to be equally respected. The challenge is to devise a means to prepare high quality samples with sufficient surface area to provide the "movie film" for either electrons or X-ray probes with the above constraint. New methods of sample preparation need to be developed. Present methods involve microtoming, arrested crystal growth, ion milling, spin coating. All have to be refined to make sample preparation more routine. The machine physics for atomically resolving reaction dynamics has been solved. Now, it all comes down to samples.

In terms of sample selection, we need to systematically explore different types of chemistry. We are looking for a sample preparation platform that will meet the above limits. In this respect, the use of polymer chemistry to introduce photoactive systems would provide a generalized approach. It is straightforward to spin coat polymers or drop cast polymers to the desired dimensions with nm precision over large areas. The film could be in endless supply. The problem is that the polymer host must not constitute more than 90% of the scattering centres so not to obscure the atomic motions of interest. Also the amorphous nature of polymers reduces greatly the sharpness of the diffraction patterns. There are means to create oriented polymer films to help in promoting constructive interference for reciprocal space imaging. There are also a number of other materials that can be processed as thin films to provide the necessary large area films for both electron and X-ray diffraction experiments. The advent of numerous nanofabrication methods have opened up a number of different means for obtaining the necessary sample geometries.

In principle, the sample preparation problem can be solved. We really need to enlist the help of synthetic chemists to provide systems poised to undergo a specific type of chemistry. To date, all the problems studied have involved well known photochemistry and standard model problems. We need to move beyond the standard problems and start to design molecular systems specifically to test various reaction mechanisms. The inspiration for specific molecular targets can be gained from theoretical calculations. The enormous success in the design of a synthetic analogue to mimic the primary photoisomerization steps involved in vision is a case in point.[96] The system was synthesized based on the truncated model used in *ab initio* calculations to treat the photoisomerization observed in rhodopsin. The molecule was designed *in silico* and was found to nearly exactly replicate the primary processes, even the observed dynamics, of the primary step of vision.[96–99] (This greatly truncated molecular system and its success in replicating the photochemistry of retinal is just another example of the key chemistry involving highly localized motions.) Given this success and recent advances in theory, I can see in the not too distant future that *ab initio* theories using a refined localized basis (to be discussed later) generate *in silico* structures to test various aspects of different reaction mechanisms. It is at this point that we need synthetic chemists to make these constructs and to play an active role in helping drive the field.

This effort will be largely frustrated if we do not develop standard templates for sample preparation and delivery to meet the requirements for atomically resolved structural dynamics. There are a few approaches that could be amenable to systematic synthetic strategies such as polymers discussed above and metal organic frameworks that are naturally 2D materials for which different active groups could be introduced. These are just two examples of many different support structures that could be used as templates for targeted synthesis. A solution to sample delivery in some form of common template is crucial to advancing the field.

Towards the quantum limit in imaging chemistry

The major issues associated with this challenge have been largely discussed in the Introduction. The most important technological advance was the development of ultrabright electron and X-ray sources of sufficient brightness to literally light up atomic motions under single shot or framing conditions. Of the two sources, electrons have more than an order of magnitude shorter de Broglie wavelength and amplitude stability than the corresponding XFEL sources. The highest space–time resolution has been achieved with electrons with time resolution on the 100 fs times cale and 0.01 Å spatial resolution.[57] In addition, the phasing problem in inverting diffraction observables to real space motions has been solved as discussed above by explicitly taking advantage of conserved atomic positions or information to provide phase references for the inversion process. This space–time resolution has been achieved without any models or refinement using theoretical inputs. These are the first full atom resolved chemical reactions. It is notable that the spatial resolution of 0.01 Å is comparable to zero point motions of small Z (<10) atoms. In this regard, the fundamental space–time requirements for imaging chemistry have been achieved. What are the ultimate limits? The reason this question is important is that the net displacement of the atoms can be directly related to the energetics of the reaction coordinate. From the dynamics, the reaction forces can be determined and the net displacements give the energetics. For example, the photoinduced electron transfer for $Pt(dmit_2)_2$ involves a net reorganization energy on the order of 1 eV typical for electron transfer processes[30] and has fully relaxed displacements of 0.1 Å. Spatial resolution of 0.01 Å translates to following motions corresponding to relaxation processes dissipating less than 100 meV. The higher the spatial resolution, the more accurate the measurements will be in comparing to theory and could achieve energy accuracy requirements comparable to spectroscopy. The important difference in benchmarking theory against prediction of specific nuclear motions as opposed to describing electronic states is that one is observing the actual motions (absolute measures) as opposed to relative changes in electronic states. The spatial resolution is determined by the SNR in the image reconstruction, which in turn depends on the source brightness, source noise, and sampling noise. For a given sample, the problem reduces to source issues. The brighter and more stable the imaging source, the higher the spatial resolution and dynamic range for contrast. Electrons have the edge over X-rays in this regard due to the much shorter wavelength and lower amplitude noise. As good as the current electron sources are, there is room for at least a factor of 100 increase in electron source brightness. The fundamental quantum limit has been addressed by the Luiten and Scholten groups.[100,101] New electron source technology based on ultracold atoms will approach this quantum limit. The other approach is to develop new photocathodes with lower emittance. The work of Bazarov *et al.*[102,103] has led to the development of trialkali photocathodes with an order of magnitude lower emittance and can be lowered another factor of 10 by going to lower temperature, for an overall factor of 100 brighter source. These developments along with improved sample preparation to increase the amount of signal averaging possible for image reconstruction will enable at least an order of magnitude higher spatial resolution with normal assumptions on scaling with increased SNR. Gas phase electron diffraction routinely obtains 0.001 Å resolution or better (*e.g.* ref. 104). I expect we will soon achieve similar resolution limits with fs atomically resolved reaction dynamics, which then puts our effective resolution to the reaction energetics below 10 meV or thermal noise. This is the true fundamental limit. It should also be pointed out that higher space–time resolution will also greatly enhance the ability to better resolve probability distributions as pointed out by Ischenko *et al.* using tomographic reconstruction[35] and also in the paper of Centurion *et al.* (DOI: 10.1039/C6FD00071A). This feature of diffractive imaging will be extremely revealing in the study of reaction dynamics involving multiple pathways and could resolve a number of issues with respect to competing ideas of dominant relaxation channels.

In addition to source issues, there is also considerable room for improvement in detection and theory of image reconstruction. As shown in the work of Ishikawa *et al.*,[57] prior knowledge of the system can provide important mathematical constraints in the image reconstruction. What is the fundamental limit with respect to detected signal counts that we need to image chemistry for a given resolution? This question has to be posed in relation to the background noise of the detector. For both XFELs and electron sources, there has been tremendous progress made in detector technology such that detection can approach noise free conditions. Each ping on the detector is a real signal count. This seemingly trivial technical detail means that it is possible to do high order spatial correlations of the accumulated signal with known constraints to reconstruct an image with far less signal. I would hazard a guess that within the next few years with the development of new algorithms for image reconstruction that we will be able to obtain atomically resolved reaction dynamics for a given resolution with an order of magnitude less signal counts. This development will be comparable to increasing the source brightness by yet another order of magnitude.

It is exciting to think of the level of detail we will soon be able to tease out of direct observations of chemical reactions. We will soon (I hope) be able to directly observe subtle quantum effects such as the Berry phase in passing through conical intersections, quantum interference effects involved in the reduction of phase space during barrier crossing *etc.* (*vide infra*). Higher space–time resolution will equate to more information and a higher level of understanding of reaction dynamics.

The theory challenge: the critical link to a new conceptual basis for chemistry

As stated above, it will not be possible to experimentally solve the phototrigger problem for all classes of chemical reactions. It is conceivable we will be able to achieve laser selective chemistry by driving key vibrational modes along ground state potentials. This development will open up a significant number of chemical processes but still for the most part unimolecular processes. We will still not have similar information on second order collision induced reaction dynamics that constitutes the bulk of chemistry. It may be possible to create molecular constructs such as the diarylethene case discussed above to approximate binary collision complexes and new perturbation schemes may be developed but it will only be possible to study a small fraction of chemical processes. In order for this field to have a broader impact in chemistry, it will be essential for there to be a close marriage between experiment and theory in which higher space–time resolution in imaging chemical processes is used as a critical benchmark for refining theoretical approaches within the fixed number of test cases. Here, I would add that the goal does not mean that we need higher and higher accuracy in predicting precise motions or energetics. We need to be able to better predict scaling relations for different reaction mechanisms and the critical points, almost like nucleation sites, where the chemistry occurs with highest proficiency – for higher and higher levels of complexity. The primary directive of this field should be to develop the theoretical tools to the point that it will be possible to extend what has been learned from experimental studies of a few model systems to all chemical processes.

The main point of this summary paper is to point out that there was a fundamental bit of physics that was missing. Chemistry is only transferrable from one molecular moiety to another if there is some self-similar process repeated over and over again for what are otherwise distinctly different many body systems. The first atomically resolved chemical processes have directly shown that there is an enormous reduction in dimensionality that occurs in the barrier crossing regions. The strong anharmonic mixing between high frequency and low frequency modes within a normal mode basis leads to these localized motions. The idea of key reaction modes is not new but to see it first hand to establish this as fact is. It is also inspiring to see the simplicity and beauty of this virtual orchestrated chemical dance. The very fact that we can now directly observe the key modes directing the chemistry establishes this phenomena as the missing bit of physics.

Here I wish to emphasize my most important point of this article. If one can identify the key modes, then it is possible to conceptually displace these modes with proper weighting to better approximate the nuclear configuration of the transition state region or critical point. This highly simplified view will empower chemists to better design molecular scaffolds to control barrier heights to achieve the desired chemistry. Chemistry is by definition a dynamical process by which matter interconverts from one form to another. The development of a reaction mode basis would properly capture the operating physics and would be no more complicated to implement than current methods based on "chemical intuition" and static structural end points. The more accurate depiction of the reaction saddle point within the context of the right physical picture for sampling this critical point would represent a major advance in conceptualizing chemistry. This in turn would translate to better control over barriers to direct the desired chemistry and scaling chemistry to higher levels of complexity. The current limits for *de novo* synthesis and imparting functions are well represented by the efforts in developing molecular motors.[105] Chemistry is stuck on the nm scale at present. The proposed dynamic structural basis may provide the improved level of understanding needed to scale chemistry to greater complexity and the *de novo* creation of active matter with functionalities rivalling biological systems.

The problem with this prospect, as pointed out in the discussion at the meeting, is that even nominally very similar molecules sharing a common parent structure show significantly different excited state dynamics. There have been attempts to find general trends in molecular dynamics such as the energy gap law that ultimately have proven not to hold in general (see discussion by Dave Townsend, DOI: 10.1039/C6FD00092D). The basic argument, and one with which I agree, is that the best means to refine theory is to be able to account for the very details that make each molecule unique. Different molecules have different structures and compositions and hence different many body potentials and properties. The excitation of one molecule in reference to another leads to a different spatial distribution of electrons in the excited state and as a direct consequence a different excited state potential energy surface leading to different forces acting on the atoms. The excited state dynamics are never the same for two different molecules. Each excited state lifetime, each nonradiative relaxation pathway, represents a molecular finger print of the unique many body potential. The ultimate goal for theory is to be able to accurately treat these excited state surfaces to the point of being able to predict the dynamics for any given system. The problem, however, is that in the search for quantifying uniqueness we lose the generalities and the simple physics that guides chemistry and makes chemical concepts transferable. The full problem also requires significantly more resources and time to execute than theoretical methods that encompass more general trends. To help guide chemistry, we need to find these general features that will compete with "arrow pushing" now in common use. We need to reveal the trends that will help to conceptualize chemistry more generally.

Let me take a few cases in point to illustrate the approach I think is needed. Let's examine the breakdown in the energy gap law, given as an example in the discussion. This generalization was formulated based on experimental studies of nonradiative relaxation in which trends were easiest to observe and do controls for the annulene system.[106–108] The larger the energy gap, the poorer the Franck–Condon (FC) overlap to assist internal conversion from the excited state to the ground state. This general observation for a given class of molecules helped formulate the need to conserve momentum in undergoing nonradiative relaxation. The momentum conservation condition is properly captured in the form of the overlap of the nuclear wavefunction probability distribution between different electronic surfaces (*i.e.*, FC factors). This level of understanding heavily influenced synthetic chemistry in the development of photostable dyes for photography and dye laser applications,[109] and solar energy conversion applications.[110] It was realized by Drexhage that substituents used to shift the absorption band for colour tuning had to employ a means to keep a rigid in-plane structure and avoid low frequency modes that lead to FC resonances. The longer the lifetime, the higher the chance of radiative relaxation, as opposed to nonradiative relaxation leading to hot ground state chemistry and degradation. The result was a very large class of photostable dyes that were found through a better understanding of the physics leading to the energy gap law. This "law" does not hold in general with the most glaring exception being the nucleobases of DNA. The bases absorb well into the UV and by the energy gap law should be long lived excited states, prone to UV photochemical degradation. DNA is known to have short lived excited states based on fluorescence quantum yields and oscillator strengths.[111] This breakdown in the energy gap law could be attributed to a high density of resonant FC modes as a result of a high DOS of low frequency modes to create resonant combinations. It is only now with much improved theory and experiment that we have found that this class of molecular systems have strongly coupled out of plane motions that create not just 1 but 2 conical intersections as a general feature in assisting very fast nonradiative relaxation.[112] We see new excited state relaxation pathways yet again when these same nucleobases are incorporated into DNA. The specific details of the molecules matter in terms of the excited state dynamics. However, I would argue that within a given class of molecules it is possible to make very accurate predictions based on observed trends. The nucleobases all show excited state surfaces that involve out of plane coupling modes and formation of CIs. This can now be predicted as a general feature knowing something about the coupling of modes conserved in the nucleobase structure – and not present in rigidized annulenes. With a mode specific basis for understanding coupling to electronic surfaces, the basic trends controlling the excited state nonradiative relaxation dynamics could be predicted.

Let me take a more relevant example with respect to developing a reaction mode basis. Consider the case of the photoisomerization of acetylene as treated by Stolow *et al.* (DOI: 10.1039/C6FD00117C) in the Discussion. This theoretical study connected the calculated reaction pathway to the XAS observable, pointing out that the reaction involves bond elongation, and torsional motion of the H(D) atoms and pyrimidalization *en route* to the photoisomer. This system is the smallest possible construct for isomerization. The primary step of vision involves the same photochemical process. The chromophore is retinal with 140 vibrational modes or degrees of freedom, which is bound within the opsin protein comprising over 350 amino acids. This system has well over 1000 dimensions in terms of possible arrangements of atoms potentially involved in the photoisomerization. How can this photoisomerization work within such a high degree of complexity? Here the beauty and power of reducing chemistry to its key reaction modes becomes apparent. For retinal itself there are 4 labile double bonds. The absorption of a photon excites an electron to an antibonding character excited state, which will lead to bond softening along any of the formal double bonds. For the protonated Schiff base of retinal in solution, all possible photoproducts are observed with quantum yields of only a few % to >10%.[113] The nonrigid nature of retinal leads to a high density of low frequency resonances for large FC overlap between electronic states, which as we learned from the physics giving rise to the energy gap law, leads to very fast nonradiative relaxation back to the ground state. Vision had to evolve a means to direct the primary step of vision to compete with this inherently ultrafast process. Recent studies have shown that the excited state potential is so highly optimized that the initially excited FC modes are coupled to the three key modes involved in isomerization, *i.e.*, bond elongation specifically at the C11=C12 position, the localized torsion at this same bond, and hydrogen out of plane motion.[99] The surrounding protein must create local fields that localize the node in the excited state wavefunction of the S_1 state at the C11, C12 position, as well as provide repulsive terms to couple to the torsion at this location. Please note, the reaction is highly localized and reduces to the very same motions involved in the photoisomerization of acetylene. A 1000+ dimensional problem with proper engineering of the system has been reduced to 3 key modes to direct the chemistry faster than any other competing process. There is only one photoproduct with an overall quantum yield of 65%. The reaction mechanism involving isomerization along polyenes, of which acetylene is the simplest possible construct, all involve 3–4 key reaction modes in the barrier crossing region. The barrier can be manipulated by proper arrangement of surrounding atoms to create spatially distinct "hot" reaction sites for which the chemistry occurs at this site orders of magnitude faster than competing processes. It needs to borne in mind that the primary step of photoisomerization involved in vision is one of the most highly optimized reaction coordinates known, with exceptional evolution pressure to optimize the quantum yield. The primary step of vision had to be optimized against the inherently ultrafast competing channel of nonradiative relaxation in polyenes that occurs on 100 fs timescales. This reaction has been so highly optimized that the motion through the CI occurs within a half period of the principle torsion mode in a concerted fashion involving all three key modes, *i.e.* occurs ballistically.[99] The reaction could not occur faster. This finding is marvellous to behold as there is no means to differentially lower the barrier of the desired process relative to the competing processes. The competing channel of nonradiative relaxation in this case is an intrinsic property of this class of molecules. The process had to be optimized to outrun nonradiative relaxation in polyenes. In conventional chemistry, the barriers of competing processes can be equally manipulated by increasing the barrier height in combination with the means used to lower the barrier for the desired process. Even relatively small differences in barriers between competing channels on the order of 0.1 eV can result in acceptable yields for synthetic chemistry.

I could cite other examples of optimized reaction coordinates in biological systems that involve complexity well beyond that encountered in current synthetic challenges. Irrespective of complexity, it is possible to have "hot" sites in the system for which the local barrier to a desired process can be made to be lowered relative to other competing processes. Theoretical methods need to be developed that focus on casting out the modes most strongly coupled to the reaction coordinate of interest. For example, the work of Olivucci *et al.* that led to an *in silico* analogue for vision was based on very detailed *ab initio* calculations.[98] From the experimental studies, this same analogue could have been deduced by consideration of the key modes and molecular scaffold to minimize the barrier to these motions as discussed above, *i.e.* by taking a page out of the book written by nature.

The challenge is to develop theoretical tools that enable projecting out the principle reaction modes from within a critical volume element of a system of arbitrary complexity. Even in the limit we can rigorously solve the full problem, the methodology will not lend itself to a simple form that will enable synthetic chemists to readily plan reactions. The generalities will be lost in favour of rigour. Theoretical approaches need to be developed that take advantage of the fact that we have additional information on the reaction coordinate. From the atomically resolved reaction dynamics, it has now been established that there are only a few dominant reaction modes that direct the chemistry. We need to focus on the few systems that will be forthcoming from the different experimental methods discussed at this meeting to test more coarse grain theories that fully exploit this information. Mike Robb (DOI: 10.1039/C6FD00067C) made an interesting suggestion to explore valence bond methods within this spirit to reduce the computation time to find the key modes and then refine from there. This theoretical approach can be benchmarked against the forthcoming atom resolved reaction dynamics and more rigorous theories in terms of properly predicting the key reaction modes and magnitude of displacements.

Here I have to emphasize the degree of accuracy does not have to conform to the accuracy of comparison to spectroscopic observables. Order of magnitude predictions of barrier distributions to project out preferred pathways is all that is needed to guide synthesis. We know this approach will work in seeing how far synthetic strategies have gone with comparatively no theory to guide them but so called chemical intuition and a highly localized depiction of reaction pathways based on static structures. This approach has led to the discovery of a wide variety of reaction mechanisms. The explicit inclusion of dynamic sampling of multiple pathways, even if not exhaustive, will represent a major advance in strategic planning reaction routes. We need to focus not on the absolute dynamics but relative dynamics. Otherwise, we risk "not being able to see the forest for the trees". As long as we can identify the key modes most strongly coupled to a desired reaction coordinate, we will know *a priori* that the displacement of these modes will open up seams to the product surface. The systematic displacement of these modes within a given molecular construct will give a reasonable estimate of the critical point in the reaction surface. Chemical means to reduce this barrier, just as nature discovered in the case of vision, can then be employed to better direct chemistry. Here it has to be stressed that the absolute rates or dynamics may be well off from the most rigorous calculations but the effect on the barrier heights for the competing processes will similarly be off. It is the control of the differential barrier for the desired process in relation to competing processes that is most important. This approach would give synthetic chemistry new tools for manipulating barrier heights and potentially extend chemistry to new length scales and complexity.

I can well see a day in which we teach chemistry using a reaction mode basis for different classifications of chemical reactions and molecular systems. These reaction modes may need to be classified according to the type of molecular architecture that facilitates motion along these coordinates as in the case for the energy gap law for annulenes or isomerization modes for polyenes. The specific critical structures could be projected to enable chemists the means to think about lowering or raising this barrier. The pedagogical approach would be to explore different displacements of these modes using real time theoretical tools to approximate the potential energy surface. This approach would allow a more accurate conceptualization of the nuclear configurations comprising the transition state region and means to control barrier heights and thereby chemistry. For a given synthetic strategy, the desired pathway could be ranked according to predicted reaction rates/barriers for other possible competing pathways. At this point, higher level, more costly, computations could be justified.

If the above can be realized, we will have formulated a new conceptual basis for chemistry, one that bridges both the structural and dynamical aspects that form the foundation of chemistry. We will have finally resolved our long standing space-time conundrum in conceptualizing chemistry. It is here where I see a very bright future for atomically resolved reaction dynamics, the theme of this *Faraday Discussion*.

4. Summary and personal perspective of the status of the field

I would like to close with one last personal observation. I recall as a student first studying chemistry thinking that I had missed the most exciting times in science. I suspect many reading this article share this sentiment. In the Introduction, I highlighted the historical developments with respect to formulating a structural basis for chemistry and our fascination with going to higher and higher spatial and temporal resolution to understand chemistry. The development of quantum mechanics and extension to chemistry was occurring in parallel. It is hard to imagine the enormous clarity that emerged once the dual particle–wave nature of matter was discovered and could be so elegantly treated mathematically within quantum theory. However, there is never just one defining moment. Science is constantly going through major advances that oscillate between fields (the quantum revolution, the structure of DNA and molecular genetics, *etc.*). I feel we are at a similar turning point in chemistry. With the recent developments on both experimental and theoretical fronts, we are beginning to see with much greater clarity the primary processes directing chemistry. Most significantly, we have arrived at the fundamental space–time resolution needed to directly observe atomic motions during the very act of chemistry – the dream of every chemist.

We now have the opportunity to solve one of the most perplexing problems in science. Chemistry involves highly nonlinear, high dimensional, problems; yet of the myriad of possible rearrangements to interconvert from one form of matter to another, there seem to be a small subset of solutions. There is an enormous reduction in dimensionality in which only a few dominant modes define the barrier crossing region and direct the chemistry during a reactive crossing. It is this reduction in effective dimensionality that provides this "magic" of chemistry and makes chemical reaction mechanisms transferable to nearly any molecule of arbitrary complexity. We can now directly observe this process and as such have an opportunity to develop a reaction mode basis for chemistry, which explicitly bridges the importance of structure and dynamics to understanding chemical processes. The study of physical chemistry has progressed from simple systems with limited information to the study of systems as complex as proteins. We can now follow the chemistry at the atomic level of detail for systems relevant to chemistry and biology. We are no longer talking about few atom systems with little consequences for the practice of chemistry. These developments will undoubtedly influence how we conceptualize and execute chemistry.

At this point, my feeling is that we will still be thinking in terms of "balls-and-springs" in picturing chemical processes. We need to continually force ourselves to include the wave particle duality to the problem. If one considers the quantum aspects of the problem, the anharmonic coupling between high frequency and low frequency modes leads to localized motions precisely because of interference terms in the vibrational wavefunctions.

In the spirit of trying to summarize the operating physics in the simplest possible terms, Fig. 6 shows the distinctive difference in the motion of a heavy effective mass for equilibrium and far from equilibrium motions in sampling the highly anharmonic region of a barrier crossing. Only two reduced reaction coordinates are shown, which is not so different than the case found for many complex systems. The executed motion of this classical ball sampling the highly anharmonic region leads naturally to additional Fourier components in its motion. It is the curvature of the potential at this point that leads to coupling between high and low frequency modes—and which modes. Now, we need to include the wave properties into this picture. The balls really represent vibrational wavepackets for which you have to consider the relative phases of the mixed

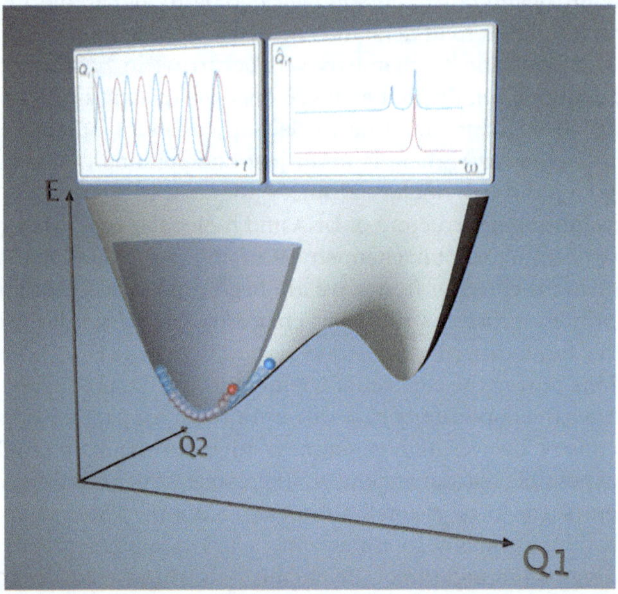

Fig. 6 Anharmonic mode coupling and reduced dimensions. The red ball shown is executing motion (time trace in inset) within harmonic limits (dark grey potential). There are 3*N*-6 orthogonal eigenfunctions that describe all possible equilibrium fluctuations using a normal mode basis. The eigenfunctions are stationary states by definition and there is no prospect of chemistry. Anharmonicity is the spice of life in that it is this aspect of statistical sampling anharmonic regions that leads to chemistry. The blue ball shows anharmonic motion (inset: time traces of motion and Fourier transform to frequencies). It is the superposition of these motions that couple to the electronic states to create the Barrier region, which acts like a doorway to the product state. The interference between high and low frequency modes leads to localized motions, and thereby reduced dimensions to the problem. The lowest frequency modes act to both propagate the system and *via* the high DOS in this spectral region to dissipatively trap the motion on the product half space. It is the coupling of these motions to the electronic states that determines the potential energy surface to create these passageways by their very motion – which will be similar for locally similar many body potentials.

wavefunction at any instant in time. There will be constructive and destructive interference of the vibrational modes of this superposition state and this will lead to localized motions. The enormous reduction in complexity and localization of atomic motions is exactly what is observed for vibrational overtones,[114–116] precisely at the typical vibrational energies corresponding to barrier positions. Given the enormous anharmonicity in the barrier region, it is difficult to imagine the wild interference effects that lead to chemistry at this point in nuclear configuration space. It is this basic physics that harbours promise to recast chemistry in a reduced reaction mode basis.

Returning to the objective to bring the wave picture into our thinking, it is extremely difficult to imagine all the interference terms that must be occurring under the highly anharmonic conditions directing atomic motions in the barrier crossing region (local maximum in Fig. 6). It must be a wild ride. (I have spent countless hours observing interference of water waves with unusual landscapes and scattering obstacles to make this statement.) Most of this information is lost anyway in statistical averaging and classic pictures provide similar results. However, if the last technological hurdles to laser driven mode selective chemistry can be solved, the theoretically predicted reaction mode basis can be used to guide experiment. This will reduce the enormous experimental parameter space to lock onto the key modes to drive the chemistry in a form of dynamic refinement. This information is directly relevant to ground state chemistry and would provide precise information on mode coupling needed to deterministically manipulate barriers. In this scenario, it will be possible to explicitly play with the relative phase relationships to master barrier transmission – a purely quantum effect. There are also important problems in directing wavepackets near conical intersections, as a scattering problem, in which the accumulated phase changes lead to destructive and constructive inference through the CI, or "spooky" disappearance of reaction probability for certain momentum states.[117,118] In short, we will be able to directly observe the "quantumness" of chemistry. We will enjoy the interference effects in how we observe the dynamics of matter waves. We are in for exciting times ahead.

Acknowledgements

I would like to thank the Max Planck Society for support. I would like to thank in particular Marcos Alta, Xinxin Chen, Gastón Corthey, Stuart A. Hayes, and María S. Grigera for help in preparing this manuscript and providing the figures and movies.

References

1 R. J. D. Miller, *Annu. Rev. Phys. Chem.*, 2014, **65**, 583–604.

2 R. J. D. Miller, *Science*, 2014, **343**, 1108–1116.

3 R. J. D. Miller, *Acc. Chem. Res.*, 1994, **27**, 145–150.

4 R. D. Miller, *Can. J. Chem.*, 2002, **80**, 1–24.

5 J. R. Dwyer, C. T. Hebeisen, R. Ernstorfer, M. Harb, V. B. Deyirmenjian, R. E. Jordan and R. J. D. Miller, *Philos. Trans. R. Soc. London, Ser. A*, 2006, **364**, 741–778.

6 M. R. Armstrong, J. P. Ogilvie, M. L. Cowan, A. M. Nagy and R. J. D. Miller, *Proc. Natl. Acad. Sci. U. S. A.*, 2003, **100**, 4990–4994.

7 M. Walther, V. Raicu, J. P. Ogilvie, R. Phillips, R. Kluger and R. J. D. Miller, *J. Phys. Chem. B*, 2005, **109**, 20605–20611.
8 M. Eckert, *Ann. Phys.*, 2012, **524**, A83–A85.
9 W. H. Bragg and W. L. Bragg, *Proc. R. Soc. London, Ser. A*, 1913, **88**, 428–438.
10 P. Atkins and J. de Paula, *Atkins' Physical Chemistry*, OUP, Oxford, 2010.
11 K. Lonsdale, *Nature*, 1928, **122**, 810.
12 A. J. Rocke, *Ann. Sci.*, 1985, **42**, 355–381.
13 L. Pauling, *J. Am. Chem. Soc.*, 1931, **53**, 1367–1400.
14 R. J. Gillespie and R. S. Nyholm, *Q. Rev. Chem. Soc.*, 1957, **11**, 339.
15 R. J. Gillespie, *J. Chem. Educ.*, 2004, **81**, 298.
16 F. Scholz, D. Himmel, F. W. Heinemann, P. v. R. Schleyer, K. Meyer and I. Krossing, *Science*, 2013, **341**, 62–64.
17 P. Hänggi and M. Borkovec, *Rev. Mod. Phys.*, 1990, **62**, 251–341.
18 D. G. Truhlar, B. C. Garrett and S. J. Klippenstein, *J. Phys. Chem.*, 1996, **100**, 12771–12800.
19 H. Frauenfelder, G. Chen, J. Berendzen, P. W. Fenimore, H. Jansson, B. H. McMahon, I. R. Stroe, J. Swenson and R. D. Young, *Proc. Natl. Acad. Sci. U. S. A.*, 2009, **106**, 5129–5134.
20 H. J. Morowitz, *Beginnings of Cellular Life: Metabolism Recapitulates Biogenesis*, Yale University Press, 1st edn, 2004.
21 K. A. Dill and H. S. Chan, *Nat. Struct. Biol.*, 1997, **4**, 10–19.
22 W. A. Smit, A. F. Bochkov and R. Caple, *Organic Synthesis: The Science Behind the Art*, Royal Society of Chemistry, 1998.
23 G. Porter, *Science*, 1968, **160**, 1299–1307.
24 M. Eigen, *Nobel Lecture*, 1967, **11**, 1963–1979.
25 M. P. Minitti, J. M. Budarz, A. Kirrander, J. S. Robinson, D. Ratner, T. J. Lane, D. Zhu, J. M. Glownia, M. Kozina, H. T. Lemke, M. Sikorski, Y. Feng, S. Nelson, K. Saita, B. Stankus, T. Northey, J. B. Hastings and P. M. Weber, *Phys. Rev. Lett.*, 2015, **114**, 255501.
26 D. D. Dlott, *Annu. Rev. Phys. Chem.*, 1986, **37**, 157–187.
27 R. Pearman and M. Gruebele, *J. Chem. Phys.*, 1998, **108**, 6561.
28 J. M. Lanzafame, R. J. D. Miller, A. A. Muenter and B. A. Parkinson, *J. Phys. Chem.*, 1992, **96**, 2820–2826.
29 J. M. Lanzafame, S. Palese, D. Wang, R. J. D. Miller and A. A. Muenter, *J. Phys. Chem.*, 1994, **98**, 11020–11033.
30 *Surface electron transfer processes*, ed. R. J. D. Miller, VCH, New York, 1995.
31 J. Schnadt, P. A. Brühwiler, L. Patthey, J. N. O'Shea, S. Södergren, M. Odelius, R. Ahuja, O. Karis, M. Bässler, P. Persson, H. Siegbahn, S. Lunell and N. Mårtensson, *Nature*, 2002, **418**, 620–623.
32 D. J. Nesbitt and R. W. Field, *J. Phys. Chem.*, 1996, **100**, 12735–12756.
33 R. Henderson, *Q. Rev. Biophys.*, 1995, **28**, 171–193.
34 A. A. Ischenko, V. V. Golubkov, V. P. Spiridonov, A. V. Zgurskii, A. S. Akhmanov, M. G. Vabischevich and V. N. Bagratashvili, *Appl. Phys. B: Photophys. Laser Chem.*, 1983, **32**, 161–163.
35 A. A. Ischenko, L. Schaefer and J. D. Ewbank, in *Proceedings SPIE* 1998, 3516, 580-587, ed. V. P. Degtyareva, M. A. Monastyrski, M. Y. Schelev and A. V. Smirnov, 1999.

36 A. A. Ischenko, L. Schafer and J. Ewbank, in *Time-resolved Diffraction*, ed. J. R. Helliwell and P. M. Rentzepis, Oxford University Press, New York, 1997, p. 323.
37 S. Williamson, G. Mourou and J. C. M. Li, *Phys. Rev. Lett.*, 1984, **52**, 2364–2367.
38 V. V. Golubkov, A. V. Zgursky, V. I. Petrov and A. A. Ischenko, Pulsed picosecond electron source, *Proceedings of the 8th European Conference on Electron Microscopy*, Budapest, 1984, pp. 61–62.
39 S. A. Akhmanov, V. N. Bagratashvili, V. V. Golubkov, A. V. Zgurskii, A. A. Ischenko and V. G. Tunkin, Production of picosecond pulses of fast electrons by laser-induced photoemission in an EMR-100 electron diffractometer, *Soviet J. Tech. Phys. Lett.*, 1985, **11**, 63–64.
40 J. C. Williamson, M. Dantus, S. B. Kim and A. H. Zewail, *Chem. Phys. Lett.*, 1992, **196**, 529–534.
41 H. Ihee, V. A. Lobastov, U. M. Gomez, B. M. Goodson, R. Srinivasan, C.-Y. Ruan and A. H. Zewail, *Science*, 2001, **291**, 458–462.
42 C.-Y. Ruan, F. Vigliotti, V. A. Lobastov, S. Chen and A. H. Zewail, *Proc. Natl. Acad. Sci. U. S. A.*, 2004, **101**, 1123–1128.
43 R. C. Dudek and P. M. Weber, *J. Phys. Chem. A*, 2001, **105**, 4167–4171.
44 S. Deb and P. M. Weber, *Annu. Rev. Phys. Chem.*, 2011, **62**, 19–39.
45 J. Charles Williamson and A. H. Zewail, *Chem. Phys. Lett.*, 1993, **209**, 10–16.
46 G. D. Goodno, G. Dadusc and R. J. D. Miller, *J. Opt. Soc. Am. B*, 1998, **15**, 1791.
47 A. A. Maznev, T. F. Crimmins and K. A. Nelson, *Opt. Lett.*, 1998, **23**, 1378.
48 T. Frigge, B. Hafke, V. Tinnemann, T. Witte and M. Horn-von Hoegen, *Struct. Dyn.*, 2015, **2**, 35101.
49 B. J. Siwick, J. R. Dwyer, R. E. Jordan and R. J. D. Miller, *Science*, 2003, **302**, 1382–1385.
50 B. J. Siwick, J. R. Dwyer, R. E. Jordan and R. J. D. Miller, *J. Appl. Phys.*, 2002, **92**, 1643–1648.
51 T. van Oudheusden, E. F. de Jong, S. B. van der Geer, W. P. E. M. O. 't Root, O. J. Luiten and B. J. Siwick, *J. Appl. Phys.*, 2007, **102**, 93501.
52 G. J. H. Brussaard, A. Lassise, P. L. E. M. Pasmans, P. H. A. Mutsaers, M. J. van der Wiel and O. J. Luiten, *Appl. Phys. Lett.*, 2013, **103**, 141105.
53 C. T. Hebeisen, R. Ernstorfer, M. Harb, T. Dartigalongue, R. E. Jordan and R. J. Dwayne Miller, *Opt. Lett.*, 2006, **31**, 3517–3519.
54 M. Hada, D. Zhang, K. Pichugin, J. Hirscht, M. A. Kochman, S. A. Hayes, S. Manz, R. Y. N. Gengler, D. A. Wann, T. Seki, G. Moriena, C. A. Morrison, J. Matsuo, G. Sciaini and R. J. D. Miller, *Nat. Commun.*, 2014, **5**, 3863.
55 C. Gerbig, A. Senftleben, S. Morgenstern, C. Sarpe and T. Baumert, *New J. Phys.*, 2015, **17**, 43050.
56 J. Hirscht, Dissertation, Universität Hamburg, 2015, p. 2015.
57 T. Ishikawa, S. A. Hayes, S. Keskin, G. Corthey, M. Hada, K. Pichugin, A. Marx, J. Hirscht, K. Shionuma, K. Onda, Y. Okimoto, S. Koshihara, T. Yamamoto, H. Cui, M. Nomura, Y. Oshima, M. Abdel-Jawad, R. Kato and R. J. D. Miller, *Science*, 2015, **350**, 1501–1505.
58 M. Gao, Y. Jiang, G. H. Kassier and R. J. Dwayne Miller, *Appl. Phys. Lett.*, 2013, **103**, 33503.
59 J. Cao, Z. Hao, H. Park, C. Tao, D. Kau and L. Blaszczyk, *Appl. Phys. Lett.*, 2003, **83**, 1044.

60 C. Guo and A. J. Taylor, *Phys. Rev. B: Condens. Matter*, 2000, **62**, R11921–R11924.
61 M. Kandyla, T. Shih and E. Mazur, *Phys. Rev. B: Condens. Matter Mater. Phys.*, 2007, **75**, 214107.
62 K. Franjic, M. L. Cowan, D. Kraemer and R. J. Miller, *Opt. Express*, 2009, **17**, 22937–22959.
63 S. Amini-Nik, D. Kraemer, M. L. Cowan, K. Gunaratne, P. Nadesan, B. A. Alman and R. J. D. Miller, *PLoS One*, 2010, **5**, e13053.
64 H. Petersen, F. Tavakoli, S. Kruber, A. Münscher, A. Gliese, N.-O. Hansen, S. Uschold, D. Eggert, W. D. Robertson, T. Gosau, S. Sehner, M. Kwiatkowski, H. Schlüter, U. Schumacher, R. Knecht and R. J. D. Miller, *Lasers Surg. Med.*, 2016, **48**, 385–391.
65 G. Sciaini and R. J. D. Miller, *Rep. Prog. Phys.*, 2011, **74**, 96101.
66 R. Ernstorfer, M. Harb, C. T. Hebeisen, G. Sciaini, T. Dartigalongue and R. J. D. Miller, *Science*, 2009, **323**, 1033–1037.
67 G. Sciaini, M. Harb, S. G. Kruglik, T. Payer, C. T. Hebeisen, F.-J. M. zu Heringdorf, M. Yamaguchi, M. H. Hoegen, R. Ernstorfer and R. J. D. Miller, *Nature*, 2009, **458**, 56–59.
68 M. Eichberger, H. Schaefer, M. Krumova, M. Beyer, J. Demsar, H. Berger, G. Moriena, G. Sciaini and R. J. D. Miller, *Nature*, 2010, **468**, 799–802.
69 T.-R. T. Han, Z. Tao, S. D. Mahanti, K. Chang, C.-Y. Ruan, C. D. Malliakas and M. G. Kanatzidis, *Phys. Rev. B: Condens. Matter Mater. Phys.*, 2012, **86**, 075145.
70 N. Erasmus, M. Eichberger, K. Haupt, I. Boshoff, G. Kassier, R. Birmurske, H. Berger, J. Demsar and H. Schwoerer, *Phys. Rev. Lett.*, 2012, **109**, 167402.
71 G. Sciaini and R. J. D. Miller, *Rep. Prog. Phys.*, 2011, **74**, 96101.
72 B. J. Siwick, J. R. Dwyer, R. E. Jordan and R. J. D. Miller, *Chem. Phys.*, 2004, **299**, 285–305.
73 M. Gao, C. Lu, H. Jean-Ruel, L. C. Liu, A. Marx, K. Onda, S. Koshihara, Y. Nakano, X. Shao, T. Hiramatsu, G. Saito, H. Yamochi, R. R. Cooney, G. Moriena, G. Sciaini and R. J. D. Miller, *Nature*, 2013, **496**, 343–346.
74 H. Jean-Ruel, M. Gao, M. A. Kochman, C. Lu, L. C. Liu, R. R. Cooney, C. A. Morrison and R. J. D. Miller, *J. Phys. Chem. B*, 2013, 131031130346003.
75 M. Irie, T. Fukaminato, K. Matsuda and S. Kobatake, *Chem. Rev.*, 2014, **114**, 12174–12277.
76 H. Jean-Ruel, R. R. Cooney, M. Gao, C. Lu, M. A. Kochman, C. A. Morrison and R. J. D. Miller, *J. Phys. Chem. A*, 2011, **115**, 13158–13168.
77 M. Woerner, F. Zamponi, Z. Ansari, J. Dreyer, B. Freyer, M. Prémont-Schwarz and T. Elsaesser, *J. Chem. Phys.*, 2010, **133**, 64509.
78 T. Elsaesser and M. Woerner, *Acta Crystallogr., Sect. A: Found. Crystallogr.*, 2010, **66**, 168–178.
79 V. Juvé, M. Holtz, F. Zamponi, M. Woerner, T. Elsaesser and A. Borgschulte, *Phys. Rev. Lett.*, 2013, **111**, 217401.
80 F. Zamponi, P. Rothhardt, J. Stingl, M. Woerner and T. Elsaesser, *Proc. Natl. Acad. Sci. U. S. A.*, 2012, **109**, 5207–5212.
81 P. Emma, R. Akre, J. Arthur, R. Bionta, C. Bostedt, J. Bozek, A. Brachmann, P. Bucksbaum, R. Coffee, F.-J. Decker, Y. Ding, D. Dowell, S. Edstrom, A. Fisher, J. Frisch, S. Gilevich, J. Hastings, G. Hays, P. Hering, Z. Huang, R. Iverson, H. Loos, M. Messerschmidt, A. Miahnahri, S. Moeller, H.-D. Nuhn, G. Pile, D. Ratner, J. Rzepiela, D. Schultz, T. Smith, P. Stefan,

H. Tompkins, J. Turner, J. Welch, W. White, J. Wu, G. Yocky and J. Galayda, *Nat. Photonics*, 2010, **4**, 641–647.

82 V. I. Prokhorenko, A. Halpin, P. J. M. Johnson, R. J. D. Miller and L. S. Brown, *J. Chem. Phys.*, 2011, **134**, 85105.

83 T. R. M. Barends, L. Foucar, A. Ardevol, K. Nass, A. Aquila, S. Botha, R. B. Doak, K. Falahati, E. Hartmann, M. Hilpert, M. Heinz, M. C. Hoffmann, J. Köfinger, J. E. Koglin, G. Kovacsova, M. Liang, D. Milathianaki, H. T. Lemke, J. Reinstein, C. M. Roome, R. L. Shoeman, G. J. Williams, I. Burghardt, G. Hummer, S. Boutet and I. Schlichting, *Science*, 2015, **350**, 445–450.

84 R. Miller, *Annu. Rev. Phys. Chem.*, 1991, **42**, 581–614.

85 G. D. Goodno, V. Astinov and R. J. D. Miller, *J. Phys. Chem. A*, 1999, **103**, 10630–10643.

86 K. Pande, C. D. M. Hutchison, G. Groenhof, A. Aquila, J. S. Robinson, J. Tenboer, S. Basu, S. Boutet, D. P. DePonte, M. Liang, T. A. White, N. A. Zatsepin, O. Yefanov, D. Morozov, D. Oberthuer, C. Gati, G. Subramanian, D. James, Y. Zhao, J. Koralek, J. Brayshaw, C. Kupitz, C. Conrad, S. Roy-Chowdhury, J. D. Coe, M. Metz, P. L. Xavier, T. D. Grant, J. E. Koglin, G. Ketawala, R. Fromme, V. rajer, R. Henning, J. C. H. Spence, A. Ourmazd, P. Schwander, U. Weierstall, M. Frank, P. Fromme, A. Barty, H. N. Chapman, K. Moffat, J. J. van Thor and M. Schmidt, *Science*, 2016, **352**, 725–729.

87 P. B. Corkum and F. Krausz, *Nat. Phys.*, 2007, **3**, 381–387.

88 N. Kuthirummal and P. M. Weber, *Chem. Phys. Lett.*, 2003, **378**, 647–653.

89 T. Zuo, A. D. Bandrauk and P. B. Corkum, *Chem. Phys. Lett.*, 1996, **259**, 313–320.

90 B. Wolter, M. G. Pullen, A.-T. Le, M. Baudisch, K. Doblhoff-Dier, A. Senftleben, M. Hemmer, C. D. Schröter, J. Ullrich, T. Pfeifer, R. Moshammer, S. Gräfe, O. Vendrell, C. D. Lin and J. Biegert, *Science*, 2016, **354**, 308–312.

91 B. C. Arruda and R. J. Sension, *Phys. Chem. Chem. Phys.*, 2014, **16**, 4439–4455.

92 S. Manz, A. Casandruc, D. Zhang, Y. Zhong, R. A. Loch, A. Marx, T. Hasegawa, L. C. Liu, S. Bayesteh, H. Delsim-Hashemi, M. Hoffmann, M. Felber, M. Hachmann, F. Mayet, J. Hirscht, S. Keskin, M. Hada, S. W. Epp, K. Flöttmann and R. J. D. Miller, *Faraday Discuss.*, 2015, **177**, 467–491.

93 L. Chen, Y. Zhao and Y. Tanimura, *J. Phys. Chem. Lett.*, 2015, **6**, 3110–3115.

94 P. Kroetz, A. Ruehl, G. Chatterjee, A.-L. Calendron, K. Muraria, H. Cankaya, F. X. Kärtnera, I. Hartl and R. J. D. Miller, High energetic and highly stable pulses from a Ho:YLF regenerative amplifier, *Proc. SPIE 9726, Solid State Lasers XXV*, ed. W. A. Clarkson and R. K. Shori, 2016, p. 97260SDOI: doi:10.1117/12.2212236.

95 M. Hada, D. Zhang, A. Casandruc, R. J. D. Miller, Y. Hontani, J. Matsuo, R. E. Marvel and R. F. Haglund, *Phys. Rev. B: Condens. Matter Mater. Phys.*, 2012, **86**, 134101.

96 I. Schapiro, S. Fusi, M. Olivucci, T. Andruniów, S. Sasidharanpillai and G. R. Loppnow, *J. Phys. Chem. B*, 2014, **118**, 12243–12250.

97 J. Briand, O. Bräm, J. Réhault, J. Léonard, A. Cannizzo, M. Chergui, V. Zanirato, M. Olivucci, J. Helbing and S. Haacke, *Phys. Chem. Chem. Phys.*, 2010, **12**, 3178.

98 M. Manathunga, X. Yang, H. L. Luk, S. Gozem, L. M. Frutos, A. Valentini, N. Ferrè and M. Olivucci, *J. Chem. Theory Comput.*, 2016, **12**, 839–850.
99 P. J. M. Johnson, A. Halpin, T. Morizumi, V. I. Prokhorenko, O. P. Ernst and R. J. D. Miller, *Nat. Chem.*, 2015, **7**, 980–986.
100 B. J. Claessens, S. B. van der Geer, G. Taban, E. J. D. Vredenbregt and O. J. Luiten, *Phys. Rev. Lett.*, 2005, **95**, 164801.
101 S. D. Saliba, C. T. Putkunz, D. V. Sheludko, A. J. McCulloch, K. A. Nugent and R. E. Scholten, *Opt. Express*, 2012, **20**, 3967–3974.
102 H. Lee, S. Karkare, L. Cultrera, A. Kim and I. V. Bazarov, *Rev. Sci. Instrum.*, 2015, **86**, 73309.
103 L. Cultrera, S. Karkare, H. Lee, X. Liu, I. Bazarov and B. Dunham, *Phys. Rev. Spec. Top.–Accel. Beams*, 2015, **18**, 113401.
104 D. A. Wann, A. V. Zakharov, A. M. Reilly, P. D. McCaffrey and D. W. H. Rankin, *J. Phys. Chem. A*, 2009, **113**, 9511–9520.
105 J. Conyard, K. Addison, I. A. Heisler, A. Cnossen, W. R. Browne, B. L. Feringa and S. R. Meech, *Nat. Chem.*, 2012, **4**, 547–551.
106 M. Bixon and J. Jortner, *J. Chem. Phys.*, 1968, **48**, 715–726.
107 R. Englman and J. Jortner, *Mol. Phys.*, 1970, **18**, 145–164.
108 W. Siebrand, *J. Chem. Phys.*, 1967, **46**, 403.
109 K. Drexhage, *IEEE J. Quantum Electron.*, 1972, **8**, 526.
110 S. Kim, J. K. Lee, S. O. Kang, J. Ko, J.-H. Yum, S. Fantacci, F. De Angelis, D. Di Censo, M. K. Nazeeruddin and M. Grätzel, *J. Am. Chem. Soc.*, 2006, **128**, 16701–16707.
111 C. E. Crespo-Hernández, B. Cohen, P. M. Hare and B. Kohler, *Chem. Rev.*, 2004, **104**, 1977–2020.
112 V. I. Prokhorenko, A. Picchiotti, M. Pola, A. G. Dijkstra and R. J. D. Miller, *J. Phys. Chem. Lett.*, 2016, **7**, 4445–4450.
113 Y. Koyama, K. Kubo, M. Komori, H. Yasuda and Y. Mukai, *Photochem. Photobiol.*, 1991, **54**, 433–443.
114 B. R. Henry and H. G. Kjaergaard, *Can. J. Chem.*, 2002, **80**, 1635–1642.
115 P. Jensen, *Mol. Phys.*, 2000, **98**, 1253–1285.
116 B. R. Henry, *Acc. Chem. Res.*, 1977, **10**, 207–213.
117 I. G. Ryabinkin and A. F. Izmaylov, *Phys. Rev. Lett.*, 2013, **111**, 220406.
118 I. G. Ryabinkin, L. Joubert-Doriol and A. F. Izmaylov, *J. Chem. Phys.*, 2014, **140**, 214116.

Poster titles

A double velocity map covariance imaging scheme for experiments with high intensity XUV and X-ray sources, **J. Lahl, F. Campi, H. Coudert-Alteirac, S. Maclot, J. Peschel, L. Rading, P. Rudawski, H. Wikmark, A. L'Huillier and P. Johnsson**, *Lund University, Sweden*

A multichannel least-squares B-spline approach to molecular photoionization. Theory, implementation and applications within the CIS approximation, **D. Toffoli and P. Decleva**, *University of Trieste, Italy*

Analysis of quantum molecular dynamics trajectories, **N. Zotev**, *University of Edinburgh, UK*

Interference effects in time-delayed nonsequential double ionization, **A. Maxwell and C. Figueira de Morisson Faria**, *University College London, UK*

Controlling the photoexcited decay of Fe(II)-*N*-heterocyclic carbene complexes through structural modifications, **M. Papai, K. B Møller, T. Rozgonyi, György Vankó and T. J. Penfold**, *Technical University of Denmark, Denmark*

Controlling two-photon ionization in H_2^+ using VUV pulses, **D. Jelovina, J. Feist, F. Martín and A. Palacios**, *Universidad Autonoma de Madrid, Spain*

Directionally controllable electron transfer in a chemically symmetrical D-B-A-B-D pentad, **T. Keane, M. Delor, S. A. Archer, J. A. Weinstein and A. J. H. M. Meijer**, *University of Sheffield, UK*

Diabatic strategies for photochemical quantum dynamics, **A. Perveaux, F. Gatti, D. Lauvergrat, H.-D. Mayer, P. J. Castro, M. Reguero, D. Mendiver Tapia, M. A. Robb, M. J. Bearpark, G. A. Worth and B. Lasorne**, *CNRS- Université de Montpellier, France*

Early-time photochemical dynamics of 1,4-diiodobenzene and methyl iodide investigated *via* ultrafast photoelectron spectroscopy, **B. Stankus, Y. Gao, D. M. Rodgers, A. Kirrander, M. P. Minitti, J. M. Budarz, C. Pemberton and P. M. Weber**, *Brown University, USA*

Ultrafast charge migration in glycine, **M. Lara-Astiaso, D. Ayuso, I. Tavernelli, P. Decleva, A. Palacios and F. Martín**, *Universidad Autonoma de Madrid, Spain*

Electron spin polarization in strong-field ionization of xenon atoms, **F. Morales Moreno, J. Kaushal, A. Hartung, M. Kunitski, K. Henrichs, A. Lauke, M. Richter, T. Jahnke, A. Kalikin, M. Schöffler, L. Schmidt, R. Dörner, M. Ivanov and O. Smirnova**, *Max Born Institute, Germany*

Enhancing high-order harmonic generation in light molecules by using chirped pulses, **M. Lara-Astiaso, R. E. F. Silva, A. Gubaydullin, P. Riviere, C. Meier and F. Martín**, *Universidad Autonoma de Madrid, Spain*

Elastic X-ray scattering from state-selected molecules, **A. Moreno, T. Northey, S. Schäfer and A. Kirrander**, *University of Edinburgh, UK*

Exploration of new techniques for precise attosecond pulse characterization, **J. M. Dahlström, J.-A. You, N. Rohringer, S. Pabst and E. Lindroth**, *Stockholm University, Sweden*

Extraction of the information on valence orbitals from the carrier envelope phase dependence of high-order harmonic generation, **S. Ohmura, H. Kono, T. Oyamada, T. Kato, K. Nakai and S. Koseki**, *Tohoku University, Japan*

Femtosecond charge and spin dynamics in the V–Cr prussian blue molecular magnet, **J. O. Johansson, J.-W. Kim, E. Allwright, D. M. Rogers, N. Robertson and J.-Y. Bigot**, *University of Edinburgh, UK*

Fixed-target for serial crystallography at synchrotrons and XFEL, **S. Oghbaey, A. Sarracini, H. M. Ginn, O. Pare-Labrosse, A. Kou, H. M. Mueller-Werkmeister, O. P. Ernst and R. J. D. Miller**, *University of Toronto, Canada*

Generation and control of elliptically polarized attosecond pulses with bicircular fields, **Á. Jiménez-Galán, D. Ayuso, F. Morales, O. Smirnova and M. Ivanov**, *Max Born Institute, Germany*

Hard-X-ray-induced multistep ultrafast dissociation: multistep core-hole-clock approach, **O. Travnikova, T. Marchenko, G. Goldsztejn, N. Sisourat, S. Carniato, R. Guillemin, L. Journel, D. Céolin, K. Jänkälä, R. Püttner, H. Iwayama, E. Shigemasa, M. Novella Piancastelli and M. Simon**, *Sorbonne Universités, UPMC Université Paris 06, CNRS, UMR 7614, France*

HHG in the 1D Fermi-Hubbard model, **R. E. Ferreira da Silva, I. Blinov, O. Smirnova, A. Rubtsov and M. Ivanov** , *Max Born Institute, Germany*

A compact X-ray laser source for attosecond science, **A. S. Johnson, D. Wood, S. Jarosch, L. Miseikis, C. Brahms, S. Parker, A. Gregory, P. Ye, C. Strüber, D. R. Austin and J. P. Marangos**, *Imperial College London, UK*

Intense XUV attosecond physics at the Lund Laser Centre, **S. Maclot, L. Rading, J. Lahl, H. Wikmark, F. Campi, J. Peschel, H. Courdert-Alteirac, P. Rudawski, A. L'Huillier and P. Johnsson**, *Lund University, Sweden*

Investigation of CH_2IBr photo-dissociation dynamics using time-dependent site-selective inner-shell ionisation on a femtosecond timescale, **H. Köckert, M. Burt, K. Amini, J. W. L. Lee, A. Gentleman, S. R. MacKenzie, C. Vallance, M. Brouard, R. Boll, C. Bomme, B. Erk, E. Savelyev, D. Rompotis, S. Bari, B. Manschwetus, D. Rolles, J. Wiese, J. Küpper, A. Rudenko, F. Brauße, A. Rouzée, P. Johnson, S. Maclot, J. Lahl, P. Olshin, A. Mereshscenko, J. Underwood, D. Holland, T. Ruchon, R. Geneaux, M. Geleoc, J. Palaudoux, F. Penent, P. Lablanquie, G. Goldsztejn, B. Cunha de Miranda, I. Ismail, M. N. Piancastelli, O. Travnikova, L. Journel, R. Guillemin, M. Simon and T. Marchenko**, *University of Oxford, UK*

Local control approach to electron transfer with ultrafast pulses, **P. Vindel Zandbergen, C. Meier, B. Y. Chang and I. R. Sola,** *Universidad Complutense de Madrid, Spain*

Machine learning for ultrafast X-ray spectroscopy at high repetition rate XFELs, **A. Sanchez-Gonzalez, P. Micaelli, C. Olivier, T. Barillot, R. Coffee, J. Marangos *et al.***, *Imperial College London, UK*

Many-body calculations on attosecond delays in photoionization, **E. Lindroth, M. Dahlström and J. Vinbladh,** *Stockholm University, Sweden*

Transient absorption spectroscopy, **A. C. Brown and H. W. van der Hart,** *Queen's University Belfast, UK*

Oscillating charge migration between glycine lone pairs silenced by nuclear delocalisation: CASSCF Ehrenfest dynamics study, **L. Polyak, A. J. Jenkins, M. Vacher, M. J. Bearpark and M. A. Robb,** *Imperial College London, UK*

Chiral states in atomic hydrogen. A clear picture of circular dichroism in photoelectron angular distributions, **A. F. Ordonez Lasso and O. Smirnova,** *Max Born Institute, Germany*

Manipulating ultrafast non-adiabatic dynamics in optical cavities, **M. Kowalewski, K. Bennett and S. Mukamel,** *University of California, Irvine, USA*

R-matrix calculations of photoelectron circular dichroism for propylene oxide, **Z. Masin, A. G. Harvey and O. Smirnova,** *Max Born Institute, Germany*

Recent developments towards time-resolved gas electron diffraction at York, **J. P. Figueira Nunes, C. D. Rankine, M. S. Robinson and P. D. Lane,** *University of York, UK*

Recollision with spin-polarized electrons in tailored laser fields, **D. Ayuso, Á. Jiménez-Galán, F. Morales, M. Ivanov and O. Smirnova,** *Max Born Institute, Germany*

Resonant multiphoton ionisation probe of the photodissociation of ammonia, **A. R. Smith, H. M. Watts, E. Jager, D. Horke, E. Springate, O. Alexander, C. Cacho, R. T. Chapman and R. S. Minns,** *University of Southampton, UK*

Sub-picosecond energy transfer by ultrashort and highly intense THz pulses, **P. K. Mishra, O. Vendrell and R. Santra,** *University of Hamburg, Germany*

The charge separated excited state of F-H-Cl as a prototype of bidirectional proton-coupled electron transfer processes, **H. An and K. Koo Baeck,** *Gangneung-Wonju National University, South Korea*

The ultrafast photodynamics of the sunscreen molecule methyl anthranilate, **N. D. N. Rodrigues, M. Staniforth, N. C. Cole-Filipiak, Y. Peperstraete and V. G. Stavros,** *University of Warwick, UK*

Photo-protective decay pathways: functionality, directionality and sensitivity, **N. Kotsina, M. Zawadzki, M. Candelaresi, L. Saalbach, D. Faccio, M. Paterson and D. Townsend,** *Heriot-Watt University, UK*

Time-resolved photoion yield spectroscopy of hydroxyindoles, **O. Ghafur, S. W. Crane, S. De Camillis, T. Y. Cowie, M. W. P. Bebbington, J. B. Greenwood and D. Townsend**, *Heriot-Watt University, UK*

Time-resolved X-ray scattering of electronic wave packets, **M. Simmermacher, N. E. Henriksen and K. B. Møller**, *Technical University of Denmark, Denmark*

Toward fully quantum modelling of ultrafast photodissociation imaging experiments. Treating tunnelling in the *ab initio* Multiple Cloning approach, **D. Makhov, T. Martinez and D. Shalashilin**, *University of Leeds, UK*

High-harmonic generation by two-color (2) field-attosecond pulse production, **M. Tudorovskaya**, *University of Edinburgh, UK*

Towards the quantum dynamics of small molecules, **D. Bellshaw and A. Kirrander**, *University of Edinburgh, UK*

Towards the ultrafast pump-probe near edge X-ray absorption spectroscopy in thin film organic polymer targets, **L. Miseikis, C. Struber, S. Jarosch, D. Austin, P. Ye, J. Marangos, A. S. Johnson and D. Wood**, *Imperial College London, UK*

Break-down of the Woodward-Hoffmann rules in cycloocta-1,3-diene and valence to rydberg transitions in pyrrole, **O. Schalk, T. Thompson, T. Geng, N. Baluyot, S. Neville, R. Thomas, T. Hansson and E. Tapavicza**, *Stockholm University, Sweden*

Ultrafast electron dynamics in nonadiabatic transitions: Correlation between the spatial change and energy exchange in time-dependent molecular orbitals, **M. Kanno, Y. Sato, S. Ohmura, H. Kono, S. Koseki, N. Shimakura and Y. Fujimura**, *Tohoku University, Japan*

Ultrafast photobiology and photochemistry at SwissFEL's Experimental Station A, **C. Milne, J. Szlachetko, G. Knopp, J. Schneider, J. Stettler, T. Humar, J. Czapla-Masztafiak, J. Réhault and R. Abela**, *Paul Scherrer Institute, Switzerland*

Ultrafast time-resolved, small angle scattering of the electrocyclic ring-opening reaction of 1,3-cyclohexadiene, **J. Ruddock, A. Kirrander, A. Moreno, B. Stankus, S. Boutet, J. Robinson, S. Carbajo, M. Hunter, A. Aquila, J. Koglin, T. Lane, M. Minitti and P. Weber**, *Brown University, USA*

Ultraviolet absorption induces hydrogen-atom transfer in Watson-Crick DNA base pairs, **K. Röttger, H. J. B. Marroux, M. P. Grubb, P. M. Coulter, A. S. Henderson, H. Böhnke, M. C. Galan, A. J. Orr-Ewing, F. Temps and G. M. Roberts**, *University of Bristol, UK*

XUV transient absorption spectroscopy of iodomethane and iodobenzene, **L. Drescher, M. Galbraith, G. Reitsma, J. Dura, M. Zhavoronkov, S. Patchkovskii, M. Vrakking and J. Mikosch**, *Max Born Institute, Berlin, Germany*

Quantum control at a conical intersection, **C. Liekhus-Schmaltz, Z. Li, A. Kaldun, G. McCracken and P. Bucksbaum**, *Stanford University, USA*

Ultrafast dynamics of resonances in radical anion clusters, **J. Verlet and J. Bull**, *Durham University, UK*

Monitoring the birth of an electronic wavepacket in a molecule with photoelectron spectroscopy, **N. Quadri, P. Decleva, D. Lauvergnat, A. Perveaux, F. Gatti, B. Lasorne, G. J. Halász and Á. Vibók**, *University of Debrecen, Hungary*

Probing ultrafast chemical dynamics by time-resolved ion imaging using the pixel imaging mass spectrometry (PImMS) camera, **K. Amini, R. Boll, A. Lauer, M. Burt, E. Savelyev, L. Christensen, F. Brauβe, B. Erk, C. Bomme, Stefan Düsterer, H. Höppner, P. Johnsson, J. W. Lee, E. Muller, N. Schirmel, C. Vallance, H. Stapelfeldt, J. Kupper, A. Rouzee, A. Rudenko, M. Brouard and D. Rolles**, *University of Oxford, UK*

Towards controlling the dissociation probability by light-induced conical intersections, **A. Vibók, G. J. Halász, A. Csehi, N. Moiseyev and L. S. Cederbaum**, *University of Debrecen, Hungary*

Coherence transfer in rhodium dicarbonyl, **H. J. B. Marroux and A. J. Orr-Ewing**, *University of Bristol, UK*

Time-resolved X-ray absorption spectroscopy with a water-window high-harmonic source, **Y. Pertot, C. Schmidt, M. Matthews, A. Chauvet, M. Huppert, V. Svoboda, A. von Conta, A. Tehlar, D. Baykusheva, J. P. Wolf and H. J. Wörner**, *Eidgenössische Technische Hochschule Zuerich, Switzerland*

Excited-state dynamics of 2-thiouracil: Simulation of neutral dynamics and time-resolved photoelectron spectrum, **S. Mai, H. Yu, J. A. Sanchez-Rodriguez, S. Ullrich, P. Marquetand and L. González**, *University of Vienna, Austria*

Interference stabilization of autoionizing states in molecular N_2 studied by time- and angular-resolved photoelectron spectroscopy, **N. Mayer, M. Eckstein, C.-H. Yang, G. Sansone, M. J. J. Vrakking, M. Ivanov and O. Kornilov**, *Max Born Institute, Germany*

Exploring highly excited electronic states of fullerenes using femtosecond laser photoelectron spectroscopy, **E. Bohl, J. O. Johansson, B. Mignolet, F. Remacle and E. E. B. Compbell**, *University of Edinburgh, UK*

High throughput femtosecond time-resolved serial crystallography using the photochip concept, **A. Sarracini, S. Oghbaey, H. M. Ginn, O. Paré-Labrosee, R. L. Owen, D. A. Sherrell, B. T. Eger, H. Duyvesteyn, A. Kohl, A. Marx, H. M. Mueller-Werkmeister, O. Ernst and R. J. D. Miller**, *University of Toronto, Canada*

The Faraday Discussions Poster Prize for the best poster was awarded to João Pedro Figueira Nunes, University of York, UK, for his poster on Recent developments towards time-resolved gas electron diffraction at York.

List of participants

Mr Fabio Albertani, *Imperial College London, United Kingdom*
Dr Jeremy Allen, *Royal Society of Chemistry, United Kingdom*
Dr Dane Austin, *Imperial College London, United Kingdom*
Dr David Ayuso, *Max Born Institute, Germany*
Prof Dr Kyoung Koo Baeck, *Gangneung-Wonju National University, South Korea*
Professor Michael Bearpark, *Imperial College London, United Kingdom*
Mr Darren Bellshaw, *University of Edinburgh, United Kingdom*
Dr Filippo Bencivenga, *Elettra-Sincrotrone Trieste S.C.p.A., Italy*
Dr Justine Bentley, *Laser Quantum Ltd, United Kingdom*
Dr Andrew Brown, *Queen's University Belfast, United Kingdom*
Mr Christian Burger, *Ludwig-Maximilians-Universität München, Germany*
Professor Eleanor Campbell, *University of Edinburgh, United Kingdom*
Professor Martin Centurion, *University of Nebraska, USA*
Dr Raluca Cireasa, *Institut des Sciences Moleculaires d'Orsay, France*
Dr Marie Cote, *Royal Society of Chemistry, United Kingdom*
Mr Stuart Crane, *Institute of Photonics and Quantum Sciences, United Kingdom*
Dr Jan Marcus Dahlström, *Stockholm University, Sweden*
Miss Natercia Das Neves Rodrigues, *University of Warwick, United Kingdom*
Professor Piero Decleva, *University of Trieste, Italy*
Professor Wolfgang Domcke, *Technische Universität München, Germany*
Dr Danielle Dowek, *Institut des Sciences Moléculaires d'Orsay, France*
Mr Lorenz Drescher, *Max Born Institute, Germany*
Professor Nirit Dudovich, *Weizmann Institute of Science, Israel*
Dr Rui Emanuel Ferreira da Silva, *Max Born Institute, Germany*
Dr Carla Figueira de Morisson Faria, *University College London, United Kingdom*
Mr João Pedro Figueira Nunes, *University of York, United Kingdom*
Ms Rachel Fisher, *University of Edinburgh, United Kingdom*
Mr Ruaridh Forbes, *University College London, United Kingdom*
Dr Lepine Franck, *CNRS ILM Lyon, France*
Dr Maxim Gelin, *Technical University of Munich, Germany*
Dr Oliver Gessner, *Lawrence Berkeley National Laboratory, USA*
Dr Omair Ghafurf, *Heriot Watt University, United Kingdom*
Mr Denis Jelovina, *Universidad Autónoma de Madrid, Spain*
Dr Andrew Jenkins, *Imperial College London, United Kingdom*
Dr Álvaro Jiménez Galán, *Max Born Institute, Germany*
Dr Olof Johansson, *University of Edinburgh, United Kingdom*
Mr Allan S. Johnson, *Imperial College London, United Kingdom*
Dr Manabu Kanno, *Tohoku University, Japan*
Mr Theo Keane, *University of Sheffield, United Kingdom*
Professor Anatoli Kheifets, *Australian National University, Australia*
Dr Victor Kimberg, *Royal Institute of Technology, Sweden*
Dr Adam Kirrander, *University of Edinburgh, United Kingdom*
Mr Hansjochen Koeckert, *University of Oxford, United Kingdom*

Dr Oleg Kornilov, *Max Born Institute*, *Germany*
Miss Nikoleta Kotsina, *Institute of Photonics and Quantum Sciences*, *United Kingdom*
Dr Markus Kowalewski, *University of California, Irvine*, *USA*
Mr Jan Lahl, *Lund University*, *Sweden*
Mr Manuel Lara-Astiaso, *Universidad Autonoma de Madrid*, *Spain*
Dr Benjamin Lasorne, *CNRS-Université de Montpellier*, *France*
Miss Andrea Lidberg, *University of Edinburgh*, *United Kingdom*
Ms Chelsea Liekhus-Schmaltz, *Stanford University*, *USA*
Professor Eva Lindroth, *Stockholm University*, *Sweden*
Dr Helen Lunn, *Royal Society of Chemistry*, *United Kingdom*
Dr Sylvain Maclot, *Lund University*, *Sweden*
Mr Sebastian Mai, *University of Vienna*, *Austria*
Dr Dmitry Makhov, *University of Leeds*, *United Kingdom*
Professor Jonathan Marangos, *Imperial College*, *United Kingdom*
Mr Hugo Marroux, *University of Bristol*, *United Kingdom*
Prof Dr Fernando Martín, *Universidad Autonoma de Madrid*, *Spain*
Dr Zdenek Masin, *Max Born Institute*, *Germany*
Ms Paloma Matia Hernando, *Imperial College London*, *United Kingdom*
Mr Nicola Mayer, *Max Born Institute*, *Germany*
Prof Dr R. J. Dwayne Miller, *Max Planck Institute for the Structure and Dynamics of Matter*, *Germany*
Dr Chris Milne, *Paul Scherrer Institute*, *Switzerland*
Dr Michael Minitti, *SLAC National Accelerator Laboratory*, *USA*
Dr Russell Minns, *University of Southampton*, *United Kingdom*
Mr Lukas Miseikis, *Imperial College London*, *United Kingdom*
Dr Pankaj Kumar Mishra, *University of Hamburg*, *Germany*
Dr Felipe Morales Moreno, *Max Born Institute*, *Germany*
Mr Andres Moreno Carrascosa, *University of Edinburgh*, *United Kingdom*
Professor Shaul Mukamel, *University of California, Irvine*, *USA*
Professor Daniel Neumark, *University of California, Berkeley*, *USA*
Dr Simon Neville, *University of Ottawa*, *Canada*
Mr Saeed Oghbaey, *University of Toronto*, *Canada*
Mr Shu Ohmura, *Tohoku University*, *Japan*
Mr Andres Felipe Ordonez Lasso, *Max Born Institute*, *Germany*
Professor Andrew Orr-Ewing, *University of Bristol*, *United Kingdom*
Dr Matyas Papai, *Technical University of Denmark*, *Denmark*
Dr Tom Penfold, *Newcastle University*, *United Kingdom*
Dr Iakov Polyak, *Imperial College London*, *United Kingdom*
Professor Michael A. Robb, *Imperial College London*, *United Kingdom*
Dr Gareth Roberts, *University of Bristol*, *United Kingdom*
Professor Daniel Rolles, *Kansas State University*, *USA*
Ms Jennifer Ruddock, *Brown University*, *USA*
Professor Artem Rudenko, *Kansas State University*, *USA*
Dr Katharina Röttger, *University of Bristol*, *United Kingdom*
Miss Lisa Saalbach, *Heriot-Watt University*, *United Kingdom*
Mr Alvaro Sanchez-Gonzalez, *Imperial College London*, *United Kingdom*
Mr Antoine Sarracini, *University of Toronto*, *Canada*
Dr Oliver Schalk, *Stockholm University*, *Sweden*
Prof Dr Taro Sekikawa, *Hokkaido University*, *Japan*
Professor Dmitry Shalashilin, *University of Leeds*, *United Kingdom*
Mr Mats Simmermacher, *Technical University of Denmark*, *Denmark*
Mr Adam Smith, *University of Southampton*, *United Kingdom*
Professor Theis Sølling, *University of Copenhagen*, *Denmark*

Ms Claire Springett, *Royal Society of Chemistry*, *United Kingdom*
Ms Katie Springham, *Royal Society of Chemistry*, *United Kingdom*
Mr Brian Stankus, *Brown University*, *USA*
Dr Vasilios Stavros, *University of Warwick*, *United Kingdom*
Miss Anne Stephansen, *University of Copenhagen*, *Denmark*
Professor Albert Stolow, *University of Ottawa*, *Canada*
Dr Daniele Toffoli, *University of Trieste*, *Italy*
Dr Dave Townsend, *Heriot-Watt University*, *United Kingdom*
Dr Oksana Travnikova, *Sorbonne Universités*, *France*
Dr Maria Tudorovskaya, *University of Edinburgh*, *United Kingdom*
Prof Dr Kiyoshi Ueda, *Tohoku University*, *Japan*
Dr Morgane Vacher, *Uppsala Universitet*, *Sweden*
Dr Jan Verlet, *Durham University*, *United Kingdom*
Professor Ágnes Vibók, *ELI-HU Non-Profit Ltd.*, *Hungary*
Miss Patricia Vindel Zandbergen, *Universidad Complutense de Madrid*, *Spain*
Dr Zhehui Wang, *Los Alamos National Laboratory*, *USA*
Professor Peter M. Weber, *Brown University*, *USA*
Professor Dr Hans Jakob Wörner, *Eidgenössische Technische Hochschule Zuerich*, *Switzerland*
Dr Junko Yano, *Lawrence Berkeley National Laboratory*, *USA*
Miss Magdalena Zawadzki, *Institute of Photonics and Quantum Sciences*, *United Kingdom*
Mr Nikola Zotev, *University of Edinburgh*, *United Kingdom*